T0111860

Farmer Innovations and Best Practices by Shifting Cultivators in Asia-Pacific

FSC
www.fsc.org
MIX
Paper | Supporting
responsible forestry
FSC® C013604

CABI is a trading name of CAB International

CABI
Nosworthy Way
Wallingford
Oxfordshire OX10 8DE
UK

CABI
200 Portland Street
Boston
MA 02111
USA

Tel: +44 (0)1491 832111
E-mail: info@cabi.org
Website: www.cabi.org

Tel: +1 (617)682-9015
E-mail: cabi-nao@cabi.org

©2023 CAB International. [Farmer Innovations and Best Practices by Shifting Cultivators in Asia-Pacific] is licensed under a Creative Commons Attribution-NonCommercial-NoDerivatives 4.0 International License

A catalogue record for this book is available from the British Library, London, UK.

The views expressed in this publication are those of the author(s) and do not necessarily represent those of, and should not be attributed to, CAB International (CABI). Any images, figures and tables not otherwise attributed are the author(s)' own. References to internet websites (URLs) were accurate at the time of writing.

CAB International and, where different, the copyright owner shall not be liable for technical or other errors or omissions contained herein. The information is supplied without obligation and on the understanding that any person who acts upon it, or otherwise changes their position in reliance thereon, does so entirely at their own risk. Information supplied is neither intended nor implied to be a substitute for professional advice. The reader/user accepts all risks and responsibility for losses, damages, costs and other consequences resulting directly or indirectly from using this information.

CABI's Terms and Conditions, including its full disclaimer, may be found at https://www.cabi.org/terms-and-conditions/.

Library of Congress Cataloging-in-Publication Data

Names: Cairns, Malcolm, editor. | Hill, Bob (Science writer), contributor. | Kurupunya, Tossaporn, contributor.
Title: Farmer innovations and best practices by shifting cultivators in Asia-Pacific / edited by Malcolm Cairns with the assistance of Bob Hill and Tossaporn Kurupunya.
Description: Boston, MA, USA : CAB International, [2023] | Includes bibliographical references and index. | Summary: "This book, the third of a series, shows how shifting cultivators from the Himalayan foothills to the Pacific Islands have devised ways to improve their farming systems. It considers the importance of swidden agriculture to food security and livelihoods, and its environmental significance across multiple cultures, crops and forest systems"-- Provided by publisher.
Identifiers: LCCN 2022055734 (print) | LCCN 2022055735 (ebook) | ISBN 9781800620094 (hardback) | ISBN 9781800620100 (ebook) | ISBN 9781800620117 (epub)
Subjects: LCSH: Shifting cultivation--Pacific Area.
Classification: LCC S602.87 .F372 2023 (print) | LCC S602.87 (ebook) | DDC 631.5/818091823--dc23/eng/20230412
LC record available at https://lccn.loc.gov/2022055734
LC ebook record available at https://lccn.loc.gov/2022055735

ISBN-13: 9781800620094 (hardback)
9781800620100 (ePDF)
9781800620117 (epub)

DOI: 10.1079/9781800620117.0000

Commissioning editor: David Hemming
Production editor: James Bishop

Printed and bound in the UK from copy supplied by the authors by CPI Group (UK) Ltd, Croydon, CR0 4YY

HAROLD C. BROOKFIELD giving a speech at a feast at the conclusion of fieldwork, in Mintima village, Chimbu province, Papua New Guinea in 1965.

Photo provided by Muriel Brookfield.

In remembrance of Harold C. Brookfield (1926-2022)

It is fitting that the final work of the late Harold Brookfield is published in this volume*, edited by one of his former doctoral students. Harold's core focus remained the relationship between rural societies and their natural environments, which he believed should be at the core of geography - a principle he maintained to the end of his life, sometimes in the face of criticism from those in geographical fashions that have since disappeared.

Harold's professional career extended over 70 years. After teaching and research in England, Ireland, South Africa, Mauritius and Australia, he came to the Research School of Pacific Studies at the Australian National University (ANU) in 1957 to undertake research in Papua New Guinea (PNG). He pioneered multidisciplinary fieldwork in Chimbu province in PNG, with anthropologist Dr. Paula Brown. As well as producing a number of books, this work led him to postulate that farmers respond to population increase by intensifying and innovating in their land management to increase productivity. During this period, he supervised postgraduate students who carried out fieldwork in rural PNG, Tonga, Fiji and Vanuatu. With Harold, they became authors of chapters in his 1973 book, *The Pacific in Transition*. He later held positions in the USA, Canada, the Caribbean and Fiji.

In 1982 Harold returned to the ANU and established the Land Management Project which involved the work of an agricultural scientist, an anthropologist, a geomorphologist/soil scientist and a geographer. During this time, he worked closely with Professor Piers Blaikie at the University of East Anglia, and together they edited *Land Degradation and Society* (1987). Harold was the Professor of Human Geography and first Convenor of the Division of Environment and Society at ANU until 1991. In 'retirement' he established the UN University-funded People, Land Management and Environmental Change project with participants in Brazil, China, Ghana, Guinea, Papua New Guinea, Uganda and Tanzania.

Harold was a prolific author, producing influential books and significant papers. A glance at his publications list indicates the breadth of his studies and changes in his focus over time. His huge contribution to geography and development studies was recognized with many honours. Harold has gone, but his influence will long endure.

R.M. (Mike) Bourke and Bryant J. Allen
College of Asia and the Pacific, The Australian National University,
Canberra, Australia

---

*
Brookfield, H., 'Raised fields, under shifting and permanent cultivation: An essay', synthesis chapter in section X, *Mounding Technologies*, in the supplementary chapters to this volume.

Wise words heard in passing

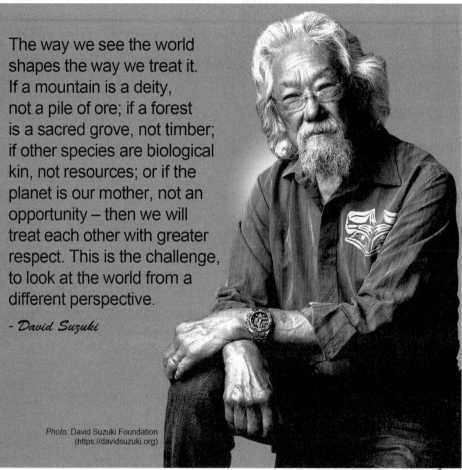

The way we see the world shapes the way we treat it. If a mountain is a deity, not a pile of ore; if a forest is a sacred grove, not timber; if other species are biological kin, not resources; or if the planet is our mother, not an opportunity – then we will treat each other with greater respect. This is the challenge, to look at the world from a different perspective.

- David Suzuki

Photo: David Suzuki Foundation
(https://davidsuzuki.org)

'I, too, approached the island in a boat in 1998, spotting from a distance a canoe in which two figures stood, fishing; others on the beach observed the encounter. Seeing us, the fishers moved back toward their shore, whereupon we left. I regret that visit; even if for some minutes, I violated their privacy and tranquility. Unlike Chau, however, it did not occur to me that I had any wisdom to impart to them. What can I, a representative of a civilization that, within the span of a few hundred years, has destabilized the biosphere of an entire planet, have to offer to a people who have thrived since the dawn of humankind on these tiny islands? Is it we who have something to teach the Sentinelese, or they us?'

MADHUSREE MUKERJEE, Senior Editor at *Scientific American* and author of *The Land of Naked People: Encounters with Stone Age Islanders* (Houghton Mifflin, 2003)

'Once we have shed the erroneous idea that shifting cultivation is necessarily historically prior to, more primitive than, and less efficient than fixed-field cultivation, there remains one further illusion to shed. That illusion is that it is a relatively static technique that has not changed much in the past millennium. On the contrary, one could argue that swiddening and, for that matter, foraging have undergone far more transformation in that period than has wet-rice cultivation.'

JAMES C. SCOTT (2009), *The Art of Not Being Governed: An Anarchist History of Upland Southeast Asia*, Yale University Press, New Haven, CT and London, pp.196–197

'Permaculture is a modern form of organic farming and agroforestry based on the criteria of keeping the soil covered at all times, imitating nature by intercropping a wide variety of species, and optimizing space by maintaining multiple vertical layers. All of these criteria are matched by shifting cultivation practices.'

DR. RAMAKRISHNA AKKINAPALLY, formerly Deputy Director General, Papua New Guinea National Agricultural Research Institute (NARI)

'My swidden experiences derive from a specific group of Karen in western Thailand I have visited over 47 years. Originally practising impressive, high-yielding swidden technologies, they have been subject to physical relocation and restrictions in their agricultural systems. Whatever innovative accommodations they have applied have been "run over" by "innovations" more or less imposed by Thai government agencies and civil society organisations. Therefore, real farmers' innovations have not been able to survive the various practices forced upon the Karen...'

ANDERS BALTZER JORGENSEN, Anthropologist, retired Chief Technical Adviser and Counsellor Development, Danish Ministry of Foreign Affairs.

'We keep providing local politicians and decision-makers with articles, press releases, research documents on traditional shifting cultivation…but the sad reality is: **they do not read them and probably would not understand them**. *Dealing with this short-sighted and narrow-minded Filipino political class is absolutely frustrating. They are the ones asking us for new evidences, but this is only part of their "delay tactics"…they do not read the work of scholars and academicians, nor do they pay any attention to the explanations given by the impacted indigenous communities. Having said this, the struggle goes on.'*

DARIO NOVELLINO, Centre for Biocultural Diversity (CBCD), School of Anthropology and Conservation, University of Kent, Winner of the Ostrom Award for 2021

'Indigenous peoples' values and knowledge provide insights for reciprocal human-nature relationships amidst the crisis of biodiversity loss and climate change.'

JOJI CARINO, Senior Policy Advisor and former Director of Forest Peoples Programme (FPP)

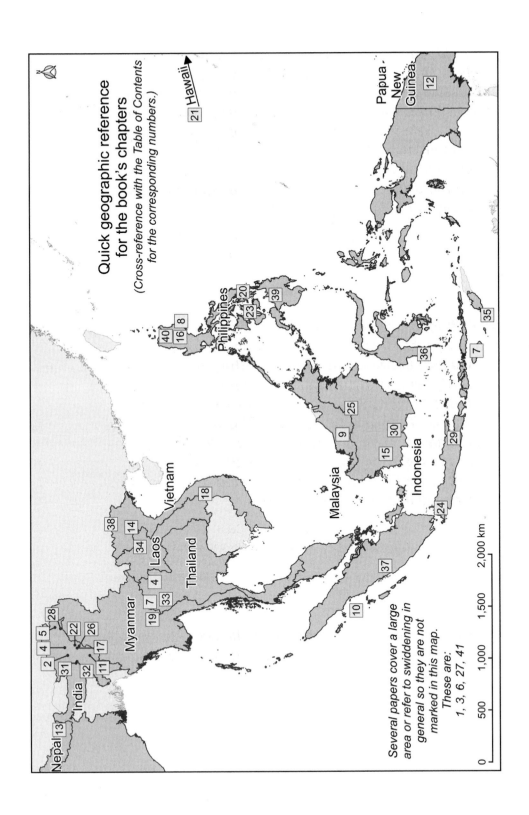

Quick geographic reference
for the book's chapters
*(Cross-reference with the Table of Contents
for the corresponding numbers.)*

21 Hawaii

Papua
New
Guinea

12

Philippines

20

23

39

8

40

16

35

7

36

25

9

15

30

Malaysia

29

Indonesia

24

Vietnam

18

Laos

38

14

34

Thailand

37

Myanmar

4

7

19

33

India

5

28

2

4

22

26

31

32

11

17

10

Nepal

13

Several papers cover a large
area or refer to swiddening in
general so they are not
marked in this map.
These are:
1, 3, 6, 27, 41

0   500   1,000   1,500   2,000 km

CONTENTS

II. Evolution of food production systems to support increasing populations

III. Fireless shifting cultivation

IV. Innovations in fallow management

A. When shifting cultivators expand forest cover

B. Favouring soil-building trees to accelerate soil recovery (biologically improved fallows)

V. Conclusions

Postscript

Indexes

Supplementary chapters

The following chapters add substantial support to the learned arguments presented in the chapters above. They are therefore recommended as additional reading on the broad subject of farmer innovations and best practices by shifting cultivators in the Asia-Pacific region. That these chapters are offered for digital download rather than published within these pages is simply a result of overwhelming contributions and limited space. A regional map on the next page shows the research locations of these supplementary chapters. The chapters themselves are available at:

https://www.cabidigitallibrary.org

CONTENTS
of the supplementary material accessible from CABI's Digital Library

Quick reference map for the supplementary chapters

Foreword: **A neophyte's tale: Sound and fury, and signifying nothing**
    Jack D. Ives

Preface - **The future of swidden agriculture in SE Asia**
    Jeff Sayer, Shintia Arwida and Agni Klintuni Boedhihartono

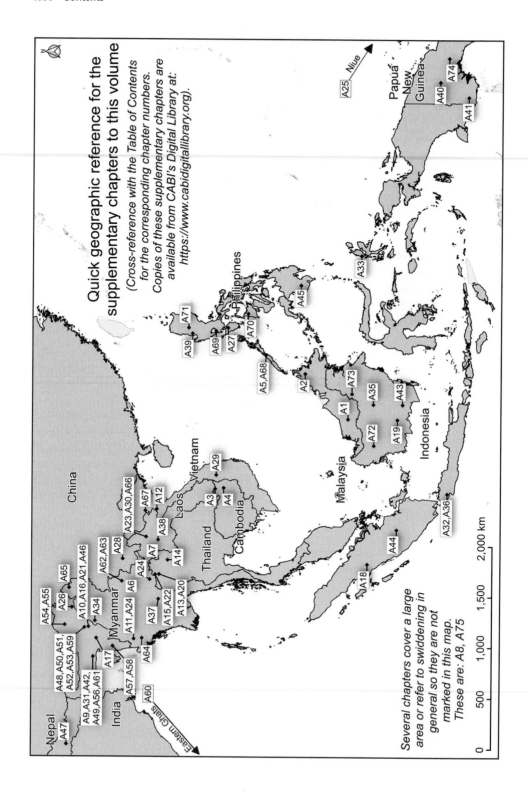

Quick geographic reference for the supplementary chapters to this volume

(Cross-reference with the Table of Contents for the corresponding chapter numbers. Copies of these supplementary chapters are available from CABI's Digital Library at: https://www.cabidigitallibrary.org).

Several chapters cover a large area or refer to swiddening in general so they are not marked in this map. These are: A8, A75

VI. Managing the factors of production: The seeds, soils and tools that farmers use

The ebb and flow of cash crops in swidden fields: The dichotomy between food security and generating cash

VII. Strategies to assure food security: A bedrock of security in an insecure world

VIII. Social innovations

IX. Bamboo-based fallow management

X. Mounding technologies

XI. Swiddening in the swamps

XII. Gender analysis

XIII. Overviews of specific places and peoples

Nepal

India

Myanmar

Lao PDR

Vietnam

The Philippines

Indonesia

Papua New Guinea

XIV. The comic section

A75. Closing with a satirical view of some of the issues
Malcolm Cairns and Paradorn Threemake

Indexes for supplementary chapters

ACRONYMS

AANCB	Average annual net cash benefit per hectare
AATB	Annual average total benefits per hectare
ACIAR	Australian Centre for International Agricultural Research
ACF	*Action Contre la Faim*, a French NGO
ADAB	Australian Development Assistance Bureau (now renamed)
ADB	Asian Development Bank
ADSPP	Ancestral Domain Sustainable Development and Protection Plan, Philippines
AERF	Applied Environmental Research Foundation, Pune, India
AFTSEMU	Agricultural Field Trials, Systems, Evaluation and Monitoring Unit, PNG
AGB	Above-ground biomass
AGC	Above-ground carbon
AIPP	Asia Indigenous Peoples' Pact
ALA	Atlas of Living Australia
AMA-JK	Alliance of Indigenous Peoples of Jalai Sekayuq and Kendawangan Siakaran, West Kalimantan
AMAN	Indonesian Alliance of Indigenous Peoples of the Archipelago
AMAP	French laboratory conducting basic research on plants and plant communities
ANR	Assisted natural regeneration
ANU	Australian National University, Canberra
ArcGIS	Geographical information system software
ARCinfo	Geographical information system software
ARF	Agriculture revolving fund
ASEAN	Association of Southeast Asian Nations
ASB	Alternatives to Slash and Burn project (ICRAF)
ASTI Council	Assam Science, Technology and Environmental Council, India
ATREE	Ashoka Trust for Research in Ecology and the Environment, Bangalore
AusAID	Former name for Australia's international aid agency
AVSI Foundation	An international NGO based in Italy specializing in cooperation and humanitarian aid projects
BATAN	National Nuclear Energy Agency of Indonesia
BD	Bulk density (of soil)
BITO	Bakun Indigenous Tribes Organization, Philippines
BNPB	National Board for Disaster Management, Indonesia
BPS	Central Bureau of Statistics, Indonesia
BRWA	Customary Land Registration Body, Indonesia
BRIN	National Research and Innovation Agency, Indonesia
BSTID	Board on Science and Technology for International Development (US)
BSU	Benguet State University, Philippines
CABI	Centre for Agriculture and Bioscience International
CAD	Community Agency for Rural Development, Myanmar
CADT	Certificate of Ancestral Domain Title, Philippines
CAR	Cordillera Administrative Region, Philippines
CARSR	Center for Agricultural Research System Research
CCD	Colony Collapse Disorder (honeybees)
CDM	Clean Development Mechanisms
CEAPRED	Centre for Environmental and Agricultural Policy Research, Extension and Development, Nepal

CEC	Cation exchange capacity
CENDI	Community Entrepreneur Development Institute, Vietnam
CESD	Center for Ethnic Studies and Development, Thailand
CF	Community forestry
CGED-N	Centre for Green Ecology Development, Nepal
CGIAR	Consultative Group for International Agricultural Research
CHARM	Cordillera Highland Agriculture Resource Management, Philippines
CHRO	Chin Human Rights Organization, Canada
CHTDB	Chittagong Hill Tracts Development Board
CIAT	International Centre for Tropical Agriculture, Cali, Colombia
CICES	Common International Classification for Ecosystem Services
CIFOR	Center for International Forestry Research
CIIFAD	Cornell International Institute for Food, Agriculture and Development
CIKARD	Center for Indigenous Knowledge for Agriculture and Rural Development
CIP-ESEAP	International Potato Centre for East, Southeast Asia and the Pacific
CIRAD	Agricultural Research Centre for International Development, France
CLMV	The countries Cambodia, Laos, Myanmar and Vietnam
CMES	Center of Mountain Ecosystem Studies (former name), Kunming, China
CMF	Center for Mountain Futures (present name), Kunming, China
CNRS	French National Centre for Scientific Research
COP	Conference of Parties to the United Nations Framework Convention on Climate Change
CRIFC	Central Research Institute for Food Crops
CRS	Catholic Relief Services
CSA	Climate-smart agriculture
CSIRO	Commonwealth Scientific and Industrial Research Organization, Australia
CSEAS	Centre for Southeast Asian Studies, Japan
CSOs	Civil-society organizations
CVPED	Cagayan Valley Program on Environment and Development, Philippines
CVPD	Citrus vein phloem degeneration virus
DAL	Department of Agriculture and Livestock, Papua New Guinea
dbh	Diameter at breast height
DENR	Department of Environment and Natural Resources, Philippines
DEQP	Department of Environmental Quality Promotion, Thailand
DILG	Department of Interior and Local Government, Philippines
DNP	Department of National Parks, Wildlife and Plant Conservation, Ministry of Natural Resources and Environment, Thailand
DoP	Department of Population, Ministry of Immigration and Population, Myanmar
EAPI	Environment and Policy Institute, East-West Center, Honolulu
EC	European Commission, also electrical conductivity (soil)
ECDF	Ethnic Communities' Development Forum
ECEC	Effective cation exchange capacity
ECHO	A global non-profit organization focused on overcoming hunger
EED	*Planète Enfants et Développement*, a French NGO
EFEO	The French School of the Far East
ENSO	El Niño Southern Oscillation (climate pattern)
ENVIS	Environmental Information System, India

EPA	United States Environmental Protection Agency
ES	Ecosystem services
ESRC	Economic and Social Research Council, UK
ETM+	Enhanced Thematic Mapper Plus (Landsat)
FALUPAM	Forest and Agricultural Land Use Planning and Management
FAO	Food and Agricultural Organization of the United Nations
FCEC	Flora of China Editorial Committee
FCPF	Forest Carbon Partnership Facility, Indonesia
FFS	Farmer field schools, Myanmar
FGD	Focus-group discussion
FHI 360	Family Health International
FLR	Forest landscape restoration
FMC	Field moisture capacity
FMNR	Farmer-managed Natural Regeneration
FOCUS	Fostering Climate Resilient Upland Farming Systems (northeast India)
FOERDIA	Forestry and Environmental Research, Development and Innovation Agency, Indonesia
FPAs	Fire Protection Associations
FPP	Forest Peoples' Programme (UK and the Netherlands)
FREM	Forest Resources and Environment Management, Vietnam
FSI	Forest Survey of India
GBH	Girth at breast height
GDP	Gross Domestic Product
GEF	Global Environment Facility
GGI-Myanmar	A manufacturer specializing in teak and hardwood
GHG	Greenhouse gasses
GHSNP	Gunung Halimun Salak National Park, Indonesia
GIAN	Global Initiative of Academic Networks, India
GIAHS	Globally Important Agricultural Heritage Systems
GIS	Geographic information system
GIZ	*Deutsche Gesellschaft für Internationale Zusammenarbeit*, a German federal enterprise for international cooperation
Gm/cc	Green manure/cover crop
GMS	Greater Mekong Subregion
GOI	Government of India
GONGO	Government-Organized Non-Governmental Organization
GoV	Government of Vietnam
GPS	Global positioning system
GPPK	Pancur Kasih Empowerment Movement, West Kalimantan
GRET	*Groupe de Recherche et d'Exchanges Technologique*, French NGO
GSBI	Global Soil Biodiversity Initiative
GTZ	*Gesellschaftfürtechnische Zusammenarbeit*, the former German technical cooperation agency, now known as GIZ
HCVRN	High Conservation Value Resource Network
HHs	Households
HI-LIFE	The Landscape Initiative for Far Eastern Himalayas
HKH	The Hindu Kush-Himalayan region
HuMa	Association for Ecology and Community-based Law Reform, Indonesia

IABI	Indonesia Disaster Expert Association
IAS	Invasive alien species
IASER	Institute of Applied Social and Economic Research, Papua New Guinea
IASSI	Indian Association of Social Science Institutions
ICAR	Indian Council of Agricultural Research
ICCC	International Conference on Climate Change
ICEF	India-Canada Environment Facility
ICFRE	Indian Council of Forestry Research and Education
ICIMOD	International Centre for Integrated Mountain Development
ICRAF	International Center for Research in Agroforestry (World Agroforestry Centre)
IDP	Internally displaced persons
IDR	Indonesian rupiah
IDMC	Internal Displacement Monitoring Centre
IDRC	International Development Research Centre, Canada
IFAD	International Fund for Agricultural Development
IFD	Integrated farm development
IFM	Indigenous fallow management
IFOAM	International Federation of Organic Agriculture Movements
IFRCRCS	International Federation of Red Cross and Red Crescent Societies
IIRR	International Institute of Rural Reconstruction, an NGO based in the Philippines
IITA	International Institute of Tropical Agriculture, Nigeria
IK	Indigenous knowledge (see also TEK)
IKAP	Indigenous Knowledge and Peoples' Network
IKSP	Indigenous Knowledge Systems and Practices
ILCAA	Research Institute for Languages and Culture of Asia and Africa, Tokyo
ILEIA	A public benefit organization in the Netherlands focused on food sovereignty
ILO	International Labour Organization
ILRI	Indian Lac Research Institute
IMF	International Monetary Fund
INBO	Research Institute for Nature and Forest, Belgium
INDUFOR	Multinational forest consulting group
InFEWS	Integrated Forest Ecosystem and Watershed Services (ICIMOD)
INGO	International Non-Governmental Organization
INNOVATION	A public scientific and technical research organization, France
INRA	*Institut National de la Recherche Agronomique*, France
IOSR	International Organization of Scientific Research
IPBES	Intergovernmental Science-Policy Platform on Biodiversity and Ecosystem Services
IPCC	Intergovernmental Panel on Climate Change
IPHHBK	In Indonesia, a permit for extracting NTFPs
IPK	Leibniz Institute of Plant Genetics and Crop Plant Research, Germany
IPO	Indigenous People's Organization
IPR	Intellectual property rights
IRD	*Institut de Recherche pour le Développement*, a French government research organization that replaced the former ORSTOM.
IRRI	International Rice Research Institute

ITPS	Intergovernmental Technical Panel on Soils
IUCN	International Union for Conservation of Nature
IV	Importance value, an index for traditional medicinal remedies
IWGIA	International Working Group for Indigenous Affairs
IWMI	International Water Management Institute
JERS	Japanese Earth-resources satellite
JICA	Japan International Cooperation Agency
JSTOR	A digital library founded in 1995 in New York
KITLV	Royal Netherlands Institute of Southeast Asian and Caribbean Studies
KMSS-Loikaw	Karuna Mission Social Solidarity-Loikaw (Myanmar)
KMT	Kuomintang – the Chinese Nationalist Party
Komnas HAM	National Human Rights Commission, Jakarta
KPE	Kasepuhan Pasir Eurih community, Indonesia
KS	Department of Statistics, Lao PDR
KSO	Khatarshnong Socio Organization, East Khasi Hills, Meghalaya
LANDSAT	US satellite-based Earth observation programme
Lao PDR	Lao People's Democratic Republic
LASS	Lao Academy of Social Science
LDTSPs	Land Degradation Target-Setting Programmes
LEAD India	Indian TV programme aiming to develop new leaders
LECS	Lao Expenditure and Consumption Survey
LEISA India	Low External Input Sustainable Agriculture (a magazine)
LFNC	Lao Front for National Construction
LGU	Local government unit, Philippines
LI-BIRD	Local Initiatives for Biodiversity, Research and Development, Nepal
LIFT	Livelihoods and Food Security Fund, Myanmar
LIPI	Indonesian Institute of Sciences, recently dissolved and replaced by BRIN
LISO	Livelihood Sovereignty Alliance
LLDASP	Inland Water Transportation Office, Pasaman district, West Sumatra
LSD	Least significant differences
LUC	Land-use committee
LUI	Land-use intensity
MA	Millennium Ecosystem Assessment
MAPs	Medicinal and aromatic plants
MARD	Ministry of Agriculture and Rural Development, Vietnam
MASP	Mapping Agricultural Systems of Papua New Guinea
MBC	Microbial biomass carbon (in soil)
MDI-Nepal	Manahari Development Institute, Nepal
Mg	Megagram (equaling 1000 kilograms)
MISEREOR	German Catholic Bishops' Organization for Development Cooperation
MNRE	Ministry of Natural Resources and Environment, Thailand
MOAC	Ministry of Agriculture and Agricultural Cooperatives, Thailand
MONRE	Ministry of Natural Resources and Environment, Vietnam
MONREC	Ministry of Natural Resources and Environmental Conservation, Myanmar
MMIID	Myanmar Institution for Integrated Development
MMSEA	Mainland Montane Southeast Asia
MOALI	Ministry of Agriculture, Livestock and Irrigation, Myanmar
MoALMC	Ministry of Agriculture, Land Management and Cooperatives, Nepal

MOECAF	Ministry of Environment Conservation and Forestry, Myanmar (outdated, see MONREC)
MODIS	Moderate Resolution Imaging Spectroradiometer
MoEF	Ministry of Environment and Forestry, India
MONREC	Ministry of Natural Resources and Environmental Conservation, Myanmar
MoSPI	Ministry of Statistics and Programme Implementation, India
MRDC	Montañosa Research and Development Center, Philippines
MSEC	Management of Soil Erosion Consortium
NAFRI	National Agriculture and Forestry Research Institute, Lao PDR
NARI	National Agricultural Research Institute, Papua New Guinea
NASA	National Aeronautics and Space Administration
NBP	Newly burnt swidden plots
NBR	Normalized Burn Ratio (soil moisture)
NBSAPs	National Biodiversity Strategy and Action Plans
NCF	National Commission on Farmers, Ministry of Agriculture, India
NCIP	National Commission on Indigenous Peoples, Philippines
NCP	Nature's Contributions to People
NDCs	Nationally Determined Contributions (to greenhouse-gas emissions)
NDF	Northern Development Foundation, Thailand
NDMI	Normalized Difference Moisture Index (soil moisture)
NDRRMC	National Disaster Risk Reduction and Management Council, Philippines
NDSP	National Data Summary Page, Reserve Bank of India
NDVI	Normalized Difference Vegetation Index
NEH	North-Eastern Hill region, India
NEHU	North-Eastern Hill University, Shillong, Meghalaya
NEPED	Nagaland Environmental Protection and Economic Development project, India
NERCORMP	North-Eastern Region Community Resource Management Project, India
NERLIP	North-East Rural Livelihood Project, India
NESFAS	North-East Slow Food and Agrobiodiversity Society
NESRC	North-Eastern Social Research Centre, Guwahati, Assam
NGO	Non-governmental organization
NIC	National Informatics Centre, Meghalaya
NIE	National Institute of Education, Singapore
NIF	National Innovation Foundation, India
NIRD	National Institute of Rural Development, India
NIRDPR-NERC	The North-Eastern Regional Centre of the National Institute of Rural Development and Panchayati Raj, India
NITI Aayog	National Institution for Transforming India
NLD	National League for Democracy, Myanmar
NLUP	New Land-use Policy, Mizoram, northeast India
NOVOA	Native Okinawan Village and Omoro Arboretum
NPK	Nitrogen, phosphorus and potassium, plant nutrients
NPV	Net present value
NRC	National Research Council (US)
NRM	Natural resources management
NRSA	National Remote Sensing Agency, India
NSO	National Statistics Office, Thailand or Philippines

NSO	National Statistical Office, Papua New Guinea
NSDMA	Nagaland State Disaster Management Authority
NTFPs	Non-timber forest products
NUoL	National University of Laos
NVA	Net value added
OCHA	United Nations Office for the Coordination of Humanitarian Affairs
OKI	Ogan Komering Ilir district, which has the largest area of peatland in South Sumatra, Indonesia
OLI	Operational Land Imager (Landsat 8 sensor)
OM	Organic matter
ORSTOM	Former name of French government research organization. See IRD
OTOP	One Tambon One Product, Thailand
P3DM	Participatory three-dimensional modelling
PAR	Platform for Agrobiodiversity Research, Rome
PASD	*Pgaz K' Nyau* (Karen) Association for Sustainable Development
PCAIV	Principal Component Analysis with Instrumental Variables
PCAARRD	Philippine Council for Agriculture and Aquatic Resources Research and Development
PCAMRD	Philippines Council for Aquatic and Marine Research and Development, now part of PCAARRD
PCARRD	Former acronym for PCAARRD
PFAF	Plants for a Future (UK)
PFE	Permanent forest estate, a Myanmar government land designation
pH	A measure of acidity or alkalinity
PIR-Bun	People's Estate Plantation Company programme, Indonesia
PKMT	Prosperity Improvement Programme for Alien Society, Indonesia
PLUP	Participatory Land-use Plans
PNG	Papua New Guinea
PNGMASP	Papua New Guinea Mapping Agriculture Systems Project
PNGRIS	Papua New Guinea Resource Information System
POINT	Promotion of Indigenous and Nature Together, Myanmar
POU	Project Operational Unit
PRA	Participatory Rural Appraisal, Participatory Rapid Appraisal, Participatory Research Appraisal
PRB	Population Reference Bureau, Washington DC
PROSEA	Plant Resources of Southeast Asia, a project documenting plant resources, Wageningen, The Netherlands
PSA	Philippine Statistics Authority
PSB	Phosphate solubilizing bacteria (biofertilizers)
PTFPP	Palawan Tropical Forestry Protection Programme
PWDs	Person-work days
PWP	Permanent wilting percentage
QGIS	Geographical information system software
RCSD	Regional Center for Social Science and Sustainable Development, Thailand
RDCC1	Rural Development and Construction Company no. 1, Ministry of Agriculture and Rural Development, Vietnam
RECOFTC	Regional Community Forestry Training Centre, Bangkok

REDD/REDD+	Reducing Emissions from Deforestation and forest Degradation, plus conservation/sustainable management/enhancement of carbon stocks
RePPProT	Regional Physical Planning Programme for Transmigration (Indonesia)
RFD	Royal Forest Department, Thailand (now part of the Ministry of Natural Resources and Environment)
RMI	Indonesia Institute for Forest and Environment
RMUs	Resource Mapping Units, Papua New Guinea
RMV	Resilient Mountain Village, a joint initiative of ICIMOD and CEAPRED
RRI	Rights and Resources Institute, Washington, DC
RRtIP	Resource Rights for the Indigenous Peoples, Myanmar
RTBV	Rice tungro bacilliform virus
SALT	Sloping agricultural land technology
SAO	Subdistrict Administrative Organization, Thailand
SAVI	Soil Adjusted Vegetation Index (soil moisture)
SCBD	Secretariat of the Convention on Biological Diversity
SEACOW	School for Ecology, Agriculture and Community Works, Kathmandu
SEAMEO	Southeast Asian Ministers of Education Organization
SEARCA	Southeast Asian Regional Center for Graduate Study and Research in Agriculture
SELAF	*Société des Études Linguistiques et Anthropologiques de France*
SES	Socioecological systems
SFCP	Soil Fertility Conservation Project, Thailand
SFM	Sustainable forest management
SHG	Self-help group
SHK Kaltim	An Indonesian NGO
SHRF	Shan Human Rights Foundation
SHRF & SWAN	Shan Human Rights Foundation and Shan Women's Action Network
SI	Solidarités International, an NGO working in areas of conflict and natural disaster
SIDA	Swedish International Development Agency
SLEM	Sustainable Land and Ecosystem Management in Shifting Cultivation Areas of Nagaland for Ecological and Livelihood Security
SMC	Soil moisture content
SME	Small- and medium-sized enterprise, Thailand
SNV	*Stichting Nederlandse Vrijwilligers*, a Dutch NGO
SOC	Soil organic carbon
SPOT-VGT	Spot vegetation satellite imagery
SPSS	Statistical Package for Social Science (computing)
SRISTI	Society for Research and Initiatives for Sustainable Technologies and Institutions, India
SSRN	Social Science Research Network, an electronic journal
STEPS	Social, Technological and Environmental Pathways to Sustainability, hosted by Sussex University, UK
SWOT	Analysis of strengths, weaknesses, opportunities, and threats
TABI	The Agro-Biodiversity Initiative (project) of the Lao PDR
TBC	The Border Consortium
TEK	Traditional Ecological Knowledge
TeROPONG	An Indonesian NGO

TESP	Tropical Ecology Support Programme, Germany
TGHK	New Forest Plan Agreements, Indonesia
TM	Landsat Thematic Mapper
TNI	The Transnational Institute
TPRI	Tropical Pesticides Research Institute
TPTI	Indonesian state regulation requiring sustainable forest utilization through selective logging
UAA	Utilized agrarian area
UHDP	Upland Holistic Development Project, Thailand
UNDCP	United Nations Drug Control Programme
UNDP	United Nations Development Programme
UNEP	United Nations Environment Programme
UNESCO	United Nations Educational, Scientific and Cultural Organization
UNFCCC	United Nations Framework Convention on Climate Change
UNFPA	United Nations Population Fund
UNPFII	United Nations Permanent Forum on Indigenous Issues
UNHCR	United Nations High Commissioner for Refugees
UN-REDD	The United Nations Collaborative Programme on Reducing Emissions from Deforestation and forest Degradation in Developing Countries, Geneva
UNU	United Nations University
UNWMP	Upper Nan Watershed Management Project, Thailand
UPCSC	University of the Philippines Cordillera Studies Center
UPLB	University of the Philippines at Los Baños
USAID	United States Agency for International Development
USAID-IUWASH	United States Agency for International Development – Indonesia Urban Water, Sanitation and Hygiene project
USDA	United States Department of Agriculture
VC	Village council
VDBs	Village Development Boards, Nagaland
VDCs	Village Development Committees, Meghalaya
VFV	Vacant, fallow or virgin land, a classification for swidden land in Myanmar
VMIs	Vegetation moisture indices
WaNuLCAS	A model of water, nutrient and light capture in agroforestry systems
WCMC	World Conservation Monitoring Centre
WWF	World Wildlife Fund

William Cairns
(12 October, 1864 - 25 June, 1932)

DEDICATION

This is the third book in a planned trilogy of volumes devoted to shifting cultivation in the Asia-Pacific region. As with the previous volumes, in this book, the Editor continues his tradition of dedicating the volume to a male ancestor from his patrilineage.

This last volume in the trilogy is dedicated to William Cairns (1864-1932) – who was the Editor's paternal great grandfather (see photo at the top of this page). In reaching so far back into the mists of time, there are to be found neither many photographic images, nor even personal stories relating to this gentleman, that are very clear.[1] William (1864-1932) was the fourth generation of this particular Cairns lineage to live on Prince Edward Island (PEI), Canada, after his ancestors had emigrated from Scotland.

The first Cairns ancestor to venture from Dumfriesshire, Scotland, to Prince Edward Island, Canada, was John Glen Cairns (1788-1871), voyaging aboard the *Isabella* (Jardine, 1985), and arriving on Island shores in 1832 (see Map 1). Travelling with John Glen Cairns was his son, William Cairns (1816-1856).[2] This William Cairns (1816-1856) was to later marry Ellen Stewart (1816-1878)(see Box 1).

---

[1] Since William Cairns (1864-1932) left this world 91 years ago, few people alive today knew him personally and can recall experiences with him. The Editor's father knew him (see Plate 14) but was not yet four years old when his grandfather died.

[2] It can be confusing to follow the Cairns genealogy because, for many generations, the family held tightly to the Scottish tradition of naming the first-born son after his paternal grandfather – and so the same names keep repeating throughout the family's history. To counter such confusion, the years of the lifespan of the ancestor in question are indicated in brackets – to distinguish him from others having the same name, but who lived at different times.

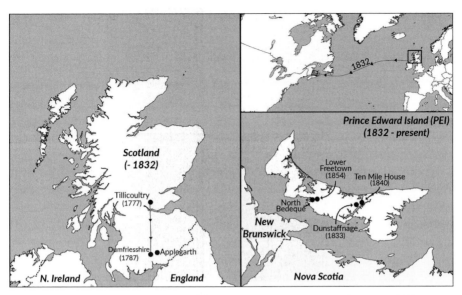

MAP 1: The migration route that brought the John Glen Cairns lineage of the Cairns clan from Scotland to Prince Edward Island, Canada.

BOX 1: Quick reference to the Cairns patrilineage

John Glen Cairns* (1788-1871)	married Jean Richardson
	Elizabeth McCaull
William Cairns (1816-1856)	married Ellen Stewart
John Cairns (1841-1899)	married Maria Crawford
William Cairns (1864-1932)	married Thirza Picketts
John Scott Cairns (1895-1973)	married Georgie Cairns
William Cecil Cairns (1928-)	married Helen Pendleton

* John Glen Cairns was a widower at the time that he journeyed to Prince Edward Island, but he married again on March 24, 1853.

In 1854, this same William Cairns (1816-1856) first bought land (100 acres purchased for the sum of £60) in Lower Freetown, and his family put down roots there (see Maps 2 and 3). This land was bought from William's uncle, Robert Cairns Sr. [1784-1878] (To be clear, this William Cairns [1816-1856] who first purchased land in Lower Freetown, was actually the paternal grandfather of the later William [1864-1932] to whom this book is dedicated.)

As already noted, the aforementioned Robert Cairns Sr. (1784–1878), who sold this land, was not one of the absentee landlords living in the British Isles, to whom the subdivided Island (see Map 2) had been granted. Rather, this Robert Cairns (1784–1878) was a brother of John Glen Cairns (1788-1871), and so was an uncle to young William Cairns (1816-1856). John Glen Cairns had been the pioneer in 'breaking a path' to Prince Edward Island in 1832 (see Map 1), and his brother, Robert, had followed that same path to PEI in 1840.[3]

---

[3] This Robert (1784–1878) was the great grandfather of the R.L. Cairns (1899-1978) shown ploughing his field in Plate 2.

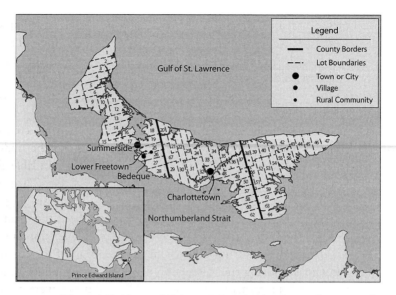

MAP 2: The subdivision of Prince Edward Island into lots, initially controlled by absentee landlords in the British Isles.

Source: Adapted from Blakeley and Vernon, 1963.

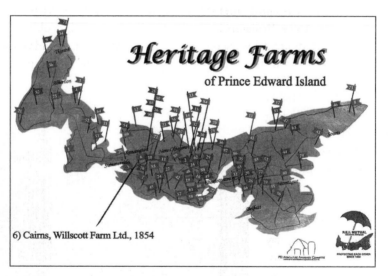

MAP 3: Heritage farms[4] of Prince Edward Island (see #6).

Source: PEI Agriculture Awareness Committee.

---

[4] Hertitage Farms" (see Map 3) are defined as farms that have continued to provide an active livelihood to members of the same family for the past 150 years (since 1864). Although the farm was begun in 1854, as explained later, the name 'Willscott Farm' wasn't coined until the early 1940s, when the farm began to milk purebred Holsteins.

Note that the oldest farms tend to be concentrated nearby PEI's main harbours, because the early settlers were all arriving by sea. Similarly, the main towns – Summerside and Charlottetown – were established on the Island's main harbours.

There was a particular reason why some of the Cairns immigrants ended up settling in Freetown (as opposed to elsewhere on the Island) at the time they did. Unlike most of the rest of North America, the colony of Prince Edward Island had been established with a leasehold system of land tenure. In the 1760s, the Island was surveyed into 67 lots (or townships) of approximately 20,000 acres each (Map 2) (Blakeley and Vernon, 1963).

These lots were subsequently granted to important and well-connected people in the British Isles, who became the proprietors of these tracts. By the mid-1800s, many Island farmers were tenants and there was very little land available for sale to individual farmers, even if they could afford to buy it. When the Cairns settlers came from Scotland, they brought some money with them from the old country. (Robert Cairns Sr.'s wife, Mary Carruthers, is thought to have been the main source (Cairns, G. A., 1999)). The Cairns arrivals were interested in owning, not renting, the land they farmed, and in also ensuring that the next generation of sons could own their own farms freehold. During the 1840s and '50s, when the younger generation of Cairns arrivals was seeking farms, sections of Lot (Township) 25 were coming on the market in parcels that consisted of viable-sized farm units. In fact, this area of Lot 25 became known as Freetown because the land in that section of the township was available as freehold. Because land in Freetown could actually be bought and owned by farmers, Cairns family members settled in that area (see Map 3).

In that same year of 1854, William Cairns (1816-1856) added to his holdings by buying another 102 acres that adjoined the first property. This second block of land was purchased for the sum of £86.10 from the Estate of Gilbert Henderson of Liverpool, England (Cairns, R. L., 1971). This gave William Cairns (1816–1856) a sizeable land holding of 202 acres. Since 100 acres was the usual viable farm size of the time (and probably the maximum amount that one farmer could actually farm, given the techniques and machinery of the day (see Plates 2 and 3 for examples of labour-intensive technologies)), we may surmise that William may have acquired the second hundred acres not so much for his own use, but to ensure there would be a farm available for more than one son, when the time came for such a need. These

PLATE 2:   Before the arrival of tractors (1920s to 1950s), heavy draught horses provided traction on Prince Edward Island's farms. Pictured above (ca 1950), R. Louis Cairns (1899-1978) ploughs one of his back fields that is now part of Willscott Farm, using a gang plow with a three-point hitch.

Photo: Mary E. Cairns.
Source: Public Archives of Prince Edward Island.

people thought in the long term. After all, they had most likely left the old country because it offered no opportunities for owning their own land. William (1816-1856) had been on PEI for 22 years before he realized his dream of owning his own farm, so he may have been quick to grab the opportunity to ensure that his two sons were looked after when an opportunity arose, to make sure there were farms for both of them long before there was an actual need.

PLATE 3: Like their Asian counterparts, PEI farmers also joined together in cooperative work groups to accomplish labour-intensive jobs. This group of Freetown farmers had joined together to saw firewood into stove-length blocks, sometime before September, 1930.

Source: Cairns family photo album.

Of course, in the story of the Freetown land acquisition, a big unknown is whether any of the land had been cleared and farmed (possibly even with buildings erected) before 1854, or whether William was arriving on virgin forest.[5] Another unknown is whether some or all of this land might have been cleared by fire. There is an old story that, in the early days, an extensive tract of forest in the Freetown area had been burnt-over by an out-of-control forest fire and the district was subsequently known to early settlers as 'the burnt lands' (Cairns, G.A., 1999). Given the subject matter of this book, it seems particularly relevant to mention this. We have no indication of whether the land occupied by William Cairns (1816-1856) was part of the burnt-over land.

After consolidating this land base, William was to farm it for only two years before going the way of his ancestors in 1856, at the age of only 40. In 1926, 72 years after the original purchase, his grandson, also named Willam Cairns (1864-1932) – the man to whom this book is dedicated – continued to build on the farm's land holdings by buying adjacent properties owned by Vernon Malone and a man named MacDonald. This type of expansion seemed to happen about every second generation, when the farm was being managed by a William namesake. Many years later, the Editor's own father, William Cairns (1928-), joined together with his father, J. Scott Cairns (1895-1973), in buying the adjacent properties of Basil Taylor (1960) and R. Louis Cairns (1970). A third adjacent farm, that of Howard

---

[5] Information in the 1861 census (Prince Edward Island Census, 1861) raises the strong possibility that at least part of William's (1816-1856) landholding was cleared at the time of purchase. The census indicated that 140 of the 202 acres were arable, an unexpectedly large quantity of land to have cleared from virgin forest and brought into cultivation in just seven years and a substantially higher clearance rate than that reported from neighouring farms settled at approximately the same time. Note that William had been dead for five of those seven years and that his two sons were aged 15 and 8 at the time of his death. Some interesting aspects of land use on the farm at that time have recently been analyzed in a paper by MacFadyen and Maddison-MacFadyen (2022).

Cairns, was purchased by William (1928-) in 1987. Today, Willscott Farm stands at about 500 acres and grows most of the feed needs for its milking herd of about 90 purebred Holsteins plus young stock (see Plates 7 and 8). The farm ships fluid milk to Amalgamated Dairies Ltd. (ADL) in Summerside. It was this herd of Holsteins that has been the main supporter of this entire trilogy of books on shifting cultivation in Asia-Pacific.

Although there is much that we don't know about the second William Cairns (1864-1932) - to whose memory this volume is dedicated - we do know that he was born in 1864 in Lower Freetown, Prince Edward Island, as the eldest child (see Plate 4) of John (1841-1899) and Maria Cairns (nee Crawford). (As far as we are aware, William didn't seem to have a middle name, as was often the custom up until about the 1850s or later.) This second William (1864-1932) was the great grandson of the John Glen Cairns[6] (1788-1871) — who was the first Cairns family member to emigrate from Scotland to PEI in 1832 (see Map 1).

PLATE 4: John and Maria Cairns raised a large family of 11 children. Their first-born was William (1864-1932) - standing in the middle of the back row in this rare family portrait. The photo was probably taken around 1890.

---

[6] Although known to later generations of the family as 'John Glen Cairns', his middle name may have been added informally later to distinguish him from others with the same name. 'Glen' doesn't appear on any official documents referring to him, or on his gravestone. Somewhere along the line (probably well into the 20th century and certainly long after he was dead) the family started referring to him as 'John Glen', and that continues to be the case. Although unofficial, it is certainly convenient to have 'Glen' tacked onto his name, as it so readily distinguishes him from all the other people also named John Cairns.

John and Maria Cairns parented a large family. After William was born, he was followed by nine girls in a row, before he finally gained a brother, Elton, 20 years later. So both the oldest and youngest children were boys, with nine girls born in between (see Plate 4). It would be logical to speculate that, in those days, male children were probably hoped for, both to carry on the family name, and to help with the farm work.[7] This is probably not different from the perspective of many of the farmers discussed in this volume.

In 1894, at the age of 30, William married Thirza Picketts (see Plate 5) from up on the Blue Shank Road, only a few miles away, 'as the crow flies' across the fields and the Wilmot River. Their first child was J. Scott Cairns (1895-1973), the Editor's grandfather and the man to whom volume II was dedicated (see Plates 12 and 13). And, in turn, the first-born (and only) son of J. Scott Cairns was the Editor's own father – the man to whom volume I was dedicated (see Plates 6 and 7). True to Scottish tradition, he was also named William (Cecil) Cairns (1928-), after his paternal grandfather.

In about 1900, when Scott was five years old, this young family splintered from the main family homestead and moved a quarter of a mile up the road to settle a new farm that they had been clearing from the forest. This new farm was to become Willscott Farm (Barde, 1977), the same ancestral farm on which the Editor grew up (Plate 8).

After Scott was born in 1895, he was quickly followed by siblings Ida May (1896), William Cecil (1898), and Amy Belle (1903) (see Plate 13 for a photo of the three surviving siblings in their later years). Tragedy struck the family in 1911, when young William Cecil (1898-1911) succumbed to a deadly combination of measles and scarlet fever and passed away at the age of 12. It was the senior William's (1864-1932) habit to keep a daily diary; later readers were to note that after young William Cecil's death, his father's journal entries took on the distinctly sad tone of a mourning parent (Cairns, W. C., personal communication, 2014).

The fact that the Editor spent his childhood wandering this land, and that so many generations of his ancestors have farmed it, have given the

PLATE 5: This photo would have been taken around 1897. William (1864-1932) and Thirza (1868-1931) Cairns must have been kept exceptionally busy, starting a new family and a new farm in unison. The toddler sitting on the table was their first-born, J. Scott Cairns (1895-1973), the Editor's grandfather and the man to whom volume II was dedicated.

---

[7] The ability of John and Maria's daughters to contribute to farm work should not be underestimated, however. The Editor's father recalls that at least some of them had a reputation for being very skilled at binding and stooking grain, in the days before it was done mechanically.

PLATE 6: William Cairns (1928–) flashes a grin at the camera whilst milking his cows, probably around 1958. William recalls that they started milking cows by machine around 1942 – but electricity didn't arrive until 1949. In those early years, before electricity was available, the milking machine was powered by an air-cooled Briggs & Stratton petrol engine. Today (2023), Willscott Holsteins are milked by a DeLaval robot, backed up by a 100 kw, 5.9 litre turbo-charged Cummins diesel generator that automatically cuts in when there is an interruption in the electricity. Through the years, it was mainly labour constraints and the economics of specialization that pushed the farm to move away from mixed farming and increasingly focus on dairy production.

PLATE 7: Probably taken in the late 1970s, William Cairns (1928–) pilots his Massey Ferguson 410 combine, whilst harvesting a mixed stand of barley and oats (often undersown with a red clover mix that provided a fodder crop the following year). The growing season is relatively short on Prince Edward Island, and farmers have to work hard to get their crops harvested before they become covered with snow. This means that grain is often combined on the cusp of winter and in the era before heated cabs, dressing warmly was a must.

BOX 2: How Willscott Farm got its name

The farm name had its origins in the early 1940s when Scott (1895-1973) and William (1928-) began adding purebred Holsteins to their barns. William (1928-) explains that their entry into dairy farming began when they bought a pregnant Holstein heifer from the Profitt family down the road, for what then seemed the astronomical price of $250. This heifer gave birth to a heifer calf and, since she was purebred, they wanted to register her with the Holstein Friesian Association of Canada.

The registration papers called for a farm name, that is used as a prefix in the purebred animal's name. As William (1928-) and Scott (1895-1973) puzzled over what to call their farm, another family member who was in the room and listening to the conversation, suggested the name of 'Willscott Farm', as an amalgamation of the names of the father and son team that was managing the farm. When registering the official name of an animal, the prefix 'Willscott' would then be followed by a given name that the farmer would generally use in recognizing the animal.

The prefix that preceded the given name of a registered animal would tell any outsider the farm on which the animal was bred. If that farm was on official test and had a reputation for high milk production, then the farm prefix that appeared on a registered animal's registration paper would provide a strong indication of the cow's likely milking ability.[8] If an animal was to be sold in an auction, a farm prefix recognized for good breeding would generally ensure that it commanded a premium price.

So it was when Scott and William were filling in the paperwork to register their first purebred Holstein calf in the early 1940s, that they first encountered the need to have a name for their farm. It was as thus that they began the foundation for their dairy herd, and that Willscott Farm got its name in the process.

---

farm that William and Thirza Cairns began,[9] a near spiritual importance to the Editor, far beyond its economic value. Throughout the Editor's career, he has always felt that his special connection to this land gave him a very personal understanding when the indigenous people with whom he worked described their strong attachments to their ancestral land. The Editor also feels this and knows that his father does too. He has followed in the footsteps of many of his ancestors in investing his life in that soil. Many parts of the landscape are connected with stories that have been passed down through the generations. During a phone call just last week, Dad had recalled that his own father (see dedication for volume II) had ploughed a corn field the day before he died in 1973 and such was his love of tractor work, had spoken of his wish that he had another field to plough. He clearly wasn't ready to leave the land that he had farmed his entire life.

---

[8] There was no guarantee of this, of course, because milk production is not only a function of genetics, but also of management. But if the right genetics are present, then the cow will have the capacity to respond to improved management and nutrition by putting more milk in the pail.

[9] With then three Cairns farms working side-by-side, this road eventually became known as the Cairns Road (Route #110), that runs perpendicularly between the Freetown Road (Route #109) and the Blue Shank Road (Route #107) (see Plate 9). A third Cairns farm was in operation from 1854 to 1930, adjacent to and just south of what became Willscott Farm. This farm had been owned by descendants of John Glen's brother, Robert [1784-1878] (first by Robert Jr. and then by Robert Jr.'s son, Alexander). It was operated by Wendell Bernard during the Editor's youth, and continues to be managed today by Wendell's son, Mark.

PLATE 8: It is now over a century since William (1864–1932) and Thirza (1868–1931) Cairns moved northward up the road in 1900 to settle a new homestead. They would probably shake their heads in amazement if they could see how Willscott Farm has developed under the care and hard work of their descendants. This has been the ancestral land of this lineage of the Cairns clan since shortly after their arrival from Scotland. By now, seven generations of the Editor's family have belonged to this land, several of whom have invested their entire working lives there, and eventually dying there. Today (2023), Willscott Farm is capably managed by James Cairns and his wife, Janet (nee Beaton). This aerial photo was taken in 2015.

Source: Bird's View Photos, Montreal.

PLATE 9: Although the number of Cairns members living on Route #110 are now far fewer than in former times, this roadside sign still harkens back to the time that the original Cairns settlers bought land in this area in 1854, and began to turn forests into fields.

PLATE 10: Thirza Cairns (1868–1931) seems lost in thought as she sits on a doorstep, probably sometime in the 1920s. Although relatively small in stature, Thirza was known as a very capable woman.

In the early 1930s, William and Thirza's family began to run into more hard times. The first casualty was William's beloved wife, Thirza (see Plate 10). In those times, preparations to wash clothes were made by first boiling water, and then pouring it into a large washing tub, with a tap at the bottom for releasing dirty water after the clothes had already been washed. In 1931, in an absurdly simple accident, Thirza poured boiling water into such a washing tub without noticing that the tap was open. This meant that the boiling water immediately exited the tap and landed on Thirza's feet, positioned beneath. Badly scalded, poor Thirza took to her bed to try to recover. But without the antibiotics that we have so easily available today, she developed pneumonia and passed away within a few days (see Plate 11).

In the last few years of his life, William (1864–1932) was afflicted by some kind of muscular dystrophy that began to render him increasingly handicapped, and unable to participate in farm work. His condition grew so debilitating that, after Thirza passed away suddenly in 1931, his daughter-in-law, Mrs. J. Scott Cairns (see Plate 12), had to feed him most of his meals until his own death came suddenly the following year, in 1932 (Plate 11). The Editor's grandfather, J. Scott Cairns (1895–1973), was left as the only adult male remaining in the family (see Plate 13), and had to learn responsibility and hard work more than most.[10] He carried an exceptionally heavy load in caring for his ailing father while he was still alive, running the farm, and parenting his own

PLATE 11: Final resting place of William and Thirza Cairns in the North Bedeque cemetery, PEI.

---

[10] See the dedication of volume II for more about J. Scott Cairns [1895–1973] (Cairns, M. F., 2017).

PLATE 12: This photo would have been taken around 1944. Like his own grandfather, J. Scott Cairns (1895-1973) began fatherhood with a son, before welcoming a string of successive daughters into his family. As the only son of a farmer and before the age of heavy mechanization, William (1928-) would have shouldered a heavy workload.

young brood (see Plate 12). His wife was to later add an explanatory footnote to Scott's diary:

In the diaries, Scott never mentions that he looked after his father day and night for months. For 13 months from Grandma's death until Grandpa passed away, I fed him

PLATE 13: By then seniors themselves, the three surviving offspring of William and Thirza Picketts were photographed at some kind of social gathering, probably around 1962. They are (from left to right), Amy Pearson, J. Scott Cairns and Ida Sanborn.

PLATE 14: This photo was likely taken in the fall of 1930. It shows the three generations of male ancestors to whom this trilogy of books has been dedicated. A young William Cairns [1928-] (volume I) sits on his father's (J. Scott Cairns [1895-1973]) (volume II) lap. Young William's paternal grandfather [1864-1932] (volume III), also named William, stands behind, to the far right.

most meals. Scott's sister, Amy (Mrs. Earle Pearson) and Grandpa's sisters visited us often and were a great help. Scott's other sister, Ida (Mrs. Eber Sanborn), lived in western Canada. His only brother, Cecil, died at age 12 (measles and scarlet fever).

Grandpa could not move a hand or foot, but never complained and appreciated everything done for him. (Mrs. J. Scott Cairns [Georgie], January 22, 1979)

We know very few personal stories about William (1864-1932) except that he was apparently fond of keeping peppermints in his pockets – a habit that his eldest son, Scott (1895-1973), was to continue throughout his own life. William was also reported to be a rather tall man, and it was undoubtedly from him that the Editor's father inherited his own considerable height (see Plate 12).

William Cairns (1864-1932) was a pioneering farmer in a new land. His legacy was in laying the foundations for an exceptionally fine farm (Plate 8) and in descending a lineage of particularly capable farmers[11] to manage it over successive generations. They are the outstanding farmers to whom this trilogy of books is dedicated (Plate 14).

---

[11] The Editor's own father's contributions to agriculture were formerly recognized in October of 2011 when he was inducted into the Atlantic Agricultural Hall of Fame (Dalhousie University, 2011).

Acknowledgements

I am grateful to my two key informants in preparing this dedication, my father, William Cairns (1928-), and Winifred Wake (nee Cairns). William is a descendant of the first Cairns member, John Glen Cairns (1788-1871), who arrived on PEI's shores in 1832, and Winifred is descended from John Glen's brother, Robert Cairns Sr. (1784-1878), who followed in 1840. Both are experts in our family's genealogy and I thank them both for sharing their time and expertise in helping me develop this dedication to my great grandfather. Warm thanks are also owed to my mother, the late Helen Ann Cairns (1933-2017), for allowing me to raid her photo albums, and even framed photos hanging on her parlour wall, for the purpose of scanning them for inclusion in this dedication.

References

Barde, B. (Producer) (1977) 'Like no other place, P.E.I.: The million acre farm' (film), *TV Ontario*, The Ontario Communications Authority

Blakeley, P. R. and Vernon, M. C. (1963) *The Story of Prince Edward Island*, J. M. Dent & Sons (Canada) Limited

Cairns, G. (1979) A hand-written note later added to her husband's 1932 diary by Mrs J. Scott Cairns.

Cairns, G. A. (1999) *Memories of My Life and Times*, privately published

Cairns, M. F. (ed.)(2017) *Shifting Cultivation Policies: Balancing Environmental and Social Sustainability*, CAB International, Wallingford, UK

Cairns R. L. (1971) *A History of Lower Freetown, Prince Edward Island*, Centennial project, Lower Freetown Women's Institute, Prince Edward Island

Cairns, W. C. (2014) Personal communication between the Editor and his father

Dalhousie University (2011) *Atlantic Agricultural Hall of Fame, Inductee: William Cecil Cairns*, Dalhousie University, Halifax, Nova Scotia, available at https://www.dal.ca/diff/aahf/inductees/william-cairns.html#:~:text=Nominated%20by%20the%20Prince%20Edward,Atlantic%20Agricultural%20Hall%20of%20Fame, accessed 8 April 2022

Jardine, R. (1985) *Freetown, Past and Present*, Freetown Historical Society, Williams and Crue, Summerside, Prince Edward Island

MacFadyen, J. and Maddison-MacFadyen, M. (2022) *Energy on the John Cairns Jr. and John Glen Cairns Sr. Farms and Lot 25, Prince County, Prince Edward Island, in 1861*, unpublished draft

Prince Edward Island Census (1861) Records found in search of microfilmed archives, Provincial Archives and Records Office, Charlottetown, Prince Edward Island, mid-1990s

FOREWORD

*Tony Simons**

Shifting cultivation or swidden agriculture is a 'catch all' term for many different types of alternating land use on the same parcel of land, over time. Typically, cycles of perennials and annuals are involved with the objectives of restoring land productivity and ecological functioning, thus underlining the strong links between agroforestry and swidden agriculture. Whilst not significant in global food production that is marketed, and almost completely undocumented in global annual statistics, shifting cultivation is significant for subsistence food production for about 250 million people worldwide. Amongst research and policy audiences, it tends to evoke either strong support or attract much criticism. Many supporters of shifting cultivators often romanticize the practice and the deep indigenous knowledge surrounding it. Detractors cite it as an antithesis to modern agriculture, perhaps because of a bias that agricultural landscapes need to be homogenized, rather than understanding and respecting the local context of many of the practices and innovations that shifting cultivators have developed.

Interestingly, at the digital visibility level, although 'swidden agriculture' does not even have enough popularity on the internet to warrant a Google Trends analysis, many aspects of swidden agriculture are synonymous with the recent and more popular terms of agroecology, regenerative agriculture and nature-based solutions – although one of these focuses more on crop rotations and the other on plot rotations. For the swidden-agriculture cognoscenti, including land stewards, forest dwellers, scientists and development practitioners, this may seem both enigmatic and rewarding. In this regard, and with hindsight, the authors and editors may wish to have been more deliberate in making these linkages to widen the popularity of shifting cultivation and its history, practices and future.

Whatever the readers' views and desires may be, to update their knowledge and perspectives on shifting cultivation this third volume in the series is the most comprehensive tome of the topic available anywhere. Malcolm Cairns and his co-editors have nurtured a long-standing practice into a very contemporary narrative. With 276 specialist authors sharing their theories, evidence and insights in 148 chapters and syntheses in this volume and its extensive supplementary collection, covering more than 20 Asia-Pacific countries – what else can hold a candle to this impressive body of work? Given the breadth and depth of this 1100-page epic and its

---

* PROFESSOR TONY SIMONS is Executive Director, Centre for International Forestry Research-World Agroforestry (CIFOR-ICRAF) and Director-General, World Agroforestry (ICRAF), Nairobi, Kenya.

supplementary chapters, it is understandable that a full glossary was not included. It would also be fitting if an enlightened donor or foundation might want to support open access to each of the individual chapters as well as the 27 'overview' synthesis papers.

The editors and authors are targeting a broad collection of audiences with this superb work. It seeks to appeal to readers interested in geography, vegetation cover, ethnicity, farming systems, agricultural ecology and political influences on shifting cultivation. It seeks not only to document much that is unpublished, but also to give syntheses of existing studies where extensive datasets and metadata are not readily accessible. As a collection of thematic and country-based chapters it is also bold in its framing of conundrums and in asking challenging questions. Questions of sustainability, productivity, social inclusion and equity, good governance and profitability. It is interesting to note the greater attention paid to income-generating activities of shifting cultivators, although this is largely limited to agricultural produce, rather than ecosystem services. However, it would be incorrect to conclude a radical shift in cultivators seeking to primarily monetize the landscapes they occupy – in stark contrast to many unsustainable logging operations and unplanned forest conversions, which typically seek to extract financial value and liquidate natural capital, rather than demonstrate any green economy or sustainability motives.

When, in 1993, World Agroforestry (ICRAF) convened the Alternatives to Slash-and-Burn (ASB) programme throughout the developing tropics, it was guided by a simple narrative involving the need to eliminate burning in shifting cultivation and to improve fallow practices. Data and analyses presented in this volume show many comparative advantages of shifting cultivation over modern agriculture practices. Interestingly, the chapter authors avoided use of the words 'burn' or 'burning', with only three mentions in the titles. They greatly preferred the term 'swidden', with 32 title-word mentions. This is very timely with the current intense international focus on the broken global food system, the foolishness of avoiding complexity in food production and the link between changing land use and management to the global climate crisis. Notwithstanding the historical and contemporary relevance of the book, it is fascinating and intriguing to read about the latest gender analyses and perspectives.

Perhaps one gap in this impressive body of work is the overall light treatment of geospatial research. The general lack of geo-references in the chapters not only seems asymmetric, but also misses capitalizing on the tremendous interest in digital analytics. However, this also offers promise to the geospatially inclined to connect more with the authors in this book and use the location-based work as a living laboratory for this important topic and other related topics.

Underlying the complexity and diversity of the topics and farming systems covered in the book is the complexity of the subject. Perhaps counter-intuitive is that shifting agriculture is relatively knowledge-intensive and context specific. This presents difficulties in research design, appeal to funding agencies and comparative analyses. It also calls for a shifting cultivation data repository to ensure that knowledge and ideas are not lost.

One highly promising new area related to shifting cultivation that CIFOR-ICRAF and partners are working on in the Pacific (Papua New Guinea), South America (Peru and Brazil), Africa (Cote d'Ivoire) and Asia (Sri Lanka, India and Indonesia), is shifts in cultivation inside long-colonized plots – even deploying cash crops such as oil palm, rubber, fruits, cocoa, and vanilla, as well as timber trees in temporal and spatial mixes. Whilst collectively such enterprises are unlikely to displace the more prevalent food, fibre, fodder, fuel and medicinal local species, they do offer new livelihood options for shifting cultivators.

An important question that this book evokes is 'what will shifting cultivation look like in 2050, as populations, urbanization and atmospheric greenhouses gases continue to rise?' Sadly, it is not a question we see raised in high-level United Nations conventions, nor in related national submissions to such conventions, including Nationally Determined Contributions (NDCs), Land Degradation Target-Setting Programmes (LDTSPs) and National Biodiversity Strategy and Action Plans (NBSAPs).

The prevalence of shifting cultivation is apparently a complex function of location, historical occurrence, biodiversity, ecosystem functioning, social institutions, social equity and political acceptance, among other parameters. Its justified recognition will likely only come when it is taken into account more in land-use, demographic, economic and social analyses. This book will be a highly influential resource when that day hopefully eventuates. Although the documented knowledge presented here is infinitesimally small compared to the indigenous knowledge of land stewards in Asia-Pacific, it greatly expands our formal knowledge on this topic.

Finally, we salute the editors and the authors and their institutions for producing this *opus magnum* and look forward to a spike in interest, and with it an increased appreciation of, and elevated priorities for, shifting cultivation – not just for its 250 million practitioners, but also for everyone in the world who benefits from their amazing stewardship of the land and its associated knowledge.

PREFACE

*Malcolm Cairns**

The origin of this series on shifting cultivation

It is with considerable relief that I sit down to write the preface for this third volume in a planned trilogy on shifting cultivation in the Asia-Pacific region. I had not been at all sure if my health would hold up long enough to complete this ambitious project. For those readers who have followed this series from the beginning, you may be aware that these volumes had their genesis in the near death of the Editor. A devastating stroke in 2008 left him lying half-paralyzed and half-conscious on his condo floor for the next four days, before he was found and taken to a hospital for medical attention. I still remember the sinking feeling in the pit of my stomach as a doctor later showed me a CT scan of my brain and outlined the large black area that had died during my cerebral haemorrhage. If brain function is used as a measure of life, then I had only half-survived that stroke. The other half had died on the floor, alone. I realized that life, as I knew it, was over. But each of us has no choice but to play the hand of cards that we are dealt. We all have challenges to overcome. Left paralyzed on my left side, I had to accept that my days of field work were finished for the time being, and that I needed to find another way to contribute to my chosen field. It was from these inauspicious beginnings that this series on shifting cultivation in the Asia-Pacific region was born.

In fact, the project began with the intent of producing a single volume – but the response to our call for papers was so overwhelming, that it soon developed into a more ambitious series of volumes that encompassed the main themes of the papers that had been proposed to us. The thematic focus of each of these volumes in the

---

* DR. MALCOLM CAIRNS is the Editor of this series of volumes. His academic background ranges from Animal Science through a Master's Degree in Environmental Studies, to a Doctorate in Anthropology from the Australian National University, awarded for a study of the cultural ecology of the Angami Nagas in northeast India (Cairns, 2013). The first volume bearing his name as Editor was *Voices from the Forest: Integrating Indigenous Knowledge into Sustainable Upland Farming* (2007, RFF Press, Washington, DC). It arose from a ground-breaking conference held at Bogor, Indonesia in 1997, and marked his publishing debut in a scientific field that has become his signature and passion. This volume is the third in a planned series that he is editing. The first, *Shifting Cultivation and Environmental Change: Indigenous People, Agriculture and Forest Conservation* was published by Earthscan in January, 2015. The second, *Shifting Cultivation Policies: Balancing Environmental and Social Sustainability*, was published by CABI in 2017. Dr. Cairns is a freelance researcher who was most recently a Fellow at the Centre for Southeast Asian Studies (CSEAS), Kyoto University, Kyoto, Japan.

trilogy follow a logical sequence. Volume I began by presenting an overwhelming tsunami of evidence that shifting cultivation is not nearly the environmental threat that its detractors would have us believe (Cairns, 2015a). Volume II built on this more favourable assessment by asking, '*Given that shifting cultivation is not the dire threat that many had thought – and that it is still of immense importance to the livelihoods of hundreds of millions of mostly indigenous peoples – what does this mean in terms of the need for policy reform?*' (Cairns, 2017). This third volume of the trilogy acknowledges that shifting cultivation supports an important sector of the rural population in Asia-Pacific, and posits, '*We accept that shifting cultivation is immensely important and the adversarial attitude towards its practitioners needs to be reversed to an attitude of support. Given that, are there ways that the system can be improved? Rather than try to impose top-down interventions on shifting cultivators* (see, for example, Enters and Lee, ch. 7, this volume)*, have some pockets of shifting cultivators themselves developed useful innovations that have potential for wider dissemination?*' (Cairns, this volume). Taken together, the three volumes guide the reader through an odyssey of learning about shifting cultivation in the Asia-Pacific region that attempts to correct a century of misunderstanding. To increase our efficiency, we've worked on these volumes much as a farmer relay-plants crops in his field, 'sowing the seeds' for the next volume, while the present work was still in progress.

Several years after my mishap with the ruptured blood vessel in my brain, I had the pleasure of meeting a highly respected colleague from the Philippines. I hadn't met Professor Percy Sajise[1] for many years, since the old days when we used to meet at SUAN[2] conferences, and he had come to Chiang Mai for meetings. Picking Professor Sajise up for dinner one evening, we were catching up on the way to the restaurant and discussing Volume I in the shifting cultivation series that was coming off-press at about that time. To explain my poor mobility and dependence on a cane, I also briefed Percy on my narrow escape from the stroke. After listening quietly to my tale, Professor Sajise exclaimed, '*Malcolm, it's amazing that you survived that! God must really have a plan for you!*' I recalled that a Naga missionary friend had said virtually the same thing. '*Well, if he does*', I replied, '*I wish that he would tell me what it is!*' '*You are already doing it!*' Percy had assured me.

I'm not sure of Percy's reading that our work on this shifting cultivation series has had divine direction, but it is true that my team members and I have devoted our lives to this work since beginning the project. Maybe Percy was right and it was our calling. Some of us have worked without pay. Each volume demanded a significant chunk from each of our lives, usually two to four years. It consumed us, as we spared no effort to develop the volumes to their highest potential. This became my life's work, something I wanted to leave behind as a legacy. When the road to completing this series seemed impossibly long and full of potholes, it was comforting to remember Percy's words. In fact, I oftentimes mused that I wouldn't have been able to complete this work without the support of two loving fathers – one in Heaven and one on Prince Edward Island. I was blessed that both had my back. To them, full credit for this work is owed.

In the midst of a climate crisis, does it make sense to continue studying shifting cultivation?

With the weather becoming alarmingly erratic and daily newscasts broadcasting increasingly shrill warnings about the need to curtail any further carbon emissions into the atmosphere, it might seem like an odd time to be publishing a volume that looks favourably upon shifting cultivation. There is a sense that we are akin to lemmings thundering dangerously close to the 1.5° Celsius abyss and likely to plunge into nothingness within our next few strides.

And yet, hundreds of millions of the world's poor continue to rely on some form of shifting cultivation for their survival. Long experience has shown that state prohibitions against shifting cultivation have scarcely worked (see Volume II), and alternative land-use systems introduced as replacements to shifting cultivation have brought their own sets of problems. Therefore, if shifting cultivation is going to remain as a bedrock of subsistence for indigenous communities living in the forest margins, the most effective strategy is probably not a continuation of the last century or so of efforts to legislate shifting cultivation out of existence, but rather, to learn from past failures and make a 180-degrees u-turn so that we can work closely with shifting cultivators in identifying ways that they can improve their traditional methods of subsistence farming. That is essentially the starting point of this volume.

A personal experience that taught me the importance of indigenous knowledge

I have to smile with some embarrassment as I recall a personal experience that highlighted the importance of local knowledge. During the writing stage of my PhD programme, I had hunkered down in Tossaporn's rubber plantation (see below) as a quiet place to work. This was at a time when Tossaporn was enrichment-planting many kinds of useful plants into her rubber plantation, and was gradually developing it into a diverse agroforest. Keen to escape my thesis for a few minutes and to get some exercise, I often volunteered for these planting assignments. As a graduate from an agricultural college, a longtime member of the Gardening Club in 4-H during my youth, and as someone with a farming background, I probably fancied that I knew a thing or two about planting things. On the day in question, some lengths of the vine, *Tinospora cordifolia* (Willd.) Miers (Menispermaceae) (known locally as *boraphet*) had been collected for me to plant nearby my office. Determined to do the job properly, I had dug a short trench and, after throwing aside any rocks and roots, had then laid the vine cutting into the trench, and carefully covered it with fine soil. After watering, I was feeling that I had done about the best job possible.

As I was admiring my handiwork, a neighbouring farmer called Na-mud wandered over to see what I was doing, undoubtedly wondering why the white man was on his knees and mucking about in the soil. Quickly realizing my objective, Na-mud reached down to pick up another length of the *boraphet* vine that was surplus. Glancing to the left and right, he then casually tossed the vine upward, where it lodged in the rubber canopy overhead. I didn't say anything – but wondered briefly what he was trying to achieve. There was considerable loss of face some time later when my carefully planted cutting withered up and died – but Na-mud's carelessly tossed vine began to sprout leaves and grow! It's been a source of good-natured ribbing in the years since.

Imagining the dawn of agriculture

Humanity's first tentative steps away from hunting and gathering towards agriculture would have been motivated by the need to find a better way of procuring food. There are several chapters (ch. 8-9) in this volume that document forest-dwelling peoples who have made this transition only relatively recently. Quite possibly, humanity's first steps towards agriculture could have begun with hunters using fire to burn old grassy areas to encourage regrowth of succulent new grass that would attract deer and other wildlife.[3] They would have discovered that localized burning in forested areas didn't produce the grass that they wanted – but instead created open plots that would take a few years for the forest to regenerate (Figure 1).[4] Remnants of fruits, grains and tubers, that had been gathered in the wild, may have been discarded near field huts in these plots, and taken root as 'accidental plantings' on the exposed soil. The same thing still happens in swidden fields today, and a sudden proliferation of fruit trees and other food plants in secondary forest often marks where a field hut had once stood in a swidden field, until it was overtaken by the regenerating forest.

FIGURE 1: Early hunters may have used fire to burn patches of old, dry grass to encourage the regrowth of succulent young grass that would attract deer and other game animals. An extension of this localized burning into the forest margins could have created the first forest openings that would have been suitable for cropping.

As human population pressures increased, and natural stocks of food plants were over-harvested, these early hunters and foragers would have felt increasing pressure to find better ways of feeding themselves. They may have been walking further and further to harvest less and less. And the foraging parties themselves would have been subject to attack by large predators searching for their own meal in the forest.

Returning from their hunting and foraging trips each day, these pre-agriculturalists would have watched their 'accidental plantings' grow and bear fruit - the same fruit that they were possibly having to walk longer and longer distances to find growing in the wild. Perhaps after an unsuccessful foraging trip one day, the returning hunter-gatherers would have realized that they had a much more accessible source of food growing just nearby their field huts. As limited as it may have been, it would have been better to harvest and cook it, than to go to bed that night with empty stomachs. Walking those long distances in search of dwindling wild food stocks was no longer necessary. The idea of intentional propagation of useful plants would have been born.

From there, it would have been a small step for folk to realize that they could intentionally distribute those same fruits, grains and tuber remnants over a wider area of the burned plot and in a more systematic fashion. All of this imagined scenario is completely hypothetical, of course – but it is easy for us to imagine our ancestors to have begun domesticating food plants in such a way, and the evolution of the first farmer. In fact, there are several contemporary examples documented in this volume, of villagers taking food plants that they have traditionally gathered from the forest, and transplanting them into their swidden fields (see Chan and Takeda, in the supplementary chapters to this volume, for example). So, this imagined scenario is still continuing today, and helps to account for both the domestication of many indigenous crops and the impressive agrobiodiversity of swidden fields. In this way, the labour of having to walk increasing distances in search of wild food stocks would have been gradually exchanged for the labour of cultivating many of those same plants in plots that they opened from the forest.[5] The spear that villagers once habitually reached for as they left their huts each morning would have been gradually replaced by a hoe. The Agricultural Revolution had begun, and its first farmers were likely to have been shifting cultivators.

Farmers have been experimenting (Figure 2) and innovating throughout the history of agriculture and their continual search for better ways of doing things is undoubtedly responsible for many of the impressive advances that humanity has made in learning how to feed a burgeoning population. It is important for each generation of farmers to pass their knowledge and traditions on to the next – but to doggedly persist in repeating things exactly as our ancestors had done would be to remain stagnant. It would have been the innovative farmers, who wanted to try something new, who would have led the way to identifying better farming techniques. The tradition of experimenting in search of better ways of managing crops and fields has persisted for millennia and continues within the advanced agricultural systems in the industrialized West today. Farmers are continually trialing new crop varieties or rotations, adjusting the amendments that they apply to their soils, or perhaps testing new machinery that their local dealership has begun offering. I recall that when my brother and I graduated from the Nova Scotia Agricultural College (NSAC) and returned home to farm, one of the ideas that we

FIGURE 2: Farmers have always tried new things, often in adjacent plots, and compared the results with their standard practices to gauge if the new innovations offered any benefits.

brought back with us was that we wanted to grow alfalfa (*Medicago sativa*) and ensile it for our dairy herd (Figure 3). We thought that it would provide a higher quality and more palatable forage than the red clover (*Trifolium pratense*, Leguminosae) mixes that we had been baling for hay up until that point. Luckily, our father (see the dedication for volume I in this series (Cairns, 2015a)) was progressive enough that he approved of our experiments. I recall the first field that James and I sowed to alfalfa, with a Brillion

FIGURE 3: A member of the Willscott Holsteins herd, whose hard work supported work on shifting cultivation half a world away.

seeder that the local Soil and Crop Improvement Association was circulating for use amongst interested farmers. It was a relatively new design of seeder that few farmers had yet bought for themselves – but that we thought would provide a more precise

seed placement than the Massey Ferguson 33 grain drill that we had traditionally used for undersowing red clover mixes with our barley crops. So, we were actually trialing several new variables at once – a new fodder crop, a new planting technology, and a new method for preserving the crop (ensiling instead of dry hay). James and I were unsure of which alfalfa varieties would grow well in our climate and soils, and so we did the same as farmers do all over the world when faced with this question. We experimented to find out. We had ordered seeds of 6 to 8 alfalfa varieties from our local seed dealer and planted them in strips, side-by-side, to see which performed best in our conditions – much as you often see swidden fields in Asia planted in strips of multiple varieties of upland rice. I still remember the field in which we undertook our first trials

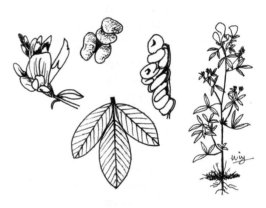

Medicago sativa L. [Leguminosae]

First cultivated in ancient Iran about 2500 years ago to feed livestock, alfalfa, also widely known as lucerne, is the world's most cultivated forage legume. It is used for grazing, hay, or silage and as a green manure/cover crop. Its primary use is feeding high-producing dairy cows, but its outstanding protein content and highly digestible fibre make it an important fodder for beef cattle, horses, sheep, goats, meat rabbits and poultry. As well, alfalfa sprouts are often part of the human diet.

in planting alfalfa. I also recall that Dad was annoyed that we planted it on a Sunday afternoon. It was always family custom that the farm should fall silent on Sundays, out of respect for the Sabbath[6], and that everyone would attend church service. In our haste to get the crops in the ground, we had violated the religious traditions of our ancestors – and this wasn't accepted any more readily than if a young swidden cultivator in Asia had opened a field in the forest without first performing a ritual to seek permission from the resident spirits (see, for example, Barber et al. and Chazée, both in the supplementary chapters to this volume). Luckily, Dad probably chalked our transgression up to youthful recklessness, and the farm continues to grow alfalfa as a main fodder crop to this day.

One of the criticisms most often hurled at shifting cultivators is that they are stagnant and resistant to change. If this was true, then this would have remained a very thin volume indeed, a pamphlet even. But as you heft this volume in your hands, its weight should convince you that shifting cultivators are keen innovators and often develop ingenious ways of farming under very difficult conditions. The fact that they have persisted over many generations and even millennia is testimony to their success. And this probably only 'scratches the surface' of the reality of what shifting cultivators have truly innovated. As Dr. K. B. Roberts points out in her chapter in the digital supplement to this volume, shifting cultivation is in itself an innovation that allows mountain dwellers to feed themselves under very challenging conditions. I sometimes wonder how well those of us who tout a 'Dr.' or 'Professor' in front of our names, and write critiques of shifting cultivators from the comfort of our 'ivory towers', would perform if forced to feed ourselves under similar conditions.

The fact that shifting cultivators all over the world have developed the same general pattern of farming system suggests that they have undergone similar kinds of experimentation and arrived at similar conclusions as to what works best. It is under extreme conditions, such as torrential rainfall, precipitous slopes, insufficient land, failing soil fertility, or more recently, market opportunities, that farmers have deviated from those standard practices and developed improved ways to manage their resources. The fireless systems of shifting cultivation, typically developed in areas of extreme rainfall (see ch. 10–12, this volume), are a case in point. Necessity was truly the 'mother of invention'.

Why hasn't research been more successful in improving the lives of shifting cultivators?

As researchers, we frequently evaluate shifting cultivators and their practices. It is interesting to ponder, if the tables were reversed, what kind of grades shifting cultivators would give us for our work in analyzing their farming systems and identifying potential improvements. If they used improvements in their living standards as their criteria for evaluation, it's unlikely that we would receive a passing grade.

The typical remoteness of shifting cultivators, on the slopes of distant mountains, creates a physical barrier to the outside world, and as part of that outside world, to the researchers who might seek to study them and their systems of land use.[7] Researchers who want to drive to their research sites and be home in time for supper, are not likely to be attracted to research on shifting cultivation. But conversely, for those who like hiking, camping and immersion in different cultures, it is likely to be the most interesting research in the world! For these same reasons of remoteness and poor accessibility, many high-ranking individuals who represent themselves as experts on shifting cultivation, and who claim to speak for its practitioners, have only distant and superficial exposure to swiddening peoples and their problems. At the very least, they are unlikely to be familiar with the lives and struggles of shifting cultivators, or be able to represent their interests at the highest levels.

The CGIAR (formerly the Consultative Group on International Agricultural Research), based in Montpellier, France, receives donor funding from various nations to support the work of 15 research centres around the world – the so-called CG centres – with an annual research portfolio of just over US$900 million (CGIAR, 2021). Under a banner declaring 'Science for Humanity's Greatest Challenges', the CGIAR appears to have abdicated its role in support of shifting cultivation – which, by various estimates made within the pages of this volume, still feeds hundreds of millions of the world's poorest people. None of the CG centres give priority to shifting cultivation as a research issue. A glance through the Table of Contents of this volume confirms that the research has been undertaken by a wide array of university professors and students, projects and institutions – but there is no one CG centre that is leading the way.

The indigenous communities who practise shifting cultivation often dress colourfully and are photogenic, so their images may often be displayed on websites or in promotional material to attract reader interest and convince donors that research is undertaken to assist the 'poorest of the poor'. But when it comes to work on the ground, the reality is probably far different. In fact, shifting cultivation is probably neglected by the research community for the very same reason that it has historically been discouraged by state governments. Money talks. Shifting cultivation is largely a subsistence-oriented system and doesn't generate either taxation revenue or large amounts of surplus cash that can be tipped into the coffers of others. Shifting cultivators are likely to be pushed aside by large agro-conglomerates that generate huge profits from monoculture plantations and are willing to donate tens of millions of dollars to research institutions if they undertake research that is useful to their business interests. This is not philanthropy; they will definitely expect a return on their investment. In this way, research centres that are meant to produce public goods for the benefit of wider humanity may be co-opted by private business interests for their own benefit. Shifting cultivators are not likely to receive so much as a second glance in this competition for research attention. *Au contraire*, if research can help to make these agro-industries even more profitable, they are likely to expand their acreage further and push even more shifting cultivators off their land, as we presently see happening with oil palm plantations in Indonesia (Potter, 2015) and Malaysia, rubber in Lao PDR (Vongpaphane and Cramb, 2015), and maize in Nan province of northern Thailand (Gypmantasiri and Limnirankul, in the supplementary chapters to this volume). This will be abetted by state policies eager to encourage agro-industries that generate tax revenues. In these cases, shifting cultivators may be reduced to poorly-paid labourers (van der Ploeg and Persoon, 2017), perhaps on the very plantations that have usurped their ancestral land.

Aside from having researchers to undertake the fieldwork, another vital component to maintaining a sustained research focus on shifting cultivation is that funding agencies have a long-term vision of the type of research issues that they wish to support, and continue to invest in them until they pay dividends. Nothing happens without money. We have been immensely fortunate, in that the Australian Centre for International Agricultural Research (ACIAR) has been a loyal supporter of our work throughout this series of volumes on shifting cultivation. This was in stark contrast to Canada's International Development Research Centre (IDRC), which was subjected to the changing policies of successive Canadian governments, and seemed, lately, to be reduced to a shadow of its former self. IDRC seemed hardly able to retain its own identity as a research institution, let alone continue its support for any specific research thrusts. Although IDRC initially 'planted the seeds' that led to this work, it was unable to continue its support long enough to see it through to full fruition. Credit for that belongs to ACIAR.

The continued success of our species has meant that land use pressures have increased unabated

Humanity's success in cultivating our food needs allowed us to reproduce prolifically, with the relative assurance that we could feed the extra mouths. With an abundance of land, the extra labour provided by large families could be directed to agricultural fields and the increased harvest would have offset the extra mouths to feed. The expansion of humanity across the planet's land mass obviously led to inexorable use pressures on the limited land suitable for cultivation. The concentration of large populations in megacities added to the need for farmers to produce large food surpluses. This has caused a major threat to the millions of other plant and animal species that are trying to coexist with us on this planet, as so much of the wildlands that they depended on for food and habitat, have been converted to agricultural use for the sole benefit of one dominant primate species – us.

As human populations increased, the expanding numbers were initially able to spill over into frontier areas. In the case of shifting cultivators, these frontiers were usually unclaimed forests on a nearby mountain. In the case of my own ancestors (see dedication for this volume), the new land into which they spilled over was across the Atlantic Ocean in unknown lands. Both groups were searching for good land to farm. But as vast as these new frontiers must have seemed, they've steadily filled with expanding populations, and agriculture has had to intensify to meet increasing food needs. There are no more 'new worlds' across the oceans, nor unclaimed new forests across the mountain ridges. Our present search for habitable planets, however, suggests that we haven't yet given up on our frontier mentality and that we are now searching quite literally for 'New Worlds' into which we can expand – not across a mountain ridge or an ocean – but possibly across a galaxy. There is a distinct danger, however, that if humanity feels that we have a fallback option of another habitable planet that we can settle, it may render us less determined to protect this extraordinary planet that we already call 'home'.[8] On Earth, there is little further scope for agriculture to expand into new areas. With the reality of more

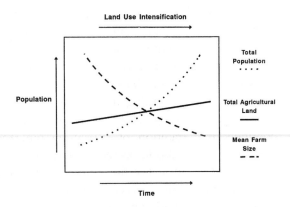

FIGURE 4: As more people needed to be fed from a stagnant land base, agriculture had to intensify if famine was to be avoided.

people to feed from a stagnant land base, there was no alternative but for agriculture to intensify and try to produce more food per land unit (Figure 4). The same population pressures that pushed our ancestors to evolve from hunters and gatherers to farmers in the first place have multiplied and continued to push for intensification of farming systems.

In the present discourse about the many serious environmental problems that confront our planet, the 'elephant in the room,' that seems to be receiving scant attention, is over-population. As a species, we seem to have not yet come to grips with the reality that this planet has a finite capacity to sustain us, and we must limit our numbers to within that capacity. There are too many of us, and we consume far too much, especially in the affluent West. At our present rate of consumption, it would take several planet Earths to sustain us all. Every farmer knows that if you put too many head of livestock in too small a field area, it will be damaging to both the field and the livestock. The 'stocking density' of humanity on this planet isn't any different.[9] If governments can't achieve something as simple as convincing their citizenry to get vaccinated or wear masks to protect themselves during a pandemic, perhaps they feel that there is simply no hope in trying to intervene in something that is far more invasive, personal and politically explosive, such as the need to limit our fertility.[10]

Continuing to focus on the fallow as an opportunity for useful interventions

Shifting cultivation is a land-use system designed for conditions of land abundance and low population pressures. Of course, it is the scarcity of these preconditions that now generates pessimism about its future. But, as testified by the case studies in this volume, shifting cultivators have been amazingly resilient in adapting to new pressures. It is not surprising that many of their innovations have focussed on the fallow – as the least intensively managed phase of the swidden cycle. It was the ingenuity of many of these indigenous fallow-management (IFM) systems that originally captured the Editor's attention during the early years of his career, and provided the forerunner (Cairns, 2007) for this series of volumes. That early work categorized farmer strategies in fallow management into a continuum of distinct typologies (Figure 5).[11] The land use under some of these systems has evolved to appear so unlike anything that we usually associate with shifting cultivation, that we barely even recognize them as variations of fallow management. This volume continues that tradition of examining indigenous fallow-management strategies (Section IV), but also looks more widely at a more complete menu of options that farmers have identified to fine-tune their swidden management.

Unfortunately, the physical limits of how large a volume can be securely bound allowed us to publish only a fraction of the contributions received in this print edition. We've therefore focused this print edition mainly on innovations to enhance the fallow phase of the swidden cycle, and have allowed the rest to spill over into a large collection of supplementary chapters that can be found in CABI's Digital Library (https://www.cabidigitallibrary.org). These supplementary chapters are listed at the end of the Table of Contents.

The kind of fallow-management strategy chosen by farmers will depend heavily on the cultivation pressures on their land and subsequently, the length of time that they can afford to keep their fields in 'fallow' (Figure 6). This next schematic clearly builds on the previous figure – but simply turns it on its end, and shows roughly what the cycle of cropping and fallow might look like under each category of fallow management. Clearly, the fallow-management strategies that farmers choose to use will have a large impact on the appearance of the landscape and the environmental services that it provides. In Indonesia, for example, the damar agroforests of Krui, in Lampung province (see Herawati et al., 2017 and plates 60 and 61 in Coloured Plates section, this volume) and the candlenut forests in South Sulawesi (see Supratman et al., ch. 36, this volume and Plates 57 to 59) both demonstrate that when farmers develop their 'fallows' into long-term agroforests, the landscape can remain

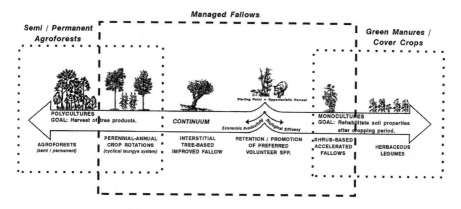

FIGURE 5: Most of the fallow-management strategies found in the Asia-Pacific region fall somewhere within this continuum of fallow-management typologies.

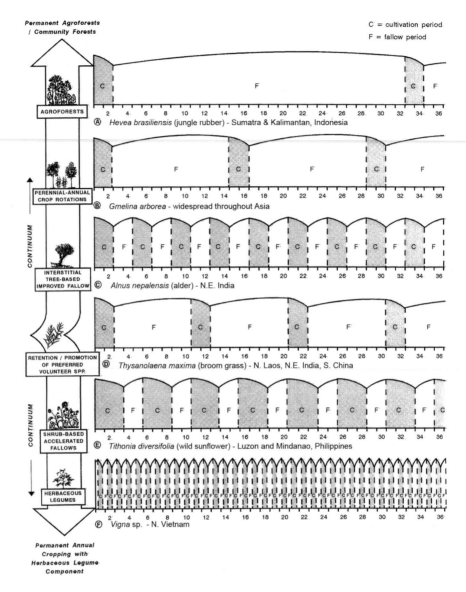

FIGURE 6:   The tempo of the swidden cycle accelerates dramatically as land-use pressures force farmers to cultivate their land more intensively. The length of time available for the fallow will largely determine what kind of fallow-management strategies might fit each farmer's circumstances.

heavily forested. But as a contrast, in the foothills of the Mt. Kitanglad Range Nature Park, in Mindanao, the Philippines (Cairns, 2015b), where the Editor undertook most of his field research on farmer management of *Tithonia diversifolia* as a preferred species for short-term fallows (see van Noordwijk et al., ch. 39, and Plates 42-44 in the Coloured Plates section, all in this volume), the forests had largely disappeared, and were replaced by a patchwork of cultivated fields and short bush fallows. This landscape was clearly well on its way to permanent cultivation and whenever farmers

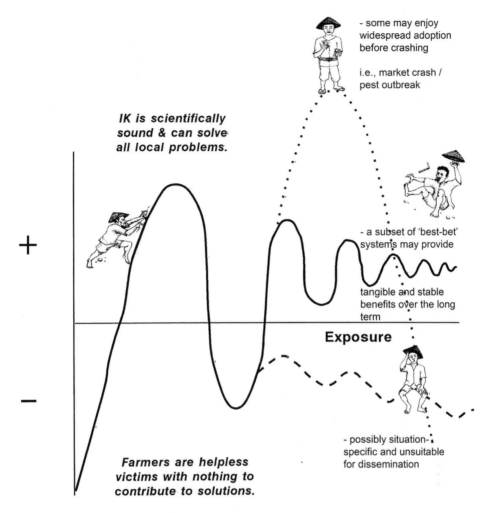

- some may enjoy widespread adoption before crashing

i.e., market crash / pest outbreak

IK is scientifically sound & can solve all local problems.

- a subset of 'best-bet' systems may provide

tangible and stable benefits over the long term

Exposure

+

–

- possibly situation-specific and unsuitable for dissemination

Farmers are helpless victims with nothing to contribute to solutions.

FIGURE 7: Although preliminary analysis may view many fallow-management strategies as promising, they will have to be trialed by farmers for some time before it becomes clear which offer tangible benefits over the long term and are worthy of further investment in research and extension. (IK = Indigenous Knowledge)

are able to buy chemical fertilizers, it is likely that even the short-term *Tithonia* fallows will also disappear. Figure 7 shows the oscillating range of opinions that vary between the extremes of believing that indigenous knowledge and practices are sufficient to guide farmers in solving most of their own problems – to the more condescending view that sees shifting cultivators as helpless victims awaiting outside researchers and projects to come and offer them solutions. In the context of these extremes, while researchers may initially become quite excited about the potential of apparent '*Cinderella*' fallow species or systems, it is likely that prolonged experience will prove some to be situation-specific and unsuitable for diffusion, others may enjoy rapid and widespread adoption before crashing, while only a subset of 'best bet' systems may provide tangible and stable benefits over the long term that qualify them for wider research and dissemination.

The movement of people and power to the cities

Until quite recently, you wouldn't have to dig too deeply into the ancestry of most people to find a farmer. Humanity's familiarity with agriculture has become increasingly distant, however, as more of us have moved from the countryside to cities. This has left new generations of children so far removed from farming that, for example, infamously, school children in Toronto need to be taught that the milk that they drink actually originates from cows, and not a bottle! As Garry Oughton notes in his chapter in the digital supplement to this volume, this migration from rural to urban, has recently progressed to the point that, for the first time in history, the majority (55%) of humanity now lives in cities. In democracies, this of course means that the majority of political power now resides in the cities, and governments elected by cities will naturally craft policies that are advantageous to city dwellers (the consumers) and are often at the expense of those who farm the land in the countryside (the primary producers).[12]

FIGURE 8: Reports of farmers cutting down trees to open spaces for planting crops, and then the further burning of the slash, are enough to convince most members of the general public that shifting cultivation must be a regressive practice.

Although people may prefer the conveniences of city life, many nonetheless wish to influence what happens with the countryside and its resources. With the demographic, economic and political power increasingly concentrated in urban centres, these urban powers increasingly try to wrest resources from rural areas.[13] Politicians, acutely aware of where the votes are[14], craft policies aimed at aiding and abetting urban interference in rural interests. For an urban observer, it probably seems obvious to conclude that any farming system that entails the cutting and burning of forest (Figure 8) is environmentally indefensible and to be condemned. It should therefore be no surprise that a land-use system as complex as shifting cultivation is equally little understood by the policy-makers who reside in the capital cities that are the seats of government power. Most recently, right-wing governments, such as those of India and Brazil, have been particularly aggressive in trying to legislate indigenous peoples off their traditional lands. Here in Thailand, from the vantage point of their high-rise penthouses, it is not hard to understand how Thailand's urban elite would hold a dim view of the shifting cultivators that they see as despoiling their favorite holidaying spots in the mountains.

This tilt of power towards the cities is not limited to the so-called 'developing world'. With the increasing size of farms and high efficiency of Western farmers in producing food, farmers have become a smaller and smaller percentage of the population, and a dwindling political force.[15] Most of the political power has gravitated to the cities. As part of this, federal and provincial governments enact legislation and regulations that impose restrictions on how farmers manage their resources, whilst having very little understanding of agriculture and its problems. A shifting cultivator may only have to worry about a wild boar or monkeys emerging from the forest and damaging his crops. For a Western farmer, being hard-working and skilled in managing his crops and livestock are no longer enough to assure success. An unsupportive politician, negotiating a free-trade deal half a world away, can destroy the economic viability of his farm overnight.[16] It is absolutely not hard to understand why shifting cultivators would want to distance themselves from government and reliance on markets.

Policy interventions in how farmers manage their resources, seem often to be born from arrogance and ignorance, rather than be informed by familiarity or research findings. They were probably drafted by urban dwellers, confident that they know more than their country cousins.

My own home province of Prince Edward Island, in Canada, is largely agricultural (Barde, 1977), but is not immune to urban interference in farmers' affairs.[17] PEI is overwhelmingly rural, with just a few small municipalities scattered across the province. In this case, the tactic is a process known as 'amalgamation', in which these municipalities unilaterally expand their boundaries so that they can encircle nearby farming communities into their tax base. They hope to continue this process of amalgamation until every farm and every field has been absorbed into a municipality and is generating tax revenues for that municipality. This tactic is, of course, attractive to the municipalities because it provides them with higher tax revenues. But for encircled farmers, it means that they are suddenly paying an added municipal tax without receiving anything in return.[18] The municipalities simply have no extra services to offer that farmers actually want.[19] Farmers have no use for the extension of streetlights, concrete sidewalks, water mains or communal sewage systems into the countryside.

Of course, rural families patronize businesses in the municipalities and would doubtlessly like to avail offers of home delivery and servicing of some of their purchases. A farm wife busy helping her husband with the harvest might want to just order pizza for supper some evening, rather than leave the tractor idle for the time that it would take her to cook a meal. A retired farmer, whose wife has already passed away, might like to order his meal from the local 'Meals on Wheels' service. But regardless of their tax bill, it is likely that in every case, they would simply be told that home delivery and servicing are available only within the actual city limits. So what exactly are the extra services that farmers would receive in exchange for their extra tax dollars?[20] The entire amalgamation movement seems to be a blatant attempt by municipalities to coerce hard-working farmers to subsidize their more comfortable urban lifestyles.[21] It is as if cities are trying to become the new colonial powers of the 21st century, and the rural areas around them are the colonies to be conquered and subjugated.

Equally worrying, farmers would be loath to give city administrators any jurisdiction over their affairs[22] and endure the harassment that would inevitably come with it.[23] At its most extreme, in some countries, urban-based organizations such as People for the Ethical Treatment of Animals (PETA), are known to actually raid farms and interfere with their livestock.

Close to my birthplace on Prince Edward Island, the nearby community of Bedeque and Area was acting aggressively expansionist back in 2015, and in a move that would have made Vladimir Putin or Xi Jinping proud, was trying to extend its boundaries to encircle a large area of nearby countryside that included my father's farm in Lower Freetown (CBC News, 2015). Some of the more resilient residents of this targeted area decided that they weren't going to accept Bedeque's expansionist agenda and began to actively petition against the proposed amalgamation. On the strength of the many rural signatures gathered on their opposing petition, the protesters were successful in driving the municipal invaders from their doors. But it is likely that they are biding their time and waiting for their next chance to 'slip their hands into farmers' pockets in search of extra tax dollars'.[24]

Referendums held elsewhere on Prince Edward Island on the same question were less successful. Although rural residents voted overwhelmingly against being amalgamated by nearby municipalities, their wishes were ultimately ignored and the amalgamations went ahead anyway. So, it seems that while human rights in Canada don't allow the state to compel its citizens to get vaccinated against a raging pandemic, new taxes can be imposed on farmers anytime.[25] This seems exemplary of the reasons that Professor James Scott (2009) hypothesized had motivated shifting cultivators to historically retreat into the mountains in the region that Professor Scott now calls 'Zomia' and distance themselves from state governments that would control and tax them.

The injustice of this proposition of amalgamation was so outrageous that my elderly father (see dedication in volume I (Cairns, 2015a)), then in his late 80s, became a leader in protesting against it[26] (Figure 9). Farmers all over the world are a similar 'breed' of people, and I often think that Dad would feel a great deal of solidarity with the shifting cultivators who are also so often treated unfairly by urban policy-makers.

Whether it is the urban masses in Asia disapproving of shifting cultivation, or the municipalities on Prince Edward Island trying to lasso farming communities into their tax base, it seems clear that the world over, farmers have an uneasy relationship with their urban neighbours.[27] Regardless of how distant urban dwellers have grown

FIGURE 9:   Standing up to Government. At a time when he should have been enjoying a well-earned retirement, William Cairns (left) decided that he needed to stand up and make his voice heard in protesting against the municipalities wanting to encircle his farm and charge him more taxes. He is pictured here with fellow protester, Mary Webster (right).

Source: SaltWire Newsletters (2015).

from the agricultural systems that feed them, or how little they actually know about them, these armchair critics seemingly don't feel it necessary to be well-informed as a precondition to hurling a lot of criticism at farmers. Ironically, the farming community that I was born into on Prince Edward Island is being harassed because they are seen as a potential source of increased taxation revenues. But the shifting cultivators that I work with in Asia are marginalized largely because they are not seen as a source of taxation revenue for the state.

Acknowledgements

A volume of this scope is only possible with the efforts, talents and resources of a great number of people – so there are many to thank. The danger, of course, is that a section of this nature will never be truly complete and will almost certainly miss important names. Nonetheless, we must try our best to acknowledge most of the names of those whose contributions have been so invaluable in developing this volume.

I want to begin by clearly acknowledging that the knowledge and practices documented in this volume are the property of the region's shifting cultivators.[28] It was their ancestors who, over generations of experimentation, developed the innovations described in the chapters that follow. They didn't publish any papers on their trials – but their findings were carefully handed down to the next generations of farmers. Like all life, shifting cultivators have co-evolved with their environments through the millennia. The innovations documented in this volume describe some of what they've learned during this time about how to farm successfully under very difficult conditions.

The next major group of contributors that was key to developing this volume was the several hundred authors and researchers who documented these farmer innovations and have been so generous in sharing their work in this volume. Writing these chapters was generally unpaid work that they undertook alongside their normal duties. Without their sacrifices and enthusiastic participation, this volume would never have been written.

Within our group of contributing authors, two deserve special mention for taking on especially demanding tasks. Professor Dietrich Schmidt-Vogt has written the introductory chapter and Professor Roy Ellen has written the concluding chapter. Both authors read almost every other chapter in the volume in preparation for writing their own respective chapters, requiring quite a large time commitment. We feel honoured to have the volume opened and closed by two such eminent experts in the field.

These authors had to remain extraordinarily patient with us during the several years that it took us to process their contributions and prepare the volume for press. A project that spans this many volumes and this many years necessitates a long-term view. Regrettably, as indicated by the tributes that appear at the bottom of the opening page of several chapters, a few of our contributing authors became grievously ill along the way and were not able to wait long enough to see their work in print.

Thais have an old custom, that you have to offer something three times and have it declined three times before you can truly be certain that it is not wanted. This custom usually applies to something like the offer of a cup of tea. But having lived in Thailand for a long time, I adopted something close to this practice in soliciting papers, and probably sorely tested the patience of a good few of my colleagues! Some of them probably wondered what part of 'No' I didn't understand! We tried to leave no stone unturned in our relentless search for useful material. For my part, I often felt much like a Border Collie, but instead of dashing about, nose to the ground, trying to herd sheep into a paddock, I was trying to herd case studies into a Table of Contents.

We are gradually narrowing our focus on the main groups whose efforts have been pivotal in the development of this volume. The next key group that deserves thunderous accolades is the small team that supported me so capably in developing the volume that you now hold in your hands. I have been extremely fortunate in that the same dedicated and talented individuals have stuck with me throughout working on this series of volumes. I regard their contributions with a combination of awe and immense gratitude! The two key team members, whose work was so central to assembling this volume, were our Copyeditor, Bob Hill, and our Design Expert, Tossaporn Kurupunya. Their dedication to these volumes has been equal to my own. The combination of talents and dedication of Tossaporn and Bob were complementary and the stuff of dreams for anyone embarking on a demanding book project of this nature. The calibre of their work was such that I'm sure that either Bob or Tossaporn could have worked for any professional publisher in the world. But this was a project amongst friends. The collegial nature of our work was perhaps best indicated in that no written contracts existed anywhere within our team. Everything operated on the basis of trust and spoken commitment.

An 'Australian Kiwi', Bob Hill (Figure 10) did all of the copyediting. His philosophy was that the volume should be easily readable by all, and not just specialists who work in the field. The chapters were, of course, often written by authors for whom English was a second or third language. That the entire volume has now been transformed into a uniform and easily-understandable writing style is testimony to Bob's hard work. As he rolled up his sleeves to begin work on each new volume, Bob must have felt much like a climber at base camp below Mt. Everest, preparing for the long climb ahead.

FIGURE 10: Bob Hill did much of the 'heavy lifting' in the creation of these volumes, by doing all of the copyediting and assisting the overall Editor with a myriad of other tasks as well. He is owed a great deal of recognition for his painstaking work on these books.

As Bob completed his first editorial pass through each chapter and took on board any revisions requested by the author/s, he would then send the file down south to Tossaporn Kurupunya, in Trang province, so that she could format each chapter according to CABI's style guidelines.

A Thai national, Tossaporn (Figure 11) was our Design Expert. She divided her time between working in her rubber plantation during the day (Figure 12), and working on this volume during the quiet of the night, with the frogs serenading her in the background. But the silence of Tossaporn's nocturnal working environment was sometimes shattered by reminders that her rubber plantation is home to a fair degree of unseen biodiversity. A small flock of poultry that roams freely under the rubber canopy by day instinctively knows that the darkness of night can bring danger and they routinely fly up into the lower branches of fruit trees near the house, to roost there for the night. Some local wildlife have apparently learnt of this.

FIGURE 11: The immense talents and patience of Tossaporn Kurupunya are on display on each page of this volume. She was our Design Expert whose careful attention to detail created the attractive formatting seen throughout this series of volumes.

As Tossaporn bent over her computer, focusing intently on formatting the chapters of this volume, the sudden loud and panicked squawking of one of these chickens usually signalled that a local python had slithered up one of these roosting trees in search of a chicken snack. Although Tossaporn seems to take all of this in stride, she probably isn't very keen on sharing her living environment with snakes, having lost a younger brother to a cobra bite many years ago.

We've tried to make these volumes more attractive and reader-friendly by including a generous amount of artwork in them. We've relied on two

FIGURE 12: When she wasn't busy at her computer, Tossaporn managed her rubber plantation to pay the bills. Although not a shifting cultivator, her feet were firmly rooted in farming.

talented artists for this purpose. The striking charcoal sketches of Paradorn Threemake (Figure 13), that we've used in all the cover designs and to open each section within the volumes, have become one of the hallmarks of this series of volumes. He has added to his artistic contribution to this volume by also preparing ink sketches to be used at the end of each section. All sketches were planned to resonate with the theme of each section in which they appear.

Another artistic hallmark of these volumes has been the intricate botanical sketches that we've scattered throughout the chapters to help familiarize readers with some of the key plants discussed in each chapter. These are the creations of Wiyono (Figure 14), an old artist friend from the days when I worked at ICRAF's regional office in Bogor, Indonesia. Wiyono[29] is a talented artist and a valued friend. He has suffered from deteriorating health in recent years and is to be commended for his perseverance in completing the many botanical sketches needed for this volume. As Wiyono seemed to struggle more in the latter stages of the project, his loyal daughter, Karina, came to work beside him and help him across the finish line.

In a volume that covers a region as vast and varied as Asia-Pacific, it is important to provide clear maps that guide the reader in easily locating the areas of research that each chapter discusses. We have been immensely blessed in having Peter Elstner (Figure 15) working with us as our cartographer in this project. Almost

FIGURE 13: Artist Paradorn Threemake at work in his studio, formerly in the basement floor of Chiang Mai's Night Bazaar. It was Paradorn's talented hand that created the charcoal sketches that appear on the cover design and open each section within the volume.

FIGURE 14: Wiyono is the talented Indonesian artist, whose sharp eye and steady hand created all the botanical sketches that help familiarize readers with some of the key plants discussed in each chapter.

all the many maps scattered through this volume are his creations. Peter has long experience working in Thailand, and has been a real pleasure to have on our team. We really learned to appreciate his reliability. Map requests were routinely submitted to Peter with the certainty that within a few days, he would get back to us with exactly what we needed. He is a German national and adds to the international diversity of our team.

Since I worked on this volume as a hemiplegic, I obviously needed more assistance in living and working than someone able-bodied. These people also played a vital role in the volume by keeping me alive and able to work. The healthcare worker who was most pivotal in keeping me functioning was Khun Suthida Chantamanas (widely and affectionately known by her nickname, Khun Ae). Not long after suffering a massive stroke in 2008, I was fortunate to meet Khun Suthida during a visit to the McKean Hospital, and to recruit her as my physiotherapist.

That was about 15 years ago, and Khun Suthida has been working with me ever since, both as a health professional and a friend. Khun Suthida has changed jobs several times during that time – but has always made time to continue to help me in whatever way she could, looking after me like a member of her

FIGURE 15: Peter Elstner was the cartographer who prepared all the maps scattered throughout this volume, that guide readers in easily locating the sites of the research discussed in each chapter.

own family. She gave birth to her second daughter, Faith, in 2015, and I've had the pleasure of watching both her girls growing up. Khun Suthida would often bring her children with her when she came to help me. In the preface of the preceding volume, I had briefly described how Faith, then a couple of months old, would tug on my beard, as if trying to stimulate new thoughts on how the second volume could be improved. As I write this preface for the third volume, Faith has already celebrated her sixth birthday, but continues to keep me grounded on what is important in life. Amongst other pearls of wisdom, Faith has taught me that it is always important

to keep some crayons and a colouring book on hand and to make time for some colouring every now and then (Figure 16)! Faith has a wonderful sense of humour and never fails to brighten up a dark mood. As I marvel at the innocence and playfulness of this beautiful child, I sometimes can't help but wonder what we've done to the world that she will inherit and whether it will be able to provide her with a healthy and happy future.

The other member of my support team is my driver, Khun Sa-ard Wangkaew. With my loss of mobility, I could hardly cope with public transportation and a driver became essential to get around town to the places that I needed to go. Khun Sa-ard has now played this role for several years, and has always proven to be an attentive driver and a good friend. Between driving jobs, Khun Sa-ard likes to check if the

FIGURE 16: Faith checks out her newest colouring book while Uncle Malcolm tries to make progress on his next swidden volume.

fish are biting and to catch frogs in the local rice fields, both hobbies that put added food on his table.

Funding support

As frugal as we were in managing this project's expenses, a project of this ambition obviously requires a substantial amount of funding to ensure that the work can continue. These funding requirements can be broadly divided into two categories: 1) those paid directly to the publisher as publishing subsidies, and 2) those used to cover our team's expenses incurred in preparing the volume for publication. The funding used for these two main purposes has come from entirely different sources.

The major cost invested in a project of this nature was, of course, the vast cost of undertaking all the original research reported on in the volume. The total price tag for this would have been many millions of dollars. But these costs were spread across considerable time and countries, and a large number of project budgets. Through the generous sharing of our contributing authors, we have been able to take advantage of this existing work and bring it together as a *tour de force* exploration of farmer innovations in shifting cultivation.

Although Canada's IDRC is no longer amongst our funding supporters, I should perhaps preface my remarks by noting that it was funding support from IDRC that had initially allowed the Editor to wander the region's swidden landscapes and develop some of the insights that have led to this series of volumes.

For funding support of this present volume, our team wishes to express our most heartfelt thanks to Australia's ACIAR, ICIMOD and ICRAF, both its Indonesian country program and its head office in Nairobi, Kenya. The full grants from all these institutions were paid directly to CABI as publishing subsidies. The generous support of these institutions is acknowledged not only here, but also by the appearance of their logos on this volume's back cover.

The day-to-day work of our team has been funded entirely from personal resources. In fact, this entire series of volumes has been supported by two agricultural products, both of which are white liquids, rubber from southern Thailand and milk from Prince Edward Island. The viability of both industries are continuously under serious threat from the predatory policies of bullying neighbour countries. This helps us to appreciate the relative security of shifting cultivators' subsistence economies.

A brief explanation is needed on how milk and rubber helped to cover the costs of producing these volumes. As I've documented in the acknowledgements sections of previous volumes in this series, the main source of funding, that allowed us to operate on a daily basis, came from the dairy farm operated by the Editor's family on Prince Edward Island, Canada (Figure 17). This funding came in the form of monthly bank transfers from the Editor's father, William Cairns (see dedication in volume I), and was used to cover the costs mainly of copyediting and artwork, as well as the Editor's own expenses.

The support from the rubber sector came through our Design Expert, Tossaporn Kurupunya, based in Trang province, southern Thailand. Tossaporn's work on these volumes has been absolutely indispensable. And yet, she claimed no salary for the long hours that she has spent working at her computer, literally for years on end. Instead, she managed to sustain herself from the modest income that she earned from her rubber plantation (Figure 18). But as many of our authors have noted in these

FIGURE 17: The green pastures of Willscott Farm were the major source of funding that supported the creation of this volume.

FIGURE 18: Another important source of support for work on this volume came from this rubber plantation in southern Thailand. Tossaporn contributed freely of her time and talents in creating this volume, whilst supporting herself by latex sales from her rubber plantation.

volumes, the flooding of rubber supply well beyond market demand has kept rubber prices depressed for many years. This meant that Tossaporn's own farmgate income was substantially reduced, and she undoubtedly had to 'tighten her belt' and minimize her personal expenses, in order to be able to continue devoting her time free-of-charge to the production of these volumes. In this way, Tossaporn's rubber plantation subsidized the production of these volumes, much as my father's dairy farm did from Prince Edward Island. The funding that produced these volumes was thus all private, and it all came from farmers. As far as I can recall, there was not one cent of public funding used in assembling this volume. As already mentioned, all grants of institutional funding were paid directly to CABI as publishing subsidies. This source of funding, directly from farmers, and the volunteer nature of much of the work invested in it, are amongst the unique aspects of this volume – setting it in stark contrast to the bloated projects that are more typically supported by UN agencies, and in which a large part of the budget is typically paid to hotels, restaurants and airlines. These funding arrangements also underline an unusual degree of personal commitment by those of us who worked on the project.

It was through these various sources that we were able to cobble together enough support to produce this series of volumes on shifting cultivation.

Looking back ...

As I think back on the journey travelled in publishing this series of volumes, I can't help but think about the many times that it almost came to an abrupt end.

For example, the stroke that felled me in 2008 left my left foot swollen, paralyzed and feeling cold. To treat these symptoms, somebody had suggested that I should soak my affected foot in hot water. I was following this soaking routine one evening, with the assistance of a hilltribe girl who worked with me as a care-giver at that time. I laid across the width of my bed, with my left foot dangling down in a bucket of hot water. Immediately next to the bucket was the small desk on which I used my laptop and the various USB devices that were connected to it. What nobody had noticed was that the electrical power bar, into which my lamp, laptop and various accessories were plugged, was sitting precariously on the edge of the desk, and just above the bucket of water in which I was soaking my foot. It should have been safely on the floor, but somebody must have picked it up and placed it on the desk, probably to make more room on the floor for the bucket of hot water. I was reading something during this soaking treatment and not paying much attention to what was happening around me.

I don't know what caused me to glance up from the papers that I was reading when I did – but I was startled to see the power bar teetering at the edge of the desk, and threatening to fall off into the bucket of water in which I was soaking my foot. With a surprised yelp, I lunged forward to grab the power bar. I don't recall if it had already fallen and I grabbed it mid-air, or if it was still teetering at the edge of the desk. My care-giver looked at me blankly, wondering what the fuss was all about. She clearly had no clue that electricity and water don't make a healthy combination. If I hadn't chanced to glance up from the papers that I was reading when I did, the work on this series would have ended then and there.

Another incident occurred late one afternoon when I was working at that same computer. I used a floor lamp to light my work area. When that lamp began to sway back and forth, I realized that the entire condominium was also swaying. I could hear excited voices in the hallway as residents shouted exchanges about the earthquake and began to evacuate. I wondered momentarily if I should join them – but half paralyzed, it didn't seem like a good idea to join a mob stampeding down six flights of stairs. I quickly gave up on the idea as impossible and settled back into my chair.

My thoughts then went to my book files, and I pondered whether, if the ceiling came crashing down and my computer was destroyed, if copies of all the book files would be safely stored on Bob's and Tossaporn's computers. I quickly realized that duplicate copies of all files would be held by my colleagues, and that even if my computer disappeared under a pile of rubble, it would not be a fatal blow to the volume's progress towards press – and so I continued on with my work, as the lamp continued to sway overhead. The earthquake wasn't very strong, and I think that I heard later that its epicentre was around Chiang Rai somewhere.

And of course, the COVID-19 pandemic has added another layer of difficulty to the project.[30] The threat was not so great to our production team because our work was mostly done in solitude anyway, so the social distancing that was being recommended by health authorities was pretty much business as usual for us.

But even our high degree of solitude was no guarantee of safety. My care-giver's first task, every morning on her way to work, was to stop off at a coffee shop and pick up a couple of cups of cappuccino for me to drink that day whilst working on this volume. One day, she mentioned that she was no longer able to buy my coffee at the same shop that we usually patronized, because the barista who usually prepared my cappuccino every morning was sick with COVID-19. This alerted me that the person who brewed my cappuccino every morning, just before it reached my hands, was harbouring the COVID-19 virus, and that those cups of coffee provided a direct line of transmission to me. The same was true of my care-giver, who probably engaged in daily banter with this sick barista as she paid for my daily caffeine fix.

The pandemic did become a problem, in that quite a number of our contributing authors contracted COVID-19 and either became too ill to write what they had planned, or fell so far behind in their regular work duties that they felt that they could no longer afford the time to write a chapter, and so were forced to withdraw their participation. There was an unfortunate confluence of circumstances in that the heaviest participation in this volume was to come from India, and as we've all watched on our television screens, COVID-19's deadly rampage across the Indian sub-continent developed into a nightmarish scenario. Crossing state lines in India required a three-week quarantine, travelling in each direction. For some of our Indian collaborators, the time that they had originally budgeted for preparing their chapters was instead wasted in successive quarantines, often without their laptops and files, that would have allowed them to use their time productively. Other collaborators were under lockdown and delayed in returning to their research sites to collect soil samples or other final pieces of data that they needed to complete their studies.

More generally, many of the authors contributing to this volume are university professors. When COVID-19 regulations forced their universities to stop classroom teaching, these professors then had to invest considerable extra time in reformatting their lectures so that they could be taught online. This added preparation time delayed some of our teaching authors from working on the chapters that they had proposed. Others were asked to substitute teach for COVID-stricken colleagues, with similar effects.

Although our team kept our heads down and worked steadily through the pandemic, COVID-19 did exact a price on this volume by sickening some of our contributing authors and causing the loss or delay of their chapters.

Despite all these challenges, our small team has pushed forward and tried to complete this volume in the best way possible. We felt that we owed this to the many farmers and researchers whose work is shared in this volume. Any deficiencies found in this work are the sole responsibility of the Editor, and not of his supporting team or sponsors.

We feel honoured to have had the privilege of working closely with the scientific community that studies shifting cultivation in developing this series of volumes, which we hope will prove useful to future generations of scholars and researchers. Assembling these volumes has been akin to a virtual conference, in which we tried to use the internet to pass the 'microphone' around as widely as possible, and make sure that every participant was given an opportunity to express his or her views. We have organized this volume a bit differently than preceding volumes, in that we have included a synthesis paper at the beginning of every section. This was our attempt to pass the 'microphone' back to senior members of our research community and ask them to offer critical perspectives of the chapters presented in each section.

The changing composition of the volumes in this series is indicative of the changing distribution of shifting cultivation across the Asia-Pacific region. The first couple of volumes that we published were dominated by case studies from Lao PDR. We are now seeing fewer contributions from authoritarian states such as China and Lao PDR, where strict state policies have sharply cut back on shifting cultivation (see Li et al., ch. 6, this volume). As the world's largest democracy, India has been more conciliatory in dealing with its shifting cultivators and has been the largest contributor of case studies to this volume.

Shifting cultivation and I have something in common, in that the demise of both of us has been prematurely reported. We both continue to survive – although neither of us is functioning as before. As a spirit passing briefly through this mortal world, this trilogy of books on shifting cultivation is something that I wanted to leave behind. It is my modest attempt to leave the world a slightly better place than I found it.

I am reminded of a dairy farmer who I used to know back on Prince Edward Island. Struggling to get to their feet, dairy cows can sometimes be clumsy and injure their teats, usually with their sharp dew claws. If the injury is serious enough, the teat can no longer be milked properly and that quarter of the udder is often lost to mastitis.

The farmer that I am recalling had a cow that had suffered such injuries twice already and so had only two remaining teats that still functioned properly. On a day that I chanced to visit this farmer, this two-teated Holstein had stepped on one of her two remaining healthy teats and took it out of comission too. Gesturing towards this now single-teated cow and explaining what had happened, Harry had deadpanned, '*If she loses many more, I may have to get rid of her!*' Like that injured cow, it will soon be time for me to 'be put out to pasture' – but hopefully not before I can finish some more work with shifting cultivators.

References

Anonymous (2022) '2001 United Kingdom foot-and-mouth outbreak', *Wikipedia*, available at https://en.wikipedia.org/wiki/2001_United_Kingdom_foot-and-mouth_outbreak, accessed 14 February 2022

Barde, B. (Producer) (1977) 'Like no other place, P.E.I.: The million acre farm' (film), *TV Ontario*, The Ontario Communications Authority

Cairns, M.F. (ed.) (2007) *Voices from the Forest: Integrating Indigenous Knowledge into Sustainable Upland Farming*, Proceedings of regional workshop held in Bogor, Indonesia on 23-27 June, 1997, Resources for the Future Press, Washington, DC

Cairns, M. F. (2013) *The Alder Managers: The Cultural Ecology of a Village in Nagaland, N.E. India*, Published online as part of the Digital Himalayas Rare Books and Manuscripts collection, at: http://www.digitalhimalaya.com/collections/rarebooks/

Cairns, M. F. (ed.) (2015a) *Shifting Cultivation and Environmental Change: Indigenous People, Agriculture and Forest Conservation*, Earthscan from Routledge, London

Cairns, M. F. (2015b) 'Ancestral domain and national park protection: Mutually supportive paradigms? A case study of the Mt. Kitanglad Range Nature Park, Bukidnon, Philippines', in M. F. Cairns (ed.) *Shifting Cultivation and Environmental Change: Indigenous People, Agriculture, and Forest Conservation*, Routledge, London, pp.597-634

Cairns, M. F. (ed.) (2017) *Shifting Cultivation Policies: Balancing Environmental and Social Sustainability*. CAB International, Wallingford, UK

CBC Digital Archives (1985) *N.B. residents raise a stink over a nearby hog farm*, available at https://www.cbc.ca/archives/entry/nb-residents-raise-a-stink-over-a-nearby-hog-farm, accessed 26 November, 2021

CBC News (2015) *Bedeque's amalgamation plan dismays many residents*, available at https://www.saltwire.com/prince-edward-island/news/group-opposing-bedeque-amalgamation-holds-meeting-in-freetown-100531/, accessed 1 February, 2022

CGIAR (2021) *The CGIAR Portfolio Prospectus*, available at https://www.cgiar.org/, accessed 12 September 2021

Chang, A. (2022) 'Researchers test growing white lupin for food on P.E.I.', *CBC News*, 7 March, 2022, retrieved from https://www.cbc.ca/news/canada/prince-edward-island/pei-white-lupins-crops-agriculture-1.6375332, accessed 26 May 2023

Garrity, D. (2007) 'Challenges for research and development on improving shifting cultivation systems', in M. F. Cairns (ed.) *Voices from the Forest: Integrating Indigenous Knowledge into Sustainable Upland Farming*, Resources for the Future Press, Washington, DC pp.3-7

Herawati, T., de Foresta, H., Rohadi, D., Banjade, M., and Fay, C. (2017) 'Negotiating for community forestry policy: The recognition of damar agroforests in Indonesia', in M. F. Cairns (ed.) *Shifting Cultivation Policies: Balancing Environmental and Social Sustainability*, CAB International, Wallingford, UK, pp.837-856

King, D. (1995) 'Rails to Agriculture group wants gov't to stop playing head games', *Island Farmer*, 11 December 1995, Montague, Prince Edward Island, p.8

Potter, L. (2015) 'Where are the swidden fallows now? An overview of oil-palm and Dayak agriculture across Kalimantan, with case studies from Sanggau, in West Kalimantan', in M. F. Cairns (ed.) *Shifting Cultivation and Environmental Change: Indigenous People, Agriculture, and Forest Conservation*, Routledge, London, pp.742-769

SaltWire Newsletters (2015) *Group opposing Bedeque amalgamation holds meeting in Freetown*, SaltWire Newsletters, Halifax, Nova Scotia, available at https://www.saltwire.com/prince-edward-island/news/group-opposing-bedeque-amalgamation-holds-meeting-in-freetown-100531/, accessed 1 February, 2022

Scott, J. C. (2009) *The Art of Not Being Governed: An Anarchist History of Upland Southeast Asia* (Yale Agrarian Studies Series), Yale University Press, New Haven, CT

van der Ploeg, J. and Persoon, G. A. (2017) 'Figments of fire and forest: Shifting cultivation policy in the Philippines and Indonesia', in M. F. Cairns (ed.) *Shifting Cultivation Policies: Balancing Environmental and Social Sustainability*, CAB International, Wallingford, UK, pp.3-26

Vongpaphane, M. and Cramb, R. (2015) 'Impacts of smallholder rubber on shifting cultivation and rural livelihoods in northern Laos', in M. F. Cairns (ed.) *Shifting Cultivation and Environmental Change: Indigenous People, Agriculture, and Forest Conservation*, Routledge, London, pp.826-840

Webber, Justice L. K. (1999) Judgement in the matter of an injunction sought by the Attorney General of Prince Edward Island against Donald Thompson, Heather Thompson, Ivan Boswall, Kenneth McNally, James McNally, Douglas Jenkins, John Doran and Lucille Boswall, GSC-16815, delivered 3 March 1999, Supreme Court, Charlottetown, Prince Edward Island

Notes

1. See the synthesis paper for Section I '*Why Farmer Innovations Should Interest Us*', this volume.
2. Southeast Asian Universities Agroecosystem Network (SUAN)
3. There are other reasons, of course, why villagers may have been burning the land. A contemporary example, here in the mountains of northern Thailand, happens near the end of the rainy season every year, when farmers often burn the forest floor to encourage the growth of *Astraeus hygrometricus* mushrooms (widely known as '*het pho*' in the local dialect). These mushrooms are in strong demand and fetch a high price at the local markets. In my own home province in Canada, fire is similarly reported to stimulate the growth of wild blueberries in burnt forest areas.
4. Even if they didn't cut down many trees, some of the systems documented by Bourke and Allen (ch. 12, this volume) in Papua New Guinea show that by simply clearing the underbrush, this alone may have opened enough ground between the trees to cultivate shade-tolerant crops.

5. It is, of course, entirely rational for farmers to resist investing scarce resources in producing products that nature provides freely. I encountered this during my early days working in dairy extension at the Prachuabkirikhan Dairy Cooperative in southern Thailand.

    One of the thrusts of my work there was to encourage small dairy farmers to establish improved pastures for their cattle. At that time, many farmers were able to cut freely-available grass (primarily *Brachiaria mutica* (Poaceae)) that grew abundantly along the roadside ditches - without needing to invest any land, labour or capital in growing their own fodder crops. Even though it was of poorer quality than they could have grown, the availability of this roadside grass was the most-often cited reason why farmers were less interested in planting improved pastures.

6. It was not only farm work that was frowned upon on Sundays. I recall my mother telling me that her uncles (Fosters), such was their piety, felt it important to polish their shoes on Saturday evenings, so that they would be ready to wear for church the next day. The reason was that even a routine task such as shining one's shoes, was considered unacceptable labour to undertake on a Sunday, even if it was in preparation to attend church service. Some farm tasks, such as milking the cows, feeding the livestock, or cleaning out the stables, obviously needed to be done daily, and were not forbidden.

7. Prior to his tenure as Director General of ICRAF, Dr. Dennis Garrity (2007) wrote about some of the challenges that are specific to undertaking research on shifting cultivation.

8. That is why you often see signs loudly proclaiming, '*There is no Planet B!*' amidst protests about climate change.

9. If farmers kept their livestock in overly-crowded conditions, amongst the first problems that they would expect would be the outbreak of fighting and disease. Witness the COVID-19 pandemic that recently rampaged across the planet. The fighting prediction is equally applicable, as more people are squabbling over fewer resources.

    When I first came to Thailand as a CUSO volunteer, I remember hearing about an innovative Thai farmer, who was able to make use of this principle of overcrowding precipitating the outbreak of disease as an alternative to spraying toxic pesticides on his crops. He would walk around his fields and manually capture as many of the problem insects as he could. These captured insects would then be kept in a very confined cage and fed little. These stresses would quickly lead to the outbreak of disease within the population of confined insects. When the disease was at its height, the farmer would pulverize the diseased insects and mix them into a solution. When this solution was then sprayed on the field, it effectively dispersed the disease pathogen so that it would infect the wider population of problem insects.

10. The 2006 movie, '*Idiocracy*' comes strongly to mind here.

11. To be clear, this schematic does not imply that if land-use pressures continue to increase, farmers' strategies for managing their fallowed lands will necessarily move horizontally along this continuum, from one typology to the next. It means only that most indigenous strategies for managing their fallow lands more effectively can be fit somewhere along this continuum.

12. Perhaps because of their expectations that politicians will work on their behalf, urban dwellers may have higher expectations of the rights and privileges that they should enjoy, and are quick to protest when they feel that their rights have been trampled upon. The 'freedom protests' against COVID restrictions that began in Canada and spread internationally may be symptomatic of this sense of entitlement and refusal to make any personal compromise for the wider public good. These protesters seem to feel that being asked to wear a mask, get a vaccine or limit their social interactions is just too much of an infringement on their personal rights and freedoms to tolerate.

    It is hard not to contrast this behaviour with that of farmers when their livestock are threatened by highly contagious diseases. For example, if British sheep and cattle farmers had acted half as unreasonably as the Canadian protesters, it is unlikely if they would have ever wrestled the 2001 outbreak of foot-and-mouth disease back under control. As it was, more than six million head of cattle and sheep had to be culled in the effort to stop the disease (Anonymous, 2022). These British farmers were not asked to do anything as paltry as wear a mask or refrain from attending the footy matches that season. Efforts to break the chain of transmission of foot-and-mouth disease meant that entire flocks and herds of livestock had to be culled. The livestock on many of these farms would have been built up over multiple generations of careful breeding, gradually improving their genetics. These farmers were looking at the overnight loss of several lifetimes of work! And yet, they understood the science behind the disease's spread and without protest, made the personal sacrifices that were necessary for the wider good of their industry.

In comparison, the Canadians protesting against COVID restrictions appeared as spoiled children engaged in a temper tantrum.

13. There are arguably some parallels between former Brazilian President Bolsonaro's determination to exploit the Amazon rainforest in Brazil, and the process of 'amalgamation' that I describe as being imposed on rural residents in my home province of Prince Edward Island, Canada. In both cases, politicians are trying to improve their economies by trampling over the rights of the proven custodians of the land, to extract resources and money. Both are trying to funnel rural resources into Government coffers.

14. A politician campaigning for votes is like a child trying to fill his bag with candy on Halloween evening. Both know that there are many more treats to be had in the cities.

15. My father has only to look out his kitchen window to confirm this reduction in farming families. He recalls that when he used to collect school taxes about 50 years ago, there were about 35 farms in the Lower Freetown area. Today, he counts only 5 farms that are fully based in Lower Freetown. That all the land continues to be cultivated testifies to the general increase in farm size. But equally as clearly, it shows the declining voting power of farmers on Prince Edward Island.

Moreover, as these families left their farms, their younger generations often migrated to the cities and became strangers to agriculture and how their food is produced. As part of this transformation, their primary interest will have changed from concern that farmers receive a fair price for their products to a sense of entitlement that they should have to spend a decreasing portion of their income for their food needs. This shift in priorities would have been part of their metamorphosis from primary producer to consumer.

16. Although farmers probably need to be unionized and have a strong collective voice more than most professions, the nature of their lifestyle seems to work against this. Farmers are widely spread across the countryside and getting essential work completed within a short growing season keeps them extremely busy. They also tend to be an independent lot, accustomed to taking care of their own affairs.

The likelihood of using strikes for negotiating better conditions is another stark contrast between urban and rural. Whilst their urban counterparts seem ready to strike 'at the drop of a hat', farmers tend not to be strongly unionized and seldom strike, even for the most critical issues. Governments count on it. In sum, probably few farmers feel that they can afford to stop working. Crops and livestock need daily care, and farmers can't let meddling governments interfere with getting their work done.

That the farmer protest in India was recently successful in forcing the Modhi government to capitulate was an outcome not only from their impressive organization and determination – but also that farmers in India remain a relatively high portion of the national population and that politicians must still covet their votes, if they want to win elections.

17. Worryingly, the Green Party that has recently begun to garner more support on Prince Edward Island and is now the official opposition in the PEI Legislative Assembly, seems quite hostile to agriculture in its policies, and tends to run urban candidates that think that they know better than farmers how their land should be managed.

18. Profit margins in farming are already slim enough without the tax collector trying to claim a larger share. This sentiment is doubly strong when farmers see successive Governments waste their tax dollars, as if they were easily earned.

19. The flow of services may actually go in the opposite direction. One example of this is that when newly-qualified veterinarians graduate from vet school these days, many will prefer the easier and more lucrative work of treating small pets in the cities than treating farmers' livestock in the countryside. Again, the cities are receiving priority, at the expense of farmers.

20. Given the much longer hours that farmers universally work than 'nine-to-five' city dwellers, if rural tax dollars are used, for example, to build more recreational facilities in the municipalities, after their day's work is finished, it is doubtful if farming families would have the time or energy left to drive into town and use the new facilities anyway.

21. Rural Prince Edward Island has had plenty of bitter experience to make it doubtful of receiving fair treatment from its municipal neighbours. A recent example of this can be found in the aftermath of the closing of the railroads on PEI in 1989. Back in 1871, before PEI joined the Canadian Confederation (in 1873), the legislature of Prince Edward Island passed the Railroad Act of 1871, authorizing construction of a railroad across the Island. Many of PEI's farmers had to surrender a corridor of their land for the construction of this railway line in the 1880s. But Section 10 of Chapter 13 in the Railroad Act of 1871 specified that if and when the railroad no longer needed this land, it would be returned to the farmers who had previously owned it, or their heirs. Contrary to this agreement, when PEI's rail system was closed down in 1989, this commitment to return the land to the farmers was broken and the land was instead repurposed for a 'Rails to Trails' project that converted the old rail lines into a long recreational corridor, primarily for the use of city folk and tourists (King, 1995). These newly purposed trails can provide hikers, bikers or snowmobilers with direct access to private farmland, without the farmers' knowledge or consent. The prospect of outsiders wandering around in farmers' back fields created the distinct danger that paddock gates could be thoughtlessly opened, allowing livestock to escape. Or that carelessly flicked cigarette butts could set barns or dry grain fields on fire. At the very least, garbage would almost certainly be strewn around by the unwelcome visitors and would run the risk of becoming caught in the throats of grazing livestock. Clearly, farm land is private land, on which the farmers pay taxes every year, and the public should have no right to trespass on it.

    When rural residents, most of whom were probably impacted farmers, protested this denial of their land by blockading sections of the trails, the Attorney General of Prince Edward Island took the dispute to the Supreme Court in Charlottetown in September, 1998 to seek a permanent injunction against any further such blockades. A group of farmers calling themselves 'Rails to Agriculture' represented farmers in the case, in opposition to the 'Rails to Trails' lobby – but a 41-page decision handed down in March of 1999 ruled against them (Webber, Justice L. K., 1999). This was yet another example of unjust urban domination over the countryside.

22. Another example of the divergent views of rural farmers and their non-farming neighbours on Prince Edward Island can be found in the wild lupins (*Lupinus polyphyllus*, Leguminosae) that bloom around mid-summer every year. The pink and purple hues of blooming lupins add to the aesthetic beauty of the Island landscape, and are commonly featured in scenic photos published in wall calendars and tourist publications.

    Folk who earn their living from the tourist industry view wild lupins as an asset that helps to augment the Island's beauty and attract tourists. Some therefore try to expand the lupin population on the Island by driving around and throwing lupin seeds into the ditches in rural areas. But to the farmers who manage the fields along those ditches, lupins are an invasive weed and not welcome.

    These invasive lupins (*Lupinus polyphyllus*, Leguminosae) found growing wild in roadside ditches, are toxic and should not be confused with the white lupins (*Lupinus albus*) that are sometimes planted as a pulse crop (Chang, 2022). Although *Lupinus albus* is a traditional pulse cultivated in the Mediterranean region, its use as a crop on Prince Edward Island remains at an experimental stage.

    But in fact, it is precisely the work of farmers that gives Prince Edward Island its well-manicured, pastoral landscape that is at the heart of PEI's charm and character, and attracts so many tourists desperate to escape their overly-frenetic lives in the cities. In this way, agriculture is central to PEI's success as a tourist destination. Farmers are thus holding up two of the main pillars of the Island's economy, agriculture and arguably, tourism.

23. Another kind of conflict arises when city folk decide that they want to move out into the country to enjoy the quiet and clean living of life in the countryside. In keeping with the behaviour of the governments that they elect, these city transplants often begin to harass their farming neighbours.

    I recall one particularly appalling case in Charlo, New Brunswick, in the 1980s, in which an established swine farmer suddenly found himself surrounded by city transplants who had built new homes around his farm and then decided that the smell of pig manure offended their delicate city noses (CBC Digital Archives, 1985). These newcomers litigated against the hog farmer to try to force him to stop the smell. The harassment against this particular farmer became so serious that he was finally driven into bankruptcy, and eventually to suicide in 1990. The case of this pork producer, bullied to an early grave by litigating neighbours, strongly highlighted the need for Right to Farm legislation in the Maritime provinces.

This is to make the point that urban interference with farming is not limited to shifting cultivators but is part of a wider global trend. It is akin to biting the hand that feeds them.

Indeed, the Right to Farm is likely to become an increasingly important issue as the world tries to grapple with some of the most threatening environmental problems confronting it. Farmers are increasingly in the crosshairs for criticism as agriculture is blamed as a major contributor to some of those major environmental problems.

For example, under the sensationalist accusation that 'cows are the new coal', pressure to reduce nitrogen emissions in The Netherlands is threatening the future of dairy farming there. Hence, the bullying of the hog farmer in Charlo, New Brunswick, now seems to be upscaled from the farm level to the industry level. It is tempting to question how solid the science is behind these interventions – or is it simply a case of the least sustainable form of human habitation – the cities – trying to distract from their own blame by trying to scapegoat farmers that they see as easy targets? Shifting cultivators are also extremely vulnerable to this same type of environmental politics. If analysts are as badly wrong about western farming systems as they have historically been mistaken about shifting cultivation, then you wouldn't want to give much credence to their advice.

24. Farmers suddenly finding themselves paying a municipal tax would not be instead of paying a provincial tax. It would be in addition to paying the provincial tax, as a totally new layer of added taxation. But at least tax dollars paid to the provincial government could possibly be spent in ways that support farmers – but the chances of this from a municipality are virtually nil. Farmers' tax dollars paid to a municipality would more likely be spent on building a new sports arena that does nothing for farmers.

25. Most of the agricultural community on Prince Edward Island trace their origins back to one of the British Isles. Some of their farming ancestors may have been tenant farmers, forced off their land in the 'old country' by the unfairness of taxes and rents levied against them by wealthy landlords. Island farmers could be forgiven if they feel that neighbouring municipalities are now trying to play the same role that the wealthy landlords once played and that history is repeating itself.

26. With cows to feed and milk both morning and evening, 365 days a year, nobody works longer hours than a dairy farmer. If most farmers were to work the 40 hours a week that is typical of an urban job, they would probably feel that they were doing so little that they were essentially on vacation. But, of course, neither dairy cows, nor most of the farmers who look after them, take vacations or know what they would feel like.

Try visiting one of PEI's beautiful beaches on a hot summer day and see how many farmers you can find relaxing there. Virtually none; they will all be in their fields working, whilst their municipal neighbours enjoy barbeques or work on their tans. If there are farmers on the beach anywhere, they will be instantly recognizable by what is known humorously as their 'farmer tans' (lily-white legs, with deeply bronzed forearms). Given this wide disparity in working hours, it seems grotesquely unfair to expect those hard-working farmers to pay added taxes so that their urban neighbours can live even more leisurely lifestyles.

27. The urban-rural divide on Prince Edward Island is not limited to space, but extends into time as well. As with the land, the roots of this discord lie within the priorities that each side holds for the use of their time. PEI's location in eastern Canada places it within the Atlantic Standard Time (AST) Zone, four hours behind Greenwich Mean Time (GMT).

Like agrarian societies everywhere, the work of PEI's farmers is scheduled around sunlight, waking up at sunrise to look after their livestock and toil in the fields, and heading home as the sun begins to dip behind the western horizon. As Earth makes its yearly revolution around the sun and winter fades into spring, the tilt of the planet's imaginary axis (23.5°) bathes the northern hemisphere with increasing daylight. As winter fades on PEI, for example, the day length increases at a rate of roughly an additional three minutes per day. To optimize use of this natural daylight, many countries, chiefly in North America and western Europe, set their clocks forward one hour during the summer months, and back again in the autumn, in a practice known as Daylight Savings Time (DST).

Like farmers in most places, PEI's farmers didn't welcome DST when it was initially being proposed, and lobbied hard against its imposition. It is the sun and the seasons that determine their farming schedules, not the clock. Having to change the clock twice a year disrupted their schedules, confused their livestock, robbed them of precious daylight in the mornings and delayed them from beginning crucial fieldwork that relied on the drying influence of full sunlight, amongst other disadvantages. In fact, until the election of Walter Shaw's Government in 1959, farmers' wishes had been respected and the use of DST had been prohibited on PEI by the previous Liberal Governments. But this law apparently 'lacked teeth' and the main municipalities of Charlottetown and Summerside ignored it and adopted DST regardless.

In contrast to the farmers, folk in the towns and cities liked the idea of having more daylight later in the day. It meant that after returning home from work, they still had plenty of daylight left to run a few errands, fire up the barbeque or play a few holes of golf.

By the time of the general election of 1959, the nonfarm vote was gaining size on PEI, and the Walter Shaw Government legislated against the farmer lobby and imposed DST on Islanders. This was yet another example that when the interests of urban and rural are contested, the urban lobby will be given its way almost every time.

28. Since shifting cultivators are often oral societies and not given to recording their farming practices in writing, it seems doubly important to have outside researchers working closely with farmers in documenting their indigenous knowledge and practices before they are lost in the mists of time. This entire volume is the outcome of that kind of close farmer-scientist collaboration.

29. Like many Indonesians, Wiyono is widely known by only one name – but it was only recently that I learned that his full name is actually Tajam Wiyono.

30. Whilst working on a Master's degree at York University, Toronto, I recall traveling off-campus one evening to attend a lecture by Dr. James Lovelock, about his Gaia theory that essentially views the world as one large and self-regulating super-organism. I remember that a member of the audience had asked Dr. Lovelock about the damage that humanity is inflicting on Earth. If memory serves me correctly, whilst acknowledging the formidable damage that we are doing to our planetary home, Dr. Lovelock had assured us that human destructiveness wouldn't be likely to actually kill Gaia, because she had 'ways of protecting herself'. I've sometimes pondered what Dr. Lovelock's view would be of the coronavirus pandemic that has plagued the world over the past several years, and whether it might simply have been an example of Gaia protecting herself?

The North American beaver (*Castor canadensis*) is widely viewed as emblematic of hard-working, thus giving rise to the old expression, '*as busy as a beaver*'. An image of a beaver is included here, because it accurately represents the work ethic of the small team that produced this volume.

I. INTRODUCTORY SECTION

Village elders are the custodians of much of the indigenous knowledge that underlies the farmer innovations documented in this volume. This old farmer looks thoughtful as she enjoys a hand-rolled cigarette. She is of the Kri people, a very small Austroasiatic minority group living in remote forest areas of Nakai district, in Khammouane province, Lao PDR.

Sketch based on a photo by Laurent Chazée in April, 2000.

Synthesis

UNDERSTANDING CHANGE AND INNOVATION IN SHIFTING CULTIVATION

Ole Mertz[*]

In an earlier volume of the Cairns series on shifting cultivation in Southeast Asia, I, together with my colleague Thilde Bech Bruun, opened a chapter by quoting a shifting cultivator in Sarawak, who had asked for ideas for improving their farming practices (Mertz and Bruun, 2017). The aim of that volume was to show that anti-shifting cultivation policies in Southeast Asia were taking a flawed approach, by working against, rather than with shifting cultivators, and thus potentially missing out on important agricultural and social-development opportunities. As if in response to the Sarawak farmer and countless others, the search for innovations in shifting cultivation – by shifting cultivators themselves – has now become the topic of this entire volume, the third in the same series, bringing to the fore the entrepreneurship and eagerness of shifting cultivators for development on their own terms.

The three chapters introducing the volume all take a point of departure in the negative views on shifting cultivation that have been pervasive in government and development circles for decades, and which appear to have prevented a thorough science on technological development of shifting cultivation (Nair, ch. 3, this volume). For example, while there has been considerable research on development of upland rice varieties, a very small part of this research has been directly aimed at shifting cultivation systems, which are the prime users and developers of upland rice varieties. The wish to eradicate shifting cultivation and to push shifting cultivators into new approaches to farming, or even completely out of agriculture, also runs the risk of losing local knowledge. For example, in one small community of just 30 to 40 households in Borneo, more than 100 names of rice varieties were recorded in the 1990s, while all development assistance was focused on cash crops like pepper, rubber, and cocoa (Christensen, 2002; Mertz and Christensen, 1997; Mertz et al., 2013;

[*] Professor Ole Mertz, Department of Geosciences and Natural Resource Management, University of Copenhagen, Denmark.

Wadley and Mertz, 2005). Several other studies have pointed to the importance of the genetic diversity and valuable traits in local varieties (Pandey et al., 2019; Rana et al., 2007), which despite relatively low yields, represent important nutritional diversity and are necessary for risk avoidance. The tenacity of shifting cultivation pointed out by Maithani (ch. 2, this volume) is very likely a function of innovative efforts in the land system, such as experimentation with local crop varieties and exchanges.

It is important to recognize that besides the push from the government and development sector to abandon shifting cultivation, shifting cultivators also react to strong pull factors. Shifting cultivation might be energetically efficient under conditions of abundant land, but it is also hard work and often insufficient to cover cash needs, tempting many to seek alternatives such as cash crops and labour migration (Cramb et al., 2009; Ornetsmüller et al., 2018; Wadley and Mertz, 2005). This is not new, of course, and as several of the introductory chapters point out, shifting cultivation was always a hybrid of land uses constantly adapting to the prevailing contexts and opportunities (Schmidt-Vogt, ch. 1, this volume) and in most of these mosaic landscapes, there are elements of both permanent farming and shifting cultivation (Nair, ch. 3, this volume). In fact, it is likely that much of the innovation in shifting cultivation stems from the exploration of introduced farming practices and crops. Examples are more than a century of cultivation of rubber in Malaysia and Indonesia (Cramb, 1988; Dove, 1993; Schmidt-Vogt, ch. 1, this volume), the more recent integration of smallholder cardamom in shifting cultivation in Laos (Ducourtieux et al., 2006), multiple tree crops in Indonesia (Burgers, ch. 37, this volume) and mixed oil palm and upland rice, as observed in Sarawak (Mertz, 2015).

In the first chapter of the introductory section, Schmidt-Vogt (ch. 1) outlines numerous other examples, many of which come from the chapters in this volume, of how shifting cultivators have been taking charge of their own development and innovations. He challenges the classical core-periphery theory, arguing that it is in the 'peripheries' that real change and economic activities are now occurring and that the 'peripheries' are central for connections across border regions, thus essentially discarding the original theory as a myopic concept of how regions develop. By drawing on a review of shifting cultivation (van Vliet et al., 2012), which has been corroborated by later reviews (Dressler et al., 2017), he also urges a change to the rhetoric of shifting cultivation, from 'conversion' to 'transformation'. There has been a vast focus on how shifting cultivation is being converted to other forms of land use, which is indeed an important trend shown in the reviews, but Schmidt-Vogt points in chapter 1 to examples in this volume to argue that there are an equally large number of examples where shifting cultivators have transformed their farming practices without the need to fully convert to monocultures.

Maithani (ch. 2, this volume) takes a more historical look at how shifting cultivation in India since colonial time has been seen as wasteful of resources and unproductive. This, he points out, is despite important evidence that upland shifting cultivation areas could be equally or even more productive than lowland settled areas, probably due to innovations in farming practices making them more resilient to irregular weather patterns and insufficient access to fertilizers and other inputs. The importance of shifting cultivation landscapes as cultural landscapes is also highlighted, and this is indeed an issue that is receiving increasing attention. For example, the UNESCO Ciletuh-Palabuhanratu Geopark in West Java, Indonesia, protects a landscape that incorporates shifting cultivation at a small scale (UNESCO, 2018; Wulandari et al., 2021), and the Globally Important Agricultural Heritage Systems (GIAHS) initiative of the Food and Agriculture Organization of the United Nations has proposed the Milpa system in Mexico (shifting cultivation with a focus on maize) as a GIAHS site. Such initiatives can be important for recognizing the role of shifting cultivation, not as a museum piece, but as an evolving and dynamic farming system based on solid local knowledge.

In the third introductory chapter of this volume, Nair addresses the linkages between shifting cultivation, agroforestry and ecosystem services by taking a point of departure in the conspicuous absence of shifting cultivation in agronomic research. Various agroforestry options have been proposed as alternatives to shifting cultivation, but many of the technologies and interventions have not lived up to expectations. Alley cropping, for example, has only been successful in areas where the rainfall is greater than 1000 mm, the soils are relatively fertile, and there are no labour shortages. The latter condition often represents a major limitation because labour is frequently in short supply in farming communities. Similar to alley cropping, the use of fast-growing nitrogen-fixing species to improve fallows has also seen difficulties in sustained adoption, especially after development and technology packages were withdrawn (Nair, ch. 3). However, fallow improvement is not an intervention developed solely by science. The first volume in this series (Cairns, 2007) is essentially a compilation of case studies of fallow improvement partly driven by local knowledge and initiatives and partly by development interventions. That volume – as well as this one – also shows that the productive element of fallow improvement is important as farmers are reluctant to plant woody species into harvested fields unless there is a clear productive value during the fallow period.

In the last part of chapter 3, Nair sees a problem in the lack of recognition of ecosystem services produced by shifting cultivation systems. The core issue is that shifting cultivation is all too often compared to old-growth forests rather than to other forms of annual or plantation agriculture. If compared to the latter, shifting cultivation often provides a broader set of ecosystem services (besides provisioning), for example, by providing more time-averaged carbon storage and scoring higher on biodiversity indicators (Mertz et al., 2021; Padoch and Pinedo-Vasquez, 2010). Nair (ch. 3) also points to the lack of knowledge on trees and woody cover in shifting cultivation areas, since many of the trees may be located on fields, or not in sufficient numbers to constitute a forest. New research will very soon remedy this gap as high-resolution satellite products combined with machine learning and artificial intelligence are already able to detect, count and measure individual trees over large areas, such as the western Sahel and Sahara (Brandt et al., 2020), and the technology is being developed to also enable counting and measuring all trees, whether outside or inside forests (Reiner et al., 2022). Cultural services are only addressed to a limited extent in chapter 3, and these are indeed studied to a much smaller extent. There is, however, an emerging literature on cultural ecosystem services (Milcu et al., 2013) and, as mentioned above, there are also new initiatives to conserve the cultural landscapes generated by shifting cultivation.

Overall, the three chapters provide a diverse and exciting introduction to the volume and set the stage for further discussions on shifting cultivation and debates on how the future research agenda should be shaped. Despite decades, if not centuries, of efforts by governments and land developers to get rid of shifting cultivation, it is not going away any time soon, and science needs to consider how shifting cultivators in Southeast Asia can be best supported for development on their own terms and in directions they choose as landscape managers.

In conclusion, I would also like to reiterate the global nature of shifting cultivation and concur with several of the authors of the introductory chapters that shifting cultivation change and innovation is not just a Southeast Asian phenomenon (Nair, ch. 3; Schmidt-Vogt, ch. 1). Quite the contrary, similar processes are found in Africa and Latin America, and the fact that the African systems, in particular, remain largely understudied has been noted in most global reviews (Mertz et al., 2021; Mukul and Herbohn, 2016; van Vliet et al., 2012). I therefore hope that the next large volume on shifting cultivation will focus on Africa, given that shifting cultivation there is most likely still on the increase (Heinimann et al., 2017). Another reason is that recent studies have pointed to shifting cultivation as the main driver of forest loss in Africa and the results are presented in high profile maps now being used by development organizations. This is unfortunate because the research acknowledges a conflation of shifting cultivation and other smallholder cash-crop farming in the analyses and mentions a higher degree of inaccuracy in identifying these drivers when compared with other drivers (Curtis et al., 2018). However, maps are powerfully influential and are often used without reading the full text. Much more and better knowledge at local level is needed to ensure that the innovative capacities of African shifting cultivators are acknowledged and that they are supported in realizing the development potential of their farming systems.

References

Brandt, M., Tucker, C. J., Kariryaa, A., Rasmussen, K., Abel, C., Small, J., Chave, J., Rasmussen, L. V., Hiernaux, P., Diouf, A. A., Kergoat, L., Mertz, O., Igel, C., Gieseke, F., Schöning, J., Li, S., Melocik, K., Meyer, J., Sinno, S., Romero, E., Glennie, E., Montagu, A., Dendoncker, M. and Fensholt, R. (2020) 'An unexpectedly large count of trees in the West African Sahara and Sahel', *Nature* 587(7832), pp.78-82, doi 10.1038/s41586-020-2824-5

Cairns, M. F. (ed.) (2007) *Voices from the Forest. Integrating Indigenous Knowledge into Sustainable Upland Farming.* Resources for the Future Press, Washington, DC

Christensen, H. (2002) *Ethnobotany of the Iban and the Kelabit*, Forest Department, Sarawak, NEPCon and University of Aarhus, Kuching and Aarhus

Cramb, R. A. (1988) 'The commercialization of Iban agriculture', in R. A. Cramb and R. H. W. Reece (eds) *Development in Sarawak: Historical and Contemporary Perspectives*, Centre of Southeast

Asian Studies, Monash University, Melbourne, Australia, pp.105–134

Cramb, R. A., Colfer, C. J. P., Dressler, W., Laungaramsri, P., Le, Q. T., Mulyoutami, E., Peluso, N. L. and Wadley, R. L. (2009) 'Swidden transformations and rural livelihoods in Southeast Asia', *Human Ecology* 37, pp.323–346

Curtis, P. G., Slay, C. M., Harris, N. L., Tyukavina, A. and Hansen, M. C. (2018) 'Classifying drivers of global forest loss', *Science* 361, pp.1108–1111

Dove, M. R. (1993) 'Smallholder rubber and swidden agriculture in Borneo: A sustainable adaptation to the ecology and economy of the tropical forest', *Economic Botany* 47, pp.136–147

Dressler, W. H., Wilson, D., Clendenning, J., Cramb, R., Keenan, R., Mahanty, S., Bruun, T. B., Mertz, O. and Lasco, R. D. (2017) 'The impact of swidden decline on livelihoods and ecosystem services in Southeast Asia: A review of the evidence from 1990 to 2015', *Ambio* 46, pp.291–310

Ducourtieux, O., Visonnavong, P. and Rossard, J. (2006) 'Introducing cash crops in shifting cultivation regions: The experience with cardamom in Laos', *Agroforestry Systems* 66, pp.65–76

Heinimann, A., Mertz, O., Frolking, S., Egelund Christensen, A., Hurni, K., Sedano, F., Parsons Chini, L., Sahajpal, R., Hansen, M. and Hurtt, G. (2017) 'A global view of shifting cultivation: Recent, current, and future extent', *PLoS ONE* 12, e0184479

Mertz, O. (2015) 'Oil palm as a productive fallow? Swidden change and new opportunities in smallholder land management', in M. F. Cairns (ed.) *Shifting Cultivation and Environmental Change: Indigenous People, Agriculture and Forest Conservation*, Earthscan from Routledge, London, pp.731–741

Mertz, O. and Bruun, T. B. (2017) 'Shifting cultivation policies in Southeast Asia: A need to work with, rather than against, smallholder farmers', in M. F. Cairns (ed.) *Shifting Cultivation Policies: Balancing Environmental and Social Sustainability*, CABI, Wallingford, UK, pp.81–96

Mertz, O., Bruun, T. B., Jepsen, M. R., Ryan, C. M., Zaehringer, J. G., Hinrup, J. S. and Heinimann, A. (2021) 'Ecosystem service provision by secondary forests in shifting cultivation areas remains poorly understood', *Human Ecology* 49, pp.271–283

Mertz, O. and Christensen, H. (1997) 'Land use and crop diversity in two Iban communities, Sarawak, Malaysia', *Geografisk Tidsskrift-Danish Journal of Geography* 97, pp.98–110

Mertz, O., Egay, K., Bruun, T. B. and Colding, T. S. (2013) 'The last swiddens of Sarawak', *Human Ecology* 41, pp.109–118

Milcu, A. I., Hanspach, J., Abson, D. and Fischer, J. (2013) 'Cultural ecosystem services: A literature review and prospects for future research', *Ecology and Society* 18(3), p.44, doi 10.5751/ES-05790-180344

Mukul, S. A. and Herbohn, J. (2016) 'The impacts of shifting cultivation on secondary forests dynamics in tropics: A synthesis of the key findings and spatio-temporal distribution of research', *Environmental Science and Policy* 55 (part 1), pp.167–177

Ornetsmüller, C., Castella, J-C. and Verburg, P. H. (2018) 'A multiscale gaming approach to understand farmers' decision making in the boom of maize cultivation in Laos', *Ecology and Society* 23(2), p.35, doi 10.5751/ES-10104-230235

Padoch, C. and Pinedo-Vasquez, M. (2010) 'Saving slash-and-burn to save biodiversity', *Biotropica* 42, pp.550–552

Pandey, D. K., Adhiguru, P., Devi, S. V., Dobhal, S., Dubey, S. K. and Mehra, T. S. (2019) 'Quantitative assessment of crop species diversity in shifting cultivation systems of Eastern Himalaya', *Current Science* 117, pp.1357–1363

Rana, R. B., Garforth, C., Sthapit, B. and Jarvis, D. (2007) 'Influence of socio-economic and cultural factors in rice varietal diversity management on-farm in Nepal', *Agriculture and Human Values* 24, pp.461–472

Reiner, F., Brandt, M., Tong, X., Skole, D., Kariryaa, A., Ciais, P., Davies, A., Hiernaux, P., Chave, J., Mugabowindekwe, M., Igel, C., Oehmcke, S., Gieseke, F., Li, S., Liu, S., Saatchi, S., Boucher, P., Singh, J., Taugourdeau, S., Dendoncker, M., Song, X-P., Mertz, O. and Fensholt, R. (2022) 'A consistent assessment of trees inside and outside forests improves estimates of total tree cover in Africa', pre-print published at Research Square, not yet peer-reviewed, doi.org/10.21203/rs.3.rs-1816495/v1

UNESCO (2018) 'Ciletuh-Palabuhanratu UNESCO Global Geopark (Indonesia)', United Nations Educational, Scientific and Cultural Organization, Indonesia, available at https://en.unesco.org/

global-geoparks/ciletuh-palabuhanratu, accessed 14 July 2022

van Vliet, N., Mertz, O., Heinimann, A., Langanke, T., Pascual, U., Schmook, B., Adams, C., Schmidt-Vogt, D., Messerli, P., Leisz, S., Castella, J-C., Jørgensen, L., Birch-Thomsen, T., Hett, C., Bruun, T. B., Ickowitz, A., Vu, K. C., Fox, J., Cramb, R. A., Padoch, C., Dressler, W. and Ziegler, A. (2012) 'Trends, drivers and impacts of changes in swidden cultivation in tropical forest-agriculture frontiers: A global assessment.', *Global Environmental Change* 22, pp.418-429

Wadley, R. L. and Mertz, O. (2005) Pepper in a time of crisis: Smallholder buffering strategies in Sarawak, Malaysia and West Kalimantan, Indonesia', *Agricultural Systems* 85, pp.289-305

Wulandari, I., Iskandar, B. S., Parikesit, P., Hudoso, T., Iskandar, J., Megantara, E. N., Gunawan, E. F. and Shanida, S. S. (2021) 'Ethnoecological study on the utilization of plants in Ciletuh-Palabuhanratu Geopark, Sukabumi, West Java, Indonesia', *Biodiversitas: Journal of Biological Diversity* 22, pp.659-667

1

METAMORPHOSIS IN THE MOUNTAINS

Swiddeners as a force for change in a changing context

*Dietrich Schmidt-Vogt**

'Pathways of metamorphosis leading up and down the mountains make being an issue of becoming'. (Campbell, 2013)

One of the catchwords of contemporary society is 'hybrid'. It is used, these days, to describe specific forms of innovation that are partly forced upon us by changing circumstances, e.g. the COVID-19 pandemic, and are partly made possible by new or adapted technologies. These include hybrid forms of working, both at home and in the office; of education, by combining classroom with online teaching; of communicating, with some people in a room and others on a screen; and also hybrid forms of transportation and motorization. We even speak of hybrid warfare, which combines conventional warfare with cyber-attacks and disinformation. The principle of hybridity in this context is to combine heterogeneous or even discordant elements for the purpose of adapting to novel challenges and opportunities as well as to changing circumstances.

Swidden farming is a hybrid form of land use that combines forest and farming in a dynamic and sequential manner and thereby creates landscapes in which fields morph into forests or other types of vegetation and vice versa. In this sense, swidden is not just a hybrid but also a metamorphic form of land use which shifts shape, often in cycles, and which keeps in motion a continuous and iterative process of land-cover change. Although many variants and even some exceptions to the rule exist, the basic principle of swidden farming is to cut and burn trees on a plot of land, to cultivate crops on this plot for a period of one or a few years and then to let the plot revert to semi-natural vegetation during a fallow period of variable length, through a process of secondary succession. The type of vegetation that becomes established

* PROFESSOR DR. DIETRICH SCHMIDT-VOGT, Faculty of Environment and Natural Resources, Albert-Ludwigs-Universität, Freiburg, Germany; formerly from the Centre for Mountain Ecosystem Studies, Kunming Institute of Botany, Chinese Academy of Sciences, and World Agroforestry Centre (ICRAF), East Asia node, Kunming, China.

depends, among other factors, on the length of the fallow period. Ideally, succession leads to the development of secondary or fallow forests that are capable of fulfilling a broad range of functions, both in the landscape and for the livelihoods of people (Chokkalingam et al., 2001; Schmidt-Vogt, 1998, 1999, 2001; Heinimann et al., 2007; Wangpakapattanawong et al., 2010; Nyein Chan and Takeda, 2016).

The hybridity and volatility of swidden farming has, in the past and up to the present, been held against it. As a form of land use that is 'neither fish nor flesh' when compared to permanent farming or forestry, swiddening has been and still is held in low esteem, especially by policy-makers and land-use planners who prefer the perspicuity of permanent land-use systems to the ambiguity of swidden farming. The fact that swidden landscapes are 'invisible' (Schmidt-Vogt et al., 2009), or 'illegible' (Scott, 2009), because the fluidity of land use and land-cover patterns cannot easily be captured on

Borassus flabellifer L. [Arecaceae]

Growing in more arid parts of South and Southeast Asia, this lofty palm has long supported shifting cultivators battling extended dry seasons (Fox, ch. 35, this volume). Known as the Palmyra Palm or Toddy Tree, its juice can be made into jaggery sugar, allowed to ferment to make toddy, or distilled to make *arak*. Other products include timber, fruit, and leaves which can be used as thatching or plaited into a wide variety of baskets and utensils.

maps, has led map-makers to zero in on the fallow phase of the swidden cycle and to represent swidden landscapes as barren land, fallow land, virgin or unused land. This practice, in combination with policies and legislation, has facilitated the conversion of swidden land into other land uses, e.g. permanent cropping and commercial tree plantations, as, for instance, through the Vacant, Fallow and Virgin Lands Management Law of 2012 in Myanmar (Lim et al., 2017) or the Land and Forest Allocation Policy in Laos (Fujita and Phanvilay, 2008). What is illegible can be erased without much ado. Methods of mapping swidden landscapes, or landscapes that contain swidden agriculture in a way that makes swiddening visible, have been developed only quite recently (Messerli et al., 2009; Hett et al., 2011; Hurni et al., 2013; Kurien et al., 2019, Li et al., ch. 6, this volume).

Conversion policies and practices led to the segregation of rural landscapes into land allocated to farming and land allocated to forest (Castella et al., 2013). In a similar vein – though in a different context – the emergence of modern or scientific forestry in Europe has led to such a segregation by severing previously existing links between forest and farming, for example in the form of forest grazing and collecting leaves as stall bedding. This segregation is manifested in the landscape by the clearcut and rectangular boundaries between forest and farmland that are, for instance, such a telling feature of the rural landscape of Germany, when seen from an airplane, and which contrast so markedly with the fuzzy forest fringes of traditional rural landscapes in the tropics, where fields may be studded with trees, such as the 'toddy palm', *Borassus flabellifer*, in more arid parts of Asia. Such feature changes can be seen as emblematic of the loss of ambiguity which Bauer (2018) deplored in his essay on the decline of diversity and the trend towards unequivocal homogeneity in so many domains.

Another feature of swidden farming that has contributed to the low esteem in which it is held is the application of fire. Although this volume contains some interesting exceptions to this rule and presents the exclusion of fire from swidden farming as an important innovation (see chapters in Section III 'Fireless Shifting Cultivation'), fire is a central element of most swidden farming. The use of fire in agriculture has been regarded as a 'stigma of primitivism' (Pyne, 2021), especially by European agronomists, and therefore modernizing land use in Europe found expression not only in the segregation of farming and forestry but also in the banishment of fire from the landscape. In countries where swidden cultivation is still practised – to the objection of the authorities – fire is also a central element in a narrative that condemns its use (Dressler, ch. A5, in the supplementary chapters to this volume) and portrays swidden cultivators as primitive peasants on the fringes of civilized states where, far away from prevailing trends and developments, they pursue their archaic livelihoods.

Shifting cultivators in a shifting context

This view has probably never been true and is especially ill-fitting in the current context of the highly dynamic Asia-Pacific region. Rapid, global and unpredictable change is the prominent feature of our age. Ernst Jünger, the German author of utopian novels (among other genres in his oeuvre), predicted in 1993 that there would be 'historical spring tides' for the first century of the new millennium, and *The Economist*, in its 2021 Christmas double issue, described the current time as the age of predictable unpredictability. This is especially true for the Asia-Pacific region, which is characterized by rapid and sustained economic development and related changes in agricultural practices, as well as in economic structure (from primary to secondary and tertiary sectors) (Heinimann et al., 2017, p.15). At present we are also learning lessons in unpredictability through our experiences with the COVID-19 pandemic and with unanticipated armed conflicts. Closely related to economic development are demographic changes, especially population growth, with implications for food security. These changes are augmented by geopolitical shifts such as the growing influence of China in the Asia-Pacific region and its implications for land use. They translate into rapid and large-scale land-use changes in response to changes in markets, to government policies and, not least, in response to the changing aspirations of the land users themselves.

The view of swiddeners as archaic farmers on the fringes of more advanced societies is, moreover, based on a rather one-dimensional and outdated view of fringes or peripheries as backwaters or realms of arrested development.

Fringes can be borderlands, where influences from adjoining countries intersect and hybridize, creating changes that are specific to transition zones such as the borderlands of mainland Southeast Asia, where the lands occupied by ethnic groups such as the Akha are dissected by international boundaries (Sturgeon, 2005, 2011). The mutability of farming practices over space and time in such borderlands, which sit astride international boundaries and can transgress administrative principles, has been referred to as landscape plasticity by Sturgeon (2005, 2021). In a completely different geographical setting, the German geographer Hermann Kreutzman (2020) has shown how apparently remote and isolated regions such as the Hunza valley in the Karakoram mountains of northern Pakistan can attain prominence because of their location at the intersection of the spheres of influence of larger political entities such as, in this case, British India and Tsarist Russia in the past, and Pakistan, India and China today.

Fringes can also be frontiers with land-use dynamics typical of frontier zones such as forest–agriculture frontiers, where forest and farmland exist in a state of flux, with farmland encroaching upon forest and forests expanding into abandoned or fallow farmland. Such changes can happen along a linear gradient, as in the case of forest transition curves leading from net deforestation to net reforestation (Mather, 1992), or as pendulum swings between agricultural expansion and ecosystem restoration (Seijger et al., 2021). In short, fringes and peripheries, instead of being seams where stagnation prevails, can be dynamic regions where changes are happening.

This is especially true for the mountainous peripheries of Asia-Pacific in the 20th and 21st centuries. Into the 20th century, swidden or shifting cultivation was the dominant land use in these mountains. From about the mid-20th century onwards, the prevailing trend in this region has been a decline of shifting cultivation and transition to other land-use systems (Padoch et al., 2007; van Vliet et al., 2012; Dressler et al., 2016; Heinimann et al., 2017).

This is by no means a global trend. The 22 authors of van Vliet et al. (2012) carried out a global assessment of the dynamics of shifting cultivation based on a systematic survey of case studies in the literature. They found significant differences between the main regions in which shifting cultivation was practised, in the tropical zones of America, Africa, and Asia. In Latin America and Africa, shifting cultivation was found to persist or, in some regions, even to be on the increase. But tropical Asia, and especially mainland Southeast Asia, was characterized by an overall decline. This trend in mainland Southeast Asia was so prominent that the authors used inset maps to highlight the situation. Portions of those maps are reproduced here (Figure 1-1).

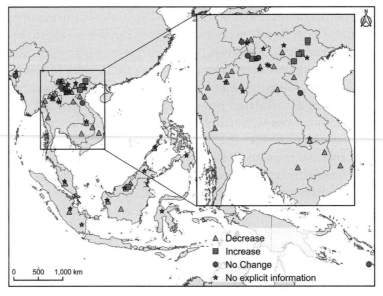

FIGURE 1-1: Change (or no change) in swidden areas of Southeast Asia, according to case studies.

Source: van Vliet et al. (2012).

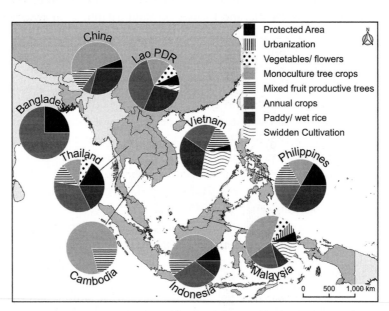

FIGURE 1-2: Transitions of swidden landscapes (the portions in each pie represent the number of case studies reporting a transition towards the various land-use types).

Source: van Vliet et al. (2012).

The second map sourced from van Vliet et al. (2012) shows the trends of land-use change after the demise of swidden (Figure 1-2). Focusing on Southeast Asia, a highly diverse picture can be seen, with trends differing from one country to another. For instance, the land uses following swidden in China and Cambodia were dominated by industrial tree crops, while field crops were more important in Thailand. However, there was a common denominator: land-use change was a process of conversion in which swidden farming was replaced by land uses that were fundamentally different.

The third map from van Vliet et al. (2012) (Figure 1-3) shows that in Southeast Asia, official policies were the predominant drivers of the decrease in swidden areas.

The trends detected by van Vliet et al. (2012) have been corroborated by the most recent global assessment of the extent and dynamics of swidden agriculture, in which Heinimann et al. (2017) predicted the end of swiddening by the end of the century – in Southeast Asia sooner than in Africa and tropical America – if current trends prevail.

That the change dynamics of swidden in Southeast Asia are in such stark contrast to those in other regions, and that the forces driving these changes seem to be so much more effective in Asia than elsewhere, justifies the regional focus of Malcolm Cairns´ swidden trilogy on the Asia-Pacific. While the first volume assesses more broadly the role of swidden in the context of environmental change (Cairns, 2015), the second volume, *Shifting Cultivation Policies: Balancing Environmental and Social Sustainability* (Cairns, 2017) recognizes the importance of policies as drivers of such change, especially in Asia-Pacific, and focuses on their influence on swiddeners and

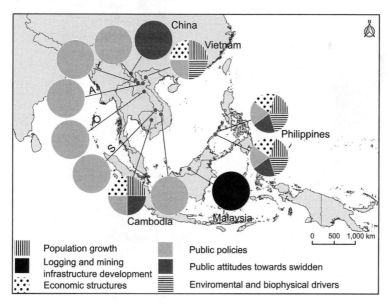

FIGURE 1-3: Drivers of decrease in swidden areas in Southeast Asia (each pie shows the drivers mentioned for each of the case studies appearing on the map).

Source: van Vliet et al. (2012).

on the response of swiddeners to these policies. The theme of this third volume is the assertion that swiddeners are not hapless victims of government policies and other contextual factors, but that they can be agents of change in their own right.

Swiddeners as agents of change

Changes affecting swiddeners are presented in the literature in predominantly negative terms: displacement, loss of tradition and indigenous knowledge, vulnerable livelihoods, exploitation, and so on. Even innovations, which in this volume are presented as resulting from the agency of farmers, can be foisted on swiddeners at their expense and for the benefit of others. A famous historical antecedent is the case of *taungya* plantation forestry in colonial Burma, which modified swiddening and instrumentalized swiddeners for the purpose of establishing teak plantations, ultimately to be owned and managed by the Forest Department (Bryant, 1997; Win and Kumazaki, 1998).

While it is true that changes can cause hardship and loss, swiddeners have proven resilient and capable of bringing about changes by themselves over the long history of this land-use system. Some well-documented examples are the development of short-rotation, stationary land-use systems based on the retention of *Alnus nepalensis* in Nagaland (Cairns et al., 2007; Cairns and Brookfield, ch. 22, this volume), and the incorporation of the newly introduced rubber tree *Hevea brasiliensis* into the swidden systems of Sumatra as 'jungle rubber' at around the beginning of the 20th century (Gouyon et al., 1993; Penot, 2007). These examples show that swiddeners are able to adapt their land use not only to pressures and constraints, but also to opportunities that they have detected themselves. While there are many case studies such as these, of how

Alnus nepalensis D. Don
[Betulaceae]

This tree enabled Angami Naga farmers to reduce their shifting cultivation system to a fallow length as short as 2 to 4 years without serious reduction in crop yields (Cairns and Brookfield, ch. 22, this volume). These nitrogen-fixing, heavy littering trees restore soil fertility in just a few years, sufficient to support a further two years of cropping. They are pollarded, becoming semi-permanent features of the swidden landscape, and in the process, produce valuable supplies of firewood.

swiddeners are responding to pressures or adapting to change, there has never been a comprehensive treatment of this subject. This is the niche that this third volume of Malcolm Cairns' swidden trilogy will occupy. In order to emphasize that swidden farmers can be proactive and take the initiative in such situations, the editor has carefully avoided speaking of farmer adaptations and has, instead, chosen the term 'farmer innovations' for the book title.

Swidden farming, instead of being the fossilized land use of unsophisticated peasants – as it has been portrayed by hostile observers – may, by its very nature, be more amenable to innovations than other systems. In the context of this chapter, I would like to define 'its very nature' in terms of the hybridity and metamorphic qualities of swidden systems that I pointed out at the outset. That metamorphosis and hybridity can also be closely intertwined is a theme of Ovid´s classical narrative poem *Metamorphoses*, in which people and demigods morph into plants, animals or stars while retaining some semblance of their prior manifestation – such as Daphne, whose arms and fingers become branches and twigs, and whose legs and feet turn into roots when she is transformed from a human into a laurel tree (Ovid, 2009).

As a hybrid of forest and farming, swidden systems can morph into new manifestations in which elements of either one or the other can acquire greater prominence or be recombined, depending on what is required in a changing context, while still retaining certain elements of the prior system and the overall hybrid and ambiguous nature of swiddening. Swidden farming is not only an ambiguous, but also an opportunistic, approach to land use. It is capable of making optimum use of synergies that already exist in nature, by maintaining diversity, and by applying a low level of manipulation, input, or creation of permanent structures such as terraces and irrigation facilities (Schmidt-Vogt, 2015).

Metamorphosis in the mountains

The transition from swidden cultivation to some form of post-swidden land use has generally been framed in terms of conversion or replacement. In the language of policy-makers, swidden agriculture is to be 'eradicated' or 'replaced' by more productive or more sustainable systems. These replacement systems are in most cases radically different from swidden systems. Famous and well-documented examples are the conversion of swidden into crop-based monocultures such as those with maize in Laos (Thanichanon et al., 2018) and northern Thailand (Gypmantasiri and Limnirankul, ch. A14 of the supplementary chapters to this volume), and the conversion of swidden systems into tree-based monocultures such as rubber plantations in Yunnan, China (Sturgeon 2011, Chen et al., 2016) or oil-palm plantations in Malaysia and Indonesia (Santika et al., 2019; Bamba and Munandar, ch. 15, and Sakuntaladewi et al., ch. 25, both in this volume). Innovation was, in this context, understood as adoption by local land users of new and externally designed and promoted land-use systems.

The main message of this volume is that swidden transitions can be transformations rather than conversions. In other words, they can take place in the form of a metamorphosis of existing models rather than in the wholesale adoption or imposition of imported models. Another message is that farmer-based transformations are not a new phenomenon and that some of them were put into practice long ago. Transformability and mutability thus appear as inherent features of swidden agriculture.

This volume is a vast repository of case-study experiences and reflections on this theme. The wealth of material that has been made available for this book by leading experts on this subject is indicated by the fact that a large number of contributions have 'spilled over' into a supplementary section.

Not surprisingly for a book that highlights the agency of farmers in a context of change, the Introductory Section has a strong focus on traditional knowledge and practices in swidden cultivation as a foundation for farmer innovations.

As population increase and attendant problems of maintaining food security are major challenges

Albizia chinensis (Osbeck) Merr. [Leguminosae]

This fast-growing multipurpose tree often appears where traditional shifting cultivation is intensifying towards settled agroforestry (see Bouan et al., ch. A64 in the supplementary chapters to this volume). It improves soil by fixing nitrogen. It also has an extensive root system and is planted for slope stabilization and reforestation of degraded land. Its dense spreading crown suppresses weeds and provides shade in tea and coffee plantations. The light, soft timber is not very durable and is used mainly for building houses and light furniture.

arising from change in Asia-Pacific, Section II on the evolution of food production systems, and Sections VI and VII in the supplementary chapters, on managing seeds, soils and tools as factors of production and strategies to assure food security, focus on the agricultural component of swidden farming, though not exclusively so.

Section IV on innovations in fallow management explores the role of swidden farmers in expanding forest cover, the role of soil-building trees in accelerating soil recovery, and the economic improvement of fallows. Specific examples of fallow management include the role of Asteraceae shrubs (van Noordwijk et al., ch. 39 and Daguitan et al., ch. 40, both this volume) and the role of bamboo, dealt with in section IX of the supplementary chapters.

The importance of innovations that are aimed at removing fire from swidden agriculture and from the landscape in an age of climate change is indicated by the fact that Section III has been exclusively dedicated to the theme of fireless shifting cultivation.

The printed volume ends with a concluding essay by Roy Ellen and a postscript by Michael Dove. Topics that are taken up in the supplementary chapters include social innovations (Section VIII), mounding technologies (Section X), swiddening in swamps (Section XI), gender analysis (Section XII) and overviews of specific places and peoples (Section XIII), which contains many contributions from case studies in northeast India, Myanmar, Laos, Vietnam, the Philippines, Indonesia and Papua New Guinea.

It is, of course, impossible to do justice to more than 114 contributions within the confines of a short introductory chapter. Therefore, I want to limit myself to highlighting those aspects that help to illustrate my statement that farmer-initiated swidden innovations can build on the inherent hybridity and mutability of swidden agriculture and to refer to just a small sample of contributions that provide supporting evidence. In this very brief conspectus, I want to focus on fallow and fire. Fire, when perceived as an agent of forest destruction (and more recently as a contributor to global warming) and fallow, when perceived as a second-rate replacement of natural forests, are elements that have been instrumental in creating the negative reputation of swiddening. They are, on the other hand, essential or even defining elements of swidden agriculture which bind forest and farming together. If fire and fallow were eliminated from the swidden complex by converting swidden lands to permanent land uses such as cropping or tree plantations, or by restoring them to forest for conservation purposes, these links would be ruptured and the result would be segregated landscapes in which farms and forests would co-exist with no -- or hardly any -- bond between them. Many innovations presented in this volume are characterized by retaining the linkages between trees and agricultural components, either in the landscape or by transferring them to stationary land-use systems, such as agroforestry.

The central element that links forest and farm in swidden systems is the fallow. It is also that part of the swidden complex that provides the greatest scope for interventions and innovations, because fallows can be managed to enhance the ecological or economic performance of swidden systems. Section IV is full of examples of how this can be achieved by drawing on the hybrid nature of swidden cultivation. A common practice of fallow management is to add desired components. These can be trees such as *Macaranga* spp. (Changkija et al., ch. 26, this volume) or '*Albizia*' (*Paraserianthes falcataria*) (Iskandar, ch. 24, this volume)[1] for the improvement of soil fertility, or non-timber plants such as broom grass (*Thysanolaena latifolia*), which can serve the dual purposes of providing income and improving the site (Bora et al., ch. 31, this volume) or even exotics like wild sunflower (van Noordwijk et al., ch. 39, Daguitan et al., ch. 40, both this volume). While management practices such as adding *Albizia* trees[1] to the fallow can assist the survival of swidden in a modified form in a context of constraint (Sakuntaladewi et al., ch. 25, this volume), there are also instances of how fallows can provide the entry point for transforming swidden into an almost entirely new land-use system. An excellent example for such a metamorphosis is provided from the island of Sumatra in the chapter 'Rotating agroforests: Using shifting

Macaranga denticulata (Blume) Müll. Arg. [Euphorbiaceae]

One of several *Macaranga* species that are used by Konyak shifting cultivators in Nagaland to improve their fallows (Changkija et al., ch. 26, this volume). The trees are either planted or establish naturally during the first year's swidden rice crop. They are nurtured through the cropping phase and then allowed to dominate the ensuing fallow. These trees have a symbiotic relationship with fungus species growing in and among their roots which help them to capture nutrients from the soil.

cultivation practices to construct a sustainable livelihood' (Burgers, ch. 37, this volume). This shows how the principle of rotation can be transferred from fields to crops so as to become an intrinsic standard of agroforestry. The opportunities offered by the introduction of cash crops such as pepper and coffee and their combination with native species such as cinnamon were utilized by farmers to develop stationary agroforests that can be managed flexibly by adjusting the rotation of crops within these forests in response to changes in market demand, fluctuations in crop yield and other contextual factors.

Fire is the driver of swidden cycles. As discussed above, it is also that element of swidden farming that has attracted the most criticism as being both archaic and a destroyer of forests. In addition, the negative perception of swidden fires has been fueled in our age of climate change by concerns about the contribution of swiddening to

Thysanolaena latifolia (Roxb. ex Hornem.) Honda [Poaceae]

Broom grass is a large, perennial, reed-like grass that grows to about 3 m on a wide range of habitats in South and Southeast Asia. Its mature flowering heads are harvested to make brooms, which earn high prices. Heavy harvest pressure on naturally-growing plants has led innovative farmers to cultivate broom grass to meet demand (see Bora et al., ch. 31, this volume). This species can thrive on degraded land, providing a source of income for poor farmers while helping to hold the soil in place.

global emissions (see an argument to refute this claim in Erni and KMSS-Loikaw, ch. 19, this volume). The removal of fire would thus be an innovation that could dramatically increase the acceptability of swidden transformations. In their chapter 'Without fire: Turning forests into agroforests on Siberut, Indonesia', Darmanto and Persoon (ch. 10, this volume) show how such a transformation can be achieved 'from scratch', by transforming a forest into a forest garden without fire by thinning the primary forest canopy and adding root crops, fruit trees and other useful trees in a cyclical process. This points out an option for a future hybrid, metamorphic, flexible and resilient land use without swidden farming – a development predicted by Heinimann et al. (2017). On the other hand, Singh et al. (ch. 11, this volume) present the example of the *yamkui* system in northeast India, in which the basic principles of swidden agriculture are retained, but the agency of fire is replaced by mulching during the fallow phase.

These few paragraphs must suffice to convince readers that this volume provides ample evidence for the statement that farmer innovations and best practices by shifting cultivators are capable of providing a 'continuing rationale for swidden farming in the 21st century' (van Vliet et al., 2013) by transforming swidden farming into systems that retain the hybridity and flexibility of the original.

Scaling mountains and upscaling a book project

No one would have been better suited to tackle the task of producing such a volume as this than Malcolm Cairns. As a researcher, he has studied land-use systems that are the products of the innovative capacities of farmers, most notably the *Alnus nepalensis*-based short-rotation swidden system of the Angami Naga in Nagaland, northeast India, already mentioned above. As a consummate networker and prolific writer, Malcolm soon found the role that suited him best: as a convener of swidden researchers in Asia-Pacific and chronicler of their research. The first volume, *Voices from the Forest: Integrating Indigenous Knowledge into Sustainable Upland Farming* (Cairns, 2007), set him on a path of exploring farmer-based approaches to land use, and this is now culminating in the present volume – the fourth. The first book was produced in a signature style that Malcolm has maintained and refined in succeeding volumes, and which is defined by an emphasis on visualizing and synthesizing. The art works of Paradorn Threemake and Wiyono make *Farmer innovations* and its predecessors instantly recognizable as 'Malcolm Cairns products'. The photographs that are assembled into coloured plates sections, accompanied by detailed captions, are of documentary value, especially in a period of accelerated change when land-use practices can pass into history in a relatively short time.

Malcolm Cairns and his team have planned and produced these volumes so carefully that they are, in structure, closer to a monograph than to the jumble of chapters that are so often offered as an edited volume. Moreover, edited volumes are too often left without synthesis chapters that pull together the strings of the many individual chapters. Malcolm has, from the beginning, made it a point to commission such chapters and has reached new heights with the production of this volume:

each section opens with a synthesis of its content, written by a leading expert on the respective topic. In the case of section IV, 'Innovations in fallow management', which on account of its bulk and complexity is divided into subsections, a synthesis even precedes each subsection. The book opens not just with an introductory chapter but with an entire Introductory Section and ends with a concluding chapter and a Postscript.

Farmer Innovations and Best Practices by Shifting Cultivators in Asia-Pacific is the capstone of an enterprise that I had the good luck to follow and accompany almost from the beginning, i.e. from 2010, when Malcolm Cairns spent a few months at the Center of Mountain Ecosystems Studies (CMES) – now Centre for Mountain Futures (CMF) – at the Kunming Institute of Botany in Kunming, China, where I was stationed at that time. Malcolm used his time in Kunming to develop the structure and outline of an edited book that would follow his landmark *Voices from the Forest* and become the first volume of his swidden trilogy. At that time, he was still recovering from a massive stroke. Watching him work and working with him, I admired, first, his courage in taking on such a massive project against the odds of physical impediment, and then the sheer doggedness with which he pursued his goal. At that time, it would have been inconceivable to many of us at Kunming that the goal would be not just one book, but a set of three books: a truly mountainous task. Throughout the years of scaling these mountains, Malcolm has kept me 'in the loop' through an unceasing flow of emails which, if printed and bound, would themselves add up to a sizeable book. I am still in awe of the fact that this venture has finally come to its completion and hope that this introductory chapter will do justice to what Malcolm and his team – Bob Hill and Tossaporn Kurupunya - have accomplished.

References

Bauer, T. (2018) *Die Vereindeutigung der Welt. Über den Verlust an Mehrdeutigkeit und Vielfalt* (the Disambiguation of the World: About the Loss of Ambiguity and Diversity), Reclam, Stuttgart, Germany

Bryant, R. L. (1997) The Political Ecology of Forestry in Burma, 1824-1994, Hurst and Company, London

Cairns, M. F. (ed.) (2007) *Voices from the Forest: Integrating Indigenous Knowledge into Sustainable Upland Farming*, Resources for the Future Press, Washington, DC

Cairns, M. F. (ed.) (2015) *Shifting Cultivation and Environmental Change: Indigenous People, Agriculture and Forest Conservation*, Earthscan from Routledge, London and New York

Cairns, M.F. (ed.) (2017) *Shifting Cultivation Policies: Balancing Environmental and Social Sustainability*, CABI, Wallingford, UK

Cairns, M. F., Keitzar, S. and Yaden, T. A. (2007) 'Shifting forests in northeast India: Management of *Alnus nepalensis* as an improved fallow in Nagaland', in M. F. Cairns (ed.), *Voices from the Forest: Integrating Indigenous Knowledge into Sustainable Upland Farming*, Resources for the Future Press, Washington, DC, pp.341-378

Campbell, B. (2013) *Living Between Juniper and Palm: Nature, Culture and Power in the Himalayas*, Oxford University Press, New Delhi, p.354

Castella, J-C., Lestrelin, G., Hett, C., Bourgoin, J., Fitriana, Y. Z., Heinimann, A. and Pfund, J-L. (2013) 'Effects of landscape segregation on livelihood vulnerability: Moving from extensive shifting cultivation to rotational agriculture and natural forests in Northern Laos', *Human Ecology* 41, pp.63-76

Chen, H., Yi., Z., Schmidt-Vogt, D., Ahrends, A., Beckschaefer, P., Kleinn, C., Ranjitkar, S. and Xu, J. (2016) 'Pushing the limits: The pattern and dynamics of rubber monoculture expansion in Xishuangbanna, SW China', *PLoS ONE* 11(2), e0150062, doi:10.1371/journal.pone.0150062.

Chokkalingam, U., Smith, J. and de Jong, W. (2001) 'A conceptual framework for the assessment of tropical secondary forest dynamics and sustainable development potential in Asia', *Journal of Tropical Forest Science* 13(4), pp.577-600

Dressler, W. H., Wilson, D., Clendenning, J., Cramb, R., Keenan, R., Mahanty, S., Bruun, T. B., Mertz, O. and Lasco, R. D. (2016) 'The impact of swidden decline on livelihoods and ecosystem services in Southeast-Asia: A review of the evidence from 1990 to 2015', *Ambio* 46, pp.291-310

Fujita, Y. and Phanvilay, K. (2008) 'Land and forest allocation in Lao People's Democratic Republic: Comparison of case studies from community-based natural resource management research', *Society and Natural Resources* 21(2), pp.120-133

Gouyon, A., de Foresta, H. and Levang, P. (1993) 'Does "jungle rubber" deserve its name? An analysis of rubber agroforestry systems in Southeast Sumatra', *Agroforestry Systems* 22, pp.181-206

Heinimann, A., Messerli, P., Schmidt-Vogt, D. and Wiesmann, U. (2007) 'The dynamics of secondary forest landscapes in the Lower Mekong Basin: A regional scale analysis', in *Mountain Research and Development* 27(3), pp.232-241

Heinimann, A., Mertz, O., Frolking, S., Christensen, A. E., Hurni, K., Sedano, F., Chini, L. P., Sahajpal, R., Hansen, M. and Hurtt, G. (2017) 'A global view of shifting cultivation: Recent, current, and future extent', *PLoS ONE* 12(9), e0184479, https://doi.org/10.1371/journal.pone.0184479.

Hett, C., Castella, J-C., Heinimann, A., Messerli, P. and Pfund, J-L. (2011) 'A landscape mosaics approach for characterizing swidden systems from a REDD+ perspective', *Applied Geography* 32, pp.608-618

Hurni, K., Hett, C., Heinimann, A., Messerli, P. and Wiesmann, U. (2013) Dynamics of shifting cultivation landscapes in Northern Lao PDR between 2000 and 2009 based on an analysis of MODIS Time Series and Landsat images', *Human Ecology* 41, pp.21-36

Jünger, E. (1993) *Siebzig verweht* 3, Klett-Cotta, Stuttgart, Germany

Kreutzmann, H. (2020) *Hunza Matters: Bordering and Ordering between Ancient and New Silk Roads*, Harrasowitz, Wiesbaden, Germany

Kurien, A., Lele, S. and Nagendra, H. (2019) 'Farms or forests? Understanding and mapping shifting cultivation using the case study of the Garo Hills, India', *Land* 8(9) 133

Lim, C. L., Prescott, G. W., De Alban, J. D. T., Ziegler, A. D. and Webb, E. L. (2017) 'Untangling the proximate causes and underlying drivers of deforestation and forest degradation in Myanmar', *Conservation Biology* 31(6), pp.1362-1372, DOI 10.1111/cobi.12984

Mather, S. (1992) 'The forest transition', *Area* 24, pp.367-379

Messerli, P., Heinimann, A. and Epprecht, M. (2009) 'Finding homogeneity in heterogeneity: A new approach to quantifying landscape mosaics developed for the Lao PDR', *Human Ecology* 37, pp.291-304

Nyein Chan and Takeda, S. (2016) 'The transition away from swidden agriculture and trends in biomass accumulation in fallow forests: Case studies in the southern Chin Hills of Myanmar', *Mountain Research and Development* 36(3), pp.320-331

Ovid (Publius Ovidius Naso) (2009) *Metamorphoses*, translated by A. D. Melville, Oxford University Press, Oxford, UK

Padoch, C., Coffey, K., Mertz, O., Leisz, S., Fox, J. and Wadley, R. L. (2007) 'The demise of swidden in Southeast Asia? Local realities and regional ambiguities', *Geografisk Tidskrift - Danish Journal of Geography* 107, pp.29-41

Penot, E. (2007) 'From shifting cultivation to sustainable rubber: A history of innovations in Indonesia', In M. F. Cairns (ed.) *Voices from the Forest: Integrating Indigenous Knowledge into Sustainable Upland Farming*, Resources for the Future Press, Washington, DC, pp.577-599

Pyne, S. (2021) *The Pyrocene: How We Created an Age of Fire and What Happens Next*, University of California Press, Oakland, CA

Santika, T., Wilson, K. A., Meijard, E., Budiharta, S., Law, E. A., Sabri, M., Struebig, M., Ancenarz, M. and Poh, T-M. (2019) 'Changing landscapes, livelihoods and village welfare in the context of oil palm development', *Land Use Policy* 87, 104073, https://doi.org/10.1016/j.landusepol.2019.104073

Schmidt-Vogt, D. (1998) 'Defining degradation: The impacts of swidden on forests in northern Thailand', *Mountain Research and Development* 18, pp.135-149

Schmidt-Vogt, D. (1999) 'Swidden farming and fallow vegetation in northern Thailand', *Geoecological Research* 8, Franz Steiner Verlag, Stuttgart, Germany

Schmidt-Vogt, D. (2001) 'Secondary forests in swidden agriculture in the highlands of Thailand', in U. Chokkalingam, W. de Jong and C. Sabogal (eds.) *Secondary Forests in Asia: Their Diversity, Importance, and Role in Future Environmental Management*, special issue, Journal of Tropical Forest Science 13, pp.748-767

Schmidt-Vogt, D., Leisz, S., Mertz, O., Heinimann, A., Thiha, Messerli, P., Epprecht, M., Cu, P. V., Chi, V. K., Hardiono, M. and Truong, D. M. (2009) 'An assessment of trends in the extent of swidden in Southeast Asia', *Human Ecology* 37(3), pp.269-280

Schmidt-Vogt, D. (2015) 'Second thoughts on secondary forests: Can swidden cultivation be compatible with conservation?', in M. F. Cairns (ed.), Shifting Cultivation and Environmental Change: Indigenous People, Agriculture and Forest Conservation, Earthscan from Routledge, London, pp.388-400

Scott, J. C. (2009) *The Art of Not Being Governed: An Anarchist History of Upland Southeast Asia*, Yale University Press, New Haven, CT and London

Seijger, C., Kleinschmit, D., Schmidt-Vogt, D., Mehmood-Ul-Hassan, M. and Martius, C. (2021) 'Water and sectoral policies in agriculture-forest frontiers: An expanded interdisciplinary research approach', *Ambio* 50, pp.2311-2321, https://doi.org/10.1007/s13280-021-01555-5.

Sturgeon, J. C. (2005) *Border Landscapes: The Politics of Akha Land Use in China and Thailand*, Silkworm Books, Chiang Mai, Thailand

Sturgeon, J. C. (2011) 'Rubber transformations: Post-socialist livelihoods and identities for Akha and Tai Lue farmers in Xishuangbanna, China', in J. Michaud and T. Forsyth (eds.) *Moving Mountains: Ethnicity and Livelihoods in Highland China, Vietnam and Laos*, UBC Press, Vancouver and Toronto, pp.193-214

Sturgeon, J. C. (2021) 'Landscape plasticity and its erasure', *Environment and Planning E: Nature and Space*, https://doi.org/10.1177/25148486211062004

Thanichanon, P., Schmidt-Vogt, D., Epprecht, M., Heinimann, A. and Wiesmann, U. (2018) 'Balancing cash and food: The impacts of agrarian change on rural land use and wellbeing in Northern Laos', *PLoS ONE* 13(12), e0209166, https://doi.org/10.1371/journal.pone.0209166

The Economist (2021) 'The new normal: The era of predictable unpredictability is not going away', Christmas double issue December 18 to 31, p.9

van Vliet, N., Mertz, O., Heinimann, A., Langake, T., Pascual, U., Schmook, B., Adams, C., Schmidt-Vogt, D., Messerli, P., Leisz, S., Castella, J-C., Joergensen, L., Birch-Thomsen, T., Hett, C., Bech-Bruun, T., Ickowitz, A., Vu., K. C., Yasuyuki, K., Fox, J., Padoch, C., Dressler, W. and Ziegler, A. D. (2012) 'Trends, drivers and impacts of changes in swidden cultivation in tropical forest-agriculture frontiers: A global assessment', *Global Environmental Change* 22, pp.418-429

van Vliet, N., Mertz, O., Birch-Thomsen, T. and Schmook, B. (2013) 'Is there a continuing rationale for swidden farming in the 21st century?', *Human Ecology* 41, pp.1-5

Wangpakapattanawong, P., Kavinchan, N., Vaidhayakarn, C., Schmidt-Vogt, D. and Elliott, S. (2010) 'Fallow to forest: Applying indigenous and scientific knowledge to tropical forest restoration', *Forest Ecology and Management* 260, pp.1399-1406

Win, S. and Kumazaki, M. (1998) 'The history of taungya plantation forestry and its rise and fall in the Tharrawaddy forest division of Burma', *Journal of Forest Planning* 4, pp.17-26

Note

1. Both chapters 24 and 25 in this volume refer to a tree known as *albizia* or *albasiah* in the local language of the study sites. This would seem certain to have arisen from its earlier identification as one of the many species in the genus *Albizia*. The name *Albizia falcataria* (L.) Fosberg was commonly replaced by the name *Paraserianthes falcataria* (L.) I. C. Nielsen – the name given to the species in this chapter. However, even that name is problematic. At the time of writing, recognized

authorities on botanical names disagreed on how it should be properly named: www.theplantlist. org said that both *Albizia falcataria* and *Paraserianthes falcataria* were synonyms of the accepted name *Falcataria moluccana* (Miq.) Barneby & J.W. Grimes. However, Kewscience's www.plantsoftheworldonline.org regarded all three of these names as synonyms of the accepted name *Falcataria falcata* (L.) Greuter & R. Rankin. All names relate to members of the plant family alternately named Leguminosae or Fabaceae.

In this chapter, there is also a botanical sketch of *Albizia chinensis*, a tree species that is also important in the context of fallow-based soil improvement, since it is also a nitrogen-fixing member of the Leguminosae family. However, it is not the species featured in chapters 24 and 25 of this volume.

2

PERSPECTIVES ON SHIFTING CULTIVATION

The northeast India experience

*B. P. Maithani**

Shifting cultivation, which is locally called *jhum*, is widespread in the hilly tracts of northeastern India. It is proving to be enigmatic to agricultural scientists, administrators and policy-makers alike. The mission of transforming shifting cultivation into a settled, 'modern' and broadly acceptable form of agriculture is an intractable task. At the macro level, shifting cultivation is reported to be on the decline. For instance, the Indian Council of Forestry Research and Education reported in its 2014 statistical year book that there had been an incredible 67% decline in the area under shifting cultivation between 2000 and 2010 (ICFRE, 2014). Similarly, the Ministry of Rural Development's 2010 Wastelands Atlas of India shows a decrease in shifting cultivation area by 46% between 2003 and 2006. Yet micro-level studies show that despite the adoption of multiple farming systems by *jhumias*, 70% of households in Manipur's Ukhrul district and 90% of households in the West Garo Hills district of Meghalaya (see Figure 2-1) still continue to practise shifting cultivation (GOI, 2018). The first attempt to estimate the extent of shifting cultivation in the northeast was made by the Scheduled Caste and Scheduled Tribe Commissioner in a report in 1960–1961. According to that estimate, about 1.86 million people were engaged in shifting cultivation on 332 million hectares of land in the region (Sharma, 1984).

The age-old mainstay of subsistence livelihoods, also known as swidden farming, or pejoratively, 'slash-and-burn', has been the subject of intense debate and discussion in official and academic circles for decades, reaching back into the era of colonial administrations. The debate, however, seems to have generated more heat than light on the issue. There are broadly two schools of thought on shifting cultivation: one blames it for deforestation and resource degradation and the other believes it is a rational and sustainable agricultural system, given the situation in which it is practised. The problem lies in the perspective. The popular view – which is also the official

* PROFESSOR DR. B. P. MAITHANI is Director of the North East India Centre, National Institute of Rural Development (NIRD), Hyderabad, India.

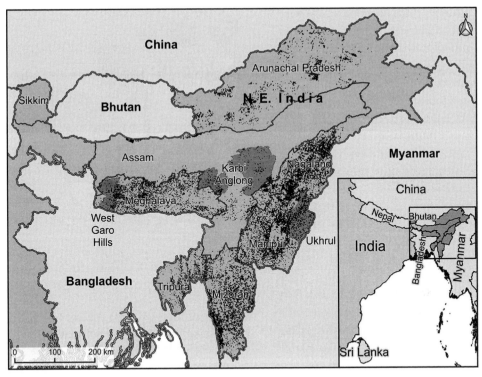

FIGURE 2-1: The states of northeast India showing the incidence and location of shifting cultivation. Repeated use of land for shifting cultivation between the years 1975 and 2018 is shown by the scattering of black spots on the uniform background of the states. Several districts discussed in this chapter are also shown with darker shading.

Source: Based on Das et al., 2021.

view the world over – considers shifting cultivation to be a primitive practice and a wasteful method of land use. Local officers appear to be 'positioning themselves' on the 'right side' of the popular professional viewpoint instead of using their experiential insights to confront and question the stereotype thinking on the issue. The familiar refrain is that it is responsible for excessive soil erosion and deforestation in fragile hill and mountain environments, causing land degradation and progressive decline in both agricultural productivity and the income of shifting cultivators. Being an antiquated form of agriculture, shifting cultivation is seen as inefficient because it uses rudimentary tools and implements, resulting in low yields that produce no surpluses to meet the needs of its practitioners beyond subsistence (Borthakur, 1983). The practices of shifting cultivation first came under administrative review in the first forest policy, published in 1894. It clearly stated that 'a system of shifting cultivation costs more to the community than it is worth and can only be permitted under due regulation'. Following the same approach, the Indian Forest Act 1927 declared that shifting cultivation was 'subject to control, restriction and abolition by the State Government' (GOI, 1983).

After independence, state policies in India have always viewed shifting cultivation as an inefficient and environmentally unsustainable practice and have offered incentives for adoption of settled agriculture and cultivation of perennial cash crops. *Jhum*-land regulations and forest policies aim at controlling shifting cultivation and promoting plantations of horticultural crops, such as rubber, teak, tea, coffee, palm oil, and so on. For example, the Working Group on Development of the North Eastern Region recommended that very high priority should be accorded to controlling shifting cultivation – 'not only for improving the quality of life of tribals, but also for checking land degradation' (GOI, 1985). It was expected that shifting cultivation would gradually disappear with the passage of time, as populations grew and farmers were exposed to modern agricultural practices. But this has not happened, despite the concerted efforts of state and central governments, and the practice remains entrenched over

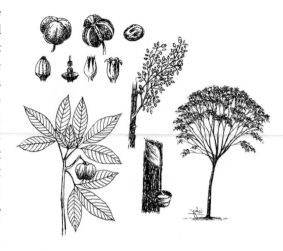

Hevea brasiliensis (Willd. ex A.Juss.) Müll. Arg. [Euphorbiaceae]

The South American para rubber tree has become a ubiquitous part of plantation monocropping throughout much of Southeast Asia. It has also been widely introduced into swidden crops as seedlings, to become a dominant part of the ensuing fallow, known as 'jungle rubber' (see Yonariza, ch. A18 of the supplementary chapters to this volume). The trees depend upon a tropical climate: a few frosts and the rubber from an entire plantation may become brittle and break once it has been refined.

large areas. In the northeastern region, for example, an area of 850,000 hectares is still being used for shifting cultivation (GOI, 2018, op. cit) (see Figure 2-1). At the global level, shifting cultivation is currently estimated to cover roughly 280 million ha, including both cultivated fields and fallows (Heinimann et al., 2017). This means that shifting cultivation must make some sense, since millions of farmers continue to practise it despite all the incentives to stop (Kerkhoff and Sharma, 2006).

The intentions of policy-makers and development personnel cannot be doubted, as their prescriptions seem to be motivated by a desire to improve the lot of shifting cultivators. However, difficulties arise when their recommendations are based more on perceptions than on sound scientific investigation of the problem. Scientific studies have been consistent in suggesting that, in many ways, there is optimum utilization of natural resources in shifting cultivation regions, and this is helpful for the stability and sustainability of agriculture in areas characterized by fragile ecosystems

(Maithani, 1997). It has been found that while modern agricultural technologies may be efficient in terms of time, labour and output per unit area, they are deficient in respect of energy input and output considerations (Ramakrishnan, 1993). In modern agriculture, 5 to 10 units of fuel energy are required to produce a single unit of food energy (Steinhart and Steinhart, 1974). On the other hand, shifting cultivators have been managing traditional agriculture for centuries with optimum yields on a long-term basis, rather than trying to maximize production on a short-term basis. In this respect, shifting cultivation is a model of productive efficiency, in which 5 to 50 units of food energy are obtained for each unit of energy expended (Ramakrishnan, 1993, op. cit.).

As mentioned above, studies have revealed how this productive efficiency is achieved, through optimum utilization of natural resources. First, the mix of crops is

Oryza sativa L. [Poaceae]

Rice is commonly the main crop in both settled and shifting cultivation in northeast India. However, studies have shown that the share of secondary crops is more balanced in shifting cultivation areas. Shifting cultivators are also known to avoid the use of seeds from 'modern' high-yielding varieties for fear of diminishing soil fertility and suffering poor crops in succeeding cropping seasons. Instead, they continue to plant their own indigenous varieties of rice.

so evolved that the root systems of different plants reach down to varying depths. In this way, different crops are able to use the nutrients of different layers of soil. Second, the variety of plants is arranged in a multi-storeyed pattern so that the total leaf area of all vegetation is extraordinarily large. This helps in collecting solar energy much more efficiently than in monocrop farming. It also provides better protection against soil erosion. Third, different plants have different growth habits and requirements. Maize, for example, is harvested at a time when upland rice plants are reaching the peak of growth and require more space. Harvested crops also leave plant residues that decay and provide organic matter and nutrients for the standing crops. Fourth, it is significant that the labour required for sowing and harvesting a variety of crops is evenly spread over a long period and is not concentrated in a seasonal peak. This helps in intensive use of manpower in sparsely populated areas. Fifth, shifting cultivation is based on the experience of practitioners over hundreds of years and its practices are ecologically compatible. For example, intercropping of different crops not only helps to optimize the use of natural processes, but also contributes significantly to soil conservation and enrichment, as well as conjunctive use of water. Sixth, the multiple-choice principle can also be seen in the management of shifting cultivation. Sowing and harvesting of crops are not one-time operations. The choice of crops is also well considered. Their gestation period may be anywhere between two and eight months. An early-maturing variety of upland rice is ready by about August, whereas some late varieties may mature in November or even December. Maize is harvested early, in July and August. Thus, harvesting of crops under shifting cultivation continues from July to January. Seventh, the wide stretches of sowing and harvesting seasons and the variety of crops grown not only give employment round the year, but they also provide an insurance against uncertainty and aberrations in the weather – particularly in rainfall. Eighth, shifting cultivators generally do not use the seeds of 'modern' high-yielding crop varieties for fear of depleting soil fertility and suffering low productivity in the second cropping period. Use of indigenous seeds is a measure of self reliance in food security (Ramakrishnan, 1993, op. cit.).

Bambusa tulda Roxb. [Poaceae]

Bamboo is one of the most important of northeast India's forest products. This species – one of the most common – grows in clumps with stems reaching 25 m tall.
As well as providing food from young shoots, the stems are used for construction, scaffolding, furniture, boxes, basketry, mats, household utensils and handicrafts. It is also used to produce paper pulp, and is commonly grown by shifting cultivators, both on the borders of swiddens or as a dominant fallow species, to help control soil erosion.

Studies by the Agro-Economic Research Centre for North East India showed for the first time that the productivity of shifting cultivation per unit area compared favourably with the productivity of wet rice under settled cultivation (Saikia and Bora, 1981). Yet another study, which focused on land-utilization patterns, production growth rates, productivity and yield stability of different crops under both shifting and settled agriculture systems, produced more interesting findings. It found although rice was commonly the main crop under both shifting and settled cultivation systems, the share of different crops was found to be more balanced in shifting cultivation areas. Similarly, production growth rates and yield per ha were found to be significantly higher in shifting cultivation areas; the mean yields of rice, maize, wheat, millets and other cereal crops were found to be considerably higher;

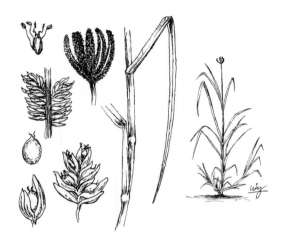

Eleusine coracana (L.) Gaertn. [Poaceae]

Finger millet is a nutritious grain crop which, among others such as rice and maize, has been shown to provide considerably higher yields in northeast India's shifting cultivation areas, rather than in settled cultivation. Despite its tiny seeds creating problems in crop establishment and competition with weeds, finger millet is being promoted in shifting cultivation areas for its benefits to human nutrition.

and the yields of all pulses and oil-seed crops were also substantially higher under shifting cultivation than in settled agriculture (Divakar, 1990).

Similar observations were made by yet another region-wide study of cropping patterns, productivity differentials and the levels and structure of agriculture development in northeast India (Srivastava, 1996). The study found that the gross value of agricultural produce per hectare and per capita was higher in the hills districts than in valley districts, where intensive settled cultivation was prevalent. These findings throw up an important message for sustainable agricultural planning in the region: in order to maximize returns from scarce land to cover food deficits, a balanced allocation of land between food grains and non-food-grain crops needs to be devised. Another message to policy-makers is that tribal communities in hilly areas with a scarcity of cultivable land are bestowed with a wisdom that suggests a centuries-long evolution of indigenous knowledge and innovativeness, contrary to the myth that people practising shifting cultivation are primitive and inefficient. The study also found that there was a very close relationship between the per-ha gross value of agricultural produce in the hill districts and the share of this value held by pulses, oil seeds, fruits and spices. It also found that the area under non-food crops was negatively related to the net cropped area as a percentage of the total area of the hill districts, suggesting that land was put to a greater variety of uses in the land-scarce hill districts and to more intensive uses in the form of monocropping in valley districts.

The innovativeness of the hill tribes, working with their indigenous knowledge, is perhaps at its best when fighting the hostility of nature caught in the grips of climate change. It was discovered in another study by the National Institute of Rural Development (NIRD) that most shifting cultivation plots were being used to grow vegetables and fruit, among other things, indicating a shift towards commercialization of shifting cultivation to ensure income security to deal with uncertainties arising from the vagaries of nature (Maithani, 1995). Thus, the indigenous response to the

Ananas comosus (L.) Merr.
[Bromeliaceae]

Tough, thorny pineapples have found multiple uses in swidden systems. They are among the crops often promoted as horticultural replacements for shifting cultivation. They are also used to create vegetative strips along the contours of swidden slopes to mitigate soil erosion. Far from these roles, pineapples were a symbol of wealth in late 18th century Europe because of the expense of importing them and the enormous cost of growing them in a temperate climate.

food crisis arising from changing climatic conditions and declining productivity in subsistence agriculture displays a process of mature, rational economic decision-making. In yet another study, shifting cultivation and, in particular, farmers' innovations, were found to contribute to forest cover and biodiversity conservation while, at the same time, maintaining agricultural and forest productivity (Kerkhoff and Sharma, 2006, op. cit.). Thus, indigenous agriculture not only shows a welcome pattern of development, but also demonstrates how scarce land resources can be used most efficiently. It is clear that the indigenous practices of shifting cultivators have evolved through experiential learning over many generations. Hence, they are better adapted to the limitations and potentials of their habitat. These factors perhaps explain the tenacity of shifting cultivation.

However, there is a persistent policy- and project-driven attempt to transform shifting cultivation by promoting cash-crop plantations (GOI, 2018, op. cit.). This policy is yielding both positive and negative effects. In a study of changing patterns of shifting cultivation in Karbi Anglong hill district of Assam, it was found that the shifting cultivation landscape was changing rapidly. *Jhum* fields and fallows were giving way to settled agriculture, terraced wet-rice fields and horticultural crops such as pineapples, bay leaf, jackfruit, lychees, bananas, plums and tree beans. Similarly, *jhum* lands in other states have been converted to horticulture and agroforestry, with tree crops like rubber and oil palm in government-sponsored plantations. Community perceptions of these changes have been mixed. Although the farmers acknowledge increased cash income, reduced drudgery and improved social status – particularly for women, through alternative livelihood options – there is also apprehension about the adverse impact of settled agriculture on agrobiodiversity, loss of traditional seed varieties, increasing input costs, increasing dependence on fluctuating markets, likely food and nutritional insecurity and drastic depletion of ecosystem services from *jhum* fallows (Aryal et al., 2019).

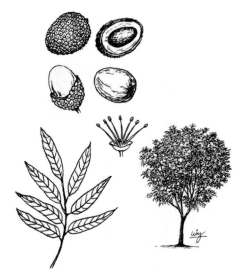

Litchi chinensis (L.) Sonn. [Sapindaceae]

Lychees are among the horticultural crops appearing on former swidden land in parts of India like Assam's Karbi Anglong district. These evergreen trees have been cultivated in their native southern China for more than 2000 years and are now grown for their fruit in many parts of South and Southeast Asia. However, thorough knowledge of the trees' nutrient requirements is said to be indispensable for profitable lychee production (Yao et al., 2020).

The change from shifting to settled cultivation is also accompanied by changes in property regimes affecting access to land and productive resources. This often leads to increasing marginalization of the poor, who are unable to invest in the conversion of their *jhum* plots or their right to access and ownership of land (Maithani, 2000). Although village land is communally owned, *jhum* land is divided between clans. Clan lands are held almost permanently by families belonging to the clans. Custom allows individual families to convert their *jhum* plots into permanently held property if they invest in developing their plots into terraced fields or orchards, or by raising a crop on the plots continuously for more than three years. With the decline of traditional institutions and increasing authority of modern state institutions, the tendency to acquire and accumulate private property has increased as families take advantage of loopholes in customary laws (Das, 1989; Karna, 1989; Roy Burman, 1989; Aryal et al., 2019).

Parkia timoriana (DC.) Merr. [Leguminosae]

Growing to 30 metres or more, the tree bean provides timber, firewood, food, folk medicines and aesthetic value. The seed pods, or beans, are eaten as a vegetable and, when mature, the hard black seeds are powdered and eaten to treat colic, flatulence and stomach ache. The powder is also applied externally to wounds and ulcers. These trees are commonly grown for their decorative value, with dangling flowers that are pollinated by fruit bats.

Government policy continues to promote the transformation of shifting cultivation lands into settled wet-rice, horticulture and agroforestry farming systems. It is argued that distortions in the traditional shifting cultivation regime and increasing populations are forcing a shorter shifting cultivation cycle, leading to declining productivity, so there is a need for change to intensive settled cultivation. However, it is not that easy, because shifting cultivation landscapes are also cultural landscapes and unless the communities adjust and adapt to a new cultural environment, shifting cultivation will continue to be practised alongside settled cultivation. It is unfortunate that the issues facing shifting cultivation have not been addressed by the people who have first-hand knowledge of practising it. Only a few of the tribal officials who have expressed themselves on the subject appear to be using their experiential insights to confront and question stereotype thinking on the issues. The greatest flaw in

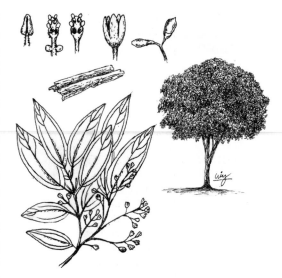

Cinnamomum tamala (Buch.-Ham.) T. Nees & Eberm. [Lauraceae]

Indian bay leaf, also known as *tejpat* or *malabathrum*, is an ancient spice that should not be mistaken for that from the Mediterranean bay laurel. Indian bay leaf is widely used in the cuisines of India, Nepal and Bhutan, approximating its native territory. But references to *malabathrum* appeared in first-century Greek texts and, its strong cinnamon-like aroma was appreciated in early Egypt and Rome. The modern market for Indian bay leaf has made these trees an alternative crop on swidden land.

our conception and perception of shifting cultivation lies in our use of theoretical deduction rather than empirical observation: the system is considered to be inferior and wasteful when viewed and compared with our notion of what hill agriculture should be. If only our way of using land in the hills was the right way! Then, Nepal would not be one of the poorest countries in the world and young people from Uttarakhand, in the western Himalaya, who have perfected the technology of terraced cultivation, would not be fleeing from their homes to suffer the indignities of menial jobs in the cities of the plains (Maithani, 2005).

When the question is asked: 'why have shifting cultivation control measures failed to achieve their stated objectives?', the simple answer is that they have so far been inadequate in effort and inappropriate in design. While planning and implementing the transformation of shifting cultivation, we have not asked ourselves some basic questions: who are the shifting cultivators, and why do they continue to practise shifting cultivation? After all, shifting cultivation is an arduous lifestyle, and when, in spite of that hardship, people continue to bear it, then there must be some genuine reasons for that.

References

Aryal, K., Thomas, S. and Choudhury, D. (2019) *Shifting Cultivation Landscapes in Transition*, Agriculture and Environment Series, Mongabay, India

Borthakur, D. N. (ed.) (1983) *Shifting Cultivation in North East India*, Research Complex for NEH Region, Indian Council for Agricultural Research, Shillong, Meghalaya.

Das, P., Mudi, S., Behera, M. D., Barik, S. K., Mishra, D. R. and Roy, P. S. (2021) 'Automated mapping for long-term analysis of shifting cultivation in Northeast India', *Remote Sensing* 13, 1066, available at https://doi.org/10.3390/rs1306106

Das, J. N. (1989) *Land System of Arunachal Pradesh*, NM Tripathy Ltd, Bombay

Divakar G. D. (1990) 'Cropping patterns, growth rates and yield stability of different crops under Ri-riad and Kynti and Roytiwari land systems', Paper presented at a National Seminar on Agrarian relations in NE India, North-eastern Regional Centre, National Institute of Rural Development, Guwahati, Assam (mimeo)

GOI (1983) *Task Force Report on Shifting Cultivation in India*, Ministry of Agriculture and Cooperation, Government of India, New Delhi

GOI (1985) *Report of the Working Group on Development of North Eastern Region during the Seventh Five Year Plan*, North Eastern Council, Government of India, Shillong, Meghalaya

GOI (2018) *Report of Working Group III; Shifting Cultivation: Towards A Transformational Approach*, National Institute for Transforming India (NITI Aayog), Government of India, New Delhi

ICFRE (2014) *Statistical Year Book*, Indian Council of Forestry Research and Education, Dehradun, Uttarakhand

Jodha N. S. (1997) 'Mountain agriculture', in B. Messerli and J. D. Ives (eds) *Mountains Of The World: A Global Priority*, The Parthenon Publishing Group, London

Karna, M. N. (1989) 'The Agrarian Scene in the North East', paper presented at a seminar on `Agrarian Structure and Land Relations in North-East India, National Institute of Rural Development, Guwahati, Assam (mimeo)

Kerkhoff, E. E. and Sharma, E. (2006) *Debating Shifting Cultivation in the Eastern Himalayas: Farmers Innovations as Lessons for Policy*, International Centre for Integrated Mountain Development (ICIMOD), Kathmandu, Nepal

Maithani, B. P. (1995) *Socio-Economic Production Studies of Tribal Communities of North East India*, North-eastern Regional Centre, National Institute of Rural Development, Guwahati, Assam (mimeo)

Maithani, B. P. (1997) 'Indigenous Agricultural Practices in NE Region: Perspectives on Shifting Cultivation', *The Administrator* XLII, LBS Academy of Administration, Mussoorie, Uttarakhand, pp.177-183

Maithani, B. P. (2000) 'Changing land relations and poverty in the eastern Himalayas', in M. Banskota, T. S. Papola and J. Richer (eds) *Growth, Poverty Alleviation and Sustainable Resource Management in the Mountain Areas of South Asia*, International Centre for Integrated Mountain Development (ICIMOD) and Zentralstelle für Ernährung und Landwirtschaft, Kathmandu, Nepal, pp.433-443

Ramakrishnan, P. S. (1993) *Shifting Cultivation and Sustainable Development: An Interdisciplinary Study from North East India*, Oxford University Press, UK

Roy Burman, B. K. (1989) 'Issues in land reforms in NE tribal areas', *Mainstream*, August 1-8, 1989, pp. 1-10

Saikia P. D. and Bora, D. (1981) *Socio-Economic Aspects of Soil Conservation Activities in the Hill Areas of North East India*, Government of Meghalaya, Shillong

Sharma, B. D. (1984) 'Shifting cultivators and their development', *Journal of Social Science and Humanities* vol. II, North Eastern Hill University, Shillong, Meghalaya

Srivastava, S. C. (1999) *Level and Structure of Development: An Inter-district Study of North East India*, National Institute of Rural Development, Hyderabad, India

Steinhart, C. E. and Steinhart, J. S. (1974) *Energy: Sources, Use and Role in Human Affairs*, Duxbury Press, North Scituate, MA

Yao, L., Bai, C. and Luo, D. (2020) 'Diagnosis and management of nutrient constraints in Litchi', chapter 45 in A. K. Srivastava and Chengxiao Hu (eds) *Fruit Crops: Diagnosis and Management of Nutrient Constraints*, Elsevier, Amsterdam, pp.661-679

3

A SYNTHESIS OF SCIENTIFIC PERSPECTIVES

on the shifting cultivation-agroforestry interface

*P. K. Ramachandran Nair**

Introduction: Historical developments

Innovations in land- and soil-management practices introduced into farming systems have resulted in evolutionary changes, leading to enhanced productivity the world over during the past three centuries. The introduction of the mechanical seed drill by Jethro Tull in the early 1700s, the initiation of agricultural research in the mid-1850s with the establishment of the Rothamsted Experimental Station in England, and the invention of the Haber process, which led to the manufacture of chemical fertilizers in the early 1900s, were three significant developments of that nature during the pre-World War II era. The Green Revolution of the 1960s, made possible by the development of 'miracle varieties' of cereal crops and supported by a confluence of scientific advances in soil and plant management, together with the formulation and implementation of enabling policies at various levels of administration from local to global levels, led to miraculous gains in food production during the latter part of the 20th century.

The introduction of all of these major innovations was preceded by periods of inefficient management of resources as well as depletion and degradation of non-renewable resources such as soil. Jethro Tull's use of a horse-drawn seed drill that helped to sow the seed economically in neat rows was a major improvement over the common practice of broadcasting seeds by hand that prevailed in England at that time. Although the method initially met with considerable resistance from large landowners, it was eventually accepted and adopted, leading to more efficient and modern farming in Britain. The second major innovation mentioned above involved a fundamental understanding of essential plant nutrients and the development and use of chemical fertilizers, which opened up opportunities for phenomenal increases in

* PROFESSOR P. K. RAMACHANDRAN NAIR is Distinguished Professor (Emeritus), School of Forest, Fisheries, and Geomatics Sciences, University of Florida, Gainesville, FL.

crop yields. The remarkable experience of the Green Revolution still reverberates in the minds of many. The use of 'miracle varieties' of crops and adoption of a package of soil- and water-management practices in many parts of the tropics led to a three-fold increase in global production of cereal crops in the last three decades of the 1900s.

Paradoxically, no such 'exciting' developments have happened in shifting cultivation, which is considered to be the oldest farming system. Known as swidden farming, slash-and-burn agriculture, and several other terms (Nair et al., 2021), numerous types of shifting cultivation are still practised in various regions around the world. Rigorous estimates of the area under shifting cultivation and the number of people dependent on it are not available, primarily because the practice encompasses a variety of land-management traditions of poor, often migratory farmers. According to estimates by the Food and Agriculture Organization of the United Nations (FAO), which have traditionally become authentic by repeated citations, the practice covered approximately 360 million hectares, or 30% of the world's exploitable soils, and supported more than 250 million people in the early 1980s. Other estimates of the area and number of practitioners include 300 million to 500 million hectares involving 200 million farmers in the tropics (Crutzen and Andreae, 1990), 300 million to 500 million shifting cultivators in 40 to 50 countries (Mertz et al., 2009), and 280 million hectares worldwide (Heinimann et al., 2017). Over the past several decades, substantial amounts of money and effort have been spent around the world on projects small and large with appealing and exciting statements of objectives and action plans for improving or replacing shifting cultivation. Yet, the core practice continues mostly in its traditional cyclic pattern: land clearing for growing food crops (cropping phase), abandoning the site after a few years and letting the vegetation regrow (fallow phase), then returning to the original site when the farmer is convinced that the land has regained its fertility (Figure 3-1). Shifting cultivation is said to have been widespread in Europe until a few centuries ago (Nye and Greenland, 1960; Greenland, 1976). Under resource-sufficient conditions, as in Europe, shifting cultivation was slowly replaced by more technologically oriented and profitable land-use systems with little or no resemblance to the original system. On the other hand, shifting cultivation and its variants are still a major form of a traditional farming system over vast areas of the tropics and subtropics (Nair et al., 2021).

Numerous publications of various types (scientific articles, descriptive and synthesis reports, proceedings of conferences and focused group meetings, and so on) are available on different aspects of shifting cultivation. The most voluminous among them are Malcolm Cairn's series of volumes (Cairns, 2007, 2015, 2017, and this volume), each involving contributions from an average of 100 scholars in more than 50 chapters and 1100 pages of descriptions and analyses of numerous examples of the practice in South and Southeast Asia and the Pacific. The practice has been prevalent in other parts of the developing world, including Africa and Latin America, although not as extensively as in Asia, but nothing comparable or as voluminous as Cairns' publications have been produced from those regions. The efforts of the numerous

FIGURE 3-1: This sketch of a swidden landscape in Luang Namtha province, northern Lao PDR, provides a glimpse of the rotational land use of Rmet shifting cultivators. The field in front of the field huts was harvested in the previous year and tree stumps are already regenerating amidst the standing rice straw. The farmers would have spent most of the rainy season living in the field huts. Behind the field huts, a new field has been cleared, left to dry, and is ready to be burnt. On the ridges and near the thalweg, older forests have been left untouched, often protecting water springs or a community burial ground.

Sketch based on a photograph by Olivier Evrard in February, 1994.

scholars who contributed to these volumes, as well as studies from other parts of the world over the years, have brought to light documented evidence of the innumerable examples of a variety of fallow and tree-cropping systems and their management by indigenous people. Most of these reports are heavily focused on social and anthropological attributes and descriptions of how they have been managed over long periods, with relatively little on the biophysical underpinnings that are essential for suggesting science-based alternatives and improvements. Because of the remoteness of the areas where such systems are practised and the lack of any discernible social and political power for the practitioners, who are predominantly poor and hapless, such systems have not attracted any systematic and sustained research investment and attention. Indeed, there has been an enduring perspective since colonial times within academic and development circles to look down upon and denigrate shifting cultivation as an unproductive, disorderly, resource-depleting, and environmentally destructive practice of the past, leading to calls for its abandonment and prohibition, as alluded to by Colfer (2017). For a student of agroforestry, however, these traditional systems offer valuable insights into how trees have been an integral part of low-input agricultural systems; what roles trees and other woody perennials play during the fallow phase of the shifting cultivation cycle in recouping the soil fertility that is depleted during the cropping phase, and to what extent these systems adapt to changing land-use patterns and market economics. Based on numerous efforts, starting with the classical work of Nye and Greenland (1960), fallow-improvement efforts under the banner of agroforestry in the recent past became the closest to an organized alternative or improvement to traditional shifting cultivation. With this background, the objectives of this short chapter are to evaluate:

1) The experience of the major agroforestry initiatives to improve shifting cultivation and synthesize the scientific principles learned from them, and
2) The nature and extent of ecosystem services provided by shifting cultivation.

The importance of the fallow phase of shifting cultivation

Early studies of soil fertility under shifting cultivation have been evaluated in various studies (Nye and Greenland, 1960; Newton, 1960; FAO/SIDA, 1974; and Sanchez, 2019; to list a few). Several 'expert consultations' were organized by the FAO during the 1980s, in some of which this author had the privilege of participating. Many of the approaches suggested as improvements and/or alternatives to shifting cultivation in those meetings emphasized the importance of retaining or incorporating the woody vegetation into the fallow phase – and even in the cultivation phase – as a key to maintaining soil productivity. The length of the fallow phase was recognized as a critical factor in terms of the system's sustainability (Nair, 1993). In various attempts to classify shifting cultivation by FAO/SIDA (1974) and reviewed by Ruthenberg (1980), the different categories designate different degrees of intensification of cultivation which can best be evaluated based on the land-use factor (L):

$$L = \frac{C + F}{F} \qquad \begin{array}{l} C = \text{length of the cropping phase (years)} \\ F = \text{length of the fallow phase (years)} \end{array}$$

Note: A related term used in some literature is the cultivation factor (R), which is the inverse of L, whereby C = R / (C+F) with C and F having the same meanings as in the land-use factor.

Despite recognition of the length of the fallow phase as the most critical factor in terms of the system's sustainability, there is a misleading tendency to present shifting cultivation as consisting of a cycle of cropping and fallow without indicating the length of the cycle's different phases (Falvey, 2017). During the early stages of shifting cultivation, the fallow periods are long, with values of L in the above equation exceeding 10. When a sedentary and permanent cultivation stage is reached, L =1. In most areas where shifting cultivation is practised, its various stages are interwoven in the agricultural landscape, such that traditional shifting cultivation and permanent production systems exist together in the same locality.

Traditional shifting cultivation with adequately long fallow periods has been considered a sound method of soil management, well adapted to the local ecological and social environment. Before the forest is cleared, a closed nutrient cycle exists in the soil-forest system. Within this system, most nutrients are stored in the biomass and topsoil, and a constant cycle of nutrient transfer from one compartment of the system to another operates through the physical and biological processes of rain-wash (i.e., foliage leaching), litterfall, root decomposition, and plant uptake. The amount of nutrients lost from such a system is negligible. Clearing and burning the vegetation leads to a disruption of this closed nutrient cycle. During the burning operation the soil temperature increases, and afterward, more solar radiation falling on the bare soil surface results in higher soil and air temperatures. This shift in the temperature regime causes changes in the biological activity of the soil. The addition of ash to the soil through burning causes important changes in soil chemical properties and organic-matter content. In general, exchangeable bases and available phosphorus increase slightly after burning; pH values also increase, but usually only temporarily. These changes in the soil after clearing and burning result in a sharp increase of available nutrients (the so-called 'fertilization effect' following a fire).

The cycle of land clearing-cropping-fallow-cropping has been repeated indefinitely in many regions where shifting cultivation has continued for centuries, though at low productivity levels. Over a long period, as population pressure has steadily increased, fallow periods have become shorter and shorter; consequently, farmers have returned to abandoned fields before they have had enough time for fertility to be sufficiently restored. The introduction of industrial crops and modern methods of crop production has also caused a diminished emphasis on the importance of the fallow period in traditional farming practices.

The agroforestry approach to fallow improvement

Consequent upon efforts to improve shifting cultivation, the nature of the practice itself has been shifting. The traditional situation of long fallows interrupted by short cropping phases has rapidly been replaced by shorter fallows over the years. Such unmanaged shorter fallows have become known as the root cause of disastrous consequences of shifting cultivation such as soil-fertility decline and accelerated soil erosion. Therefore, it was a logical consequence to accept that managed permanent (sedentary rather than shifting) cultivation systems encompassing some of the advantages of traditional shifting cultivation would be preferable to fallow-depleted forms of the traditional system. Under the leadership of two institutions of the Consultative Group for International Agricultural Research (CGIAR) based in Africa – IITA (the International Institute of Tropical Agriculture) in Nigeria, established in the 1960s, and ICRAF (the International Centre for Research in Agroforestry, now known as World Agroforestry), established in Kenya in the late 1970s – major efforts were undertaken for about three decades, beginning in the 1970s, to integrate fast-growing, nitrogen-fixing trees and shrubs with crop production on farmlands.

Various names were proposed for different categories of fallows to emphasize the major distinguishing features of each; natural, improved, sequential, rotational, enriched, managed, and mixed are adjectives commonly used to indicate familiar types (Nair et al., 2021). Depending on how the woody species were integrated with crops, these fallow practices became known as Alley Cropping (Kang and Wilson, 1987) and Improved Fallow systems (Buresh and Cooper, 1999). The experience with these major fallow-improvement initiatives will be briefly reviewed here. Some prototype farming systems that would allow farmers to have continuing access to and dependence on land, even during the 'no-cropping' (rather than the fallow) phase, as alternatives to shifting cultivation, have been proposed in the northeastern parts of India. But the extent to which such alternatives are adopted by shifting cultivators will depend more on the social, economic and anthropological conditions than on the biological merits of the suggested alternatives (FAO, 1985; 1989).

The rise and fall of alley cropping

It became evident in the early 1990s, after about two decades of experience with alley cropping in different parts of the tropics, that the potential benefits of the technology had been overblown. Among several factors that were identified as critically important for its success, the major one was soil moisture. In many semiarid regions, the rainfall follows a unimodal pattern, and extends over only four to five months. In these conditions, nitrogen-fixing tree legumes that were planted as hedgerows were unable to develop an adequate number of new branches (foliage) to allow more than two or three pruning events during a year – one before planting the alley-cropped cereal (or other short-duration crop) and one or a maximum of two during the crop's growth. The mulch yield and, therefore, nitrogen contributions were also correspondingly lower, to the extent of being inadequate for producing a satisfactory crop yield. Additionally, shade effects were caused by the hedgerows as well as a reduction in the area of land available for crop production (in a square configuration, 20 hedgerows, each casting severe shade over an area 1 m wide and 100 m long, will cover 2000 sq. m per hectare, or 25% of the total area). The additional labour that was required to maintain and prune the hedges was another limitation. Furthermore, farmers could choose to remove the mulch for use as animal fodder, rather than adding it to the soil, as was the case in Haiti (Bannister and Nair, 1990). While all factors related to the biological advantages of alley cropping were important, the social acceptability and adoption potential of the practice were equally – or even more – important. In addition to common difficulties in popularizing an improved agricultural technology developed by research stations among target farmers, some features of alley cropping counterbalanced its advantages and hindered its widespread adoption (see Enters and Lee, ch. 7, this volume). These included the need for additional labour and skills that were required for hedgerow pruning and mulch application, loss of cropping area to the hedgerows, difficulty in mechanizing agricultural operations, and potential for the hedgerow species to become a weed, an alternate host for pests and pathogens, a shelter for grain-eating birds, and the possible cause of increased termite activity, especially under dry conditions.

In retrospect, alley cropping was no exception to the all-too-common experience in agricultural development initiatives in the tropics: excessive expectations and euphoria that accompany the introduction of any new initiative, followed by the disappointment that ensues when the expectations are not fully met. The reasons are several and well-known from past experiences: the craze and race for finding immediate solutions to long-standing, complex, and multifaceted problems; the simplistic and trivial nature of proposed solutions; popularization of the solution (technology) without adequate testing; and so on. An exaggerated emphasis was placed (the 'panacea' syndrome) on the advantages and expectations from agroforestry, and researchers and the development community were under severe pressure to bring out some 'magic' that could be used to erase all the massive problems of deforestation, land degradation, food scarcity and poverty, and all the related issues. Alley cropping, being one of the early technologies of agroforestry, was welcomed with a wave of extreme enthusiasm. Although it was based on the sound land-management principle of a biological means of maintaining and improving soil fertility, its limitations became evident when introduced to areas that were unfavourable and would later be acknowledged as beyond its limits (for example, dry areas). Some proponents of alley cropping took extreme positions and went to great lengths, using only positive results and ignoring the not-so-positive ones, while others just denigrated and dismissed it. Some 'played it safe' by joining the bandwagon at first, and later (when it became clear that the going was rough) trying to 'become smart' by criticizing it as having a basis in weak science (Sanchez, 2019). Others, however, argued that the results of tropical alley-cropping datasets needed further analyses before there was a rush to 'throw out the baby with the bathwater' (Vandermeer, 1998).

To sum up the alley cropping experience, the consensus was that the practice could be promising under conditions where the annual rainfall during cropping seasons was more than 1000 mm and the soils were reasonably fertile, with no serious nutrient deficiencies or extreme soil reactions. Under these conditions, the results of alley cropping would even be better if the land was gently sloping (less than 10% slope) and there was no shortage of labour during cropping seasons. Under such conditions, the advantages of soil health and environmental sustainability arising from reduced use of chemical fertilizers would make alley cropping a winner. An important point was that under conditions where alley cropping was appropriate, such as in the lowland humid tropics, the technology could be adapted for both low and high levels of productivity.

The rise and fall of improved fallows

From the early 1980s, alley cropping was the most-talked-about and researched topic in tropical agroforestry. But by the mid-1990s, it was replaced by an ICRAF-promoted technology called 'improved fallows'. This referred to 'deliberate planting of fast-growing, usually leguminous, species for rapid replenishment of soil fertility, and implies the use of improved tree and shrub species during the fallow phase' (Sanchez, 1995) (Figure 3-2). The technology attained prominence when ICRAF began to focus its institutional efforts on improved fallows as an approach to soil-fertility management for enhancing crop production in the nutrient-depleted soils of sub-Saharan Africa. Faced with the frustration of the failure of alley cropping to deliver expected benefits from the soil-improvement potential of trees and shrubs, ICRAF began to promote improved fallows as a breakthrough in improving crop production and alleviating hunger and poverty (Sanchez, 1999). It became the centre's flagship programme and, understandably, dominated the tropical agroforestry scene and generated a lot of expectations. Numerous publications (research reports, journal articles, conference proceedings, and so on) became available during the ensuing 10 to 15 years; a notable one was *The Science and Practice of Improved Fallows*, a book-length compilation of mostly experiential descriptions of improved fallows from various countries, published as a special issue of the journal *Agroforestry Systems* (Buresh and Cooper, 1999). The flow of publications then slowed down gradually, except for a few summaries and reviews (Ajayi et al., 2007; Sileshi et al., 2008).

The driving force behind the development of the improved-fallow technology was the same as that of alley cropping, i.e., the search for new approaches to respond to soil-fertility problems, primarily in sub-Saharan Africa, resulting from the breakdown of traditional farming systems that had once enjoyed the benefit of long fallow periods. Scientifically, the improved-fallow concept was founded on well-known principles – not very different from the foundations of tropical alley cropping – that planting fast-growing tree species, especially those that fix nitrogen, produces easily

FIGURE 3-2: Forest clearance leads to a rapid decline in soil fertility, and one approach to creating improved fallows in shifting cultivation is the *taungya*-planting of valued trees into swiddens along with the annual crops. After a few seasons of annual cropping, the trees 'take over' to become the dominant fallow species. Here, a luxuriant-looking rice (*Oryza sativa*) crop in the first year of establishment (left) of a teak (*Tectona grandis*) and *Eucalyptus camaldulensis* (not shown) plantation. A year later (right), the trees have grown taller, but the rice crop is poorer than that of the first year, with rice plants showing signs of yellowing, a typical symptom of nitrogen deficiency.

Photos: P. K. R. Nair, lowland Thailand, 1992 and 1993, respectively.

decomposable biomass to provide nitrogen for food crops growing together with or following the tree species, increases soil organic matter, and improves the physical conditions of the soil. In retrospect, the rather disappointing performance of tropical alley cropping provided a good background and incentive for presenting the soil-improvement potential of trees as a new approach to the issue.

Since improved fallows was a new technology in sub-Saharan Africa, it had to be tested for various technical issues, such as screening and selection of species and fallow establishment and management. While these time-consuming procedures continued, the technology was taken prematurely to the dissemination stage based on its assumed promise and potential for success. A noteworthy aspect of the technology development, however, was the realization that the success of such technologies crucially depended on their suitability to local conditions, and that this could best be realized by farmers' participation in technology development and adaptation. Consequently, farmers were involved in assessing the technology and making modifications based on their experiences. In order to assess the extent of farmers' adoption of the improved-fallow technology, those who planted trees for a second cycle were identified as 'adopters', while those who were still in the first cycle of tree fallows were designated as 'users'. Presenting the case study of Zambia, Ajayi et al. (2007) describe how the scaling up of the technology to different parts of the country was coordinated by a network comprising representatives of ICRAF, government research-and-extension services, farmer organizations and non-governmental organizations (NGOs). From fewer than a hundred planters in the early 1990s, the number of farmers who planted improved-fallow trees reportedly increased each year by tens of thousands. Subsequent reports indicated, however, that the initial euphoria fizzled out and the number of adopters declined gradually to the extent that out of the nearly 700,000 smallholder farmers who were reportedly planting improved fallows in East and Southern Africa, only 7% of them – fewer than 50,000 – continued to do so after four years. The major reason for the lack of adoption of the technology was later attributed to the lack of a financial package to offset opportunity costs for one or two years (Sanchez, 2019). Strangely, however, while the improved-fallows technology was being promoted as 'rock-solid', the failure of tropical alley cropping was attributed to the weakness of its science, although the scientific foundations of both were similar.

The lessons learned from fallow improvement technologies

The lessons to be learned from three decades of efforts to develop and popularize fallow-improvement technologies involving soil-improving trees and shrubs can be summarized as follows:

- Both alley cropping and improved fallows are two sides of the same coin; the scientific principles of both are fundamentally the same (although this is not unanimously accepted).
- Both technologies performed well under conditions of adequate water availability during crop growing seasons, but were unsuitable for dry areas.

- The dichotomy between the two technologies perpetuated by some researchers turned out to be counterproductive; opportunities were lost for critical efforts focused on finding the best ways of incorporating the proven benefits of including fast-growing woody legumes and other species in smallholder farming systems on nutrient-poor tropical soils.
- The 'post-mortem' analysis identified two major reasons for the low level of adoption of these technologies: (1) administrative failures to create an enabling environment for providing credit and financial support, seeds and other planting materials, fertilizer amendments as needed, and so on, and (2) strategic failures in pushing the boundaries of testing way beyond the ecological regions that were the 'safe' zones for these technologies.

Irrespective of the technical merits (and demerits) of these two fallow-improvement technologies, the important factors that determined their suitability as alternatives to or improvements of shifting cultivation were socio-economic considerations. Being management-intensive practices, these technologies demanded skilful management of resources and constant attention to various operational details of both the crops and the hedgerow species. On the other hand, shifting cultivation in its traditional form is a low-intensity activity; once the main initial operations of land clearing and crop sowing are done (which, of course, are extremely labour-intensive operations), the crop does not receive much attention until it is ready for harvest. These differences in management intensity could have been a major hindrance to the extensive adoption of the two technologies. The various socio-economic reasons, in general, that impede the successful adoption of new and improved agricultural technologies are relevant here; however, they are too well known to be included in this discussion.

Shifting cultivation and ecosystem services

Ecosystem services

Until about the turn of the century, soil-related issues (fertility decline, erosion, and so on) were cited as the major aspects of ecosystem degradation associated with shifting cultivation. Then, the concept of ecosystem services (ES), i.e., *the benefits that people obtain from ecosystems*, became prominent following the publication of the Millennium Ecosystem Assessment (MA, 2005), the massive United Nations report that grouped ecosystem services into four major categories: provisioning, regulating, supporting and cultural. Soon, that categorization came under criticism for 'category overlap', or too many services fitting into more than one of the four categories of the MA framework (Notte et al., 2017). For example, nutrient cycling, water cycling and some other services included under supporting services could be counted as provisioning services as well. To address this issue, the Intergovernmental Science-Policy Platform on Biodiversity and Ecosystem Services (IPBES), an 'intergovernmental body which assesses the state of biodiversity and of the ecosystem services it provides to society', introduced the concept of *nature's contributions to people* (NCP: initially termed *Nature's Benefits to People*), a term embracing 'all contributions of nature to the quality of life of humans (both positive and negative) as NCP' (Torralba et al., 2016; IPBES, 2017; Diaz et al., 2018). Another important frame of reference in ecosystem research is the Common International Classification for Ecosystem Services (CICES), proposed by the European Environment Agency (Maes et al., 2016). The CICES categorizes ecosystem services into three categories: provisioning, regulating and cultural. Despite all of these developments, ES and NCP are widely used synonymously, and the MA framework continues to be the most recognized classification of ecosystem services.

Shifting cultivation has traditionally been denigrated as an unproductive, disorderly, resource-depleting and environmentally destructive practice of the past, and considered more as a provider of ecosystem disservices than services. Because of this strong bias against the practice, no serious effort has been made to examine whether it indeed provides any ES at all. Moreover, as stated earlier, until about the turn of the century, the numerous attempts that were made to evaluate the effect of shifting cultivation were concentrated on the extent of damage caused in terms of soil loss and degradation. As pointed out by Mertz et al. (2021), the focus of such evaluations was on the extent of impacts on ecosystems and livelihoods that could be consequent upon changes to traditional practices (Mertz and Magid, 2001; Bruun et al., 2009; Rerkasem et al., 2009; van Vliet et al., 2012; Dressler et al., 2017; Rasmussen et al., 2018). With increasing interest in studies of the cultural heritage and traditions of shifting cultivators, as reported in Cairns' massive volumes (Cairns, 2007, 2015, 2017 and this volume), and the broadening of the concept of ecosystem services from formerly narrow perceptions to also include biodiversity and cultural services, the value of shifting cultivation in ecosystem services is receiving new attention. Mukul and Herborn (2016) reported on a comprehensive meta-analysis of results from 412 studies on shifting cultivation conducted during the period from 1950 to 2014. The outcome showed high degrees of variability in research emphasis at various times (temporal heterogeneity) and research locations (spatial heterogeneity) across tropical regions. The analysis involved studies from 45 countries; 52.2% of them representing tropical Asia and the Pacific, 31.8% tropical America and the Caribbean, and 16% Africa. Of the 412 studies, 52% were conducted since 2001. The major areas of research were anthropology/human ecology (52%), soil-related (17%), plant ecology (15%), agricultural production/management (13%) and agroforestry (8%), with noticeable shifts in the research themes across the regions and study periods. Despite the limitations of the analysis based on arbitrary grouping of the studies into various categories with overlapping boundaries (especially among soil quality, plant ecology, agricultural production/management and agroforestry), it is clear that, in the past, anthropology/human ecology was the predominant theme of research on shifting cultivation. However, the authors of the study noted that in recent years, environmental impacts had received more attention.

A perfunctory review (with no detailed analysis) of the increasing volume of such literature emerging from research on shifting cultivation since 2001 shows that:

1) studies on ecosystem services of shifting cultivation are on the increase;
2) soil-carbon dynamics related to climate-change mitigation and biodiversity conservation are two of the major themes of such studies; and
3) although some reports on provisioning services (dealing with items other than food materials) and cultural services are also available, they are mostly descriptive.

Carbon storage and sequestration

Just as in many other aspects of shifting cultivation, studies on soil-carbon dynamics and sequestration have been few, superficial, uncoordinated and inconclusive; therefore, the knowledge base is poor and inadequate for arriving at meaningful conclusions. Moreover, the reported studies on carbon (and organic matter) under shifting cultivation, starting with the classical one by Nye and Greenland (1960) and followed by many others – some of which have been cited in this chapter – focused on the rate of decline of carbon. Overall, studies that quantify soil organic carbon (SOC) stocks in shifting cultivation systems are not only particularly few, but are also inconclusive, as documented by Mukul and Herbohn (2016). Chan et al. (2016) found no relationships between fallow age and SOC stocks in a chronosequence of fallows (1 to 53 years) in Myanmar, nor did Bruun et al. (2009) in a study involving newly cleared fields that had been fallowed for 5 to 38 years in Sarawak, Malaysia. In terms of carbon storage, studies have primarily focused on old-growth forests, and while the knowledge base on secondary forests is rudimentary (Mertz et al. 2012), evidence of their carbon-storage potential is emerging from several case studies (Jepsen, 2006; Kho and Jepsen, 2015; Mukul et al., 2016; Bruun et al., 2018; Matos et al., 2020). Bruun et al. (2021) reported from a study in northern Laos that there was no evidence that intensification of shifting cultivation led to a decline in total soil-carbon stock; however, the reasoning behind this observation was unfortunately not elucidated. McNicol et al. (2015) showed that allometric determination of net biomass in shifting cultivation fallows did not account for the extensive root biomass and stump resprouting of trees in fallows, leading to underestimates of root biomass by 50% to 60% and significantly lower total carbon-storage estimates in shifting cultivation areas. Several other studies have also highlighted different weaknesses in the current level of understanding of the topic (van Vliet et al., 2012; Mukul and Herbohn, 2016; Dressler et al., 2017; Rasmussen et al., 2018).

Several methodological problems that are associated with carbon sequestration studies in agroforestry systems have been described by Nair et al. (2021, ch. 20, section 20.4.3); these apply to shifting cultivation as well. An additional issue is that many studies on shifting cultivation compare the conditions between old-growth (natural) forests and secondary (and successional) forests, and consider shifting cultivation as a form of secondary forest. But the fallow and cropping cycles of shifting cultivation do not represent secondary forest. As Mertz et al. (2021) and several others have noted, there is an abundance of oversimplified assumptions in studies of carbon under shifting cultivation. Without going too deeply into the details here, this author's perception is that although the reported studies on soil carbon dynamics and sequestration under shifting cultivation are both too few and superficial, overall, the practice negatively impacts soil carbon sequestration.

Biodiversity conservation

Along with climate change, loss of biodiversity is the other major environmental and ecosystem-service issue that has captivated global attention over the past two decades. It is now clearly recognized that the narrow range of species that provide food and other products for human use cannot be sustained without the optimal presence and activities of a great diversity of other plants, animals, and organisms. The IPBES (2017) assessment of the state of global biodiversity was a significant document. It presented an alarming picture of the rapid decline of biodiversity in the immediate past and painted seriously grave scenarios for the immediate future.

The information base for conservation of biodiversity under shifting cultivation is arbitrary and observational rather than experimental or scientific. The prevailing impression of shifting cultivation as an undesirable and destructive practice is so negative that a substantial body of strong evidence is needed to counter it; unfortunately, no such evidence is available. Another major issue is the rather superficial nature of studies on biodiversity conservation in agroforestry, shifting cultivation and other inadequately researched land-use systems. Often, such reports are one-shot exercises that report the numbers of various species of components (mostly plants, especially trees) found in a particular location when the counting was done. For example, 'it is estimated that there are approximately 3.04 trillion trees in the world' (https://www.geographyrealm.com). Not even wild estimates are available for the number of trees in the approximately 300 million ha where shifting cultivation is practised, but it is plausible that many trees (of all ages and forms) do exist in shifting cultivation areas. Such numbers have no significance unless the dynamics of the tree species over time and the effects of their interactions, both with other components of the systems of which they are a part, and with the overall system, are carefully assessed. Biodiversity in and of itself is not the issue; assessment and exploitation of biodiversity conservation for the overall performance of a land-use system are the critical issues. These cannot be assessed by a one-time count or estimate of species composition. In such discussions, secondary forests are the environmental feature that comes closest to shifting cultivation, but as mentioned earlier, specific reviews on the role of secondary forests have had a relatively narrow focus. Moreover, such reviews have often considered logged forests rather than shifting cultivation systems (Edwards et al., 2010). Nevertheless, some reviews have reported that secondary forests can be important for biodiversity if old-growth forests are nearby, and that both old-growth and 'older secondary' forests play an important role in species conservation (Dent and Wright, 2009; Sayer et al., 2017). Others have focused on the recovery of forest ecosystems after shifting cultivation (Teegalapalli et al., 2009; Martin et al., 2013); but they do not reflect on the continued use of shifting cultivation and fallow cycles in the landscapes. The bottom line is that even if shifting cultivation supports high biodiversity, it cannot be equated with biodiversity conservation.

Summary: Ecosystem services of shifting cultivation

The brief discussion above suggests that the terms ecosystem services and shifting cultivation do not fit together well; using them together sounds like an oxymoron in the conventional understanding of the terms. Ecosystem services are *the benefits that people obtain from ecosystems*; in the conventional sense, these include mainly the provisioning, regulating, and supporting services categories. As discussed earlier, shifting cultivation is not credited with any significant provision of non-food services. Comparing the ecosystem services in secondary forests and other land uses in shifting cultivation landscapes, Mertz et al. (2021) found that carbon storage, biodiversity, and soil variables dominated such studies, with very few on provisioning services and none available in the literature on cultural services. They concluded that the narrow focus on specific ecosystem categories strongly limited the understanding of secondary forests and other fallow vegetation in shifting cultivation areas and that it was more relevant to compare such secondary forests with other agricultural systems rather than with old-growth forests. On the other hand, shifting cultivation scores highly in terms of the cultural and social services it provides to the indigenous people who practise it. Unfortunately, there are no standard procedures for assessing the value of such services in quantitative terms. As the old dictum goes, anything that cannot be counted tends to get ignored. Thus, even if shifting cultivation provides significant ecosystem services, they are largely unappreciated.

Conclusions

The major scientific issue that has dominated discussions on the shifting cultivation-agroforestry interface is the importance of fallow management. It is well appreciated that in the crop-fallow-crop-fallow cycles of traditional shifting cultivation, the soil fertility that is lost during the cropping phase is recouped during the fallow phase – if the fallow phase is long enough, depending on the local soil and climatic conditions. Various types of fallow improvements have been tried, all of them involving the integration of fast-growing trees and shrub species, preferably nitrogen-fixing ones, with crop production. Depending on how the trees/shrubs were integrated with crops, the technologies were known as alley cropping (simultaneous) or improved fallow (sequential, with two or more years under the fallow). A major characteristic of both technologies was that they were less suitable for semiarid and arid regions with low or limited water availability. Despite several efforts by local and international agencies, farmer-adoption rates for both technologies were disappointingly low. Reasons given for their low performance include administrative failures to create an enabling environment for providing credit and financial support to ensure timely availability of inputs, and strategic failures in pushing the boundaries of testing to ecological regions that were way beyond the 'safe' zones for the technologies. Given that the fallow-improvement technologies were well-grounded on solid scientific principles, it may not be unrealistic to expect that tweaking the technologies by introducing appropriate management interventions and revamping the administrative and policy support might overcome the initial difficulties in the course of time.

Regarding the provision of ecosystem services by shifting cultivation, there is an age-old general perception that shifting cultivation contributes to ecosystem disservices rather than services. In the ecosystem-service categories of the Millennium Ecosystem Assessment, shifting cultivation could impact services in the provisioning, regulating, and supporting categories. However, even if fallow-improvement technologies were adopted by farming communities in shifting cultivation areas, it is unlikely that those technologies would become net providers of such services. Social and cultural services are the categories in which shifting cultivation could be a net provider of ecosystem services, but the absence of practical measures to assess the extent of such services realized through shifting cultivation remains a handicap. Thus, it is doubtful that shifting cultivation will become recognized as a provider of ecosystem services.

References

Ajayi, O. C., Place, F., Kwesiga, F. and Mafongoya, P. (2007) 'Impacts of improved tree fallow technology in Zambia', in H. Waibel and D. Zilberman (eds) *International Research on Natural Resource Management: Advances in Impact Assessment*, CABI, Wallingford, UK, and Science Council, CGIAR, Rome, pp.147-168

Bannister, M. E. and Nair, P. K. R. (1990) 'Alley cropping as a sustainable agricultural technology for the hillsides of Haiti: Experience of an agroforestry outreach project', *American Journal of Alternative Agriculture* 5, pp.51-59

Bruun, T. B., de Neergaard, A., Lawrence, D. and Ziegler, A. (2009) 'Environmental consequences of the demise of swidden agriculture in Southeast Asia: Carbon storage and soil quality', *Human Ecology* 37, pp.375-388

Bruun, T. B., Berry, N., de Neergaard, A., Xaphokahme, P., McNicol, I. and Ryan, C. M. (2018) 'Long rotation swidden systems maintain higher carbon stocks than rubber plantations', *Agricuture, Ecosystems and Environment* 256, pp.239-249

Bruun, T. B., Ryan, C. M., de Neergaard, A. and Berry, N. J. (2021) 'Soil organic carbon stocks maintained despite intensification of shifting cultivation', *Geoderma* 388, 114804, https://doi.org/10.1016/j.geoderma.2020.114804, accessed 13 October 2021

Buresh, R. J. and Cooper, P. J. M. (eds) (1999) 'The science and practice of short-term improved fallows', special issue, *Agroforestry Systems* 47, pp.1-356

Cairns, M. F. (ed.) (2007) *Voices from the Forest: Integrating Indigenous Knowledge into Sustainable Upland Farming*, Resources for the Future, Washington, DC

Cairns, M. F. (ed.) (2015) *Shifting Cultivation and Environmental Change: Indigenous People, Agriculture and Forest Conservation*, Earthscan, London

Cairns M. F. (ed.) (2017) *Shifting Cultivation Policies: Balancing Environmental and Social Sustainability*, CABI, Wallingford, UK

Chan, K. M. A., Balvanera, P., Benessaiah, K., Chapman, M., Díaz, S., Gómez-Baggethun, E., Gould, R., Hannahs, N., Jax, K., Klain, S., Luck, G. W., Martín-López, B., Muraca, B., Norton, B., Ott, K., Pascual, U., Satterfield, T., Tadaki, M., Taggart, J. and Turner, N. (2016) 'Opinion: Why protect nature? Rethinking values and the environment', *Proceedings of the National Academy of Sciences* 113, pp.1462-1465

Colfer, C. J. P. (2017) 'Foreword', in M. F. Cairns (ed.) *Shifting Cultivation Policies: Balancing Environmental and Social Sustainability*, CABI, Wallingford, UK, pp.xxvi-xxxi

Crutzen, P. J. and Andreae, M. O. (1990) 'Biomass burning in the tropics: Impact on atmosphere chemistry and biogeochemical cycles', *Science* 250, pp.1669-1678

Dent, D. H. and Wright, J. S. (2009) 'The future of tropical species in secondary forests: A quantitative review', *Biological Conservation* 142, pp.2833-284

Díaz, S., Pascual, U., Stenseke, M., Martín-López, B., Watson, R. T., Molnár, Z., Hill, R., Chan, K. M. A., Baste, I. A., Brauman, K. A., Polasky, S., Church, A., Lonsdale, M., Larigauderie, A., Leadley, P. W., van Oudenhoven, A. P. E., van der Plaat, F., Schröter, M., Lavorel, S., Aumeeruddy-Thomas, Y., Bukvareva, E., Davies, K., Demissew, S., Erpul, G., Failler, P., Guerra, C. A., Hewitt, C. L., Keune, H., Lindley, S. and Shirayama, Y. (2018) 'Assessing nature's contributions to people', *Science* 359(6373), pp.270-272, doi: 10.1126/science.aap8826

Dressler, W. H., Wilson, D., Clendenning, J., Cramb, R., Keenan, R., Mahanty, S., Bruun, T. B., Mertz, O. and Lasco, R. D. (2017) 'The impact of swidden decline on livelihoods and ecosystem services in Southeast Asia: A review of the evidence from 1990 to 2015', *Ambio* 46, pp.291-310

Edwards, D. P., Hodgson, J. A., Hamer, K. C., Mitchell, S. L., Ahmad, A. H., Cornell, S. J. and Wilcove, D. S. (2010) 'Wildlife-friendly oil palm plantations fail to protect biodiversity effectively', *Conservation Letters* 3, pp.236-242

Falvey, J. L. (2017) 'Policies impacting shifting cultivation', in M. F. Cairns (ed.) *Shifting Cultivation Policies: Balancing Environmental and Social Sustainability*, CABI, Wallingford, UK, pp.43-63

FAO (1985) *Changes in Shifting Cultivation in Africa: Seven Case Studies*, FAO Forestry Paper 50/1, Food and Agriculture Organization of the United Nations, Rome

FAO (1989) *Household Food Security and Forestry: An Analysis of Socioeconomic Issues*, Food and Agriculture Organization of the United Nations, Rome

FAO/SIDA (1974) *Shifting Cultivation and Soil Conservation in Africa*, FAO Soils Bulletin 24, Food and Agriculture Organization of the United Nations, Rome

Greenland, D. J. (1976) 'Bringing green revolution to the shifting cultivator', *Science* 190, pp.841-844

Heinimann, A., Mertz, O., Frolking, S., Egelund Christensen, A., Hurni, K., Sedano, F., Chini, L. P., Sahajpal, R., Hansen, M. and Hurtt, G. (2017) 'A global view of shifting cultivation: Recent, current, and future extent', *PLoS ONE* 12(9), e0184479, https://doi.org/10.1371/journal.pone.0184479

IPBES (2017) *Update on the Classification of Nature's Contributions to People by the Intergovernmental Science-Policy Platform on Biodiversity and Ecosystem Services*, IPBES, Bonn, Germany, available at https://www.ipbes.net/system/tdf/downloads/pdf/ipbes-5-inf-24.pdf, accessed 13 October 2021

Jepsen, M. R. (2006) 'Above-ground carbon stocks in tropical fallows, Sarawak, Malaysia', *Forest Ecology and Management* 225, pp.287-295

Kang, B. T. and Wilson, G. F. (1987) 'The development of alley cropping as a promising agroforestry technology', in H. A. Steppler and P. K. R. Nair (eds) *Agroforestry: A Decade of Development*, ICRAF, Nairobi, Kenya, pp.227-243

Kho, L. K. and Jepsen, M. R. (2015) 'Carbon stocks of oil palm plantations and tropical forests in Malaysia: A review', *Singapore Journal of Tropical Geography* 36, pp.249-266

MA (2005) *Millennium Ecosystem Assessment: Ecosystems and Human Well-being*, Synthesis, Island Press, Washington, DC

Maes, J., Liquete, C., Teller, A., Erhard, M., Paracchini, M. L., Barredo, J. I., Grizzetti, B., Cardoso, A., Somma, F. and Petersen, J. (2016) 'An indicator framework for assessing ecosystem services in support of the EU Biodiversity Strategy to 2020', *Ecosystem Services* 17, pp.14-23, doi: 10.1016/j.ecoser.2015.10.023

Martin, P. A., Newton, A. C. and Bullock, J. M. (2013) 'Carbon pools recover more quickly than plant biodiversity in tropical secondary forests', *Proceedings of the Royal Society B: Biological Sciences* 280, 20132236

Matos, F. A. R., Magnago, L. F. S., Aquila Chan Miranda, C., de Menezes, L. F. T., Gastauer, M., Safar, N. V. H., Schaefer, C. E. G. R., da Silva, M. P., Simonelli, M., Edwards, F. A., Martins, S. V., Meira-Neto, J. A. A. and Edwards, D. P. (2020) 'Secondary forest fragments offer important carbon and biodiversity co-benefts', *Global Change Biology* 26

McNicol, I. M., Berry, N. J., Bruun, T. B., Hergoualc'h, K., Mertz, O., de Neergaard, A. and Ryan, C. M. (2015) 'Development of allometric models for above and belowground biomass in swidden cultivation fallows of Northern Laos', *Forest Ecology and Management* 304, pp.104–116

Mertz, O. and Magid, J. (2001) 'Shifting cultivation as conservation farming for humid tropical areas', in: L. García-Torres, J. Benites and A. Martínez-Vilela (eds) *Conservation Agriculture, a Worldwide Challenge, First World Congress on Conservation Agriculture*, vol. II, Offered Contributions, Environment, Farmers Experiences, Innovations, Socio-economy, Policy, XUL, Córdoba, Spain

Mertz, O., Padoch, C., Fox, J., Cramb, R. A., Leisz, S. J., Nguyen, T. L. and Vien, T. D. (2009) 'Swidden change in Southeast Asia: Understanding causes and consequences', *Human Ecology* 37, pp.259–264

Mertz, O., Müller, D., Sikor, T., Hett, C., Heinimann, A., Castella, J. C., Lestrelin, G., Ryan, C. M., Reay, D., Schmidt-Vogt, D., Danielsen, F., Theilade, I., van Noordwijk, M., Verchot, L. V., Burgess, N. D., Berry, N. J., Pham, T. T., Messerli, P., Xu, J., Fensholt, R., Hostert, P., Pfugmacher, D., Bruun, T. B., de Neergaard, A., Dons, K., Dewi, S., Rutishauer, E. and Sun, Z. (2012) The forgotten D: challenges of addressing forest degradation in complex mosaic landscapes under REDD+, *Geografisk Tidsskrift-Danish Journal of Geography* 112, pp.63–76

Mertz, O., Bruun, T. B., Jepsen, M. R., Ryan, C. M., Zaehringer, J. G., Hinrup, J. S. and Heinimann, A (2021) 'Ecosytem service provision by secondary forest in shifting cultivation areas remains poorly understood', *Human Ecology* 49, pp.271–283, https://doi.org/10.1007/s10745-021-00236-x

Mukul, S. A. and Herbohn, J. (2016) 'The impacts of shifting cultivation on secondary forest dynamics in tropics: A synthesis of the key findings and spatio-temporal distribution of research', *Environmental Science and Policy* 55, Part 1, pp.167–177

Mukul, S. A., Herbohn, J. and Firn, J. (2016) 'Cobenefits of biodiversity and carbon sequestration from regenerating secondary forests in the Philippine uplands: Implications for forest landscape restoration', *Biotropica* 48, pp.882–889

Nair, P. K. R. (1993) *An Introduction to Agroforestry*, Springer, Dordrecht, The Netherlands

Nair, P. K. R., Kumar, B. M. and Nair, V. D. (2021) *An Introduction to Agroforestry: Second edition: Four Decades of Scientific Developments*, Springer, Dordrecht, The Netherlands

Newton, K. (1960) 'Shifting cultivation and crop rotation in the tropics', *Papua New Guinea Agricultural Journal* 13, pp.81–118

Notte, A. L., D'Amato, D., Mäkinen, H, Paracchini, M. L., Liquete, C., Egoh, B., Geneletti, D. and Crossman, N. D. (2017) 'Ecosystem services classification: A systems ecology perspective of the cascade framework', *Ecological Indicators* 74, pp.392–402, https://doi.org/10.1016/j.ecolind.2016.11.030

Nye, P. H. and Greenland, D. J. (1960) *The Soil Under Shifting Cultivation*, Commonwealth Bureau of Soils, Harpenden, UK

Rasmussen, L. V., Coolsaet, B., Martin, A., Mertz, O., Pascual, U., Corbera, E., Dawson, N., Fisher, J. A., Franks, P. and Ryan, C. M. (2018) 'Social-ecological outcomes of agricultural intensification', *Nature Sustainability* 1, pp.275–282

Rerkasem, K., Lawrence, D., Padoch, C., Schmidt-Vogt, D., Ziegler, A. and Bruun, T. B. (2009) 'Consequences of swidden transitions for crop and fallow biodiversity in Southeast Asia', *Human Ecology* 37, pp.347–360

Ruthenberg, H. (1980) *Farming Systems in the Tropics*, second ed., Oxford University Press, London, UK

Sanchez, P. A. (1995) 'Science in agroforestry', *Agroforestry Systems* 30, pp.5–55

Sanchez, P. A. (1999) 'Improved fallows come of ages in the tropics', *Agroforestry Systems* 47, pp.3–12

Sanchez, P. A. (2019) *Properties and Management of Soils in the Tropics*. John Wiley, New York

Sayer, C. A., Bullock, J. M. and Martin, P. A. (2017) 'Dynamics of avian species and functional diversity in secondary tropical forests', *Biological Conservation* 211, pp.1–9

Sileshi, G., Akinifesi, F. K., Ajayi, O. C. and Place, F. (2008) 'Meta-analysis of maize yield response to woody and herbaceous legumes in sub-Saharan Africa', *Plant and Soil* 307, pp.1–19

Teegalapalli, K., Veeraswami, G. G. and Samal, P. K. (2009) 'Forest recovery following shifting cultivation: An overview of existing research', *Tropical Conservation Science* 2, pp.374–387

Torralba, M., Fagerholm, N., Burgess, P. J., Moreno, G. and Plieninger, T. (2016) 'Do European agroforestry systems enhance biodiversity and ecosystem services? A meta-analysis', *Agriculure, Ecosystems and Environment* 230, pp.150-161, http://dx.doi.org/10.1016/j.agee.2016.06.002

van Vliet, N., Mertz, O., Heinimann, A., Langanke, T., Pascual, U., Schmook, B., Adams, C., Schmidt-Vogt, D., Messerli, P., Leisz, S., Castella, J.C., Jørgensen, L., Birch-Thomsen, T., Hett, C., Bruun, T. B., Ickowitz, A., Vu, K. C., Fox, J., Cramb, R. A., Padoch, C., Dressler, W. and Ziegler, A. (2012) 'Trends, drivers and impacts of changes in swidden cultivation in tropical forest-agriculture frontiers: A global assessment', *Global Environmental Change* 22, pp.418-429

Vandermeer, J. H. (1998) 'Maximizing crop yield in alley crops', *Agroforestry Systems* 40, pp.199-208

A swidden hut sits unused in Phongsaly district, northern Laos, after the Phunoy shifting cultivators had finished harvesting their rice crop.

Sketch based on a photo by Olivier Evrard in November, 1998.

Why farmer innovations should interest us...

An Angami Naga farmer pollards an alder tree as he reopens his *jhum* field for cultivation. The management of alder trees scattered across swidden fields reached its zenith in Khonoma village, Nagaland, in northeast India, allowing farmers to shorten fallow periods to as little as two or three years without substantial loss of productivity (see Cairns and Brookfield, ch. 22 and plates 11 to 13 in Coloured Plates section I, this volume). We should be learning from such farmer innovations.

Sketch based on a photo by Malcolm Cairns in 2001.

Synthesis

TRADITIONAL KNOWLEDGE AND PRACTICES IN SWIDDEN AGRICULTURE

A window on sustainability

*Percy E. Sajise**

Introduction

Sustainability, in its simplest meaning, is the ability to maintain or prolong. It could refer to the capacity of a biological system to remain functional and productive indefinitely. This is reflected in its endurance or resilience to absorb disturbances and still retain its basic structure and functions, in order to provide basic goods and services needed by human society. The general characteristics and principles of sustainability seem simple and non-debatable. However, it is complex, multi-dimensional and highly contextual, making it difficult to apply in operational terms.

How, then, do traditional knowledge and practices in swidden agriculture provide a window or guidance in promoting sustainability? The chapters in this introductory section, as well as many others in this volume, draw attention to why swidden-farmer innovations should be of interest to us. They provide some of the needed lessons on sustainability, given that swidden agriculture, as an age-old practice, has been generally described as sustainable. These lessons are:

a) Sustainability mainly comprises three major and interacting pillars: (1) ecological; (2) economic; and (3) socio-cultural. These three major elements must work in a symbiotic and complementary manner in a particular context, so that goods and services generated by the natural resource base and needed by human society are produced on a sustainable basis. This is well described in the chapters of Rosado-May et al. (ch. 4) and Riba (ch. 5), in this section, and it will become more apparent in following chapters of this volume. For example, in shifting

---

* Dr. Percy E. Sajise, Senior Fellow, Southeast Asian Ministers of Education Organization (SEAMEO)'s Regional Center for Graduate Study and Research in Agriculture (SEARCA), College, Laguna, Philippines; Adjunct Professor, School of Environmental Science and Management, University of the Philippines at Los Baños, Philippines.

cultivation, the growing of diverse crops and raising of animals are tied up to ecological considerations pertaining to various types of niches, in terms of soil fertility, exposure to sunlight, soil-moisture conditions and issues of conservation, in respect of natural-resource regeneration. The same practices are also linked to the need of households for food, fuel, energy, building materials, religious rituals and other traditional practices. Prevailing social organizations and norms are also responsible for governance that ensures allocation of resources and resolution of conflicts.

b) A knowledge system comprising both indigenous knowledge (IK) and formal knowledge emanating from research institutions provides the integrating element that promotes symbiotic interactions among the three sustainability pillars. IK is the main driving force that brings this synergy and complementarity to shifting cultivation. It is an accumulated and integrated knowledge system honed by generations of proven practices, and is embedded in the individuals who manage, benefit from, and maintain the natural resource base involved in swidden agriculture. The depth of this inherited knowledge invokes a greater investment in sustainability and generates reasons for making sure that the three pillars are promoted in order to enhance it. In transformed and intensive agricultural systems, on the other hand, the roles are assumed separately by different sectors, e.g. production, processing and marketing, where benefits are divided, more often than not, inequitably. This results in a lack of complementarity among the three pillars and a reduced or total absence of sustainability. This was well described in the chapters of Riba (ch. 5, this volume) and Enters and Lee (ch. 7, this volume). Riba, for example, clearly indicates that IK related to people's reactions and perceptions of ecological processes and events becomes part of their culture, applying directly to cultivation, food habits, healing practices, hunting and fishing, and other economic activities.

This equilibrium of endogenous and exogenous factors and processes for promoting sustainability in shifting cultivation – as it has been traditionally practised over many generations in various parts of the world – is undergoing rapid transformations, as described by Li et al. (ch. 6, this volume). This is manifested in shortening of fallow periods and increasing intensity of production. The driving forces are mainly population growth, alternative land uses, forest-conservation policies, and market and infrastructure development. The more common consequences of these changes are enhanced soil erosion, declining water quality, reduced biodiversity and diminished levels of carbon stocks, both above the ground and in soil organic carbon. These result in overall reduction of sustainability and capacity to buffer climate change in areas where traditional shifting cultivation was once practised. These changes are inevitable and a reality. The challenge is, how do we respond to this reality and restore some level of sustainability to these transformed swidden landscapes? How do we draw lessons from the sustainability characteristics of traditional shifting cultivation and apply them to rapidly changing landscapes that are being transformed by more intensive but less sustainable agriculture and resource use? Going back to lessons learned on the sustainability of shifting cultivation practices, the following are some suggestions for the application of key elements:

a) Analyze the nature of the three pillars of sustainability and their interactions. For example, is the technology used for intensive production promoting over-extraction and non-regenerative practices that are affecting the natural resource base (soil, biodiversity, nutrient cycling, the balance of trophic levels and others)? What are the impacts of policies on land ownership and security of tenure vis-à-vis the practices of those engaged in agricultural production? What are the impacts of markets on the increasing cash needs of farming households? This analysis should be able to identify which of the three pillars is a stronger driver of negative interactions that are promoting non-sustainability, and what interventions are needed to reverse the trajectory.

b) The knowledge system, and how it provides the integrating and synergistic 'glue' among the three sustainability pillars, should also be analyzed. The rapid transformation process and changes driven by both exogenous and endogenous factors involved in traditional shifting agriculture create a knowledge gap that makes IK inadequate as an integrating factor to bring about a sustainable production system. These factors may include introduction of new crops and breeds of animals with different requirements and new pests and diseases, the need for more external inputs, climate change, and many other changes. This will require the addition of a complementary formal-knowledge system, which in many cases will also be inadequate because commonly, it will have been derived from monodisciplinary approaches. Such approaches may further promote negative interactions among the three pillars of sustainability. This type of knowledge system could be derived from intercultural education as described in the chapter of Rosado-May et al. (ch. 4, this volume).

In the face of this rapid transformation in shifting agriculture landscapes, a knowledge system must have the following characteristics:

a) Relevant IK combined with a formal holistic-knowledge system generated from interdisciplinary research, and

b) Inter-generational continuity of this combined and evolving knowledge system.

Enters and Lee (ch. 7, this volume) have provided insights into how this kind of knowledge system can be generated, to bring about synergy in the interactions of the three pillars of sustainability. This will require that transformations taking place in shifting cultivation landscapes should be more process-based and 'bottom-up', rather than 'top-down'; closely involving the farmers themselves and with less emphasis on physical targets and outcomes. Another big challenge is a demographic backdrop where more young people are leaving the farm and being drawn into urban centres, away from rural areas, creating an increasing divide between the land and its caretakers. This divide is reflected in the varying interests of young and old farmers, as indicated by Rosado-May et al. (ch. 4, this volume). Policies are another important exogenous factor. An important consideration is the type of policies that enable farmers to increase their capacity to adopt practices that promote sustainability, such as those related to land tenure security, transfer of appropriate formal knowledge systems, recognition of gender roles and links to markets.

Case studies of successful climate-change adaptations in Southeast Asia have demonstrated the importance of the '3Ps' (Participatory assessment, Partnership building and Process-based). These are directly relevant to the process of transforming shifting cultivation landscapes and practices (Sajise et al., 2016). The highly contextual nature of the pillars of sustainability requires participatory assessment, which requires both the identification of key stakeholders and provision of space for their meaningful involvement in the actual process of assessing the nature and relationships of the three pillars. Partnership-building is important, given that new knowledge and relationships are needed to ensure that the process of transforming shifting agriculture practices will be sustainable. Such partnerships may involve key actors at various hierarchical levels, with resource users at the core. This will allow the resource users to assert their needs and assume ownership of the whole process. In the final analysis, the resource users are the ultimate determinants and gatekeepers of sustainability.

References

Sajise, P. E, Cadiz, M. C. H. and Bantayan, R. B. (eds) (2016) *Learning and Coping with Change: Case Stories of Climate Change Adaptation in Southeast Asia*, Southeast Asia Regional Center for Graduate Study and Research in Agriculture (SEARCA), College, Laguna, Philippines

4

MAKING THE INVISIBLE VISIBLE

The role of indigenous knowledge in the sustainability of shifting cultivation

Francisco J. Rosado-May, Prasert Trakansuphakon, Seno Tsuhah and Phrang Roy*

Introduction

Shifting cultivation is a generic name for a variety of food-production systems that have been practised for centuries across continents and cultures (Harrison and Turner, 1978; Majumdar, 1990; Palm et al., 2005; Ickowitz, 2006; Cairns, 2015). Heinimann et al. (2017) calculated the global area devoted to shifting cultivation on the basis of existing Landsat-based deforestation data covering the years from 2000 to 2014, using very high-resolution satellite imagery with an accuracy level in their reckoning of higher than 87%. With 62% of the investigated one-degree cells in the humid and sub-humid tropics currently showing signs of shifting cultivation – the majority in the Americas (41%) and Africa (37%) – the authors were confident in saying that this form of cultivation remains widespread, and it would be wrong to speak of its general global demise over recent decades. However, they had to accept that their estimate of 280 million hectares devoted to shifting cultivation around the world was clearly smaller than the areas mentioned in literature they had reviewed. These ranged up to 1 billion hectares, mostly due to differences in the methodologies used in measuring the area covered by slash-and-burn farming systems. Regardless of whether the area under shifting cultivation (also known as swidden farming) is decreasing or increasing, the future of these farming systems is raising important environmental issues as well as questions related to the livelihood security and resilience of the people currently dependent upon them.

There have been many attempts to estimate the number of people involved in shifting cultivation. Notable among them, the Director of the Land and Water

---
* Dr. Francisco J. Rosado-May, Professor, Intercultural Maya University of Quintana Roo, Mexico; Dr. Prasert Trakansuphakon, President, Pgakenyaw Association for Sustainable Development, Thailand; Seno Tsuhah, Project Team Leader, North East Network, India; and Phrang Roy, Coordinator, The Indigenous Partnership, India/Italy.

Development Division of the Food and Agriculture Organization of the United Nations, R. Dudal, wrote in the Preface of FAO Soils Bulletin 53 (1984) that between 200 and 300 million people were dependent on this system of cultivation, and they were facing failing yields, more poverty and reduced opportunities to subsist and improve their living standards. Osman (2013), citing a publication by ICRAF in 1998, estimated that about 250 million people practised shifting cultivation around the world. In a report for the World Bank, Chomitz et al. (2007) estimated that about 70 million people lived in remote tropical forests and about 800 million rural people lived in or near tropical forests and savannas; areas in which shifting cultivation was

Cnidoscolus chayamansa McVaugh
Synonym of *Cnidoscolus aconitifolius* (Mill.) I.M.Johnst. [Euphorbiaceae]

Sometimes known as tropical or Maya spinach, this plant is an important swidden vegetable on Mexico's Yucatán Peninsula. Its edible leaves have high iron and vitamin contents and are known as a remedy for kidney stones.

practised. Literature reviewed by Ribeiro Filho et al. (2013) indicated that between 35 million and 1 billion people were dependent on shifting cultivation. On the other hand, Mertz et al. (2009) were sceptical about the estimated number of farmers practising shifting cultivation because of a lack of solid information due to differences in methodologies for collecting data.

Shifting cultivation has been a part of human history for centuries. Dove (2015, pp.3-24), for instance, in an excellent review of the epistemology of swidden agriculture, discussed the contributions of Carl Nilsson Linnaeus, in the mid-18th century, towards understanding the role of this type of agriculture in the development of Europe. Morley (1946, p.141), a classic author on Maya culture, published the following description of shifting cultivation:

> The modern Maya method of raising maize is the same as it has been for the past three thousand years or more – a simple process of felling the forest, burning the dried trees and bush, of planting, and changing the location of corn fields every few years. This is practically the system of agriculture practised in the American wet tropics even today, and indeed is the only method available to a primitive people (Morley, 1946).

This account is not very different from the historical descriptions of shifting cultivation in India and Sri Lanka by Kingwell-Banham and Fuller (2012); by Schmidt-Vogt

(2001) for Thailand; or by Spencer (1977) for Southeast Asia, only the main crop is not maize, but rice (Figure 4-1).

Even with doubts about the accuracy of data regarding the extent of land used for shifting cultivation and the number of people dependent upon it, the historical and cultural roots of this food-production system, attributed to indigenous people, are acknowledged and well documented (e.g. Erni, 2015, for South Asia; Watters, 1971, for the Americas).

FIGURE 4-1: Karen women harvesting rice from a swidden in northern Thailand.

Photo: Prasert Trakansuphakon.

Shifting cultivation has been the topic of research by different scientific disciplines over the past five decades. According to Angelsen (1994), it is possible from the literature to identify different driving forces behind studies of shifting cultivation. One is its association with deforestation. For instance, Rahman et al. (2011) suggest that shifting cultivation, still prevalent in the uplands of eastern Bangladesh, contributes significantly to forest loss and is the main cause of land degradation. Garbyal (1999) makes the assumption that shifting cultivation, which he calls 'a primitive agricultural practice with a devastating impact on the environment', can be improved by providing alternative land-based permanent occupations. He describes a programme launched by the Mizoram (India) government in 1984, called New Land Use Policy, which basically failed, and he calls for its restructuring, but not in favor of shifting cultivation. Swidden farming has been blamed for deforestation in

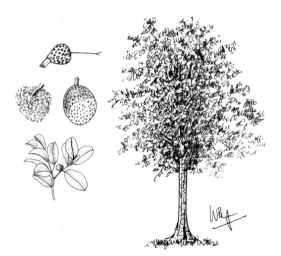

Brosimum alicastrum Sw. [Moraceae]

The *Ramon* tree is a significant forest species in Mexico and Central America. The history of its use reaches back to the ancient Maya people. The nuts have high nutritional value. In times of food shortage they are ground into flour to supplement maize flour. When stewed, they taste like mashed potato; when roasted the flavour is like coffee, or chocolate.

the Amazon (Fearnside, 2005), in Asia (Fox, et al., 2000; Seidler and Bawa, 2001) and elsewhere (Palm et al., 2005). Another driving force is the association of shifting cultivation with climate change, one example of which was a study by Kauffman et al. (2009), which showed that carbon emissions resulting from swidden agriculture in the tropics exceeded the ability of secondary forest regrowth to sequester carbon. This was related to habitat fragmentation, an increase in the area burned and the total quantity of biomass consumed by fire per unit of area (Figure 4-2). Using a sophisticated method, Silva et al. (2011) estimated the annual emissions of carbon dioxide, methane, carbon monoxide, nitrous oxide and other oxides of nitrogen from biomass burning in shifting cultivation systems in tropical Asia, Africa and America. Their results revealed large uncertainties in emission estimates for all five gases, and their mean estimates were lower than those found in previous studies of atmospheric emissions from biomass burning in swidden systems. In addressing the debate over the sustainability of shifting cultivation and its impact on the soil, Ribeiro Filho et al. (2013) reviewed the literature published on shifting cultivation over a 30-year period and reported that the nature of the impact depended on the phase of the swidden system and on the soil properties (physical, chemical and biological). The idea that shifting cultivation is not a single system with just one phase is hardly surprising. Spencer (1977) described 18 distinct types of shifting cultivation within Southeast Asia alone, and Brookfield and Padoch (1994) argued that swidden agriculture was actually many hundreds, or even thousands, of systems.

The assumption that small-scale farming and shifting cultivation are major causes of deforestation has been challenged and the sustainability of swidden systems has now been recognized, so different driving forces have been created for shifting cultivation research. For instance, Fox et al. (2000) concluded that the Vietnam government's policies on shifting cultivation would fail if there was no solid understanding that the system rested on the temporary removal of trees, but not of the forest, and that if it was left alone to work the result would be conservation of biodiversity, soil and land cover, as well as carbon sequestration. Other studies have found that when public policy sets conditions for shifting cultivation to be replaced by commercial agriculture or oil palm or rubber plantations and the like, there are higher environmental impacts, including deforestation (Geist and Lambin, 2002; Rudel et al., 2009; DeFries et al., 2010; van Vliet et al., 2012, 2013). A study supported by Oxfam-GB and reported by the Northern

FIGURE 4-2: A Karen community group plants crops on a recently burnt swidden in northern Thailand.

Photo: Prasert Trakansuphakon.

Alnus nepalensis D. Don [Betulaceae]

One of the most effective fallow trees. This species is used in its eastern Himalayan range to enrich depleted soil with heavy leaf litter as well as fixing nitrogen from the atmosphere. It also provides generous quantities of firewood when a fallow plot is reopened for cultivation.

Colocasia esculenta (L.) Schott [Araceae]

Taro is widely cultivated as a swidden food crop for its corms, leaves and leaf stems. It is one of the few food crops, along with rice and lotus, that can be grown under flooded conditions.

Development Foundation and the Huay Hin Lad community in Northern Thailand (NDF, 2012), showed that shifting cultivation as developed by the Karen community was both sustainable and contributed to capturing carbon, contrary to popular perceptions that shifting cultivation causes loss of carbon into the atmosphere.

Angelsen (1994) has proposed another driving force in the quest to understand shifting cultivation: economic analysis and modelling of the system. The main argument is the need to better understand the nature of economic decision-making and farmers' responses to exogenous changes, including land-ownership issues, in order to design effective policy instruments. Angelsen's (1994) analysis adopts two economic approaches: open and subsistence. It identifies the effects of exogenous changes due to population growth, technological progress and increased risk. The driving concept is based on literature reported in the study (e.g. Ruthenberg, 1980), and assumes the negative impacts of shifting cultivation, which is considered to be an agricultural practice from an early stage in the evolution of agricultural systems, practised by indigenous people. The same driving concept is discussed by Mai (1999), who studied shifting cultivation in the Central Highlands of Vietnam and concluded that poverty associated with permanent food shortages was the main cause of the continuation of shifting cultivation. Contrary to the assumptions that guide the research exemplified by that of Angelsen (1994) and Mai (1999) – which basically involves blaming the victims – Arifin and Hudoyo (1998) contend that there are other approaches to understanding the socio-economic and political elements present in shifting cultivation. They say the assumption that shifting cultivation is associated with negative environmental and social issues is a simple-deterministic paradigm aligned with Neo-Malthusian or Neo-Marxian thinking (Arifin and Hudoyo, 1998). In emphasizing the urgent need for effective farming methods that do not pollute the environment or affect farmers' health, Arifin and Hudoyo (1998) argue that in an open economy, farmers are trying to adopt more permanent and more intensive land-use practices to adjust to market forces. These authors advise against a fallow system for lowland rice cropping, arguing that while a long fallow system is profitable and efficient, it is subject to increased pressure over time.

The challenges of connecting the indigenous knowledge in shifting cultivation with education

There is a driving force missing in the research surrounding shifting cultivation. It is the need to connect the neglected, underutilized, invisible, but highly complex and sophisticated traditional knowledge, upon which shifting cultivation systems have rested for centuries, with present-day education systems and new young generations. To discuss this, it is vital to present the following premise: *the thinking that sustains shifting cultivation, in its different forms, was not developed with the idea of destroying the environment; it was with the idea of working with the environment.* In the light of this principle, it would be important to learn the age and knowledge of farmers on whose plots studies are carried out and the number of years of experience they have in shifting cultivation. It would be appropriate, also, to assume that examples of shifting cultivation that have been studied and show significant degradation of nature are in direct correlation with varying degrees of failure to transfer traditional knowledge, and the inability of the farmers involved to learn it.

Indigenous cultures pass on knowledge generation after generation, by creating a system in which there is participation by all family members, of all ages, and observation and practising lie at the core of the process (Figure 4-3). This has been documented extensively in the Americas (e.g. Correa-Chávez et al., 2015; UNESCO, 2017a), and also in Asia-Pacific and Africa (e.g. Kapoor and Shizha, 2010).

Unfortunately for the transmission of indigenous knowledge to new generations in the formal school system, there is a strong influence of Western thinking in the education of indigenous and non-indigenous people in Asia and elsewhere. Altbach (1989) documented the impact of Western academic models, practices and orientations that not only shaped the nature of higher education systems in India, Malaysia, Indonesia and Singapore, but also excluded local knowledge. In a study for the Asian

FIGURE 4-3: A Maya *milpero* (shifting cultivator) in Mexico teaches his grandson how the *ibes* bean (*Phaseolus lunatus*) can be intercropped with maize and both harvested at the same time.

Photo: Francisco J. Rosado–May.

in international research centres such as the International Rice Research Institute and the Center for International Forestry Research, the school system in each country has embraced the Green Revolution and developed teaching resources, training and research programmes to advance permanent intensive agriculture instead of shifting cultivation and other traditional farming systems. This pervasive focus has prevented young generations of indigenous people from accessing their own traditional knowledge systems or learning from them, and has ignored opportunities to use indigenous ways of learning. Attending school

Development Bank, of new challenges for educational development and cooperation in Asia for the 21st century, Hirosato (2001) does not even mention the existence of conspicuous indigenous knowledge in the region, much less suggest how it could be incorporated into the formal school system.

Using the assumption that shifting cultivation has negative effects, and under pressure to conserve natural resources, countries have designed policies aimed at stopping shifting cultivation and promoting more permanent and intensive land uses. To do so, they have established two main strategies: one is based on Green Revolution technologies for annual crops (Farmer, 1981; Herath, 1985; Pingali, 2012) and the other on tree plantations (Ball, 1995; INDUFOR, 2012; Fox et al., 2014). With science and technology produced

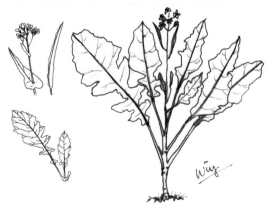

Brassica campestris (L.)
Synonym of *Brassica rapa* L.
[Brassicaceae]

This species, which is closely related to turnip, napa cabbage and *bok choy*, is often known as field mustard. It grows naturally in many conditions (see Figure 4-4) and its seed is harvested to extract oil. It produces fields of bright yellow flowers that attract large numbers of butterflies.

has meant that children and young adults have been unable to spend enough time with their parents and grandparents, or with anyone else from their communities, in order to learn about shifting cultivation or any other community endeavour. In these circumstances, how could the younger generations have learned the indigenous knowledge that is fundamental to shifting cultivation? It is a surprise to many indigenous observers that the negative effects detected on some shifting cultivation systems are not greater than they presently are (Figure 4-4).

FIGURE 4-4: A *jhumia* (shifting cultivator) harvests vegetables from the rich plant diversity of a mountainside swidden in Nagaland, northeast India.

Photo: North-East Slow Food and Agrobiodiversity Society (NESFAS).

However, there are attempts to make the value and importance of indigenous knowledge more 'visible'. For instance, Thaman (2003) proposes decolonizing the field of Pacific studies and reclaiming indigenous perspectives, knowledge and wisdom that have been devalued or suppressed. Realizing the importance of language for indigenous education in the context of several United Nations goals for sustainable development, the United Nations' Educational, Scientific and Cultural Organisation (UNESCO) has been working in Cambodia, the northern Philippines, Malaysia, Myanmar and Thailand to encourage learning and the use of mother tongues (UNESCO, 2017b). According to Hallinger (1998), the rapid economic growth of Asian countries has impacted on education systems in a multicultural setting, with countries needing a clear approach based on their own social fabric, rather than simply a capacity to implement reform policies.

The exclusion from the formal school system of indigenous

Artemisia absinthium L. [Asteraceae]

A common swidden species known in Western countries as Wormwood, a bitter extract of this plant is used in the spirit Absinthe. It is also used as a medicine in cases of dyspepsia, for various infectious diseases and to treat internal parasites.

knowledge and indigenous ways of learning about shifting cultivation can be explained by the way in which that knowledge is regarded in Western thinking. Most, if not all, of the early literature about shifting cultivation uses terms like 'primitive' (Morley, 1946, Garbyal, 1999), or the Marxist view in Vietnam that considers shifting cultivation to be a primitive stage in the cultural evolution of agriculture (Rambo, 1995). There is little wonder that the general perception is that there is no valuable knowledge in shifting cultivation! However, is it not logic to assume that there is a wealth of sophisticated knowledge and information supporting a system that has prevailed for centuries, overcoming many types of pressures, and still has evidence of solid bases that provide elements of sustainability?

Case studies from Mexico, India and Thailand

The following information was obtained from three indigenous communities in each of three different countries: Yucatec Maya communities (X-Yatil, Chancah Veracruz and Xhazil) in Quintana Roo, Mexico; Naga (Chizami and Enhulumi) and Khasi (Nongtraw) communities in India; and Karen communities (Hin Lad Nai, Mae Yod and Mae Umpai) in Thailand. All of these communities maintain a strong tradition of shifting cultivation and are located in biodiversity hotspots known as Mesoamerica and Indo-Burma (Myers et al., 2000) (Figure 4-5). Based on examples of successful sustainable farming systems (Everton, 2012, p.16, pp.57-85; NDF, 2012), as well as access and the confidence of the informants, the people were grouped into two age categories: 30 to 60 years old and 20 to 29 years old in Mexico and Thailand, and 40 to 60 years old and 20 to 39 years old in India. Five farmers were randomly selected as informants from each age group in each community. The objective was to detect differences in their understanding of, and performance in, shifting cultivation.

The information sought to provide answers to the following questions:

1. Based on the experience accumulated through their years in farming, what is their perception of the relative importance of factors limiting the success of shifting cultivation? Are there differences according to age group or ethnic group?
2. Are younger generations of farmers losing knowledge about shifting cultivation systems developed in their communities?
3. How many decisions that demand knowledge do experienced farmers make in order to perform successfully under a shifting cultivation system?

When asked to rank their main concerns regarding their farming systems, there was a clear consensus, except for the younger Maya group. Their main concern was to prevent the erosion of knowledge about shifting cultivation from generation to generation. Land security was the second most important concern for all groups. The younger groups seemed to be more concerned about marketing their products than the older groups. On issues of soil fertility and pest and weed management as limiting factors, there was basically no trend detected among the various groups (Table 4-1).

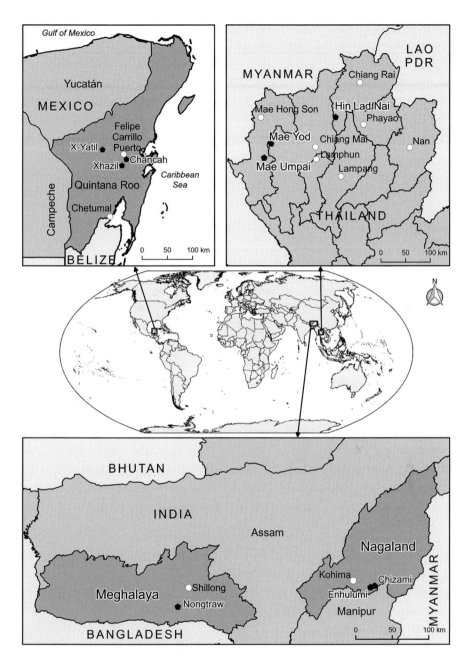

FIGURE 4-5: Locations of the study communities in Mexico, Thailand and northeast India.

TABLE 4-1: Perception of the relative importance of six limiting factors affecting shifting cultivation systems in three indigenous groups from each of Mexico, India and Thailand.

| Limiting factor | Origin/Age group | | | | | |
| | Older group | | | Younger group | | |
	Maya	N.E. India	Karen	Maya	N.E. India	Karen
Passing on traditional knowledge to younger generations	1	1	1	3	1	1
Land security	2	2	2	1	2	2
Soil fertility	4	5	3	2	5	3
Pest management	5	3	4	5	3	4
Weed management	6	4	4	4	4	4
Marketing	4	4	3	2	3	2

Note: Each figure represents a consensus of relative-importance ranking provided by 15 farmers, with number 1 regarded as the factor of greatest concern and number 6, the least. The older group is composed of farmers from 30 or 40 to 60 years old, whereas the younger group is composed of farmers from 20 to 29 or 39 years old. The communities are X-Yatil, Chancah Veracruz and Xhazil, Yucatec Maya villages in Quintana Roo, Mexico; Chizami and Enhulumi, Naga communities and Nongtraw, a Khasi village, in northeast India; and Hin Lad Nai, Mae Yod, and Mae Umpai, Karen communities in northern Thailand.

To illustrate erosion in the process of transmitting traditional knowledge to younger generations, nine parameters where chosen and studied in all nine villages, with the same informants from both age groups (Table 4-2). While 87 to 100% of the older group of shifting cultivators said they did not use external chemical inputs, 7 to 87% of the younger generation said they did (Figure 4-6). Although the average number of species planted in Mayan swiddens was noticeably smaller than that in northeast India and Thailand, in all cases the older-generation farmers planted more species (6 to 22) in their swiddens than the younger farmers (1 to 16 species). The range (minimum to maximum) of wild species identified by the older group before clearing their shifting cultivation fields was about double the number identified by their younger counterparts. The number of edible plants that were not planted, but were identified by the older group before the first weeding of their swiddens, was almost three times higher than the younger group of Mayan farmers and almost double the number identified by the younger groups in northeast India and Thailand.

FIGURE 4-6: A farmer in northeast India applies chemicals to a swidden crop.

Photo: North-East Slow Food and Agrobiodiversity Society (NESFAS).

TABLE 4-2: Farmer's ages and experience in shifting cultivation systems.

Parameters	Older group			Younger group		
	Maya	N.E. India	Karen	Maya	N.E. India	Karen
Farmers not using external chemical inputs (%).	90	100	87	13	93	53
Farmers using external chemical inputs (%).	10	0	13	87	7	47
Average number of species planted after burning the field.	6	20	22	1	16	16
Number of wild species recognized in a 50 m x 50 m plot before clearing for shifting cultivation.*	31-37	37-47	34-50	8-12	20-28	22-27
Number of edible plants recognized growing naturally on a swidden before the first weeding.	6-8	8-13	9-11	1-2	5-8	6-8
Average number of years of formal schooling.	1	4	4	7	9	12
Average number of years working with father or grandfather on shifting farming before becoming independent.	20	10	17	8	15	12
Average age of beginning to accompany parents or grandparents to swiddens.	8	10	7	13	14	10
Average age of becoming fully independent farmers.	30	21	25	21	19	26

Note: Each figure represents the consensus ranking provided by 15 farmers – five from each of the three indigenous communities chosen in each of the three countries. The older group is comprised of farmers from 30 or 40 to 60 years old, whereas the younger group is comprised of farmers from 20 years old to 29 or 39. *These figures relate to plots with very similar soil and climatic conditions. The communities are X-Yatil, Chancah Veracruz and Xhazil, Yucatec Maya villages in Quintana Roo, Mexico; Chizami and Enhulumi, Naga communities and Nongtraw, a Khasi village, in northeast India; and Hin Lad Nai, Mae Yod, and Mae Umpai, Karen communities in northern Thailand.

The younger groups had 7 to 12 years of formal schooling, whereas the older group had between 1 and 4 years. Thus, the time spent by the younger group with their parents or grandparents, learning about shifting cultivation or other community endeavours, was considerably less (8 to 15 years) than the time spent by the older generation (10 to 20 years). The older group started accompanying their parents or grandparents to their swiddens when they were 7 to 10 years old, compared with 10 to 14 years old for the younger group. The older group became fully independent farmers at an older age – 21 to 30 years old – than the younger group, who became independent when aged 19 to 26. All three indigenous groups, from different parts of the world, showed the same trends (Table 4-2).

The above information assumes that indigenous farmers have a wealth of knowledge that guides their decisions in managing their traditional and sustainable swidden systems (Figure 4-7). To understand this process, it was important to find out how many issues and what kind of questions weighed on the minds of those farmers considered by their communities to be the 'holders of the secrets' to successful shifting cultivation. The assumption was that the success and sustainability of their farming systems rested on the knowledge they held and their ability to correctly answer those questions, and that younger generations, with little or no contact with their elders, would have fewer proper answers.

Over the course of two years, participatory field visits and days of conversations with the 'older group' of experienced indigenous farmers allowed us to measure the depth and breadth of knowledge required for successful swidden farming (Figure 4-8). These experienced farmers, recognized by their communities as 'holding the secrets' for successful shifting cultivation, did not use external chemical inputs, lived on the outputs of their farms, had great biodiversity in their fields and provided advice and seeds to other community members.

The complete shifting cultivation process was divided into nine different stages and from these arose a total of 193 questions requiring a thorough knowledge of the system in order to make appropriate decisions (Table 4-3). For each of the nine stages, the questions were classified

FIGURE 4-7: A farmer harvests millet from his swidden in Nagaland, northeast India.

Photo: Seno Tsuhah.

FIGURE 4-8: A *jhumia* (shifting cultivator) rests after a bountiful harvest from her swidden in Nagaland, northeast India.

Photo: Seno Tsuhah.

into three aspects of shifting cultivation that are always present and interconnected: spiritual, community/social and technical issues. The questions represented an extraordinary agricultural system and a wealth of knowledge that has been tested for centuries, and has prevailed. The list of questions, and the knowledge needed to answer them, was basically the same for all indigenous groups in this study. One important aspect of shifting cultivation that is not illustrated by this exercise is the dynamics of the system; indigenous knowledge is constantly being innovated, adapted and created. Even when an external eye cannot see changes, they are still happening. It has been called 'dynamics without motion', a description coined by

TABLE 4-3: Questions for which a traditional shifting cultivator must have answers for development of a successful farming strategy.

Stage	Origin/Age group Classification				Examples of questions/decisions that demand sound knowledge for successful shifting cultivation.
	Spir	Com	Tech	Total	
Selecting the field	2	3	13	18	What spiritual ceremonies are performed before and after selecting the field?
					Who selects the plot?
					What indicators should be used to select the plot (types and conditions of plants, insects, animals, soil)?
Clearing the field	3	8	5	16	What ceremonies and prayers are performed when asking permission of supernatural beings to clear the forest and use the soil?
					What is the role of the men, women or children who are helping?
					What criteria should be used in deciding when to start clearing the field?
Fire management	2	11	13	26	Who participates in the spiritual ceremony before burning?
					What are the factors to consider for good coordination between participants in the burning?
					How does a farmer read natural indicators of weather conditions to determine good conditions for burning?
Planting	3	8	21	32	When exactly is a spiritual ceremony performed before planting?
					Are seeds shared within the community or exchanged with other communities?
					Once the date for planting is set, what species and varieties should be planted?

TABLE 4-3 (cont.): Questions for which a traditional shifting cultivator must have answers for development of a successful farming strategy.

Stage	Origin/Age group Classification				Examples of questions/decisions that demand sound knowledge for successful shifting cultivation.
	Spir	Com	Tech	Total	
Pre-harvest management.	8	7	31	46	When should the spiritual ceremony be performed? Who can teach how weeds, pests and diseases should be managed, and what ceremonies are needed to ask for rain, if it is needed? There are various types of pests, like worms, both above and below the ground, ladybugs, birds, animals, and so on. What are the various techniques used to handle and manage these pests?
Harvest.	6	3	9	18	Is there a ceremony and prayers to perform before harvesting, to give thanks for the harvest? Have there been changes in the community decision process over the years in terms of getting help for harvesting? How does a farmer know that the moisture content of the seeds is suitable for harvesting?
Post-harvest management.	2	3	12	17	Are there any prayers and ritual practices for the storage of grain? Who can teach the techniques of efficient post-harvest management, and how can this be taught? Is there a need for any special measures to care for both the seeds and the storage place, before the harvest is placed in storage?
Trade/ marketing.	3	3	8	14	Is there a special ritual to ask for good trading/selling? Is trading or selling based on climatic or social factors, or both? How does a farmer set minimum and maximum prices for sales?

TABLE 4-3 (cont.): Questions for which a traditional shifting cultivator must have answers for development of a successful farming strategy.

Stage	Origin/Age group Classification				Examples of questions/decisions that demand sound knowledge for successful shifting cultivation.
	Spir	Com	Tech	Total	
Designing an agro-ecosystem: Long term visions.	2	3	1	6	What is the role of beliefs in a long-term vision for food production? What factors are considered in decisions about soil conditions, planning, diversification, accumulated knowledge and innovations, and the community decision-making process? Should a farmer keep clearing new plots every other year or develop a system that does not depend solely on shifting cultivation every year?

Note: Spir = Spiritual; *Com* = Community/social; *Tech* = Technical. Information was collected from three successful and highly experienced farmers from the indigenous groups in each of the three countries: Yucatec Maya in Mexico, Naga and Khasi from NE India, and Karen from Thailand.

an eminent expert on North American native studies, Professor Leroy Little Bear, himself a Blackfoot Indian from Canada (Little Bear, 2011). Without this element, it would be very hard to explain both the successful resistance to outside pressures and the resilience of shifting cultivation in many parts of the world.

Having the list of 193 questions – examples of which appear in Table 4-3 – represented a baseline of knowledge about shifting cultivation. The next step was to put these questions to young farmers in the study communities to determine how much knowledge they had about their farming system when compared to that of experienced farmers. Following suggestions from the elders in each community, five young farmers who were known for their dedication to farming were selected and asked each of the 193 questions relating to the nine stages of shifting cultivation identified in Table 4-3. Their responses were rated either good or bad by the researchers and these judgments were confirmed by the elders when necessary. The assumption was that if young farmers who were dedicated to their job did not provide proper answers, then those who were less than dedicated would know much less.

Table 4-4 presents the percentage of appropriate answers given by young farmers to questions grouped according to each of the nine stages of shifting cultivation. Based on these results, the younger Yucatec Maya farmers are losing traditional knowledge on shifting cultivation much more rapidly than their counterparts in northeast India and northern Thailand. With the exception of their responses related to post-harvest management, marketing and visions for the future, the young Yucatec Maya farmers seem to be retaining traditional knowledge related to the spiritual and communal aspects of shifting cultivation better than they are the technical knowledge. In India and Thailand, the younger farmers seem to have more concerns about the technical knowledge than the other two aspects of shifting cultivation, once again with the

TABLE 4-4: Percentage of young indigenous farmers from each region who provided correct answers to questions regarding traditional knowledge on their shifting cultivation system.

| Indigenous groups | Stage of the system / Aspect of management | | | | | | | | |
| | Field selection | | | Clearing the field | | | Fire management | | |
	Spir.	Tech.	Com.	Spir.	Tech.	Com.	Spir.	Tech.	Com.
Yucatec Maya, Mexico	53	19	26	53	38	35	39	16	20
Naga and Khasi, NE India	100	50	66	100	73	72	100	70	76
Karen, Thailand	96	96	100	68	100	100	33	84	100

| Indigenous groups | Stage of the system / Aspect of management | | | | | | | | |
| | Planting | | | Pre-harvestment | | | Harvest | | |
	Spir.	Tech.	Com.	Spir.	Tech.	Com.	Spir.	Tech.	Com.
Yucatec Maya, Mexico	51	49	49	43	45	53	47	49	33
Naga and Khasi, NE India	75	83	100	72	95	69	93	98	77
Karen, Thailand	51	49	49	43	45	53	47	49	33

| Indigenous groups | Stage of the system / Aspect of management | | | | | | | | |
| | Post-harvest management | | | Trade/Marketing | | | Designing the agroecosystem. Long term vision | | |
	Spir.	Tech.	Com.	Spir.	Tech.	Com.	Spir.	Tech.	Com.
Yucatec Maya, Mexico	36	51	53	28	82	55	22	20	15
Naga and Khasi, NE India	100	61	75	100	64	53	50	60	44
Karen, Thailand	100	96	100	100	99	27	0	32	66

Note: Spir. = Spiritual; Tech. = Technical; Com. = Community

exception of their responses related to post-harvest management, marketing and visions for the future. It is interesting to note that responses to questions about the design of shifting cultivation agroecosystems – the ninth stage of the farming system and a crucial one for its future and sustainability – is the stage where younger farmers show the least knowledge.

Passing on traditional knowledge to new generations is a matter of great concern to all indigenous communities (see Table 4-1). The substitution of traditional methods

FIGURE 4-9: Having built his field hut, a Maya *milpero* in Quintana Roo, Mexico, prepares to plant his swidden.

Photo: Francisco J. Rosado-May.

of farming with external chemical inputs is growing as new generations take over. Relatively young shifting cultivators do not have the same depth of knowledge about their natural resources as older farmers, so the management of their food systems has lost much of its traditional knowledge and the farmers rely more and more on different, 'outside' technologies (see Table 4-2). There seems to be a correlation, or at least a perception, that school systems are not facilitating the transmission of community-developed knowledge, so there is not only erosion of both traditional knowledge and the age-old system through which it is transferred to new generations,

Sabal mexicana Mart. [Arecaceae]

Known locally as *Huano,* this is a common plant in the *milpas* (swiddens) of Mexico's Yucatán Peninsula (see Figure 4-9). It is grown mainly for roofing materials and for weaving hats and baskets. The fruit and leaves at the 'palm heart' are also used as food.

but the school systems also hinder the creation of new community knowledge (see Table 4-2). Table 4-3 shows that in order to successfully sustain a shifting cultivation system, a farmer needs extensive, high-quality, sophisticated knowledge – a body of traditional knowledge that can only come from years of accumulated hands-on experience, advice from knowledgeable elders and participatory observation skills (Figure 4-9).

In differing degrees, Table 4-4 confirms the process by which traditional knowledge is being lost to new generations. However, these studies have provided a baseline list of questions related to essential traditional knowledge needed for success as a shifting cultivator, and this list may be improved as changes in critical knowledge for the sustainability of traditional shifting cultivation systems is monitored. The social fabric and farming systems that produced this traditional knowledge over countless years provide conditions under which it is passed on efficiently from one generation to the next. Therefore, there is a need to understand not only the traditional knowledge itself, but also the conditions and processes under which successful shifting cultivation is perpetuated. Only then can that knowledge lead to innovation, encouraged within the social fabric of a community that for many generations has been either buoyed by success or suffered from the failure of its farming systems.

Intercultural joint creation of knowledge: A viable alternative for the future of shifting cultivation

The near future of shifting cultivation is dependent on many factors, and the lives of shifting cultivators and their families seem destined to follow one of several scenarios. Some will keep farming in a sustainable manner; others will continue to farm, but in unsustainable ways; some will farm more permanently on one piece of land; others will move away from their communities and forget about farming; and some will have children who will undertake formal studies of viable alternative farming systems, both sustainable and unsustainable. As we have argued, the permanent farming scenario could lead to greater social, economic and environmental problems, so there is a need to work on new approaches

Ricinus communis L. [Euphorbiaceae]

Often grown in the swiddens of Southeast Asia, castor oil is extracted from the seeds of this species for medicinal purposes, particularly to expel intestinal parasites. This species is also regarded as the world's most poisonous common plant, producing the toxin ricin. Although cases of poisoning are rare, the lethal dose in adults is said to be four to eight raw seeds.

to sustainable farming and food production. Such approaches could learn from, as well as support, the sustainability of shifting cultivation (Siahaya et al., 2016).

Assuming that governments will contribute significantly to the achievement of the United Nations' Sustainable Development Goals (United Nations, 2015), as a framework for attending to the many issues related to shifting cultivation, it is important to recommend the development of viable alternatives in the fields of education, indigenous knowledge and ways of learning. Common sense dictates that new generations should acknowledge traditional indigenous knowledge and provide for its enhancement, improvement and adaptation to meet the new and unprecedented challenges of our time.

Research on incorporating indigenous knowledge into mainstream education in Asia is very scarce, but some localized and successful experiences have been recorded, such as using a community-based approach to sustainable forest management (e.g. Colchester, 1994). It seems that the attitude which considers shifting cultivation 'primitive' is reflected in a similar regard for the value of indigenous knowledge. For instance, Fien and Tilbury (1996) presented a report to UNESCO on learning for sustainable development without once considering the importance of indigenous

Hibiscus cannabinus L. [Malvaceae]

A plant that has spread beyond its swidden origins, kenaf produces one of the world's most important natural fibres. Its uses range from rope and coarse cloth through to fine paper. The leaves are also edible and the seeds yield an edible vegetable oil.

knowledge in Asia and the Pacific. However, there is solid recent research to support new approaches in education that incorporate indigenous knowledge and ways of learning, even in government school systems (Jacob et al., 2015). For example, in a study of the differences between the underlying principles of Western science and the knowledge and wisdom of indigenous people from different countries, Morgan (2003) identifies a process of shifts from appropriation to appreciation and then accommodation, which goes both ways and presents challenges to the design of new teaching systems. In further examples, Nogueira Soares et al. (2017) demonstrated the importance of culture in the construction of knowledge for the use of woody flora in Brazil, and after extensive research on indigenous ways of learning, Rogoff (2014) developed a model of the processes involved, called 'Learning by Observing and Pitching In (LOPI)'. These findings have been very helpful in designing the institutional settings and teaching systems of an Intercultural University at Quintana Roo, in Mexico. More than 90% of the university's students are of indigenous origin, and it has achieved top performance indicators (Rosado-May, 2017a).

Intercultural education is present at different levels on all continents (Cushner 1998). It opens spaces for local knowledge and languages to be incorporated into the curricula of both basic and higher education. In 2010, UNESCO completely changed its approach to learning for sustainable development after sponsoring new research that upended the 1996 findings of Fien and Tilbury. The new study, by de Leo (2010), recommended an approach to sustainable development through intercultural education. However, the research emphasis on intercultural education in Asia has not yet concerned itself with how shifting cultivation knowledge, and ways of learning shifting cultivation, should be incorporated into school curricula. This is a challenge still to be overcome.

The Asia Education Foundation (AEF, undated) acknowledges interculturality as a dynamic concept and embraces UNESCO's (undated) definition:

Alpinia officinarum Hance
[Zingiberaceae]

Lesser Galangal is cultivated throughout Southeast Asia as a spicy food additive and a herbal remedy with stimulant and digestive effects. It should not be confused with Greater Galangal (*Alpinia galanga*), which is used extensively in Thai and Indonesian cuisines.

(T)he existence and equitable interaction of diverse cultures and the possibility of generating shared cultural expressions through dialogue and mutual respect.

Using different words and focusing more on cognitive processes, Rosado–May (2015) writes:

(I)ntercultural education is the result of a process in which different systems of constructing knowledge (e.g. local and scientific) interact in a safe environment, providing conditions for synergies to occur and thus creating new knowledge; intercultural knowledge.

This definition opens new avenues for indigenous knowledge to share with scientific methods and learn from them, under equal conditions. This was how the concept of agroecology was first adopted in Latin America (Rosado-May, 2015), as a result of interaction between indigenous Maya knowledge and Western concepts of ecology. Indigenous knowledge is thus recognized as being of great value. Rosado-May's definition was the basis for an approach called intercultural agroecology, which creates conditions for scientists and local farmers to jointly create intercultural knowledge with both deserving acknowledgment for their contributions on the bases of equity and cognitive justice (Rosado-May, 2017b).

For the authors of this chapter, all of whom are of indigenous origin and born to shifting cultivators, reference to shifting cultivation as a primitive food-production system by some authors reflects a huge potential for the field of intercultural agroecology. There exists an enormous wealth of sophisticated, yet 'primitive', knowledge that explains the lately acknowledged sustainability of shifting cultivation. We can only imagine the great potential that this represents, if there is an openness to jointly create new intercultural knowledge, hand-in-hand with 'modern' science. This process is needed more urgently than ever before.

References

AEF (undated) *Intercultural Understanding Toolkit*, Asia Education Foundation, Melbourne, available at: http://www.asiaeducation.edu.au/professional-learning/pathways-and-toolkits/intercultural-understanding/definitions/multicultural-to-intercultural (accessed 13 November 2017)

Altbach, P. G. (1989) 'Twisted roots: The Western impact on Asian higher education', *Higher Education* 18(1), pp.9-29

Angelsen, A. (1994) *Shifting Cultivation Expansion and Intensity of Production: The Open Economy Case*, working paper 3, Development Studies and Human Rights, CHR Michelsen Institute, Bergen, Norway

Arifin, B. and Hudoyo, A. (1998) *An Economic Analysis of Shifting Cultivation and Bush-fallow in Lowland Sumatra*, Southeast Asia Policy Research working paper no. 1, Alternatives to Slash-and-Burn Indonesia Consortium, University of Lampung and ICRAF Southeast Asia, Available at: https://www.researchgate.net/publication/228610236_An_economic_analysis_of_shifting_cultivation_and_Bush-Fallow_in_lowland_Sumatra (accessed 12 November 2017)

Ball, J. B. (1995) *Development of Eucalyptus Plantations – An Overview*, proceedings of the Regional Expert Consultation on Eucalyptus, Bangkok, Thailand, October 4-8, 1993, vol. 1, pp.15-27, Available at http://www.fao.org/forestry/25872-0c162e555bb41fe8fe9cf076acda4216d.pdf (accessed 14 November 2017)

Brookfield, H. and Padoch, C. (1994) 'Appreciating agrodiversity: A look at the dynamism and diversity of indigenous farming practices', *Environment, Science and Policy for Sustainable Development* 36(5), pp.6-45

Cairns, M. F. (ed.) (2015) *Shifting Cultivation and Environmental Change: Indigenous People, Agriculture and Forest Conservation*, Earthscan by Routledge, London and New York

Chomitz, K. M., Buys, P., de Luca, G., Thomas, T. S. and Wertz-Kanounniko, S. (2007) *At Loggerheads? Agricultural Expansion, Poverty Reduction, and Environment in the Tropical Forests*, World Bank, Washington, DC

Colchester, M. (1994) 'Sustaining the forest: The community-based approach in South and South-East Asia', *Development and Change* 25(1), pp.69-100

Correa-Chávez, M., Mejía-Arauz, R. and Rogoff, B. (eds) (2015) *Advances in Child Development and Behavior. Children Learn by Observing and Contributing to Family and Community Endeavors: A Cultural Paradigm*, vol. 49, Elsevier-Academic Press, San Diego, CA

Cushner, K. (ed.) (1998) *International Perspectives on Intercultural Education*, Routledge, New York

DeFries, R. S., Rudel, T., Uriarte M. and Hansen, M. (2010) 'Deforestation driven by urban population growth and agricultural trade in the twenty-first century', *Nature Geoscience* 3(3), pp.178-181

de Leo, J. (2010) *Reorienting Teacher Education to Address Sustainable Development: Guidelines and Tools. Education for Intercultural Understanding*, UNESCO, Asia and Pacific Regional Bureau for Education, Bangkok

Dove, M. R. (2015) 'The view of swidden agriculture by the early naturalists Linnaeus and Wallace', in M. F. Cairns (ed.) *Shifting Cultivation and Environmental Change: Indigenous People, Agriculture and Forest Conservation*, Earthscan by Routledge, London, pp.3-24

Dudal, R. (1984) 'Preface', in *Improved Production Systems as an Alternative to Shifting Cultivation*, FAO Soils Bulletin 53, Food and Agriculture Organization of the United Nations, Rome, available at: http://www.fao.org/docrep/018/ar128e/ar128e.pdf (accessed 10 November 2017)

Erni, C. (ed.) (2015) *Shifting Cultivation, Livelihood and Food Security: New and Old Challenges for Indigenous People in Asia*, Food and Agriculture Organization of the United Nations (FAO), International Work Group for Indigenous Affairs (IWGIA), and Asia Indigenous People's Pact (AIPP), Bangkok

Everton, M. (2012) *The Modern Maya: Incidents of Travel and Friendship in Yucatan*, University of Texas Press, Austin, TX

Farmer, B. H. (1981) 'The "green revolution" in Asia', *Geography* 66(3), pp.202-207

Fearnside, P. M. (2005) 'Deforestation in Brazilian Amazonia: History, rates and consequences', *Conservation Biology* 19(3), pp.680-688

Fien, J. and Tilbury, D. (1996) *Learning for a Sustainable Environment: An Agenda for Teacher Education in Asia and the Pacific*, UNESCO Regional Office for Asia and the Pacific, Bangkok

Fox, J., Truong, D. M., Rambo, A. T., Tuyen, N. P., Cuc, L. T. and Leisz, S. (2000) 'Shifting cultivation: A new old paradigm for managing tropical forests', *Bioscience* 50(6), pp.521-528

Fox, J. M., Castella, J-C., Ziegler, A. D. and Wesley, S. B. (2014) 'Rubber plantations expand in mountainous Southeast Asia: What are the consequences for the environment?', *Asia Pacific Issues* 114, East-West Center, Honolulu, HI

Garbyal, S. S. (1999) '"Jhuming" (shifting cultivation) in Mizoram (India) and new land use policy: How far it has succeeded in containing this primitive agriculture practice', *The Indian Forester* 125(2), pp.137-148

Geist, H. and Lambin, E. (2002) 'Proximate causes and underlying driving forces of tropical deforestation', *BioScience* 52(2), pp.143-150

Hallinger, P. (1998) 'Educational change in Southeast Asia: The challenge of creating learning systems', *Journal of Educational Administration* 36(5), pp.492-509

Harrison, P. D. and Turner, B. L. II (eds) (1978) *Pre-Hispanic Maya Agriculture*, University of New Mexico Press, Albuquerque, NM

Heinimann, A., Mertz, O., Frolking, S., Christensen, E. I, Hurni, K., Sedano, F., Chini, L. P., Sahajpal, R., Hansen, M. and Hurtt, G. (2017) 'A global view of shifting cultivation: recent, current, and future extent', *PLoS One* 12(9), DOI: 10.1371/journal.pone0184479 (accessed 7 November 2017)

Herath, H. M. G. (1985) 'The green revolution in Asia: Productivity, employment and the role of policies', *Oxford Agrarian Studies* 14 (1), pp.52-71

Hirosato, Y. (2001) 'New challenges for educational development and cooperation in Asia in the 21st century: Building indigenous capacity for education reforms', *Journal of International Cooperation in Education* 4(2), pp.1-24

Ickowitz, A. (2006) 'Shifting cultivation and deforestation in tropical Africa: Critical reflections', *Development and Change* 37(3), pp.599-626

INDUFOR (2012) *Strategic Review on the Future of Forest Plantations*, INDUFOR, Helsinki, Finland

Jacob, W. J., Cheng, S. Y. and Porter, M. K. (eds) (2015) *Indigenous Education: Language, Culture and Identity*, Springer, New York

Kapoor, D. and Shizha, E. (2010) *Indigenous Knowledge and Learning in Asia/Pacific and Africa*, Palgrave Macmillan, New York

Kauffman, J. B., Hughes, R. F. and Heider, C. (2009) 'Carbon pool and biomass dynamics associated with deforestation, land use, and agricultural abandonment in the Neotropics', *Ecological Applications* 19(5), pp.1221-1222

Kingwell-Banham, E. and Fuller, D. Q. (2012) 'Shifting cultivators in South Asia: Expansion, marginalization and specialization over the long term', *Quaternary International* 249, pp.84-95

Little Bear, L. (2011) *Native Science and Western Science*, The Library Channel, ASU Library, Arizona State University, available at: https://lib.asu.edu/librarychannel/2011/05/16/ep114_littlebear (accessed 1 December 2017)

Mai, P.T. (1999) 'Socio-economic analysis of shifting cultivation versus agroforestry system in the upper stream of Lower Mekong watershed in Dak Lak province', M.A. thesis in economics of development, National University – HCMC, College of Economics, Vietnam and Institute of Social Studies, The Hague, The Netherlands

Majumdar, D. N. (ed.) (1990) *Shifting Cultivation in Northeast India*, Omsons Publications, New Delhi

Mertz, O., Leisz, S. J., Heinimann, A., Rerkasem, K., Thiha, Dressler, W., Pham, V. C., Vu, K. C., Schmidt-Vogt, D., Colfer, C. J. P., Epprecht, M., Padoch, C. and Potter, L. (2009) 'Who counts? Demography of swidden cultivators in Southeast Asia', *Human Ecology* 37, pp.281–289

Morgan, D. L. (2003) 'Appropriation, appreciation, accommodation: Indigenous wisdom and knowledge in higher education', *International Review in Education* 49(1-2), pp.35–49

Morley, S. (1946) *The Ancient Maya*, Stanford University Press, Stanford, CA

Myers, N., Mittermeier, R. A., Mittermeier, C. G., da Fonseca, G. A. B. and Kents, J. (2000) 'Biodiversity hotspots for conservation priorities', *Nature* 403, pp.853-858

NDF (2012) *Climate Change, Trees and Livelihood: A Case Study on the Carbon Footprint of a Karen Community in Northern Thailand*, Northern Development Foundation and the Huay Hin Lad community, with support from Oxfam-GB, Chiang Mai, Thailand, available at http://unfccc.int/resource/docs/2012/smsn/ngo/240.pdf (accessed 12 November 2017)

Nogueira Soares, D. T., Sfair, J. C., Reyes-García, V. and Baldauf, C. (2017) 'Plant knowledge and current uses of woody flora in three cultural groups of the Brazilian semi-arid region: Does culture matter?', *Economic Botany* 20(10), pp.1-16

Osman, K. T. (2013) *Forest Soils: Properties and Management*, Springer, London

Palm, C. A., Vosti, S. A., Sanchez, P. A. and Ericksen, P. J. (eds) (2005) *Slash-and-burn Agriculture: The Search for Alternatives*, Columbia University Press, New York

Pingali, P. L. (2012) 'Green revolution: Impacts, limits, and the path ahead', *Proceedings of the National Academy of Sciences of the United States of America* 109(31), pp.12302-12308

Rahman, S. A., Rahman, F. and Sunderland, T. (2011) 'Causes and consequenses of shifting cultivation and its alternative in the hill tracts of eastern Bangladesh', *Agroforestry Systems* 84(2), pp.141-155

Rambo, A. T. (1995) 'Perspectives on defining highland development challenges in Vietnam: New frontier or cul-de-sac?', in A. T. Rambo, R. R. Reed, L. T. Cuc and M. R. DiGregorio (eds) *The Challenges of Highland Development in Vietnam*, East-West Center, Honolulu, HI, pp.21-30

Ribeiro Filho, A. A., Adams, C. and Sereni Murrieta, R. S. (2013) 'The impacts of shifting cultivation on tropical forest soil: A review', *Boletim do Museu Paraense Emílio Goeldi. Ciências Humanas* 8(3), pp.693-727

Rogoff, B. (2014) 'Learning by observing and pitching in to family and community endeavors: An orientation', *Human Development* 57(2-3), pp.69-81

Rosado-May, F. J. (2015) 'The intercultural origin of agroecology: Contributions from Mexico', in V. E. Méndez, C. M. Bacon, R. Cohen and S. R. Gliessman (eds) *Agroecology: A Transdisciplinary, Participatory and Action-oriented Approach*, Advances in Agroecology Series, CRC Press, Boca Raton, FL, pp.123-138

Rosado-May, F. J. (2017a) 'Challenges and opportunities in guiding intelligence with intelligence: The intercultural academic model in Quintana Roo, Mexico', in F. González González, F. J. Rosado-May and G. Dietz (coords) *La gestión de la educación superior intercultural en México. Retos y perspectivas de las universidades interculturales* (The Management of Intercultural Higher Education in Mexico: Challenges and Perspectives of Intercultural Universities), Universidad Autónoma de Guerrero and El Colegio de Guerrero A.C., Chilpancingo, Guerrero, México (Spanish language)

Rosado-May, F. J. (2017b) 'Intercultural alternatives to challenges for the glocal advancement of Agroecology', *Proceedings of the Sixth International Symposium on Agroecology*, Universidad Autónoma Juárez de Tabasco, Villehermosa, Tabasco, Mexico (Spanish language with English summary)

Rudel, T. K., DeFries, R. S., Asner, G. P., and Laurance, W. F. (2009) 'Changing drivers of deforestation and new opportunities for conservation', *Conservation Biology* 23(6), pp.1396-1405

Ruthenberg, H. (1980) *Farming Systems in the Tropics*, third edition, Oxford University Press. Oxford, UK

Schmidt-Vogt, D. (2001) 'Secondary forest in swidden agriculture in the highlands of Thailand', *Journal of Tropical Forest Science* 13(4), pp.748-767

Seidler, R. and Bawa, K. S. (2001) 'Logged forests', *Encyclopedia of Biodiversity* 3, pp.747-760

Siahaya, M. E., Hutauruk, T. R., Aponno, H. S. E. S., Hatulesila, J. W. and Mardhanie, A. B. (2016) 'Traditional ecological knowledge on shifting cultivation and forest management in East Borneo, Indonesia', *International Journal of Biodiversity Science, Ecosystem Services and Management* 12(1-2), pp.14-23

Silva, J. M. N., Carreiras, J. M. B., Rosa, I. and Pereira, J. M. C. (2011) 'Greenhouse gas emissions from shifting cultivation in the tropics, including uncertainty and sensitive analysis', *Journal of Geophysical Research* 116, D20304, doi:10.1029/2011JD016056 (accessed 3 December 2017)

Spencer, J. (1977) *Shifting Cultivation in Southeast Asia*, University of California Publications in Geography, vol. 19, University of California Press, Berkeley, CA

Thaman, K. H. (2003) 'Decolonizing Pacific studies: Indigenous perspectives, knowledge and wisdom in higher education', *The Contemporary Pacific* 15(1), pp.1-17

UNESCO (undated) *UNESCO Guidelines on Intercultural Education*, Section of Education for Peace and Human Rights, Division for the Promotion of Quality Education, Education Sector, United Nations Educational, Scientific and Cultural Organization, Paris, available at: http://unesdoc.unesco.org/images/0014/001478/147878e.pdf (accessed 13 November 2017)

UNESCO (2017a) *Indigenous Knowledge and Practices in Education in Latin America: Exploratory analysis of How Indigenous Cultural Worldviews and Concepts Influence Regional Educational Policy*, Regional Office for Education in Latin America and the Caribbean, United Nations Educational, Scientific and Cultural Organization, Santiago, Chile

UNESCO (2017b) *Mother Tongue Matters for Sustainable Futures: Three Case Studies*, United Nations Educational, Scientific and Cultural Organization, Bangkok, available at http://bangkok.unesco.org/content/mother-tongue-matters-sustainable-futures-three-case-studies (accessed 12 November 2017)

United Nations (2015) *Sustainable Development Goals*, available at: http://www.un.org/sustainabledevelopment/sustainable-development-goals/ (accessed 13 November 2017)

van Vliet, N., Mertz, O., Heinimann, A., Langanke, T., Pascual, U., Schmook, B., Adams, C., Schmidt-Vogt, D., Messerli P., Leisz, S., Castella, J-C., Jørgensen, L., Birch-Thomsen, T., Hett, C., Bech-Bruun, T., Ickowitz, A., Vu, K.C., Yasuyuki, K. and Ziegler, A. D. (2012) 'Trends, drivers and impacts of changes in swidden cultivation in tropical forest-agriculture frontiers: A global assessment', *Global Environmental Change* 22(11), pp.418–429

van Vliet, N., Adams, C., Guimarães Vieira, I. C. and Mertz, O. (2013) '"Slash and burn" and "shifting" cultivation system in forest agriculture frontiers from the Brazilian Amazon', *Society and Natural Resources* 26(12), pp.1454-1467

Watters, R. F. (1971) *Shifting Cultivation in Latin America*, Forestry Development Paper no. 17, Food and Agriculture Organization of the United Nations, Rome

5

TRADITIONAL ECOLOGICAL KNOWLEDGE

among the shifting cultivators of Arunachal Pradesh, India

*Tomo Riba**

A reminiscence

> I was lucky to be born the son of a shifting cultivator in a small village, amid lush tropical evergreen forest and with the Kidi river flowing by, just a few metres from my house. The forest and the river had enough biotic resources to meet our simple needs. Although my hectic teaching profession now gives me very little opportunity to maintain my youthful connection with nature, the hidden instincts that I acquired as a child still stimulate an awareness of plants and animals in the forest and rivers, and responses to the changing seasons. The sharpness of pebbles beneath bare feet; the biting chill of the river in winter; the mindless sucking of leeches as we hurried through the break-of-day to reach the field before the marauding birds – such sweet memories have enriched my adult life. Still, I long for the umami taste of roasted field mouse, grasshoppers and crickets, such as my father used to feed me. And the shrill call of cicadas in June, heralding relief from the sultry heat. How I miss them all!

Introduction

People who know the least about shifting cultivation are those who talk about it the most. In their eyes, shifting cultivation is a primitive form of cultivation that destroys forest ecosystems, leads to carbon dioxide emissions, increases soil erosion and is generally unproductive.

It is blamed for most of the world's ecological and environmental problems, such as the extinction of biotic life and global climatic change. This negative bias has

* Dr. Tomo Riba, Professor, Department of Geography, Faculty of Environmental Sciences, Rajiv Gandhi University, Itanagar, Arunachal Pradesh, India.

blinded them to the fact that scholars from all over the world are interested in documenting the practice of shifting cultivation. For example, Japan – among the most scientifically developed of nations – wants to revive the lost practice of shifting cultivation in their country, on a demonstration plot. After studying shifting cultivation practices all over the world, Kyoto University cultivated a demonstration plot in 2010, and your author was fortunate to take part in burning the field (Figure 5-1).

On another notable occasion, the English economist and Emeritus Professor of Development Studies at Oxford University, Professor Barbara Harris-White, visited the author's village during the harvest season (Figure 5-2). She was studying carbon dioxide emissions from the burning of biomass in shifting cultivation fields. After observing the cultivation techniques in detail, she commented: 'There is science in this cultivation.'

A significant portion of the population in Arunachal Pradesh is engaged in shifting cultivation (Figure 5-3). Were it not for this fact, these people would be unemployed and causing social problems. Shifting cultivators are not unemployed, nor are they homeless and landless. They keep themselves engaged and live a meaningful life. Although they may not be affluent, there are no beggars among them. Unlike urban societies, social problems are very rare in communities of shifting

FIGURE 5-1: Your author (centre) joins in the burning of a demonstration swidden plot by Kyoto University in Japan.

Photo: Yashiyuki Kosak.

FIGURE 5-2: Professor Barbara Harris-White (left) with an elderly villager at harvest time in the author's village.

Photo: Tomo Riba.

cultivators, because they are too busy to have idle minds. The widespread discourse of non-participants, who have never observed shifting cultivation closely, has distorted the reputation and human value of shifting cultivation. They see no further than slashing and burning a chosen area of forest and planting seeds.

Shifting cultivators are acutely perceptive of every aspect of their ecosystem, and they behave accordingly. Centuries of repetition of the same agricultural system has helped them to retain the best practices and continue to refine their farming system. They carry out their economic activities with a clear sense of the ecological functioning of their surrounding ecosystem, and their ability to 'read nature' is one of their unique talents. A farmer can stand in the forest with his or her eyes closed,

FIGURE 5-3: Arunachal Pradesh shares a long border with China, and is India's northeastern-most state. In 2011 it had a population of 1.38 million.

and can read nature with their nose and ears. They can identify the smell of different animals and their movements; even as far as which animal has taken different fruit. They can identify the favoured places of different plants. For them, the forest is a source of food, medicine, raw materials and recreation.

A printed calendar hanging on a wall may fail to accurately predict the arrival or end of seasons for various economic activities, but the natural calendar, based on the behaviour of animals, birds, and insects; the appearance of new leaves, flowers and fruit; the smell of the soil, and the direction of seasonal winds and the movement of clouds, guide shifting cultivators in their performance of annual activities.

All of these knowledge systems are popularly known as Traditional Ecological Knowledge (TEK) – a description that is growing in importance. In the management of natural resources, TEK is very eco-friendly. It is a way in which human activities are adjusted according to the ecological norms of a physical location. Different societies have different levels of reaction to the stimuli of nature. Over time, these reactions become part of their culture, in the form of guiding principles when dealing with nature. These subjective ideas are reflected in their objective world, applying directly to cultivation, house construction, food habits, healing practices, hunting and fishing, and so on.

Shifting cultivators clear the forest to grow crops, and wild vegetation is replaced by domesticated plants. They are maintainers of crop diversity. Shifting cultivation is part of their environment; so is the forest. When the cropping is finished, shifting cultivation will help the formation of secondary forest, which will have a high potential for supporting more organisms than primary forest.

In every stage of shifting cultivation, farmers follow Traditional Ecological Knowledge in order to avoid mistakes and accidents. It is therefore in the practice of shifting cultivation that TEK achieves its ultimate expression. It involves the use of different sharp tools, felling big trees, climbing slopes, crossing rivers and carrying huge loads. In the past there were ferocious animals; nowadays there are still poisonous snakes and insects and poisonous plants and fruits. Without TEK, human survival in the forest would not be possible.

Every plant and animal has a best season for collection, and wild items have different best seasons for their use. Shifting cultivators do not have separate vegetable and fruit gardens. These things are collected from the forest. Thus, Traditional Ecological Knowledge is the guiding principle for these people in choosing suitable times for carrying out any economic activity.

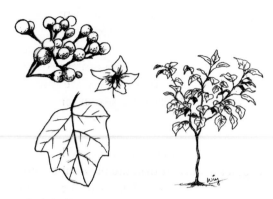

Solanum indicum L. [Solanaceae]

This stiff, prickly shrub grows to 1.5 m and is attracting increasing interest as a medicinal crop. Known variously as brihati, poison berry or Indian nightshade, it is highly regarded in Ayurvedic medicine. The fruit and roots are said to be one of the finest herbal remedies for coughs, colds, sore throats and asthma. It is also a good herbal appetizer, a digestive and diuretic, and is beneficial to the heart.

TEK and local sayings

Traditional Ecological Knowledge is not in written form because its practitioners do not have a written language. Its myriad details are memorized, to be recalled and refreshed through constant repetition. Sometimes, there can be confusion in reading the messages of nature. In this case, old sayings are brought to mind, to match and compare what is happening in nature, with the plants and animals. The feelings of people and the 'inward understanding' of these old sayings defy precise translation; many of them relate to bodily sensations, such as touching, listening, and sensations of the skin. There are sayings about the seasonal behaviour of plants and animals and the human response to these changes, especially in the field of shifting cultivation. The following are a few of the main messages of nature, as they relate to the changing seasons.

Deechi (winter): This is the last part of December and the first half of January, when the weather is very cold. Sometimes a weather disturbance from the west makes it even worse. This is the 'off' season from major agricultural activities. Outdoor activities related to the forest become very difficult if it rains. Low-growing plants on rocks begin to dry out. The damp and cold may persist for five days, then there follows another 10 days without change. Painful cracks form on the toes and palms of the hands. Fish become inactive and remain confined to riverbed holes, where they are easy to catch. There is silence in the forest because the animals and birds are not moving. Sometimes, during the cold and damp winter days, piglets, puppies, and even birds, die. This is the season for clearing the forest for early fields, especially for maize.

`*Agloo*: This occupies the second half of January and the first half of February. The weather is warmer than the previous month. The severely cold days are over. Snakes and lizards basking in the sun are a common sight. Birds and animals begin to move about. The soil has a mouldy smell. This is the time for clearing the forest for the main swidden and sowing crops in the early field that was cleared in the depths of winter.

`*Lukww and Lumii*: This is the second half of February and the first half of April. Spring is coming to the forest and the days are warmer, but occasional rain brings back the cold. The sound of migratory birds returning from the south to the north is heard in the sky '*Huu.. haa, huu..haa!*' They never miss the time of their migration. Birds and animals are seen more as they begin to range further. It is the time of nesting. In the river, holes can be seen where sand has been shifted aside by eels, and batches of *danio* fish begin moving downstream at night. Tadpoles are seen near the bank.

The leaves of wild yam plants grow to the size of frogs' palms. Roof-palm trees begin to flower, and different seasonal bird calls can be heard in the nearby forest. This is the season for sowing crops in the main field.

Luchwwr, Ilo and Tenlo: This extends from the second half of April to the first half of July. It is midsummer, and the long, hot days are sultry and oppressive. This is the hardest part of the year for

Zanthoxylum rhetsa DC. [Rutaceae]

This tree grows to 35 m, and is valued as a source of spice and medicinal materials. The bark is used as a lime-pepper-flavoured food seasoning. The immature fruit has the taste and aroma of orange peel, and the ripe seeds taste at first like lemons, but leave a burning sensation. The fruit is used as a treatment for dyspepsia, asthma and bronchitis, heart troubles, toothache and rheumatism.

farmers. It is the season of storms, heavy rainfall, landslides and floods. Even small streams can become difficult to cross. It is also the season of sicknesses like fever, dysentery, and diarrhoea. There is constant uncertainty about the weather, and lazy and lethargic farmers find it difficult to undertake strenuous work.

It is said that in this season, wild boars in their dens fail to sense the movement of hunters because of the sound of heavy rain. As the floods begin, *gara* fish move into small streams to lay their eggs. The long, shrill call of cicadas heralds an imminent start to the flowering and fruiting of different trees in the forest. The cicadas give solace and hope to farmers, that the days of food scarcity will soon be over. The days are long, the granaries are empty and wild fruit is not yet ready. Meanwhile, there is the heavy work of weeding, and in this season, farmers become physically very weak.

`**Hoo:** From the second half of July to the first half of August. The period of acute food scarcity is over. The forest is now full of wild fruit, like different varieties of cardamom, figs and bananas. Some yams can also be harvested. Even the maize from the early field is ready to eat.

During this month, the sub-surface soil is saturated because of the long monsoon rain, and even small downpours lead rapidly to surface runoff, causing landslides and flooding. Migratory birds start moving from north to south. By now, the cold season has come to Tibet, to the north. Towards the end of August, temperatures begin to fall in the mornings and evenings. At this time, the weeding is almost over. Crops like maize, cucumbers, melons, pumpkins, vegetables, and mushrooms are available. The long-awaited rice is forming panicles and setting grain, bringing great solace to farmers, although they have been able to predict the level of production all along, based on the fertility of the soil and the health of the plants.

`**Hwwtv:** This extends from the second half of August into the first half of September. Floods may still be frequent, although the temperature has fallen considerably, especially in the morning and evening. On the horizon, the bright flowers of rattan adorn the sunlit mountainsides. The farmers

Clerodendrum colebrookianum Walp., a synonym of *Clerodendrum glandulosum* Lindl. [Lamiaceae]

The end of the monsoon is the start of the flowering season for this shrub or small tree, commonly known as the East Indian glory bower. It is characterized by a foetid smell, and is best known for its medicinal properties. The leaf juice is widely used as a treatment for high blood pressure, and the smell of the wood is believed to relieve many medical complaints, including coughs and dysentery.

have a saying: 'If you have any confusion about the month *hwwtv*, observe the flowers of rattan'. This is the time for harvesting to begin in the main field.

`**Pwra:** This is the last half of September and the first half of October. Streams may still be swollen, but the water is falling. Wild boars become fat from feasting on chestnuts, and this is the best season for hunting them. All of the animals, birds and even the fish have eaten well, and are delicious for being oily. The harvesting is almost over.

Lub: The last half of October and the first half of November, when days are pleasant and the weather is cool. The summer is over, and dryness prevails. Some trees start shedding their leaves, and some winter fruits begin to appear in the forest. By now, major flooding is over. There is a scene of damage and scattered riparian debris. Logs, big and small, are found stuck between boulders along river banks. This is the season for large-scale fishing; the river is partially dry and a bund is built to divert the main channel. After the fish are collected, the bund is removed. The shrimp and crabs are delicious to eat. The crabs make holes that are easy to see from the soil thrown out as they dig. Peacocks in the forest change their sound and call '*oiiik…oiiik!*' This is the start of the 'off' period for agriculture, but root crops like yam and taro are still harvested. Snakes are often seen on the surface of the earth, before they go into hibernation, and sometimes, people are bitten. The mornings and evenings become cold.

Ratv: This is the second half of November and the first half of December. Everywhere, there is dryness and silence in the forest; the sounds of birds and animals are seldom heard. Creepers and climbers start drying. From the drying leaves, farmers can find wild yams, and collect them. In the old field, there are still residual items like yams, taro and ginger to be collected. The cold weather has driven field mice into their holes, with heaps of debris, and the farmers catch them by blowing smoke into their holes. By this time, the abandoned field has been fully colonized by plants and

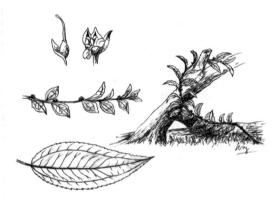

Pouzolzia sanguinea (Blume) Merr. [Urticaceae]

This weak-stemmed evergreen shrub grows up to 3 m tall, and is commonly found in fallowed swiddens in dense communities offering mutual support. Its tender young leaves and shoots are gathered and cooked as a vegetable. A fibre obtained from the bark is useful as cordage, and to make rope and nets. A paste made from the roots is used medicinally, to treat boils.

regeneration has begun. Plant succession in the shifting cultivation field will be complete within a few years. Although this is the 'off' season, a few farmers will still go to the old field to collect firewood or wild vegetables. Wild vegetables collected from newly fallowed fields are very tender and tasty. The farmers are also busy collecting construction materials.

TEK at different stages of cultivation

Every stage of shifting cultivation, from the demarcation of field boundaries to storing the grain and preserving seeds for the next crop, is systematically managed following traditional practices that have been passed down by countless earlier generations. Failure to adhere to these ecological norms of the place will mean the failure of crops. Therefore, farmers are the best decision-makers and planners. The various stages of shifting cultivation are guided by the following examples of traditional ecological knowledge:

Marking field boundaries

Previously, recording field boundaries on paper was not possible because the farmers were illiterate. Even today, there is no cadastral map showing the extent, value and ownership of land. Farmers find the traditional system of demarcating field boundaries to be more convenient because it is commonplace for paper records to be damaged by termites or accidental fires. Therefore, they consider prominent physical features like ridges, depressions, streams or rivers, cliffs, slopes, rocks and big trees. Because most locations have been repeatedly cleared for shifting cultivation, the field boundaries of the past are remembered by the farmers; even for remote fields. Sometimes, they plant certain valuable trees to mark the boundaries. But care must be taken when the field is burnt, in case the fire escapes. In the case of disputes regarding field boundaries, these will be settled by villagers according to evidence at the field location.

TEK in selection of a site

Once a field site is decided and clearing has begun, the decision cannot easily be changed. Therefore, site selection is one of the main factors in determining productivity, and farmers have learned well the skills of judging forest areas where crop production will be good or bad. For example, crop productivity will always be better on a slope that gets abundant sunshine, compared to an area where the duration of sunshine is not so high. As a result, some areas of forest are never cleared for agriculture because they don't receive enough sunshine. The most important factor in selecting a patch of forest for cultivation is the fertility of the soil. This is determined by the age of the forest, i.e. the length of the fallow period since it was last cultivated. The age of the forest not only determines the fertility, but also the likelihood of weed infestations. Thus, the farmers consider forest that is between 10 and 15 years old to be ideal, because the dormancy period of weed seeds on the forest floor will have been exceeded. If younger forest is cleared it will be infested with weeds that will inhibit the growth of crops. Sometimes, farmers have to abandon a field because they are unable to cope with fast-growing weeds. On the other hand, forest that has been fallowed for too long is unsuitable for shifting cultivation because of thick deposits of humus. The soil becomes spongy and crops 'over-grow' in size and set less grain. Such quick-growing crops are weak and fall to the ground. Moreover, forest that has been

FIGURE 5-4: An old swidden farmer in his traditional attire, entering his field to begin a day's work.

Photo: Tomo Riba.

fallowed for a long time has big trees that are difficult to clear and the field is likely to be littered with big logs.

Generally, farmers avoid clearing isolated fields because crops are more likely to be attacked by wild animals and birds. The same can also be said for fields very close to the settlement, but the damage in this case comes from domestic animals like gaurs, cows, pigs and goats. To protect crops from domestic animals, a farmer has to build a fence around the field, and this is a very arduous task requiring huge quantities of wood, bamboo, rope and manpower (Figure 5-4).

TEK during clearing of the forest

The common view of the clearing of forest for shifting cultivation is limited to slashing and burning. This simplistic view does not present an accurate picture of forest clearing for shifting cultivation in Arunachal Pradesh. The drying of fallen trees, the successful burning of the dry slash, growth of weeds and finally the growth of crops are all dependent on how herbs and shrubs are cleared from the site. Banana trunks and large-sized creepers and climbers require longer to dry than other slashed debris. Tall climbers also interfere with the free fall of big trees as they topple. Therefore, farmers cut banana plants and creepers well in advance; even a year before they begin to clear the rest. The forest-clearing task begins from the bottom of the field, so that the farmer does not have to bend too much. It is also easier to move upward. A farmer carries two machetes, one big and one small. The small one is most frequently used; the big one comes out to fell big trees. Dead fallen leaves on the ground are tossed, creepers are pulled out and climbers are pulled down, and all are cut into pieces for early drying. Herbs and shrubs are cut down systematically and branches are cut off. Not a single plant is allowed to remain alive. First, the herbs and shrubs are cleared and big trees remain standing until the fallen undergrowth is dry. If the big trees are felled at the same time as the undergrowth, leaves and branches lying against the ground will not dry properly.

On a selected day, during sunny weather with a dark moon at night, the big trees are felled. Most accidents in forest clearing occur at this time. Generally, more than two men fell the trees. First, they study the bend and balance of a tree and free it of entangling creepers and other trees that might disrupt its fall. They work out the 'safe side', where they can run when the tree begins to fall. First, a man will cut into the bending side of the tree, and then the other side, a little bit above the first cut. Another man will watch the top of the tree and warn the axeman when it begins to move. Great caution is taken to ensure that the falling tree does not split and injure the workers. After the tree is down, the branches are lopped from the main trunk so that the fire catches everything.

Burning the field

Burning is the most important part of shifting cultivation. The flames will clear the fallen trees and open the soil to the light. Burning not only clears the debris but also kills weed seeds that remain in the topsoil as well as other pests. It also enriches the soil by adding potash – the water-soluble part of the ash from the burnt vegetation, which is a plant nutrient. A field that is not properly burnt increases the farmers' workload because the remaining debris and half-burnt logs need to be cleared (Figure 5-5). Crop productivity becomes very low in poorly-burnt fields because of increased weed growth, pests, and lesser enrichment with ash. Sometimes, because of an unsuccessful burn, a farmer will find he or she is unable to clear the debris and will abandon the field. Thus, great care is always taken, from the first day of clearing the forest, to maximize the chances of a successful burn. Big trees are felled carefully and the branches of fallen trees are lopped.

The proper selection of a day and time for burning the field is also essential. After felling the trees, farmers wait anxiously for sunny days in which to burn the field. Sometimes, untimely rain delays the fire and subsequently leads to delays in sowing crops. On rare occasions, farmers have been forced to abandon a field because burning it has proven to be impossible.

Burning is done around noon, after a long and continuous run of sunny days. Careful preparation precedes the setting of fires. Bundles of split bamboo are prepared and kept dry, for use as torches. The farmers will study the plot carefully, note the direction of the wind, and find a place where they can take shelter after the flames are firmly established. Normally, the fire is set to burn uphill, so a path is cleared at the bottom of the field from which the farmers can light several points simultaneously, to create a broad fire front moving up the field. If the field is near to a settlement and there is chance of the flames spreading, the fire may be started at the top of the slope, so that it burns gradually downwards. Meanwhile, villagers are on full alert, with water ready to avert any accidents.

Clearing remaining half-burnt logs and debris

Work on clearing the remaining logs and debris begins on the day after the fire. Half-burnt logs remain soft for a few days, and are easy to cut, but when they become completely dry, cutting them is very difficult. First, big logs are cut into convenient sizes and pegged across the slope to prevent soil erosion. Pieces of timber that are firm and straight are set aside for construction of a field hut. The rest are either kept for use as future firewood, or are burnt. At different points, heaps of debris are formed, and these will be used later for dumping uprooted weeds. However, care is taken to avoid creating safe havens for

FIGURE 5-5: An elderly woman farmer returns home after a day in her swidden, carrying a load of firewood.

Photo: Tomo Riba.

mice. These places are also used for growing cucumbers and melons, and during wet and damp winter, field mice are caught when they shelter in the heaps of debris.

Location of the field hut

Generally, the field hut is located at centre of the field, from where the farmer has a view of the entire field and can watch for marauding birds, which tend to attack the crops in depressed spots near the edge of the field, close to the forest (Figure 5-6). A central location is also convenient when carrying harvested grain from different parts of the field for storage.

TEK in sowing the crops

Sowing the crops is the most important stage in any type of agriculture. It requires careful decision-making regarding the crops to be sown and the parts of the field that are set aside for different crops and crop combinations. Care must always be taken to avoid one crop becoming a weed that inhibits the growth of another. A combination of millet and rice is suitable, as is a combination of rice and cassava. Maize and rice can also be sown together, as long as the maize is sown sparsely and is harvested early. Climbers and creepers like cucumbers, gourds, melons, climbing beans, and so on, are grown near the edge of the field. Taro and ginger are grown where there is a generous deposit of ash. Mustard is also broadcast on to areas where ash deposits are thicker and is harvested early, so as to allow space for rice. Soybeans are grown on either side of the field path, or at the field boundaries. Onions, chillies and eggplants are grown around the field hut within easy reach.

Sowing techniques

First of all, mustard seeds are broadcast by an experienced hand on gentle slopes with a good ash deposit. This is done immediately after the burn, when half-burnt debris is still being cleared. Millet seed is also sown by broadcasting. Yams are planted near to tree stumps on the

FIGURE 5-6:   After the field is burnt, a field hut is built with a view of the entire swidden.

Photo: Tomo Riba.

slope so that they get climbing support and are easy to harvest by digging.

Seeds sown on slopes are easily blown away by strong winds, washed away by rain, or simply tumble downslope. Sowing of the upland rice and maize is therefore a meticulous and highly technical operation (Figure 5-7). Women carry the seeds in small baskets on their backs, along with a dibbling stick about the size of a walking stick, in their right hand. The left hand holds a handful of grain. Working quickly and rhythmically, they dig holes in the soil with the dibble sticks and instantly throw an equal number of seeds into each hole with unerring accuracy. With smooth movements, they sweep soil over the holes with one leg. Sometimes men use rake-style bamboo brooms to cover the holes. An expert woman can make more than 60 holes per minute, while maintaining an equal distance of about 30 cm between the holes. All the time, the planters are gossiping happily with their friends. The advantages of dibbling the seed are quick germination due to soil moisture and deep rooting. The grains cannot easily be removed by wind and rain, or eaten by birds. Germination is thus guaranteed and the plants become healthy.

The sowing of rice and maize is begun at a corner of the field while other workers are still clearing the half-burnt debris. Many activities, like cutting the logs into pieces and re-burning the debris, help to prevent attacks by birds while the seed is newly planted or most exposed. Some birds may even dig the seeds, especially maize, out of the holes. Such birds are trapped. The farmers know how long germination will take, and they monitor the field closely.

FIGURE 5-7:   A group of women planting rice with dibbling sticks in a newly burnt swidden.

Photo: Bomchak Riba.

Shifting cultivation is rainfed agriculture, and the crops are grown only in the summer season. The clearing of the forest starts in winter and harvesting ends in the autumn. Thus, for most of the year, certain crops remain in the field. Upland rice is the principal crop. Other crops include millet, maize, pulses, gourds, cucumbers and melons, varieties of beans, onions, ginger, sesame, chillies, yams, taro, eggplants, and so on. The farmers have been growing these crops and saving the best seeds for generations. They are risk averse, and don't readily accept new seeds. They have full faith only in their own seeds, from the plant varieties they have been using for centuries.

TEK in weeding

Weeds are the greatest enemies of crops. Weeding is also the toughest stage of shifting cultivation. During this season, the days are long and hot. At the same time, the granaries are empty and there are no crops ready to feed hungry families. Even in the forest, wild fruit is not yet ready. Warm and humid weather with a long growing season favours weeds, giving them faster growth than the crops. The moment the surface of the soil is cleared, the long-waiting dormant weed seeds begin to sprout. Some weeds, like bananas, germinate within a week. In a field that is not weeded properly, the growth of crops is retarded due to a lack of sunlight and soil nutrients. Such fields are also easily attacked by other pests, such as rodents, grasshoppers, crickets and bugs.

Normally, farmers will weed a field three times in the course of one cropping season. The first is a selective weeding of very fast-growing plants like bananas, *Setaria megaphylla* – a grass species with very sharp leaves that cut hands during weeding – the emergence of numerous wild cardamom plants, and so on. These plants are either pulled by hand or are dug out. The second weeding is the most systematic. Soil around crop plants is loosened with a trowel while weeding. Hoes are not used for fear of damaging the crops. In some cases, excessive crop growth is thinned. Fallen plants are given support and soil is added to their roots. Weeds are collected and

Setaria megaphylla (Steud.) T. Durand & Schinz [Poaceae]

Broad-leaved bristle grass is a native of Southeastern Africa that is now found in many tropical and subtropical areas – including Arunachal Pradesh, where it flourishes as an early weed in swiddens and its broad leaves inflict painful cuts. It favours low-lying and shady areas near streams, and a popular use in its native Africa is to roll the leaves into a bundle and use them to clean dishes.

dumped at different designated spots. Some weeds do not readily die, so they are laid out in the sun to dry. The final weeding is another selective procedure in which leaves sprouting from tree stumps and remaining fast-growing weeds are removed with machetes or knives.

In all three cases, weeding starts at the bottom of the slope, for easy bending. The farmers carry small cane baskets on their backs to collect wild vegetables, mushrooms and grasshoppers. If the field is far from the settlement, the weary farmers will spend their nights in the field hut, to save time wasted in walking back and forth. While weeding, the farmers usually avoid killing certain insects and birds that live in the field.

The farmers usually carry burning rice husks held in a collection of broad leaves stitched into a conical shape. Smoke from the smouldering husks keeps flies and other insects away and hot ash is used to start fires at the field hut. Biting insects can cause a lot of problems. Being close to a fire also gives a psychological sense of security.

TEK in protecting the crops

Protection of their crops is a challenging task for shifting cultivators. Since the field is in the forest, crops are attacked by birds, rodents, wild animals and insects. Farmers do not have effective means of protecting crops from diseases and pests.

First and foremost, the crops must be protected from competing weeds, as mentioned above. Most farmers have their own 'best techniques', and manage weed control effectively. They also have many indigenous techniques for protecting crops from birds and wild animals. First, they have to know the behaviour of the birds and animals of the forest, such as their feeding habits, the times they can be expected to visit, the prints of their feet in the soil, and their calls and smells. In the case of fields that are distant from settlements, the most damage is done by birds, all the time from sowing to harvesting. Wild animals are drawn to the crops only during fruiting time, and they come mostly at night, which makes it difficult to stop the damage they cause. Birds come to the field early in the morning and in the evening, before sunset. After the rice and maize are sown, one member of a farming family – most often a child – will go to field early in the morning, before sunrise, to scare away the marauding birds. After reaching the field, he or she will beat a big dry empty bamboo or an empty tin while shouting, to scare the birds away. A fire is lit in the field hut so smoke will further dissuade the birds. He or she will then move around the periphery of the field, to ensure that birds don't settle. In the evening, the farmer will light fires at different spots in the field and remain there, until late.

During sowing, or just after sowing, the protection concentrates only on birds. But during fruiting, the crops are not only attacked by birds, but also by many animals, such as rodents, bears, boars, stags, monkeys, porcupines and civets. Different protection methods are used for different animals. To scare the birds, a network of bamboo ropes are strung across the field, connecting the far corners of the field with the farm hut. At different points in this network, partially split bamboo clappers are fixed, so that when the ropes are pulled, the clapping sound drives away both birds and animals. Sometimes, a bamboo rope that is connected to the field network is stretched across a nearby river, and from that point, two ropes are dropped into the river, attached to a big floating object. The current draws the floating object back and forth, pulling the ropes on the field network of bamboo clappers, creating a constant noise. Another, simpler method of scaring a flock of birds is to shoot arrows into their midst.

In order to scare-off monkeys and other big animals, noise-generating instruments that are driven by gusts of wind are placed at several points in the field. Scarecrows are also erected among the crops. Rodents and porcupines are trapped in various snares that are fixed around the field. Big animals like bears and boars are ambushed and killed. The farmers use many eco-friendly traps to catch unwanted intruders.

Another method, called *palu-pale*, is where farmers cut down trees around the field to deter big animals, which are very sensitive to freshly cut trees. However, monkeys are a different matter. They are very difficult to keep out of the crops, and have even been known to steal grain from the field hut.

The best technique for preventing domestic animals, such as goats, pigs, cows and mithun (*Bos frontalis*) from plundering the crops is to raise a fence around the field. Fields that are some distance from the settlement are not bothered by domestic animals (Figure 5-8), except for mithun, and the common way of preventing mithun from entering the field is to block various entry points. There is also another method, where all the mithun are driven to a large enclosed patch of forest until the harvest is over. Care is taken to ensure that the animals are safe from wild beasts like tigers. The enclosed patch of forest is visited regularly, either by a person engaged for the task or by the owner of the mithun. Special care is taken of mithun cows nearing calving. If certain mithun repeatedly enter a field, the owner is asked to keep the animals tethered. If it still occurs, the angry farmer may burn a rug to chase the mithun away. A burning rug smells similar to a tiger, and the

FIGURE 5-8: A farmer returns from the field with food for pigs. The long pieces are heart-wood from tree ferns.

Photo: Tomo Riba.

mithun scatter away through the forest in blind panic, They have even been known to tumble off cliffs in the effort to get away from the smell of a tiger. But if the owner is informed and his mithun continue to damage crops, then fines may be imposed to compensate the farmer for lost crops.

If there is a river alongside the field, fencing is difficult, and a hanging fence may be built over the water. In the case of uncontrollable domestic animals, a ring of wood may be fixed around an animal's neck to prevent it from forcing its way through a fence and into the crops.

Harvesting

Harvesting time is the best and most festive time of year for farmers, especially if the crop is a good one. There are different times for harvesting different crops. The cultivation has been planned so that the first crops harvested create space for the rice, which is still maturing. Crops that were grown around the field hut are also harvested early to make space for threshing the grain. Generally, harvesting begins either from the corner of the field that is under the greatest threat from marauding animals, or from that side of the field where crops have ripened first.

During the harvest, a farmer carries a large basket for the ripened grain. As well, he or she will carry a small basket at their hip, in which to place grasshoppers, crickets and wild vegetables. When harvesting rice, the farmers cut only the panicles and leave the rest of the plant in the field, to cover the soil from the sun and rain. Partially ripened rice is collected separately for preparation of flattened rice. Maize is harvested before the rice. Yams and taro are harvested much later, after their leaves have dried. Harvesting of any crop at an immature stage leads to spoilage, so the farmers always wait until crops are properly ripened.

Preservation of seeds

Farmers have a very simple method of preserving seeds for the next cropping season. They always collect seed from fruit that appear first, and have healthy growth. Fruit or grains that have been marked for the gathering of seeds for the next cultivation are not harvested quickly and are allowed to remain on the tree or in the field till they reach complete maturity. Such plants are tied with rope so that others do not pick them by mistake. Maximum care is taken to protect these seeds from becoming damp, and they are

FIGURE 5-9: The granary has a two-storeyed design to prevent the entry of rats.

Photo: Tomo Riba.

stored after drying properly in the sun. Occasionally, after drying in the sun, care must be taken that the seeds are not damaged by powder post beetles or fungus. Seeds like maize, yams and taro are kept in the farmer's main house, where they are smoked over the fireplace. Some seeds are kept in the old farm hut. Many items like eggplants, sugar cane and chillies, are either grown from seeds or old plants are transplanted from the old field during rainy days, to prevent them from drying out. Seeds of pulses, mustard and millet are stored in dry bamboo tubes.

The granary has a special two-storeyed design (Figure 5-9). In between the ground and the first floors, discs are fitted to the pillars so that rats cannot climb up to the grain. A ladder to gain entry to the granary is used only when the farmer visits. For the rest of the time, it is kept elsewhere.

Shifting cultivators do not

Lyctus brunneus [Bostrichidae]

One of several species of small beetles known for tunneling through wood and producing a fine flour-like powder – hence the common name powder post beetle. In Arunachal Pradesh's swidden communities, these insects can wreak havoc among stored seeds if a farmer is not vigilant, or the seeds are not protected by the smoke above a kitchen fireplace.

readily accept new seeds from others, due to fear of uncertainty. They do not want to experiment with new crop varieties, because in the case of failure, there is no escape from disaster. They have faith only in their own seeds, that have been handed down to them from generation to generation. They know all about their old seeds, and can predict the production of their crops from the very beginning.

Fallowing

Farmers are acutely aware of the importance of fallowing fields. Generally, a field is cultivated for one year only. Sometimes, a portion of a field is cultivated twice or even three times, due to certain circumstances, especially for the sake of old women (Figure 5-10). But farmers do not normally allow cultivation of the same field for more than one year because it affects the timely attainment of maturity of the forest that rises up from the cultivated field. During the fallow period, large-scale felling of trees is not allowed. Collection of minor items like wild fruit, raw materials for craft work, and firewood, as well as trapping of animals, are not restricted. Sometimes, farmers will cut big climbers that form a heavy canopy in the crown of trees and prevent forest growth by obstructing the sunlight. Such climbers also break tree branches because of their weight. Big plantains are also cut in fallow growth.

Preservation of agricultural tools

Shifting cultivators use very simple tools. Most tools are designed out of locally available raw materials collected from the forest, except for those that are made from iron or steel, like machetes, axes, knives and scrapers. The farmers are very possessive of their tools and hardly ever share them with others. They always keep them away from children. Most of the men get angry when a knife or machete is used by a woman because they may not be careful and the tools may be blunted. They know that moisture is the enemy of any tool, whether it is made of cane, wood, bamboo or metal. Thus, they always keep them in dry places. If baskets woven from cane or bamboo are used in the rain, they are then kept over the hearth to dry. Ash is considered to be the best preservative of such items. Immature materials from the forest, or broken plants, are never used to make tools because these materials are soft and easily damaged by termites and powder post beetles. Storage in the smoke above the hearth is a general rule. Even items such as meat for future use are roasted over the household fire.

Conclusions

The use of Traditional Ecological Knowledge among the people of Arunachal Pradesh is gradually losing its importance, even in rural areas. This is due to changes of occupation towards a non-forest-based economy, migration of educated people to urban centres, conversion to Christianity, and availability of factory-made goods. Diminishing contact with the natural world has led to a decline in the people's attachment to the forest. Change is necessary for the betterment of human existence, but many hurry to make easy changes without realizing the potential they have for harm. One of the main causes of a rapid decline in biotic life in the forest is replacement of traditional bows and arrows and wood, cane and rope traps, with guns and steel traps that have greater accuracy and are more effective. The use of gillnets, chemicals and blasting have led to a depletion of aquatic life. Some farmers have even started using chemicals, such as herbicides, for fishing without having much knowledge of the side effects.

Traditional Ecological Knowledge in the protection of crops from birds and wild animals is rarely used nowadays because of a reduction in the number of birds and animals. After the introduction of modern hunting techniques with guns and other effective tools, the depletion of biotic life has gained pace. In many places near to urban centres, wild animals like boars, bears, monkeys and deer have become things of the past.

There has been a decline in the popularity of Traditional Ecological Knowledge, and one of the factors responsible for this is compulsory elementary education in every village. Children do not get enough time to interact with nature and with lessons delivered by their parents. The education system has become very competitive, and students spend most of their time indoors. Gradually, in pursuit of higher learning, they move away from their villages and ultimately settle in urban centres.

While many changes have come to the rural people of Arunachal Pradesh, the food habits have remained almost unchanged. Better cooking techniques for wild foods from the forest have improved, along traditional lines. Consumption of wild vegetables and medicinal herbs has increased, mainly due to the sharing of Traditional Ecological Knowledge via electronic communications, such as e-mail, Facebook, WhatsApp, and so on.

FIGURE 5-10: They cannot while away their time sitting at home. These old farmers still visit nearby fallowed swiddens where they once worked among the crops, gathering small contributions to family kitchens.

Photo: Tomo Riba.

Someday, most of the forest where shifting cultivation is practised today will have become primary forest, because the shifting cultivators will be no more. But unlike the secondary forests that follow shifting cultivation, these new primary forests will have relatively little plant diversity. The canopy layers will remain, but there will be no herbs and shrubs in the darkness beneath, due to the absence of sunlight.

6

CHANGES IN THE LENGTH OF FALLOW PERIODS AND LAND-USE INTENSITY OF SWIDDEN AGRICULTURE

In montane mainland Southeast Asia from 1988 to 2016

*Peng Li, Zhiming Feng and Chiwei Xiao**

Introduction

In more than 60 developing countries in the tropics, between 300 and 500 million upland farmers practise swidden agriculture, or shifting cultivation – a system pejoratively known as slash-and-burn farming. They are mostly from ethnic-minority groups that are scattered throughout the mountainous and hilly regions of Latin America, central Africa, and South and Southeast Asia, at elevations above 200 metres above sea level (masl) (Goldammer, 1988; Brady, 1996; van Vliet et al., 2012; Li et al., 2014). Currently, the only available map of the global coverage of swidden agriculture – largely speculative and conspicuously old-fashioned – is in a book on economic geography (Butler, 1980). In addition, the highly-dynamic, annually-changing and spatially-random nature of swidden agriculture means that it is seldom included in land-cover classification maps, with the exception of the SPOT-VGT-based Global Land Cover 2000 (1 km), covering shifting cultivation, but only in South and Southeast Asia, China, Africa and South America (Silva et al., 2011).

Information with which to update the global extent of shifting cultivation is very limited, especially data based on remote sensing (Heinimann et al., 2017). This lack of data occurs at a time when the length of fallow periods in swidden systems has been greatly shortened over recent decades, due to population growth, alternative land uses (e.g. cash-crop intensification), forest conservation policies and infrastructure development across the mountainous regions of the tropics (van Vliet et al., 2012; Ziegler et al., 2012).

In Southeast Asia, the practice of swidden agriculture is an ages-old tradition (van Vliet et al., 2012; Li et al., 2014) which can be traced back to the Neolithic

*   PROFESSOR PENG LI, PROFESSOR ZIHMING FENG and DR. CHIWEI XIAO are all from the Key Lab for Resources Use and Environmental Remediation, Institute of Geographic Sciences and Natural Resources Research, Chinese Academy of Sciences.

FIGURE 6-1:   A mosaic of swiddens (the grey patches) and fallows of different ages, typical of a shifting cultivation landscape, seen here in Huameuang district, Houaphanh Province, Lao PDR. Accurately measuring the extent of shifting cultivation at provincial or national levels, with a view to analyzing trends in sustenance of upland communities, conservation of forests and use of available land – among other things – has hitherto been almost impossible. This chapter presents an innovative approach to gathering remotely sensed data on the changing status of shifting cultivation.

Photo: Peng Li, November 2016.

era. Because it generally involves quasi-periodic slashing-and-burning of natural and secondary vegetation in the same place, there are major concerns over the environmental consequences, including deforestation and forest degradation. When damage from illegal logging is also considered, Southeast Asia has seen very high levels of deforestation and forest degradation in the past (Achard et al., 2002: Corlett, 2005). The prevailing slash-and-burn practice has impacted tropical rainforest and dry biomes in mainland Southeast Asia in particular, raising serious issues of soil erosion and decline in water quality (Gupta, 2005a; Ziegler et al., 2009). Active fires induced by the burning of biomass in other tropical areas have also led to serious environmental effects (Li et al., 2020). Additionally, recent developments in agriculture in mainland Southeast Asia have seen shortening of the restorative fallow phase of shifting cultivation, or replacement of swidden systems with other intensified land uses, particularly industrial tree plantations such as rubber, teak, and *Acacia mangium*. These transformations have had different effects on biological diversity (Rerkasem et al., 2009a), and have resulted in substantial reductions of aboveground and soil organic carbon stocks (Bruun et al., 2009). Therefore, detection of the historical extent of swidden agriculture and monitoring of fallow-period variations and land-use intensity using available satellite data are very important for assessing environmental sustainability and estimating carbon sequestration (Figure 6-1).

The diversity, complexity and dynamics of tropical swidden cultivation result in considerable difficulties in monitoring it with satellite data. This may well explain the scarcity of reports on this traditional farming system that have used remote sensing techniques. Since the 2000s, free access to remotely-sensed products, such as Moderate Resolution Imaging Spectroradiometer (MODIS, 1999 and later) and Landsat-family sensor data, has greatly promoted investigations of shifting cultivation using remote sensing. For instance, a landscape-mosaics approach based on existing land-cover inventories, obtained from SPOT satellite data, was developed to quantify swidden agriculture in the Lao PDR at an above-pixel scale (Messerli et al., 2009; Hett et al., 2012). Similarly, a landscape-metrics approach based on MODIS time-series products was also developed to overcome dependency on available land-cover data (Hurni et al., 2013a). However, the use of coarser spatial–resolution data usually leads to misclassifications and omissions, because swidden fields normally have an area less than one hectare. In addition, the rotational nature of swidden agriculture poses a substantial challenge to its detection with single-date satellite data (Hurni et al., 2013b; Li

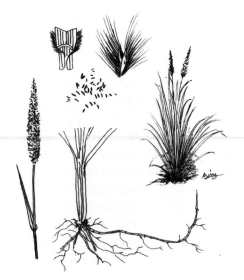

Imperata cylindrica (L.) Raeusch. [Poaceae]

This perennial grass, perhaps more than any other species, marks the degradation of forest land. Known as cogon grass or thatch grass, it covers large areas of upland Southeast Asia, growing in dense swards that deny other species the ability to regenerate the forest. It is a highly flammable fire-adapted species prone to burn annually, killing almost everything but its mat of underground rhizomes, from which it rises again to dominate its territory.

and Feng, 2016). Therefore, it requires timely and updated geospatial databases of newly-opened swidden fields at improved spatial and temporal resolutions.

Recently, we developed a Landsat-based multi-step threshold algorithm that takes advantage of multiple spectral bands (visible, near-infrared, and shortwave-infrared) of Landsat-8 Operational Land Imager (OLI) sensors to generate the first map of swidden agriculture (including fallows at varied stages) in mainland Southeast Asia and Yunnan province, China (Li and Feng, 2016). This approach is based on the unique phenological features of swidden agriculture, i.e. the quasi-periodic slashing-and-burning of natural and secondary vegetation in the same place between March and April in certain years. At that time, the newly-opened swidden fields are a mixture of felled, dried, and burnt vegetation and exposed soil, at the peak of the dry season (from March to April), in particular. Therefore, an algorithm was developed on the basis of four vegetation-moisture indices (VMIs): the Normalized Difference Vegetation Index (NDVI) (Tucker et al., 1986), the Soil Adjusted Vegetation Index (SAVI), the Normalized Difference Moisture Index (NDMI) (Vogelmann et al., 1988), and the Normalized Burn Ratio (NBR). The intention was to identify and track those target pixels that resulted from slashing, drying and burning over time within dry seasons. Unlike other landscape ecology-based approaches that primarily use existing land-cover data or single-date imagery, our Landsat-based multi-step threshold algorithm combined VMIs that were sensitive to changes in the vegetation canopy, land-surface water moisture, and exposed soils at the pixel scale. We used this algorithm to map newly-opened swidden fields and swidden fallows at different stages of forest recovery.

Acacia mangium Willd.
[Leguminosae]

This species is noted in this chapter for its use in industrial tree plantations, which replace swidden systems as an intensified form of land use. The trees grow to 30 m, often with a straight trunk. The timber is heavy, hard, and very strong, and is often used for furniture, doors and window frames. The tree's ability to fix nitrogen in the soil makes it popular for agroforestry uses and rehabilitation of degraded land.

More recently, we used our algorithm to generate longitudinal maps showing 29 years of annual newly burnt swidden plots in montane mainland Southeast Asia, using Landsat historical data products from the years 1988 to 2016. Therefore, there are two main pathways for monitoring the dynamics of swidden agriculture and its associated fallows (Li et al., 2018a). Using a 'space- time conversion' method, they can be called 'space-for-time (S-T) substitution' and 'time-for-space (T-S) substitution', respectively. One aims to detect newly burnt plots and swidden fallows of various ages with single-year remote sensing data. It can be named 'S-T' substitution because it shows temporal changes in swidden agriculture through the spatial distribution of variously aged fallows, from one to five and even more than 20 years old. Swidden fallows of various ages indirectly reflect the temporal development of newly opened swiddens. The other pathway aims to detect newly burnt swidden plots with inter-annual remotely sensed data. This can be regarded as 'T-S' substitution because it shows the spatial differences in the landscape of swidden agriculture by examining the distribution of annual newly burnt plots over nearly three decades. The annual extraction of newly burnt swidden plots indirectly reflects the general spatial pattern of swidden agriculture.

The annual dynamics are closely related to land-use intensity and the varying length of fallow periods in traditional farming. However, changes in the fallow phase of the swidden cycle are typically analyzed on a local scale, and for a particular moment in time, depending on where in the world the analysis is made. Our investigation of fallow-period variations contributes to fully understanding the intensity of swidden farming as well as its socio-economic and biophysical effects. In this study, we attempt to reveal the dynamic changes in fallow length and land-use intensity in mainland Southeast Asia over the 29 years from 1988 to 2016, by using probability comparative analysis of the 'for-this-moment' pixels recorded on each of those years.

Saccharum officinarum L. [Poaceae]

Sugar cane has long been a traditional swidden crop. Its emergence as a plantation cash crop to replace shifting cultivation has followed population growth, economic development and environmental-protection policies. About 70% of the world's sugar production comes from this species and related hybrids. The sweet juice is extracted from the stems to make sugar, and the by-products are used as livestock feed.

Materials and methods

The study area

Mainland Southeast Asia is the continental part of Southeast Asia (Figure 6-2) comprising Cambodia, the Lao PDR (Laos), Myanmar, Thailand, and Vietnam (Chuan, 2005). The region is located within tropical and subtropical climatic zones, with the latter normally referring to the northern mountainous regions of Myanmar. According to the Köppen climate classification, mainland Southeast Asia primarily has a tropical savanna climate, but also has a tropical monsoon climate, especially in the western Myanmar hills, the southern borderland of Myanmar and Thailand, and a long belt from Cambodia's Elephant and Cardamom hills to the Mekong delta and the southern end of the Annamite mountain chain. This part belongs to an extensive monsoon climate system in which the prevailing winds reverse directions every six months. The climate is characterized by greater seasonality, higher extremes in both temperature and precipitation, and more pronounced dry spells (Chuan, 2005). As for seasonality, mainland Southeast Asia has a cool dry season from November to February, a hot dry season from March to April, and a rainy season from May to

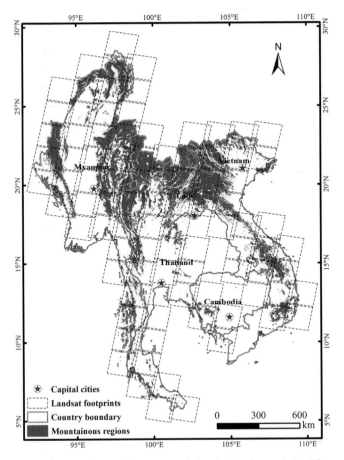

FIGURE 6-2: The study area, mainland Southeast Asia. The shaded area is the redefined mountainous area (or redefined montane mainland Southeast Asia) that is covered by 91 Landsat footprints (path and row).

October. The dry season, typically controlled by the northeast monsoon, is characterized by a low cloud cover with less than 20 mm of rainfall per month. This seasonal condition not only facilitates felling, drying, and burning of household-based swidden fields, but also contributes to the acquisition of remote-sensing images from Landsat-family sensors that are either free of clouds, or have a cloud coverage of less than 30% (Li et al., 2018b).

Mainland Southeast Asia's geomorphology consists mainly of three intermingled landform units: mountain ranges, plateaux and plains. Quasi-parallel mountain ranges lying in a north-south axis feed five major river systems, from west to east: the Ayeyarwady (Irrawaddy), Thanlwin (Salween), Chao Phrya, Mekong, and Red rivers. The rivers either cut through or bypass the plateaux and form a number of valleys, plains and deltas of various sizes in the middle and lower basins. Montane mainland Southeast Asia – the mountainous regions of the mainland – usually refers to upland areas more than 300 m above sea level (Ziegler et al., 2009; Fox et al., 2014). In this study, the mountainous area of mainland Southeast Asia is redefined. It comprises land with different parameters of elevation, slope gradient and local relief, based on the Mountain Research Initiative typology, in order to reduce the spectral noise effects of permanent farmland and corresponding human settlements on the detection of swidden practices. More information about this redefining process can be found in our previous study (Li and Feng, 2016). Montane mainland Southeast Asia normally comprises most of the northern mountainous region of Myanmar, the western Myanmar hills, the central range of hills (including the Shan highlands, the hills of northern Thailand and Laos, the Tenasserim hills and the central highlands of the Malay Peninsula), the Elephant and Cardamom Hills of Cambodia and the Annamite range, extending from Laos and Vietnam to northern Cambodia (Gupta, 2005b). In the uplands, acrisols are the dominant soil types, limiting agricultural output due to high acidity, low fertility and susceptibility to erosion (Dudal, 2005). Hence, swidden agriculture with a short-term fallow is still very common, but is transforming rapidly into other acid-tolerant commercial plantations, such as rubber and *Acacia mangium* (Dudal, 2005).

There are two broad categories of vegetation in mainland Southeast Asia (Corlett, 2005): a small proportion of lowland equatorial evergreen rainforest in the southern part and a vast area of seasonal tropical forest because of seasonal drought. The mountainous part was once home to large, continuous tracts of natural forest (Corlett, 2005), but much of this has been modified by human activities, including slashing-and-burning and illegal logging. Because of topographical differences, the forests can be grouped into four types: evergreen mountain forests (above 1000 masl), evergreen lowland forests (below 1000 masl), mixed deciduous forests and fragmented and degraded

Hevea brasiliensis (Willd. ex A.Juss.) Müll.Arg. [Euphorbiaceae]

Prominent among the tree crops replacing shifting cultivation, rubber has grown in importance as the natural altitude and latitude ranges of this lowland species have been stretched by plantations that provide cash income for upland farmers. However, latex tapping can only begin when the trees are five to seven years old or more. The need to wait for income and poor market prices for rubber have detracted from this tree's appeal.

evergreen forests (Stibig et al., 2004, 2007), with the latter two making up the largest remaining tropical forests in mainland Southeast Asia (Rerkasem et al., 2009b). The peoples living in these forests include various and diverse ethnic minority groups (e.g. Akha, Chin, H'mong-Mien and Shan) that have been practising subsistence-level swidden agriculture for countless generations. This traditional farming system remains a dominant land-use category in montane mainland Southeast Asia (Fox and Vogler, 2005; Schmidt-Vogt et al., 2009). In recent decades, under the pressure of population growth, economic development and national ecological-conservation policies, swidden agriculture has undergone extensive transformations, including conversion into cash-crop cultivation (rubber, sugar cane, maize, and so on) and shortened fallow periods (Cairns, 2015).

Annual data of newly burnt swidden plots from the years 1988 to 2016

With the Landsat-based multi-step threshold algorithm of four VMIs, we were able to prepare 29 longitudinal maps of newly burnt swidden plots in mainland Southeast Asia – the first showing the situation in 1988 and one for every subsequent year until 2016 (Li et al., 2018a) (Figure 6-3). Over that period of nearly three decades, an annual average area of 60,800 sq. km, or 3.15% of the total land area in the uplands, was opened each year for new swidden plots. For more information about extraction of data for newly burnt swidden plots, as well as national analyses, please refer to our previous study (Li et al., 2018a).

Calculation of land-use intensity based on the occurrence of newly burnt plots

In this study, land-use intensity of swidden farming refers to the frequency of occurrence of newly burnt plots at the pixel level within a given period. For instance, a frequency of 12, measured by counting the occurrence and reoccurrence of swidden plots being burnt over the years from 1988 to 2016, indicated that the same plots were repeatedly slashed and burnt 12 times by local farmers. We used all of the maps of newly burnt plots between a given year (beginning with 1988) and 2016, e.g. from 1988 to 2016, 1998 to 2016, 1999 to 2016, and so on, and calculated the occurrence frequency. The occurrence of newly burnt plots ranged from just once to 29 times over the study period. Analyses of the area of newly burnt plots with various occurrence frequencies, within different temporal intervals, e.g. from 1988 to 2016, 1998 to 2016 or 2008 to 2016, were able to help in understanding changes in swiddening intensity.

Calculating the length of fallow periods in swidden agriculture, based on newly burnt plots

To understand changes in the length of fallow periods, we calculated the time that passed between the first occurrence of a newly burnt plot and the next occasion on which burning occurred at the same spot, at pixel level. This process began with 1988,

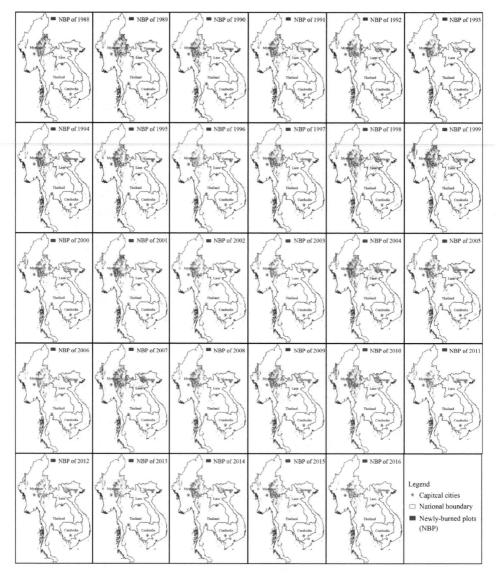

FIGURE 6-3: The 29 longitudinal maps of newly burnt swidden plots (NBP) in redefined montane mainland Southeast Asia from 1988 to 2016, derived from a Landsat-based multi-step threshold algorithm of four vegetation-moisture indices (NDVI, SAVI, NDMI and NBR).

and then progressed by examining the 29 annual maps up to 2015. The calculation processes, which covered two major procedures, were finished with Python language using ArcGIS 10.x software. The first step was to compute the years that passed between the first burning and subsequent burnings at a pixel scale. For example, if a newly burnt plot was detected on the 1988 map, a repeated burning of the same plot might be detected in any subsequent year, from 1989 to 2016. The temporal gaps between the initial burning and those that followed on the same pixel were given values in years, ranging from 1 to 28. When the pixels showed the first appearance of a newly burnt plot in 1989, it was given the value of one – the same as a one-year difference between successive burnings. A value of two was given to those that first emerged in 1990, and a value of 28 was given to those cleared exclusively in 2016. For cases with more than two successive burnings, we counted the individual year gaps between the first burning in 1988 and the second burning, then between the second and the third, the third and the fourth, and so on. Based on the analyses, we generated 28 maps of temporal gaps, or years passing, between the first occurrence in 1988 and subsequent re-appearances of newly burnt plots. The maps reflected the temporal gaps between the first in 1988 and the second, third, fourth and fifth re-occurrences, and so on up to the 28th and 29th. Then the same calculating procedures were carried out for newly burnt plots that first occurred in 1989, 1990, 1991, and so on up to 2015. The number of year-gap maps decreased from 27 for the situation in 1989 to just one for 2015, since the end year for all of the maps was 2016. The second step was to take the resulting total of 406 thematic maps showing the temporal gaps between two occurrences of burning and explore the changes in the length of swidden fallow periods in montane mainland Southeast Asia in the years from 1988 to 2016.

Results and analysis

Changes in the land-use intensity of swidden agriculture, based on the annual occurrence of newly burnt plots in the years from 1988 to 2016

Figures 6-4 and 6-5 clearly show that there was a consistent trend of distribution and change in the area of newly burnt plots under different occurrence frequencies. As the durations got shorter, e.g. from a start in 1988 to a start in 1997, both finishing in 2016, the maximum occurrence frequency became smaller accordingly. Hence, we generated 18 bar graphs (Figures 6-4 and 6-5) that clearly show the changes in the total area of newly burnt plots at various occurrence frequencies (or different fallow lengths) over three periods: 29 years (1988 to 2016); 20 years (1997 to 2016); and 12 years (2005 to 2016).

The data showed that the total area of plots that were burnt fewer than five times (i.e. had a lesser frequency of occurrence), was typically more than one-million ha in montane mainland Southeast Asia. Plots that were burnt only once in periods that began in 1993, and in every subsequent year until 2005, and ended in 2016, covered a maximum area of six-million ha. However, the maximum areas for plots that were burnt only once fell to figures consistently below five-million ha in periods that began in 1988, 1989, 1990 and 1991, and ended in 2016. It exceeded five-million ha in the period from 1992 to 2016. To the contrary, plots that were newly burnt between six and 28 times generally covered a total area less than one-million ha, although the area showed a slightly incremental trend. As the frequency of occurrence increased (i.e. the fallow period was shortened) the total area covered decreased rapidly, from a high of about five- to seven-million ha to nearly one-million ha, followed by a very slow

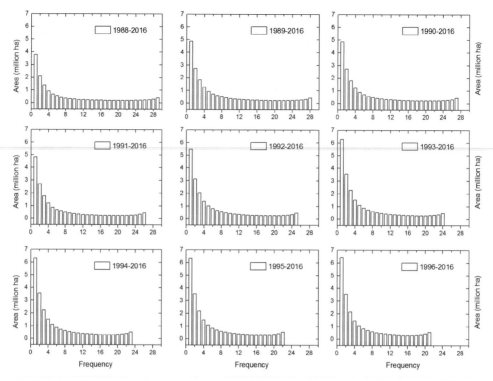

FIGURE 6-4: Changes in the total area covered by newly burnt swidden plots at different occurrence frequencies (or different fallow lengths) between each of the years from 1988 to 1996, and 2016.

reducing trend to a minimum area of around 500,000 ha, and then showed a slightly increasing trend back to roughly one-million ha.

It may be concluded that farmers paid high attention to the intensity of their cultivation and let the swidden plots recover with an acceptable fallow phase that was as long as possible. Generally speaking, more than 60% of newly burnt plots had an average reoccurrence frequency of five times over the study period, or about five to six years of fallow. However, other newly burnt plots occurred at shorter fallow cycles of less than six years. The correlation between the total area covered by newly burnt plots and their fallow length clearly showed that swidden agriculture in montane mainland Southeast Asia was practised to the fullest extent possible. The relationship between total area and frequency is clearly shown in Figures 6-4 and 6-5. The figures for the remaining years of the study, i.e. from 2006 to 2016 (not shown here), were almost the same. Overall, the intensity of shifting cultivation in montane mainland Southeast Asia was low, although shortened fallow lengths were reported at a local level, with swiddens covering a smaller area.

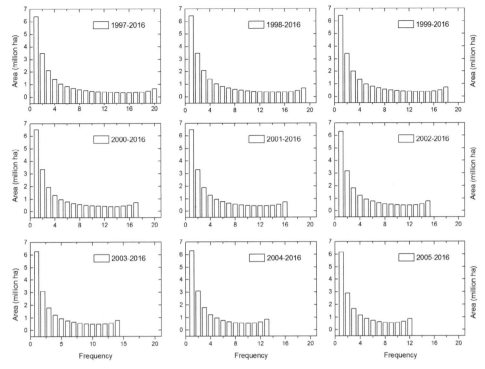

FIGURE 6-5: Changes in the total area covered by newly burnt swidden plots at different occurrence frequencies (or different fallow lengths) between each of the years from 1997 to 2005, and 2016.

National differences in the total area of newly burnt plots of varied swiddening intensity

The above analyses clearly show the changes in swiddening intensity for montane mainland Southeast Asia as a whole. However, they do not provide an understanding of national variations in the total area covered by newly burnt plots with different occurrence frequencies (or fallow lengths). Therefore, we took all of the maps of newly burnt plots between 1988 and 2016 and calculated the land-use intensity of swidden agriculture in the five countries covered by the study, by comparing the cumulative area of newly burnt plots with different occurrence frequencies. We first extracted all of the pixels showing the opening of new swidden plots in the years from 1988 to 2016. Then we calculated their related area at provincial level. As there are huge differences in the area of provincial units in Cambodia, Laos, Myanmar, Thailand and Vietnam, we also computed the ratio of the cumulative area covered by newly burnt plots at different occurrence frequencies (or fallow lengths) and the total land area of each province in the five countries.

Figure 6-6(a) shows the differences in the cumulative total area of newly burnt plots in each province of the five countries in montane mainland Southeast Asia. In terms of area used over the 29 years of the study, three provinces in Myanmar ranked as the top three. Shan state had the largest total area of newly burnt plots with

7.01-million ha, followed by Kachin state with 3.49-million ha and Sagaing Region with 1.62-million ha, mainly because of the vast land area in these provinces comprised of mountains and plateaux. The fourth-largest province for newly opened swiddens in mainland Southeast Asia was Chiang Mai province in the north of Thailand – that country's second-largest province – with about 1.22-million ha. The fifth-largest province, and the largest in the Lao PDR, was Luang Prabang province in the north of the country, with nearly 1.11-million ha. The total areas used for swidden farming over the study period in all of the other provinces in the five countries of montane mainland Southeast Asia were consistently less than one-million ha. Six provinces – Oudomxay, Luang Namtha and Bolikhamxai in the Lao PDR, Mae Hong Son and Tak in Thailand and Chin state in Myanmar, in descending order – had more than half-a-million ha. The provinces with the largest areas of newly burnt plots in Vietnam and Cambodia were Nghe An province in the north-central coastal region of Vietnam – that country's largest province by area – and Pursat province in the western part of Cambodia, with about 370,000 ha and 80,000 ha, respectively. Thus, it can be concluded that those provinces with large areas of mountains and plateaux tended to have large tracts of shifting cultivation. Spatially, newly burnt swidden plots were mainly opened and reopened in the northern parts of Myanmar and Laos, as well as in the western or northern parts of Vietnam, near their borders with China, and in the northern provinces of Thailand (Figure 6-6(a)). These remote "hot spots" of swidden activity usually happened in the agricultural-forest frontiers where upland ethnic groups relied heavily on relatively well-preserved primary forest.

Figure 6-6(b) shows the different ratios of cumulative swiddening area to provincial land area in each country over the past nearly three decades. Further, it clearly shows that the spatial distribution of newly opened swiddens was more concentrated in the northern borderlands of mainland-Southeast-Asian countries, with the exception of Cambodia. The ratios of total swidden land to provincial land area better reveals the spatial differences in newly burnt plots with a variety of swiddening intensities.

Although the three northern provinces of Myanmar (Shan State, Kachin State and Sagaing region), Chiang Mai in Thailand and Luang Prabang in Laos were the top-five provinces for swidden area, with more than one-million ha each, it is very interesting to note that the highest ratio of cumulative swidden land to provincial land area (up to 80.56%) was in Luang Namtha province, in northern Lao PDR. This was followed by Oudomxay province, another mountainous province in northern Laos with elevations between 300 and 1800 masl, which had a ratio of 75.88%. Then there were four provinces with ratios ranging between 53% and 56%. They were Mae Hong Son and Chiang Mai in Thailand and Bokeo and Luang Prabang in the Lao PDR. Next came Shan state in Myanmar (43.11%), Cao Bang province in Vietnam (40.84%), and Tak province in Thailand (40.13%). They were followed by Kachin state (Myanmar), Bolikhamxai (Laos) and two Vietnamese provinces (Dien Bien and Ha Giang) with ratios of cumulative swiddening area to provincial land area ranging between 31% and 38%. More interesting was the comparison between Shan state, which had the largest area of newly burnt swidden plots in mainland Southeast

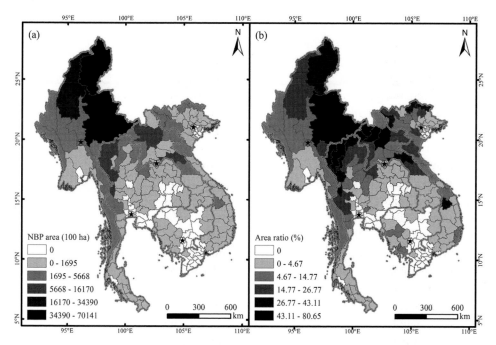

FIGURE 6-6:   The provincial distribution of newly burnt swidden plots (NBP) with very low, low, moderate, high and very high intensity between 1988 and 2016.

Note: Blank provincial units were not included in the study area.

Asia, and the two provinces of Laos, Luang Namtha and Oudomxay, which had the highest ratio of swidden land to total area, while they had only 810,000 ha and 910,000 ha of newly burnt plots. This clearly indicated that swidden agriculture was quite commonly and intensively practised in these provinces during the years from 1988 to 2016. This was supported by the additional fact that four other provinces in Laos and Thailand had swidden land to total land ratios of more than 50%.

The actuality of dense and prevalent distribution of newly burnt plots for this traditional livelihood farming system was brought home to us during two field trips, in late November, 2016 and mid-May, 2018. We travelled along route 13 in the Lao PDR between Boten checkpoint in Luang Namtha province and Pak Mong, in Luang Prabang. Analogously, we witnessed the wide distribution of newly burnt plots in Chiang Mai, Thailand, and in Bokeo and Luang Namtha provinces, on a field trip in late March, 2013, along the Kunming to Bangkok highway. A similar situation could be seen in the uplands along the Tengchong-Myitkyina Road (also well known as Stilwell Road) in eastern Myitkyina in Kachin State, Myanmar, in early February, 2015.

The above analyses clearly illustrate the differences in the overall area of newly burnt plots over the 29-year study period. However, one question remains unanswered: how, and/or where, was the diversity and intensity of all of the newly burnt swidden plots distributed or located in the uplands of montane mainland Southeast Asia over the past three decades? To further compare the provincial differences in swiddening intensity, we adopted the Jenks 'natural breaks' classification method, which is also called the Jenks Optimal Classification Algorithm (Jenks and Coulson, 1963). The advantage of this method lay in determining the best arrangement of values into different contiguous classes, so as to minimize the squared deviation within each class. The aim was to divide the data presented above into five major classes of swiddening intensity: very low, low, moderate, high and very high. The very low level corresponded to swidden plots being newly burnt once or twice; the low level with occurrence frequencies ranging from three to eight times; the moderate level with plots burnt nine to 16 times; the high level with 17 to 23 times; and the very high level with plots burnt 24 to 29

Manihot esculenta Crantz [Euphorbiaceae]

Cassava is a woody shrub that commonly appears in swiddens because it tolerates poor soil, resists dry conditions, and can be harvested any time. It is the third-largest source of food carbohydrates in the tropics, but it is a poor source of protein and vitamins. The starchy tubers contain toxic compounds that must be released by cooking. A powdery extract from dried cassava is known as tapioca.

times over the study period. Figures 6-7 through 6-11 present the spatial differences in swiddening intensity at provincial level in montane mainland Southeast Asia over the years 1988 to 2016. In all cases, the figures labeled (a), in the left panel, show differences in intensity within the total area of newly opened swiddens, while figures labelled (b), in the right panel, show the ratio of cumulative swidden areas, regardless of different land-use intensities, to total land area at provincial level. We then further examined the related spatial characteristics with different frequencies of occurrence and reoccurrence at country level.

Area of newly burnt plots at different swiddening intensities in Myanmar

From 1988 to 2016, newly burnt swidden plots in Myanmar at a very low level of land-use intensity added up to a large area of 9.13-million ha (Figure 6-7(a)). The top three provincial units, Kachin and Shan states and Sagaing region, accounted for 81.32% of swiddening in Myanmar at this very low level of intensity over the three decades. Specifically, the country's northernmost Kachin state opened and reopened a cumulative swidden area of 3.39-million ha, equalling 36.46% of the hilly state's

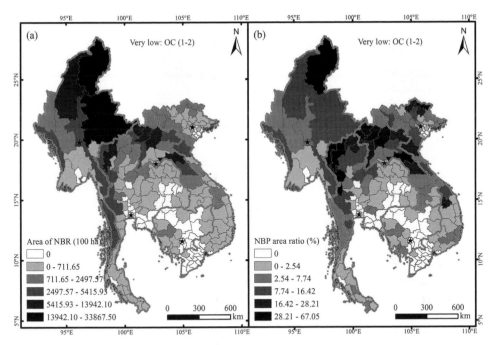

FIGURE 6-7: Provincial distribution of newly burnt plots (NBP) at a very low level of swiddening intensity during the years from 1988 to 2016.

Note: Blank provincial units were not included in this study; OC = occurrence frequency.

land area. It was followed by Shan state and Sagaing region, with 2.64-million ha (16.24% of state area) and 1.39-million ha (13.70% of area), respectively (Figure 6-7(b)). Rakhine state, on Myanmar's western coast, and Tanintharyi region, in the long narrow southern part of the country, both had more than 300,000 ha of swiddens at very low intensity. However, Kayin state and other states or regions in Myanmar had much smaller areas of less than 300,000 ha at this very low intensity.

In the low-intensity swiddening category, Myanmar had newly burnt plots that added up to a large total area of 2.12-million ha over the 29-year study period (Figure 6-8(a)). The top three provinces in this category, Shan state, Chin state and Sagaing region, together accounted for 87.78% of Myanmar's low-intensity shifting cultivation over the three decades. Specifically, northeastern Shan state had 1.56-million ha of swidden fields opened and reopened, amounting to 9.68% of the plateau state's total area. It was followed by Chin state and Sagaing region, both with about 140,000 ha of low-intensity swiddens, covering areas amounting to 3.62% and 1.41% of each administrative unit, respectively (Figure 6-8(b)). Another six administrative units, Kachin state, Mandalay region, Magway region, Rakhine state, Kayin state and Kayah state, had low-intensity areas decreasing in that order from 90,000 to 10,000 ha, to equal about 1% of the land area of each unit. Tanintharyi and other states or regions had much smaller areas of less than 10,000 ha of newly burnt low-intensity plots, amounting to less than 0.1% of their total area.

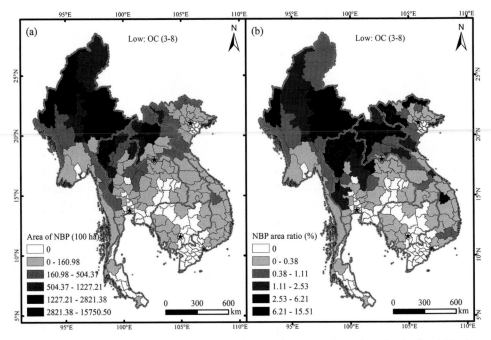

FIGURE 6-8: Provincial distribution of newly burnt plots (NBP) at a low level of swiddening intensity during the years from 1988 to 2016.

Note: Blank provincial units were not included in this study; OC = occurrence frequency.

In terms of the moderate (Figure 6-9(a)), high (Figure 6-10(a)) and very high (Figure 6-11(a)) levels of swiddening intensity, the cumulative area of newly burnt plots in Myanmar from 1988 to 2016 was 1.14-million ha, 930,000 ha and 1.43-million ha, respectively. Myanmar was the only country in mainland Southeast Asia with large areas of newly burnt plots in the moderate to high categories. Among Myanmar's provinces and regions, the prevailing swidden practices in Shan state were a major contributor to the overall figures. In addition, Chin state, Mandalay region, Sagaing region, Magway region and Kayah state were important provincial units for the practice of swidden agriculture over the past three decades. Of these, Shan state, Mandalay region and Magway region had larger areas of newly burnt plots in the high and very high levels of swiddening intensity than those at a moderate level, for example, equal to about 11.49% of total land area for high and very high intensities (Figures 6-10(b) and 6-11(b)) compared with merely 5.70% in the moderate category (Figure 6-9(b)). Of greater interest, Kayah state was the only provincial unit with a total area of high to very high swiddening intensity, amounting to 12.46% of the provincial land area, that was larger than the percentage (about 4.83%) of very low to moderate levels. This clearly indicated that Kayah state had very high land-use intensity of swidden agriculture in the study period from 1988 to 2016. In summary, because very low to low levels of swiddening intensity accounted for about 76.3% of shifting cultivation in Myanmar, the overall intensity of swidden agriculture in the country was basically at a low level. However, newly burnt plots with moderate to very high levels of land–use intensity added up to 3.49% of the country's total land area. This, also, was a large percentage.

Area of newly burnt plots at different land-use intensities in the provinces of the Lao PDR

Laos, the only landlocked country in Southeast Asia, had a cumulative area of newly burnt plots at the very low level of swiddening intensity amounting to 4.56-million ha over the three-decade study period (Figure 6-7(a)). These plots were burnt only once or twice in that time. Moreover, they were scattered widely in every province across the entire mountainous area of the country. Luang Prabang, Oudomxay, Luang Namtha and Bolikhamxai were the top four provinces with the largest areas of newly opened swiddens, with approximately 790,000 ha, 710,000 ha, 670,000 ha and 540,000 ha, respectively. Four other provinces, Houaphan, Bokeo, Khammouane and Xayabury, had about 300,000 ha of newly burnt swidden plots. The remaining provinces, which, with the exception of Phongsaly, are in the country's central and southern regions, each had less than 100,000 ha of newly burnt plots. The total area of newly burnt plots in Luang Namtha added up to equal 67.06% of the province's total land area – the largest area ratio in Laos. It was followed by Oudomxay and Bokeo, which were both above 50%, and Luang Prabang and Bolikhamxai, where the ratios were over 30% (Figure 6-7(b)). In the remaining provinces, the area of newly burnt plots amounted to a much lower proportion of the land area, less than about 20%.

In the category for low-level swiddening intensity, Laos had a cumulative area of about 850,000 ha of newly burnt plots (Figure 6-8(a)). In this category, the plots were burnt between three and eight times in the 29 years of the study. As in the previous category, they were widely scattered across the whole mountainous area of the country, making Laos the only country with newly burnt plots at low-level intensity occurring and reoccurring in all provinces. Among them, Luang Prabang, Luang Namtha and Oudomxay were the top three provinces with the largest areas of low-level swiddening intensity: approximately 280,000 ha, equal to 13.87% of the total land area, 190,000 ha or 15.51% of the total land area, and 120,000 ha or 12.22% of the total land area, respectively (Figure 6-8(b)). The area of newly burnt plots at low-level intensity in another six provinces – Xayabury, Houaphan, Bokeo, Phongsaly, Bolikhamxai and Xiangkhouang – were between 80,000 ha (equal to 4.94% of total land area) and 10,000 ha (1.10% of total land area). Finally, Khammouane and the remaining provinces all had less than 10,000 ha (equal to 1% of total land area) committed to low-level intensity swiddening.

The area covered by newly burnt plots at moderate and high levels of land-use intensity in Laos was approximately 120,000 ha (Figure 6-9(a)) and 20,000 ha (Figure 6-10(a)), respectively. The moderate-intensity plots occurred and reoccurred mainly in four northern provinces: Luang Prabang, Oudomxay, Sayabury and Luang

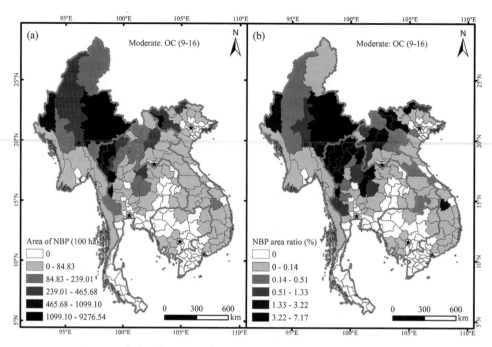

FIGURE 6-9: Provincial distribution of newly burnt plots (NBP) at a moderate level of swiddening intensity during the years from 1988 to 2016.

Note: Blank provincial units were not included in this study; OC = occurrence frequency.

Namtha. However, these amounted to merely 10,000 ha to 40,000 ha per province, equal to 1% or 2% of the land area of each (Figure 6-9(b)). Phongsaly and the remaining provinces all had less than 10,000 ha of newly burnt plots at moderate intensity. This pattern was also seen in the distribution of newly burnt plots at high swiddening intensity, but the percentage of total land area covered by these plots fell below 0.29% in all provinces (Figure 6-10(b)). The area of land occupied by newly burnt plots at very high levels of swiddening intensity was much smaller, amounting to only 1.39-million ha in the entire country (Figure 6-11(a)), with proportions of land area used for this purpose in individual provinces falling below 0.04% of total area (Figure 6-11(b)). Very high intensity swiddening was mainly confined to Sayabury, Khammouane and Luang Prabang provinces. In summary, since newly burnt plots with a very low to low level of land-use intensity amounted to 97.65% of all shifting cultivation detected in Laos, the country's swiddening intensity was found to be extremely low.

Area of newly burnt plots at different land-use intensities in the provinces of Thailand

Over the three study decades, Thailand had a total area of 3.69-million ha of newly burnt plots at a very low level of swiddening intensity, i.e. plots that were burned only once or twice in the study period (Figure 6-7(a)). Newly burnt plots detected by

Landsat in the years from 1988 to 2016 were widely distributed in six geographical regions delineated by the National Research Council of Thailand on the basis of biophysical features, such as landforms and drainage. They were the country's north, northeast, central, east, west and south regions. However, at a provincial level there were considerable differences in the area of newly cleared swiddens at a very low level of intensity. Chiang Mai province, in the north region, had the largest area, covering nearly 780,000 ha and giving the province the country's highest ratio of cumulative swidden area to total land area of 33.78% (Figure 6-7(b)). The second-, third-, and fourth-largest areas of very low intensity swidden were in Tak province (west region), with about 510,000 ha, equal to 28.21% of total land area; Mae Hong Son province (north region), approximately 440,000 ha or 33.16% of total land area; and Kanchanaburi province (west region), about 310,000 ha or 15.62% of total land area. The provinces of Lampang, Chiang Rai, Nan and Phayao in the north region, Phitsanulok in the central region, Loei in the northeast and Uthai Thani in the west had moderate areas of between 100,000 ha and 200,000 ha of newly burnt plots at very low intensity over the course of the study, making up between 10% and 18% of the total land area of the provinces. Finally, Phetchabun and the remaining provinces in Thailand accounted for 10% or less of the country's newly burnt plots at very low swiddening intensity, totalling less than 70,000 ha in area.

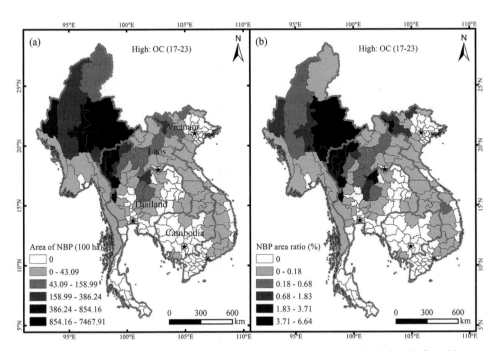

FIGURE 6-10:   Provincial distribution of newly burnt plots (NBP) at a high level of swiddening intensity during the years from 1988 to 2016.

Note: Blank provincial units were not included in this study; OC = occurrence frequency.

With regard to the low level of swiddening intensity (plots burnt three to eight times from 1988 to 2016), Thailand had a total area of only 870,000 ha of newly burnt plots (Figure 6-8(a)). Once again, these plots were widely distributed in the six geographical regions. However, there were considerable differences at provincial level. Chiang Mai and Mae Hong Son provinces in the north region had the largest areas of newly opened swiddens at low land-use intensity, covering about 210,000 ha and 150,000 ha respectively, with a ratio of cumulative swidden-land area to total land area of 9.04% and 11.33%, respectively (Figure 6-8(b)). The area of newly burnt plots at low intensity in Tak, Chiang Rai, Nan, Loei, Lampang, Kanchanaburi, Lamphun, Phayao, Phetchabun, Phitsanulok, Uttaradit, Uthai Thani and Phrae provinces ranged in descending order from 100,000 ha to 10,000 ha, equalling 1.54% to 6.13% of total land area. Finally, Kamphaeng Phet and all remaining provinces accounted for less than 1% of all newly burnt plots in the low-intensity category, altogether covering less than 10,000 ha.

Thailand had a total area of about 490,000 ha of newly burnt plots in the moderate swiddening intensity category, i.e. plots that were burnt nine to 16 times in the course of the 29-year study period (Figure 6-9(a)). With around 110,000 ha, Chiang Mai was the only province with a cumulative number of newly burnt plots at this land-use intensity covering more than 100,000 ha. The province's ratio of swidden land in the moderate intensity category to total land area was 4.77% (Figure 6-9(b)). Eleven provinces − Mae Hong Son, Tak, Loei, Chiang Rai, Lamphun, Lampang, Phetchabun, Nan, Kanchanaburi, Phayao and Chaiyaphum − had newly burnt plots with cumulative areas ranging in decreasing order from 60,000 ha to 10,000 ha, with percentages of total land area ranging from 6.54% to 0.66%. All of the remaining provinces in the country had a combined total of less than 10,000 ha, indicating that swidden agriculture was not commonly practised there.

In the categories for high and very high swiddening intensity, in which swiddens were burnt between 17 and 29 times between 1988 and 2016, Thailand had plots covering approximately 300,000 ha (Figure 6-10(a)) and 140,000 ha (Figure 6-11(a)), respectively. Three provinces − Chiang Mai, Mae Hong Son and Tak − had newly burnt plots with a total area between 20,000 and 90,000 ha, equal to 1.13% to 3.71% of the total land area (Figure 6-11(b)). In addition, Loei, Lamphun, Chiang Rai and Phetchabun provinces all had between 20,000 ha and 50,000 ha of newly burnt plots at the high level of swiddening intensity, amounting to between 1.77% and 6.54% of the total land area (Figure 6-10(b)). In Thailand's remaining provinces, swidden agriculture has not been commonly practiced over recent decades. In summary, because shifting cultivation at the very low to low levels of land-use intensity represented 83.10% of all newly burnt plots over the 29 years of the study, the overall intensity of swidden agriculture in Thailand was low.

Area of newly burnt plots at different land-use intensities in the provinces of Vietnam

Vietnam had only 2.16-million ha of swidden plots that were burnt just once or twice – the lowest level of swiddening intensity – over the years from 1988 to 2016 (Figure 6-7(a)). These were distributed to differing degrees throughout the country's eight administrative regions, the northeast, northwest, Red river delta, north central coast, south central coast, central highlands, southeast and Mekong river delta (Boateng, 2012). However, differences in swidden area were very evident among the country's 58 provinces and five centrally controlled municipalities. Nghe An province in the north central coast region – the largest province by land area – had the largest swidden area of about 340,000 ha in the very low category of land-use intensity, equal to 20.56% of the province's land area (Figure 6-7(b)). Cao Bang province in the northeast region ranked second, with about 230,000 ha, but this represented the largest percentage of provincial land area, at 33.86%. Another eight provinces, Ha Giang and Bac Kan in the northeast, Kon Tum in the central highlands, Thanh Hoa and Quang Binh in the north central coast region, Quang Nam in the south central coast, and Dien Bien and Son La in the Northwest, had very low intensity swiddens covering 100,000 ha to 180,000 ha, or between 7% and 26% of provincial land area. In the northwest province of Lai Chau and all remaining provinces in Vietnam, very low intensity swiddening was not the dominant land use and areas of newly burnt plots covered less than 70,000 ha per province, equal to less than 10% of land area.

Only a small area of approximately 400,000 ha of swidden plots was produced in Vietnam by burning between three and eight times over the study period – the low level of land-use intensity (Figure 6-8(a)). These plots were in all eight administrative regions, with the exception of the Mekong river delta (Boateng, 2012), but once more, differences between individual provinces were large. In the northwest region, Dien Bien province, which borders with China's Yunnan province and Phongsaly province in the Lao PDR, topped the order of newly burnt plots at this level of intensity, with about 100,000 ha, amounting to 10.73% of the province's land area (Figure 6-8(b)). This was followed by Kon Tum in the central highlands, Lai Chau and Son La in the northwest, Ha Giang and Cao Bang in the northeast and Nghe An and Thanh Hoa in the north central coast region, which had between 10,000 ha and 50,000 ha of newly burnt plots, equal to 1.06% to 6.05% of their total area. In Quang Tri province, in the north central coastal region, as well as in all other provinces in Vietnam, newly burnt plots at a low level of swiddening intensity covered less than 10,000 ha per province, amounting to less than 1% of their land area.

In the moderate, high and very high categories of swiddening intensity, Vietnam had total areas of approximately 200,000 ha (Figure 6-9(a)), 60,000 ha (Figure 6-10(a)) and 10,000 ha (Figure 6-11(a)), respectively, over the 29 years of the study. Newly burnt plots in these categories were detected mainly in northern provinces bordering China, including Dien Bien, Lai Chau, Son La, Ha Giang and Cao Bang, and in the central highlands province of Kon Tum, which is bordered by Laos and Cambodia. Dien Bien had approximately 69,400 ha, equal to 7.17% of its land

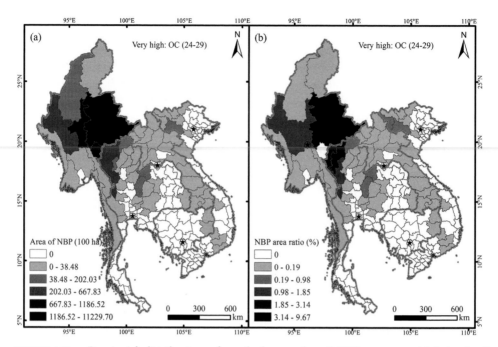

FIGURE 6-11: Provincial distribution of newly burnt plots (NBP) at a very high level of swiddening intensity during the years from 1988 to 2016.

Note: Blank provincial units were not included in this study; OC = occurrence frequency.

area (Figure 6-9(b)); about 25,500 ha (2.64%) (Figure 6-10(b)); and about 3800 ha (0.40%) (Figure 6-11(b)) at the three intensity levels, respectively. Lai Chau had about 34,800 ha, equal to 3.78% of its land area; approximately 12,100 ha (more than 1.31%); and about 1400 ha, (0.16%), at the three intensity levels, respectively. Son La had approximately 30,400 ha (2.12%); 13,400 ha (0.94%); and 5700 ha (0.39%) of newly burnt plots at the three levels, respectively, and Ha Giang had about 29,800 ha (3.71%); 10,000 ha (1.24%); and 1100 ha (0.13%) at the moderate, high, and very high levels, respectively. Kon Tum and Cao Bang provinces had approximately 19,000 ha (equal to 2.34% of their land area) and 2300 ha (0.29%) of newly burnt plots at the moderate and high levels of swiddening intensity, and about 6200 ha (0.92% of land area) at a very high swiddening intensity. In summary, since the very low to low levels of swiddening intensity applied to 90.39% of all newly burnt plots detected in Vietnam over the study period, it could be concluded that the overall intensity of swidden agriculture in Vietnam was low.

Area of newly burnt plots at different land-use intensities in the provinces of Cambodia

Cambodia, with its low-lying central plain of the Tonle Sap and low mountains and uplands, had only a small area of newly burnt shifting cultivation plots, even at the very low and low occurrence frequencies (long fallow periods). Pursat province, in the western part of the country, including the mountains of the Cardamom range, had the largest area of newly opened plots, covering nearly 70,000 ha (Figure 6-7(a)), as well as the second-largest swidden-area to total-area ratio, of 6.31% (Figure 6-7(b)). Four other provinces, Kampot, Preah Sihanouk, Battambang, and Kampong Speu, had very small swidden areas of 10,000 ha to 30,000 ha, although the area of swidden detected in Preah Sihanouk equalled 7.61% of its total area – the country's largest such percentage because of the province's small land area. It should be pointed out that both Mondulkiri and Ratanakiri provinces in the northeastern part of the country showed only very small areas of swidden agriculture at the very low level of swiddening intensity. This result contradicts the commonly held view that the two provinces are a principal region of swidden agriculture (Schmidt-Vogt et al., 2009). With regard to other higher levels of swiddening intensity, only Mondulkiri province showed a very small area, which decreased over the study years from 2660 ha to 610 ha of newly burnt plots (Figures 6-9(a), 6-10(a), 6-11(a)). Overall, 96.28% of all newly burnt plots in the country were at the lowest level of swiddening intensity. This indicated a very low level of swidden agriculture in Cambodia, because shifting cultivation at low, moderate, high and very high levels of intensity was seldom detected.

Changes in the area of newly burnt plots with short-, medium-, and long-term fallow periods during 1988 to 2016

In this study, the fallow period in swidden agriculture refers to the interval of years between any two consecutive occurrences of newly burnt plots in situ, detected by Landsat sensors at the pixel level. Therefore, the length of fallow periods is negatively related to the land-use intensity of swidden agriculture. A short-fallow system usually has high swiddening intensity, while medium-to-long fallow systems have low swiddening intensity. However, the relationship between the length of fallow periods and the intensity of swiddening may become much more complex and dynamic in the case of fallow periods that are deliberately manipulated by the farmers. For example, higher-intensity swidden systems may occur synchronously with short-, medium- and long-term fallow systems occurring over a long duration of perhaps 20 to 29 years.

The fallow period normally follows cultivation, or cropping. It is a time when the soil is 'rested' and the forest regenerates; plant nutrients are restored and weeds are suppressed. The effectiveness of the fallow depends largely on how long it endures before the plot is cleared and burned for the next period of cultivation, so the length of the fallow is fundamentally important. The shorter the fallow, the less capable is the soil of supporting a new crop. At the end of the fallow phase, a plot is likely to be cleared and burnt for new cultivation. At this point, the Landsat sensors captured the newly burnt plot and the start of a new swidden cycle.

There is no fixed classification system for the length of fallows. There are all sorts of shifting cultivation systems classified as having short-, medium- or long-term fallows, based on previously published studies in tropical mountainous regions. According to our previous field investigations and commonly applied methods of categorizing the length of fallow periods, there are three such categories in montane mainland Southeast Asia (Mertz et al., 2009; Ziegler et al., 2012). These are short-term or 'young' fallows with a duration of about five years or less (Sarmiento, 2000; Metzger, 2002, 2003), medium-term fallows ranging in duration from five to 10 years (Pelzer, 1945; Nyerges, 1989), and long-term fallows or long bush-fallow systems lasting more than 10 years (Pelzer, 1945; Arifin and Hudoyo, 1998; Metzger, 2002, 2003; Fox et al., 2014).

During the study period from 1988 to 2016, the frequency at which newly burnt plots were detected related directly to the intensity with which swiddening land was being used. The frequency with which swiddens were being burnt varied from once (very low intensity) to 29 times (very high intensity) over the study period. To understand the changes in the length of different fallow periods, we first selected newly burnt plots with time spans ranging from three to 12 years. Within each of these time spans, we identified those pixels that revealed a given occurrence frequency of new burning. Targetting these pixels, we

Eucalyptus deglupta Blume
[Myrtaceae]

Another industrial timber crop, these trees can grow to 75 m, with massive buttressed trunks up to two and a half metres in diameter. Known as rainbow gums, they have smooth, orange-tinted bark that sheds in strips, revealing streaks of pale green, red, orange, grey, and purplish brown. Plantations are harvested mainly for pulpwood, to make white paper, although they are also popular ornamental trees.

calculated the land area they represented by multiplying their number by the constant ground-area value of a 30m pixel – about 0.09 hectare. We then examined and compared the relationship between the given occurrence frequency of new burning and the area represented by the pixels that detected it, so as to reveal spatial changes in the length of fallow periods. More specifically, time intervals of three, four and five years were used to delineate variations in short-term fallows (for example, from 1988 to 1991, 1992 and 1993, or from 1990 to 1993, 1994 and 1995). Periods of seven, eight and nine years were used for the counterpart analysis of medium-term fallows (e.g. from 1988 to 1995, 1996 and 1997, or 2000 to 2007, 2008 and 2009). Long-term fallows were compared at intervals of 10, 11 and 12 years (e.g. 1988 to 1998, 1999 and 2000, or from 2000 to 2010, 2011 and 2012).

On the whole, the area of newly burnt plots with smaller frequencies of occurrence, i.e. longer fallow periods, was greater than those with higher frequencies of occurrence. In considering these results, it should be remembered that the area of newly burnt swiddens was cumulative – every time a swidden was reburnt, its area was added to the total, whereas the total land area subject to recurrent burning remained static. In the case of the shorter time intervals, this may appear at first impressions to give extra weight to the short-fallow systems. Thus, the cumulative area of plots burnt again after three years and four years added up to more than 50% of the total land area subject to recurrent burning. At the time intervals of three and four years, swiddens that were burnt again after short fallow periods had a cumulative area equal to 63.71% of the total land area. Swiddens with medium-term fallows added up to 55.16% of the total land area, and long-fallow systems equalled 51.68%. The equivalent area percentages of short-, medium- and long-term fallows increased to 76.46%, 65.10% and 61.25%, respectively, when the first three occurrence frequencies (e.g. three, four and five years) were all taken into consideration. Thus, even at the shorter time intervals with greater accumulation of short-fallow swiddening, long-fallow systems maintained their percentage. In other words, the land-use intensity of newly burnt swidden plots was generally low to moderate, although between 23.54% and 38.75% of newly burnt plots had frequencies of occurrence larger than three, and greater swiddening intensities. Swiddens with short-fallow periods covered about 2.07- to 3.94-million ha; those with medium-term fallows, 5.17- to 6.12-million ha; and those with long fallows covered 6.53- to 7.24-million ha.

Taking the shortest time interval of three years in the study of short-term fallows, the average areas of swiddens that were burnt once, twice, thrice and four times in that period were approximately 5.5-million ha, 2.4-million ha, 1.73-million ha and 2.07-million ha, respectively. When the time interval – which we called the 'step length' – was increased to four years, the average areas of swiddens burnt from one to five times became approximately 5.69-million ha, 2.42-million ha, 1.58-million ha, 1.38-million ha, and 1.76-million ha, respectively. Similarly, when the step length was increased to five years, the average areas of swiddens burnt from one to five times were about 5.87-million ha, 2.49-million ha, 1.54-million ha, 1.22-million ha and 1.17-million ha, respectively.

Studies of the medium-term fallow period, beginning with a step length of seven years, showed that the average areas of swiddens burnt once, twice, thrice, four and five times over that seven years were approximately 6.13-million ha, 2.65-million ha, 1.58-million ha, 1.15-million ha and 950,000 ha, respectively. When the step length was eight years, the average areas of swiddens burnt from one to five times became approximately 6.20-million ha, 2.73-million ha, 1.61-million ha, 1.15-million ha and 920,000 ha, respectively. With a step length of nine years, the average areas of swiddens burnt from one to five times became approximately 6.24-million ha, 2.81-million ha, 1.64-million ha, 1.16-million ha and 910,000 ha, respectively.

Taking the long-term fallow period, with a step length of 10 years, the average areas of swiddens burnt once, twice, thrice, four and five times were approximately 6.25-million ha, 2.88-million ha, 1.67-million ha, 1.17-million ha and 910,000 ha, respectively. Extending the step length to 11 years, the average swidden areas that were burnt from one to five times became approximately 6.24-million ha, 2.95-million ha, 1.7-million ha, 1.18-million ha and 910,000 ha, respectively. At a step length of twelve years, swiddens that were burnt from one to five times covered an average of approximately 6.29-million ha, 3.08-million ha, 1.78-million ha, 1.22-million ha and 930,000 ha, respectively.

It should be noted that there was an increasing trend in the average area covered by swiddens that were burnt just once or twice, and were therefore at the lowest end of swiddening intensity. There was a continuous expansion of newly burnt plots in the 1990s, a slow decline in the 2000s, and a sharp reduction in the 2010s, with two key time nodes – 1992 and 2011 – for changes in the average area of swiddens at different burning frequencies. The primary reasons for these changes are not clearly understood and require further research, including field surveys and household interviews. However, to our knowledge, the frequent launch of geo-economic cooperation mechanisms since the 1990s, including the Greater Mekong Subregion (GMS) economic cooperation programme may have been a major factor in both the continuous increment of newly burnt swidden plots and plantations of industrial trees (e.g. rubber, teak and eucalypts). Geo-economic cooperation under the GMS, involving Cambodia, Laos, Myanmar, Thailand, Vietnam and neighbouring China and India, may have been a strong driver of extensive and rapid land cover and land-use changes, including the practice of swidden agriculture. The sharp reduction since 2011 may have been closely related to stricter environmental governance imposed by each of the countries of mainland Southeast Asia, particularly since they became partner members of the United Nations Collaborative Partnership on Reducing Emissions from Deforestation and Forest Degradation (UN-REDD), in about 2011.

Changes in the length of shifting cultivation fallow periods from 1988 to 2016

So far, our analyses of swiddening in the period from 1988 to 2016 have revealed the overall land-use intensity of shifting cultivation in the mountainous regions of mainland Southeast Asia. What remains to be examined are changes in the time intervals between the first occurrence of burning and following re-occurrences of burning in the same swidden plots. For example, new occurrences of burning may have been detected five times on particular plots between 1988 and 2016 – the first in 1988, the second in 1990 (or 1992), the third in 2000 (or some other year after 1992), the fourth in 2009, and the final in 2016. But while we may know that a plot has been reclaimed for new cultivation five times in this period, we have yet to explore changes in the time between the first and subsequent burnings. Were the times between burnings equally separated, i.e. every five to six years, or were they longer, say, before 2000, and shorter after 2000, or just the opposite? And were there other change patterns over the three-decade study period?

Considering the long duration of 28 years between 1988 and 2016, we added a longer-term fallow type to help reveal changes in the length of fallows over this period. We therefore separated fallow periods into four categories: short-term

(1 to 5 years); medium-term (6 to 10 years); long-term (11 to 20 years) and longer-term (21 to 28 years). After calculating the area of newly burnt plots between the first occurrence (beginning with 1988) and subsequent burnings over the study period, we were able to separate newly burnt swidden plots into the four types. We further compared changes in the length of fallows with the average area of newly burnt plots in each fallow type. To keep the data comparable, the calculation of fallow durations all began from 1988 and ended in 1996, 2006, 2011 and 2016 for the four fallow lengths. Figure 6-12 shows changes in the fallow periods affecting plots of average area within the short-, medium-, long- and longer-term fallow categories, respectively.

With regard to short-term fallows, in Figure 6-12(a) each column refers to an average area size over five consecutive years, for example, 1988 to 1993, 1989 to

Tectona grandis L.f. [Lamiaceae]

Teak produces one of the most important timbers in the world. Hence, this deciduous species is increasingly grown as a plantation timber crop. It can grow to 13 m tall, with a straight bole 10 cm in diameter, in just five years, but can eventually reach 45 m and live for more than 80 years. It is a pioneer species, but unlike others, is able to persist, dominate and naturally regenerate towards the climax phase of forest succession.

1994, 1990 to 1995, and 2011 to 2016. The highest and lowest five-year-average area sizes were approximately 11.41-million ha in the period from 2004 to 2009 and about 2.75-million ha from 1988-1993. When the five-year average areas were then averaged to create decadal averages, the figures from the 1990s and the 2000s were approximately 8.63-million ha and about 9.26-million ha, respectively. Comparative approximate figures from the late 1980s and the early 2010s were a meagre 5.35-million ha and 6.56-million ha, respectively. It can be concluded that short-term fallows were the dominant form of swidden agriculture. The average areas increased continuously from the late 1980s to the 2000s, then began to decline in the early 2010s.

With regard to medium-term fallows, each column in Figure 6-12(b) refers to an average area size in a maximum of 10 consecutive years after 1988, by which time swiddens first burnt in 1988 could be identified as belonging in this category. Examples of periods lasting more than five years are 1988 to 1998, 1989 to 1999, 1990 to 2000, and 2006 to 2016. The highest and lowest five-year average sizes were

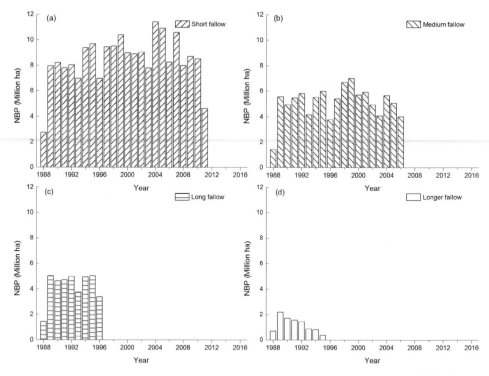

FIGURE 6-12: Temporal changes and average areas covered by newly burnt swidden plots in short-, medium-, long- and longer-term fallow categories during 1988 and 2016.

Note: Each column represents an average area covered by swiddens over consecutive years within the fallow length of each category, beginning with the year shown on the horizontal axis.

approximately 6.98-million ha in 1999 to 2009, with a first occurrence of burning in 1999, and approximately 1.4-million ha in the period from 1988 to 1998, for a first occurrence in 1988. The quasi-decadal averages of five-year average areas in the late 1980s, the 1990s and the early 2000s were approximately 3.48-million ha, 5.46-million ha and 5.04-million ha, respectively.

With respect to long-term fallows, each column in Figure 6-12(c) refers to an average area size over a maximum of 20 consecutive years after 1988, for example, 1988 to 2008, 1989 to 2009, 1990 to 2010 and 1996 to 2016. The highest and lowest average area sizes were approximately 5.03-million ha in the period from 1995 to 2015, for the first occurrence of burning in 1995, and approximately 1.43-million ha during 1988 to 2008, for the first occurrence in 1988. Quasi-decadal averages in the late 1980s and the early 1990s were approximately 3.23-million ha and 4.48-million ha, respectively. Finally, in terms of the longer-term fallows, each column in Figure 6-12(d) refers to an average area in a period of more than 20 consecutive years from 1988 to 2016. The highest and lowest five-year average areas were approximately 2.19-million ha in the period from 1989 to 2016, from a first occurrence in 1989; and about 380,000 ha from 1995 to 2016, from a first occurrence in 1995. The quasi-decadal averages of average areas over more than 20 years in the late 1980s and

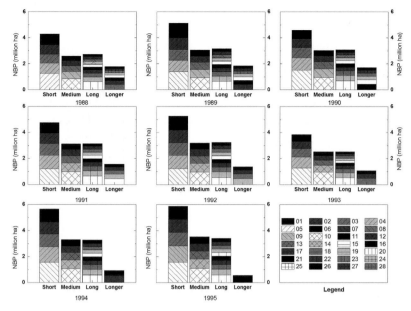

FIGURE 6-13:   Changes in the area of newly burnt plots in the short-, medium-, long- and longer-term swidden-cycle categories in periods starting with every year from 1988 until 1995), and ending in 2016.

Note: Legend shows years of swidden cycle.

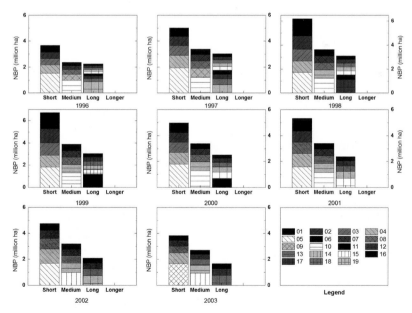

FIGURE 6-14:   Changes in the area of newly burnt plots in the short-, medium-, long- and longer-term swidden-cycle categories in periods starting with every year from 1996 to 2003, and ending in 2016.

Note: Legend shows years of swidden cycle.

the early 1990s were approximately 1.44-million ha and 1.13-million ha, respectively. It can be concluded that the longer-term fallows occurred or re-occurred more frequently in the 1980s than in the 1990s, which is quite different from the short-, medium- and long-term fallows, although the longer-term fallows were the least obvious type of swidden agriculture in mainland Southeast Asia in recent decades.

Further, we calculated changes in area sizes between the first detection of new burning and consecutive re-occurrences of burning at the pixel level, starting in 1988 and on every following year until 2011, and finishing in 2016 (Figures 6-13, 6-14 and 6-15). This was able to accurately detect the length of swidden cycles, i.e. both the cultivation and fallow phases, between recurrent burnings of the plot. However, it generally follows that a long swidden cycle involves a long fallow phase.

Conclusions

The length of fallow periods in shifting cultivation has been greatly shortened by a combination of factors, including population growth, alternative land uses and infrastructure development across the mountainous regions of the tropics in the years from 1988 to 2016. Knowledge of the dynamic processes of natural vegetation regeneration under swiddening – including management of the fallow phase – is critical for much ongoing research, such as that involving nutrient cycling, fertility

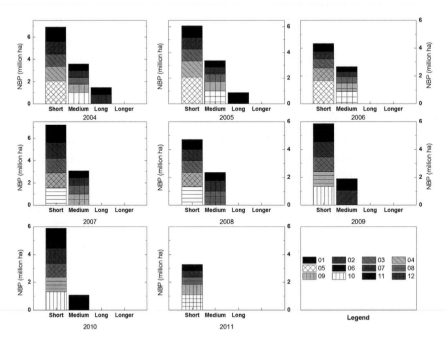

FIGURE 6-15: Changes in the area of newly burnt plots in the short-, medium-, long- and longer-term swidden-cycle categories in periods starting with every year from 2004 to 2011, and ending in 2016.

Note: Legend shows years of swidden cycle.

restoration and carbon sequestration. However, apart from some sporadic and scattered studies at a local scale, longitudinal and large-scale analyses of swidden-fallow dynamics in the tropical uplands are seldom reported, let alone the regeneration processes of natural vegetation. For ages, montane mainland Southeast Asia has been the home of traditional swiddening and fallow management, yet the views of its necessity, sustainability and future development remain controversial and distant (Li et al., 2014; Pham et al., 2020). Analogously with the development of this region, swidden agriculture has undergone a rapid and dramatic evolution, or transformation, with sharp reductions in fallow length and increased intensity of land use. In this study, maps of annually detected newly burnt swidden plots were generated from dense Landsat Thematic Mapper (TM), Enhanced Thematic Mapper Plus (ETM+) and Operational Land Imager (OLI) images acquired during the peak of dry seasons (March, April and May) in each year from 1988 to 2016. The maps were applied to analyze swidden intensity and the length of fallow periods in swidden agriculture over nearly 30 years of shifting cultivation practice in montane mainland Southeast Asia.

We drew the following conclusions: First, the area covered by newly burnt plots with a lesser frequency of occurrence (less than five recurrent burnings), was typically greater than one million ha in montane mainland Southeast Asia. Second, the area covered by newly-burned plots at these lesser occurrence frequencies was much greater than that of plots that were burnt at larger frequencies of occurrence, i.e. those with shorter fallow periods. Finally, the dominant form of shifting cultivation was that using short-term fallows. The corresponding areas covered by shifting cultivation plots in short-, medium- and long-term fallow systems increased continuously from the late 1980s to the 2000s, and began to decline in the early 2010s.

Acknowledgements

This research work was partially supported by the National Natural Science Foundation of China (nos. 41971242 and 41301090), Youth Innovation Promotion Association of the Chinese Academy of Sciences (no. CAS2020055) and Program for BINGWEI Excellent Young Talents of Institute of Geographic Sciences and Natural Resources Research, Chinese Academy of Sciences (no. 2018RC201). We would like to thank Dr. Malcolm Cairns and Bob Hill for their helpful comments and suggestions on this chapter.

References

Achard, F., Eva, H. D., Stibig, H. J., Mayaux, P., Gallego, J., Richards, T. and Malingreau,J-P. (2002) 'Determination of deforestation rates of the world's humid tropical forests', *Science* 297(5583), pp.999-1002

Arifin, B. and Hudoyo, A. (1998) 'An economic analysis of shifting cultivation and bush-fallow in lowland Sumatra', research report submitted to Alternatives to Slash-and-Burn Indonesia Consortium, ICRAF Southeast Asia, Bogor, Indonesia

Boateng, I. (2012) GIS assessment of coastal vulnerability to climate change and coastal adaption planning in Vietnam. *Journal of Coastal Conservation* 16(1), pp.25-36

Brady, N. C. (1996) 'Alternatives to slash-and-burn: A global imperative', *Agriculture, Ecosystems and Environment* 58(1), pp.3-11

Bruun, T. B., de Neergaard, A., Lawrence, D. and Ziegler, A. D. (2009) 'Environmental consequences of the demise in swidden cultivation in Southeast Asia: Carbon storage and soil quality', *Human Ecology* 37(3), pp.375-388

Butler, J. H. (1980) *Economic Geography: Spatial and Environmental Aspects of Economic Activity*, John Wiley & Sons, New York

Cairns, M. F. (ed.) (2015) *Shifting Cultivation and Environmental Change: Indigenous People, Agriculture and Forest Conservation*, Earthscan from Routledge, London

Chuan, G. K. (2005) 'The climate of Southeast Asia', in A. Gupta (ed.) *The Physical Geography of Southeast Asia*, Oxford University Press, Oxford, UK, pp.80-93

Corlett, R. T. (2005) 'Vegetation', in A. Gupta (ed.) *The Physical Geography of Southeast Asia*, Oxford University Press, Oxford, UK, pp.105-119

Dudal, R. (2005) 'Soil of Southeast Asia', in A. Gupta (ed.) *The Physical Geography of Southeast Asia*. Oxford University Press, Oxford, UK, pp.94-104

Fox, J. and Vogler, J. B. (2005) 'Land use and land-cover change in montane mainland Southeast Asia', *Environmental Management* 36(3), pp.394-403

Fox, J., Castella, J-C. and Ziegler, A. D. (2014) 'Swidden, rubber and carbon: Can REDD+ work for people and the environment in montane mainland Southeast Asia?', *Global Environmental Change* 29, pp.318-326

Goldammer, J. G. (1988) 'Rural land use and wildland fires in the tropics', *Agroforestry Systems* 6(3), pp.235-252

Gupta, A. (2005a) 'Accelerated erosion and sedimentation in Southeast Asia', in A. Gupta (ed.) *The Physical Geography of Southeast Asia*, Oxford University Press, Oxford, UK, pp.239-249

Gupta, A. (2005b) 'Landforms of Southeast Asia', in A. Gupta (ed.) *The Physical Geography of Southeast Asia, Oxford University Press*, Oxford, UK, pp.38-64

Heinimann, A., Mertz, O., Frolking, S., Christensen, A. E., Hurni, K., Sedano, F., Chini, L. P., Sahajpal, R., Hansen, M. and Hurtt, G. (2017) 'A global view of shifting cultivation: Recent, current, and future extent', *PLoS ONE* 12(9), e184479, https://doi.org/10.1371/journal.pone.0184479, accessed 13 October 2020

Hett, C., Castella, J-C., Heinimann, A., Messerli, P. and Pfund, J-L. (2012) 'A landscape mosaics approach for characterizing swidden systems from a REDD+ perspective', *Applied Geography* 32(2), pp.608-618

Hurni, K., Hett, C., Epprecht, M., Messerli, P. and Heinimann, A. (2013a) 'A texture-based land cover classification for the delineation of a shifting cultivation landscape in the Lao PDR using landscape metrics', *Remote Sensing* 5(7), pp.3377-3396

Hurni, K., Hett, C., Heinimann, A., Messerli, P. and Wiesmann, U. (2013b) 'Dynamics of shifting cultivation landscapes in Northern Lao PDR between 2000 and 2009, based on an analysis of MODIS time series and Landsat images', *Human Ecology* 41(1), pp.21-36

Jenks, G. F. and Coulson, M. R. (1963) 'Class intervals for statistical maps', *International Yearbook of Cartography* 3, pp.119-134

Li, P. and Feng, Z. M. (2016) 'Extent and area of swidden in montane mainland Southeast Asia: Estimation by multi-step thresholds with Landsat-8 OLI data', *Remote Sensing* 8(1), 44. https://doi.org/10.3390/rs8010044, accessed 14 October 2020

Li, P., Feng, Z. M, Jiang, L. G., Liao, C. and Zhang, J. (2014) 'A review of swidden agriculture in Southeast Asia', *Remote Sensing* 6(2), pp.1654-1683

Li, P., Feng, Z., Xiao, C., Boudmyxay, K. and Liu, Y. (2018a) 'Detecting and mapping annual newly-burnt plots (NBP) of swiddening using historical Landsat data in montane mainland Southeast Asia (MMSEA) during 1988-2016', *Journal of Geographical Sciences* 28(9), pp.1307-1328

Li, P., Feng, Z. and Xiao, C. (2018b) 'Acquisition probability differences in cloud coverage of the available Landsat observations over mainland Southeast Asia from 1986 to 2015', *International Journal of Digital Earth* 11(5), pp.437-450

Li, P., Xiao, C., Feng, Z., Li, W. and Zhang, X. (2020) 'Occurrence frequencies and regional variations in VIIRS global active fires', *Global Change Biology* 26(5), pp.2970-2987

Mertz, O., Padoch, C., Fox, J., Cramb, R., Leisz, S. J., Nguyen, T. L. and Tran, D.V. (2009) 'Swidden change in Southeast Asia: Understanding causes and consequences', *Human Ecology* 37(3), pp.259-264

Messerli, P., Heinimann, A. and Epprecht, M. (2009) 'Finding homogeneity in heterogeneity: A new approach to quantifying landscape mosaics developed for the Lao PDR, *Human Ecology* 37(3), pp.291-304

Metzger, J. P. (2002) Landscape dynamics and equilibrium in areas of slash-and-burn agriculture with short and long fallow periods (Bragantina region, NE Brazilian Amazon)', *Landscape Ecology* 17(5), pp.419-431

Metzger, J. P. (2003) 'Effects of slash-and-burn fallow periods on landscape structure', *Environmental Conservation* 30(4), pp.325-333

Nyerges, A. E. (1989) 'Coppice swidden fallows in tropical deciduous forest: Biological, technological, and sociocultural determinants of secondary forest successions, *Human Ecology* 17(4), pp.379-400

Pham, T. T., Moeliono, M., Brockhaus, M., Wong, G. Y. and Le, N. D. (2020) 'The politics of swidden: A case study from Nghe An and Son La in Vietnam', *Land Use Policy 99,* 103050, https://doi.org/10.1016/j.landusepol.2017.10.057, accessed 24 September 2021

Pelzer, K. J. (1945) *Pioneer Settlement in the Asiatic Tropics: Studies in Land Utilisation and Agricultural Colonization in Southeastern Asia,* American Geographical Society, New York

Rerkasem, K., Lawrence, D., Padoch, C., Schmidt-Vogt, D., Ziegler, A. D. and Bruun, T. B. (2009a) 'Consequences of swidden transitions for crop and fallow biodiversity in Southeast Asia'. *Human Ecology* 37(3), pp.347-360

Rerkasem, K., Yimyam, N. and Rerkasem, B. (2009b) 'Land use transformation in the mountainous mainland Southeast Asia region and the role of indigenous knowledge and skills in forest management', *Forest Ecology and Management* 257(10), pp.2035-2043

Sarmiento, L. (2000) 'Water balance and soil loss under long fallow agriculture in the Venezuelan Andes', *Mountain Research and Development* 20(3), pp.246-253

Schmidt-Vogt, D., Leisz, S. J., Mertz, O., Heinimann, A., Thiha, T., Messerli, P., Epprecht, M., Cu, P.V., Vu, K. C., Hardiono, M. and Dao, T. M. (2009) 'An assessment of trends in the extent of swidden in Southeast Asia', *Human Ecology* 37(3), pp.269-280

Silva, J. M. N., Carreiras, J. M. B., Rosa, I. and Pereira, J. M. C. (2011) 'Greenhouse gas emissions from shifting cultivation in the tropics, including uncertainty and sensitivity analysis', *Journal of Geophysical Research* 116(D20), doi:10.1029/2011JD016056

Stibig, H. J., Achard, F. and Fritz, S. (2004) 'A new forest cover map of continental southeast Asia derived from SPOT-VEGETATION satellite imagery', *Applied Vegetation Science* 7(2), pp.153-162

Stibig H. J., Belward, A. S., Roy, P. S., Rosalina-Waasrin, U., Agrawal, S., Joshi, P. K., Beuchle, R., Fritz, S., Mubareka, S. and Giri, C. (2007) 'A land-cover map for South and Southeast Asia derived from SPOT-VEGETATION data,' *Journal of Biogeography* 34(4), pp.625-637

Tucker, C. J., Justice, C. O. and Prince, S. D. (1986) 'Monitoring the grasslands of the Sahel, 1984-1985', *International Journal of Remote Sensing* 7(11), pp.1571-1581

van Vliet, N., Mertz, O., Heinimann, A., Langanke, T., Adams, C., Messerli, P., Leisz, S. J., Pascual, U., Schmook, B., Schmidt-Vogt, D., Castella, J-C., Jorgensen, L., Birch-Thomsen, T., Hett, C., Bruun, T. B., Ickowitz, A., Vu., K. C., Yasuyuki, K., Fox, J. M., Dressler, W., Padoch, C. and Ziegler, A. D. (2012) 'Trends, drivers and impacts of changes in swidden cultivation in tropical forest-agriculture frontiers: A global assessment', *Global Environmental Change* 22(2), pp.418-429

Vogelmann, J. E. and Rock, B. N. (1988) 'Assessing forest damage in high-elevation coniferous forests in Vermont and New Hampshire using Thematic Mapper data', *Remote Sensing of Environment* 24(2), pp.227-246

Ziegler, A. D., Bruun, T. B., Guardiola-Claramonte, M., Giambelluca, T. W., Lawrence, D. and Than, L. N. (2009) 'Environmental consequences of the demise in swidden cultivation in montane mainland Southeast Asia: Hydrology and geomorphology', *Human Ecology* 37(3), pp.361-373

Ziegler, A. D., Phelps, J., Yuen, J. Q., Webb, E. L., Lawrence, D., Fox, J. M., Bruun, T. B., Leisz, S. J., Ryan, C. M., Dressler, W., Mertz, O., Pascual, U., Padoch, C. and Koh, L. P. (2012) 'Carbon outcomes of major land-cover transitions in SE Asia: Great uncertainties and REDD+ policy implications', *Global Change Biology* 18(10), pp.3087-3099

7

STRIPPING THE STRIPS

The failure of soil- and water-conservation technologies in northern Thailand and eastern Indonesia

Thomas Enters and Justin Lee[*]

> We believe that development practice cannot be critically examined independently of the social and political environment in which it occurs. Yet formal and technical project appraisal techniques are frequently taught and written about in textbooks in a disembodied way which gives rise to misleading conclusions about their appropriateness and the risks associated with their use. The debate about issues in development practice: sustainability, participation, land degradation are in a similar manner commonly treated in a rhetorical fashion, or informed only by highly generalized information gleaned from overviews at a national level.
>
> (Porter et al., 1991, pxvi)

Introduction

Sloping agricultural land technology (SALT) and its many manifestations, notably alley cropping, arrived on the rural-development scene in the 1980s with the promise of improving soil fertility and agricultural productivity in upland farming systems. Thirty years later, SALT technology is still being put forward as a sustainable rural-development solution, especially for smallholders in the developing world (e.g.

---

[*] Dr. Thomas Enters conducted field studies in Thailand as a postgraduate student at the Australian National University in Canberra, Australia. He worked formerly for the Center for International Forestry Research (CIFOR), the Food and Agriculture Organization of the United Nations (FAO), the Center for People and Forests (RECOFTC) and the UN Environment Program (UNEP). He is now retired. Dr Justin Lee conducted field studies on Sumba island in Indonesia as a postgraduate student at the University of Adelaide in South Australia. He joined Australia's Department of Foreign Affairs and Trade in 1995, was posted to diplomatic missions in Indonesia and Papua New Guinea and served as Australia's High Commissioner to Bangladesh before becoming Australia's High Commissioner for Climate Change between 2012 and 2014. He is currently Australia's High Commissioner to Malaysia.

Ogunlana et al., 2010; Wilson and Lovell, 2016). Surprisingly, despite the time since SALT's initial development and thousands of publications on SALT technologies (Wolz and DeLucia, 2018), one of the persistent reasons given for its slow and low adoption rates, according to the authors above, is that smallholders are not aware of its benefits and they lack experience in managing trees. For those who may be seeking to promote SALT technologies to solve the problems of the 21st century, the following are lessons learned from two studies into SALT adoption conducted in rural Thailand and Indonesia in the early 1990s (Figure 7-1). It is a story that confirms the view of Hoare (2017), who, as a veteran of more than 20 years' work on international development projects, wrote:

> For all the successes that have been achieved by international development projects ... there has been an astounding waste of money and effort on projects doomed to failure because they told the 'beneficiaries' what they were getting and how they should adapt, rather than starting from their perceived agricultural and natural-resource problems and their ideas on how to solve their problems.

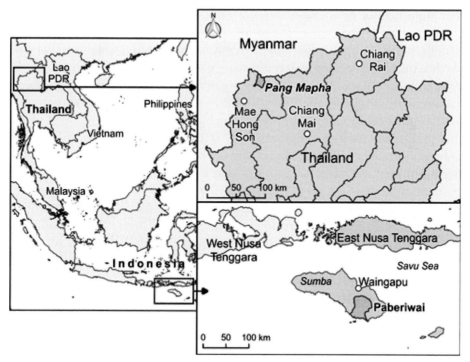

FIGURE 7-1: Locations of the study sites in Mae Hong Son province, northern Thailand and Paberiwai district, East Sumba regency, East Nusa Tengarra province, Indonesia.

Northern Thailand[1]

In June 1995, a group of about 150 soil-conservation specialists from 25 countries took part in a two-day field trip in northern Thailand, organized by the Soil and Water Conservation Society of Thailand. On previous days, they had debated issues related to soil-conservation extension in a meeting room in Chiang Mai, far from agricultural fields and farmers. The field trip was to discuss *in situ* challenges and opportunities and to pose questions to farmers participating in various soil- and water-conservation projects on sloping lands. The trip was to open the eyes of participants (including those of your first author) as to what was happening on the ground.

The specialists were looking forward to the trip to the countryside. Their questions and discussions at the project sites told of their enthusiasm about what various projects had achieved and how they were personally contributing to soil conservation, the fight against soil erosion and land degradation, and improving the livelihoods of millions of poor farmers. Each stop confirmed project successes and the participants clearly enjoyed the landscape and the upland environment. However, few recognized that in between the project areas and sites, there were neither contour strips nor alley cropping; there were no soil- and water-conservation technologies to be observed. They were blinded by localized successes and were thus unable to see the bigger picture of why farmers were not adopting innovations designed for their benefit.

In 1978, the authors and editors Peter Kunstadter, E. C. Chapman and Sanga Sabhasri called the people practising marginal agriculture in northern Thailand *Farmers in the Forest*. Much has been written about these ethnic minorities. In Thailand, they are usually referred to as 'hilltribes', although they call themselves *chao khao*, meaning hill or mountain people or highlanders. They are also found in many other parts of mountainous Southeast Asia. While some migrated to the mountains in recent years, usually fleeing conflict and persecution, many have lived there for as long as 200 years. As James C. Scott tells us in his 2009 *Anarchist History of Upland Southeast Asia*, they are people practising *The Art of Not Being Governed*. By that he means that they are people who have fled the policies and upheavals of the organized state societies that surround them, including slavery, conscription, taxes, corvée labour, epidemics and warfare. This may seem irrelevant to the adoption of soil-conservation technologies and reducing deforestation, but as we will show, it is not. It is a disposition that has contributed to other project failures, not only to those relating to soil and water conservation and natural-resource interventions.

Until the mid-1950s, there was little government concern in Thailand for the highlands and their inhabitants (Hoare, 2017). The first contacts between officials and hilltribe communities started slowly in the more accessible areas, through social welfare programmes. While concern for border security was the basis for Thai government interventions during the 1970s, 20 years later it was the poor economic conditions of most communities, opium production, narcotics trafficking, soil erosion and deforestation. Hence, starting in the mid-1980s, donor-assisted projects focused for about 20 years on national watershed protection and resource conservation, opium eradication and the attainment of basic human needs. Originally, the principal agencies involved in highland development were the Office of the Narcotics Control Board and the Department of Public Welfare. However, a Cabinet resolution on watershed classification in 1985 effectively increased the power of the Royal Forest Department (RFD), which sought to strengthen the State's rights to most of the highlands.

For decades, if not centuries, the hilltribes employed several strategies to avoid state-based 'civilization'. As well as dispersing into the highlands, they adopted cropping practices that enhanced mobility, i.e. they practised shifting cultivation (also called swiddening or, pejoratively, slash-and-burn). They were usually described as subsistence farmers, although most of them had, for many decades, enjoyed extensive trading relationships with lowland merchants – and not only for opium. Besides rice (Figure 7-2), maize and opium, by the mid-1980s many communities were also growing cash crops such as sesame or red kidney beans, and even green beans, baby carrots, cabbages, coffee and many other crops. To produce upland rice, the shifting cultivation fields were used for one or two years and then left fallow or used for other crops such as maize. New fields were cut out of primary or secondary forests for further rice production, which the lowland bureaucracies, especially the RFD, regarded as an unacceptable practice as it led to the slow destruction of the nation's northern watershed forests.

Shifting cultivation is a perfect agricultural system in situations of low population density, which allow for long fallow periods to rejuvenate the land and restore plant nutrients. However, it is commonly accepted that higher population densities lead to shorter fallow periods, which in turn lead to intensified weed growth and lower agricultural yields. On steep slopes this was exacerbated by soil erosion (Figure 7-3), especially the loss of fertile topsoil. Hence, in tropical and subtropical countries, shifting cultivation in mountainous environments has been identified as a major cause of productivity losses as well as negative external consequences that are believed to threaten both the livelihoods of lowland communities and national economies. In 1984, Brown and Wolf, for the Worldwatch Institute, called soil erosion 'the quiet crisis in the world economy', and many soil scientists, specialists for donor-funded projects and government departments agreed with that assessment. Sloping Agricultural Land Technology (SALT) was proposed as a solution to the

FIGURE 7-2: Typical upland rice swiddens in northern Mae Hong Son province, northern Thailand.

Photo: Thomas Enters

problem, by keeping soil in place on sloping lands through the establishment of contour strips of grass or other plants and growing crops in the alleys between the strips (Figure 7-4). As the Worldwatch Institute's Brown and Wolf (1984) and many other publications hailed the virtues of SALT technology, policy-makers felt they had a solution to the predicament of the 'hilltribes'. Alas, the 'Farmers in the Forest' knew little about the furore – and nothing about SALT.

Explaining the failure

Extending SALT to shifting cultivators in northern Thailand turned out to be a failure for a variety of reasons. The first was of a technical nature: shifting cultivators shift to new plots because productivity declines. However, productivity declines are not related solely to reductions in soil fertility induced by soil erosion. As early as 1960, Nye and Greenland had informed us that continuous cropping, i.e. not shifting, had the following consequences:

* multiplication of pests and diseases;
* an increase in weeds;
* deterioration in the physical condition of the soil;
* erosion of the topsoil;
* deterioration in the nutrient status of the soil; and
* changes in the number and composition of soil fauna and flora.

FIGURE 7-3: Serious and easily visible soil erosion on a field in northern Thailand, early 1990s.

Photo: Thomas Enters

FIGURE 7-4: Contour strips in a field in northern Mae Hong Son province, northern Thailand, in the early 1990s.

Photo: Thomas Enters

Since then, many others have pointed out that declining productivity should not always be ascribed to nutrient depletion due to soil erosion. In 1994, Ong of the World Agroforestry Centre (ICRAF) called alley cropping an 'Ecological Pie in the Sky', although he was careful and posed the issue as a question. In 1991, the late Sam Fujisaka (of the International Rice Research Institute) identified 13 reasons why farmers failed to adopt innovations intended to improve the sustainability of upland agriculture. It was the same year, during his field work, that the first author met Fujisaka in Chiang Mai and was inspired by his work. Thirteen reasons appeared to be too many, so in a 1994 publication, Fujisaka focused on only six reasons:

- farmers have not encountered the problem the innovation sets out to solve;
- farmer practices are equal to or better than the innovation;
- the innovation does not work;
- extension of the new technology fails to inform farmers;
- the innovation costs too much; and/or
- 'social factors' hinder adoption of the innovation.

From a purely technical perspective, the SALT technologies extended in northern Thailand simply did not work. They did not address several of the aspects identified by Nye and Greenland in 1960. Between the contour strips, soil erosion was still happening, which meant that the most fertile soil ended up in the strips, where in many cases, no crops were grown. In addition, some of the grasses used for the strips turned out to be invasive, and spread across entire fields, reducing productivity or making crop cultivation impossible. Other crops such as *Leucaena leucocephala* and pigeon pea (*Cajanus cajan*) attracted livestock, and the farmers certainly did not want animals walking through their fields. The first author's research showed that farmers adopted SALT (at least initially), because they received positive incentives, including seeds, fertilizers and payments. However, over time the incentives were replaced by threats, including warnings of resettlement and confiscation of Thai identity (ID) cards. The farmers' initial behaviour of welcoming SALT did not last. They soon began to pretend adoption of SALT to please government officials and extension workers by establishing 'token strips' in exposed areas (Figure 7-5). As a result, challenging issues related to the adoption of SALT were neither properly discussed nor addressed.

FIGURE 7-5: A 'token strip' in an exposed field in northern Thailand, intended to feign adoption of SALT technologies to please government officials.

Photo: Thomas Enters

In addition to these issues, SALT made life more difficult for farmers. More labour had to be used for weeding. Fields were usually worked up and down the slope, but with SALT, farmers had to walk around the contour strips. Harvesting *L. leucocephala* and pigeon pea as livestock feed also required additional labour, which was often in short supply, particularly during the rainy season. What might have worked on research plots did not work in a real situation because labourers were paid to do all the work on the numerous research plots in northern Thailand in the 1980s and 1990s. Hence, in the north of Mae Hong Son province, what appeared to be a promising innovation in 1990 looked like a failure only four years later. Meanwhile, deforestation had continued because the farmers were still cutting primary forests to make new fields.

To refer again to Scott's (2009) work *The Art of Not Being Governed*, in the early 1990s hilltribe people were still placing distance between their communities and state appropriation. This is probably no longer the case. In 1990, four of the five case-study villages in Pang Mapha subdistrict, Mae Hong Son province, were no older than six years and one was 19 years old (Figure 7-6).

It is questionable whether these people (they were all Lahu) were ready to settle and invest in the long term. In late 2010, the villages were still there and more of the households had pickup trucks and satellite dishes, something that was beyond imagination in the early 1990s, when the people were threatened with resettlement and the revocation of their Thai ID cards. Hence, the timing of interventions depends on the perceptions that people have of their future. People on the move will not invest and intimidating people is not a recipe for success. Instead, it invites failure.

Finally, we need to ask to what extent these so-called subsistence farmers viewed themselves as farmers in perpetuity. This question has been addressed by Rigg (1997) – among many others – in his book *Southeast Asia: The Human Landscape of Modernization and Development*, in a chapter entitled *New Rural Worlds: More than the Soil*. Visit Thailand's mountainous north today and you will be struck by changes of all sorts. Such changes were touched upon by Preston in the case of central Java, Indonesia, as long ago as 1989. Southeast Asia is experiencing 'deagrarianization' – in other words agrarian transformation or change in its rural areas – although there has

FIGURE 7-6: Luk Khao Lam, one of the study villages in Pang Mapha subdistrict, Mae Hong Son province, in December 1990.

Photo: Thomas Enters

been a massive increase in the production of some crops, such as maize. When Preston published his work in 1989, this was not an issue in remote areas of Mae Hong Son province. But since then there have been significant changes. The tourism industry has boomed and provides employment for many young people who have moved away from the villages. Even people who have stayed behind gather at scenic viewpoints to sell their wares and ask to be paid for the right to take photographs of them in traditional costumes that were common in the early 1990s, but have since been replaced by T-shirts and jeans. This shows how versatile the people are, or perhaps how versatile they want to be. Hence, clinging to the past and not expecting farmers to leave the forest is certainly a mistake. Perhaps this was not an easy expectation in the 1980s and 1990s, although Kunstadter and his co-authors were hinting at it as early as 1978. Apparently, there was no room and no time to take them seriously, because there were timelines and targets that had to be met.

Eastern Indonesia

In the early 1990s, a similar story unfolded on the island of Sumba in Indonesia's East Nusa Tenggara province. For about a decade, an international non-governmental organization (NGO), in cooperation with a local proxy and the regional government, had been working to address land degradation and diminishing agricultural productivity in East Sumba's remote Paberiwai district (Lee, 1995).[2] The technologies being promoted in this campaign were alley cropping using tree legumes, and instigation of slow-forming terraces on sloping fields that required little labour. The alley cropping involved the use of mostly *Calliandra* species in contour strips, following the successful use of *Leucaena leucocephala* as livestock fodder in nearby West Timor (Jones, 1983) and similar experiences elsewhere in semi-arid tropical uplands.

In eight villages where the NGO had been working for more than five years, more than 60% of all farmers had commenced legume terracing. Farmer extensionists, demonstration plots and peer-to-peer learning were said to be responsible for this rapid and widespread adoption. International and domestic visitors would come to the villages for rapid assessments of legume terraces that had been established mainly in home gardens around the centre of villages. Farmers receiving support from the NGO would reaffirm to outsiders the benefits of this innovation, including reduced erosion and protection of water sources, higher crop yields, more fodder and more firewood.

The reality in Paberiwai was more complex. Of more than 100 farmers who planted tree legumes and attempted intercropping, more than half abandoned their legume gardens, usually after three to five years. While 'exhibition' alley cropping gardens were maintained in prominent locations around the centre of villages (Figure 7-7), the landscape elsewhere was littered with overgrown and neglected *Calliandra* groves (Figure 7-8). Almost all households, including members of the village elite, continued to have much larger traditional swiddens at the edges of the forest (East Sumba's largest moist forest and now the Laiwangi Wanggameti National Park). Farmers also preferred to harvest specialist firewood and construction timber from the forest rather than use tree legumes for these purposes. Livestock were left to roam freely on surrounding grasslands, and the livelihoods of the farmers remained dependent on traditional shifting cultivation and gathering of forest products.

Paberiwai, in Sumba's mountainous forested interior, had long been remote and neglected by successive administrations. The Dutch 'pacified' the island's interior as recently as the early 20th century and then left the mountain Sumbanese largely alone (Kuipers, 1998). This changed under the Suharto government in Indonesia, with its emphasis on modernization and development. As described by Dove (1985), shifting cultivation was seen by the centrist government as an inferior agricultural system and a major cause of deforestation. Much of the traditional territory of

FIGURE 7-7: Demonstration alley cropping near the centre of a village in Paberiwai, Sumba.

Photo: Justin Lee

FIGURE 7-8: 'Landscapes were littered with overgrown and neglected groves of *Calliandra calothyrsus*'.

Photo: Justin Lee

Paberiwai's villages was rezoned by the government as protection forest and water catchment and villagers were threatened with relocation. In some cases, this was forcibly carried out.

Villagers in Paberiwai were given the opportunity to remain in their home territory if they took up permanent settled agriculture along the main roads on the ridge tops, away from the fertile forest lands. The provincial government introduced agricultural-intensification programmes in the 1980s, including planting improved dryland rice varieties, and in the early 1990s it launched further programmes to raise agricultural productivity and enhance rural livelihoods. NGOs were enlisted to provide service delivery and to support these initiatives. Across the province, the number of small-scale NGOs doubled in the late 1980s and by 1991 there were more than 100 such organizations. In Paberiwai, tree-legume intercropping was seen as a way in which villagers could intensify their agricultural systems, become permanent farmers and avoid resettlement. While technical support for this programme was delivered by an NGO and its farmer extensionists, it was also actively promoted under government programmes and village elites used peer-group pressure and coercion. It was in this flawed inception of alley cropping and the approach to the programme that the seeds for its failure were sown.

Explaining the failure

In Paberiwai, the widespread adoption and then abandonment of alley cropping was foremost a failure of policy and approach. Alley cropping was introduced to address the externally perceived problem of forest and land degradation. However, villagers were more interested in marketable crops and trees, prices at local markets, improving poor transport infrastructure and reducing shortages of labour and draught power. Of course, coercion and the prospect of resettlement were also concerns, but these could be avoided by conceding the establishment of token legume gardens around village centres.

At its core, alley cropping, as presented, was an unappealing technology to farmers. It was too labour intensive, since it required preparation earlier in the season and constant pruning. Whether a real problem or not, farmers also believed that the roots of tree legumes interfered with food crops (especially cassava), and that pests hid in the contour hedgerows. Moreover, they felt that the technology brought no noticeable improvement in soil fertility. Matters were made worse by the fiction that alley cropping could increase the productivity of permanent gardens on the poor soils of ridge tops, which were especially infertile and unattractive.

The opportunity costs for farmers were also high. They preferred to devote their excess labour to profitable off-farm activities such as collecting forest products (eaglewood, cinnamon and birds) or earning food from work programmes. There were also more attractive farming options. The invasive weed *Chromolaena odorata* had spread rapidly across East Sumba in preceding decades and in many areas had replaced the rhizomatous grass invader *Imperata cylindrica* at the forest margins. *C. odorata* was a more useful plant to succeed crops in shifting cultivation, as a fallow species. It was perceived to fertilize the soil, was easy to clear and could be slashed and burned without upsetting the forest department (see McWilliam (2000) for details of farmers' relationships with *C. odorata* in nearby Timor). All of these activities earned higher returns for less work, and left more time for the highly ritualistic Sumbanese to fulfil their ceremonial commitments.

The bigger failure was the inability of the NGO and the authorities promoting these technologies to recognise these weaknesses and instead to continue devoting resources to essentially the same programmes for more than a decade. Some adaptations were made: bench terraces were replaced with slow-forming terraces and an initial reliance on *Leucaena leucocephala* was diversified to other species after it was attacked by a psyllid pest. But these changes were largely cosmetic.

The NGO model was a central problem. From the outset, it was set up to deal with an environmental issue of greatest concern to international donors and the central government, and not an issue of concern to the local people. The NGO was designed to roll out a pre-determined technology as quickly and broadly as possible, but there was little spontaneous adoption. The NGO approached farmers and offered alley cropping as a rigid 'take it or leave it' package. Horticultural and estate-crop trees, which were more desired by farmers, were only offered as a reward after tree legumes had been planted and worked for some years.

Most concerning was the inadequacy of assessment, monitoring and evaluation, to identify problems with the technology. A range of factors contributed to this. Farmer-to-farmer extension made sense in a remote location where almost 90% of the people had never been visited by a government agricultural-extension officer, but it also created perverse incentives and self-fulfilling feedback loops. Many young farmers adopted alley cropping in order to become NGO extensionists, so they could access the relatively attractive career structure it offered in an environment with few opportunities to earn wages. They also became what was essentially a paid cadre of farmer advocates for the technology, who could parrot its benefits to any external assessors.

These international and national assessors did themselves no favours; they stayed on average less than two days in the field. Well-intentioned and looking to avoid offence and gain the trust of their hosts, they were ripe for mystification. Local communities preyed upon this, experienced as they were in passive resistance and telling external observers what they wanted to hear about a range of government food-for-work and land-revegetation projects. Gatekeepers were used to keep short-term visitors focused on the exhibition gardens at the centre of villages, rather than allowing them to examine the range of practices at the forest margin that subverted these well-intentioned projects.

Failure was also due to a mistaken sense of what constituted success. Success was target-oriented on the numbers of farmers adopting alley cropping and planting tree legumes, rather than in-depth assessments of improvements in agricultural productivity, farmer incomes or farmer livelihoods. It was never assumed that there could be non-agricultural socio-political reasons for the initial adoption, and farmers' eventual abandonment of the technology was able to be conveniently ignored as the trees continued to stand, albeit unused.

Eventually, to meet its targets and to sustain both its organizational structure and the accolades of national and international assessors, the NGO maintained the programme's momentum by becoming increasingly dependent on the coercive support of local government officials. Government directives made it compulsory for farmers to alley-crop terraces on certifiable land or face relocation to Sumba's southern coastal plain, away from the mountainous interior. Local officials used various forms of coercion to follow up on these edicts. The NGO itself adopted such measures, threatening to withdraw from villages that did not adopt the technology as proposed and relied heavily on peer-group pressure. As a result, tree legumes were planted in highly visible areas to provide a facade of development and success.

What can be done better?

So what have we learnt? How can we improve – if we think that we can improve at all?

First, quite a few things have improved over the years. As Hoare (2017) pointed out with regard to adoption of a 'bottom-up' approach in northern Thailand:

> In terms of achieving successful implementation of a project and widespread adoption of its changes and innovations, involvement of the beneficiaries, using 'bottom-up' approaches, is now widely regarded as the only way to proceed. It was not always like that.

However, effective engagement of people (and we should probably not always call them 'beneficiaries') is much more than ticking boxes (something that Hoare certainly did not suggest). We need to realize that our thoughts of conserving soils and forests may clash with aspirations to have better roads, health services, education and ID cards.

Second, we should think more about what 'bottom-up' means. It is not just listening to people and involving them, but genuinely empowering people to help themselves. Regardless of how empathetic we as outsiders may be, we can never fully put ourselves in their situation. We have to strive to give them their own voice, even if that means devolving our own power and privilege, and the responsibility to identify and implement ways of meeting their own needs.

Third, we should put less emphasis on meeting physical targets and outcomes, and more on process. Genuine innovation and sustainable behavioural change takes time and requires suitable processes to generate support from all stakeholders. This is especially important in forestry projects where excluded or disenfranchised groups or individuals can impact the success of the whole project through burning or using other forms of destructive behaviour. And those motivated for the wrong reasons will not help to bring about change in the long term.

Fourth, we should probably be more open about failures, learn from mistakes and, at times, hold accountable those people who are responsible instead of trying to cover up failures, as discussed by Porter et al. (1991), with regard to a project in Kenya. There is plenty of good advice available for reducing the incidence of failure, such as the advice provided by Brandon and Wells (2009) about integrated conservation and development projects.

Fifth, any research that is not multi-disciplinary in nature should be questioned. It is incomprehensible why we have human geographers and anthropologists publishing in their journals and soil scientists and hydrologists in theirs. Results obtained in small research plots are meaningless, but they continue to be presented as conclusive. At the same time, social scientists often do not properly understand the biophysical sciences. Why do these very educated and smart people get away with reductionist approaches when we have known for decades that they contribute to failure more often than to success?

Sixth, a serious effort must be made to understand where people – in this case 'The Farmers in the Forest' – come from. They have evaded the State for 200 years. Can we bring them around with a few development projects, whatever their nature? Again, more time should be spent in formulating projects.

Finally, people should never be frozen in time. Subsistence farmers today may not be, or want to be, subsistence farmers tomorrow. To avoid failures, an effort must be made to engage with people (and nobody suggests that this is easy), to see where they see themselves in the near or more distant future. They may not want to be farmers anymore. The history of Southeast Asia (and other parts of the world) tells of dramatic change, not always for the better, even in remote corners such as northern Mae Hong Son province, a beautiful part of Thailand, and the Mist Kingdom of Sumba in Indonesia.

None of the recommendations above is probably new or innovative. We have mainly built on what others have said or written before. Perhaps the biggest failure we have witnessed is our weak capacity to listen, learn and adapt, and most importantly, to stand up and say: 'I was wrong!'

And that should probably be the first, rather than the last, point that we make.

References

Brandon, K. and Wells, M. (2009) 'Lessons for REDD+ from protected areas and integrated conservation and development projects', in A. Angelsen (ed.) *Realising REDD+: National Strategy and Policy Options*, Center for International Forestry Research, Bogor, Indonesia, pp.225-235

Brown, L. R. and Wolf, E. C. (1984) *Soil Erosion: Quiet Crisis in the World Economy*, Worldwatch Paper 60, Worldwatch Institute, Washington, DC

Dove, M. R. (1985) 'The Agroecological mythology of the Javanese and the political economy of Indonesia', *Indonesia* 39, pp.1-36

Fujisaka, S. (1991) 'Thirteen reasons why farmers do not adopt innovations intended to improve the sustainability of upland agriculture', in *Proceedings of an International Workshop on Evaluation for Sustainable Land Management in the Developing World*, 15-21 September 1991, Chiang Rai, Thailand

Fujisaka, S. (1994) 'Learning from six reasons why farmers do not adopt innovations intended to improve sustainability of upland agriculture', *Agricultural Systems* 46 (4), pp.409-425

Hoare, P. (2017) 'Lessons learned in northern Thailand: Twenty years of implementation of highland agricultural-development and natural-resource management projects', in M. F. Cairns (ed.) *Shifting Cultivation Policies: Balancing Environmental and Social Sustainability*, CABI, UK, pp.799-818

Jones P. H. (1983) 'Lamtoro and the Amarasi model from Timor', *Bulletin of Indonesian Economic Studies* 19(3), pp.106-112

Kuipers, J. C. (1998) *Language, Identity and Marginality in Indonesia: The Changing Nature of Ritual Speech on the Island of Sumba*, Cambridge University Press, Cambridge, UK

Kunstadter, P., Chapman, E. C. and Sabhasri, S. (eds) (1978) *Farmers in the forest: Economic development and marginal agriculture in Northern Thailand*, University Press of Hawaii for the East-West Center, Honolulu, HI

Lee, J. L. (1995) 'Participation and pressure in the Mist Kingdom of Sumba', PhD dissertation to the University of Adelaide, Australia

McWilliam, A. (2000) 'A plague on your house? Some impacts of *Chromolaena odorata* on Timorese livelihoods', *Human Ecology* 28(3), pp.451-469

Nye, P. H. and Greenland, D. J. (1960) *The Soil under Shifting Cultivation*, Technical Communication No. 51, Commonwealth Bureau of Soils, Harpenden, UK

Ogunlana, E. A., Noomhorm, A. and Silakul, L. (2010) 'Alley farming in Thailand', *Sustainability 2010*, 2, pp.2523-2540

Ong, C. K. (1994) 'Alley cropping: Ecological pie in the sky?', *Agroforestry Today* 6(3), pp.8-10

Porter, D., Allen, B. and Thompson, G. (1991) *Development in Practice: Paved with Good Intentions*, Routledge, London and New York

Preston, D. A. (1989) 'Too busy to farm: Under-utilization of farm land in Central Java', *Journal of Development Studies* 26(1), pp.43-57

Rigg, J. (1997) *Southeast Asia: The Human Landscape of Modernization and Development*, Routledge, London and New York

Scott, J. C. (2009) *The Art of Not Being Governed: An Anarchist History of Upland Southeast Asia*, Yale University Press, New Haven, CT

Wilson, M. H. and Lovell, S. T. (2016) 'Agroforestry: The next step in sustainable and resilient agriculture', *Sustainability 2016* 8(6), p.574, doi.org/10.3390/su8060574

Wolz. K. J. and DeLucia, E. H. (2018) 'Alley cropping: Global patterns of species composition and function', *Agriculture, Ecosystems and Environment* 252, pp.61-68

Notes:

1. The northern Thailand part of this chapter is based on a presentation made by the first author at an Asia-Pacific Forest Policy Think Tank Expert Consultation entitled *Snatching Success from the Jaws of Failure*, on 7-8 December 2016 in Manila, Philippines, organized by the Regional Office of the Food and Agriculture Organization of the United Nations in Bangkok and the Institute for Environmental Science for Social Change.

2. In 2000, Paberiwai was subdivided into two districts, Paberiwai and Matawai Lapau.

As seen in this field, some shifting cultivators in Mongar district, Bhutan, are in the habit of retaining *Ficus roxburghii* trees in their swidden fields, as a preferred source of fodder for their cattle.

Sketch based on a photo by Malcolm Cairns, ca. 1996.

II. EVOLUTION OF FOOD PRODUCTION SYSTEMS TO SUPPORT INCREASING POPULATIONS

An Agta hunter draws his bow against an unseen target in the Philippines province of Aurora (see Minter and Headland, ch. 8, this volume). Hunting deer, wild pigs and monkeys with bows and arrows was a main economic activity of Agta men in the 1960s. Home-made shotguns came into use in the 1970s and by 2010, bows and arrows had disappeared.

Sketch based on a photo by Thomas Headland in 1965.

Synthesis

FORAGING IS NOT ANTITHETICAL TO FARMING

Swidden cultivation, following the perspective of present-day hunter-gatherers

*Edmond Dounias**

It is a truism to say that governments, whatever their latitude, distrust nomadic people. Better access to education, health services, markets and job opportunities are among the recurrent promises brandished by authorities in their efforts to persuade the last of the present-day hunter-gatherer societies to settle down.

Present-day hunter-gatherers comprise around 10 million people. They represent 2.1% of the world's indigenous peoples. This small fraction of humankind – approximately 0.12% of the worldwide population – speaks nearly 5% of the 7700 languages that are still spoken around the world. But languages are receding dramatically, and so is the cultural diversity that is related to them (Hays et al., 2022) and the biological diversity that is hosted in the territories of these indigenous peoples (IPBES, 2019). Despite their small numbers, then, hunter-gatherer groups maintain a critical repository of human cultural and linguistic diversity, and play an important role in maintaining the biodiversity of the lands on which they live (Garnett et al., 2018).

Hunter-gatherer groups are not only recognizable by their high mobility along linear territorial paths that are used on a seasonal basis, but they also have socio-political features in common, which can be summarized as (1) egalitarian social structures; (2) sociality based on inclusion and non-coercive relations; (3) an emphasis on individual autonomy; and (4) environment-oriented ontologies that contribute to sustainable forms of natural-resource use (Bird-David, 2015). They are finally distinguishable by their far-end position along the spectrum of marginalization, making them most vulnerable to its pernicious effects. Almost everywhere, they experience rapid dispossession of their lands, forced sedentarization and targeted assimilation processes. Furthermore – like most indigenous peoples – they also face

* Dr. EDMOND DOUNIAS, Research Director at the French National Research Institute for Sustainable Development (IRD), CEFE, Univ Montpellier, CNRS, EPHE, IRD, Montpellier, France.

imminent threats that include climate change and environmental deterioration, extractive industries, land insecurity, and social unrest (Mamo, 2020). Most hunter-gatherer societies now confront both exclusion and assimilation simultaneously. They are no longer able to survive solely from their traditional subsistence techniques, and now need to engage in diversified economic activities.

A large majority of persisting hunter-gatherer societies, whose livelihoods used to depend mainly on foraging activities, have now adopted swidden cultivation (Dounias et al., 2007). Compared to the swiddens of full-time farmers, those of hunter-gatherer societies are much smaller in size and less carefully prepared, cleaned, and maintained. Tree cutting and burning is minimal, or replaced by slash and mulch; there is less weeding and elimination of vegetal debris and there are neither fences nor surveillance. The crops are poorer in diversity, with fewer cultigens and fewer cultivars of each cultigen. In the end, crop yields appear ridiculously low in comparison to yields obtained by neighbouring traditional farmers.

Nevertheless, it is important to adopt the foragers' point of view in order to better capture the logic of their swiddening activities, and not to be misled by the false mediocrity of their practices.

First of all, by practising shifting cultivation, hunter-gatherers wish to reassure the outside world of their efforts to conform to a way of life deemed decent by the authorities, and thus to deserve the consideration due to them. Their swiddens are seen as evidence of their efforts to settle down and to request the socio-political advantages granted to farmers. As hunter-gatherers, they have always been deprived of such advantages. Comforting the administrative authorities allows them to legitimately claim basic civil rights, including identity cards, the right to vote, and access to primary healthcare. At the very least, they can claim to be regarded as human beings worthy of consideration.

The apparent paucity of hunter-gatherers' farming practices should not be interpreted as a lack of skill, but rather as a deliberate choice. Accordingly, action plans carried out by authorities and many non-governmental organizations (NGOs) in order to improve their capability and autonomy in farming are a waste of time and reveal a lack of foresight. In most cases, swidden agriculture constitutes only a secondary component of the diversified land-use system of hunter-gatherers. Their skills in swiddening are long learned and constantly practised when they work as labourers in the fields of more experienced farmers. Many hunter-gatherers prefer to hire out their labour and receive in-kind payments in return. Full-time farmers pay them out of their agricultural surpluses, and this compensates hunter-gatherers for the shortcomings of their agriculture. Thus, everyone reaps a reciprocal benefit.

The low-yielding swiddens of present-day foragers reveal a deliberate strategy to invest minimally in crop production and a well-reasoned decision to attribute a marginal value to farming and to redirect their efforts to other components of a diversified livelihood strategy. This translates into an intention to devote as little time as possible to their farms, including swiddens, plantations and home gardens, and when it is necessary to increase this time, they prefer to accept the tasks of salaried agricultural labourers.

Most importantly, the historical ecology approach reveals a connivance between hunter-gatherers and their natural environment, based on a co-construction of ecosystems whose dynamics result from the combined interventions of humans and natural engineers. Through generations that have succeeded one another over millennia, hunter-gatherers have profoundly oriented the distribution of spontaneous resources in space and time. They have shaped the mosaic-like structure of the ecosystems upon which they rely for their subsistence.

Unless they are fully domesticated, the ecosystems that sustain hunter-gatherers are no longer completely 'natural'. The shaping of these ecosystems by formerly nomadic hunter-gatherers involves very diverse forms of intervention along a continuum between two idealized extremes referred to as 'wild' versus 'domesticated'. The swiddening practices of these peoples are fully integrated along this continuum; they participate in a cosmogony that aims to accompany natural ecological dynamics and is without any pretention of eventually taming nature.

Hunter-gatherers show a strong preference for cultivating less toxic varieties of staple crops; their intention is clearly meant to reduce the food-processing time when these crops are harvested. In line with this same intention, hunter-gatherers prefer to be paid 'in kind' in exchange for their labour. The direct conversion of 'pay' into immediate consumption of meals allows a substantial saving of time that would otherwise be spent in harvesting, storage, and tedious processing of foods that contain toxins.

Another peculiarity of the swiddens and home gardens owned by hunter-gatherers is their relative lack of spice plants. As well as acting upon the sensory organs, spices are known for their antimicrobial properties, which are particularly necessary when food processing and consumption are deferred from the moment when the food resources are obtained. Hunter-gatherers are characterized by their 'immediate-return economy'; they obtain a direct return from their labour and their food is neither elaborately processed nor stored (Woodburn, 1982). Thus, they enjoy consuming very quickly their freshly acquired food products. This propensity for immediate consumption makes the recourse to spices and their anti-bacterial properties unnecessary (Dounias et al., 2006).

The swiddens form pools of cultivars that are accessible when needed, without restrictions, to all members of the community, especially when fields have turned into fallows. Because they are neither protected by fences nor safeguarded, the swiddens also attract wildlife. In essence, swiddening hunter-gatherers accept a loss of crop production because the loss is advantageously compensated by the capture of game. Furthermore, wild animals are attracted to the cultivated plots by fruit and seeds produced by trees that have not been felled and that purposely persist in the swiddens. This wildlife has a major part to play in dissemination processes and significantly contributes to the fast regeneration and enrichment of forest regrowth.

The small, sparsely planted and poorly controlled crop fields contribute to a much faster recovery of the forest. Hunter-gatherers' swiddens then mimic spontaneous and ephemeral forest gaps and windfalls. These clearings provide optimal conditions for the spontaneous propagation of useful plants that are light demanding for their establishment, such as wild yams, sago palms and rattans. These will be exploited later, during foraging activities. Clearings created for swiddening thus indirectly contribute to the overall dynamics of forests and their enrichment with valuable resources, and they participate in the perpetuation of foraging activities. Over time, these pervasive yet non-invasive interventions of hunter-gatherers, referred to as 'paradomestication' (Dounias, 2014), tend to exert a control over the spatial and temporal distribution of spontaneous resources and to mitigate the uncertainty in food supply during foraging activities.

It is worth mentioning that foraging is not an exclusive practice of hunter-gatherers alone, and has persisted among many farming societies. Foraging was, and still is, an integrative component of farming systems, rendering the classical dichotomy between eating from the wild versus cultivating crops and herding cattle inaccurate.

Lastly, forest is commonly presented as a safety net for forest peoples. Contemporary hunter-gatherers have adopted a meaningful opposite view, in that they consider their swiddens as a safety net for their foraging activities.

References

Bird-David, N. (2015) 'Hunting and gathering societies: Anthropology', in J. D. Wright (ed.) *International Encyclopedia of Social and Behavioural Sciences,* second edition, Elsevier, Oxford, UK, pp.428-431

Dounias, E. (2014) 'From foraging to … foraging', *Non-wood Forest Products Newsletter* 4, Food and Agriculture Organization of the United Nations, Rome

Dounias, E., Selzner, A., Koppert, G. J. A. and McKey, D. (2006) 'Hot recipes and evolution: Do hunter-gatherers eat more spicy foods than their sedentarized neighbours?' paper presented at the annual meeting of the Society of Anthropology of Paris, 18-20 January 2006, Paris

Dounias, E., Selzner, A., Koizumi, M. and Levang, P. (2007) 'From sago to rice, from forest to town: The consequences of sedentarization on the nutritional ecology of Punan former hunter-gatherers of Borneo', *Food and Nutrition Bulletin* 28(2), pp.S294-S302

Garnett, S. T., Burgess, N. D., Fa, J. E., Fernández-Llamazares, A., Molnár, Z., Robinson, C. J., Watson, J. E. M., Zander, K. K., Austin, B., Brondizio, E. S., Collier, N. F., Duncan, T., Ellis, E., Geyle, H., Jackson, M. V., Jonas, H., Malmer, P., McGowan, B., Sivongxay, A. and Leiper, I. (2018) 'A spatial overview of the global importance of indigenous lands for conservation', *Nature Sustainability* 1, pp.369-374

Hays, J., Ninkova, V. and Dounias, E. (2022) 'Hunter-gatherers and education: Towards a recognition of extreme local diversity and common global challenges', *Hunter-Gatherer Research* 5(1-2) (2022 [for 2019])

IPBES (2019) *Summary for Policymakers of the Global Assessment Report on Biodiversity and Ecosystem Services,* Intergovernmental Science-Policy Platform on Biodiversity and Ecosystem Services (IPBES) secretariat, Bonn

Mamo, D. (ed.) (2020) *The Indigenous World 2020,* International Work Group for Indigenous Affairs (IWGIA), Copenhagen

Woodburn, J. (1982) 'Egalitarian societies', *Man* 17, pp.431-451

8

WHY PHILIPPINE FORAGERS HAVE NOT BECOME FARMERS

Two decades of research on Agta swidden cultivation (1983 to 2004)

Tessa Minter and Thomas N. Headland[*]

Introduction

'We have to farm, because the forest can no longer support our children.'

'The good thing about farming is that after planting, you can just wait for the harvest.'

These statements, or very similar ones, were made repeatedly over past decades by our Agta informants living in various parts of the Sierra Madre mountain range, in the northeastern Philippines. They are a good indication of why many Agta feel pressure to adjust from foraging to farming, and they allude to the actual position that cultivation has within their overall hunting and gathering mode of existence. Moreover, they summon up questions about the circumstances that may either help or hinder foragers on the path to becoming farmers (Bellwood, 2005, pp.28, 39; Headland, 1986).

The Agta are descendants of populations that first settled in what is now the Philippines between 30,000 and 60,000 years ago. Related groups in the region include the Semang and Batek of peninsular Malaysia and the Onge, Jarawa and Sentinelese of the Andaman Islands (Endicott, 1999; Pandya, 1999). Within the Philippines, 32 groups share a common ancestry with the Agta, among them the Aeta (or Ayta) of western Luzon, the Batak of Palawan, the Ati of Panay, the Ata of Negros and the Mamanua of Mindanao (Headland, 2003, p.9, 2010, p.112; Eder, 1987). Only over the past 4000 years have these original populations encountered Austronesian agriculturalists (Bellwood, 1999, p.284, 2005, p.135).

---

[*] DR. TESSA MINTER, Assistant Professor, Leiden Institute of Cultural Anthropology and Development Sociology, Leiden, the Netherlands, who worked with the Agta of Isabela province in various periods between 2003 and 2014; and DR. THOMAS HEADLAND, Summer Institute of Linguistics, Dallas, Texas, who, along with his family, lived among the Agta people of Casiguran for most of the 48 years from 1962 to 2010.

While the arrival of farming populations in their territories has resulted in a range of socio-economic relationships with these peoples (Headland and Reid, 1989, 1991; Minter, 2009), it has not led to a full shift of foraging to farming. Of the various Philippine foraging groups mentioned above, only the Aeta of Zambales have transformed into shifting cultivators, or swiddeners, although the time and the circumstances of this transformation remain unknown (Brosius, 1990, pp.19-23).

Nevertheless, most contemporary hunter-gatherers have long been growing some of their own food (Kelly, 2013, p.2). While it is impossible to know when the Agta began to practise cultivation, it is well-established that during the (late) Pre-Hispanic period in the Philippines they were already involved in small-scale, part-time agriculture (Headland, 1986, p.227, 1987; Headland and Headland, 1988; Headland and Reid, 1989, 1991, p.45). The earliest known historical reference to this comes from Father Santa Rosa, who wrote in 1745 that the Agta in the Casiguran area (see Figure 8-1) made their own fields (Headland, 1986, p.216). From this point

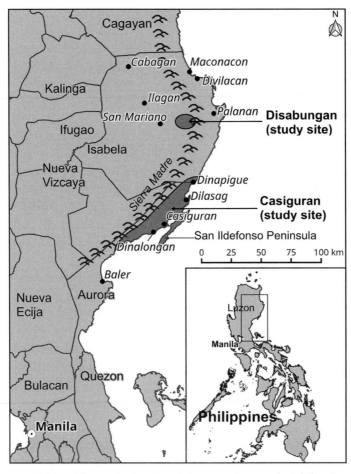

FIGURE 8-1: The study sites in northeastern Luzon, the Philippines.

in time, references to Agta cultivation appear more regularly in historical records, albeit always accompanied by comments on its marginal nature, and by accounts of attempts (usually by the authors themselves) to improve it (Headland, 1986, pp.217-226).

Indeed, the prime objective of consecutive colonial and post-colonial agencies has been to turn the Agta from mobile foragers into sedentary farmers. There are many examples of such attempts, along with their poor success (see Headland, 1986, pp.216, 229-236, 593-597; Minter, 2017, pp.131-135). But even if this 'campaign against the mobile hunter-gatherer lifestyle' (Rai, 1982, p.7) hasn't been successful in terms of its actual outcomes, it has made the Agta deeply aware of what mainstream society expects of them.

Government and civil-society organizations have so persistently delivered the message that the Agta have no future without farming that many of them at least keep up appearances, mostly by proclaiming the importance of farming. Yet, despite such statements being partly socially motivated, the Agta are genuine in their concern that the forest can no longer support their foraging way of life. Over the past 200 years, and especially in the second half of the 20th century, the Agta's natural and social environment has undergone tremendous changes as a result of two factors: population growth and deforestation (see van der Ploeg and Persoon, 2017). The impacts of these factors will become clear from our case descriptions.

Around the year 2000, there were about 11,000 Agta (Headland, 2003, p.9) living in hamlets scattered along the coasts and rivers of eastern Luzon. Within the population there was, and still is, considerable variation in the emphasis on foraging vis-à-vis other livelihoods, mobility, and in the extent to which Agta lived in proximity to non-Agta. Much of this variation was determined by the extent of deforestation and commercial dynamite fishing, and the speed and nature of agricultural expansion. However, the majority of Agta occupied a niche of 'commercial hunting and gathering' (Headland, 1986, p.77), a term coined by Hayden (1981, p.346). That is, they subsisted on a combination of fishing, hunting and gathering and barter trading of forest and marine products with neighbouring farmers. In addition, they engaged in extensive cultivation and various forms of paid labour, ranging from farm labour to wage work in the extractive industries (Minter et al., 2012). A common trait across the Agta population was, and is, their poor health and high mortality rate, resulting from nutritional and infectious disease, poor delivery of health services, and alcoholism (Early and Headland, 1998; Headland et al., 2011; Minter, 2010, p.58; Page et al., 2018).

Case studies

In the context of population pressure, resource competition, cultural change and social marginalization, what does Agta swidden cultivation look like? And what is its position within the wider livelihood strategy of commercial hunting and gathering? These are the questions that we will explore in this chapter. Before outlining our methodology and describing the Agta's swiddens in detail, we will first sketch the situation of the two Agta groups whose cultivation practices we studied. The cases derive from Headland's (1986) and Minter's (2010) PhD dissertations.

Casiguran

The Casiguran Agta live in the foothills and along the coasts of the eastern side of the Sierra Madre mountains (Figure 8-1). Prior to World War Two, the overall population density in this area was only about 1.4 people per sq. km. The non-Agta population consisted mostly of wet-rice farmers, with whom the Agta maintained trading relations. By 1984, the population density had grown to 51 people per sq. km and was increasing daily. The total population of Casiguran Agta numbered just over 600, while there were about 35,000 non-Agta residents of various ethnic backgrounds (Headland, 1986, pp.15-16). The fast growth of the non-Agta population resulted from various processes, notably the steady development of the logging industry since 1928, with a peak in the 1970s; the opening of an open-pit magnesium mine at Dinapigue (Figure 8-1) which operated from 1960 to the early 1970s; the construction of a government road connecting the southern town of Baler with Casiguran in 1977, and a boom in rattan gathering which, from 1978 onwards, attracted thousands of collectors (Headland, 1986, pp.250-256, 290, 297). The cumulative effect of these developments on the population and on the forests meant that by the mid-1980s, the Casiguran Agta were no longer able to live by hunting and gathering. Commercial rattan gathering and wage labour in coconut plantations became their major activities, while cultivation gained importance among some Agta groups.

Disabungan

The Disabungan Agta live some 60 km to the north of Casiguran, on the western side of the mountain range (Figure 8-1). Their hamlets are at the forest fringe, at an altitude of around 500 m above sea level (masl). The two nearest non-Agta villages are a three-hour hike downstream. These two villages and surrounding farms arose as a result of logging-related migration in the 1950s, and continue to have very fast-growing populations. They went from around 1000 inhabitants each in the year 2000 (NSO, 2000), to more than 1500 and 1600 inhabitants respectively in the year 2015 (PSA, 2015). Both villages are located in areas that were once the Agta's fishing and hunting grounds. When industrial logging ended in the early 1990s, it was followed by clandestine logging, operated by numerous small groups of men from villages and towns, financed by politically influential individuals (van der Ploeg et al., 2011). The Agta hamlets became a 'gateway' to the forest, and the logging teams passed-by or camped there on a daily basis. As a result, the Disabungan Agta came to depend on heavily disturbed secondary forest for their foraging activities, which they complemented with income from logging labour, barter trade and, as we will see, a bit of swidden cultivation.

Methods

The data on swiddening that we present here were collected among the Casiguran Agta (in Aurora province) by Headland in 1983 and 1984, and two decades later among the Disabungan Agta (in Isabela province) by Minter in 2004 and 2005 (see Figure 8-1). In studying Agta cultivation practices, and livelihoods more generally, both of us combined the classic ethnographic methods of participant observation and various forms of interviewing with a more quantitative approach, consisting of records on swiddens, time allocation and diet. Detailed descriptions of these can be found in Headland (1986, pp.32-60) and Minter (2010, pp.300-317). We will summarize our approach below, after a short note on the strengths and limitations of our data.

A major strength is that our data were collected as part of long-term fieldwork among two Agta populations: Headland since 1962, with his latest visit in 2010, and Minter since 2002, with her most recent visit in 2014. The foremost limitation of our present contribution is that the data we present here were collected decades ago. However, as no recent work has studied Agta farming practices as systematically as we have, ours is the best available material to date. Moreover, our data remain relevant in shedding light on the role of swidden cultivation in hunter-gatherers' adaptations to rapid social and environmental change. We do not wish to suggest that Agta swiddening systems are static, so we therefore present and discuss our findings in the past tense.

Another note must be made on comparability. As Minter largely followed Headland's methodology, our data are in many ways suitable for comparison, but the Agta groups whose swidden cultivation practices we studied differ considerably in size, namely 224 people in Casiguran and 60 people in Disabungan, although the mobility of the Agta meant that group size and composition were always fluid.

Swiddens

We studied the characteristics of Agta swiddening by observing and measuring fields and what was growing in them. Using a measuring tape and a notebook (and, in Minter's case, a GPS), we manually measured and recorded the outer boundaries of Agta swiddens; the different areas within them planted in certain crops; and the unplanted area. We measured and listed the different crops found in each swidden, made notes on the soils and areas surrounding the swiddens and estimated the swiddens' hillside slopes (Headland used a protractor for this). Headland measured and mapped all 43 Agta swiddens cleared in the Casiguran area in 1983 and 1984, while Minter did the same for the 35 swiddens made in the Disabungan area in 2004 and 2005. We further used semi-structured interviews to gather information on who cleared and cultivated the land, its history, and the source of the seeds and cuttings used.

Time allocation

Both Headland and Minter conducted time allocation studies with the purpose of knowing how the Agta were dividing their time between different activities, with a focus on livelihood activities. This was done by listing the major activity an individual was involved in during a certain day. The information was obtained through direct observation and interviewing. The many different activities that people undertook were reduced to several larger categories of core activities. Time spent in these activities was expressed as a percentage of the total number of person-work-days (PWDs) recorded. These numbered 3283 PWDs for Casiguran, recorded by Headland, and 281 PWDs for Disabungan, recorded by Minter.[1]

Diets

Knowing what people ate and where that food came from provided another indication of the relative importance of different livelihood activities, including swiddening. Headland and his wife compiled a record of 558 meals eaten in the first half of 1984, by interviewing Agta on the main starch and the main side dish that they had eaten in the previous three meals. Minter kept records based on direct observation and measurements of meal components as they were prepared, combined with interviews on where the ingredients came from. She did this for 43 meals from the Disabungan area, which were recorded among four households in two five-day periods in August 2004 and March 2005.

Agta swidden cultivation examined across time and space

We now turn to a description of Agta swiddening activities, consisting of successive stages of land selection and preparation, planting, guarding and weeding and finally, harvesting.

Land selection and preparation

As is typical of swidden cultivation, land once cultivated is left fallow for a longer period than it has spent under crops (Conklin, 1961, p.27). In the case of the Casiguran Agta, 57% of the swiddens they cultivated in 1983 were left to lie fallow in as little as seven or eight months after they were burnt. However, fields planted with root crops were abandoned to fallow within 20 months after burning. The Agta generally only cultivate a certain area of land for one season, so the cycle begins with the selection and marking of new land.

Towards the end of the wet season – any time between November and February – the Agta marked their land by cutting a boundary path around the area to be cleared. The Casiguran Agta made most of their new swiddens in 1983 at elevations below 100 masl, in both secondary and primary forest (Table 8-1). In contrast, the Disabungan Agta cleared their swiddens at higher elevations, around 500 masl. In

TABLE 8-1: Areas from which Agta cleared swiddens (as % of total area cleared).

Location and swidden size	Primary forest	Secondary forest	Primary and secondary forest	Brushland	Total
Casiguran (1983) (Total area cleared: 7.7 ha)	27.4	32.8	23.6	16.2	100%
Disabungan (2004) (Total area cleared: 1.8 ha)	0	39.3	0	60.7	100%

2004, they slashed almost two thirds of their swidden area in flat brushland along river banks, which had been cleared by a logging company in the 1970s. The rest was cut from secondary forest, as there was no longer any primary forest in Disabungan (Figure 8-2).

Clearing took place around February and was a male-dominated activity, although women were usually involved as well. Both Headland (1986, p.335) and Rai (1982, p.168) observed the Agta helping each other in clearing with reciprocal exchange groups. These sometimes involved non-Agta workers, as well. By the early 2000s, this practice had disappeared among the Disabungan Agta. Older informants lamented the decreasing interest in exchange labour by the younger generation, saying that individual households tended to work for themselves. Indeed, the 2004–2005 Disabungan swiddens were cleared either by members from within one nuclear family or by the nuclear families of siblings, who often clustered their swiddens together (see also Rai, 1982, p.167).

Clearing involved several substages, the first of which was cutting small trees and underbrush. This was done around March, usually with axes and machetes. Women often helped men in this task. Next, men felled the big trees. The Casiguran Agta used axes to do this, but in the 1980s they often hired experienced non-Agta, who owned chainsaws, for the job, paying in cash or labour. The Disabungan Agta, being surrounded by loggers, either borrowed a chainsaw to cut down the big trees themselves, or requested loggers to do so, usually in exchange for logs, meat or labour.

The Agta tended to completely clear-cut their fields, with the exception of strangler fig trees (*Ficus* spp.), which were not cut down for fear of upsetting supernatural beings that were thought to inhabit them. When felling

FIGURE 8-2:   Disabungan swiddens and hamlet in the midst of secondary forest in 2004.

Photo: Tessa Minter.

was complete, men and women worked through the field lopping off branches from fallen trees (Figure 8-3). This was aimed at hastening the drying of the debris in preparation of burning, which took place at the height of the dry season, in April or May (Figure 8-4).

Swidden size

Agta swiddens were 'tiny' (Headland, 1986, p.339; Rai, 1982, p.170). However, there was a wide range of swidden sizes both within and between the two groups: from 50 to 4000 sq. m and 2 to 2700 sq. m at Casiguran and Disabungan respectively. However, when the average planted area per capita was calculated (Table 8-2), the difference between the two groups was reduced to a negligible 20 sq. m. On average, Agta in both areas had a per capita swidden area of almost 280 sq. m.

Crop variety

In line with their size, Agta swiddens were planted with a modest variety of crops (Table 8-3). The majority of swiddens in both groups were planted with between one and three crops, but there were considerable differences between them. In Casiguran, the overall number of crops found was 47, while in Disabungan it was only 18. Moreover, in Casiguran, the largest number of different crops found in one swidden was 47. In Disabungan, this was only 11.

Staples like rice and root crops, including cassava, sweet potatoes and taro, occupied the highest percentage of swiddens in both areas (Table 8-3). At Disabungan, the total area covered by each crop confirmed the importance of

FIGURE 8-3: After the trees were felled at Casiguran, the branches were lopped and prepared for burning.

Photo: Thomas Headland.

FIGURE 8-4: The fires were lit at the height of the dry season, in April or May.

Photo: Thomas Headland.

TABLE 8-2: Planted area per capita in Casiguran (1983) and Disabungan (2004).

Location and year	Agta population	Total planted area (sq. m)	Planted area per capita (sq. m)
Casiguran (1983)	224	59,508	266
Disabungan (2004)	60	17,300	288

these staples: 45% of swidden area was planted with upland rice and 19% with root crops.

There were also interesting differences between the two areas. Bananas, for instance, were a major crop in Casiguran, but were relatively unimportant in Disabungan. Rai (1982, p.168) had already noted this in the late 1970s, but the situation in the early 2000s is best explained by the impact of the banana 'bunchy top' virus, which affected bananas throughout the region from the 1990s (Masipiqueña, 2003, p.110), and super typhoon Harurot, which destroyed 98% of bananas in the region in July 2003 (Huigen and Jens, 2006, p.2124). Another difference concerned white corn, which occurred in one third of Casiguran swiddens, but in more than half of the Disabungan swiddens, where it occupied 14% of the total swidden area in 2004.

The most striking difference between the two areas related to fruit trees. While these occurred regularly in Casiguran swiddens, they were absent in Disabungan. The most logical explanations for this are found in the location and history of the gardens. The Casiguran swiddens were situated at low elevations, near the coast, so there was high occurrence of coconuts. Moreover, in many of these swiddens, non-Agta cultivators were involved at some point, often even as original owners and planters. In contrast, the Disabungan Agta had only recently begun to cultivate swiddens, and they were the first to do so at their location. They had formerly lived at lower elevations, and these areas had been taken over by farmers. Logging and the subsequent expansion of agriculture had pushed them uphill and upstream.

Ipomoea batatas (L.) Lam.
[Convolvulaceae]

Sweet potatoes have a higher nutrient density than cereals and are the most efficient staple crop in terms of land use, yielding about 70,000 kilocalories per hectare per day. They were introduced to the Philippines in the 16th century by Spanish galleons plying the Pacific.

TABLE 8-3: Crops found in Casiguran and Disabungan swiddens in 1983 and 2004.

Crop common name	Scientific name	% of swiddens in which crop was found (rounded)	
		Casiguran 1983 (n= 43)	Disabungan 2004 (n=35)
Staples			
Upland rice	Oryza sativa	74	29
Cassava	Manihot esculenta	72	63
Sweet potatoes	Ipomoea batatas	53	63
Taro 1	Colocasia esculenta	37	29
Taro 2	Colocasia sp.	37	–
White corn	Zea mays	33	51
Yams	Dioscorea alata	9	–
Legumes			
String beans	Phaseolus vulgaris	23	26
Mung beans	Vigna radiata	2	3
Winged beans	Psophocarpus tetragonolobus	2	–
Gourds			
Squash	Cucurbita maxima	7	11
Loofa gourd	Luffa sp.	5	–
Common gourd	Lagenaria leucantha	2	–
Vegetables			
Eggplants	Solanum melongena	21	6
Okra	Abelmoschus esculentus	16	9
Pechay	Brassica rapa	–	3
Fruit trees			
Coconuts	Cocos nucifera	37	3
Calamansi	Citrus microcarpa	9	–
Balsem apples	Momordica charantia	7	–
Jackfruit	Artocarpus heterophyllus	7	–
Avocados	Persea americana	5	–
Breadfruit	Artocarpus altilis	5	–
Pomelos	Citrus maxima	2	–
Fruits			
Bananas	Musa acuminata	54	9
Pineapples	Ananas comosus	19	12
Papaya	Carica papaya	12	6
Tomatoes	Lycopersicon esculentum	5	6
Watermelons	Citrullus lanatus	2	–
Herbs and spices			
Mustard	Brassica juncea	16	–
Onions	Allium sp.	16	–
Ginger	Zingiber officinale	7	–
Lemon grass	Andropogon citratus	7	–
Garlic	Allium sativum	5	–
Kaempferia	Kaempferia galanga	2	–
Oregano	Oreganum vulgare	2	–

TABLE 8-3 (cont.): Crops found in Casiguran and Disabungan swiddens in 1983 and 2004.

Crop common name	Scientific name	% of swiddens in which crop was found (rounded)	
		Casiguran 1983 (n= 43)	Disabungan 2004 (n=35)
Others			
Sugar cane	*Saccharum officinarum*	47	–
Chillies	*Capsicum* sp.	9	3
Peanuts	*Arachis hypogaea*	5	3
Marigolds	*Tagetes* sp.	5	–

Planting

Swidden farmers are known to select rice seed and other planting materials from their own fields and save it for the next planting season. Agta generally didn't do this. For instance, for the 2004 swidden cycle in Disabungan, cassava and sweet potatoes were the only crops that were replanted from previous swiddens. Most planting materials in both Casiguran and Disabungan were obtained from non-Agta farmers, either as gifts or through exchanges.

The time of planting differed per crop, per area and per year (see also Rai, 1982, pp.169; 336; Griffin, 1989, pp.64-66), but most of it took place between April and July. At least in Disabungan, the task of obtaining planting materials, as well as planting the crops, was predominantly performed by women, sometimes aided by children and elderly men. Planting root crops and vegetables was done individually or in pairs of related women (Figures 8-5 and 8-6), and was spread out over various days, weeks or even months.

The situation was different for rice, which was planted in one concerted effort, and by a group, consisting of either a nuclear family or of members from within an extended family. Most of this happened in June. One person

Colocasia esculenta (L.) Schott
[Araceae]

Taro is one of the earliest of cultivated plants. Traces of its use have been found in many archaeological sites, dating as far back as 20,000 years. It was a popular crop planted by the Agta, particularly at Disabungan. Elsewhere in the Philippines, taro is a common vegetable, with the corms, leaves and stems forming part of the local cuisine.

FIGURES 8-5 and 8-6: Jennalyn Matias (left) and Cecille Wagi planting taro at Disabungan in January 2010.

Photo: Tessa Minter.

made holes in the soil with a dibble stick while rice seeds were sown in the holes by the other planters (Figures 8-7 and 8-8). In Casiguran, a total area of 5.2 ha of swidden land was planted with upland rice in 1983, but the Disabungan Agta planted just under 0.8 ha of land with upland rice in 2004. Thus, the Disabungan Agta had roughly 100 sq. m less swidden land planted in rice per capita than the Casiguran Agta.

Guarding and weeding

As the second statement at the opening of this chapter suggests, most Agta paid little attention to their swiddens in the time between planting and harvesting. Fields were generally unguarded, although a few had tin cans or leaves hanging from rattan strings to scare off birds, while the occasional wild pig that raided fields was shot or killed by burying small, baited explosive charges around the edges of fields. Weeding was done irregularly (see also Griffin, 1989, p.64). The only swiddens weeded

FIGURES 8-7 and 8-8: Dibbling rice seeds (top) and planting root crops (below) at Casiguran.

Photo: Thomas Headland.

by the Disabungan Agta were those cleared from brushland (Figure 8-9). They experienced less weed-growth in fields cut from the forest.[2] Of all the time they spent in agriculture, the Casiguran and Disabungan Agta, respectively, gave 13% and 38% to weeding, most of which was done by (elderly) women. (Figure 8-10).

Returns from swidden cultivation and contribution to diets

As with clearing and planting, rice harvesting was mostly done by closely related nuclear families, with those who planted the rice usually harvesting it. This happened between late September and early November, and took one to several days per swidden, depending on its size and the number of harvesters. Vegetables and root crops were harvested throughout the year, and this was usually done by children, or their mothers or grandmothers. This often happened little by little: sweet potatoes, cassava, corn and vegetables were usually gathered in quantities just large enough for one meal (see also Rai, 1982, p.171).

FIGURE 8-9:   Mario Matias weeding his rice field at Disabungan in June 2011.

FIGURE 8-10:   An Agta mother, Carmen Matias, weeding a field of sweet potatoes at Disabungan in January 2004.

Photos: Tessa Minter.

Because of the way root crops were harvested, it was difficult to gain a systematic insight into yields. This was a serious limitation in our studies of Agta agricultural production; possibly resulting in underestimation of the Agta's agricultural productivity, or at least of its nutritional importance. In Casiguran, root crops were eaten at 5% of meals (Headland, 1991, p.6). Minter observed that the Disabungan Agta ate root crops on four out of 20 observation days in 2004 and 2005. They ate them mostly as 'snacks' outside main meals, especially on days that meals were skipped. This confirmed the idea that the Agta depended on root crops as 'famine food', when rice was unavailable. Indeed, as Headland (1986, p.348) noted for the small (13 sq. m) per-capita plots planted in cassava and sweet potato, these were probably best considered as a 'highly adaptive custom'.

Clearly, rice was the preferred staple in both areas: the Casiguran and Disabungan Agta ate rice at 92% and 74% of their meals respectively. But the level of their dependence on non-Agta farmers for their rice is clear from Table 8-4, which shows the rice yields of both groups. Even though the Casiguran Agta did better than the Disabungan Agta, their net rice yield per capita was still only 13 kg, against a negligible 1.8 kg for the Disabungan Agta. Headland (1986, pp.349, 500) calculated that if the Casiguran Agta had not shared their rice with the wider Agta population of Casiguran (some 600 people), it would have fed them for 43 days. However, refusing to share was, and still is, probably the strongest taboo within Agta society. And so, sharing their rice with all other Agta, the entire 1983 rice harvest (9.1 metric tons of unhusked rice) was calculated to last only about 15 days (Headland, 1986, pp.349, 500, 554, 1991, p.4; Headland and Headland, 1988). Similarly, Minter observed that the Disabungan Agta shared their extremely poor rice harvest among at least nine households. It was eaten up within a week.

Manihot esculenta Crantz
[Euphorbiaceae]

Cassava is the third-largest source of carbohydrates in tropical regions, after rice and maize. It was found in most Agta swiddens at Casiguran and Disabungan, where it was used as a 'snack' food, or to avoid famine when rice was unavailable. Elsewhere in the developing world, cassava is a major staple food, providing a basic diet for an estimated half a billion people.

According to the little available information on upland rice harvests in adjacent areas, the rice yields of neighbouring non–Agta swiddeners ranged from about 1800 to 2000 kg of unhusked rice per hectare (Romero, 2006, p.115). At this rate, the rice fields of the Casiguran and Disabungan Agta yielded only about 50% and 10%, respectively, of what their non–Agta neighbours harvested per hectare.

Interestingly, the Casiguran Agta considered 1983 to be a successful agricultural year. Weather conditions were favourable throughout the agricultural cycle and there had been no significant pest problems (Headland, 1986, p.349). Yet, from the perspective of most farmers, their efforts could hardly be called successful. One explanation for the poor yields may be found in the fact that the Casiguran Agta were

TABLE 8-4: Per capita area planted in upland rice and estimated yields of unhusked and milled rice.

Location and year	Population	Area planted in rice		Harvested rice (kg)			
		Total (ha)	Per capita (sq. m)	Unhusked total	Unhusked yield/ha	Milled total	Milled yield/capita
Casiguran (1983)	224	5.2	231	4700	900	3000	13
Disabungan (2004)	60	0.8	133	169	200	110	1.8

heavily involved in paid labour on the farms of non-Agta neighbours during the clearing, planting and harvesting seasons of the 1983 agricultural cycle. Indeed, overall investment of time in their own agricultural activities (4.4% of total time) was only half of that invested in paid farm labour (8.7%). They were even busier working in rattan collection, on which the Casiguran Agta spent as much as 25% of their time (men 33% and women 17%) (Headland, 1986, pp.468, 485).

The situation was similar for the Disabungan Agta. They, too, and men in particular, invested minimal time in their swiddens compared to the time they spent in paid labour. In their case, farm labour was unimportant, but logging labour and rattan work took as much as 32% and 14%, respectively, of men's time, against 4% in swiddening. This was partly compensated by the women, who spent 13% of their time in swiddening (Minter, 2010, pp.149; 178).

An additional explanation for the poor rice harvests of the Disabungan Agta in 2004 rested in the adverse effects of logging in the area. First, deforestation in upstream areas and along river banks had resulted in an increased risk of flooding. Almost all of the Disabungan rice swiddens that were created along river banks were washed out in the rainy season of 2004. In addition, these and other fields suffered from damage by free-roaming *carabao* (water buffaloes, *Bubalis bubalis*) that were brought in to haul timber from the forest. This was despite the Agta's repeated requests to loggers that the animals be tethered.

The Disabungan Agta responded by clearing all of their swiddens for the following 2005 cycle in secondary forest on steeper and higher slopes. However, they abandoned these in the next rainy season and moved to a previous hamlet an hour's hike upstream. When visiting them there in 2007, they explained that the continued presence of loggers and their *carabao* had led them to move. They had reopened old swiddens, and while the results appeared to be more rewarding, they were concerned about the future. An old logging road was being reconstructed, and it was expected to be only a matter of weeks before the loggers and their animals caught up with them.

Discussion: Agta swiddening in perspective

We have used data from two different Agta groups, in two different time periods, and on two sides of the northern Sierra Madre mountain range. The Casiguran Agta (in the early 1980s) made their swiddens in lowland areas, in proximity to both the coast and neighbouring farming communities. In contrast, the Disabungan Agta (in the early 2000s) were settled in the forest interior, on higher elevations and far away from farming communities.

Neither of these two Agta groups showed signs of being on a trajectory that led from foraging to farming (Headland and Headland, 1988). Indeed, from our descriptions of Agta swiddening practices, it is clear that these practices were atypical in many ways. Based on Headland's detailed comparison of 13 swidden variables with those of 19 other Southeast Asian swidden societies (Headland, 1986, pp.349-353, 501-504),[3] some of the most striking differences are discussed below.

Agta swiddens were much smaller than those of full-time swidden farmers. Even the Aeta of Zambales (also in the Philippines) were reported to make significantly larger swiddens, with each one covering between 0.15 and 1.5 ha (Brosius, 1990, p.88). Agta swiddens also had relatively large unplanted areas, a feature we will get back to below. Moreover, the Agta grew far fewer different crops per field and often planted extremely small quantities of each of them. Rice, although it took up the largest share of the swidden area, held much less importance than it did for other swiddeners. This was obvious from the relatively small areas planted in rice, the small number of rice varieties planted, and the low yields. Some crops (notably coconuts, herbs, spices and fruit trees), were more commonly a part of Casiguran swidden farming than they were in Disabungan. We have

Dioscorea alata L. [Dioscoreaceae]

Purple yams were found in Agta swiddens at Casiguran in 1983, but 21 years later at Disabungan, they were not grown. Originating in Island Southeast Asia and Papua New Guinea, yams were carried both east and west by Austronesian voyagers. They remain an important crop in the Philippines, where purple yams are called *ube*, and are used in various traditional and modern desserts.

suggested that this may have been due to the different histories and locations of the swiddens: these crops may have more frequently occurred in Casiguran swiddens due to their closer proximity to non–Agta farmers (who, in some cases, were involved in them as original planters).

Overall time invested in swiddening was very low in both groups. This was especially true for weeding, a task undertaken repeatedly and extensively among other Southeast Asian shifting cultivators (Schlegel, 1979, p.52), including, again, the above mentioned Aeta (Brosius, 1990, p.91). Furthermore, the Agta allowed their swiddens to revert to fallow much sooner than other swiddeners. Finally, perhaps the most remarkable characteristic of Agta swiddening was its 'complete lack of any religious ritual'. That is, while most other swiddeners marked each stage in the swiddening process with '[…] a great deal of ritual [involving] magic, augury, divination, sacrifices, ceremony, etc.' (Hires and Headland, 1977; Headland, 1986, p.338), throughout our years with the Agta, we saw none of this.

However anomalous Agta cultivation practices may seem to have been in comparison to those of other swiddeners, some of their key characteristics are reflected in the scant literature on cultivation among other hunter-gatherers. Kitanishi (2003), for instance, describes how the Baka of Cameroon refuse to cultivate anything except plantains, which they grow (rather satisfactorily) in small, unguarded and unweeded fields. Within the Philippines, Balilla et al. (2012, p.697) found that only a quarter of Aeta families at Mariveles, in southwestern Luzon, planted swiddens (exclusively with sweet potatoes, taro and bananas), while the remainder did not engage in cultivation at all. Thus, as we will argue in the remainder of this chapter, the deviance of Agta swiddening should not lead to its denunciation. Instead, it should be assessed in the light of the Agta's overall mode of subsistence, as well as in the context of the rapid social and environmental changes faced by the Agta.

Agta swiddening from a forager's perspective

Understanding Agta swiddening requires that it be seen from the perspective of foragers, and not from that of farmers. This allows us to see that within their subsistence mode, cultivation takes an instrumental (not necessarily marginal) position. It creates the view that '[…] Agta economic emphases are placed elsewhere than on farming' (Headland, 1986, p.350).

This is evident primarily from the little time that was invested in cultivation relative to the many other things that Agta did as part of their diversified livelihood package. But some interesting differences within groups were obscured by looking at overall time investment, as we did. That is, time spent in cultivation, as in other activities, varied between individuals. While a few Agta families were quite committed to farming, most others were less so, while some didn't spend any time in cultivation at all. For instance, almost half of the 94 adults in Headland's sample never spent a single day in cultivation. As one of Minter's informants remarked in the early 2000s: "Why grow rice if you can buy it?"

Indeed, being 'commercial hunter-gatherers' (Headland, 1986; Hayden, 1981, p.346), Agta have, for a very long time, had many different ways of 'buying rice' and other cultivated food: through trading it for meat, fish, rattan and other non-timber forest products. In addition, they obtained much of their rice through providing labour, whether it was on non-Agta farms, as was the case in Casiguran, or in logging, like in Disabungan. This economic diversification had an important advantage: by keeping different options open, the Agta were spreading risk in a volatile environment.

Aside from diversification, there was another key principle underlying the commercial hunting-gathering strategy. The Agta are a classic example of what Woodburn (1982) called 'immediate-return foragers': they '[...] obtain a direct [...] return from their labour' and 'Food is neither elaborately processed nor stored.' (Woodburn, 1982, p.432). In other words, in such systems, energy invested in a certain activity should generate results immediately – not four months later. And the other way around: food generated on one day was to be consumed immediately. Closely related to the necessity for immediate consumption was the strong social pressure to share whatever one had (Woodburn, 1982, pp.440-442).

Thus, as opposed to 'delayed-return' foragers – such as the Inuit and indigenous peoples of the northwest coast of Canada and the United States, who store some of the food they produce – there was only one place where the Agta saved food, and that was in other people's stomachs. Immediate-return foragers like the Agta, the Baka in Central Africa (Kitanishi, 2003) and the !Kung in southern Africa show marked disdain for anyone who attempts anything, including saving food, that looks like they are trying to 'get ahead' of the rest (see Lee, 1969). This partly explains the fact that the Agta generally did not keep seeds or other planting material for next season. The taboos on storage and the ethical obligation to share also imply that the gains from agriculture do not outweigh the effort invested (Sahlins, 1972, p.32; Bellwood, 2005, pp.31-32; Woodburn, 1988, p.57; Headland, 1986, p.349).

But there is a final characteristic of immediate-return systems that compromises a full commitment to cultivation: mobility (Woodburn, 1982, p.435). While the level of mobility differed between groups, households and even individuals, for many Agta it continued to be important as a means to ease social tension, maximize economic opportunity, escape unhealthy conditions and mediate spiritual danger (Minter, 2017; Page et al., 2018). Committing to agriculture would require giving up on mobility, which from the perspective of immediate-return hunter-gatherers, had far-reaching social consequences.

So, given these obvious mismatches between the delayed-return nature of agriculture and the immediate-return nature of hunting-gathering, why did the Agta have swiddens at all? While 'off-farm' activities (i.e. commercial hunting-gathering) were in fact far more reliable sources of rice supply than their own fields, having fields of their own was still important for the Agta in at least two ways. First, small as they were, Agta swiddens were an important source of 'famine food', notably in the form of root crops. Second, as we will discuss below, the Agta's swiddening practices also had important symbolic value in the current social-environmental constellation of the rural Philippines.

Adapting to social and environmental change

Since 1950, the Philippine population as a whole increased from 20 million to 107 million people (PRB, 2018). At the same time, the Agta population hardly grew, so they became a small minority in areas that they once shared with only a handful of farmers. Moreover, while non-Agta people were previously confined to villages and town centres, they spread along rivers and into the foothills, where they competed with the Agta for game, fish and other forest products. This resource pressure came on top of massive forest loss and degradation as a result of industrial logging that took place throughout the Sierra Madre from the early 20th century, and most intensively from the 1970s to the 1990s. After the logging companies withdrew, clandestine logging continued and secondary forest was converted into farmland (van den Top, 2003). Thus, the forest on which the Agta depended was drastically reduced in quantity and quality, a process that continues today.

Moreover, the Agta lack tenurial security over the remaining forest land. Despite a legal framework for providing Philippine indigenous groups with collective legal ownership of their 'ancestral domains' having been in place since 1997, this has not benefited the Agta (see van der Ploeg et al., 2016; Headland, 1999, 2013). This has resulted in an alarming situation, as Agta land is not only under pressure from population increase, forest exploitation and agricultural expansion, but Agta are also being displaced by tourism and infrastructure development (Headland and Headland, 1999; Cruz et al., 2013; Hagen and Minter, 2020).

Against this backdrop, we may view the Agta's increasing emphasis on cultivation as a symbolic act – specifically their previously mentioned tendency to clear far more land than they can actually plant. Given land ownership traditions in the Philippine uplands, which prescribe that land belongs to whoever has cleared it, Agta swiddeners may have felt that they needed to lay claim to as much land as possible. Even if this didn't have *de jure* legitimacy, it did at least give more *de facto* security than leaving forest land intact (see van der Ploeg and Persoon, 2017, p.3). This is because forest land is seen as idle land; land in need of 'development', which is a broader reflection of the widely held idea that farming is superior to foraging.

Conclusion

In this chapter we have documented the swidden farming practices of the Agta, a people who have long been pressured to become farmers, while, in reality, farmers they are not. We have argued that Agta swiddening can only be understood in the context of their niche of commercial hunting and gathering under circumstances of rapid social and environmental change. Extensive cultivation has long been part of their diversified livelihood strategy, but government and non-governmental organizations, and the individuals leading them, have misjudged it as marginal and inferior and therefore as being in need of improvement.

In the design of programmes to enhance the Agta's farming practices, change agents have invariably focused on sedentary rice cultivation. This reflects the cultural preference for rice in mainstream Filipino society, rather than its nutritional value. In fact, a high occurrence of vitamin B1 deficiency among the Agta can be linked to the dominance of milled rice in their diets (Early and Headland, 1998, p.104). From a health perspective, it would therefore be much wiser to encourage the continuation of root-crop cultivation. Moreover, a full switch from foraging to rice farming is incompatible with the cultural values, social organization and economic principles underlying the Agta's hunting and gathering livelihood. Even if the Agta aspired to such a switch, they would be unlikely to succeed because they would be competing for arable land with a large population that has lived as sedentary farmers for generations.

Thus, instead of launching yet another programme to turn foragers into farmers, development policy ought to focus on strengthening the Agta's niche as commercial hunter-gatherers. We therefore call for an end to the centuries-old campaign against the Agta's way of life, and for protection of what remains of the Philippine forest and the Agta's rights to it.

References

Balilla, V. S., Anwar-McHenry, J., McHenry, M. P., Marris Parkinson, R. and Banal, D. T. (2012) 'Aeta Magbukún of Mariveles: Traditional indigenous forest resource use practices and the sustainable economic development challenge in remote Philippine regions.' *Journal of Sustainable Forestry* 31, pp.687-709

Bellwood, P. (1999) 'Archaeology of Southeast Asian hunters and gatherers', in R.B. Lee and R. Daly (eds), *The Cambridge Encyclopaedia of Hunters and Gatherers*, Cambridge University Press, Cambridge, UK, pp.284-288

Bellwood, P. (2005) *First Farmers: The Origins of Agricultural Societies*, Blackwell Publishing, Oxford, UK

Brosius, J. P. (1990) *After Duwagan; Deforestation, Succession and Adaptation in Upland Luzon, Philippines*, Center for South and Southeast Asian Studies, University of Michigan, Ann Arbor, MI

Conklin, H. C. (1961) 'The study of shifting cultivation', *Current Anthropology* 2, pp.27-61

Cruz, J. P., Headland, T. N., Minter, T., Grig, S., Phil, M. and Aparentado, M. G. (2013) 'Land rights and inclusive development: The struggle of marginalized peoples against APECO and the pursuit of alternative development pathways', *Langscape* 2(12), pp.54-63

Dove, M. R. (1985) *Swidden agriculture in Indonesia; The subsistence strategies of the Kalimantan Kantu*, Mouton publishers, Berlin/New York/Amsterdam

Early, J. D. and Headland, T. N. (1998) *Population Dynamics of a Philippine Rainforest People; The San Ildefonso Agta*, University Press of Florida, Gainesville, FL

Eder, J. F. (1987) *On the Road to Tribal Extinction: Depopulation, Deculturation, and Adaptive Well-being among the Batak of the Philippines*, University of California Press, Berkeley, CA

Endicott, K. (1999) 'The Batek of Peninsular Malaysia', in R. B. Lee and R. Daly (eds), *The Cambridge Encyclopaedia of Hunters and Gatherers*, Cambridge University Press, Cambridge, UK, pp.298-302

Griffin, P. B. (1989) 'Hunting, farming and sedentism in a rainforest foraging society', in S. Kent (ed.) *Farmers as Hunters; The Implications of Sedentism*, Cambridge University Press, Cambridge, UK, pp.60-70

Hagen, R. V. and Minter, T. (2020) 'Displacement in the name of development. How indigenous rights legislation fails to protect Philippine hunter-gatherers', *Society and Natural Resources* 33(1), pp.65-82, DOI: 10.1080/08941920.2019.1677970

Hayden, B. (1981) 'Subsistence and ecological adaptations of modern hunter/gatherers', in: R.S.O. Harding and G. Teleki (eds) *Omnivorous primates: Gathering and Hunting in Human Evolution*, Columbia University Press, New York, pp.344-421

Headland, T.N. (1986) *Why Foragers Do Not Become Farmers: A Historical Study of a Changing Ecosystem and its Effect on a Negrito Hunter-Gatherer Group in the Philippines*, University Microfilms International, Ann Arbor. MI

Headland, T. N. (1987) 'The wild yam question: How well could independent hunter-gatherers live in a tropical rain forest ecosystem?' *Human Ecology* 15, pp.463-491

Headland, T. N. (1991) 'How Negrito foragers live in a Philippine rainforest: What they eat and what they don't eat', paper presented at an international symposium, 'Food and Nutrition in the Tropical Forest: Biocultural Interactions and Applications to Development', sponsored by the United Nations Educational, Scientific and Cultural Organization (UNESCO), September 10 to 13, 1991, Paris, available online at ResearchGate

Headland, T. N. (1999) 'The Indigenous Peoples' Rights Act: A triumph of political will', *SIL Electronic Working Papers 1999-004*, (August), Summer Institute of Linguistics, Dallas, TX, https://www.researchgate.net/publication/283645835_The_Indigenous_Peoples%27_Rights_Act_A_Triumph_of_Political_Will

Headland, T. N. (2003) 'Thirty endangered languages in the Philippines', working paper, the Summer Institute of Linguistics, University of North Dakota Session, vol. 47, available online at ResearchGate

Headland, T. N. (2010) 'Why the Philippine Negrito languages are endangered', in M. Florey (ed.) *Endangered Languages of Austronesia*', Oxford University Press, Osford, UK, pp.110-118

Headland, T. N. (compiler) (2013) 'Indigenous peoples' rights and inclusive development: An open letter on APECO to President Aquino and the Filipino people from an international community of scientists', available online at https://www.facebook.com/notes/letter-on-apeco-to-pres-aquino-from-international-scientific-community/casiguran-agta-people-in-trouble-please-sign-this-statement/180758425412857

Headland, T. N. and Headland, J. D. (1988) 'Rice cultivation practices in a Negrito foraging society in northeastern Luzon, Philippines', *International Rice Research Newsletter* 13(5), pp.38

Headland, T. N. and Headland, J. D. (1999) *Agta Human Rights Violations: Why Southeast Asian Negritos are a Disappearing People*, Summer Institute of Linguistics, Dallas, TX, available at https://scholars.sil.org/thomas_n_headland/controversies/agta_human_rights_violations

Headland, T. N. and Reid, L. A. (1989) 'Hunter-gatherers and their neighbours from prehistory to the present', *Current Anthropology* 30, pp.43-66

Headland, T. N. and Reid, L. A. (1991) 'Holocene foragers and interethnic trade: A critique of the myth of isolated independent hunter-gatherers', in S. A. Gregg (ed.) *Between Bands and States: Interaction in Small-scale Societies*, Southern Illinois University, Carbondale, IL, pp.333-340

Headland, T. N., Headland, J. D. and Uehara, R. (2011) *Agta Demographic Database: Chronicle of a Hunter-Gatherer Community in Transition*, SIL Language and Culture Documentation and Description, version 2.0., Summer Institute of Linguistics, Dallas, TX, available online at www.sil.org/resources/publications/entry/9299

Hires, G. A., and Headland, T. N. (1977) 'A sketch of western Bukidnon Manobo farming practices, past and present', *Philippine Quarterly of Culture and Society* 5, pp.65-75

Huigen, M. G. A. and Jens, I. C. (2006) 'Socio-economic impact of super typhoon Harurot in San Mariano, Isabela, the Philippines', *World Development* 34, pp.2116-2136

Kelly, R. (2013) *The lifeways of hunter-gatherers: The foraging spectrum*, Cambridge University Press, Cambridge, UK

Kitanishi, K. (2003) 'Cultivation by the Baka hunter-gatherers in the tropical rainforest of Central Africa.' *African Study Monograph* 28, pp.143-157

Lee, R. B. (1969) 'Eating Christmas in the Kalahari', *Natural History* (December), pp.14-22, 60-63

Masipiqueña, M. D. (2003) 'Upland food production systems in the Sierra Madre: Realities and prospects', in J. van der Ploeg, A. B. Masipiqueña and E. C. Bernardo (eds) *The Sierra Madre Mountain Range: Global Relevance, Local Realities*, CVPED/Golden Press, Tuguegarao, Cagayan, Philippines, pp.103-118

Minter, T. (2009) 'Contemporary relations between the Agta and their farming neighbours in the northern Sierra Madre, the Philippines', in K. Ikeya, H. Ogawa and P. Mitchell (eds) *Interactions between Hunter-gatherers and Farmers from Prehistory to Present*, SENRI, Osaka, Japan, pp.205-228

Minter, T. (2010) 'The Agta of the Northern Sierra Madre Natural Park: Livelihood strategies and resilience among Philippine hunter-gatherers', PhD dissertation to Leiden University, Leiden, The Netherlands, https://openaccess.leidenuniv.nl/handle/1887/15549

Minter, T. (2017) 'Mobility and sedentarization among the Philippine Agta', *Senri Ethnological Studies* 95, pp.119-150

Minter, T., de Brabander, V., van der Ploeg, J., Persoon G. A. and Sunderland, T. (2012) 'Whose Consent? Hunter-Gatherers and Extractive Industries in the Northeastern Philippines', *Society and Natural Resources* 25(12), pp.1241-1257, DOI: 10.1080/08941920.2012.676160

NSO (2000) '2000 Census of population and housing, Isabela'. Report no. 2-48b, vol. 1, National Statistics Office, Manila

Page, A. E., Minter, T., Viguier, S. and Migliano, A. B. (2018) 'Hunter-gatherer health and development policy: How the promotion of sedentism worsens the Agta's health outcomes', *Social Science and Medicine* 197, pp.39-48

Pandya, V. (1999) 'The Andaman islanders of the Bay of Bengal', in R. B. Lee and R. Daly (eds) *The Cambridge Encyclopaedia of Hunters and Gatherers*, Cambridge University Press, Cambridge, UK, pp.243-247

PSA (2015) *Census of Population 2015*, Region II (Cagayan Valley), Total Population by Province, City, Municipality and Barangay, Philippine Statistics Authority, Manila

PRB (2018) *2018 World Population Data Sheet*, Population Reference Bureau, Washington, DC, http://www.worldpopdata.org/map

Rai, N. K. (1982) 'From forest to field: A study of Philippine Negrito foragers in transition', PhD dissertation to the University of Hawaii, Manoa, HI

Romero, M. R. (2006) 'Investing in the land; Agricultural transition towards sustainable land use in the Philippines forest fringe', PhD dissertation to Leiden University, Leiden, The Netherlands

Sahlins, M. (1972) *Stone Age Economics*, Aldine, Chicago, IL

Schlegel, S. A. (1979) *Tiruray Subsistence: From Shifting Cultivation to Plow Agriculture*, Ateneo de Manila University Press, Quezon City, Philippines

van den Top, G. (2003) '*The Social Dynamics of Deforestation in the Philippines: Actions, Options and Motivations*' NIAS press, Copenhagen

van der Ploeg, J., Masipiqueña, A. B., van Weerd, M. and Persoon, G. A. (2011) 'Illegal logging in the Northern Sierra Madre Natural Park', *Conservation and Society* 9(3), pp.202-215

van der Ploeg J., Aquino, D. M., Minter, T. and van Weerd, M. (2016) 'Recognizing land rights for conservation? Tenure reforms in the northern Sierra Madre, The Philippines', *Conservation and Society* 14(2), pp.146-160

van der Ploeg, J. and Persoon G. A. (2017) 'Figments of fire and forest. Shifting cultivation policy in the Philippines and Indonesia.' in M. F. Cairns (ed.) *Shifting Cultivation Policies: Balancing Environmental and Social Sustainability*, CABI, Wallingford, UK, pp.3-26

Woodburn, J. (1982) 'Egalitarian societies'. *Man* 17, pp.431–451

Woodburn, J. (1988) 'African hunter-gatherer social organization: Is it best understood as a product of encapsulation?', in T. Ingold, D. Riches and J. Woodburn (eds) *Hunters and Gatherers: History, Evolution and Social Change*, vol. 1, Berg, Oxford, New York and Hamburg, pp.31-64

Notes

1. The sampling methods for time-allocation records at Casiguran and Disabungan were different. Headland recorded the activities of 331 adults during monthly visits to different hamlets in 1983 and 1984, by listing the primary activity on the day of the visit and that of the previous day. This resulted in 3283 PWDs (1709 for men and 1574 for women). Minter's sample for the Disabungan Agta consisted of the activities of all individuals aged six and over that were present in the watershed during three five-day periods of observation in January and August 2004 and April 2005. This covered the main activities of 37 individuals, resulting in 281 PWDs (145 for boys/men and 136 for girls/women).

2. Dove (1985, p.223) observed the same approach to weeding among Kalimantan Kantu' swiddeners.
3. The other swidden societies included the Hanunoo, Tiruray, Dayak, Kenyah, Pwo Karen, Lamet, Mandaya, Gaddang, Lahu Nyi, Lawa', Cuyunon, Skaw Karen, Tagbanwa, Iban, Kantu', Bangon, Napsaan, Mandaya and Thai lowland.

9

OUR FARMS ARE AS GOOD AS THEIRS

From hunters and gatherers to shifting cultivators in
Sarawak, East Malaysia

*Valerie Mashman**

Introduction

This chapter describes the factors that encouraged a semi-nomadic people, the Penan,
who were living as hunter-gatherers in the forests of Borneo, to build a settlement
at Long Beruang in northern Sarawak, Malaysia, and become shifting cultivators of
upland rice. It was a process encouraged and supported by a nearby community of
swidden farmers who simply wanted the Penan 'to be like them and to eat like them'.
So while many recent studies of shifting cultivation tell of an ancient practice that is
in decline, this chapter supports its relevance as a recent source of food security for a
community of forest dwellers who were willing to transform their lifestyle. Although
the shifting cultivators of the study area are beset by the wider issues affecting swidden
farming across the region, such as the penetration of a capitalist economy, logging
and out-migration of younger generations, these matters are beyond the scope of
this chapter. The narratives of how the Penan came to settle at Long Beruang, in the
Kelapang river area of the upper Baram, recur in the recollections of the elders of the
Kelabit people, in a longhouse at Long Peluan, downriver.

The Penan settlement of Long Beruang was established in about 1961 above the
confluence of the Bale and Beruang rivers by five enterprising pioneers, Lake' Loi,
his brother, Lake' Gar, and his brothers-in-law Lake' Pa', Lake' Juman and Tama
Bawang. The area is in the upper reaches of the Baram river system of northern
Sarawak, about 20 miles from the border with Indonesian Borneo (Figure 9-1). This
is an hour's journey upriver by boat from the settlement of their benefactors at the
Kelabit longhouse of Long Peluan, which was established about 50 years previously.
There is no intermarriage between the two settlements. Although the leaders of
the Long Peluan longhouse are mainly Kelabit, their pioneering community also

* DR. VALERIE MASHMAN, has a PhD in anthropology and is a research fellow at the Institute of Borneo
Studies, Universiti Malaysia, Sarawak.

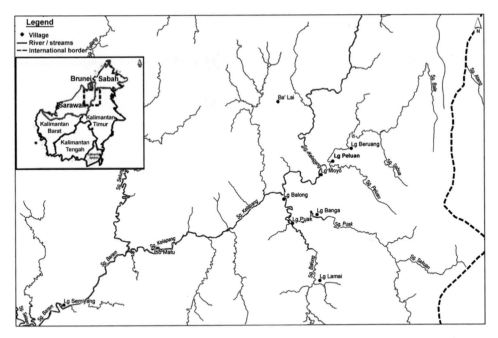

FIGURE 9-1: The settlements of Long Beruang and Long Peluan in northern Sarawak, Malaysia.

Map: Lee Guan Heng.

has Ngurek and Sa'ban ancestry, so the mixing of backgrounds and social equality characterize the outlook of the people. Indeed, district Christian elders often refer to people simply as coming from a particular settlement, rather than belonging to a specific ethnic group. This enables everyone to relate to one another and removes any previous overtones of superiority or inferiority based on ethnicity.

The early Penan settlers at Long Beruang came from another settlement, which was established in 1958. They moved because of a community disagreement and a desire to find better hunting grounds (Murang, 1993, p.91). At that time, the Kelabit of Long Peluan were farming in the Long Beruang area and they suggested that the Penan should stay there. Eventually, the Penan began to build huts and established their settlement at Long Beruang, where they live today.

There are now about 70 families living in Long Beruang, in wooden houses (Figure 9-2) and a longhouse built for them by a logging company. The Penan have forgone the animist beliefs of their former nomadic lifestyle and have adopted Christianity. Their community has a village church run by the Borneo Evangelical Mission. The area was selectively logged in the 1990s and local community leaders negotiated for an area of forest to be conserved across the watersheds of the Kelapang and Bale rivers.[1] They also negotiated for construction of a road connecting their area with the coast. Today, there is a road from Long Beruang to Long Peluan and Long Banga, and many households own motorcycles.

Much of the material for this chapter derives from visits to Long Peluan from 1984 to the late 1990s and then more recently in the years from 2010 to 2015. Interviews were recorded with the Kelabit elders of Long Peluan, who formed close relationships with the first Penan to settle in the area, and with Penan elders in Long Beruang. Today, many of the Penan speak Malay and Kelabit and interviews were conducted in a combination of these languages and Penan.

FIGURE 9-2: Present-day houses in the Penan settlement of Long Beruang.

All photographs in this chapter were taken by Valerie Mashman.

This chapter builds on the account of Murang (1993) who described the role of a Sa'ban Christian missionary, Lahang Apoi, in encouraging the Penan to settle at Long Lamai – well to the south of their eventual settlement at Long Beruang. The literature that describes the processes of the Penan settling down (Nicolaisen, 1976; Sellato, 1994; Langub, 2003; Chan, 2007; Sercombe and Sellato, 2007; Zawawi and NoorShah, 2012) does not cover the role played by the Kelabit who, in many cases, live nearby, in the transition of the Penan to a settled life. However, Janowski (1997) discusses the attitude of the Kelabit towards their nomadic neighbours and suggests that the Kelabit thought of the Penan as being 'forever children'. This was because the way of life of hunters and gatherers was considered by the Kelabit as 'play', whereas swidden cultivation of rice was perceived as 'work', which brought prestige to 'adults'. Janowski's (1997) position highlights a view that the two ways of life stood in opposition to each other. Her work provoked an immediate critical response from the Kelabit and later from a Kelabit anthropologist, Poline Bala (2002), who argued that Janowski's work was marred by a theoretical framework which illustrated binary oppositions that did not exist in day-to-day life.

It has been acknowledged that settled groups played a crucial role in the processes of the Penan settling down, even though the Penan were both diffident about the relationships and suspicious of settled people for fear of being cheated (Langub, 2003, p.137; Murang, 1993, p.94). These apprehensions were borne out in observations by colonial officers who described the Penan being treated as the slaves or subjects of settled people, who were their 'overlords' (Sellato, 1994, p.169). There were also reports that the economic relationships between the Penan and settled groups were exploitative, with unfair exchanges being made for trading jungle produce (Brosius, 2007, p.304). This chapter will highlight a more nuanced and complex relationship between the Kelabit and the Penan. I will argue that for the Penan in Long Beruang, the Kelabit of the upper Baram became a point of reference, and they exemplified a different way of life as Christians and shifting cultivators for the Penan to follow.

The Kelabit and the Penan

Although the Kelabit of the upper Baram river traded forest products with the Penan, they were less dependent on the Penan to obtain jungle produce to trade than were other river-based groups such as the Kayan and the Kenyah. The Kelabit were very familiar with headwater trails and also collected resins and bezoar stones and searched for rhinoceroses to trade.[2] They practised shifting cultivation and were known as 'the people who were on the move' (*lun rupu*). Like the Penan, they hunted and shared their meat with the rest of their community and they depended on the forest for foraged vegetables, rattan and timber for their houses and boats (Figure 9-3).

The main differences between the two groups were that the Penan relied on sago, harvested from *Eugeissona utilis* palms, as their staple food, whereas the Kelabit farmed

FIGURE 9-3: A Kelabit hunter.

upland rice; and the Penan lived in temporary nomadic shelters whereas the Kelabit lived in longhouses that were rebuilt every five years or so. There had been close relationships between the Penan, the Kelabit and the Sa'ban in the upper Baram area for three or four generations and the Kelabit played an important role in the Penan process of settlement. Indeed, the Kelabit of Long Peluan fondly refer to the Penan in their vicinity as 'our Penan' (*Penan tauh*). This is often used in an inclusive context that conveys openness and affection and does not have the connotations of possession or monopoly that Brosius (2007, p.303) identified when he described how 'longhouse aristocrats were proprietary about "their" Penan'. In turn, the Penan use the word *pade*', meaning brother, when talking to their counterparts in Long Peluan, which conveys deep acquaintance and familiarity (Murang, 1993, p.96).

Such feelings of inclusion and relatedness are very much built upon today. All villages in the district are considered to be on a par with each other, and hold district-wide church services in rotation. Everyone enjoys reciprocal hospitality, with overnight stays and all meals provided. But first, it is important to consider how the Penan remember their past, before they settled down.

Life before settling down

The Penan understand their identity as being closely linked to life in the forest, which makes them different from other people. Lake' Isak from Long Beruang defines his identity as a Penan:

Eugeissona utilis
[Arecaceae]

Wild Bornean sago was the staple food of the Penan before they ended their nomadic lifestyle. They shifted around the forest seeking out new stands of this palm.

Ever since I remember, I was called Penan. There was no other name but Penan. Penan are people who depend on the resources of the forest. Yes, we call ourselves *kelunan* (human beings). *Va'e* are other people; strangers (Figure 9-4).

Lake' Isak remembers the camps (*lamin*) where he lived a nomadic existence as a child with his parents. They never camped by big rivers; only by small rivers 'because of the spirits that live in the big rivers'. They did not use boats, and were not used to fishing with nets. They lived in small shelters raised a few feet off the ground. Each family had its own shelter, carefully built to avoid the prevailing wind blowing smoke inside the house from the hearth, which always faced the sleeping area. The shelters had no walls, but had generous roofs made of layers of wild ginger or palm leaves (Figure 9-5). A whole camp of 10 to 20 related families (*panen*) lived in shelters together. The men would leave the camp to hunt for wild boar, deer and smaller animals. Some women might hunt for smaller animals, as they were also proficient users of blowpipes. On longer hunting trips, the whole family would move together. The families harvested sago (*Eugeissona utilis*) together. Sago flour was most often consumed on the go, before it went mouldy, and was stored in a basket above the fireplace at the camp.

FIGURE 9-4: Lake' Isak whittles blowpipe darts.

There are mixed recollections of nomadic life. Lake' Isak remembers a place where they stayed that he hated because it was very unhealthy. He was also very afraid of thunder and lightning, and held tight to his father because of the loud sounds. Now he likes living in a sturdy house where he feels protected from storms, rain and the hot sun. However, he is also nostalgic about the ease with which the Penan were once able to harvest sago (*uvut*). 'People were happy. Everything was there and they could get it anytime they wanted'. Missionaries recall how the Penan would eat all day in times of plenty, but there were also times when food was scarce and people went very hungry. Children were taught not to wail when they were starving (Malone, 2014, pp.104, 112). Lake' Isak's older brother, Lake' Tawan, recalled such hardships when asked about his childhood as a nomad:

FIGURE 9-5: A Penan temporary shelter in a forest camp.

> Life in the forest was bad. We were hungry. If there is no food we just stayed put. Life now is very good. There is enough to eat. Before, we lived on shoots, in the forest…we had nothing. Everything had to be done by hand. Life was very tough. We did not have any clothes, and it was very cold. In the forest, we would die of cold (*matei genin*).

Clearly, the cold was something that Lake' Tawan remembered. In the highlands of Sarawak the temperature can drop as low as 12° Celsius at night. All he had to protect him from the cold was the warmth of a fire.

Sometimes, the Penan nomads had to move camp suddenly because of a death in the community. The custom was to cover the body of the deceased with the roof from the shelter and leave it behind while the rest of the camp moved to another place.[3] The elders remembered that when Lake' Loi's mother died, they wanted to get rid of the memory of her passing as fast as possible. In Lake' Isak's words: 'We don't want to remember them, we want to get them out of our mind'.

Another reason for moving camp was to find more sago. Once the mature *Eugeissona utilis* palms in one area had been harvested and processed, the group moved on to find new sources of their staple food. Moving to a new camp was done with great care. For instance, before the people moved, sufficient sago flour had to be prepared to last them for a while. Lake' Isak's nephew, Tabare, explained:

Some of the most able-bodied people go ahead to choose the next stopping place and to make sure it is safe and there is enough food. They will wait until they have good signs (*amen*). They have to survey the area to make sure there are no enemies around. When they come back, they make sure there are no footmarks left there; no rotten wood. They make sure there is nothing left for the enemy. It's a special mission – if they decide to stay for quite some time they have to make sure it's a safe place. They have to be alert; to make sure there are no enemies, human or animal, and they also need to observe the signs (*amen*) for spiritual protection.

The need to be watchful and alert originated in times of warfare, when the Penan had to be wary of headhunters. Fear of outsiders (*v'ae*) was a distinct aspect of Penan behavior until the late 1930s and 1940s. Lake'Tawan remembers that Balare's brother, Labong, was abducted by a dominant chief on the Bahau river in the 1940s. The Penan did not mix with other people, so they were wary and frightened when they came into contact with outsiders.

The first meetings

The headman of Long Peluan, Malian Tepun, recalled:

The Penan would be very shy; they lived in the forest. If they saw you in the forest, they would move away, they would not want to see you. In the olden days, the Penan could be encouraged to come close us with the offer of tobacco (*sigup*). We would call out, '*sigup, sigup!*'.

If the Kelabit were hunting in the forest, they might sense that there were Penan in the vicinity. Ose Murang, a Kelabit elder in his sixties, remembered how it was possible to sense the presence of Penan in the forest: 'I was out hunting with my brother at Arur Telingan, in the headwaters of the Kelapang. "There are Penan close by," my brother said, and I asked him how he knew that. "I can smell them," he replied. I couldn't smell anything, but my brother told me the Penan didn't like the river and wouldn't bathe very often.' The Penan were known to be reluctant to bathe in rivers as they believed that there were spirits who took the lives of people who dared to swim (Malone, 2014, p.108).

At this time, the Kelabit and the Penan were not so familiar with each other, and Lake'Tawan related how one Kelabit boy in Long Peluan had been scared of him when, some years earlier, he had visited the boy's parents at their farm. Lake' Tawan had brought a gift of deer meat with him, and was sitting with the parents of the small boy, who was just learning to talk. He screamed 'Penan, Penan!' as if the Penan were to be feared. Their appearance was very different to that of the Kelabit because they wore very little clothing, their hair was long and wild-looking and they carried blowpipes and sometimes had pet monkeys with them. Lake'Tawan said he reassured the boy by saying, 'I am not Penan. I'm not going to harm anyone'. He meant by this that he was no longer a Penan who lived far away and apart from the Kelabit, but was a friend. At that time the Penan became anxious when they encountered the Kelabit in the forest, and Kelabit children were not accustomed to seeing them. Eventually, the younger Kelabit boys learned how to use Penan blowpipes for hunting birds and squirrels around the longhouse.

As time went on, a rapport of accepting and giving hospitality developed between the Penan and the Kelabit. This was illustrated by an encounter described by the Kelabit elder Ose Murang:

Manihot esculenta Crantz
[Euphorbiaceae]

> Another time, we chanced upon Tama Bawang at his temporary shelter in the forest, by the head-waters of the Beruang river. He made us feel very welcome and shared his food with us. He nonchalantly emptied and scraped out his pot where the dog had been eating and used it to prepare sago for us to eat. We accepted this graciously and we weren't any the worse for it.

Tapioca, or cassava, was one of the first food sources cultivated by the Penan, eventually becoming a substitute for sago as their staple food.

Settled life

In the years following the Japanese occupation in World War Two, the Penan began to meet the Kelabit on the Bale river, upriver from their present settlement, and traded items such as resin (from *Agathis kinabaluensis*), wild rubber (from *Leuconotis anceps*), wild meat and woven baskets and mats in exchange for cloth, axes and bush knives. The Kelabit had become Christian and the current headman of Long Peluan, Malian Tepun, recalled how the Penan were invited to become Christian by his father, Soh Tepun, and to settle so they would no longer be hungry.

> After we had been Christian for a long time, we had big celebrations at Easter called *irau*, and Penan who lived far away in the jungle would come a long way to visit us. Those who were less shy and who were ready to talk to us were Lake'Loi, Juman and Belare Jabu, who were the fathers of today's [Penan] leaders. Over time, the *irau* were very well attended. The Penan became familiar with us and later my father, Soh Tepun, made an offer to Lake' Loi and Belare:

Leuconotis anceps Jack
[Apocynaceae]

The source of wild rubber traded by the Penan before they settled, in exchange for cloth, axes and bush knives.

'If you want to believe in the Christian god, then we will send someone to teach you.'

That's how it came about that we asked Tama Lawai Lahang to go and teach them in the forest.[4] At that time, the Penan were nomads in the headwaters of the Ano river, between Long Peluan and Long Banga. Later, as they got familiar with us, they came closer to the longhouse and eventually came to church in the village. So that's how Soh Tepun asked them to settle down and build houses in a place where they would be happy to live. He told them: 'If you set up houses, there will be no animals, no forest shoots as vegetables. You'll make gardens and plant tapioca.'

Agathis kinabaluensis de Laub.
[Araucariaceae]

When they were still forest nomads, the Penan harvested resin from these trees to use in trade with the Kelabit.

The Penan took to planting tapioca (cassava – *Manihot esculenta*) because it required little attention and grew easily close to their houses. It could be harvested easily and both the leaves and the tubers could be eaten. Eventually, it became a substitute for their staple food, sago (Mashman, 2011).

Lake' Isak explained how Christianity became attractive to the Penan:

> The problems in the past were the omens (*amen*). An omen like a bird calling, or a fallen tree, would tell you that you couldn't go out and look for food. So you became hungry as you ran out of food. Now we believe in God and have religion. It is much better than in the past.

Christianity gave them the freedom to search for food whenever they wanted, without heeding the omens of their old animistic beliefs. They also began to bury their dead, so they no longer had a reason to move to a new location. Eventually, they lost their fear of the river spirits and began to make boats and fishing nets and became avid fishermen. When they began to settle and planted tapioca and bananas, the men still went out to the forest to hunt and collect forest products. And while the founders made their way to Long Beruang through settlements at Long Belaka and Long Lamei, they learnt from Tama Lawai Lahang to read in Penan and later continued their education with Kenyah teachers and Australian missionaries.

Kelabit elder Ose Murang continued the story:

> The Penan settled as our neighbours in Long Beruang in 1961. I remember visiting and seeing them in makeshift huts, without walls, made of small round logs, built under the shade of big trees close to where the Beruang river flows into the Bale. The huts were small – about ten feet square. They used round logs because they had no equipment or expertise in making planks like us. The roofs consisted of layers of big leaves from fan palms, wild ginger and *Macaranga* trees. They didn't sew the leaves together in layers to make a more leak-proof roof like we did. In one corner of the hut was the hearth.

For the Penan living at Long Beruang, the experience of participating in shifting cultivation with the Kelabit was the beginning of a period of very gradual change. Lake' Tawan's wife, Pahin, shared her recollections of this time:

> We followed Soh Tepun (the Kelabit leader, also known as Tama Tuloi) and his people here. They were then farming at Ba Beruang, where there is

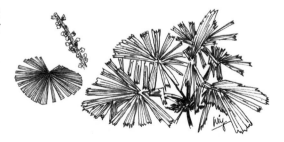

Licuala valida Becc. [Arecaceae]

Fan palm leaves were used to create shelter for Penan families when they lived in temporary forest huts.

a type of jackfruit (*bua' kiran* – *Artocarpus odoratissimus*) by the bank. There is a stream there known after him – Ba' Tama Tuloi.[5] It was difficult work to farm when we started. We didn't have the tools. Tama Gerawat gave Jalong a chainsaw. So it was easy to fell the trees. We went with the *irah lebo*, the settled people, to their farms. They taught us to plant tapioca near Pa' Sebua'. They eventually gave us bush knives and axes. We learnt by working with *irah lebo* (the settled people). But now, we do it like the people of Long Peluan.

It is interesting that the name *Tuloi* was given to a stream, *Ba Tuloi*, which ran through Tuloi's farm at Pa Beruang. Thus, the names of Kelabit elders began to feature in what was to become the Penan landscape, reflecting the association between the two peoples. Jalong, a Penan elder now living at Long Lamai, recollected the time when the Penan still ate sago starch and sago shoots:

> We ate sago shoots. We didn't eat shoots from the rattan palms. It's only now that we eat them, having learned from you. Later on, I planted clumps of sago in the grounds at Long Beruang. I got the seeds from Long Dano. We had enough to eat. There was a lot of sago in the hills.

The Penan still went to forage for sago and to hunt, but there was a slow building of relationships between families at Long Peluan and Penan families. Ose Murang continued his account:

> At that time, the Long Peluan people had been farming in the Beruang river area for about three years, from about 1958.[6] All of the Long Peluan families farmed in the same area. It made our working together for the *baya'* system of exchange labour much easier. The Penan began to join in, to help with farming of hill *padi*, as the Kelabit were working nearby on their farms (Figure 9-6). Food, consisting of rice, vegetables – both planted and foraged – fish and wild meat, was shared freely among the farming

Artocarpus odoratissimus Blanco
[Moraceae]

The 'kind of jackfruit' that grew near the river bank, marking the place now called Long Beruang, where the Penan and Kelabit first interacted.

groups, including the Penan. Most of the Kelabit families lived in temporary farm huts on their farms at Long Beruang, close to the Penan, and would return to the long-house only at the weekends for the Sunday church service. We would

FIGURE 9-6:   A group of Kelabit farmers harvesting upland rice in a swidden at Long Peluan.

punt and paddle longboats to get to the area.

Jalong, a Penan elder, remembered visiting Long Peluan: 'As young boys, we joined in the farming and harvesting. We used to go fishing and hunting together'. There was an initial empathy between the two groups. The Kelabit elders recalled that they had a genuine desire to feed the Penan and wanted them 'to be like us and eat like us'. The Penan visits became more frequent. They would come to the longhouse at Long Peluan and sit in groups on the verandah, waiting for mealtimes, because they were hungry.[7] The Kelabit hospitality was sometimes stretched when they found that they were feeding more mouths, the whole year round. The burden fell on the Kelabit women, who were juggling a heavy workload, having to look after the farms, fetch water from the river, pound rice by hand, cook dinner and care for babies at the same time. Then the Penan often turned up at the longhouse. Over time, closer relationships were forged between certain families. The Kelabit spoke Penan and, eventually, the Penan spoke Kelabit. The Kelabit asked the Penan to bring meat or rattan (*Calamus optimus*) to the longhouse and this gave the Penan a sense of being able to reciprocate and contribute to the Kelabit families who offered them food.

The fact that the Kelabit had abundant food all the year round and were never hungry must have been the trigger that motivated the Penan to start farming, despite the hardship of working on swiddens in the hot sun. The Penan complained about it, saying 'We can't go into the sun, we will die!' To begin with, their swiddens were smaller than the Kelabit farms and Penan farmers worked only in the early morning and late afternoon. Ose Murang recalled passing by Penan farms at Long Beruang:

I remember that in 1965, when I was on my way back from school, I passed by a Penan farm when they were staying at Pa Beruang. It didn't look like a *padi* [rice] field. It hadn't burnt well. This might have been because they were unaware of the importance of selecting the right time and the right weather for burning and planting. Some stages of land preparation had been left out, such as slashing the undergrowth and lopping branches from the felled trees before the first burning of the land. They dibbled the holes for the *padi* too far apart, so the ripening stalks were thinly distributed. They didn't spend much time weeding either. This was because they still needed to go into the forest to collect sago and to hunt as their *padi*-framing was hardly giving them any yield. They only gave a little time to their *padi* farming at this stage. They were just trying out farming. That's why their farms were often untended. They were away a lot in the forest. You could say their farms were experimental.

Calamus optimus Becc. [Arecaceae]

Rattan found in the forest enabled the Penan to contribute to Kelabit households that offered them food.

This was a familiar story. In the early stages of their settled life it was usual for the Penan to spend less time on cultivation than on continued hunting and gathering. Consequently, insufficient attention was paid to their crops once the seeds had germinated, and the resulting grain yields provided sustenance for only three to four months (Langub, 2003, p.144) (Figure 9-7). The Penan were combining their survival strategies and were giving priority to those that gave more immediate economic returns. Hence, they combined farming with hunting and gathering, selling wild meat, fish and handicrafts and working as wage laborers farther afield (Sellato and Sercombe, 2007, p.26).

Ose Murang recalled how support from the Kelabit was crucial in the early days, when the Penan were beginning to farm in the forest:

FIGURE 9-7: Upland rice pushes its way through the unburnt debris of a Penan swidden.

The Kelabit gave the Penan *padi* [rice] seeds, banana suckers and tapioca sticks to plant. My parents would talk to them about farming and encourage them not to abandon their farms. I remember overhearing my parents saying that the Penan would pull the tapioca roots out of the ground before the tubers were mature.

The developing relationship between the Kelabit and the Penan was not all plain sailing; there was one bone of contention: the disappearance of sugar cane and tapioca tubers planted by the Kelabit. It happened years ago, when the Penan were still 'experimental' swidden farmers, but it was unexpectedly recalled by a bedridden Penan elder, Tama Bawang, when a Kelabit group from Long Peluan visited him recently. He admitted that he and his kinsmen might have had something to do with it. He continued: 'I told the Kelabit, when I come here there are people who take your tapioca. So I told them, "ask for it back, let us not quarrel over it. When I'm not good, not behaving well, I take your tapioca, your sugar cane; there's nothing wrong if you get angry with me because I'm doing the wrong thing".'

Looking back, Ose Murang believes that the problem over the disappearance of the tapioca, bananas and sugar cane planted by the Kelabit on their outlying farms in Long Beruang came about because the Penan did not realize that the Kelabit, by planting the crops, considered them to be their property. The differing perceptions of ownership of planted crops reflected a wider gap in understanding between the Kelabit and the Penan with regard to land ownership, and this led to further tensions. When the Penan began to settle in Long Beruang as hunters and gatherers, they did so with the agreement of the Kelabit community at Long Peluan. Gradually, they began to farm fallow land that had previously been cleared by the farmers of Long Peluan, on the understanding that the land was being loaned. This agreement has been forgotten over time, and in Long Peluan there is now some resentment that the younger generation of Penan and new migrants to Long Beruang do not acknowledge the previous history of the land. These issues have become more acute for the people of Long Peluan as some families seek more land to pass on to their children.

Despite the tensions and resentment regarding land issues, new kinds of partnerships have evolved between the farmers of

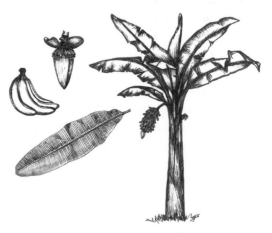

Musa acuminata Colla [Musaceae]

Banana suckers were among the first planting materials offered to the Penan. Later, filched bananas stirred up trouble with their benefactors.

Long Peluan and Long Beruang. The Penan farmers at Long Beruang have become increasingly efficient. This is reflected in the way they have developed their own Penan terms for the various phases of shifting cultivation that are practised by the Kelabit, e.g. slashing the undergrowth (*memara*); felling the trees (*nebeng*); cutting the branches so there will be a good burn (*meto'*), and so on.[8] Every Penan family is now practising shifting cultivation and there are a few wet-rice fields located near the river. The average yield per household is about 24.7 sacks of rice per hectare,[9] but 82.6 per cent of recently interviewed farmers said their harvest was not enough to last them throughout the year (Sarawak Development Institute, 2014). However, some households farm bigger fields and have surpluses sufficient to sell rice to families in Long Peluan.

In a phenomenon that is affecting shifting cultivation communities across the region, the Kelabit community at Long Peluan is becoming increasingly depopulated. Younger generations are moving away for employment and education, leaving the older folk to cope as best they can. This means the farmers have problems finding sufficient labour to work their fields. The traditional method of reciprocal labour (*baya'*) places a heavy burden on those who are too busy with child-rearing and other work to reciprocate. The alternative is to pay others to work, and there is a standard rate of payment for labour in the fields, including meals for the day and often transport as well. With these payments, Kelabit families find support from the Penan as well as their Sa'ban neighbours in Long Banga. 'The Penan can work just like us,' the Kelabit farmers say, and the Penan farmers agree: 'Now, our farms are just as good as those of the Kelabit.' The Kelabit think the Penan are becoming just like them. 'They aren't like they used to be; they are just like us.' Earlier in this chapter, Lake' Isak was quoted as defining the Penan as a people dependent on the forest. There are doubts that this is still strictly relevant for those who have become today's shifting cultivators: perhaps the Penan are becoming something different. To outside observers, it is difficult to tell a young Penan woman from a young Kelabit woman. Both have lightly tanned complexions from being outdoors, they converse fluently in Penan, Malay and Kelabit, and both have a confident demeanour when meeting new people. Significantly, the Penan woman have become experts in weaving the sort of mats and baskets the Kelabit need for processing their rice harvests and the Kelabit, who have less time and inclination to collect rattan and weave, readily buy them from the Penan (Mashman, 2012). The Penan farmers often supply neighbouring villages with vegetables grown in their swiddens and they regularly have extra rice for sale, particularly a variety of sticky rice that fetches higher prices.

The starkest change for the Penan may be in overall lifestyle. From the forest shelters of living memory, their wooden houses now have solar-powered electricity provided by the Malaysian government and the community has a service centre provided by the Sarawak state government which houses a kindergarten, a computer center that provides internet access, and a room for handicraft activities, where craft items are collected by handicraft extension workers to be taken to markets. There is a small general shop in Long Beruang that is run by a villager who stocks and sells subsidized essential goods. However, many Penan also shop around for the best prices in Long Peluan and Long Banga. Men work periodically in logging camps or in the offshore oil and gas industry based in the coastal town of Miri, close to the mouth of the Baram river. Despite logging in the area, forest resources are still a significant source of income. This is derived mainly from selling rattan mats and baskets to neighbouring villages for use in processing rice, collecting agarwood (*Aquilaria beccariana*) and selling wild meat, fish and vegetables to other villages in the vicinity. Children attend primary school at Long Banga – 20 minutes away by road – and secondary school at Bario, the district centre, three hours away by road, or at Marudi, which is a full day's journey away.

Aquilaria beccariana Tiegh.
[Thymelaeaceae]

Agarwood remains an important trading product for the Penan, although the trees are a threatened species. The fungi-infected heartwood has high value for making incense, perfume and traditional medicine.

Conclusion

The Kelabit people of Long Peluan talk about the Penan as being an integral part of their own history of settlement in the upper Baram area and their stories reflect the encouragement and support they gave to the Penan as they came to settle and become shifting cultivators of upland rice. The Penan narratives are, for the most part, positive about the experience of settling down as farmers. Furthermore, it is difficult to find elders who express the wish to return to the realities of life as forest nomads, despite nostalgic memories of the forest and the seasonal abundance of food. The people of Long Peluan now rely on the Penan to help them labour in their swiddens, as well as providing wild meat, fish and vegetables in return for cash payments. They begin to look on the Penan differently, and say 'the Penan are like us'. Gone are the times

FIGURE 9-8: Brothers in the Sarawak forest: Lake' Tawan of Long Beruang (Penan – standing) and Malian Tepun of Long Peluan (Kelabit).

when the Penan were outsiders, reluctant to meet the Kelabit because the mere sight of them brought screams of fear from Kelabit children. Nowadays, when elders from Long Beruang and Long Peluan come together they reflect on their shared history and acknowledge common experiences that have created strong bonds akin to brotherhood. In the words of Lake' Tawan:

> Our parents worked together to start life here. They [the Kelabit] gave us clothes and bush knives … Malian [Tepun] said we are brothers, we have the same soul, the same mind, the same understanding (*kua kenin kua seruh kua jam*). That's what our parents did to start life here (Figure 9-8).

References

Bala, P. (2002) *Changing Borders and Identities in the Kelabit Highlands*, Dayak Studies Contemporary Society Series no. 1, The Institute of East Asian Studies, Unit Penerbitan Universiti Malaysia Sarawak, Kuching, Malaysia.

Brosius, J. P. (2007) 'Prior transcripts, divergent paths: Resistance and acquiescence to logging in Sarawak, East Malaysia', in P. Sercombe and B. Sellato (eds) *Beyond the Green Myth*, NIAS Press, Copenhagen, pp.289-333

Chan, H. (2007) 'History and the Punan Vuhang: Response to economic and resource change', in P. Sercombe and B. Sellato (eds) *Beyond the Green Myth*, NIAS Press, Copenhagen, pp.199-226

Janowski, M. (1997) 'The Kelabit attitude to the Penan: Forever children', *La Ricerca Folklorica* 34, pp.55-58

Langub, J. (2003) 'Penan response to change and development', in C. Padoch and N. L. Peluso (eds) *Borneo in Transition: People, Forests, Conservation and Development*, Oxford University Press, Kuala Lumpur, pp.131-150

Malone, P. (2014) *The Peaceful People: The Penan and their Fight for the Forest*, Strategic Information and Research Development Centre (SIRD), Petaling Jaya, Malaysia

Mashman, V. (2011) 'Tapioca for breakfast with the Penan in Long Beruang', *The Sarawak Museum Journal Centenary Issue* LXIX(90), pp.43-52

Mashman, V. (2012) 'The baskets of the Kelabit of Long Peluan and their neighbours', in B. Sellato (ed.) *Plaited Arts and Crafts from Borneo*, Lontar Foundation, Jakarta, and NIAS Press, Copenhagen, pp.176-192

Murang, O. (1993) 'Kelabits, Sa'bans, Kenyah and their Penan neighbours: Partners in development', in V. H. Sutlive Jr (ed.) *Change and Development in Borneo*, Borneo Research Council Inc., Williamsburg, VA, pp.83-100

Nicolaisen, J. (1976) 'The Penan of the Seventh Division of Sarawak: Past, present and future', *Sarawak Museum Journal* 24(45), New Series, pp.35-61

Rahman (2012) 'Long Beruang', in I. Zawawi and M. S. NoorShah (eds) *Masyarakat Penan dan Impian Pembangunan* (Penan Society and Imagined Development: Narratives of Marginalization and Identity), Strategic Information and Research Development Centre (SIRD), Petaling Jaya, Malaysia, pp.99-107

Sarawak Development Institute (2014) *Socio-Economic Survey for Kedaya Telang Usan Miri Agropolitan Project*, Sarawak Development Institute, Kuching

Sellato, B. (1994) *Nomads of the Borneo Rainforest*, University of Hawaii Press, Honolulu

Sellato, B. and Sercombe, P. (2007) 'Introduction: Borneo, Hunter-Gatherers and Change', in P. Sercombe and B. Sellato (eds) *Beyond the Green Myth*, NIAS Press, Copenhagen, pp.1-49

Sercombe, P. and Sellato, B. (2007) *Beyond the Green Myth*, NIAS Press, Copenhagen

Uda, E. (2012) 'Penanaman padi bukit dan ekonomi di Long Lamai (Cultivation of Hill Rice and Economy in Long Lamai)', in I. Zawawi and M. S. NoorShah (eds) *Masyarakat Penan dan Impian*

Pembangunan (Penan Society and Imagined Development: Narratives of Marginalization and Identity), Strategic Information and Research Development Centre (SIRD), Petaling Jaya, Malaysia, pp.71–78

Zawawi, I. and NoorShah, M. S. (2012) *Masyarakat Penan dan Impian Pembangunan* (Penan Society and Imagined Development: Narratives of Marginalization and Identity), Strategic Information and Research Development Centre (SIRD), Petaling Jaya, Malaysia

Notes

1. For more details see Rahman (2012, p.102) and Malone (2014, p.158).
2. The Kelabit say that the last of their forefathers to catch a rhinoceros in the headwaters of the Bale river were named Tama Bun and Araya Ewat, and the event occurred about three generations ago. When I refer to the Kelabit of Long Peluan, I include the Sa'ban, Ngurek and Kenyah who have married into the longhouse and their mixed-blood descendants.
3. At other times a body would be wrapped in a palm leaf cover or a mat and placed between the buttress roots at the base of a tree, or left in the hollow of a tree.
4. Also known as Lahang Apoi, Tama Lawai Lahang gained literacy skills at school in Indonesia, and at a mission school in the Sarawak town of Lawas. He taught Kelabit elders literacy for several years at Long Peluan.
5. Tama Tuloi was Soh Tepun's name, taken on the birth of his first child.
6. They continued farming in the Beruang river area until 1969.
7. Sellato (1994, p.179) notes that the Penan did not like the taste of rice. In the case of the Penan at Long Beruang, there was little evidence of this.
8. See Uda (2012) for further descriptions of Penan terms for the processes of shifting cultivation.
9. One 'sack' contains about 30 kilograms of unhusked rice.

The Batak of Palawan are another ethnic minority in the Philippines that hasn't yet fully made the transition from hunting and gathering to farming. The Batak hunter shown here had strung a vine across the forest canopy, near Tina village, and was crossing it to gain access to the lair of a flying squirrel that he hoped to capture.

Sketch based on a photo by Malcolm Cairns in 1993.

III. FIRELESS SHIFTING CULTIVATION

A Mentawaian farmer clears a forest plot for cultivation of food crops in South Siberut, in the Mentawai archipelago, off West Sumatra, Indonesia. These farmers have developed a shifting cultivation system that does not use fire and transforms forests into agroforests (see Darmanto and Persoon, ch. 10, this volume).

Sketch based on a photo by Gerard Persoon in August, 1985.

Synthesis

CAN SHIFTING CULTIVATION BE TRANSFORMED BY GOING FIRELESS?

*Eklabya Sharma**

Introduction

Shifting cultivation, negatively described as slash–and–burn, has long been viewed by state and development agencies as an outdated and destructive practice. Generally, the traditional practice is accused of causing air pollution and contributing to climate change (Tinker, 1996). It is also blamed for causing other adverse environmental impacts such as deforestation, loss of forest biodiversity, soil erosion and nutrient leaching (Kerkhoff and Sharma, 2006). Efforts to eradicate or transform the practice into settled or more benign forms of agriculture have largely been unsuccessful. Shifting cultivation still persists worldwide as a dynamic mix of cultivated and natural land cover spread over 280 million hectares worldwide (Heinimann et al., 2017).

In some pockets across the eastern Himalaya, Southeast Asia and Oceania, farmers have introduced fireless systems that offer a positive feature within the diversity of shifting cultivation practices. In the following section, the authors of three chapters on fireless shifting cultivation have described variations in these innovative practices from northeast India, Indonesia and Papua New Guinea. These no-burn systems seem to have evolved in response to specific local needs in the three study areas. The cultivation of a variety of crops and fallowing, without burning, can therefore be viewed as location-specific adaptation strategies. Fireless shifting cultivation is a movement towards a semi-sedentary form of agriculture, and commercialization of harvests from these fields holds promise of transformation towards sustainable practices. However, the issues of large areas of land required for this practice and land-tenure rights will need to be addressed first.

In the Papua New Guinea case study, regular slash-and-burn is most common (Bourke et al., 1998). However, 25% of the rural population follow a practice

---

* Dr. Eklabya Sharma was formerly Deputy Director-General of the International Centre for Integrated Mountain Development (ICIMOD), Kathmandu, Nepal.

involving little or no burning after the vegetation is slashed. This study shows unique innovations by farmers to meet the challenges of high rainfall, leaching and low soil fertility (Bourke and Allen, ch. 12, this volume). The no-burning practice reduces soil erosion, decreases nutrient leaching and improves soil fertility. In this new form of shifting cultivation, undergrowth is cleared without fire, the crops are seeded, and trees are slashed for crops to grow between the fallen branches. Interestingly, this agricultural system involves slashing trees and allowing them to fall onto the growing food crops. Soil fertility is managed by creating compost mounds, rotating root and legume crops, and transferring large volumes of organic matter from swamp taro growing in ditches to the cultivated fields. Litter decomposition happens quickly, within a span of two months, making nutrients easily available to crops in the field. Another strategy employed by these same farmers is deriving bulk food from tree crops, commercializing the products and achieving food security by trading. The key innovations in this study are therefore sustainable agriculture and the commercialization of commodity tree crops.

The case study from northeast India shows that the majority of farmers follow a combination of three forms of agriculture – shifting cultivation, terrace cultivation and home gardens. In the pursuit of shifting cultivation there are three variants in the study area (Singh and Choudhury, 2015). The first is conventional slash-and-burn with one or two years of cropping before moving to a new plot; the second is similar to the first, but with extended years of cropping termed 'two-year-plus'; and, the third variant is the practice of not using fire. The fireless system was observed only in small sections or by a few households, but in one village 75% of farmers followed this practice (Singh et al., ch. 11, this volume). It was found in areas where vegetation was predominantly shrubs or grasses interspersed with a few nitrogen-fixing alder trees. In these areas, which produce less biomass, the farmers innovated the 'no burning' practice to conserve organic matter for mulching and increasing soil fertility. Cultivation of legume crops in the first year of cropping enhanced soil fertility in subsequent years. The need for labour was substantially decreased. Farmers also commercialized commodity crops for higher income. This system shows ecological stability and traditional farmers shifting to entrepreneurship to attain economic gains.

The Indonesian case study describes the conversion of forests into agroforestry without the use of fire (Darmanto and Persoon, ch. 10, this volume). In these landscapes, the transition from settlement to agriculture fields to vegetation of secondary- and primary-forest is gradual, and in some cases hardly noticeable (Persoon, 2001). This is mainly because fire has not been used in the conversion of forests into agroforestry. This comes from the traditional belief that fire is a 'hot' element that will cause sickness and even death, and communities fear the use of fire in opening up forests. The innovation in this case is a cycle beginning with the clearing of forests, followed by the cultivation of food crops like tubers and bananas, and a final stage involving the return of forest characteristics. This practice actually reduces run-off and leaching of nutrients, and leftovers from the original vegetation help the growth of cultivated crops. This system mainly produces fruit from tree crops and tubers. Coconuts are the most noticeable fruit crop in forest gardens, but there are also three species of durian, jackfruit, mangoes, *Garcinia* sp., *Lansium* sp., *Nephelium* sp. and many others. Local trees used for construction are *Shorea* and *Dipterocarpus* sp. Farmers follow territory ownership by adopting a patrilineal group lineage. In this case, they opted for transformation by adapting to new crops and animals. However, migrant farmers arriving in the study area are introducing the use of fire for clearing forests. These external influences reveal persistent pressure on indigenous communities and their ways of life. Aspirations are changing; future forest cultivation is seen to have perennial crops, to ensure sustained income against the alternative of collecting forest products and trading.

All three chapters describe unique fireless practices where the cleared biomass is used to protect the soil and grow food crops in high rainfall areas, thereby reducing nutrient leaching, helping with weed control and increasing soil fertility. Biodiversity loss, deforestation and infestation by noxious weeds are relatively less in areas with these practices, compared to conventional shifting cultivation areas. These fireless practices and the use of fruit trees have resulted in a semi-sedentary form of agriculture providing permanency to its inhabitants. In all the case studies, farmers showed signs of developing entrepreneurship and an inclination for trading produce for better economic return. This emerging transformation of agricultural practices seen in these three studies demonstrates ecological stability and economic sustainability, which may be the way forward for many shifting cultivation farmers and landscapes. Fireless shifting cultivation also fits well with climate action to meet the challenges of global climate change and ensuring food security among indigenous communities who follow traditional livelihood practices.

References

Bourke, R. M., Allen, B. J., Hobsbawn, P. and Conway, J. (1998) *Agricultural Systems of Papua New Guinea*, Working Paper No. 1, Papua New Guinea: text summaries (two volumes), Australian National University, Canberra, available at https://openresearch- repository.anu.edu.au/bitstream/1885/110392/1/01_Bourke_Agricultural_Systems_paper%20 1_2002.pdf, accessed 1 March 2019

Heinimann, A., Mertz, O., Frolking, S., Egelund Christensen, A., Hurni, K., Sedano, F, Parsons Chini, L., Sahajpal, R., Hansen, M. and Hurtt, G. (2017) 'A global view of shifting cultivation: Recent, current, and future extent', *PLoS ONE* 12(9), e0184479, doi:10.1371/journal.pone.0184479, accessed 8 December 2021

Kerkhoff, E. and Sharma, E. (2006) *Debating Shifting Cultivation in the Eastern Himalayas: Farmers' Innovations as Lessons for Policy*. International Centre for Integrated Mountain Development, Kathmandu, Nepal

Persoon, G. A. (2001) 'The management of wild and domesticated forest resources on Siberut, West Sumatra', *Antropologi Indonesia* 25(64), pp.68-83

Singh, L. J. and Choudhury, D. (2015) 'Fallow management practices among the Tangkhuls of Manipur: Safeguarding provisioning and regulatory services from shifting cultivation fallows', in M. F. Cairns (ed.) *Shifting Cultivation and Environmental Change: Indigenous People, Agriculture and Forest Conservation*, Earthscan from Routledge, London, pp.449-467

Tinker, P. B., Ingram, J. S. I. and Struwe, S. (1996) 'Effects of slash-and-burn agriculture and deforestation on climate change', *Agriculture, Ecosystems and Environment* 58, pp.12-22

10

WITHOUT FIRE

Turning forests into agroforests on Siberut, Indonesia

*Darmanto and Gerard A. Persoon**

Introduction

One of the striking surprises during biodiversity surveys on the island of Siberut, in the Mentawai archipelago, West Sumatra, Indonesia, is the relatively high number of fruit trees and other useful plant species in what initially seem to be dense 'virgin' forests. In the course of the island's recent history, numerous inventories have been made of its biodiversity and the frequent occurrence of perennial fruit trees has been a persistent 'discovery'. Closer examination has revealed that the forests are full of traces of human activity and indications of migration across the island's various watersheds.

In most areas of the tropics that are, or once were, dominated by rainforests, there used to be a quite obvious transition when moving beyond the village zone and its various agricultural activities into the forest: the vegetation types changed, tall trees became dominant while the undergrowth became thinner. Often the transition zone was easily recognizable, with newly cleared fields and the remains of recent fires. But the island of Siberut, off the west coast of Sumatra, is different. There are no signs of burnt vegetation and the transition from settlements to intensely used agricultural fields to secondary and primary forest types of vegetation is gradual or in some cases even hardly noticeable. In a comparative perspective, this appears to be a rather rare phenomenon (Spencer, 1966; Schieffelin, 1975).

* Dr. Darmanto is a research fellow at the Oriental Institute, Czech Academy of Science, Prague. He graduated from Leiden University in The Netherlands in 2020, writing a PhD thesis on the social values of food and food-related-activities on Siberut island. From 2006 to 2011, he worked for the UNESCO Man-and-Biosphere Siberut project. Dr. Gerard A. Persoon is Emeritus Professor of Environment and Development at Leiden University in The Netherlands. He undertook initial fieldwork on Siberut from 1979 to 1982 and spent many periods there in later years for various research projects or in the context of conservation projects.

In this chapter we will discuss the traditional system of land use on the island, which is characterized by its lack of fire in the process of transforming the rainforest into an agroforest. This process will be described in detail, with reasons why we consider this to be a sustainable tradition in the wide range of indigenous practices of shifting cultivation in Southeast Asia.

The island of Siberut

The island of Siberut is the largest of the four main islands of the Mentawai archipelago, off the west coast of Sumatra (Figure 10-1). Its total land mass is about 4030 sq. km and the present population is about 35,000. Most of them are indigenous Mentawaians, but there are also substantial groups of migrants who have come to the island in search of arable land, or as fishermen, traders, civil servants or teachers, either in governmental schools or those of the catholic mission. These migrants are mainly of Minangkabau, Javanese or Batak descent.

Traditionally, the population lived in large communal houses called *uma*, along the banks of the rivers that cut through the dense tropical lowland rainforest in the hilly landscape. These rivers were also used for human transport and to deliver forest products to the small harbour villages at the mouths of the main rivers. People obtained food from various agricultural activities. Sago has always been the staple food

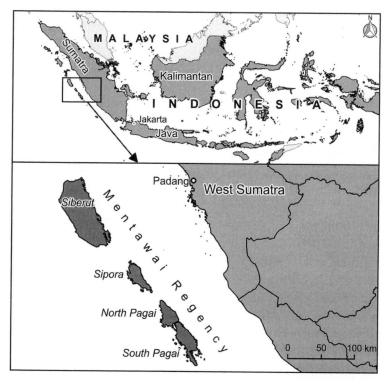

FIGURE 10-1:   The island of Siberut, in Indonesia's Mentawai archipelago.

and additional food has been obtained from hunting, gathering and fishing. Trade in non-timber forest products such as rattan provided exchange products like ironware, tobacco, cloth, coffee, sugar and other goods. For many years, copra has been the main product from the island's coconut crop. Socially, Mentawaian society is egalitarian. There are patrilineal groups, also called *uma*, but there is no hierarchy between their members, apart from variations based on skills and age. The voices of women and younger males may not be equal to those of adult males in public gatherings, but they have autonomy in the domestic space. The groups are exogamous: women leave the group in which they were born and marry into another group. Basically, every man and woman is free to perform the tasks needed to make a living on the island; there is no division of labour, apart from that between men and women. The only exception is the medicine man, who performs specific tasks related to people and the way they deal with the environment.[1]

It was not until the beginning of the 20th century that Siberut became a part of the Dutch East Indies. Until that time the Dutch colonial powers were not really interested in the Mentawaian islands: the population was considered to be rather primitive and there were no interesting resources that were considered valuable by the colonial government. The area was also believed to be unsuitable for the establishment of plantations of commercial crops like rubber. Once in a while, an inspection visit was made to the islands by military officials stationed at Padang, the capital of West Sumatra. Protestant missionaries from the Rheinische Mission in Wuppertal, Germany, were invited by the colonial government to 'start the civilization process' through conversion to Christianity. The first colonial administrators were stationed on the Mentawaian islands at the end of 19th century, with a military outpost. In 1909, the first missionary, Lett, was killed by the locals. Soon afterwards, the colonial government decided to expand its military presence on the islands, and suppression of headhunting was one of the first aims. A number of violent clashes followed, between soldiers and local communities on the islands (Van Beukering, 1947; Schefold, 1990, 2017). Officials and missionaries also tried to persuade the local population to begin cultivating rice, which was regarded as a superior food to the traditional staple, sago. Because of the limited need for labour in producing sago, it was also considered to be a 'lazy man's food' (Persoon, 1992).

The Mentawaians believe that all elements in the natural world, including animals, plants, rivers, the sea and natural forces like the wind, have spirits and souls. Every creature can also radiate and emanate power (*bajou*) that can do harm to human beings and cause sickness. In their daily activities, men and women routinely disturb the environment by engaging in agricultural activities or going fishing or hunting. As souls and spirits are invisible, but can produce *bajou*, extensive rituals have to be performed and special offerings made to restore the balance between humans and the environment. Acts of offering (*pasibulu*) are the core activity in the local religion, called *sabulungan*. Extensive taboo periods, in which no productive labour may be performed, are an important part of the ritual cycle. Local medicine men, or *kerei*, play a key role in the performance of the necessary rituals.

Officialdom perceived the *sabulungan* religion as the core of a primitive culture that prevented the Mentawaians from advancing along with the majority of Indonesia's peoples. After the country declared independence in August 1945, local administrators maintained colonial policies that suppressed some local religious practices and promoted regular village life, so as to enable children to go to school. In 1954, local people were forced to abandon the practice of *sabulungan* and to choose one of the officially recognized religions. In the 1970s, under the regime of President Suharto, the forests of Siberut were granted to several logging companies. At the same time, the Department of Social Affairs began a civilization and development programme, based on resettlement policies. New villages, with houses for nuclear families, were built along the coast and people were forced to leave their traditional communal houses. Cultivation of rice and commercial crops like cloves was strongly promoted and pig raising was banned in the new villages. Schools, churches and polyclinics were built in the settlements to bring mainstream Indonesian village life to isolated Siberut (Departemen Sosial, 1994). As a result of this government interference, the population of Siberut became more concentrated in resettlement villages along the coast, whereas in the past its more than 250 patrilineal groups were scattered all over the island.

As this was happening, Siberut began to attract the attention of conservation agencies and organizations for the protection of indigenous peoples. The publication of a report entitled 'Saving Siberut', followed by its implementation in the field, was the first serious attempt by the World Wildlife Fund (WWF) to protect the island's wildlife and its traditional culture (WWF, 1980). Many more such attempts were made in following years by various conservation and donor agencies, including the Asian Development Bank and Conservation International. Then, in the 1990s, Siberut was 'discovered' as a tourist destination for Western backpackers in search of a 'Stone Age culture', and government interference became more relaxed. In fact, Siberut began to appear on touristic maps of Indonesia with illustrations of tattooed medicine men wearing loin cloths and glass beads, in a bid to attract more people (Figure 10-2). This was in sharp contrast to the repressive cultural policies of earlier decades (Persoon, 1994). Another

FIGURE 10-2: A present-day medicine man searches for special plants in the Mentawaian forest.

Photo: Gerard Persoon.

recent 'discovery' was the giant waves for surfing around the smaller Mentawaian islands. While the local people had been afraid of the breaking waves, professional surfers were attracted by them. Now, Mentawai is classified as one of the best surfing sites in the world, and it attracts large numbers of skilled surfers. This new attraction is likely to have a major impact on the islands in the future (Sentosa Group, 2017).

In terms of administrative and governmental structure, the Mentawai archipelago used to be a part of Padang-Pariaman district (*kabupaten*), in the province of West Sumatra. However, in 1999, soon after the fall of President Suharto, the Mentawai archipelago gained the status of an autonomous *kabupaten*. This implied, among other things, that the islanders could elect their own leaders. But it also implied that, more than in the past, the area had to generate its own financial resources. Since 1999, the district has built up its own bureaucracy with local staff, while the number of administrative subdistricts (*kecamatan*) and village units (*desa*) has been expanded to 10 and 43, respectively (Eindhoven, 2019).

Man-made ecosystems and productive zones

The people of Siberut have vocabularies for different types of ecosystems and productive zones. Understanding how they perceive and manage natural ecosystems will enable us to examine the whole cycle of shifting cultivation, particularly the absence of fire in any stage of cultivation.

The people of Siberut claim that real 'empty land' never existed on their island (Persoon, 2001). They differentiate between various productive zones that can be exploited in a number of ways. In the past, unoccupied forest land was recognized among and between the patrilineal groups (*uma*) on the island, and every man and woman – as a member of one of these groups – had access to the island's natural resources within a particular part of a watershed.

Generally, Mentawaians divide the entire space on the island into various types of ecosystems. The first is the domesticated and semi-domesticated ecosystem, containing several productive zones: dwelling places (*pulaggaijat*); home and forest gardens (*tinungglu* and *mone*); sago gardens (*pusaguat*); and taro fields (*pugettekat*). These spaces are characterized by intensive activities, and each zone has its own values. *Pulaggaijat* is the most domesticated space. It refers to the flat area in a valley that is occupied by human settlement. Derived from the word *laggai* meaning 'trunk' or 'stone', *pulaggaijat* has a figurative meaning as the place of origin and is associated with a particular *uma* that discovered and claims the land. Nowadays, *pulaggaijat* means hamlet, or village, and is defined by intense social interaction, houses, and daily human activities such as cooking and eating.

In the area near riverbanks, or in the wettest parts of gardens, sago palms (*pusaguat*) and taro fields (*pugettekat*) are cultivated (Figure 10-3). While sago is strongly associated with men, taro is exclusively planted by women. Despite these gendered resources, sago and taro gardens are sometimes literally located side-by-side, where the work of men and women is complementary. Sago stands provide the staple

FIGURE 10-3: A stand of sago palms on Siberut, with a taro field in the foreground.

Photo: Gerard Persoon.

food, but apart from the starch harvested from the palms, they also yield many other valuable products, such as sago grubs, leaves for roofing material and bark for walls. Unprocessed sago is also fed to chickens and pigs. Taro fields are commonly fenced to keep pigs out. Taro is mainly used to produce dumplings called *subbet*, made up from mashed taro and banana rolled in grated coconut. These are eaten for breakfast and are an important food during rituals. The taro fields also provide a source of protein as the ponds give women an exclusive area in which to catch frogs and little fish (Figure 10-4).

The forest gardens are usually found at some distance from the settlements. There are two types, or cyclical stages, of forest gardens: *tinungglu* and *mone*. These will be described in detail in the next section. Beyond the forest gardens is the real forest, or *leleu*. The forest is regarded as prime undomesticated space with its own 'hidden culture' (Schefold, 2017) – a world of autochthonous and ancestral spirits whose relationships with the living are maintained through rituals. The forest provides game animals (primates, deer, birds, lizards and many smaller animals), edible and medicinal plants and timber and bamboo for constructing houses or canoes. Rattan is also an important product from the forest, mainly for commercial purposes, but also for domestic basket-weaving.

FIGURE 10-4: A woman catches frogs and small fish in a taro field.

Photo: Gerard Persoon.

Finally, there are the rivers and creeks and the coastal area, including mangrove forests. Rivers and small lakes (*bat oinan*, bodies of fresh water) are part of undomesticated space. Their main use is for transportation, but they are also a source of daily protein, such as small fish, shrimps and molluscs. *Bat oinan* tend to be associated with women, as a wide range of fishing techniques is used mainly, but not exclusively, by women to catch these animals with various nets and traps. Another undomesticated ecosystem is *nusa*, a local name for the broader coastal zone. It is specifically associated with small islets scattered off the east coast of Siberut, but also includes mangrove forests, coral reefs, beaches and tidal flats. Some islets have been domesticated and converted into coconut groves. Mangrove forests produce stems and poles for house construction while coral reefs in shallow water are important for both men and women to catch fish, hunt turtles and collect crabs and molluscs.

Metroxylon sagu Rottb. [Arecaceae]

The sago palm is the source of Siberut's staple food, and has a long history of supporting ancient populations in lowland Southeast Asia. It is a suckering or multi-stemmed palm with each stem flowering only once. The sago starch is harvested from the stems by washing it out of the pulverized pith with water.

The *tinungglu-mone* cycle: transforming primary forest into agroforest

The establishment of forest gardens involves two complementary processes: *tinungglu* and *mone*. *Tinungglu* is an opening stage, associated with the clearing of forest and the cultivation of food crops, especially tubers and bananas, while the *mone* is the later, final stage, associated with the return of forest characteristics and the dominance of fruit trees. *Tinungglu-mone* is a cycle that produces a kind of integrated shifting cultivation system within the tropical rainforest of Siberut (cf. Conklin, 1957). As we will describe, the two stages of the cycle are strongly linked to each other.

The opening of the *tinungglu* begins with the clearing of a small patch of forest or an old forest garden. Once a man has decided that it is time to clear a new field, he goes into the forest that is part of the territory of his patrilineal group, within a particular watershed. This may be a substantial distance from his settlement. While inspecting the forest, he looks for indicators that tell him something about the fertility of the soil. He also looks for beehives and useful wild plants that can be harvested now or in the future, such as rattan. Steep slopes are usually avoided, but if there is a swamp in that part of the forest, it may be used to plant sago palms. He makes a rough calculation of the area he needs to clear to serve his needs for the coming years. The fields are usually between half a hectare and one-and-a-half hectares. In some cases, the man's relatives (brothers or sons) may decide to clear an adjacent field at the same time.

Before starting to clear the undergrowth, the man performs a ritual in which he asks the forest spirits for permission to cause the harm that he is going to inflict on the trees and the land. The ritual involves the sacrificing of one or more chickens as an offering to the spirits of the land and trees. After a few days the actual clearing can start. First, with the use of a sharp machete, all of the undergrowth, including small trees and shrubs, is cut.

One of the interesting aspects of the *tinungglu* phase is that the new fields are cleared without the use of fire. Fallen trees, weeds, grass, wild vegetation and other debris are not burnt. Mentawaians believe that felled trees contain and emanate an enormously dangerous power (*bajou*) that can harm people, and the 'hot' element from the fire can trigger the release of *bajou* from the slashed plants. *Bajou* from fallen trees is believed not only to do harm, but also to cause a serious sickness called *oringen*, which can only be cured by a shaman in a healing ritual called *pabetei*. From an ecological perspective, there is a clear reason why fire is unwanted. Mentawaians are not seed cultivators; their diet relies mostly on sago, tubers and bananas. These staple food plants are capable of competing with the weeds and grasses that grow quickly in the high rainfall and wet climate of an opened area. These plants also lack serious natural diseases. So, instead of burning the forest, the slashed vegetation is left as a mulch.

Once the cleared vegetation has withered somewhat, the forest floor is further cleared to enable the planting of seeds and seedlings. A large variety of tubers and bananas is cultivated, along with vegetables, annual fruits and various useful species of bamboo, rattan, sugar cane, pepper, medicinal and ornamental plants, and plants that produce various poisons that are used in hunting and fishing. The seedlings are taken from other fields or from relatives, and are raised near the houses or in other locations in the weeks or even months before the new field is cleared. Both men and women contribute to the work. Men tend to be associated with sago, bananas and other useful species, while women bear prime responsibility for root crops, especially taro, cassava and sweet potatoes.

After the planting is finished, the most difficult part begins: the cutting of all the remaining mature trees, some of which may be forest giants with impressive trunks. The trees are felled with axes. In some cases, a platform two or three metres high is constructed around the trunk of a tree in order to avoid the need to cut extensive buttresses (Figure 10-5). More recently, chainsaws have been used in this clearing process. Before the larger trees are cut, some of the smaller trees are partially cut. Then, when the larger tree falls, it takes down the smaller trees with it – provided the direction of the large falling tree has been accurately calculated. Only skilful cutting can achieve this labour-saving trick. Some useful trees may be left standing. When the felling of the large trees is complete, the forest field looks absolutely chaotic (Figure 10-6). A layer of tree trunks lies some metres high, the trees having fallen over each other, leaving branches thick with leaves as a tangled barrier to anyone trying to cross the field. All in all, the preparation of the field may take more than two months of hard work. Even the fallen forest giants are not burnt, but are allowed to gradually decompose. By this method of forest clearance, the forest floor is never directly exposed to rain, sunlight or wind. The withering vegetation is gradually replaced by a new layer of vegetation rising from the planted seeds and seedlings. Topsoil run-off and leaching of nutrients from the forest are greatly reduced. In fact, the 'leftovers' from the original vegetation help the growth of both the cultivated crops and the wild but useful plants in the next stage.

FIGURE 10-5: After the seeds are planted, it's time to cut down the mature trees. An axeman begins the Herculean task of felling a forest giant.

Photo: Gerard Persoon.

FIGURE 10-6: Bananas and cassava force their way through the tangled debris after the opening of a new forest garden on Siberut.

Photo: Gerard Persoon.

After a number of weeks, when the foliage has shrivelled, the owner of the field may return to do some additional trimming of tree branches to make it easier to walk around amid the forest debris. Meanwhile, many of the newly planted seedlings will have found their way in between the tree trunks and shrivelling vegetation. Gradually, fresh cassava and banana leaves begin to appear, and with regular pruning and cutting of regrowing vegetation, the cultivated food plants are able to grow well. The tubers and bananas mature quickly and after about four to six months, they can be harvested. The first harvest marks the fresh garden (*tinungglu*) phase. A small ritual is commonly enacted as part of the harvest, to mark the cycle of cultivation. After the first year, the *tinungglu* phase may continue to provide good harvests of some crops for about three years, depending on the quality of the land and the ability of the plants to continue producing food.

When production from the tubers and bananas begins to decline, the owner of the field may do some enrichment planting of seeds and seedlings. The cultivation of fruit trees is, for the most part, neither seasonally determined nor rigorously planned. The cultivators occasionally collect fruit-tree seeds from mature fields or nearby sites. Slowly, crops like cassava and bananas are replaced by fruit trees and other crops that grow well in the shadow of the other plants and begin to dominate the vegetation. Tubers and bananas no longer do well because the spreading canopy of the maturing fruit trees creates too much shade.

Across all fruit species, coconuts (*Cocos nucifera*) are perhaps the most noticeable feature of forest gardens, providing a stable source of food and by-products for many purposes, such as hut walls, skirts, mats, drinking cups and tobacco storage as well as yielding copra for commercial purposes. However, the most socioculturally valuable and desirable plant in the forest garden is durian. Mentawaians cultivate three species of durian: *toktuk* (*Durio oxleyanus*), *posinoso* (*Durio graveolens*) and *doriat* (*Durio zibethinus*). The presence of all three durian species defines, and is something indicative

of, a forest garden. By definition, a *pumonean* (mature garden) is plural for many *mone* (durian trees). Alongside durian is a broad range of fruit trees such as *langsat* (*Lansium parasiticum*), *rambutan* (*Nephelium lappaceum*), jackfruit (*Artocarpus heterophyllus*), mango (*Mangifera indica*), mangosteen (*Garcinia mangostana*) and many others (Table 10-1).

The forest gardens are gradually filled not only with fruit trees, but also with other useful species (Figure 10-7). Young shoots of rattan and forest-tree species are sometimes taken from the wild and planted alongside other introduced cultivars. Local trees that are important for construction, such as *Shorea* and *Dipterocarpus* species, are cultivated not only to mark out the garden but also as future sources of timber. Special *raggi* trees (*Derris elliptica*), which produce

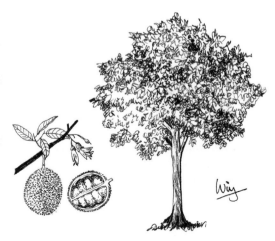

Durio zibethinus L. [Malvaceae]

Durian is the most valued plant in Siberut's forest gardens. This is the most popular species, although two other *Durio* species can also be found in the gardens. Famous for its powerful odour, regard for durian is split between eager devotion and intense disgust. Many different durian cultivars produce fruit for global markets.

TABLE 10-1: Productive and useful tree species planted in Mentawaian forest gardens (after Darmanto, 2006).

Species	Local and common names
Fruit trees	
Lansium parasiticum	Siamung (langsat)
Artocarpus integer	Peigu (cempedak)
Mangifera sp.	Sipeu (wild mango)
Mangifera odorata	Kueni (Kuini)
Nephelium sp.	Bairabbit (wild rambutan)
Garcinia dioica	Kairiggi (button mangosteen)
Syzygium sp.	Ailuluppa (water apple)
Durio zibethinus	Doriat (durian)
Durio graveolens	Toktuk (red durian)
Durio oxleyanus	Pusinoso (wild durian)
Garcinia x mangostana	Lakkopak (purple mangosteen)
Cocos nucifera	Toitet (coconut)
Gnetum gnemon	Tojet (paddy oats)
Baccaurea lanceolata	Teggeiluk (wild langsat)
Citrus maxima	Muntei (pomelo)
Mangifera macrocarpa	Abbangan (wild mango)
Syzygium aqueum	Ailuluppa (water bell)
Syzygium pycnanthum	Ailuluppa leleu (wild rose apple)
Psidium guajava	Sabbui (common guava)
Forest/Wild trees	
Shorea lamellatta	Katuka (white meranti)
Shorea sp.	Sirikdik (red meranti)
Shorea sp.	Karai (red meranti)
Pentace triptera	Kaboi (melunak)
Hopea dryobalanoides	Maencemi (merawan)
Aquilaria malaccensis	Simoitek (agarwood)
Campnosperma auriculatum	Tumu (madang)
Artocarpus scortechinii	Baiko (terap)
Oncosperma sp.	Ariribuk (nibung)
Radermachera gigantea	Elagmata (purple tree jasmine)
Alstonia sp.	Kakaddut (pulai)
Arenga pinnata	Poula (sugar palm)
Derris elliptica	Raggi (tuba)
Bamboo species	
Schizostachyum zollingeri	Manggeak (clumping bamboo)
Gigantochloa apus	Okbuk (string bamboo)
Schizostachyum brachycladum	Sisurat (yellow bamboo)
Dendrocalamus asper	Betuk (giant bamboo)
Introduced tree species	
Myristica fragrans	Pala (nutmeg)
Cinnamomum burmanni	Kayu manis (cinnamon)
Coffea Robusta	Kopi (coffee)

TABLE 10-1 (cont.): Productive and useful tree species planted in Mentawaian forest gardens (after Darmanto, 2006).

Species	Local and common names
	Introduced tree species
Citrus aurantiifolia	Arimau sareu (key lime)
Citrus nobilis	Arimau simananam (common orange)
Citrus hystrix	Arimau boitok (kaffir lime)
Syzygium aromaticum	Cengkeh (clove)
Hevea brasiliensis	Karet (para rubber)
Areca catechu	Pinang (betel nut)
Syzygium malaccense	Ailuluppa sareu (malay water apple)
Nephelium lappaceum	Rambutan sareu (rambutan)
Manilkara zapota	Sau (sapodilla)
Theobroma cacao	Coklat (cocoa)
Mangifera indica	Abbangan sareu (mango)
Artocarpus heterophyllus	Peigu sareu (jackfruit)
Annona muricata	Sirsak (soursop)

poison, are planted alongside introduced species such as nutmeg (*Myristica fagrans*), cinnamon and various citrus species. However, most useful species in the gardens are not cultivated. More than half of the plants used for various domestic purposes grow without – or with very little – human intervention (Darmanto, 2006). About 63% of wild plants used for medicinal purposes (Ave and Sunito, 1990) – to be drunk as herbs (*koiloket*) or used for magic rituals (*gaud*) – are found in the gardens. The absence of burning allows most tree seeds and shoots in a previously forested area to sprout without much disturbance, and the wild species from the forest spontaneously occupy the gardens alongside the mature fruit trees. In the following years, the forest gardens gradually fill with a combination of cultivated and wild species in which fruit trees, particularly durian, dominate the canopy and the vegetation. Enrichment planting may be done whenever there is space on the forest floor or when particular seedlings are available. Wild plants that have no particular function may be cut whenever this is believed to be necessary. So an agroforest slowly replaces the original forest. Together, all of the fruit trees and uncultivated but useful species make up and mark the *mone* cycle.

The growing gardens are visited and tended regularly. Since they are scattered or situated several hours by foot or canoe from settlements, huts may be built to ensure effective care and attention. These range from simple shelters from the frequent rain to regular houses, and they signify the completion of the *mone* cycle. Although temporary, the huts serve as pivot-points for other subsistence activities revolving around forest gardening, such as fishing, hunting, seeking rattan, gathering wild food, making canoes and other related projects. The huts are also important for animal husbandry of pigs and chickens, which is an integral part of forest cultivation and, more broadly, for social exchange. The animals are not only important as a source of protein, consumed mainly in communal rituals, but they also play a role in basic

FIGURE 10-7: The different stages of the *Tinungglu-pumonean* cycle. **(1)** Cleaning the forest floor of shrubs, climbers and small trees; **(2)** Planting of seeds and seedlings of cassava, bananas and numerous other crops; **(3)** Slashing of major trees. Useful trees are left standing. **(4)** After six months, harvesting of first root crops and fruit, enrichment planting of seeds and seedlings. Construction of field hut and introduction of chickens to the field hut. **(5)** Maturing fruit and other trees gradually overshadow the root crops and bananas. Use of some trees for memorial purposes (*kirekat*); **(6)** Some fruit trees gradually dominate the vegetation until it develops into a true agroforest. Wild tree species may be cut.

social relations such as marriage and mortuary ceremonies. Semi-domesticated animal husbandry allows the pigs to roam around the mature *mone*, where they can do little harm because most of the fruit is beyond their reach. The huts enable people to stay in the gardens and take care of their pigs. For older people, they allow them to avoid the noise of life in a settlement and to maintain traditional practices.

The mature forest gardens (*pumonean*) can yield fruit and other useful products for more than three generations. In some cases, they may last for up to five or six generations, as long as the durian trees bear good fruit. When their productivity declines, another site is selected and the process is repeated. However, there is no definitive date on which a mature garden should be abandoned and a new one created. Often, the cycle is repeated, and a mature garden in decline becomes a new *tinungglu*, or opening stage, for the cultivation of food crops and the addition of useful forest tree species. The only requirement is the clearing of old and unproductive fruit trees. New gardens are also commonly created when a couple marries and starts a family, or when a family moves to a new location.

Codiaeum variegatum (L.) Rumph. ex A.Juss. [Euphorbiaceae]

Commonly known as crotons, these shrubs with multi-coloured foliage grow to 3 m, and on Siberut are often used in rituals and may be planted around memorial trees for the deceased (*kirekat*). These shrubs contain an oil that is violently purgative.

The forest gardens are used not only for harvesting fruit and other non-timber forest products, but they also have an important ritual function. Mentawaians believe that when a person dies, his or her soul (*simagere*) leaves the body to wander in the environment. Such wandering souls may bring harm to people, causing illness or even death. Therefore, soon after a person dies, a ritual is performed in which a new body is offered to the soul of the deceased. Usually a big fruit tree is selected for this purpose, and representations of the hands and feet of the person are carved into the bark (Figure 10-8). The identity of the person may be indicated in other ways, such as holes to indicate the size of the person, or in some cases even a face is carved in the tree. These memorials are called *kirekat*, and a ring of ornamental plants, usually crotons, is planted around such trees. From that moment onwards, the tree performs a particular function in the deceased person's community. Under no circumstances can it be cut or sold, or become an element in the payment of bride price or the payment of a fine for a violation of local rules related to theft, adultery, or other offenses. It is no wonder that, in retrospect, it is difficult to clearly establish distinctions between mature

FIGURE 10-8: A *kirekat* in a Siberut forest. This tree has been given to the soul of a deceased person as a 'new body'. The person is identified by the markings on the trunk. Under no circumstances can it be cut or sold.

Photo: Gerard Persoon.

forests, different kinds and degrees of secondary regrowth and new gardens, domesticated or non-domesticated plants, and farming or gathering activities.

Over several centuries, the complementary cycle of *tinungglu* and *mone* has created a patchy forest garden mosaic covering a vast area. The practices of forest cultivation and movement of the people have altered the character of Siberut's primary forest in measurable ways. Cultivation increases the number of useful species, conserves the number of stands of particular tree species, decreases wild species that are not frequently used, creates patches of culturally productive sites in more accessible areas and produces numerous sites of densely planted fruit trees. The distribution of many other useful trees throughout the lowlands of Siberut reflects patterns of human modification and serves as a convenient botanical indicator of settlement and migration history (Tulius, 2012). It is a vivid example of the co-evolution of humans and nature that can be found throughout the Indonesia archipelago (Ellen, 1999).

Transformation and persistence: Introduction and integration of new crops and animals

While Mentawaian forest cultivation is often regarded as an ecologically-sound and sustainable system (WWF, 1980; ADB, 1995), it has attracted little appreciation from the government. Over the decades,

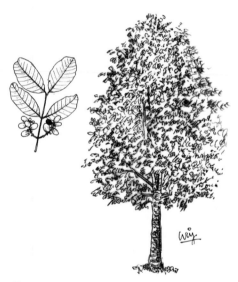

Gnetum gnemon L. [Gnetaceae]

Growing to 20 m, these evergreens are often planted in Mentawaian forest gardens. Their nut-like seeds are used to make a sour soup, are crushed and deep fried to make crackers, and form the basis of many traditional Indonesian dishes. Extracts are being studied for their chemical and medicinal qualities.

the introduction of rice cultivation has been a recurring theme for colonial and post-colonial governments and missionaries seeking to increase productivity and replace sago and tubers, which they considered to be inferior foods. Post-colonial governments enthusiastically introduced rice to replace a culture that was strongly dependent on 'nature' and 'animistic religion'. Cultivating rice was seen as a way to encourage people to become a part of Indonesia's national culture and to accept state intervention, so from the early 1960s, rice cultivation was a core feature of government development objectives (Persoon, 2001). Distributing rice seeds and encouraging people to plant rice were standard development activities.

Decades of encouragement have not seen Mentawaians, particularly those on Siberut, fully embrace rice cultivation. This does not mean that the local population has no interest in cultivating rice; often it is planted on flood plains and in swampy areas along riverbanks, slightly separated from the forest gardens. Principally, rice cultivation does not greatly alter existing cultivation practices, but it does require additional work in clearing fairly low vegetation and digging drainage canals in wet areas. However, the slashed debris is not burned, but mulched to slowly release its nutrients into the soil. Fire is unnecessary, as burning slashed debris in wet areas is considered a waste of energy and also dangerous from a religious standpoint. Over the four-month growing period of the rice, the field is given little attention before being animatedly harvested. While it is hard to say whether rice has been successfully introduced, one regular pattern emerged in the early years of development projects: rice was planted only one or two times. Tools and seeds were initially subsidized, making cultivation attractive. However, as soon as the nutrients in the swamp were depleted and weeds and grass became uncontrollable, rice cultivation lost its appeal. When the government took a more relaxed attitude, rice fields were abandoned and became swamps again.

There are many reasons why rice cultivation has barely disrupted the *tinungglu-pumonean* cycle on Siberut. Ecologically, rice suffers in competition with weeds and grass. Stands of sago and tuber crops enjoy only limited attacks from pests and diseases, whereas rice is susceptible to plant diseases and irregular rainfall. It is also easily destroyed by birds, rats, mice, deer and wild pigs. There are storage problems in a humid climate, especially when storing seed for the following year. Furthermore, successful and sustainable rice cultivation in both hilly and irrigated areas requires a certain level of social organization (Boomgaard, 2003). The collective labour exchanges and hierarchal social structure required for rice cultivation are unknown to the Mentawaians, and much of the work involved in the rice cultivation cycle would keep them away from many social and religious obligations. All rice-cultivation activities must be carried out consistently and quickly, as delays result in lower yields, and the constant planting, tending, guarding and harvesting would break various taboos and rituals. Rice also has a symbolic difference: it lacks the 'richness' of local food, such as fish, shrimp, sago worms and hunted meat and pork. Rice consumption also challenges the Mentawaian core identity of being sago eaters, compared to the majority of Indonesians who are considered to be rice eaters (Persoon, 1992).

Some introduced species have been more successful than rice. The expansion and fluctuations of clove cultivation since the 1970s have been a part of forest cultivation, mainly as a cash crop for both indigenous and in-migrating people. Cloves (*Syzygium aromaticum*) were first introduced in the 1960s by Minangkabau officials with the general view that the trees would become a stable and permanent crop, while being less destructive than local agricultural practices. Clove cultivation was met with little resistance and quickly spread all over the island, because it did not disrupt the forest gardens and was easily planted among mature fruit trees. The plant was not at risk of any serious diseases, needed no intensive labour and bore fruit just a few years after cultivation. The absence of taboos or other related cultural attributes meant the crop was also easily harvested. The only problem was that cloves grew well mainly along the coast and on islets east and south of Siberut that had regular dry winds from the sea. In

Syzygium aromaticum (L.) Merr. & L.M.Perry [Myrtaceae]

The clove tree comes from the fabled 'spice islands' (Maluku) in eastern Indonesia, and is regularly found in Siberut's forest gardens. Its dried flower buds are used as an aromatic food additive, are found in some Indonesian cigarettes, and are also used to treat a wide range of ailments, from toothache to fever and high blood sugar.

these areas, cloves had a more serious effect on forest gardens than other crops. Along the east coast, clove trees dominated – and in some places entirely replaced – fruit trees. In the interior, the effect on existing gardens was less destructive because the trees produced less than regular yields due to the high humidity. In the early 1980s, monopolization of the clove trade by the New Order (Suharto) regime contributed to the downfall of clove prices. The clove trees were soon abandoned, but existing forest gardens remained intact (Persoon, 1997).

When clove was largely abandoned, patchouli (*Pogostemon cablin*) started to boom, with a slight effect on forest cultivation. Locally, this plant is known as *patokkailo* or *nilam*, and is used mainly for medicinal purposes. For world markets, however, it contains a valuable oil used in the perfume and incense industries. An alkaloid compound is obtained through a distillation process involving its fresh leaves. For the Mentawaians, the lack of other cash-earning activities encouraged the intensive incorporation of this crop into existing forest cultivation systems. As patchouli could be harvested within a relatively short period, it became popular with those who needed quick cash. Patchouli required fresh and fertile soil and shade. In the forest gardens, decomposing vegetation provided nutrients, shade and protection against heavy rains during the initial growth phase. The first harvest could be obtained after just six months and subsequent harvests at intervals of a few months. However, planting it with other crops was found to reduce the amount of oil in the leaves and branches, and replanting in the same field failed to improve the yield. The Mentawaian farmers found that after the first harvest, the plants were no longer productive, so it became usual practice to replace them at that stage with tubers, bananas or fruit trees. Various reports have suggested that patchouli generated some deforestation and soil destruction (Ave and Sunito, 1990; Göltenboth and Timotius, 1996). However, it can safely be said that the crop has generally had little impact on the *tinungglu-pumonean* cycle because it is planted in newly cleared fields on which trees and branches have not been burned, and it simply replaces the cultivation of tubers and sago for one or two years. The *tinungglu* phase is a little bit longer as a result and the cultivation of food species delayed. However, patchouli does not fundamentally disrupt the forest cultivation cycle. When the patchouli is exhausted in a year, the opened forest can be filled with tubers, banana seeds and fruit trees, starting a new forest garden cycle.

The forest gardens have not only seen introduced plants, but also animals. It began with a government viewpoint that pigs were dirty animals that spread disease, damaged crops and were a nuisance to villagers. Moreover, raising pigs meant following religious taboos and specific rituals that were seen as irrational and superstitious. Mainland livestock, such as cattle, buffaloes and goats, were considered superior and were introduced through various development programmes to replace pigs. These animals were initially met with some reluctance, but over the years, the Mentawaians found ways to integrate them into their cultivation systems. Goats were the only exception. Cattle, buffaloes and ducks, although they were introduced in order to stimulate modern animal husbandry, were accepted and rather successfully 'indigenized', in that they were raised in the same way as pigs. They were taken upstream to the forest gardens once the fruit trees were sufficiently mature, and were allowed to roam freely without regular fences or tethers. When they were no longer in daily contact with humans, the animals gradually became shy and even semi-wild. Nowadays, they are only caught and slaughtered for occasional communal ceremonies.

The Mentawaians have always had an interest in new opportunities and have never rejected any introduced species, even when they knew it was not suitable for cultivation. All species alien to their culture have been adopted and cultivated in the same way that they have nurtured native species over the centuries. When these introduced plants and animals provide social, cultural and, importantly, economic advantages, the people are keen to adopt them, ferociously exploit them, and gradually integrate them into their forest gardens. When such species do not provide much by way of advantages, the people simply abandon them, return to the old plants, or search for new ones. Introduced plants with little ability to adapt to the island's ecology, such as coffee, cinnamon, nutmeg and areca palm, are found only sparsely and do not receive intensive care in any forest garden. The Mentawaian form of shifting cultivation or forest conversion is not static, unproductive or backward. The histories of various introduced species show that external influences are tested and, to some extent, incorporated into existing forest cultivation systems.

Artocarpus integer (Thunb.) Merr. [Moraceae]

A recent development: introducing the use of fire

Widely known by its Malay name *cempedak*, this large evergreen fruit tree is a close relative of jackfruit and breadfruit. Within the tough-skinned fruit, fleshy, edible arils surround the large seeds. The fruit can be eaten fresh or cooked, and the seeds fried or roasted like chestnuts.

Historically, migrants to Siberut from the mainland paid little attention to agricultural activities. A small number of Minangkabau bought patches of land on small islands off the east coast of Siberut for coconut and clove cultivation. However, the majority of migrants happily filled particular social niches, becoming traders or civil servants. Others depended on fishing for their livelihood and did not compete with Mentawaians for land. Migrants were reluctant to move inland because of the presence of pigs and because growing mainland crops like rice and corn, with their agricultural techniques, would have been almost impossible.

More recently, the picture has changed slightly. It started with 'cocoa fever' in the mid-2000s. Cocoa (*Theobroma cacao*) was voluntarily brought to the island in 1996 by a school teacher who sold cocoa seeds to supplement his low salary. Some families bought and planted the crop in the shade of fruit trees, but failed to give it special care. The landscape of Siberut is suitable for cocoa, as the crop benefits from a 'forest rent', i.e. the soil of a newly-cleared forest plot does not require fertilizer and the disease load is initially low (Ruf, 2011). The opportunity cost is limited because it can be intercropped with annuals such as bananas, pineapples and tubers, and it begins to yield after just two to three years. Thus, the island's ecology was suitable – at least initially – since forested land was relatively abundant. Moreover, entry costs for cocoa cultivation were minimal.

To begin with, there were few signs that cocoa would generate a fortune. However, the collapse of cocoa production in Sulawesi and West Africa due to disease and civil war, respectively (Ruf et al., 1996, Neilson, 2007) changed the fate of the crop. A succession of harvests coincided with good cocoa prices in global markets. A handful of Siberut pioneers sold their first harvest, earned decent profits, and began to enjoy significantly improved lifestyles. Inspired, their neighbours soon took up cocoa cultivation. The main attraction was that cocoa was a perennial, enabling them to adopt a more positive view as 'real farmers' building up a long-term investment. There was even a running joke that cocoa trees could provide growers with a 'pension' in the future.

As cocoa required new agricultural arrangements, it changed the perception of both Mentawaians and migrants regarding land, while significantly transforming the *tinungglu-mone* cycle. For maximum yields, cocoa required an open space larger than the usual forest garden. The crop transformed both forests and what was a diverse semi-domesticated landscape. Farming families favoured the establishment of new cocoa gardens either in a newly opened lowland forest or a freshly cleared sago garden. Conversion of swampy areas to cocoa involved clearing all of the vegetation in a wetland and draining the water.

Unlike clove or patchouli, cocoa could not be easily integrated into the *tinungglu-mone* cycle without disrupting it. As it failed to produce the expected yield when cultivated in combination with fruit trees, cocoa was mostly planted as a monocrop, requiring more intensive labour than a traditional forest garden. A good cocoa garden had to be dry, young sprouts needed to be regularly pruned and grass and shrubs had to be routinely weeded. Thus, the arrival of cocoa transformed the natural undergrowth and shrubs, which were important for preventing erosion, into weeds. Useful but uncultivated species were similarly unwanted. The monocrop trees disrupted the *tinungglu* cycle because they required longer cultivation. Unlike patchouli cultivation, cocoa made impossible the planting of tubers and sago, followed by fruit trees, because each cocoa crop was in place for more than two decades.

The arrival of cocoa was accompanied by new farming techniques. Migrant farmers and a few elite Mentawaians introduced fire in the opening stages of cocoa cultivation. They roughly cultivated plots of between two and 10 hectares – larger than the local standard. In a record sale, an 80-hectare piece of land was sold to a group of migrants and a 60-hectare plot to a high-ranking Mentawaian officer. They hired and instructed wage labour to clear and burn the forest, so the forest and soil were destroyed through unplanned 'slash-and-burn'. Their traditional religion had taught the people that such use of fire was spiritually dangerous and ecologically ineffective, and that if they burnt the forest they would be stricken by a sickness called *oringen*. The proponents of the new techniques pointed out that their burning had not resulted in anyone falling sick, so burning became regarded as an effective way to clear debris and prevent uncontrollable grass. But cultivation of tubers and sago was no longer possible. Wild species were cleared regularly and fruit and forest trees were no longer expected to dominate the vegetation.

There is no doubt that this rapid clearing of forests on Siberut was damaging. Most importantly, it began an erosion of long-standing forest cultivation practices that modified the forest, increased its genetic diversity and usefulness, and permitted extraction on a sustainable basis.

The future

Political, ecological and economic changes have led to a revision of local aspirations for future forest cultivation on Siberut. Across the island, both men and women now express the desire to own perennial cash crops that will enable them to generate a regular income. Moving to the coastal zone and planting monocrops like cloves and cocoa, while catching fish from the sea, is preferable to living in the interior and earning cash solely from collecting forest products and selling them to traders. Many no longer desire a future for their children in cultivating the forest in a traditional setting. Instead, they want education for their children, which will enable them to leave the village in a bid to find permanent employment elsewhere.

Annona muricata L. [Annonaceae]

Soursop is a fruit tree regularly found in Siberut's forest gardens. A native of tropical America and the Caribbean, the fruit is popular around the world as a fruit-juice drink and as candies, or sorbets. Soursop is also widely promoted as an alternative cancer treatment, although there is no medical evidence to support this claim.

The future of existing forest cultivation on Siberut is difficult to predict, although the case of a proposed oil-palm plantation is perhaps enlightening. Between 2010 and 2016, five companies sought a permit to secure 73,000 hectares of land across the Mentawai islands for an oil-palm plantation and forest estate. One or two land claimants were ready to sign an agreement to support the companies and lease their land, but few people welcomed the proposal and the majority strongly opposed it (Puailiggoubat, 2013, 2015). They were worried that they would permanently lose their land after the plantation and the issue of how the companies would, in the future, deal with claimants to the land became a subject for heated discussion. More importantly, the implication that a vast area of forest would be burnt, including sago palms and fruit trees, and replaced with monotonous ranks of oil palms, was beyond the imagination of the people. They frequently remarked that they could not understand how they would survive without food from their *tinungglu* and continued social exchanges without sago, durian trees and pigs from the *mone* (Darmanto, 2016).

Meanwhile, in anticipation of the need for economic crops, some people have tried to intensively domesticate a native rattan species and to plant para rubber; both are seen as a suitable 'fit' for their mature forest gardens (*pumonean*). *Manau* rattan (*Calamus manan*) is a major exported forest product that has been exploited for centuries. The depletion of *manau* in the wild has encouraged people to begin domesticating it. As a native species, there have been few ecological problems integrating *manau* into the gardens and it has gradually become a favourite crop. However, such is not the case for para rubber. Development projects had brought it to the island on several previous occasions, so it was not totally unknown. It was a simple matter to incorporate it into the mature forest gardens. But, until recently, sales of latex have been unheard of, because rubber produces little latex on the island. It is believed that heavy rainfall may be the main reason for this.

Mentawaians may have different degrees of attachment to their current shifting cultivation system and commitment to an agrarian future. However, their attitudes have always been tactical and dependent on their perceived material interests, existing agricultural practices and

Artocarpus heterophyllus Lam. [Moraceae]

Jackfruit is the largest known tree-borne fruit, capable of weighing in at 55 kg each. A mature tree can also produce as many as 200 fruits per year. It can be eaten ripe or unripe, and the seeds are also eaten. The wood is termite-proof and is used for house construction and making furniture.

religious beliefs, in a physical environment that is, at the same time, also a spiritual environment. Therefore, it is not surprising that their modifications of shifting cultivation systems reflect this attitude. The only lesson to be learned from the history of Mentawaian forest cultivation is the emphasis on sustainable reproductive cycles for a combination of food crops, animal husbandry and cash–crop production, including experimentation and the possible incorporation of new crops.

Conclusion: why is there no fire on Siberut?

Why has fire been absent in Mentawaian forest cultivation? From a local viewpoint, fire is a 'hot' element that can activate emanations of power (*bajou*) from felled forest trees and other cleared plants. The collision of fire and the *bajou* of the vegetation can produce an enormous danger that may cause sickness (*oringen*), even death. From an ecological perspective, there are clear reasons: the excessive rainfall on the island, and the absence of a clearly marked dry season, make burning a poor choice. The absence of fire ensures that the shielding properties of cut debris prevents the leaching of nutrients from surface humus and soil, and these can then be used by newly-planted tubers and sago in the *tinungglu* phase (e.g. Thurston, 1997). Twigs, leaves and timber decompose slowly, releasing nutrients at different rates. In this way, the soil is never directly exposed to the sun or rain; it is always covered by vegetation at various stages of growth or decay, thereby reducing susceptibility to erosion. Later, the nutrients can be utilized by fruit and spice trees that are planted while the first crop is maturing in the *mone* cycle. By the time the debris cover has lost its protective properties, the surface soil is already held together by a layer of low grasses, shrubs and fairly low trees. This form of shifting cultivation also prevents the converted forest from being covered by invasive grasses, such as *Imperata cylindrica*, that are usually associated with slash-and-burn agriculture and which may render large areas useless for agriculture. These kinds of 'wastelands' or 'idle grasslands' do not occur on Siberut.

Local food regimes and man-made ecosystems also contribute to the absence of need to use fire. Ecologically, tubers, bananas and sago compete successfully with weeds and grasses that grow quickly in open areas with a high rainfall and wet climate. Moreover, they do not require additional nutrients from ash. Burning to clear diseases is also unnecessary because tubers and bananas, unlike grains, do not have serious natural diseases. The people only permit the slashing of vegetation for mulching, to release biomass into the land. The variety of productive zones, producing diverse sources of food, means the people do not have to rely solely on forest gardens. The distinctive feature of Mentawaian forest gardens lies in the role played by fruit trees in delivering the main product. In terms of diet, the fruit is a kind of snack; an edible item that is complementary to the staple food and meat, both of which are available from other productive zones. For daily consumption, forest gardens are of secondary importance. Staple foods are already provided by sago palms and roots and tubers such as yams and taro, while daily meat is gathered from rivers, ponds and the coastal area, and includes sago grubs. The people of Siberut have traditionally relied on gathering and collecting, as well as cultivating semi-domesticated resources. They are less dependent on fruit from the forest gardens.

Forest gardens are less crucial in terms of diet because, at most, they are harvested only once a year. 'Great fruit seasons' (*rura*), when all the fruit trees simultaneously produce great yields, generally occur only about once every three years. Although some cultivated trees routinely produce fruit each year, it is not really significant for daily consumption. The lesser dependency on forest gardens contributes to the lack of cultivation techniques. Mentawaians do not burn fallen trees, weed-out grasses and wild vegetation, or mark the boundaries of cultivated areas. Socially, cultivation without fire and reliance on tubers does not require the cooperative working arrangements and labour exchanges commonly featured in many types of shifting cultivation that depend on rice or maize. Guarding against escaping fire, planting grains, protecting the rice and collective harvesting require a complicated and often rather hierarchical social structure. The absence of complicated labour arrangements certainly reflects the egalitarian ethos of Mentawaian social organization.

The importance of the forest gardens, or agroforests, lies not only in the value of fruit trees as a source of food, but also in their social and religious functions. The presence of fruit trees and, in particular, trees that are living memorials (*kirekat*) to deceased people, is important to the maintenance of claims over land and forests. The agroforests, with all their inherent signs of human activity, are instrumental in 'reading the landscape'. In addition to their function as a source of food and a wealth of other materials, the forests reflect the movements of people, both distant and recent, and the ways in which they have been used for social and religious purposes.

References

ADB (1995) *Siberut National Park Integrated Conservation and Development Management Plan*, vol. I/III, Asian Development Bank and the Republic of Indonesia Ministry of Forestry, Jakarta

Ave, W. and Sunito, S. (1990) *Medical Plants of Siberut*, WWF International, Gland, Switzerland

Boomgaard, P. (2003) ´In the shadow of rice: Roots and tubers in Indonesian history, 1500–1950´, *Agricultural History* 77(4), pp.582–610

Conklin, H. C. (1957) *Hanunoo agriculture: A report on an integral system of shifting cultivation in the Philippines*, Food and Agriculture Organization of the United Nations, Rome

Darmanto (2006) 'Studi ekologi perladangan hutan tradisional masyarakat mentawai (pumonean) di pulau Siberut, Sumatra barat', in H. Soedjito (ed.) *Kearifan Tradisional Dan Cagar Biosfer di Indonesia, Prosiding Piagam MAB 2005 Untuk Peneliti Muda dan Praktisi Lingkungan di Indonesia*, Komite Nasional MAB Indonesia, Lembaga Ilmu Pengetahuan Indonesia dan UNESCO Kantor Jakarta, Bogor, pp.57-118

Darmanto (2016) *Maintaining Fluidity, Demanding Clarity: The Dynamics of Customary Land Relations among Indigenous People of Siberut Island, West Sumatra*, Department of Asian Studies, Murdoch University, Perth, Australia

Departemen Sosial (1994) *Pembinaan kesejahteraan masyarakat terasing di Indonesia* (Development of Welfare for Isolated Communities in Indonesia), Departemen Sosial, Jakarta

Eindhoven, M. (2019) 'Products and producers of social and political change. Elite activism and politicking in the Mentawai Archipelago, Indonesia', PhD dissertation to Leiden University, The Netherlands

Ellen, R. (1999) 'Forest knowledge, forest transformation: Political contingency, historical ecology, and the renegotiation of nature in Central Seram', in T. M. Li (ed.) *Transforming the Indonesian Uplands: Marginality, Power, Production*, Harwood Academic Publishers, Amsterdam, pp.131-156

Göltenboth, F. and Timotius, K. H. (1996) ´Impact of rainforest destruction: The Siberut Island case, Sumatra, Indonesia', in D. S. Edwards (ed.) *Tropical Rainforest Research: Current Issues*, Kluwer Academic Publishers, Dordrecht, The Netherlands, pp.425-433

Neilson, J. (2007) 'Global market, farmers, and the state: Sustaining profits in the Indonesian cocoa sector', *Bulletin of Indonesia Economic Studies* 43(2), pp.227-250

Persoon, G. A. (1992) ´From sago to rice: Changes in cultivation in Siberut, Indonesia´, in E. Croll and D. Parkin (eds) *Bush Base, Forest Farm: Culture, Environment and Development*, Routledge, London, pp.187-199

Persoon, G. A. (1994) 'Vluchten of veranderen. Processen van verandering en ontwikkeling bij tribale groepen in Indonesië (Flights of change: Processes of change and development among tribal groups in Indonesia), PhD dissertation to Leiden University, the Netherlands

Persoon, G. A. (1997) *Defining wildness and wilderness: Minangkabau images and actions on Siberut* (*West Sumatra*), The Future of Tropical Forest Peoples (APFT), ULB Brussels and Canterbury, UK

Persoon G. A. (2001) 'The management of wild and domesticated forest resources on Siberut, West Sumatra', *Antropologi Indonesia* 25(64), pp.68-83

Puailiggoubat (2013) Ramai-Ramai Tolak Sawit (Busy refusing oil palms)', *Puailiggoubat* 11(268), 15-31 July, Citra Mandiri Foundation, Padang, Indonesia

Puailiggoubat (2015) 'Masyarakat dan Pemda Mentawai Tolak HTI (The community and the Mentawai district government refuse industrial timber estate)', *Puailiggoubat* 13(320), 15-30 September, Citra Mandiri Foundation, Padang, Indonesia

Roth, R. B. (1985) ´Simeuluë - Nias - Mentawai - Enggano: eine bibliografische Ergänzung und Erweiterung (1959-1984) zu Suzuki's "Critical survey of studies on the anthropology of Nias, Mentawei and Enggano" (1958), *Anthropos* 19/1, pp.421-470

Ruf, F. (2011) 'The myth of complex cocoa agroforests: The case of Ghana', *Human Ecology* 39, pp.373–388

Ruf, F., Ehret, P. and Yoddang (1996) 'Smallholder cocoa in Indonesia: Why a cocoa boom in Sulawesi?', in W. G. Clarence-Smith (ed.) *Cocoa Pioneer Fronts Since 1800: The Role of Smallholders, Planters and Merchants*, Macmillan Publishers, London, pp.212-234

Schefold, R. (1990) 'Amiable savage at the doors of paradise: Missionary narratives about the Mentawai Islands (Indonesia)', in P. Kloos (ed.) *True Fiction: Artistic and Scientific Representation of Reality*, VU University Press, Amsterdam, pp.21-35

Schefold, R. (2017) *Toys for the souls. Life and Art on the Mentawai Islands*, Premedia, Bornival, Belgium

Schefold, R. and Persoon, G. A. (eds.) (2002) 'Nias-Mentawai-Enggano: Diversity and commonality within an island chain in western Indonesia', Special issue, *Indonesia and the Malay World* 30(88), pp. 221-378

Schieffelin, E. L. (1975) 'Felling trees on top of the crop: European contact and the subsistence ecology of the Great Papuan Plateau', *Oceania* 46(1), pp.25-39

Sentosa Group (2017) *Integrated Tourism Area, Mentawai Bay, Indonesia*, Sentosa Group, Singapore

Spencer, J. E. (1965) *Shifting Cultivation in Southeastern Asia*, University of California Press, Berkeley, CA

Suzuki, P. (1958) *Critical Survey of Studies in the Anthropology of Nias, Mentawai and Enggano*, KITLV bibliographic series 3, Martinus Nijhoff, The Hague, The Netherlands

Thurston, H. D. (1997) *Slash/Mulch Systems. Sustainable Methods for Tropical Agriculture*, Westview Press, Boulder, CO

Tulius, J. (2012) *Family Stories: Oral Tradition, Memories of the Past and Contemporary Conflict Over Land in Mentawai, Indonesia*, Leiden University Press, The Netherlands

van Beukering, J. A. (1947) *Bijdrage tot de anthropologie der Mentaweiers* (Contribution to the Anthropology of the Mentawaians), Kemink & Zoon, Utrecht, The Netherlands

WWF (1980) *Saving Siberut: A Conservation Master Plan*, World Wildlife Fund Indonesia Programme, Bogor

Note

1. There is an extensive ethnography about the Mentawaian islands and Siberut in particular. Three complementary bibliographies have been published by Suzuki (1958), Roth (1985) and Schefold and Persoon (2002).

11

FIRELESS SHIFTING CULTIVATION

A lesser-known form of swiddening practised by
Tangkhuls in parts of Ukhrul district, Manipur, India

*Loushambam Jitendro Singh, Thingreiphi Lungharwo and Dhrupad
Choudhury**

Introduction

Shifting cultivation has been practised throughout the tropics and subtropics for
countless generations (Whitmore, 1984). It has been blamed historically for large-
scale deforestation and environmental destruction. Scientific literature has described
the practice as one in which the clearing and burning of vegetation is an integral part.
This has led to a common perception that shifting cultivation is guilty of all that its
detractors claim, and the practice has been given the label 'slash-and-burn' (however,
see Spencer, 1966). Of late, shifting cultivation has also been accused of polluting
the atmosphere and releasing significant quantities of carbon, thereby contributing to
climate change. It has thus drawn renewed criticism from foresters, environmentalists
and governments alike, while reinforcing calls for its eradication and replacement
with settled forms of agriculture.

While calls for the eradication of shifting cultivation have gathered momentum,
efforts to transform it into more benign forms of agriculture have not been
encouraging. The management of shifting cultivation has proven to be enigmatic
for governments and development practitioners alike, despite concerted efforts
and substantial expense. Shifting cultivation continues to persist, particularly across
much of the uplands of the eastern Himalayas and Southeast Asia. A comprehensive
account of the reasons for the persistence of shifting cultivation is given by Leduc
and Choudhury (2012). They suggest that limited access to markets and support
programmes, coupled with a lack of viable and appropriate technical options, underlie

* DR. L. JITENDRO SINGH is Assistant Project Manager for Natural Resource Management and
Environment, North East Rural Livelihood Project (NERLIP), Guwahati, Assam; THINGREIPHI
LUNGHARWO is a Global Indigenous Fellow for Biodiversity at the GEF Small Grants Programme, United
Nations Development Programme; and DR. DHRUPAD CHOUDHURY is Chief of Scaling Operations at the
International Centre for Integrated Mountain Development (ICIMOD), Kathmandu, Nepal.

the reasons why it continues. Meanwhile, governments and communities alike explore options and affordable alternatives that may help to bring shifting cultivation to an end.

It has been suggested that an acceptable compromise may be available, involving a more benign form of shifting cultivation that may represent a transitory phase while governments and development agencies search for long-term solutions. The compromise dispenses with burning and reduces emissions, while increasing the regenerative fallow phase that allows forests to regrow. An extended cultivation phase, together with the longer fallow, would represent a 'semi-sedentary' form of agriculture. While it is generally true that the use of fire is an indispensable part of shifting cultivation in most parts of the world, more benign forms which dispense with the use of fire have been reported by Kabu (2001) from the Solomon Islands and Garrity and Chun (2001) from parts of Southeast Asia, particularly Indonesia and Papua New Guinea (see also Darmanto and Persoon, ch. 10, and Bourke and Allen, ch. 12, this volume).

This chapter will describe an agricultural practice found in some of the villages in the northern parts of Ukhrul district, in Manipur state, northeast India. It is called 'fireless shifting cultivation', and may offer the basis for developing a viable model for a transitory phase between traditional swidden farming and sedentary, or permanent, agriculture. This practice not only dispenses with the need for burning, but also offers a cultivation phase that spans at least four to five years, thereby offering a 'semi-sedentary' state, such as that being sought by the agents of transformation.

Shifting cultivation has long been central to the livelihood and food-security pursuits of the farmers in the uplands of Ukhrul district. What sets their version of shifting cultivation apart from that of their counterparts in northeast India and elsewhere is its longer cultivation phase, which lasts for at least four or five years and, in past times, extended for as long as seven years. While this effectively translates into a state of 'semi-sedentary' farming, it also allows fallow periods that are sufficiently long to achieve forest regeneration that is more effective than that seen elsewhere in the region.

A promising version of this form of shifting cultivation, and one of direct relevance to this discourse, is that locally known as *yamkui*. It is practised by a small proportion of upland farmers in parts of northern Ukhrul, in particular, in the village of Kalhang and its neighbouring villages (Figure 11-1). In contrast to the traditional form of shifting cultivation practised elsewhere, burning and fire play no part in *yamkui*. It also has a cultivation phase that is at least four years long.

Study site

The study of fireless shifting cultivation, which was focused around Kalhang and its neighbouring villages, was part of a larger study of 'two-year-plus' shifting cultivation, conducted across eight villages in Ukhrul district of Manipur. The village of Kalhang lies approximately between latitude 24° to 25° 41′ north and longitude 94° to 94°

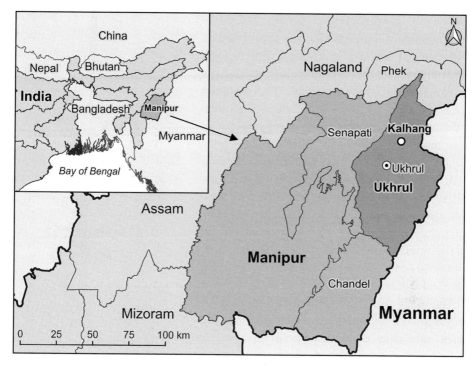

FIGURE 11-1: Manipur state, in Northeast India, showing the study village, Kalhang, in the northern part of Ukhrul district.

47′ east, with an altitudinal range of 1823 to 1845 m above sea level (masl). It is 45 km from the district headquarters, also named Ukhrul. The district ranges in elevation from 913 to 3114 masl and is bound on the north by Phek district of Nagaland, and by Manipur's Senapati district to the west and Chandel district to the south. The entire eastern flank of the district is bordered by Myanmar. The district is mountainous, with deep, narrow riverine valleys. The climate is temperate, with temperatures ranging from 20° to 33° Celsius, and annual rainfall of between 1600 and 2100 mm spread over the year, with the heaviest downpours in June and July. The people inhabiting the study villages are Tangkhuls, who depend mainly on shifting cultivation for their sustenance.

Agricultural land use in the study villages

The main occupation of the study-village inhabitants is agriculture, and shifting cultivation, or *jhum*, locally known as *Ahanglui/Khayailui*, is its predominant form. About 24% of households also practice terrace cultivation, and home gardens are kept by 82% of households (Table 11-1). There are at least three forms of swidden farming practised in the study villages. The first is conventional shifting cultivation, as practised by most of the upland communities in the region and elsewhere, with one or two years of cropping before the farmers move to a new plot, the land is left fallow and the forest is allowed to regenerate for a period presently varying from 4 to 12 years. In our survey, 29.23% of the respondent households said they pursued this form of shifting cultivation. The second type varies from the first only in terms of the duration of the cropping phase. The cropping period is more than two years, and so we called it the 'two-years-plus' practice. Cropping ranges from three to five years, and in exceptional cases, can go beyond this, with a fallow phase ranging from 9 to 15 years. New plots are cleared only at the end of the cropping phase of five years and farmers cultivate only a single plot. The 'two-years-plus' practice was reported by 43.57% of the total households surveyed, but by more than 60% of the households in four of the eight surveyed villages.

The third variation, although reported by only 8.37% of the total households surveyed and practised by only a handful of households in five of the eight villages studied, is a novel variation of shifting cultivation that does not use fire. Although the overall percentage of households practising this variation may seem insignificant, within one of the villages – Kalhang – 73.55% of the households followed this form of shifting cultivation (Table 11-1). We called it 'fireless shifting cultivation'. It is locally known as *yamkui*.

The figures show that shifting cultivation is the predominant agricultural land use in the study villages. However, they also throw up some intriguing questions: why do the upland farmers in these villages practice three different versions of shifting cultivation? Why do we find this heterogeneity when the farmers belong to the same ethnic group? What prevents the farmers from having a uniform practice? And, of even more interest, *yamkui* doesn't seem to be a very popular practice within the study villages (as reflected by the percentage of farmers practising this system), so why is it so widespread among the farmers in Kalhang village?

Imperata cylindrica (L.) Raeusch. [Poaceae]

The bane of many upland cultivators, thatch grass, cogon grass, or blady grass is widely regarded as an invasive species. It spreads aggressively via countless wind-blown seeds and dense mats of rhizomes, which account for 60% of its biomass. It is highly flammable and depends on regular fires that destroy competing vegetation while the rhizomes survive to maintain its ecological dominance.

Why do farmers in Ukhrul practise different forms of shifting cultivation?

Discussions with farmers in Kalhang and surrounding villages, as well as with elders in Ukhrul, suggested that the choice of adopting a particular form of shifting

TABLE 11-1: Agricultural land use in the study villages.

Village	Distance from HQ (km)	Total households	Households with terraces	Home Gardens	Households practising: 1–2 yr Jhum	2-yrs-plus jhum	Fireless jhum	With paddy in jhum
Luirishimphung	65	38	7 (18.4)	28 (66.67)	8 (21.05)	23 (60.53)	2 (5.26)	–
Kalhang	45	121	24 (19.83)	121 (97.58)	–	101 (83.47)	89 (73.55)	–
Nungbi Khullen	40	237	74 (31.22)	221 (90.95)	43 (18.14)	62 (26.16)	3 (1.27)	–
Nungbi Kajui	39	267	71 (26.59)	187 (68)	109 (40.82)	87 (32.58)	38 (14.23)	30
Paoyi	46	282	52 (18.44)	176 (60.9)	31 (10.99)	209 (74.11)	–	–
Paorei	40	216	55 (25.46)	204 (91.89)	33 (15.28)	141 (65.28)	8 (3.7)	23
Phungcham	36	207	69 (33.33)	177 (83.1)	106 (51.21)	30 (14.49)	–	52
Halang	19	305	41 (13.44)	294 (95.15)	159 (52.13)	76 (24.92)	–	–
Total		1673	393 (23.49)	1408 (82)	489 (29.33)	729 (43.57)	140 (8.37)	105 (6.28)

Note: Figures in brackets signify the percentage of households within each category.
Source: Singh (2009); Singh and Choudhury (2015).

cultivation is determined by a variety of factors: the primary drivers seem to be the terrain in which a plot is located; the soil characteristics of an individual plot and its surroundings; the density of woody vegetation at the location; the options open to a farmer for income generation from agriculture; and the family labour that is available. All of these factors are considered when a farmer decides whether to opt for conventional shifting cultivation with one or two years of cropping, 'two-years-plus' shifting cultivation, or *yamkui* – the fireless version. The discussions also revealed that households may adopt two versions of shifting cultivation at any given time, but never a combination of conventional shifting cultivation and the 'two-years-plus' version.

Conventional shifting cultivation with a cropping phase of one or two years is practised in plots with steep slopes and poor soil that usually has a poor capacity to retain moisture and is often pebbly. Crop yields tend to fall in the second year, compelling the farmer to move to a new plot. The two-years-plus version, on the other hand, occupies those plots that have good vegetation that is dominated by woody species, irrespective of slopes. Dense woody vegetation provides sufficient litter, thereby improving both soil fertility and the ability to retain moisture. Such plots allow cultivation to continue for more than two years, sometimes for as long as five years, until weeds become difficult to control, forcing the farmer to move to a new plot. It is important to point out that none of the farmers felt that there

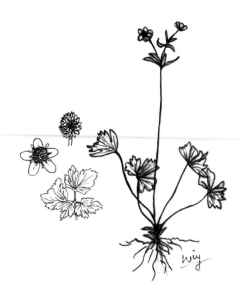

Ranunculus cantoniensis DC.
[Ranunculaceae]

This perennial herb, which displays bright yellow flowers and is valued for its medicinal qualities, is very particular about the conditions in which it grows. It prefers wet or marshy locations, which are uncommon in upland areas. The whole plant is gathered, washed and sun-dried or is used fresh for external treatments of eye complaints, jaundice, rheumatic pain or malaria.

was a decline in soil fertility, and the compulsion to move was forced solely by an uncontrollable increase in weed growth. If weeds could be controlled effectively, farmers felt that they could continue cultivation for a few more years before leaving the land fallow.

Yamkui, or fireless shifting cultivation, is practised in those areas where the vegetation is predominantly composed of shrubs and grasses interspersed with a few trees. Alder (*Alnus nepalensis*) is the dominant tree species in such plots, but the density is low, with only a few individuals dispersed across the plot. When these plots are cleared for cultivation (Figure 11-2), the vegetation fails to yield sufficient biomass. If burned, the yield of ashes and char would be inadequate and it could quickly blow away in the prevailing winds or be washed downslope in rain, leaving the soil deprived of nutrients and incapable of delivering a satisfactory crop yield. Such plots do not offer the option of clearing by fire, and farmers must mulch the vegetation in order to grow their crops. Such plots are usually farmed by the more marginalized members of the community and are located in areas characterized by grassy patches along ridges.

Crop diversity in shifting cultivation plots in Ukhrul,

FIGURE 11-2: The first step in *yamkui* is clearing the vegetation.

All photographs in this chapter were taken by the author L. Jitendro Singh.

irrespective of the version of the practice, differs from that found elsewhere because of the absence of rice. In the uplands of Ukhrul, rice cultivation occurs only in flooded terraces. The traditional staple diet of Tangkhuls once consisted of maize and wheat, and rice was an uncommon food. However, with time and acculturation, diets changed and rice became a local staple. Nowadays, households that do not have terraces and depend solely on shifting cultivation have no choice but to buy or barter rice, and this has given rise to the need to grow cash crops.

Farmers practising *yamkui* grow a variety of legumes in their fields, starting from the first year, and many of these have a good market value. Of particular importance are rice beans (*Vigna umbellata*), known locally as *Naga dal*, and flat beans, for which there is a healthy demand locally as well as in nearby markets. As demand for their crops increased steadily over the years, *yamkui* farmers diversified and introduced other commodity vegetables such as cabbages and peas, thus increasing their income.

Farmers in Kalhang were quick to see this opportunity, and realized that while traditional crops could be grown in their two-year-plus shifting cultivation fields, *yamkui* offered a good opportunity to grow commodity crops and commercialize shifting cultivation with a

Alnus nepalensis D. Don [Betulaceae]

Himalayan alder is the predominant tree species in *yamkui* plots, although the density is low. Frankia bacteria living in root nodules enable these trees to fix nitrogen in the soil. They also shed copious quantities of leaf litter, ranking them among the most important of fallow soil improvers.

negligible need for inputs other than seeds and labour. This offered to conserve traditional and local crops, on one hand, while simultaneously moving towards commercial agriculture on the other. Even the labour requirement was much less than that demanded by other forms of shifting cultivation or terrace farming. Over the years, more and more farmers in Kalhang adopted this form of shifting cultivation. At the time of this study, more than two-thirds of the households in the village had adopted *yamkui* as one of their agricultural systems, dedicated primarily to commercial agriculture, while a two-year-plus swidden provided their household food and nutritional needs. In a sense, the farmers of Kalhang have been entrepreneurial, harnessing a neglected production system to grasp commercial opportunities. In doing so, they have optimized production from available labour and inputs, within the menu of options open to them.

Land preparation in *yamkui* or fireless shifting cultivation

For villagers in Kalhang, the beginning of July is the time for preparation of fields for *yamkui*. After site selection and field allocation, the farmers slash the vegetation and heap it in bunds along the contours of their fields (Figure 11-3). Branches of standing trees are lopped and tree trunks and larger branches are laid along the contours on top of the vegetative bunds (Figures 11-4 and 11-5). Excess logs are removed for use as firewood or for fencing. The farmers believe that the absence of fire allows the survival of underground parts of plants and seeds, along with soil micro-fauna and microbial activity, while it provides nutrients to the germinating crops once they sprout and start growing.

The vegetation is slashed in July as opposed to October or November in conventional shifting cultivation (Table 11-2). The undergrowth, particularly herbs and shrubs, are uprooted by light digging and clodding and removing plant parts from the soil. The slashed vegetation is piled along the contours as mulch (Figure 11-6), thus slowly giving rise to 'vegetative bunds' along the contours. Logs from felled trees are also laid along the contours to strengthen the bunds, while smaller branches, twigs and foliage are piled on to the bunds. As the biomass decomposes over time, the bunds become 'nutrient-access banks', with nutrients leaching out slowly, to be immediately available to the growing crops (Figure 11-7).

Crop cultivation

After the completion of land preparation, seeds are sown immediately and the whole process is completed by end of July. Crop diversity in the first year is limited, and depends on the capability of the land and soil type. The majority of farmers grow flat beans and a short variety of rice beans in the first year, while others also intercrop

FIGURE 11-3: After the vegetation is slashed, it is gathered into bunds across the contours of the plot.

FIGURES 11-4 and 11-5: Trees on the plot are cut down (left) and the logs laid on top of the freshly made vegetative bunds (right).

TABLE 11-2: Calendar of activities in fireless shifting cultivation.

Calendar months	Activities in each cropping year			
	Year 1	Year 2	Years 3 and 4	Year 5
January	–	–	–	–
February	–	Cleaning, sowing	Cleaning, sowing	Cleaning, sowing
March	–	Cleaning, sowing	Cleaning, sowing	Cleaning, sowing
April	–	Sowing, first weeding	Sowing, first weeding	Sowing, first weeding
May	–	–	Second weeding	Second weeding
June	–	Second weeding	Third weeding	Third weeding
July	Slashing, cleaning*, planting	First harvest, second sowing	First harvest, second sowing	First harvest, final harvest
August	First weeding	First harvest, third weeding	First harvest, fourth weeding	First harvest, final harvest
September	Second weeding	Fourth weeding	Fifth weeding	–
October	–	–	Sixth weeding	–
November	Harvesting	Second harvest	Second harvest	–
December	Harvesting	Second harvest, final harvest	Second harvest, final harvest	–
January	–	Final harvest	Final harvest	–

Note: *Clearing and mulching.
Source: Singh (2009).

with peanuts, depending on how soft or hard the soil is. A few farmers also grow a variety of maize, some spices and vegetables. The whole process, from slashing to sowing, can be completed in a day or two, depending on available labour. After sowing in July, the next major activity is in the last week of August, when the field is weeded for the first time. A second weeding occurs in the second week of September. Harvesting begins in December and, depending on yields and crop mixes, could continue into January.

The subsequent agricultural activities and calendar, as outlined in Table 11-2, mimic those for two-year-plus shifting cultivation, including the length of the cultivation phase (Table 11-3). Crop diversity, crop mixes, sowing and sequential harvesting in the *yamkui* from the second year onward, show no difference from the two-year-plus system, except that cereals are grown sparsely, are restricted to varieties of maize and Job's tears, and exclude wheat.

FIGURE 11-6: After the vegetative bunds are formed, the plot is ready for planting.

Farmers' perceptions of fireless shifting cultivation

Farmers' perceptions of *yamkui* are extremely positive. In contrast to conventional shifting cultivation or even the two–year–plus system, both of which require substantial amounts of labour for forest clearing, burning and land development, plot preparation for *yamkui* is less labour intensive, since it is restricted to plots that lack dense woody vegetation. Because there is no need for burning, fire management or clearing of unburnt debris, the demands for labour are considerably reduced (Table 11-4). Land preparation activities in *yamkui* begin in July, in the midst of the rainy season, when the soil is softer and easier to manage. The entire operation of clearing, land development and sowing can be completed in a day or two. In contrast, burning in conventional shifting cultivation makes the predominantly clayey-shale soils in the study sites harder, and given that burning and subsequent land development are done during the dry season in February or March, these activities are more labour intensive than those in *yamkui*. Therefore, the farmers feel that fireless shifting cultivation has major advantages in terms of savings on labour. In addition, with the deliberate cultivation of legumes as primary crops in the first year and nutrients being leached from the mulched vegetation presumably allowing cropping for more than two years, most of the crops cultivated in the two–year–plus system can also be grown in *yamkui* (see Table 11-3). All of these factors lead farmers to believe

FIGURE 11-7: From one season, the bunds form a decomposing 'nutrient bank'.

Arachis hypogaea L. [Leguminosae]

Kalhang farmers may grow peanuts in the first year of *yamkui* cultivation – if the soil is soft enough to allow these legumes to penetrate the upper layers so their seed pods can develop under the ground. As well as improving the soil, peanuts are classified both as a grain legume and an oil crop. They are rich in human dietary needs.

TABLE 11-3: Fireless shifting cultivation: Crop diversity in different cropping years.

Crops grown	Crop diversity across different years (n = 50 plots)				
	Year 1	Year 2	Year 3	Year 4	Year 5
Cereals					
1 Coix lacryma-jobi L.		x	x	x	
2 Setaria italica (L.) P. Beauv.*		x	x		
3 Zea mays L. (red kernel)		x	x		
4 Zea mays L. (variegated kernel)	x	x	x	x	
5 Zea mays L. (white kernel)		x	x	x	x
6 Zea mays L. (yellow kernel)		x	x		
Legumes					
1 Arachis hypogaea L.	x	x	x		
2 Lablab purpureus (L.) Sweet		x	x	x	
3 Glycine max (L.) Merr. (big grain)	x	x	x		
4 Glycine max (L.) Merr. (small grain)	x	x			
5 Phaseolus vulgaris L.		x	x	x	
6 Pisum sativum L.	x	x			
7 Vicia faba L.	x	x	x		
8 Vigna umbellata (Thunb.) Ohwi & H.Ohashi	x	x	x	x	
Tubers/rhizomes/roots					
1 Colocasia esculenta (L.) Schott		x	x	x	
2 Colocasia antiquorum Schott		x	x	x	
3 Dioscorea bulbifera L.		x	x	x	
4 Ipomoea batatas (L.) Lam.		x	x	x	
5 Manihot esculenta Crantz.		x	x	x	
6 Solanum tuberosum L.		x	x		
Spices					
1 Allium cepa (M).		x	x	x	
2 Allium hookeri Thwaites		x	x	x	
3 Allium ramosum L.		x	x		
4 Allium ampeloprasum L.		x	x	x	
5 Allium sativum L.		x	x	x	
6 Apium graveolens L.		x	x	x	
7 Coriandrum sativum L.	x	x	x		
8 Elsholtzia blanda (Benth.) Benth		x	x	x	
9 Ocimum americanum L.		x	x		
10 Zingiber officinale Roscoe		x	x	x	
Vegetables					
1 Benincasa hispida (Thunb.) Cogn.		x	x	x	
2 Brassica napus L. var.	x	x	x	x	
3 Brassica nigra (L.) K.Koch			x	x	
4 Cucumis sativus L.		x	x	x	
5 Cucurbita maxima Duchesne		x	x	x	
6 Lycopersicon esculentum Mill.		x	x	x	
7 Solanum melongena L.		x	x		
8 Solanum sp.		x	x	x	

Note: *Cultivated in two-year-plus system.
Source: Singh (2009).

TABLE 11-3 (cont.) : Fireless shifting cultivation: Crop diversity in different cropping years.

Crops grown	Crop diversity across different years (n = 50 plots)				
	Year 1	Year 2	Year 3	Year 4	Year 5
Others					
1 *Helianthus annuus* L.		x	x		
2 *Nicotiana tabacum* L.		x	x	x	
3 *Passiflora edulis* Sims		x	x		
4 *Anisomeles indica* (L.) Kuntze		x	x	x	
Total no. of crops	9	41	40	27	1

Note: *Cultivated in two-year-plus system.
Source: Singh (2009).

TABLE 11-4 : Comparative labour requirements for plot development in conventional and fireless shifting cultivation.

Conventional shifting cultivation	Fireless shifting cultivation (Yamkui)
1 Slashing the vegetation.	Clearing the shrubs and herbs (by shallow digging and clodding).
2 Piling dried vegetation.	Laying vegetation in bunds along plot contours.
3 Burning the slashed vegetation (except for bigger branches and logs that are used for erosion control).	Branches and logs are laid in the contour bunds, apart from some that are kept for firewood.
4 Second burning, after piling the unburnt debris.	–
5 Plot cleaning (shallow digging to soften the soil and remove roots).	–
6 Contour barriers to erosion created by unburnt logs and bigger branches.	–
Plot development completed with activity no. 6. Planting of crops begins	*Plot development completed with activity no. 3. Planting of crops begins.*
Soil quality, and its implications for labour requirements	
» Soil becomes hard due to burning and requires more labour to dig and remove roots.	» The soil is soft because it is the rainy season; less labour is required, compared to other forms of shifting cultivation.
Constraints to expansion of *Yamkui*	
» Labour shortages as plot clearing and development for *yamkui* clash with wet-rice transplanting in flooded terraces.	
» Lean season for rice, making it difficult to afford food as well as wages for labourers (a customary requirement among the Tangkhuls).	

that their returns to labour from *yamkui* are much better than those from other forms of shifting cultivation. In addition, since *yamkui* is carried out in areas where woody vegetation is sparse, farmers can reduce the fallow phase and return to the plot earlier if necessary. Finally, with *yamkui* being practised mainly for cultivation of cash crops, farmers feel that this system may be a better option for moving towards sedentary or permanent farming than other forms of shifting cultivation.

Why is fireless shifting cultivation not predominant and widespread?

Given the perception of farmers that *yamkui* has advantages over the practice of conventional shifting cultivation, it is surprising that its popularity seems to be disproportionate to its adoption, and the system is not practised more widely. We sought to understand the reasons underlying this situation, in order to scale this practice. Consultations with the farmers and village elders provided some insights, and while the reasons seemed plausible and simple, they were not insurmountable.

Yamkui, or fireless shifting cultivation, is practised in areas where woody vegetation is sparse, but shrubs and grass offer sufficient biomass. Such plots must also have a good layer of topsoil to support crops. Large patches with the required vegetation and soil conditions are not always available, because dense vegetation, often with a concentration of woody species, regenerates quite quickly on most fallow land. This regeneration is aided by the Tangkhuls' robust fallow-management practices (see Singh and Choudhury, 2015). Therefore, there is not always sufficient suitable land available in the villages to allow an expansion of *yamkui*, and this constrains the wider adoption of the practice.

Moreover, the primary activities for *yamkui* – plot clearing, land development and sowing – begin in July, and they coincide with the major activities for rice cultivation in terraces, particularly transplantation. There is an immediate conflict of interests, resulting in a shortage of labour for *yamkui*. Agricultural labour in terraces offers households a good opportunity to earn much-needed cash, and given a choice between working in wet terraces or in *yamkui*, villagers will opt for working in the terraces without hesitation. A custom in the uplands also requires that a household engaging labour must provide a midday meal for the labourers, and in today's context, that means a meal based on rice. This represents a challenge for poor farmers and marginal households because, in the months preceding July, the food savings of most families are exhausted and the availability of rice dwindles. This makes it extremely difficult for households to afford the food and wages for extra labour, so poor and marginal families are unable to

Vigna umbellata (Thunb.) Ohwi & H.Ohashi [Leguminosae]

Rice beans are one of the first crops grown in freshly cleared *yamkui* plots. Perhaps best known as an intercrop with maize or sorghum, this short-lived warm-season annual produces high-quality dried grain and nutritious livestock fodder, as well as being an effective green manure. It is one of several legume species grown in the fireless system.

afford *yamkui*, despite all of its advantages. In combination, these factors prevent *yamkui* from expanding to become a widely adopted form of shifting cultivation. In the given context, this is regrettable and solutions need to be sought urgently.

Lessons from *yamkui* for managing change in shifting cultivation

Fireless shifting cultivation, or *yamkui*, as practised by the Tangkhul *jhumias* of Ukhrul district, Manipur, seems to offer significant benefits and holds potential for the development of a model for managing a transition from shifting cultivation to sedentary, or permanent, agriculture. The upland farmers of Ukhrul provide important lessons in the management of shifting cultivation for researchers, governments and development agencies involved in managing change in upland agriculture. The farmers have demonstrated that shifting cultivation can be practised without the need for burning. Of course, this is contextual and applicable only in those situations where

the slashed vegetation is not predominantly woody. However, with fallow cycles rapidly declining in most areas, shifting cultivation in the uplands of South and Southeast Asia is nowadays practised in areas where fallow vegetation is dominated by shrubs and sparse trees, so *yamkui* could hold important lessons relevant to the management of short-cycle shifting cultivation. In such situations, *yamkui* – or a model based on *yamkui* – becomes eminently pertinent. The most important and attractive attribute of *yamkui*, and one that strengthens advocacy for its wider dissemination and promotion, is its lack of fire; its use of mulching, rather than burning. This neutralizes one of the key criticisms of shifting cultivation – that the use of fire gives rise to environmental pollution.

A positive strength of the practice is its central dependence on mulching. Scientific studies over many years have established the importance of leaf litter, prunings and crop residue in maintaining soil–nutrient status, as an important aspect of agricultural sustainability (Lal, 1989; Handayanto et al. 1994; Carsky et al., 1998; Whitbread et al., 1999; Erenstein, 2002). The mulched vegetation in *yamkui*,

Pisum sativum L.
[Leguminosae]

Peas were a leguminous crop introduced to the *yamkui* system to both boost farmers' returns from market sales and to help to enrich the soil by fixing nitrogen. With a proven history of more than 6000 years as a human food, peas grow well in cool, high-altitude tropical areas. Their dry weight is about one-quarter protein and one-quarter sugar.

together with the deliberate cultivation of a variety of legumes in the first year, seem to provide farmers with the nutrient inputs required to support satisfactory crop yields without the need for external inputs. Mulching of vegetation is also known to positively affect soil ecology. It reduces run–off and evaporative losses, thereby conserving water and improving soil moisture retention (Erenstein, 2003). Through the process of smothering and/or allelopathic effects, mulch is also reported to control weed growth (Akobundu, 1987). Although scientific studies are required to assess these effects in *yamkui*, it is safe to assume that such benefits may accrue, leading to positive outcomes for the soil and subsequently for crop growth.

The *yamkui* system in Ukhrul also demonstrates that the cropping phase in shifting cultivation can be increased to more than two years, thus introducing a semi-sedentary form of agriculture. Although this attribute may be subject to local soil capacities, the deliberate predominance of legumes (and hence the probable nitrogen improvement of the soil) suggests that such an intensification of the cropping phase

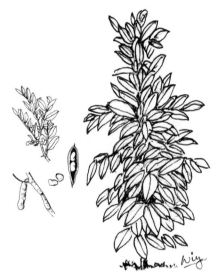

Vicia faba L. [Leguminosae]

Broad beans are another legume species used by the farmers of Kalhang in their fireless swidden farming. These are among the most ancient plants in cultivation. The broad, leathery pods each contain three to eight seeds that are highly nutritious as a human food. The plants are also hardy, can grow in harsh conditions, and are often used as a cover crop to prevent erosion.

Phaseolus vulgaris L. [Leguminosae]

The common bean has a long history of cultivation and is grown around the world. In the *yamkui* system, it is regularly cultivated in the second, third and fourth years of cropping. It is yet another legume contributing to soil enrichment. Its most popular forms are dried beans, or seeds harvested at full maturity, and snap beans – tender pods harvested as a green vegetable before the seed-development stage.

may be possible elsewhere, as legumes are known to improve soil nutrient content (Blair et al. 1990). This not only extends the cropping phase in shifting cultivation, but also results in an extended fallow period, thereby assisting soil recovery. These are positive attributes that offer a basis for agricultural-development models that are desperately required for managing change in shifting cultivation.

Labour requirements in *yamkui* are also substantially less than those for conventional shifting cultivation. Labour migration out of rural areas is a phenomenon affecting many parts of upland Asia-Pacific, so any agricultural system that requires a reduced labour intensity would be beneficial to the farmers who stay back. In such situations, *yamkui*, with its lower labour requirements, seems a viable option. However, as pointed out earlier, a scarcity of labour resulting from the demands of terraced wet-rice cultivation is a formidable challenge, especially for poor and marginalized households. Government support in the form of labour subsidies, targeted specifically at poorer households, could provide a solution to this problem. A second option could be the revival of community labour groups to help out the needy households. Such groups were once a common social feature of upland communities. If governments and development programmes were to extend support to meet the labour requirements – by extending cash and food grants so the poor could engage and feed labourers – perhaps smallholders and marginal farming households could overcome the hurdles and adopt fireless shifting cultivation, thus providing a broader solution to concerns about the practice of shifting cultivation in its traditional forms.

The upland farmers of Kalhang have amply demonstrated their entrepreneurial skills in harnessing a neglected agricultural system to their advantage. Thus, they provide a lesson that should be studied in depth, not just for its technical value, but also as an appropriate case for analysis as a prototype business model. Finding a viable answer to the management of change in shifting cultivation areas is an issue that has vexed governments and development agencies for decades. The *yamkui* system of Ukhrul needs detailed assessment, for it may offer a solution to this long-standing problem.

Triticum aestivum L.
[Poaceae]

Wheat was once a staple food of the Tangkhuls, along with maize. However, rice is now their staple of choice and wheat is excluded from the *yamkui* system, which is focused on growing cash crops to support purchases of rice. Contrarily, wheat is now the world's most widely grown crop, and the one earning the highest monetary yield.

Acknowledgements

The authors acknowledge the warmth and close cooperation extended by the villagers of Kalhang, who, during this research, openly accepted the research team as close members of their community. The authors also acknowledge the support of the North Eastern Region Community Resource Management Project (NERCORMP) team in Ukhrul district. This study would not have been possible without the financial support of the NERCORMP project, funded by the Government of India and the International Fund for Agricultural Development (IFAD), Rome. The authors are also grateful for the help of Dr K. Haridasan, formerly of the State Forest Research Institute, Itanagar, Arunachal Pradesh, in identifying plant species used for mulching.

References

Akobundu, I. O. (1987) *Weed Science in the Tropics: Principles and Practices*, Wiley, Chichester, UK

Blair, G., Catchpoole, D. and Horne, P. (1990) 'Forage tree legumes: Their management and contribution to the nitrogen economy of wet and humid environments', *Advances in Agronomy* 44, pp.27-54

Carsky, J. R., Hayashi, Y. and Tian, G. (1998) *Benefits of Mulching in the Sub-humid Savanna Zone: Research Needs and Technology Targeting*, Resource and Crop Management Research Monograph, International Institute of Tropical Agriculture, Ibadan, Nigeria

Erenstein, O. (2002) 'Crop residue mulching in tropical and semi-tropical countries: An evaluation of residue availability and other technological implications', *Soil and Tillage Research* 67(2), pp.115-133, DOI: 10.1016/S0167-1987(02)00062-4

Erenstein, O. (2003) 'Smallholder conservation farming in the tropics and sub-tropics: A guide to the development and dissemination of mulching with crop residues and cover crops', *Agriculture, Ecosystems and Environment* 100(1), pp.17-37, DOI: 10.1016/S0167-8809(03)00150-6

Garrity, D. and Chun, K. L. (2001) 'Shifting cultivation in Asia: Diversity, change, indigenous knowledge and strategies', in *Shifting Cultivation: Towards Sustainability and Resource Conservation in Asia*, International Institute of Rural Reconstruction, Cavite, Philippines, pp.3-13

Handayanto, E., Cadisch, G. and Giller, K. E. (1994) 'Nitrogen release from prunings of legume hedgerow trees in relation to quality of the prunings and incubation method', *Plant and Soil* 160(2), pp.237-248

Kabu, R. (2001) 'Traditional slash and mulch systems in the Solomon Islands', in *Shifting Cultivation: Towards Sustainability and Resource Conservation in Asia*, International Institute of Rural Reconstruction, Cavite, Philippines, pp.196-201

Lal, R. (1989) Conservation tillage for sustainable agriculture: Tropics versus temperate environments. *Advances in Agronomy* 42, pp.85-197, DOI:10.1016/S0065-2113(08)60524-6

Leduc, B. and Choudhury, D. (2012) 'Agricultural transformations in shifting cultivation areas of Northeast India: Implications for land management, gender and institutions', in D. Nathan and V. Xaxa (eds) *Social Exclusion and Adverse Inclusion: Development and Deprivation of Adivasis in India*, Oxford University Press, New Delhi, pp.237-258

Singh, L. J. (2009) 'A case study of shifting cultivation practices among the Tangkhuls of Ukhrul district, Manipur', PhD dissertation to the Department of Ecology and Environmental Sciences, Assam University, Silchar, Assam, India

Singh, L. J. and Choudhury, D. (2015) 'Fallow management practices among the Tangkhuls of Manipur: Safeguarding provisioning and regulatory services from shifting cultivation fallows', in M. F. Cairns (ed.) *Shifting Cultivation and Environmental Change: Indigenous People, Agriculture and Forest Conservation*, Earthscan from Routledge, London, pp.449-467

Spencer, J. E. (1966) *Shifting Cultivation in Southeastern Asia*, University of California Press, Berkeley and Los Angeles, CA

Whitbread, A., Blair, G. J., Naklang, K., Lefroy, R. D. B., Wonpasaid, S., Konboon, Y. and Suriyaarunroj, D. (1999) 'The management of rice straw, fertilizers and leaf litter in rice cropping systems in Northeast Thailand: Rice yields and nutrient balances', *Plant and Soil* 209, pp.29-36

Whitmore, T. C. (1984) *Tropical Rain Forest of The Far East*, Oxford Science Publications, (second edition), Clarendon Press, Oxfordshire, UK

12

NO BURNING

Some unusual elements of agricultural systems in Papua New Guinea

*R. Michael Bourke and Bryant Allen**

Introduction

Shifting cultivation is the basis for production of most arable food crops in Papua New Guinea (PNG), with a fallow phase dominated by self-sown trees, grass and other vegetation to restore soil fertility. New food gardens are formed by clearing fallow vegetation, which is generally burnt. However, for 25% of rural villagers, or about 1.7 million people, burning the fallow vegetation is a minor practice. In this chapter we describe two of these situations. In the first, undergrowth is cleared with minimal burning and food crops are then planted under the forest trees. Sometime later, the trees are cut down on top of the growing crops and the food plants continue to grow between the fallen timber. The second practice is where the undergrowth is cleared and none of it, or only a small proportion, is burnt. Commonly, timber is left standing or is removed for firewood, while most of the smaller fallow vegetation is left on the soil surface without burning or removal from the plot.

Most of the areas where these practices occur share a number of characteristics: rainfall is very high to extremely high (4000 to 7000 mm/year) with no regular dry periods; soil fertility is low; the climate is hot (sea level to 1400 m altitude); and population density is very low. The high rainfall, humidity and temperatures result in a mass of leaves and small branches decomposing quickly and contributing significant amounts of nutrients to the growing food plants. The innovations described in this chapter have been developed by villagers over the years to enhance productivity in marginal environments. The often widely dispersed locations of these practices suggest independent discovery rather than adoption from neighbouring communities.

* Dr. R. Michael Bourke and Dr. Bryant Allen are both honorary associate professors at the College of Asia and the Pacific, The Australian National University, Canberra.

Background to PNG agriculture and research methods

The population of Papua New Guinea in 2018 was about 8.4 million, with the majority (6.8 million) being rural villagers who grew most of their own food.[1] An estimated 83% of food energy and 75% of protein was derived from locally grown produce. The most common sources of food energy were the staple foods. Sweet potatoes were by far the most important of these. Other staple foods included bananas, sago (*Metroxylon sagu*), cassava, various yams (*Dioscorea* spp.), *Colocasia* taro and *Xanthosoma* taro (Bourke et al., 2009, pp.130-144).

The physical environment used for agriculture in PNG is complex and varied. Rainfall ranges from 1000 mm to more than 9000 mm per year, with the greatest concentration of people living and farming in the 1800 mm to 3500 mm rainfall zone. People live and farm from sea level to an altitude of 2800 metres above sea level (masl). The following altitudinal zones are commonly recognized: lowlands (0 to 600 masl); intermediate zone (600 to 1200 masl); highlands (1200 to 1800 masl); high altitude (1800 to 2200 masl); and very high altitude (2200 to 2800 masl). The physical environment may be divided into five basic landforms: mountains and hills (not of volcanic origin); landforms of volcanic origin; plains and plateaux; floodplains; and raised coral reefs and littoral areas. More than half (52%) of the total land area consists of mountains and hills. Almost 19% is plains or plateaux and 18% is floodplains, leaving volcanic landforms, raised coral reefs and littoral areas to make up the rest (Allen and Bourke, 2009, pp.56-61, 87-94; McAlpine et al., 1975; McAlpine and Quigley, c. 1995).

In the 1990s, the authors, together with colleagues from Australia and Papua New Guinea, mapped and described village agriculture over the entire country in some detail (Allen et al., 1995; Bourke et al., 1998)(Figure 12-1). Based on extensive fieldwork over a six-year period, along with literature reviews, we generated a comprehensive database on more than 100 aspects of PNG village agriculture, known as the Mapping Agricultural Systems of PNG (MASP) database. We delineated 287 discrete 'agricultural systems' based on the following criteria:[2]

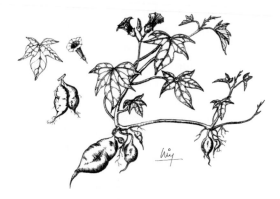

Ipomoea batatas (L.) Lam.
[Convolvulaceae]

Sweet potatoes are by far the most important staple crop in Papua New Guinea, and are grown in most of the study sites for this chapter. While sweet potato provides less edible energy and protein per unit weight than cereals, it has higher nutrient density than cereals. It is a rich source of beta-carotene, a precursor to vitamin A.

1. Fallow type (the vegetation that is cleared from a garden site before cultivation);
2. Fallow period (the length of time for which a garden site is left unused between periods of cultivation);
3. Cultivation intensity (the number of consecutive crops before a garden is left fallow);
4. The staple, or most important, food crops;
5. Soil-fertility maintenance techniques, other than natural-regrowth fallows; and
6. Sources of cash income from agricultural activities.

FIGURE 12-1: Author Mike Bourke takes notes among the felled trees in a food garden in Honinabi village, Western province, PNG, in July 2013. Bananas, the staple food in the area, are growing through the fallen fallow vegetation.

Photo: Joe Saferius.

Data presented in this chapter draws heavily on the MASP database and some of the publications arising from this research, including monographs devoted to the (then) 19 provinces in PNG. In this paper we draw heavily on the MASP monographs that describe agricultural systems in Gulf province (Hide et al., 2002b), Southern Highlands province (Bourke et al., 2002c), West Sepik province (Bourke et al., 2002d) and Western province (Allen et al., 2002a) and, to a lesser extent, on those describing systems in East New Britain province (Bourke et al., 2002b), East Sepik province (Allen et al., 2002b), Manus province (Hide et al., 2002a) and Morobe province (Bourke et al., 2002a).

There is some variability in most of the agricultural systems discussed here. However, the analysis focuses on dominant crops and activities and glosses over some variations within a defined system. For example, in System 1405 in the Yangoru and Wingei area in East Sepik province, the fallow vegetation on about 75% of the agricultural land is short grass, and on the rest it is short woody regrowth (Allen et al., 2002b). Fallow vegetation is not burnt in the grassland areas, but it is burnt in forested parts of the system. The description here refers to the grassland area only. Similarly, where swamp taro (*Cyrtosperma merkusii*) is grown,[3] there is no fallow vegetation and hence no burning. This crop is the most important staple food in only one agricultural system: that on the atolls east of Bougainville (Bourke and Betitis, 2003), so this is the only system where swamp taro is grown that is noted in Table 12-1. In other locations, swamp taro is a minor food crop and land management reported in this chapter is that used for more important food sources.

Managing fallow vegetation

Shifting cultivation is the basis for production of most arable food crops in Papua New Guinea. Soil fertility is restored after a fallow phase involving self-sown trees, grass and other vegetation. The most common type of fallow vegetation, used by 37% of the rural population, is tall woody regrowth; that is, trees more than 10 m high. Other fallow types are tall grass (used by 19% of the rural population), low woody regrowth (17%), a mix of grass and woody vegetation (16%), short grasses (11%) and savannah (0.4%) (Bourke and Allen, 2009, pp.235–241, 534).

New food gardens are formed by clearing fallow vegetation, which is generally burnt. However, for 25% of rural villagers (1.7 million people) burning the fallow vegetation is a minor practice (Table 12-1). In this chapter we describe two of these situations. In the first, villagers clear

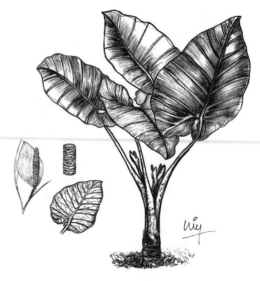

Cyrtosperma merkusii (Hassk.) Schott [Araceae]

Swamp taro is the largest of the plants known collectively as taro. The corms can reach 80 kg in weight. One of very few carbohydrate crops that can be grown on low-lying coral atolls, its cultivation is increasingly threatened by rising sea levels caused by climate change.

the undergrowth with minimal burning and then plant food crops under the forest trees. Sometime later, the trees are cut down on top of the growing crop and the food plants continue to grow between the fallen timber. The second practice is where the undergrowth is cleared and none of this cut vegetation, or only a small proportion of it, is burnt. Commonly, trees are left standing in the plots or are removed for firewood, while most of the smaller fallow vegetation is left on the soil surface without burning or removal.

There are a number of other situations where there is minimal burning of fallow vegetation (Table 12-1), but we do not consider them in detail here. The first is where there is minimal fallow material because cultivation of food gardens is semi-permanent or permanent. In the western part of the Central Highlands, soil fertility is maintained by incorporating organic matter, including some grasses that grow during short-duration fallows, into mounds of soil to form 'composted mounds' (Bourke et al., ch. 40, in the supplementary chapters to this volume). In other highland locations, soil fertility is maintained by a combination of transferring organic matter from drainage ditches to the soil surface and rotation of a root crop (usually sweet potatoes) and a leguminous food crop, most commonly peanuts. In the very specialized agriculture practised on some atolls, soil fertility is maintained by transferring large

TABLE 12-1: Agricultural systems in Papua New Guinea where fallow vegetation is not burnt, or burning is only a minor practice.

Main reason that burning is absent or is a minor practice	Environmental zone (rainfall and altitude) and locations	Number of agricultural systems	Estimated population, 2018
Undergrowth is cleared; food crops planted; trees felled onto growing crops.	Very high rainfall areas in inland lowlands, Western and Gulf provinces; adjacent areas in Southern Highlands; Baining Mountains of New Britain; Oro/Morobe provincial border. Up to 1400 masl.	12	140,928
Undergrowth cleared; some vegetation burnt; most is heaped in garden or left on soil surface or along garden edge.	Very high rainfall highlands and intermediate altitude zone in Telefomin and Oksapmin area, Sandaun province. Mostly 1400 to 2200 masl, but as low as 400 m.	5	40,524
Undergrowth cleared; trees pollarded, killed or removed. Some fallow vegetation is burnt; most is left on soil surface.	Variable environments with high or medium rainfall; lowland, intermediate and highland altitudes. Kaintiba, Gulf province; Bema-Aseki area, Gulf and Morobe provinces; Yangoru area, East Sepik; southwest of Leron Plains, Morobe province; western Manus island. (100 to 1900 masl).	5	86,371
Semi-permanent or permanent gardens. Soil fertility maintained by incorporating organic matter into mounds.	High rainfall highlands in Enga, Southern Highlands and Western Highlands. Mostly 1600 to 2800 masl, but as low as 600 m.	14	920,084
Semi-permanent or permanent gardens. Soil fertility maintained by transfer of organic matter from drainage ditches onto beds and peanut/sweet potato rotation.	Seasonally dry and high rainfall highland valleys in Western Highlands, Simbu and Eastern Highlands provinces (1400 to 2700 masl).	5	428,310
Permanent plots, atolls. Soil fertility maintained by transfer of organic matter to plots.	High rainfall atolls east of Bougainville Island. Sea level.	1	1,098

TABLE 12-1 (cont.): Agricultural systems in Papua New Guinea where fallow vegetation is not burnt, or burning is only a minor practice.

Main reason that burning is absent or is a minor practice	Environmental zone (rainfall and altitude) and locations	Number of agricultural systems	Estimated population, 2018
Few or no food gardens.	High rainfall lowlands. Middle Fly river; Sepik river mouth; small islands off East Sepik north coast; small islands near Manus island (0 to 50 masl).	4	29,123
Fallow vegetation is removed or incorporated.	High rainfall lowlands. Beach ridges near Kupiano station, Central province; peri–urban gardens near Madang town. (0 to 30 masl).	2	30,530
Total		48	1,676,968

Note: Population estimates for 2018 calculated using 2000 census data and a population growth rate of 2.7% per annum (NSO, 2002).
Data source: Mapping Agricultural Systems of PNG (MASP) database (Allen et al., 1995; Bourke et al., 1998).

volumes of organic matter to plots of swamp taro (*Cyrtosperma merkusii*) and some *Colocasia* taro grown in pits of fresh water.

In a limited number of locations in PNG, villagers derive the bulk of their food from tree crops, particularly sago and leaves from *Gnetum gnemon*, as well as from small garden plots. In other locations, most food comes from trading. In these circumstances, villagers clear only small areas for arable food gardens, so there is little or no burning of fallow vegetation. Finally, in several other locations, fallow vegetation is removed from garden plots with minimal or no burning.

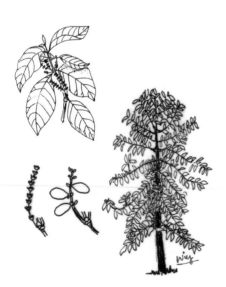

Gnetum gnemon L. [Gnetaceae]

This small- to medium sized tree is unlike others in its genus, most of which are lianas. The young leaves, nut-like seeds and male and female cone-like structures (strobili) are all eaten. The strobili are rich in a kind of natural phenol that has antibacterial and antioxidative qualities.

Agricultural systems where trees are felled on to growing food crops

In 12 agricultural systems, the undergrowth in forest fallows is cleared and food crops are planted beneath the standing trees. When the new food plants are established, around 8 to 12 weeks after planting, the trees are cut down on top of them. The estimated number of people using this practice in PNG in 2018 was 141,000 (Table 12-1).

There are two regions where this practice is common. One is in the Kiunga to Nomad area of Western province and extending eastwards into Hela and Southern Highlands provinces (Figure 12-2). The altitude range is 50 to 700 masl. Rainfall is extremely high, with mean annual rainfall between 6000 and 7000 mm. The most important foods are combinations of bananas and sago, except for the Mt Sisa area in Southern Highlands – at a higher altitude of 600 to 1200 masl – where sweet potatoes are the most important food.

The other region is in the northeast of Gulf province and nearby areas in Morobe and Central provinces (Figure 12-2). In this case, the altitude range is 200 to 1400 masl, while mean annual rainfall is very high at 4000 to 6000 mm. The most important foods in this sub-region are combinations of bananas and sweet potatoes.

The following is a brief summary of an agricultural system where trees are dropped on to growing crops. It describes Agricultural System 0105, an area of 658 sq. km in Western province, with an estimated population of 3900 in 2018. The system is located near the Tomu river, west of Mt Bosavi, and east of Nomad station, between the Nomad and Rentoul rivers.

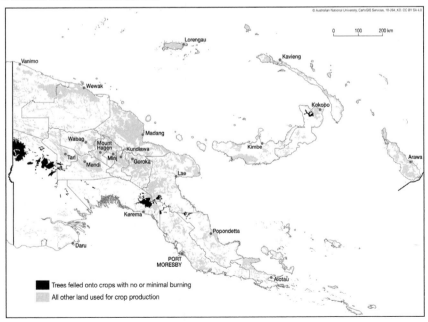

FIGURE 12-2: Locations in Papua New Guinea where trees are felled on to growing food crops.

Source: MASP database.

Bananas are the most important crop; sago is an important food; other crops are *Colocasia* taro, yams (*Dioscorea alata*), *Xanthosoma* taro and sweet potatoes. Tall woody regrowth, more than 25 years old, is cultivated using two methods. The first involves clearing beneath the trees and planting bananas with some *Colocasaia* taro. The trees are then felled on top of the crops. There is no burning and little fencing. In the second method, trees are felled and burnt and the gardens are strongly fenced. Root crops are planted in these gardens. The fallows used for these two types of gardens are similar in age and type. Only one planting is made before fallowing. Breadfruit (*Artocarpus altilis*) (grown for its edible nuts) and *marita* (*Pandanus conoideus*) (grown for its fruit, which is rich in pro-Vitamin A and oil) are very common in fallows. (Adapted from Allen et al., 2002a, pp.35-36) (Figure 12-3).

The second example of an agricultural system where trees are dropped on to growing crops is at a higher altitude (600 to 1200 masl), located on the southern slopes of Mt Sisa in Southern Highlands province and extending into Western province. The following passage describes Agricultural System 0707/0109, in an area of 131 sq. km, with an estimated population of just over 1000 in 2018.

Sweet potatoes are the most important crop; sago is an important food;

Musa cvs. [Musaceae]

There are literally hundreds of cultivated varieties (cvs) of bananas in Papua New Guinea. Most of the edible varieties are interspecific crosses between *M. acuminata* and *M. balbisiana*. Collectively, they are one of the most important food crops in the study areas, and grow towards sunlight through a tangle of felled trees in the study areas.

FIGURE 12-3: *Colocasia* taro and bananas rise through the tangled and decomposing fallow debris in a food garden at Kwomenai, in Western province, August 2013.

Photo: Timothy Sharp.

other crops are *Colocasia* taro, cassava, bananas, *Xanthosoma* taro and yams (*Dioscorea alata*). The undergrowth is cleared beneath tall woody regrowth that is more than 15 years old. Two types of gardens are made. In the first, covering about 80% of the cultivated land, the site is strongly fenced. Sweet potatoes are planted by dibbling and the trees are then felled on top of the crop. There is no burning. Cut vegetation is placed in large heaps within the fenced area. In the second type of garden, bananas and *Colocasia* taro are planted and the trees felled on to the crops. There is no burning or heaping of vegetation. Only one planting is made before fallowing. Household gardens are common. *Marita* (*Pandanus conoideus*) is grown extensively in fallows. (Adapted from Bourke et al., 2002c, pp.39-40).

The practice of dropping trees on to growing food crops is also followed in two other parts of PNG, but only on plots of *Xanthosoma* taro. One is in the mountainous region on the border of Oro and Morobe provinces. The other is in the western Baining mountains on the island of New Britain (Figure 12-2). In the Oro/Morobe border area, sweet potatoes are the most important crop and *Colocasia* taro is an important crop. On land to be planted with either of these crops, fallow vegetation is felled and burnt. However, *Xanthosoma* taro is planted beneath standing trees and the trees are felled onto the crop two or three months after planting (Bourke et al., 2002a).

In the western Baining mountains, *Xanthosoma* taro is the most important crop and *Colocasia* taro is an important crop. Most fallow vegetation is not burnt. In taro gardens, fallow vegetation is cut and tree branches are removed from the garden site for firewood. Some leaves and small branches are burnt in locations intended for planting tobacco and certain vegetables.[4] For *Xanthosoma* taro gardens, the undergrowth is removed from the site and trees are ringbarked. After the *Xanthosoma* taro has been planted and becomes established, some trees are felled on top of the growing crop. A thick mulch of leaves and twigs covers the soil surface (Bourke et al., 2002b).

Agricultural systems where most fallow vegetation is left on the soil surface

In 10 agricultural systems, some of the fallow vegetation is burnt, but most of it is left on the soil surface (or in some locations, it is moved to the edge of the garden plot).

Telefomin-Oksapmin area

Five of these agricultural systems are in the Telefomin and Oksapmin area in the far west of the highlands, near the border with Indonesian New Guinea (West Papua) (Figure 12-4). In this area, the practice is most common in mountains at altitudes of 1400 to 2200 masl, but it is also found in some places as low as 400 masl. The rainfall in this sub-region is a high 3500 to 5000 mm per annum. Historically, *Colocasia* taro was the most important crop in this sub-region, but over the past 60 years, sweet potatoes have been adopted. The most important food crop is either *Colocasia* taro or sweet potatoes, or a combination of the two, with some *Xanthosoma* taro also grown. An estimated 40,500 people employ this practice in the Telefomin-Oksapmin area of West Sepik province (Table 12-1).

In the Telefomin-Oksapmin area, *Colocasia* taro is planted beneath the forest trees (Figure 12-5). As the taro crop matures, trees are thinned and eventually all trees may be either cut down or killed and left standing, so that direct sunlight is allowed to reach the crop. Some of the cleared undergrowth is burnt, but much is heaped within the gardens or along the garden edge. A thick layer of mulch is left on the soil surface at planting (Figure 12-6). (Bourke et al., 2002d). The more detailed description that follows comes from the Yapsei area north of Telefomin:

Colocasia esculenta (L.) Schott [Araceae]

Taro has long been a ubiquitous food crop in Papua New Guinea. Traces of its use have been found at Kuk Swamp, in the Wahgi valley of Western Highlands province, where an ancient agricultural system is believed to be about 10,000 years old. It is grown for its edible corms. They are toxic when raw, but made edible by cooking.

> Most of the trees are not felled, but the undergrowth beneath the tall woody regrowth is cleared and planted with taro. As the *Colocasia* taro matures, more and more trees are killed by ringbarking or, if they are on the edge of a garden, felled outwards, to allow more light into the garden. This process is extended progressively across a garden site in a strip, exposing the soil to the heavy rainfall for only a short period. There is minimal burning of fallow vegetation. An associated technique in this area is the felling of all the trees on a garden site some years before it is to be planted in taro. The felled forest decomposes rapidly and provides the soil with additional nutrients. A low secondary forest quickly colonises the site again but, when the site is planted in taro a few years later, the size of the trees that have to be thinned is much reduced.

Other lowland and intermediate-altitude locations

In five other agricultural systems, most of the fallow vegetation is left on the soil surface or is removed with minimal burning. Four of the five locations are in different subregions in PNG and the agricultural systems vary considerably between these locations. The estimated 2018 population in these five locations was 86,400 (Table 12-1).

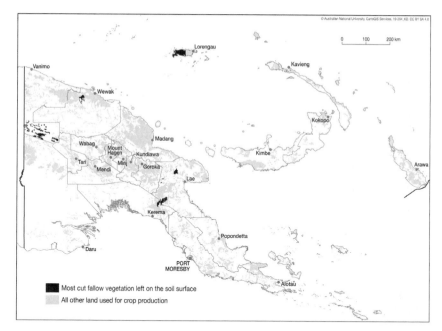

FIGURE 12-4: Locations in Papua New Guinea where most fallow vegetation is left on the soil surface.

Source: MASP database.

FIGURE 12-5: A crop of *Colocasia* taro planted under a forest near Telefomin, in West Sepik province, June 1991.

Photo: Mike Bourke.

FIGURE 12-6: Small *Colocasia* taro plants in a deep layer of organic matter typical of the fireless forest gardens in the Telefomin area of West Sepik province, June 1991.

Photo: Mike Bourke.

The first location is in the Kaintiba area in northeast Gulf Province (Figure 12-4). The altitude range is 600 to 1000 masl and the mean annual rainfall is 4500 to 5000 mm. Sweet potatoes are the most important crop; bananas an important crop and *Xanthosoma* taro, *Colocasia* taro, cassava and sago are also grown. Villagers clear the undergrowth beneath tall woody regrowth and plant food crops beneath the trees. The trees are then pollarded and the branches are burnt in heaps or removed from the gardens. Standing trees are killed by burning the base (Hide et al., 2002b, pp.47–48).

The second location is northeast of Kaintiba in the Bema–Aseki area of Gulf and Morobe provinces (Figure 12-4). This location is higher than the Kaintiba area, at 1400 to 1900 masl, but the mean annual rainfall is similar (4000 to 5000 mm). Sweet potatoes are the most important crop, with smaller areas of *Colocasia* taro, bananas and *Xanthosoma* taro. Here, the tall woody regrowth is cleared, with many trees left standing. Felled trees are cut into pieces and stacked in low heaps over the garden site. In other places, short hurdles, about 3 m long and 1 m high, are constructed and branches and leaves heaped over them. The heaps are distinctive and are a form of incipient composting. *Colocasia* taro and bananas are planted in and around the heaps. There is very little burning, except for small piles of leaf litter. Green leafy vegetables are planted in the ash. Leaves and other material are also spread over the surface as mulch (Hide et al., 2002b, pp.49-50).

The third agricultural system is practiced in two small valleys in the mountains south of Leron Plains in Morobe province and just east of the Eastern Highlands border (Figure 12-4). The altitude is 1000 to 1100 masl and mean annual rainfall is 2000 to 2500 mm. A mix of tall cane grass and low woody regrowth is cleared and mostly removed from the garden site. The fine litter remaining is left for two months to decompose, with very little burning. Sweet potatoes (the most important food), taro, bananas, corn and green vegetables are planted into the decomposing litter (Bourke et al., 2002a, pp.135-136).

The fourth agricultural system in this group is located west of Yangoru and north of Wingei in hilly grasslands and patches of low woody regrowth in East Sepik province (Figure 12-4). This is a lowland (100 to 300

Xanthosoma sagittifolium (L.) Schott [Araceae]

Xanthosoma taro is grown in some of the study areas, where trees are felled on to the growing crop a few months after it is established. The cormels are rich in carbohydrates, calcium, iron and phosphorus.

masl) area with a seasonally dry rainfall pattern (1700 to 2000 mm/year). Yams (*Dioscorea esculenta*) and bananas are the most important crops. The grasslands are dominated by *Imperata cylindrica*. The grass is pulled up and either laid on the surface or along 10 cm high 'fences' that divide the garden into smaller plots about 10 to 15 m apart. There is minimal burning in the predominantly grassland areas, although short woody regrowth is cleared and burnt elsewhere in the system (Allen et al., 2002b, pp.29–30).

The final agricultural system in this diverse group is at the western end of Manus island (Figure 12-4), where agriculture is practised from sea level to 600 masl and the mean annual rainfall is 3500 to 4000 mm. Sago is the staple food and agriculture is generally of minor importance. Small food gardens are made in tall woody regrowth. The fallow vegetation is cut and either dried and burnt or left as mulch and not burnt (Hide et al., 2002a, pp.41–44).

Discussion

Metroxylon sagu Rottb. [Arecaceae]

Sago is an important food – and in some parts, a staple – throughout the study areas. The starch is harvested from the felled trunks of the palms by washing it out of the pulverized pith with water. The palms are multi-stemmed, and each stem flowers once and then dies. The stems are cut for sago extraction just before they flower.

Allowing the land to lie fallow is the most common method of restoring soil fertility in PNG agricultural systems. When these fallows are reopened for cultivation, the villagers generally dispose of the vegetation by slashing and burning it. However, 25% of the rural population do little or no burning. They adopt this approach for a number of reasons, the most common being that there is little fallow vegetation and soil fertility is maintained by incorporating green organic matter into mounds (known locally as 'composting'), transferring organic matter from drainage ditches to the soil surface or growing a leguminous food crop in rotation with the staple sweet potatoes. In a limited number of locations, people have few or no food gardens and depend on sago as their staple food, so there is little or no fallow vegetation to manage. Elsewhere, it is either removed from the garden site or incorporated into the soil.

Our focus here is on the two other situations where little or no fallow vegetation is burnt. Firstly, where villagers clear the undergrowth and plant food crops under the trees. After 8 to 12 weeks, men cut the trees onto the growing crops. Leaves and other cut vegetation decompose and provide nutrients for the growing crops. This practice occupies a number of discrete locations in parts of Western, Gulf, Southern Highlands and Gulf provinces and smaller adjoining areas in Morobe and Northern provinces (Figure 12-2).

Felling large trees on to the developing food plants results in much less damage than might be expected. As the big trees fall, the food plants disappear beneath a mass of twisted branches and tree trunks, but within a few weeks, the bananas, taro, sweet potatoes and other food crops are growing through the mass of fallen vegetation (Figure 12-1).

These areas share a number of characteristics: rainfall is very high to extremely high (4000 to 7000 mm/year) with no regular dry periods; soil fertility is low; the climate is hot from sea level to an altitude of 1400 masl; and the population density is generally very low. The high rainfall, humidity and high temperatures result in the mass of leaves and small branches decomposing rapidly and contributing significant quantities of plant nutrients to the growing food crops. Within a few months, all but the trunks of the large trees have decomposed. This technique reduces exposure of the soil to the very high rainfall and reduces the amount of weeding required.

The second agricultural practice involving a low burning rate is where villagers clear the undergrowth in the fallow, remove or kill large trees and leave most of the smaller fallow vegetation on the soil surface. In the Telefomin-Oksapmin area, *Colocasia* taro is planted under the trees, which are progressively removed or killed to allow more sunlight to reach the growing crop (Figure 12-7). For both taro and sweet potato gardens in this sub-region, much of the fallow vegetation is left on the soil surface with minimal or no burning. As in the zone where trees are felled on to growing crops, the rainfall is very high and soil fertility is low.

In three of the other five locations where villagers leave most of the fallow vegetation on the soil surface, the rainfall is very high. However, in the two remaining locations, rainfall is not excessive and measures only 1700 to 2500 mm/year. It is not obvious why this practice has been adopted in these locations where rainfall is lower and presumably there is a lesser loss of nutrients by leaching and oxidation of organic matter.

The innovations described in this chapter have been developed by villagers in generations past to enhance food-crop productivity in marginal environments. The widely dispersed locations of these practices suggest independent discovery rather than adoption from neighbouring communities. Indeed, similar practices have been reported

FIGURE 12-7: Trees have been cleared to let sunlight into this taro garden in the Telefomin area of West Sepik province.

Photo: Mike Bourke.

from elsewhere in the humid tropics. In the first published description of the practice of felling trees on to growing crops in PNG, Schieffelin (1975) notes that this practice may be more widespread. He cites Conklin (1957, p.52), who mentioned that the Hanunoo people in the Philippines planted some of their crops in slashed gardens before the trees were felled and the debris burnt. Denevan (2001, p.69) summarizes a variation of shifting cultivation in parts of South America which he describes as 'slash/mulch'. Fallow vegetation is not burnt after cutting, but rather allowed to decompose, with a large portion of the organic material allowed to return to the soil. He notes that this practice is found in very wet locations (as in PNG), in the Choco department of Pacific Colombia, near the city of Guanare in Venezuela's Portuguesa state, in coastal Ecuador and in some parts of Amazonia. In ch. 10 of this volume, Darmanto and Persoon describe land use on Siberut Island, West Sumatra, Indonesia, where fire is not used in the transformation of rainforest into an agroforest.

Acknowledgements

Much of the information presented in this chapter was based on discussions in villages and food gardens with numerous rural villagers in all parts of PNG. People freely gave their comprehensive knowledge about agriculture in their area, generally with no prospect of any recompense. We acknowledge that the innovations and practices featured in this chapter were developed by them and their ancestors so they could survive and thrive in environments that were often marginal for agricultural production. This research was funded by the Australian government through AusAID and the Australian National University.

References

Allen, B. J. and Bourke, R. M. (2009) 'Part 1: People, land and environment', in R. M. Bourke and T. Harwood (eds) *Food and Agriculture in Papua New Guinea*, ANU E Press, The Australian National University, Canberra, pp.27-127, available at http://press-files.anu.edu.au/downloads/press/p53311/pdf/part11.pdf, and http://web.archive.org/web/20110225075407/http://ips.cap.anu.edu.au/ssgm/resource_documents/lmg/png_ag_tables/, both accessed 1 March 2019

Allen, B. J., Bourke, R. M. and Hide, R. L. (1995) 'The sustainability of Papua New Guinea agricultural systems: The conceptual background', *Global Environmental Change* 5(4), pp.297-312, available at https: //www.researchgate.net/publication/223761209_The_sustainability_of_Papua_New_Guinea_agricultural_systems_The_conceptual_background, accessed 1 March 2019

Allen, B. J., Hide, R. L., Bourke, R. M., Akus, W., Fritsch, D., Grau, R., Ling, G. and Lowes, E. (2002a) *Agricultural Systems of Papua New Guinea Working Paper No 4. Western Province: Text Summaries, Maps, Code Lists and Village Identification*, Department of Human Geography, Australian National University, Canberra, available at https://openresearch-repository.anu.edu.au/bitstream/1885/13639/1/Agricultural%20systems%20of%20Papua%20New%20Guinea%20Working%20Paper%20No%204.pdf, accessed 1 March 2019

Allen, B. J., Hide, R. L., Bourke, R. M., Fritsch, D., Grau, R., Lowes, E., Nen, T., Nirsie, E., Risimeri, J. and Woruba, M. (2002b) *Agricultural Systems of Papua New Guinea Working Paper No 2. East Sepik Province: Text Summaries, Maps, Code Lists and Village Identification*, Department of Human Geography, Australian National University, Canberra, available at https://openresearch-repository.anu.edu.au/bitstream/1885/13637/1/Agricultural%20systems%20of%20Papua%20New%20Guinea%20Working%20Paper%20No%202.pdf, accessed 1 March 2019

Bourke, R. M., Allen, B. J., Hobsbawn, P. and Conway, J. (1998) *Agricultural Systems of Papua New Guinea Working Paper No 1. Papua New Guinea: Text Summaries* (two volumes), Department of Human Geography, Australian National University, Canberra, available at https://openresearch-repository.anu.edu.au/bitstream/1885/110392/1/01_Bourke_Agricultural_Systems_paper%20 1_2002.pdf, accessed 1 March 2019

Bourke, R. M., Allen, B. J., Hide, R. L., Fereday, N., Fritsch, D., Gaupu, B., Grau, R., Hobsbawn, P., Levett, M. P., Lyon, S., Mangi, V. and Sem, G. (2002a) *Agricultural Systems of Papua New Guinea Working Paper No 19. Morobe Province: Text Summaries, Maps, Code Lists and Village Identification*, Department of Human Geography, Australian National University, Canberra, available at https:// openresearch-repository.anu.edu.au/bitstream/1885/13673/1/Agricultural%20systems%20of%20 Papua%20New%20Guinea%20Working%20Paper%20No%2019.pdf, accessed 1 March 2019

Bourke, R. M., Allen, B. J., Hide, R. L., Fritsch, D., Geob, T., Grau, R., Heai, S., Hobsbawn, P., Ling, G., Lyon, S. and Poienou, M. (2002b) *Agricultural Systems of Papua New Guinea Working Paper No 14. East New Britain Province: Text Summaries, Maps, Code Lists and Village Identification*, Department of Human Geography, Australian National University, Canberra, available at https:// openresearch-repository.anu.edu.au/bitstream/1885/13658/1/Agricultural%20systems%20of%20 Papua%20New%20Guinea%20Working%20Paper%20No%2014.pdf, accessed 1 March 2019

Bourke, R. M., Allen, B. J., Hide, R. L., Fritsch, D., Grau, R., Hobsbawn, P. Konabe, B., Levett, M. P., Lyon, S. and Varvaliu, A. (2002c) *Agricultural Systems of Papua New Guinea Working Paper No 11. Southern Highlands Province: Text Summaries, Maps, Code Lists and Village Identification*, Department of Human Geography, Australian National University, Canberra, available at https:// openresearch-repository.anu.edu.au/bitstream/1885/13655/1/Agricultural%20systems%20of%20 Papua%20New%20Guinea%20Working%20Paper%20No%2011.pdf, accessed 1 March 2019

Bourke, R. M., Allen, B. J., Hide, R. L., Fritsch, D., Grau, R., Lowes, E., Nen, T., Nirsie, E., Risimeri, J. and Woruba, M. (2002d) Agricultural Systems of Papua New Guinea Working Paper No 3. West Sepik Province: Text Summaries, Maps, Code Lists and Village Identification, Department of Human Geography, Australian National University, Canberra, https://openresearch-repository.anu.edu. au/bitstream/1885/13638/1/Agricultural%20systems%20of%20Papua%20New%20Guinea%20 Working%20Paper%20No%203.pdf, accessed 1 March 2019

Bourke, R. M. and Betitis, T. (2003) *Sustainability of Agriculture in Bougainville Province, Papua New Guinea*, Land Management Group, Australian National University, Canberra, available at https:// www.researchgate.net/publication/320297253_Sustainability_of_Agriculture_in_Bougainville_ Province_Papua_New_Guinea, accessed 1 March 2019

Bourke, R. M. and Allen, B. (2009) 'Part 3. Village food production systems', in R. M. Bourke and T. Harwood (eds) *Food and Agriculture in Papua New Guinea*, ANU E Press, The Australian National University, Canberra, pp.193-269, available at http://press-files.anu.edu.au/downloads/ press/53311/pdf/part31.pdf, and http://web.archive.org/web/20110225075407/http://ips.cap. anu.edu.au/ssgm/resource_documents/lmg/png_ag_tables/, both accessed 1 March 2019

Bourke, R. M., Gibson, J., Quartermain, A., Barclay, K., Allen, B. and Kennedy, J. (2009) 'Part 2. Food production, consumption and imports', in R. M. Bourke and T. Harwood (eds) *Food and Agriculture in Papua New Guinea*, ANU E Press, The Australian National University, Canberra, pp.129-192, available at http://press-files.anu.edu.au/downloads/press/p53311/pdf/part21.pdf, and http://web. archive.org/web/20110225075407/http://ips.cap.anu.edu.au/ssgm/resource_documents/lmg/ png_ag_tables/, both accessed 1 March 2019

Conklin, H. (1957) *Hanunoo Agriculture*, Food and Agriculture Organization of the United Nations, Rome

Denevan, W. (2001) *Cultivated Landscapes of Native Amazonia and the Andes*, Oxford University Press, Oxford, UK

Hide, R. L., Allen, B. J., Bourke, R. M., Fritsch, D., Grau, R., Helepet, J. L., Hobsbawn, P., Lyon, S., Poienou, M., Pondrilei, S., Pouru, K., Sem, G. and Tewi, B. (2002a) *Agricultural Systems of Papua New Guinea Working Paper No 18. Manus Province: Text Summaries, Maps, Code Lists and Village Identification*, Department of Human Geography, Australian National University, Canberra, available at https://openresearch-repository.anu.edu.au/bitstream/1885/13672/1/Agricultural%20

systems%20of%20Papua%20New%20Guinea%20Working%20Paper%20No%2018.pdf, accessed 1 March 2019

Hide, R. L., Bourke, R. M., Allen, B. J., Fereday, N., Fritsch, D., Grau, R., Lowes, E. and Woruba, M. (2002b) *Agricultural Systems of Papua New Guinea Working Paper No 5. Gulf Province: Text Summaries, Maps, Code Lists and Village Identification*, Department of Human Geography, Australian National University, Canberra, available at https://openresearch-repository.anu.edu.au/bitstream/1885/13640/1/Agricultural%20systems%20of%20Papua%20New%20Guinea%20Working%20Paper%20No%205.pdf, accessed 1 March 2019

Humphreys, G. (1991) Personal communication between author Mike Bourke and the late Dr. Geoff Humphreys, a soil scientist at the Australian National University and later Macquarie University, Sydney.

McAlpine, J. R., Keig, G. and Short, K. (1975) *Climatic Tables for Papua New Guinea. Division of Land Use Research Technical Paper No 37*, Commonwealth Scientific and Industrial Research Organization (CSIRO), Canberra

McAlpine, J. R. and Quigley, J. (c. 1995) *Natural Resources, Land Use and Population Distribution of Papua New Guinea: Summary Statistics from PNGRIS*, PNGRIS Report No. 7, Papua New Guinea Resource Information System, Australian Agency for International Development, Canberra

NSO (2002) *Papua New Guinea 2000 Census: Final Figures*, National Statistical Office of Papua New Guinea, Port Moresby

Schieffelin, E. L. (1975) 'Felling the trees on top of the crop: European contact and the subsistence ecology of the Great Papuan Plateau', *Oceania* 46(1), pp.25-39

Notes

1. The population has increased by more than 50% since 2000. This figure was extrapolated from the 2000 national census at an assumed growth rate of 2.7% per year, which was the population growth rate recommended by the National Statistical Office in 2002 (NSO, 2002). There is considerable doubt about the accuracy of data from the more recent 2011 national census. Hence we have extrapolated from the more reliable 2000 census and the 1980 to 2000 inter-census growth rate to generate 2018 estimates of populations using different agricultural techniques (Table 12-1).

2. Originally, garden and crop segregation were among the criteria used to define an agricultural system. However, these parameters were not used in practice to define systems because they were defined by the first five criteria listed on page 245. When the database was standardized after completion of field mapping, sources of cash income from agricultural activities was added as a criterion to help to define agricultural systems.

3. Swamp taro, with the accepted botanical name *Cyrtosperma merkusii*, is also known by many synonyms, among them *C. chamissonis*, *C. edule*, and *C. lasioides*.

4. In many parts of PNG, leafy vegetables, tobacco and other crops that have a high nitrogen requirement are commonly planted in small areas of gardens where fallow vegetation has been burnt. In some locations in Enga province, at an altitude of 1600 to 2000 masl, villagers burn heaps of small branches and twigs of she-oak trees (*Casuarina oligodon*) and later plant them with vegetables. The scientific explanation for this practice is that the heat of the fire kills soil microflora and this releases nitrogen and other plant nutrients that are used by the growing crops (Humphreys, 1991).

Some innovative farmers in a high-rainfall area of West Sepik province, Papua New Guinea, have learned how to farm under a partial forest canopy with *Colocasia* taro and a few other less light-demanding crops (see Bourke and Allen, ch. 12, this volume). Only the undergrowth has been cleared before planting, and a thick layer of mulch is left on the soil surface. Under the high-rainfall conditions where this system is practised, much of this organic matter decomposes during the life of the garden and provides nutrients to the growing crops. As the crops mature, the overstorey of trees is progressively removed or killed by ringbarking to allow more sunlight to reach the crops growing beneath.

Sketch based on a photo by R. Michael Bourke in June, 1991.

IV. INNOVATIONS IN FALLOW MANAGEMENT

Fallow enrichment with rattan – an innovation that is well developed in Kalimantan, Indonesia - is a compelling example of an economically improved fallow (see Schreer, ch. 30, this volume). This Bukidnon farmer in Mindanao, the Philippines, was actually harvesting wild rattan, but depletion of natural stocks and increasing prices may soon persuade him to begin cultivating the forest palm himself (see plates 54 to 56 in Coloured Plates section II, this volume).

Sketch based on a photo by Malcolm Cairns, ca. 1996.

Synthesis

FALLOW MANAGEMENT

Indigenous knowledge, vegetative succession, practical experience and planting commercial trees

*Do Dinh Sam**

Fallow is one of the key stages in a chain of traditional swidden-farming practices pursued by farmers in mountainous regions, typically covered by tropical forest, in many countries of Asia, Africa and Latin America. The main stages in rotating shifting cultivation begin with selecting the land to cultivate, then proceed to clearing the land of vegetation and burning it, sowing seeds using dibble sticks, weeding and tending the growing crop, harvesting, and then leaving the land fallow, so that after a period of rest and recovery, cultivation may be resumed.

The fallow phase is essentially intended to restore soil fertility through reforestation after years of cultivation, so that farming can be continued in later cycles. Fallow-land management plays a key role in traditional shifting cultivation and is a stage in the cycle where alternatives may be sought to continued 'slash-and-burn' farming.

Fallow-land management is a process with two aspects: on one hand, technical, and on the other, socio-economic and environmental. The technical aspect is based on the process of reforestation and follows the initiatives and practical experience of the farmers. The socio-economic and environmental aspects relate to the need of farmers to improve their livelihoods and forest managers to enforce forest protection, especially in the current era of global climate change and the important role of forests in absorbing the greenhouse gas carbon dioxide.

The following section of this volume has two chapters, the first from Nepal and entitled *Fallow management in transforming shifting cultivation systems*, by authors Kamal P. Aryal, Keshab Thapha, Rajan Kotru and Karma Phuntsho (ch. 13), and the second from authors Mai Van Thanh and Tran Duc Vien in Vietnam, with the title *Fallowed swidden fields in Vietnam: Floral composition, successional dynamics and farmer management* (ch. 14). Although the research was conducted in two countries and had differing in-depth aspects, the findings were similar. The discussion of

---

* Professor Dr. Do Dinh Sam is a senior researcher in forest ecology and environment and Vice President of the Vietnam Soil Science Association.

Aryal et al. is directed towards transforming traditional shifting cultivation and analysing the major drivers of change in fallow-land management in particular and shifting cultivation in general, while the research in Vietnam concerns traditional management of fallow land based on the restoration of forest vegetation and continued cultivation in the next cycle. Both studies pay attention to the innovations of famers in land-use practices, and it is clear that the management of fallow land has two different goals: first, the continuation of traditional shifting cultivation and second, gradual replacement of shifting cultivation with more sustainable practices.

In considering the management of fallow land, the crucial aspect is the length of the fallow. Research in both countries concluded that in the past, the fallow period had been quite long and depended on the natural conditions for forest rehabilitation. This could require 15 or 20 years of fallow before the soil fertility increased, allowing farmers to continue to cultivate.

At present, the trend is towards shorter fallow periods. The maximum length of fallow may be from five to nine years, with a minimum period of only two to three-and-a-half years. The causes of the shorter fallow period in Nepal, as identified by Aryal et al., (ch. 13, this volume), are (a) increasing population leading to the shrinking and subdivision of arable land; (b) the state denying people the ability to freely cut and burn, especially when the forest has recovered after a long fallow period; (c) the food needs of the people and the cultivation of cash crops, such as medicinal herbs, which can be harvested after three to six years and are easy to market, even though farmers understand that fallow periods should be as long as possible, to ensure better soil fertility and better harvests.

Shortening fallow periods and the causes of their reduction are not found only in Nepal and Vietnam, but also in other countries. As a consequence, soil fertility is not maintained, forest rehabilitation is slowed down, and farmers do not have enough fertile land for the next cycle of cultivation. They are thus forced to go further into the forest to practise shifting cultivation. Alternatively, they need initiatives to effectively and sustainably manage fallow land, matched by suitable state policies.

The two following chapters deal with fallow-management conditions in their study areas in the light of these trends. The results are common to many countries in which shifting cultivation is practised, and before the findings are discussed, it is necessary to better understand the views on fallowing land held by both farmers and policy-makers. Farmers' views are not unanimous and depend on the types of shifting cultivation they practise. In Vietnam, for example, there are three types of shifting cultivation: (a) pioneer shifting cultivation is practised mainly by farmers of the Mong ethnic minority living between 600 and 1000 metres above sea level (masl). They make maximum use of soil fertility and then abandon the exhausted land without any intention to use it again. They pay no attention to fallow management and must relocate entire hamlets, sometimes travelling great distances to cultivate new areas of forest. (b) Supplementary shifting cultivation is practised by Nung and Muong farmers living at lower altitudes. They also have wet-rice fields, so shifting cultivation provides further support for the family economy. They are therefore interested in fallowing their land and manage it effectively, trending towards fixed cultivation. (c) The most common and more typical form is rotational shifting cultivation, following different cycles after the fallow period. This is practised by many ethnic groups who are very interested in the management of fallow land, restoring forest and rejuvenating soil fertility. Many of their practices demonstrate deep experience and traditional knowledge. However, their traditional views on fallow land and its management also change over time due to many influencing factors.

Policy-makers commonly consider that fallowed land is not 'owned' on a long-term basis. In Vietnam, fallows are considered to be unused land, even if people have made use of it. An 'unused land' category can be found in current land-use statistics, and this official attitude affects farmers' approach to management of fallow land.

The studies in the two following chapters are related to rotational swidden cultivation. They detail many initiatives and approaches by farmers to managing fallow land – most of them involving indigenous knowledge that has accumulated over many generations. The traditional approach to restoring soil fertility is forest rehabilitation. Shifting cultivation creates a 'gap' in the forest, and after cropping the forest is restored in a period of fallow which increases the fertility of the soil. In this process, the farmers select useful species for growth in the regenerating forest. An in-depth study of the forest restoration process in Vietnam over 5, 10 and 20 years of fallow has produced statistics on the composition of the vegetation in each period. Many species of trees are reported in the recovering forest, including useful species such as woody trees and bamboo, and particularly those that produce non-timber forest products, such as medicinal plants, palm leaves, fruit trees, vegetables, bamboo shoots, mushrooms and spices.

Forest restoration usually follows four stages of vegetative succession. In the first five years the dominant species include grass and light-tolerant shrubs. From five to 10 years, many fast-growing tree species with medium light demand emerge, followed by slower-growing, shade-demanding species. From 10 to 15 years, the forest gradually recovers. The number of species increases, and after 20 years it is relatively fixed. Experimental formulae have calculated that the above- and below-ground biomass of a forest after 20 years of rehabilitation reaches 64 tons per hectare. In Vietnam, there are abundant studies of forest rehabilitation after shifting cultivation and the capacity of those forests to absorb carbon dioxide. The precise quantity of CO_2 that can be absorbed begins with the above calculation of biomass of rehabilitated forests. The quantity of carbon in the wood is estimated to be about half of the total biomass, so the quantity of carbon represented as carbon dioxide can be calculated by the formula: CO_2 = quantity of C x 44/12. Thus, the amount of CO_2 absorbed by regeneration of the forest is 64 tons/ha x 0.5 x 44/12 = 117 tons of CO_2/ha.

This capacity invites participation in the United Nations' REDD+ programme (Reducing Emissions from Deforestation and Forest Degradation), through the sale of carbon credits. Therefore, forest rehabilitation in shifting cultivation fallows not only improves soil fertility but also has great value in the use of forests to absorb greenhouse gases and contribute to the mitigation of climate change by taking part in the carbon market.

Indigenous knowledge has been applied to the management of fallow land in both of the study countries, most commonly involving legumes and multipurpose plants. Legumes provide food products while increasing soil fertility by fixing atmospheric nitrogen in the soil. Vegetative residues are also used as mulch, adding nutrients to the soil as they decompose. In Nepal the most common fallow–management species are the legumes *Leucaena leucocephala*, *Cajanus cajan* and *Flemingia congesta*, the multipurpose tree *Diploknema butyraceae* and the medicinal plant *Swertia chirayita*. In Vietnam, farmers of the Tay ethnic minority sow seeds of the fast-growing multi-purpose trees species *Melia azedarach* among their corn and cassava crops. After five years of fallow period, timber from the trees is sold for use in making furniture and building houses. Other studies in Vietnam show that ethnic-minority farmers in the Central Highlands, where the country's largest area of natural forest is found, often sow green beans in fallowed land. These produce both household food and a marketable cash crop while boosting soil fertility through leguminous nitrogen fixation.

It should be emphasized that the traditional knowledge and experience of farmers is best exercised in rotational shifting cultivation systems where there are only low levels of population pressure and large areas of natural forest.

In fact, in most countries the pressure on shifting cultivation is so great that farmers are changing their approach to fallow-land management with a clear trend towards planting trees as cash crops. A typical study in Nepal showed that after three to six years of cultivation, medicinal plants or cardamom are planted. About 75% to 85% of interviewed villagers wanted to change to planting commercial crops, especially local medicinal plants, on their fallow land. The integration of horticultural crops like bananas, pineapples and vegetables is also being used as a fallow-management strategy in study sites in central Nepal. The study also shows the potential for development of non-timber forest products, especially many species of medicinal plants, and the establishment of medicinal cooperatives to contribute to the development, conservation and processing of these species. Research in Vietnam also shows that a trend towards planting multi-purpose commercial trees on fallow land is becoming common. Such trees include *Styrax tonkinensis* for paper, fibre materials and resin for export to perfume industries in Europe, the bamboo species *Dendrocalamus barbatus* for house-building materials and paper pulp, and some palm species for thatched roofs and cash sales. Other studies in Vietnam have also found a great diversity of cash-crop systems on fallow land. In different areas, farmers cultivate *Cinnamomum cassia* for the value of its bark and essential oils, native trees for wood and seeds for spices, and *Michelia tonkinensis* for its medicinal value. In many places, farmers are cultivating exotic *Acacia mangium* for chipping wood or paper pulp. Agroforestry models are also applied in some localities, such as growing native plants with rattan or the medicinal herb *Morinda officinalis*.

After discussing commercial cultivation models on fallow land, the researchers in Nepal put the question: Is the traditional shifting cultivation system changing? They identify five major drivers of change: (a) population growth; (b) government policy;

(c) migration; (d) the introduction of commercial crops; and (e) social pressures. It can be said that the changing face of shifting cultivation and the five major drivers of change are true not only in Nepal, but also in other countries, especially with the current pressures of climate change. The question should not remain; the facts of the matter are affirmed. In Vietnam, for example, policies and trends in migration since 1968 have followed a campaign supporting fixed cultivation and sedentarization. In 2007 the Ministry of Agriculture and Rural Development launched a project to support upland people who implement sustainable agriculture and forestry cultivation in former shifting cultivation areas.

Many policies have been introduced, such as planning shifting cultivation areas, allocating residential land, cultivated land and forests to households for protection, mobilizing them to participate in afforestation, to take part in sustainable agroforestry projects, and convert to economically valuable crops on shifting cultivation land. Vietnam has also achieved the United Nations' Millennium Development Goal of eradicating hunger and is now implementing a sustainable poverty-reduction strategy. All of those policies strongly affect the change away from traditional shifting cultivation. However, free migration has become a major problem in Vietnam that needs to be addressed, especially ethnic minorities moving from the north of the country to the Central Highlands, where fertile land and large areas of natural forest are open for exploitation. Pioneer shifting cultivation, with farmers venturing deeper and deeper into the forest, is common. Fallow land becomes privately owned, and many areas are transferred or sold for planting industrial tree crops such as coffee and rubber.

In conclusion, both chapters in this section show that the management of fallow land is playing an important role, not only in the cycle of rotating shifting cultivation, but also in creating alternatives to traditional swiddening. The fallow period is growing shorter, but diverse traditional indigenous knowledge is being applied to this land, to retain its value in sustainable land use and organic agriculture. Amid the current high pressures on shifting cultivation, the management of fallow land is gradually changing to permanent systems, particularly those involving commercial tree crops, and traditional swidden cultivation is trending towards fixed agriculture.

COLOURED PLATES* ...Part I (plates 1-33)

Taking advantage of a multi-purpose tree. Finding himself without a tripod in the field, the Editor uses a pollarded *Alnus nepalensis* stump to steady a long zoom lens and keep it sufficiently still as he composes a shot of the swidden fields on a distant slope. See Cairns and Brookfield, ch. 22, this volume, for more agronomic uses of the Himalayan alder.

Sketch based on a photo by Carolin Meru (2002).

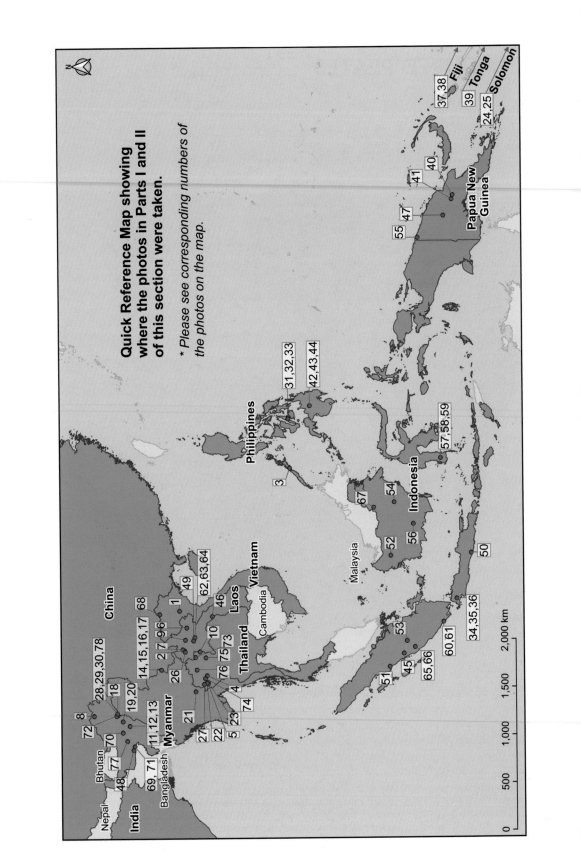

Quick Reference Map showing where the photos in Parts I and II of this section were taken.

Please see corresponding numbers of the photos on the map.

LANDSCAPE-LEVEL INNOVATIONS

1. Spreading the risk of rice production across both wet and dry fields.

When valley bottoms are available, shifting cultivators often also construct wet rice terraces, so that they have a dual system for growing rice. This has become known as 'composite swiddening' (see Cairns and Brookfield, ch. 22, this volume). It provides increased food security, in that if one system fails, the other probably won't.

1. This village was at about 300 metres above sea level (masl) in Tat hamlet, Tan Minh commune, Da Bac district, Hoa Binh province, northern Vietnam. The Tay Da Bac ethnic group that farms these lands shows that shifting cultivation and wet rice farming can co-exist side-by-side on the landscape in a 'composite' farming system. Although farmers were planting hybrid rice varieties in their terraces in an effort to increase rice production, the terraces in the valley bottom were too small to produce the village's entire food needs, so swiddens were opened on adjacent hillsides as a supplementary food source.

Photo: Nguyen Thanh Lam (October, 2003).

2. Farming at higher altitudes provides a healthier environment for both farmers and their livestock.

2. Probably between 1000 and 1100 masl, clouds commonly drifted across this swidden field in Ban Konkud, Naelae district, in Luang Namtha province, Lao PDR. In taking this photo, the photographer wanted to illustrate the point that the Khmu farmers who cultivate these lands (see Chazée, ch. A66, in supplementary chapters to this volume) spend about six months of the year living in their swidden fields, and so – as shown here – build quite elaborate field huts to live in during this time.

Photo: Olivier Evrard (2006).

When Lao PDR implemented a policy to resettle shifting cultivators out of the uplands to lower altitudes near roads and other infrastructure, one of the unintended outcomes was a serious rise in morbidity and mortality amongst both the resettled shifting cultivators and their livestock (see Chazée, 2017).

Villages established at high altitudes may have placed themselves above the altitudinal belt within which malarial mosquitoes are a serious threat. If these villagers open swiddens that are downslope from their village and enter the malarial zone, then farmers working in these fields during the day will try to retreat back upslope to the relative safety of their village by dusk before the mosquitoes begin to bite (Yaden, personal communication, 2001).

3. Small swidden fields cleared from large forests are only a temporary disturbance to the forest ecology.

3. This Batak village in Palawan, the Philippines, was carefully rotating its swiddens around the periphery of its residential area, cutting down on time spent walking to and from their fields. All fields were surrounded by forest patches that acted as seed sources, and accelerated recovery of secondary forest after the swiddens were left fallow.

Photo: Malcolm Cairns (1993).

4. Designation of 'spirit forests' near the village encourage forest conservation.

Local guardian spirits are often enlisted to ensure that the environment is treated with respect (see Barber et al., ch. 29, and Chazée, ch. A30, in supplementary chapters to this volume). Another conservation-oriented practice is maintaining a sacred forest adjacent to the village, where any cutting is strictly forbidden, on penalty of a traditionally imposed fine. These spirit forests can often be identified as uncut forests adjacent to the village residential area.

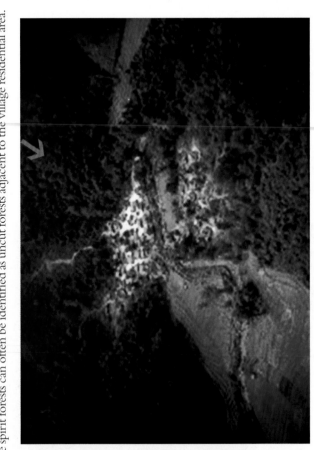

4. Karen elders of the Pa Pae community carry out a forest worship ceremony (*luj pgaj*) in a sacred forest on Doi Chang in Lamphun province, northern Thailand.

Photo: Bancha Mu Hae (2019).

5. Almost 58 years after he took this photo, Dr. Peter Kunstadter recalls that Ban Pa Pae, in Mae Sariang district, northern Thailand, retained a sacred forest. The red arrow indicates where it was, just above and slightly to the right of the village residential area. The trees look to be noticeably taller in that section of the photo. The rules in those days prohibited the cutting or taking of anything from sacred forests, on penalty of a fine and a ceremony to the spirits. Since then, the community has undergone development that has ignored those traditions.

Photo: Peter Kunstadter (1965).

5. Managing land use to try to accommodate wildlife populations.

6. This photograph was taken in a high-altitude area of Houaphan province in northeastern Laos. The area had been occupied relatively recently by H'mong shifting cultivators, probably only 10 to 30 years before this photo was taken. The photographer's attention was drawn to a small uncut patch of forest (red arrow) that divided two swidden blocks and connected two forested areas. He was convinced that it had been left intentionally by the H'mong farmers as a wildlife transit corridor. Although this conviction was not subjected to interview inquiries, if true, it would have demonstrated a relatively sophisticated approach to upland forest management.

The photographer further speculated that as pioneer shifting cultivators, the H'mong were probably more reliant on wildlife as a supplementary food source than most Lao Theung or Lao Loum ethnic groups, and so would have had a strong incentive for trying to accommodate wildlife on their landscape.

Photo: Alan Potkin (1997).

6. The mosaic of land use created by shifting cultivation supports increased biodiversity.

The main point to make here is that shifting cultivation creates a patchwork landscape, ranging from uncut forests to cultivated fields, and all stages in-between – and thus creates diverse ecological conditions that provide habitat for a wide range of biodiversity. Some of the plants that find a home within this mosaic may prove to have value to farmers as food or medicine. As documented by some of the chapters in this volume (e.g., Tran and Pham, ch. 38, this volume, and Nyein Chan and Takeda, ch. A11, in supplementary chapters to this volume), some of the most promising of these may be introduced into swidden fields or fallows on a trial basis and eventually become domesticated.

Alternatively, if shifting cultivation is pushed too hard, and goes beyond its ecological resilience, the result may be entire landscapes covered by *Imperata* grasslands that are often referred to as 'green deserts' in reference to their impoverished biodiversity.

8. These swidden patches were newly opened, probably by Adi shifting cultivators, in Jengging circle, Upper Siang district, in Arunachal Pradesh, northeast India.

Photo: Anirban Datta-Roy (May 9, 2013).

7. These swidden fields were being managed by Rmet shifting cultivators in Ban Daklueng, Nalae district, in Luang Namtha province of northern Lao PDR. The rice appears ripe and harvest will begin soon.

Photo: Olivier Evrard (2013).

7. Retaining trees on ridges and the upper boundary of fields provides a seed source, and accelerates regeneration of secondary forest when the field is left fallow.

Many shifting cultivators deliberately maintain trees and other vegetation on ridges and on the highest side of swidden fields, with the intention that the trees will reseed downhill during the fallow years, and thus accelerate regeneration of secondary forest.

9. This swidden, photographed along Route 13 in Ban Xong Ja, Nambak district, in Luang Prabang province, Lao PDR, was opened in 2018 (see Li et al., ch. 6, this volume), and shows the typical farmer practice of retaining trees along the highest perimeter of swidden fields, so that they can 'rain their seeds down' on the field below, and accelerate the transition from the cultivation phase back to secondary forest. The farmers have also fenced this field to prevent predation by both wildlife and free-roaming livestock.

Photo: Li Peng (May 19, 2018).

Alnus nepalensis-based fallows in southwest Yunnan province, China (see Guo et al., 2007).

15. It is quite possible that the Wa farmers may not even be fully cognizant of the strength of the coppicing ability of *Alnus nepalensis*. By cutting their alder trees young and low, most of the stumps simply die out and there is very little stump survival or coppicing, as is evident in the photo above.

Photo: Malcolm Cairns (2001).

14. On the southwest border of Yunnan province in China, Wa shifting cultivators have also learnt the benefits of *Alnus nepalensis* as a fallow species that will accelerate soil rejuvenation during the fallow period. But unlike their counterparts in Nagaland (see Cairns and Brookfield, ch. 22, this volume), the Wa have not learned how to pollard their alder trees. Instead, each fallow stand of alder trees is cut low and mostly killed out at the time of reopening the fallow (see next plate), and must then be totally re-established by replanting alder seedlings into the swidden field (see Plate 17). These photos were taken in Amo village, Xinchang town, Ximen county, Yunnan province, China.

Photo: Malcolm Cairns (2001).

16. There appears to have been some stump survival in this field. A Wa farmer hoes weeds in her swidden, with some regenerating alder trees visible behind her. As she passes each alder tree, she appears to be pausing to break off some of the side branches and place them at the base of the tree as a mulch.

*Photo:*Malcolm Cairns (2001).

17. With the previous stand of alder trees almost completely killed out at the time of reopening the fallow, the farmer then has to set about replanting a new crop of alder seedlings into his swidden. An established alder fallow can be seen on a slope in the background.

*Photo:*Malcolm Cairns (2001).

Retention of *Schima wallichii* as a soil-building tree.

18. Experience leads us to expect that when fallows are shortened to only a few years, the trees disappear from the fallow vegetation and are replaced by grasses, shrubs and bamboos. Yet this photo of a Konyak fallow in Mon village of Nagaland, northeast India, shows that farmers can effectively intervene in retaining the tree component of their fallow regrowth. Although the fallow shown in this photo is only two years old, it already has an impressive number of trees that will provide an effective fallow. Most of these trees were *Schima wallichii*, either retained as relict emergents from the previous fallow vegetation, or regenerated from the stumps of trees cut when the fallow was last reopened. Managing *S. wallichii* in this way is particularly common amongst the Konyak tribe of Nagaland, at elevations around 1000 masl. Shifting cultivators' preference for this tree species in their *jhum* fields is based on two main traits: 1) farmers believe that the leaves of *S. wallichii* increase soil fertility, and 2) the trees provide superior firewood.

Photo: Ango Konyak (July, 2006).

19. At Settsu village in Nagaland's Mokokchung district, the swidden fields display numerous *Schima wallichii* trees dispersed across the landscape. *Schima wallichii* is commonly retained by shifting cultivators at warmer altitudes (300–1000 masl) to assist soil rejuvenation, very similar to the role of *Alnus nepalensis* at cooler altitudes (900–2000 masl). Both species are well suited for their roles in that both have thick bark that helps them to survive swidden fires, and both coppice profusely when cut. Shifting cultivators strongly associate both species with soil improvement and seek to retain them in their swidden fields. Interestingly, a fallow stand comprised of half *Schima wallichii* or less is viewed as a positive indicator of fertile soil – but a higher presence of *S. wallichii* is viewed less favorably (Yaden, personal communication, 2020).

Photo: Amenba Yaden (ca 2001–2006).

20. When retained in a swidden field, *S. wallichii* trees are not pollarded, but as seen from these trees towering over a swidden in Longkum village of Mokokchung district, Nagaland, they have clearly had only their side branches lopped off to reduce shading on nearby crops, and left with only a few branches at the top. That leaves them as tall, thin sentinels standing over the ripening rice crop. Although the trunks of many have obviously been charred by fire, most have survived and are coppicing profusely.

Photo: Temjen Toy (2009).

More general retention of relict emergents in swidden fields (see Schmidt-Vogt, 2007).

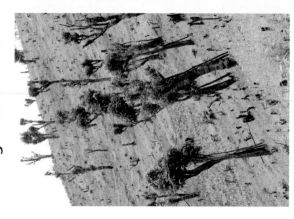

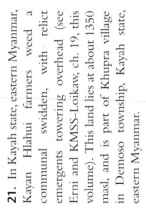

21. In Kayah state, eastern Myanmar, Kayan Hlahui farmers weed a communal swidden, with relict emergents towering overhead (see Erni and KMSS-Loikaw, ch. 19, this volume). This land lies at about 1350 masl, and is part of Khupra village in Demoso township, Kayah state, eastern Myanmar.

Photo: Christian Erni (July, 2016).

22. When reopening their fallows, the Lawa men of Tun village, in Thailand's Mae Hong Son province, climb selected trees to trim their crowns and reduce shading on crops that will later be planted nearby. A few branches are usually left near the top of the tree to help ensure its survival. This is how relict emergents are created.

Photo: Dietrich Schmidt-Vogt (October 2, 1991).

23. Members of a *PgazK'Nyau* (Karen) community at Dok Daeng village in Hot district, in Thailand's Chiang Mai province, say that a few branches were left at the top of this relict emergent to *'provide a perch for the scarlet minivet bird'*. They don't rationalize it as a way to increase the chances of the relict emergent surviving, but that is likely to be its main benefit. This bird has no particular significance to the villagers, but seems to be used as a metaphor referring to the accommodation of nature within their farming system more generally.

Photo: Bue Paw Thaworn Kamponkun (2004).

Working under a partial canopy on South Pacific islands, shifting cultivators commonly retain trees in their gardens to provide support for climbing yams (*Dioscorea esculenta*).

24-25. Communal labour was being used to establish this yam garden in a relatively old fallow, probably 10–15 years old, on Bellona in the Solomon Islands (see Mertz et al., 2012). Whilst opening the fallow, larger trees have been retained to provide support for the climbing yam vines. When the undergrowth is slashed, it is left to dry and then piled around the retained trees. When these piles of slash are burned, the fire stresses the trees and causes them to shed their leaves, but not die. The shed leaves decompose and both augment the soil fertility provided by the ash, and admit more light to the crops that will be planted below. These trees serve as climbing poles for the yams (*Dioscorea esculenta*) that are one of the most important crops. In both photos, the farmers can be seen using sticks to loosen the soil and prepare small mounds for planting yams. Banana and taro will probably be planted as secondary crops.

Photo: Thilde Bech Bruun (August, 2006).

When opening a fallow, felled trees are often cut high to increase their chances of survival.

26. A Khmu shifting cultivator approaches his field hut in a swidden field at Ban Konkud, Naelae district, in the Lao PDR's Luang Namtha province. The tree stumps that he is passing illustrate the practice of not cutting trees too low, as a measure to improve their survival rate. Although regrowth was not yet visible on these stumps, the photographer assures that they were still alive and would soon be resprouting new foliage. That these trees were small and multi-stemmed is probably a vestige from previous times when they had been cut and resprouted multiple shoots that competed with each other, and so did not grow large.

Photo: Olivier Evrard (August, 2006).

27. Although the stump was not cut very high and is badly charred from the swidden fire about 2 weeks earlier, this *Castanopsis acuminatissima* (white oak) stump was already resprouting vigorously. Although the bole of the tree had been cut and burnt, its roots remain intact and continue to anchor the soil in place, preventing land slippage. In view of this, the photographer who took this shot argues that when *Pgaz K'Nyau* (Karen) shifting cultivators open a fallow for cultivation, they are not destroying the forest, but rather, renewing it. Farmers commonly eat the acorns from this tree, and use the wood for firewood. This photo was taken in Mae Omphai village of Maelanoi district in Mae Hong Son province of northern Thailand.

Photo: Bue Paw Thaworn Kamponkun (2002).

Macaranga-based fallows in Mon district, Nagaland, northeast India (see Changkija et al., ch. 26, this volume).

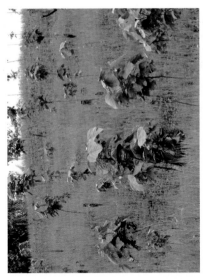

28. In Nagaland's Mon district, the Konyak Nagas have learnt to take advantage of *Macaranga* trees that self-seed in their swidden fields and establish in tandem with the rice crop.

*Photo:*Malcolm Cairns (2002)

30. A passer-by unfamiliar with the system, could easily assume that these *Macaranga* trees have been *taungya*-planted into this swidden field. But no – this is truly a case of the Konyak *'hitching a ride with nature'*, and these trees are all self-sown. The farmer has only to do some thinning and transplanting to achieve an even distribution of trees across the field.

*Photo:*Sapu Changkija (May, 2016).

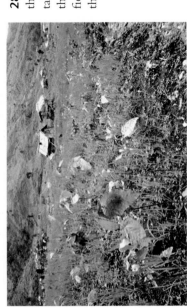

29. The *Macaranga* trees develop into an even-aged, almost pure stand during the fallow period. The fallow pictured next to this closest field hut was eight to nine years old and was photographed in Tangya village, under Phomching division of Mon district, Nagaland.

*Photo:*Sapu Changkija (May, 2016).

Leucaena leucocephala-based fallows in Barangay Naalad, Cebu, the Philippines (see Suson and Lasco, ch. 23, this volume, and Lasco, 2007).

31. The hillsides of Barangay Naalad, on the island of Cebu, are clothed in *Leucaena* forests that have been planted by local shifting cultivators as an improved fallow species. Just as they would have done with natural forests, farmers carve fields out of these *Leucaena* forests on a cyclical basis to plant arable crops.

Photo: Malcolm Cairns (ca 1995).

32. Not yet planted, this newly-opened field offers a clear view of the fascine structures that farmers stake across their fields to reduce erosion, similar to the log bunds discussed by Lotha et al., ch. A21, in the supplementary chapters to this volume.

Photo: Malcolm Cairns (ca 1995).

33. This Sugbuanon farmer was transplanting tobacco seedlings into the alleyways of his field that was managed under the 'Naalad system', that combines soil improvement and conservation. In this system, farmers are able to plant two croppings per year, the first maize, and then tobacco during the drier part of the year. Note the fascine structures that form the alleyway within which he is working.

Photo: Jose Dadan (November, 1992).

Continued in Coloured Plates Part II ...

13

FALLOW MANAGEMENT

In transforming shifting cultivation systems

Kamal P. Aryal, Keshab Thapa, Rajan Kotru and Karma Phuntsho[*]

Introduction

Shifting cultivation, locally known as *khoria kheti* or *bhasme*, is a traditional farming system used by many ethnic groups in Nepal. However, the practice is currently undergoing unprecedented transformations due to various socio–economic and political changes in the country (Aryal et al., 2010). In the course of these transformations, fallow management has become a key challenge, providing a means by which farmers can maintain their production and productivity. Fallows, locally known as *lhose,* are an integral part of the shifting cultivation farming system, and are essential to enable the soil to recover its fertility and structure after the cultivation phase. In the course of this recuperation, a range of products meeting both household requirements and socio–cultural functions can be gathered from fallows (Bruun et al., 2006; Kerkhoff and Sharma, 2006; Aryal and Kerkhoff, 2008). The fallow phase is a period of forest growth that is left undisturbed, between short periods of cultivation on the same plot of land. It supports optimum forest growth after annual crops are harvested. Land used for shifting cultivation is often too steep for annual cropping, and this often persuades farmers to leave it fallow for several years, depending on the size of the land and the availability of human resources to work on it.

Shifting cultivators maintain a fallow for a considerable time to restore the fertility of the previously cultivated soil in order to provide a bumper harvest in the following cultivation phase (Prinz, 1987; Reijntjes et al., 1992; Cairns, 2007; Aryal and Kerkhoff, 2008). In addition to restoring soil fertility, fallows maintain a

[*] KAMAL P. ARYAL, natural-resources management (NRM) analyst, International Centre for Integrated Mountain Development (ICIMOD), Kathmandu, Nepal; KESHAB THAPA, Programme Officer, Local Initiatives for Biodiversity, Research and Development (LI-BIRD), Pokhara, Nepal; DR RAJAN KOTRU, Senior NRM Specialist and Team Leader, Integrated Forest Ecosystem and Watershed Services (InFEWS), ICIMOD; and KARMA PHUNTSHO, NRM Specialist, ICIMOD.

diversity of plants and animals and provide for their interaction, helping to maintain an ecological balance and support conservation of biodiversity (Prinz, 1987; Kerkhoff and Sharma, 2006; Schmook, 2010). In the face of climate change and consequent climatic variability, fallows minimize the loss of soil from sloping land during heavy precipitation as well as maintaining soil moisture, which helps to resist prolonged drought. These services reflect an autonomous adaptation strategy by swidden farmers to deal with extremes of weather in the changing climatic context. More importantly, fallow vegetation can sequester carbon from carbon dioxide to help to mitigate climate change and this has the potential for involvement in international carbon trading, allowing the shifting cultivation cycle to be compared to a permanent cropping system.

However, the period of fallow depends on several factors, such as access to land for shifting cultivation, its location, aspect and ownership, as well as vegetation growth. In developing countries such as Nepal, fallow periods in shifting cultivation systems have been as long as 12 years or more (Aryal and Kerkhoff, 2008). However, in recent decades, the pressure on these lands has increased, causing a considerable shortening of the fallow period. Population growth and social change has brought increasing pressure to bear on shifting cultivators to produce more on less land. If one considers the cropping phase as the sole source of production to meet rising demand, then intensification pressures result in extension of the cropping phase and shortening of the fallow phase. But, since fallow forests are the main source of soil fertility, such a shortening greatly jeopardises short-term food security as well as long-term sustainability.

This study is part of a regional research project on shifting cultivation in Bangladesh, Bhutan and Nepal. However, the findings presented here focus only on Nepal. This chapter looks at how indigenous peoples are managing the fallow phase to transform shifting cultivation systems, and what strategies farmers in the study villages are pursuing to compensate for the loss of the products and services that were provided by the longer fallows of past times. Furthermore, the study focuses on medicinal plants that are harvested from the fallows and surrounding forests. These are not only important for the household medicine cabinet and for treating local ailments, but also for generating income.

Materials and methods

Study site and study population

The study was conducted in four Village Development Committee areas (hereafter called villages) in four districts of Nepal's Central and Eastern Development Regions. Jogimara and Siddhi villages are in Dhading and Chitwan districts of central Nepal, while Lelep and Pawakhola villages are in Taplejung and Sankhuwasabha districts of eastern Nepal (Figure 13-1). The major criterion used in selecting the study sites was their ability to represent areas where shifting cultivation is a major farming system with varying fallow lengths and practices. Other supplementary criteria were shifting

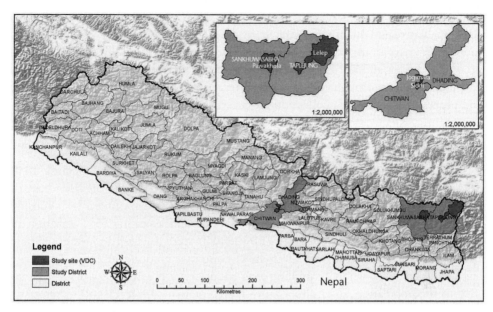

FIGURE 13-1:   Study sites and districts.

cultivators of multiple ethnic origin and areas having less and more road and market access, in order to get diverse field-level information on the intensity of shifting cultivation. Within the villages, shifting cultivators for the study were identified through stakeholder consultation and with the help of key informants.

The study sites in central Nepal have a tropical to warm-temperate climate with an altitude range of 250 to 1250 metres above sea level (masl), whereas in eastern Nepal the study villages are situated at an altitude of 1680 to 2200 masl, with a temperate to alpine climate. Because of the steepness of the terrain, altitudes can vary hundreds of metres within the same village. The study sites in eastern Nepal are one-and-a-half to two days' walk away from the district headquarters, in a northeasterly direction. In this case, one day equals 12 hours of walking. In contrast, the study villages in central Nepal require only three to six hours of walking. There are a number of ethnic groups in each of the study villages. Limbu, Sherpa and Rai are the predominant inhabitants of eastern Nepal, whereas Chepang and Magar farmers practise shifting cultivation in central Nepal. Agriculture (shifting cultivation) is the main source of livelihood for the majority of households in the study villages, and it is largely subsistence-oriented. However, land use and livelihood strategies have changed in recent times and farmers have diversified their options. The geographical, socio-economic and biophysical conditions, the farming systems and the alternative options in shifting cultivation are similar at the two sites in eastern Nepal, just as they are similar at the two sites in central Nepal. Hence, the research findings and analysis in this chapter will be presented by comparing one pair of sites with the other: essentially, eastern versus central.

Methods

Both qualitative and quantitative data were collected through primary and secondary sources. Primary information was collected through household surveys, focus-group discussions and key-informant interviews, whereas secondary information was collected from a thorough review of the literature. A brief description of methods and processes used for data collection are presented below.

Household survey

A survey was conducted in 276 households from a sampling frame totalling 1448 households. A total of 137 respondents from central Nepal and 139 from eastern Nepal were chosen for the survey. The sample size was determined by choosing a 90% confidence level with 5% margin of error and assuming 50% variation in the population (Bartlett et al., 2001). Households were selected by using a simple random-sampling method with the help of computer random numbers. At village level, different proportional sampling was adopted, with a sample size of 14% to 30% of the population, to ensure a more representative sample. Details of sampling are presented in Table 13-1. A semi-structured questionnaire was developed for the household survey. It was focused on information about farming practices and particularly on shifting cultivation systems, changes to these systems and to fallow periods and management of fallows, as well as introduced options in shifting cultivation and management of non-timber forest products (NTFPs).

Focus group discussions

Focus-group discussions were held for collection of information on specific topics of interest. Altogether, eight of these discussions (two at each study site; one a mixed group and one with women only), involving 20 male and 20 female shifting cultivators, were conducted in order to understand the various strategies that farmers were adopting for the management of fallows in order to maintain past levels of production from the same pieces of land. The first group discussion involved five male and five female participants and these were later separated into men only and women only. The gender disaggregation was aimed at discovering the role of gender

TABLE 13-1: Sampling details for the household survey.

Districts	Villages	Households practising shifting cultivation	Sample % of households	No. of households sampled or surveyed
Dhading	Jogimara	560	14	76
Chitwan	Siddhi	266	23	61
Sakhuwasabha	Pawakhola	372	17	64
Taplejung	Lelep	250	30	75
Total		1448	21*	276

Note: *This figure is a mean percentage of the sampled households.

in management of NTFPs. The discussion was further focused on different coping strategies that farmers were applying in order to manage soil fertility. Perceptions of changes in shifting cultivation systems, drivers of these changes and conservation and management of NTFPs were also discussed.

Key informant interviews

This method was chosen to gather special information on some topics of interest from key informants with specialized knowledge that other members of the community lacked. Members of each community helped to identify two local healers with knowledge about the use of medicinal plants, so a total of eight local healers were interviewed for their knowledge and practices in using medicinal plants for household health care.

Desk review

Secondary information related to the research was derived from a review of literature on shifting cultivation and alternative land-use options, village profiles, district profiles, national statistics and national land-use-related policy documents.

Data analysis

The quantitative information was entered into MS Excel sheets and was analysed through descriptive statistics simply by computing mean, standard deviation and percentage with the 2011 version of Statistical Package for Social Science (SPSS). The qualitative information was analysed through the content and theme of the study.

Results

Is the traditional shifting cultivation system changing?

An overwhelming 85% of respondents at the study sites said they had experienced changes in the traditional shifting cultivation system in general and the fallow period in particular. Changes in the shifting cultivation system were visible due to changes in external drivers and internal societal change. About 75% of respondents in eastern Nepal reported that traditional shifting cultivation land was being converted to cardamom and *chiraita* (*Swertia chirayita*)-based farming (the latter being a pharmacological crop).[1] They had experienced changes in their traditional shifting cultivation system that had led it to become a more cash–oriented and cash–productive system, while maintaining their traditional knowledge and innovation. However, 85% of respondents in central Nepal said that traditional shifting cultivation systems now had the barest minimum fallow period and had almost been converted into annual-farming systems because of population pressures that had resulted in land fragmentation. Demographic pressures, cash-cropping and market influences and socio-economic pressures, together with the policy environment, were forcing shifting cultivation, as a land use, to change, and shifting cultivators were being left with no option but to modify their traditional systems.

Figure 13-2 shows that the drivers of change are more or less different in eastern and central Nepal. At the eastern study sites, participants named the introduction of cash crops (N=115), especially *chiraita* and cardamom, followed by government policies that fail to recognize the practice of shifting cultivation (N=105) and migration of the entire household or seasonal migration of family members (N=100) as major drivers of change in land-use practices. In central Nepal, the major factors affecting traditional shifting cultivation land use were reported to be government policies (N=135), followed by population growth leading to land fragmentation (N=125). The average household size among shifting cultivators is quite high compared to Nepal's national average. The average family size in eastern Nepal is six and in central Nepal, seven, compared to the national average of

Swertia chirayita [Roxb.] H. Karst. [Gentianaceae]

Despite growing importance as a cash crop in Nepal, this medicinal herb is critically endangered. Its widespread uses in traditional medicine to treat ailments such as liver disorders, malaria and diabetes (see endnote 1) have resulted in over-exploitation and it is on the verge of extinction in the wild.

4.7. The swelling population not only creates more mouths to feed, but also the number of hands claiming inherited land. As a result, the land is fragmented as it is apportioned out to each new generation. Farmers are obliged to intensify cropping to meet growing food requirements from a rising population, so they reduce the fallow period to increase available cropping land from plots that diminish in size with every generation.

More than 75% of the respondents in both regions mentioned that government policy was the leading cause of transformation of shifting cultivation. This, they said, was mainly because the government had not recognized shifting cultivation as a land-use option, and farmers could not register their land for this purpose. None of the policies of the government of Nepal recognize shifting cultivation as a farming system. Under the government's forest and biodiversity-conservation related policies, shifting cultivation is looked upon with unconstructive and negative acuteness. This negative attitude, as a premeditated accompaniment to the policy environment, is intended to daunt shifting cultivators. Furthermore, social pressures are also playing an important role in changing the system. In eastern Nepal, Sherpa and Rai people

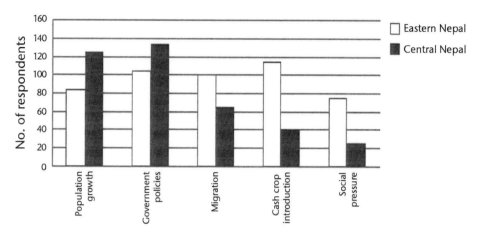

FIGURE 13-2:   Major drivers of change to traditional shifting cultivation systems.

are relatively rich compared to the Chepang and Magar of central Nepal, where there is a general belief that shifting cultivators are among the poorer members of society who do not have enough food for the whole year. Outsiders look upon them with extreme negativity; as encroachers. Consequently, shifting cultivators feel some degree of social pressure, and this is persuading them away from farming and into non-agricultural sectors.

Despite these changes, shifting cultivation (traditional as well as modified) continues to be a major contributor to livelihoods in the study villages, providing 60% of household needs; whereas the mainstream population has sufficient *khet* (irrigated lowland where paddy rice is the main crop) and *bari* (dry upland, where maize is the major crop) to produce significantly more food than the shifting cultivators. People in the study villages in the central region believe that traditional shifting cultivation is a part of their ethnic culture and identity, and they feel that they should not be discouraged from expressing their cultural values. In addition, they perceive shifting cultivation as a system that provides various ecosystem services to the mainstream population, in terms of water conservation, soil conservation and ecological balance, as well as providing products for their daily needs, such as food and firewood, while encouraging tourism.

Current status of fallow periods

Due to changes in the shifting cultivation systems, the status of fallows has also changed, over time. In the eastern part of Nepal, land used to be fallowed for 12 to 15 years, depending on the family size and the land holdings. That was 20 to 25 years ago. Fallow periods have tumbled in this time to between 7 and 12 years, and now they have fallen even further. The study found that the average maximum fallow period in eastern Nepal is now just over nine years, compared with a little more than three years in central Nepal. On the other hand, the average minimum fallow period is down to five years in eastern Nepal, against less than two years in central Nepal (Figure 13-3). The fallow period in eastern Nepal has not fallen as dramatically as that in central districts, even though some swidden farmers have given priority to cash cropping the medicinal plant *chiraita*. This cash–cropping trend followed high market prices for *chiraita* in 1994–1995. Nevertheless, the farmers said fallow periods had been kept as high as possible to maintain soil fertility for good yields in the succeeding cultivation phase.

The story is a different one in central Nepal, where the cropping-and-fallow cycle of shifting cultivation has in many cases been transformed into a bi-annual cropping cycle. In some areas it has been abridged to two to three years, compared to the 10 to 15 years of fallow that was commonplace two decades ago. There are two main reasons given by farmer-respondents for this abrupt transition of fallows in the central area: fears of government policy and less land from which to support an increasing population. Most of the shifting cultivation land at the study sites in central Nepal is unregistered. Farmers believe that if they allow the forest to regrow in long periods of fallow, the land will be incorporated into nearby community forests or captured by the state. According to the definition of forest in the National Forest Act, unregistered land lying around forested areas falls under the general category of government forest. At the same time, because of increasing population pressures, shifting cultivators have shortened their fallows in order to grow food to sustain their livelihood. This does not alleviate their food insecurity because reduced fallow periods provide insufficient time for recovery of soil fertility, resulting in poor harvests.

When asked about their reasons for reducing the fallow period, more than half (N=160) of all respondents (N=276) reported that the major reasons were shrinking land size per family and land fragmentation due to ever-increasing population growth. When this number was split according to study sites, it was found that 80% of the respondents who gave shrinking land size as a main reason for change were from central Nepal. This showed that fallow length was directly linked to the size of a family's land holdings. Eighty respondents out of the total of 276 gave use of land for cash crops and other innovative land-use options as major reasons for shortening

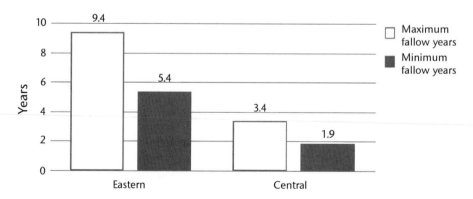

FIGURE 13-3:   Status of fallow periods at study sites.

fallows, making this reason second only to shrinking land size. Once again, when these 80 respondents were split according to study sites, it was found that 55 of them were from eastern Nepal. They explained that *chiraita* was the major cash crop occupying shifting cultivation land, and it could only be harvested from three to six years after planting, and this interfered with the fallow period. Other respondents (N=36) reported that intensification of the cropping pattern and lengthening of the cropping phase were the major reasons they saw for shortening fallow periods.

Fallow forests are an integral part of shifting cultivation systems and are essential for recovering soil fertility and providing a range of products to meet household requirements. Hence, it is important to manage fallows to meet household demands amid the changing circumstances.

Fallow-management practices at the study sites

The general perception of participants in the focus-group discussions was that the challenges of managing shorter fallows lay either in maintaining an optimum fallow by integrating high-value crops that grew well in fallowed land or integrating practices that restored soil fertility and land productivity. The study found some management practices in short fallows that aimed to deliver the same products and achieve the same production levels as previously recorded from longer fallows. For example, to enrich soil fertility in conditions of reduced fallow length or non-existent fallows, farmers in central Nepal are introducing alley cropping with fast-growing legumes such as *Leucaena leucocephala*, *Cajanus cajan* and *Flemingia congesta*. *F. congesta* is not an introduced species, but was found growing in a fallowed field. When its added value as a fallow species was recognized it was planted intensively as a hedgerow species. The three species have multiple uses: they produce fodder, they are deep rooted, they fix nitrogen and adapt easily to the local environment. Ninety per cent of the respondent farmers in

Flemingia congesta W.T. Aiton, a synonym of *Flemingia macrophylla* (Willd.) Merr. [Leguminosae]

This woody, deep-rooting shrub grows to about two-and-a-half metres, and is commonly used for mulching, weed control and soil protection, particularly in contour hedgerows for erosion control. A powder made from the fruit is used for dyeing silk to a brilliant orange colour.

central Nepal and 65% in eastern Nepal said they had begun intercropping legumes in between other crops – mostly with maize. Legumes that are commonly grown at the study sites in central Nepal include black gram, horse gram, cowpeas and green beans, whereas beans, cowpeas and soybeans are the major leguminous crops grown alone or intercropped with other crops in eastern Nepal.

One interesting example comes from the Chepang people, an indigenous group in central Nepal that depends heavily on shifting cultivation. The Chepang raise a traditional fruit tree that they call *yosi,* for a variety of purposes. The tree, *Diploknema butyracea,* is called *chiuri* in Nepali or Indian butter tree in English. The ripened fruit can be eaten, the kernels are collected for oil, and cakes made from the kernels after processing are used for fishing, as well as being a very good fertilizer and insecticide. The Chepang people use the cakes as fertilizer, mainly with introduced cash crops like banana and ginger, in the *khoria* (swidden) land.

Strip cropping is also practised in order to conserve soil and soil fertility, and thereby to increase productivity. Farmers are not aware of the capacity of leguminous crops and plants to fix atmospheric nitrogen into the soil. However, they have observed the ways these plants enhance production from intercrops or from crops that are planted after the legumes are harvested. A number of farmers in the study villages in central Nepal have also ceased burning crop residues and are instead using the slashed material for mulching. The idea for this came from the farmers themselves, even though the benefits of mulching are well known to agronomists. Covering the soil during the rains helps prevent erosion, but more importantly, the nutrients in the mulch are released more slowly, during a longer part of the cropping phase. Controlled burning is still practised in eastern Nepal, the main reasons being preparation of fields for *chiraita* cultivation and nursery beds for finger millet. In the household survey, 95% of respondents said that germination of seed was very good in burnt fields. Furthermore, farmers also leave uprooted weeds around the base of maize stems in swiddens, helping to retain soil moisture and

Diploknema butyracea (Roxb.) H. J. Lam [Sapotaceae]

The seeds of the Indian butter tree yield a fat that is used in cooking, as fuel for lamps, and as a body lotion. The fruit is eaten fresh or is used to distil alcohol, oil-cakes made from the processed kernels are used as fertilizer, and the tree itself is used as firewood.

adding soil nutrients after decomposition. The main sources of soil nutrients in the shifting cultivation fields of eastern Nepal are those built up naturally during fallows, controlled burning and natural decomposition of leaf litter. Farmers usually retain preferred species such as *gogan* (*Sauraria nepalensis*) and alder (*Alnus nepalensis*) in their swiddens, thereby influencing species composition in the fallow forests. None of the people in the eastern study villages use farmyard manure or other fertilizers on their shifting cultivation plots. In contrast, about 35% of the farmers in central Nepal began using chemical fertiliser on their maize crops more than eight years ago. The fertilizer, mainly urea, which releases nitrogen into the soil, is an attempt to compensate for the poor level of soil nutrients because of short fallow periods.

Rotation of plots for crop cultivation is also practised for increased yields, and livestock and crop production in shifting cultivation are strongly linked. Fallowed fields are important grazing areas and a major source of grass and fodder, and the presence of grazing animals supplies *in-situ* manuring for crop production. Farmers in eastern Nepal raise livestock as a means of managing soil fertility. Local traditional practices such as terracing, water channelling, fire management and stone walling are also pursued in the east, but they are not applied everywhere. Leaving trees in swiddens during the cropping phase (Figure 13-4) is a common practice for soil and water conservation as well as soil-nutrient management. Seventy per cent of respondents in central Nepal mentioned this practice, but only 25% of respondents in eastern Nepal spoke about it.

FIGURE 13-4:   Useful trees retained here in the cropping phase give forest restoration a boost in the fallow.

Photo: Kamal Aryal.

In order to respond to the drivers of change and to enhance their livelihoods, shifting cultivators have adopted some alternative uses of their swidden land. These include horticulture-based agroforestry in central Nepal and cardamom–alder agroforestry and *chiraita*-based swidden fallows in eastern Nepal. In focusing on cash–earning non–timber forest products (NTFPs) such as cardamom (*Amomum subulatum*) and *chiraita* in their fallows, the farmers of eastern Nepal believe that *chiraita*, for one, is an important crop for fallow management. It can be harvested twice during a five-year fallow. These NTFPs have a high market value and are being marketed mainly by middlemen and local businessmen. Besides these, other NTFPs found in fallow regrowth have similar markets. They include:

- *Allo* (*Girardinia diversifolia*), a perennial herb that produces a fibre from the stem bark that is used to make threads, ropes and strings. It is also used to make rough clothes. Besides this, tender parts of the plant are thoroughly boiled and eaten as a vegetable.
- *Argeli* (*Edgeworthia gardneri*), a shrub that is commonly used for making a local Nepali paper. Furthermore, juice from its roots is used for eye disorders and its bark and leaves are used to poison fish.
- *Budo Okhati* (*Astilbe rivularis*), a perennial rhizomatous herb that is used to treat diarrhoea and dysentery.
- *Majitho* (*Rubia cordifolia*), a climbing herb, the root of which is used to cure coughs, diseases of the uterus and jaundice problems. This plant is also used for dyeing coarse cotton fabrics, blankets and carpets.
- *Pakhanbed* (*Bergenia ciliata*), a perennial rhizomatous herb commonly used to treat coughs, fever and headaches. The dried rhizomes constitute the drug *Paashaanabheda*, which has astringent, tonic, anti-scorbutic and laxative properties.

The integration of horticultural crops like bananas, pineapples and vegetables is being used as a fallow-management strategy in the study sites in central Nepal. Such options contribute to cash income, support family nutrition and enhance or sustain the productivity of land. Cultivation of broom grass (*Thysanolaena maxima*) has also been adopted by a majority of households in the central Nepal study sites. These adoptions have significantly improved their economic and nutritional status as well as contributing to their livelihood options.

Fallows and forests as a source of medicinal plants for household health care

The use of a wide range of medicinal plants was reported in the study villages. The most important 21 species prescribed by local healers for household-level health care are listed in Table 13-2. Most of the medicinal plants are gathered from fallow fields and forests. Interviews with local healers at the study sites found that the plants were used in the form of juice, pastes, infusions or powders. For instance, the bark of *Lyonia ovalifolia* is used to make a paste to treat dislocated bones of both humans

TABLE 13-2: Medicinal species used for household health care in the study areas.

Local name	Scientific name	Plant type	Uses
Eastern Nepal			
Timur	*Zanthoxylum armatum*	Small Tree	Stomach problems
Chiraita	*Swertia chirayita*	Perennial herb	Fever, cough, headache, cold, throat problem, cuts and wounds
Kutki	*Neopicrorhiza scrophulariiflora*	Perennial herb	Bile disorders, blood and lung fevers, high blood pressure, sore throat, cough, intestinal pain
Jatamansi	*Nardostachys jatamansi*	Perennial herb	Heart disease, headache
Hadachur	*Viscum articulatum*	Herb	Fractures, dislocation of bones
Nirmasi	*Aconitum ferox*	Flowering plant	Anti-poison
Pakhanbed	*Bergenia ciliata*	Perennial herb	Fever, cold and cough
Central Nepal			
Rudilo	*Pogostemon glaber*	Shrub	Used for indigestion
Gurjo	*Tinospora sinensis*	Climber	Fever, stomach upset
Harro	*Terminalia chebula*	Deciduous tree	Cough and cold, astringent, laxative
Bhakyamlo	*Brucea javanica*	Deciduous tree	Dysentery, stomach upset
Bhimsen pati	*Buddleja asiatica*	Small tree	Skin disease
Bhyakur	*Dioscorea deltoidea*	Herbaceous climber	Lice control and bush poison
Amala	*Phyllanthus emblica*	Small deciduous tree	Cooling, haemorrhage, diarrhea
Kafal	*Myrica esculenta*	Evergreen tree	Bark-astringent, asthma and cough
Timur	*Zanthoxylum armatum*	Small tree	Toothache, cholera
Barro	*Terminalia bellirica*	Deciduous tree	Astringent, diarrhoea
Hadachur	*Viscum articulatum*	Herb	Fractures, dislocation of bones
Ghod Tapre	*Centella asiatica*	Perennial herb	Skin disease, nerves and blood purifier
Pipla	*Piper longum*	Slender aromatic climber	Diseases of respiratory tract, such as coughs, bronchitis, asthma
Kurilo	*Asparagus racemosus*	Twining herb	Cooling and indigestion, tender shoots provide vital tonic for recovering patient

and livestock at both of the study sites. Similarly, the medicinal plant that has become a popular fallow species, *Swertia chirayita,* is used to cure fever, coughs, headaches, colds and throat problems. The whole plant is used to create these medicines. During the household survey, we asked the general question: do people use medicinal plants for household-level health care? An overwhelming 80% of households said yes, they did use medicinal plants. The dependency on local medicine was higher in central Nepal (75%) than at the eastern study sites (60%). However, modern medicines are used as well, particularly by those with sufficient cash or those who have little faith in local treatments.

Medicinal plants diversify income opportunities

Besides household health care, a number of species are being harvested from fallow fields or gathered from forests for commercial sales. Gurjo (*Tinospora sinensis*), kurilo (*Asparagus racemosus*), harro (*Terminalia chebula*), bhakyamlo (*Brucea javanica*), barro (*Terminalia bellirica*), amala (*Phyllanthus emblica*) and pipla (*Piper longum*) are some of the common species being traded in central Nepal. Praja Cooperative, a cooperative run by Chepang people based in Chitwan district of central Nepal, has been promoting the marketing of plant products. These plant parts are being collected from both fallows and forests on a seasonal basis. In recent years, some of the more important species have been cultivated on fallow land as well as in home gardens. About 40% of respondent households at the study sites in central Nepal said they were involved in making income from medicinal plants. They were supplying the NTFPs to the Praja Cooperative (more recently known simply as Chepang Cooperative) in Shaktikhor village (see Box 13-1).

In the study sites in eastern Nepal, the main traded medicinal plants are the perennial herb chiraita (*Swertia chirayita*), the shrub budo okhati (*Astilbe rivularis*) and another perennial herb, pakhanbed (*Bergenia ciliata*). *S. chirayita*, is valued for its analgesic qualities; *B. ciliata*, which constitutes the drug *Paashaanabheda*, has astringent, tonic, anti-scorbutic and laxative properties; and *A. rivularis* is used to treat diarrhoea and dysentery. These plants have a high market value and are being marketed mainly by middlemen and local businessmen. About 65% of households in the study villages earn income from chiraita. One farmer, Laxman Rai, of Lelep village in Taplejung district, eastern Nepal, reported that he was earning about 50,000 Nepalese rupees (about US$564) per year from sales of chiraita. However, it can be harvested only after three years of cultivation, and for only three years after that. After four years of cultivation, when the plants have been harvested for one year, production declines. Laxman Rai explained that he was expanding his chiraita crop by planting it in more of his fallowed land.

In eastern Nepal there are four systems for marketing chiraita. Some who gather it as an NTFP sell it themselves in local markets, while some local traders also buy the chiraita to trade it on wholesale markets. Chiraita can also be sold to businessmen by individual farmers or local middlemen, and finally, there are wholesalers who collect chiraita in bulk and sell it in big markets within Nepal or export it to India and China (Figure 13-5). Despite its high value, 45% of household respondents mentioned that

---

BOX 13-1: Chepang Medicinal Plant Cooperative

Chepang Medicinal Plant Cooperative was established in 1995 to increase the income of the Chepang people, who depend on natural resources, especially non-timber forest products, by promoting local business entrepreneurship and sustainable management of natural resources. Situated in Shaktikhor village in Chitwan district, central Nepal, the cooperative has a collection centre, storage facilities, a processing plant and a market outlet. Its most interesting aspect is that it is based on niche-products – local medicinal and aromatic plants – that are specific to the Chepang. When it started, the cooperative had 25 shareholders. The number reached 360 by the end of 2011.

The cooperative has been supplying non-timber forest products to private companies in Narayangarh, a nearby market town, and in the Nepalese capital Kathmandu, through contracts with traders. The NTFPs include *Terminalia chebula*, *Terminalia bellirica*, *Phyllanthus emblica*, *Diploknema butyracea* and *Tinospora sinensis* (dry vines worth US$0.16/kg), and *Asparagus racemosus* (at US$3.39/kg). The products are sold to Life Care (a company producing medicines), Gorkha Ayurved (a company producing homeopathic medicine) and the Alternative Herbal Company. On average, the farmers earn about 7500 Nepalese rupees per annum (about US$85) from selling their medicinal-plant products. The product with highest demand in the market is *Tinospora sinensis*. This species is generally used in the treatment of fever and to prevent recurrence of disease symptoms. It contains glucoside, alkaloids, essential oils and a mixture of fatty oils. *Asparagus racemosus*, a twining herb, is generally sold as a table vegetable in urban markets. Tender shoots are cooked and eaten as a vital tonic for recovering patients. The roots and leaves are also used in the manufacture of pharmaceutical drugs.

Initially, farmers collected these NTFPs from fallows and forests, but an increasing volume being taken from forests and the conversion of national forests to community forests has meant that free collection is no longer permitted. Responding to this, farmers are cultivating NTFPs such as *gurjo* (*Tinospora sinensis*), *tejpat* (*Cinnamomum tamala*), *kurilo* (*Asparagus racemosus*), *harro* (*Terminalia chebula*), *barro* (*Terminalia bellirica*) and *amla* (*Phyllanthus emblica*) in their fallow fields. They have also begun domestication of these species in their home gardens.

Linking local products to markets through the cooperative has contributed to livelihoods in the Chepang community. In one season, a farmer can sell at least 50 to 100kg of *gurjo*, the NTFP in highest demand. In addition to NTFPs, the cooperative is also marketing agricultural products grown by the farmers, including the legumes black gram, horse gram, beans and cowpeas, along with spices and honey. The main challenge has been to match supply with market demand.

---

the marketing system for *chiraita* was not well established. It relies on just a few businessmen and heavy fluctuations in prices are a major challenge for farmers, they said. Hence, improving the marketing system would benefit both farmers and traders.

Discussion

Shifting cultivation, as a farming system, is undergoing an unprecedented transformation due to various socio–economic and political changes in Nepal (Aryal et al., 2010). In the course of this transformation, fallow management has become a key challenge for farmers, in order to maintain production at the same level as they used to enjoy when fallow periods were longer. The shortening of fallow periods

FIGURE 13-5: Author Kamal Aryal (right) interviews a farmer carrying a bundle of dried *Chiraita* for sale.

Photo: Elisabeth Kerkhoff.

in shifting cultivation is a widespread phenomenon occurring in many countries (Dalle and De Blois, 2006; Kerkhoff and Sharma, 2006). In general, the fallow period in shifting cultivation systems in Nepal has decreased over time. Recent studies have shown that in some places the fallow period has ceased to exist, and the farming system has been transformed into continuous cropping (Aryal and Kerkhoff, 2008; Kafle et al., 2009). Since one of the principal functions of the fallow period is to restore soil fertility following cropping, the shortening of fallow periods greatly jeopardises short-term food security as well as long-term sustainability. The reasons for this change range from household-bound factors such as food security, available human labour, livelihood options and choices of crops to policy-bound factors,

processes and decisions such as the declaration of conservation areas, enactment of national laws and regulations, access to and ownership of the land, and so on. Families having alternative livelihood options may choose to maintain a long-fallow system, or they may give up shifting cultivation altogether; but families with very little land for cultivation may struggle to continue shifting cultivation with relatively short fallows (Coomes et al., 2000). In such circumstances, the challenge for researchers and development workers is to identify and promote cultivation practices and crop combinations on shifting cultivation land that will enhance food security, restore soil fertility and adapt to the shortened fallow.

Managing the fallow in these transforming shifting cultivation systems should mean promoting and adopting sustainable management options to respond to shortened periods of fallow. This study, involving 276 shifting cultivation-dependent households in central and eastern Nepal, revealed some constructive options for managing fallowed land in transforming shifting cultivation systems. These included cash cropping, horticultural cropping and conservation practices in agricultural production. Such options contribute to cash income, support family nutrition and enhance or sustain the productivity of land. However, management options and interventions also need to increase soil fertility, crop productivity and food security as well as reducing soil losses and human pressures on the land (Jakobsen, 2006).

Farmers at the study sites reported that multipurpose tree species that were usually also capable of fixing atmospheric nitrogen were generally protected during land clearing and were managed during the cropping phase. Intercropping of maize with a variety of legumes, including cowpeas, black gram, common beans, soybeans and horse gram, is a long-standing practice in central Nepal. Rather than pulling the beans at harvest time, the stalks are cut and the stumps and roots are left in the ground to decompose and add nutrients. Similar practices were also documented by Kerkhoff and Sharma (2006). However, the challenge in managing shortened fallows is opting for either maintaining an optimum fallow by integrating high-value crops that grow well in a fallow environment, or integrating practices that restore soil fertility and land productivity. Such strategies should focus on promoting or integrating alternative options that are socially acceptable, environmentally friendly, gender friendly, economically viable and profitable and adaptable to climatic variability. Shifting cultivators in the study villages have adopted some alternative land uses, including horticulture-based agroforestry in central Nepal and cardamom-alder agroforestry and *chiraita*-based shifting cultivation in eastern Nepal. These findings are also documented in Aryal and Kerkhoff (2008) and Aryal et al. (2010).

An important aspect of this transformation of shifting cultivation is the management of fallows by conserving and cultivating medicinal plants. Many studies have documented the harvesting of a wide range of medicinal plants from fallows and forests for household health care, and have shown that the use of medicinal plants remains prevalent in the study area. The 21 most important and most frequently used of these medicinal species are listed in Table 13-2. Of these, 14 are from central Nepal and seven from eastern Nepal. The dependency on local medicine was higher in central Nepal (about 75%) than in the eastern districts (60%). However, the use of such plants as medicines is declining, and the decline is more pronounced in eastern Nepal than in the central study sites The capacity to afford modern health care and medications is a major reason for the decline in use of local treatments in eastern Nepal, despite the area having poor access to good hospitals. However, people of older generations in the area prefer to use local medications – as far as they work. The declining use of local medicines and treatment has also been noted by similar studies carried out by Shrestha and Dhillion (2006) and Aryal et al. (2009). However, people in central Nepal have no alternative but to use local treatments, although members of the younger generation will try to go to a hospital for treatment if they can arrange enough money. An important reason for this change is that traditional healers, who used to treat village people, are now very few and far between, and the transfer of knowledge about treatment practices and preparation of traditional medicines is very poor. Moreover, the younger generation is no longer interested in studying traditional healing (Ladio and Lozada, 2004).

Despite local trends in the use of traditional medicines, medicinal plants have become a product of major economic importance in recent decades. The commercial cultivation and domestication of important plant species have begun in many parts of the country, and a number of species are gathered from forests and fallows and traded in markets for economic gain. *Gurjo* (*Tinospora sinensis*), *kurilo* (*Asparagus racemosus*), *harro* (*Terminalia chebula*), *bhakyamlo* (*Brucea javanica*), *barro* (*Terminalia bellirica*), *amla* (*Phyllanthus emblica*) and *pipla* (*Piper longum*) are some of the common species being traded in central Nepal. In eastern Nepal, the main traded medicinal plants are *chiraita* (*Swertia chirayita*), *budo okhati* (*Astilbe rivularis*) and *pakhanbed* (*Bergenia ciliata*). These plants have high market value and are being marketed mainly by middlemen and local businessmen (Ghimire et al., 2008). Another study by Manandhar (2002) found that the collection and marketing of *S. chirayita* for medicinal purposes in eastern Nepal was showing good profitability. However, over-exploitation and illegal harvesting and trading have threatened a number of species (Shrestha and Dhillion 2006; Chaudhary, 1999). Unsustainable harvesting from the wild, to meet both household and commercial demand, could lead to genetic erosion. In contrast, as market access has increased, farmers have been adjusting their crop selection more and more, to incorporate commercial species into their fallows and home gardens. This study has found that farmers in the eastern study sites are commercially cultivating *chiraita* in their fallowed shifting cultivation fields as well as in other land, when in the past it was harvested only from natural forests. Similar initiatives have also been reported with regard to cultivation of *kurilo* (*A. racemosus*) in central Nepal.

Despite the high importance of these medicinal crops for household health care and income generation, they have been overlooked by the government's development and research programme (Kerkhoff and Sharma, 2006; Aryal et al., 2010). It is now important for us to be asking: Are we managing these resources so that they will be available for future generations? Can we manage these resources in changing socio-economic and biophysical conditions? To some extent, this management process has already been initiated at local level; farmers at the study sites as well as in other parts of Nepal have begun cultivating these plants in fallow fields and domesticating them in home gardens, to both conserve and utilize these species (Aryal, 2010). Furthermore, organizations like the Praja Medicinal Plant Cooperative in central Nepal are training farmers in the cultivation and management of economically important medicinal plants in their fallow fields and home gardens.

Conclusions

Management of fallows in transforming shifting cultivation systems is important to sustain the food security of millions of shifting cultivators. Farmers are well aware of the importance of fallows for restoring soil fertility, and the need to achieve the same production levels when fallows are shorter. They have responded to this need with a number of innovative options under different circumstances that are socially, economically and environmentally feasible. However, the many good practices and local innovations that farmers are adopting for the effective management of their fallows are yet to be documented and up-scaled. Understanding and recognizing these innovative fallow-management practices is crucial to addressing the management of both shifting cultivation and the natural resources involved in transforming shifting cultivation systems.

Acknowledgements

The authors wish to thank all the respondents and community members of the study villages who patiently shared their time, insights and views about shifting cultivation systems, their management and associated changes. Special thanks go to Canada's International Development Research Centre (IDRC) for financial support for this study, which is part of a regional project on shifting cultivation. The International Centre for Integrated Mountain Development (ICIMOD) and Local Initiatives for Biodiversity Research and Development (LI-BIRD) are acknowledged for their management support during the entire study period.

References

Aryal, K. P. (2010) 'Conservation through utilisation: An analysis of indigenous people's dependency on forest resources for livelihood support in the middle hills of Nepal', in *Proceedings of the 15th International Forestry and Environment Symposium*, part II: full papers, Department of Forestry and Environment Science, University of Sri Jayewardenepura, Sri Lanka, pp.164-171

Aryal, K. P. and Kerkhoff, E. (2008) *The Right to Practice Shifting Cultivation as a Traditional Occupation in Nepal: A Case Study to apply ILO Conventions No. 111 (Employment and Occupation) and 169 (Indigenous and Tribal peoples),* International Labour Organization, Kathmandu

Aryal, K. P., Berg, A. and Ogle, B. (2009) 'Uncultivated plants and livelihood support: A case study from the Chepang people of Nepal', *Ethnobotany Research and Applications* 7, pp.409-422

Aryal, K. P., Kerkhoff, E. E., Maskey, N. and Sherchan, R. (2010) 'Shifting Cultivation in a Sacred Himalayan Landscape', in *A Case study from Kangchenjunga Conservation Area*, First Edition, WWF Nepal, Kathmandu

Bartlett, J. E., Kotrlik, J. W. and Higgins, C. C. (2001). 'Organisational research: Determining appropriate sample size in survey research', *Information Technology, Learning and Performance Journal* 19 (1), pp.43-50

Bruun, T. B., Mertz, O. and Elberling, B. (2006) 'Linking yields of upland rice in shifting cultivation to fallow length and soil properties', *Agriculture Ecosystems and Environment,* vol 113, no 1-4, pp.139-149

Cairns, M. F. (2007) 'Conceptualizing indigenous approaches to fallow management: A roadmap to this volume', in Malcolm Cairns (ed.) *Voices from the Forest: Integrating Indigenous Knowledge into Sustainable Upland Farming,* Resources for the Future, Washington, DC, pp.16–36

Chaudhary, R. P. (1999) 'Biodiversity prospecting in Nepal: Constraints and opportunities', in *Proceedings of the 8th International Workshop on BIO-REFOR, 28 November – 2 December, 1999,* Kathmandu, Nepal, Biotechnology assisted Reforestation Project (BIO-REFOR), International Union of Forest Research Organisation (IUFROs) Special Programme for Developing Countries, Kathmandu, pp.256-260

Coomes, O. T., Grimard, F. and Burt, G. J. (2000) 'Tropical forests and shifting cultivation: Secondary forest fallow dynamics among traditional farmers of the Peruvian Amazon', *Ecological Economics* 32, pp.109–124

Dalle, S. P. and De Blois, S. (2006) 'Shorter fallow cycles affect the availability of noncrop plant resources in a shifting cultivation system', *Ecology and Society* 11(2), http://www.ecologyandsociety.org/vol11/iss2/art2/, accessed 24 October 2019

Ghimire, S. K., Sapkota, I. B., Oli, B. R. and Parajuli, R. R. (2008) *Non-timber Forest Products of Nepal Himalaya: Database of Some Important Species Found in the Mountain Protected Areas and Surrounding Regions,* WWF Nepal, Kathmandu

Himalaya Drug Company. (no date) *Herbal Monograph (Swertia Chirayita),* https://herbfinder. himalayawellness.in/swertia-chirayita.htm, accessed 24 October 2019

Jakobsen, J. (2006) 'The role of NTFPs in a shifting cultivation system in transition: A village case study from the uplands of North Central Vietnam', *Danish Journal of Geography* 106(2), pp.103-114

Kafle, G., Regmi, B. R., Pradhan, B., Lama, T. L., Shrestha, P. K., Limbu, P. and Karki, S. (2009) *Land Use Change and Human Health in Eastern Himalayas: An Adaptive Ecosystem Approach,* report for Local Initiatives for Biodiversity, Research and Development, submitted to the International Centre for Integrated Mountain Development (ICIMOD), Kathmandu

Kerkhoff, E. and Sharma, E. (2006) *Debating Shifting Cultivation in the Eastern Himalayas: Farmers' Innovations as Lessons for Policy,* International Centre for Integrated Mountain Development (ICIMOD), Kathmandu

Ladio, A. H. and Lozada, M. (2004) 'Patterns of use and knowledge of wild edible plants in distinct ecological environments: A case study of a Mapuche community from northwestern Patagonia', *Biodiversity and Conservation* 13, pp.1153-1173

Manandhar, N.P. (with the assistance of Sanjaya Manandhar) (2002) *Plants and People of Nepal,* Timber Press, Portland, OR, ISBN 0-88192-527-6

Prinz, D. (1987) 'Improved fallow: Increasing the productivity of small holder farming systems', *ILEIA Newsletter,* 3(1), pp.4-7

Reijntjes, C., Haverkort, B. and Waters-Bayer, A. (1992) *Farming for the Future: An Introduction to Low-External Input and Sustainable Agriculture,* Macmillan, London, http://www.ciesin.org/ docs/004-176a/004-176a.html, accessed 24 October 2019

Schmook, B. (2010) 'Shifting maize cultivation and secondary vegetation in the Southern Yucatan: Successional forest impacts of temporal intensification', *Regional Environmental Change* 10, pp.233–246, http://sypr.asu.edu/pubs/REC_Schmook.pdf, accessed 24 October 2019

Shrestha, P. M. and Dhillion, S. S. (2006) 'Diversity and traditional knowledge concerning wild food species in locally managed forest in Nepal', *Agroforestry Systems* 66, pp.55-63

Notes

1. The medicinal plant *Swertia chirayita* is known by the name *charaita* in Nepal. The drug chiretta is obtained from the dried plant. The ethanolic extract of *S. chirayita* exhibits hypoglycaemic activity, and the compound may have clinical application in control of diabetes. It is also a herbal antiseptic and antifungal agent and an extract of the plant exhibits significant anti-inflammatory activity. (Himalaya Drug Company, n.d.)

14

FALLOWED SWIDDEN FIELDS IN VIETNAM

Floral composition, successional dynamics and farmer management

Mai Van Thanh and Tran Duc Vien[*]

Introduction

Swidden cultivation is the foremost agricultural technique used by indigenous peoples and one of the most important land-use systems in tropical areas of the world (Karthik et al., 2009). Moreover, it is a practice that has existed in montane mainland Southeast Asia for centuries and remains widely practised in the region to this day (Fox and Vogler, 2005; Cramb et al., 2009; Schmidt-Vogt et al., 2009; Mertz et al., 2013). In many countries, governments have seen swidden cultivation as a backward agricultural system that is destructive to the environment (McElwee, 2004; Hoang, 2011; Chan et al., 2013). However, researchers are agreed that this traditional method will continue to be an important practice in the future (van Vliet et al., 2012, 2013).

In Vietnam, swidden cultivation remains a prominent means of subsistence agriculture. Hai et al. (2003) estimated that about 50 ethnic groups (out of the country's total of 54 ethnic groups) were practising swidden agriculture, involving about nine million swiddeners working an area of 3.5 million hectares (ha) in the country's uplands. Earlier, the government estimated that about 50% of the 110,000ha of forest lost per year during the 1980s and 1990s was lost because of swidden cultivation (Sam, 1994). Government data goes on to show that swidden cultivation accounted for 54% of total forest loss in the country in 2012, and 71.6% of

An earlier version of this chapter is in Tran Duc Vien, A. Terry Rambo and Nguyen Thanh Lam (eds) *Farming with Fire and Water: The Human Ecology of a Composite Swiddening Community in Vietnam's Northern Mountains* (2009), Kyoto University Press and Trans Pacific Press (https://www.kyoto-up.or.jp/books/9784876984688.html?lang=en). It is republished here with the kind permission of Kyoto University Press.

[*] DR. MAI VAN THANH, Research Fellow, Centre for Applied Climate Sciences, University of Southern Queensland, Australia; and DR. TRAN DUC VIEN, Professor and former Director of the Center for Agricultural Research and Ecological Studies at Hanoi University of Agriculture, Vietnam.

total forest area lost in the years 2008 to 2011 (Forest Protection Department, 2013). The government has also noted that hundreds of thousands of villagers in the uplands still engage in swidden cultivation on an area of more than one million ha of land (MARD, 2007; MONRE, 2011).

Despite the official attitude towards swidden cultivation, it is a highly dynamic land-use system in terms of both space and time (Hett et al., 2012). Generally, it undergoes cyclical land-cover changes from crop to fallow, and contrary to a widespread belief that fallow land is simply lying idle, the fallow stage plays a significant socio-economic and environmental role. This is because the fallow not only restores the fertility of the soil and accumulates biomass in swidden fields (Kass and Somarriba, 1999; Scott et al., 1999), but it also enhances the livelihood of swidden communities through the provision of non-timber forest products (Quang and Anh, 2006; Cunningham, 2011; Ghorbani et al., 2012). Vegetation cover during the fallow period, when former swiddens are cloaked in young forest, has been considered as an instrument for future agreements to reduce emissions from deforestation and forest degradation (REDD) (Chan et al., 2013). In Vietnam, the fallow period – the time allowed for the forest to recover – has been growing shorter over the past few decades. This trend is engendered by a bundle of factors, including growing population, improved access to markets for swidden crops and stringent forest-land allocation policies imposed by the government (Tran, 2004). As a consequence, the structure and patterns of the vegetation and the way it is managed by local villagers have changed substantially. In this chapter, we identify species diversity, describe successional dynamics and investigate local knowledge in managing the fallow stage of the swidden rotation cycle. We do this though the lens of a case study in Ban Tat, a hamlet located in the country's mountainous northwest.

Methods

The study area

This study was conducted in 2004, in Ban Tat, sometimes referred to here as Tat hamlet, a part of Tan Minh commune in Da Bac district of Hoa Binh province (Figure 14-1). The village is in the Da river watershed, near the Hoa Binh hydropower plant, which supplies most of the country's electricity. The village lies at an elevation of about 360 m above sea level (masl) and is surrounded by mountains. Its topography is undulating, with hill slopes ranging from 7 to 45 degrees. Prior to the 1960s, the entire area surrounding Ban Tat was covered by primary forest (Rambo and Tran, 2002). The forest cover has since been dramatically reduced by excessive logging of timber and swidden cultivation (Sikor, 1998; Sikor and Phuc, 2011). The area surrounding the village is now regarded as barren land with a poor forest cover. The hill tops and ridge lines around the village are covered with mature secondary forest with a slight degree of canopy differentiation. Most slopes are covered with swiddens or recently fallowed plots growing grasses, herbs and scattered patches of bamboo and small trees. Lower valley slopes and storm-drainage courses remain surrounded by woody vegetation and secondary forest, although in some cases these patches have been scorched by fires escaping from the swiddens. Near-slopes and hillocks in non-protected areas around the hamlet have mostly been cleared for planting canna or corn, cassava swiddens or tree gardens. There is very little fallow land. Valley bottoms are covered by wet-rice fields.

Most of Ban Tat's people are of the Tay ethnic minority and have been living in the area for many generations, with the forest and swidden cultivation as their primary sources of livelihood.[1] Most houses are located near a main road that runs through the village. On average, each household in Ban Tat has about four swidden plots, often four to five kilometres away from their homes, requiring three to five hours of walking. Some households have plots 10 km from

Canna indica L. [Cannaceae]

The vibrant colours of *Canna* flowers are well known throughout Southeast Asia. For thousands of years, the plants have also been a minor food crop and, as such, they are cultivated around the hamlet of Ban Tat. The large tubers can be eaten raw, but are usually baked, and the leaves are used to wrap food. In China, *Canna* starch has been combined with polyethylene to produce biodegradable plastics.

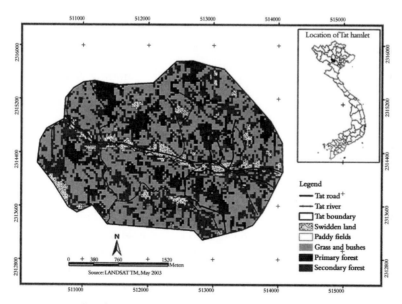

FIGURE 14-1: Land use in Ban Tat and the location of the hamlet in northern Vietnam (inset).

the village, requiring an entire day of walking just to reach the swidden. According to the villagers, the traditional swidden cycle once involved one or two years of cultivation, followed by 10 to 15 years of fallow. However, at the time of the study, the usual fallow period lasted only one or two years.

Site selection

The swidden–fallow sites for this study were chosen on the bases of fallow age, soil type, slope gradient, position of the fields and cultivation history. The selection involved two steps: discussion with the villagers and characterization in the field. In the first step, we organized a focus-group discussion with knowledgeable famers in Ban Tat. This helped in gathering information about social aspects of the households and the village, swidden cultivation, and changes associated with swiddening over time as well as the drivers of these changes. We also asked villagers about the types of fallow management that were practised in the village and evaluated these systems by employing a simple matrix-ranking exercise (Toai et al., 1998). Discussions and consultations with the famers also helped us to identify sample fallow plots in the field. The second step involved transect walks with villagers who were deeply knowledgeable about the flora in the area and identifying fallow plots to make sure that they met our selection criteria.

Field measurements

The flora of the study plots was identified through four transect walks that cut across different habitats in Ban Tat's territory, including fallow plots of different ages, representing different stages of successional regrowth (very early, early, intermediate and late). The fallow ages chosen for the study were one to two years, five years, 10 years and 20 years. A specimen of each of the flora species encountered was collected and the villagers were asked to give local names and explain how each species was used (i.e., for timber, food or medicine). The scientific names of the species were determined by morphological comparison and from secondary data on flora in Vietnam (Ho, 1993). Botanical phyla were listed in order of vegetation evolution, i.e from plants that scattered spores (Psilotophyta, Lycopodiophyta and Polipodiophyta) to plants with seeds (Pinophyta and Magnoliophyta). The species were listed in alphabetical order based on phylum, family, genus and species – except for the Magnoliophyta phylum, which was listed in class order.

Structure and composition

In total, 12 sample plots were established (each fallow age repeated three times) using a nested sampling design (Figure 14-2). Vegetation at different fallow ages was measured in the different plots for structure and composition. The A plots measured 50 m x 50 m (0.25 ha) and these were used in habitats following 20 years of fallow. All timber trees having a diameter at breast height (dbh) larger than 10 cm were measured for dbh, height of crown and height of stem.[2] Trees shorter than 10 metres were measured directly, while taller trees were measured by using the similar-triangles method (Hinh and Giao, 1998). Standard B plots measured 31.5 m x 31.5 m (0.1 ha) and were applied in plots that had been fallowed for periods from four or five years to 10 years. Measurement of tree species was similar to that in the A plots. The C plots were 10 m x 10 m in size and were used to measure the number of species of bushes, secondary-timber species and bamboo. D plots were 2 m x 2 m and were used to measure grass biomass under the woody canopy.

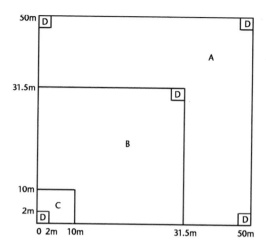

FIGURE 14-2: Comparison of standard plots A, B, C, and D.

Estimation of biomass

The biomass of vegetation in plots of different fallow ages was measured using methods appropriate to each class of vegetation. For trees, the volume of all the specimens growing inside an A plot was determined by the following equation:

$$V_0 = \frac{\Pi \times dbh_{1.3}{}^2}{4} \times H_{crown} \times f_{1.3}$$

Where V_0 = volume of timber; $f_{1.3}$ = coefficient ($f_{1.3}$ = 0.42 for natural forest and 0.45 for planted forest) (Hinh and Giao, 1998). The tree biomass was estimated from their total volume using the following formulas of the Nissho Iwai Research Institute (Que et al., 2003):

$$M = 0.5 \times V$$
$$V = V_1 \times 1.2$$
$$V_1 = 1.33 \times V_0$$

Where M = biomass of trees; V_0 = volume of above-ground biomass of timber; V_1 = total volume of above-ground biomass of trees (leaves, timber, and stems); and V = total volume of above- and below-ground biomass of trees.

For bushes, bamboo, and lianas (vines), the average wet weight per plot was directly determined and then converted to dry weight of the specific vegetation (i.e., bushes, bamboo or vines) per hectare. The biomass of grass was determined by picking, washing and weighing all of the specimens growing inside D plots (2 m x 2 m).

Results and Discussion

Diversity of floristic species

A total of 419 species belonging to 283 genera, 113 families and five phyla (Psilotophyta, Lycopodiophyta, Polipodiophyta, Pinophyta and Magnoliophyta) were found in Ban Tat's territory. It was clear that the flora in the hamlet was very diverse. These categories of flora fell into five life forms: grass, trees, lianas (vines), bushes and palms. Grass species were the most numerous (Table 14-1). The composition of the flora at each fallow age was different, as it was affected by both the monsoon climate and topographical factors. Variations in sunlight regimes at the different sites led to further differentiation. The canopy stratification was relatively complex due to a high frequency of human intervention. Almost all ecological floral groups were represented at different successional stages of fallow. These included grasses (very early stage), annual and perennial light-demanding pioneer flora (early stage), flora with medium light demand (intermediate stage) and shade-tolerant flora (late stage). Interestingly, there was a wide variation in the number of species belonging to each family or genus. The Fabaceae (Leguminosae) family had the largest number of species (28), followed by Euphorbiaceae (25 species); Poaceae (18); Asteraceae (16); Rubiaceae (15); and Moraceae (13).

Remarkably, many useful plants were found in the study-area fallows. Villagers identified 398 out of the 419 species of plants (95%) as being useful for some purpose. These species were categorized according to eight different uses: medicine, fruit, timber, vegetables, ornamentals, spices, dye, and for chewing with betel. As shown in Table 14-2, plants with medicinal properties were most numerous, followed by fruit, timber and vegetables.

TABLE 14-1: Distribution of flora life forms in fallows in Tat hamlet.

Type	Number of species	Percentage (%)
Grasses	178	42.5
Trees	137	32.7
Lianas	45	10.7
Bushes	50	11.9
Palms	9	2.2
Total	419	100.0

TABLE 14-2: Useful fallow–plant species in Tat hamlet.

Uses/functions	Number of species	Percentage (%)
Medicine	225	53.7
Fruit	59	14.1
Timber	39	9.3
Vegetables	44	10.5
Ornamental	8	1.9
Spice	17	4.1
Dye	3	0.7
Chewed with betel	3	0.7

Successional dynamics of fallow flora

Changes in composition

The number and type of species within plots changed in the process of succession, beginning from a very early fallow age. The number of tree species increased markedly from the second year of fallow to the 20th year. The economic value of the tree species also increased. Lower-value *Styrax tonkinesis* and *Rhus javanica* were found from the second year to the fifth year of fallow; *Neonauclea calycina* and *Saurauia roxburghii* were present from the fifth to the 10th year, at which time *Ormosia balansae* was also identified; and high-value *Elaeocarpus petiolatus* and *Desmos chinensis* were finally found in the 20th year, along with several wild fruit trees with market value, such as *Garcinia cochinchinensis* and wild litchi.

Changes in life forms

The importance of different floral life forms also changed in the course of succession (Table 14-3). Although the species structures of the second and fifth years of fallow were varied, almost all species were light-tolerant plants, and tree species were outnumbered by grasses and bush

Styrax tonkinensis Craib ex Hartwich
[Styracaceae]

This fast-growing woody tree, called *bo de* at the study site, can grow to 25 metres in 10 years. A benzoin resin tapped from the trunk after six or seven years of age is widely used in pharmaceuticals and to make fragrances for household products such as soaps and shampoos. The wood is light and soft, and the species is an important source of pulpwood for paper production.

Table 14-6 shows that diverse plant products were collected from fallows during the cooperative period. Bamboo stems, palm fruit and leaves and bamboo shoots were considered important products during this period. Fallow products were used mainly for home consumption, such as in house construction and for food, livestock fodder and firewood. After the cooperative period, however, the relative importance and functions of these products changed. Bamboo is still the most important product, followed by bamboo shoots, timber, firewood and broom grass (Table 14-7). Today, many more products are collected for sale in markets, although the villagers still use them for livestock fodder, house construction, fuel, fencing and their own food.

Dendrocalamus brandisii (Munro) Kurz [Poaceae]

Sometimes known as velvet-leaved bamboo or teddy-bear bamboo, this is one of the largest clumping bamboo species. The stems grow to 33 metres and are up to 20 cm in diameter. They are known for their strength, and uses range from boat masts to construction and handicrafts. The shoots are eaten as a vegetable. The stems grow to their full height in just one year, sometimes growing 30 cm in a day.

Fallow-management systems

Farmer strategies for managing fallow land have changed over time in response to increasing population, market demand and government policies. During the cooperative period, population density was still low and villagers did not face land shortages. Natural fallows and the augmentation of natural-fallow regeneration by planting *Melia* trees and palm trees in the fallow were the most popular systems of fallow management. More recently, villagers have faced greater constraints on land for cultivation, and in response to these constraints they are using several adapted fallow-management systems, as follows:

Natural fallow

This is a traditional practice that allows vegetation to regenerate naturally without any human intervention. Its advantage is that soil fertility is naturally restored without any investment of capital or labour. However, because there is no intervention, soil erosion can occur during the early years of fallow, so a longer period might be required in order for soil fertility to recover. Consequently, this method is only applied to swidden plots located a long distance from the settlement.

TABLE 14-7: Uses for plants collected from fallows at the present time.

Use	Bamboo shoots	Bamboo	Firewood	Timber	Animals	Palm leaves	Medicinal plants	Broom grass	Mushrooms	Total
						Product				
Food	3	–	–	–	5	–	–	–	2	10
Cash sales	10	8	5	4	2	5	5	5	2	46
Medicine	–	–	–	–	1	–	3	–	–	4
Fuel	–	5	10	–	–	–	–	–	–	15
Housing	–	5	–	4	–	10	–	–	–	19
Fencing	–	7	–	5	–	–	–	–	–	12
Fodder	10	2	–	3	–	–	–	10	–	25
Total	23	27	15	16	8	15	8	15	4	131

Melia-*planting on fallow land*

This simple agroforestry model has long been practised in Tat hamlet (Tran, 2007). Farmers usually sow *Melia azedarach* seeds into their swiddens before burning the fields, because *Melia* seed only germinates after exposure to high temperatures. Both crops and *Melia* trees grow together in the swidden field. After several crops are harvested, the field is left fallow while the *Melia* grows to maturity. *Melia* grows quickly and its timber is used for domestic purposes. *Melia* leaf has a bitter taste and is used as green manure and as a method of biological pest control in paddy fields. The harvesting period for *Melia* usually begins 8 to 10 years after planting. After the Melia is harvested, the next swidden cycle begins. This model is still popular in Tat hamlet.

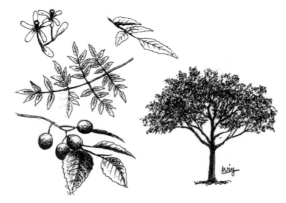

Melia azedarach L. [Meliaceae]

Often known as chinaberry and grown in many countries as an ornamental, this species is commonly around 12 metres tall but can grow much taller. The flowers are small, fragrant, and pale purple. Its high-quality timber is often mistaken for teak, but is generally under-utilized. The fruit, which are poisonous to humans, are eaten by birds and large animals without risking damage to the toxic inner seed.

Bamboo-planting on fallow land

Most households now plant bamboo on their fallow land. This land-use system has been promoted by various development projects in order to restore areas considered by the government to be 'barren land'. The projects provide farmers with subsidized seedlings, fertilizer and training. In addition, bamboo (*luong*) has high economic value and there is strong market demand. It grows very quickly and suits the land and weather conditions in Ban Tat. However, this model has several disadvantages: bamboo is most suitable for growing in lowland areas with moist soils; it requires high initial investment and removes a large quantity of nutrients from the soil when harvested; and the leaves of bamboo are highly resistant to decomposition. The two latter factors adversely affect following swidden cycles.

Styrax-planting on fallow land

This land-use system is commonly practised in many parts of Vietnam's northern mountain region, including Ban Tat. The villagers plant the fast-growing native *Styrax* (*Bo de*) on their fallow land. This species occurs as a volunteer but can also be planted by the famers. It is later sold for paper pulp, for timber, or to make matches. While this land-use system is not popular in Ban Tat, several households plant the species in their fallows so it can later be used for firewood and fencing. It also speeds-up the vegetation cover in the early stages of fallow.

Palm-planting on fallow land

This model is relatively popular in Ban Tat. At its beginning, crops such as corn, cassava and rice are interplanted with palms for several years. When the palms reach the closed-canopy stage, cropping is discontinued. Palm leaves are used for roof thatching and for cash sales. In recent years, this income source has significantly enhanced local   liivelihoods and created jobs for villagers, especially during the off-farm season from November to January.

Coppicing trees in swiddens

This method leaves stumps of trees with a height of 30 cm to 40 cm above the ground when trees are cut to open new swidden fields. The stumps help to prevent soil erosion and landslides. During the cropping phase, the stumps begin to regenerate buds and shoots, which promote rapid regrowth of secondary forest when the field is fallowed. The regenerating trees are a source of firewood and also encourage quick recovery of soil fertility for the next swidden cultivation cycle.

Farmer perceptions of fallow-management systems

This section sets out to examine the relative value of different fallow-management systems used in Ban Tat in the eyes of the villagers. Seven criteria were identified in discussions with a group of villagers, including both male and female farmers from different age groups. They were: 1) generates income; 2) requires little investment; 3) offers effective protection against soil erosion; 4) improves soil fertility; 5) requires simple techniques; 6) suits the local situation; and 7) minimizes demands for labour. The matrix used a scoring system of 0 to 10, with 0 meaning the value was not important for the system, while 10 meant that it was very important. The results are presented in Table 14-8.

As can be seen, the farmers generally favoured systems that require less investment, use simple technologies and are most suitable to their specific situations. Generation of income, improving soil fertility and controlling soil erosion are all lesser concerns. Labour demand is not seen as important. Consequently, the locally developed models of planting *Melia* and palms in swidden fallows are the most preferred.

Conclusions

This chapter has shown that fallow land is far from being the useless wasteland it is commonly portrayed to be in the Vietnamese official mythology of swidden cultivation (Jamieson et al., 1998; Rambo and Tran, 2002). Fallow swidden fields play vital ecological and economic roles in the functioning of Tat hamlet's composite swidden agroecosystem.

The flora found on swidden fallows in this study was highly diverse. The mosaic of fallow fields in various stages of succession provided habitats for species with a wide range of ecological requirements, from early pioneers to those adapted to life

TABLE 14-8: Farmer perceptions of the value of different fallow-management systems.

	Systems							
Criteria	*Natural fallow*	*Melia planting*	*Bamboo planting*	*Styrax planting*	*Palm planting*	*Coppicing*	*Total*	*Ranking*
Generates income	3	8	10	4	8	3	33	4
Requires little investment	10	8	3	8	8	10	47	1
Regulates soil erosion	3	6	4	4	8	8	29	6
Improves soil fertility	5	7	2	6	5	7	32	5
Requires simple techniques	10	8	3	8	8	8	45	2
Suits the local situation	7	8	4	7	6	6	37	3
Minimizes demand for labour	0	6	8	6	7	0	27	7
Total score	38	51	34	43	50	42		
Ranking	6	1	5	3	2	4		
% households using	20	50	90	50	45	30		

under the conditions of secondary forest. This habitat diversity contributed to the maintenance of overall floristic biodiversity.

The fallow flora was found to be composed of many species with high economic value, especially the timber species and medicinal plants. Products collected from fallow land have always contributed considerably to local livelihoods and social activities, although the significance of specific products has changed considerably, particularly following Vietnam's economic reforms. Specifically, products from the fallows have become important commodities used by villagers to earn cash income from markets.

In terms of successional dynamics, we found that species and species composition differed significantly according to the age of the fallows, especially between the early and intermediate and the early and late stages. However, between the intermediate and later stages, the only difference we found was in structure.

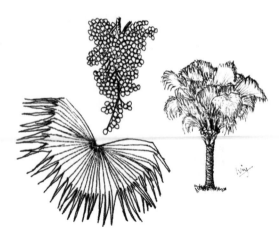

Livistona chinensis (Jacq.) R.Br. ex Mart. [Arecaceae]

The Chinese fan palm grows up to 15 m on a solitary stem and is topped by a dense crown of fan-shaped leaves. The leaves are harvested to make hats, brooms, raincoats and – at the study site – roof thatching. The palms are also a popular fallow species at Ban Tat for earning cash from market sales. Despite this popularity – or perhaps because of it – this species is regarded as threatened.

We also documented the changes in species composition, dominance, and above-ground biomass that were due to human intervention. The above-ground biomass changed significantly over the course of the fallow. The greatest change occurred in large-tree species, while the change in biomass of small trees, bushes and grasses did not present a clear trend. The villagers have experimented with a number of innovative management systems on their fallow land in an effort to adapt to increasing shortages of land for swidden cultivation. These include planting fallows with *Melia*, palms, bamboo and *Styrax*. Of these, the two locally developed systems – planting of *Melia* and palm – have been most successful in the eyes of the villagers. These systems would seem to have the most potential for expansion in Tat hamlet and beyond.

References

Chan, N., Takeda, S., Suzuki, R. and Yamamoto, S. (2013) 'Establishment of allometric models and estimation of biomass recovery of swidden cultivation fallows in mixed deciduous forests of the Bago Mountains, Myanmar', *Forest Ecology and Management* 304, pp.427-436

Cramb, R. A., Colfer, C. J. P., Dressler, W., Laungaramsri, P., Trang Le, Q., Mulyoutami, E., Peluso, N. L. and Wadley, R. L. (2009) 'Swidden transformations and rural livelihoods in Southeast Asia', *Human Ecology* 37(3), pp.323-346

Cunningham, A. B. (2011) 'Non-timber products and markets: Lessons for export-oriented enterprise development from Africa', Tropical Forestry no 7, in S. Shackleton, C. Shackleton and P. Shanley (eds) *Non-Timber Forest Products in the Global Context*, Springer-Verlag. Heidelberg, Germany, pp.83-106

Fox, J., and Vogler, J. B. (2005) 'Land-use and land-cover change in montane mainland Southeast Asia', *Environmental Management* 36(3), pp.394-403

Forest Protection Department (2013) Official Government of Vietnam website, www.kiemlam.org.vn, accessed on 29 October 2013 (Vietnamese language)

Ghorbani, A., Langenberger, G., Liu, J-X., Wehner, S. and Sauerborn, J. (2012) 'Diversity of medicinal and food plants as non-timber forest products in Naban River Watershed National Nature Reserve (China): Implications for livelihood improvement and biodiversity conservation', *Economic Botany* 66(2), pp.178-191

Hai, V. D., Que, N. D. and Thuong, P. N. (2003) *Shifting Cultivation and Forest Regeneration during the Fallow Period in Vietnam*, Nghe An Publishing House, Nghe An, Vietnam

Hett, C., Castella, J-C., Heinimann, A., Messerli, P. and Pfund, J. L. (2012) 'A landscape mosaic approach for characterizing swidden systems from a REDD+ perspective', *Applied Geography* 32(2), pp.608-618

Hinh, V. T. and Giao, P. N. (1998) *Forest Inventory and Planning*, Agricultural Publishing House, Hanoi

Ho, P. H. (1993) *Flora in Vietnam*, Agricultural Publishing House, Hanoi

Hoang, C. (2011) 'Forest thieves: State resource policies, market forces, struggles over livelihoods and meanings of nature in a northwestern valley of Vietnam', in T. Sikor, P. T. Nghiem, J. Sowerwine and J. Romm (eds) *Upland Transformation in Vietnam*, National University of Singapore Press, Singapore

Jamieson, N. L., Le, T. C. and Rambo, A. T. (1998) *The Development Crisis in Vietnam's Mountains*, East-West Center, Honolulu

Karthik, T., Veeraswami, G. G. and Samal, P. K. (2009) 'Forest recovery following shifting cultivation: An overview of existing research', *Tropical Conservation Science* 2(4), pp.374-387

Kass, D. C. L. and Somarriba, E. (1999) 'Traditional fallow in Latin America', *Agroforestry Systems* 47, pp.13-36

MARD (2007) *Proposal on Supporting Upland People in Sustainable Agroforestry Cultivation on Swidden Land in the period of 2008-2012*, Ministry of Agriculture and Rural Development, Hanoi

McElwee, P (2004) 'Becoming socialist or becoming Kinh? Government policies for ethnic minorities in the Socialist Republic of Vietnam', in C. Duncan (ed.) *Civilizing the Margins*, Cornell University, New York, pp.182-213

Mertz, O., Egay, K., Bruun, T. B. and Colding, T. S. (2013) 'The Last Swiddens of Sarawak, Malaysia', *Human Ecology* 41(1), pp.109-118

MONRE (2011) *Decision 2282 dated 8 December 2011 on Approval and Announcement of Statistics on Land Areas as of 1 January 2011*, Ministry of Natural Resources and Environment, Hanoi

Quang, D. V. and Anh, T. N. (2006) 'Commercial collection of NTFPs and households living in or near the forests: Case study in Que, Con Cuong and Ma, Tuong Duong, Nghe An, Vietnam', *Ecological Economics* 60(1), pp.65-74

Que, N. D., Phuong, V. T., Huy, L. Q., Giang, D. T., Tung, N. T. and Thang, N. V. (2003) *Identifying Indicators and Criteria for Forest Plantation in Vietnam towards the Clean Development Mechanism*, Forest Science Institute of Vietnam, Hanoi

Rambo, A. T. and Tran, D. V. (2002) 'Social organization and the management of natural resources: A case study of Tat hamlet, a Da Bac Tay ethnic minority settlement in Vietnam's northwestern mountains', *Southeast Asian Studies* 39, pp.299-324

Sam, D. D. (1994) *Shifting Cultivation in Vietnam: Its Social, Economic and Environmental Values Relative to Alternative Land Use*, International Institute for Environment and Development, London

Schmidt-Vogt, D., Leisz, S., Mertz, O., Heinimann, A., Thiha, T., Messerli, P., Epprecht, M., Cu, P., Chi, V., Hardiono, M. and Dao, T. (2009) 'An assessment of trends in the extent of swidden in Southeast Asia', *Human Ecology* 37(3), pp.269-280

Scott, L. T., Palm, C. A. and Buresh, R. J. (1999) 'Ecosystem fertility and fallow function in the humid and subhumid tropics', *Agroforestry Systems* 47, pp.183–196

Sikor, T. (1998) 'Forest policy reform: From state to household forestry', in M. Poffenberger (ed) *Stewards of Vietnam's Upland Forest*, Asia Forest Network, Tagbilaran City, Philippines

Sikor, T. and Phuc, X. T. (2011) 'Illegal logging in Vietnam: Lam tac (Forest Hijackers) in practice and talk', *Society and Natural Resources* 24(7), pp.688–701

Toai, B. D., Ngai, N. B. and Hong, N. N. (1998) *Participatory Rural Appraisal in Agricultural and Forest Extension Activities*, Agriculture Publishing House, Vietnam

Tran, D. V. (2004) 'Changes in the composite swiddening system in Tat hamlet in Vietnam's northern mountains in response to integration into the market system', in H. Furukawa, M. Nishibuchi, Y. Kono and Y. Kaida (eds) *Ecological Destruction, Health, and Development: Advancing Asian Paradigms*, Kyoto University Press, Kyoto, pp.453–465

Tran. D. V. (2007) 'Indigenous fallow management with *Melia azedarach* Linn in northern Vietnam', in M. F. Cairns (ed.) *Voices from the Forest: Integrating Indigenous Knowledge into Sustainable Upland Farming*, Resources for the Future, Washington, DC, pp.435–443

van Vliet, N., Mertz, O., Heinimann, A., Langanke, T., Pascual, U., Schmook, B., Adams, C., Schmidt-Vogt, D., Messerli, P., Leisz, S., Castella, J-C., Jørgensen, L., Birch-Thomsen, T., Hett, C., Bech-Bruun, T., Ickowitz, A., Vu, K. C., Yasuyuki, K., Fox, J., Padoch, C., Dressler, W. and Ziegler, A. D. (2012) 'Trends, drivers and impacts of changes in swidden cultivation in tropical forest-agriculture frontiers: A global assessment', *Global Environmental Change* 22(2), pp.418–429

van Vliet, N., Mertz, O., Birch-Thomsen, T. and Schmook, B. (2013) 'Is there a continuing rationale for swidden cultivation in the 21st Century?', *Human Ecology* 41(1), pp.1–5

Notes

1. The Tay ethnic group is the largest of Vietnam's 54 ethnic minorities.
2. Diameter at breast height (dbh) was measured 1.3 m from the ground; H_{crown} was measured from the ground to the top of the tree, and H_{stem} was measured from the ground up to the first main branch.

A. When shifting cultivators expand forest cover

A Bugis woman in Southeast Sulawesi, Indonesia, plants valued perennial seedlings to enhance her swidden production and family income. The introduced perennials will be nurtured among her annual crops to become a major and perennial part of the fallow vegetation when annual cropping in this field ends. This mimics the *taungya* system, in which usually timber trees are interplanted with annual crops in swidden systems (see Roshetko et al., ch. 29, and plates 49 and 50 in Coloured Plates section II, this volume).

Sketch based on a photo by Yusuf Ahmad/AgFor project, ICRAF Indonesia, ca. 2014.

Synthesis

SHIFTING CULTIVATION AND FOREST CONSERVATION

Historic legacies and future possibilities

*Stephen F. Siebert**

Shifting cultivation entails the cultivation of crops *and* the regulated management of vegetation in a secondary-forest fallow. The chapters in this section illustrate the diversity, sophistication and ecological importance of indigenous swidden/fallow systems. They also attest to the resilience, adaptability and potential of some swidden practices to conserve forests, maintain ecosystem functions and enhance household livelihoods in a future that will entail rapid and profound social, economic and environmental change.

It is well known that indigenous swidden systems can be compatible with forest conservation and contribute to ecological services and functions (Fox et al., 2000; Xu et al., 2009; Padock and Pinedo-Vasquez, 2010). These case studies illustrate how regulated and sophisticated management by swidden practitioners can foster forest conservation, restoration of degraded lands and maintenance of cultural diversity, while enhancing the value of traditional ecological knowledge and practice and protecting the historic rights of indigenous communities. The chapter by Huy, for example (ch. 18, this volume), documents the means by which long-fallow swidden systems maintain forest structure and soil fertility in Vietnam. Daguitan et al. (ch. 16, this volume) describe how Ifugao practices, operating at landscape scale, facilitate natural forest regeneration and watershed management in support of irrigated rice cultivation. Erni and KMSS-Loikaw (ch. 19, this volume) describe sophisticated fallow management by swidden farmers in Myanmar, which includes retaining and pruning forest trees to facilitate secondary succession and the maintenance of soil-carbon levels. Similarly, Mukul et al. (ch. 20, this volume) document how fallow management increases tree species diversity and conserves soil-carbon in the Philippines. Nakro and Konyak (ch. 17, this volume) describe how swidden

* Dr. Stephen F. Siebert is Professor Emeritus of Tropical Forest Conservation and Management, W.A. Franke College of Forestry and Conservation, University of Montana, Missoula, MT.

farmers in Nagaland manage trees on a site-specific basis to sustain a diversity of forest products, accelerate forest regeneration and minimize soil erosion. Bamba and Munandar (ch. 15, this volume) detail an inspiring effort to revive customary swidden/agroforestry practices in Kalimantan to address household livelihood needs, maintain cultural traditions, secure state recognition of indigenous tenure rights, and fight expansion of agro-industrial plantations and logging concessions.

Generations of experimentation, accumulated traditional knowledge and practice, and customary governance by swidden farmers and communities has sustained and, in some cases, enhanced soil fertility, hydrologic functions and biological diversity. In a classic paper, Connell (1978) noted that intermediate-scale ecological disturbances increase functional heterogeneity, thereby increasing opportunities for coexistence among species assemblages, which correlates with biodiversity. While Connell and many other ecologists do not consider anthropogenic disturbances 'natural', some indigenous communities have regulated and managed swidden-associated ecological disturbances for centuries (Xu et al., 2009; Balee, 2013; Siebert and Belsky, 2014). In fact, recent analysis indicates that the vast majority of all terrestrial environments have been inhabited, used and managed for at least 12,000 years (Ellis et al., 2021). These practices not only increased landscape heterogeneity and the availability of preferred plant and animal species, but also created and maintained desired edaphic and hydrologic conditions and sustained household needs and cultural traditions.

Swidden/fallow management practices also offer insights, means and pathways to restoring and maintaining ecological services and functions in the future. As documented by these chapters, fallow management practices can: 1) increase household livelihood security and reduce economic and environmental risks; 2) restore degraded lands, including those dominated by *Imperata cylindrica*, which are widespread throughout tropical Asia, and 3) assist marginalized communities to secure customary rights to historic land and forest resources. Payment for ecosystem services, such as carbon sequestration and biodiversity/forest conservation, are suggested as means to encourage more widespread adoption of sustainable-management practices. However, it is important to recognize that customary swidden systems were developed over time to address site-specific social, economic and environmental needs. While some practices might inform economic and environmental concerns elsewhere, most notably regarding household food security and climate change, they are ecologically and culturally site-specific (i.e. indigenous to particular places, times and conditions) and thus cannot be uncritically applied in other locations or social contexts. Efforts to legitimize and expand historic practices also have socio-economic and political implications, including increased influence by non-governmental organizations and/ or nation-states over forest and governance practices, and deeper integration of (neo-liberal) capitalist values and institutions in traditional practices (Fletcher and Büscher, 2017).

Indigenous swidden systems and the environments in which they once thrived face profound challenges throughout the tropical world. These case studies argue that the most significant challenges include: 1) government ignorance of, and often purposeful disregard for, indigenous rights to forests and land, 2) macroeconomic and state policies that favour privatization and short-term profit maximization through logging, mining, industrial agriculture and plantation development, and 3) cultural change and associated loss of traditional ecological knowledge and practices. To this list I would add the global embrace of neo-liberal capitalism that privileges Western perspectives and technology and values only those social, cultural and biological attributes that can be monetized. To paraphrase Einstein: not all values can be monetized and not all that is monetized has value.[1] Indigenous swidden/fallow management practices were developed and co-evolved in particular human and environmental contexts to create and maintain site-specific socio-ecological systems. As such, they reflect context-specific ways of seeing, living and relating to the world that remain only partially understood, and that have both utilitarian and intrinsic value. As these chapters illustrate, building upon historic swidden fallow-management practices can also contribute to a more productive, sustainable and just future.

References

Balee, W. (2013) *Cultural Forests of the Amazon: A Historical Ecology of People and Their Landscapes*, University of Alabama Press, Tuscaloosa, AL

Connell, J. (1978) 'Diversity in tropical rain forests and coral reefs', *Science* 199, pp.1302-1310

Ellis, E. C., Gauthier, N., Goldewijk, K., Bird, R. B., Boivin, N., Díaz, S., Fuller, D. Q., Gill, J. L., Kaplan, J. O., Kingston, N., Locke, H., McMichael, C. N. H., Ranco, D., Rick, T. C., Shaw, M. R., Stephens, L., Svenning, J-C. and Watson, J. E. M. (2021) 'People have shaped most of terrestrial nature for at least 12,000 years', Proceedings of the National Academy of Sciences of the United States 118(17), e2023483118, available at https://doi.org/10.1073/pnas.2023483118, accessed 3 January 2022

Fletcher, R. and Büscher, B. (2017) 'The PES conceit: Revisiting the relationship between payments for environmental services and neoliberal conservation', *Ecological Economics* 132, pp.224-231

Fox, J., Truong, D. M., Rambo, A. T., Tuyen, N. P., Cuc, L. T. and Leisz, S. (2000) 'Shifting cultivation: A new old paradigm for managing tropical forests', *BioScience* 50, pp.521-528

Padoch, C. and Pinedo-Vasquez, M. (2010) 'Saving slash and burn to save biodiversity', *Biotropica* 42, pp.550-552

Siebert, S. F. and Belsky, J. M. (2014) 'Historic livelihoods and land uses as ecological disturbances and their role in enhancing biodiversity: An example from Bhutan', *Biological Conservation* 177, pp.82-89

Xu, J., Lebel, L. and Sturgeon, J. (2009) 'Functional links between biodiversity, livelihoods, and culture in a Hani swidden landscape in southwest China', *Ecology and Society* 14(2)

Note

1. 'Not everything that counts can be counted, and not everything that can be counted counts – attributed to Albert Einstein.

15

DAHAS

Innovations in shifting cultivation by the Dayak of West Kalimantan to fight deforestation and climate change

*John Bamba and Aries Munandar**

Introduction

In the 1980s, land was a bountiful resource in Kalimantan. It is now in short supply and many indigenous communities live in precarious circumstances. In 1979, Sanggau district in West Kalimantan was selected as the site for the first oil-palm plantations in Indonesian Borneo (Potter, 2015). Since then, the explosive spread of both government and private estates across all five provinces of Kalimantan has replaced forests and fallows with tidily ranked and seemingly boundless landscapes of oil palm, with immense consequences for the people and the environment. The inexorable growth of plantation monocrops in Kalimantan, the disappearance of swidden fallows and the demise of shifting cultivation, along with the effects on the livelihoods of Dayak communities in particular, has been thoroughly reported in scientific literature (e.g. Dove, 1985; Potter and Lee, 1998; Colchester et al., 2006; Potter and Badcott, 2007; Gillespie, 2010; McCarthy, 2010; Fortin, 2011; Potter, 2011; Carlson et al., 2012a, 2012b; Julia and White, 2012).

In about 1985, West Kalimantan had an estimated 338,200 shifting cultivators, or about 12% of the population (Weinstock, 1989, citing RePPProT, 1987, and Department of Forestry, 1983). Indigenous swidden farming has now been pushed back to the remote boundaries of the oil palm areas, and there are real doubts that what remains will survive for long (Potter, 2015). A 2010 study found that if all concession leases then in existence were planted to oil palm, 34% of Kalimantan's lowlands outside protected areas would be devoted to this crop. Making matters worse, if logging concessions and pulp-wood plantations were added, this would set the scene for 54% of Kalimantan's lowlands being under the control of extractive land-based industries, not including mining (Potter, 2015).

---

* JOHN BAMBA was a co-founder and director of Institut Dayakologi from 1999 to 2015 and is currently chairman of the Pancur Kasih Empowerment Movement (GPPK), Pontianak, West Kalimantan; ARIES MUNANDAR is a freelance journalist living in Kubu Raya, West Kalimantan.

In the seeming hopelessness of this situation, this chapter tells of an innovative renaissance of indigenous farming by dispossessed Dayak communities who have used their indigenous knowledge to create permanently cultivated communal gardens on the fringes of remaining forest patches. This has brought a resurgence of Dayak pride and interest in self-sufficient agriculture. It is perhaps more surprising for the fact that the dispossessed farmers have been wrongly accused of causing some of the major environmental disasters that have emerged from the transformation of land use in Kalimantan, and have brought complaints from neighbouring countries.

A background of disasters

Smog has been a recurring disaster in Indonesia almost every year since 1997. In a 2015 report, the United States National Aeronautics and Space Administration stated that the 1997 event produced the worst smog ever recorded (NASA, 2015). That disaster was the result of forest fires and burning for land clearance which were concentrated in two areas of Indonesia: Sumatra and Kalimantan.

Indonesia's National Board for Disaster Management (BNPB) estimated that the fires in 1997 swept across nine million hectares of land (BNPB, 2015). The economic, social and ecological losses were incalculable. Air travel was impossible as airports were closed for day after day. Travel by road was also affected; thick smoke reduced traffic to a standstill in major cities of Kalimantan and Sumatra, the two islands mostly affected by the fires, because of poor visibility. Thousands of people suffered respiratory-tract infections due to smoke inhalation. The majority were babies, children and the elderly. Schools, colleges and universities were forced to close as students became victims of the air pollution. The government advised members of the public to reduce their outdoor activities.

Cordyline fruticosa (L.) A.Chev. [Asparagaceae]

This palm-like plant grows up to 4 m tall with broad, elongated leaves at the end of a thin trunk. The leaves can range from red to green and variegated. It is one of the most important plants related to the animist religions of Austronesians. Among the Dayak, red-leaved plants are used to ward against evil spirits and to mark boundaries. They are also used in healing rituals and at funerals.

The Center for International Forestry Research (CIFOR), based at Bogor in Indonesia, estimated that financial losses in the country's forestry and agriculture sectors resulting from the 1997-1998 forest fires amounted to more than US$8.5 billion. Tourism was reckoned to have suffered losses of $4.5 billion (CIFOR, 2015). These costs do not include those of the health sector or negative impacts caused by reductions in productivity. Similarly, it is difficult to estimate the economic costs of ecological damage and recovery.

The smog disasters seem to have become an annual event, coinciding with the arrival of every dry season, despite intensive government measures to prevent and control them. No effort is spared in land and air operations, including cloud seeding, but the fires can only be brought under control by the eventual arrival of the rains.

In 2015 there was nearly a repeat of 1997's worst-ever smog disaster when the El Niño weather event affected the islands of Indonesia and exacerbated the situation. According to the National Board for Disaster Management (BNPB), the area affected by fires in 2015 was 'only' 2.6 million hectares – far less than in 1997. Of this area, 1.74 m ha (67%) had mineral soils and 0.87 m ha (33%) was peatland. Even so, the fires were widespread and affected 16 of Indonesia's 34 provinces (BNPB, 2015). Nineteen deaths were directly attributed to smoke inhalation, and an estimated 500,000 people suffered from respiratory complaints (Kusuma, 2017).

Impacts on investment

It is hard to avoid the conclusion that the smog has resulted from uncontrolled land clearing for oil-palm plantations: the late 1990s marked the beginning of Indonesia's palm-oil boom (Figure 15-1). It was hoped that palm oil would compensate for the collapse of the country's timber industry, which was declining due to a critical shortage of raw material.

In the 1990s, world demand for crude palm oil escalated steeply. An environmental impact assessment in 1997 pointed out that global demand rose from 9.39 million tonnes in 1993 to 12.26 million tonnes just four years later (Prasetijo, 2016). The Indonesian government seized the tempting market opportunity to rescue the country from an economic crisis. In 1997, the government launched a drive to

FIGURE 15-1: An oil-palm plantation in West Kalimantan, showing the processing plant in the background. Oil-palm concessions in the province are reported to cover 4.9 million hectares.

Photo: © Douglas Sheil.

increase the area of oil-palm plantations to 5.5 million ha by the year 2000, with an estimated production of 10.6 million tonnes of palm oil by 2015 (Casson, 1999).

This policy decision was supported by the International Monetary Fund (IMF) and the World Bank. The IMF granted a US$43-billion loan to support Indonesia's agribusiness sector, and this included oil-palm plantations (Prasetijo, 2016).

Palm oil was regarded as a gift from heaven: it not only fetched high prices and became Indonesia's biggest source of tax revenue, but it was also acclaimed as the country's biggest employer at a time of job shortages and high unemployment. In 1998, more than two million people were reported to be dependent upon it economically (Casson, 1999).

Such large-scale expansion triggered unhealthy competition and massive ecological destruction. As land was in short supply, the palm-oil industry looted protection forest and conservation areas, including deep peat-soil areas. It also invaded communities' agricultural and customary lands and forests, as well as their sacred sites.

Much of West Kalimantan's land, covering a total of 14.68 million ha, has now been divided up between concession holders. According to the Civil Society Coalition for Just and Sustainable Spatial Planning (*Koalisi Masyarakat Sipil untuk Tata Ruang yang Adil dan Berkelanjutan*), oil-palm concessions in West Kalimantan cover 4.9 million ha. Around 17% of these (841,610 ha) are operating within the forest estate (Pahlevy, 2016).

The expansion of oil palm also impacts on social conditions and provokes land conflicts. Data from the National Geodatabase team recorded 165 conflicts over agricultural land in Kalimantan in 2012. Of these, 108 cases resulted from the expansion of oil-palm plantations (Munandar, 2013).

Increased emissions

It is believed that oil-palm plantations usually begin with the clearing of land by burning. This is both practical and very cheap. The head of the Data and Information Centre at *Badan Nasional Penanggulangan Bencana* (BNPB -- the National Board of Disaster Management), Dr. Sutopo Purwo Nugroho, is quoted as saying that land clearing costs only 600,000 to 800,000 rupiah per hectare when owners use fire. Other methods of clearing the land cost as much as 3.4 million rupiah per hectare (Wahyuni, 2015).

Evidence that plantation companies are burning to clear land is frequently presented by environmentalists and environmental organizations. Fires are traced from satellite images or direct field observations. Sutopo of BNPB has made such information available through his Twitter account and this has been cited by the media. He revealed that oil palm had been planted on land cleared by burning at Nyaru Menteng, in Central Kalimantan. Some companies have been prosecuted for suspected land clearing by burning, but there have been no convictions.

The international community has subjected Indonesia to scrutiny and pressure. Some neighbouring countries have complained about the smog polluting their air space. They point out that carbon dioxide released by the fires accumulates in the atmosphere as a greenhouse gas and this is a principal cause of climate change (Figure 15-2).

Research by the Center for International Forestry Research estimates that land clearance by burning has added over 1 million tonnes of carbon to Indonesian emissions, despite the government's target of reducing emissions by 29%

FIGURE 15-2: Fires in Indonesia, lit to clear land, have led to complaints of air pollution from neighbouring countries.

Photo: David Gaveau.

by 2030 (CIFOR, 2015). The issue has been the focus of heated debate at several international forums, including the Conference of the Parties to the United Nations' Framework Convention on Climate Change.

Climate change

The effects of climate change are already noticeable in West Kalimantan. Several areas have reported crop failures due to extreme weather conditions and unpredictable seasons. Major storms and floods are also becoming increasingly common.

Data analysis by a climatology station at Mempawah shows year-on-year temperature increases in West Kalimantan. Examples of these increases in average temperatures are 0.024° Celsius in Ketapang district; 0.012°C in Mempawah and Kubu Raya; and 0.001°C in Kapuas Hulu (Munandar, 2014).

Agricultural and horticultural crops are sensitive to changes in weather patterns. Studies by the International Rice Research Institute, as quoted by Widiarta (2016), have concluded that every 1°C rise in air temperature can produce an 8% to 10% fall in rice harvests. According to Tschirley (2007), crop production can fall by more than 20% if the temperature rises by more than 4°C (Tschirley, 2007, cited by the Indonesian Agricultural Research and Development Board, 2011).

Plagues of plant pests and diseases have also increased in recent years, particularly in areas that have suffered from deforestation and forest degradation. Temperature fluctuations and increases in humidity trigger explosions in certain insect populations. Conversely, food sources and natural predators are scarce as a consequence of changes in climate and vegetation.

Locusts (*Locusta migratoria*) have become one of the most serious pests attacking crops in West Kalimantan. Plagues of these insects were officially designated 'natural disasters' in Ketapang district in 1990. A single colony of locusts can multiply to thousands or even millions of individuals very rapidly, and rice, vegetables and other village crops can be destroyed in a single attack.

Controls generally focus on large-scale use of insecticides, rather than on natural methods or improving environmental quality. The control procedures tend to follow methods used by the government and the Indonesian army as long ago as 1967, when locusts invaded most of the farmland in the Marau and Jelai Hulu subdistricts of Ketapang (Bamba, 2013).

Some local communities have tried to use local knowledge to control locust attacks, such as a ceremony called *beniat*. However, these have yet to show positive results because they have not been applied systematically and some members of communities fail to observe the traditional restrictions involved.

Locust plagues are still common in Ketapang district, and similar invasions have occurred in Sintang and Melawi districts. Forests that once formed a natural barrier to pest attacks have been cleared and the land is now laid bare to exploitation and land-use changes.

Criminalization of local farmers

The 'smog season' can also create anxiety in local farming communities because they are often accused of being the main causes of the smog. Farmers are persecuted, arrested, face criminal charges and are imprisoned. Yet land clearing for indigenous farming usually starts around May, or the middle of the year. The smog often appears both before and after this (Figure 15-3).

The Police Chief of West Kalimantan issued a public statement in 2014 (Hayat, 2014) and again four years later (Irawan, 2018), in which people who used fire to clear land were threatened with 10 years in prison and fines of up to 10 billion rupiah. The police argued that the statement was based on the Indonesian Criminal Code, Environmental Law No, 32, 2009; Forestry Law No. 41, 1999; and the Plantation Law No. 39, 2014.

This formal declaration remains in force today. In practice, it is most often used against small-scale farmers and ordinary people rather than corporations. Local farmers have been intimidated by fear of arrest and their lives made more difficult by aerial water bombing to douse forest fires. Their fields are mistakenly targeted by water bombs from helicopters because they are thought to be the sites of uncontrolled fires.

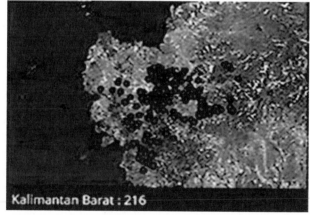

Kalimantan Barat : 216

FIGURE 15-3: An image of West Kalimantan taken by the US NOAA-18 satellite showing 216 'hotspots' on 14 September 2016.

Source: Indonesian Meteorology, Climatology and Geophysics Agency (*BMKG*), accessed via tribunepontianak.co.id.

The release of 5,000 litres of water in each 'bomb' devastates farmers' fields and creates deep craters. Apart from damage to their land, the farmers also fear for their lives, although there have been no deaths recorded from 'bomb' strikes. However, fears of prosecution or 'water bombing' have forced some farmers to cease local farming, and this has created more problems, including the threat of malnutrition and other social issues. The government has failed to propose any practical solutions.

Local knowledge

Despite the official action against it, indigenous shifting cultivators still use fire to create new fields for their crops. This burning is permitted under the law. Tree trunks and branches, plus any remaining shrubs, are burnt during land clearance. Indigenous shifting cultivation uses no chemicals and is heavily dependent on natural sources of fertilizer. The ash that is produced by the fires provides nutrients for crops and reduces soil acidity. Burning by farmers is not done randomly: the use of fire is tightly controlled by customary law to prevent it from spreading into adjacent forests or fields. The farmers must create natural fire breaks around a field before it is burnt.

These rules have been in force since ancient times. An example of this is found among the Dayak Jalai communities in the Jelai Hulu subdistrict of Ketapang. These indigenous farmers always engage dozens of people to set a fire and to control it. Customary sanctions (*sanksi adat*) are imposed on any farmer who lets a fire get out of control, so that it damages other peoples' fields or forests (Munandar, 2015).

This type of land clearance by burning does not violate official regulations. Provisions that recognize

Celosia argentea L. [Amaranthaceae]

Plumed or silver cockscomb is an important annual plant found in Dayak swiddens and *dahas*. It has deep meaning in local shifting cultivation rituals. However, in China and India it is often regarded as a troublesome weed. The flowers range from pink to light violet, and can last up to eight weeks. The seeds are tiny – more than 1500 to the gram. The flowers and leaves may be eaten as a vegetable.

'local knowledge' are included in the Environmental Protection and Management Law No. 32, 2009. While clause 1h of this law states that no-one can clear land by burning, the following clause affirms complete respect for indigenous practices. According to the explanatory notes, the interpretation of 'local knowledge' includes burning to clear a maximum of two hectares of land per family for growing local crop varieties, with a surrounding barrier to prevent the fire spreading to adjacent areas.

Most farming households in West Kalimantan cultivate only half a hectare to one hectare of land. Many have less than this and should therefore be categorized as small-scale farmers or gardeners (*petani guram*). These numbered 81,287 households in 2013, or 13.18% of the population (Central Bureau of Statistics, 2013) (Table 15-1).

At the very least, the census figures in Table 15-1 illustrate the extent of land ownership in West Kalimantan, including dry-land cultivation. They support the premise that the use of burning by indigenous communities to clear land does not break the law. They also show how the area of land cultivated by small-scale farmers has been reduced as result of massive land clearance for oil palm plantations and mining in West Kalimantan.

TABLE 15-1: Number of farming households with land and small-scale farming households by district/city, West Kalimantan, 2003 to 2013.

Location	Farming households				Small-scale farming households			
	2003	2013	Growth		2003	2013	Growth	
			Number	%			Number	%
Sambas	84,061	91,066	7,005	8.33	25,226	25,197	−29	−0.11
Bengkayang	32,940	37,310	4,370	13.27	7,462	3,497	−3,965	−53.14
Landak	54,932	60,213	5,281	9.61	3,114	1,476	−1,638	−52.6
Pontianak	29,039	26,310	−2,729	−9.4	11,259	8,203	−3,056	−27.14
Sanggau	64,016	70,021	6,005	9.38	6,151	2,963	−3,188	−51.83
Ketapang	64,283	61,875	−2,408	−3.75	16,625	8,314	−8,311	−49.99
Sintang	63,193	68,963	5,770	9.13	5,621	2,220	−3,401	−60.51
Kapuas Hulu	37,301	39,061	1,760	4.72	5,079	2,821	−2,258	−44.46
Sekadau	33,351	35,183	1,832	5.49	3,611	984	−2,627	−72.75
Melawi	32,221	32,250	29	0.09	2,720	1,024	−1,696	−62.35
Kayong Utara	14,934	16,246	1,312	8.79	2,896	2,598	−298	−10.29
Kubu Raya	59,721	62,091	2,370	3.97	14,151	13,374	−777	−5.49
Kota Pontianak	11,771	3,856	−7,915	−67.24	9,425	2,959	−6,466	−68.6
Kota Singkawang	12,720	12,450	−270	−2.12	7,235	5,657	−1,578	−21.81
West Kalimantan total	594,483	616,895	22,412	3.77	120,575	81,287	−39,288	−32.58

Source: Agriculture Census, West Kalimantan, Central Bureau of Statistics, 2013.

Community land management

As well as using fire to clear their land, indigenous shifting cultivators practise rotational cultivation: old fields are left fallow, allowing the regrowth of scrub and forest. They can be cleared again for new cultivation after at least three to five years.

This rotational land use – otherwise known as shifting cultivation – allows 'resting time' so the nutrient content and fertility of the land can be restored. It also prevents over-exploitation, which could threaten the physical, biological and chemical structure and functions of the soil.

Shifting cultivation also prevents large-scale land clearance at any particular location. Importantly, it provides a way for indigenous peoples to mitigate potentially disastrous impacts from climate change by avoiding the release of carbon into the atmosphere that is associated with land-use change.

For indigenous farmers, deciding where to clear land and establish fields is not done randomly: customary law and other considerations must first be taken into account. For example, it is forbidden to create fields on sacred sites, where there are protective forests (*hutan lindung*), or where there are sources of clean water. As well, land clearance may only be carried out at particular times.

When practised properly, indigenous farming shows that these peoples have long been capable of implementing best practices in sustainable land use. Unfortunately, their farming systems are too often blamed for environmental damage, including the smog problem. Their land-management systems are stigmatized when identified as shifting cultivation, because this title has long been wrongly burdened with negative connotations.

Permanent agroforestry plots (*pedahasan*)

For indigenous Dayak communities, fields and farms are much more than simply spaces or places to till the soil to meet their families' needs. Fields are the economic focus of their lives and the epicentre of their customs and communities. This is certainly the case for the Dayak Jalai, the people whose innovations lie at the heart of this chapter.

For these people, many traditions, customs, cultural traits, customary laws, community rituals and aspects of everyday behaviour are related to, or inspired by, farming. The *menyandam* ritual starts the farming cycle; *menimbung* is when seeds are taken out of the grain store; *baabuang hulat* is a ritual seeking protection against pests and diseases of the rice crop; *menjulang atuq* is when the harvested rice is placed in the grain store; and there are many other rituals and traditions.

The cultivation by indigenous farmers of many local varieties of rice also indirectly conserves genetic diversity. The varieties they choose depend on the growing conditions and the lie of the land. Each variety produces rice with a different texture and flavour (Munandar, 2017).

The *dahas* system

Natural-resource management, as practised by Dayak communities, provides for settled, or permanent cultivation, as well as shifting cultivation. This usually involves a plot of land on which the farmers plant rice, vegetables and rubber trees, in addition to various local fruit trees. Aspects of the system borrow from the principles of indigenous forest gardens, called *tembawang*, involving collections of fruit and nut trees, and other useful species.

In the Jalai, Kendawangan and Pesaguan Dayak communities of Ketapang district, this permanent land-use system is known as *dahas* or *pedahasan*. The fields are generally located on the edge of the forest, far away from the main settlement – so far, in fact, that farming families often build a second home there. Villagers stay in their 'field settlement' (*dahas*) during the growing season, while working the land, planting vegetables, caring for and harvesting their fruit crops, and collecting non-timber forest products such as rattan and medicinal plants from the nearby forest.

Each *dahas* has five to seven basic homes for farming families. Each has the right and the choice to cultivate land collectively or as individual families. The usual customary regulations and principles apply within the *dahas*, just as in the main settlement. Beside each 'field home', there is a *jurung* or rice store, and farming families usually keep livestock there as well, such as pigs and chickens.

There are similarities with agroforestry systems that combine elements of agriculture and forest management. However, a *dahas* does not involve logging or using timber for commercial purposes.

A member of a community may manage a *dahas* all of his or her life, and even pass it on to the next generation. A *dahas* often forms the basis for a new settlement as the community expands.

But back to the circumstances that saw these indigenous principles revive the independence and spirit of present-day Dayak communities. The customary lands

Manilkara zapota (L.) P.Royen
[Sapotaceae]

Sapodilla is a fruit tree cultivated in *dahas*. Growing to 30 m tall, it produces two crops of fruit per year. The fruit are up to 8 cm in diameter, with pale yellow to earthy brown flesh with a sweet, malty flavour. The trees die easily if the temperature falls to frosty, and the bark yields a white gummy substance traditionally used as chewing gum in its native Mexico and Central America.

of Kampung Tanjung, in the Jelai Hulu subdistrict of Ketapang district (Figure 15-4) had long been neglected and deserted by the community. For year after year, the land had been swept by fires. The people had no idea who was setting the fires, which totally destroyed all of their planted land and forest. The people had lost hope, and the whole area had become a sea of coarse grass (*ilalang - Imperata cylindrica*).

At the same time, corporate investors were eager to turn this land into oil-palm plantations and mines. They included PT Andes Sawit Mas (ASM) of the Poliplant Group and PT Kwam (Harita Group). Since then, the PT ASM concession has been taken over by Cargil (Gunui', 2018).

This pressure from extractive industries triggered conflicts over land and the community's boundaries. Everyone claimed to own land that was to be swallowed by concessions. The invasion of investors reduced the community's customary lands and agricultural areas, including the *dahas.*

However, a 70-year-old member of the Kampung Tanjung community named Daniel decided to stand up to the companies and to defend his lands against the investors' expansion. As his campaign strategy, he chose to use a peaceful approach based on local wisdom. Daniel donated eight hectares of his land to AMA-JK (Aliansi Masyaraat Adat Jalai Sekayuq and Kendawangan Siakaran – the Alliance of Indigenous Peoples of Jalai Sekayuq and Kendawangan Siakaran), so that it could become a model *dahas.* AMA-JK was the only indigenous community organization

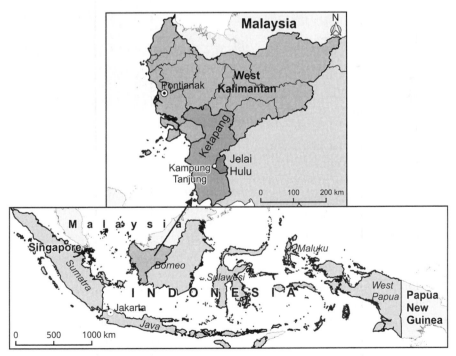

FIGURE 15-4: The site of *Dahas Pancing*, at Kampung Tanjung, Jelai Hulu subdistrict, Ketapang district, West Kalimantan province.

that had been consistently advocating indigenous peoples' rights in this area. The organization then approved the planning, management and care of the land involved in the scheme.

The scheme also attracted the support of Institut Dayakologi, a non-governmental organization based in the West Kalimantan capital, Pontianak, that is working to revitalize and support the restitution of Dayak identity and culture. The institute assumed the role of 'parent body', and the AMA-JK organization revived and re-established the *dahas* model of collective customary farming. The main aims were to protect and restore the community's land and to make the *dahas* the heart of the community's economic development.

Creating the *Dahas Pancing*

Daniel donated eight hectares of his land to AMA-JK for the following reasons: First, he was deeply frustrated by the fires that burnt his land almost every year. Second, despite the fact that his land was covered by *imperata* grass (*Imperata cylindrica*), he believed that a forest could be regrown on it by using local wisdom plus hard work and patience. Third, a *dahas* system could be effective in strengthening the local community's rights to land as well as protecting it from both commercial encroachment and forest fires, as advocated by AMA-JK and Institut Dayakologi. Fourth, as he was a member of the alliance formed by the Dayak Jalai and Kendawangan communities, he believed that AMA-JK had the capacity and resources to create a *dahas* in that area.

Soon after receiving the land, AMA-JK began planning the creation and management of *Dahas Pancing*. The project attracted the participation of people from 10 villages in the Jalai and Kendawangan areas. Several workshops and informal discussions resulted in a plan for developing the degraded land into a well-managed *dahas*. A management team consisting of some activists from AMA-JK and farmers appointed by local people's representatives was set

Mangifera indica L. [Anacardiaceae]

Mangoes are among the fruit trees being cultivated in *dahas*. They produce what is reputedly the world's most popular fruit. Although native to the Indian subcontinent, hundreds of cultivated varieties have been introduced to other warm regions. These are large trees, capable of growing to 45 m tall with a crown width of 30 m. They can produce fruit for more than 40 years and live for a century or more.

up to take charge. The team agreed to start with shifting cultivation and gradually develop the area into a *dahas* within three to five years. One of the first tasks was to create fire breaks around the area to minimize the risk of forest fires. The creation of the *Dahas Pancing* then followed a step-by-step process:

Year one: Defeating the *imperata* grass was the first big task. *Imperata cylindrica* is well-known for its fast growth, long, tough leaves that cut upon contact, and strong rhizomatous roots that make an area almost impossible to cultivate. These characteristics have led to the perception among farmers that *imperata* grass indicates infertile land that has been over-exploited and destroyed by regular fires. Clearing an area covered by *imperata* grass by cutting or burning does little more than encourage more growth. Usually, the only option is to use a glyphosate-based herbicide, but using a chemical herbicide was contrary to the values of the local culture. *Imperata* grass requires a lot of sunlight, and it can hardly survive in the shade of a well-forested area. To 'defeat' the grass, its growth would have to be weakened, and this would take time and patience. Heavy objects, such as wooden poles and tree branches, were placed on the grass to force it to bend and lie flat on the ground. Thus, the bottom layer of the grass died and decayed due to a lack of sunlight. This process was maintained over a period of time, and the grass was gradually killed. Meanwhile, as many fast-growing trees as possible were planted to block the sunlight. The entire first year was spent on these processes, and after this time, the *imperata* grass had been significantly reduced.

Year two: The plot became a shifting cultivation field, or swidden. Following local traditions, the land was cleared, burnt, and crops planted, including ritual observances of *menyandam, menimbung, baabuang hulat* and *menjulang atuq.* About half of a hectare was allocated specifically for rice. The land was cleared by cutting down the trees and clearing any remaining grass. It was left to dry for a few weeks, then it was burnt. Villagers representing the 10 village members of AMA-JK gathered to plant the rice (Figure 15-5), along with other local species of vegetables, spices, medicinal plants and flowers. A simple hut was also built near the rice field, as a shelter.

It is important to note that the procedures of shifting cultivation were employed only as a means to launch the process of creating the *dahas;* the rice crop was not an end goal. The main purpose was the gradual reforestation of the degraded area covered by *imperata* grass (Figure 15-6). The productivity

FIGURE 15-5: Villagers plant the first crop of upland rice as the degraded site at *Dahas Pancing* becomes a productive swidden.

Photo: Institut Dayakologi.

of the rice field was very low, reflecting the degraded nature of the area. Nevertheless, growing the rice crop was very important both culturally and technically, as the land moved towards the creation of a *dahas* (Figure 15-7).

Year three: The third year was focused on further management of the land. Areas were established for a farm house, a fruit garden, a rubber garden, a fish pond, animals – including pigs, ducks, chickens and grazing livestock – and a forest area. All were initiated in the third year. Every village member of AMA-JK was involved in the processes and they worked several days in each month based on an agreed schedule. AMA-JK also hired two local people to work full time at the *dahas*. They lived in the new farm house.

Year four: This was a year of intensive maintenance and enrichment planting. The results of initial planning were carefully observed and evaluated by the management team. The focus was on making sure that various plants and trees were growing as planned

FIGURE 15-6:   A woman and other farmers clear wild plants from a rice field.

Photo: Hendrikus Adam.

FIGURE 15-7:   A woman working in a rice field at *Dahas Pancing*, Kampung Tanjung.

Photo: Institut Dayakologi.

and the area was developing into a forest based on the *dahas* concept. Sick or dying plants or trees were immediately replaced with healthy new ones and wild grasses – including the last remnants of the *imperata* grass – were cleared. At this stage, wild animals, including certain birds and squirrels began to appear in the *dahas*, as various trees had grown significantly.

Year five: New vegetation, comprising a variety of trees, had grown at the *Dahas Pancing* to replace the *imperata* grass. A new environment with a vigorous forest area was subject to integrated management based on local knowledge. The *Dahas Pancing* was also generating income for AMA-JK through the sale of pigs, chickens, fish and vegetables.

More than a decade has now passed since the *Dahas Pancing* was created. The severely degraded area, formerly covered by *imperata* grass, has become a productive, well-managed agroforest where biodiversity of flora and fauna have been regained. Rubber tapping activities have begun and fruit trees have started to bear fruit. The *Dahas Pancing* has become an example of how local knowledge and wisdom can provide solutions to environmental destruction and provide sustainable incomes for local people.

Customary law

The *pedahasan* system of settled farming communities is a manifestation of local knowledge and sustainable natural-resource management. Collectivity, family bonds and social solidarity are built up between *dahas* members and this supports the principle of sustainability.

The *pedahasan* also reflect how customary communities are established and managed according to principles of collective ownership and inheritance. These principles form the structures or means of protecting and sustaining the sovereignty of customary law and the collective and private ownership of land.

Parkia speciosa Hassk. [Leguminosae]

Stink beans, as they are known, are grown throughout Southeast Asia and northeastern India and contribute to many regional cuisines. Thus, they are valuable trees in *dahas*. The trees grow to 30 m, bearing bunches of long, flat pods containing green beans, which have a strong and pervasive smell. The pods can also be eaten when young, and the mature beans dried for later use.

According to Krissusandi Gunui', from the local indigenous community organization AMA-JK, there are six major problems associated with sustaining the *dahas* system:

- The *pedahasan* system is stigmatized by the belief that it is old-fashioned and outmoded.
- Problems arise from the privatization of land through certification by groups or individuals belonging to a *dahas*.
- The government may define areas as forest estate or conservation forest, when this includes *pedahasan*.
- The expansion of land concessions and investment in rural areas, which is afflicting traditional land uses.
- Information about the *dahas* system is not included within the 'local content' component of the school curriculum.

- Advocacy and campaigning in support of the system are still minimal (Gunui', 2018).

Institut Dayakologi is taking the lead in addressing the problems facing the *dahas* system, and since 2009, it has revitalized the *dahas* by strengthening and developing the system from the inside out. It has established two model locations: *Dahas Pancing* and *Dahas Sangkuh Pinang Lakaq.* These *dahas* are both located in areas that have been affected by fires.

Community strengthening and independence

As well as strengthening the solidarity of the Jalai community, *Dahas Pancing* was a collective farming initiative that was planned to focus on the community's needs, priorities, management, efficiency and effectiveness. It has thus generated creativity and self-empowerment initiatives among the people.

The development of the *dahas* gradually stimulated community resistance and desire to protect the sovereignty of their customary rights against the onslaught of external investors (Figure 15-8). Although they were unable to save all of their customary land, the rediscovery of the *dahas* system encouraged the community to revive its indigenous practices in other locations.

Dahas Pancing, which was established on the customary lands of Kampung Tanjung between 2012 and 2013, has since become

Artocarpus heterophyllus Lam. [Moraceae]

Jackfruit are among the many tree species cultivated in the recovered land of *dahas* systems. Individual fruit on these trees can reach a weight of 55 kg, and the trees are prolific producers. The fruit are formed from the fusion of the ovaries of multiple flowers, and the succulent, edible parts are fleshy coats surrounding the seeds, which are also edible. The unripe fruit are also eaten as a meat substitute.

FIGURE 15-8: A man feeds pigs in a *dahas* at Pasir Mayang village, nearby Kampung Tanjung, in Jelai Hulu subdistrict. Establishment of this *dahas* followed the pattern created at *Dahas Pancing.*

Photo: Aries Munandar.

the pattern for the development of other collective farming areas. As these have increased in number and area, organizing the communities in support of the *dahas* system has become progressively easier. In the original case of Kampung Tanjung, the community's customary lands were better protected because the people were prepared to stand up for their customary rights. During the initial phase of the development of the *dahas* at Kampung Tanjung, the community confronted the companies that were seeking land concessions on five separate occasions to protect their land rights (Gunui', 2018).

Development of *Dahas Pancing* has also promoted some productive and sustainable enterprises. These include planting rubber trees, raising livestock such as chickens, geese and pigs, building fish ponds, setting up seedling nurseries and developing horticulture.

Rubber cultivation is now in its eighth year at Kampung Tanjung, with productive trees covering three to four hectares. Raising laying hens is another venture, with a direct connection to markets. The *dahas* raises enough pigs, geese and chickens to meet the community's needs.

Raising fish in a pond was not as successful as hoped because of problems with water supply and flooding in the wet season. Vegetable production has also decreased, mainly because efforts were diverted into planting trees in a reforestation drive.

The selective breeding of plants and livestock at the *dahas* has been directed towards increasing productivity in line with community needs. Seed selection has been directed towards improving the yields of rubber and fruit trees, such as durian and other local species. Priorities in the vegetable fields have been given to runner beans, aubergines, chillies and other local vegetables, while priorities for livestock improvement have gone to laying hens, tilapia fish (*Oreochromis niloticus*) and snakehead fish (*Channa striata*).

Durio zibethinus L. [Malvaceae]

The natural range of the durian includes Kalimantan and the *dahas* study site. Sometimes claimed to be the King of Fruits, the pungent durian evokes reactions ranging from deep appreciation to intense revulsion. It is a species propagated mainly by cloning, layering, and grafting, producing hundreds of cultivars with widespread commercial benefits for those with a milder smell or creamier flesh.

Climate change: mitigation and adaptation

The programme which began in Kampung Tanjung has restored the land and revived the spirit of the *dahas*, both of which were top priorities. It has focused on selecting and cultivating seedlings of productive trees such as durian, jackfruit, stink beans (*Parkia speciosa*), mangoes, *sawo* (sapodilla – *Manilkara zapota*) and *lengkeng* (longan – *Dimocarpus longan* subsp. *malesianus*). Timber species have also been planted.

So far, more than 2000 trees of 19 different species have been planted in the *Dahas Pancing*. Serious efforts have also gone into growing rubber trees, for their economic value. To date, 3,573 rubber trees have been planted in an area of four hectares (Gunui', 2018).

Now that many types of trees have reached a good height, the land that was once in a critical condition is covered by lush, green vegetation. The agroforest (*pedahasan*) atmosphere and identity of the area has been restored. Once again, the trees are recycling the air; producing oxygen and absorbing carbon dioxide. They are also circulating water, which protects the area from drought.

Dimocarpus longan subsp. malesianus Leenh. [Sapindaceae]

This is a relative of the better-known longan, but is better-adapted to the lowland tropics. It grows up to 30 m tall, with a bole about 30 cm in diameter and buttresses up to two metres. This is one of the many fruit trees planted in *dahas*, producing seasonal crops of succulent fruit. The timber of this species is also strong, tough and very durable.

Greenhouse gases are the generally accepted cause of global warming and climate change. The United States Agency for International Development – Indonesia Urban Water, Sanitation and Hygiene project predicts an average temperature rise of 0.8 to 1.0°C between 2020 and 2050 (USAID-IUWASH, 2013, cited by Laksono and Damayanti, 2014). Trees play an important role in reducing levels of greenhouse gases, particularly levels of carbon dioxide, which are greatly increased by land clearance. This is a small part of the contribution made by trees to life on earth. Each tree is capable of producing around 1.2 kg of oxygen per day. It could therefore be claimed that planting a tree could save the lives of at least two people, since – on average – each of us uses about 0.5 kg of oxygen a day (Febrinastri, 2015).

Carbon–absorption rates depend on the type of vegetation, the species composition, its age and its rate of growth and development, as well as the condition of the soil. According to Hiratsuka et al. (2006), the amount of carbon stored in the surface layers of soil in secondary forest that has regrown on a field after it has been burnt is between 7.5 and 55.3 tonnes per hectare (Hiratsuka et al., 2006, cited by Forestry Research and Development Board, 2010).

The last stronghold

The *Dahas Pancing* agroforestry system represents an innovative approach that is based primarily on local wisdom. It has revived the community's ancestral role in protecting the environment. Natural resources are managed, not exploited; their utilization is not destructive and their usefulness is protected for future generations.

Such local wisdom is not related solely to management techniques, but involves nurturing and passing on sentiments of communality and solidarity. Throughout modern societies, the values of ancestors are fading, and indigenous communities are not immune to the diminution of indigenous wisdom. However, the *Dahas Pancing* has had the effect of binding the community of several villages together and reviving their spirit to protect their sovereignty. Moreover, this has happened without neglecting the very real economic priorities in peoples' everyday lives.

The *Dahas Pancing* has become the green economic movement of the Dayak Jalai. Utilization of natural resources is set at a level that balances people's basic needs with the carrying capacity of the environment. Economic demands are not set above or given higher priority than ecological demands. Indeed, we could say that economic benefits are merely a kind of bonus resulting from respectful treatment of the environment.

The rebirth of *Dahas Pancing* still leaves the Dayak peoples of West Kalimantan a long way from solving all the complexities of the area's environmental problems. However, the social movement that has arisen with development of the *dahas* is a part of the solution, and creates space for shared learning. Local communities have found that such involvement frees them from feeling trapped; the victims of the capitalist drive that has engulfed their land.

The long struggle to restore the *pedahasan* agroforestry system could become a model to be replicated elsewhere. It is sorely needed, given the increasing destruction associated with the stranglehold of concessionaires on the natural environment. Replication would help the movement to become bigger and stronger, enabling it to defend indigenous sovereignty and indigenous forms of community life. At the very least, it would bolster the bargaining position of local communities vis-à-vis the outside world.

The *pedahasan* system is a simple one, but it is capable of making a big contribution to the global community, by helping to mitigate the effects of climate change. It is a difficult task, but not a hopeless one, especially if the kind of initiative seen in the *Dahas Pancing* of Kampung Tanjung can ignite the spirit of indigenous communities on a wider scale.

Acknowledgement

The authors are grateful for the work of Liz Chidley, who translated the original draft of this chapter from Bahasa Indonesia to English.

References

Agricultural Research and Development Board (2011) *Pedoman Umum Perubahan Iklim Sektor Pertanian* (General Guidelines for Climate Change in Agricultural Sector), Badan Penelitian dan Pengembangan Pertanian (Agricultural Research and Development Board), Jakarta

Bamba, J. (2013) *Dayak Jalai di Persimpangan* (Dayak Jalai at the Crossroads), Institut Dayakologi, Pontianak, West Kalimantan

BNPB (2015) *Evaluasi Bencana Asap: Tahun 1997 Lebih Parah dari 2015* (Smog Disaster Evaluation: 1997 was worse than 2015), Badan Nasional Penanggulangan Bencana (National Board for Disaster Management), Jakarta

Carlson, K. M., Curran, L. M., Ratnasari, D., Pittman, A. M., Soares-Filho, B. S., Asner, G. P., Trigg, S. N., Gaveau, D. A., Lawrence, D. and Rodrigues, H. O. (2012a) 'Committed carbon emissions, deforestation, and community land conversion from oil palm plantation expansion in West Kalimantan, Indonesia', *Proceedings of the National Academy of Sciences of the United States of America* (PNAS) 109(19), pp.7559-7564

Carlson, K. M., Curran, L. M., Asner, G. P., Pittman, A. M., Trigg, S. N. and Adeney, J. M. (2012b) 'Carbon emissions from forest conversion by Kalimantan oil palm plantations', *Nature Climate Change Letters*, published online 7 October 2012

Casson, A. (1999) *The Hesitant Boom: Indonesia Oil Palm Sub-Sector in an Era of Economic Crisis and Political Change*, Center for International Forestry Research (CIFOR), Bogor, Indonesia

Central Bureau of Statistics (2013) *Laporan Hasil Sensus Pertanian* (Report of Census Result on Agriculture), Badan Pusat Statistik Kalimantan Barat (West Kalimantan Central Bureau of Statictics), Pontianak, West Kalimantan

CIFOR (2015) *Kabar Hutan. Menghalau Asap: Sebab dan Dampak Kebakaran lahan* (Forest News. Getting Rid of Smog: Causes and Impacts of Forest Fires), Center for International Forestry Research, Bogor, available at Forestnews.cifor.org, 10 November 2015

Colchester, M., Jiwan, N., Andiko, Sirait, M., Firdaus, A. Y., Surambo, A. and Pane, H. (2006) *Promised Land. Palm Oil and Land Acquisition in Indonesia: Implications for Local Communities and Indigenous Peoples*, Forest Peoples Programme, Perkumpulan Sawit (group watching oil palm), Association for Legal Reform and Community-Based Ecology (HuMA) and the World Agroforestry Centre (ICRAF)

Department of Forestry (1983) *Data Pokok Peladang Berpindah* (Basic Data on Shifting Cultivation), Department of Forestry, Jakarta

Dove, M. R. (1985b) 'Plantation development in West Kalimantan, I: Extant population/labour balances and II: The Perceptions of the Indigenous Population' *Borneo Research Bulletin* 17, pp.95-105 and 18, pp.3-27

Febrinastri, N. (2015) '*Tanam 1 Pohon Selamatkan 2 Nyawa* (Planting 1 Tree is Saving 2 Souls)', *beritasatu.com* 12 April 2015, available at https://www.beritasatu.com/lingkungan/264755/tanam-1-pohon-selamatkan-2-nyawa, accessed 6 July 2020

Forestry Research and Development Board (2010) *Cadangan Karbon pada Berbagai Tipe Hutan, dan Jenis Tanaman di Indonesia* (Carbon Reserves in Various Types of Forests and Plants in Indonesia), Tim Perubahan Badan Litbang Kehutanan, (Team of the Forestry Research and Development Board), Jakarta

Fortin, C. J. (2011) 'The biofuel boom and Indonesia's oil palm industry: The twin processes of peasant dispossession and adverse incorporation in West Kalimantan', paper presented at an International Conference on Global Land Grabbing, 6-8 April 2011, Institute of Development Studies, University of Sussex, UK

Gillespie, P. A. (2010) 'Politics, power and participation: A political economy of oil palm in the Sanggau district of West Kalimantan,' PhD dissertation, Australian National University, Canberra

Gunui', K. (2018) *Napak Tilas Dahas Pancing dan Dinamikanya* (Flashback to *Dahas Pancing* and its Dynamics), Institut Dayakologi, Pontianak (unpublished)

Hayat, N. (2014) '*Polda Kalbar Keluarkan Maklumat Penanggulangan Pembakaran Hutan*' (West Kalimantan Police Issue Notice on Overcoming Forest Fires), Antaranews.com, 24 February 2014, available at https://kalbar.antaranews.com/berita/320583/polda-kalbar-keluarkan-maklumat-penanggulangan-pembakaran-hutan, accessed 5 July 2020 (Indonesian language)

Irawan, Y. K. (2018) '*Maklumat Kapolda Kalbar: Pelaku Pembakaran Lahan Didenda Rp 10 Miliar* (Notice of West Kalimantan Police Chief: Perpetrators of Land Burning Fined Rp 10 Million), *kompas.com*, available at https://regional.kompas.com/read/2018/02/23/13185111/maklumat-kapolda-kalbar-pelaku-pembakaran-lahan-didenda-rp-10-miliar, accessed 17 August 2019 (Indonesian language)

Julia and White, B. (2012) 'Gendered experiences of dispossession: Oil palm expansion in a Dayak Hibun community in West Kalimantan', *Journal of Peasant Studies* 39(3-4), pp.995-1016

Kusuma, H. (2017) '*Sri Mulyani: Sawit Ada di Dimensi Negatif Karena Kebakaran Hutan*' (Sri Mulyani: There are Negative Dimensions of Palm Oil Due to Forest Fires), *Detikfinance*, 2 February 2017, available at https://finance.detik.com/industri/d-3412185/sri-mulyani-sawit-ada-di-dimensi-negatif-karena-kebakaran-hutan, accessed 5 July 2020 (Indonesian language)

Laksono, B. A. and Damayanti, A. (2014) *Analisis Kecukupan Jumlah Vegetasi dalam Menyerap Karbon Monoksida dari Aktivitas Kendaraan Bermotor di Jalan Ahmad Yani Surabaya* (Analysis of Vegetation Sufficiency in Absorbing Carbon Monoxide from Motor Vehicles in Ahmad Yani Street, Surabaya), Sepuluh Nopember Institute of Technology, Surabaya, Indonesia

McCarthy, J. F. (2010) 'Processes of inclusion and adverse incorporation: Oil palm and agrarian change in Sumatra, Indonesia', *Journal of Peasant Studies* 37 (4), pp.821-850

Munandar, A. (2013) '*Terjerat Kemudahan Izin Baru* (Trapped by New Licence Incentives)', *Media Indonesia*, 10 June 2013, Jakarta

Munandar, A. (2014) '*Tanaman Padi, cuma Dapat Benih* (Planting Rice to Harvest Seeds)', Media Indonesia, 15 October 2014, Jakarta

Munandar, A. (2015) '*Tanam Padi di Dayak Jalai* (Planting Rice in Dayak Jalai)', *Media Indonesia*, 3 May 2015, Jakarta

Munandar, A. (2017) '*Tuah Rumpun Padi di Ladang Dayun* (Paddy Fortunes at Dayun's Ricefield)', *Media Indonesia*, 19 February 2017, Jakarta

NASA (2015) *Indonesian Smog the Worst on Record*, US National Aeronautics and Space Administration, News Office, DW, 2 October 2015.

Pahlevy, A. (2016) '*Korsup KPK di Kalimantan Barat, Akankah Masalah Perkebunan Sawit Terselesaikan?* (Coordinator and Supervisor of Corruption Eradication Commission in West Kalimantan: Will the Oil Palm Plantation Problem be Resolved?)', *mongabay.co.id*, 9 April 2016, Pontianak, available at https://www.mongabay.co.id/2016/04/09/korsup-kpk-di-kalimantan-barat-akankah-masalah-perkebunan-sawit-terselesaikan/, accessed 6 July 2020, Indonesian language

Potter, L. (2011) 'Agrarian transitions in Kalimantan: Characteristics, limitations and accommodations', in R. de Koninck, S. Bernard and J-F. Bissonnette (eds) *Borneo Transformed: Agricultural Expansion on the Southeast Asian Frontier,* NUS Press, Singapore, pp.152-202

Potter, L. (2015) 'Where are the swidden fallows now? An overview of oil palm and Dayak agriculture across Kalimantan, with case studies from Sanggau, in West Kalimantan', in M. F. Cairns (ed.) *Shifting Cultivation and Environmental Change: Indigenous People,* Agriculture and Forest Conservation, Earthscan from Routledge, London, pp.742-769

Potter, L. and Lee, J. (1998) *Tree planting in Indonesia: Trends, impacts and directions*, Occasional Paper No 18, Centre for International Forestry Research (CIFOR), Bogor, Indonesia

Potter, L. and Badcock, S. (2007) 'Can Indonesia's complex agroforests survive globalisation and decentralisation? Sanggau District, West Kalimantan' in J. Connell and E. Waddell (eds) *Environment, Development and Change in Rural Asia-Pacific: Between Local and Global*, Pacific Rim Geographies, Routledge, London and New York, pp.167-185

Prasetijo, A. (2016) '*Mengapa Sawit Begitu Perkasa* (Why Palm Oil is so Strong)', *Etnobudaya.net* 10 May 2016, available at https://etnobudaya.net/2016/05/10/mengapa-sawit-begitu-perkasa/, accessed 5 July 2020 (Indonesian language)

RePPProT (1987) *Review of Phase I Results, West Kalimantan*, Regional Physical Planning Programme for Transmigration (RePPProT), Land Resources Development Centre, Overseas Development Administration (ODA), Foreign and Commonwealth Office, London, and Department of Transmigration, Government of Indonesia, Jakarta

Wahyuni, T. (2015) '*BNPB Bongkar Motif dan Modus Kebakaran Hutan dan Lahan* (The National Board for Disaster Management (BNPB) reveals the Motives and Modus of Forest Fires), *CNN Indonesia.com*, 29 July 2015, available at https://www.cnnindonesia.com/nasion al/20150729182700-20-68935/bnpb-bongkar-motif-dan-modus-kebakaran-hutan-dan-lahan, accessed 5 July 2020 (Indonesian language)

Weinstock, J. A. (1989) 'Study on shifting cultivation in Indonesia', phase 1 report for FAO project 'UTF/INS/065/INS', unpublished background paper accompanying *Situation and Outlook of the Forestry Sector in Indonesia* (5 vols) Food and Agricultural Organization of the United Nations and the Government of Indonesia, Jakarta

Widiarta, I. N. (2016) *Teknologi Pengelolaan Tanaman Pangan dalam Beradaptasi terhadap Perubahan Iklim pada Lahan Sawah* (Food Crop Management Technology of Paddy Field Adaptive to Climate Change), Center for Research and Development of Food Crops, Bogor

16

THE *MUYONG* SYSTEM

Assisting forest regrowth to protect water supplies and food security

*Florence Daguitan, Robert T. Ngidlo and Irish P. Baguilat**

Introduction

Using forests to capture and store water, as well as supporting agroforestry systems, can significantly contribute to food security for highland communities. Alongside this idealized role for forests, there is another reality, and a significant threat: a lack of food can lead to deforestation as farmers try to grow more food on increasingly limited land resources, setting off a vicious cycle of degradation that steeply reduces food security. Care and protection of forests by indigenous people, according to age-old traditional knowledge, has therefore become a subject for increasing scientific attention. One such group of forest farmers is the Ifugao, in the Central Cordillera mountain ranges of northern Luzon, in the Philippines, who, from ancient times, have developed a unique way of growing and tending to forests. It is known as the *muyong,* a name from the local language meaning forest or woodlot. *Muyong* are

Professor Robert T. Ngidlo passed away suddenly in 2019, aged 63, during the development of this chapter. His peers, colleagues and students at Ifugao State University at Lamut, Ifugao, have recognized his significant contribution to forestry education and research. Dr. Ngidlo served as both Dean of the university's Institute of Forestry and as director for research and development. He also authored several publications on Ifugao indigenous forest-management systems, lauding the ingenuity and resilience of the Ifugao people. Prior to joining the university, he was involved in several watershed projects in Ifugao province. He earned his PhD in Forestry at the University of the Philippines, Los Baños. – Irish P. Baguilat.

* FLORENCE DAGUITAN, Indigenous Peoples' International Centre for Policy Research and Education (Tebtebba Foundation), Baguio, Benguet, Philippines; PROFESSOR ROBERT T. NGIDLO, Professor of Silviculture, Department of Forestry, Ifugao State College of Agriculture and Forestry, Nayon, Lamut, Ifugao, Philippines; and IRISH P. BAGUILAT, Program Manager, Food Security and Nutrition, Philippine Country Program, International Institute of Rural Reconstruction, Cavite, Philippines.

privately owned, and are managed according to a system that is deeply deeply ingrained in the culture of the Ifugao people, and which has been recognized internationally as ideal.

The *muyong* system can be viewed from different perspectives, either as a forest conservation strategy, a watershed rehabilitation technique, a farming system or a strategy of assisted natural regeneration (ANR). The Ifugao were successfully practising ANR long before its recognition in the forestry sector as a strategy for forest regeneration. The system is living proof of Ifugao knowledge of silviculture, agroforestry, horticulture and soil and water conservation. It involves an intricate web of relationships between the Ifugao and their natural resources, extending even into their spirit world (Ngidlo, 1998), which sustain the conservation of the physical environment and different ecosystems. The system is strengthened by customary laws and cultural beliefs that are mostly expressed as taboos, prescribing 'no harm' to the natural world.

In 2009, a study of the *muyong* system was initiated in Hungduan municipality, in Ifugao – a province with the same name as its people (Figure 16-1). Its aim was to document the establishment, management and maintenance of the *muyong*; to study their influence on local food and water security; and to make recommendations for support or enhancement of the system.

It was intended to be a research project involving community action, but various limitations led to it becoming an exercise of documenting the community's indigenous knowledge, supported by a literature review and *in situ* studies of the *muyong* system in various parts of the province. It culminated in key-informant interviews and observations in January and February, 2020, which updated the system's recent condition. This chapter draws upon the experiences, knowledge and data recorded in that study.

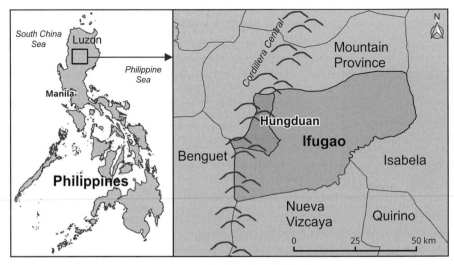

FIGURE 16-1: The location of Hungduan municipality in the central Cordillera mountain range, north Luzon, Philippines.

The study site

The location is one of the rugged and precipitous sites of the famed rice terraces of the Philippine Cordilleras, declared a World Heritage site by the United Nations Educational, Scientific and Cultural Organization (UNESCO) in 1995. The subsistence livelihood of the Ifugao people is dependent upon rice grown in the terraces. Alongside this system is shifting cultivation, providing a safety net if one or the other system fails. The terraces and groups of *muyong* are located at high elevations of 500 to 1500 metres above sea level (masl), on slopes of 50% and more, providing a wide range of forest products and ecosystem services and, essentially, to supply life-giving water to the terraces. In some parts of Ifugao, the *muyong* can be found higher than the rice terraces, at 1600 to 2000 masl.[1]

Hungduan is one of 11 municipalities of Ifugao province.[1] It has a total land area of 26,030 hectares. It is peopled by the Tuwali, one of the three major ethnolinguistic groups in the province that are collectively known as Ifugao. Hungduan's population in 2015 was 9400, living in 1641 households. The municipality is characterized by rugged terrain, steep river valleys and highland forests. It has five of the 10 highest mountains in the Philippines.

The name of the province is derived from the term *i-pugo*, meaning 'people of the earth'. It now refers to the province, the people and their languages. Aside from their spectacular rice terraces, the Ifugao people are also known for their woodcarvings – which are of export quality – and their forests, especially the privately-owned secondary forests or forest gardens known as *muyong*, which exist in nine of the province's 11 municipalities.

The Ifugao landscape is made up of eight partially interdependent components, based on elevation, terrain and vegetation. These include the *muyong* (private forests), *inalahan* (communal forests), *payoh* (rice terraces), *habal* or *uma* (swiddens), *boble* (settlements), *magulun* (communal grasslands, mainly covered with *cogon* grass (*Imperata cylindrica*)), *mabilau* (communal lands covered with cane-like grasses (*Miscanthus* sp), and *wael* or *wangwang* (creeks and rivers). Most of these components, and the role they play in maintaining the watershed, can be seen in Figure 16-2. Each of the land components has its own distinct biodiversity.

FIGURE 16-2: A schematic showing some of the principal features of an Ifugao lanscape, and how various components interconnect. The areas labelled 'woodlots' are *muyong*, which are interspersed within the rice-terraces landscape.

Source: Adapted from Castonguay et al. (2016).

Establishment and ownership of *muyong*

Muyong are privately owned patches of forest that are managed by individuals, families, or clans according to traditional practices. Covering between half a hectare and three hectares of land, they are established primarily as a source of firewood, construction materials, wooden farm tools, food, medicines and, most importantly, water.

The subsistence economy in Ifugao revolves around the production of rice in terraced paddies (Klock and Tindungan, 1995), which is highly dependent upon stable supplies of water. The *muyong* are a major component of the overall agricultural-production system and serve as the primary 'recharge' zone. Water flowing out of the *muyong*, located at the upper fringe of the rice terraces, dictates the overall physical soundness of terrace cultivation and the condition of the whole watershed unit.

The *muyong* system is the result of a learning process based on interaction between the Ifugao people and their environment, which is manifest in their profound appreciation for the multiple values of forests. The decision to develop a *muyong* usually comes from the male head of a family. It may be established in newly growing areas of forest, in shifting cultivation fallows, or above or adjacent to areas of rice terraces. If a principal reason is to support rice growing in terraces, *muyong* may be established where there are natural springs, but this is not an essential criterion for selecting a *muyong* site, as the people are also aware that a forest serves to 'recharge' water resources. They may also be established after an area has been cultivated as a swidden. After several seasons of cultivation, the swidden is left fallow and a rich biodiversity of seeds invade the area on the wind, along with wildlife. The process of natural succession may take 20 years or more, depending on the condition of the soil. However, the owner may accelerate the process by planting preferred tree species or by transplanting wildlings from nearby forest and removing undesirable weeds. In areas where stray and wild animals threaten the *muyong*, fences are built.

Management of the *muyong* is done with overriding concern for conserving water, rice production and the stability of the rice terraces. Thus, *muyong* have become a storehouse of both flora and fauna (Ngidlo, 1998). A study by Rondolo (2001) involved creating plots measuring 25 m by 25 m in the centre of 67 *muyong*. A total of 264 plant species, many of them indigenous, were found in the plots. They belonged to 71 plant families. Euphorbiaceae was the most dominant family, followed by Moraceae, Meliaceae, Leguminosae, Poaceae, Anacardiaceae and Rubiaceae. Of the 264 species, 234 were considered useful – with many having more than one use (Table 16-1). The number of species per individual woodlot ranged from 13 to 47, with an average of 30 species. Most of them were endemic to the region (Rondolo, 2001).

While *muyong* are often established near the rice terraces or in fallowed swidden fields, in some cases, the house or houses of the owners are within the *muyong*. Such an arrangement facilitates the allocation of labour and time for maintenance and enhancement of the *muyong*. The owner must continue to remove unwanted plants, on the basis of his judgment of which species may be useful. Shrubs, herbs, vines

TABLE 16-1: Important trees in Ifugao *muyong* and their uses.

Ifugao name	Common Name	Scientific Name	Uses
Alimit	Hagimit	*Ficus minahassae*	Indicator of water/stores water
Anablon	Amublit/Balunti	*Macaranga dipterocarpifolia*	Firewood; fast-growing endemic tree, the seeds of which are sown in swiddens, to grow later in the fallow.
Arnoh/Arnus	Japanese alder	*Alnus japonica*	Wood carving; firewood
Bangtinon		*Toona calantas*	Wood carving
Bayakot	Daha	*Macaranga caudatifolia*	Firewood
Bitaor	Pamintaogan	*Calophyllum soulattri*	Firewood
Bulon	Bili tree	*Garcinia vidalii*	Timber for house construction
Buluh	Copper leaf	*Acalypha angatensis*	Leaves used to hasten fruit ripening
Buwa	Areca palm	*Areca catechu*	Betel nuts for chewing
Gutmo	Katmo	*Vaccinium whitfordii*	Firewood, charcoal for blacksmith
Halinghingon	Japanese eurya	*Eurya japonica*	Firewood
Hawili	Kalumbaya	*Elaeocarpus bontocensis*	Timber for housing
Palayon	Acorn	*Quercus subsericea*	Timber for house construction
Piwis	Dungarug	*Ficus ribes*	Water-bearing trees
Tabangawen	Bani/Itangan	*Weinmannia luzoniensis*	Firewood
Talanak	Kalaingai	*Astronia cumingiana*	Firewood
Tikom/Ti-om	Kalamansanai	*Neonauclea calycina*	Timber for house construction
Tipanglan	Giant fern	*Cyathea sp.*	Ornamental, for posts
Towol/Tower	Tuai tree	*Bischofia javanica*	Timber for house construction
Umog	Ayusan	*Clethra tomentella*	Firewood, charcoal for blacksmith

Notes: Sources of most of these common and botanical names: Rabena et al. (2015). Uses described in key-informant interviews.

or other plants that are useful are left to grow. Borders with adjacent *muyong* are regularly cleared of vegetation to demarcate boundaries and prevent conflicts. Trees or other plants that are absent, but deemed to be important and useful, are taken from other *muyong* or from the natural forest and are transplanted in the *muyong*. Some trees are regularly pruned of their branches, and the prunings used as firewood. When they reach harvestable size, trees may be cut for use as round posts for houses or as timber for walls and flooring.

Labour invested in developing a *muyong* is a key factor in claiming ownership. An entire community will recognize the ownership of a family that maintains, enriches and constantly cares for its woodlot. The developers and owners do not have title deeds to prove their ownership, but their continual maintenance, enrichment and care are proof enough for the community. However, while the *muyong* are recognized as privately-owned or owned by a clan, the owners usually share the resources with other villagers – as long as they seek permission. Resources shared may include saplings or remnants of harvested trees that are not needed by the owner. Community members who do not own a *muyong* are also allowed to cut saplings for firewood or harvest whole trees, with the owner's permission. When harvestable trees are available, requests to take two or three of them are usually granted. This enables the conservation of main communal watershed forests. Hence, while they are regarded as private property, to a large extent *muyong* contribute to the common good.

FIGURE 16-3: A mature *muyong* supplies water to downslope rice paddies.

Photo: Florence Daguitan.

The *muyong* are therefore both productive and protective. As production sites, local people derive their fruit and medicinal plants from the *muyong* (Hayama 2003), along with raw materials for construction and wood carving. During heavy rainfall, water is stored in the soil mantle, aided by the canopy and roots of the trees. The water eventually finds its way to the slopes in the form of springs, and these become a permanent source of water for irrigation and drinking and provide vital water and nutrients for the rice terraces (Figure 16-3). Their position along the slopes allows for the optimal use of water and nutrients and minimizes erosion despite

Ficus minahassae (Teijsm. & Vriese) Miq. [Moraceae]

This widely spreading tree grows in *muyong* at elevations up to 1350 masl. The Ifugao believe that it indicates, or stores, water. This tree depends on a highly-specialized wasp species for pollination. The edible fruit develop in spherical bunches on long aerial roots hanging from the trunk and major branches. In local folk medicine, the leaves can be applied to the body as an antirheumatic treatment.

the steepness of the terrain. While being the primary production area for rice, the terraces also harbour a rich aquatic biodiversity, including fish, shellfish, molluscs, insects and other species. To sustain this aquatic population, the rice terraces need to be inundated the whole year round. The *muyong* make a significant contribution to the continuous availability of water. They stabilize hydrological conditions and sustain the microclimate in their local area (Sajise, 1979). Therefore, a synergy exists between forest cover, water output and rice production. The larger the area of forest cover, the greater the output of water and the larger the rice harvest. The biodiversity supported by the main components of the Ifugao landscape is detailed in Table 16-2.

The Ifugao as farmers and agroforesters

Butic and Ngidlo (2002) reported that in the past, the Ifugao migrated to areas where forests existed and from there began to transform the landscape into other productive uses. It was only recently that they adopted tree planting in sparsely vegetated woodlots as a measure to restore depleted wood cover. However, this does not deny the wealth of indigenous knowledge employed by *muyong* owners in the management of their forests. Forest protection is a common traditional concern of all villagers in Ifugao (Figure 16-4). In traditional tourism areas nearby Hungduan, hamlets of 10 to 30 families have reportedly banded together to protect whole hillsides of community forest and *muyong* to ensure that their rice terraces have water in perpetuity. Intrusion in *muyong* plots is dealt with severely. A person caught cutting trees without permission is fined and required to pay the equivalent value of the trees in pigs and chickens. Until recently, failure to pay such fines led to the possibility of banishment or death (Butic and Ngidlo, 2002).

TABLE 16-2: Key species and associated biodiversity found in various components of the Ifugao landscape.

Landscape component	Key species	Associated biodiversity
Muyong (private woodlots)	264 plant species, 234 of which are being used by the community.	10 varieties of climbing rattan 45 species of medicinal plants 41 bird species
Inalahan (communal forests)	Assumed to have only about 200 species due to commercial timber extraction in relatively accessible public forests.	6 indigenous mammals, 2 of them endemic
Payo (rice terraces)	10 or more varieties of rice 4 species of fish 7 species of edible molluscs	Crabs, crayfish, frogs and insects that keep the paddy terraces active year round
Wael/Wangwang (streams and rivers)	Populated by a broad range of fish and molluscs of the same species as those found in the rice terraces	Larger fish such as eels, as well as monitor lizards and other vertebrates

The Ifugao are considered to be traditional practitioners of agroforestry. They adopted agroforestry in woodlots and multiple cropping in shifting cultivation fields as an economic insurance in case of crop failure in the rice terraces. They have successfully been using assisted natural regeneration (ANR) in their *muyongs*, without professional intervention, for many years (Butic and Ngidlo, 2002). Implicit in the application of ANR is an array of silvicultural activities including thinning, cleaning, pruning and salvage cutting, to enhance the growth and development of natural stands.

FIGURE 16-4: Members of the Hungduan community at work in the heart of a mature *muyong*.

Photo: Yolanda W. Buyago.

Generally, harvesting of timber crops is highly selective and is based on the *muyong* owner's extensive knowledge of the various tree species and their use. Harvesting trees is seasonal in nature, or as needed. During the summer months, owners select and cut trees, then split, dry and store them as firewood to be used during the cold monsoon months. *Muyong* owners follow an efficient system of timber utilization referred to as whole-tree harvesting. Roots and buttresses are often excavated along with a few feet of remaining trunks, which may have been cut to length for use as vertical posts to support single-roomed houses; branches are cut to length for general use, and small branches and twigs gathered and bundled for firewood or fencing. Only the leaves are left in the forest to decompose. *Muyong* owners can name any tree and its specific uses, as well as when it can be harvested.

Investigating the *muyong* system

As mentioned earlier, a six-months-long study of the *muyong* system was launched in 2009, to investigate the establishment, management and maintenance of the *muyong*; to document the customary laws governing them; to study the influence of *muyong* on local food and water security; and to make recommendations for support or enhancement of the system. It soon became obvious that, despite the diligence of the owners and the fiercely protective attitude of Ifugao towards their woodlots, the *muyong* had long been under threat in a landscape where forest exploitation was rampant, populations and household needs were increasing and economic considerations were turning the tide of subsistence towards the cash economy.

The *muyong* system is regarded as a traditional forest-management practice, developed as a result of a learning process involving many years of human interaction with the environment. Most *muyong* in Hungduan municipality are joined to the rice terraces, and management of both the *muyong* and the terraces is not only interconnected, but also remains the sole responsibility of individual owners, familes, or clans.

There is no formal organization binding the Ifugao; they generally attribute value to their environment on the basis of their long-standing cultural ways and practices (Butic and Ngidlo, 2002). Nevertheless, while benefits from the *muyong* flow mainly to the owners, establishment and management of the woodlots are still activities with major community value. It is done with an overriding concern for conserving water and rice production and bringing stability to the rice terraces. There is also an informal arrangement for *muyong* owners to help distressed neighbours during the performance of rituals or during times of emergency. So

Macaranga dipterocarpifolia Merr. [Euphorbiaceae]

A prominent link between shifting cultivation and the *muyong*, the seeds of this fast-growing tree are commonly planted in swiddens, to develop as saplings within the crop, later to mature in the fallow phase. It is at this time that swidden fallows can be developed as *muyong*, and the trees are then nurtured for their use as firewood.

while maintenance of a *muyong* is the responsibility of the individual, family or clan that owns it, the sustainability of the system is a collective responsibility of the whole community, whose members subject themselves to various customary laws, including the following:

- When a *muyong* is owned by a clan, any member of that clan is free to cut trees, as long as the other members of the clan are informed, and only mature and harvestable trees are taken. Non-owners can also cut trees with the permission of the *muyong* owner, but the owner will identify the trees and limit the number that can be cut.
- Community members who do not own a *muyong* are allowed to cut trees for firewood, provided they seek permission. A person who cuts a sapling for fuel without permission may be subjected to verbal reproach, but anyone who cuts a full-grown tree without permission will be treated as a thief and ordered to pay for any damage done.
- Intrusion into *muyong* is dealt with severely. A person caught cutting trees without permission is fined and required to pay the equivalent value of the trees in pigs and chickens.

- If swiddens are opened near a *muyong*, the shifting cultivators must build fire breaks before burning slashed vegetation to prevent fires from spreading into the *muyong*. This is a non-negotiable requirement.
- When a tree is cut for timber, it is felled so as not to damage other growing plants. For every tree that is cut, a minimum of two are planted to replace it.
- Cutting big trees that are close to springs is forbidden, in the belief that they are the homes of unseen spirits.
- Some settlements are in the middle of *muyong* clusters. Not only are houses built within a *muyong*, but the forefathers of the residents are also buried there, so the *muyong* is consequently recognized as an abode of ancestral spirits and treated as a sacred site.
- Ifugao who own a *muyong* usually share the bounty by allowing fellow villagers to gather the remnants of harvested trees for firewood. It is common practice among *muyong* owners to allow poorer members of their community to harvest one or two trees for house construction, after permission has been given.

Key informants recall a time when entire communities held strong respect for the customary laws and management procedures related to the *muyong*, and followed them closely. Under these conditions, the maintenance and protection of the *muyong* system involved the whole community in a collective responsibility (Jang and Salcedo, 2013). Such cultural norms recognized poor behaviour in the past and provided the necessary 'plugs' to prevent a recurrence of past abuses and conflicts in the use of natural resources. Nowadays, communities still recognize the crucial role of *muyong* in protecting the forest, stabilizing water supplies and conserving biodiversity, thus lessening pressures on primary vegetation in natural forests. However, social and economic pressures pose a grave threat to the survival of the system. Among the early problems to beset the stability of the *muyong* system was a steady increase in population, which resulted, among other things, in expansion of settlements and a boom in the woodcarving industry. Both resulted in the clearing or partial clearing of *muyong*.

Woodcarving

Wood for carving is one of the major raw materials extracted from the *muyong*. Woodcarving has long been a major cultural and creative activity of the Ifugao. Many of their carvings are said to closely reflect their relationship with nature and with their gods, and there was a time when their carving was done only as gifts for their families. The advent of the cash economy meant that the number of trees cut for carving increased with rising market demand, but government restrictions eventually led to a significant decrease in the carving trade.

During the boom years, in the 1960s and 1970s (Sajor, 1999), large numbers of carvings were transported to Manila and other cities and even reached international markets. Raw materials were initially drawn from *muyong* and communal forests. However, supplies of wood dwindled with the introduction of mechanized tools such as chainsaws and the forests of the area were directly threatened. This led to government intervention. Wood carvings and other forest products were confiscated in large quantities at highway checkpoints set up by the Department of Environment and Natural Resources, even when the producers had permits from their local municipalities or the wood came from their own *muyong*. In 1996, the department introduced the *Muyong* Resource Permit, under laws prohibiting the cutting of trees, even within privately owned *muyong*. The reason, the department said, was its recognition of 'the biological and physical significance of *muyong* for the life sustenance [of the Ifugao] and the [desire to] manage, protect and conserve these areas against destructive human exploitation.' For a fee of 792 pesos

Alnus japonica (Thunb.) Steud. [Betulaceae]

This fast-growing tree is grown in *muyong* as a source of wood to be used in woodcarving or as firewood. It also benefits surrounding plants because bacteria in nodules on its extensive lateral root system fix nitrogen from the atmosphere. It is often used to stabilize slopes and soil in landslides. The bark and leaves are used locally to treat cancer, gastric disorders, and hepatitis.

(about US$30.27), an applicant offering proof of *muyong* ownership could obtain a *Muyong* Resource Permit, allowing all carved wood to be transported without fear of confiscation. However, as suitable wood supplies declined further, carvers went further afield, to buy wood from localities not involved in the woodcarving industry, and used their *Muyong* Resource Permits to clear the way for its transportation. This stopped when international prices for wood carvings plummeted. Nowadays, the local woodcarving industry continues, but it caters to a smaller market. Wood for carving is still extracted from *muyong* and community forests, but in sustainable volumes.

Land-use change

Another problem is land-use change. The richness of plant diversity in the *muyong* has long played an important role in conserving plant resources. However, according to Rondolo (2001) the *muyong* are increasingly being converted into other land uses as their owners seek cash income. In her 2001 study, Rondolo found that almost all of the 67 *muyong* that she studied contained commercial plantings of coffee (88%), bananas (66%) and citrus trees (49%). Moreover, seven *muyong* were being cleared for residential development to accommodate a growing upland population.

At the height of the coffee boom in the 1980s, *muyong* owners planted vast amounts of coffee, and many Ifugao families found a new affluence from the sale of coffee beans. However, a decline in the coffee industry in the years before the new millennium led many of them to abandon coffee and replace it with other fruit-bearing trees. Prominent among these was *santol* (*Sandoricum koetjape*) and other introduced species that significantly diminished the number of forest-tree species. The *muyong* of today resemble an agroforestry system made up of forest trees (80%), fruit trees (10%) and bananas (5%), with rattan, betel nut and citrus making up the rest. The integration of commercial tree crops and herbs into *muyong* occurred at the same time as bananas, taro and *cadios* (*Cajanus cajan*) were introduced into the Ifugao shifting cultivation system. Edible rattan (*Calamus manillensis*) is now included in almost all woodlots. Rattan also provides poles and canes for handicrafts, while betel palm (*Areca catechu*) and *ikmo* (*Piper betle*) are cultivated in *muyong* to cater for the habit of betel-nut chewing, which is popular among the Ifugao people and is used in some rituals. Other plants include wild fruit berries, guavas, mountain tea and herbal medicines.

Changes to the character of the *muyong*, as repositories of local plant diversity, are also being hastened by 'enrichment planting', introduced with the intention of rejuvenating depleted *muyong* areas. While many *muyong* owners would prefer to use indigenous tree species, seedlings for these species are difficult to find, so they are left with no option but to use exotic fast-growing species like *Gmelina arborea*, *Swietenia macrophylla* and *Cassia spectabilis*, all of which are promoted by the government. These offer the benefits of quick turnaround and profits (Butic and Ngidlo, 2002), but

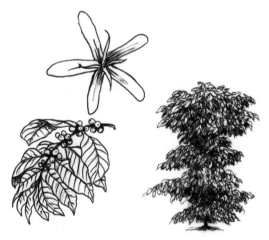

Coffea arabica L. [Rubiaceae]

A 'coffee boom' in the 1980s saw large areas of *muyong* planted to *Coffea arabica*. As long ago as the 16th century, Ethiopia, the country of its origin, saw a world market for coffee developing from plantations in Yemen, across the Red Sea, and banned the export of plants and seeds. But a group of Dutchmen smuggled a pocketful of seeds to Indonesia, and from this meagre genetic beginning grew the world's huge coffee crop. It is believed that nearly 95% of *Coffea arabica's* genetic diversity remains in Ethiopia.

they threaten the biodiversity of the *muyong*. It has been widely observed that there is no understorey growth beneath *Gmelina* trees. These species provide excellent raw materials for the wood carving and furniture industries. However, the invasive impact of alien tree species is a major concern (Joshi et al., 2000), and further studies are needed to determine their impacts on indigenous species. Butic and Ngidlo (2002) have recommended research into the mass propagation of *muyong* species for future reforestation.

Butic and Ngidlo (2002) also reported instances of inappropriate approaches being made to *muyong* owners by development projects, ostensibly to help the woodlot owners cope with problems besetting the woodcarving industry. The owners were enticed to clear portions of their *muyong* for replacement with exotic species. In some cases, local people were taught to clear *muyong* and to plant tiger grass or broom grass (*Thysanolaena latifolia*), or to implement sloping agricultural land technology (SALT), both of which were alien to their established practices.

Loss of biodiversity

The woodcarving industry, which has long been fuelled by tourism and caters for both local and export markets, has resulted in a loss of biodiversity. At the same time, the local construction industry has expanded, leading to over-exploitation of premium wood species in both the *muyong* and public or communal forests. Overcutting over the past 50 years has reduced the number of species in *muyong* to about 200 (DENR, 2006). Declining forest cover and species composition affects the water-holding capacity of the *muyong*, thereby reducing their ability to provide water to the terraces. Climate change is adding to this problem, with lower rainfall leading to a lack of water for the terraces, and productivity is declining.

The 1980's also saw the intrusion of modern agricultural technologies into the rice terraces. Foremost among them was replacement of traditional rice varieties with new high-yielding varieties. Although the high-yielding varieties substantially increased rice yields, the associated use of pesticides and other harmful synthetic inputs threatened the extinction of aquatic plants and mollusc species and changed the face of biodiversity in the terraces.

The interdependence of the multi-functional components of the Ifugao landscape was highlighted in studies by the Philippine Agriculture Department's Bureau of Soils and Water Management in 2005. The studies concluded that the Ifugao rice terraces had the capacity to supply basic rice needs and provide a variety of environmental services and rural amenities, combining to maintain a viable rural environment (Concepcion, 2005). However, the capacity of the terraces to provide environmental services has been deteriorating, owing to problems like low rice productivity, outmigration, overcutting of trees in the *muyong*, erosion and landslides, unmaintained irrigation systems and pests like golden apple snails (*Pomacea canaliculata*) and giant earthworms. The earthworms burrow deep into the soil, disrupting the water-supply function of the terrace walls, resulting in the walls drying up, cracking and eroding (Calderon et al., 2009).

Built-up areas and pollution

As already emphasized, the Ifugao *muyong* are part of an interdependent mosaic of land uses in a spectacularly mountainous location. It follows that if one aspect of the landscape is challenged or disturbed, the other parts are also affected. Owners of the Ifugao rice terraces are facing increasing difficulties in making a living from their mountainside paddies. Natural disasters such as typhoons, droughts, landslides and pests and diseases have brought significant problems for Ifugao farmers. Integration into the market economy has also brought new pressures for change, and terrace farmers are increasingly exposed to uncoordinated development, such as conversion of rice terraces to residential and commercial uses, loss of biodiversity from application

Cyathea contaminans (Wall. ex Hook.) Copel. [Cyatheaceae]

Variously called the giant fern and the mountain tree fern, this species thrives in the moist mountain environment of the *muyong*. It grows to seven metres, with leaf fronds up to 2.5 metres long, with spiny stems. The trunks can be used as fence posts or as raw material for handicrafts such as vases or statuettes. The fibrous material is used as a growing medium for orchids.

of pesticides and inorganic fertilizers (DENR, 2006), and pollution. The growth of backyard pig-raising has made rivers and creeks unsuitable for bathing and washing, and polluted water has found its way into terraces on lower slopes. Uncoordinated infrastructure development now threatens to reduce the aesthetic value of the rice terraces as a tourist attraction.

Bringing the situation up-to-date

Key-informant interviews conducted in Hungduan, Ifugao, in January and February, 2020,[2] produced the following points of view regarding the current state of the *muyong*:

*   *Muyong* all over Hungduan are reported to be robust. There is a significant view that this robustness results from people preferring a cash-based system that leaves shifting cultivation areas to grow into *muyong*. Almost all sloping land has a forest cover, except for a few slashed areas that have been converted to vegetable gardens. There is also a significant increase in the number of people using liquefied petroleum gas for fuel, rather than firewood. A former employee of the Department of Environment and Natural Resources (DENR) attributed

the robustness of the *muyong* to the department's Integrated Natural Resources and Environment Management Programme. Under this programme, about 665 hectares of *muyong* have been enhanced between 2011 and 2017 by the provision of cash to *muyong* owners to enrich their *muyong* with native forest species. These plantings have seen an average 85% survival rate.

- A major present-day threat to the *muyong* is the weakening of customary governance by community elders, and changes in the style of housing. Increasingly, houses are being built of concrete, and their construction requires hundreds of round logs. These are being cut randomly, with disregard for the sustainable cutting of trees in the *muyong*. Another threat is the conversion of forest to commercial vegetable production, and the possibility that this will lead to *muyong* being used for this purpose.

- Most key informants believe that the rice terraces will deteriorate if the multi-faceted challenges being faced by rice farmers are not addressed. However, they are confident that the *muyong* will be retained. They agree that younger generations should be educated about the importance of the rice landscape as a production unit that embodies the spirituality and sound cultural farming practices of the people, and which is a habitat for insects, fish, snails, and edible weeds that enrich the local diet.

- They conclude that the rice landscape is a key component of the tourism industry in their province, with potential for further development. Therefore, both the *muyong* and the rice terraces must be sustained.

Conclusions

The multi-functional character of the Ifugao landscape has not changed for many generations. The local people continue to produce rice and gather firewood, timber, fruit, vegetables, medicinal plants and other resources from its bountiful production systems, ranging from the rice terraces to communal forests, shifting cultivation systems and uniquely, the *muyong*. The landscape is therefore a sustainable production system, meeting both socio-economic and ecological objectives. But there is a vital bottom line to its future viability: water must continue to flow to the terraces. In this regard, the *muyong* are the crucial 'recharge' zone, gathering water and directing it into irrigation channels.

Yet the *muyong* are increasingly under threat; their once proud concentration of diverse indigenous species is being changed to trees selected for their high commercial value and fast-growing exotics, and use of the land is changing as owners pursue cash incomes. Unless such conversion is stopped, the number of *muyong* will dwindle, and as they disappear, so too will their vital environmental services and their protection of a substantial part of the region's biodiversity.

The best approach may be regulation of timber harvesting coupled with enrichment planting to maintain a close canopy cover and improve the water-holding capacity of the *muyong*. Any future activities in the *muyong* should be conditional upon close consideration of the continued capacity of the forest ecosystem to capture water and supply it to the terraces.

The *muyong* need to be in a constantly healthy condition, to yield sufficient water for irrigation. Any action undertaken in the interests of improving just one aspect of this vast and varied production system, such as the use of pesticides in rice, must first be examined for its possible effects on the functioning of the entire system and its biodiversity.

Local government intervention, through local legislation, is needed to bolster the capacity of *muyong* owners to improve and strengthen local governance of the landscape. However, a bigger portion of this responsibility should remain with the owners of the *muyong* and the rice terraces – as has been the case for countless years.

The *muyong* system exemplifies the depth of traditional knowledge related to forest ecosystems, the sustainable use of resources and the application of wise customary laws. These privately-owned forests deliver a large measure of benefit to their human dependents, while conserving the watershed and contributing to the sustainability of the rice terraces. Yet they flounder in a conflict of perspectives related to land ownership, between the Ifugao people and the State. In official regard, the *muyong* of the Ifugao are the property of the Philippine government.

Areca catechu L. [Cyatheaceae]

This medium-sized palm grows to about 20 m and is important in the muyong for its seed crop, the areca nut or betel nut. Accompanied by a leaf from the vine *Piper betle*, the nuts are habitually chewed by countless thousands of people in East and Southeast Asia and the Pacific. They contain alkaloids that are intoxicating and addictive. Moreover, chewing the nuts has been directly linked to oral cancer.

But according to customary law, the Ifugao are assured of rightful ownership of their *muyong* and adjoining lands. They claim the rice terraces and the *muyong* will remain, 'until the end of time'. Despite their problems, present indications suggest that this may well be the case.

References

Butic, M. and Ngidlo, R. T. (2002) *Muyong Forest of Ifugao: Assisted Natural Regeneration in Traditional Forest Management*, Food and Agriculture Organization of the United Nations, Regional Office for Asia and the Pacific, Bangkok, Thailand

Castonguay, A. C., Burkhard, B., Müller, F., Horgan, F. G. and Settele, J. (2016) 'Resilience and adaptability of rice terrace social-ecological systems: A case study of a local community's perception in Banaue, Philippines', *Ecology and Society* 21(2):15, available at http://dx.doi.org/10.5751/ES-08348-210215, accessed 22 September 2020

Calderon, M. M., Dizon, J. T., Sajise, A. J. U., Bantayan, N. C. and Salvador, M. G. (2009) *Towards the Development of a Sustainable Financing Mechanism for the Conservation of the Ifugao Rice Terraces in the Philippines*, College of Forestry and Natural Resources, University of the Philippines Los Banos, Philippines

Concepcion, R. N. (2005) *Multi-functionality of the Ifugao Rice Terraces*, Bureau of Soils and Water Management, Department of Agriculture, Manila

DENR (2006) *The Ifugao Rice Terraces Project Framework*, Department of Environment and Natural Resources, Manila, Philippines

Hayama, A. (2003) 'Local forest management in the rice terraces area of Banaue, The Philippines', in *People and Forest-policy and Local Reality in Southeast Asia*, Institute for Global Environmental Strategies. Kanagawa, Japan

Jang, J. W. and Platt-Salcedo, S. (2013) 'The socio-political structure that regulates the Ifugao forest maintenance', *Proceedings of the Fourth International Conference on Biology, Environment and Chemistry* 58, pp.85-94

Joshi, R., Matchoc, O. R. O., Bahatan, R. G and Pena, F. A-D. (2000) 'Farmers' knowledge, attitudes and practices of rice crop and pest management at Ifugao rice terraces, Philippines,' *International Journal of Pest Management* 46(1), pp.43-48

Klock, J. and Tindungan, M. (1995) 'The past and the present: A meeting of forces for a sustainable future', *Forest, Trees and People Newsletter* no. 29, FTP Programme Network, International Rural Development Centre, Sweden, and the Community Forestry Unit, Food and Agriculture Organization of the United Nations, Rome

Ngidlo, R. T. (1998) *Conserving Biodiversity: The Case of the Ifugao Farming System*, Philippine Council for Agriculture, Aquatic and Natural Resources Research and Development (PCARRD), Los Baños, Philippines

Rabena, M. F., Macandog, D. M., Cuevas, V. V. and Espaldon, M. O. (2015) 'A vegetation inventory of a traditional secondary forest (*muyong*) in Kinakin, Banaue, Ifugao, Northern Luzon, Philippines', *Philippine Journal of Systematic Biology* 9, pp.16-24

Rondolo, M. T. (2001) 'Fellowship Report', *Tropical Forest Update* 11(4), International Tropical Timber Organization, Japan

Sajise, P. (1979) 'Some ecological considerations for agroforestry (for whom?)', paper presented at a PCARRD symposium-workshop on agro-forestry, Los Baños, Laguna, Philippines

Sajor, E. (1999) *Cutting Trees and the Dynamics of Social Change: The Case of the Ifugao Muyong in the Philippine Uplands*, Institute of Social Studies Working Papers No. 294. The Hague, Netherlands

Notes

1. A municipality is the second administrative unit in the Philippines. A *barangay* (village) is the smallest unit.
2. Key informants:
   Mateo and Linda Gano (60+ years old) of Bokiawan, Hungduan
   Mario Ibay, 68 years old, Bokiawan, Hungduan
   Yolanda W. Buyago, mid-30s, Hapao, Hungduan
   Elena Madrid, early-40s, Hapao, Hungduan
   Yolanda Dulnuan, Poblacion, Hungduan

17

AN INVENTORY OF TREE STANDS

In forest fallows of Nagaland

Vengota Nakro and Ango Konyak*

Introduction

Shifting cultivators are aware of the role played by trees during the fallow phase of the swidden cycle. They 'pump' soil nutrients up from deeper layers of soil and make them available to agricultural crops, and they improve the soil structure by adding humus, which supports the formation of clay–humus complexes that improve the soil structure. Shade from trees helps to maintain a favourable soil temperature, and a canopy of trees breaks the impact of raindrops, making the soil less vulnerable to erosion. The shifting cultivators of Nagaland therefore take great care to populate their fallow fields with trees, in a process that begins in the cropping period.

Panoramic views of the landscape in the Nagaland districts of Mon, Tuensang, Longleng and Kiphere, where shifting cultivation is the main occupation, are deeply impressive as *jhum* (swidden) fields enter the fallow period and natural regeneration of secondary forest begins. Outsiders – even trained agronomists – may be deceived

This is a tribute to my dear colleague, Vengota Nakro, who passed away in June 2020. He dedicated more than 15 years of his life to the Nagaland Empowerment of People through Economic Development (NEPED) project, and was respected and admired for his vast experience, commitment and contributions to sustainable *jhum* practices and community-based biodiversity conservation in Nagaland. Despite a well-deserved reputation as an expert in this field, he remained a humble, hardworking, insightful and dedicated man who, to me and many others, was a mentor and a friend. He retired as Additional Director of Nagaland's Department of Soil and Water Conservation in 2018, but continued to serve as Project Support Specialist for the IFAD-funded Fostering Climate Resilient Upland Farming Systems in the Northeast (FOCUS) Project. He is dearly missed and will always be remembered for his service to the people of Nagaland. – Ango Konyak.

* VENGOTA NAKRO, Project Support Specialist, FOCUS-IFAD Project, Nagaland, India; and ANGO KONYAK, Assistant conservator of Forests and Deputy Project Director, Nagaland Forest Management Project, Nagaland, India

into believing that these areas are secondary succession forests, and not shifting cultivation fields. The question of how swidden farmers have been able to master the management of these fallow processes is one that is often asked. However, instead of acknowledging the vast knowledge employed by farmers in the course of managing their resources, government officials instead induce them to plant 'economically viable' tree species, using perceptions of economic viability that are not those of the farmers, but those of the persuasive officials themselves.

The area that is annually under crops in the cultivation phase of the swidden cycle in Nagaland is about 1221 sq. km, out of a total geographical area of 16,579 sq. km (Department of Land Resources, 2011). The average length of the shifting cultivation cycle is nine years. Therefore, the computed total area involved in shifting cultivation (cultivated land plus fallowed land) is about 10,993 sq. km. A conventional understanding of shifting cultivation is that — due to population pressure — the *jhum* cycle is becoming shorter and farmers are encroaching more and more into pristine forest areas. As a result, there is a loss of the vegetative cover that facilitates a recharging of perennial springs; there is not enough time for natural regeneration to take place, there is an accelerated loss of biodiversity, and so on. If these assumptions were based on facts, the area under forest cover would have been decreasing over the years. But that is not the case in Nagaland. Assessments by the Forest Survey of India from 2001 to 2009 (See Table 17-1) show that there was an increase of 448 sq. km in forest area between 2001 and 2003. And while the total forest area decreased by 266 sq. km between 2005 and 2007, there was an increase of 464 sq. km in the area of dense forest between 2003 and 2007. The forest area remained constant between 2007 and 2009. Thus, the misconception that shifting cultivation is responsible for degradation of pristine forest in recent times needs to be reviewed.

Recognizing the fact that shifting cultivation brings about accelerated soil erosion, runoff and nutrient losses, leading to a decline in crop production per unit of area, swidden cultivators make use of various measures to prevent these adverse effects. These include

Clerodendrum glandulosum Lindl.
[Lamiaceae]

This flowering shrub is commonly known as East Indian Glory Bower and is often found in swidden fallows. It is a species with vulnerable status in northeast India for its loss of habitat. Leaf extracts are widely used medicinally to treat rheumatic pains, high blood pressure, diabetes and obesity. The leaves can also be eaten as a vegetable.

TABLE 17-1: Area under different types of forest in Nagaland (sq. km).

Forest type	Assessment of Forest Survey of India				
	2001	*2003*	*2005*	*2007*	*2009*
Dense forest	5393	5707	5838	6171	1274
Moderately dense					4897
Open forest	7952	7902	7881	7293	7293
Scrub forest	47	231	13	2	2
Total	13392	13840	13732	13466	13466

Note: Tree cover outside forests is 300 sq. km.
Source: Ministry of Environment and Forest, Government of India.

placing tree poles along the contours of slopes to check soil erosion and reduce runoff (see Lotha et al., ch. A21, in the supplementary chapters to this volume), cultivation of legume crops to restore soil fertility and, most importantly, populating the field with trees by planting, preserving stumps and nurturing natural regeneration during weeding operations.

In order to quantify the standing trees in fallows in Nagaland, an inventory was carried out in 21 villages in nine districts across the state. Case studies are described below.

Methodology

The survey used the general principle of Minimum Representative Area by adopting the following steps:

1. Systematic selection of sampling areas. Sampling plots were placed at regular distances along a baseline;
2. Sampling design: A stand analysis was made covering all strata, from seedlings to pole-sized trees to mature trees. These procedures were followed:

   a) The main parameters of shifting-cultivation practices prevalent in the study villages were obtained in consultation with village elders. These included the number of years in a *jhum* cycle, the length of the cropping period and the fallow period in years, the names of locations where shifting cultivation was taking place and the number of people who would be involved in cultivation in the study area. We then identified which particular locations would be cultivated in the survey year – in this case, 2007.

   b) A compartment measuring 20 m x 20 m was established as a baseline. Within this compartment, diameters at breast height (dbh), useable heights and local names of all strata of trees were recorded. Botanical names were also recorded when identified. Twenty-five such compartments were established in all directions from the baseline compartment, making up a total area of 1 hectare, and all information was recorded.

c)   The distance between one compartment and the next was determined by the following formula:

$$D = \sqrt{(c \times N \times 10000/n)}$$

where D = the distance from one compartment to the next; N = the total population that is going to cultivate that location; n = the sample size (25 in this survey); and c = approximate area cultivated per person in 1 hectare

The study showed that variation in the tree population and the number of species per hectare depended largely on the farmers' management of the field during cropping, and not necessarily on the length of the following fallow period. Two case studies were then identified for comparison by inventory of trees. The first, in a village called Phuktong, was in an area with a 10-year *jhum* cycle involving two years of cropping followed by eight years of fallow. The second, in a village called Tuophema, had a 24-year *jhum* cycle with four years of cropping followed by 20 years of fallow.

Phuktong village

A 10-year jhum cycle: Two years of cropping followed by eight years of fallow

Phuktong village is located 13 km north of Mon, the headquarters of Nagaland's northernmost district, which is also called Mon, at an altitude of 990 metres above sea level (masl) (Figure 17-1). The village has a total population of 1897, living in 179 households. Shifting cultivation is the main occupation of the village and all

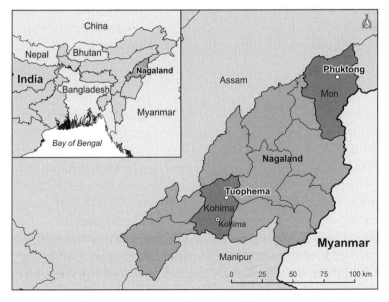

FIGURE 17-1:   The study sites in Mon and Kohima districts, Nagaland, northeast India.

villagers are involved. The *jhum* cycle takes 10 years – two years for cropping and eight years of fallow. Trend analysis conducted in the village showed that the last time the villagers opened primary forest to expand the area under cultivation was in 1948. Since then, the primary forest has remained intact in terms of conversion to agricultural purposes and the total area under *jhum* has remained the same.

To populate their *jhum* fields with trees, the farmers preserve tree seedlings that emerge amid crops by protecting them during weeding operations (Figure 17-2). In subsequent weeding processes, space adjustments are made by uprooting the saplings when they are too dense. Because the leaves of the tree saplings tend to shade the crops being grown, the farmers trim the branches. This helps the trees to grow straight. The branches are trimmed again when the field is slashed for the second year of cropping. The slash is burned in a controlled fashion so as to avoid harming the trees.

FIGURE 17-2: A *Macaranga* sp. seedling emerges in a rice crop in Phuktong village. This tree species is popular among Naga farmers for its soil-enrichment properties.

Photo: Ango Konyak.

The Konyak farmers at Phuktong manage the trees as though they are one of the crops. From the first slash for cultivation, a farmer is aware of the need for trees, to be used mainly for firewood, for poles to be placed across slopes as a measure to control soil erosion (see Lotha et al., ch. A21, in the supplementary chapters to this volume), thence left to rot as an aid to soil fertility, and to be used in house construction, particularly as roof rafters.

Standing trees that are preserved from previous *jhum* cycles are a common sight. The dominant species in this case is *Schima wallichii*, as shown in Table 17-2. The farmers lop the branches and leave only the tops, so the standing trees do not shade the crops. *Schima wallichii* thrives on pollarding and coppicing; it can withstand and survive lopping and tree topping. The farmers find this species useful for house construction as posts and it also makes good firewood. But in the farmers' view, any tree that can be preserved in their *jhum* field is good for future use.

Although there is something special about the manner in which Konyak farmers deal with trees as a 'crop' and nurture them for the next *jhum* cycle, the Phuktong site was chosen for the inventory study purely by random selection, and not for any particular speciality (Figure 17-3).

The total number of trees per hectare in the Phuktong inventory was 2210, with 95 species. The high tree population was the result of a conscious effort by farmers

TABLE 17-2: Main stand parameters of common woody species (≥ 2 cm dbh) in *jhum* plots at Phuktong village.

Local name	Scientific name	No. of trees per hectare	Avg dbh (cm)	Avg height (m)	Avg basal area (sq. m/ha)	Volume (cu. m/ha)
Pok	*Macaranga* spp.	335	11.40	8.93	4.66	36.39
Lak	*Schima wallichii*	294	14.10	9.73	6.83	64.47
Aomah	*Brucea javanica*	145	9.01	5.66	1.06	4.32
Yam	Unidentified	134	11.09	10.10	1.78	13.65
Meant	Unidentified	88	12.24	10.88	1.57	15.34
Pangpoi	*Duabanga grandiflora*	81	16.47	13.59	2.25	23.15
Aling	*Balakata baccata*	77	12.04	10.03	1.32	12.19
Hen	*Stereospermum* sp.	63	19.31	12.06	2.55	27.31
Hengmah	Unidentified	51	12.93	8.73	0.92	7.78
Leak	*Terminalia myriocarpa*	47	13.70	11.87	0.89	7.66
85 other species		895	12.85	10.15	15.41	140.30
Total		2210	13840		39.23	352.57

Notes: The inventory followed eight years of fallow, which was supported by planting and sowing of pioneer tree species. Some of the larger trees originated from previous jhum cycles. The average height is the average useable height. Volume was calculated according to the formula: B x H x 0.6. Volume figures are the sum of individual stands. The form factor of individual trees is 0.6.

to preserve naturally occurring tree seedlings in individual *jhum* plots during the cropping period, in order to accelerate the recovery of soil fertility during the fallow period.

FIGURE 17-3: Before and after. Left: shifting cultivation (*jhum*) fields in Phuktong village after clearing, at the very beginning of the cultivation phase in March, with heavily pruned trees standing starkly in the swiddens. Right: the same swidden area just five months later, in August, with flourishing regrowth on the trees amid a healthy crop, with saplings rising through the upland rice.

Photos: Ango Konyak.

The farmers have developed a system of preserving naturally regenerated saplings, mostly *Macaranga denticulata*, and preserving poles from trees of previous *jhum* cycles, especially those of *Schima wallichii*. The combined stand of these two species numbered 629, or about 29% of the total. *Duabanga grandiflora* is a species that villagers preserve from the previous *jhum* cycle fields for its use as a timber species. Saplings of this species can be preserved during burning by providing firebreaks and covering them with sheaths cut from banana stems. *Brucea javanica* is preserved for its medicinal properties.

Grouping by diameter class shows that trees up to 19.9 cm in diameter at breast height have regenerated following clear felling for cultivation 10 years earlier (Table 17-3; Figure 17-4). The number of trees per hectare in this

Schima wallichii Choisy [Theaceae]

This medium to large evergreen tree is often saved from previous *jhum* cycles by farmers opening swiddens. They are trimmed of branches and stand like poles with a tuft of foliage at the top, but they coppice vigorously. The wood is often traded or used for construction. The flower petals are used medicinally, to treat uterine disorders or hysteria, and the bark is used as an antiseptic.

FIGURE 17-4: Before and after photographs from Angjangyang village, south of the study village of Phuktong, but still in Mon district, Nagaland. Left: after the village's *jhums* were cleared for cultivation in 2009. Right: two years later, and the same swiddens area in a second-year *jhum* fallow.

Photos: Ango Konyak.

TABLE 17-3: Basal area and volume by diameter class of woody species (≥ 2 cm dbh) in a *jhum* plot at Phuktong village.

Diameter class (cm)	No./ha	dbh (cm)	Height (m)	Basal area (sq. m/ha)	Volume (cu. m/ha)
2 – 4.9	61	4.11	6.08	0.07	0.26
5 – 9.9	1152	6.96	6.99	4.59	20.24
10 – 14.9	387	12.13	9.11	4.62	25.12
15 – 19.9	239	16.96	11.71	5.74	38.81
20 – 24.9	119	22.51	15.45	4.93	44.24
25 – 29.9	125	27.70	16.83	7.57	76.80
30 – 34.9	84	32.21	19.33	6.86	79.58
35 – 39.9	29	36.74	21.24	3.08	39.24
40 – 44.9	5	42.02	17.40	0.69	7.41
45 – 49.9	7	47.34	21.29	1.23	15.77
50 – 54.9	2	52.52	20.00	0.43	5.11
Total	2210			39.81	352.58

Notes: The inventory followed eight years of fallow, which was supported by planting and sowing of pioneer tree species. Some of the larger trees originated from previous *jhum* cycles. The average height is the average useable height. Volume was calculated according to the formula: B x H x 0.6. Volume figures are the sum of individual stands. The form factor of individual trees is 0.6.

class is 1839, which is 83% of the total, but with a volume of only 84.43 cu. m/ha they represent only 24% of the total volume.

The dominant presence of trees with a dbh of 2 to 9.9 cm indicated in Table 17-3 that early succession processes were in progress. The trees in this class will be felled during subsequent cultivation, and will either be collected for firewood, placed across slopes as soil-conservation measures, or burnt. The presence of trees with a dbh of 2 to 4.9 cm indicated that some recalcitrant tree species such as *Melia azedarach* had recently been germinating.

The number of trees preserved from previous cycles was only 371, but their volume was 268.15 cu. m/ha, or 76% of the total volume. Trees in this class are used for specific purposes. Species such

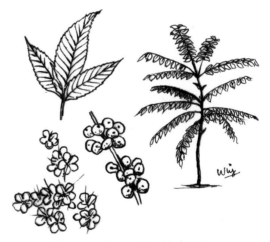

Brucea javanica (L.) Merr. [Simaroubaceae]

This shrub or small tree is often saved from swidden fires for is medicinal qualities. All parts of the tree, but most often the seeds and roots, are used, mainly in the treatment of amoebic dysentery, diarrhoea and malaria. Compounds isolated from the fruit are under study for their anti-cancer qualities. The fruit and roots of this plant are often found in local markets.

as *Schima wallichii* will be felled in times of need, for house construction, and species like *Duabanga grandiflora* will be preserved as support for cultivation of betel leaves, which are sold when mature.

In Table 17-4, the projected population of standing trees ranges from as low as 575 to as many as 3700 per hectare. As stated in the methodology, the stand parameters were taken at random, and consisted of no fewer than 25 *jhum* farmers' fields, with various tree densities and volumes for optimum coverage and analysis.

In compartment 19 of Table 17-4, there are only 23 trees, but a volume of 5.56 cu. m/ha. This indicates that some of the trees there have a large dbh, which also implies that these trees are taller, and with wider canopy coverage. For this reason, the farmers avoided growing more trees in this compartment during the cultivation period because they would have interfered with the *jhum* crops.

TABLE 17-4: Projected tree population ($\geq$ 2cm dbh) in 25 compartments in *jhum* fields at Phuktong village.

Compartment number (each 20 m x 20 m)	Number of trees (per ha)	Volume (cu. m/ha)	Estimated tree to tree distance in compartment (m)	Projected tree population of compartment
1	100	14.45	4	2500
2	98	12.64	4	2450
3	97	8.98	4	2425
4	149	24.38	3	3725
5	50	6.47	8	1250
6	50	9.18	8	1250
7	82	8.21	5	2050
8	78	14.87	5	1950
9	88	12.05	5	2200
10	108	6.51	4	2700
11	100	10.76	4	2500
12	100	3.21	4	2500
13	99	2.59	4	2475
14	124	9.57	3	3100
15	121	5.94	3	3025
16	68	20.54	6	1700
17	88	30.89	5	2200
18	29	4.15	14	725
19	23	5.56	17	575
20	148	33.81	3	3700
21	100	23.38	4	2500
22	100	29.65	4	2500
23	99	28.89	4	2475
24	50	11.47	8	1250
25	61	14.60	7	1525
Total	2210	352.74		

In comparison, there are 99 trees in compartment 13, but with a volume of only 2.59 cu. m/ha, which indicates that the trees have a smaller dbh. In the first year of cultivation, the farmers select roughly a third of these naturally regenerating saplings, on the basis of preference and spacing, to grow with the *jhum* crops. In the second year of cultivation, these saplings are protected during burning and allowed to grow, since they will not affect the crops. After the second-year harvest, more naturally regenerating tree saplings are allowed to grow, with the objective of utilizing them during the next *jhum* cycle. This may be as poles for soil conservation or as rafters for roofing. This shows that the farmers take care to repopulate their fields with trees for the fallow period and later reap the benefits from having done so.

Tuophema village

A 24-year jhum cycle: four years of cropping followed by 20 years of fallow

Tuophema village is in Kohima district, 44 km from the district headquarters, at an altitude of 1380 masl. It is believed to be the oldest, but the fastest-developing village in the Angami region. There are 370 households in Tuophema, of which 205 are actively engaged in *jhum* cultivation. In this village, rice is produced mostly in irrigated terraces and maize and other vegetables are grown in *jhum* fields. The area opened for *jhum* cultivation each year is about 90 hectares, and the normal *jhum* cycle is 10 years.

Although the normal shifting cultivation cycle is 10 years, in some locations the villagers subject their fields to cultivation for four to five years, then leave them fallow for more than 20 years. In the first two years of cropping the farmers pursue normal *jhum* operations, but at the start of the third year, the field is ploughed deeply and the exposed roots are collected along with crop residues and these are burnt. A layer of soil as thick as 10 cm or more is placed on top of the burning roots and crop residues and the fire is left to slowly burn out overnight. This procedure is repeated all over the field, at distances of two to five metres, depending on the availability of combustible debris. The next day, the farmers spread the burnt soil all over the field and seeds are sown. Because of this intensive burning of soil and a cultivation procedure that is not accompanied by planting of tree seedlings or sowing tree seeds into the cultivated *jhum*, the population of trees in these locations is lower than that in nearby *jhums* that are cultivated on a normal 10-year cycle, despite the practice having a 20-year fallow.

As can be seen from Table 17-5, a total of only 28 species were found in the Tuophema *jhum* field, with *Quercus* spp. being the most dominant because of its importance as firewood in household use. The economic value of this wood has increased because of heavy demand for firewood in the district capital, Kohima. As a result, farmers have started to nurture saplings that germinate in their fields. The total number of trees per hectare in the Tuophema field was 407, even though the area had been in fallow for more than 20 years. The low tree-population density and species diversity was attributed to the farmers' practice of cultivating the land continuously for four years without supporting forest regeneration by planting or preserving trees that would have seeded the area. Another reason was the intensive burning of crop residue beneath a layer of soil – a process in which even the recalcitrant seeds were destroyed. The 407 trees per hectare standing in the area were largely the result of germination from natural seed dispersal processes and growth from old stumps that were resistant to fire.

Table 17-6 shows that there were 286 trees in the Tuophema *jhum* with a dbh of 2 to 9.9 cm. These constituted over 70% of the total stand. This indicates that the establishment of trees in the area started recently, and not immediately after the fallow phase began, such as happens following a

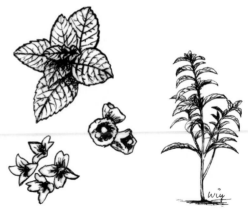

Mentha arvensis L. [Lamiaceae]

Wild mint is a perennial herb native to temperate regions of Europe and western and central Asia. The species name *arvensis* means to grow in a cultivated field, attesting to its ease of propagation. The leaves are commonly brewed into a tea to treat colds and aid digestion. The leaves contain menthol, which is used in food, drinks, cough medicines, creams and cigarettes.

TABLE 17-5: Main stand parameters of common woody species (≥ 2 cm dbh) in a *jhum* field at Tuophema village, following 20 years of fallow.

Local name of tree species	Botanical name	Number. of trees per hectare	Avg dbh (cm)	Avg total height (m)	Basal area (sq. m/ha)	Volume (cu. m/ha)
Seitho	*Quercus* spp.	83	8.82	3.97	0.62	1.84
Kenyhu	*Phyllanthus emblica*	75	10.66	5.31	1.00	3.84
Ze	*Albizia* spp.	48	9.57	5.84	0.58	2.36
Mege	Unidentified	30	7.19	3.65	0.15	0.46
Sotsi	*Callicarpa arborea*	22	5.76	3.93	0.08	0.31
Zomhu	*Rhus chinensis*	21	2.73	2.81	0.01	0.02
Tekha vohei	*Ficus* spp.	18	2.88	3.17	0.01	0.02
Chiede	*Ficus* spp.	16	10.45	3.81	0.22	0.66
Mecho	*Schima wallichii*	16	9.67	6.31	0.13	0.57
Paga-kemeluo	*Stereospermum* spp.	9	14.54	5.67	0.21	0.97
18 other species		69	14.33	6.43	2.39	19.37
Total		407			5.41	30.43

Notes: This inventory followed 20 years of fallow, without sowing or planting tree species. Some of the larger trees originated from previous *jhum* cycles. The average height is the average useable height. Volume was calculated according to the formula: B x H x 0.6. Volume figures are the sum of individual stands. The form factor of individual trees is 0.6.

TABLE 17-6: Basal area and volume-by-diameter class of woody species (≥ 2cm dbh) in a *jhum* plot at Tuophema village.

dbh class (cm)	No. of trees per ha	Avg dbh (cm)	Avg useable height (m)	Basal area (sq. m/ha)	Volume (cu. m/ha)
2 – 4.9	177	3.48	2.95	0.18	0.35
5 – 9.9	109	8.67	4.69	0.66	1.94
10 – 14.9	67	13.08	6.56	0.91	3.53
15 – 19.9	25	17.61	8.00	0.61	2.92
20 – 24.9	13	23.76	11.85	0.58	4.06
25 – 29.9	4	28.34	5.50	0.25	0.83
30 – 34.9	0	0.00	0.00	0.00	0.00
35 – 39.9	4	39.01	4.00	0.48	1.16
40 – 44.9	2	42.36	14.00	0.28	2.37
above 45	6	55.57	15.00	1.46	13.27
Total	407			5.41	30.43

Notes: This inventory followed 20 years of fallow, without sowing or planting tree species. Some of the larger trees originated from previous *jhum* cycles. The average height is the average useable height. Volume was calculated according to the formula: B x H x 0.6. Volume figures are the sum of individual stands. The form factor of individual trees is 0.6.

normal *jhum* cycle with two years of cropping. There were 391 trees in the 2-to-24.9 cm dbh classes, representing regrowth during the fallow period. They had attained a volume of 12.80 cu. m/ha, or 42% of the total volume of 30.43 cu. m/ha. A significant contribution to the stand volume of 13.27cu. m/ha was made by six trees with a dbh of more than 45 cm. These were identified as *Bombax ceiba*.

Phyllanthus emblica L. [Phyllanthaceae]

This small to medium-sized tree, commonly known as Indian gooseberry, is arguably the most important plant in traditional folk medicine in Asia-Pacific. The edible fruit are sour, bitter, and astringent, but Ayurvedic medicine says they are potent rejuvenators that stall the degenerative processes of old age. All parts of the plant are used to combat many ailments, from liver complaints and anaemia to hair loss.

TABLE 17-7: Projected tree population (≥ 2cm dbh) in 25 compartments in *jhum* fields at Tuophema village.

Compartment number (each 20 m x 20 m)	Number of trees (per ha)	Volume (cu. m/ha)	Estimated tree to tree distance in compartment (m)	Projected tree population of compartment (per ha)
1	34	1.39	12	850
2	15	0.51	27	375
3	7	0.26	57	175
4	11	0.66	36	275
5	5	0.02	80	125
6	26	0.35	15	650
7	11	0.55	36	275
8	23	1.76	17	575
9	18	1.01	22	450
10	5	0.17	80	125
11	20	5.20	20	500
12	34	1.39	12	850
13	15	0.51	27	375
14	7	0.26	57	175
15	11	0.66	36	275
16	5	0.02	80	125
17	26	0.36	15	650
18	11	0.55	36	275
19	23	1.76	17	575
20	18	1.01	22	450
21	5	0.17	80	125
22	20	5.20	20	500
23	18	1.01	22	450
24	5	0.17	80	125
25	34	5.47	12	850
Total	407	30.43		

Note: This inventory followed 20 years of fallow without sowing or planting tree species. Some of the larger trees originated from previous *jhum* cycles.

Table 17-7 shows that the projected tree-stand density per hectare in most of the compartments in the Tuophema *jhum* field was below 500. In only three compartments does the projected number of trees rise as high as 850. In five of the compartments, the projected number of trees is as low as 125 per hectare, which is very low.

Conclusion

A short cropping phase followed by a long fallow period is favourable for regeneration of trees, when compared to a *jhum* cycle with a longer cropping period followed by long fallow period, as can be seen in Table 17-8. In order to populate fallowed *jhum* fields with trees, it is necessary to nurture naturally occurring tree seedlings,

TABLE 17-8: Comparison of main parameters of tree stands in fallowed *jhum* fields following different cropping and fallow cycles.

Stand parameter	Tuophema village: 4 yrs cropping followed by 20 yrs fallow	Phuktong Village: 2 yrs cropping followed by 8 yrs fallow
Number of trees per ha	407	2210
Total basal area (sq. m/ha)	5.41	39.81
Total volume (cu. m/ha)	30.43	352.58

preserve fire-resistant species with branches lopped and carefully manage regrowth from coppiced stumps during the cropping phase.

With the exception of a few pockets where the cycle has shortened to five or six years, the average length of the *jhum* cycle in Nagaland is nine years, including two years of cropping. This study has shown that trees are regenerating in the fallow period of shifting cultivation, and this is contributing to the maintenance of a consistent total forest area in the State of Nagaland, contrary to the common belief that *jhum* cultivation is responsible for forest degradation.

References

Department of Land Resources (2011) *State Perspective and Strategic Plan (SPSP) of Nagaland*, Department of Land Resources, State Government of Nagaland, Kohima

18

SHIFTING CULTIVATION AND REHABILITATION OF NATURAL FOREST ECOSYSTEMS

In the Central Highlands of Vietnam

*Bao Huy**

Introduction

Shifting cultivation, which is also known as swidden agriculture, or pejoratively as slash–and–burn farming, is an ancient farming practice that remains the dominant and traditional land use in the mountainous regions of Southeast Asia, providing a livelihood for most of the ethnic minorities that inhabit these upland regions (Cherrier et al., 2018; Li et al., 2014). It has been estimated that there are about 100 million shifting cultivators in this region (Christanty, 1986).

In Vietnam, more than 50 ethnic minorities practise shifting cultivation, in numbers second only to Indonesia in Asia (Sam, 1996). In Vietnam's Central Highlands, which are home to many ethnic minorities, shifting cultivation is the farming system of necessity because the groundwater table is about 20 metres below the soil surface, making the irrigation of wet-rice fields impracticable (Huy et al., 1998). Throughout the tropics, most of the stages of shifting cultivation are similar (Christanty, 1986). The upland farmers apply indigenous knowledge passed down by their ancestors in selecting forest sites, clearing the vegetation and burning it, intercropping, protecting their crops, harvesting and fallowing, as described by many authors (e.g. Thinh and Son, 1998; Nhung, 1998; Dao, 1999; Dung, 2000; Hung, 2004). The size of each swidden, or farming plot, is dependent on the size of a household and the number of labourers it can call upon. The main crops are upland rice, mixed with chillies, cucumbers and melons, and the schedule of cultivation is drawn from long experience. Rice yields are unstable because the crop is highly dependent on the forces of nature (Huy et al., 1998). The choice of land to cultivate is important because the indigenous farmers aim to protect upstream watershed areas, and when clearing the forest, they often retain big trees and the roots of others to

---

* DR. BAO HUY, Professor, Consultancy for Forest Resources and Environment Management (FREM), No. 06 Nguyen Hong, Buon Ma Thuot, Dak Lak, Vietnam.

avoid soil erosion (Huy et al., 1998). The length of fallow periods depends on the area of arable land that is available to each village, and the time it takes for the soil to recover its fertility after cultivation. Land is usually allowed to lie fallow for between eight and 15 years. Thus, indigenous people apply a sustainable 'land-use plan' based on a cycle of forest-farm-fallow-forest, within the boundaries of each traditional village in the Central Highlands (Huy et al., 1998; Hung, 2004).

In recent decades, the traditional practice of upland cultivation in the Central Highlands has been disrupted by the introduction of industrial farming systems, using chemicals, and foreign exploitation of forest and land resources (Cuc, 1995; Huy et al., 1998, 2018). Specialized industrial monoculture crops such as coffee and rubber have replaced traditional shifting cultivation on sloping land, and these systems are facing problems from lack of water and soil erosion, as well as environmental pollution due to the use of chemicals and inorganic fertilizers (Huy et al., 2018). This transition has had a profound effect, as the increasing economic value of industrial crops such as coffee, cashews and rubber have seen these systems replacing the fallowed upland fields of shifting cultivators. The government has favoured and supported the march of these specialized monoculture systems rather than opting for a gradual evolution (van Noordwijk et al., 2008; Schmidt-Vogt et al., 2009), such as waiting for forest restoration, or using agroforestry to restore the natural forest ecosystem.

In an environment where agricultural production is being driven by globalization, shifting cultivation still plays an important role for indigenous people in terms of culture, economy, society and livelihoods, mainly in the subtropical and tropical zones of Southeast Asia, from the Chittagong Hill Tracts of Bangladesh, through Myanmar, Nepal and Bhutan to southwest China, northeast India, northern Thailand, the Lao PDR, Cambodia and Vietnam (Cherrier et al., 2018).

This study examines the recovery of forest structure and soil fertility and changes that occur over different fallow periods, as a basis for proposing improved upland-farming practices by ethnic minorities in the Central Highlands of Vietnam.

Lithocarpus balansae (Drake) A.Camus [Fagaceae]

A lofty broad-leaved evergreen of the beech family, this tree prefers to grow alongside streams at elevations from 400 to 1900 masl. Native to southeast China, Myanmar, Laos and Vietnam, it grows up to 30 m and dominates forest regrowth in Vietnam's Central Highlands after about eight years of fallow. It is one of the more dominant species in the area's primary forest.

Materials and methods

The study site

This study was carried out in Vietnam's Central Highlands, an eco-region with eight ecological zones. The Central Highlands (Figure 18-1) have the country's highest cover of tropical forests. Our study focused on the region's main forest type: evergreen broadleaf forest, and fallow lands within these forests, which are structurally complex, with mixed-species composition. The common species are from the Fagaceae, Myrtaceae and Lauraceae plant families

The elevation of evergreen broadleaf forests in the study area ranges from 700 to 1000 metres above sea level, with slopes of up to 30° in some areas. Mean annual rainfall is between 2000 and 2600 mm, with a dry season lasting for three months and mean annual temperatures ranging from 22 to 25°C. The forests grow on a main soil type of sedimentary rock (Chien, 1986; Hijmans et al., 2005; Fischer et al., 2008).

Shifting cultivators in the study area are of the native M'nong ethnic minority.

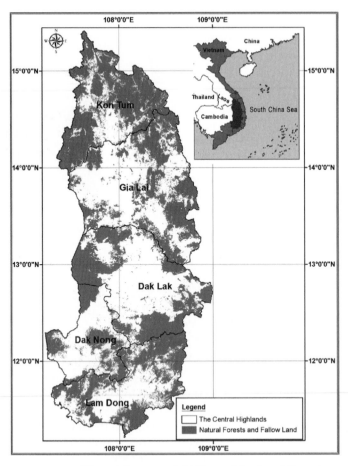

FIGURE 18-1:   Natural forests and shifting cultivation fallows in the Central Highlands ecoregion of Vietnam.

Data collection and analysis

A total of 15 nested circular plots were established in five fallow and forest areas. These areas consisted of fallows aged 2, 4, 8 and 10 years, and an area of primary forest. Three plots were established in each area, comprising two concentric circles, or subplots. Within a wider subplot, with a radius of 17.4 m and covering 1000 sq. m, all trees with a diameter at breast height (dbh) of 10 cm or more and a tree height (H) of 2 m or more were measured. Within a smaller subplot, with a radius of 5.64 m and an area of 100 sq. m, regeneration trees with a dbh of less than 10 cm and a tree height of 2 m or more were measured. Within each subplot, species, dbh and H were recorded.

We arranged distributions of height with intervals of 4 m and identified dominant forest-tree species by using an Importance Value Index (IV_i %) (Curtis and McIntosh, 1950; Cottam and Curtis, 1956; Narayan and Anshumali, 2015) as follows:

$$(1) \qquad IV_i\% = \frac{N_i\% + BA_i\%}{2}$$

$$(2) \qquad N_i\% = \frac{N_i}{N}$$

$$(3) \qquad BA_i\% = \frac{BA_i}{BA}$$

Where $IV_i\%$ is the IV % indicator for the ith species; $N_i\%$ and $BA_i\%$ are density and basal area percentage for the ith species; N_i (tree) and BA_i (sq. m) are the density and basal area of the ith species; and N (trees) and BA (sq. m) are total tree density and basal area of the stand, respectively.

Three soil samples (0.5 kg each) were collected at a depth of 0 to 30 cm at representative locations within each plot. The soil properties were analysed, including humus and N% in the soil using the Tyurin and Kononova method (Kononova, 1966); pH_{KCl}, using the Meter method (Huang and Summer, 2012); P_2O_5 (mg/100 g soil), using the Oniani method, (Oniani et al., 1973); and K_2O (mg/100 g soil) using the flame photometer method (Estefan et al., 2013).

For an overview and discussion of shifting cultivation in Vietnam, as well as that in Southeast Asia more generally, we undertook a literature review. We also surveyed and interviewed local people to describe their swidden practices and record their indigenous knowledge pertaining to the composition of integrated crop species, different durations of fallow, soil-fertility changes and forest restoration during different fallow periods, and trends of changing fallow-land use in conditions of market-driven agricultural development.

Results

The farming systems

There are common patterns to the shifting cultivation systems used by the ethnic minority groups of Vietnam. These include:

Rotational shifting cultivation: This form of swidden farming is common to most of the ethnic minorities (Figure 18-2). It involves two to four years of cultivating crops on a forest plot. When the productivity of the crops begins to decline, farmers clear new fields from the forest and the old fields are abandoned for natural regeneration of the forest and replenishment of soil fertility. The old fields can be returned to cultivation after a fallow period of seven to 15 years. In this way, cultivation is periodically rotated, or 'shifted', and land use is stable and lasting (Sam, 1994, 1996,1998; Huy et al., 1998; Hung, 2004). This type of shifting cultivation has many variations, with some farmers in remote areas distant from markets still using the same traditional methods (Hung, 2004, Huy et al., 1998). In some areas where fallowing is not observed because the government does not recognize farmers' land-tenure rights (Jong, 2007), there is partial application of agroforestry to prolong the period of cultivation (Huy, 2014, De Royer et al., 2016). In many other areas, where there is easy market access and high demand, the government has encouraged the transformation of shifting cultivation land into monoculture cropping systems (Huy et al., 2018; De Royer et al., 2016; Schmidt-Vogt et al., 2009; van Noorwijk et al., 2008).

Pioneer shifting cultivation: Farmers practise slash–and–burn cultivation until the land is exhausted and the productivity of crops decreases sharply. They then abandon the old fields and enter the forest to develop new fields. In some cases, when forest land is no longer available near a village, the entire village may move to a new site (Sam, 1994; Huy et al., 1998; Hung, 2004). This type of farming has been adopted by communities that use traditional shifting cultivation methods in areas where there is limited access to markets. Nowadays, this form of shifting cultivation is seldom applied because most ethnic minorities are expected to settle. At the same time, there is insufficient forest land available to support this form of shifting cultivation.

FIGURE 18-2: Landscape in the Central Highlands of Vietnam showing various phases of shifting cultivation.

Photo: Bao Huy.

Shifting cultivation in combination with wet-rice cultivation: This composite form of farming has been adopted by ethnic-minority communities who cultivate wet-rice fields in valleys and coordinate this with farming of upland fields in the surrounding hills. Many ethnic-minority groups, such as the Tay, Nung, Thai and Muong, practise this pattern of cultivation in Vietnam's Northern Mountainous Region and the practice is being adopted in the Central Highlands (Sam, 1994; Huy et al., 1998; Hung, 2004). The government of Vietnam has promoted this pattern of cultivation in the Central Highlands by building irrigation systems for indigenous communities to grow wet rice, while others have been trying to maintain shifting cultivation. This approach has not only decreased fallow areas, but has also had a deep impact on the customs and culture of indigenous shifting cultivators in the Central Highlands.

The common and traditional process of swidden farming followed by most ethnic-minority communities begins with choice of a forest site. The site is then cleared by slashing the vegetation and felling trees. After being allowed to dry, the slash is then burnt, and soon afterwards, the crops are sown. The crops are then tended and weeded as they grow and measures are taken to prevent attacks by animals and birds. Finally, the crops are harvested. After two or three years of this cultivation, the plot is allowed to lie fallow for 10 to 15 years (Huy et al., 1998). Vien (2007) described an alternative form of shifting cultivation typical of farmers in the north of Vietnam. In this case, rice was grown in swiddens for two or three years, followed by two years of cassava cultivation. The plots were then fallowed for five to 10 years. The entire shifting cultivation cycle of ethnic-minority farmers in the uplands of Vietnam covers about 10 to 12 years, on average. In this respect, it is similar to shifting cultivation systems practised in other Asian countries (Cherrier et al., 2018).

The swidden-farming process in the uplands follows a calendar of events that is highly dependent on the weather, and climatic conditions prevailing in mountainous areas within ecoregions. In the Central Highlands there are two distinct seasons: rainy and

Cratoxylum cochinchinense (Lour.) Blume [Hypericaceae]

This small deciduous tree grows to 18 m and, in the study site, is often found in the first four years of fallow. It is harvested as a source of food, medicine, dyes and wood. The young fruit is used as a spice for cooking and young shoots are eaten raw as a vegetable. The roots, bark and twigs are used as a treatment for colds and diarrhoea. The wood is very hard and highly valued.

dry. In the dry season, from January to April, there is strong sunshine, strong winds and almost no rain, so the burning of new fields takes place at the end of the dry season. Cultivation occurs only during the rainy season, from May to December (Dao, 1999; Huy et al., 1998; Dung, 2000).

Up until the 1970s, most indigenous farmers in the Central Highlands were engaged in rotational shifting cultivation. The farming system was stable because natural forest cover was extensive, population density was low, and the communities adhered strictly to customary rules of land tenure (Jong, 2007). There was no destruction of forests or environmental degradation (Hung, 2004). However, the population density accelerated rapidly because of migration – both legal and illegal – from the north of the country. Thus, demand for food production and industrial tree planting increased the pressure on dwindling land resources, and the time for which land was allowed to lie fallow after cultivation grew shorter. Without the opportunity for restored soil fertility, the land was burnt again for cultivation, leading to a vicious cycle in which the forest retreated and increasing areas of land became degraded (Thinh and Son, 1998; Sam, 1996, 1998; Huy et al., 1998; Dung, 2000; Hung, 2004).

The recovery of soil fertility after cultivation increases hand-in-hand with rehabilitation of forest vegetation in the fallow period. However, the length of time for which the land must lie fallow depends on the specific situation. Sam (1996) and Hung (2004) found that where swidden fields were in small clusters in the forest, the natural vegetation was restored very quickly in the fallow, taking from seven to 10 years for full regeneration. In cases where swiddens were cleared from bamboo-dominated forests, it took a long time – often more than 15 years – to restore a stable forest, and if swiddens were cleared in areas of low forest cover, then forest restoration was slow and fallows needed to last more than 20 years.

Indigenous knowledge of shifting cultivation and forest fallows offers a sound basis for solutions for forest ecosystem rehabilitation and improving livelihoods in different human-ecology contexts.

Aporosa octandra var. *malesiana* Schot [Phyllanthaceae]

A fast growing tree that is commonly found in the first eight years of fallow. Along with others in its genus, this tree accumulates aluminium and is restricted to acidic soils, which prevail in rainforests (Schot, 2004). The wood is hard and dark brown, suitable for house construction or furniture and the leaves can be used to dye cloth black.

Changes in soil properties during different fallow periods

Shifting cultivation often occurs on sloping land, so after a few years of cultivating food crops, the soil deteriorates and the farmers fallow the land to restore its soil fertility for the next cultivation cycle. This study sought to evaluate the resilience of soil fertility over different fallow periods, and to compare this with the fertility of primary forest land. With three soil samples taken from each of 15 fallow and forest plots (as described earlier), 45 samples were analysed and their main properties were averaged. These properties included pH_{KCl}, Humus %, N %, P_2O_5 %, K_2O %, P_2O_5 (mg/100 g soil) and K_2O (mg/100 g soil) (Table 18-1).

Figure 18-3 shows that the accumulation of phosphorus (P) and potassium (K) content in soils taken from eight-year-old fallow land were nearly as good as those in primary forest. At the same time, the percentage of nitrogen (N), P and K in the soils approached the percentages in primary forest only after a fallow lasting more than 10 years. However, after 10 years of fallow, the pH_{KCl} and humus content were still low compared to primary forest. The pH_{KCl} value for 10 years of fallow was 4.21 compared to 5.50 in primary forest; and humus content was 5.35% in the 10-year-old fallow, while in primary forest it was 6.50%. Figure 18-4 shows how these differing fallow plots, and the primary forest, appeared to the research team.

The restoration of forest structure according to fallow time

The dynamics of the distribution of tree height in fallows of different ages are shown in Table 18-2 and Figures 18-5 and 18-6. It is easy to see that the natural regenerative capacity of fallow land is very strong. After two and four years of fallow, the density of regenerating trees was between 14,000 and 23,000 trees/ha with a height (H) equal to or over 2 m. The density of trees with a diameter at breast height (dbh) equal to or over 10 cm then decreased sharply when the fallow period exceeded 10 years and gradually approached the density of primary forest (Table 18-2). The

TABLE 18-1: Soil-property indicators at a depth of 0 to 30 cm in fallow plots of different age, compared to primary forest.

Years of fallow	pH_{KCl}	Humus %	N %	P_2O_5 %	K_2O %	P_2O_5 mg/100 g soil	K_2O mg/100 g soil
2	3.86	3.67	0.13	0.12	0.02	2.50	8.82
4	3.90	4.49	0.17	0.09	0.02	3.30	11.24
8	3.92	5.04	0.22	0.11	0.03	3.55	12.01
10	4.21	5.35	0.23	0.06	0.05	3.89	12.35
Primary Forest	5.50	6.50	0.25	0.13	0.09	3.90	12.40

Note: Indicators were averaged from nine soil samples taken from three plots in each of the four fallow ages, plus primary forest.

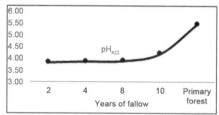

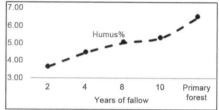

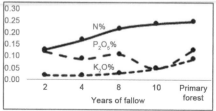

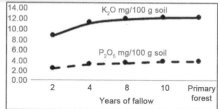

FIGURE 18-3:   Trends of soil properties in different fallow periods, compared to those in primary forest.

FIGURE 18-4:   Forest fallows of different age in shifting cultivation areas of Vietnam's Central Highlands, compared to primary forest. Top left: two-years-old fallow; top right: four-years-old fallow; above left: eight-years-old fallow; above right: more than 10-years-old fallow; left, primary forest.

Photos: Bao Huy.

TABLE 18-2: Distribution of tree height after different fallow periods compared to primary forest.

Tree height (m)	Number of trees/ha				
	Two years of fallow	Four years of fallow	Eight years of fallow	Ten years of fallow	Primary forest
4	14012	18005	509	121	5
8		5012	303	223	61
12			242	125	142
16				97	123
20				78	99
24					87
28					27
32					11
Total trees/ha	14012	23017	1054	644	555
Average height ($\bar{H}$) (m)	2.2	4.3	9.5	11.4	18.3

Notes: Numbers of trees per ha were averaged from data from three plots for each fallow period. At two and four years of fallow, trees with height equal to or greater than 2 m and a diameter at breast height less than 10 cm were counted; at eight and 10 years of fallow and primary forest, trees with height equal to or greater than 2 m and a diameter at breast height equal to or greater than 10 cm were counted.

average height ($\bar{H}$) of regenerating trees increased rapidly in the first stage and after 10 years of fallow, $\bar{H}$ was about 0.6 $\bar{H}$ of primary forest (Figure 18-6).

After 10 years of fallow, the height structure of the regenerating forest was approaching that of primary forest. Expressed in a graph (Figure 18-5), the height distribution in the 10-year-old fallow formed an apex to the left of that representing primary forest. The structure of the evergreen broad-leaved forest consisted of three layers of tree canopy: a lower canopy with a height of less than 12 m, an ecological layer with a height of 12 to 28 m and a canopy overpass with a height of more than 28 m. Figure 18-5 shows that after 10 years of fallow, the regenerating forest had formed two main forest layers: an understorey and an ecological layer, and the ecological circumstances of the forest were thus gradually rehabilitated.

Restoration of tree-species component after different periods of fallow

There were significant changes to the tree-species component of the fallows as they became older. The dominant tree species, with an importance value index (IV) greater than 5%, in the two-year-old fallow were mostly pioneer trees that were small, soft, sunlight-demanding and fast-growing. These included species such as *Aporosa octandra* var. *malesiana*, *Cratoxylum cochinchinense*, *Dillenia ovata* and *Grewia paniculata*. After four years of fallow, large timber species with economic value emerged among the dominant tree species, including *Dalbergia cochinchinensis*,

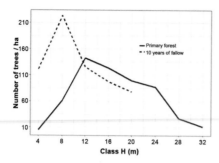

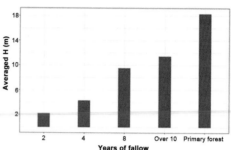

FIGURE 18-5: Distribution of tree height after 10 years of fallow and in primary forest.

FIGURE 18-6: Average tree height of stands in fallows of different age, compared to primary forest.

Melicope pteleifolia and *Lagerstroemia calyculata*. These species were able to regenerate because of the forest ecological environment created by the earlier pioneer species. At this stage of fallow, the dominant-tree-species component included both the sunlight-demanding pioneer species group and large valuable trees. In the period from eight to 10 years of fallow, most of the sunlight-demanding pioneer species had been excluded from the dominant-tree-species component by the large timber trees, and by the time the fallow reached 10 years of age, the dominant tree species came closer to the dominant-species component in the primary forest, and included *Lithocarpus* spp., *Persea odoratissimus*, *Schima crenata*, *Syzygium odoratum*, and *Walsura* sp. At this stage, the dominant-species component was stable (Table 18-3).

As the duration of the various fallows grew longer, the number of dominant tree species increased and reached a peak concentration in the eight-years-old fallow. The number of dominant tree species then decreased, due to space competition and the change in ecological conditions, which became unsuitable for most of the sunlight-demanding pioneer species. When the fallow exceeded the age of 10 years, the number of dominant tree species with an importance value index (IV) greater than 5% decreased to the lowest point and came close to that of primary forest (Figure 18-7).

Discussion

Restoration of soil fertility and natural forest during periods of fallow

The restoration of soil fertility during the fallow period was due to the regeneration of forest vegetation, and the organic relationship between these two elements. This study was conducted in areas of evergreen broadleaved forest with periods of drought not exceeding two to three months – an environment that facilitates the rapid restoration of forest vegetation, thereby promoting improvement of soil fertility during the fallow.

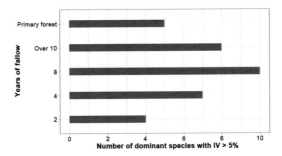

FIGURE 18-7:   Distribution of dominant tree species after different periods of fallow, compared with the dominant tree species in primary forest.

TABLE 18-3: Changes to the dominant tree species after different fallow periods, and comparing these with primary forest.

Dominant tree species with IV > 5%		*IV% in different fallow periods and in primary forest*				
Botanical names	*Vietnamese name*	*2 years*	*4 years*	*8 years*	*10 years*	*Primary forest*
Albizia lebbekoides	song ran			6.34		
Aporosa octandra var. *malesiana*	thau tau	16.07	10.87			
Cratoxylum cochinchinense	lanh nganh do ngon	7.14	5.43			
Dalbergia cochinchinensis	trac		8.70	5.12		
Dillenia ovata	so	7.14				
Melicope pteleifolia	dau dau 3 la		5.43	7.23		
Microcos tomentosa	co ke	5.36	7.61			
Lagerstroemia calyculata	bang lang		5.43	5.82	5.12	
Lithocarpus spp.	de			32.92	20.34	16.17
Persea odoratissima	khao			5.17	12.45	15.23
Schima crenata	cho xot			10.34	9.03	12.45
Semecarpus annamensis	sung				5.88	
Syzygium odoratum	tram la nho			5.65	5.45	6.00
Vitex pinnata	binh linh			5.23	5.12	
Walsura sp.	nhan rừng		6.52	6.23	7.34	5.14
Total of dominant tree species		4	7	10	8	5

Notes: IV = importance value index. Average IV percentage values were calculated from data from three plots for each fallow period.

Christanty (1986) pointed out that the length of the fallow period was the most important factor to ensure the success of shifting cultivation; and that there were some additional ways of improving the rehabilitation of soil fertility and prevention of soil erosion on sloping land that could be incorporated at different stages of the shifting cultivation cycle. This study found that with a fallow period of at least 10 and preferably 15 years, the soil properties recovered and the structure and dominant-tree-species composition of the fallow forest almost stabilized, and the fallow was gradually approaching the status of primary forest. These findings support the indigenous knowledge of shifting cultivators in the Central Highlands. Along with the application of rotational cultivation, they use upland fields and forests sustainably and, after the cultivation phase, close their swiddens for an appropriate fallow period of 10 to 15 years (Figure 18-8). These results are also consistent with those of Vien (2007), Dao (1999), Huy et al. (1998) and Cherrier et al. (2018), although they differ from those of Christanty (1986), who found that in the tropics generally, a fallow period of 20 to 25 years was needed for forest rehabilitation.

It follows that the original forest ecosystem in the Central Highlands could be rehabilitated through the adoption of simple silvicultural measures such as zoning and protection of forest fallows. Some substantial alternatives should be considered to improve soil fertility and forest rehabilitation alongside the continued practice of shifting cultivation. These might include controlling the crop/fallow ratio in forest corridors that are about 100 m wide and oriented in an east-west direction, replacing natural fallows with planted fallows, and practising agroforestry, or maintaining continuous ground cover through minimum tillage during fallow periods (Christanty, 1986).

Sustainable management of fallow land and application of agroforestry for rehabilitation of forest ecosystems

Although fallow land occupies millions of hectares in Vietnam, as in many other

Lagerstroemia calyculata Kurz [Lythraceae]

Crepe Myrtle is a common forest species at low altitudes in Vietnam, and the trees are widely planted in many countries as ornamentals, showing vivid displays of pink or pale violet flowers. The trees grow up to 30 m, and are thought to have gained dominance in parts of Vietnam's Cát Tiên National Park simply because the relatively poor quality of the wood allowed them to survive repeated logging.

countries in Southeast Asia, fallow land is not recognized by either the forest-land classification system (Jong, 2007) or the forestry sector's system of measures for forest restoration (De Royer et al., 2016). Driven by misconceptions about its impacts on forests, most governments in Asian countries have long been trying to ban traditional shifting cultivation, while indigenous peoples fight to keep their ancestral fallow lands. In many cases, they must farm their swiddens illegally (Cherrier et al., 2018). Therefore, communities that rely on customary fallow land for their

FIGURE 18-8: A M'nong farmer cultivating his swidden in Dak Nong province, Central Highlands, Vietnam, in 2018.

Photo: Bao Huy.

livelihood are less likely to receive official land-use rights from their governments (Jong, 2007). Challenged by an increasing population due to in-migration from the north of the country as people seek better-quality land to cultivate, and pressure from market forces that drive the development of commodity cropping, indigenous people are hard-pressed to protect their customary fallow land. For these reasons, fallow lands are not effectively managed.

Consistent with the findings of Watters (1971), shifting cultivation remains a realistic solution for sustainable farming in the humid tropics, provided there is sufficient forest and forest land; where the population pressure is not high; and where

Syzygium odoratum (Lour.) DC. [Myrtaceae]

An evergreen species common in mountainous broad-leaved forests in China and Vietnam, growing to 20 m tall with a diameter at breast height up to 30cm. This study reported its appearance as a dominant species in eight-year-old fallows, with an importance value close to that in primary forests. Better known species from the same genus include cloves (*S. aromaticum*) and roseapples (various *Syzygium* species).

fallow periods are either long enough to restore soil fertility or upland shifting cultivation systems can be improved by adopting agroforestry practices (Huy and Hung, 2011; De Royer et al., 2016). This could be the basis for sustainable forest land use and livelihoods, while preserving the culture of ethnic minorities in the uplands. These findings are consistent with those of Cherrier et al. (2018), who urged the generation of a common understanding about the significance of shifting cultivation for safeguarding the livelihoods of local people and protecting the forest environment. At the same time, these authors called for efforts to minimize the misconceptions of governments and researchers regarding shifting cultivation and the need for lengthy fallows. Adding to these voices, Kleinman et al. (1995) argued that slash-and-burn agriculture was ecologically sustainable because it did not depend on outside inputs, such as fossil-fuelled energy for fertilizers, pesticides and irrigation.

Over recent decades, a major challenge to the sustainable use of fallow land has been the Vietnam government's willingness to support the rapid expansion of monoculture cropping systems such as coffee, pepper, rubber, cashews and acacia into areas of fallow land – even the conversion of natural forests into industrial cropping systems (De Royer et al., 2016; Huy et al., 2018). The use of fallow-forest land, or conversion of dipterocarp forests, for industrial monocultures has been a demonstration of environmental degradation and economic failure in the Central Highlands. In many cases, monoculture systems failed to survive because of ecological hardship, but the conversion destroyed the forest ecosystem by robbing the soil of its vegetative cover and exposing it to severe conditions of drought and fire in the dry season and waterlogging in the rainy seasons. The soil nutrients were thus eroded (Huy et al., 2018). Cruz (2015) also pointed out that converting shifting cultivation land to other uses negatively affected biodiversity and soil fertility because of excessive land use.

The replacement of shifting cultivation with highly intensive monoculture systems generally increases the income of farmers in the short term, but it has negative effects on society and indigenous cultures at different levels. From an environmental point of view, the shift to other land-use systems results in deforestation, long-term biodiversity loss, increased weeds, reduced soil fertility and accelerated soil erosion (van Vliet et al., 2012). In addition, the expansion of monocultures reduces the carbon stocks of the forest (Ziegler et al., 2011). Therefore, despite the global trend towards intensive land uses, shifting cultivation still plays an important role in mountainous regions as a safeguard of diversity that avoids the risks of monocropping systems.

Other experiences in Vietnam – and elsewhere in Southeast Asia with tree-rich fallow lands – have suggested that crop species, when used in agroforestry systems, along with forest-enrichment planting (Huy et al., 2018), can help to rehabilitate the forest ecological environment (Sabogal, 2007). Agroforestry systems also demonstrate economic resilience, thanks to a diversity of short- and long-term species, along with the creation of effective carbon pools and provision of essential environmental services such as watershed protection and measures against soil erosion on sloping hills (De Royer et al., 2016; Huy, 2014; Huy et al., 2018). Swidden areas could thus be recognized as a forest type attracting payments for environmental services and from schemes such as Reducing Emissions from Deforestation and Forest Degradation (REDD+) in Vietnam (Moeliono et al., 2016).

Conclusions

Following the cultivation of swidden crops in evergreen broad-leaved forests in the Central Highlands of Vietnam, fallow periods that last for more than 10 years are likely to recover and stabilize soil fertility, forest structure and the composition of dominant tree species. If, after 10 to 15 years of fallow, there is no further use for the fallowed land, a measure of forest protection would ensure successful rehabilitation of the original forest ecosystem. Alternatively, swidden-based agroforestry systems would be an appropriate approach to harmonizing the maintenance of traditional shifting cultivation with the current tendency to intensify the use of fallow land by converting it to monocultures of commodity crops.

Schima crenata Korth. [Theaceae]

This forest species is widespread throughout China and Southeast Asia at altitudes between 700 and 1000 masl. Growing to 20 m, it is harvested for its timber. It has a strong importance value in eight- and ten-year-old fallows and a dominant place in primary forest. There have been calls for research into its life history, population and harvest trends to avert possible habitat threats from agricultural expansion.

The economic significance of monocultures of global commodity crops is seriously compromising the important role of shifting cultivation systems in the uplands of Vietnam. The expansion of monoculture cropping poses a serious threat to swidden farming systems that ensure the livelihoods and well-being of local people while conserving their traditional knowledge and cultures.

Diversification of food crops that improve the use of fallow land, combined with income from forest trees, can improve the livelihoods of indigenous people while simultaneously rehabilitating the forest environment. This is an important current action for the Central Highlands of Vietnam. However, it can only be achieved by incorporating local ecological knowledge in the management of shifting cultivation fallows and using forest-agricultural science and technology advances to rehabilitate the forest ecosystem and develop agroforestry systems on degraded land.

Acknowledgements

This study was partly funded by the Department of Science and Technology, Dak Lak province, Vietnam, with the participation of colleagues from the Faculty of Agriculture and Forestry, Tay Nguyen University, Vietnam.

References

Cherrier, J., Maharjan, S. K. and Maharjan, K. L. (2018) *Shifting cultivation: Misconception of the Asian Governments*, Reviews, the Food and Agriculture Organization of the United Nations, Bangkok

Chien, N. V. (1986) *Sub-ecoregions in the Central Highlands of Vietnam*, Scientific and Technical Publishing House, Hanoi

Christanty, L. (1986) 'Shifting cultivation and tropical soils: Patterns, problems and possible improvements', in G. G. Marten (ed.) *Traditional Agriculture in Southeast Asia: A Human Ecology Perspective*, Westview Press, Boulder, CO

Cottam, G. and Curtis, J. T. (1956) 'The use of distance measures in phytosociological sampling', *Ecology* 37(3), pp.451–460

Cruz, A., (2015) *The Past, Present and Future of Swidden Agriculture*, World Agroforestry Center (ICRAF), Nairobi, available at blog.worldagroforestry.org/index.php/2015/05/07/the-past-present-and-future-of-swidden-agriculture/, accessed 1 May 2018

Cuc, L. T. (1995) 'Rehabilitation of degraded land in the midlands of Northern Viet Nam', in L. T. Cuc and A. T. Rambo (eds) *Some Issues in Human Ecology in Viet Nam*, Agricultural Publishing House, Hanoi

Curtis, J. T. and McIntosh, R. P. (1950) 'The interrelations of certain analytic and synthetic phytosociological characters', *Ecology* 31(3), pp.434–455

Dao, B. M. (1999) *Traditional Cultivation of Ethnic Minority Groups in Central Highlands*, Social Science Publishing House, Hanoi

De Royer, S., Ratnamhin, A. and Wangpakapattanawong, P. (2016) *Swidden-fallow Agroforestry for Sustainable Land Use*, Policy brief No. 68, Agroforestry Options for Asean series No. 2, Bogor, Indonesia; World Agroforestry Center (ICRAF), Southeast Asia Regional Program, Jakarta, and Asean-Swiss Partnership on Social Forestry and Climate Change

Dung, T. T. (2000) 'Assessment of the status of shifting cultivation on the Buon Ma Thuot plateau and recommendations', PhD dissertation to the University of Natural Science, Hanoi

Estefan, G., Sommer, R. and Ryan, J. (2013) *Methods of Soil, Plant, and Water Analysis. A Manual for the West Asia and North Africa Region*, third edition, International Center for Agricultural Research in the Dry Areas (ICARDA), Beirut, Lebanon

Fischer, G., Nachtergaele, F. O., Prieler, S., Teixeira, E., Toth, G., van Velthuizen, H., Verelst, L. and Wiberg, D. (2008) *Global Agro-Ecological Zones Assessment for Agriculture* (GAEZ, 2008), International Institute for Applied Systems Analysis (IIASA), Laxenburg, Austria and the Food and Agriculture Organization of the United Nations, Rome

Hijmans, R. J., Cameron, S. E., Parra, J. L., Jones, P. G. and Jarvis, A. (2005) 'Very high resolution interpolated climate surfaces for global land areas', *International Journal of Climatology* 25 (1965–1978)

Huang, P. M., Li, Y. and Summer, M. E. (2012) *Handbook of Soil Sciences: Resource Management and Environmental Impacts*, second edition, CRC Press, New York

Hung, V. (2004) 'Management of fallow lands of shifting cultivation in Dak Lak province, Viet Nam', Ph.D. Dissertation to Viet Nam Academy of Forest Sciences, Hanoi

Huy, B. (2014) 'CO_2 sequestration estimation for the Litsea-Cassava agroforestry model in the Central Highlands of Vietnam' in *Compendium of Abstracts of World Congress on Agroforestry*, 10–13 Feb 2014, New Delhi, India, Indian Council for Agricultural Research and World Agroforestry Center (ICRAF) Global Initiatives.

Huy, B. and Hung, V. (2011) 'State of agroforestry research and development in Vietnam', *Asia-Pacific Agroforestry Newsletter (APANews)* 38(2011), pp.7-10

Huy, B., Hoa, N.V. and Dinh, N.D. (1998) *Assessment of current forest land management to recommend sustainable forest land use in Dak Lak province, Viet Nam*, Technical Report, Department of Science and Technology, Dak Lak province, Vietnam.

Huy, B, Tri, P. C. and Triet, T. (2018) 'Assessment of enrichment planting of teak (*Tectona grandis*) in degraded dry deciduous dipterocarp forest in the Central Highlands, Vietnam', *Southern Forests: a Journal of Forest Science* 80(1), pp.75-84, DOI: 10.2989/20702620.2017.1286560

Jong, W. D. (2007) 'Understanding forest landscape dynamics', in J. Rietbergen-McCracken, S. Maginnis and A. Sarre (eds) *The Forest Landscape Restoration Handbook*, Earthscan, London, pp.49-56

Kleinman, P. J. A., Pimentel, D. and Bryant, R. B. (1995) 'The ecological sustainability of slash-and-burn agriculture', *Agriculture, Ecosystems and Environment* 2(52), pp.235-249

Kononova, M. M. (1966) *Soil Organic Matter: Its Nature, Its Role in Soil Formation and Soil Fertility*, second English edition, Pergamon Press, London

Li, P., Feng, Z., Jiang, L., Liao, C., and Zhang, J. (2014) 'A review of swidden agriculture in Southeast Asia', *Remote Sensing* 2014 6(2), pp.1654-1683, doi:10.3390/rs6021654.

Moeliono, M., Thuy, P. T., Dung. L. N., Brockhaus. M., Wong., G. Y., Kallio. M. and Tien, N. D. (2016) 'Local governance, social networks and REDD+: Lessons from swidden communities in Vietnam', *Human Ecology* 44(4), pp.435-448

Narayan, C. and Anshumali (2015) 'Diversity indices and importance values of a tropical deciduous forest of Chhotanagpur plateau, India', *Journal of Biodiversity and Environmental Sciences* 7(1), pp.358-367

Nhung, M. L. T. (1998) *Indigenous Knowledge in the Central Highlands of Viet Nam*, Tay Nguyen University, Buon Ma Thuot, Dak Lak province, Vietnam

Oniani, O. G., Chater, M. and Mattingly, G. E. G. (1973) 'Some effects of fertilizers and farmyard manure on the organic phosphorus in soils', *Journal of Soil Science* 24, pp.1-9

Sabogal, C. (2007) 'Site-level strategies for managing secondary forests', in J. Rietbergen-McCracken, S. Maginnis and A. Sarre (eds) *The Forest Landscape Restoration Handbook*, Earthscan, London, pp.97-107

Sam, D. D. (1994) 'Shifting cultivation in Vietnam: Its social, economic, and environmental values relative to alternative land use', *Forestry and Land Use*, No. 3, International Institute for Environment and Development, London

Sam, D. D. (1996) *Overview of Shifting Agriculture in Vietnam*, Agricultural Publishing House, Hanoi, Vietnam

Sam, D. D. (1998) 'Swidden agriculture and sustainable forest management in Viet Nam', in *Proceedings of a Workshop on Sustainable Forest Management and Forest Certification*, Department of Forestry, Ministry of Agriculture and Rural Development, Vietnam

Schmidt-Vogt, D., Leisz, S. J., Mertz, O., Heinemann, A., Thiha, T., Messerli, P. and Dao, T. M. (2009) 'An assessment of trends in the extent of swidden in Southeast Asia', *Human Ecology* 37(3), pp.269-280

Schot, A. M. (2004) *Systematics of Aporosa (Euphorbiaceae)*, Blumea supplement 17, Leiden University branch, National Herbarium of the Netherlands, p.53

Thinh, N. D. and Son, C.T. (1998) *Traditional Regulations of Ede and M'nong Ethnic Minority People*, National Political Publishing House, Hanoi, Vietnam

van Noordwijk, M., Mulyoutami, E., Sakuntalakewi, N. and Agus, F. (2008) *Swiddens in Transition: Shifted Perceptions on Shifting Cultivators in Indonesia*, World Agroforestry Center (ICRAF) Southeast Asia Regional Program, Bogor, Indonesia

van Vliet, N., Mertz, O., Heinimann, A., Langanke, T., Pascual, U., Schmook, B., Adams, C., Schmidt-Vogt, D., Messerli, P., Leisz, S., Castella, J. C., Jørgensen, L., Thomsen, T. B., Hett, C., Bruun, T. B., Ickowitz, A., Vu, K. C., Yasuyuki, K. and Ziegler, A. D. (2012) 'Trends, drivers and impacts of changes in swidden cultivation in tropical forest-agriculture frontiers: A global assessment', *Global Environmental Change* 2(22), pp.418-429, doi.org/10.1016/j.gloenvcha.2011.10.009

Vien, T. D. (2007) 'Indigenous fallow management with *Melia azedarach* Linn. in Northern Vietnam', in M. F. Cairns (ed.) *Voices from the Forest. Integrating Indigenous Knowledge into Sustainable Upland Farming*, Resources for the Future Press, Washington, DC, pp.435–444

Watters, R. F. (1971) *Shifting Cultivation in Latin America*, Food and Agriculture Organization of the United Nations, Rome

Ziegler, A. D., Fox, J. M., Webb, E. L., Padoch, C., Leisz, S. J., Cramb, R. and Vien, T. D. (2011) 'Recognizing contemporary roles of swidden agriculture in transforming landscapes of Southeast Asia', *Conservation Biology* 25(4)

19

CALCULATING THE CARBON BALANCE

Questioning the contribution of shifting cultivation to climate change in Kayah state, Myanmar

*Christian Erni and KMSS-Loikaw**

Introduction

Throughout Asia, governments from the colonial period to the present day have devised policies and laws seeking to eradicate the practice of shifting cultivation (see, for example, Cairns (2017), Fox et al. (2009) and Padoch et al. (2007) for a general view; IWGIA (2007) for Laos; Pulhin et al. (2005) for the Philippines; Laungaramsri (2005) and Forsyth (1999) for Thailand; Phuc (2008) for Vietnam; and Dove (1985) for Indonesia). Many of the arguments against this land use – that it is economically inefficient, ecologically harmful, 'backward' and primitive – have been proven inaccurate or completely wrong (See, for example, Dove, 1983, 1985, 1996; Padoch, 1985; Forsyth, 1999, Laungaramsri, 2005; Nielsen et al., 2006; Forsyth and Walker, 2008). Notwithstanding all of the evidence, the attitudes of decision-makers, and consequently state policies, have hardly changed in more than a century.

The menace of global climate change has recently led to a new charge being levelled against shifting cultivators, with a full measure of assumption and prejudice: that the practice of burning new swidden fields releases significant quantities of carbon dioxide (CO_2) into the atmosphere, thus contributing to climate change. There is no argument against the fact that fires release CO_2 into the atmosphere. But the carbon-emissions accusation overlooks the fact that carbon is cyclical, and fails to tell the whole story. This chapter joins an increasing number of recent studies that have found that long-fallow systems of shifting cultivation do not cause net carbon emissions, but are carbon neutral or even carbon positive, i.e. they sequester more carbon than they emit.

---

* Dr. CHRISTIAN ERNI holds a Ph.D. in Social Anthropology from Zurich University, Switzerland. He worked for nearly 20 years for the Copenhagen-based International Work Group for Indigenous Affairs (IWGIA) and is now an independent researcher and consultant. KMSS-LOIKAW, or Karuna Mission Social Solidarity-Loikaw (the capital of Kayah state in Myanmar), is one of 16 Diocese-level offices of a faith-based social network of the Catholic Church of Myanmar.

These efforts to once again prove the inaccuracy of policy-sensitive dialogues about shifting cultivation come at a time when economic transformations, population increases and policy interventions that promote large-scale agro-industrial plantations, on one hand, and forest conservation and thus the restriction of access to land for upland farmers on the other, have already led to a rapid decline of shifting cultivation throughout Southeast Asia (Fox et al., 2009). Yet this centuries-old practice still feeds millions of people, and with its decline a treasury of indigenous wisdom is slipping through the fingers of a world facing environmental disaster. Recent global analyses show that while transitions from shifting cultivation to other forms of land use lead, in many cases, to increased incomes, other expected benefits, above all those with respect to forest and biodiversity conservation or increased carbon sequestration, are not at all certain (van Vliet et al., 2012; Ziegler et al., 2012; Dressler et al., 2016).

In the words of van Vliet et al. (2012, p.418) 'swidden remains important in many frontier areas where farmers have unequal or insecure access to investment and market opportunities, or where multifunctionality of land uses has been preserved as a strategy to adapt to current ecological, economic and political circumstances.'

In Myanmar, many upland communities depend for their livelihood on shifting cultivation. Reliable figures on the number of shifting cultivators in the country are non-existent and estimates vary considerably, from 2 million to 20 million people. However, up to half of the country's upland population is believed to make a living from shifting cultivation (Springate-Baginski, 2013). Yet, like elsewhere in the region, negative prejudices against shifting cultivation prevail among the country's decision-makers. In the 2015 election manifesto of the National League for Democracy, a chapter entitled 'The Freedom and Security to Prosper' includes the statement: 'The NLD will carry out the following activities in order to reduce the current levels of pollution and environmental harm, and to create a better environment: ... 2. Farming: We will provide education and practical assistance in order to eradicate shifting cultivation practices.'

In 2011, Myanmar joined the United Nations Collaborative Programme on Reducing Emissions from Deforestation and Forest Degradation in Developing Countries (UN-REDD). The draft National REDD+ Strategy,[1] written by the UN-REDD programme and the Ministry of Natural Resources and Environmental Conservation (MONREC), mentions shifting cultivation as one of the direct drivers of deforestation, with an estimated impact on 6 to 7 million hectares in upland areas. It states that the area has decreased since 2000 and that it is 'expected to decrease further, if conflicts can be resolved' (REDD+ Myanmar, 2018, p.18). The strategy does not propose any direct action to limit shifting cultivation, but encourages the promotion and support of other forms of land use, including improved tenure security through implementation of the Land Use Policy of 2016. Importantly, the draft strategy departs from the approach of other UN-REDD documents (FAO, UNDP, UNEP, 2008), which say that shifting cultivation is 'associated with' deforestation, rather than being 'a driver of deforestation'.

In 2015, Myanmar submitted a list of Intended Nationally Determined Contributions to the United Nations' Framework Convention on Climate Change, prominent among them the aim to have 30% of its land area as a permanent forest estate and 10% of its land area as protected areas by 2030.

Concern about the country's continuing negative attitude towards shifting cultivation and its likely effects on upland communities led to a unique research effort, and ultimately, to this chapter. A faith-based social network of the Catholic church in Myanmar called Karuna Mission Social Solidarity-Loikaw has been working with shifting cultivation communities for many years. Mandated by the Catholic Bishops' Conference of Myanmar, KMSS-Loikaw became concerned about the negative impact on shifting cultivators arising from policies on climate change and environmental conservation. It decided to have a say in discussions about the impact of shifting cultivation on climate change by conducting a study in conjunction with one of its partner communities. The aim was to measure the 'carbon footprint' of a swidden farming community[2] – in reality, a task of such complexity it would daunt the best equipped and funded of scientific inquiries.

The chosen community was Khupra, near the western border of Kayah, Myanmar's smallest state, which shares a border with northern Thailand (Figure 19-1). The research team comprised 10 members of the Khupra community and five members from KMSS-Loikaw's Livelihood Programme, with three 'trainers', including the author. The participatory action research (below) was conducted between October 2015 and July 2016. The results were published as a report, entitled *Livelihood, Land Use and Carbon: A Study on the Carbon Footprint of a Shifting Cultivation Community in Kayah State* (Erni and KMSS-Loikaw, 2017). This chapter is drawn from that report.

In the following sections, we will describe the study village and its people, its agricultural systems, the methodology used in the study and finally the results, with discussion and conclusions. The focus eventually falls on the sophisticated fallow management of Khupra villagers and their not uncommon practice of trimming and retaining mature trees in their swiddens, and the impact these have on carbon sequestration – and therefore the village's carbon footprint – throughout the shifting cultivation cycle. Finally, the level of carbon stocks found in Khupra's shifting cultivation lands are compared with those in alternative land uses.

Khupra community

Khupra is a community of the Kayan Hlahui, one of the indigenous peoples of Kayah State. The three villages that comprise Khupra community are located in Demoso township at around 1350 metres above sea level (masl). The highest peak in Khupra's territory, Paikana, rises to 1803 masl. The community's territory covers about 19.4 square kilometres (1940 hectares), consisting mainly of mountainous land, with very little flat land along the Kanklo river and its tributaries. Most of the uplands are covered by secondary forest, grass and shrub land and some residual old-growth forest. Most of the secondary forests are shifting cultivation fallows at various stages of regeneration.

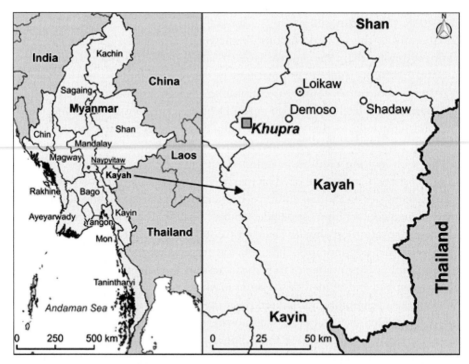

FIGURE 19-1:   Location of the study village of Khupra, in Kayah state, Myanmar.

The three villages are *Dou Du* (Big Village in Burmese and usually referred to in English as Upper Khupra) with 62 households; *Dou Lû* (Old Village or Old Khupra) with 26 households; and *Dou Thû* (New Village or New Khupra) with 24 households. Those who live in Upper Khupra are baptists, while residents of Old Khupra and New Khupra are catholics. A household census conducted for this study found that the total population of the 112 households was 612, of which 323 were males and 289 females.

According to oral history, Khupra was founded around 1360 A.D. Since then, it has had 22 village leaders, whose names are all remembered and recorded by Khupra villagers. Over the centuries, there has been considerable movement of people in and out of Khupra and the village site has shifted several times. Village history says that the people split into baptists and catholics following a power struggle between two village chiefs. Twenty families living in today's Upper Khupra supported one of the chiefs and became baptists in 1903, while around 30 families in Old and New Khupra supported his opponent and became catholics in 1906.

The entire Khupra community, which then totalled 66 households, was forced to relocate to Aungmingala in Demoso township during 1991 and 1992 because of conflict between the armed forces of Myanmar and supporters of the Karenni National Progressive Party. Due in part to outmigration, the population of Khupra grew by only 38% over the 85 years between 1906 and 1991. However, over the past 25 years the population has grown by 70%, or an average 2.8% per year.

Livelihood and land use

With the exception of just one household, everyone in Khupra is engaged in farming. However, most households are not able to cover all their needs from farming alone and pursue a broad range of other livelihood activities to support food security by generating cash income. These include the sale of handicrafts and forest products, doing carpentry or masonry work for others, daily labour in their own and neighbouring villages and seasonal labour migration. Small retail shops provide additional income for a few families, but only one household lives entirely from the operation of a retail shop.

Most households own some land for wet-rice cultivation, although the size of the holdings differ considerably, ranging from 1215 sq. m to 1.4 hectares (14,000 sq. m). Due to the limited area of flat land for wet rice, terraces have been cut in suitable areas with technical training provided in 2010 and 2011 by the French non-governmental organization Action Contre la Faim. The terraces rely on rainfall and water from creeks and streams. Some families have begun transforming their swidden and grazing land into rice terraces, although not all families can afford the intensive labour involved, and continue to rely on shifting cultivation to make ends meet.

All households in Khupra grow upland rice in swiddens, with 99 out of the community's 111 farming households either owning or co-owning shifting cultivation land. Around half of the shifting cultivation plots are owned by individuals (in most cases men), and the rest are owned collectively by clans. The size of land holdings for shifting cultivation ranges from 0.4 to 8 ha, with an average size of 4.25 ha. A few people have established permanent gardens and orchards on individually-owned shifting cultivation land and have started growing cardamom, coffee and bamboo as cash crops.

Since land ownership is unevenly distributed, almost half of all households in Khupra were renting land in 2016. In some cases they were allowed free use of the land, but in most cases rent was paid either in cash (10,000 kyat, or about US$8.51 per field), a share of the rice harvest (usually one third) or in labour (10 days).

Only 13 of the 111 rice-growing households were able to produce enough rice to cover their annual needs. Nineteen had enough for about eight months; 26 for about six months, and the rest – almost half of all households – did not have enough rice to last for six months. Various income-generating activities were therefore of critical importance for the livelihood of the Khupra villagers, since most of them needed cash to buy rice and other necessities.

Cash crops such as cucumbers, pumpkins, beans, lack root (a kind of vegetable) and, above all, chillies, are grown in shifting cultivation and paddy fields, but only about 10% of all households plant these cash crops. The most important source of cash income is casual labouring, either in the local area or through seasonal migration. About 70% of all households earn money in this way. The second most important source of cash is forest products, above all the sale of baskets and brooms made from bamboo and broom grass, respectively. A broad range of other forest products are collected for cash income: firewood, timber for construction, fibres for weaving mats, herbal medicines, honey, flowers and stems of wild bananas, orchids, mushrooms, wild vegetables and the meat of wild animals. More than 90% of all households gather forest products for their own use and consumption, while almost 40% depend partly on the sale of forest products for their cash income.

Increased mobility, thanks to affordable Chinese motorcycles and itinerant petty traders, has given Khupra villagers better access to manufactured goods, as well as processed and fresh food like fish, meat and vegetables. Even though most households regularly buy vegetables in the market, especially in the dry season, kitchen gardens, swidden fields, fallows and forest are still important food sources.

More than 70% of Khupra's land area is covered by forest. However,

Thysanolaena latifolia (Roxb. ex Hornem.) Honda [Poaceae]

Making brooms from wild broom grass provides an important source of income for villagers at Khupra. This species grows in dense clumps up to four metres high. As well as making brooms from the flowers, the leaves can be used as fodder for livestock and the stems can be used as firewood.

most of it is shifting cultivation fallow and is therefore subject to forest-succession dynamics. Thus, more than half of the village territory is covered by a mosaic of secondary forest between 1 and 14 years old. The mosaic is constantly changing as new areas are cleared and old fields are left fallow every year. About 17% of Khupra's land is under permanent forest (Table 19-1). The community has declared an area of 7.2 hectares around the spring that supplies their water as a protected forest. There are three small spirit forests and five old and new cemetery areas. In these forests, cutting of trees is forbidden, so they are the only remnants of old-growth forest.

Forest land is owned collectively, mainly by four clans, members of which live in all three villages. However, in New Khupra, the people decided that grazing and forest land near Paikana mountain and Mo stream belonged to the whole community. Clan ownership implies that clan members have use rights over these lands, but all village members can access and use forest resources on clan-owned forest land and all community members and even people from neighbouring communities can let their cattle and buffaloes graze on land that is owned by the clans.

TABLE 19-1: Land use in Khupra.

Land use	ha	%
Agricultural land		
Wet-rice fields (*sūena*)	69.07	3.56
Shifting cultivation (*sue khu sue kar*)	1058.46	54.55
Gardens (*thoū tar phar*)	12.21	0.63
Forest land		
Use forests (*thoūphar*)	302.36	15.58
Protected and spirit forests (*kan du*)	11.20	0.58
Cemeteries (*lu khu*)	13.58	0.70
Other land		
Grazing land (*bon panan kalan ansar gan*)	430.92	22.21
Settlements (*doū*)	38.49	1.98
Caves (*lōn Ku*)	2.65	0.14
KNPLF border guard camp	1.25	0.06
Total land area	**1940.19**	**100.00**

Two thirds of all households own cattle and/or buffaloes, in total 121 head of cattle and 35 buffaloes. Most of these animals are allowed to graze on pastures that are collectively owned by the clans. Land that is used mainly for grazing covers about 20% of the village's territory. However, in some parts of this grazing land the forest is recovering and a few families have begun opening small plots for shifting cultivation.

Shifting cultivation

It has been estimated that shifting cultivation contributes to the livelihoods of between 15 and 35 million people in Southeast Asia (Dressler et al., 2016). Most of these people belong to ethnic groups that are generally subsumed under categories such as ethnic minorities, tribal people, hilltribes, aboriginal or indigenous peoples.

Shifting cultivation can be defined as 'a land-use system that employs a natural or improved fallow phase that is longer than the cultivation phase of annual crops – sufficiently long to be dominated by woody vegetation, and cleared by means of fire [in order for the cycle to be repeated]' (Mertz et al., 2009, p.261). The precise term for shifting cultivation in the Burmese language is *shwe pyaung taungya* (Ennion, 2015).

The cycle

The people of Khupra practice a long-fallow system of shifting cultivation (Figure 19-2). The new cycle starts early in the year when the area to be cleared is identified by clan elders. Khupra villagers prefer to have their swiddens next to each other, since less labour is required for tasks like making fire breaks and building fences. Therefore, most of a year's new fields are usually cleared in two or three large blocks – a practice that leads to the creation of contiguous areas of fallow land of the same age.

The main criteria for site selection are ownership and the age of the fallow. People prefer to clear a new field on their own land simply because borrowing land from others implies paying rent. However, since they only clear fields in forest that has been fallow for at least 10 years, some families face the unavoidable need to borrow land from others.

New swiddens are cleared through a system of labour exchange involving all of the families owning plots in a large block. By April, the

FIGURE 19-2: A family group weeds a recently planted swidden, which has been cleared from secondary forest after at least 10 and up to 14 years of fallow regeneration.

Photo: Ingatio, KMSS-Loikaw.

fields are burnt. All families are once again obliged to contribute labour for making a fire break around the block and helping to control the fires. After the fires die down, the collective workforce erects a fence around the block to keep animals out of the swiddens.

When the first rains of the wet season are about to fall, each family plants its own plot. Again, exchanges of labour among the families sharing a block are common. Along with the main crop of upland rice, other cereals like maize or millet and various tubers and vegetables are intercropped, such as taro, yams, cucumbers, pumpkins, mustard, lettuce, coriander and beans. Some families may plant part of a field with cash crops like chillies.

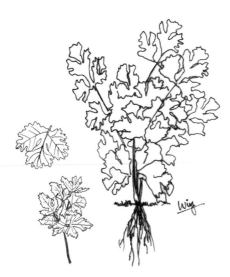

Coriandrum sativum L. [Apiaceae]

Coriander is one of the vegetable species intercropped with upland rice in Khupra swiddens. The entirety of this plant is edible; the seeds are used as a spice, the leaves as herbs, and the roots as flavouring ingredients. Its history reaches back to Neolithic times.

Seeds are sown by dibbling, in small teams of a man and a woman. The man usually makes a shallow hole with a dibble stick, into which the woman drops the seeds. Compared to harrowing or ploughing, this causes only minimal disturbance of the topsoil and helps to reduce soil erosion.

During the growing season, the fields are weeded at least two or three times. No chemical inputs are used in shifting cultivation in Khupra, although they are not formally prohibited. The first vegetables can be harvested a few weeks after planting. Maize is harvested in July and August, rice in November and millet in December.

Only rarely is a field used for a second year; it is usually left fallow after one growing season. Young fallows continue to provide tubers and some vegetables until the secondary forest takes over. On the rare occasions when a field is replanted for a second year, it is with maize and other less demanding crops like chillies.

A vital aspect of the Khupra shifting cultivation system is that the villagers never cut all of the trees when they open a new field. They prefer to retain some larger trees and only lop their branches, leaving the cultivated swiddens studded with large trees with recovering foliage (Figure 19-3). Some smaller trees are also left standing and lopped, to provide support for beans or yams, and many trees are cut a foot or two above the ground, so they quickly grow shoots again from the root stock. Many of the large trees in Khupra swiddens have obviously survived many swidden cycles and have grown to a considerable size despite repeated lopping. The practice of retaining so-called 'relict emergents' has long been observed as one of the hallmarks of sustainable long-fallow shifting cultivation (Schmidt-Vogt, 2007). The practice drew mention in reports of sustainable shifting cultivation by the Lawa people of northern Thailand (Kunstadter, 1974, 1978a, 1978b; Kunstadter et al., 1978; Sabhasri, 1978; Schmidt-Vogt, 1997, 1999), and was singled out for studies of its contribution to the 'astonishing floristic and structural complexity of secondary forests' in Lawa fallows (Schmidt-Vogt, 2007). Not only does the retention of large trees – 'relict emergents' or, in this chapter, called

FIGURE 19-3: A vital aspect of Khupra's shifting cultivation system is the retention of large trees – a practice that boosts both forest recovery and the ability to sequester carbon.

Photo: Christian Erni.

lopped or residual trees – help the forest fallow to establish itself very quickly after the cultivation phase, but this study has found that it also plays an important role in the shifting cultivation system's ability to sequester carbon.

In Khupra, fields are left fallow for at least 10, and usually for 12 to 14 years before they are cleared again for new cultivation. The fallow-management practices of Khupra's farmers, i.e. keeping large trees in the swiddens, coppicing trees so that the roots remain alive to grow new shoots soon after burning, and minimal disturbance of the soil, allow a rapid growth of secondary forest. After 12 to 14 years, the field is free of weeds and the soil fertility has recovered sufficiently for a new cycle to begin.

Scope of the study

This study was never intended to be a comprehensive and detailed quantitative assessment of carbon emissions and sequestrations in the Khupra community. With a territory of about 19.4 square kilometres covered by a complex mosaic of vegetation involved in diverse land uses and livelihood practices, such a study would have required time, financial and human resources well beyond those at our disposal.

Nevertheless, the study still tried to get a basic idea of the magnitude of the greenhouse-gas impact of livelihood and land uses in Khupra, and thus of the community's 'carbon footprint'. But it was largely based on rough estimates helped by external data and the limited data the team was able to collect in the field. Direct data collection mainly involved land use under shifting cultivation, which was the focus of the study.

While the study hoped to provide insights into underlying processes and general trends in the carbon dynamics of land use in Khupra community, technical limitations made the quantitative assessments indicative, at best, and the study's main value may therefore lie in raising questions and identifying needs for further research.

Bombax ceiba L. [Malvaceae]

This species is commonly among those left standing when a Khupra swidden is cleared. It is one of the tree species that produces kapok – fine white fibres like cotton – in its seed capsules. The dry cores of its bright red flowers are also used in a spicy soup that is popular in neighbouring Shan state.

Methods

As an exercise in participatory-action research, the study was conducted by a team composed of members of the Khupra community, staff of Karuna Mission Social Solidarity, and advisers.[3] The participatory–action research was supported by the author, along with Prawit Nikornuaychai and Gam Angkang Shimray. This support group acted as trainers during field testing of the research methodology, gave technical support during data consolidation and analysis, and wrote the eventual reports.

A combination of quantitative and qualitative research methods were applied. Basic demographic and socio-economic data was collected in a household census conducted by the research team. This data was complemented by findings from a range of standard qualitative research methods such as group interviews, individual interviews, focal group discussions and various other commonly used research tools and methods that followed the participatory learning and action (PLA) approach, such as community mapping, timelines, transect walks, land-use sketch mapping and priority ranking.

Spatial information on land use was collected through participatory mapping, with the help of both satellite imagery and a GPS ground survey. This was processed with Geographical Information System software (ArcGIS and QGIS).

Carbon-stock assessments were conducted in six areas: a newly cut swidden field and fallows that were two, five, six, nine and 12 years old. In each of these areas between five and 10 sample plots measuring 10 m by 10 m were chosen randomly, in proportion to the size of each area. The tree and bamboo biomass (and the corresponding carbon stock) was estimated using a standard forest-biomass assessment method that involved measuring the diameter of trees at breast height, and bamboo according to the species, i.e. per individual pole or whole clumps. The assessment method used allometric equations for trees and bamboo in dry evergreen forests developed by The Centre for People and Forests (RECOFTC) for forests in Thailand. The hills of Kayah state fall within the same bioregion and the allometric equations were therefore considered applicable to the forests of Khupra village. Undergrowth was cut and litter collected from four 1 m by 1 m subplots in each of the sample plots. They were dried and both wet and dry weights were recorded for calculation of their carbon content.

Soil samples were taken from each of the four subplots within each sample plot, and a combined sample for each sample plot was analysed for pH value, organic matter and carbon content at Mae Jo University in Chiang Mai.

Land use and carbon

Sequestration of atmospheric carbon dioxide (CO_2) in vegetation and soil organic matter is an important factor affecting the concentration of greenhouse gases in the atmosphere. Any form of human land use in which the vegetation cover is changed has an impact on the exchange of greenhouse gases between the atmosphere and terrestrial ecosystems. As land use changes, so does the balance between emissions and sequestration of greenhouse gases, depending on the kind of vegetation cover resulting from land-use change. This is most obvious when a forest is cleared, as described in the 2000 report of the Intergovernmental Panel on Climate Change:

...the carbon stocks in aboveground biomass are either removed as products, released by combustion, or decay back to the atmosphere through microbial decomposition. Stocks of carbon in the soil will also be affected, although this effect will depend on the subsequent treatment of the land. Following clearing, carbon stocks in aboveground biomass may again increase, depending on the type of land cover associated with the new land use. During the time required for the growth of the new land cover – which can be decades for trees – the aboveground carbon stocks will be smaller than their original value (IPCC, 2000).

As has been found in most parts of the world, almost all of the original vegetation cover in Khupra's village territory has been altered by centuries of human land use. The most drastic transformation, implying the largest loss of carbon stock, is where forest has been turned into permanent agricultural land, such as wet–rice fields or grassland for grazing.

Estimates of aboveground carbon stocks in two of Khupra's cemetery forests showed a very high level of more than 400 tons of carbon per hectare (t C/ha), due to the high density of very large trees. These small residual forests may not be representative of the area's original forests; thus a somewhat lower carbon stock value should be assumed, as a benchmark against which carbon losses can be calculated. The aboveground carbon in Southeast Asian forests is reported to range from 40 to 400 t C/ha, with a median value of 220 t C/ha (Ziegler et al., 2012, p.3092). The Guidelines for National Greenhouse Gas Inventories, published by the Intergovernmental Panel on Climate Change, give a range of 105 to 169 t C/ha for tropical seasonal forests in Asia, like those found at Khupra (Eggleston et al., 2006).[4]

The aboveground carbon stock of grasslands, pastures and shrubs ranges from 3 to 35 t/ha (Ziegler et al., 2012). Our carbon-stock estimates of grazing land in Khupra ranged from 3.5 to 15.3 t C/ha, with an average of 7.25 t C/ha. So if an original forest carbon stock of 137 t C/ha is assumed, the conversion to grasslands means a reduction of 94.7% of carbon

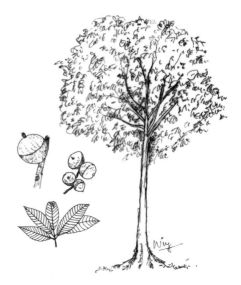

Lithocarpus elegans (Blume)
Hatus. ex Soepadmo
[Fagaceae]

This species can grow to 30 metres and is a regular 'relict emergent' in swiddens in neighbouring northern Thailand (Schmidt-Vogt, 2007). Its acorns are edible, either raw or cooked, and its timber finds many uses, among them providing firewood and making charcoal.

stock. Continuous annual cropping systems, like wet-rice fields, have a carbon stock of 1 to 4 t C/ha (Bruun et al., 2009, p.380), which implies a carbon-stock loss of 97.1 to 99.3% as forests are turned into paddy fields.

Land under shifting cultivation has a much smaller carbon-stock loss because most of the land is kept under fallow secondary forest. Based on our own field measurements, we estimated that aboveground carbon stocks in 9- to 12-years-old fallow forests were between 90 and 95 t C/ha (Figure 19-4). This means that after the first clearing of the forest for a swidden field, when almost all of the original carbon stock is lost, there is a recovery of 65.7 to 69.3% of the original carbon stock by the time the fallow forest on that plot of land is old enough and ready for a new cropping cycle. At the landscape level, long-fallow systems of shifting cultivation, such as that practised by the people of Khupra, have been reported to have an aboveground carbon stock of between 25 and 110 t C/ha (Ziegler et al., 2012). In Khupra, the time-averaged aboveground carbon stock for all shifting cultivation land is probably around 60 t C/ha, which would be about 44% of the original carbon stock.

There is no doubt that the amount of carbon stored in the vegetation covering Khupra's territory is much less today than what it was before human settlement. However, the transformation of the landscape has been nowhere near as drastic as that which occurred in the densely settled rice-growing areas of Southeast Asia's lowlands, where forests have almost completely disappeared. In the highlands, forests are still part of the landscape, despite hundreds of years of human settlement. In a landscape created by shifting cultivation, there is a complex mosaic of forest, shrub and grassland that is constantly changing, and along with it, the carbon stocks of its vegetation (Figure 19-5).

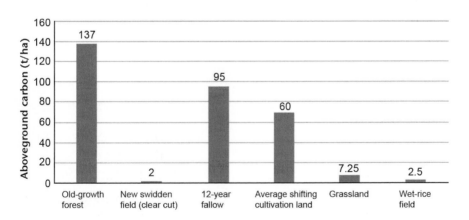

FIGURE 19-4: Comparison of aboveground carbon stocks as a result of different land-use changes at Khupra.

Carbon-stock change

Aboveground and belowground biomass carbon

When a secondary forest is cleared for shifting cultivation, a large part of the aboveground biomass is burnt and its carbon is directly emitted as carbon dioxide into the atmosphere. After cultivation the field is left fallow and the aboveground biomass, consisting of trees, bamboo and undergrowth, and thus the aboveground carbon, is rapidly restored. The carbon stock in leaf litter and dead wood is also restored. Studies of shifting cultivation in Southeast Asian forests have shown that biomass restoration in natural forest regrowth is very rapid during the first 10 years of fallow (4 to 20 t C/ha per year), which is followed by a period of slower growth (Bruun et al., 2009). As forests mature after 50 to 100 years, there is very little net accumulation of biomass.

Data for fallow biomass recovery in the upland tropical seasonal forests of Myanmar, including those at Khupra, were not available at the time of writing this chapter. However, there has been one recent study in Myanmar's Bago mountains, in an evergreen lowland forest at 250 to 450 masl (Chan et al., 2016). The range of biomass increments during the first 10 years of fallow was 3.9 to 22.38 t C/ha per year – quite similar to the figures reported in the study mentioned above.

In Khupra, recovery of tree biomass is fast, above all because the root stock of many trees remains intact and new shoots emerge soon after a field has been cleared (Table 19-2). Furthermore, because large trees are often not cut but only lopped, a considerable proportion of the biomass, and thus carbon stock, is left, even in a newly cut field. As well, there is partly charred dead wood in these fields, most of which is later collected and used as firewood.

The biomass – and thus the carbon content – of the lopped trees that remain in a swidden field can be estimated with the standard method used in assessing forest carbon stocks.[5] For a large field cleared at Khupra in 2016, the estimated carbon stocks in residual trees amounted to 28.82 t C/ha (Table 19-3).

It was estimated that the belowground biomass of lopped trees was at least the standard 20% of the aboveground biomass, although the percentage was probably much higher than that because many of the coppiced trees remained alive, with their root systems intact. It was very difficult to estimate the belowground biomass because it was impossible to predict,

FIGURE 19-5: Khupra's shifting cultivation landscape is a complex mosaic of forest, shrub and grassland that is constantly changing, and along with it, the carbon stocks of its vegetation.

Photo: Christian Erni.

TABLE 19-2: Aboveground carbon stocks in new fields and fallows of different ages at Khupra (kg C/ha).

Swidden stage	Aboveground carbon				
	trees	bamboo	undergrowth*	litter and dead wood**	total
New field	28,820		2000	1489	32,309
2 years fallow	17,601		6622	4709	28,932
5 years fallow	111,706		2329	4292	118,327
6 years fallow	108,493		2259	5062	115,814
9 years fallow	52,706	37,565	1671	3031	94,973
12 years fallow	90,300		418	1506	92,224

Notes: *In a new swidden field the 'undergrowth' is the standing crop (rice, and so on); ** In a new swidden field there is no litter, but there is some dead wood, while in fallow fields there is hardly any dead wood. Litter tends to decrease in old fallows, probably because of decomposition into the topsoil.

TABLE 19-3: Aboveground carbon in residual trees as a percentage of total aboveground carbon in new fields and fallowed forests at Khupra (kg C/ha).

Swidden stage	Aboveground carbon		
	Total	Residual trees	% of total
New field	32,309	28,820	89.2%
2 years fallow	28,932	10,773	37.2%
5 years fallow	118,326	58,115	49.1%
6 years fallow	115,814	47,075	40.6%
9 years fallow	94,973	17,008	17.9%
12 years fallow	92,224	22,932	24.9%

immediately after the field was cleared, which coppiced trees would survive and how much of the root biomass would remain intact after coppicing. In any case, swiddens with as many lopped and coppiced trees as those measured at Khupra would have a combined aboveground and belowground carbon stock in the remaining tree biomass that was well over 30 t C/ha. This was much higher than the aboveground and belowground carbon stocks of the planted crop (rice, tubers and vegetables) plus the small amount of dead wood that remained in the field. The aboveground carbon stock of the standing crop was not likely to exceed 2 t C/ha (Bruun et al., 2009) and the corresponding belowground stock would not be more than 0.4 t C/ha. Dead wood was estimated to be around 1.5 t C/ha, although some of this would still be collected and used as firewood (Table 19-3).

The presence of residual trees also explained the high carbon stocks in some of the young fallow forests. The total carbon stock measured in five- and six-years-old fallows exceeded the stocks in 9- and 12-years-old fallows because they contained more large residual trees, some of them over 40 cm in diameter at breast height. In the two-years-old fallow, residual trees made up 37% of the biomass and thus carbon stock, and in the five- and six-years-old fallows this rose to 40% and 50%. In nine-

and 12-years-old fallows residual trees accounted for only about 18% and 25% of the biomass, respectively (Figure 19-6).[6]

Due to the presence of residual trees in the newly cleared swidden, the aboveground carbon stock was still about 30% of that in a mature fallow forest. This figure may not be as high in all cleared fields because others may not contain as many residual trees. In any case, were it not for the conscious fallow management of Khupra's farmers in preserving trees in their fields, the aboveground carbon stock in new fields and, subsequently, the fallow, would have been much lower. A new field without residual trees would probably have not more than 2 t C/ha (Bruun et al., 2009), and a mature fallow around 70 t C/ha, instead of the measured 90 to 95 t C/ha.

The significance of residual trees for biomass and carbon storage throughout the shifting cultivation cycle becomes more evident when Khupra's shifting cultivation system is compared with another in which the preservation of residual trees is not such a methodical practice. A study of Karen shifting cultivation in the Bago mountains area of central Myanmar by Chan et al. (2016) focused on bamboo forest as a preferred vegetation type. Fallow management in the Bago system was similar to that practised by Khupra villagers; trees were cut in a similar manner, keeping the root stock of most trees intact so that they could sprout new shoots soon after burning. However, much fewer trees were left standing and lopped. Table 19-4 compares the aboveground carbon stocks in fallows of different ages in Khupra with those in the Bago mountains study area of Chan et al. (2016). It shows that fallows at Khupra, with a larger number of residual trees, had much larger carbon stocks. However, if the carbon stocks of residual trees had been excluded, the aboveground carbon in the respective shifting cultivation systems would have been quite similar.

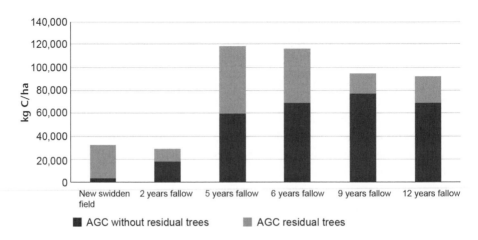

FIGURE 19-6: Aboveground carbon (AGC) stocks at various stages of Khupra's shifting cultivation cycle.

TABLE 19-4: Comparison of aboveground carbon stocks in shifting cultivation fields with (Khupra) and without (Bago) many residual trees (t C/ha).

Fallow year	Khupra	Khupra excluding residual trees	Bago mountains site of Chan et al. (2016)
0	32.309	3.489	
1			19.53
2	28.932	18.159	12.13
4			89.51
5	118.326	60.211	42.79
6	115.814	68.739	
8			67.14
9	94.973	77.965	35.11
10			69.93
12	92.224	69.292	
15			89.87

Soil carbon

The analysis of soil samples collected from the top 30 cm in the six swidden areas at Khupra showed an organic-matter content of between 4.16% (the new swidden field) and 5.69%. An organic-matter content of 3% to 5% is considered 'relatively high' for tropical upland soils (Dang and Klinnert, 2001). Consequently, the carbon content of soils on Khupra's shifting cultivation land was also fairly high, ranging from 47 t C/ha to more than 64 t C/ha (Table 19-5, Figure 19-7). These are similar to soil-carbon stocks measured in shifting cultivation fallows elsewhere in the region (Chan et al., 2016, p.581).

As with the aboveground carbon stocks, we were unable to directly measure and document changes in the soil-carbon stocks of a particular field over an entire swidden cycle. We were only able to analyse soils in fallow fields of different ages and compare them. This method had obvious limitations, since the properties of soils could vary greatly, even within a small area, particularly in the mountains due to differing geological and microclimatic conditions. This may, for example, explain the exceptionally high carbon content in the soil of the two-years-old fallow.

The results of the soil analysis for the six areas did not show great differences between soils from five-years-old and nine-years-old fallows. Also, organic matter and carbon stocks of the soil from the new swidden field, which had been burnt just three months before the soil samples were taken, were still high – only about 10% less than in the soils of older fallows and 17% less than the exceptionally high content measured in the 12-years-old fallow.

Soil carbon makes up more than 50% of the total carbon in the newly burnt field. Even in fallow fields with higher aboveground biomass, soil carbon is still around 30% of the total carbon stock (Table 19-6, Figure 19-8).

TABLE 19-5: Soil organic matter and soil carbon at different stages of shifting cultivation at Khupra.

Swidden stage	Organic matter (%)	Soil carbon (kg/ha)
New field	4.16	47,012
2 years fallow	5.66	63,997
5 years fallow	4.65	50,630
6 years fallow	4.66	52,741
9 years fallow	4.63	52,405
12 years fallow	5.69	64,365

As in all long-fallow systems of shifting cultivation, carbon stocks on Khupra's swidden land undergo rapid changes: a radical reduction when the field is cleared, followed by rapid recovery during biomass regrowth in the fallow period. Due to the effective fallow management of Khupra's swidden farmers, a fairly large permanent stock of carbon is retained throughout the cycle in the form of residual trees, root stocks and soil carbon. It can therefore be concluded that total carbon stocks on Khupra's shifting cultivation are fairly high.

We estimate that the time-averaged aboveground carbon stock on Khupra's swidden land, under a 12-year cycle, is about 60 t C/ha. This is assuming a moderate to low occurrence of residual trees, thus an initial stock of 10 t C/ha on a newly cleared swidden and 90 t C/ha for a 9- to 12-year-old fallow. With a corresponding belowground stock of biomass carbon equal to one-fifth of the aboveground stock, plus an average 50 t C/ha of soil carbon, the average total carbon stocks on Khupra's swidden land would amount to roughly 120 t C/ha. This means that Khupra's 1057 ha of shifting cultivation land stores at least 127,000 tons of carbon, and probably much more. According to the United States Environmental Protection Agency's Greenhouse Gas Equivalencies Calculator (EPA, 2017), this corresponds to 470,000 tons of carbon dioxide.

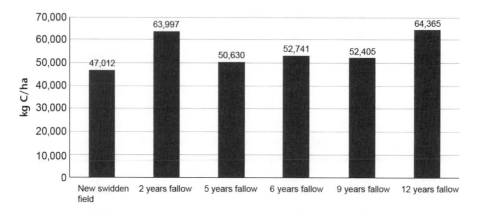

FIGURE 19-7: Belowground carbon stocks at different stages of the shifting cultivation cycle at Khupra.

TABLE 19-6: Soil carbon stocks at various stages of the shifting cultivation cycle at Khupra, as a percentage of total carbon stocks at each stage (kg C/ha).

Swidden stage	Soil carbon	Total carbon stocks	% of total carbon
New field	47,012	85,484	55.0
2 years fallow	63,997	102,498	62.4
5 years fallow	50,630	191,763	26.4
6 years fallow	52,741	190,706	27.7
9 years fallow	52,405	165,767	31.6
12 years fallow	64,365	174,728	36.8

Carbon emission and sequestration

In shifting cultivation, most carbon emissions occur when the slashed vegetation of a newly cleared field is burnt. The actual amount of biomass burnt is difficult to estimate. It varies in a particular year according to the quantity of vegetation that has to be slashed and how well the field burns. The completeness of the burn is always a major concern for shifting cultivators. A clean and comprehensive burn means less work in cleaning the field before sowing and more ash to enrich the soil with plant nutrients and to help to balance its pH level, since tropical soils tend to be acidic. Unsuccessful burning of the slashed biomass causes a lot of additional work as the remaining debris has to be heaped up and burnt again. Large stems and branches never burn completely, but are a valuable source of firewood. In Khupra, three months after a new swidden field was burnt, there was hardly any wood left lying in the field; almost all of it had been carried to the village.

As already pointed out, many large trees are left standing when a field is cleared, and only their branches are lopped. In the case of the new swidden field cleared at Khupra in 2016, this meant that its aboveground carbon stock was still 28 t C/ha, even after clearing and burning. Assuming that prior to clearing, this same field had an aboveground carbon stock of about 95 t C/ha (the stock measurement made in a 9- to 12-years-old fallow), and that all the rest of the newly cut swidden's carbon

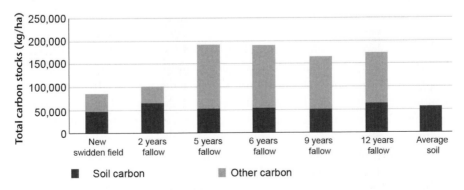

FIGURE 19-8: Total carbon stocks at various stages of the shifting cultivation cycle and soil carbon as a percentage of total carbon.

stock is emitted through burning and decomposition, then the carbon emissions from shifting cultivation at Khupra would be between 67 t C/ha for fields with residual trees left standing and up to 95 t C/ha for fields that were clear cut. This would be equivalent to between 245.89 and 348.65 tons of CO_2 per ha.

The total area cleared for swidden fields at Khupra varies from year to year. With the help of satellite images, villagers tried to identify the areas cleared in particular years and made rough estimates covering the 10 years prior to 2016. The total area cleared each year varied considerably, according to the farmers' engagement in other income-generating activities. Estimates of areas cleared for swiddens over the preceding 10 years ranged from 48 to 114 ha, with an average of 88.7 ha. This figure was just about what we expected for a fallow cycle of 12 years, since one-twelfth of Khupra's total shifting cultivation area is 88 ha.

Castanopsis armata (Roxb.) Spach [Fagaceae]

This tree species is native to the eastern Himalayas, Bangladesh, Myanmar, Thailand and Vietnam. Swidden cultivators in Myanmar and northern Thailand are known to retain these trees when clearing plots for cultivation. They help the forest to recover during the fallow phase.

Therefore, if we assumed that on average one-twelfth of the shifting cultivation area was cleared every year, this would generate emissions equivalent to 21,638 t CO_2 if the fields had many residual trees, or 30,681 t CO_2 if the fields were clear cut. This would amount to about 35 to 50 t CO_2 per capita per year in the Khupra community.

However, all of this carbon is sequestered again over the ensuing 10- to 14-years-long fallow period. This observation is supported by the findings of other recent studies, which conclude that long-fallow shifting cultivation can be carbon neutral or even carbon positive, i.e. it can sequester more carbon than it emits (Dressler at al., 2016).

A similar scenario is very likely with regard to belowground carbon, particularly soil carbon. Carbon emissions from decomposing organic matter in the soil during the cultivation period seem to be quite limited, as soil carbon levels in new fields were found to be only about 10% lower than those in the soil of an old fallow. It is therefore likely that the soil carbon lost during cultivation is recovered during the fallow. This observation is in line with the findings of other recent studies:

Several studies looking at the whole fallow–cropping cycle of swidden systems in Borneo have found SOC [soil organic carbon] contents to be unaffected by swiddening (Bruun et al., 2006; Kleinman et al., 1996; Mertz et al., 2008). In another study from Borneo, de Neergaard et al. (2008) found that neither SOC concentration (%) nor total content was affected by swiddening in soil layers down to 90 cm depth. Similar findings are reported by Sommer et al. (2000), who examined the upper 6 m of an ultisol in Brazil (Bruun et al., 2009, p.379).

Schima wallichii Choisy [Theaceae]

This species is valued for its timber, which is hard, with a straight grain and a fine texture, so it finds a multitude of uses. The trees, which are often retained when swiddens are cleared, are also used to provide shade to crops such as cardamom. They are also valued for their ability to help restore forests during the fallow phase of the swidden cycle.

The length of the fallow is critical for maintenance of soil carbon levels. A comprehensive review of research on the impacts of shifting cultivation on tropical forest soil found that shorter fallow time did not allow for the entrance of new carbon into the soil organic carbon and that the system began to emit carbon into the atmosphere. However, in longer fallows 'time allows for the entrance of new C into the SOC. The system begins to capture C from the atmosphere' (Ribeiro Filho et al., 2013, p.714).

The review concludes that shifting cultivation cannot be considered 'to be unsustainable *per se* in relation to the soil system dynamics' and that the causes that may lead to unsustainability 'are related to the social, economic, political, and cultural changes that affect the communities that subsist on this agricultural practice', that may 'cause the resilience capacity of the soils to be surpassed' (Ribeiro Filho et al., 2013, p.720).

A study among the Dayaks in West Kalimantan, Indonesia, found that long-fallow shifting cultivation, as practised by the Dayaks, 'makes soils more fertile than those in undisturbed primary forest'. The researcher found her findings 'surprising, […] because they contradict negative perceptions of 'slash and burn' methods that the Dayaks employ to clear land' (Science Daily, 1997).

Shifting cultivation is the main source of carbon emissions in land use by the villagers of Khupra, but under the long-fallow system that they practise, all of this carbon – and maybe, as other research suggests, even more – is sequestered by the fallow vegetation, i.e. in land use under long–fallow shifting cultivation there is a direct offset of carbon emissions.

Grasslands and fire management

Khupra villagers keep livestock on free-range pastures. These grasslands are on former shifting cultivation land and are maintained through regular burning. The farmers sometimes intentionally burn pastures in order to maintain open grasslands, or they are burned by hunters in order to flush out game. Sometimes, fires are caused accidentally or occur naturally. However, the vegetation recovers quickly when the monsoon rains begin. Thus, while these annual fires emit a considerable amount of CO_2, it is sequestered again as the vegetation grows back.

Bauhinia variegata L. [Leguminosae]

Mountain ebony – as this species is commonly known - is often retained in freshly-cleared swiddens in Khupra. Growing to only about 12 metres, these trees are known for their eye-catching displays of pink, purple or white flowers. Their buds and fruit are edible. Many parts of these trees also have medicinal uses.

Uncontrolled fires, either from burning grasslands or swiddens, occasionally escape into fallow areas, and this delays the start of a new swidden cycle in the damaged plot. Hunters causing such fires are expected to apologize to the owners of the fallow land. If unintended fires occur too frequently on fallow land, it may become permanent grassland – a change that has been reported in shifting cultivation areas throughout the region. However, the reverse also occurs: when fire is absent for long enough, forest regrows on grasslands as a result of natural succession. At Khupra, villagers report that grass and shrub lands have not expanded and, in some places, natural regrowth of forest has reached a point where land can once again be used for shifting cultivation.

Controlling fire and natural reforestation of grass and shrub land are two factors with the highest potential for increasing carbon sequestration in Khupra's territory. Aboveground carbon stocks measured in six sample plots in five different grazing areas ranged from 4.7 to 15.3 t C/ha, with an average of 7.2 t C/ha. The plot with the highest value contained more scrub and young trees and had carbon stocks similar to those of some two-years-old fallow plots. Allowing the forest in these areas to recover for a further 10 years would probably lead to an eight- to 10-fold increase in biomass.

There is an increasing awareness among Khupra villagers of the need to conserve forest, which is reflected, for example, in a decision by the people of New Khupra to declare part of former gazing and swidden land as community forest. Biomass, and thus carbon stocks, in these areas will certainly increase over coming years.

Shifting cultivation and carbon in Khupra: Some conclusions

Our study shows that annual emissions of CO_2 from burning new shifting cultivation fields are considerable, but under a long-fallow system such as that practised in Khupra, carbon is rapidly sequestered again by secondary forest regrowth during the fallow period. We also conclude that CO_2 emissions from burning grasslands are similarly sequestered by the regrowth of grasses and shrubs.

Our findings are supported by other recent studies that have concluded that long-fallow systems of shifting cultivation do not cause net carbon emissions, but are carbon neutral or even carbon positive, i.e. they sequester more carbon than they emit.

In Khupra's long-fallow system of shifting cultivation, fields are left fallow for 12 to 14 years. Conscious fallow management, including the preservation of lopped residual trees and coppicing of trees, thus keeping their roots alive, ensures a rapid recovery of fallow forest after the cultivation phase ends.

Since more than 90% of shifting cultivation land is covered by fallow forest at any given point in time; since there are relatively high carbon stocks in cultivated fields due to the presence of residual trees and high belowground carbon due to the survival of the roots of coppiced trees; and since there are high levels of soil carbon throughout the swidden cycle, this form of land use preserves considerable stocks of carbon. Our rough estimates of average total carbon stocks, including aboveground and belowground biomass carbon and soil carbon, suggest that Khupra's 1058 ha of swidden land holds an average 120 t C/ha. The total area, therefore, must store at least 127,000 tons of carbon, equivalent to about 466,000 tons of CO_2.

Comparison with alternative land uses

Several other studies in recent years have also shown that land under long fallow shifting cultivation can store a considerable amount of carbon. It is certainly less than might be stored in an undisturbed natural forest, but comparing the environmental aspects of shifting cultivation with those of primary forests is problematic, 'most fundamentally because a primary forest is not a production system, thus for the farmers, forests do not represent an alternative to swidden cultivation' (Bruun et al., 2009, p.377).

Some of these studies have also concluded that it is rather doubtful that other forms of land use, such as plantations, which are widely promoted by policy-makers, are a better alternative for sequestering carbon. A meta-analysis of 250 studies reporting aboveground and belowground carbon estimates for different land-use types indicated 'great uncertainty in the net total ecosystem carbon changes that can be expected from many transitions, including the replacement of various types of swidden agriculture with oil palm, rubber, or some other types of agroforestry systems' (Ziegler et al., 2012, p.3087). The study concluded that 'there is little evidence to suggest that transitions from swidden agriculture to many other land uses will directly or reliably produce positive carbon gains, e.g., from intermediate- or long-fallow swidden systems to oil palm and rubber plantations' (Ziegler et al. 2012, p.3095).

Ficus racemosa L. [Moraceae]

The many uses of cluster fig – as it is known – tempt farmers to retain these trees when clearing swiddens. The fruit, which grows in clusters on the trunk, can be eaten raw, cooked, or dried and ground. The leaves, bark, fruit, roots and sap are all used medicinally, to treat ailments ranging from tonsillitis and diarrhoea to mumps and coughing blood.

The study also found that with an aboveground carbon stock of 25 to 143 t/ha for rubber plantations and 25 to 110 t/ha for long-fallow swidden systems, there was no great difference between the two (Ziegler et al., 2012, p.3092). Other forms of land use that are often encouraged or actively promoted seem, as well, to be no better alternatives (Table 19-7).

Impacts other than carbon sequestration and storage also need to be taken into account when considering the adoption of alternative forms of land use. While a transition from shifting cultivation to cash cropping may lead to an increase in income for many farmers, is has been shown that such changes may also have long-term negative impacts on the environment and local livelihoods. As well as reduced carbon sequestration, environmental impacts include 'a permanent decrease in forest cover at the landscape scale, combined with substantial losses of wild biodiversity and agro-biodiversity, increases in weed pressure, decreases in soil fertility, accelerated erosion and declines in stream water quality' (van Vliet et al., 2012, p.426). The impact on local livelihoods includes increased economic vulnerability and reduced food security (Dressler et al., 2016).

TABLE 19-7: Aboveground carbon stocks in vegetation under different forms of land use (t C/ha).

Land use	Aboveground carbon stocks	Studies
Shifting cultivation		
Long fallow-systems (>10 years fallow)	80 (24-160)	Bruun et al., 2009
8-years fallow forest	74	van Noordwijk et al., 1995
4-years fallow system	8-9	Bruun et al., 2009
Agroforests		
Rubber agroforest (Indonesia)	90	Bruun et al., 2009
Rubber (agro)forest (Indonesia)	116	van Noordwijk et al., 1995
Permanent agriculture seasonal crops		
Continuous annual cropping	1-4	Bruun et al., 2009
Annual cropping vegetables	2	van Noordwijk et al., 1995
Annual cropping cassava	4	van Noordwijk et al., 1995
Monoculture tree plantations		
Casuarina tree monoculture plantation	21-55	Bruun et al., 2009
Rubber plantation	50	Bruun et al., 2009
Oil palm		
Indonesia, 20-25 years rotation	48-91	Bruun et al., 2009 van Noordwijk et al., 1995
Malaysia	36	Bruun et al., 2009

Therefore, it is not only doubtful that replacing shifting cultivation with other forms of land use will have any carbon benefits, but possible economic benefits may also be outweighed by negative environmental and social impacts. Instead of pushing for the blanket eradication of shifting cultivation, land-use policies need to be informed by a better understanding of this form of land use. These policies also need to be sensitive and responsive, not only to ecological and economic issues, but also to the social and cultural conditions of the communities involved.

The way ahead

Shifting cultivators are innovative and adaptive; they have been quick to respond to both pressures and opportunities and have often transformed their land-use systems in ways that differ considerably from what is conventionally considered to be shifting cultivation (Padoch et al., 2007). Building on and improving long-fallow shifting cultivation systems may not only strengthen local livelihoods but further increase carbon sequestration and other environmental services.

From the farmers of Khupra we can learn that good fallow management, which preserves belowground biomass, soil carbon and aboveground biomass carbon in the form of residual trees, has a considerable potential for maintaining large carbon stocks, supporting rapid fallow regrowth and, overall, increasing carbon sequestration in shifting cultivation.

What we have also learned in this study is that good fire management, and above all the prevention of unwanted fires in grasslands and fallow, is critical, and that there is considerable potential for increasing overall carbon stocks by allowing natural reforestation of parts of grasslands. Some villagers in Khupra are already using some reforested grasslands for shifting cultivation, and others have been experimenting with agroforestry gardens on fallow land.

External support in the form of research and experimentation, conducted in cooperation with farmers, together with technical support, credit and market access for agroforestry products, along with other measures that take the existing form of land use as a point of departure, can lead to new and innovative forms of land use that increase incomes and offer alternatives to urban migration. Keeping people on the land by providing decent and sustainable livelihoods can help to prevent the spread of lifestyles that squander stored carbon and create heavy carbon footprints that humanity simply cannot afford.

Acknowledgements

Financial support for this study and publication of its main report was generously provided by the German Catholic Bishops' Organisation for Development Cooperation, MISEREOR, and the International Work Group for Indigenous Affairs (IWGIA), with funding from the Danish Ministry of Foreign Affairs. We would also like to express our gratitude to Dr. Pathipan Sutigoolabud, at the Faculty of Agricultural Production, Mae Jo University, Chiang Mai, Thailand, for his support and advice.

References

Cairns, M. F. (ed.) with Hill, B. and Kurupunya, T. (2017) *Shifting Cultivation Policies: Balancing Environmental and Social Sustainability,* CABI, Wallingford, UK

Chan, N., Takeda, S., Suzuki, R. and Yamamoto, S. (2016) 'Assessment of biomass recovery and soil carbon storage of fallow forests after swidden cultivation in the Bago Mountains, Myanmar', *New Forests* 47(4), pp.565-585, DOI 10.1007/s11056-016-9531-y

Bruun, T. B., Mertz, O. and Elberling, B. (2006) 'Linking yields of upland rice in shifting cultivation to fallow length and soil properties', *Agriculture, Ecosystems and Environment* 113, pp.139-149, doi: 10.1016/j.agee.2005.09.012

Bruun, T. B., de Neergaard, A., Lawrence, D. and Ziegler, A. D. (2009) 'Environmental consequences of the demise in swidden cultivation in Southeast Asia: Carbon storage and soil quality,' *Human Ecology* 37, pp.375-388, DOI 10.1007/s10745-009-9257-y

Dang, N. T. and Klinnert, C. (2001) 'Problems with and local solutions for organic matter management in Vietnam', in C. Martius, H. Tiessen and P. Vlek (eds) *Managing Organic Matter in Tropical Soils: Scope and Limitations*, Kluwer Academic Publishers, Dordrecht, The Netherlands, pp.89-97

Dove, M. R. (1983) 'Theories of swidden agriculture and the political economy of ignorance', *Agroforestry Systems* 1, pp.85-99

Dove, M. R. (1985) 'The agroecological mythology of the Javanese and the political economy of Indonesia', *Indonesia* 39 (April), Cornell Southeast Asia Program, pp.1-36

Dove, M. R. (1996) 'So far from power, so near to the forest: A structural analysis of gain and blame in tropical forest development', in C. Padoch and N. Peluso (eds) *Borneo in Transition. People, Forests, Conservation and Development*, Oxford University Press. Kuala Lumpur, pp.41-58

Dressler, W. H., Wilson, D., Clendenning, J., Cramb, R., Keenan, R., Mahanty, S., Bruun, T. B., Mertz, O. and Lasco, R. (2016) The impact of swidden decline on livelihoods and ecosystem services in

Southeast Asia: A review of the evidence from 1990 to 2015', *Ambio* 46(3), pp.291-310, doi:10.1007/s13280-016-0836-z

de Neergaard, A., Magid, J. and Mertz, O. (2008) 'Soil erosion from shifting cultivation and other smallholder land uses in Sarawak, Malaysia, *Agriculture, Ecosystems and Environment* 125, pp.182-190, doi: 10.1016/j.agee.2007.12.013

Eggleston, H. S., Buendia, L., Miwa, K., Ngara, T. and Tanabe, K. (eds) (2006) *IPCC Guidelines for National Greenhouse Gas Inventories*, National Greenhouse Gas Inventories Programme, Intergovernmental Panel on Climate Change, Geneva and Institute For Global Environmental Strategies, Kanagawa, Japan

Ennion, J. D. (2015) 'From conflicting to complementing: The formalisation of customary land management systems governing swidden cultivation in Myanmar', Master of Law thesis to Victoria University, Wellington, New Zealand, available at http://www.burmalibrary.org/docs21/Ennion-2015-Formalisation_of_swidden_land_use_rights_in_Myanmar-red.pdf, accessed 9 September 2018.

EPA (2017) *Greenhouse Gas Equivalencies Calculator*, available at https://www.epa.gov/energy/ghg-equivalencies-calculator-calculations-and-references, accessed 12 September 2018

Erni, C. (ed.) (2017) *Livelihood, Land Use and Carbon: A Study on the Carbon Footprint of a Shifting Cultivation Community in Kayah State*, Karuna Mission Social Solidarity-Loikaw, Loikaw, Kayah state, Myanmar

FAO, UNDP, UNEP (2008) *UN Collaborative Programme on Reducing Emissions from Deforestation and Forest Degradation in Developing Countries, Framework Document*, Food and Agriculture Organization of the United Nations, Rome; United Nations Development Programme, New York; and United Nations Environment Programme, Nairobi

Forsyth, T. J. (1999) 'Questioning the impacts of shifting cultivation', *Watershed* 5(1), pp.23-29

Forsyth, T. and Walker, A. (2008) *Forest Guardians, Forest Destroyers: The Politics of Environmental Knowledge in Northern Thailand*, Silkworm Books, Chiang Mai

Fox, J. M., Fujita, Y., Ngidang, D., Peluso, N., Potter, L., Sakuntaladewi, N., Sturgeon, J. and Thomas, D. (2009) 'Policies, political-economy, and swidden in Southeast Asia', *Human Ecology* 37(3), pp.305-322

Gibbs, H. K. and Brown, S. (2007a) *Geographical Distribution of Woody Biomass Carbon Stocks in Tropical Africa: An Updated Database for 2000*, Carbon Dioxide Information Center, Oak Ridge National Laboratory, Oak Ridge, TN

Gibbs, H. K. and Brown, S. (2007b) *Geographical Distribution of Biomass Carbon in Tropical Southeast Asian Forests: An Updated Database for 2000*, Carbon Dioxide Information Center, Oak Ridge National Laboratory, Oak Ridge, TN, https://cdiac.ess-dive.lbl.gov/epubs/ndp/ndp068/ndp068b.html accessed 2 April 2023

Gibbs, H. K., Brown, S., Niles, J. O. and Foley, J. A. (2007) 'Monitoring and estimating tropical forest carbon stocks: Making REDD a reality', *Environmental Research Letters* 2(4), doi:10.1088/1748-9326/2/4/045023

IPCC (2000) *Land Use, Land-Use Change and Forestry*, Intergovernmental Panel on Climate Change, Geneva

IWGIA (2007) *The Indigenous World 2007*, International Work Group for Indigenous Affairs, Copenhagen

Kleinman, P. J. A., Bryant, R. B. and Pimentel, D. (1996) 'Assessing ecological sustainability of slash-and-burn agriculture through soil fertility indicators', *Agronomy Journal* 88, pp.122-127

Kunstadter, P. (1974) 'Usage et Tenure des Terres Chez les Lua' (Thailande) (Land Tenure and Use among the Lua', Thailand)', *Etudes Rurales* 53-56, pp.449-466

Kunstadter, P. (1978a) 'Ecological modification and adaptation: An ethnobotanical view of the Lua' swiddeners in northwestern Thailand', in R. I. Ford (ed.) 'The nature and status of ethnobotany', *Anthropological Papers* 67, pp.168-200

Kunstadter, P. (1978b) 'Subsistence agricutural economies of Lua' and Karen hill farmers, Mae Sariang district, northwestern Thailand', in P. Kunstadter, E. C. Chapman and S. Sabhasri (eds) *Farmers in the Forest*, the University Press of Hawaii, Honolulu, pp.74-133

Kunstadter, P., Sabhasri, S. and Smitinand, T. (1978) 'Flora of a forest fallow environment in northwestern Thailand', *Journal of the National Research Council of Thailand* 10(1), pp.1-45

Lasco, R. D. (2002) 'Forest carbon budgets in Southeast Asia following harvesting and land cover change', *Science in China* 45, pp.55-64

Laungaramsri, P. (2005) 'Swidden agriculture in Thailand: Myths, realities and challenges', in: C. Erni (ed.) 'Shifting cultivation', *Indigenous Affairs* 2/05, International Work Group for Indigenous Affairs (IWGIA), Copenhagen

Mertz, O., Wadley, R. L., Nielsen, U., Bruun, T. B., Colfer, C. J. P., de Neergaard, A., Jepsen, M. R., Martinussen, T., Zhao, Q., Noweg, G. T. and Magid, J. (2008) 'A fresh look at shifting cultivation: Fallow length an uncertain indicator of productivity', *Agricultural Systems* 96, pp.75-84, doi: 10.1016/j.agsy.2007.06.002

Mertz, O., Padoch, C., Fox, J., Cramb, R. A., Leisz, S. J., Nguyen, T. L. and Tran, D. V. (2009) 'Swidden change in Southeast Asia: Understanding causes and consequences', *Human Ecology* 37(3), pp.259-264

Nielsen, U., Mertz, O. and Noweg, G. T. (2006) 'The rationality of shifting cultivation systems: Labor productivity revisited', *Human Ecology* 34(2), pp.201-218

Padoch, C. (1985) 'Labor efficiency and intensity of land use in rice production: An example from Kalimantan', *Human Ecology* 13(3), pp.271-289

Padoch, C., Coffey, K., Mertz, O., Leisz, S. J., Fox, J. and Wadley, R. L. (2007) 'The demise of swidden in Southeast Asia? Local realities and regional ambiguities', *Geografisk Tidsskrift-Danish Journal of Geography* 107(1), pp.29-41

Phuc, T. X. (2008) *Does Forest Devolution Benefit the Upland Poor? An Ethnography of Forest Access and Control in Vietnam*, Resource Politics and Cultural Transformation in the Mekong Region (working paper series), Regional Center for Social Science and Sustainable Development (RCSD), Faculty of Social Sciences, University of Chiang Mai

Pulhin, J. M., Amaro, M. C. Jr. and Bacalla, D. (2005) 'Philippines community-based forest management 2005', in *First Regional Community Forestry Forum: Regulatory Frameworks for Community Forestry in Asia*, proceedings of a regional form, 24-25 August 2005, The Centre for People and Forests (RECOFTC), Bangkok

REDD+ Myanmar (2018) *National REDD+ Strategy Myanmar* (draft) Ministry of Natural Resources and Environmental Conservation, Naypyidaw, available at http://www.myanmar-redd.org/wp-content/uploads/2019/04/Document-TF-4-5-Draft-National-REDD-Strategy.pdf, accessed 2 April 2023

Ribeiro Filho, A. A., Adams, C. and Murrieta, R. S. S. (2013) 'The impacts of shifting cultivation on tropical forest soil: A review', *Boletim do Museu Paraense Emílio Goeldi. Ciências Humanas* 8(3), pp.693-727

Sabhasri, S. (1978) 'Effects of forest fallow cultivation on forest production and soil', in P. Kunstadter, E. C. Chapman and S. Sabhasri (eds) *Farmers in the Forest*, the University Press of Hawaii, Honolulu, pp.160-184

Schmidt-Vogt, D. (1997) 'Forests and trees in the cultural landscape of Lawa swidden farmers in northern Thailand', in K. Seeland (ed.) *Nature is Culture: Indigenous Knowledge and Socio-Cultural Aspects of Trees and Forests in Non-European Cultures*, Intermediate Technology Publications, London, pp.44-50

Schmidt-Vogt, D. (1999) 'Swidden farming and fallow vegetation in northern Thailand' *Geoecological Research* 8, Franz Steiner Verlag, Stuttgart

Schmidt-Vogt, D. (2007) 'Relict emergents in swidden fallows of the Lawa in northern Thailand: Ecology and economic potential', in M. F. Cairns (ed.) *Voices from the Forest: Integrating Indigenous Knowledge into Sustainable Upland Farming*, Resources for the Future, Washington, DC, pp.37-54

Science Daily (1997) 'Duke researcher finds some tropical farming practices have surprising consequences', *Science Daily*, 16 August 1997, available at https://www.sciencedaily.com/releases/1997/08/970816100758.htm, accessed 13 September 2018

Sommer, R., Denich, M. and Vlek, P. L. G. (2000) 'Carbon storage and root penetration in deep soils under small-farmer land-use systems in the eastern Amazon region, Brazil', *Plant and Soil* 219, pp.231-241, doi: 10.1023/A:1004772301158

Springate-Baginski, O. (2013) *Rethinking Shifting Cultivation in Myanmar: Policies for Sustainable Livelihoods and Food Security*, University of East Anglia, UK and Pyoe Ping, Myanmar, available at http://www.burmalibrary.org/docs20/Springate-Baginski-2013-Rethinking_Swidden_Cultivation_in_Myanmar-en-red.pdf, accessed 5 September 2018

van Vliet, N., Mertz, O., Heinimann, A., Langanke, T., Pascual, U., Schmook, B., Adams, C., Schmidt-Vogt, D., Messerli, P., Leisz, S. J., Castella, J. C., Jørgensen, L., Birch-Thomsen, T., Hett, C., Bech-Bruun, T., Ickowitz, A., Vu, K. C., Yasuyuki, K., Fox, J., Padoch, C., Dressler, W. and Ziegler, A. D. (2012) Trends, drivers and impacts of changes in swidden cultivation in tropical forest-agriculture frontiers: A global assessment', *Global Environmental Change* 22(2), pp.418-429

Wright, L. A., Kemp, S. and Williams, I. M. (2011) '"Carbon footprinting": Towards a universally accepted definition', *Carbon Management* 2(1), pp.61-72, doi:10.4155/cmt.10.39

Ziegler, A. D., Phelps, J., Yuen, J. Q., Webb, E. L., Lawrence, D., Fox, J. M., Bruun, T. B., Leisz, S. J., Ryan, C. M., Dressler, W., Mertz, O., Pascual, U., Padoch, C. and Koh, L. P. (2012) 'Carbon outcomes of major land-cover transitions in SE Asia: Great uncertainties and REDD+ policy implications', *Global Change Biology* 18(10), pp.3087-3099

Notes

1. REDD+ (or REDD-plus) replaces the previously used acronym REDD. It refers to 'reducing emissions from deforestation and forest degradation in developing countries, and the role of conservation, sustainable management of forests and enhancement of forest carbon stocks in developing countries'. This is the terminology used by the Conference of Parties (COP) of the United Nations' Framework Convention on Climate Change (UNFCCC).

2. A 'carbon footprint' is commonly understood as the total amount of greenhouse gases produced to directly and indirectly support human activities, usually expressed in equivalent tons of carbon dioxide (CO_2). According to Wright et al. (2011), a 'carbon footprint' is a measure of the total amount of carbon dioxide and methane (CH_4) emissions of a defined population, system or activity, considering all relevant sources, sinks and storage within the spatial and temporal boundary of the population, system or activity of interest. It is calculated as carbon dioxide equivalent, using the relevant 100-year global warming potential.

3. The community members who were part of the research team were: From Upper Khupra, Khu Ka Bwe, Saw Saw Htoo, Htoo Dayl, Poe Po and Day Khaing; from Old Khupra, Giovanni Shwerino, Leone and Marta Ma Oo; and from New Khupra, Pilatio and Anthony. Research team members from KMSS-Loikaw were Christina Ti Myar, Elena, Anne Mary Ne Ne, Alesio Ngairi and Ignatio.

4. The terminology used in classifying forest types in Southeast Asia differs from author to author. Seasonally dry evergreen upland forests like those at Khupra (above 1000 masl) may fall into categories like evergreen moist forest (distinguished from evergreen rainforest), hill evergreen forest, evergreen mountain forest, tropical seasonal forest, or mixed evergreen upland forest.

   In this chapter, Khupra's forests are regarded as 'tropical seasonal forest' in accordance with the IPCC's guidelines (Eggleston et al., 2006) and recommendations for carbon-stock assessment, i.e. the median figure between 106 and 169 tC/ha, which is 137 tC/ha. This corresponds more or less with 138 tC/ha recommended by the International Conference on Climate Change (ICCC) in 1997 (Lasco, 2002), and the 142 tC/ha of Gibbs and Brown (2007a, 2007b), cited in Gibbs et al. (2007).

5. In the case of freshly lopped trees in newly cut fields, the biomass and carbon stocks were calculated using the formula for tree trunks only, while for residual trees in fallow fields on which the branches had grown back, the formula for whole trees was used.

6. In our stock assessments, in young (2-years-old) fallows, trees with a diameter at breast height (dbh) of more than 20 cm were considered residual trees. For five- to six-years-old fallows, residual trees were those with a dbh greater than 25 cm and in older (nine- to 12-years-old) fallows, residual trees were those with a dbh of more than 30 cm.

20

CARBON AND BIODIVERSITY OUTCOMES UNDER DIVERGENT MANAGEMENT SCENARIOS

Lessons from upland Philippines shifting cultivation landscapes

Sharif A. Mukul, John Herbohn, Jennifer Firn and Nestor Gregorio[*]

Introduction

The Philippines is both a biodiversity hotspot and a megadiverse country (Myers et al., 2000; Posa et al., 2008). It has also experienced one of the highest rates of deforestation in Southeast Asia and was among the first countries to introduce a massive reforestation programme to address its rapid loss of forest and biodiversity (Chokkalingam et al., 2006; Pulhin, J. M. et al., 2007). Shifting cultivation, which is known locally as *kaingin*, is a widespread land use in the Philippines (Kummer, 1992). To many smallholder farmers living in remote rural areas of the country, it is also a major livelihood strategy (Herbohn et al., 2014; Mukul, 2016). However, as in many other tropical countries, forestry policies in the Philippines have attempted to restrict *kaingin*, based on the assumption that it has detrimental impacts on the environment (Lawrence, 1997; Suarez and Sajise, 2010).

In the Philippines, 53% of all land is considered to be forest, based on the national land classification system which regards all areas with a slope of more than 18% as 'forest', irrespective of forest cover (Jahn and Asio, 2001). Upland areas cover approximately 55% of the country's total land area, and are important mostly because they contain most of the country's remaining forests and have been subject to intensive use by humans as well as suffering severe land degradation (Cramb, 1998) (Figure 20-1). In these upland areas, *kaingin* is a prominent land use. It can be categorized into three distinct types, based on the sites where it is practised: the *tubigan* system, the *katihan* system and the *dahilig* system (Olofson, 1980). The *tubigan* and *katihan*

[*] Dr. Sharif Ahmed Mukul, Research Fellow, Tropical Forests and People Research Centre, University of the Sunshine Coast, Queensland, Australia; Professor John Herbohn, Tropical Forests and People Research Centre, University of the Sunshine Coast, Queensland, Australia; Professor Jennifer Firn, School of Biology and Environmental Science, Queensland University of Technology, Brisbane, Australia; and Dr. Nestor Gregorio, Research Fellow, Tropical Forests and People Research Centre, University of the Sunshine Coast, Queensland, Australia.

systems are practised in areas of lower elevation or on gently sloping land with limited irrigation facilities. The *dahilig* system is widely practised in heavily forested areas and on steeper slopes (Olofson, 1980).

Although the Philippines is a pioneer in large-scale forest landscape restoration (FLR), access by smallholders and subsistence farmers to such efforts remains very limited (Le et al., 2014). Therefore, *kaingin* will continue to be an imperative land-use in

FIGURE 20-1: A typical upland landscape in the Philippines, with complex mosaics of secondary regrowth, terraced rice fields, coconut plantation, disturbed and old-growth forests.

All photographs in this chapter were taken by the author, Sharif A. Mukul.

the country's upland areas until greater access to such state-regulated reforestation programmes is accessible by local communities under community forestry or other participatory schemes (Pulhin, J. M. et al., 2007; Mukul et al., 2016a). Moreover, secondary fallow forests regrowing after *kaingin* are generally not viewed as suitable targets for biodiversity conservation and carbon retention in the upland Philippines (Mukul et al., 2020).

Drawing upon an empirical study conducted on the island of Leyte (Figure 20-2) and other relevant case studies in the Philippines, we will demonstrate in this chapter that secondary forests that are recovering after *kaingin* have high potential for biodiversity and carbon co-benefits. In our empirical study, we also found that secondary forests regrowing after *kaingin* have the potential for use as a cost-effective reforestation measure with multiple benefits for the people and environment of upland areas of the Philippines. We also discuss measures that we believe are essential for such programmes to succeed.

The island of Leyte

We conducted this study on the island of Leyte, in the Philippines (Figure 20-2). The island is the eighth largest in the country, covering an area of about 800,000 hectares (ha). The major cash crops of the island include coconut (*Cocos nucifera*), *abaca* (Manila hemp – *Musa textilis*) and maize (*Zea mays*). Geographically, the island is located between 124°17′ and 125°18′ East longitude and between 9°55′ and 11°48′ North latitude. It receives relatively even distribution of rainfall throughout the year with an annual rainfall of about 4000 mm (Jahn and Asio, 2001). The mean annual temperature is 28°C, which remains constant throughout the year (Navarrete et al., 2013).

Cocos nucifera L. [Arecaceae]

The coconut palm is one of the most useful and widely grown tropical tree crops. The seed (or nut), the flowers, the sap, the roots and pith from the stem are all popular foods; various extracts, particularly oil from the nut, have numerous medicinal qualities and industrial applications; and the leaves, midribs, fibre, nut shells and timber from the stem all have important uses. The palms are also ideal for combining with other crops.

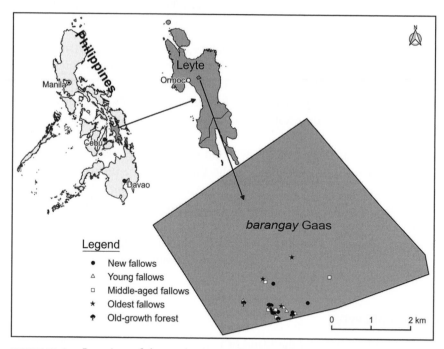

FIGURE 20-2: Location of the study sites at *barangay* Gaas, on the island of Leyte, the Philippines.

For the ecological survey, we purposively selected *barangay* Gaas, situated within the territory of the city of Ormoc. A *barangay* is the smallest administrative unit in the Philippines, similar to a village. Gaas has a comparatively high altitudinal range with a relatively greater extent of undisturbed forests and low population density. These factors favour the regeneration of *kaingin* fallows into secondary forests (Chazdon,

2014). Smallholders living in the area usually grow *abaca* or coconut in their *kaingin* fallows after the end of cultivation in order to generate some financial gain (Figure 20-3).

Ecological surveys

Our survey was confined to *kaingin* fallows that were at least one ha in size. We categorized our fallow sites into four different categories: new fallow sites (up to 5 years old); young fallow sites (6 to 10 years old); middle-aged fallow sites (11 to 20 years old); and oldest fallow sites (21 to 30 years old) (Figure 20-4). We took five replicates from each fallow class with old–growth natural forest as our control. In each study site we identified tree species with a diameter at breast height (dbh) of at least 5 cm, and measured both the diameter and height. More about our survey design and data collection can be found in Mukul (2016).

Musa textilis Née [Musaceae]

Known as *abaca* in its native Philippines, this banana species produces inedible fruit but the strongest-known natural fibre. Often called Manila hemp, it was originally used for making rope, but fibre from an extensive *abaca* industry is now used for making tea bags, banknotes, decorative papers and a wide variety of textiles and handicrafts. The fibre comes from the leaf stems of mature plants, and world demand still outstrips supply.

Biodiversity, carbon measurements and trade-off analysis

We used tree species diversity, i.e. number of unique tree species per unit area, as a measure of biodiversity (see Mukul et al., 2020). Aboveground biomass was estimated on a per-ha basis following the generic allometric equation developed by Chave et al. (2014) and this was expressed in Megagrams (Mg).[1] We used localized wood-density estimates from our study sites and carbon content was assumed to be 50% of the dry woody biomass (see Mukul et al., 2016b).

FIGURE 20-3: A *kaingin* fallow being used to grow *abaca* (*Musa textilis*) in the study area.

FIGURE 20-4: Different sites in the study area. Clockwise from the top left, (a) a newly opened *kaingin*; (b) a middle-aged fallow; (c) the oldest *kaingin* fallow; and (d) a control area of old-growth forest.

For biodiversity and carbon trade-off analysis (measured as the additionality (Δ)) of different related land uses or land covers, we used the median values for tree-species diversity and biomass carbon for each land use or land cover (such as fallows of different ages, old-growth control forest, plantation forest). Additionality was measured as the difference in biodiversity and carbon between our control old-growth forest sites and other related land uses or land covers, including regrowing *kaingin* fallows, reported in the Philippines. The value could be either positive (+) or negative (-).

Biodiversity and carbon co-benefits and trade-offs associated with different land uses and land covers in upland Philippines

We found the highest biodiversity (i.e. tree species density) in our oldest *kaingin* fallow sites, followed by control old-growth forests, middle-aged fallows and young fallows (Figure 20-5). Aboveground biomass carbon was significantly higher (P < 0.01) in our control old-growth forest sites than in all of the *kaingin* fallow sites (Figure 20-6).

In Table 20-1 we present the potential biodiversity and carbon trade-offs associated with land uses or land covers that may replace *kaingin* landscapes in upland Philippines after cultivation has ended and sites have been abandoned. Apparently, old-growth

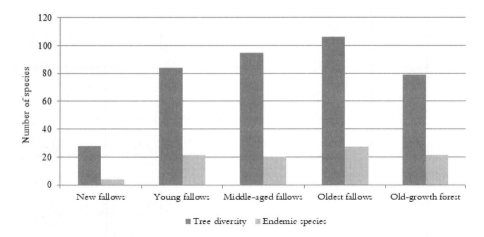

FIGURE 20-5:   Tree diversity in regrowing secondary forest sites (after *kaingin* abandonment) and reference old-growth forest on Leyte island, the Philippines.

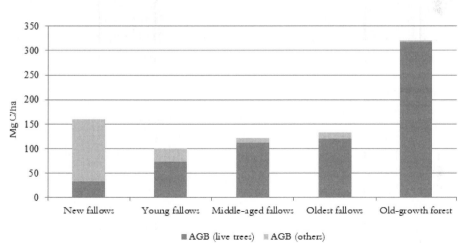

FIGURE 20-6: Aboveground biomass (AGB) carbon in regrowing secondary forest sites (after *kaingin* abandonment) and reference old-growth forest on Leyte island, the Philippines.

forest always provides the highest carbon benefits (380.49 Mg C/ha) when compared with other common land uses or land covers. Although biodiversity was highest in our oldest *kaingin* fallows (21 to 30 years old), the number of endemic species was highest in the old-growth forest sites. An 80-year-old plantation of *Swietenia macrophylla* came closest (264 Mg C/ha) to equalling the carbon stock of the old-growth forest, followed by a mixed plantation of *Parashorea malaanonan* and *Anisoptera thurifera* of a similar age, and *kaingin* fallows of different ages (Table 20-1). The high level of aboveground biomass carbon in new fallow sites was attributed mainly to high levels of coarse dead wood remaining after forest clearing.

The conservation values of different plantations in upland Philippines were very low compared to *kaingin*-fallow secondary forests (see Table 20-1). Only our oldest *kaingin* fallows had the potential to provide biodiversity and carbon co-benefits comparable with those of old-growth forests. As expected, agricultural land uses, such as rice fields, sugar cane and plantations of commercially important species such as oil palm, coconuts, and fast-growing *Gmelina arborea*, *Albizia falcataria*, and *Acacia* sp., had lower conservation values and less importance for carbon sequestration than fallow forests regrowing after *kaingin* cultivation or native dipterocarp forest (Table 20-1). Monocultures of fast-growing timber species were found to have lower carbon benefits than mixed plantations.

Conclusion and management implications

Parashorea malaanonan Merr. [Dipterocarpaceae]

This critically endangered species was part of an 80-year-old plantation that showed relatively high carbon stocks, but low biodiversity, when research studies compared it with natural *kaingin* fallows of different ages. This tree grows to 60 m, and is sought-after for its timber, for use in carpentry and furniture. In the carbon-stock research it was growing along with *Anisoptera thurifera*, another dipterocarp species that is used for harvesting resin.

Our study revealed that secondary forest that is regrown after *kaingin* abandonment can provide substantial biodiversity and carbon co-benefits when compared with old-growth forest, fast-growing timber species, and other commercially important species that may grow on land after it has been used for *kaingin*. While some plantations may also hold superior potential for carbon storage and sequestration in tropical regions (see Erskine et al., 2006, for example), their conservation value is not comparable to that of old-growth forests and secondary forests regrowing after shifting cultivation. Moreover, such plantations require high management and maintenance costs (Gregorio et al., 2015).

In tropical regions, uncertainties about forest-carbon levels and the distribution and recovery rate of secondary forests are the main constraints to including forests that are regrowing after shifting cultivation in the global voluntary carbon market (Mertz et al., 2012; Ziegler et al., 2012; Mertz, 2009). Because of the dynamic nature of secondary-forest regrowth after shifting cultivation, the instability of biodiversity and biomass carbon in such landscapes can be an issue (Mukul and Herbohn, 2016). In such circumstances, biodiversity and carbon co-benefits can be achieved

TABLE 20-1: Biodiversity and carbon stocks and associated trade-offs in common land uses and land covers in upland Philippines.

Land-use/cover	Tree diversity[1]	Carbon[2] (Mg C/ha)	Additionality[3] (Δ) Δ Biodiversity	Additionality[3] (Δ) Δ Carbon	Age[4]	Sources
Old-growth forest	45	380.49	–	–	–	Mukul et al. (2016a,b, 2020)
Post-*kaingin* forests						
New fallow	5	155.1	-40	-225.39	5	As above
Young fallow	39	87.99	-6	-292.5	10	As above
Middle-aged fallow	42	135.73	-3	-244.76	20	As above
Oldest fallow	47	155.71	+2	-224.78	30	As above
Dipterocarp forest	NA	221.0	–	-159.49	NA	Lasco and Pulhin (2009)
Grasslands						
Imperata sp.	0	8.5	-45	-371.99	1	Lasco and Pulhin (2009)
Sacharrum sp.	0	13.1	-45	-367.39	1	Lasco and Pulhin (2009)
Plantations						
Swietenia macrophylla	1	264.0	-44	-116.49	NA	Racelis et al. (2008)
Acacia sp.	1	81.0	-44	-299.49	NA	Lasco and Pulhin (2009)
Albizia falcataria	1	48.69	-44	-331.8	9	Lasco (2002)
Gmelina arborea	1	54.32	-44	-326.17	9	Lasco (2002)
Parashorea malaanonan + Anisoptera thurifera	2	241.25	-43	-139.24	80	Lasco and Pulhin (2009)
Parashorea malaanonan + Dipterocarpus grandiflorus	2	125.61	-43	-254.88	80	Lasco and Pulhin (2009)
Coconut	1	86.0	-44	-294.49	30	Lasco (2002)
Oil palm	1	55.0	-44	-325.49	9	Pulhin, F. B. et al. (2014)
Agriculture						
Rice	0	3.1	-45	-377.39	1	Lasco and Pulhin (2009)
Abaca	0	5.7	-45	-374.79	1	Lasco (2002)
Sugar cane	0	12.5	-45	-367.99	1	Lasco and Pulhin (2009)

Notes: [1] Only the main and/or characteristic plant diversity of a particular land use or land cover is considered here; [2] aboveground carbon in tree biomass; [3] the difference, either positive or negative, between control old-growth forest and respective land uses or land covers; [4] stand age; NA, not available.

by either avoiding further intensification of land use and landscape degradation, or by promoting natural regeneration. Intensification can be avoided by allowing longer fallow cycles or by growing multipurpose species that are also common in the forest. Enhancement of natural regeneration, on the other hand, can be achieved by preventing further use of the area for shifting cultivation and by assisted natural regeneration (Chazdon and Guariguata, 2016).

Our trade-off analyses, in all cases, found that regrowing secondary forests after *kaingin* abandonment outperformed other land uses and available reforestation measures in upland Philippines, with regard to biodiversity and carbon co-benefits. However, the density and growth of population are major drivers of intensification of shifting cultivation systems in most of the tropics (van Vliet et al., 2012), and this may reduce carbon benefits due to shorter fallow periods and more frequent cultivation (Lawrence et al., 2010).

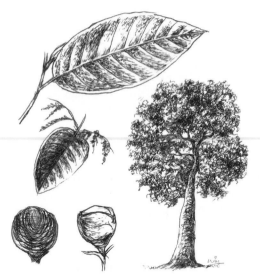

Shorea polysperma Merr.
(an unresolved name)
[Dipterocarpaceae]

Known in its native Philippines as red meranti or 'Philippines mahogany', this tree has a straight, buttressed bole and grows to 40 m. It is regarded as critically endangered because of loss of habitat and heavy exploitation from the wild for its timber, which is a lucrative export item. Uses for the timber include interior and exterior panelling, joinery, furniture and cabinetwork.

Presently, global forest-carbon credits are valued at more than US$100 billion per year, making this an emerging, fast-growing sector (Petrokofsky et al., 2011; Peters-Stanley and Yin, 2013). In addition, policy-makers have recently committed to the Bonn Challenge, an initiative to restore 150 million ha of degraded forests globally by 2020 and 350 million ha by 2030 (Locatelli et al., 2015). However, the prospect of including regrowing secondary forests in the emerging global carbon markets depends largely on reliable estimates of carbon and their biodiversity benefits (Maron et al., 2013; Law et al., 2015).

Forest conservation in the Philippines has clearly visible benefits to local livelihoods and climate-change mitigation (Sheeran, 2006; Lasco et al., 2013). Due to the large areas involved and their importance to smallholders, secondary forest regrowth after shifting cultivation could provide important benefits to both the environment and local communities, if properly incorporated in REDD+ (Reducing Emissions from Deforestation and Forest Degradation) and CDM (Clean Development Mechanisms) schemes (Mukul and Byg, 2020). We found that allowing forests to regenerate naturally after *kaingin* abandonment could be a cost-effective restoration measure in the Philippines, with a high potential for integration into REDD+ and CDM schemes. However, it is critical that members of local communities be involved in such activities, with clearly defined rights and responsibilities (Mukul et al., 2014). Improving environmental governance through legal and regulatory reform, better land allocation, land tenure and management, law enforcement and monitoring are also crucial (Le et al., 2012; Chazdon, 2013; Baynes et al., 2015)

Lithocarpus celebicus (Miq.) Rehder [Fagaceae]

This tree grows up to 40 m tall with a straight bole free of branches up to 20 m and buttresses up to 1 m high. It is commonly found in lowland and montane forests, occasionally up to 3000 m in elevation. The wood is strong, hard and heavy and is suitable for construction. It is used locally for fence posts, mining props, shingles and boat building. It is a good firewood and can be used to make charcoal.

References

Baynes, J., Herbohn, J., Smith, C., Fisher, R. and Bray, D. (2015) 'Key factors which influence the success of community forestry in developing countries', *Global Environmental Change* 35, pp.226-238

Chave, J., Réjou-Méchain, M., Búrquez, A., Chidumayo, E., Colgan, M. S., Delitti, W. B. C., Duque, A., Eid, T., Fearnside, P. M., Goodman, R. C., Henry, M., Martínez-Yrízar, A., Mugasha, W. A., Muller-Landau, H. C., Mencuccini, M., Nelson, B. W., Ngomanda, A., Nogueira, E. M., Ortiz-Malavassi, E., Pélissier, R., Ploton, P., Ryan, C. M., Saldarriaga, J. G. and Vieilledent, G. (2014) 'Improved allometric models to estimate the aboveground biomass of tropical trees', *Global Change Biology* 20(10), pp.3177-3190

Chazdon, R. L. (2013) 'Making tropical succession and landscape reforestation successful', *Journal of Sustainable Forestry* 32, pp.649-658

Chazdon, R. L. (2014) *Second Growth: The Promise of Tropical Forest Regeneration in an Age of Deforestation*, University of Chicago Press, Chicago, IL

Chazdon, R. L. and Guariguata, M. R. (2016) 'Natural regeneration as a tool for large-scale forest restoration in the tropics: Prospects and challenges', *Biotropica* 48, pp.844-855

Chokkalingam, U., Carandang, A. P., Pulhin, J. M., Lasco, R. D., Peras, R. J. J. and Toma, T. (eds) (2006) *One Century of Forest Rehabilitation in the Philippines: Approaches, Outcomes and Lessons*, Centre for International Forestry Research (CIFOR), Bogor, Indonesia

Cramb, R. (1998) 'Environment and development in the Philippine uplands: The problem of agricultural land degradation', *Asian Studies Review* 22, pp.289-308

Erskine, P. D., Lamb, D. and Bristow, M. (2006) 'Tree species diversity and ecosystem function: Can tropical multi-species plantations generate greater productivity?', *Forest Ecology and Management* 233, pp.205-210

Gregorio, N., Herbohn, J., Harrison, S. and Smith, C. (2015) 'A systems approach to improving the quality of tree seedlings for agroforestry, tree farming and reforestation in the Philippines', *Land Use Policy* 47, pp.29-41

Herbohn, J. L., Vanclay, J. K., Ngyuen, H., Le, H. D., Harrison, S. R., Cedamon, E., Smith, C., Firn, J., Gregorio, N. O., Mangaoang, E. and Lamarre, E. (2014) 'Inventory procedures for smallholder and community woodlots in the Philippines: Methods, initial findings and insights', *Small-scale Forestry* 13, pp.79-100

Jahn, R. and Asio, V. B. (2001) 'Climate, geology, geomorphology and soils of the tropics, with special reference to Leyte island (Philippines), in *Proceedings of the 8th International Seminar and Workshop on Tropical Ecology,* Visayas State College of Agriculture, Baybay, Leyte, pp.25-43

Kummer, D. M. (1992) 'Upland agriculture, the land frontier and forest decline in the Philippines', *Agroforestry Systems* 18, pp.31-46

Lasco, R. D. (2002) 'Forest carbon budgets in Southeast Asia following harvesting and land cover change', *Science in China* 45, pp.55-64

Lasco, R. D. and Pulhin, F. B. (2009) 'Carbon budgets of forest ecosystems in the Philippines', *Journal of Environmental Science and Management* 12, pp.1-13

Lasco, R. D., Veridiano, R. K. A., Habito, M. and Pulhin, F. B. (2013) 'Reducing emissions from deforestation and forest degradation plus (REDD+) in the Philippines: Will it make a difference in financing forest development?', *Mitigation and Adaptation Strategies for Global Change* 18, pp.1109-1124

Law, E. A., Bryan, B. A., Torabi, N., Bekessy, S. A., McAlpine, C. A. and Wilson, K. A. (2015) 'Measurement matters in managing landscape carbon', *Ecosystem Services* 13, pp.6-15

Lawrence, A. (1997) Kaingin *in the Philippines: Is it the End of the Forest?,* Rural Development Forestry Network Paper 21, London, UK

Lawrence, D., Radel, C., Tully, K., Schmook, B. and Schneider, L. (2010) 'Untangling a decline in tropical forest resilience: Constraints on the sustainability of shifting cultivation across the globe', *Biotropica* 42, pp.21-30

Le, H. D., Smith, C., Herbohn, J. and Harrison, S. (2012) 'More than just trees: Assessing reforestation success in tropical developing countries', *Journal of Rural Studies* 28, pp.5-19

Le, H. D., Smith, C. and Herbohn, J. L. (2014) 'What drives the success of reforestation projects in tropical developing countries? The case of the Philippines', *Global Environmental Change* 24, pp.334-348

Locatelli, B., Catterall, C. P., Imbach, P., Kumar, C., Lasco, R., Marín-Spiotta, E., Mercer, B., Powers, J. S., Schwartz, N. and Uriarte, M. (2015) 'Tropical reforestation and climate change: Beyond carbon', *Restoration Ecology* 23, pp.337-343

Maron, M., Rhodes, J. R. and Gibbons, P. (2013) 'Calculating the benefit of conservation actions', *Conservation Letters* 6, pp.359-367

Mertz, O. (2009) 'Trends of shifting cultivation and the REDD mechanism', *Current Opinion in Environmental Sustainability* 1, pp.156-160

Mertz, O., Müller, D., Sikor, T., Hett, C., Heinimann, A., Castella, J-C., Lestrelin, G., Ryan, C. M., Reay, D. S., Schmidt-Vogt, D., Danielsen, F., Theilade, I., van Noordwijk, M., Verchot, L. V., Burgess, N. D., Berry, N. J., Pham, T. T., Messerli, P., Xu, J., Fensholt, R., Hostert, P., Pflugmacher, D., Bruun, T. B., de Neergaard, A., Dons, K., Dewi, S., Rutishauser, E. and Sun, Z. (2012) 'The forgotten D: Challenges of addressing forest degradation in complex mosaic landscapes under REDD+'. *Danish Journal of Geography* 112, pp.63-76

Mukul, S. A. (2016) 'Shifting cultivation in the upland secondary forests of the Philippines: Biodiversity and carbon stock assessment, and ecosystem services trade-offs in land-use decisions', PhD dissertation to the University of Queensland, Australia

Mukul, S. A. and Herbohn, J. (2016) 'The impacts of shifting cultivation on secondary forest dynamics in the tropics: A synthesis of the key findings and spatio-temporal distribution of research', *Environmental Science and Policy* 55, pp.167-177

Mukul, S. A. and Byg, A. (2020) 'What determines indigenous Chepang farmers swidden land-use decisions in the central hill districts of Nepal?', *Sustainability* 12, article number 5326

Mukul, S. A., Herbohn, J., Rashid, A. Z. M. M. and Uddin, M. B. (2014) 'Comparing the effectiveness of forest law enforcement and economic incentives to prevent illegal logging in Bangladesh', *International Forestry Review* 16, pp.363-375

Mukul, S. A., Herbohn, J. and Firn, J. (2016a) 'Co-benefits of biodiversity and carbon sequestration from regenerating secondary forests in the Philippine uplands: Implications for forest landscape restoration', *Biotropica* 48, pp.882-889

Mukul, S. A., Herbohn, J. and Firn, J. (2016b) 'Tropical secondary forests after shifting cultivation are important carbon sources and sinks', *Scientific Reports* 6, article number 22483

Mukul, S. A., Herbohn, J. and Firn, J. (2020) 'Rapid recovery of tropical forest diversity and structure after shifting cultivation in the Philippine uplands', *Ecology and Evolution* 10, pp.7189-7211

Myers, N., Mittermeier, R. A., Mittermeier, C. G., da Fonseca, G. A. B. and Kent, J. (2000) 'Biodiversity hotspots for conservation priorities', *Nature* 403, pp.853-858

Navarrete, I. A., Tsutsuki, K. and Asio, V. B. (2013) 'Characteristics and fertility constraints of degraded soils in Leyte, Philippines', *Archives of Agronomy and Soil Science* 59, pp.625-639

Olofson, H. (1980) 'Swidden and *kaingin* among the southern Tagalog: A problem in Philippine upland ethno-agriculture', *Philippine Quarterly of Culture and Society* 8, pp.168-180

Peters-Stanley, M. and Yin, D. (2013) 'Maneuvering the mosaic: State of the voluntary carbon markets, 2013', *Forest Trends*, Washington, DC and Bloomberg New Energy Finance, New York

Petrokofsky, G., Holmgren, P. and Brown, N. D. (2011) 'Reliable forest carbon monitoring: Systematic reviews as a tool for validating the knowledge base', *International Forestry Review* 13, pp.56-66

Posa, M. R., Diesmos, A. C., Sodhi, N. S. and Brooks, T. M. (2008) 'Hope for threatened tropical biodiversity: Lessons from the Philippines', *BioScience* 58, pp.231-240

Pulhin, F. B., Lasco, R. D. and Urquiola, J. P. (2014) 'Carbon sequestration potential of oil palm in Bohol, Philippines', *Ecosystems and Development Journal* 4, pp.14-19

Pulhin, J. M., Inoue, M. and Enters, T. (2007) 'Three decades of community-based forest management in the Philippines: Emerging lessons for sustainable and equitable forest management', *International Forestry Review* 9, pp.865-883

Racelis, E. L., Carandang, W. M., Lasco, R. D., Racelis, D. A., Castillo, A. S. A. and Pulhin, J. M. (2008) 'Assessing the carbon budgets of large leaf mahogany (*Swietenia macrophylla* King) and Dipterocarp plantations in the Mt. Makiling Forest Reserve, Philippines', *Journal of Environmental Science and Management* 11, pp.40-55

Sheeran, K. A. (2006) 'Forest conservation in the Philippines: A cost-effective approach to mitigating climate change?', *Ecological Economics* 58, pp.338-349

Suarez, R. K. and Sajise, P. E. (2010) Deforestation, swidden agriculture and Philippine biodiversity', *Philippine Science Letters* 3, pp.91-99

van Vliet, N., Mertz, O., Heinimann, A., Langanke, T., Pascual, U., Schmook, B., Adams, C., Schmidt-Vogt, D., Messerli, P., Leisz, S., Castella, J-C., Jørgensen, L., Birchthomsen, T., Hett, C., Bech-Bruun, T., Ickowitz, A., Vu, K. C., Yasuyuki, K., Fox, J., Padoch, C., Dressler, W. and Ziegler, A. D. (2012) 'Trends, drivers and impacts of changes in swidden cultivation in tropical forest-agriculture frontiers: A global assessment,' *Global Environmental Change* 22, pp.418-429

Ziegler, A. D., Phelps, J., Yuen, J. Q., Webb, E. L., Lawrence, D., Fox, J. M., Bruun, T. B., Leisz, S. J., Ryan, C. M., Padoch, C. and Koh, L. P. (2012) 'Carbon outcomes of major land-cover transitions in SE Asia: Great uncertainties and REDD+ policy implications', *Global Change Biology* 18, pp.3087-3099

Note

1. One Megagram is equal to 1000 kg, and is otherwise known as one tonne, or one metric ton.

This sketch shows the concept of *taungya*-planting valued trees into swidden crops, in this case, showing a farmer's field in Mokokchung district, Nagaland, northeast India. This reflects the work of the local NEPED project, which was based on the premise of promoting and building on traditional *taungya*-planting practices to introduce valued trees into swidden fields. This sketch shows a field after the first year's rice crop had already been harvested, and although distant, you can see that the trees are already robustly established by the end of this first cropping year. These trees will be starting to form a canopy by the end of the second cropping year, and will then be left to become an almost monospecific fallow that the farmer can later harvest for sale – rather than just 'slash-and-burn' it for its ash content - when he opens the field for his next phase of arable cropping.

Sketch based on a photo by Malcolm Cairns, ca. 1998.

B. Favouring soil-building trees to accelerate soil recovery (biologically improved fallows)

A Wa farmer in China's Yunnan province prunes a soil-enriching alder (*Alnus nepalensis*) tree in her swidden field, in the knowledge that it will grow to improve the fallow (see plates 14 to 17 in Coloured Plates section I, this volume). The Wa farmers would benefit from learning the skill of pollarding alder trees, which is used to great effect by farmers in nearby Nagaland (see Cairns and Brookfield, ch. 22, this volume).

Sketch based on a photo by Malcolm Cairns in April, 2001.

Synthesis

SOIL-BUILDING TREES

Back to the future

*Dennis Garrity**

There is a great wave of concern sweeping over the world right now that is questioning the basis for conventional models of industrial, input-based farming. There is also intensive exploration of new bases for a transformation towards much more agroecological approaches to agriculture.

There are numerous reasons for this shift in emphasis. Farmers themselves are increasingly searching for ways to reduce their costs by using practices that take better advantage of soil ecology to feed their crops, rather than using out-of-the-bag inputs, and to enhance the supply of soil organic matter that can be produced on the farm.

Massive pressure is building up from consumers for healthier foods, and for producers to drastically reduce their greenhouse-gas emissions to address the accelerating climate emergency.

For these, and many other reasons, there is renewed interest in learning about less conventional ways to diversify agricultural production and use integrated and lower-cost ways to provide necessary nutrients and crop protection. This trend is opening up new scope for appreciating the knowledge possessed by current practitioners of more traditional farming systems, and also for learning from the experiences of farmers in times past.

Such a perspective makes the seven chapters in this section of the volume particularly edifying. It is a fascinating collection that demonstrates the various ways that soil-building trees can be effectively incorporated into agricultural systems. The context-specific options that are examined include:

---

* DR. DENNIS GARRITY is Chair of the Global EverGreening Alliance; Distinguished Research Fellow and Former Director-General of the World Agroforestry Centre (ICRAF); Senior Fellow of the World Resources Institute; former Drylands Ambassador for the United Nations' Convention to Combat Desertification; and a board member of Global Landcare.

- Trees that are permanently grown in fields of annual crops and are pollarded at the beginning of each cropping cycle (ch. 22 on the alder system in Nagaland).
- Trees that are planted during the cropping cycle to regenerate the soil and produce a harvestable timber crop during the fallow period, before being removed and sold when cropping is resumed (ch. 26 on *Macaranga* systems in Nagaland; ch. 24 on *Albizia* in West Java; and ch. 25 on *Albizia* in Kalimantan).
- Trees that are regenerated after a long cropping cycle, that rebuild the soils during the fallow period, and are then removed at the beginning of the next cropping cycle (ch. 23 on *Leucaena* in Naalad, Cebu).
- Trees that are nurtured in long fallows and are integrated with the crops to provide abundant leaf compost for the cropping cycle (ch. 21 on the *pakukui* system in Hawaii).
- Trees that are nurtured for many ecological, cultural and livelihood benefits (ch. 27 on Java cedar across the Asian region).

These stories build a compelling case for the diverse ways that the incorporation of trees into farming systems can not only contribute to more sustainable farming, but also become a source of more diverse and increased family income. One system that is not covered in this collection is the permanent establishment of trees in continuously cropped fields that are coppiced back to near the ground surface each year, building up soil organic matter and organic nitrogen supplies sufficient to sustain continuous cropping. These systems were featured in the first volume in this three-volume series (Garrity, 2015).

Each of these examples has important insights and practical experiences that provide lessons for future efforts to assist smallholders to enhance their productivity and sustainability and, on a broader scale, to contribute to the current process of rethinking and recreating agricultural systems across the tropical world. They are particularly relevant to smallholders who are currently practising fallow-rotation systems, or who may be gradually transforming these systems towards more permanent field agriculture. But they are also pertinent to farms that are already continuously cropped and in danger of being severely degraded and losing their production potential.

The question is: which practices being described here might hold the greatest promise for extrapolation and adaptation to agriculture elsewhere in Southeast Asia, or to other places in the tropics?

A prevalent theme of the chapters is adaptation to changing circumstances. The commercialization of agriculture is happening rapidly in the region, even in the more remote upland communities that provide many of the cases described here. Since the cost of labour is on the rise almost everywhere, how can these practices be further adjusted to reduce the labour inputs required, while continuing to provide their productive and environmental benefits?

The first chapter (ch. 21) is Dr. Noa Kekuewa Lincoln's fascinating excursion into traditional Hawaiian farming systems, to both document them and revive them for modern usage. The abundant body of literature that Lincoln has produced on this topic, over decades of work, speaks elegantly of his dedication to this vision and mission.

History was the starting point for his efforts. Lincoln notes that precolonial Hawaiians 'intensified virtually every arable habitat, made extremely marginal environments productive, utilized innovation to develop unique farming methods, and sustained production for hundreds of years without the use of external inputs.' This inspired his efforts to better understand how this was done.

He learned that pre-colonial Hawaiians planted trees widely to build up soil fertility, particularly the species called candlenut (*Aleurites moluccanus*). These trees grew rapidly and their foliage decomposed quickly, making an excellent growing environment for food crops, which were produced in intimate association with the trees. They also established agroforestry gardens in more steeply sloping and less fertile zones and niches.

The chapter focuses on the extensive candlenut forests that were used to practise shifting cultivation. Small taro gardens were prepared by felling a number of the trees in a concentrated area. The trees were allowed to decompose, creating a rich humus for planting the staple crop of taro. Lincoln shows that this particular mulching system has outstanding performance, producing enormous crop yields. At the same time, the farmers harvested the candlenuts for many other uses.

This work has been instrumental in reinforcing a revival and extension of traditional Hawaiian agriculture. Many Hawaiians, and their local producer organizations, are becoming involved in the further development of these systems, buttressed by a modern surge of interest in ecological farming on the islands. The movement has recently been documented by Schwartz (2020) in her survey of unique landscape restoration experiences. One wonders whether this effort to build on traditional 'agri-cultures' may become the harbinger and inspiration for a broader movement to do likewise, not only in Hawaii, but also in many other parts of Asia and around the world.

In the next chapter (Cairns and Brookfield, ch. 22), Malcolm Cairns describes how his interest was piqued by the unique way in which nitrogen-fixing Himalayan alders (*Alnus nepalensis*) were established and managed in the crop fields of a village in Nagaland. He admired the long-term success of this alder-based shifting cultivation system, which had served the community well for many generations. Moreover, he saw it as a model that others might copy or adapt in the transition from swiddening to more continuous farming, using this or other tree species that improve soil and earn cash.

The farmers in the Cairns and Brookfield study actively propagated *Alnus nepalensis* so that it became an almost pure stand in their fallows. When the land was to be cultivated again, the trees were not cut down or burnt, but rather they were skillfully pollarded. By nurturing *Alnus* trees permanently in their fields, the Naga farmers could sustain their soil fertility even with very short fallow periods – an innovation that should be of wide interest to many upland communities that are experiencing a transition from shifting cultivation to continuous cropping because of intensifying land shortages.

Interestingly, there is an analogue between the Naga *Alnus* system and the integration of a sister species, *Alnus acuminata*, into continuously cropped farmlands at similar elevations in the African country, Rwanda. There, however, *Alnus* is not pollarded, but rather it is allowed to grow to produce timber. Because it is highly compatible with food crops, especially when it is moderately pruned, it is possible to produce the tree as a cash crop within the continuously cropped fields. This is another variation that can have wide application in smallholder tropical agriculture. In Kenya, for instance, the intercropping of *Grevillea robusta* and *Melia volkensii* with maize and other annual crops is becoming a popular practice, and these systems are now being promoted on a large scale.

But back to the Naga village where the *Alnus nepalensis* system is practised. Times there are changing and the availability of off-farm work both in and outside the village has focused farmer interest on farming systems that require less labour. They are now replacing the *Alnus* cropping systems with tree-crop systems that reduce their labour investment: agroforestry systems that feature cardamom, with *Alnus* as an overstorey tree for timber production. Transitions towards similar permanent agroforestry systems are occurring in many other upland areas in the region, where opportunities for off-farm work are beckoning farmers away from their fields.

The elegant *Alnus* pollarding practice that was developed by the Nagas should stimulate much more investigation for direct extrapolation to other remote communities in the region at elevations from 1000 to 2000 masl. The basic principles should also be directly applied to the vast number of smallholder farming systems where continuous or near-continuous cropping is practised.

For example, how might insights from these systems reinforce current efforts to scale-up alley-cropping systems with leguminous shrubs in other parts of the tropics, especially in the East African countries of Kenya, Malawi and Zambia, and in South Asia, particularly in Sri Lanka?

Currently, the most common recommendation in these systems is to coppice the shrubs down close to the ground level. But there is also experience to suggest that pollarding of the trees at a height of three metres or more, to sustain a canopy coverage of 10% to 15% of the field, may better protect annual crops from drought stress and increase crop yields.

Faidherbia albida is a unique acacia species that is nurtured by millions of farmers across the African continent (Garrity et al., 2010). It is a legume that can be grown at densities of 100 trees per hectare, because it sheds its foliage at the beginning of the rainy season. It remains dormant and leafless while the crops are growing. Given that drought damage is becoming increasingly severe in many areas, overstorey trees creating a light shade should become a more widespread climate-smart practice.

Regrettably, there appears to have been few attempts to embed the principles and practices of the Naga system into more formal agricultural extension, in Nagaland or beyond. This should now become an important mission.

Chapter 26 discusses an alternative practice that has been developed in the more remote northeastern corner of Nagaland, at lower elevations with a more tropical or subtropical climate than that where the *Alnus nepalensis* system is found. Professor Sapu Changkija and colleagues examine how three similar species of *Macaranga* trees are commonly used there as dominant fallow vegetation in shifting cultivation systems to intensify production and sustain soil fertility.

They observe that swiddens cleared from fallows that were dominated by *Macaranga* trees have richer soil and the crop harvests are better than those from fallowed fields without the trees. During the cropping year, *Macaranga* seedlings emerge naturally among the upland rice and other crops. The young trees are then protected during the crop-weeding operations – a form of farmer-managed natural regeneration of trees in the fields (Garrity et al., 2010). Additional *Macaranga* seeds are also broadcast into the fields to provide a more uniform tree density.

This is similar to the practice of actively seeding the favoured tree species *Albizia* (*Paraserianthes falcataria*) in the shifting cultivation systems of the Baduy people in Java (Iskandar, ch. 24, this volume), and in Kalimantan (Sakuntaladewi et al., ch. 25, this volume), and the seeding of *Leucaena leucocephala* into swidden fields before they are fallowed in the Naalad system in the Philippines (Suson and Lasco, ch. 23, this volume). Farmers nurture the young trees throughout the cultivation phase. They grow alongside the annual crops, and then the trees dominate the site during the years of fallow recovery. Although the primary use of the *Macaranga* trees is in fallow enrichment to increase crop yields, they are also an important source of firewood and construction timber, so they serve multiple purposes.

The authors found that the ability of the *Macaranga* species to boost recovery in secondary-forest fallows lies in their effect on the soil, among the roots. There is a complex symbiotic relationship between the trees and fungal species – a mycorrhizal association. At the heart of this relationship is greater efficiency in accessing soil nutrients. The innovative use of fallow-enriching species, such as *Macaranga*, maintains soil quality, even with shorter fallows. Upland rice yields of 3.5 t/ha and higher are obtained in short-fallow systems associated with high-density stands of *Macaranga denticulata* (4200 trees per hectare) on the poorest of mountain soils. In an environment in which fallow periods have been reduced to six years and less, the dense cultivation of this species has maintained a rich diversity of swidden crops.

The authors note that a better understanding of the role and contribution of *Macaranga* in traditional shifting cultivation could provide solutions to the restoration of subsistence crop yields in shortening fallow systems elsewhere. *Macaranga* species are distributed across a very wide range in South and Southeast Asia, suggesting that their incorporation into shifting cultivation systems, and perhaps permanent-field farming as well, could be scaled-up in many countries. The cultivation of trees in dense stands is also characteristic of the fallow-management systems promoted in Eastern Zambia and western Kenya (Garrity et al., 2010).

Changkija and his colleagues further note that 'The whole gamut of indigenous knowledge involved in shifting cultivation could be adopted as theoretical bases for practical and field-based approaches to more intensive agricultural systems elsewhere. Technologies arising from shifting cultivation could thus have a great bearing on other farming systems'. I agree.

In chapter 24, Professor Johan Iskandar describes an evolving system of swidden cultivation in western Java that follows the same trajectory as the *Macaranga* system in Nagaland. In this case, the Baduy people cultivate the leguminous *Albizia* tree (*Paraserianthes falcataria*) as a means of accelerating the regeneration of soil fertility in their swidden fields. They also cultivate a great diversity of annual legume crops and establish several other valuable leguminous trees during the cropping cycle, and retain them in their fields through successive cropping periods for food, other products, and soil enhancement.

The seeds of the *Albizia* trees are mixed with rice seeds and are planted in freshly opened swiddens. They become a dense stand of trees during the fallow period, which may exceed a population of 400 trees per hectare, enabling the rapid recovery of soil fertility. This innovation, introduced several decades ago, has been of critical success in sustaining annual crop productivity as the fallow periods declined over the years. The fallows are now typically only four years in duration, due to intensifying population pressure.

This practice has also contributed in another important way to the sustainability and livelihood outcomes of the Baduy swidden system: the sale of the *Albizia* logs at the end of the fallow period generates a substantial additional cash income of more than US$2500 per hectare from the timber crop. Increasingly, we see cash benefits produced by fallow-tree enterprises becoming a crucial factor in the stability of swidden farming systems. In fact, tree fallows may, under some conditions, replace annual cropping entirely. This is the direction in which the Naga *Alnus* system is moving (Cairns and Brookfield, ch. 22, this volume).

Chapter 25 describes another successful Indonesian case in which *Albizia* (*Paraserianthes falcataria*) has been incorporated into a swidden system to provide both accelerated soil fertility and attractive cash returns from timber production. Niken Sakuntaladewi and his colleagues preface this story with a broad assessment of the general circumstances and challenges facing shifting cultivators in Indonesia. They note that swidden systems there are in a period of tremendous change. Traditional farming systems are either evolving or being spontaneously replaced by mixed gardens, agroforests, plantations (rubber or cacao) or rice fields.

They then describe the case of a community in Kalimantan that recently confronted its difficulties in sustaining its swidden systems by adopting the *Albizia*-fallow solution. This enabled them to surmount their challenges and to thrive. Key factors were that the life cycle of the tree species fitted comfortably within the shortening shifting cultivation cycle, and there were lucrative cash returns from the timber. This situation closely mirrors the case of the Baduy in West Java (Iskandar, ch. 24, this volume).

The practices for establishing the *Albizia* trees in Kalimantan are quite similar to those described by Iskandar in West Java. The tree seedlings are planted in the first year of rice cultivation, about one month after the rice has been planted. They are nurtured along with the annual crop, and in the second year they are thinned to the desired population. The trees grow to become a dense and dominant stand during the fallow period. They are harvested for their timber after about seven years, and then the swidden plot is returned to annual cropping.

The farmers emphasize that the silvicultural treatment of their *Albizia* plantations is very easy and quite inexpensive, while the land regains its fertility in a relatively short time under the *Albizia* fallow. Thus, it appears that the *Albizia* solution has quite wide extrapolation potential for fallow rotation systems across Indonesia and beyond.

In chapter 23, Suson and Lasco document another highly sustainable farming system that evolved on steeply sloping land on Cebu Island in the central Philippines. It has been practised for the past 170 years. The system is based on the culture of another leguminous tree species, *Leucaena leucocephala*, which is grown as a dense plantation during the fallow period. This enables quite productive and indefinitely sustained cropping on agriculturally marginal and extremely erosive land.

Surprisingly, it allows the land to be cultivated for annual crops (maize and tobacco) for a continuous stretch of seven years. It is then fallowed under a plantation of *Leucaena* for an additional seven years before the *Leucaena* is cut down for the next cropping cycle. During this period, the *Leucaena* becomes an abundant source of high-protein fodder for ruminant livestock. The associated animal enterprise provides an important source of income generation, analogous to the income-generating timber of the tree fallows in the two Indonesian cases.

During the seven-year cropping period, two successive crops are cultivated in each of the seven years, for a total cycle of 14 crops. This is a land-use intensity that greatly exceeds those of the other systems reported in this section – suggesting that its fertility regeneration potential is quite extraordinary in the annals of Asian shifting cultivation. This seems to well justify the authors' contention that it 'offers a technology capable of beneficial replication in many parts of Asia-Pacific'.

It so happens that the location where this system was developed and is practised is quite close to a major industrial city; an urban center that has seen explosive growth over recent decades. The availability of abundant opportunities for off-farm labour has inflated local wage rates, and this is now threatening the economic viability of this productive, but labour-intensive, agricultural system.

It would seem that the natural tendency of the farmers will be to shift toward more permanent tree crops, so as to drastically reduce the amount of labour demanded by the system. This is happening under somewhat analogous circumstances to the evolution of the *Alnus* system in Nagaland (Cairns and Brookfield, ch. 22, this volume), and similar trends are observed in the peri-urban areas of many other large cities in Asia and Africa. Nevertheless, the Naalad system may be quite suitable for communities that are more remote from centres of population, have very limited land availability per family, and need annual-crop production for food security.

There is increasing pressure to increase returns to labour almost everywhere, and this is a major factor in adopting such systems. Therefore, extrapolation of the Naalad system might be more successful if the system could create more permanent terraces, rather than the temporary soil-erosion structures (*balabags*) that have been deployed until now.

The Naalad case provides such an interesting example of farmer innovation for sustainable steep-land cultivation that it deserves serious attention for adaptation and extension. As the authors emphasize, 'there is a widespread need for a farming system that uses only two plots of land, delivers two crops every year, supports livestock, provides firewood, and has an impressive proven record'.

Chapter 27 discusses a process of agrodeforestation in Southeast Asia and the Pacific islands. Dr. Randolph Thaman discusses the case of Java cedar (*Bischofia javanica*), one of the most important agroforestry trees in the region, with quite wide adaptation. He calls these trees 'sentinels of the past', noting that they remain an important component in shifting cultivation in some areas, due to their wide cultural utility and perceived role in soil enrichment.

Although it may be severely pollarded, ring-barked or coppiced, the tree rarely dies, and is allowed to regenerate in gardens and fallows because of its cultural and ecological importance. For example, Java cedar (*koka*) was a basis for the traditional Tongan polycultural shifting-cultivation system, which was focused on yam culture. However, although *koka* systems are apparently sustainable, they are declining in many areas. The disappearance of these trees from the agricultural landscape is due to a failure to protect or replant them in agricultural land-use systems. This is another example of a theme covered in the Naga *Alnus* and Naalad *Leucaena* chapters: the recent decline of tree-based systems for regeneration of soil fertility resulting from agricultural commercialization.

This chapter brings us full circle in this *tour de table*, as it evokes a vision for a possible renaissance of traditional agricultural systems in many Pacific nations, based on Java cedar fallows or intercropping with traditional food crops – such as those being developed now in Hawaii, described in the candlenut chapter (Lincoln, ch. 21, this volume). Thaman advocates for the conservation and deliberate replanting of Java cedar and other appropriate trees, to reverse the process of agrodeforestation and the spread of monoculture cropping. This draws attention to the need to adapt new ways of managing such species, so they can play a more important role in evolving agroecological farming systems.

Summing up

One theme that is present in all of the chapters in this section is that innovative swidden cultivators are finding many ways to creatively adapt and incorporate specific tree species into their farming systems, to enable them to thrive despite changing circumstances. Insights gained from more deeply assessing these models will be of great value in assisting further development of swidden agriculture, or in making the transition to more permanent-field farming, or in restoring productivity to land that is degrading under continuous cultivation.

Global concern is intensifying around the imperative that agriculture must transform; to reduce unsustainable and polluting practices, adapt to a rapidly changing climate, and to sequester vast quantities of carbon dioxide from the atmosphere. I am confident that these concerns will shift attention dramatically towards ways in which we can 'perennialize' all of our agricultural systems by making the use of trees and shrubs a universal practice. This trend will demand that researchers and practitioners turn their interest to understanding and applying the principles and practices that shifting cultivators use today and have been using for countless generations past.

Therefore, it is encouraging to have such a rich collection of articles from Southeast Asia, showing how farmers everywhere might benefit from soil-building trees, and thus bring us effectively 'back to a more agroecologically sound future'.

References

Garrity, D. P. (2015) 'Learning to cope with rapid change: Evergreen agriculture transformations and insights between Africa and Asia', in M. F. Cairns (ed.) *Shifting Cultivation and Environmental Change: Indigenous People, Agriculture and Forest Conservation*, Earthscan from Routledge, London, pp.235-258

Garrity, D., Akinnifesi, F., Ajayi, O., Sileshi, G. W., Mowo, J. G., Kalinganire, A., Larwanou, M. and Bayala, J. (2010) 'Evergreen agriculture: A robust approach to sustainable food security in Africa, *Food Security* 2(3), pp.197-214

Schwartz, J. (2020) *The Reindeer Chronicles and Other Inspiring Stories of Working with Nature to Heal the Earth*, Chelsea Green Publishing, White River Junction, VT, pp.125-152

COLOURED PLATES ...Part II (plates 34-78)

During the time of his PhD fieldwork, the villagers of Khonoma village, in Nagaland, grew accustomed to the Editor often lurking in the background, trying to document their farming practices on film. It was Khonoma's ingenious management of the Himalayan alder tree that had initially piqued the Editor's interest and brought him to their fields (see Cairns and Brookfield, ch. 22, this volume). Although this sketch doesn't show much of the alder trees (but see plates 11 to 13 in Coloured Plates Part I, this volume), it does show how farmers had, over generations, cleared the stones from their fields and used them to build walls along the slope contours that would hold the soil in place and reduce erosion. These walls were commonly used in tandem with the alder trees, and both were part of the landesque capital (Blaikie and Brookfield, 1987) developed by Khonoma's farmers over the centuries.

Sketch based on a photo by Carolin Meru (2002).

SYSTEMS-LEVEL INNOVATIONS: BIOLOGICALLY IMPROVED FALLOWS (continued …)

Paraserianthes falcataria-based fallows in Indonesia (see Iskandar, ch. 24 and Sakuntaladewi et al., ch. 25, both this volume).

34. *Paraserianthes falcataria* seedlings have been prepared by Outer Baduy farmers and are awaiting transplanting into their rice swiddens. This nursery was photographed in Cisiment village of Leuwidamar district in Lebak Regency, Banten province, Indonesia. These seedlings are the start of the *Paraserianthes falcataria* fallows that will both improve the soil and earn the farmers additional income through wood sales.

Photo: Johan Iskandar (2018).

35. * An Outer Baduy farmer works between the young *Paraserianthes falcataria* trees, as she slashes dried rice stems and weeds, after the first year's rice crop has been harvested. The slashed material is left to rot and becomes organic fertilizer for the young *Paraserianthes falcataria* trees. The field pictured here was at the end of the first cropping year.

Photo: Johan Iskandar (1995).

36. This photo, taken in the same village as plate 34 (above), shows what a *Paraserianthes falcataria* fallow looks like at about three years of age. Fallow length under the Outer Baduy system of management usually ranges between three and five years, so these trees are likely to be harvested in the next year or two, and the land will then return to the next phase of arable cropping in the swidden cycle. By *taungya*-planting N-fixing *Paraserianthes falcataria* as a preferred fallow species, farmers are able to earn dual benefits of both soil improvement and income through wood sales.

Photo: Johan Iskandar (2016).

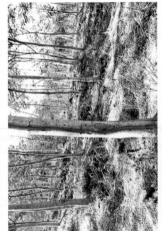

* Plate 35 (above) was earlier published by Berghahn Books in *"Modern Crises and Traditional Strategies: Local Ecological Knowledge in Island Southeast Asia"* (2007).

Retention of *Bischofia javanica* in swiddens in the South Pacific (see Thaman, ch. 27, this volume).

38. Semi-deciduous, trifoliate compound leaves of *koka* or Java cedar (*Bischofia javanica*). Shifting cultivators in the South Pacific consider that leaf fall from this tree creates the best soil for cultivation of yams and other staple root crops. This photo was taken at Suva, Fiji.

Photo: Randolph Thaman (2004).

39. These *Bischofia javanica* trees had been protected in a shifting sweet potato garden on Tongatapu island, Tonga.

Photo: Randolph Thaman (2016).

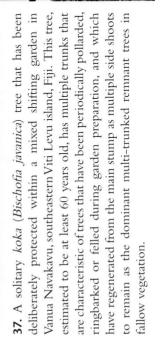

37. A solitary *koka* (*Bischofia javanica*) tree that has been deliberately protected within a mixed shifting garden in Vanua Navakavu, southeastern Viti Levu island, Fiji. This tree, estimated to be at least 60 years old, has multiple trunks that are characteristic of trees that have been periodically pollarded, ringbarked or felled during garden preparation, and which have regenerated from the main stump as multiple side shoots to remain as the dominant multi-trunked remnant trees in fallow vegetation.

Photo: Randolph Thaman (April, 2013).

Management of *Casuarina oligodon* as an improved fallow in the highlands of Papua New Guinea (see Bourke, 2007).

40. From a low-flying aircraft, over the western part of Eastern Highlands province in Papua New Guinea (PNG), a large food garden can be seen being cleared from what was primarily a *Casuarina oligodon* fallow, mixed with small amounts of other self-sown woody vegetation.

Most of the *Casuarina* trees that can be seen left standing would have been ring-barked and killed. Others will have had most of their side branches cut off without the trees being killed. Most of the trees have not been felled – but some have been cut down and used to construct a fence around the garden to keep out domestic pigs. Other trunks may have been laid on the ground to delineate individual plots belonging to different women gardeners. The garden will be planted mostly with sweet potatoes as the staple food crop.

Photo: R. Michael Bourke (October, 1990).

41. Similar to the reputation of *Schima wallichii* in northeast India (see plates 18 to 20 in Coloured Plates Part I, this volume), shifting cultivators in PNG associate *Casuarina oligodon* with improved soil fertility and work to increase its presence in their fields. In a system analogous to that described for *S. wallichii* in India, PNG farmers retain *C. oligodon* trees in their fields as relict emergents, and lop off most of the side branches, primarily to reduce shading on nearby crops. The farmer pictured here was pruning the side branches from a *C. oligodon* tree in Chimbu province, PNG, in the course of opening a *Casuarina* fallow for cultivation. It is interesting to see shifting cultivators from very different cultures and separated by great distances, managing different soil-building trees in similar ways and for similar purposes.

Photo: R. Michael Bourke (ca 1999).

Tithonia diversifolia-based fallows in the Philippines (see van Noordwijk et al., ch. 39, and Daguitan et al., ch. 40, both this volume).

43. This photo of a farmer standing next to a *Tithonia diversifolia* fallow in Bukidnon, the Philippines, shows just how dense and woody this shrub can become.

Photo: Malcolm Cairns (1995).

44. The soil under a *Tithonia* fallow is usually black and rich with biotic activity.

Photo: Malcolm Cairns (1995).

42. *Tithonia diversifolia* is commonly known as 'wild sunflower'. It is likely that this shrub owes its wide distribution across Asia to its aesthetic qualities as an attractive flower, rather than any recognition of its agronomic properties. Native to Mexico, it is thought to have been brought to Asia by missionaries as an ornamental flower.

Photo: Malcolm Cairns (1995).

Other Asteraceae-based fallows across the Asia-Pacific region (see van Noordwijk et al., ch. 39, this volume).

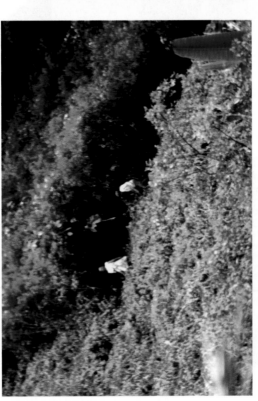

45. These Minangkabau shifting cultivators show that *Tithonia diversifolia* isn't the only Asteraceae shrub that can provide fallow functions when fallows are shortened to only a few years. The Minang couple pictured above were busy opening an *Austroeupitorium inulaefolium* fallow on the slopes of Dusun Sungai Manau Atas, in the central rift valley of West Sumatra, Indonesia (see Cairns, 2007).

Photo: Malcolm Cairns (1994).

46. Although the photographer estimated that this fallow was likely to have been only one-and-a-half years old, the abandoned field hut was already disappearing under a green thicket of fast-growing shrubs, most likely *Chromolaena odorata*, Asteraceae (see Roder et al., 2007). This land was swiddened by Tai Bo farmers in Hinbonn district of Khammouane province, in the Lao P.D.R.

Photo: Keith Barney (May, 2006).

When land use intensifies beyond the scope of standard fallowing techniques, some farmers reshape their land into mounds or raised beds (see Bourke et al., ch. A40, Hitchcock, ch. A41, and Tiwari, ch. A42, all in supplementary chapters to this volume).

Shifting cultivation relies on the biomass built up over the fallow period to replenish soil fertility and keep the system sustainable. As land use intensifies and fallows shorten, the amount of fallow biomass drops precipitously, and is no longer enough to rejuvenate soil fertility. Both systems pictured below appear to maximize use of the limited amount of biomass available by concentrating its application to smaller seedbeds.

47. In an extreme example of how farmers have responded to pressures to intensify their land use, farmers in the Upper Lai Valley, in Enga province of Papua New Guinea, shape their fields into very large mounds (see Bourke et al., ch. A40 and Hitchcock, ch. A41, both in the supplementary chapters to this volume). Available organic matter is composted in these mounds. By far the most important crop planted in these mounds is sweet potatoes, the staple food for this area.

As intensive as this land use is, strictly speaking, it is still a shifting cultivation system. After perhaps 30 to 40 years of this intensive cultivation, the whole area covered by mounds in this photo will be left uncultivated and will grow tall cane grasses and scattered shrubs. The fences that surround it to keep pigs out will be opened and and the pigs allowed in.

Photo: Bryant Allen (1981).

48. Khasi farmers, in the East Khasi Hills district of Meghalaya, almost always plant their potatoes on raised beds (see Tiwari, ch. A42, in supplementary chapters to this volume). Slashed vegetation is placed on the raised beds, covered with soil and then burnt under oxygen-limited conditions. Any unburnt debris is then removed from the beds in preparation for planting the potatoes. Aside from focusing fertility in the raised planting beds, the author suggests that this system offers further advantages of reducing soil density, improving its moisture retention and reducing chances of water logging on the growing crop.

This photo provides an aerial view of the raised bed system as practised in Pomlakarai village, in the East Khasi Hills district of Meghalaya, northeast India.

Photo: Evamary Diangdoh (2013).

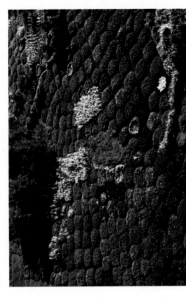

EXTENDING THE WINDOW OF PRODUCTIVITY INTO THE FALLOW PHASE

Taungya-planting valued trees into swidden crops cuts across most of the 'economically improved' fallows pictured in this section. Alternatively, a less labour-intensive strategy is, during weeding operations, to simply retain valued trees that sprout from the soil seed bank. This natural regeneration may then be supplemented by *taungya*-planting to fill in any gaps and thicken the stand. However it is established, in all cases, the central objective is to get the seedlings planted during the cropping phase, and then use the 'fallow period' to grow a marketable product.

Converting swidden fallows into valuable timber stands (see Roshetko et al., ch. 29, this volume).

We start with planting timber trees into swidden fields since this is one of the most widespread strategies used across the region to achieve a marketable product from the 'fallow period'. As natural timber became more scarce and expensive, many shifting cultivators saw an opportunity to *taungya*-plant timber trees into their swidden fields and earn added income through timber sales. The purpose of the 'fallow' thus evolved from almost solely rehabilitation to take on an important component of economic production. Farmers have long managed livestock as 'living banks' that can be converted to cash during times of urgent financial need. In a similar way, shifting cultivators can harvest some timber trees during times of financial stress, such as during sickness, or perhaps to pay for school fees, a wedding or other cultural celebrations. Arguably, the introduction of teak into swidden lands could more accurately be considered as a conversion to a new type of land use, rather than the economic improvement of a 'fallow', *per se*. In areas where fallow periods are typically five to ten years, these teak plantations may take 20 to 50 years to fully mature and be harvested, after which it is doubtful if the farmer will return to growing upland rice on this land, in a swidden rotation.

49. Teak interplanted with upland rice in Luang Prabang province of Lao PDR. Most farmers *taungya*-plant teak into the upland rice about one to two months after planting the rice. Field preparations for the rice crop provide a good growing environment for teak. But competition from the rice can reduce teak growth substantially in the first year, especially if planting the teak is delayed and if the rice crop is vigorous. Rice cropping is continued for two to four years before fallowing. If the conditions are favourable for teak, the farmer may get his first returns from teak sales after about 20 years. This long time to market is obviously a major obstacle for poor upland farmers who can't afford to wait that long.

Photo: Walter Roder (1994).

50. The teak being harvested in this photo was probably established by a Javanese farmer *taungya*-planting into a swidden field. After the trees dominate the site, annual crops might continue to be cultivated in the understorey when market prices are good and family labour is available. When harvesting a few trees, the farmer may open a gap large enough to intercrop between the trees. Otherwise, less light-demanding crops, such as ginger, can be cultivated in the understorey. This timber harvest was photographed in Gunung Kidul, Yogyakarta, Indonesia. The man harvesting the tree was a local contractor, known to have good chainsaw skills. He was hired by the timber trader, who had bought the standing tree.

Photo: Iwan Kurniawan, ICRAF (2008).

Primarily within Indonesia, another prime example of *taungya*-planting being used to create 'economically improved fallows' is the 'jungle rubber' system (see Yonariza, ch. A18, in supplementary chapters to this volume, and Penot, 2007).

51. Like so many of the systems pictured in this section, 'jungle rubber' begins by *taungya*-planting young seedlings into a swidden field. This photo shows the stage of growth of the young rubber seedlings at the end of the second cropping year.

In cases where swidden fields are not *taungya*-planted with rubber seedlings, the cropping period in Pasaman district, West Sumatra, is usually limited to a single year of upland rice. But when rubber seedlings are interplanted amongst the swidden crop, farmers usually extend their cropping phase to a second year, often planted with patchouli (*Pogostemon cablin*) or chillies. This allows the rubber seedlings a second year of favourable growing conditions before being forced to compete with regenerating fallow vegetation. After two years, the rubber seedlings are considered to be large enough to survive the fallow period.

The field shown here was managed by Minangkabau shifting cultivators in Silayang village, Pasaman district, West Sumatra, Indonesia.

Photo: Yonariza (April, 2016).

52. This photo shows a mature 'jungle rubber' fallow that is already being tapped, in Indonesia's West Kalimantan. It shows the secondary forest that regenerates between the rows, and gives the system an untidy appearance, from which the moniker 'jungle rubber' is derived. This plantation was managed by a Dayak farmer in Kopar village, Kabupaten Sanggau, in West Kalimantan, Indonesia.

Photo: Eric Penot, CIRAD, UMR Innovation (November, 2019).

53. A farmer in the Muara Bungo region of Jambi province, Sumatra, shows his technique for tapping what appear to be quite old trees in his rubber agroforest. It shows the jungle character of this agroforest, with a high plant diversity between over-aged rubber trees. The tree shown here may be 40 to 45 years old.

Photo: Hubert de Foresta (1991).

Another well-known fallow-enrichment system in Indonesia develops into rattan gardens (see Schreer, ch. 30, this volume).

54. A Dayak Benuaq farmer deftly transplants a young rattan seedling, as part of the process of establishing a rattan garden at Pepas Eheng village in Kutai Barat district, East Kalimantan. The Dayak Benuaq are one of the few ethnic groups in this area that are known to commonly plant rattan.

Photo: Wahyu Widhi, NTFP-EP Indonesia (2013).

55. A mature rattan climbs through a canopy of secondary forest regrowth in Papua New Guinea. This rattan would have established naturally. The Krisa people who live in this area do not plant rattan since there are sufficient natural stocks available from regrowth vegetation. This photo was taken in Krisa village, Sandaun province, Papua New Guinea.

Photo: Stefanie Belharte (October, 1997).

56. A Ngaju Dayak woman harvests canes from a rattan garden near the village of Baun Bango, on the bank of the Katingan river in Central Kalimantan. Depending on the species, rattan is usually ready to harvest at about seven to ten years of age. This woman worked under a share-harvesting arrangement, with 50% of the rattan value paid to the harvesters.

Photo: Gerard Persoon (June 18, 2005).

South Sulawesi provides another compelling example of fallows being transformed into agroforests (see Supratman et al., ch. 36, this volume).

57. This photo shows the community candlenut forests (*hutan kemiri rakyat*) in South Sulawesi, from which swidden fields are cyclically cleared to plant food crops for about 4 years. It largely parallels standard shifting cultivation, except that it is man-made candlenut (*Aleurites moluccanus*) forests that are cleared for cultivation, instead of natural forests. Similar to the 'jungle rubber' system discussed in plates 51–53, these candlenut 'fallows' provide farmers with an income throughout the 'fallow' phase.

The overall rotational pattern takes place in a honeycomb shape over a period of 25 years, before cultivators return to the original plot. The honeycomb pattern is divided into 8 hexagonal plots, each covering about ¼ of a hectare, totalling an average of 2 hectares per household. This landscape was photographed in Cenrana Baru village, Maros district, South Sulawesi, Indonesia.

Photo: Andy Kurniawan (April, 2021).

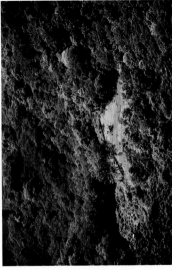

58. A handful of the candlenut fruit. Fallen fruit are simply gathered from the forest floor, so there is no need to climb the trees for harvest. The harvest season usually spans from August to October each year. The innovators of this candlenut system self-identify as belonging to the Bugis Camba ethnic group.

Photo: Nurhidayanti (August, 2020).

59. This group of Bugis Camba farmers has just completed a *makkalice* harvest of a candlenut stand. *Makkalice* is an open harvest season, in which anybody has the right to harvest, even if they have no ownership claims to the trees. Farmer groups can get together in certain locations and gather up any unharvested nuts that they find on the ground. The owners of the candlenut stands usually create boundaries to try to prevent their harvest from falling onto a neighbour's land.

Photo: Nurhidayanti (August, 2020).

As exemplified by the damar agroforests of Krui, Indonesia, the *taungya*-planting of valued trees into swidden fields can develop into permanent agroforests that remove the land from the swidden cycle (see Herawati et al., 2017, and Michon et al., 2007).

60. At first glance, this landscape appears to be covered by lush tropical forests – but more discerning observers will recognize that it is a man-made landscape, planted by the *Orang Krui* in Indonesia. Studies have found that in the natural, old growth forests in this area, damar trees (*Shorea javanica*) with a trunk diameter greater than 20 cm at breast height (dbh), are present, but only at an average density of one tree per hectare. Another proof of the planted nature of the forest is the history of these lands and the damar trees, as told by the damar agroforest farmers themselves. They confirm that all these damar trees had been planted in the typical sequential agroforestry system that they developed. Further, this impressive forest is comprised of many individual agroforest plots, averaging about one hectare in size, that are individually owned and managed. This photo was taken in the territory of Penengahaan village, near Krui in Lampung province, Sumatra, Indonesia.

Photo: Hubert de Foresta (1993).

61. This climber was harvesting damar resin in the damar agroforests of Pahmungan village, near Krui, in Sumatra's Lampung province. On average, damar trees can begin to be tapped about 20 years after they have been planted. The damar tree being tapped in this photo was probably about 60 years old.

When a damar agroforest is planted, the land is completely removed from the swidden cycle, since the agroforest will never be completely cut down in its entirety. Rather, damar and fruit trees are individually replaced when they die or when they are felled because they become unproductive.

Photo: Hubert de Foresta (1992).

Lao PDR has its own resin-producing system – but one that fits more comfortably within a swidden cycle (see Vongkhamho and Ingalls, ch. 34, this volume, and Fisher et al., 2007).

62. The Lao system of producing resin begins when seeds of *Styrax tonkinensis*, that are part of the soil seed bank, germinate naturally when a fallowed plot is cleared and burned in preparation for cultivation. Farmers recognize the young *S. tonkinensis* seedlings growing amidst their rice, and are careful to retain them when weeding. The *S. tonkinensis* sapling shown in this photo was a product of this process of selective weeding, and shows the size reached at about ten months of age, after the first crop of rice has already been harvested.

Photo: Simone Vongkhamho (March 12, 2017).

63. This photo shows the main tools used in collecting solidified benzoin resin. The small spade is used for prying the dried resin from the tree and into the bamboo basket.

Photo: Simone Vongkhamho (February 21, 2016).

64. A Khmu farmer taps a *Styrax tonkinensis* tree in his 'managed fallow'. It is the Khmu who are the main practitioners of the 'Styrax fallow system' and who have accumulated the greatest local knowledge on its management, learned over many generations of use (see also Fischer et al., 2007). The age that the *S. tonkinensis* trees are tapped is determined largely by the swidden cycle in use – but in most cases, tapping begins when the trees are six to seven years old and may continue up to the 15th year. The tapping season runs from June to November. This coincides with the season of flowering and fruiting for *S. tonkinensis*, and is the time that the tree produces the most resin. However, August is the peak time for tapping. Farmers wait four to five months after tapping before they begin to collect the solidified resin, usually from November to April. This tapping-collection sequence is usually limited to only once a year to avoid the trees becoming too weak and dying prematurely.

Photo: District Agriculture and Forestry Office (DAFO), Phonthong district, Luang Prabang (November 12, 2013).

Using the 'fallow period' to grow a spice in Jambi province, Indonesia (see Burgers, ch. 37, this volume).

65. Minangkabau farmers in Jambi province *taungya*-plant *Cinnamomum burmanni* into their swidden fields as an investment to pay for large, one-time expenditures such as going on a pilgrimage to Mecca, paying for a funeral or a wedding, to educate their children, or perhaps to cover the cost of travelling to Malaysia in search of work (Burgers, 2004). The farmer shown here, peeling the bark from a felled tree, was planning to get married, and was harvesting his cinnamon crop to finance it. The bark is considered to be the only commercially valuable part of the *Cinnamomum burmanni* tree. After the bark is removed, the remaining bole of the tree is usually just left in the field to decay and return nutrients to the soil. Since these plantations tend to be in remote upland areas, far from roads, it is usually not considered worthwhile to carry the wood out. In situations where fields are close to a road, the wood may be sold as firewood, although prices are poor.

When large-scale harvesting is not needed, farmers may gradually harvest bark from individual trees or even just from branches, to cover daily or weekly expenses. This photo was taken in Selampaung village, Gunung Raya district in Kerinci Regency, Jambi, Indonesia.

Photo: Paul Burgers (ca May, 2000).

66. This photo shows the aftermath of harvesting a cinnamon 'fallow'. With the 'fallow' vegetation cleared, the farmer continues with the cropping phase of his swidden cycle. In this case, the farmer had planted a mixed crop of tobacco and chillies among the *Cinnamomum burmanni* stumps. This phase of arable cropping usually continues for two years.

What is most notable about this photo is the profuse coppicing from the *Cinnamomum burmanni* stump, next to the farmer. When harvesting, farmers purposely cut their trees a bit high, at about 1.5 m., to increase the rate of stump survival and coppice growth. This is usually done when the price of cinnamon bark is expected to be high, and the farmer wants to accelerate when he will be able to harvest his next crop and cash in on those high prices. Another alternative would be to completely kill out the old stumps and plant an entirely new crop of *Cinnamomum burmanni* seedlings – but since the coppices grow faster, they can typically shorten the time to market by two to three years. This system of harvesting coppices can be repeated many times, but reportedly does not produce the highest quality cinnamon bark.

Photo: Paul Burgers (ca 1999–2000).

Indigenous peoples have long used forests as their pharmacies. Some shifting cultivators are seeking to bolster that role by planting medicinal plants as an understorey in their swidden fallows (see Tran and Pham, ch. 38, this volume, and Leaman, ch. A73, in supplementary chapters to this volume).

68. This Dao farmer, in Vietnam's northernmost Ha Giang province, was weeding her cardamom (*Amomum tsao-ko*) when she noticed that she was being photographed by visitors. This land had previously been used for growing upland rice and maize in a swidden pattern – but the fallow had now been planted with an understorey of shade-tolerant cardamom. The cardamom seed pods are used as both a spice and a medicine for treating a long list of ailments. Some farmers are earning high returns from growing cardamom for Chinese markets - reportedly as much as 10 to 20 times what they could earn from growing maize or upland rice.

The cardamom shown in this photo was planted in the previous year. Harvests of the seed pods can continue for many years from a single planting. October is the main month for harvesting.

This photo was taken at about 1500 masl in Lung Tao village, Cao Bo commune, Vi Xuyen district, Ha Giang province, in northern Vietnam, not far from the border with China. The Dao are recognized as among the ethnic minority groups in Vietnam with a long history of practising shifting cultivation.

Photo: Dang Van Dui (2018).

67. A Kenyah elder in East Kalimantan removes the bark of *Lansium domesticum* (langsat) to prepare a treatment for malaria (see Leaman, ch. A73, in supplementary chapters to this volume). This photo was taken in a forest fallow, possibly more than 40 years old, at Long Sungei Barang in Apo Kayan, East Kalimantan. *Lansium domesticum* is one of the valued species that is often left standing when fallows are reopened for cultivation, so this particular tree may have been even older than the rest of the fallow regrowth, of which it was a part. Langsat is native to this area, so we don't know if this tree was part of the natural vegetation, or might have been introduced by the farmer.

Photo: Danna Leaman (ca 1988).

Elevating broom grass from an occasionally collected non-timber forest product (NTFP) to an actively-propagated fallow crop (see Bora et al., ch. 31, this volume).

69. Khasi shifting cultivators in the Dawki area of Meghalaya, in northeast India, had transformed this fallowed hillside into a thick carpet of broom grass (*Thysanolaena latifolia*). Aside from providing the farmer with an added source of income, researchers also report that it has a positive effect on the environment. Broom grass usually grows up to an elevation of about 2000 masl.

Photo: Indrani P. Bora (May, 2010).

70. As part of a project promoting broom grass as a crop that can be grown on fallowed land, these Karbi farmers in the Karbi Anglong district of Assam are seen harvesting their crop. Broom grass generally requires about nine months to reach a harvestable stage. The inflorescences, complete with stems, are harvested once a year, from December to February, and sales may continue until March or even May. During the first year of growth, yields from broom grass are negligible. They reach a peak in the second and third years, after which there is a gradual decline. The project showed broom grass to be a profitable crop.

Photo: Indrani P. Bora (December, 2010).

71. This was a local collection centre in the Dawki area of Meghalaya, to which broom grass growers brought their products for sale. The only processing required for broom grass before marketing, is that it is properly dried in the sun. The flowering heads brought here are the raw materials that will be fashioned into brooms.

Photo: Indrani P. Bora (March, 2010).

A few other useful trees that have become domesticated in fallowed lands.

72. Farmers in northeast India widely value *Livistona jenkinsiana* as a roofing material. They therefore see value in intensifying its use beyond the opportunistic harvest of wild plants found growing naturally in their environment, and have actively planted it as a 'fallow crop' on *jhum* land near their villages. It is propagated by seeds and leaf harvest may begin at the age of four or five years. Individual plants may persist up to 80 years, so it is doubtful if this land will ever again be brought back under shifting cultivation.

L. jenkinsiana is well-suited to tropical and sub-tropical conditions and is cultivated by a wide range of tribal groups across Nagaland and Arunachal Pradesh.

Photo: Sapu Changkija (June, 2008).

73. Shifting cultivators in the uplands of Nan province, in northern Thailand, are able to earn extra income by domesticating a local jungle spice, *Zanthoxylum limonella* (*ma kwaen*), in their fallowed fields, as part of a wider agroforestry system (see Hoare et al., 2007). *Z. limonella* is the larger of the trees seen in this photo, planted in association with lychee and cotton. This system will eventually develop into a lychee orchard, and so take the land out of use for swiddening. *Z. limonella* grows widely in the forests of Nan, up to 700 masl. Its fruit and seeds are used as a jungle spice in traditional Thai food, in traditional medicine and in the perfume industry. *Z. limonella* regenerates naturally in swidden fields and farmers transplant additional seedlings to thicken the stand to about 80 trees per hectare. It has mostly been the Northern Thai, Thai Lue, and Khmu ethnic groups who have adopted *Z. limonella* as a 'fallow crop'.

This field was photographed in Pang San village, Song Kwae district, Nan province, Thailand.

Photo: Peter Hoare (2000).

Northern Thailand provides another example of a valued component of fallow regrowth being domesticated and transplanted into long-term plantations.

When shifting cultivation was widely practised in Nan province of northern Thailand, and fallows were lengthy, one of the valued plants that regenerated naturally from the soil seed bank was *Melientha suavis* (Opilliaceae). This small evergreen tree is native to Southeast Asia, and its young shoots, leaves and flowers are widely valued as a wild vegetable. In northern Thailand, it is locally known as *pak wan pa*, and is in high demand at local markets.

74. In the act of establishing a *Melientha suavis* plantation, this farmer was transplanting seedlings from a nursery, probably at about one or two months of age. But farmer experiments found that transplanting younger seedlings (18 days) provided a higher survival rate. This plantation was being established in Ban Muang Uang, Ban Luang subdistrict, Chom Thong district, Chiang Mai province, to demonstrate *M. suavis* cultivation as a step towards increasing the agrobiodiversity of food plants promoted by the nearby organic vegetable planting area of the Royal Project. The plant spacing appears to be about 2 X 2 meters, which would amount to about 400 trees/rai*.

Photo: Arratee Ayuttacorn (May 25, 2022).

* A rai is a Thai unit of area equal to 1,600 square metres.

75. This two-year old plantation of *Melientha suavis* had been planted by local Thai farmers, along with a companion crop of N-fixing *Sesbania grandiflora*. Such *M. sauvis* plantations can begin to provide farmers with a monthly income after two to three years, and may continue to be harvested for more than 10 years, depending on the pruning practices used. Farmers can harvest 50–100 grams of young shoots and leaves/tree/month, and sell them for about 250 Thai baht/kg. This plot was photographed in Nam Tok subdistrict, Na Noi district, Nan province.

Photo: Phrek Gypmantasiri (March 8, 2019).

Some farmers have learnt how to harness insects to convert fallow vegetation into useful products (see Chakraborty and Choudhury, ch. 32, and Trakansuphakon and Ferrari, ch. 33, both this volume).

76. The wild flowers that bloom in swidden fallows contribute to honey production (see Trakansuphakon and Ferrari, ch. 33, this volume). Shown here, a member of the *Pgaz K'Nyau* (Karen) village of Hin Lad Nai was harvesting a honey crop from his swidden fallow. The peak season for honey harvest is in May. The hive has been placed in the farmer's old swidden hut to protect it from the elements and provide a welcoming environment for the bees. Much as cattle often follow behind after shifting cultivation to graze fallow grasses and crop residues, in this case, it is the bees that follow behind shifting cultivation in search of nectar from flowering components of the fallow vegetation. This provides farmers with another saleable product from their swidden fields. Hin Lad Nai is located in Ban Pong subdistrict of Wiang Pa Pao district in Chiang Rai province, northern Thailand.

Photo: Chalit Saphaphak (ca 2018).

77. Karbi shifting cultivators in West Karbi Anglong district of Assam, northeast India, manage their swidden fallows to provide host plants for the tiny insects (*Kerria chinensis* and *K. rangoonensis*) that produce the resinous stick lac (see Chakraborty and Choudhury, ch. 32, this volume). In response to a request from one of the co-authors, this young Karbi farmer held up a branch from a *Ficus religiosa* tree to show the stick lac that he was harvesting.

Lac resin finds applications in food, dyes, textiles and pharmaceutical industries and is used by goldsmiths in making jewellery. It is also used as a surface coating in electrical and other fields.

Photo: Sanat Chakraborty (October, 2011).

Many farmers find advantage in opening swiddens communally.

78. This swiddening area in Mon district of Nagaland, northeast India, is typical of the large communal swiddens that Nagas commonly cultivate. The distribution of field huts provides a good indication of how the land has been sub-divided between individual Konyak families.

There are a number of persuasive reasons that recommend opening swiddens communally:

- Everyone shares the same access path, so it is easier to keep it cleared;
- Fencing is reduced, and it is easier to guard crops against damage by wildlife and stray livestock;
- When flocks of birds plunder the crops, the losses are spread across a wide area and no single field is completely wiped out;
- Since many farmers are working in the same general area, the arrangement is conducive for communal labour;
- Traditional village leadership finds large swidden blocks easier to manage;
- The communal nature of the fields adds a social element to working in swiddens. It is 'more fun' to work in groups;
- In Nagaland's head-hunting past, working together offered greater security. There was more safety in numbers.

Photo: Malcolm Cairns (ca 2001).

Continued in Coloured Plates Part III, in the supplementary material to this volume ...

21

PAKUKUI

The productive fallow of ancient Hawaii

Noa Kekuewa Lincoln[*]

Introduction

The Hawaiian Islands, one of the endpoints of Polynesian settlement of the Pacific, saw the development of unique agricultural advances that have not been seen anywhere else. Strong environmental differences between Hawaii and the South Pacific, explained in greater depth below, provided agricultural opportunities in Hawaii that were, at best, uncommon in other island groups. The settlers of the Hawaiian Islands inhabited Hawaii by the 11th century (Athens et al., 2014). They arrived in several waves of migration from southern island groups such as the Marquesas and Tahiti, representing the end of a several-millennia-long journey from Papua New Guinea through Samoa and Near Oceania, eventually reaching the vast eastern Pacific (Kirch, 2010). The settlers, therefore, emerged from a long history of farming the relatively smaller, steeper and less fertile landscapes of the South Pacific (Yen, 1993). Upon arriving in Hawaii and discovering new opportunities, as well as new limitations to be overcome, the settlers, over time, developed unique methods of farming and associated practices.

These practices were highly successful, and Polynesian agriculture is said to have reached its zenith within Hawaii, contributing to a complex social system of competing island chiefdoms (Kirch and Zimmerer, 2011). More than any other Polynesian islanders, Hawaiians predominantly relied on agriculture rather than marine resources for food (Handy et al., 1972). They intensified virtually every arable habitat (Ladefoged et al., 2009), made extremely marginal environments productive (e.g. Schilt, 1984), utilized innovation to develop unique farming methods (McCoy

[*] Dr. Noa Kekuewa Lincoln is an Assistant Researcher in Indigenous Crops and Cropping Systems within the Tropical Plant and Soil Sciences Department, College of Tropical Agriculture and Human Resources, University of Hawaii at Manoa and President of the Mala Kalu'ulu Cooperative, which engages in traditional agroforestry restoration and education on Hawaii Island.

and Graves, 2010), and sustained production for hundreds of years without the use of external inputs, metals, draught animals, legumes or cover crops (Handy et al., 1972; Lincoln and Vitousek, 2017). Not only are these achievements yet to be widely recognized, but in many cases the mechanisms behind them are still poorly understood (e.g. Lincoln and Vitousek, 2015). Nevertheless, the extraordinary nature of Hawaiian agriculture has not gone unnoticed. Early visitors to the islands noted that Hawaiian agricultural practices 'far exceed in point of perfection the produce of any civilized country within the tropics' (Menzies, 1920, p.81). More recently, archaeologists have marvelled that 'the [rainfed system of the Kona region] is without equal in Hawaii, and probably in the nation, in terms of the extensiveness of a prehistoric modification of the land' (Newman, 1974, p.8).

The ethno-agroecology of Hawaii Island

Hawaii boasts perhaps the densest network of ecosystem diversity on the planet. These ecological niches primarily stem from three interrelated factors. The first is the sequential age gradient of the islands, which across the eight main (youngest and largest) islands (Figure 21-1) extends from freshly emerging lava on the southern

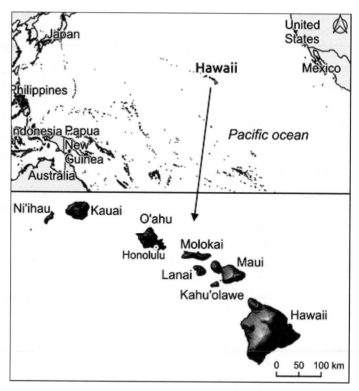

FIGURE 21-1: The eight major Hawaiian islands that are the largest and youngest of the archipelago. Hawaii Island, the southernmost, is the youngest of all with active volcanism and mountain peaks reaching over 4200 metres.

areas of Hawaii island to 5 million-year-old soils on Kauai and Ni'ihau (Juvik and Juvik, 1998). The islands are also among the highest mountainous islands in the world, which drives altitudinal changes such as temperature, solar radiation and vapour pressure (Körner, 2007). The high mountains also drive some of the strongest rainfall gradients on the planet, ranging from ~200 mm/yr to more than 10,000 mm/yr (Giambelluca et al., 2013).

The primary factors of age and rainfall interact in predictable ways, to drive secondary differences in topography and soil fertility. The myriad conditions represented by the wet-dry and old-young matrix create diverse landscapes that differ in their ecologies. Hawaii, the youngest, tallest and largest of the islands, encompasses much of the ecological diversity in the archipelago. Hawaii Island hosts 25 of the 31 life zones described by the Holdridge system of bioclimatic classification of land areas (Asner et al., 2005). In some areas, one can walk from a coastal desert to a montane cloud forest in a short three-hour hike. Along the way, a traveller would notice the change from arid to moderately moist to wet and the accompanying shifts in vegetation, but would likely be less aware of the vastly different types of soils and soil processes traversed (Vitousek and Chadwick, 2013).

The diversity of ecosystems and their drivers are the same underlying factors that drove the different application of farming techniques in ancient Hawaii. A large body of interdisciplinary literature has focused on the landscape-level distribution of these agricultural systems (e.g., Kirch, 2005; Kurashima and Kirch, 2011; Ladefoged et al., 2009; Lincoln et al., 2014; Vitousek et al., 2004, 2014). In general, Hawaiian agriculturalists employed broad strategies based on landscape parameters (Figure 21-2).

Most, if not all, of the agricultural strategies found across the archipelago were utilized at some level on Hawaii island (the largest and southernmost island) (Figure 21-3). Everywhere that surface water could be gravitationally fed, networks of terraced infrastructure were built for flooded agriculture akin to rice paddies, while areas of high natural fertility and adequate rainfall were densely planted with combinations of high-yielding crops. Fertile but dry areas were irrigated by diverting streams to support hybrid agricultural systems. Less fertile areas with adequate rainfall were converted to various forms of agroforestry, using the trees to accumulate fertility over time. Even in seemingly inhospitable conditions, Hawaiian cultivators employed techniques to farm where 'we scarcely perceive a particle of soil' (Ellis, 1917, p.337).

Hawaiian agriculture was based on agro-ecological strategies, adopting methods and timing of production that fell in line with the natural ecology of the landscape. For instance, in the Kona region there were banded planting 'zones' that mimicked the natural ecotones encountered as one moved up the mountain (Lincoln and Ladefoged, 2014), and in the Kohala region, plantings followed the seasonal rainfall and temperature shifts across the slopes (Kagawa-Viviani et al., 2018a). These self-sustaining systems needed to manage nutrients and moisture to ensure long-term maintenance without the use of external inputs (e.g. Lincoln et al., 2014, 2017; Lincoln and Vitousek, 2015).

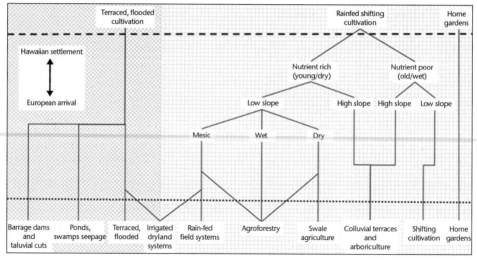

FIGURE 21-2: A rough typology for agricultural production systems in pre-contact Hawaii. The general categories are not exhaustive and represent a spectrum of practices. The dashed horizontal line is intended to demarcate systems that the Polynesian discoverers of Hawaii had in mind when they arrived; the lower dotted line indicates the range of techniques employed at the time of European arrival. There is no implication that cropping systems were static over time along a given line. To the contrary, we know that some systems expanded over time, rain-fed field systems underwent infilling and intensification, and intensive management of the fallow, as well as the cropping phase, began in shifting cultivation systems. Other systems no doubt developed as well.

Source: Lincoln et al. (2018).

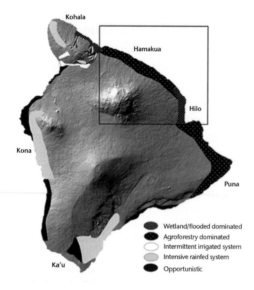

FIGURE 21-3: General patterns of agricultural reliance on Hawaii Island, estimated by ethnographic sources, archaeological surveys and bio-geochemical models. It is important to note that the forms of agriculture shown are only broad categories of dominant forms. Many nuanced variations occurred within any area, including areas that do not show any agriculture. Major regions are labelled and the box represents the study area and the extent of maps in Figures 21-8, 21-9 and 21-10.

Source: Lincoln et al. (2018).

Agroforestry in ancient Hawaii

As seen throughout much of Polynesia, arboriculture – the cultivation of trees and shrubs – played a significant role in Hawaiian agriculture (Kirch, 1994; Huebert, 2014; Maxwell et al., 2016; Dotte-Sarout and Kahn, 2017 ; Quintus et al., 2019). Tracing their ancient roots back to Southeast Asia, most of the Hawaiian crops hailed from the Indochina region. These crops were critical for the settlement of Polynesia, as the islands were virtually devoid of native fruits and tubers. Agriculture in the Hawaiian islands was, therefore, built almost exclusively upon plants introduced by the Polynesian people. Many of them survive and produce well in diversified forest conditions (Table 21-1), including breadfruit, candlenut, paper mulberry, mountain

TABLE 21-1: Crops commonly used in Hawaiian arboricultural systems.

Scientific name	Hawaiian name	Common name	Brief description of primary uses
Aleurites moluccanus	kukui	Candlenut	Dye, wood, ink, oil, medicine
Alocasia macrorrhizos	ape	Swamp taro	Food
Artocarpus artilis	ulu	Breadfruit	Food, caulking, wood, sandpaper
Broussonetia papyifera	wauke	Paper mulberry	Fibre
Calophyllum inophyllum	kamani		Wood
Cocos nucifera	niu	Coconut	Food, wood, fibre, water, medicine
Colocasia esculenta	kalo	Taro	Food
Cordia subcordata	kou		Wood
Cordyline fruticosa	ki	Ti leaf	Physical resource, food
Curcuma longa	olena	Turmeric	Medicine, food
Dioscorea alata	uhi		Food
Dioscorea bulbifera	hoi		Food
Dioscorea pentaphylla	pi'a		Food
Hibiscus tiliaceus	hau		Fiber, medicine
Ipomoea batatas	'uala	Sweet potato	Food
Morinda citrifolia	noni		Medicine, dye, food
Musa spp.	mai'a	Banana	Food
Pandanus tectorius	hala	Pandanus	Plaiting, fiber, food
Piper methysticum	'awa	Kava kava	Mild narcotic
Pipturus albidus	mamaki		Medicine
Saccharum officinarum	kō	Sugar cane	Food, medicine
Schizostachyum glaucifolium	'ohe	Bamboo	Physical resource
Syzygium mallaccense	'ohi'a ai	Mountain apple	Food, wood
Tacca leontopetaloides	pia	Arrowroot	Food
Thespesia populnea	milo		Wood
Touchardia latifolia	olonā		Fiber
Zingiber zerumbet	'awapuhi	Shampoo ginger	Medicine, cosmetic

apple, coconut, kava kava (*Piper methysticum*), banana, shampoo ginger (*Zingiber zerumbet*), noni (*Morinda citrifolia*), and pandanus. Endemic plants used for resources and medicine also thrive. Ethno-historical sources record extensive application of arboriculture in Hawaii (Quintus et al., 2019). In particular, breadfruit is known to have played a dominant role in extensive arboricultural systems across the archipelago (Lincoln and Ladefoged, 2014; Meilleur et al., 2004). However, research has largely ignored arboriculture because of a lack of archaeological infrastructure and evidence associated with it. Therefore, although a robust literature and investigation of Hawaiian agriculture exists, arboriculture is severely underrepresented. This had led to a simplified understanding of Hawaiian arboriculture with an emphasis on permanent, breadfruit-dominated arboricultural systems.

Agroforestry was particularly prevalent in areas that were too steep, too rocky, too infertile, or too salty for more intensive production systems. It was therefore common in colluvial areas and areas with very young soils, as well as those along the coast or with excessive rainfall (Lincoln and Vitousek, 2017).

As an example, valley walls were often too steep to support intensive rainfed agriculture. If they were fertile, colluvial soils were sometimes worked to form rudimentary terraces (Kirch, 1977; Kurashima and Kirch, 2011). More often, however, semi-wild tree and shrub plantings were established that would provide resources, seasonal products and unmanaged reserves against disasters that might cause the loss of intensive systems nearby. The planting of breadfruit, in particular, is evidenced by dozens of historical and prehistorical references (e.g. Meilleur et al., 2004). These valley plantings accounted for the bulk of agricultural production in some areas and were a pivotal component of agriculture throughout the islands (Allen, 2004; Kurashima and Kirch, 2011).

Very young, rocky areas with minimal soil were often augmented with forms of arboriculture. Regions such as Kona and Puna, with extensive lava flows less than a few thousand years old, are famous for their breadfruit, coconuts and pandanus groves (Lincoln and Ladefoged, 2014; Meilleur et al., 2004). The following is a description from 1778:

> [They] march[ed] … over a … track…consisting of little else than rugged porous lava and volcanic dregs…when [they] entered the breadfruit plantations whose spreading trees with beautiful foliage were scattered about that distance from the shore along the side of the mountain as far as [they] could see on both sides. (Menzies, 1920, p.74)

Coastal areas were another location where arboriculture was extensively practised. Here, trees supplied resources as well as creating pleasurable habitats. Pandanus is particularly famous for providing a pleasant ambiance in which to work, in the shade of trees open to the cool ocean breeze. Several tree crops fare well in the salt-intrusion zone, including coconuts, pandanus, *milo* (*Thespesia populnea*), noni (*Morinda citrifolia*), hau (*Hibiscus tiliaceus*), and kou (*Cordia subcordata*). The adoration of these coastal plantings is expressed in traditional sayings, such as *Puna, kai nehe i ka ulu hala* (Puna, where the sea murmurs to the *hala* [pandanus] grove) (Pukui, 1983).

Finally, and perhaps most importantly, Hawaiians engaged in arboriculture in areas too infertile for more intensive cropping. For this purpose, Hawaiians made use of both natural and novel forests. Maps of homesteads in the mid-19th century show small areas of native-canopy forest with dense understorey plantings of useful ferns and shrubs being preserved amidst a larger agricultural landscape. This practice capitalized on increased nutrient cycles and microhabitats associated with forests. In this way, larger sections of native wet forests in the Kona region were used to add a measure of resilience to the agricultural landscape:

Syzygium malaccense (L.) Merr. & L.M.Perry [Myrtaceae]

A common tree in Hawaii, mountain apple is native to island Southeast Asia and Australia, and was introduced deliberately to Remote Oceania as a 'canoe plant'. The red, pink or white fruit have been described as 'bland but refreshing'. The trees grow to 18 m and are often valued for their shade as well as their hard wood.

> …we entered the forest, the verge of which was adorned with rich and fruitful plantations of bananas and plantains, from which we supplied ourselves with a good stock for our journey (Menzies, 1920, p155).

However, in some regions, it may be that Hawaiians planted trees specifically to accumulate fertility. In these systems, very fast-growing woody plants that decomposed quickly, such as candlenut and *hau* (*Hibiscus tiliaceus*), were cultivated. In essence, these plantings were shifting cultivation systems in which both the 'fallow' and 'cropping' phases provided useful products. In this chapter I will explore the candlenut-based system in greater detail.

Kukui (candlenut) in ancient Hawaii

While the cultivation of taro as the core staple of the Hawaiian people (Winter et al., 2018; Kagawa-Viviani et al., 2018b) was the focus of a shifting cultivation system within candlenut forests, it is important to note that candlenut, known locally as *kukui*, was an important tree in the Hawaiian agricultural economy (Abbott, 1992; Lincoln, 2009).

Candlenut was first domesticated in island Southeast Asia, where remains of harvested candlenuts recovered from archeological sites in Timor and Morotai have been dated to about 13,000 years ago (Blench, 2004). The trees were widely introduced into the Pacific islands by Austronesian voyagers, to become naturalized on high volcanic islands like Hawaii (Kirch, 1989; Larrue et al., 2010; Weisler et al., 2015). As Figure 21-4 shows, the tree is now distributed throughout the New- and Old-World tropics.

Although *kukui* was not a significant food source in Hawaii, a relish known as *'inamona* frequently enlivened the Hawaiian diet. Candlenut kernels were roasted and salted to yield a condiment that tasted like toasted macadamia nuts (Abbott, 1992). The nuts, however, were a source of feed for both wild and domesticated pigs in ancient Hawaii, and extensive groves of *kukui* are associated with the feeding of *Kamapua'a*, the mischievous hog god (Handy et al., 1972).

The principal use of candlenut was as a fuel source. The extremely oily nuts were either burnt directly or used to extract a liquid oil. So universal was the application that the words for lamp/light (*lama*) and candlenut (*kukui*) are frequently used interchangeably. Traditional lighting included several types of lamps and torches. The simplest, *kālī kukui*, consisted of candlenut kernels skewered on dry coconut leaflet midribs. Lit at one end, the kernels burned one by one every 15 minutes or so, leading to their use as a measure of time. Another version was to string the kernels on longer reeds and bind several strings together before wrapping them in *ti* (*Cordyline fruticosa*) leaves to prevent the kernels from burning too rapidly (Fornander, 1920). Another disposable form was the *lama 'ohe*, in which hollow lengths of bamboo were filled with *kukui* kernels and burned, producing stronger light for a longer time, but also a large amount of smoke and soot. More sophisticated were *pōho kukui* – hollowed stones that held a reservoir of pure *kukui* oil with a wick made from cloth

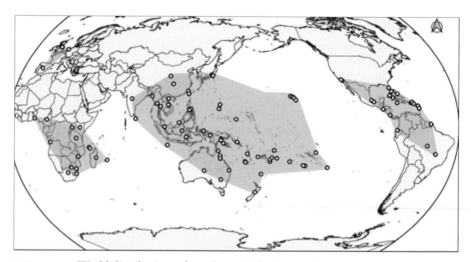

FIGURE 21-4: World distribution of candlenut (*Aleurites moluccanus*).

Source: After *Discover Life* (2019) Global Mapper.

Aleurites moluccanus (L.) Willd. [Euphorbiaceae]

So widespread and so ancient is the human use of candlenut that its precise origins are difficult to establish. The early Hawaiians made it the basis of a unique shifting cultivation system. They also used the oil-rich nuts as burning torches, made *leis* (garlands) from the silver-green leaves, used the oil as a preservative, made tattoo ink from charred nuts and a red-brown dye from the inner bark.

or cord to create a bright burning candle. Similarly, cloth was tightly bound to poles before soaking it in pure oil and lighting it to produce torches. Following the arrival of Europeans, *kukui* nut oil was shipped abroad in the early 1800s, with as much as 10,000 gallons (about 37,850 litres) exported per year (Wilcox and Thompson, 1913).

The oil was commonly applied to the skin in medicinal ointments, or in its pure form as a moisturizer and protection against the sun. The oil was also commonly applied to many material goods, to aid in waterproofing, preservation and aesthetics. *Kapa*, the common bark cloth of old, was often oiled with *kukui* oil to increase its longevity and to make the fabric softer and more comfortable against the skin. The oil reacts well with wood and provides a beautiful finish and sheen. For instance, the wood of *'ahakea* (*Bobea* spp.) is a dull orange–brown that becomes a dark gold when rubbed with *kukui* oil (Abbott, 1992). Traditionally, *kukui* oil was the final finish on all parts of a canoe, with frequent applications 'after each use of the canoe' to maintain the waterproofing (Abbott, 1992, p.81).

Another common use of *kukui* was to make dyes. The candlenut tree produces a range of dye colours but, more importantly, it is rich in tannins that serve as a mordant to help fix colours in materials. The Polynesian standard of brown is not common in Hawaii, but when applied is made from the bark of the *kukui* tree, while beige comes from the immature fruits. The inner bark of the roots provides a red colour that was often used to dye fishnets to make them less visible to fish and to increase their longevity. However, the most common application of *kukui* was for black ink. Canoes, for instance, use a black paint known as *pa'ele* for decoration and preservation. The recipe often called for a charcoal base mixed with juice from the inner bark of *kukui* roots. With a high concentration of *kukui* juice, the paint was resistant to abrasion and water (Abbott, 1992). Similarly, highly permanent stamping of traditional cloth with a thick ink made from the ash of *kukui*-nut shells mixed with *kukui* oil was a common practice. A similar application of this ink, sometimes made from charred *kukui* shells and sugar cane juice, was for tattooing (Lincoln, 2020).

Kukui was also a vital part of the Hawaiian medicinal pharmacopoeia. The oil and sap of *kukui* can be a strong laxative, and in higher quantities serves as a total purge that was standard treatment for many ailments. The sap is applied to heal mosquito bites, canker sores and other small topical ailments as both a treatment and a sealant. For larger sores and ulcers, a salve is used that has mashed roasted *kukui* nuts as a base. The leaves may be used to wrap areas of swelling, bruising and minor sprains. For this treatment, bruised leaves can be placed against the skin and these are wrapped in fresh leaves to seal the area. The oil is used to prevent scarring and stretch marks. Other applications include the use of charcoal from the shells for sore throats (Abbott, 1992), internal applications of extracts to treat ear, nose, and throat disorders (Young et al., 2005), and more complicated remedies (Chun, 1994).

Kukui also provides construction materials. Its lightweight wood is useful for canoe floats, house construction, and other applications (Malo, 1903). The hard nuts were commonly used to make children's toys, such as whistles and spinning tops. The leaves, the delicate white flowers and the polished nut shells were all used in *leis*, or garlands for adornment.

Some unique applications of *kukui* relate to fishing. Fishermen chew the slightly dried nuts and spit them into the water to break the surface tension and remove reflections, giving them greater visibility of shallow reefs. Mashed candlenut and coconut may also be used to create a chum; dipping fishhooks and spearheads into the mixture helps to attract fish. A large amount of the nuts and coconut, when fermented with seafood waste such as shrimp peels, creates an odorous mixture which, when added to a vegetarian chum, serves to attract fish better.

Kukui as a mulch

Kukui is the most commonly referenced mulch material in Hawaiian agriculture, followed by *hau* (*Hibiscus tiliaceus*), *'ama'u* fern (*Sadleria* spp.) and sugar cane. *Kukui* mulch was stamped into the mud of flooded terraces and used to fertilize rainfed gardens. Taro grown with *kukui* mulch is consistently described as producing the largest and most delicious corms. In a recent experiment we grew taro in pure mulches of candlenut, sugar cane, and *hau* (*Hibiscus tiliaceus*), and the growth in candlenut mulch was by far the largest (by ~150%), despite it having the lowest nitrogen content of the three treatments. Although *kukui* was clearly the most effective mulch, there were no clear explanations as to why. In a separate trial we found that although *kukui* had a moderately high carbon-to-nitrogen ratio (~28:1) and low leaf-nitrogen concentrations (~1.15%), it decomposed much more quickly than the other two mulches. In a wet environment (~3500 mm/yr), 95% of *kukui* leaf mass decomposed in only nine months, compared to only ~65% of sugar cane (Lincoln, 2017). During decomposition, *kukui* exhibited a moderate level of nitrogen fixation (~0.9gN per kg of dry mulch) (Lincoln, 2017).

Colocasia esculenta (L.) Schott [Araceae]

Taro – another 'canoe plant' introduced by migrants – is the core staple of the Hawaiian people. Shifting cultivation plots fertilized by litter from candlenut forests were once the basis for prodigious production of taro. In Hawaiian culture, the word for family, *'ohana*, comes from *'ohā*, the shoot that grows from a taro corm, suggesting that people similarly grow from their families.

FIGURE 21-5: A planting pit filled with organic composting material in a lava field.

Photo: Noa Kekuewa Lincoln.

Candlenut has been used for soil development in many areas across Hawaii at different scales. Small-scale application was common on very young lava flows where minimal soil development had occurred, and patches of candlenut trees served to supply a handful of plantings (Handy, 1940). In these situations, litterfall was gathered into relatively impermeable pits in the lava and composted in order to create a growing medium. Local organic waste and small amounts of soil that could be excavated nearby was added to these enclosures, or *pa*, to aid in the rotting of composts (Kalokuokamaile, 1922, translated by M. K. Pukui) (Figure 21-5). The beds would be prepared several months before the rainy season so that when the rains came the new soil would be at an optimal level of development. In such enclosures sweet potatoes, taro, and sugar cane grew luxuriantly. On the young lava flows of Hawaii Island additional soil was occasionally brought in from other areas to allow for cultivation (Handy et al., 1972). While a handful of these cultivation pits would commonly be located near households to compost local waste, within novel or native forests 'many times 40' of these pits would comprise an agricultural field. When *kukui* was the principal component, these *pa* (enclosures) were appropriately referred to as *pakukui*.

The *Pakukui* of Hamakua

While the small-scale use of *kukui* litter for mulch was common across the islands, this chapter focuses on extensive candlenut forests that were used to practice shifting cultivation. These forests of candlenut were established with the intent of felling the trees and using the mulch to grow crops, primarily taro. This differed from the use of *kukui* as mulch or for built soils, as described above, because of its extent and management. The most famous of these candlenut forests was in the Hamakua region (see Figure 21-3). The area has moderately young soils and rainfall in excess of 2000 mm/yr (Giambellucca et al., 2013). The high rainfall is enough to strip the soils of available nutrients and therefore these soils are suboptimal for intensive cropping of annuals (Vitousek et al., 2014). However, the soils are not so denuded as to make cultivation impossible (Vitousek and Chadwick, 2013). This moderate soil fertility appears to have supported extensive application of the *pakukui*:

> The broad slopes of the wet windward coast (Hamakua) of Hawaii between the abrupt eroded shore and the high forest…were, before the era of early sugar plantations, completely covered by *kukui* forest. […] This is the area which, in Hawaiian lore, was known as the domain of *Kamapua'a*, in contradistinction to the drier and lava-strewn areas…which was *Pele's* [volcano goddess] country. When the rain-drenched forests covered these slopes of Hamakua, undoubtedly great numbers of wild hogs roamed these forests, feeding on the *kukui* nuts. In a *kukui* forest there were open glades called *pakukui* in which taro was grown (Handy et al., 1972, p.231).

As an alternative definition, *pakukui* means 'shiftless, vagabond; to wander' (Pukui and Elbert, 1986). This definition conjures the image of 'wandering' taro patches through a *kukui* forest in a practice of shifting cultivation. The perspective is that the candlenut forests provided both *kukui* nuts and the nutrient accumulation necessary for taro cultivation. The application of the system is described in several historical sources:

> *Kihapai*, or small gardens, in *pakukui* were prepared by felling a number of *kukui* trees in a concentrated area, and allowing for the trees to decompose, creating rich humus for planting. Large holes, up to 10 feet (3 m) in circumference and 3 feet (0.9 m) in depth, were then dug and filled with *kukui* leaves, which, once decomposed, were turned and planted. (Handy et al., 1972, p.110)

> These were big holes, nine feet or less in circumference, according to the wish of the worker. The depth was from the fingertips to the armpit or collarbone. The hole was filled with *kukui* leaves, then covered over with soil. Four *kukui* branches were also buried in the holes with the leaves down and the stems up. These branches were used so that one knew whether the leaves had rotted away in the hole, by pulling up one of the branches to see. If, upon pulling up

a branch, the leaves had not rotted, then it was left alone until another time. If they had rotted, the taro stalks were planted. [...] There were no other leaves used except the *kukui*. (Kalokuokamaile, 1922, translated by M. K. Pukui).

In the beginning, the taro grew out of their holes like young banana plants. When they were grown, they were not like other dry garden or wet garden taro, for these surpassed them in growth. When the taro (corm) had developed both the parent stalk and the offshoots, the offshoots were pulled away from the parent and the space between stepped on with the feet to separate them from the parent. These would not be the only offshoots, for the parent would send out more, each with fully developed corms. (Kalokuokamaile, 1922, translated by M. K. Pukui).

It is an amazing thing, the growth of the *kalo* (taro) in this way; it grows seven feet high and more. And its *kalo* weighing 20 pounds or more, according to the size of the *pakukui* pit, so is the growth of the plant and corm (Kalokuokamaile, 1932, translated by N. K. Lincoln).

Estimating the *Pakukui*: An interdisciplinary venture

Collectively, the qualitative descriptions above provide a good overview of the location, method and outcome of the planting style, but leave more details untold. The descriptions of taro produced are impressive, saying they surpassed all other forms of taro cultivation in Hawaii, in terms of size and quality. Yet, we do not know how productive this method may have been at landscape scale, nor how extensive it was on the landscape. Starting in 2015, research colleagues and I used multiple methods to examine these questions.

Extent of the Pakukui

We used botanical surveys to examine the potential extent of the *pakukui* system in the Hamakua and Hilo regions of Hawaii Island (see Figure 21-3). *Kukui* has a very distinct, silvery appearance on the landscape due to short downy hairs on the leaves (Figure 21-6). This

FIGURE 21-6: Prior to conversion to sugar plantations, vast forests of *kukui* were grown for harvest and nutrient accumulation, the remnants of which can still be found withstanding the pressure of abandonment and invasive species, such as this patch in east Maui.

Photo: Forest and Kim Starr.

appearance also corresponds to a unique spectral signature that can be used to recognize *kukui* from remote images (Figure 21-7). Using remote sensing we identified more than 25,000 *kukui* trees in southern Hamakua and northern Hilo districts. We verified 500 trees on the ground and found no falsely identified trees, but we systematically underestimated the number of trees when they were growing close together. Although the landscape has been highly altered, converted first to sugar cane in the mid-1800s and subsequently to forestry and diversified agriculture starting in the mid-1900s, many traditional crops persisted in the rivers and rivulets where heavy machinery was not used. Because candlenut is not grown contemporarily for any agricultural practices, we assumed that the trees occurring in valleys were primarily descendant from remnant plantings prior to European contact. Furthermore, *kukui* has relatively large, round, and heavy fruit that do not migrate uphill well; we assume, therefore, that the modern extent of *kukui* represents, with some drift, the ancient extent of agroforestry cultivation in the region.

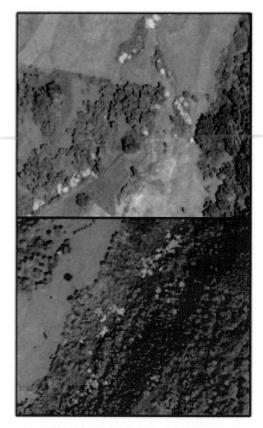

FIGURE 21-7: Examples of candlenut visibility on the landscape using remote imagery, in this case WorldView-3 data. The bright silvery leaves make for easy and distinct identification, even within dense forest.

Concurrently, we surveyed 200 m of every river and rivulet at ~50m elevation for breadfruit (*Artocarpus altilis*). Breadfruit was the dominant tree associated with arboriculture in Hawaii and is relatively easy to identify because of its large size and large, distinct leaves. The occurance of breadfruit in this environment indicated the likely presence of permanent, diversified arboricultural developments as opposed to the shifting cultivation of the *pakukui*. Again, trees occurring in overgrown or uncultivated areas were assumed to be remnant trees from traditional agroforestry plantings. The results from these two botanical surveys, collectively, tell a story of the extent and form of ancient arboriculture and the co-evolution of practices and politics in the context of the landscape properties.

Artocarpus altilis (Parkinson ex F.A.Zorn) Fosberg [Moraceae]

Like other plants in this chapter, breadfruit was a 'canoe plant', originating in New Guinea, the Maluku islands and the Philippines, and carried eastwards across the Pacific by the Austronesian migration, to become one of Hawaii's most important food sources. It has become a staple food in many of the world's tropical regions. Its name comes from the texture of the fruit when cooked.

In Figure 21-8, it can be seen that moving from south to north, breadfruit is prevalent in the gullies until about halfway up the coast, when it drops off the landscape completely. The point at which remnant breadfruit stops appearing on the landscape coincides with a dramatic shift in rainfall and soil type. Collectively, these shifts in landscape properties manifest in suitability for candlenut cultivation, which we modelled using established methodologies (Gross, 2014).

By plotting the *kukui* trees graphically (Figure 21-9), we observe more subtle changes that are not readily apparent on the maps. We again display the spatial distribution of the mapped trees, but also depict the *kukui* trees as small black dots against the lighter grey of breadfruit to distinguish those that occur in the southern political units, which contain breadfruit, and those that occur in the northern political units that do not. Two cropping patterns are evident: in the south a mixture of

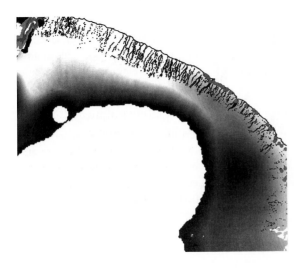

FIGURE 21-8: Map of the study area in the Hamakua and northern Hamakua regions (see box on Figure 21-3) depicting surveyed candlenut (small black dots) and breadfruit trees (larger grey dots). The mapped trees are overlayed on a model of candlenut habitat suitability ranging from 0 (black) to 100 (white), as modeled using methods in Gross (2014).

breadfruit and *kukui*, and in the north exclusive *kukui* groves that occur more densely. At the point of transition between the two systems, there is a significant increase in the upper elevation that remnant *kukui* trees are found, from ~375 m in the south to ~650 m in the north (Figure 21-9 - upper). It is also apparent that the density of *kukui* trees is much higher in the north than in the south. The underlying fertility of the soils in the north and south also differ dramatically (Figure 21-9 - lower). Using a soil-fertility index developed to map the distribution of intensive Hawaiian cultivation (Ladefoged et al., 2009; Kurashima et al., 2019), we see that a strong shift in soil fertility begins where breadfruit cultivation ceases.

A final piece of the story considers the socio-political units called *ahupua'a* – the local territories that composed the larger regions that were mainly controlled by chiefs (Gonschor and Beamer, 2014). The point where landscape properties shift and breadfruit cultivation ceases coicides with a significant socio-political point on the landscape (Figure 21-10). At this point, an *ahupua'a* occurs that extends far into the uplands. Notice that the units to the south truncate into the highlighted unit, which effectively 'caps' the southern units in the uplands.

The 'overarching' nature of this territory and its exclusive access to the mountain summit clearly indicate its political importance. Correspondingly, this territorial unit demarcates the larger regional boundary between the Hilo and Hamakua districts (Figure 21-3). The regional demarcation indicates that different family lineages and chiefly lines dominated, particularly in the deep history of these regions.

Botanical surveys may be unreliable indicators of traditional cropping systems, and there may be historical and other explanations for the distribution of candlenut and breadfruit trees to the ones we have observed. However, given the multiple alignments of two remnant tree species, the landscape-level shift in climate and soils,

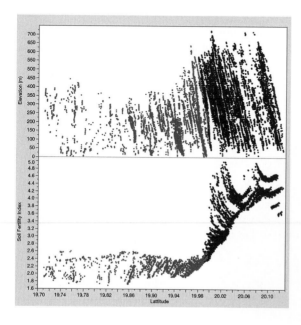

FIGURE 21-9: Graphical representation of candlenut trees, illustrating (upper) the elevation of each tree as one moves northward through the study region and (lower) underlying soil fertility as represented by a fertility index. Points representing candlenut trees are darker and breadfruit trees lighter, for visual effect to emphasize the differences between the southern area that contains breadfruit trees and the northern area that does not.

and the socio-political divisions, our interpretation is that the patterns observed accurately represent traditional arboriculture.

We believe that the northern, more fertile soils of Hamakua were home to a massive candlenut forest that supported the *pakukui* cultivation system. Here, candlenut was cultivated to an elevation of ~650 m - approaching the temperature limitation for optimal *kukui* growth (Figure 21-8). In the more infertile south, we suggest that a permanent system of arboriculture was established that used a range of perennial crops such as breadfruit, bananas, candlenuts and coconuts – all of which are evident on

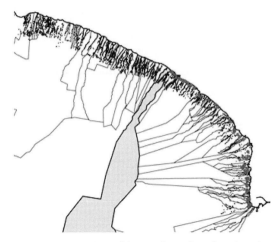

FIGURE 21-10: Map of the study region showing the candlenut (small black dots) and breadfruit trees (large grey dots) over the traditional socio-political units known as ahupua'a. The point at which the landscape properties and farming methods shift coincides with a major socio-political boundary.

the landscape today. In the southern region, soil fertility, rather than temperature, constrained the extent of agriculture and an upland limit to the expansion of arboriculture can be seen on the basis of soil fertility (Figure 21-9).

Kukui *growth and nutrient accumulation*

After better defining the extent of the *pakukui*, our research group further explored the potential productivity and fallow period of the system by looking at growth rates and nutrient accumulation in the trees. We sampled 243 trees of known ages in the region, ranging from 3 to 25 years old. Although they showed highly variable growth in the first 3 to 7 years, overall growth fitted very well to regression equations for both diameter at breast height (=-9.021135 + 16.757157 * log(Age); r^2 0.95, p<0.0001) and height (=-3.635323 + 8.5568025 * log(Age); r^2 0.90, p<0.0001).

Using a series of reported and developed figures, we generated some initial estimates of the productivity of the *pakukui* system based on nitrogen dynamics (Table 21-2). We applied a general pan-tropical allometric equation (Chave et al., 2014) and candlenut wood densities (Krisnawati et al., 2011) to estimate aboveground biomass based on the growth curves generated above, then segregated that biomass into trunks, branches and leaves based on the characteristics of broadleaf tropical trees. We used our own analyses of nutrient content, decomposition rates, and nitrogen fixation during decomposition (Lincoln, 2017) to generate a relationship between tree age and total nitrogen accumulation, using Hawaii-specific landscape-level inputs such as nitrogen deposition and mineralization. We then looked at nutrient accumulation and availability using only the leaves and stems of the trees (assuming the large branches and trunks were not composted) and conservatively assumed that 20% of the available nutrients were utilized to satisfy the demands of taro growth. Conversely, we used our own measures of taro anatomy and nutrient concentrations along with reported literature (Goenaga and Chardon, 1995) to estimate nitrogen demand in taro to be ~205 gN per 100 wet kg of taro.

Although the calculations made a range of assumptions and simplifications, this first-order estimation at least provided a picture of the patterns and potential productivity of the region. We made a best-guess for the growth, density, and nutrient accumulation, based on literature review and our own experiments, and a model total harvest per hectare based on different fallow lengths by relating the annual area of cultivation to the length of fallow (Table 21-2). While we do not suggest that the total yields are accurate, we do suggest that the general pattern of productivity is reasonable.

According to our model, the system had the potential to generate in the order of 20,000 kg of taro per hectare per year. There is a point at which the trees reach a maximum annual yield, due to their overall declining nutrient accumulation as they age, with younger trees gaining biomass

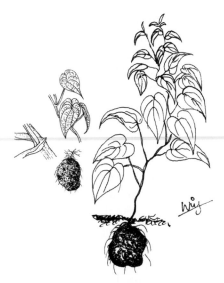

Dioscorea alata L. [Dioscoreaceae]

Yams are another important part of the Hawaiian diet. Archeological evidence suggests that yams were exploited in island Southeast Asia and Papua New Guinea long before they were carried across the Pacific in canoes. The plant, which is sterile, is known only in its cultivated form. It cannot cross bodies of water, so it has become an important food crop throughout the tropical world purely by human agency.

and nutrients much more rapidly than older trees. In our model this inflection point occurs at year 12. However, the system is within 95% of maximum productivity from years 7 to 26. It is likely than any form of fallow rotation that falls within this window would be virtually indistinguishable from the farmer's perspective. This is consistent with other traditional descriptions of shifting taro production in the Pacific:

> With dryland taro, the practice of shifting cultivation is still being followed. Normally, virgin forest would be cleared and planting holes made in the ground using a selected stick about 1.5 m long. In the first year, taro would be the main crop, but by the second and third years an intercropping pattern would have developed. The land would be fallowed after two to three years of cropping.

In some areas if the climate and vegetation allow, regeneration of forest species could proceed at a rapid rate. Soil nutrient replenishment by natural means could be achieved in a few years, and the same piece of land could then be replanted. (Vilsoni, 1993, p.86).

TABLE 21-2: Some key parameters of a production model developed for the *pakukui* system of Hamakua.

Tree age (yr)	DBH (m)	Height (m)	Crown area (sq. m)	Total biomass (kg/ha)	20% of N accumulation (kg/ha)	Potential taro yield (wet) (kg/ha/yr)	% of max yield (%)
2	0.03	2.30	2.36	0.3	2.26	11,298	56.3
3	0.09	5.77	7.50	10.0	4.24	14,135	70.5
4	0.14	8.23	10.90	31.7	6.52	16,295	81.2
5	0.18	10.14	13.45	61.3	8.85	17,699	88.2
6	0.21	11.70	15.49	95.9	11.16	18,604	92.7
7	0.24	13.02	17.19	133.4	13.43	19,191	95.7
8	0.26	14.16	18.66	172.9	15.66	19,569	97.6
9	0.28	15.17	19.93	213.5	17.83	19,808	98.7
10	0.30	16.07	21.07	254.7	19.95	19,952	99.5
11	0.31	16.88	22.09	296.2	22.03	20,029	99.9
12	0.33	17.63	23.02	337.8	24.07	20,059	100.0
13	0.34	18.31	23.87	379.3	26.07	20,055	100.0
14	0.35	18.95	24.66	420.6	28.04	20,026	99.8
15	0.36	19.54	25.38	461.6	29.97	19,980	99.6
16	0.37	20.09	26.06	502.2	31.87	19,920	99.3
17	0.38	20.61	26.70	542.5	33.75	19,851	99.0
18	0.39	21.10	27.29	582.4	35.60	19,775	98.6
19	0.40	21.56	27.86	621.8	37.42	19,694	98.2
20	0.41	22.00	28.39	660.9	39.22	19,611	97.8
21	0.42	22.42	28.90	699.4	41.00	19,525	97.3
22	0.43	22.81	29.38	737.6	42.76	19,438	96.9
23	0.44	23.19	29.84	775.2	44.51	19,350	96.5
24	0.44	23.56	30.28	812.5	46.23	19,263	96.0
25	0.45	23.91	30.70	849.3	47.94	19,175	95.6
26	0.46	24.24	31.11	885.7	49.63	19,089	95.2
27	0.46	24.57	31.50	921.7	51.31	19,003	94.7
28	0.47	24.88	31.87	957.3	52.97	18,919	94.3
29	0.47	25.18	32.23	992.4	54.62	18,836	93.9
30	0.48	25.47	32.58	1027.2	56.26	18,754	93.5

Conclusion: Reviving the *pakukui*

Following European contact in Hawaii, several forms of traditional agriculture rapidly declined, primarily due to the population crash that accompanied the introduction of foreign diseases. Among the practices that declined rapidly was the *pakukui*. Sugar plantations were established in the region starting in the mid–1800s, which were consolidated into Hawaii's second-largest sugar cane plantation – the Hamakua Sugar Company – in 1883. Since statehood in 1959, these plantations have all failed, and in their wake there has arisen a revival of Hawaiian culture, foods, and agriculture (Kagawa-Viviani et al., 2018b; Lincoln et al., 2018).

Building upon this research, colleagues and I have partnered with two non-profit organizations to begin restoring the *pakukui* system in northern Hamakua in order to relearn and experience, at first hand, this lost system. Through these efforts we aim, as we have done in other systems, to explore aspects of the cropping system through experience and experimentation (e.g. Kagawa and Vitousek, 2012; Lincoln et al., 2017, 2018; Marshall et al., 2017). These efforts are still in the early establishment stage, altering what was a long-established pasture back into a candlenut forest to reinitiate the practice of nutrient accumulation and natural fertilization to realize significant taro productivity. Through these efforts we seek to contribute to knowledge revival and the biocultural restoration of Hawaii to further inform appropriate and sustainable resource mangement (Chang et al., 2019).

References

Abbott, I. A. (1992) *Lāʻau Hawaii: Traditional Hawaiian Uses of Plants*, Bishop Museum Press, Honolulu, HI

Allen, M. S. (2004) 'Bet-hedging strategies, agricultural change, and unpredictable environments: historical development of dryland agriculture in Kona, Hawaii', *Journal of Anthropological Archeology* 23, pp.196-224

Asner, G. P., Elmore, A. J., Hughes, R. F., Warner, A. S., and Vitousek, P. M. (2005) 'Ecosystem structure along bioclimatic gradients in Hawaii from imaging spectroscopy', *Remote Sensing of Environment* 96(3-4), pp.497-508

Athens, J. S., Rieth, T. M., and Dye, T. S. (2014) 'A paleo-environmental and archaeological model-based age estimate for the colonization of Hawaii', *American Antiquity* 79(1), pp.144-155

Blench, R. (2004) 'Fruits and arboriculture in the Indo-Pacific region', *Bulletin of the Indo-Pacific Prehistory Association* 24 (The Taipei Papers (Volume 2)), pp.31-50

Chang, K., Winter, K. B. and Lincoln, N. K. (2019) 'Hawaii in Focus: Navigating Pathways in Global Biocultural Leadership', *Sustainability* 11(1), p.283

Chave, J., Réjou Méchain, M., Búrquez, A., Chidumayo, E., Colgan, M. S., Delitti, W. B., Duque, A., Eid, T., Fearnside, P. M., Goodman, R. C., Henry, M., Martínez-Yrízar, A., Mugasha, W. A., Muller-Landau, H. C., Mencuccini, M., Nelson, B. W., Ngomanda, A., Nogueira, E. M., Ortiz-Malavassi, E., Pélissier, R., Ploton, P., Ryan, C. M., Saldarriaga, J. G. and Vieilledent. G. (2014) 'Improved allometric models to estimate the aboveground biomass of tropical trees', *Global Change Biology* 20(10), pp.3177-3190

Chun, M. N. (ed.) (1994) *Native Hawaiian Medicines*, First People's Production, Honolulu, HI

Discover Life (2019) Global Mapper, *Aleurites moluccanus*, available at https://www.discoverlife.org/mp/20m?kind=Aleurites+moluccana, accessed 12 April 2019

Dotte-Sarout, E., and Kahn, J. G. (2017) 'Ancient woodlands of Polynesia: A pilot anthracological study on Maupiti Island, French Polynesia', *Quaternary International* 457, pp.6-28

Ellis, W. (1917) *A Narrative of a Tour Through Hawaii, or Owhyhee: With Remarks on the History, Traditions, Manners, Customs, and Language of the Inhabitants of the Sandwich Islands* (No. 2), Hawaiian Gazette Company, Honolulu

Fornander, A. (1920) *Fornander Collection Of Hawaiian Antiquities And Folklore*, vol. 6, Bishop Museum, Honolulu, HI

Giambelluca, T. W., Chen, Q., Frazier, A. G., Price, J. P., Chen, Y-L., Chu, P-S., Eischeid, J. K. and Delparte, D. M. (2013) 'Online rainfall atlas of Hawaii', *Bulletin of the American. Meteorological Society* 94, pp.313-316, doi: 10.1175/BAMS-D-11-00228.1.

Goenaga, R., and Chardon, U. (1995) 'Growth, yield and nutrient uptake of taro grown under upland conditions', *Journal of Plant Nutrition* 18(5), pp.1037-1048

Gonschor, L., and Beamer, K. (2014) 'Toward an inventory of Ahupua'a in the Hawaiian kingdom: A survey of 19th- and early-20th-century cartographic and archival records of the island of Hawaii, *The Hawaiian Journal of History* 48, pp.53-87

Gross, J. J. (2014) 'Assessment of future agricultural land potential using GIS and regional climate projections for Hawaii island: An application to macadamia nut and coffee', PhD dissertation to the University of Hawaii at Manoa, Honolulu

Handy, E. S. C. (1940) *Hawaiian Planter-Volume I, His Plants, Methods and Areas of Cultivation*, Bulletin 161, Bishop Museum Press, Honolulu, HI

Handy, E. S. C., Handy, E. G. and Pukui, M. K. (1972) *Native Planters in Old Hawaii: Their Life, Lore, and Environment*, Bishop Museum Press, Honolulu, HI

Huebert, J. M. (2014) 'The role of arboriculture in landscape domestication and agronomic development: A case study from the Marquesas Islands, East Polynesia', PhD dissertation to the University of Auckland, New Zealand

Juvik, S. P. and Juvik, J. O. (1998) *Atlas of Hawaii*, University of Hawaii Press, Honolulu, HI

Kagawa, A. K. and Vitousek, P. M. (2012) 'The Ahupua'a of Puanui: A resource for understanding Hawaiian rainfed agriculture', *Pacific Science* 66(2), pp.161-173

Kagawa-Viviani, A., Lincoln, N., Quintus, S., Lucas, M. and Giambelluca, T. (2018a) 'Spatial patterns of seasonal crop production suggest coordination within and across dryland agricultural systems of Hawaii Island', *Ecology and Society* 23(3)

Kagawa-Viviani, A., Levin, P., Johnston, E., Ooka, J., Baker, J., Kantar, M. and Lincoln, N. K. (2018b) 'I Ke Ewe Aino o Ke Kapuna: Hawaiian ancestral crops in perspective', *Sustainability* 10(12), p.4607

Kalokuokamaile, Z. P. K. (1922) 'Ke ano o ke Kalaiaina', in *Ka Nupepa Kuokoa*, 22 June 1922, p.4

Kalokuokamaile, Z. P. K. (1932) *Kepelino's Traditions of Hawaii*, Bulletin 95, Martha Beckwith (ed.), Bernice P. Bishop Museum, Honolulu, HI

Kirch, P. V. (1977) 'Valley agricultural systems in prehistoric Hawaii: An archaeological consideration', *Asian Perspectives* 20(2), pp.246-280

Kirch, P. V. (1989) 'Second millennium B. C. arboriculture in Melanesia: Archaeological evidence from the Mussau Islands', *Economic Botany* 43(2), pp.225-240

Kirch, P. V. (1994) *The Wet and The Dry: Irrigation and Agricultural Intensification in Polynesia*, University of Chicago Press, Chicago, IL

Kirch, P. V. (2005) *From Chiefdom to Archaic State: Social Evolution in Hawaii*, Brigham Young University, Provo, UT

Kirch, P. V. (2010) 'Peopling of the Pacific: A holistic anthropological perspective', *Annual Review of Anthropology* 39, pp.131-148

Kirch, P. V. and Zimmerer, K. S. (2011) *Roots of Conflict*, School for Advanced Research Press, Santa Fe, NM

Körner, C. (2007) 'The use of 'altitude' in ecological research', *Trends in Ecology and Evolution* 22(11), pp.569-574

Krisnawati, H., Kallio, M. and Kanninen, M. (2011) *Aleurites moluccana (L.) Willd.: Ecology, silviculture and productivity*, Center for International Forestry Research (CIFOR), Bogor, Indonesia.

Kurashima, N. and Kirch, P. V. (2011) 'Geospatial modeling of pre-contact Hawaiian production systems on Moloka'i Island, Hawaiian Islands', *Journal of Archaeological Science* 38(12), pp.3662-3674, doi:10.1016/j.jas.2011.08.037

Kurashima, N., Fortini, L. and Ticktin, T. (2019) 'The potential of indigenous agricultural food production under climate change in Hawaii,' *Nature Sustainability* 1

Ladefoged, T. N., Kirch, P. V., Gon, S. M., Chadwick, O. A., Hartshorn, A. S. and Vitousek, P. M. (2009) 'Opportunities and constraints for intensive agriculture in the Hawaiian archipelago prior to European contact', *Journal of Archeological Science* 36, pp.2374-2383, doi:10.1016/j.jas.2009.06.030

Larrue, S., Meyer, J-Y. and Chiron, T (2010) 'Anthropogenic vegetation contributions to Polynesia's social heritage: The legacy of candlenut tree (*Aleurites moluccana*) forests and bamboo (*Schizostachyum glaucifolium*) groves on the Island of Tahiti', *Economic Botany* 64(4), pp.329-339. doi:10.1007/s12231-010-9130-3.

Lincoln, N. (2009) *Ethnobotanical Guide to Native Hawaiian Plants*, Bishop Museum Press, Honolulu, HI

Lincoln, N. K. (2017) 'Nitrogen fixation in Hawaiian agroecological systems, paper presented at the annual meeting of the Ecological Society of America, Portland, Oregon

Lincoln, N. K. (2020) *Ko: An Ethnobotanical Guide to Native Hawaiian Sugarcane Varieties*, University of Hawaii Press; Honolulu, HI

Lincoln, N. and Ladefoged, T. (2014) 'Agroecology of pre-contact Hawaiian dryland farming: The spatial extent, yield and social impact of Hawaiian breadfruit groves in Kona, Hawaii', *Journal of Archaeological Science* 49, pp.192-202, doi:10.1016/j.jas.2014.05.008

Lincoln, N. K. and Vitousek, P. (2015) 'Nitrogen fixation during decomposition of sugarcane (*Saccharum officinarum*) is an important contribution to nutrient supply in traditional dryland agricultural systems of Hawaii', *International Journal of Agricultural Sustainability* 5903(September), pp.1-17, doi:10.1080/14735903.2015.1071547

Lincoln, N. K. and Vitousek, P. M. (2017) 'Indigenous Polynesian agriculture in Hawaii', *Oxford Research Encyclopedia of Environmental Science* March (2017), Oxford University Press, Oxford, UK, doi: 10.1093/acrefore/9780199389414.013.376

Lincoln, N., Chadwick, O. and Vitousek, P. (2014) 'Indicators of soil fertility and opportunities for pre-contact agriculture in Kona, Hawaii', *Ecosphere* 5(April), pp1-20, doi:10.1890/ES13-00328.1

Lincoln, N. K., Kagawa-Viviani, A., Marshall, K. and Vitousek, P. (2017) 'Observations of sugarcane in traditional Hawaiian cropping systems', in R. Murphy (ed.) *Sugarcane: Production Systems, Uses, and Economic Impact*, Nova Science Publishers, Hauppauge, NY

Lincoln, N., Rossen, J., Vitousek, P., Kahoonei, J., Shapiro, D., Kalawe, K., Pai, M., Marshall, K. and Meheula, K. (2018) 'Restoration of 'Aina Malo'o on Hawaii Island: Expanding Biocultural Relationships', *Sustainability* 10(11), p.3985

Malo, D. (1903) *Hawaiian Antiquities*, translated by N. B. Emerson, Special Publication 2, Bernice P. Bishop Museum, Honolulu, HI

Marshall, K., Koseff, C., Roberts, A., Lindsey, A., Kagawa-Viviani, A., Lincoln, N. and Vitousek, P. (2017) 'Restoring people and productivity to Puanui: Challenges and opportunities in the restoration of an intensive rain-fed Hawaiian field system', *Ecology and Society* 22(2)

Maxwell, J. J., Howarth, J. D., Vandergoes, M. J., Jacobsen, G. E. and Barber, I. G. (2016) 'The timing and importance of arboriculture and agroforestry in a temperate East Polynesia Society: the Moriori, Rekohu (Chatham Island)', *Quaternary Science Reviews* 149, pp.306-325

McCoy, M. D. and Graves, M. W. (2010) 'The role of agricultural innovation on Pacific Islands: A case study from Hawaii Island', *World Archaeology* 42(1), pp.97-107, doi:10.1080/00438240903430340

Meilleur, B. A., Jones, R. R., Tichenal, C. A. and Huang, A. S. (2004) *Hawaiian Breadfruit: Ethnobotany, Nutrition, and Human Ecology*, College of Tropical Agriculture and Human Resources, University of Hawaii at Manoa, Honolulu, HI

Menzies, A. (1920) *Hawaii Nei 128* years ago (publisher unidentified)

Newman, T. S. (1974) *Kona Field System*, Hawaii Register of Historic Places Nomination Form, Site 10-37-6601, State Historic Preservation Division, Kapolei, Honolulu, HI

Pukui, M. K. (1983) *Ōlelo No'eau: Hawaiian Proverbs & Poetical Sayings*, Bishop Museum Press, Honolulu, HI

Pukui, M. K., and Elbert, S. H. (1986) *Hawaiian Dictionary: Hawaiian-English, English-Hawaiian*, University of Hawaii Press, Honolulu, HI

Quintus, S., Huebert, J., Kirch, P.V., Lincoln, N. K. and Maxwell, J. (2019) 'Qualities and contributions of agroforestry practices and novel forests in pre-European Polynesia and the Polynesian outliers', *Human Ecology* 47(6), pp.811-825

Schilt, R. (1984) *Subsistence and Conflict in Kona, Hawaii: An Archeological Study of the Kuakini Highway Realignment Corridor*, Bishop Museum Press, Honolulu, HI

Vilsoni, F. (1993) 'Sustainable taro culture: Fiji situation', in L. Ferentinos (ed.) *Proceedings of the Sustainable Taro Culture for the Pacific Conference*, 24-25 September 1992, University of Hawaii, Honolulu, HI, pp.84-87

Vitousek, P. M. and Chadwick, O. A. (2013) 'Pedogenic thresholds and soil process domains in basalt-derived soils', *Ecosystems* 16(8), pp.1379-1395

Vitousek, P. M., Ladefoged, T. N., Kirch, P. V., Hartshorn, A. S., Graves, M. W., Hotchkiss, S. C., Tuljapurkar, S. and Chadwick, O. A. (2004) 'Soils, agriculture, and society in precontact Hawaii', *Science* 304(5677), pp.1665-1669

Vitousek, P. M., Chadwick, O. A., Hotchkiss, S. C., Ladefoged, T. N. and Stevenson, C. (2014) "Farming the rock: a biogeochemical perspective on intensive agriculture in Polynesia', *Journal of Polynesian Archaeology* 5(2), pp.51-61

Weisler, M. I., Mendes, W. P. and Hua, Q. (2015) 'A prehistoric quarry/habitation site on Moloka'i and a discussion of an anomalous early date on the Polynesian introduced candlenut (*kukui, Aleurites moluccana*)', *Journal of Pacific Archaeology* 6(1), pp.37-57

Wilcox, E.V. and Thompson, A. R. (1913) 'The Extraction and Use of Kukui Oil', in E.V. Wilcox (ed.) *Hawaii Agricultural Experiment Station*, Honolulu Press Bulletin 39, Honolulu, HI

Winter, K., Lincoln, N. and Berkes, F. (2018) 'The social-ecological keystone concept: A quantifiable metaphor for understanding the structure, function, and resilience of a biocultural system', *Sustainability* 10(9), p.3294

Yen, D. E. (1993) 'The origins of subsistence agriculture in Oceania and the potentials for future tropical food crops', *Economic Botany* 47(1), pp.3-14

Young, R. A., Cruz, L. G. and Brown, A. C. (2005) 'Indigenous Hawaiian nonmedical and medical use of the Kukui tree', *Journal of Alternative and Complementary Medicine* 11(3), pp.397-400

22

COMPOSITE FARMING SYSTEMS IN AN ERA OF CHANGE

In Nagaland, northeast India

*Malcolm Cairns and Harold Brookfield**

Introduction: composite farming systems

Composite systems, as we abbreviate the 'composite swiddening agroecosystems' definition by Rambo (1996, 2007), are those in which radically different technologies such as swidden agriculture, permanent wet-rice farming, home gardens, perennial-crop farming and enduring tree-crop cultivation are found together within the farming complex of one community. Composite systems are not new, and are not all that uncommon in Southeast Asia, but little integrated discussion occurs in the literature. Rambo, Menzies (1996), Yin (2001) and Menzies and Tapp (2007)

This is a tribute to a close friend and colleague, Khrieni Meru, who played a large role in the success of the research that produced this chapter. Khrieni was a capable farmer and a strong pillar of the Khonoma community. He and his family hosted me during my research in their village, and treated me as a member of their own family. Khrieni's enthusiasm, work ethic, and inquiring mind were all invaluable in completing this research. He was often as keen as I was, to see what the data showed! We became like brothers, and so it was with deep sadness that I learned in May 2016, that, after a brief illness, Khrieni had passed away at a hospital in Dimapur, at the age of 60. He left behind his loving wife, Gonguü, two daughters and a son. I take this opportunity to salute an old friend, who left us far too early. Rest in peace, old friend! – Malcolm Cairns.

This is a revised version of a paper originally published in April 2011 in *Asia Pacific Viewpoint*, vol. 52 (1), pages 56-84, and reproduced here with that journal's kind permission. The lead author's complete Ph.D dissertation, from which this research was taken, is available online at http://www.digitalhimalaya.com/collections/rarebooks/

* DR. MALCOLM CAIRNS is a freelance researcher who was most recently a Fellow at the Centre for Southeast Aian Studies (CSEAS), Kyoto University, Kyoto, Japan; DR. HAROLD BROOKFIELD spent the final years of his career as an Emeritus Professor of Geography and Anthropology at the College of Asia and the Pacific, Australian National University, Canberra. After a lengthy period of declining health, Professor Brookfield passed away in Canberra on May 22, 2022, at the age of 96 years. (See Remembrance page in the front matter of this volume.)

are among the few who review the interrelations that make such mixed systems successful. There have, on the other hand, been several studies on the productivity of swidden cultivation, and in some cases a comparison with wet-rice cultivation (e.g. Padoch, 1985; Cramb, 1989; Conelly, 1992; Hunt, 2000; Mertz, 2002). The most recent of those that dealt with a composite system looked only at the swidden element (Nielson et al., 2006). Until now, no-one has attempted to quantify in detail the manner in which farmers manage their inputs into the distinctive subsystems of composite farms, or has comparatively measured the total outputs thus obtained. This paper pioneers such quantitative measurement.

We present data on the operation of the principal subsystems, wet-rice terraces and swiddens, found in two villages in Nagaland, northeastern India, in the years 2000 and 2001. This was in a period of rapid economic and social change in Nagaland. The data rest principally on two years of supervised day-by-day diary-keeping by groups of men and women farmers in support of Cairns' fieldwork in Nagaland. The accompanying tables, diagrams and photographs all relate to 2000 and 2001 and all statements referring to this time of fieldwork are placed in the past tense, even though some conditions described may well remain unchanged. The compelling priority of other work and severe illness delayed the writing of this paper so that its central data are already historical, although hopefully interesting methodologically, despite the delay.[1] To discuss change, it becomes necessary to view the situation in 2000 and 2001 in perspective, involving us in challenging problems of interpreting trends both within and since the date of Cairns' field material. We tackle these problems toward the end of this chapter.

Relevant aspects of the modern debate regarding swidden cultivation

Locating our material in the literature was not easy, since most writing on wet rice concerns regions that have experienced the Green Revolution; Nagaland has not. Thus, the swiddening literature has greater relevance. Writing on swidden cultivation in Southeast Asia has also undergone a major shift. Until recently, the impending collapse of unimproved systems, due to the imposition of shorter and shorter cycles of clearing, burning and cultivation on sensitive forest ecosystems, was a recurrent theme (Cairns, 2007). On the evidence presented by Schmidt-Vogt et al. (2009), there are now few on-going shifting cultivation systems in Southeast Asia in which 10 years of fallow, the minimum suggested by Ramakrishnan (1992), is still achieved. Yet the predicted disasters have not occurred.[2]

Other forces have instead hit shifting cultivation. Writers nowadays are principally concerned with its rapid replacement by other forms of land use. Padoch et al. (2007) focused attention on this issue, and were followed by a set of papers presented at a symposium held at Hanoi, Vietnam, in 2008, and published in a special issue of the journal *Human Ecology* in 2009. While new livelihood strategies are sometimes adopted voluntarily, the reasons given for change are often in the category of external forcing: heavy advice, prohibitive legislation, land alienation or exclusion from land by conservation zoning. Fox et al. (2009) described how shifting cultivators had been marginalized as members of ethnic minorities, deprived of access to land or brought almost forcibly into the commercial economy in a subordinate status. Scott (2009) echoed this argument by describing the events of the past half-century as the 'last enclosure' in the uplands, meaning the capture by state control of the last major farming regions in which people could organize their social and productive systems in diverse ways of their own choosing.

Scott (2009) said more. Most recent writers seem at least tacitly to accept the traditional stance that shifting cultivation is the most ancient of all farming systems, and therefore that it is primitive. Yin Shaoting (2001) had earlier denied this and pointed out that farmers who used knives, fire and dibble sticks in swiddens were fully aware of the use of the hoe and plough, and of irrigated terracing; in some instances, they combined these very different technologies in their management repertoire. Scott went much further, and argued that shifting cultivation had been a political choice taken in historical times. Swidden agriculture offered fewer opportunities for appropriation of agricultural surplus by tax-gathering governments, their police and armies. Swiddening, he said, was footloose, and an 'escape agriculture'. Brookfield (2011a, 2011b) has written critically on these generalizations, but in this chapter we allow Cairns' data to tell their own different story.

Introducing the Angami Nagas, the research and the two villages

The Nagas of northeastern India and adjacent areas of Burma are a warrior-like people, famous for their headhunting (von Fürer Haimendorf, 1946; Saul, 2005). The majority practise swidden farming with a few wet-rice patches or none, but the Angami of southern Nagaland are different in that they depend primarily on wet rice grown in sometimes elaborate systems of irrigated terraces. Shifting dry fields – *jhums* as swidden fields are known in northeast India – are for them a subsidiary source of livelihood. Because the rest of this paper is about Nagaland, we use the northeast-Indian term *jhum* from this point forward. Some Angami *jhums* are roughly terraced, but not levelled, and where we use the term 'terrace' alone, it always means the wet-rice terraces only.

The Angami Nagas were first comprehensively described by Hutton (1921). Because of the events briefly recounted below, rigorous description of them suffers a long gap after his writings and those of other British administrators. The PhD thesis on which most of this paper is based (Cairns, 2009) has yet to be published in printed form, but it provides the first comprehensive ethnographic account since Hutton (1921). Cairns' first interest was agronomic. He was impressed by the use of nitrogen-fixing, heavy-littering Himalayan alders (*Alnus nepalensis*) as a sole fallow species in the *jhums* of a leading Angami village, Khonoma (Cairns et al., 2007). It was this distinctive use of soil-improving trees in the *jhum* component that attracted his attention to the village. His research there eventually expanded to become a wider analysis of the dynamics of Khonoma's cultural ecology, but from early on, it was decided to also study a second village with a composite system that lacked the use of alders in the *jhum* fields. Selection fell on Tsiesema, which turned out to be fortunate because there were substantial contrasts (Figure 22-1).

The highly elaborated alder-*jhum* system remained one of Khonoma's major claims to distinction. *A. nepalensis* is a pioneer colonist of open spaces, and Khonoma farmers actively propagated it so that it became almost a pure stand in their fallows. When the land was to be cultivated, the alders were not cut down or burnt, but were skilfully pollarded (Figure 22-2). Only the small slashed branches and debris were burnt. Pollarding, a woodsmanship practice encountered with several tree species in scattered localities all the way from western Europe to China, makes use of the self-renewing power of trees. The tree is cut two metres or so above the ground, leaving a permanent trunk from which shoots grow up to become poles that are cut for firewood or other uses at intervals of years.[3] Under conditions of severe land shortage in the recent past, a fallow period under managed alder of only several years was sufficient to restore soil fertility for a further two years of cropping, and to go on doing so, cycle after cycle (Figure 22-3). Khonoma, and to a lesser extent, a Chang Naga village called Chingmei, in present-day Noklak district, are rather unique in Nagaland in their intensity of management of *A. nepalensis*. But the management of *A. nepalensis* as a preferred fallow extends beyond Nagaland's borders into neighbouring states, most notably by Tangkhul Nagas in Manipur. Beyond this, to

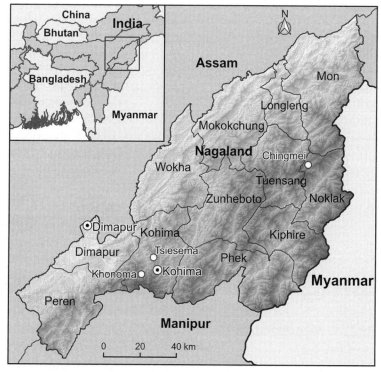

FIGURE 22-1: Nagaland and the location of the study villages, Khonoma and Tsiesema.

our knowledge, in all of south and east Asia, these pollarding methods find a close observed parallel only in the treatment of *Cassia siamea* (sometimes known as *Senna siamea*) for firewood production in southwestern China.

Yet Khonoma's *jhums* remained subsidiary to the wet-rice terraces. The village's foundation stories strongly emphasize the potential of the adjacent valley for wet-rice cultivation, and although there is no written history before the 1830s, we can be reasonably confident that wet-rice intensification came before *jhum* intensification.[4] The valley below the village is terraced and irrigated in a manner that has drawn comparison with the famous Ifugao terraces in the Philippines, and oral history indicates that the terraces were completed at least 250 years ago. The village site itself, on a sharp ridge, was eminently defensible. By the 19th century, when the population of Nagaland had grown substantially and warfare between villages had become endemic, the village was strongly fortified.

FIGURE 22-2: Pollarding alders at Khonoma.

All photographs in this chapter were taken by Malcolm Cairns.

Khonoma became powerful, influential and feared over a wide area (Hutton, 1921). It was never conquered by other Nagas, but was unable to resist the more heavily armed British, who seized the village in 1879 in the course of campaigns to pacify the Naga Hills. Nor was Khonoma able to resist the Indian army, which for a second time, destroyed and depopulated most of the village in 1956, early in a prolonged

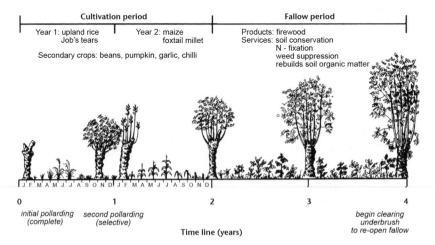

FIGURE 22-3: Historic phases of alder management through a four-year cycle at Khonoma prior to 1956.

struggle to put down a movement, based initially at Khonoma, for Naga independence. Two of Khonoma's three major clans dispersed almost completely into the forest and to other communities following this rout. Some individuals bought land in the lower hills, mostly in what is now Dimapur district (Figure 22-1), land which they still retain. Tsiesema was less directly involved in these hostilities.

The war ended in stages between 1964 and 1975, but it was not until the 1990s that peace became more or less permanent.[5] Meanwhile, in 1963 Nagaland became a state within the Indian union, with one distinctive condition that is important to this

Diplazium esculentum (Retz.) Sw. [Athyriaceae]

Believed to be the world's most commonly eaten fern, the young fronds of this species are a popular vegetable in the study villages. This fern grows to about 50 cm and it is often cultivated as a home-garden or swidden crop. When Khonoma was attacked and its people driven into the forests, this fern was a vital part of their diet, that allowed them to survive.

discussion. Eighty-eight per cent of the land, and forests in particular, remained under local control. This contrasted with only 10% of forest cover being locally controlled in India as a whole. There has been no alienation to create externally controlled forest reserves. Peace re-opened trading links with the Assam lowlands, but although many of the wartime refugees returned to Khonoma, the modern village remains smaller than it was before 1956. Trade and emigration have relieved the severe population pressure on local resources that had earlier driven a remarkable intensification of its agriculture.[6]

Alnus nepalensis D. Don [Betulaceae]

Himalayan alder is a large tree that grows rapidly to as high as 30 metres. It occurs throughout the Himalayas, from Pakistan through Nepal, Bhutan and northeast India to Yunnan, in southwest China. Nitrogen fixation by *Frankia* bacteria in root nodules and copious litterfall make this species a catalyst for a remarkable intensification of agriculture, particularly as it is managed in Khonoma.

For Nagaland as a whole, national-government funds provide much the largest part of a state budget that has risen almost year-by-year, rapidly since 1980, and in 2018-2019, it exceeded the equivalent of US$250 billion. Most of this supports a greatly enlarged civil service. Smaller funds are directed to Village Development Boards (VDBs) in each registered village, under a scheme begun in the 1980s. Sundry other national poverty-reduction supports also reach Nagaland communities. This national support is without specific 'strings'. It needs to be stressed that despite the trauma in the second half of the 20th century, the Angami Naga villages have lost neither land nor control over its use. Official and semi-official documents describe efforts to persuade Naga farmers to adopt 'modern' agricultural methods, and there have been inputs in provision of infrastructure, germplasm and (at Tsiesema) machinery. But decision-making has not been taken out of local hands. External forces have been strongly influential under both British and Indian rule, but the changes reported below were determined ultimately only by the villagers themselves.

Researching the basis for a comparison

Because of continuing low-level unrest in Nagaland, Cairns was permitted to visit only for short periods of one to three months, totalling about one year over more than three years between 1999 and 2003. Intervening periods were occupied in documentary work, data tabulation and GIS analysis. Sustained participant observation was therefore not an option. To obtain continuous data through the whole of 2000 and 2001, two groups each of 10 collaborating farming families in Khonoma and Tsiesema agreed to keep diaries of their activities on selected fields, guided by research assistants resident in the two villages. In each village a 'farmer coordinator' was selected first and, in consultation with Cairns, he then chose another nine collaborating families. While hopefully reasonably representative, therefore, the 20 families were in no sense random samples of the village populations. Diary-keeping became a family activity and children were often involved. Separate records were kept of labour inputs and of all material inputs and outputs, to facilitate cross-checking between diaries.[7]

Standard lists of field operations were developed for both wet-rice terrace and *jhum* cultivation, and divided into six sequential categories: 1. wood harvest; 2. nursery management; 3. field preparation; 4. planting; 5. plant management; and 6. crop harvest. Cairns and his assistants reviewed each family's activities with them, line-by-line, to clarify what exactly was done, then inserted the activity code that fitted best. This made possible reduction of all text into simple numbers which could be manipulated for statistical analysis. Farmers recorded their labour in hours, since much work was done in small instalments of only a few hours here and there, particularly during non-peak times.

Material inputs and outputs were all recorded not as quantities, but in terms of their local price (in Indian rupees (₹)), regardless of whether they were bought or sold. Often germplasm was saved rather than bought, and a high proportion of crops was consumed domestically and not sold. The 'values' given in the text and tables are therefore imputed values, and not necessarily either actual expenses or income. This is shown in the tables, but to avoid circumlocution in the text, they are treated as though they were real when findings from the diary records are discussed.

The selected fields, 40 in total (10 wet-rice terraces plus 10 *jhum* fields in each village), were carefully measured and located. The diaries provided data at field level. The area (sq. m) of these monitored fields had to be measured in order to be able to convert the findings to a more meaningful measure of data per hectare of land.[8] The data also needed to be examined in terms of what it meant to the average village household, since it was here that most decisions were made on how agricultural lands were managed. The Angami patrilineal system of inheritance fosters considerable disparity in land ownership. The research thus needed to look beyond the monitored fields, at the total land holdings of each of the collaborating farmers. For the wet-rice terraces, it was a straightforward matter of surveying all of the additional plots – beyond the one monitored – owned by each collaborating farmer. This showed the total wet-rice land cultivated by each farmer, and allowed the diary findings to be scaled up to household level.

The extent of *jhum* land was not so easy to measure and an indirect proxy method was used to estimate the area of additional *jhum* holdings of each of the collaborating farmers, beyond the single field that was being monitored. Villagers themselves measured *jhum* land by how much seed it took to plant it or by the expected yield. By applying these widely understood equivalences to the monitored fields that were of a known hectarage, the planting rates and expected yields reported by farmers could be calibrated on a per hectare basis.[9] This, in turn, allowed estimation of the approximate area of their additional *jhum* fields.

The organization of production

Angami villages, and even the major clans within them, governed themselves without chiefs. Important decisions were taken by an informal council consisting of elders. There was no heritable ranking, either of descent groups or individuals.[10] Each major clan held blocks of land within and around the villages. Only land beyond the normal limits of cultivation was regarded as the domain of the village as a whole. Within the clan territories, individual male-headed families held effectively private claim to regularly-utilized land, including all the wet-rice fields and most of the *jhums*.

Most agricultural labour was performed by household members working in their own fields. This was particularly true in Khonoma, where fields tended to be smaller. In Tsiesema's larger fields, groups of often related farmers banded together and worked in turn in each member's fields in preparing the seedbeds, transplanting and harvesting. Between these times of peak demand, labour reverted to household sources as each family tended its own fields. Most of the labour recorded was performed by women. Gender aspects of work are discussed in full in Cairns (2009) and in Colfer et al. (2015). Historically, the principal role of men was to defend the villages and their people from the danger of headhunting raiders. Although these dangers were long in the past, men still contributed only a minority share of the field work during the study, especially at Khonoma. Only the actual pollarding of trees was almost exclusively a male task. The data often showed village women performing two or three times as much agricultural labour as their menfolk. It was solely in field preparation and in the wood harvest at Tsiesema that male labour input exceeded that of women. In Khonoma, female labour input also preponderated in most categories of field activities.

Wage labour was found most frequently when relatively affluent households, often with more lucrative off-farm sources of income, hired their least prosperous neighbours to work in their fields. These neighbours often had little land of their own to cultivate, and some lacked even the means to rent land. It was thus usually a transaction between the most and the least affluent strata of the village. During the period of field research, the standard wage rates for hired labour were ₹100/day for men,[11] ₹80/day for women and ₹60/day for children (under 15 years of age).[12] Some of Khonoma's wet rice was produced under share-cropping arrangements. When wet-rice fields were rented, the usual terms were for the harvest to be evenly divided between the cultivator and land owner. This was also the arrangement by which Khonoma villagers used the wet-rice land that they bought down in the lower hills during the Indo-Naga war. Owners contracted lowland families to manage them on a share-cropping basis. About 15 truckloads of bagged paddy arrived in Khonoma each autumn from these outside fields.

Chickens and also pigs, though fewer than in earlier times, were kept by most households. Cattle were now little managed by Angami families. Migrant Nepali families, usually staying in ramshackle accommodations on village outskirts, tended the cattle in return for a share of the milk and calves. Their standard of living was far below that of their Angami hosts. Mithun (*Bos frontalis*) were still highly prized at Khonoma. They wandered in the forests in communal herds and received little human handling until it was time to capture them for slaughter or sale – in modern times, usually to celebrate a wedding. The little attention that they did get was from herders who camped in outlying sheds and, from time to time, blew on buffalo horns to summon the mithun for inspection, rewarding them by sprinkling salt on nearby rocks.

The two villages and their land

Khonoma occupies a rugged basin backing onto the forested Barail range of mountains, cresting at over 2700 metres above sea level (masl) (Figure 22-4). The valley is terraced up to 1800 masl, but rice was cultivated only below 1520 masl. The adjacent hillslopes carried *jhum* fields up to 2100 masl. Using GIS analysis, Cairns found that only 41.5% of the total village territory lay below 2000 masl and had a slope of less than 30%, making it readily cultivable by Angami technologies. The total

FIGURE 22-4:   The residential area of Khonoma village sits on a highly defensible spur.

population of Khonoma, not including the Nepali herders who lived on the outskirts, was 2145 in 2001, lower than before 1956. More than 3600 people who were born in Khonoma lived elsewhere. The first British observers to see Khonoma in the 19th century were struck by the shortage of agricultural land and during the period of British rule (1879 to 1946) the effective population density on cultivable land probably exceeded 250/sq. km. Disregarding the purchased land toward the Assam border, the figure was around 160/sq. km in 2000-2001. It was high population density in the two or three centuries before 1956 that forced the full elaboration of the alder-*jhum* system at Khonoma, with fallow periods as short as two years on the most intensively used *jhum* land.

TSIESEMA, by contrast, is on less mountainous but dissected terrain at lower altitude, and almost all of its land is cultivable.[13] Its full altitudinal range is from 522 to 1460 masl, and the village's residential area occupies the highest land. Tsiesema had 788 people in 2001, and there had been no major reduction as at Khonoma. Population density at the time of fieldwork was only 52/sq. km. Tsiesema's wet-rice lands were scattered in small valleys, whereas most of Khonoma's were in a single large block (Figures 22-5 and 22-6). Almost all of Khonoma's rice was grown in this block with only a minor contribution from the *jhum* fields. In Tsiesema, by contrast, a large part was grown in the *jhum* fields, making this village more representative of Nagaland as a whole. The normal fallow period at Tsiesema was around 13 years, and earlier it was longer. Tsiesema is too warm for alders to flourish, and the woody fallows were comprised of a wide range of tree species.

FIGURE 22-5: A Tsiesema flight of wet–rice terraces surrounded by secondary forest (*jhum* fallows).

FIGURE 22-6: Part of the Khonoma's wet–rice terrace system, where retained *Albizia stipulata* trees are thickly scattered across the landscape.

The wet-rice terrace systems in 2000-2001

Labour inputs

In 2000-2001, both villages grew a single crop of wet rice in each year, grown from the farmers' own seeds with no use of either high-yielding varieties or the external inputs that characterize Green Revolution systems. Tsiesema made some use of oxen and there was a single hand tractor, but most labour there, and all of it at Khonoma, was performed by hand. The monitored fields in both villages were worked only by hand.

The entire Angami agricultural calendar has been built around wet–rice cultivation and, in 2000–2001, its demands took priority at all times. The first hoeing was done individually during the winter months. Thereafter, all activities were synchronized.[14] When work was required on the wet–rice crop, everything else was set aside. Particularly at Khonoma, there was a palpable sense of competition and nobody wanted to be seen finishing last or doing a less admirable job than his or her neighbour.

KHONOMA. The very similar rhythm of labour inputs into both villages' wet-rice terraces is shown in Figure 22-7. At Khonoma, the first tillage was completed over December to February. The rice nurseries were established in March, but otherwise there was a relative lull in March and April. There was a major spike in activity over late-May and June as the fields were puddled and transplanted. This was the busiest period of the year. Then there was only light work tending the crop from July to September. A final surge at harvest time, mostly in October, ended the wet-rice cropping year. This was the sequence of activities that has kept the Angami fed for as long as they can remember.

TSIESEMA. The pattern in Tsiesema shows minor differences related mainly to the altitudinal spread of the Tsiesema fields (760 to 1355 masl) compared to those monitored at Khonoma (1331 to 1527 masl). Tsiesema began its first tillage about a month later and did less of it.[15] We see a dispersed pattern with, most notably, rice transplanting continuing into July and even August. Table 22-1 additionally shows that the collaborating farmers in Khonoma invested one-third more labour hours than their counterparts in Tsiesema, who worked a significantly larger area of wet-rice land.

Material inputs and outputs

The material flows into and out of the wet-rice terraces are presented in Figure 22-8. It is here that we first encounter the major changes that had already taken place. During the British period, government officers introduced potatoes and other cool-climate vegetables to the northeast Indian hills. These were largely lost in the wartime period after 1942, but were re-introduced around 1980 and have been adopted in both the wet-rice terraces and the *jhums*.

TABLE 22-1: Summary of data from wet-rice-terrace fields monitored over cropping seasons in 2000 and 2001, averaged over both seasons.

	Khonoma	Tsiesema	Difference K to T
Sampled fields			
Average altitude (masl)	1427	1093	+ 30.6 %
Household level			
Average area of wet rice (sq. m)	4183	5568	- 33.1 %
Labour inputs (hrs)	1705	1280	+ 33.2 %
Hectare level			
Labour inputs (hrs/ha)	4076	2299	+ 77.3 %
Planting materials (₹/ha)	1606	403	+ 298.5 %
Gross harvested outputs (₹/ha)	37,361	23,097	+ 61.8 %
Net returns to land (₹/ha)[f]	35,755	22,694	+ 57.6 %

Notes: Net harvested outputs = gross harvested outputs minus planting materials; Net returns to land = gross harvested output/ha minus planting materials/ha; ₹= Indian rupees.

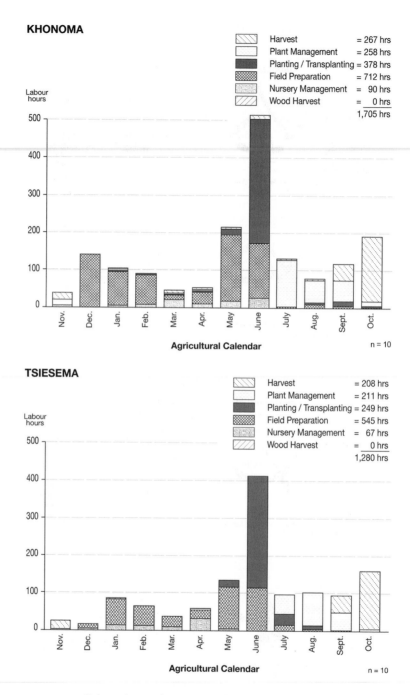

FIGURE 22-7: Labour inputs in monitored wet-rice terraces.

Note: Since, unlike *jhum* fields, there is little variation in how ponded fields are managed from one year to the next, the data collected during the 2000 and 2001 cropping years has been aggregated and is presented here as an average.

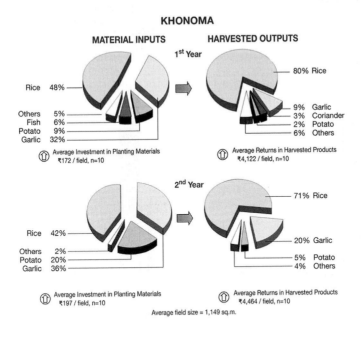

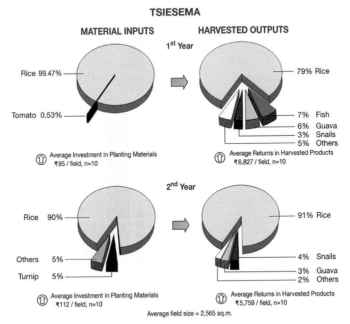

FIGURE 22-8:   Material inputs and outputs from monitored wet-rice terraces.

KHONOMA. The first thing that stands out from Figure 22-8 is that the wet-rice terraces were being planted to much more than rice. There was a seemingly expanding trend for terraces to be converted into intensively managed vegetable gardens. This was particularly so on the higher-altitude terraces. Since about 1990, terraces above 1520 masl had been completely converted to dry fields. While rice continued to be the major output from all other terraces, garlic and potatoes were gaining importance as cash crops, in addition to a wide variety of greens and other vegetables. This was happening under two distinct patterns. In the first, selected terraces were set aside from the standard flooding procedures, and vegetables planted as a summer crop instead of rice. In the second, after the rice crop was harvested, the soil was quickly turned over again and the vegetables planted as winter crops. They were harvested the following spring in time to prepare the land for the usual summer crop of rice. This second pattern made more intensive use of the land since the winter vegetables were an adjunct to the usual rice crop, not a replacement. This diversification of cropping patterns also shows up in the labour graphs and somewhat clouds the pattern of activities relating strictly to rice. Labour spent planting in the autumn months (Figure 22-7), for example, was for potatoes and garlic planted in the terraces as winter crops.

TSIESEMA. Conversion from subsistence cereals to vegetable cash crops had not happened in Tsiesema's wet-rice terraces, where hotter conditions offered fewer advantages for producing cool-climate crops.[16] Although the terraces themselves continued to be planted almost exclusively to wet rice, Tsiesema did diversify its production with subtropical fruit (guavas, oranges, bananas, peaches, pomelo) harvested from the terrace bunds or the vicinity of the field huts.

In partly real and partly imputed terms, the diary-keeping group of farmers in Khonoma was spending nearly three times as much on planting materials for their terraces as their counterparts in Tsiesema. This was primarily due to the higher costs of potato sets (₹12/kg) and garlic cloves (₹10/kg) that were increasingly supplementing rice (₹4.6 to ₹5.8/kg) in Khonoma's terraces. But these costs were

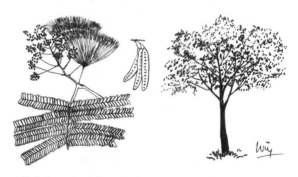

Albizia stipulata B. Boivin [Leguminosae] Syn. for *Albizia chinensis* (Osbeck) Merr.

Another soil-building species managed by Khonoma farmers, but this one in their wet-rice terraces (as seen in Figure 22-6). Growing to 30 m or more, these trees provide nitrogen fixation to improve the soil. Their extensive root system makes them suitable for stabilizing slopes. The wood is not very durable, but is resistant to attacks by termites and other insects, and is used in house building and for light furniture.

FIGURE 22-9:   Khonoma farmers transplanting rice into their terraced fields. The stone monoliths commemorate large merit feasts given in the past by wealthy individuals.

cushioned, in that most of the planting materials could be set aside from the previous year's harvest. In return for their investment, the 10 Khonoma farmers were reaping harvests averaging an imputed value of ₹37,361/ha from their terraces (Table 22-1). Of the harvested output, 71% to 80% was rice and the remainder mostly garlic, potatoes and coriander. Although they were larger in size, the Tsiesema terraces were earning somewhat less at ₹23,097/ha. Rice comprised 79% to 91% of the value and the balance was mostly fish, guavas and snails.

The diary-keeping farmers in both villages were, on average, each cultivating in the vicinity of half a hectare of wet-rice land – Khonoma somewhat less and Tsiesema a bit more. Khonoma was managing its smaller land area much more intensively, both in labour and planting materials (Figures 22-7 and 22-8). These extra inputs enabled Khonoma to coax about 18% more harvest value from its limited land. Significantly, almost 25% of the harvest value per household (₹3858) came from vegetable cash crops that were relatively new introductions. This intensified management was providing Khonoma with returns of close to 60% more harvest value from a hectare of land than was reaped in Tsiesema. But the same was not true for the productivity of labour. Despite Khonoma's increasing adoption of high-value crops, Tsiesema farmers were still managing to earn almost 13% more for every day that they spent working in their fields.[17] The picture that emerges thus fits with Khonoma's history as a village short of agricultural land. It had to prioritize returns to land as opposed to labour. This intensification was further encouraged by the closeness of Khonoma's terraces to the village residential area (walking times to and from fields averaged 22.5 min), so fields could be visited often without wasting much time in transit. But in Tsiesema, where the population had never grown so high, the terraced fields were widely scattered across the landscape (walking times to and from fields averaged 61 min), and there had been less pressure to intensify land use.

Jhum cultivation in 2000-2001

Labour inputs

With terraces supplying most of the villagers' rice needs, the role of Angami *jhums* was of a supplementary nature, providing firewood and additional grains and a wide range of dryland crops to add variety to their diet. *Jhum* management was thus fitted around the exigencies of wet-rice terracing – spatially in terms of where the fields were located, temporally in terms of the flow of labour inputs and the crops that it was expected to produce. This might also be true for most Angami villages, but we cannot say that primacy of wet rice would characterize all composite systems in Southeast Asia. Rambo's (1996) seminal report on the Tay system in Hoa Binh province, Vietnam, gives no indication of such priority. The Tay system combined wet rice with swidden rice and other elements, and obtained much higher rice yields from its small wet-rice terraces, though each system gave a closely comparable return to labour inputs. Each provided about half of the total rice production. More widely in the region, enduring crops have occupied only a small area before recent changes.

First-year jhums

Khonoma. As shown in Figure 22-10, the new *jhum* cycle began with reopening the fallow and harvesting the firewood obtained in the process of pollarding the alder trees from November to February. Field preparation followed directly, gaining momentum in February and March; the fields were then planted in March; management of the crops dominated over the next four months, interrupted by the peak season for transplanting rice into the irrigated terraces. It was only in July that a similar amount of labour again went into the *jhums* as into the wet-rice terraces. Harvest began in June and extended into November. Although these were the rough contours of the graph, it was also clear that small amounts of soil tillage and replanting continued throughout the summer months, representing Khonoma's use of intercropping, relay-planting and crop rotations as it planted faster-maturing vegetables in intensified cropping patterns.

Tsiesema. *Jhums* here continued to be rice-based. Activities in Tsiesema's *jhums* tended to be done once and were then finished for the year. There was not the continual re-tillage and replanting that was found in Khonoma's *jhums* as one crop was harvested and that piece of ground then became available for another. *Jhum* cultivation in Tsiesema began with wood harvest over the winter months of November to February; field preparation started in tandem with removal of the wood, peaked in February and March, and was completed by May. Planting began in mid-March, peaked in April, and spilled over into May; plant management got underway in April, became busiest in May (when the rice was hilled early in the month and then weeded again at the end), and continued until September. The busy time concluded with harvest in September and October. As was the case with Tsiesema's wet-rice terraces, the significant range in altitude amongst the monitored

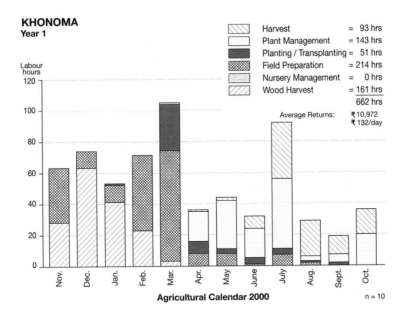

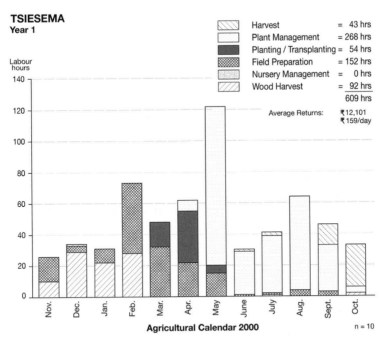

FIGURE 22-10: Labour inputs into monitored first-year *jhums*.

jhum fields (967 to 1453 masl) stretched out the time frame under which each field operation continued. Planting in the lower *jhums* could still be going on more than two months after completion of the last upper *jhum*.

Second-year jhums

KHONOMA. In the second cropping year, there was a dramatic reduction in labour inputs into the monitored *jhum* fields in both villages (Figure 22-11). Firewood now came from first-year *jhums* opened elsewhere that year. The residues from the previous year's crops needed only to be cleared away and the soil tilled lightly to get

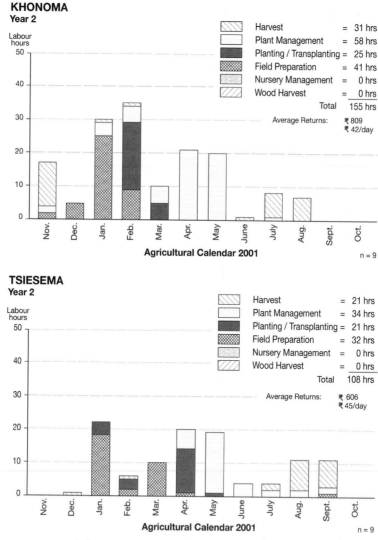

FIGURE 22-11: Labour inputs into monitored second-year *jhums*.

Note: The vertical scale (Labour hours) on this graph is double that in Figure 22-10.

the seedbed into shape for replanting. This much-reduced field preparation was done mostly in January, followed by planting in February. Modest labour inputs continued as the crop was cared for over the next four months and was finally harvested in July and August. As autumn approached in Khonoma, the alder coppices were already beginning to form a canopy and the yellow crop residues quickly disappeared under a flush of weeds that marked the first stages of fallow succession. Although the sequence of field operations over the two cropping years was still quite clear, it was more dispersed, with more activities going on concurrently than would have been the case earlier, when Job's tears were the main crop of first-year *jhums*, with a millet-perilla intercrop planted during the second year.

TSIESEMA. The second cropping year in the monitored *jhum* fields in Tsiesema had even less importance than that in Khonoma and sometimes did not happen at all. Some crops were more tolerant of second-year cropping. Chilli yields usually improved in the second year (assuming the original plants persisted), sometimes even doubling. Maize, millet, cucumber and yams would still yield satisfactorily, but not the rice staple. There was one particular non-glutinous variety that was normally used when a Tsiesema *jhum* field was replanted with rice for a second consecutive year. The farmers' attitude towards second-year fields appeared opportunistic; the soil was still sufficiently fertile and weed populations were not yet out of control, so they tried to harvest whatever they could with minimal inputs. But at Tsiesema it was not uncommon for less promising fields to be returned to fallow after only a single cropping year.

Material inputs and outputs

KHONOMA. Crop data for 2000-2001 (Figure 22-12) show a system in transition, in which traditional subsistence crops lingered alongside new introductions aimed at outside markets. This was most obvious in the first cropping year. The Job's tears that had long been the forte of first-year *jhums* had been largely pushed aside by potatoes, now constituting almost three-quarters of total planting-material costs.[18] Soybeans, taro and Job's tears each added another 4%, and another 36 minor crops combined for the balance.

The output data show that a massive 46% of the total imputed value of outputs at Khonoma, during the first cropping year in the monitored *jhums*, came from the firewood harvested when the fallow was opened. Potatoes, maize and soybeans were the main harvested food crops, but a wide range of minor crops was grown in small amounts for kitchen consumption. This latter category included small quantities of wild flora (perennial buckwheat and other wild vegetables) and fauna (birds, crabs, grasshoppers) that farmers collected as they worked in their fields.

The second-year reduction in labour inputs was accompanied by an equally dramatic narrowing of agrobiodiversity (average 17.9 crops/field [first-year]: 4.8 crops/field [second-year]). Cash crops had made fewer inroads here, and the millet-perilla intercrop that traditionally occupied all second-year *jhums* still represented

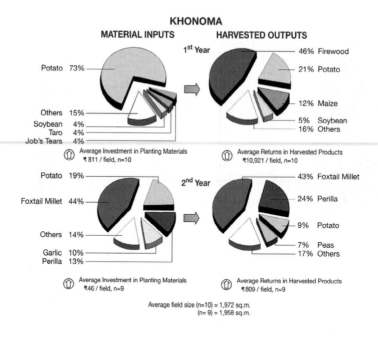

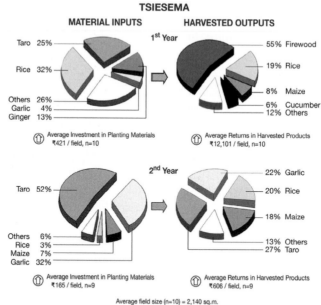

FIGURE 22-12: Material inputs and outputs from the monitored *jhums.*

more than half of expenditures on planting materials. But an emerging trend towards conversion from such subsistence crops to higher-value cash crops was evident in the shares of potato sets and garlic. The bulk of the harvest value in the second cropping year still came from foxtail millet and perilla.

FIGURE 22-13:   Khonoma pollarded-alder *jhum* planted to potatoes.

TSIESEMA. Compared with Khonoma, this village showed a lower level of agrobiodiversity in its *jhum* fields (53 species [K] : 40 species [T]),[19] but the nature of the fields was strikingly different from the intensively-managed market gardens, terraced and scattered with alders, found in Khonoma (compare Figures 22-13 and 22-14). Tsiesema's *jhums* remained more typical of the region: rice-based, with a sprinkling of maize, chillies, tomatoes and other crops throughout, with yams and cucumbers targeted at heavily-ashed spots where unburnt debris had been piled and re-burnt. Field boundaries were often marked by rows of maize or taro, and there were small patches of vegetables near the field huts for preparing noon meals. Rice was the largest single expense in planting for the first cropping year, followed by root crops (taro, ginger and garlic). The remainder comprised a wide array of minor crops.

FIGURE 22-14: A Tsiesema *jhum* planted with rice.

Even more than at Khonoma, firewood dominated the list of harvested products from Tsiesema's first-year *jhums* (55%). At Tsiesema, the older fallows contained oak and other high-value tree species. The trees were cut low rather than pollarded and firewood of higher average value was yielded.[20] In the limited cropping that Tsiesema did in the second year, emphasis was shifted from cereals to root crops that were more nutrient-efficient under conditions of declining fertility. The major planting-material costs were for taro and garlic, but the total cost per field was very small.

Cost-benefit analysis of jhum cultivation

A distillation of the findings from the *jhum* diaries is presented in Table 22-2. Angami *jhums* were small compared to those of other Naga subgroups that rely on *jhum* cultivation for virtually all of their food needs. Keitzar (1998) reported that the *jhum* area cultivated by Nagaland households varied from 0.5 to 2.5 hectares, but the diary-keeping farmers in Khonoma and Tsiesema each opened up only around one-fifth of a hectare of new *jhum* land in 2000 (Figure 22-10). There were important differences between the two sets of sampled fields. The 10 monitored *jhum* fields in

TABLE 22-2: Summary of data from *jhum* fields monitored over 2000–2001 cropping seasons.

	Khonoma	Tsiesema	Difference K to T
Per household			
First-year cropping			
Costs			
Labour inputs (hours)	662	609	+8.7%
Planting materials (₹)	811	421	+92.6%
Benefits			
Gross harvested outputs (₹)	10,921	10,058	+8.6%
Net harvested outputs (₹)	10,110	9,637	+4.9%
Net returns to labour (₹/day)	122	127	-4.1%
Second-year cropping			
Costs			
Labour inputs (hours)	155	108	+43.5%
Planting materials (₹)	46	165	-258.7%
Benefits			
Gross harvested outputs (₹)	809	608	+33.1%
Net harvested outputs (₹)	763	443	+72.2%
Net returns to labour (₹/day)	39.4	32.8	+20.1%
Per hectare			
First-year cropping			
Costs			
Labour inputs (hrs/ha)	3357	2846	+18%
Planting materials (₹/ha)	4113	1967	+109.1%
Benefits			
Gross harvested outputs (₹/ha)	55,380	47,000	+17.8%
Net harvested outputs (₹/ha)	51,268	45,033	+13.8%
Second-year cropping			
Costs			
Labour inputs (hrs/ha)	786	505	+55.6%
Planting materials (₹/ha)	233	771	-230.9%
Benefits			
Gross harvested outputs (₹/ha)	4102	2841	+44.4%
Net harvested outputs (₹/ha)	3869	2070	+86.9%

Notes: Net harvested outputs = gross harvested outputs minus planting materials; Net returns to land = gross harvested output/ha minus planting materials/ha; ₹= Indian rupees.

Khonoma had an average elevation of 1671 masl, with a northern exposure – too cold for rice, but ideal for potatoes and other cash crops grown for lowland markets. The Tsiesema fields had an average elevation of 1264 masl with a predominantly western aspect, too warm for the semi-temperate *jhum* crops that Khonoma then grew, but quite satisfactory for rice.

The *jhums* sampled in Khonoma were also more uniformly clustered, from 25 to 40 minutes' walk from the village's residential area; those in Tsiesema were strung out from the edge of the village's residential area (5 minutes' walk to 1453 masl) down the slope midway to its most distant boundaries (110 minutes' walk to 967 masl).[21] The more distant *jhums* had higher costs, most significantly in terms of daily transit times, but also from the increased danger of wildlife damage. Often, Tsiesema farmers would need to sleep over in distant *jhums,* both to cut down time spent walking and to guard the crops against wild boar and other nocturnal visitors.

The data in Table 22-2 parallel the general dynamics already seen in the terraces (Table 22-1). Khonoma was investing more, both in terms of labour and planting materials, in its smaller land area. This was earning it about 8.6% more gross harvest value than Tsiesema's fields. But the labour that Khonoma used to achieve those gains received lower returns than did Tsiesema's less intensive management. Second-year *jhums* in both villages provided meagre outputs and abysmal returns to labour. Of all activities supporting the livelihood of Angami villagers, planting second-year *jhums* was probably one of the least remunerative uses of their time.

The major role of Tsiesema's *jhums* remained subsistence – supplementing yields from the terraces to create a position of assured rice security. Khonoma farmers, on the other hand, were exploiting their comparative advantages. The role of Khonoma's cooler *jhums* thus became cash generation to pay school fees, electricity bills, buy clothing and kitchen supplies, while families relied on their wet-rice terraces (and often the additional land bought at lower altitudes) for their rice needs. Supplementary rice could easily be purchased from the market in Kohima.

The most marked contrast between the two villages was in the length and treatment of their *jhum* fallows. On the normative basis that farmers married and assumed full responsibility for their own farms before age 25 and continued

Melia azedarach L. [Meliaceae]

This member of the mahogany family appears naturally in fallow regrowth in Tsiesema's *jhums* – at a density that has much to do with grazing cattle (see endnote 20). However, the woody seeds within the copious fruit are known to be highly toxic to humans. Commonly known as Chinaberry, the high-quality timber is dark brown to red, and is often confused with teak.

active farming until almost age 70, it follows that during a working life of, say, 45 years, a Tsiesema farming family would cultivate a particular *jhum* plot three times, whereas their Khonoma counterparts would do so five times. The contrast is clear in Figure 22-15.

Condensing a detailed analysis in Cairns (2009), the averaged annual imputed product of a family's *jhums* over 45 years, on a per hectare basis, would be ₹3976 for Tsiesema, but ₹6613 for Khonoma. The contrast in long-term land productivity would have been even more marked when Khonoma's fallows were shorter, in the century or more before 1956. It remained significant in 2000-2001. Investment in managing *A. nepalensis* by pollarding and maintaining almost pure stands had converted these stands into landesque capital with on-going value that could be realized on a regular and quite frequent basis. On a scale of capital valuation over time, the Khonoma alder-*jhums* were of considerably higher worth than the more natural fallows of Tsiesema.

The alder-*jhum* system at Khonoma, with its alternating wood-crop and field-crop use of the land, was a composite element different from classic swiddening in an important quantitative as well as qualitative sense. Both of us (Brookfield, 2001; Cairns, 2007) and others cited above (e.g. Menzies, 1996; Yin Shaoting, 2001; Rambo, 2007) have insisted that shifting cultivation should be thought of not as a single system, but as a diverse set of dynamic systems. In this perspective, the alder-*jhum* system should be regarded as representing an unusually intensive form, highly sustainable as well as uncommonly productive.[22]

Complementarity and a contradiction in the composite systems in 2000-2001

We have seen how the labour demands of wet-rice terrace cultivation dominated the allocation of inputs in 2000-2001, and by all accounts, had done so for generations. Yet the technology itself had remained stagnant, benefiting from none of the innovations brought elsewhere by the Green Revolution. The improvements in income came entirely from the new crops and the markets they commanded. More dramatic changes had taken place in the *jhums,* especially when viewed in terms of returns to labour. The 2000-2001 data in Tables 22-1 and 22-2 (summarized in Table 22-3 below) show that in imputed-value terms, first-year *jhum* cultivation provided between 1.6 and 1.8 times the returns to labour invested in the wet-rice terraces. The margin of superior returns from *jhum* cultivation had grown as farmers substituted high-value commodity crops for the food grains that they had planted in the past, and in a more comprehensive manner than on the terraces.[23]

Operation of two complementary subsystems had given Angami farmers both security and scope for flexibility.[24] These strategies enabled both villages to roughly quadruple the imputed value of their first-year *jhum* harvests per hectare, compared to what Cairns (2009) estimated for the pre-market era when they planted only food grains for subsistence use. They were able to do this while using the terraces to preserve the goal of rice self-sufficiency imposed on them by persistent insecurity

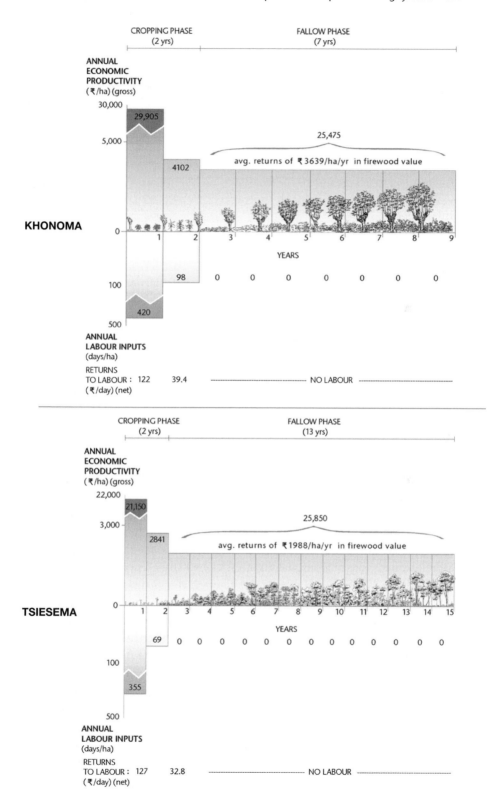

FIGURE 22-15:   Investment and returns in contrasted *jhum* systems

TABLE 22-3: The seeming contradiction in 2000-2001: Where labour was most profitably employed and where it was actually invested.

	Khonoma	*Tsiesema*
Imputed returns to labour per H/H	₹/day	₹/day
Terraces	70	79
First-year *jhums*	122	127
Second-year *jhums*	39	33
Total	231	239
First-year *jhums* as % of terrace	174%	161%
Actual inputs of labour per H/H	Hours	Hours
Terraces	1705	1280
First-year *jhums*	662	609
Second-year *jhums*	155	108
Total	2522	1997
First-year *jhums* as % of terrace	39%	48 %

Notes: One working day = eight hours; H/H=household; ₹=Indian rupees.

in the past, continuing almost to the end of the 20th century. Despite the absence of high-yielding varieties, they obtained reasonable yields, estimated by the state Department of Agriculture to average 2.5t/ha on the Khonoma terraces and 2t/ha on those of Tsiesema (Keitzar, 2010). Calculating on the basis of production reported by the two groups of collaborating farmers on all their terrace plots gave higher yield values – suspiciously high. We rejected these because of potential errors in the data, even though the same methods gave a modest and entirely credible 1.3t/ha for the Tsiesema first-year *jhum* crop of upland rice (Cairns, 2009).[25] The best we can firmly conclude about yields from the wet-rice terraces is that they remained rewarding and, as we shall see below, apparently still do.

We took a long-term perspective of the *jhums* in Figure 22-15 and the associated discussion. If we take a similar perspective of the wet-rice terraces, we recall that they require no fallowing and, with only modest inputs, mainly of green manure, can be productive every year. At 2000-2001 costs and prices, they were capable of yielding produce valued close to an annual imputed ₹12,900 for a Tsiesema family, and ₹15,000 for a family in Khonoma (Table 22-1). Though this is much less than the imputed product of first-year *jhums*, these yearly-available values were respectively more than 3.25 and 2.25 times the annual productivity of *jhum* sites averaged over 45 years of cultivation and fallow. Taking this perspective, they were easily the most productive part of the total space cultivated by the two villages. What Angami Naga composite-system farmers had done in the past was to manage different parts of their ecological environment so that each part yielded its optimal returns for subsistence objectives. But when the objectives began to change, a new pattern of responses emerged.

Signs of its emergence were visible at Khonoma in 2000-2001, but in a puzzling manner. Although vegetables had come to dominate first-year *jhum* production, maintenance of some *jhums* was lagging and by 2002, the opening of new *jhums* had become patchy. Only about a third of one particular *jhum* block that was due to be opened was actually cleared in that year. A survey of all Khonoma families revealed that fully one-third of them had opened no first-year *jhums* at all in 2002. Re-inspection of the data uncovered what seemed, on *prima facie* grounds, to be illogical behaviour.

The summary in Table 22-3 presents a stark contradiction between the advantages obtained by working on first-year *jhums* rather than on the terraces, and the fact that actual inputs still went mainly into the latter. Given that farmers have not been slow to change their cropping patterns in response to modern opportunities, it was surprising that they had not also shifted their inputs in favour of making more and larger profitable first-year *jhums*. At the end of fieldwork, directions were uncertain and in writing his thesis conclusions three years later, Cairns was inevitably ambiguous. He praised the success of the alder-*jhum* system, and saw it as a model that others might copy or — using other soil-improving and/or cash-earning trees — adapt, in the transition from swiddening to more continuous farming (cf Cairns et al., 2007). Yet at Khonoma, he feared that it might vanish within 25 years (Cairns, 2009).

Seeking to understand the pathway of change

Although we have no hard data beyond 2002, and there has been no opportunity for new fieldwork, we have a useful body of information on subsequent change. Cairns has been kept well *au courant* with events in Khonoma, though not at Tsiesema. What we know is outlined below, and it all follows from what was observable in 2002. To explain change means that we must be able to predict the present from the past, but due to the contradiction discussed above, we have a classic uncertainty situation. The determinants of what was observed in 2000-2001 were not fully understood. Logically, confident prediction is impossible in these circumstances (although one might not guess this from the many confident predictions that are written). It became essential to reduce the uncertainty. We therefore re-examined the data for clues and tested suggestions in correspondence between Cairns and some of his best Nagaland informants, learning more about events in the process.

Four suggestions could be offered. First, the high seed costs of the profitable new *jhum* crops — potatoes and garlic; might these alone be capable of excluding the less affluent from participation? This explanation appealed to certain informants, but Cairns' data did not encourage belief that this would be relevant except to the poorest of the poor. The two groups of collaborating households had additionally kept financial diaries from April 2001 to March 2002, and their records unmistakably showed that agricultural inputs were only a minor element among recorded expenses (Cairns, 2009). But might unequal access to suitable land also be an element? Many villagers laid claim to little land. Among the Tsiesema farmers who kept diaries, claims to *jhum* and fallow land ranged from more than 5 ha down to 0.06 ha, and the range was similar though smaller at Khonoma, with four of 10 farmers claiming less than 0.6 ha. But land shortage was not necessarily an obstacle - at least not in 2000-2001. Five of the 10 *jhum* fields monitored at Khonoma (48.25% of the total area monitored in 2000) were on rented or borrowed land. Two of the 10 at Tsiesema were on borrowed land.

A third possibility arose from the observed conversion of some upper wet-rice terraces to permanent commodity-crop fields in the decade before Cairns' fieldwork, discussed earlier. Had the 2000-2001 fieldwork caught an early stage of an ongoing transformation of many more wet-rice terraces into vegetable fields? Terraced land is much easier to work than *jhum* land, and the terraces lying below the unsewered village receive abundant nutrients via stormwater. Intuitively, it seemed likely that these would follow the higher-altitude terraces into conversion, especially if more families were now willing to depend on buying cheap rice imported from the plains. Could this be why many families were losing interest in opening a *jhum*? This hypothesis quickly bit the dust. In the late-summer of 2010 the very terraces in question, close to the village, were described by a highly reliable informant as producing a bountiful crop of rice (Meru, 2010). Clearly, rice production continues to flourish.

The fourth suggestion was more far-reaching. Was the emergence and growth of a non-farm economy the reason for seeming disinterest in a good agricultural opportunity? Together with education for children and young adults, non-farm activities absorbed a growing share of the potential workforce by 2000-2001, especially, but no longer only, the men.[26] Employment in the villages themselves was surveyed in 2002, and reported in appendices to Cairns (2009). The results are summarized in Table 22-4.

In terms of percentages of village populations employed, and the average wage, these jobs were more significant in Tsiesema than in Khonoma. In both villages, the

TABLE 22-4: Jobs and their value in the two research villages in 2002.

	Khonoma	Tsiesema
Wage incomes (recorded)		
Number of government jobs in village	80	82
Number of private jobs in village	36	7
Total wage earners in village	116	89
As % of village population	5%	11%
Total wages per month (₹ per month)	517,900	450,500
Average monthly wage of employees (₹ per month)	4465	5062
Imputed agricultural incomes		
Mean annual H/H wet-rice terrace production averaged to per month (₹)	1302	1072
Mean annual H/H first-year *jhum* production averaged to per month (₹)	910	838
Mean annual H/H second-year *jhum* production averaged to per month (₹)	67	51
Total imputed value per month per mean H/H (₹)	2279	1961

Notes: Job information from surveys by Cairns in 2002. Agricultural income calculated from Tables 22-1 and 22-2 by dividing annual gross values by 12; H/H = household; ₹ = Indian rupees.

mean monthly wage of these employed people amounted to fully twice the per-month imputed value of an average diary-keeping family's production from its terraces and *jhums* together. Nor was this all. In addition to these within-village employees, there were also weekly commuters to jobs in the state capital, Kohima, whose wages were not recorded. Some villagers with well-placed patrons obtained significant incomes from government contract work, and many others earned irregular incomes from skilled artisanal work in stone masonry, basketry, weaving and carpentry.[27]

The 2001-2002 financial diaries underscored the importance of these cash incomes. For these village families, most farm produce was still used mainly for consumption. The average on-farm component of total money income, even including firewood, came to only 13.3% at Khonoma and 35.9% at Tsiesema. Members of most cooperating households earned wages which, together with income from skilled artisanal work and contracts, were the principal family sources of cash (Cairns, 2009). All of these were active farming families, selected as such. In the villages as a whole, the most financially hard-pressed households were those that had not succeeded in tapping into one of the new income niches and continued to rely solely on agriculture. For those who were successful, off-farm work constituted a livelihood element in strong competition with both forms of agriculture. Nothing we have learned suggests that the significance of off-farm incomes has diminished in subsequent years. Here we seem to come closest to a satisfactory explanation of our 2000-2001 contradiction. Pluriactivity had come to southern Nagaland and was profoundly modifying family priorities. This was the more so as rewarding alternatives to *jhum* cultivation were emerging.

The events can be briefly outlined. In addition to the market-garden opportunity that people had discovered for themselves, the state agricultural authorities were promoting a small number of perennial crops among Nagaland's farmers. So also was the one major international project that had worked in Nagaland agriculture.[28] The bushy spice crop *Amomum subulatum* (commonly called 'large cardamom' , 'black

Amomum subulatum Roxb.
[Zingiberaceae]

This spice-producing species gained popularity in Khonoma as an alternative to shifting cultivation. Planted in the shade of *Alnus*-dominated fallows, the dried seed pods of large or black cardamom were sold in India for their strong camphor-like flavour, with a smoky character. The pods are also used in Chinese medicine to treat stomach disorders and malaria.

cardamom', or sometimes 'Nepalese cardamom') was the principal of these; others were tea and sometimes passionfruit. They were being planted in the *jhum* areas. In some of the mature *jhum* fallows that were not being opened in the early 2000s, these crops, mainly *A. subulatum*, were instead being planted under the alders. Some farmers had chosen timber as a cash crop. At its simplest, some were doing this just by leaving fields in fallows of indefinite length and thus allowing selected coppiced branches on the alder trees to reach more marketable girths as timber.

According to informants, this new agroforestry pattern of using alder as a shady overstorey has now become more common, permitting many now-pluriactive families to retain control over their land and its produce with quite minimal labour inputs. Sales of *A. subulatum* were being made within India, and not on volatile international markets, and prices were continuing to rise. Nevertheless prices are at risk from market fluctuations and the security that comes with diversity in subsistence cropping patterns can be lost. There are also dangers of vulnerability to pests and diseases.[29]

It would seem that new long-term crops are taking the place of the *jhums*, and we can now confirm Cairns' (2009) tentative prediction. Whether alder becomes a plantation overstorey or a timber crop, it is probable that its long historical role as a fallow improver in Khonoma's *jhums* is coming towards an end.[30] The pace of change accelerated after a long delay imposed by the period of insecurity between the 1950s and 1990s. The patterns of change include transformation of a *jhum* system that had been remarkably successful both ecologically and economically. Rice, on the other hand, has not yet been either displaced or revolutionized. The changes are comparable to, but not identical with, those experienced in most other parts of the larger Southeast Asian region, and they have happened without the negative external forcing commonly reported in other areas

Solanum betaceum Cav.
[Solanaceae]

These shrubs or small trees grow well under an alder canopy. Commonly known as tamarillos or tree-tomatoes, they were promoted in the study villages as a cash-crop. A native of South America, the name tamarillo was chosen 50 years ago by New Zealand growers for its 'exotic appeal'. A subtropical species, it thrives at high elevations in the tropics, producing large quantities of egg-shaped fruit that are highly nutritious but low in calories.

(Mertz et al., 2009). Partial transformation of the system into cash-crop farming is based on exploitation of regional-market opportunities and is thus economically less vulnerable than transformations that rely on specialization for international markets that may be more volatile. This is still a composite system, but of an increasingly different content from that observed and quantified nearly two decades ago.

Concluding remarks

Rambo (1996, 2007) described composite systems as robust by comparison with unmixed shifting cultivation systems, and we endorse this view. Composite-system farmers have an advantage that the pluriactive farmers in developed countries also have: they can shift their inputs between different forms of activity in an efficient manner (Brookfield and Parsons, 2007). The successful management of both natural and human resources was central to the old composite system in these Angami Naga villages and now the creation of a large non-farm sector creates new management challenges. Part-time farming was, in an embryonic form, already present in the days when men were primarily warriors. It was common in 2000-2002. We do not know if it will now become general or if there will instead be a growing neglect of agriculture as a whole as the presently adult generation ages and is replaced by a better-schooled generation with its known preference for non-farm work. Nagaland is now a small and peripheral part of a large country which is undergoing accelerating change, most of it in the direction of expanding its manufacturing and tertiary sectors. That is the context of the future.

Returning finally to the historical debate, a better understanding of composite systems should help dispose of notions that upland agriculture is necessarily 'footloose'. Where farming systems involve substantial investments in landesque capital, such investment tends to fix them in place. In Nagaland, Khonoma is very clearly a land-bound polity of some antiquity. Its *jhum* innovations reinforced the land-bound status initially conferred by the long flights of terraced rice fields. Khonoma had to remain where it was, building up strong defences, becoming powerful regionally and intensifying all elements of its farming. When, half a century ago, much of the village was destroyed and depopulated, many of its people returned as soon as they could, to care for their largely undamaged investments in land and trees. We suggest, in conclusion, that greater attention to the composite systems of the region might itself diversify an argument which has been too strictly focused on an over-simplified model of shifting cultivation, in regard to external forcing that has also been over-simplified (Brookfield, 2011a, 2015). The case of the Angami Nagas is not typical, involving a traumatic historical experience which, in the end, has left them with more autonomy than is enjoyed by many of the region's upland minorities. But it does demonstrate the resilience of farming people who develop multiple skills and quickly adapt to changing conditions.

Acknowledgements

The research detailed in this chapter was made possible by the generous assistance of Canada's International Development Research Centre (IDRC), while Cairns himself was supported by a PhD scholarship awarded by the Australian National University. An advisory panel comprising Drs. Harold Brookfield, Bryant Allen, Michael Bourke, Andrew Walker and Nicholas Tapp provided many useful suggestions throughout the study. Special thanks are owed to Dr. John Graham for his assistance in administering the IDRC grant, and for remaining supportive throughout the work. Local research assistants in each village, Carolin Meru in Khonoma and Koulie Mere in Tsiesema, were pivotal in undertaking fieldwork. Further support was provided by a Farmer Coordinator in each village, Khrieni Meru in Khonoma (see tribute to Khrieni on the opening page of this chapter) and Zapuvi Mere in Tsiesema. The residents of the two villages, and especially the collaborating farm families, who kept diaries, provided invaluable support. Two Ao Naga government officers deserve special mention for their exceptional support: Dr. Supong Keitzar was Director of Nagaland's Department of Agriculture at that time, and Amenba Yaden was a Deputy Conservator of Forests with Nagaland's Department of Forestry. Through them, we were guided by Naga expertise in both agriculture and forestry. Both men became personal friends who often went out of their way to assist this research.

At the fieldwork 'base camp' in Chiang Mai, Thailand, Tossaporn Kurupunya played a highly important role in collating field data and organizing them into easily-understood figures. The graphs and charts in this chapter were largely her creations. Kittima Kidarn assisted with the mapping and GIS components of the research. Jenny Sheehan (ANU Cartography) skilfully redrew all the diagrams for better presentation in greyscale. Peter Elstner redrew the map of Nagaland to show new district boundaries that were established after the fieldwork. Both authors gratefully acknowledge the assistance of Nicholas Tapp, Bob Hill and Muriel Brookfield in reading and commenting constructively on earlier drafts of this chapter. We are also grateful to two anonymous referees for valuable suggestions on better presentation of our arguments.

References

Baruah, S. (2003) 'Confronting constructionism: Ending India's Naga war', *Journal of Peace Research* (Oslo, Norway) 40, pp321–338

Boserup, E. (1965) *The Conditions of Agricultural Growth*, Earthscan, London

Brookfield, H. (2001) *Exploring Agrodiversity*, Columbia University Press, New York

Brookfield, H. (2011a) 'Farming in the Southeast Asian uplands: The trouble with generalization', in K. G. Saxena, L. Liang and K. Tanaka (eds) *Land Management in Marginal Mountain Regions: Adaptation and Vulnerability to Global Change*, Bishen Singh, Mahendra Pal Singh, Dehradun, India, pp.1–11

Brookfield, H. (2011b) 'Scott and others on history in the uplands of Southeast Asia', review article, *Asia-Pacific Journal of Anthropology* (TAPJA), pp.489–494

Brookfield, H. (2015) 'Shifting cultivators and the landscape: An essay through time', in M. F. Cairns (ed.) *Shifting Cultivation and Environmental Change: Indigenous People, Agriculture and Forest Conservation*, Earthscan, London

Brookfield, H. and Parsons, H. (2007) *Family Farms, Survival and Prospect: A World-wide Analysis*, Routledge, Abingdon, UK

Cairns, M. F. (2007) (ed. and contrib.) *Voices from the Forest: Integrating Indigenous Knowledge into Sustainable Upland Farming*, Resources for the Future, Washington, DC

Cairns, M. (2009) *The Alder Managers: The Cultural Ecology of a Village in Nagaland, N.E. India*, PhD dissertation to the Australian National University, Canberra, available at *Digital Himalaya* (Rare Books and Manuscripts Collection) University of Cambridge and Yale University, http://www.digitalhimalaya.com/collections/rarebooks/, accessed 17 October 2019

Cairns, M. F., Keitzar, S. and Yaden, T. A. (2007) 'Shifting forests in northeast India: Management of *Alnus nepalensis* as an improved fallow in Nagaland', in M. F. Cairns (ed.) *Voices from the Forest: Integrating Indigenous Knowledge into Sustainable Upland Farming*, Resources for the Future, Washington, DC, pp.341–378

Colfer, C. J. P., Minarchek, R. D., Cairns. M. F., Aier, A., Doolittle, A., Mashman, V., Odame, H. H., Roberts, M., Robinson, K. and Van Esteerik, P. (2015) 'Gender analysis: Shifting cultivation and indigenous people', in M. F. Cairns (ed.) *Shifting Cultivation and Environmental Change: Indigenous People, Agriculture and Forest Conservation*, Earthscan from Routledge, London

Conelly, W. T. (1992) 'Agricultural intensification in a Philippine frontier community: Impact on labor efficiency and farm diversity', *Human Ecology* 20, pp.203-223

Cramb, R. A. (1989) 'The use and productivity of labour in shifting cultivation: An East Malaysian case study', *Agricultural Systems* 42, pp.209-226

Fox, J., Fujita, Y., Ngidang, D., Peluso, N., Potter, L., Sakuntaladewi, N., Sturgeon, J. and Thomas, D. (2009) 'Politics, political-economy, and swidden in Southeast Asia', *Human Ecology* 37, pp.305-322

Hunt, R. C. (2000) 'Labor productivity and agricultural development: Boserup revisited', *Human Ecology* 28, pp.251-277

Hutton, J. H. (1921) *The Angami Nagas, with some Notes on Neighbouring Tribes*, Macmillan, London (reprinted 1969, Oxford University Press, Bombay [Mumbai])

Keitzar, S. (1998) *Farmer Knowledge of Shifting Cultivation in Nagaland*, project report to the International Development Research Centre (IDRC), Canada

Keitzar, S. (2010) Personal communication between author Cairns and Dr Supong Keitzar, 6 Nov 2010

Leach, E. R. (1954) *The Political Systems of Highland Burma: A Study of Kachin Social Structure*, Athlone Press and Harvard University Press, London and Cambridge, MA

Menzies, N. (1996) 'The changing dynamics of shifting cultivation practices in upland southwest China', in B. Rerkasem (ed.) *Montane Mainland Southeast Asia in Transition*, Chiang Mai University, Chiang Mai, pp.51-68

Menzies, N. and Tapp, N. (2007) 'Fallow management in the borderlands of southwest China: The case of *Cunninghamia lanceolata*', in M. F. Cairns (ed.) *Voices from the Forest: Integrating Indigenous Knowledge into Sustainable Upland Farming*, Resources for the Future, Washington, DC, pp.425-434

Mertz, O. (2002) 'The relationship between fallow length and crop yields in shifting cultivation', *Agroforestry Systems* 55, pp149-159

Mertz, O., Wadley, R. L., Nielsen, U., Bruun, T. H., Pierce Colfer, C. J., de Neergaard, A., Jepsen, M. R., Martinussen, T., Zhao, Q., Noweg, G. T. and Magid, J. (2008) 'A fresh look at shifting cultivation: Fallow length an uncertain indicator of productivity', *Agricultural Systems* 96, pp.75-84

Mertz, O., Padoch, C., Fox, J., Cramb, R. A., Leisz, S. J., Lam, N. T. and Tran, D.V. (2009) 'Swidden change in Southeast Asia: Understanding causes and consequences', *Human Ecology* 37, pp.259-264

Meru, V. (2010) Personal communication between author Cairns and Dr. V. Meru of Khonoma village, 27 August 2010

Nielson, U., Mertz, O. and Noweg, G. T. (2006) 'The rationality of shifting cultivation systems: Labour productivity revisited', *Human Ecology* 34, pp.201-218

Padoch, C. (1985) 'Labor efficiency and intensity of land use in rice production: An example from Kalimantan', *Human Ecology* 13, pp.271-287

Padoch, C., Coffey, K., Mertz, O., Leisz, S., Fox, J. and Wadley, R. L. (2007) 'The demise of swidden in Southeast Asia? Local realities and regional ambiguities', *Geografisk Tidsskrift – Danish Journal of Geography* 107, pp.29-41

Ramakrishnan, P. S. (1992) *Shifting Agriculture and Sustainable Development: An Interdisciplinary Study from North-Eastern India*, UNESCO and Parthenon, Paris and Park Ridge, NJ

Rambo, A. T. (1996) 'The composite swiddening agroecosystem of the Tay ethnic minority of the northwestern mountains of Vietnam', in B. Rerkasem (ed.) *Montane Mainland Southeast Asia in Transition*, Chiang Mai University, Chiang Mai, pp.69-89

Rambo, A. T. (2007) 'Observations on the role of improved fallow management in swidden agricultural systems', in M. F. Cairns (ed.) *Voices from the Forest: Integrating Indigenous Knowledge into Sustainable Upland Farming*, Resources for the Future, Washington, DC, pp.780-801

Saul, J. (2005) *The Naga of Burma: Their Festivals, Customs and Way of Life*, Orchid Press, Bangkok

Schmidt-Vogt, D., Leisz, S. J., Mertz, O., Heinimann, A., Thiha, T., Messerli, P., Epprecht, M., Cu, P. V., Chi, V. K., Hardiono, M. and Dao, T. M. (2009) 'An assessment of trends in the extent of swidden in Southeast Asia', *Human Ecology* 37, pp.269-280

Scott, J. C. (2009) *The Art of Not Being Governed: An Anarchic History of Upland Southeast Asia*, Yale University Press, New Haven, CT, and London

The Economist (2010) 'Isolation ward: Blockade of a Northeast Indian state', vol 395, No 8683 (22 May 2010), p.32

von Fürer Haimendorf, C. (1946) *The Naked Nagas: Headhunters of Assam in Peace and War*, Thacker, Spink & Co, Calcutta

Yin, Shaoting (2001) (translated by Magnus Fiskesjö) *People and Forests: Yunnan Swidden Agriculture in Ecological Perspective*, Yunnan Education Publishing House, Kunming, China

Notes:

1. An extended period of fieldwork created a large body of data which had to be analyzed and presented in a PhD dissertation within the limited time allowed. There was also heavy editorial work on Cairns (2007). Not long after these tasks were completed, Cairns suffered a severe stroke for which, in 2022, he remains under treatment. Brookfield, who had been one of Cairns' principal supervisors, joined him in writing this paper in 2010.

2. Finding no consistent relationship between crop yield and the length of fallow in a range of systems, Mertz et al. (2008) concluded that swidden systems do not necessarily self-destruct when fallow periods are shortened. There remains scope for fallow-improvement innovations even at a late stage in the fallow-shortening process.

3. The principle is the same as with coppicing, except that coppiced trees are cut close to the ground. Both practices rejuvenate the tree. Some Angami alders are thought to be well over 100 years old.

4. The foundation stories place Khonoma's origin in the 14th century, when a group from western Burma came to the site from a previous location on the edge of Manipur. By this period, wet rice was being managed widely in incipient valley states throughout Southeast Asia and northeastern India. Manipur was one of several small valley states in the region.

5. The Naga independence movement split into several mutually hostile factions and some low-level violence continued. The question of Naga nationalism in relation to the modern state is far from resolved (Baruah, 2003; *The Economist,* 2010).

6. The agricultural history of Khonoma is a demonstration, not just a test, of the Boserup (1965) hypothesis regarding agricultural intensification driven by population pressure.

7. A combination of close farmer supervision as the diaries were being kept, followed by careful verification at the end of each cropping year, was aimed at maximizing data accuracy to the extent possible. Irregular group meetings were held throughout the fieldwork to discuss any problems and make needed adjustments.

8. Farmer-coordinators were trained in using a compass, altimeter and clinometer so that, together with a 30 m measuring tape, they were able to survey the 40 monitored fields quickly. After

adjustment, coordinates were imported into ARCinfo to create polygons representing each field. The boundaries were also entered into a GPS and overlain on electronic base maps of the villages.

9. The measures involved a number of local containers, the capacity of which had to be measured for conversion to kg. Results confirmed that this worked well in all the monitored fields.

10. This republican form of government would seem to correspond to the 'gumlao' polity of many Kachin in adjacent Burma, as described by Leach (1954). There seems never to have been a time when Angami villages had hereditary chiefs, but the Sema Nagas, to their northeast, did until recently. Angami political equality does not imply economic equality. In the past, wealthy individuals could acquire status by giving large merit feasts that were recorded by stone monoliths in the rice fields; some are to be seen in the background in Figure 22-9. Status has more recently been shown by prominent display of prestigious possessions.

11. US$1 = ₹47.78 in September 2001.

12. These modern rates downgrade the relative value of women's work. Before the British abolished slavery in the later 19th century, a female slave was worth two or three times as much as a male slave.

13. Tsiesema is a more recent village than Khonoma and lacks the oral-history detail that is remembered at Khonoma. As 'Chiswema', Hutton (1921) placed it within the 'Kohima group', the old Kohima village itself possibly being roughly contemporary with Khonoma. Over time, the early Angami villages spawned offshoot settlements, some of which grew large themselves.

14. Rice planting began first in the cooler ecology of the uppermost terraces and then 'rippled' down to lower altitudes where warmer temperatures nurtured faster rice growth.

15. As part of its field preparations, Tsiesema often had the added labour costs of repairing terraces and irrigation canals that had been damaged by landslips. Unlike Khonoma, Tsiesema did not have stone walls to stabilize the terraces and keep them from slipping down the slope. Repairing such slips was an annual chore and damage could be so bad and so frequent that terraces were abandoned.

16. Agrobiodiversity in the Khonoma terraces was highly variable, depending on the basket of resources that each farmer had at his disposal and his strategy for managing them. Some continued the old tradition of planting only rice, but other monitored fields were planted with up to a dozen crops in addition to the main rice crop. This was not happening in Tsiesema. The warmer terraces in Tsiesema continued to be planted primarily to rice, with occasional small amounts of maize, chillies or tomatoes. Some Tsiesema farmers who were unable to produce all their rice needs planted higher-priced glutinous rice (₹20/kg) with the intention of selling it in the Kohima market (for making rice beer) and using the revenue thus earned to buy the non-glutinous rice they needed at ₹12/kg. Many households also relied on sales of sticky rice to generate the money needed to pay their children's school fees. However, glutinous rice was grown at a cost. Yields were roughly 25% to 33% lower than those from non-glutinous varieties.

17. Throughout this discussion, a working day is taken to consist of eight hours. All data were recorded in hours.

18. Root crops were relatively more expensive to plant than cereals because the planting materials cost more per kg and were planted at more kg/ha. It is likely that higher planting costs prevented some poorer households from planting potatoes, especially if the seed potatoes had to be imported from Shimla, in Himachal Pradesh. While their more prosperous neighbours earned good returns from potato sales, the poorer sector remained stuck in a subsistence economy.

19. Inventories of crops grown in each subsystem suggested that the lower agrobiodiversity of Tsiesema's *jhums* might have been because they were planting some of these other crops in small permanent home gardens close to the village. Cairns and his collaborators found a total of 75 crop species in the permanent gardens sampled in Tsiesema (n = 10), compared to a more modest 42 in Khonoma (n = 9). Tsiesema managed these gardens relatively intensively, applying cattle dung and continually planting new crops as old ones were harvested. Villagers could harvest something to eat from their permanent gardens throughout the year, as well as feedstuffs to include in the mash that they boiled for their pigs every evening. Surpluses were sold.

20. Although Tsiesema did not manage its *jhum* fallows anywhere nearly as intensively as Khonoma, it nevertheless had less obvious ways of manipulating fallow vegetation to favour desired species. *Melia azedarach*, a valued timber species, provides one such example. As Tsiesema's *jhum* fields were left

fallow, free-ranging cattle followed behind to graze the crop residues and fallow regrowth. These cattle were particularly fond of eating any *Melia* fruit that they could find, and in this case, were playing a very important role in seed dispersal. Another vital advantage of this process was that passage through the digestive systems of cattle broke the normally dormant state of *Melia azedarach* seeds. When the land was later reopened for cultivation, the *Melia* seed germinated in tandem with the *jhum* crops and was then protected during subsequent weeding operations. With no extra labour costs, this valued species thus had a high frequency in the subsequent fallow regrowth, constituting a form of improved fallow in its own right.

21. These sample biases, which they indeed are, were to some degree a product of having had to select the farmers before the fields. In Khonoma, there was a more intentional bias toward farmers in one of the three major clans, in which the alder-*jhum* system was more highly developed than among the other two major clans.

22. The share of firewood sales in this productivity needs a brief further comment. Closeness to the state capital Kohima created a particularly strong market for firewood from Khonoma and Tsie-sema, but it was an important *jhum* product even in remote areas. Chingmei village in present-day Noklak district in eastern Nagaland (Figure 22-1), also grew alder in its *jhum* fields, its methods often compared to those of Khonoma, but at a lesser intensity. Informants there explained that when they opened a fallow, they earned more from the harvest of alder wood than from the food crops that were then planted. Alder firewood was sold at ₹200/stack, each stack measuring about 0.9 m x 1.2 m x 0.9 m. One respondent noted that beyond satisfying his own firewood needs, he was able to sell another 28 stacks, earning him ₹5600.

23. Farmers participating in the research were learning from and responding to data as they became available. Analysis of their diaries, for example, confirmed that planting potatoes and other vegetable crops was far more profitable than traditional staples such as Job's tears or millet. The following year (2002) witnessed a large upsurge in potatoes planted in *jhum* fields. Better information was improving their capacity to make informed decisions.

24. In explaining why *jhum* cultivation was valued, farmers frequently pointed to the many months over which successive crops were harvested. There was flexibility in the system. *Jhums* could be managed with a minimalist strategy – such as near monocultures of maize or Job's tears – in which labour inputs were few. Alternatively, they could be managed extremely intensively, incorporating a wide menu of crops interplanted in combinations of crop rotations, intercropping and relay-cropping patterns. This, in turn, also provided flexibility in producing either enough for home consumption only or extra for the market.

25. Conversion from local containers to kg, described in endnote 9, worked well on the monitored plots, but in this calculation, there were two additional sources of potential error. Reported total production was related to total terrace area owned, including the unmonitored plots. These latter plots were surveyed but, being generally of irregular shape, there may have been errors. In addition, some farmers might have exaggerated their remembered total production of wet rice. It was therefore decided to reject what were suspiciously high results.

26. A comprehensive village census in Khonoma in 2001 found more than 1000 fewer people than did the national census of the same year. The missing people were living and working elsewhere, but still gave Khonoma as their place of residence in the national census.

27. Stonemasonry, involving the chiselling of stone blocks from the numerous sandstone boulders in stream beds, is an ancient skill at Khonoma. It built the terrace walls and the formidable fortifications of the old village. At the time of fieldwork, it earned money by supplying the building boom in the nearby state capital, Kohima.

28. NEPED, or 'Nagaland Empowerment of People through Economic Development', was at the time of fieldwork, a Canadian-funded project concerned principally with the improvement of shifting cultivation from within. It was later supported by state and national funds. Beginning in 1995, it promoted the concepts underlying Khonoma's alder-*jhum* system elsewhere in Nagaland, and more recently, has been concerned with the stabilization of shifting cultivation by plantation crops, especially large cardamom (*Amomum subulatum*). It was through NEPED that Cairns first visited Khonoma in 1995.

29. Khonoma experienced this problem when, early in its conversion of *jhum* land to *A. subulatum* plantations, an unidentified disease swept through the introduced *A. subulatum*. Diseased planting material had been brought from Sikkim in the Himalayan middle hills.

30. The expansion of roads was central to these changes. In the case of Tsiesema, the response included construction of a new village away from the hilltop site and more accessible by road. In 2000-2002, Khonoma had daily bus services not only to Kohima, but also to the larger town of Dimapur on the edge of the Assam plains. The buses were heavily used for the transportation of low-volume produce for sale in the towns. Bulkier produce, including firewood, went by truck.

Himalayan alder (*Alnus nepalensis*
D. Don [Betulaceae])

23

SUCCESSFUL FARMING ON PRECIPITOUS SLOPES

A 170-year-old indigenous improved-fallow system at Naalad in the Philippines

*Peter D. Suson and Rodel D. Lasco**

Introduction

Agroforestry has long been practised in the Philippines by smallholder farmers, including indigenous peoples living in upland areas and practising traditional systems of shifting cultivation. With a rising population and an increasing ratio of people to land, some farmers have created innovations in the form of improved fallow systems, either to boost the speed of soil rejuvenation or to enhance the productivity of the fallow. One of the most notable examples of this is an improved fallow system at Naalad, on the island of Cebu in the Philippines, which not only does away with burning, but also drastically reduces the area required for fallowing, in order that the land can recover from cropping.

This chapter aims to show that the Naalad system was developed more than a century and a half ago as an adaptation to a steep and rugged environment. Moreover, it has never been a static system, but one upon which farmers have continually improvised. We will outline the reasons why the system continues to exist, but will also explain that current challenges may result in its imminent breakdown and abandonment. Its demise will not be the result of any inherent fault in the system, which still offers a technology capable of beneficial replication in many parts of Asia-Pacific.

The Naalad system

The Naalad improved fallow system is found in the village (*barangay*) of Naalad, part of Naga City, on the island of Cebu in the Philippines (Figure 23-1). Its most remarkable feature is that the cultivation and fallow phases are roughly of the same duration, achieving a cultivation-fallow ratio of 1:1, in contrast to a typical shifting

* PROFESSOR PETER D. SUSON, Mindanao State University-Iligan Institute of Technology, Philippines; and DR. RODEL D. LASCO, Philippine Country Director, World Agroforestry Centre (ICRAF).

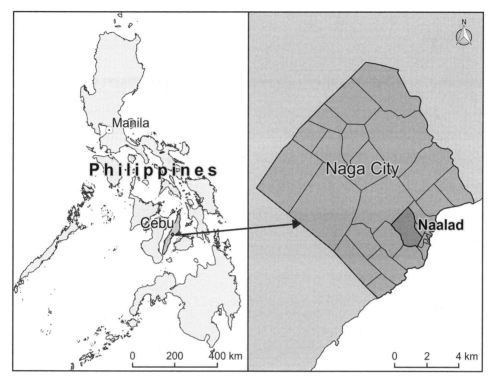

FIGURE 23-1:   The location of Naalad, in Naga City, Cebu, the Philippines.

cultivation system in which one year of cultivation would be followed by 15 to 25 years of fallow. Another outstanding aspect is that the Naalad system does not involve burning. The heavy cropping capacity of this fireless system is made possible by the use of a leguminous nitrogen-fixing tree, locally called *Ipil-ipil* (*Leucaena leucocephala*), as the exclusive fallow vegetation. This chapter will refer to the species simply as *Leucaena*.

At the end of the fallow, the *Leucaena* trees are slashed. But in contrast to traditional shifting cultivation, they are not burnt. Instead, the trees are felled and cut into stakes, poles, and smaller biomass. Stakes of *Leucaena* about 30 cm long are driven into the ground at regular intervals along the contours of the slope to be cultivated. Smaller poles and branches with a maximum diameter of about 2 cm are piled horizontally against the uphill side of the stakes and the remaining biomass is piled behind them as green manure. The main function of these structures is to control soil erosion. In fact, as sediment gathers behind them, small terraces are formed after a few years.

The fallow period in the Naalad system is about seven years, after which the plot is cleared and cropped for another seven years, with two crops per year. It then cycles back to fallow. The soil- and water-conservation structures that follow the contours of the cultivated slope are called *balabags* (Figure 23-2), which in the Visayan dialect means to obstruct or block. The *balabags* were originally spaced between one and two metres apart, based on horizontal distance, and crops were planted between them.

Leucaena leucocephala (Lam.) de Wit [Leguminosae]

The species that lies at the heart of the Naalad improved fallow system, *L. leucocephala* has been dubbed 'the miracle tree'. It is fast-growing, fixes nitrogen in the soil and provides nutritious animal fodder, along with generous quantities of firewood. However, it is also considered to be one of the tropical world's most invasive species, capable of smothering native vegetation.

But that has since changed. According to Lasco (2007), farmers say that in the early years of the system, up to five rows of maize could be planted in the alley between *balabags*. However, over the years the number of maize rows has been progressively reduced, as has the distance between the *balabags*, so that there are now only one or two rows of maize to each alley. The reason for this is that the maize plants nearest to the *balabags* reportedly grew better than the plants in the middle of the alleys, suggesting that as well as minimizing soil erosion, the *balabags* improved the quality of the soil (Lasco, 2007). This could be due to the nutrient contribution from decaying *balabags* as well as accumulation of more fertile sediments. A favourable microclimate around the *balabags* could also provide habitat for soil organisms such as earthworms, resulting in improved physical and chemical qualities.

The period of cultivation is usually seven years, because after that time the *balabag* structure deteriorates to a level where it can no longer effectively reduce soil erosion. Researchers first saw the decay of the dead *Leucaena* poles and branches as a basic flaw in the system. It was feared that with the collapse of the *balabags* there would be very high erosion rates, given the steepness of the slopes involved (Lasco, 2007). However, it was found that farmers were using the collapse of the *balabags* as a key indicator of when the field should be fallowed. In traditional

FIGURE 23-2: *Balabags*, the structures that retain soil and water and support cultivation until they deteriorate and fail after about seven years.

Photo: Peter Suson.

shifting cultivation, declining yields and weed problems are the main reasons for fallowing land and shifting elsewhere (Sanchez, 1976). In Naalad, it is possible that the decay and collapse of the *balabags* coincides with unacceptably low yields or weed problems (Lasco, 2007). So, after about seven years, the system shifts from cultivation to fallow. *Leucaena* seeds are sown into the fields, and surviving stumps are allowed to sprout (Lasco, 2007). While the shift to fallow usually occurs after seven years of cropping, it may happen after six or eight years, depending on the condition of the *balabags*.

Since the length of the fallow period is equal to the length of cultivation, farmers theoretically need only two parcels of land to make the system ecologically sustainable (Lasco, 2007). Another unusual aspect is that fallow and cultivated plots are usually adjacent to one another, minimizing the labour required in swapping cultivation for fallow, and vice versa (Figure 23-3).

History

According to a key informant in the study of the Naalad system (Perfecto Daclan), the improved fallow system was developed as a result of the displacement of farmers when, in the 1840s, during the Spanish colonial period (Juan, 1993), a rich gentleman named Don Pedro Cui claimed ownership of the flatlands of Naalad for sugar cane production. In those days it was a known practice of the government to grant vast tracts of land to a favoured subject. Don Pedro Cui was a very prominent figure: a street in Cebu City – the capital of Cebu – still bears his name.

The displaced farmers were pushed on to the steep slopes of Naalad, where more than 71% of the terrain is classified as hilly (Figure 23-4), with more than 77% of it having a slope of more than 18% (Tables 23-1 and 23-2).

Our key informant went on to claim that his grandfather was among the group of farmers that developed the improved fallow technology after observing that soil accumulated at a place where cleared vegetation had been dumped. This place was also clearly more fertile than its surroundings. In the light of those observations, they came up with the idea of building fences on contours across slopes to collect eroded soil.

Using a process of trial and error, they experimented with poles and stakes from seven different tree species, eventually finding that the locally named

FIGURE 23-3: A cultivated plot alongside a mature *Leucaena* fallow.

Photo: Malcolm Cairns.

Ipil-ipil (*Leucaena leucocephala*) was the most effective species for dominating the Naalad fallows and later for building the fences. It was known as a prolific generator of leaf litter, its foliage made good green manure, it grew rapidly and vigorously, and poles cut from the trees lasted longer and were slower to deteriorate than those of other species. Moreover, the foliage could be used as livestock fodder and the timber made good firewood.

The fallow species

The Naalad farmers were using a local variety of *Leucaena* called *kabaryo*, which is resistant to psyllid (jumping plant lice) infestation. This variety persisted when, during the late-1980s and early-1990s, such an infestation

FIGURE 23-4:   The farmers were pushed on to steep terrain. Their fields, secured by *balabags*, nestled within dense stands of fallowed *Leucaena*.

devastated huge areas of *Leucaena*. Because of its wide-ranging benefits and very rapid growth rate, *Leucaena* has been variously named 'the miracle tree' and 'the alfalfa of the tropics', and has been the subject of extensive scientific study. The value of *Leucaena* leaf litter as a fertilizer is widely recognized in tropical regions, and the species has been used for several decades to create contour hedgerows or live-barrier terraces for erosion control, but with widely recognized benefits in terms of soil fertility (Dijkman, 1950; Parera, 1983).

TABLE 23-1: Elevation classes at Naalad.

Elevation (m asl)	Area (Hectares)	% of total	% Lowland or hilly land
0–50	67.80	28.60	28.60
50–100	45.74	19.30	
100–150	43.77	18.46	71.40
150–200	45.98	19.39	
200–250	33.78	14.25	
Total	237.06	100	100

TABLE 23-2: Slope classes at Naalad.

Slope (%)	Description	Area (Hectares)	% of total	% flat or sloping
0–18	Level to rolling	53.70	22.65	22.65
18–30	Rolling to hilly	75.41	31.81	
30–50	Steep hills and mountains	74.45	31.41	77.35
>50	Very steep hills and mountains	33.50	14.13	
	Total	237.06	100	100

Leucaena leaves are fragile and decompose quickly, providing a very rapid, short-term influx of nutrients related to a low carbon-to-nitrogen ratio (Weeraratna, 1982). In most respects, *Leucaena* is also one of the highest-quality and most palatable livestock-fodder trees of the tropics (Jones, 1979, 1994). The leaf quality compares favourably with alfalfa (*Medicago sativa*) in feed value, except for its higher tannin content (Jones, 1979) and mimosine toxicity to non-ruminants (Bray, 1995). Nevertheless, *Leucaena* leaves have high nutritive value with high palatability, digestibility, intake and crude protein content, resulting in impressive animal production with 70% to 100% increases in animal live-weight gains compared with pure grass pastures (Shelton and Brewbaker, 1994; Jones, 1994).

As well as providing soil nutrients through generous litterfall, *Leucaena* also fixes atmospheric nitrogen in the soil via rhizobial bacteria in nodules that grow on the roots of the trees. Annual nitrogen accumulation rates of 600 kg/ha and as high as 1 t/ha have been recorded (Halliday and Samosegaran, 1983), but measurements of nitrogen-fixation rates vary widely.

For all of its benefits, there is a 'down-side' to *Leucaena*: it is regarded as one of the world's worst invasive species. It forms dense thickets and other vegetation is outcompeted, with a consequent reduction in species diversity (Weber, 2003). According to Henderson (2001) *Leucaena* is a potential transformer of habitats. The transformation of the Naalad slopes did not, in the first instance, arouse such concern. The crops grown between the *balabags* were originally – and still are – mainly maize and tobacco. The rural people of Cebu are one of the few groups in the Philippines whose staple food is maize (*Zea mays*), rather than rice. Maize has the ability to grow in extremely varying conditions, as evidenced by the fact that, in global terms, it is grown much more extensively than rice. On the other hand, the drier and relatively cooler upland conditions at Naalad made it a favourable environment for tobacco (*Nicotiana tabacum*).

Dynamism

It is clear that, although the Naalad farming system was originally created to enable farming of the area's steep slopes (Figure 23-5), the farmers have been continually active in fine tuning it. A major adjustment was the decreased distance between the *balabags* and the number of rows of crops in each alley.

Initially, they planted nine rows of crops behind each *balabag*. This was reduced to six rows and finally three rows or even just one. This meant the *balabags* were closer together (Figure 23-6). More *balabags* were needed, so adjustment of the system began to depend on the availability of poles.

One row of crops was preferred because the increased number of *balabags* effectively controlled soil erosion and the quality of soil in the alleys encouraged vigorous and uniform growth. This is observed to be the result of concentrating the deposition of eroded topsoil into narrower alleys and avoiding the creation of soil-fertility gradients in broader alleys where soil fertility is highest in the lowest parts of the alley where sediments gather.

As the increased number of *balabags* raised demand for wood in their construction, the farmers began to allow *Leucaena* trees to grow along the *balabag* rows to serve as braces for poles and to retain *Leucaena* biomass placed laterally across the slopes. In this way, they reduced the amount of wood needed for *balabag* construction.

Some farmers introduced a degree of precision to their construction of

FIGURE 23-5:   The steep Naalad slopes with cultivated plots interspersed with fallow plots of *Leucaena*.

Photo: Malcolm Cairns.

FIGURE 23-6:   Placing the *balabags* closer together improved soil and water conservation and led to more uniform crop growth.

Photo: Peter Suson.

balabags by using a carpenter's level to ensure that the structures followed the contours accurately. This extra effort made the control of erosion even more effective because perpendicular blocking substantially reduced the force of surface run-off and reduced the detachment of soil particles.

The farmers also added other cash crops to their system, notable among them: onions (*Allium cepa*) (Figure 23-7). As *Leucaena* fallows faced heavy pressure to meet the demands of *balabag* construction, the farmers experimented with other species, but their performance in controlling soil erosion was poorer than *Leucaena*. Nevertheless, the trial and error illustrated the farmers' keenness to adapt the system to meet changing circumstances.

Allium cepa L. [Amaryllidaceae]

Onions were a testimony to the high nutrient levels in the Naalad swiddens because they grow best in fertile soils. A biennial plant, onions are usually grown as an annual. They have been grown and selectively bred in cultivation for more than 7000 years and are now a common food crop around the world.

Incorporation of livestock

One of the biggest modifications of the system was the incorporation of livestock – particularly the raising of goats. The fallow plots of *Leucaena* took on a new purpose: as well as rejuvenating the soil and providing timber for the *balabags*, the foliage was used simultaneously as a source of fodder, on a cut-and-carry basis.

However, the livestock venture involved a considerable measure of compromise. The number of goats raised per household was limited to an average of six. Had this restriction not been imposed, the Naalad farmers would have begun raising too many goats, fodder harvesting would have intensified at the expense of soil rejuvenation and firewood production. Inordinate fodder harvesting would also have reduced the quantity of leaves to be used as green manure and lowered the amount of photosynthate translocated by the trees for production of wood biomass. Thus the production level of one product – goats – was compromised so as not to adversely affect the production of other products. Moreover, to compensate for the reduced nutrient-generation of *Leucaena* leaves, the farmers began carrying livestock manure back to the cultivated plots from the third year onward as a fertilizer to maintain soil fertility. Figure 23-8 shows the place occupied by the livestock component in the modified agroforestry system.

In a 1993 study, Lasco and Suson found that there was no significant difference between the soil chemical properties in the fallow and cultivated plots at Naalad. This raised the question of why there was a need to fallow the land when

FIGURE 23-7: Onions were introduced as a new crop, seen here growing between tobacco seedlings.

Photo: Peter Suson.

there was no significant difference in soil fertility between the two plots. Interviews with farmers revealed that cultivation was not abandoned because of reduced fertility, because livestock manure served as source of nutrients for the maize and tobacco crops from third year onwards. Rather, cultivation was abandoned because the *balabags* were deteriorating and without them the crops would literally be washed downslope by the rain.

Yield, income and profitability

A surprising aspect of the Naalad improved-fallow system is that despite the farm plots being located on very steep slopes, the yields per unit area are relatively high (Tables 23-3 and 23-4). The system yields multiple products. In a one-year period, products generated included: maize, tobacco, firewood, goats and to varying degrees other products such as fruit and vegetables.

Maize (Zea mays)

Maize yield was 3.1 tons/hectare (Table 23-5). This is more than 50% higher than the national average yield in the Philippines, which was 1.3 tons/hectare (DA, 1993). This was also prior to the advent of genetically modified maize varieties.

Tobacco (Nicotiana tabacum)

Tobacco yield on a per-hectare basis was 1.1 tons (Table 23-5). This was higher than the average yield in Cebu province, which was .65 t/ha, but lower than the average national yield of 1.3 t/ha (DA, 1993).

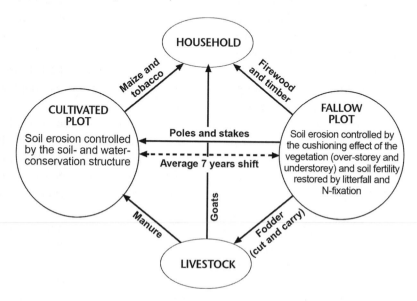

FIGURE 23-8: Interaction between elements of the Naalad agroforestry system.

TABLE 23-3: Cost of tobacco production at Naalad.

Inputs	Average cost per farm plot	Proportion (%)
Hired Labour	1057.58	38
Non–Hired Labour	1360.20	49
Fertilizer	164.40	6
Insecticide	171.33	6
Transport	31.00	1

Note: All costs are expressed in Philippine pesos (₱). Data were gathered in study interviews.

Firewood

Available data for firewood production did not reflect total productivity (Tables 23-4 and 23-5), because the amount of firewood harvested depended on family size. Although firewood is occasionally sold, the main purpose in harvesting it is to meet household needs. Furthermore, intensive harvesting of firewood is usually avoided because this could result in a shortage of materials for constructing *balabags* and indirectly influence both soil fertility and fodder production.

Goats

As mentioned earlier, goats are fed on *Leucaena* foliage harvested from fallow plots. Goat numbers are limited to about six per household out of concern that over-harvesting of fodder could jeopardize the productivity of all other elements of the system.

Profitability

Tables 23-4 and 23-5 show that, in 1993, the Naalad system was profitable. All of the produce from the system showed positive net income and percentages of returns on investment. The average return on investment for both average-sized farms and on a per-hectare basis was 149% per year (Tables 23-7 and 23-8 respectively). In other words, for every peso invested there was a corresponding return of ₱1.49 over one year. One of the reasons for the high average return on investment was the low level of production costs, indicating the cost effectiveness of the system.

At a practical level, there was a more important question than that related to returns on investment: Was the net income above or below the poverty threshold – the minimum level of income deemed adequate in provincial Cebu? Based on the average farm size, the farmers' income per year was ₱10,462.37. Since the size of cultivated plots per farmer ranged from 0.11 to 1.25 ha (Eslava, 1984), the annual income ranged from ₱3141 to ₱35,693 per year.

According to the Philippines' National Statistics and Census Office (now the Philippine Statistics Authority) the poverty threshold for a statistically average family of 5.4 people in provincial Cebu was ₱27,000 (NCSO, 1990). This meant that Naalad farms of average size (0.525 ha) or less provided an annual income that was below poverty level, and this was due purely to insufficient farm size. In fact, a farm size of one hectare was needed to provide an income that was just above the poverty threshold.

TABLE 23-4: Average production and profit values for an average-sized farm at Naalad, based on 1993 prices.

	Maize [a]	Tobacco [b]	Firewood [c]	Goats [d]	Others [e]
Yield [f]	567.7 kg per season	194.2 kg per season	215 cu. m per year or 150 bundles	6 head per year	N.A.
Commercial unit	Kilogram	Manos [g]	Bundle	Head	N.A.[h]
Price per commercial unit	9/kg [i]	N.A.[j]	7/bundle	700/head	N.A.
Gross income	5109.30	7219.20	1050	4200	500
Production cost	1013.61[k]	3722.44	438.48 [l]	1680[m]	315.00
Erosion control cost [n] (balabags)	223.30	223.30			
Net income [o]	3872.39	3273.46	611.52	2520	185.00
Return on investment [p] (%)	304.00	83.00	139.00	150.00	59.00

Notes: All monetary values are expressed in Philippine pesos. The average exchange rate in 1993 was US$1 = ₱28.05 (https://www.in-philippines.com/fx-rates-historic-from-1965-usd-php/).
[a] Maize yield = average number of maize packs per plot x average number of maize ears per pack x weight of kernels per ear.
[b] Tobacco yield = average number of manos (see g, below) per plot x average weight per manos.
[c] Firewood yield = Number of bundles per year x volume per bundle.
[d] Goats = average number of goats sold per year.
[e] This category includes minor products such as fruit and vegetables planted randomly in farm plots.
[f] This row shows the yield per average effective area of farm size, equal to 1832 sq. m. The effective area of farm size = total average farm size ÷ 2 – area occupied by balabags. The total average farm size = 0.525 ha, based on Eslava (1984). Division by two results from the farm size consisting of two plots, cultivated and fallowed. The area occupied by balabags = 793 sq. m.
[g] One manos consists of 100 tobacco leaves. The weight of a manos ranges from .65 to 1 kg.
[h] Not applicable because this involves composite data from minor products with different commercial units.
[i] Maize is planted for home consumption. To determine whether growing maize is a commercially viable enterprise it was necessary to compute it for profitability.
[j] Not applicable, because this value is an average of various prices that depend on tobacco leaf quality (see Table 23-3).
[k, l, m] No input costs other than labour (see Table 23-6 for maize)
[n] Total cost of erosion control is ₱1339.80 per cultivated plot. But since the cultivation period is six years, the cost of erosion control is distributed throughout the period. Hence, erosion control costs ₱223.30 per year.
[o] The total net income of the farming system is ₱10,462.37 for an average farm size of 0.525 ha.
[p] Return on investment (%) = Net income ÷ (Total Cost) x 100.

The wide range of farm sizes at Naalad reflects unequal land distribution. In fact, 70% of the farmers were tenants. Factors that contributed to the wide range of farm sizes included famers who began farming their plots earlier than the rest; distribution of farm lots to numerous children, leading to fragmentation across succeeding generations; and some farmers who struck financial difficulties and sold the rights to use their land.

TABLE 23-5: Average total production and profit values per hectare, based on 1993 prices.[a]

	Maize	Tobacco	Firewood	Goats [b]	Others
Yield	3.1 tons per season	1.1 tons per season	1174 cu. m per year or 819 bundles	6 head per year	N.A.
Commercial unit	Kilogram	Manos	Bundle	Head	N.A.
Price per commercial unit	9/kg	N.A.	7/bundle	700/head	N.A.
Gross income	27,900	39,406.10	5733	4,200	2729.25
Production cost	5532.80	20,319	2393.45	1680	1719.40
Erosion control cost (*balabags*)	1218.85	1218.85			
Net income [c]	21,148.35	17,868.25	3335.55	2,520	1009.85
Return on investment [d] (%)	313	183	240	250	159

Notes: [a] Values were derived from Table 23-4 using ratios and proportions. All monetary values are expressed in Philippine pesos (see Table 23-4 notes); [b] The number of goats in this case is independent of farm size. Hence, values remain the same, even on a per hectare basis; [c] Total net income for the farming system is ₱45,882; [d] Average return on investment for the farming system is 149%.

TABLE 23-6: Cost of maize production at Naalad.

Input	Average cost per farm plot	Proportion (%)
Hired labour	385.00	75
Non-hired labour	128.30	25

Note: Monetary values expressed in Philippine pesos.

TABLE 23-7: Annual and net income and average returns on investment for an average-sized farm at Naalad.

System products	Net income	Returns on investment (%)
Maize	3872.39	313
Tobacco	3273.46	83
Firewood	611.52	139
Goats	2520	150
Others	185.00	59
Total (annual)	10,462.37	
Monthly	871.86	
Average		148.80

Note: All monetary values expressed in Philippine pesos.

Much of Naalad is classified as timber land, even though the area no longer has forests. The classification is more legal in nature, in that land classified as timber land is owned by the state. According to the law, all of the farmers who were farming inside the area classified as timber land were squatters. However, this did not seem to affect their commitment to the land. This was reflected in the way most of them

TABLE 23-8: Annual and net income and average returns on investment for one hectare at Naalad.

System products	Net income	Returns on investment (%)
Maize	21,148.35	313
Tobacco	17,868.25	83
Firewood	3339.55	139
Goats	2520	150
Others	1009.85	59
Total (annual)	45,886	
Monthly	3823.83	
Average	1473.60	148.80

Note: All monetary values expressed in Philippine pesos.

practised a farming system that was concerned with conservation. In some instances, people who owned land let it to tenants. According to Eslava (1984), 70% of the farmers were tenants, 11% were both tenants and owner-operators and 19% were owner operators. The confidence found among farmers occupying plots inside the timber-land area could have reflected the level of tolerance shown by the Naga City Municipal Government, which was so clearly aware of the farms that it collected land taxes from the farmers.

Assessment

The fact that the Naalad agroforestry system has been in existence for more than 170 years is a testament to its sustainability. There are several reasons for this:

- It does not require a large area to sustain crop production. In fact it needs only two plots, one for fallow and the other for cultivation, and these roles are exchanged on average every seven years.
- The versatility of *Leucaena* as the dominant fallow vegetation. It is a nitrogen-fixing tree with heavy litterfall that produces green manure to restore soil fertility. The trees are easy to propagate and have a high reproductive capacity. They provide fodder for livestock, firewood and timber, as well as conserving soil and water on steep slopes.
- The system effectively controls soil erosion on very steep slopes. Under a *Leucaena* fallow, the dense overstorey foliage and the presence of undergrowth have a cushioning effect on heavy downpours. In the cultivation phase, soil erosion is controlled by the fence-like *balabags*, built across the contours of a slope to obstruct rainwater run-off. Modifications to the system have included building the *balabags* closer together, so that the alleys between them accommodate only one row of crops, thereby making erosion control more effective (Figure 23-6).
- As shown in Figure 23-6, it is a closed system, although an occasional external input such as insecticide is provided for tobacco. However, this accounts for only 6% of the total production cost (Table 23-3).

Zea mays L. [Poaceae]

Maize (or corn) was first domesticated by indigenous people in southern Mexico about 10,000 years ago. Total production now surpasses that of wheat and rice, although it is mostly used as animal feed, or to make ethanol. At Naalad, maize was the farmers' staple food, and between the *balabags*, its yields were 50% higher than the Philippine average.

- All elements of the system, including *Leucaena*, maize, tobacco and goats, reproduce easily, ensuring a ready supply of materials to sustain it; there is no need to purchase new materials for each new cycle. A clear example of this is the variety of maize used at Naalad. It is open pollinated and the same stock of seeds can be used from one generation to the next without compromising crop quality, as long as there is no substantial cross-pollination from other maize varieties. This is unlike hybrid varieties of maize, for which new seeds must be purchased for each and every new crop.

Challenges

For all of its effectiveness and success, the Naalad agroforestry system faces various challenges that may lead to its breakdown or abandonment by farmers. The following are among the more serious of them:

Labour input

One of the major drawbacks of the Naalad system is its very high demands for labour (Tables 23-3 and 23-6). Labour is either hired, contributed by the farmer and his family, or both. Consequently, labour is a potential source of constraint, especially when family size is small and money for hiring outside labour is scarce. These constraints are mitigated by male farmers working in the city and using part of their earnings to hire labour. This set-up is a viable option because labour costs at village level are half of those in the city. However, in the absence of male farm labour due to outside employment, most farm activities are left to women, when traditionally this was not their role. Furthermore, in the absence of employment opportunities in the city, only those with a large household labour force or an ability to hire labour can continue to meet the labour demands of the farming system. Others might attempt

Nicotiana tabacum L.
[Solanaceae]

Tobacco provided only moderate returns on investment when compared with other aspects of the Naalad system. *N. tabacum* is a herbaceous plant found only in cultivation, and is believed to be a hybrid of several other *Nicotiana* species. It is commercially cultivated around the world for its leaves, which are processed into tobacco.

to modify the system to reduce the need for labour or seek new technical alternatives to reduce labour inputs, but these efforts generally mean a steep reduction in the quality of work. In other words, in the absence of employment opportunities in the city, the farming system becomes unfeasible for many farmers.

Human predation

Villagers are reported to have over-harvested fodder from the fallowed plots, and as a result the *Leucaena* trees have become so weakened that other species have begun to take over the fallows. Contributing to this problem was the introduction of an improved cattle breed that was seen as a substitute for goats. The cattle were far more demanding of fodder than the goats, and species succession in the weakened *Leucaena* fallows began to threaten the whole system. The cultivated plots were also assailed by villagers who were unable to find enough firewood in the fallows, so they pilfered firewood from the *balabags*. Both activities threatened the farming system with adverse effects – specifically, substantial loss of nutrients and soil.

Sparse rainfall

This problem constantly cropped up in study interviews. According to the farmers, insufficient rainfall results in delays to planting of crops, which in turn exposes the crops to insect infestation, reduces the usual two crops per year to just one, and affects the growth of *Leucaena*, thereby reducing the availability of fodder and firewood and hindering Leucaena's effectiveness as a soil rejuvenator.

A 1992 study by Boling et al. found that the Cebu area had suffered what the authors called 'a significant climate change' in the middle of the 20th century, after the Naalad system had been in existence for about one hundred years. The area has what is known as a 'type three' climate, out of four climate types felt in the

Philippines. This is described as a tropical monsoon climate in which seasons are not very pronounced, with a relatively dry period from November to April and wet during the rest of the year. Annual rainfall should measure between 1600 mm and 2000 mm. In the period from 1901 to 1940, rainfall at Naalad generally fell within that range. But in the 40 years following this, from 1948 to 1988, the relatively dry months became longer and, according to Boling et al., about 70% of the rainfalls measured in those years were at the lower end of the type three range, or below it (Table 23-9). The researchers were unsure of what had caused the lower rainfall, but its result had been a decrease in farming and an increase in other subsistence activities. Some farmers even spoke of farming being reduced to a mere sideline.

The outlook for the Naalad system

Recent visits to the Naalad area have revealed that a landscape once alternating between patches of green fallow and brown plots of cultivated land (Figure 23-9) is now a virtually unbroken stand of *Leucaena* (Figure 23-10). Prior to the new millennium, more than 100 people were reportedly involved in the Naalad system of farming. That number has now fallen to just 10 farmers. Across most of the hillsides, *Leucaena* has risen above its supportive role and lived up to its notoriety as an invasive species by smothering all evidence of the cultivated plots and their *balabags*. In time, it is hoped that the permanently fallowed hillsides will revert to forest.

TABLE 23-9: Annual rainfall figures for Cebu, 1965 to 1991.

Year	Total rainfall (mm)	Comparison with 'type 3' climate rainfall range
1965	1528.80	<1600 mm
1970	1491.60	<1600 mm
1975	1887.60	Upper limit of the range
1976	1728.00	Lower limit of the range
1977	1519.20	<1600 mm
1978	1694.00	Lower limit of the range
1979	1430.40	< 1600 mm
1980	2186.40	>2000 mm
1981	1272.00	<1600 mm
1982	1104.00	<1600 mm
1983	1694.00	Lower limit of the range
1984	1732.80	Lower limit of the range
1985	1750.40	Lower limit of the range
1986	1696.80	Lower limit of the range
1987	994.60	<1600 mm
1988	1378.80	<1600 mm
1989	2352.00	>2000 mm
1990	1839.60	Upper limit of the range
1991	1473.60	<1600 mm

Note: Rainfall range for type 3 climate is 1600 to 2000 mm per annum.
Source: NSO (1992).

What appears to be an imminent end to a successful innovation in shifting cultivation is in no way indicative of an inherent weakness in the system itself. Rather, it has to do with the small and fragmented size of the farming plots and a rapidly developing neighbourhood. Naga City, of which Naalad is a part, is now known as the 'industrial city of the south'. It has two large power plants and an equally large cement factory, and all of them are expanding. Construction is in full swing, as is the development of a full range of commercial activities. The result has been a generous range of employment opportunities for the farmers of Naalad, who were struggling to rise above poverty on farm plots that were too small. The enhanced economic status of the villagers is now obvious in the improved quality of their homes.

It is believed that the economic trend will continue, and the remaining hillsides retained by *balabags* are expected soon to become a part of indigenous agricultural history. Indeed, the demise is expected by Naalad village officials, who have made a special feature of the Naalad improved fallow system and placed it on display in their museum.

The Naalad system was developed by indigenous farmers deprived of their land and in desperate need of food security for their families. For more than 170 years, it not only served that purpose, but from precipitous slopes it also out-performed farms on lowland plains, with little or no outside inputs. The improved fallow system may soon disappear from the hills of Cebu province in the Philippines, but *Leucaena* grows readily in most tropical locations within 15 to 25 degrees north or south of the equator. Within that area there is a widespread need for a farming system that uses only two plots of land, delivers two crops every year, supports livestock, provides firewood, and has an impressive proven record.

FIGURE 23-9: Cultivated plots scattered around the steep slopes of Naalad at the height of the improved fallow system.

Photo: Peter Suson.

FIGURE 23-10: The same area today. An unbroken *Leucaena* canopy hides any evidence of the farming system.

Photo: Peter Suson.

References

Boling, A. A., Garrity, D. P., Franco, D. T. and Ramirez, A. M. (1992) 'Historical trends in rainfall distribution in the Philippines', *Philippines Journal of Crop Science* 16(3)

Bray, R. A. (1995) 'Possibilities for developing low mimosine lines', in H. M. Shelton, C. M. Piggin and J. L. Brewbaker (eds) *Leucaena: Opportunities and Limitations*. Proceedings of a Workshop, Bogor, Indonesia, ACIAR Proceedings no. 57, Australian Centre for International Agricultural Research, Canberra, pp.119-124

Dijkman, M. J. (1950) '*Leucaena*: A promising soil-erosion-control plant', *Economic Botany* 4(4), pp.337-349

DA (1993) *Annual Report*, Department of Agriculture, Manila

Eslava, F. M. (1984) 'The Naalad style of upland farming in Naga, Cebu, Philippines: A case of an indigenous agroforestry scheme', country report submitted by Filipino participants to an ICRAF course in agroforestry, 1 to 20 October, 1984, University of Pertanian, Malaysia

Halliday, J. and Somasegaran, P. (1983) 'Nodulation, nitrogen fixation, and Rhizobium strain affinities in the genus Leucaena'. in *Leucaena Research in the Asian-Pacific Region*, proceedings of a workshop, Singapore, International Development Research Centre (IDRC), Ottawa, Canada

Henderson, L. (2001) *Alien Weeds and Invasive Plants*, Plant Protection Research Institute Handbook no. 12, Paarl Printers, Cape Town, South Africa

Jones. R. J. (1979) 'The value of *Leucaena leucocephala* as a feed for ruminants in the tropics', *World Animal Review* 31, pp.13-23

Jones, R. J. (1994) 'Management of anti-nutritive factors – with special reference to Leucaena', in R. C. Gutteridge and H. M. Shelton (eds) *Forage Tree Legumes in Tropical Agriculture*, CAB International, Wallingford, UK, pp.216-231

Juan, M. C. (1993) personal communication between author Suson and Maria Cristina Juan, who was then a member of a faculty at the University of the Philippines, Visayas

Lasco, R. D. (2007) 'The Naalad improved fallow system in the Philippines and its implications for global warming', in M. F. Cairns (ed.) *Voices from the Forest: Integrating Indigenous Knowledge into Sustainable Upland Farming*, Resources for the Future Press, Washington, DC

Lasco, R. D and Suson, P. D. (1993) 'On-farm evaluation of the Naalad indigenous agroforestry system: Preliminary findings', paper presented at an international symposium on multi-purpose tree species, Manila

NCSO (1990) *Annual Report*, National Census and Statistics Office, Manila

NSO (1992) *1991 Statistical Yearbook*, National Statistics Office, Manila

Parera, V. (1983) '*Leucaena* for erosion control and green manure', *Leucaena Research in the Asian-Pacific Region*, proceedings of a workshop, Singapore, International Development Research Centre (IDRC), Ottawa, Canada

Sanchez, P. A. (1976) *Properties and Management of Soils in the Tropics*, John Wiley and Sons, New York

Shelton, H. M. and Brewbaker, J. L. (1994) '*Leucaena leucocephala*: The most widely used forage tree legume', in R. C. Gutteridge and H. M. Shelton (eds) *Forage Tree Legumes in Tropical Agriculture*, CAB International, Wallingford, UK, pp.15-29

Weber, E. (2003) *Invasive Plant Species of the World: A Reference Guide to Environmental Weeds*, CAB International, Wallingford, UK

Weeraratna, C. S. (1982) Nitrogen release during decomposition of Leucaena leaves, *Leucaena Research Reports* 3, p.54

24

MAINTAINING SOIL FERTILTY AND SUSTAINABLE SWIDDEN CULTIVATION WITH *'ALBIZIA'* TREES

An innovation of the Outer Baduy community, south Banten, Indonesia

*Johan Iskandar**

Introduction

Cultivation of rice in Indonesia can be divided into two main systems: swidden farming or shifting cultivation (*huma* in Sundanese or *ladang* in Indonesian) and wet-rice farming (*sawah*). In the past, the predominant practice of the rural people of West Java and Banten – the westernmost province of Java – was swidden, rather than wet-rice farming. The cultivation of wet rice came to the Banten area in 1520, coinciding with the establishment of the sultanate of Banten (Iskandar, 1998). It was more than two centuries later, in 1750, when wet-rice farming first spread from Central to West Java (Geertz, 1963, p.44).

Nowadays, most farmers in West Java and the Banten area have ceased swidden farming and turned to wet-rice cultivation, so swiddening has all but disappeared. Notable exceptions are the Baduy community of Banten and Kasepuhan of Sukabumi, both in West Java, where traditional swidden farming continues (Iskandar, 1998). The Baduy community of Kanekes village, in Leuwidamar subdistrict, Lebak district, in the southern part of Banten province, not only continues traditional shifting cultivation, but under the management of the village farmers, the practice also flourishes. These people have been able to maintain a swiddening system with a wide variety of crops through the innovative management of soil fertility and

An earlier version of this chapter appeared in *Biodiversitas* 19(2), pp.453-464, March 2018.

* Dr. Johan Iskandar is Professor of Ethnobiology in the Department of Biology, Faculty of Mathematics and Natural Sciences and School of Postgraduate Studies on Environmental Science, Universitas Padjadjaran, Bandung, Indonesia. He is also a Senior Researcher at Universitas Padjadjaran's Centre for Environmental and Sustainability Science.

economically productive management of both swidden and fallow land, including the integration of *albizia* (*Paraserianthes falcataria*) trees into their swiddens.[1]

Culturally, the Baduy people can be divided into two groups: 'Inner Baduy' (*Urang Baduy Dalam* or *Urang Baduy Jero*) and 'Outer Baduy' (*Urang Baduy Luar* or *Urang Baduy Panamping*). The Inner Baduy people are considered to be stronger upholders of tradition than those of Outer Baduy. The Baduy territory of Kenekes village (*desa*) can be divided in to three main areas: the 'Inner Baduy area' (*Daerah Baduy Dalam*), the 'Outer Baduy area' (*Daerah Baduy Luar* or *Daerah Panamping*), and the '*Dangka* area' (*Daerah Dangka*). The Inner Baduy area, occupied by the Inner Baduy people, consists of three hamlets (*kampung*): Cibeo, Cikartawarna and Cikeusik. There are 56 hamlets in the larger Outer Baduy area, prominent among them being Cipondoh, Kaduketug, Gajeboh, Marengo, Kadujangkung, Kadukter, Cihulu and Cicakal Muhara. The *Dangka* area is adjacent territory owned by members of Indonesia's Muslim majority. It comprises the hamlets of Kamancing, Cihandam, Cibengkung and Garehong, in which reside the informal leaders of the Baduy community (Figure 24-1).

The practice of swidden cultivation is considered by the Baduy as an obligation of their religion. They believe that their territory is sacred land that must be specially managed, while the land of the neighbouring Muslim majority is considered to be non-sacred, profane, i.e., not subject to Baduy religious strictures.

According to the local regulations of Lebak district, the Baduy area of Kanekes village covers 5136.58 hectares. Of this, about 3000 hectares of land is designated as protection forest, while 2136.58 hectares is used for agriculture and settlements (Kurnia and Sahabudin, 2010). In 2017, the total population of the Baduy community was 11,699 people, living in 3413 households. Of this population, 89.64%, or 10,488 people, were Outer Baduy and the remaining 10.36%, or 1211 people, were Inner Baduy (Statistics of Kanekes village, 2017).

Swidden cultivation has long been, and still remains, the main source of

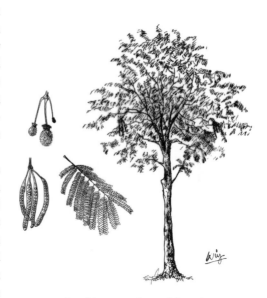

Parkia speciosa Hassk.
[Leguminosae]

This tree legume is highly valued for its edible beans, known as *petai* in Indonesia. It is grown in both Baduy swiddens and fallows as a soil improver that also provides a commercial product to generate household income. The long, flat seed pods are often known as stink beans.

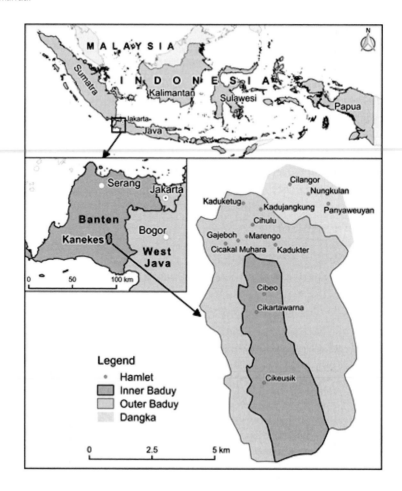

FIGURE 24-1: The long-standing borders of Baduy territory, approximately based on Kanekes village in Leuwidamar subdistrict, Lebak district, in Indonesia's Banten province. The three areas that are central to this study – Inner Baduy, Outer Baduy and the Dangka area – are shown.

Source: Adapted from Iskandar and Ellen, 2000.

Baduy subsistence. Some Baduy men are also involved in making traditional woven bags from cloth made from the bark of young *teureup* (*Artocarpus elasticus*) trees, tanned with *gintung* (*Bischofia javanica*) bark or *salam* (*Syzigium polyanthum*). These handicrafts are produced for personal use, but may also be sold to visitors or small shops in their villages and neighbouring areas. The men may also be involved in petty trading of non-rice products such as durian (*Durio zibethinus*), bananas (*Musa x paradisiaca*), *petai* (*Parkia speciosa*), and brown sugar from *Arenga pinnata* palms. Meanwhile, the women, particularly Outer Baduy women, are involved in making traditional woven cloth, and their products are sold to visitors.

Today, the area of forest in Baduy territory has decreased, the population has increased and rural areas such as Kanekes village have been targeted by very intensive market-economy penetration. Despite all of this, the Baduy community still practices intensive shiftmg cultivation, based on their traditional ecological knowledge, beliefs and world views (cf. Conklin, 1957; Iskandar, 1998; Berkes, 1999; Toledo, 2002). The Outer Baduy people, in particular, have been able to maintain a system of swiddening that produces a variety of crops through innovative management of both soil fertility and the productivity of their swidden and fallow land. Prominent among their innovations is the integration of leguminous *albizia* (*Paraserianthes falcataria*) trees into their farming system.[1]

Traditional management of swidden farming by the Outer Baduy

This chapter discusses strategies undertaken by Outer Baduy farmers to maintain the soil fertility and the sustainability of their swidden systems, and in particular three innovative practices: the cultivation of a high diversity of crops; the planting of legumes; and the use of economic *albizia* trees in the cultivation and fallow cycle.

Cultivating a high diversity of crops

Five types of anthropogenic land use have been identified in the Outer Baduy area. These are swidden fields (*huma*); secondary forest on fallow land (*reuma*); temporary gardening of annual crops after the initial rice crop has been harvested and before a swidden is left fallow (*kebon*); perennial mixed gardens (*kebon campuran*); and hamlet forests (*dukuh lembur*). Of these, the *huma* and *reuma* have seen dynamic change. All Outer Baduy households usually cultivate upland rice and other annual crops in a swidden field (*huma*) every year. After a short period of cultivation, each swidden plot is fallowed for several years before being opened again for cropping. The compulsion to practice traditional shifting cultivation comes from their religion, which is called *Sunda Wiwitan* (original Sundanese) or *Agama Baduy* (Baduy religion), and which regards swidden farming as a duty. Therefore, before the annual swidden farming calendar begins, each Outer Baduy household must find a plot of fallowed land with mature secondary forest. Generally, in order to open a new swidden, the farmers prefer land that has been fallow for four years, since this is the current maximum period for which most of the land in their territory is fallowed. In order to overcome shortages of available land in which to open a new swidden, the Outer Baduy farmers (unlike the Inner Baduy people) have developed a cultural strategy in which they may leave their home territory to find suitable fallowed land. They undertake 'temporary out-migration' to areas adjacent to Baduy territory that are owned by Muslim-majority Indonesians, where they can obtain fallowed land by providing labour, renting or sharecropping (Iskandar et al., 2018). There, they make a farm house and practise their traditional shifting cultivation. The time that they remain in non-Baduy territory is determined by the availability of appropriate fallowed land that is ready for clearing and new cultivation in their home area.

At the same time every year, the households of Outer Baduy that have claimed appropriate fallowed land in their village area make preparations to begin the swidden cycle. First, a ritual is performed that seeks permission from the spirits of the forest (*narawas*) to undertake swidden farming on the plot, allowing the family to cultivate in peace. Then, the underbrush growing on the fallowed land is cut over an area of about half a hectare. This is usually done in the month of *Kanem* (June – July). Two months later, in the month of *Kadalapan* (August – September), selected trees are cut, while others with economic value, such as arenga palm (*Arenga pinnata*), petai (*Parkia speciosa*), jengkol (*Achidendron pauciflorum*) and durian (*Durio zibethnus*) are pruned of branches. Some of the cut branches are collected for firewood and the rest of the debris is left to dry in the sun. The dry debris is collected, made into piles, burnt (*ngahuru*) and re-burnt (*ngaduruk*) to produce ash, which is used as an organic fertilizer to supplement nutrients in the topsoil for growing rice and other annual crops. By tradition, the Outer Baduy are not permitted to use inorganic fertilizers.

After the debris is burnt, the rice and other annual crops are planted before the rainy season begins in the month of *Kasalapan* (September – October). However, before this, the ritual of planting rice (*upacara ngaseuk*) is performed in the centre of the swidden field. The ritual is usually performed by the head of the household in what is regarded as a sacred place, called *pungpuhunan*, in the centre of the swidden. It measures about one metre by one metre, and is bordered by *barahulu* (*Amomum* sp.) stems. The ritual begins when the head of the household enters the *pungpuhunan* and sits down in front of a woven basket containing sacred rice grains. For each swidden plot that is at least half a hectare in size, three sacred rice varieties (landraces) must be planted in separate parts of the swidden. These varieties, called *pare koneng, pare siang* and *pare ketan langasari*, are sown at the centre, east and west of the field, respectively. Traditional practice demands that there must be no contact between the three varieties, as a measure to maintain their purity. Non-sacred rice varieties are planted along the boundaries of the sacred varieties in order to ensure their separation. This

Archidendron pauciflorum (Benth.)
I. C. Nielsen [Leguminosae]

This tree legume has commercial value for its edible beans and seeds, and it is regularly left standing but heavily pruned when Baduy swiddens are opened. The seeds contain high amounts of protein and are low in fat. Although they are mildly toxic, the seeds are often used in traditional medicine throughout the region.

traditional prohibition has an important practical function in maintaining the genetic distinctiveness of different varieties, by ensuring that each sacred variety can be harvested separately and its purity more easily guaranteed (cf. Richards, 1992; Cotton, 1996, p.246) (Figure 24-2). Thus, the rice diversity of Outer Baduy swidden farming is higher than that of wet-rice farming by Muslim farmers in paddies neighbouring the Baduy area. Based on surveys, 89 local rice varieties (landraces) are

FIGURE 24-2: The rice diversity of Outer Baduy swidden farming is higher than that in nearby wet-rice systems.

Photo: Johan Iskandar.

customarily cultivated by both Inner Baduy and Outer Baduy communities. These varieties are traditionally classified according to their maturation period, grain phenotypes, straw and panicle colour, culinary qualities and sacredness (Iskandar and Ellen, 1999). The number of rice varieties in nearby wet-rice paddies is lower because of the modernization of wet-rice farming in a Green Revolution programme introduced by the Indonesian government in the early 1970s. Farmers were told that they had to cultivate high-yielding varieties developed by the International Rice Research Institute in the Philippines. The negative impact of the Green Revolution programme was the mass extinction of local rice varieties. For example, ethnobotanical research has shown that about 88 varieties of rice were grown in wet-rice systems in villages of Majalaya subdistrict, Bandung district, West Java, before the Green Revolution programme. Following implementation of the programme, fewer than 10 varieties remained in the same area. In other areas of West Java, including the Rancakalong area of Sumedang district, 60 rice varieties were recorded before the Green Revolution programme. Later, only 20 varieties could be found (Iskandar and Iskandar, 2018).

In addition to planting rice varieties, the swidden farmers of Outer Baduy also plant at least 19 species of annual non-rice crops. These usually include *hiris* (*Cajanus cajan*), *hanjeli* (*Coix lacryma-jobi*), *taleus* (*Colocasia esculenta*), *hui manis* (*Dioscorea alata*), *mantang* (*Ipomoea batatas*) and *singkong* (*Lablab purpureus*). The seeds of these annual crops are mixed with rice seeds before planting, so that even among the non-rice varieties, the Outer Baduy swiddens have a high diversity (Table 24-1).

TABLE 24-1: Various annual non-rice crops commonly planted in the swiddens of Outer Baduy.

Species	Family	Baduy/ Sundanese name	Landraces	Cropping system*
Pigeon pea (*Cajanus cajan*)	Leguminosae	*hiris*		Planted in the same holes as rice.
Job's Tears (*Coix lacryma-jobi*)	Poaceae	*hanjeli*		Regularly planted along the swidden edges.
Taro (*Colocasia esculenta*)	Araceae	*taleus, talas*	*taleus manis, taleus ketan, taleus hejo*	Planted along with rice.
Cucumber (*Cucumis sativus*)	Cucurbitaceae	*bonteng, mentimun*		Seed mixed with rice before planting.
Butternut squash (*Cucurbita moschata*)	Cucurbitaceae	*waluh*		Seed mixed with rice before planting.
Purple yam (*Dioscorea alata*)	Dioscoreaceae	*hui manis*	*hui kalapa, hui ramo, hui ketan*	Planted so they can creep up nearby trees.
Sweet potato (*Ipomoea batatas*)	Convolvulaceae	*mantang, boled, ubi jalar*	*mantang taropong, mantang coklat, mantang hideung, mantang sinden*	Planted along with rice.
Hyacinth bean (*Lablab purpureus*)	Leguminosae	*roay*		Planted so they can creep up nearby trees.
Cassava (*Manihot esculenta*)	Euphorbiaceae	*dangdeur, sampeu, singkong*	*sampeu nangka, sampeu roti, sampeu koneng/ sampeu mantega sampeu racun*	Planted along with rice.
Velvet bean (*Mucuna pruriens*)	Leguminosae	*kepes*		Planted along with rice.

TABLE 24-1 (cont.): Various annual non-rice crops commonly planted in the swiddens of Outer Baduy.

Species	Family	Baduy/ Sundanese name	Landraces	Cropping system[*]
Banana (*Musa x paradisiaca*)	Musaceae	*cau, pisang*	*cau astroli, cau emas, cau beuleum, cau galek, cau gejloh, cau klutuk, cau muli, cau nangka, cau papan, cau raja, cau raja bulu, cau raja cere, cau rejang, cau tanduk, dan cau uli/ketan*	Planted along with rice.
Wild foxtail millet (*Setaria viridis*)	Poaceae	*degeng*		Planted in the same holes as rice.
Common bean (*Phaseolus vulgaris*)	Leguminosae	*kacang merah*		Planted along with rice.
Chinese potato (*Plectranthus rotundifolius*)	Lamiaceae	*kumili*		Planted along with rice.
Winged bean (*Psophocarpus tetragonolobus*)	Leguminosae	*jaat*		Planted along with rice.
Wild sugar cane (*Saccharum spontaneum* var. *edulis*)	Poaceae	*tiwu endog, trubus*		Planted along with rice.
Sugar cane (*Saccharum officinarum*)	Poaceae	*tiwu, tebu*		Planted along with rice.
Cowpea (*Vigna unguiculata*)	Leguminosae	*kacang penyut*	*jagong coklat, jagong manis*	Planted in the same holes as rice.
Maize (*Zea mays*)	Poaceae	*jagong, jagung*		Planted along with rice.

Note: [*]All of these crops are cultivated together with rice in a multiple cropping system.

On this basis, it can be inferred that the swidden systems of Outer Baduy, as well as having secured food supplies for the farming communities (Soedjito, 2015, p.438), have played an important role in conserving various rice and non-rice crop varieties. Throughout the year, the Outer Baduy swiddens produce a wide variety of food crops that are resistant to pests and which provide a buffer against market fluctuations (Aryal and Choudhury, 2015, pp.282-283). It may also be argued that this swidden farming system has been important for enhancing productivity, crop resilience, stability and agroecosystem functions (cf. Sajise, 2015, p.404).

Various legumes, including *hiris* (*Cajanus cajan*), *roay* (*Lablab purpureus*), *kacang merah* (*Phaseolus vulgaris*), *kepes* (*Mucuna

Cajanus cajan (L.) Millsp. [Leguminosae]

Pigeon pea is one of the world's oldest food crops, having been cultivated for about 3,500 years. Baduy farmers plant *C. cajan* seeds in the same holes as those for rice, allowing this perennial legume to grow along with a wide range of other crops.

pruriens), *jaat* (*Psophocarpus tetragonolobus*) and *kacang penyut* (*Vigna ungulata*) are regularly planted along with the rice – the perennials among them treated like annual crops. In addition, some perennial leguminous trees are commonly planted in the swiddens, including *jengkol* (*Achidendron pauciflorum*), *jeungjing* (*Albizia chinensis*), *kitoke* (*Albizia lebbeck*) and *petai* (*Parkia speciosa*) (Table 24-2).

TABLE 24-2: Leguminous species commonly planted by Outer Baduy farmers in both swiddens (*huma*) and fallowed land (*reuma*).

Species	Baduy name	Uses
Achidendron pauciflorum	*jengkol*	Vegetable
Albizia chinensis	*jeungjing*	Firewood, building material
Albizia lebbeck	*kitoke*	Firewood, building material
Erythrina variegata	*dadap*	Support for pepper vine (*rinu*), canopy for shading
Flemingia lineata	*kitambaga*	Green manure
Paraserianthes falcataria	*albasiah, kalabise*	Firewood, building material, petty trading
Parkia biglobosa	*peundeuy*	Medicine, handicrafts
Parkia speciosa	*petai*	Vegetable
Pterocarpus indicus	*angsana*	Firewood

Perennial legume trees are mostly found in swiddens because they have been grown on fallowed land (*reuma*) that is reverting to secondary forest. They are not cut down when swiddens are opened, but the branches are pruned as the land is prepared for cultivation. As a result, the perennial legume trees that grow in Outer Baduy swidden systems play an important role in providing food, building materials, firewood and green manure, which is an inexpensive source of organic fertilizer to build up or maintain soil organic matter and fertility (cf. Reijntjes et al., 1992, p.168). These traditional practices of Outer Baduy farmers, including the planting of various annual legumes mixed with the main crop of rice,

Albizia chinensis (Osbeck) Merr. [Leguminosae]

This large evergreen is another legume species planted in their swiddens by Baduy farmers. It can grow more than 40 metres tall and as well as fixing nitrogen in the recovering soil and providing timber and firewood, it offers fodder for livestock.

are considered to be ecologically sound, particularly when their traditions forbid the use of inorganic fertilizers in their swiddens. In order to maintain soil fertility, the legume crops are applied intensively as a form of green manure. The Outer Baduy swidden systems are thus believed to be economically viable, because they not only produce enough food for self-sufficiency, but various commercial legumes, including *petai* (*Parkia speciosa*) and *jengkol* (*Achidendron pauciflorum*) also produce surpluses that can generate cash income.

Planting various legumes

Each Outer Baduy farming household cultivates a swidden every year, preferably in the Baduy area when suitable fallowed land is available. Secondary forest (*reuma*) that has been fallowed for at least four years is usually considered suitable for planting rice and annual crops. The swiddens usually produce only one crop of rice, although sometimes a second crop is planted in the same plot if the soil is considered sufficiently fertile. After harvesting the rice, the land is usually fallowed and natural vegetation succession proceeds to develop an immature secondary forest. After one year this is known as '*reuma 1*'. However, in some cases, before the land is fallowed, it is planted with annual crops, such as sweet potatoes (*Ipomoea batatas*), sweet cassava (*Manihot esculenta*) and wild sugar cane (*Sacharum spontaneum* var. *edulis*). The swidden then becomes a *kebon* (garden). After the annual crops are harvested from the *kebon*, the land is then fallowed and after one year it, too, is referred to as '*reuma 1*'. Later, after the land has been fallowed for at least four years (*reuma 4*), it can be reopened for swidden cultivation.

These days – unlike past practices – the vegetation on fallowed land consists of mixed natural undergrowth and economic plants producing fruit, vegetables, building materials and firewood. These economic perennials grow over the four years of fallow, and when a new swidden is opened and prepared for rice and annual crops they are not cut, but their branches are pruned so they do not shade the crops. During the following fallow period, the economic trees regrow their branches and the natural undergrowth develops from dominant grasses into shrubs. In the process, fruit, vegetables, building materials and firewood can be continuously harvested from the

Albizia lebbeck (L.) Benth. [Leguminosae]

This native of South and Southeast Asia, New Guinea and northern Australia is planted in Baduy swiddens for its timber and firewood. Elsewhere it is valued for its medicinal properties in treating lung and eye problems. In the forest setting, it is known for the rattling sound made by the seeds in their pods.

fallowed land and surpluses can be sold to local middlemen to earn cash incomes. These products include bananas (*Musa x paradisiaca*), durian (*Durio zibethinus*), *petai* (beans) (*Parkia speciosa*), *jengkol* (*Achidendron pauciflorum*) and *aren* (sugar palm) (*Arenga pinnata*) (cf. Iskandar, 2007, pp.125-126; Weinstock, 2015, p.179). At the same time, soil fertility is improved by heavy litter produced by the trees and natural undergrowth. Indeed, the abovementioned *P. speciosa* and *A. pauciflorum* are both leguminous species, and to add to the fertility improvement, other perennial legumes such as *jeungjing* (*Albizia chinensis*) and *kitoke* (*Albizia lebbeck*) are added to the fallow vegetation (see Table 24-2). The adoption of legume crops and/or tree-fallow species as a nitrogen-fixing strategy for sustainable soil-fertility management in swidden systems is a widely recognized practice throughout the tropics (Bunch, 2015, p.224; Ramakrishnan, 2015, p.188).

The need to speed up the recovery of soil fertility by planting perennial legumes has arisen because of the shrinking availability of land in the Outer Baduy area. As long as fallowed land is still available, each Outer Baduy household will cultivate a swidden every year, moving from one swidden area to another within the Outer Baduy territory. In the past, tradition dictated that land should be fallowed for an odd number of years, i.e. at least five or seven years (Purnomohadi, 1985; Iskandar, 1992). However, the former abundance of fallowed secondary forest has dwindled as the population has increased. Consequently, the fallow period, during which the soil recovers between one cropping period and the next, has shrunk to three or four years. To speed up the recovery of soil fertility, the farmers have adopted the planting of perennial legumes. These legumes have nodules on their roots, which contain bacteria that are able to fix nitrogen from the atmosphere. Some of this nitrogen is available to the host plant, while sloughed and disintegrating nodules increase the levels of nitrogen in the surrounding soil. In return, the bacteria are supplied with carbohydrate by the host plant (Purseglove, 1987, pp.199-200). It can therefore be argued that the various legumes grown on fallowed land by Outer Baduy farmers have been important in their bid to accelerate the improvement of soil fertility (Reijntjes et al., 1992, p.168).

More recently, some other economic perennials have been introduced to the Outer Baduy swidden systems to increase household incomes. These include *coklat* (*Theobroma cacao*), *kopi* (*Coffea canephora*) and *albizia* (*Paraserianthes falcataria*). In fact, the cacao and coffee trees have occupied some of the fallowed land and this has been taken out of the swidden cycle. In some cases, cacao and coffee trees are planted along with *dadap* (*Erythrina variegata*). The *dadap* trees provide shading for the young coffee and cacao trees and, because they are legumes, they also improve soil fertility (cf. Hensleigh and Holaway, 1988, p.149). In addition, the *dadap* trees are commonly used to support creeping pepper plants (*Piper baccatum*), which are grown for sale to generate cash income. Unlike coffee and cacao, *albizia* is cultivated in an innovative multiple cropping system involving trees growing alongside rice in traditional swiddens.

Initially, the informal leaders (*puun*) of the Outer Baduy rejected the adoption and planting of coffee and cacao in the knowledge that these trees, when planted in a four-year fallow, would stop the traditional swidden cycle. Both coffee and cacao need longer than four years to reach maturity and they cannot be cut down to make way for new rice cultivation. However, the leaders were persuaded to allow the cultivation of both trees in fallowed land in the Outer Baduy area, and a paradox has arisen: on one hand, the swidden cycle is threatened with interruption by the coffee and cacao trees, while on the other, each of the increasing number of Baduy farmers is driven by a cultural obligation to practice swidden farming on a dwindling area of land (Iskandar, 1998; Iskandar and Ellen, 1999).

Erythrina variegata L. [Leguminosae]

Adding a splash of colour to Baduy swiddens, these thorny legumes – locally called *dadap* – have dense clusters of scarlet flowers. Their canopies are used to shade other crops and the trees are also used to support climbing pepper vines.

Fortunately, no such problem exists in the adoption of the *albizia* (*Paraserianthes falcataria*) innovation. The seeds of the *albizia* trees are mixed with rice and planted in freshly-opened swiddens. They grow alongside the various swidden crops and when the rice is harvested the trees are allowed to continue growing for the four years of the fallow period. If, at that stage, the land must be opened for new cultivation of rice and annual crops, the *albizia* trees can be harvested. There is no disruption to the swidden cycle. In the five years of their growth, the *albizia* trees bring ecological benefits, including accelerated recovery of soil fertility, because *P. falcataria* is categorized as a legume with the ability to share a symbiotic relationship with

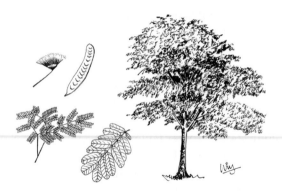

Paraserianthes falcataria (L.) I. C. Nielsen, a synonym of *Falcataria moluccana* (Miq.) Barneby and J. W. Grimes [Leguminosae]

Although this chapter uses the local name *albizia* for this tree, it doesn't actually belong to the *Albizia* genus (see endnote 1). Its rapid growth and nitrogen-fixing capacity lie at the heart of the Baduy shifting cultivation system, in which as many as 100 of these trees per quarter-hectare may grow throughout the swidden and fallow cycle.

rhizobia – bacteria that fix atmospheric nitrogen. As well, the sale of the *albizia* logs provides a substantial economic benefit.

The *albizia* innovation is a significant strategy, among many others undertaken by Outer Baduy farmers, in a bid to overcome the problems associated with a lack of available land and shrinking fallow periods. Intensive planting of other legumes in both swiddens (*huma*) and fallowed land (*reuma*) is another ploy. Most Outer Baduy people have also been involved in 'temporary migration' (*nganjor*) to areas adjacent to Baduy territory. They have obtained swidden fields in territory belonging to Muslim landowners by various means, including renting, share cropping of rice production and providing labour, and more recently, by sharing income from *albizia* harvests (Iskandar et al., 2018). Some Outer Baduy farmers have also bought land in non-Baduy areas and have obtained local government land certificates.

The use of economic *albizia* trees in swidden farming

Swidden cultivation by the Outer Baduy people has for countless generations depended on their world view and traditional ecological knowledge (cf. Iskandar, 1998; Berkes, 1999; Toledo, 2002). Their agricultural system is characterized by a rotation of short periods of cropping followed by longer periods of fallow. The fallowing of fields following cultivation is a key element in the management of soil fertility, because vital plant nutrients that are used by annual crops are returned to the soil during the fallow. Thus, many scholars hold the view that as long as population density remains low and fallow periods are long enough to restore soil fertility, swidden cultivation may be sustainable in the long run, wherever it is practised around the world (cf. Okigbo, 1984; Purnomohadi, 1985; Iskandar, 2007; Iskandar et al., 2016, 2017; Bunch, 2015, p.221; Ramakrishnan, 2015, p.187; Weinstock, 2015, p.184). Swidden cultivation in Outer Baduy has been pressured by a rapidly increasing population over recent centuries. However, the swidden practices of the Outer Baduy people have not been static; they have shown considerable dynamism over time. For example, the Outer Baduy farmers have diversified their crops, initiated 'temporary migration' to find land in non-Baduy areas, adopted the production of palm sugar, made handicrafts, involved themselves in petty trading and introduced commercial crops of coffee, cacao and *albizia* trees, all in an effort to make their swidden cultivation sustainable (Iskandar and Ellen, 2007; Iskandar, 2007; Iskandar and Iskandar, 2016; Iskandar et al., 2018). Of these, the introduction of *albizia* into the swidden-farming cycle has been most successful in maintaining soil fertility and providing cash income – indeed, in achieving sustainable swidden cultivation to support the cultural identity of both Inner and Outer Baduy communities.

Initially, a number of economic plant species were introduced to the Outer Baduy farmers. They included cloves (*Syzigium aromaticum*), coffee (*Coffea canephora*), cacao (*Theobroma cacao*), and albizia (*Paraserianthes falcataria*). However, the informal leaders (*puun*) of the Outer Baduy rejected cloves and banned cultivation of the trees in their area. Consequently, cloves are grown in adjacent non-Baduy areas, particularly by Outer Baduy people who have bought land there. Coffee was similarly banned because of the likelihood that coffee plantations would disrupt the swidden cycle. However, some Outer Baduy people planted small numbers of coffee trees on fallowed land (*reuma*). These were slashed down during annual 'clean-ups' by leadership staff, but they regrew

Durio zibethinus L. [Malvaceae]

While a number of commercial tree crops have been introduced to Baduy farmers, durian, it seems, has always been with them. The farmers protect durian trees in their swiddens and fallows for their commercial value, and trade the spiky, pungent fruit for household cash.

from the stumps of the felled trees. Meanwhile, there was intensive planting of coffee in non-Baduy areas. Eventually, the planting of both coffee and cacao trees was permitted on fallowed land in the Outer Baduy area, as an alternative to using the land for swidden farming. Coffee production has been used mainly for domestic consumption and some surpluses are sold, but cacao production, in the form of dry beans, is mainly sold to middlemen.

Since its first introduction in the 1980s, *albizia* has found a completely different role in Outer Baduy swidden systems. Closely akin to the traditional use of *kacang hiris*, or pigeon pea (*Cajanus cajan*), *albizia* is planted intensively in new swiddens by mixing its seed with that of the rice and other annual crops. *Albizia* seeds are collected and before planting are soaked in warm water for about five minutes to break the seed dormancy (cf. Soerianegara and Lemmens, 1994, p.322; Iskandar and Ellen, 2000, p.10). The seeds are plunged back and forth several times between warm water and cold water. They are then mixed with the rice and sown in the same holes in the swidden soil. In addition, some *albizia* seeds are planted between rows of rice (Figure 24-3). Later, during weeding, the spacing of the *albizia* seedlings is rearranged; those that are too close together are moved to other places where the seedlings are growing more sparsely.

Originally, the *albizia* seeds were mainly gathered from beneath mature trees growing in fallow land. In the 1990s, however, some *albizia* seeds were offered for sale by an agricultural supplier in the district capital, Rangkasbitung, at a cost of 5000 rupiah (about US$2.15) per kilogram (Iskandar and Ellen, 2000, p.10). In 2017, *albizia* seedlings about 50 cm tall could be bought for 1000 rupiah (US$0.07) each. Moreover, with a view to planting *albizia* in their swiddens, many Outer Baduy people have begun growing seedlings in plastic bags in their home gardens and elsewhere. When the seedlings reach about 20 cm in height, they are transplanted into the swiddens, in holes between the rows of rice, preferably when the rice is a few weeks old. Before planting, the root tip of each *albizia* seedling is cut, placed in the bottom of the hole and buried in soil. The seedling is then placed in the hole on top of this and packed with soil. This is a form of protection against termite

FIGURE 24-3: An Outer Baduy farmer transplants *albizia* seedlings into his swidden rice crop.

Photo: Johan Iskandar.
Source: Iskandar and Ellen, 1999.

attack. The seedlings are not planted densely, so as to allow sunlight to penetrate to the growing rice. When the swidden is weeded, some small branches and leaves of the *albizia* are pruned using a sharpened bamboo stick and are used as compost. If the same swidden is planted with rice for a second year in succession, the young *albizia* trees are composted with cut rice straw during land preparation (Iskandar and Ellen, 2000, p.10).

FIGURE 24-4: During the four-year fallow period the *albizia* trees become a dominant component of the fallow vegetation.

Photo: Johan Iskandar.
Source: Iskandar and Ellen, 1999.

After the rice is harvested, the land is fallowed for at least four years, to allow soil fertility to recover. The young *albizia* trees continue to grow in the secondary-forest fallow (Figure 24-4). After several years, improvement of soil fertility has been accelerated by nitrogen fixation and the provision of other nutrients from *albizia* biomass, mainly leaf litter and branches. Since chemical fertilizers are prohibited by the religious leaders of the Baduy, nutrient input to fallowed land is dependent entirely upon compost and nitrogen-fixing bacteria. Indeed, based on laboratory evidence, young *albizia* trees fix between 17% and 37% of the nitrogen in the atmosphere (Iskandar and Ellen, 2000), and mature trees around 55% (People et al., 1991, cited in Giller amd Wilson, 1991, p.187). After the four-year fallow, the farmer who planted the *albizia* stand will return to claim his or her harvest of timber.

Earlier, when the land is fallowed, an Outer Baduy farmer will look for another plot that has been fallow for at least four years, in order to open a new swidden. The farmer will search in both the Outer Baduy area and the adjacent non-Baduy area. If appropriate fallow land is not available in Outer Baduy area, the farmer will 'temporarily migrate' to practice swidden farming in the non–Baduy area (Figure 24-5).

Up to the end of the 1990s, the people of Outer Baduy who were unable to open new swiddens in their own territory obtained suitable land in adjacent non-Baduy areas by renting, share cropping of swidden rice, or selling their labour. Today, however, some Outer Baduy farmers have purchased land in non-Baduy areas and

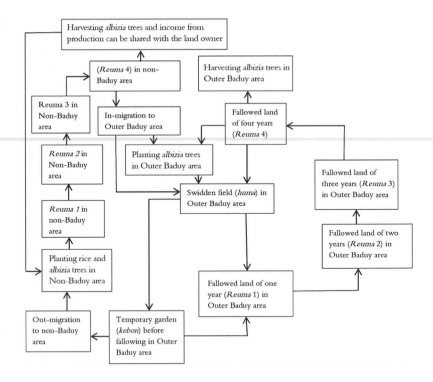

FIGURE 24-5: Use of *albizia* trees in the swidden cultivation cycles followed by Outer Baduy farmers.

they 'temporarily migrate' to practise their traditional swiddening on their own land. These owners of land in non-Baduy areas not only plant economic *albizia* trees on their own land, but also rent land or enter share-cropping arrangements to extend their *albizia* plantations (Iskandar et al., 2018). Those without their own land still 'temporarily migrate' to non-Baduy areas when there is no suitable land within their own territory and rent swiddens or use the value of *albizia* trees to sweeten share-cropping deals.

Planting *albizia* trees in non-Baduy territory follows a similar procedure to that practised in Outer Baduy swiddens: either the seeds are mixed with those of rice when the swidden is first planted, or seedlings are planted with the young rice crop. After the rice and annual crops are harvested, the *albizia* trees are allowed to grow as a dominant component of the fallow vegetation, while the farmers move on in their search for new swidden land. After the four-year fallow, farmers who own their own land in non-Baduy territory enjoy the full benefit from the harvest of timber (Figure 24-6); those who rent or share-crop usually share the income with the land owner and if they wish to begin the cycle again, must make a new agreement. After several years of 'temporary migration' they may return to their village in Outer Baduy, particularly when there is fallow land there that is ready to recultivate.

Normally, the trees are sold to local middlemen before they are harvested. The sale price is usually based on an average value per tree. In 1984 and 1985, a five-year-old *albizia* tree was worth between 3500 and 4000 rupiah, so if there were 300 trees on a fallowed plot, the Outer Baduy farmer could sell them all for between 1,050,000 and 1,200,000 rupiah. By 1995 and 1996, the value of each tree had risen to between 9000 and 10,000 rupiah, and on the basis of harvesting 300 trees from a plot of half a hectare, an Outer Baduy household could earn between 2,700,000 and 3,000,000 rupiah (US$1179 to $1285). The price continued to rise. By 2003 and 2004, each *albizia* tree was worth between 10,000 and 12,000 rupiah, earning between 3,000,000

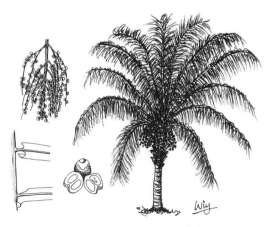

Arenga pinnata (Wurmb) Merr. [Arecaceae]

When the timber is harvested and Baduy swiddens are cleared, sugar palms are regularly left standing. The sap is harvested to produce sugar, is made into a traditional cold sweet drink, or is fermented to produce palm wine. The immature fruit are also widely consumed.

and 3,600,000 rupiah for a crop of 300 trees. (Devaluation of the rupiah made this equivalent to between US$307 and $400). More recently, in 2017 and 2018, a

harvest of 300 five-year-old *albizia* trees could earn an Outer Baduy family between 18,000,000 and 19,500,000 rupiah (US$1250 and $1353 at current exchange rates) (Table 24-3).

Thus, it can be concluded that planting economic *albizia* trees integrated with rice in a traditional swidden system can deliver significant benefits for Outer Baduy farmers. These include improving and maintaining soil fertility in both swiddens and fallows, while earning cash income to provide for daily household needs.

FIGURE 24-6: Freshly harvested *albizia* timber ready for sale as a four-year-old fallow is opened for new cultivation.

Photo: Johan Iskandar.

TABLE 24-3: Value of *albizia* harvested after four years of fallow on a half-hectare plot, between 1984 and 2016.

Year	Local market value (rupiah per tree)	Total value of 300 trees from a half-hectare plot (rupiah)	Harvest of hulled rice (beras) from an equivalent plot (kg)
1984 to 1985	3500 to 4000	1,050,000 to 1,200,000	4200 to 4800
1995 to 1996	9500 to 10,000	2,850,000 to 3,000,000	3557 to 3428
2003 to 2004	10,000 to 12,000	3,000,000 to 3,600,000	1142 to 1371
2017 to 2018	60,000 to 65,000	18,000,000 to 19,500,000	1600 to 1733

Note: In 1985, 1996, 2004 and 2018, one kilogram of hulled rice (*beras*) was worth Rp250, Rp875, Rp2625 and Rp11,250, respectively. In 2017 and 2018, US$1 was worth between 13,397 and 14,406 rupiah.

Source: Author's field notes (1985, 1996, 2004 and 2018)

Conclusions

This chapter has discussed innovations developed by the farmers of Outer Baduy to maintain soil fertility in both their swiddens and fallowed land. These innovations were driven by the need to enhance the sustainability of Outer Baduy swidden farming, which was under significant pressure from changing socio-economic and environmental conditions, including population pressure, the development of a market economy and a decreasing area of mature secondary forest in which to open new swiddens.

Swidden farming, or shifting cultivation, has been practised by the Outer Baduy for centuries, based on their traditional ecological knowledge. The obligation to cultivate swiddens has thus become strongly embedded in their culture. In the past, when the population was still low and mature secondary forest was abundant within their territory, each Outer Baduy household was free to choose plots of mature secondary forest in which to open new swiddens. Year after year, they moved from place to place. Nowadays, the much larger Outer Baduy population has led to the need to intensify and increase agricultural production from both swiddens and fallow land to meet the needs for extra food and household cash income. As the available land diminished, the farmers had to abandon their long tradition of using odd numbers of years – mainly five and seven – for the length of the fallow. It has now come down to four years – a period that under traditional circumstances would not allow adequate recovery of soil fertility and this, in turn, would cause reduced productivity and lower incomes. Faced with these changes, the Outer Baduy farmers introduced a variety of innovations, including farming a higher diversity of crops, planting various legumes and growing economic *albizia* trees as part of their traditional swidden system. They have achieved both ecological and socio-economic benefits by maintaining soil fertility in their swiddens while accelerating and improving the recovery of fallowed land. At the same time, they have made their farming system more resistant to pests and climatic changes and provided a variety of agricultural products for home consumption, while achieving surpluses to boost the economic well-being of Outer Baduy communities. In the current times of change, including limitations on available areas of mature secondary forest, more mouths to feed with an increasing population and intensive penetration of a market economy in their villages, Outer Baduy farmers have been able to maintain a viable system of traditional shifting cultivation through innovative and economically productive management of their fallowed land.

References

Aryal, K. and Choudhury, D. (2015) 'Climate change: Adaptation, mitigation and transformation of swidden landscapes: Are we throwing the baby out with the bathwater?' in M. F. Cairns (ed.) *Shifting Cultivation and Environmental Change: Indigenous People, Agriculture and Forest Conservation*, Earthscan from Routledge, London and New York, pp.281-288

Berkes, F. (1999) *Sacred Ecology: Traditional Knowledge and Resource Management*, Taylor and Francis, Philadelphia

Bunch, R. (2015) 'Learning from migratory agriculture around the world', in M. F. Cairns (ed.) *Shifting Cultivation and Environmental Change: Indigenous People, Agriculture and Forest Conservation*, Earthscan from Routledge, London and New York, pp.221-234

Conklin, H. (1957) *Hanuoo agriculture: A report on an integral system of shifting cultivation in the Philippines*, Forestry Development Paper 12, Food and Agriculture Organization of the United Nations, Rome

Cotton, C. M. (1996) *Ethnobotany: Principles and Applications*, John Willey and Son, Chichester, UK

Geertz, C. (1963) *Agricultural Involution: The Process of Ecological Change in Indonesia*, University of California Press, Oakland, CA

Giller, K. E. and Wilson, K. J. (1991) *Nitrogen Fixation in Tropical Cropping Systems*, CAB International, Wallingford, UK

Hensleigh, T. E. and Holaway, B. K. (eds) (1988) *Agroforestry Species for the Philippines*, US Peace Corps, Washington, DC

Iskandar, J. (1991) 'An evaluation of the shifting cultivation system of the Baduy society in West Java, using system modeling', Master's Thesis to Chiang Mai University, Chiang Mai, Thailand (unpublished)

Iskandar, J. (1992) *Ecology of the Swidden Farming in Indonesia: Case Study in Baduy, South Banten, West Java*, Penerbit Djambatan, Jakarta (Indonesian language)

Iskandar, J. (1998) 'Swidden cultivation as a form of cultural identity: The Baduy case', PhD dissertation to the University of Kent at Canterbury, UK (unpublished)

Iskandar, J. (2007) 'Responses to environmental stress in the Baduy swidden system, South Banten, Java', in R. F. Ellen (ed.) *Modern Crises and Traditional Strategies: Local Ecological Knowledge in Island Southeast Asia*, Berghahn Books, New York and Oxford, UK, pp.112-132

Iskandar, J. and Ellen, R. F. (1999) 'In situ conservation of rice landraces among the Baduy of West Java', *Journal of Ethnobiology* 19(1), pp.97-125

Iskandar, J. and Ellen, R. F. (2000) 'The contribution of *Paraserianthes* (*Albizia*) *falcataria* to sustainable swidden management practices among the Baduy of West Java', *Human Ecology* 28(1), p.17

Iskandar, J. and Ellen, R. F. (2007) 'Innovation, "hybrid" knowledge and the conservation of relict rainforest in Upland Banten', in R. F. Ellen (ed.) *Modern Crises and Traditional Strategies: Local Ecological Knowledge in Island Southeast Asia*, Berghahn Books, New York and Oxford, UK, pp.133-142

Iskandar, J. and Iskandar, B. S. (2016) 'Resilience of Baduy traditional agroforestry systems in responses to environmental and socio-economic changes', *Journal of Indonesian Natural History* 1(1), pp.19-24

Iskandar, J. and Iskandar, B. S. (2018) 'Ethno-ecology, rice biodiversity and modernization of rice cultivation: Case study on Baduy community and Kampung Naga', *Biodjati* 3(1), pp.36-50 (Indonesian language)

Iskandar, J., Iskandar, B. S. and Partasasmita, R. (2016) 'Responses to environmental and socio-economic changes in Karangwangi traditional agroforestry system, South Cianjur, West Java', *Biodiversitas* 17(1), pp.332-341

Iskandar, J., Iskandar, B. S. and Partasasmita, R. (2017) 'Introduction of *Paraserianthes falcataria* in the traditional agroforestry "*huma*" in Karangwangi village, Cianjur, West Java, Indonesia', *Biodiversitas* 18(1), pp.295-303

Iskandar, B. S., Iskandar, J. and Partasasmita, R. (2018) 'Strategy of the Outer Baduy community of South Banten (Indonesia) to sustain their swidden farming traditions by temporary migration to non-Baduy areas'. *Biodiversitas* 19(2), pp.453-464

Kurnia, A. and Sihabudin, A. (2010) *It is Time Baduy to Talk*, PT Bumi Aksara, Jakarta (Indonesian lamguage)

Okigbo, B. N. (1984) 'Improved permanent production systems as an alternative to shifting intermittent cultivation systems', in B. N. Okigbo (ed.) *Improved Production Systems as Alternatives to Shifting Cultivation*, FAO Soils Bulletin 53, Food and Agriculture Organization of the United Nations, Rome, pp.1-100

People, M. B., Bergersen, F. J., Turner, G. L., Sampet, C., Rerkasem, B., Bhromsiri, A., Nurhayati, M. and Herridge, D. F. (1991) 'Use of natural enrichment of N in plant available soil N for the measurement of symbiotic N2 fixation', in *Stable Isotopes in Plant Nutrition, Soil Fertility and Environmental Studies*, Proceedings of a symposium jointly organized by the International Atomic Energy Agency and the Food and Agriculture Organization of the United Nations, 1–5 October 1990, Vienna, pp.117-130

Purnomohadi, S. (1985) 'System of socio-economic interaction and management of natural resources among the Baduy community, Kanekes village, South Banten', Master's Thesis to the Postgraduate Faculty of the Department of Natural Resources and Environment, Agricultural University, Bogor, Indonesia (Indonesian language)

Purseglove, J. W. (1987) *Tropical Crops: Dicotyledons*, Longman Scientific and Technical, John Wiley and Sons, New York

Ramakrishnan, P. S. (2015) 'Shifting agriculture and fallow management options', in M. F. Cairns (ed.) *Shifting Cultivation and Environmental Change: Indigenous People, Agriculture and Forest Conservation*, Earthscan from Routledge, London and New York, pp.186-198

Reijntjes, C., Haverkort, B. and Waters-Bayer, A. (1992) *Farming for the Future: An Introduction to Low-external-input and Sustainable Agriculture*, Macmillan Press, London and Basingstoke, UK

Richards, P. (1992) 'Rural development and local knowledge: The case of rice in central Sierra Leone', *Etnwicklungsetnologie* 1, pp.26-40

Sajise, P. E. (2015) 'Biodiversity and swidden agroecosystems: An analysis and some implications', in M. F. Cairns (ed.) *Shifting Cultivation and Environmental Change: Indigenous People, Agriculture and Forest Conservation*, Earthscan from Routledge, London and New York, pp.401-419

Soedjito, H. (2015) 'Shifting cultivators, curators of forests and conservators of biodiversity: The Dayak of East Kalimantan, Indonesia', in M. F. Cairns (ed.) *Shifting Cultivation and Environmental Change: Indigenous People, Agriculture and Forest Conservation*, Earthscan from Routledge, London and New York, pp.420-448

Soerianegara, I. and Lemmens, R. H. M. J (1999) 'Timber Trees: Major Commercial Timbers', in L. T. Hong, R. H. M. J. Lemmens, S. Prawirohatmodjo, I. Soerianegara, M. S. M. Sosef and W. C. Wong (eds) *Plant Resources of Southeast Asia* 5(1), Springer-Verlag, Berlin and Heidelberg, pp.319-325

Toledo, V. M. (2002) 'Ethno-ecology: A conceptual framework for the study of indigenous knowedge of nature', in J. R. Stepp, F. S. Wyndham and R. K. Zarger (eds) *Ethnobiology and Biocultural Diversity*, The University of Georgia Press, Athens, GA

Weinstock, J. A. (2015) 'The future of swidden cultivation', in M. F. Cairns (ed.) *Shifting Cultivation and Environmental Change: Indigenous People, Agriculture and Forest Conservation*, Earthscan from Routledge, London and New York, pp.179-185

Note

1. The correct botanical name for this species is problematic. The name *albasiah*, given to the trees locally, and *albizia* in this chapter, would seem certain to arise from its earlier identification as one of many species in the genus *Albizia*. The name *Albizia falcataria* (L.) Fosberg was more recently commonly replaced by the name *Paraserianthes falcataria* (L.) I. C. Nielsen – the botanical name given to the species in this chapter. At the time of writing, recognized authorities on botanical names disagreed on how it should be properly named: www.theplantlist.org said that both *Albizia falcataria* and *Paraserianthes falcataria* were synonyms of the accepted name *Falcataria moluccana* (Miq.) Barneby & J. W. Grimes. However, Kewscience's www.plantsoftheworldonline.org regarded all three of these names as synonyms of the accepted name *Falcataria falcata* (L.) Greuter & R. Rankin. All names relate to members of the plant family alternately named Leguminosae or Fabaceae.

25

INNOVATION HELPS SWIDDEN CULTIVATION TO SURVIVE

'Albizia' fallows recover fertility and improve performance in conditions of limited land, dearth of labour and a ban on burning

*Niken Sakuntaladewi, E. P. Setyawan, S. Ekawati and A. Wibowo**

Introduction

This chapter discusses innovations by swidden cultivators, some of the world's most vulnerable people, in facing various challenges in order that they can continue their agricultural practices and survive. Swidden farming, shifting cultivation, or pejoratively, slash-and-burn, is the oldest and most primitive form of agricultural production (Weinstock, 2015), yet it is still commonly practised in many tropical regions of Asia–Pacific, Africa and South America (Ramakrishnan, 2015; Allen and Filer, 2015). This agricultural practice has existed for many hundreds of years, with a more or less similar cycle of activities: slashing, cutting, drying, burning, planting, harvesting and fallowing.

This traditional cultivation system relies entirely on nature. Slashing and cutting of vegetation is performed in the dry season, and the energy of the sun is employed to dry the shrubs and trees that have been felled so that they burn completely. Fire breaks are made around the field to ensure that the fire does not spread into nearby areas, and when the fires are lit, the wind direction is taken into account so that it helps the burn and farmers working on the fire are not threatened. The resulting ash is used to fertilize the land. Planting is done just before the onset of the rainy season because the crops will depend solely on rainwater. After the crops are harvested, and usually after just one year, the field is abandoned and the land left fallow so that natural processes will allow it to recover its fertility, preferably over 10 or more years (Weinstock,

* Dr. Niken Sakuntaladewi, National Research and Innovation Agency, Indonesia; E. P. Setyawan, formerly Social Development Coordinator, World Wide Fund for Nature, Indonesia, now Advisor on Community-based Peatland Management and Rehabilitation, GIZ, Indonesia; Dr. S. Ekawati, National Research and Innovation Agency, Indonesia; and A. Wibowo, Community Organizer, World Wide Fund for Nature, Indonesia.

2015). The farmers must clear new fields in a constant cropping-and-fallow cycle, creating a mosaic landscape of regenerating patches of forest, until they return after a decade or more to clear the plot of land where they began. Consequently, shifting cultivation in its traditional form requires a considerable area of land – that which is being cultivated, plus that lying fallow.

In Indonesia, swidden cultivation is commonly practised on outer islands, and the number of swidden cultivators is not known with certainty. Statistical data for 2014 indicate that despite the increased number of families living in forest areas, the number of swidden cultivation families has declined over the past 10 years (Table 25-1). There is no explanation of why this phenomenon has occurred.

Swidden cultivation communities are not always isolated, even though they live in and around forests. Many of them are equipped with various electronics and communication technologies, and are within reach of government development programmes. However, their lifestyle is quite vulnerable, simply because it relies heavily on nature and the availability of land.

The environmental sustainability of swidden cultivation has long been a matter for debate among scholars. Studies by the Indonesia-based Center for International Forestry Research (CIFOR) have found that when it is done properly, shifting cultivation can create natural ecosystems with high biodiversity, carbon-rich reserves and a low risk of soil erosion (Pearl and O'Connell, 2016). But in order to achieve this, there are a number of essential requirements, among them the need for a long period of fallow (Dressler et al., 2015). Given the world's growing population, Weinstock (2015) considers shifting cultivation less appropriate because it requires a lot of land, which is used only once or twice before being fallowed. Ultimately, the suitability of shifting cultivation for the 21st century remains questionable. This is particularly so because this farming system is usually practised by communities living in or around forests and involves clearing of forest areas by felling bushes and trees and burning them. Use of this technology raises questions regarding the existence of forests, the effect it has on forest cover, the emissions it generates, and the best use of forests (Weinstock, 2015).

TABLE 25-1: Number and percentage of households in areas surrounding forests who practise swidden cultivation, 2004 and 2014.

Description	2004	2014
Number of households in areas surrounding forests	7,804,970	8,643,228
Number of households in areas surrounding forests who practise swidden cultivation	259,959	242,866
Percentage of households in areas surrounding forests who practise swidden cultivation	3.3	2.8

Source: BPS (2015b).

Challenges to swidden cultivation practices

Various studies have found that population increase in general has the result of reducing forest areas (Nath and Mwchahary, 2012; Gul et al., 2014; Debel et al., 2014). Thus, with forest areas continuing to diminish, the management of forests has become a matter of forest protection and improvement, and this has brought considerable challenges for shifting cultivators.

Indonesia's Central Bureau of Statistics (BPS) defines swidden cultivation as an agricultural business activity involving the cultivation of seasonal crops or food crops that traditionally shifts inside and outside of forest areas without regard for the preservation of forest resources, land, or water (BPS, 2015b). This definition makes it clear that shifting cultivation is no longer considered to be in accord with current conditions in the country. With Indonesia's population reaching more than 258.7 million in 2016 (BPS, 2017b), along with the need for national development, conservation efforts, and the government's commitment to reduce emissions, an agricultural practice that keeps moving from one place to another, that clears forest areas by burning, and which abandons the land when it is no longer fertile, is at least inappropriate.

With all of these recent developments, swidden cultivators in Indonesia are facing substantial challenges in pursuing their livelihoods. Among other things,

a) there is limited land for cultivation;
b) increasing areas of the country are being taken by plantation commodities that promise cash incomes, such as oil palm, rubber and cocoa;
c) climate change is making it difficult for them to follow a normal calendar, and to decide when to farm;
d) government policies now include a prohibition on burning forest to clear land; and
e) agricultural production is decreasing while family needs are rising.

Limited land for farming

Limited available land for farming is now a major challenge for shifting cultivators, and to some extent, this is closely linked to government development policies. As a developing country, Indonesia needs a huge amount of money to build the country to meet the needs of a continually increasing population, and forests have become an income source for this development. With tropical forests covering 120 million hectares – making it the country with the world's third–largest forested area – the Indonesian government began commercializing its forests in the 1970s by granting timber concessions to entrepreneurs (Anonymous, 2011). In 2000 there were 600 timber-concession licenses covering a total area of more than 64 million ha. However, by 2015 that number had fallen to about 270 timber concessions covering an area of about 21 million ha.

Although the number of forest concessions has fallen significantly, the establishment of many new districts, oil-palm plantations, and housing and road developments has further shrunk forest areas. Utilization of forests for these various purposes has heavily impacted the availability of land for swidden farmers, who are no longer free to farm in forest areas where management or forest-use permits have been granted to other parties.

Development of oil palm, rubber and cocoa plantations

Swidden systems in Asia, including Indonesia, are in a period of tremendous change, largely driven by population dynamics and forward-looking economic policies (van Noordwijk et al., 2015). Traditional agricultural farming systems are either evolving or being spontaneously replaced by mixed gardens, agroforests, plantations (rubber or cocoa) or rice fields (van Noordwijk et al., 2015; Penot, 2007). Oil-palm (*Elaeis guineensis*) plantations, rubber and cocoa are growing rapidly in Indonesia. Historically, state-owned oil-palm plantations (PT Perkebunan Nusantara) began to grow in the 1970s, while smallholder oil-palm plantations grew after 1979, with the support of the World Bank. All were oriented towards economic development, in the form of creating employment opportunities, improving community welfare, and generating foreign exchange. In 1980, oil-palm plantations in Indonesia covered 294,560 ha. The rapid growth of smallholder plantations was partly due to the government's Peoples' Estate Plantation Company (PIR-Bun) programme and the development of oil-palm mills to process plantation products into crude palm oil. The government's seriousness towards expansion of oil palm could be seen from the massive land bank of more than 100,000 ha granted per company, and authority to grant permits for palm plantations resting in the hands of a minister and regional heads, as part of moves towards regional autonomy. Within 20 years, from 1990 to 2010, the land covered by oil-palm plantations in Indonesia grew from 1.1 million ha to 8.4 million ha, and by 2018 these plantations had spread over 14.3 million ha (Directorate General of Plantations, 2015; BPS, 2019).

The acquiescence of villagers to expansion of oil-palm plantations came about for a variety of reasons. Giving two typical examples: in Cempaka Mulia Barat, Central Kalimantan province, the people were originally dependent on rubber and rattan gardens. They turned to oil palm because prices for rubber and rattan fell dramatically (Rumboko et al., 2018). In the village of Tepian Buah, East Kalimantan province, the change was purely economic: the community allowed companies to plant oil palm on its farming land, and the farmers were given work as labourers by the oil-palm company for weekly or monthly salaries. Thus, at village level, the development of oil-palm plantations is often a controversial balancing act between economic benefits and undesirable social and environmental impacts.

In Indonesia, rubber (*Hevea brasiliensis*) is one of the leading agricultural products because it has an important role to play in generating foreign exchange (Fauzi, 2008). The prominent economic role of rubber has diminished with plummeting prices, and for many rubber growers – particularly smallholders – care of plantations has become a waiting game, in the hope that prices will recover. Before the market downturn, many swidden cultivators planted their fields with rubber, sometimes at a rate of more than 10 ha of rubber per household. Rubber seedlings were usually planted at the same time as upland rice was planted in the swiddens, or alternatively, after the rice was harvested. The seedlings were not planted intensively, and they were big enough to compete with regenerating secondary forest when the swidden was abandoned and left fallow. The result was a fallow forest interspersed with rubber trees – a system known as 'jungle rubber'. Currently, smallholders maintain rubber plantations located around their settlement more intensively, but those further distant tend to be 'left behind' (Setiabudi et al., 2013).

In East Kalimantan's Mahakam Ulu district, smallholder rubber plantations reached a total area of 1839 ha in 2016, with dry–rubber production of 366 tons of latex, or productivity of 199 kg/ha (BPS, 2017a, pp.142-143) (Table 25-2). This was far below the average dry-rubber productivity rate of national smallholder rubber plantations in 2016, which had reached 834 kg/ha (BPS, 2017b). Nevertheless, the area planted to rubber in Mahakam Ulu district continued to increase from 1705 ha in 2013 (BPS, 2014) to 1839 ha in 2016 (BPS, 2017a, p.274).

Rubber production is the mainstay of the district, representing about 88.48% of total production from the district's plantations. However, the existing production comes from trees grown from low-quality seeds, so potential production cannot be guaranteed. In addition, trees in existing plantations are already old and have never been maintained. Tapping is not performed well and latex production is low because the trees are tapped too often (Rusdiana et al., 2017).

Cocoa (*Theobroma cacao*) has become an agricultural product with an important role in Indonesia's agricultural development, particularly in terms of employment, regional development, increasing farmers' welfare, and increasing foreign exchange. Over the past two decades, the country's area committed to cocoa plantations has

Theobroma cacao L. [Malvaceae]

The cocoa tree is a native of South America, growing up to 8 metres tall. The fruit, which grow on the main trunk and major branches, weigh about 500 gm each and contain up to 60 seeds known as cocoa beans, which are used to produce chocolate. World production of these beans reached 5.3 million tons in 2018, the bulk of which came from millions of farmers with small plots. Indonesia was the world's third-largest producer, behind the Ivory Coast and Ghana.

TABLE 25-2: Development of community rubber area and production in Mahakam Ulu district from 2013 to 2017.

Year	2013	2014	2015	2016	2017
Smallholder rubber plantation area (ha)	1705	1709	1839	1839	1839
Smallholder rubber plantation (tons)	275	274	343	366	–

Note: BPS (2018) gave no information on rubber production in 2017.
Source: BPS (2014, 2015a, 2016, 2017a, 2018).

grown continuously and national cocoa production has grown accordingly. However, the productivity of cocoa plantations is unstable and even tends to decline.

East Kalimantan province is one of the highest producers of cocoa beans in Indonesia. The 'beans' are actually the seeds of the tree. They are used to produce chocolate. Cocoa trees were first introduced to farming communities in the 1990s, after which there was rapid development and many large cocoa plantations were established. Based on statistical data from 2015, cocoa plantations in North and East Kalimantan covered an area of 8296 ha, of which 10.3% was in East Kalimantan's Mahakam Ulu district.

Climate change

Climate change is already a part of our lives. It has resulted in environmental disasters, increased frequency of extreme weather events, and uncertainty over what the weather will bring, and when. These disruptions are having serious consequences in the agricultural sector. Studies have recorded symptoms that threaten the sustainability of food production in countries such as Indonesia, China, the Americas, Africa and Europe (Gommes, 1998; Olesen and Bindi, 2002; Naylor et al., 2007; Natawidjaja et al., 2009; Wang, 2010; Li et al., 2010; Shen et al., 2010).

The linking of forest fires, deforestation and forest degradation to climate change, and the complexity of emissions-reduction targets, have added to the challenges in swidden cultivators' lives. They do not know what climate change is; what causes the change of the seasons, or whether they contribute to the occurrence of irregular seasons. What they do understand is that the seasons are no longer regular, and if they fail to determine the precise times for land preparation, they will have extra work to do, possibly extra costs to meet, and production will be less than optimal. It is becoming an issue of survival. A few decades ago, swidden cultivators could be sure of when they could start land preparation. Now, rainy seasons and periods of drought are no longer predictable, making it difficult to determine when they should start preparing the land. Shrubs, bushes, and trees that have been slashed need to be as dry as possible for a perfect burn, and this requires hot, dry weather. They also need to be able to predict the arrival of the wet season, so they know when to start planting their fields.

There is uncertainty at village level about what the key issues are, and how to handle them appropriately. To overcome the impacts of climate-driven crop failures,

Choudhury (2012), in Aryal and Choudhury (2015), recommended planting a wide diversity of crops, so that swidden farmers could harvest different crops throughout the year and avert risks to their families' food security. Aryal and Choudury (2015) concluded that planting a diversity of crops in swidden cultivation is a strategy to spread risks and serve as a risk-insurance mechanism, including risks arising from climate change. However, van Noordwijk et al. (2015) stated that most of the agricultural alternatives to shifting cultivation have an increased sensitivity to climate variability, so that swidden transitions may well be riskier as a result of climate change.

Government prohibition of burning

Forest and land fires incur economic losses and have a negative impact on the health of local communities. Such fires occurring in Indonesia also bring protests from neighbouring countries, such as Singapore and Malaysia, as the smoke envelops their territories. Therefore, controlling forest and land fires has become one of the many concerns of the Indonesian government.

As a forested third-world country, Indonesia has recently been among the top 10 emitters of greenhouse gases. The government is committed to reducing emissions by 29% before 2030, and by up to 41% in this time if there is foreign assistance. The aim is also to boost economic growth by 7%. The country plans to address this commitment by involvement in the United Nations' Reducing Emissions from Deforestation and Forest Degradation (REDD) schemes, among other strategies.

Forest and land fires are among the drivers of deforestation and forest degradation. These fires occur almost every year in Indonesia. Medrilzam (2015) says that the number of hotspots in the years from 2011 to 2014 indicate that 72% of fires occur in non-forest areas, such as plantations, agricultural areas and bushland. One of the causes is land preparation, including forest clearance and burning. Deforestation in and outside forest areas in 2015 to 2016 remained quite high, covering 820,000 ha despite a significant decline since the 1990s (Sugardiman, 2017).

The Director of the Indonesian Disaster Expert Association (IABI), Professor H. A. Sudibyakto, believes that recent forest and land fires have become a major threat to Indonesia. In addition to damaging tropical wetland ecosystems, forest and land fires are accelerating the climate-change process. In 2009, the Government of Indonesia issued Law No. 32, on Environmental Protection and Management. Article 69 paragraph (1) of this law prohibits the use of fire to open land. It prescribes sanctions not only against anyone lighting such fires, but also against responsible officials in the area.

In this effort to prevent forest and land fires, the government gave an exception to some people who burn the forest for land-clearing purposes. Paragraph (2) of article 69 recognizes the genuine local wisdom of people who clear land with fire within their respective territories. Furthermore, a regulation issued by the State Minister of Environment in 2010 states that indigenous and tribal people should inform their village head if they plan to burn land with a maximum area of two hectares per household, in order to plant local crop varieties. However, this provision does not apply in conditions of normal rainfall, long drought, or 'dry climate', in which case, both the perpetrators of burning and local officials will be sanctioned. These sanctions are feared by the authorities, who, to be safe, opt to prohibit people from burning. The management of fires to clear land is part of the indigenous knowledge of swidden-farming communities, learned and passed down by many earlier generations, and has become a part of their culture (Talaohu, 2013). Therefore, the ban has shaken the ancestral customs of most people living in and around Indonesia's forests.

Decreasing crop production and increasing family needs

Swidden cultivators are now faced with rapidly declining soil fertility. This is due, among other things, to shorter fallow periods, as farmers resort to longer periods of cultivation in order to produce more from a decreasing area of land. The land is thus denied the restorative qualities of lengthy fallows. This phenomenon has also been reported in northeast India, where the fallow period has fallen to five years or less (Ramakrishnan, 2015). This forces a steep decline in agricultural yields.

At the same time, the cost of living for families continues to increase, along with increasingly varied necessities of life, and prices of goods for family needs are becoming more expensive. For shifting cultivators living in Kalimantan's forested interior, the cost of inland transportation is added to already expensive prices. To be able to finance the necessities of family life, shifting cultivators are turning to a range of alternatives, including:

- looking for new livelihoods, such as working for timber companies or agricultural plantations, opening small-business stalls, mining gold, or illegal logging;
- selling their land to oil-palm plantations;
- adopting new cropping systems that have good economic value, such as rubber or cocoa plantations; or
- planting nitrogen-fixing crops or tree-fallow species on their agricultural land for sustainable soil-fertility management (Ramakrishnan, 2015).

There are also those farmers who continue to practise shifting cultivation, but with some improvements.

We will now discuss the lives of Dayak tribespeople in Laham village, Mahakam Ulu district, East Kalimantan province, and their innovative approaches to overcoming challenges that have arisen mainly from the government's policy prohibiting the use of fire to clear land, the limited availability of land for farming, decreasing field production, and climate change. These innovations have come from their determination to continue shifting cultivation, so that they can protect and maintain a variety of rice that has, for hundreds of years, been their staple food, and to uphold the culture that accompanies swidden agriculture. The World Wide Fund for Nature (WWF) has assisted the villagers of Laham to improve their swidden cultivation.

The following sections of this chapter will tell of the dawning realization of the farmers of Laham village that the myriad problems confronting their livelihoods boiled down to the need to work within existing strictures; to improve their management of an increasingly brief fallow period so that soil fertility is restored in the time available, allowing them to continue to practise shifting cultivation, albeit with some innovative changes. The answer was a fast-growing leguminous tree, *Paraserianthes falcataria*, which is widely known as *albizia*. Its life cycle fits comfortably within a shorter shifting cultivation cycle, it enriches the soil, and it provides a valuable harvest of timber when a field is re-opened for new cultivation (see also Iskandar, ch. 24, this volume).

Research methods

Data were collected from December 2017 to May 2018, from WWF representatives who served as field facilitators, five key village representatives, three heads of farmer groups, and 18 out of a total of 29 swidden cultivators in Laham village (62%), who have planted *albizia* trees

Oryza sativa L. [Poaceae]

Rice is the most widely consumed staple food for more than half of the world's population. Its importance is reflected vividly at this chapter's study site. While much smallholder-farming development in Indonesia has seen traditional swiddening replaced by plantation crops, the shifting cultivators of Laham village in East Kalimantan are instead deliberately using tree crops to support their continued swidden production of a rice variety that has, for hundreds of years, been their staple diet.

as a fallow crop. Respondents included men and women. The data were collected through focus-group discussions, in-depth interviews, and field visits, which, among other things, involved measuring the diameter of sample *albizia* trees. The trees are widely known as *albizia* because of a former botanical identity.[1] In Laham village, the trees are called *sengon* or *sengon solomon*, but in this chapter we have chosen to use *albizia* because it is a name recognized in many parts of the region.

The data collected included: (1) respondents' family education levels; (2) farming area and details of fallow management; (3) innovations in swidden cultivation practices and their benefits; (4) costs involved in practising swidden cultivation; (5) challenges in implementing the innovative systems; (6) the diameter of *albizia* trees; and (7) the benefits and potential impacts – socially, economically, and environmentally –

of planting *albizia* as a fallow species. The data were analysed using qualitative and quantitative descriptive analysis, and included a cost analysis of the local farming system.

The study site

Demographics of Laham village

Laham Village, which is also the capital of Laham subdistrict, is located along the Mahakam river – the largest river in East Kalimantan – in an area that is quite remote (Figure 25-1). It is about 55 km from Ujoh Bilang, the capital of Mahakam Ulu district (about one-and-a-half hours by speed boat), or 491 km from Samarinda, the capital city of East Kalimantan province. From Samarinda, Laham village can be accessed by light aircraft (about 30 minutes), or by road to Kubar regency (taking about 11 hours), thence by speed boat up the Mahakam river for a further three hours.

The village of Laham had 800 inhabitants in 2003. It was named subdistrict capital in 2004, and by 2018, its population had grown to 1416 people, living in 351 households. Most of Laham's people are from the Bahau Dayak tribe, but other tribes represented in the population are Dayak Kayan, Dayak Penihing, Dayak Kenyah, Flores and Batak.

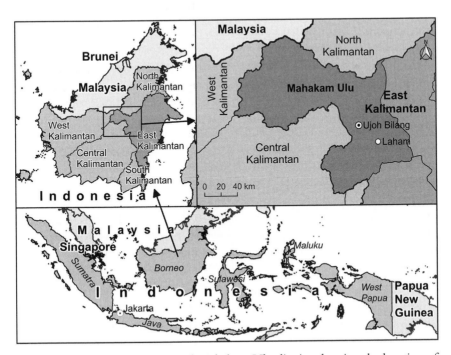

FIGURE 25-1:   East Kalimantan and Mahakam Ulu district, showing the location of the study village, Laham.

Basic education came to the village in 1911, when a Catholic missionary established a school. It was one of the milestones of education in East Kalimantan. Nowadays, as a subdistrict capital, Laham has an educational infrastructure ranging from kindergarten to elementary school and junior and senior high schools. The government's compulsory education programme up to junior high school and the presence of television in the village have created a social emphasis on the importance of education, for both adults and the younger generation. Older-generation

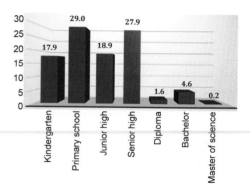

FIGURE 25-2: Percentages of various education levels among the people of Laham village.

villagers are eager to follow informal learning packages to get a better education and to send their children away for a college education. As a result, many of Laham's villagers have a high education level, even to college (Figure 25-2).

The research found that 36% of the household heads and their spouses had a high-school education. Among Laham's parents, awareness of the importance of education had led them to send their children away from the village to get a better education. School-aged children generally attended school classes and about 22% of the village's younger generation had reached college level.

However, few youngsters over 19 years old had junior high or even elementary education, due to economic limitations and disability (Figure 25-3). About 22% of these older children had been sent to the provincial capital to attend universities, at a high cost. With a higher education, there was a possibility that they would look for jobs in the city, thus reducing the family's labour force to work on the farm.

The people of Laham village are very familiar with various technologies; television, fridges and cellphones are common. Roads that were constructed by timber companies make access to their farm fields easy. Motorcycles have become a means of mobility both within the village and into the fields. Laptops, computers and printers are being used to run the village administration.

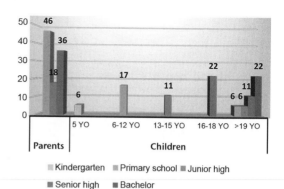

FIGURE 25-3: Education levels as a percentage of the families of respondent farmers in Laham village.

The livelihood and use of the land

Most of Laham's villagers work as swidden cultivators. Some of them have additional duties as village employees and are paid from government-allocated funds for a village-development programme. Many of the village's teachers are migrants. Thus, although the population of Laham has increased, there has been no substantial change in the number of swidden cultivators. This form of farming is part of the identity of the Dayak Bahau community, and its primary function is to grow upland rice, the staple food, for family consumption. In addition to growing subsistence rice crops, many farmers cultivate rubber or cocoa trees for their economic value. Recently, *albizia* (*Paraserianthes falcataria*) has become a popular fallow crop for the value of its timber, among other vital reasons.

Rubber is a traditional commodity that has long been cultivated by communities in East Kalimantan province. In 2013, the total area of rubber plantations in the province was 103,117 ha. Of this, community plantations covered 89,341 ha; the remainder were state and private-company plantations (Dinas Pekebunan Provinsi Kalimantan Timur, 2018). Mahakam Ulu district has only smallholder rubber plantations, and these covered 1839 ha in 2016 (see Table 25-2). The extent of rubber plantations in the wider East Kalimantan province belies the fate of rubber cultivation in Laham village.

About 60% of Laham's people own rubber plantations. In 2013, latex attracted a good price, reaching 13,000 rupiah (Rp) (about US$1.07)/kg, and at that time a government programme sought to improve the community's economy by introducing rubber planting as an alternative to shifting cultivation. When the trees reached tapping age, the community began selling unprocessed latex. Unfortunately, the price of latex declined sharply, falling to Rp2000 (about US15 cents)/kg in 2017. The low returns made the Laham villagers reluctant to tap the

Hevea brasiliensis (Willd. ex A.Juss.) Müll.Arg. [Euphorbiaceae]

Rubber is one of Indonesia's leading agricultural products. In the study area, farmers eagerly interspersed their swidden fields with rubber seedlings – developing a system known as 'jungle rubber' – to increase their household income. Then the market price for rubber plunged. While many conveniently nearby plantations are still maintained, more distant ones have been abandoned as farmers play a 'waiting game', hoping for better prices.

latex and their rubber plantations – particularly those that were distant from the village – were neglected, and are no longer productive. However, while the farmers turned to another commodity, cocoa, the rubber trees were not cut down. There is hope, still, that one day, the good prices for rubber latex will return.

Cocoa is widely cultivated in Mahakum Ulu district, with an area of 754.2 ha in 2014, when about 60% of this area had trees that were not yet in production. Laham village was one of 30 villages that each received 20,000 cocoa seedlings from the Mahakam Ulu district government. The trees were planted on community swidden land, but the growth rate was slow. In 2017, the Laham village administration decided to boost cocoa plantation by providing 105,000 certified seeds for a village programme in which 125 ha of former swidden fields was planted to cocoa. This indicated that more and more people were eager to grow cocoa to get additional family income.

When cocoa first became a plantation crop with economic potential, farmers began by allocating 1 ha to 2 ha of land (with an average of 1.4 ha) for cocoa trees. They hesitated to allocate large amounts of land because of the high costs of cocoa farming, the lack of management expertise for cocoa plantations, the limited availability of labour, and the need to retain land for shifting cultivation in order to feed their families. In Laham, prices for dry cocoa beans ranged from Rp20,000 to Rp30,000 (US$1.48 to US$2.21 in 2017)/kg, so the farmers adopted the same cultivation procedure for cocoa that they used for rubber: planting the seedlings in their swidden fields after they had finished planting rice. They took care of the growing seedlings while cultivating the rice, and after the rice was harvested, the cocoa generally grew towards maturity in a polyculture with other perennial trees such as rambutan (*Nephelium lappaceum*) and durian (*Durio* spp.)(Rusdiana et al., 2017).

Land planted with rubber or cocoa is automatically removed from the shifting cultivation cycle over many years, until the trees are no longer productive. Cocoa begins to yield fruit at five years of age and continues producing at economic levels for up to 30 years (Lebe et al., 2017). Similarly, rubber can be tapped at around five years of age and its economic life is between 20 and 25 years (Astuti et al., 2014). With plantation land taken out of the farming cycle in Laham, there was less and less land remaining for shifting cultivation. The field rotation – and hence the fallow period – grew shorter as farmers tried to cultivate crops for as long as possible. They soon faced the problem of decreasing soil fertility.

The main capital of the Laham village community in meeting the needs of its families is the village land. Swidden cultivation was expected to provide food security, while the rubber and cocoa plantations were expected to provide cash income. However, neither agricultural practice has been able to meet the needs of Laham's swidden families.

The onset of 'modernity' and the emphasis on education, particularly that away from the village, has had a profound effect on the availability of labour for shifting cultivation and the cost of farming. Interviews with research respondents led to a list of cultivation activities that were once performed by mutual or reciprocal labour, but nowadays involve the hiring of paid workers (Table 25-3). The standard cost of labour in Laham village is Rp50,000 (US$3.45)/day, in addition to the cost of food and beverages. That work which still attracts mutual or reciprocal labour still requires provision of food, drinks and snacks.

The costs incurred by respondents' families for swidden farming reportedly varied from Rp1.05 million (US$72.51)/ha to Rp13.65 million ($942.61)/ha, with an average of Rp7.52 million ($519.30). Fifty-four per cent of respondents spent more than Rp9 million ($621.50)/ha.

External pressures

The farmers' urgency to 'go where the money is', and improve their families' incomes, is an 'internal' influence on both their farming practices and the transformation from shifting cultivation to plantations. However, the transformation is being driven mainly by powerful 'external' influences, including government policy and market forces (Aryal and Choudhury, 2015).

Mahakam Ulu district, of which Laham subdistrict is a part, covers 15,315 sq. km. More than 80% of the district is covered by forest, including protected forests, production forests and customary forests (Rusdiana et al., 2017). According to the Central Bureau of Statistics of East Kalimantan province (BPS, 2017a), 14 timber companies have concessions covering 728,345 ha of forest in the province. They are licensed to utilize timber forest products, and the area they manage is no longer accessible by the farmers of local communities. Many people remain unaware of where the boundaries lie between forest areas that are open to them, and those that are closed.

Land-allocation policies favouring big companies have consequences for swidden cultivators living in and around forest areas. None of Laham's farmers has been able to open up new forest land for farming, because numerous companies have been granted permission to manage the land surrounding the village's territory. On the villagers' own land, the long production life of rubber and cocoa trees means that land allocated to these plantations is unlikely to be farmed again, so the farmers have been forced to re-use land that was planted just a few years earlier.

TABLE 25-3: Farming schedule of swidden cultivation at Laham and the need for hired labour.

Month	Activities	Undertaken by
May	Slashing	Hired labour
June – July	Tree felling, chopping, drying	Chainsaw rental; farmers perform this task
August	Burning	Farmers/mutual cooperation
September – October	Planting	Hired labour
September	Weeding	Hired labour
February – March	Harvesting	Hired labour or mutual cooperation

Source: Focus-group discussions with swidden cultivators, Laham village.

Oil palm is a pervasive commodity in Mahakam Ulu district. Oil-palm plantations are spread throughout the district and have penetrated the territories of villages around Laham village. Tempted by the lure of economic benefits promised by large oil-palm companies, neighbouring villagers allowed the companies to plant oil palm on their village land. They were then given work as labourers, earning a monthly salary to take care of oil-palm plantations growing on their land.

The Mahakam Ulu district government has identified land in Laham village's territory as a potential area for development of oil-palm plantation, but more than 90% of Laham's people have rejected the proposal. They put forward the following reasons:

- Allowing an oil-palm company to use village land would mean the loss of the community's rights over its land, and this would mean an inability to cultivate the rice varieties that constitute their staple diet. Moreover, those rice varieties would be lost without cultivation and the associated cultural and traditional values would also be lost.
- The people of Laham have learned from the experience of surrounding villages that have allowed private oil-palm companies to use their land. Those villagers no longer have land to cultivate, the air temperature is hot, and their income is limited because the amount of money they earn depends on the company. The Laham villagers point to Long Gelawang, one such neighbouring village, which now has to buy clean water for household use, costing up to Rp600,000 (US$41.43)/month/household.

Barriers to getting the most out of swidden farming

Swidden cultivation aims to meet basic subsistence family needs. It is a low-input system that relies on natural mechanisms for drying slashed vegetation and maintaining soil fertility. However, Talaohu (2013) points out that the success of shifting cultivation systems rests on natural mechanisms, so it is extremely vulnerable to the failure of these mechanisms. Laham's shifting cultivators face many challenges, not the least of which is the expense of practising swidden farming. Figure 25-4 illustrates the various internal and external factors influencing the ability of the village farmers to meet the challenges of their livelihood.

The *albizia* innovation

The Laham villagers have been unable to generate enough income from their farm land, and they have no other fixed income. With an increasing number of households and no possibility of expanding their farming area, they have been forced to shorten the fallow period and cultivate the land again before its fertility is recovered. Research respondents say the fallow periods on their land vary from as little as three years up to 10, with an average of around seven years. They admit that, for the land's fertility to be

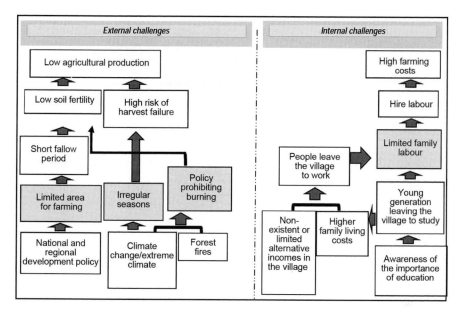

FIGURE 25-4: Challenges faced by Laham villagers in swidden cultivation.

restored, the fallow period should be at least 10 years. The consequence of a rotation of less than 10 years is a decrease in soil fertility, leading to decreased agricultural yields (Weinstock, 2015). The farmers of Laham point out that they need three cans of rice seed/ha when planting their swiddens. The yields are uncertain, ranging from 60 to 100 cans. But paying for labour takes up to 40 of those cans, leaving the farming family with only 20 to 60 cans for a year's consumption.

Slowly, the swidden cultivators became aware that there were other issues to be considered in the management of their limited land in order for them to generate additional income, as well as to achieve food security. In the traditional practice of shifting cultivation, with its wider availability of land and long fallow periods, these issues didn't arise. The clear relationship between the limited land area, the inadequate fallow period and dwindling soil fertility led the farmers to realize that they had to find some means of restoring the soil fertility within the existing limitations. This was consistent with advice from a wide body of research, which had left no doubt that swidden cultivation was reliant on the principles of fallow management (Mulyoutami et al., 2010; Ramakrishnan, 2015). The key to solving their problems was wiser use of the fallow period: the farmers had to combine the restoration of soil fertility with generation of additional income.

One solution was to plant leguminous species so that their nitrogen-fixing abilities would enrich the soil as they grew to become part of the fallow vegetation. This strategy arose from technologies involving 'fast-growing fertilizer trees', promoted by the World Agroforestry Centre (ICRAF) and other development-oriented organizations in the 1990s. Under this programme, the leguminous species *Sesbania sesban* successfully restored fertility to exhausted soils in Africa (Raintree and Warner, 2015).

Some of Laham's swidden cultivators first planted the fast-growing leguminous species *albizia* (*Paraserianthes falcataria*) in 2004, but simply as a sign of land ownership. No other purpose for the trees was seen, so no care was taken of them. Many pioneer species grew around the *albizia* and, as a result, they had no commercial value. That attitude changed when one farmer broadcast *albizia* seeds into his swidden and nurtured them as they grew big enough to dominate the fallow period. The trees grew well, and six or seven years later, he received a good price when he sold the timber. He crowned his commercial venture by buying a motorcycle to travel to and from his swidden fields, and the interest of other farmers was captured.

With the understanding that *albizia* provides economic value as well as fertilizing the soil, one by one the swidden cultivators of

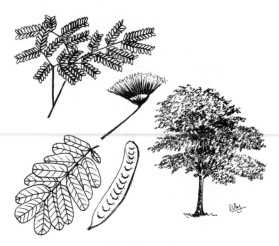

Paraserianthes falcataria (L.)
I.C.Nielsen[1] [Leguminosae]

This fast-growing species is increasingly being recognized as a valuable multipurpose tree. It is a nitrogen-fixing soil improver and produces valuable timber in a period that is often brief enough to fit within the shifting cultivation cycle. Native to a region including Indonesia's Maluku islands, Papua New Guinea and the Solomon Islands, the massive trunks of these trees were traditionally used to make dugout canoes.

Laham village began to plant this species so that it became a dominant part of their fallow vegetation. In 2017, when this data was collected, 29 swidden cultivators had planted *albizia* in their fields, with a total area of 92 ha. Figure 25-5 shows the area of land planted to *albizia* by individual farmers, as a percentage of the village's total.

The swidden cultivators of Laham village grow annual crops in their swiddens for one or two years before leaving a plot to lie fallow and move to a new site. *Albizia* seedlings are planted in the first year of rice cultivation, about one month after the rice has been planted. The seedlings are nurtured along with the annual crop, and in

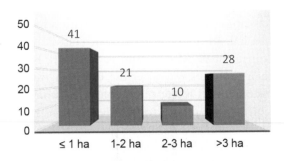

FIGURE 25-5: Areas of land planted to *albizia* by individual farmers in 2017, as a percentage of total *albizia* plantation (n = 29).

the second year, they are thinned and herbicide is used to kill encroaching grass. In the third year, the *albizia* is weeded, and is soon big enough to deprive weeds and grass of sunlight. The trees grow to become a dense and dominant stand until they are harvested for their timber and the swidden plot is returned to annual cropping.

Figure 25-6 shows an *albizia* plantation in a fallowed swidden at Laham village that was a focus for data gathering by the authors in 2018. The stand was six-and-a-half years old. The diameter at breast height of the 43 trees in the stand ranged from 18.8 cm to 58.2 cm, with an average diameter of 33 cm, calculated from circumferential measurements ranging from 59 to 183 cm (Figure 25-7). However, the *albizia* trees in this stand did not get silvicultural treatment, so that only a portion of them got enough sunlight to grow well. The trees will be harvested according to their intended use, after eight years for pulpwood production, 12 to 15 years for carpentry production, or 10 to 15 years if the trees are planted in an agroforestry system (Krisnawati et al., 2011).

FIGURE 25-6: A stand of 43 *albizia* trees in a swidden fallow at Laham village. The trees were planted in 2011 and were six-and-a half years old when this photograph was taken in 2018.

FIGURE 25-7: Each tree was measured for this study. The diagram below shows the diameter distribution of the trees in the fallow stand.

Photos: Niken Sakuntaladewi and Ari Wibowo.

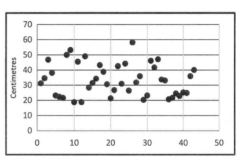

The price for *albizia* timber in Laham village is Rp300,000 (US$20.72)/cu. m, if the grower delivers the timber to the buyer at the banks of the Mahakam river. If the buyer has to fell the trees and transport the timber to the riverbank, the price is Rp150,000 ($10.36)/cu. m.

According to the farmers, buyers are happy to come to the village and harvest the trees in the fields. Interview respondents reported the successful sale of seven-year-old *albizia* trees by some swidden cultivators for between Rp15 million (US$1035.84) and Rp150 million ($10,358.40). This cannot yet be regarded as a maximum profit because of the weak bargaining position of the swidden cultivators. Nevertheless, they are exhilarated by their new earning capacity and consider that *albizia* trees are their future investment.

In interviews, the farmers said the silvicultural treatment of *albizia* plantations was very easy and quite cheap. The trees do not require specific treatment and have no need for fertilizer. Importantly, under an *albizia* fallow the land regains its fertility in a relatively short period. The land is ready for cultivation again after six to seven years; the farmers no longer have to wait for ten years or more. Therefore, the use of Laham's limited agricultural land is becoming more efficient, and the temptation to open forest areas for swidden cultivation is reduced.

Albizia trees are now in great demand among various groups in the Laham area, especially community-forest farmers. Looking to its future use of the trees, Laham village established its own nursery for producing seedlings of the *sengon solomon* variety of *albizia* in 2017 (Figures 25–8 and 25–9). It was launched with a contribution of 15,000 seedlings from the World Wide Fund for Nature. The swidden cultivators were keen to plant *sengon solomon* because of a belief that this variety grows bigger stems and the price is much higher as a result.

FIGURES 25-8 and 25-9: Laham's village nursery, showing *albizia* seedlings growing in plastic bags under a screen (top), and ready for transplanting into swiddens (bottom).

Photos: Ari Wobowo.

In the nursery, the *albizia* seeds are grown in plastic bags, and are ready for planting at about two months of age. The seedlings are planted into rice swiddens about a month after the rice is planted, at a spacing of about 3 m by 3 m, resulting in about 1000 seedlings/ha. After about a year, 30% to 40% of the seedlings are thinned to leave 600 to 700 trees to grow to maturity.

No alternative to burning

Despite their innovative approach to the fallow period in their practice of shifting cultivation, the farmers of Laham say that one aspect of the traditional system is unchangeable: there is no alternative to using fire to clear fields for cultivation.

Interviews and focus-group discussions revealed that all respondents understood that the Indonesian government had prohibited the burning of land and forest. They learned of the ban from TV broadcasts, social meetings with police officers, and many other sources. They agreed that the policy was a good one, and that if fires were not managed carefully, they could spread into the forest and the smoke could impact community health. However, they still thought that the policy was not appropriate in their case, because they took so many precautions. They were not aware that the policy was aimed at big companies, and that there was still room for swidden cultivators to burn, as long as it was no more than two hectares of land per household, and local traditional wisdom was used throughout the operation.

In 2017, a burning incident shocked the Laham community. A group of swidden cultivators cleared eight hectares of land between them and burned it all at the same time. It resulted in a phone call to the Laham headman from the police, informing him that there was a forest fire in his area and delivering a stern warning. The fire had been detected by satellite and the warning heightened local caution.

Nevertheless, interview respondents concluded that burning for land preparation was both necessary and inevitable. They concurred with the findings of Katterings et al. (1999), cited in Mulyoutami et al. (2010), that burning is the most effective and quickest way of clearing agricultural land; it suppresses the growth of weeds and other vegetation, especially after planting crops; it turns biomass into a natural fertilizer that benefits plants and soil; it soothes the soil, so that the seeds grow faster; and it is an effective way to kill pests and pathogens.

Laham's farmers say they have no other land-clearing technology to replace burning, and that no swidden fire on Laham territory has ever escaped into the forest. Moreover, they say that the use of *albizia* as a fallow species has brought improvements to land preparation:

- Land clearing is easier, faster and lighter because *albizia* trees reduce the growth of shrubs. The trees will also be felled by a buyer, so the farmer no longer needs to undertake slashing and cutting or pay for labour, chainsaw rental fees and other land-preparation costs.
- Big *albizia* trees are transported to the Mahakam riverside by the buyers, leaving only small branches, twigs, leaves and a few other plants, so the time required for drying is shorter.

- Burning is a faster process, taking less than two hours, because there is less material to burn. The land is also cleaner because there are no big timbers left in the field.
- Smoke from the fire is greatly reduced, so there are no longer 'calls from government officers'.
- After burning an *albizia* fallow, many seedlings sprout spontaneously, meaning there is no longer a need to buy *albizia* seeds and tend to a nursery. The only need is to thin the natural growth, so the costs are much lower.

The future

In one sense, the development of *albizia* fallows has solved the apparently inescapable problem of dwindling soil fertility and diminishing harvests in Laham village, because of inadequate land for practising shifting cultivation. The new system recovers soil fertility in a shorter time, delivers a generous boost to household incomes, and improves other aspects and costs of swidden farming. However, after a few decades of increasingly desperate efforts to increase their incomes, the new innovation leaves the farmers with three agricultural or plantation systems, all of which are 'up and running', on the same inadequate areas of land. In economic terms, all respondents still show interest in the three systems and their various products: rubber, cocoa, and *albizia* from shifting cultivation. However, most respondents are more interested in cocoa because they expect it will provide a weekly income. With rubber, they still expect the price of latex to improve, so they keep their trees. They are aware that *albizia* will take seven years to provide economic benefits, so they consider this 'family savings for the future'. However, when it comes to environmental benefits and ease of land preparation, the swidden cultivators of Laham show more interest in *albizia*. Aside from the potential economic benefits, they believe that with *albizia*, soil fertility will recover quickly and they can farm again, with easier land preparation. Figure 25-10 shows farmers' perceptions of the benefits of the three systems.

Conclusions

Laham's villagers are determined to continue their practice of shifting cultivation, not only to protect their food security, but to maintain the Hudoq culture that accompanies the traditional agricultural system. They point out that this culture, which plays a big role in uniting the Laham community, has been lost to villagers who surrender their land to oil-palm companies. Following

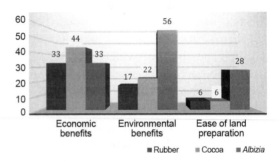

FIGURE 25-10: Farmers' perceptions of the benefits of rubber, cocoa, and *albizia* plantations.

their objectives, the villagers have managed to retain the greatest asset in their lives – the land. It is becoming more valuable and is sought-after by others, who are prepared to fight for it.

Their agricultural system relies entirely on nature, and it is facing both external and internal challenges, resulting in declining agricultural production, increasing costs for swidden farming, and increasing costs for meeting the growing needs of families. The external challenges include land constraints, climate change, and the government's ban on burning to clear land. The biggest challenge from within is reduced family labour. This is related to lifestyle changes that follow higher education, interaction with outsiders, and government development programmes.

The swidden cultivators of Laham village are fixated on two issues: (1) the ban on burning and their inability to use any other means than fire to prepare their land, and (2) the need for a fallow period of more than ten years to restore soil fertility, which implies the need for plenty of agricultural land, when that in Laham is increasingly limited.

Growing *albizia* as a fallow species is seen as a way to enable a continuation of shifting cultivation. The years in which the trees grow to maturity are seen as 'a time of saving', and the eventual fairly high economic returns bring additional income to village households. Importantly, the leguminous trees also hasten the return to soil fertility, with fallows lasting about seven years, compared with about ten years for a traditional fallow. *Albizia* also means lighter work in land preparation, as buyers will harvest and remove the trees, reducing the cost of land preparation and potentially reducing emissions from burning.

The *albizia* innovation is increasingly being adopted by the shifting cultivators of Laham village because it responds to their problems, it fits with existing biophysical and socio-economic conditions, and moreover, it is easy to do.

Acknowledgements

The authors would like to acknowledge support provided by the Forest Carbon Partnership Facility (FCPF-Indonesia) in achieving field data collection in Laham village, and also the World Wide Fund for Nature, for its assistance during the data collection and providing valuable information.

References

Allen, B. and Filer, C. (2015) 'Is the 'bogeyman' real? Shifting cultivation and the forests, Papua New Guinea', in M. F. Cairns (ed.) *Shifting Cultivation and Environmental Change: Indigenous People, Agriculture, and Forest Conservation,* Earthscan from Routledge, London, pp.517-545

Anonymous (2011) 'Hak Pengusahaan Hutan' (Forest Concession Rights), in *Green My Earth,* retrieved from https://hendryferdinan.wordpress.com/2011/03/14/hak-pengusahaan-hutan-hph/, accessed 11 July 2021

Aryal, K. and Choudhury, D. (2015) 'Climate Change: Adaptation, Mitigation and Transformation of Swidden Landscapes. Are we throwing the baby out with the bathwater?' in M. F. Cairns (ed.) *Shifting Cultivation and Environmental Change: Indigenous People, Agriculture, and Forest Conservation,* Earthscan from Routledge, London, pp.281-288

Astarika, R. (2015) 'Peremajaan karet rakyat sebagai solusi peningkatan produktivitas karet (studi kasus di provinsi Jambi)' (Rejuvenation of smallholder rubber as a solution to increase productivity: A case study in Jambi province), *Jurnal ilmiah Univarsitas Batanghari Jambi* 15(1), pp.105-109

Astuti, M., Hafiz, Yuningsih, E., Wasingun. A. R., Nasution, I. M., Mustikawati, D. (2014) *Pedoman Budidaya Karet (Hevea brasiliensis) yang Baik* (*Guidelines for Good Rubber* (Hevea brasiliensis) *Cultivation*), Kementerian Pertanian, Direktorat Jenderal Perkebunan (Directorate-General of Plantations, Ministry of Agriculture)

BPS (2014) 'Statistics of East Kalimantan province, 2014', in *Provinsi Kalimantan Timur Dalam Angka 2014* (East Kalimantan province in Figures 2014), retrieved from https://kaltim.bps.go.id/publication/2014/12/29/e1f40006068a483eca1d7659/kalimantan-timur-dalam-angka-2014.html, accessed 10 August 2021

BPS (2015a) 'Statistics of East Kalimantan province, 2015', in *Kalimantan Timur Dalam Angka 2015* (East Kalimantan in Figures 2015), retrieved from https://kaltim.bps.go.id/publication/2015/11/02/a21fa528921162ad9cdeab82/kalimantan-timur-dalam-angka-2015.html

BPS (2015b) *Statistical Year Book of Indonesia, 2015*, Badan Pusat Statistik (Central Bureau of Statistics) retrieved from https://www.bps.go.id/statictable/2015/09/22/1849/jumlah-dan-persentase-rumah-tangga-di-sekitar-kawasan-hutan-yang-melakukan-perladangan-berpindah-2004-dan-2014.html, accessed 18 March 2021

BPS (2016) 'Statistics of East Kalimantan province, 2016', in *Provinsi Kalimantan Timur Dalam Angka 2016* (East Kalimantan Province in Figures 2016), retrieved from https://kaltim.bps.go.id/publication/2016/07/15/9743adfa7b30d83ab79fb777/provinsi-kalimantan-timur-dalam-angka-2016.html

BPS (2017a) 'Statistics of East Kalimantan province, 2017', in *Provinsi Kalimantan Timur Dalam Angka 2017* (East Kalimantan Province in Figures 2017) retrieved from https://kaltim.bps.go.id/publication/2017/08/11/f39c3916581e94fc12c3fb2d/provinsi-kalimantan-timur-dalam-angka-2017.html, accessed 10 August 2021

BPS (2017b) *Statistical Year Book of Indonesia, 2017*, Badan Pusat Statistik (Central Bureau of Statistics)

BPS (2018a) 'Statistics of East Kalimantan province, 2018', in *Provinsi Kalimantan Timur Dalam Angka 2018* (East Kalimantan Province in Figures 2018), retrieved from https://kaltim.bps.go.id/publication/2018/08/16/9341dae4a1306ccfee98a393/provinsi-kalimantan-timur-dalam-angka-2018.html

BPS (2018b) *Forest Concession Rights Company Statistics*, BPS-Statistics Indonesia, retrieved from https://www.bps.go.id/publication/2019/11/29/32f23bcfc7f6c1166deabd1a/statistik-perusahaan-hak-pengusahaan-hutan-2018.html, accessed 10 August 2021

BPS (2019) *Statistik Kelapa Sawit Indonesia* 2019 (Indonesian Oil Palm Statistics 2019), BPS-Statistics Indonesia, retrieved from https://www.bps.go.id/publication/2020/11/30/36cba77a73179202def4ba14/statistik-kelapa-sawit-indonesia-2019.html

Choudhury, D. (2012) 'Why do *jhumias jhum*? Managing change in shifting cultivation areas in the uplands of northeastern India', in Sumi Khrisnan (ed.) *Agriculture and a Changing Environment in Northeastern India*, Routledge, New Delhi, pp.78-100

Debel, F., Tilahun, U. and Chimdesa, D. (2014) 'The impact of population growth on forestry development in East Wollega zone: The case of Haro Limu district', *Journal of National Sciences Research* 4(18), ISSN 2225-0921 (online)

Dressler, W., Wildon, D., Clendenning, J., Cramb, R., Mahanty, S., Lasco R. D., Keenan, R. J., Phuc, X. T. and Gevana, D. T. (2015) '*Examining how Long-fallow Swidden Systems Impact upon Livelihood and Ecosystem Services Outcomes Compared with Alternative Land Uses in the Uplands of Southeast Asia*', Working Paper 174, Center for International Forestry Research (CIFOR), Bogor, Indonesia

Direktorat Jenderal Perkebunan (Directorate-General of Plantations) (2015) *Statistik Perkebunan Indonesia 2014-2016. Kelapa Sawit* (Statistics of Indonesian Oil-Palm Plantations, 2014–2016), Ministry of Agriculture, Jakarta

Fauzi, A. (2008) 'Kesesuaian Laham tanaman karet (*Hevea brasiliensis*) berdasarkan aspek agroklimat di Sulawesi Tenggara' (Land suitability for rubber (*Hevea brasiliensis*) based on agro-climatic aspects in Southeast Sulawesi), skripsi, Departemen Geofisika dan Meteorologi, Fakultas Matematika dan Ilmu Pengetahuan Alam, Institut Pertanian Bogor (dissertation to the Department of Geophysics and Meteorology, Faculty of Mathematics and Natural Sciences, Bogor Agricultural University)

Gommes, R. (1998) *Some Aspects of Climate Variability and Food Security in Sub-Saharan Africa*, Royal Meteorological Institute and Royal Academy of Overseas Science, Brussels

Gul, S., Khan, M. A. and Khair, S. M. (2014) 'Population increase: A major cause of deforestation in Ziarat district', *Journal of Applied and Emerging Sciences* 5(2)

Krisnawati, H., Varis, E., Kallio, M. and Kanninen, M. (2011) 'Paraserianthes falcataria (L.) *I.C. Nielsen: Ekologi, silvikultur dan produktivitas (Paraserianthes falcataria* (L.) I.C. Nielsen: Ecology, Silviculture and Productivity), Center for International Forestry Research (CIFOR), Bogor, Indonesia

Lebe, D., Fadhli, Z. and Philips, M. (2017) *Akses untuk Pembiayaan: Kasus Kakao. Pelatihan Sektor Kakao untuk Lembaga Keuangan. Bagian 1: Pelatihan sektor kakao* (Access to Finance: The Case of Cocoa. Sector Training for Financial Institutions), retrieved from www.swisscontact.org/indonesia

Li, C., Ting, Z., and Rasaily, R. G. (2010) 'Farmer's adaptations to climate risk in the context of China: Research on the Jianghan plain of the Yangtze river basin', *Agriculture and Agricultural Science Procedia* 1, pp.116-125

Medrilzam (2015) *Policy and Institutional Reforms to Effectively Combat Peat Fire*, Office of the Deputy Minister for Maritime and Natural Resources. Kementerian PPN/Bappenas (Ministry of National Development Planning of Indonesia/National Development Planning Agency), Jakarta

Mulyoutami, E., van Noordwijk, M., Sakuntaladewi, N. and Agus, F. (2010) *Perubahan Pola Perladangan. Pergeseran Persepsi Mengenai Para Peladang di Indonesia* (Changing Patterns in Farming: Shifting Perceptions of Cultivators in Indonesia), World Agroforestry Centre (ICRAF), Bogor, Indonesia

Natawidjaja, R. S., Supyandi, D., Tulloh, C., Tridakusumah, A. C., Calford, E. M. and Ford, M. (2009) *Climate Change, Food Security and Income Distribution: Adaptations of Small Rice Farmers*, Crawford School of Economics and Government, The Australian National University, Canberra

Nath, D. C. and Mwchahary, D. D. (2012) 'Population increase and deforestation: A study in Kokrajhar district of Assam, India', *International Journal of Scientific and Research Publications* 2(10)

Naylor, R. L., Battisti, D. S., Vimont, D. J., Falcon, W. P. and Burke, M. B. (2007) 'Assessing risks of climate variability and climate change for Indonesian rice agriculture', *Proceedings of the National Academy of Sciences of the United States of America* 104(19), pp.7752-7757

Olesen, J. E. and Bindi, M. (2002) 'Consequences of climate change for European agricultural productivity, land use and policy', *European Journal of Agronomy* 16, pp.239-262

Pearl, H. and O'Connell, E. (2016) *Pertanian perladangan berpindah menciptakan ekosistem alami* (Swidden agriculture creates a natural ecosystem), Center for International Forestry Research (CIFOR), Bogor, Indonesia, available at https://forestsnews.cifor.org/40193/pertanian-perladangan-berpindah-menciptakan-ekosistem-alami?fnl=id, accessed 11 July 2021

Penot, E. (2007) 'From shifting cultivation to sustainable jungle rubber: A history of innovations in Indonesia', in M. F. Cairns (ed.) *Voices from the Forest: Integrating Indigenous Knowledge into Sustainable Upland Farming*, Resources for the Future Press, Washington, DC, pp.577-599

Raintree, J. and Warner, K. (2015) 'Agroforestry pathways revisited: Voices from the past', in M. F. Cairns (ed.) *Shifting Cultivation and Environmental Change: Indigenous People, Agriculture and Forest Conservation*, Earthscan from Routledge, London, pp.87-121

Ramakrishnan, P. S. (2015) 'Shifting agriculture and fallow management options: Where do we stand?' in M. F. Cairns (ed.) *Shifting Cultivation and Environmental Change: Indigenous People, Agriculture and Forest Conservation*, Earthscan from Routledge, London, pp.186-198

Rumboko, L. R., Hakim, I., Sakuntaladewi, N. and Komarudin, H. (2018) 'Politics, migration and rural transformation in oil palm locations in Sumatra and Kalimantan', paper delivered to the LANDac Annual International Conference, The Netherlands Land Academy, Utrecht, the Netherlands

Rusdiana, O., Supijatno, Ardiyanto, Y. and Widodo, C. E. (2017) 'Potensi pengembangan kehutanan dan pertanian kabupaten Mahakam Ulu, provinsi Kalimantan Timur' (Development potential for forestry and agriculture in Mahakam Ulu district, East Kalimantan province), *Journal of Regional and Rural Development Planning* 1(2), pp.144–131

Setiabudi, Budiman, A. and Hultera (2013) *Laporan Pemetaan Tutupan Lahan Kabupaten Kutai Barat dan Mahakam Ulu Provinsi Kalimantan Timur* (Land Cover Mapping Report, West Kutai and Mahakam Ulu regencies, East Kalimantan province), retrieved from https://docplayer.

info/45665205-Laporan-pemetaan-tutupan-lahan-kabupaten-kutai-barat-dan-mahakam-ulu-provinsi-kalimantan-timur-oleh-setiabudi-arif-budiman-hultera.html, accessed 12 August 2021

Shen, S., Basist. A. and Howard, A. (2010) 'Structure of a digital agriculture system and agricultural risks due to climate change', *Agriculture and Agricultural Science Procedia* 1, pp.42–51

Sugardiman, R. A. (2017) *Kebijakan Pemerintah dalam Inisiatif Pemetaan dan Data Baseline untuk Monitoring Kegiatan karbon (REDD+) pada Tingat Kabupaten/Tapak* (Government Policy on Mapping and Data Baseline Initiatives for Monitoring Carbon Activities (REDD+) at District and Site Levels), FGD Penutupan Laham dan Monitoring Karbon (REDD+) Kegiatan Forest Program II (FGD land closure and carbon monitoring (REDD+) forest programme), River Flow Management Centre and Barito protected forest, Batang Hari district, Jambi province, Indonesia

Talaohu, M. (2013) 'Perladangan berpindah: Antara masalah lingkungan dan masalah sosial' (Shifting cultivation: Between environmental and social problems), *Populis* 7(1)

Tamzil, M., Santoso, R. P., Saipul, M. and Setyawan, E. P. (2016) *Kajian Potensi Pengembangan Usaha Komoditi Kakao Kampung Long Tuyoq–Mahakam Ulu* (Studies of Potential Cocoa Commodity Business Development in Long Tuyog village, Mahakam Ulu district), WWF Indonesia and the Penabulu Foundation

van Noordwijk, M., Minang, A. and Hairiah, K. (2015) 'Swidden transitions in an era of climate-change debate', in M. F. Cairns (ed.) *Shifting Cultivation and Environmental Change: Indigenous People, Agriculture and Forest Conservation*, Earthscan from Routledge, London, pp.261-280

Wang, J. (2010) 'Food security, food prices and climate change in China: A dynamic panel data analysis', *Agriculture and Agricultural Science Procedia* 1, pp.321-324

Weinstock, J. A. (2015) 'The future of swidden agriculture', in M. F. Cairns (ed.) *Shifting Cultivation and Environmental Change: Indigenous People, Agriculture and Forest Conservation*, Earthscan from Routledge, London, pp.179-185

Wulandari, S. A. and Kemala, N. (2016) 'Kajian komoditas unggulan sub-sektor perkebunan di Provinsi Jambi' (A study of leading sub-sector plantation commodities in Jambi province), *Jurnal ilmiah Universitas Batanghari, Jambi* 16(1), pp.134-141

Note

1. The correct botanical name for this species is problematic. It was earlier identified as one of many species in the genus *Albizia* – hence its local name. The name *Albizia falcataria* (L.) Fosberg was more recently commonly replaced by the name *Paraserianthes falcataria* (L.) I.C. Nielsen – the botanical name given to the species in this chapter. At the time of writing, recognized authorities on botanical names disagreed on how it should be properly named: www.theplantlist.org and www.worldfloraonline.org said that both *Albizia falcataria* and *Paraserianthes falcataria* were synonyms of the accepted name *Falcataria moluccana* (Miq.) Barneby & J. W. Grimes. However, Kewscience's www.plantsoftheworldonline.org regarded all three of these names as synonyms of the accepted name *Falcataria falcata* (L.) Greuter & R. Rankin. All names relate to members of the plant family alternately named Leguminosae or Fabaceae.

26

TRADITIONAL USE OF *MACARANGA* TREES FOR SOIL FERTILITY

By Naga shifting cultivators in northeast India

*Sapu Changkija, Dwipendra Thakuria and Alomi Cynthia**

Introduction

Nagaland is one of India's 'frontier' states. It is flanked by Myanmar in the east, Assam state in the west and north, and Manipur state in the south. It has a total area of 16,580 sq. km, with a total population of 1.99 million (in 2001) (Figure 26-1). The Nagas are the indigenous inhabitants of the state and make up about 84% of the total population. Most of them are Christians, and their languages are part of the Tibeto-Burman language family. Nagaland is covered by young mountain ranges and hilly areas with sharp crests and deep gorges with narrow valleys. The state has a monsoon climate with average annual rainfall of 2500 mm and 85% relative humidity. The total rainfall occurs over about seven months, from April to October. Exposed as it is to this heavy rainfall, the mountainous state is endowed with a unique and rich diversity of terrestrial flora and fauna along the altitude gradient from flood plains to high mountain ranges, supporting diverse agricultural systems involving different farming practices. The main source of livelihood for the Naga peoples is agriculture, and the topography of the land has dictated that the main form of agriculture is shifting cultivation, known locally as *jhum*. This has resulted in a rich agrobiodiversity within microclimatic ecosystems. Over the centuries, the elements of this rich agro-biodiversity have evolved in intricate association with communities distributed along the mountain and hill slopes. More than being simply a means of supplying food, the practice of *jhum* has become a way of life. Interactions between the tribal people and natural systems have helped to maintain the richness of species and genetic materials in both production systems and mountain forests.

* PROFESSOR SAPU CHANGKIJA, Department of Genetics and Plant Breeding, School of Agricultural Sciences and Rural Development, Nagaland University, Mezdiphema Campus, Nagaland; DR. DWIPENDRA THAKURIA, Professor of Soil Science and Microbiology and Principal Investigator, College of Post Graduate Studies, Central Agricultural University, Meghalaya; and DR. ALOMI CYNTHIA, Researcher, Institute of Naga Studies, Tetso College, Dimapur, Nagaland.

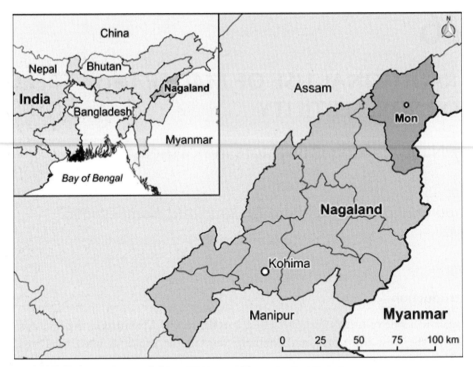

FIGURE 26-1:   Nagaland, in northeast India, showing Mon district, where Konyak Naga farmers have developed a fallow management system featuring *Macaranga* trees.

The impact of the environment on a culture is often perceived primarily through an examination of its subsistence technologies. This study sought to make a detailed survey of a traditional mountain farming system based on the use of *Macaranga* species in shifting cultivation. This system has developed over centuries of interaction between the people and their environment. With their simple technologies, the people have developed subsistence skills suited to a subtropical area characterized by hilly terrain with gentle slopes.

Shifting cultivation (*jhum*)

Traditional shifting cultivation, or *jhum*, in the highlands of Nagaland is a complex system that is sustainable if the initial cultivation phase is followed by a fallow phase that is long enough to allow secondary-forest vegetation to regrow and soil nutrients to recover. Traditional shifting cultivation is normally practised on a permanent basis, as every household in a community rotates its fields and fallow forests around the established community land of its village. In the initial process, the farmers (*jhumias*) clear a large tract of land by slashing the undergrowth, weeds, and smaller trees in secondary forests as well as lopping the branches from big trees. The slashed biomass is allowed to dry and is burnt before the onset of the wet season. Burning releases nutrients previously stored in the forest biomass on to the soil surface, making it available for crops in the cultivation phase. Burning also helps to neutralize acid soil, eradicate the seeds of weeds and drive out pests and rodents.

After being cleared and burnt, *jhum* fields may be planted with annual crops for one or two, or sometimes three, years. But to avoid severe depletion of soil fertility and increasing weed infestation, the community will make a collective decision to move to a new field, leaving the old one to lie fallow and regenerate naturally. After fallow regeneration has reached a mature stage, providing sufficient biomass to enable productive re-cultivation, the full cycle of shifting cultivation can begin again. This may take between six and 15 years, depending on how scarce new land is.

The most remarkable aspect of *jhum* cultivation is the huge variety of cereals, vegetables and fruit that can be grown in a single *jhum* field – an attribute that is beyond the capacity of farming land on the plains. Mixed cropping and a multi-layered structure are basic traditional features of shifting cultivation, making each field a miniature 'biodiversity hot spot' in the landscape of a farming village. It is common to find between 40 and 70 non-rice crops growing in a single household *jhum* of about one hectare.

The dynamics of shifting cultivation have led to the belief that as long as both population density and pressure to produce large surpluses of crops are low, this form of agriculture is a highly sustainable practice that does not result in significant degradation of the ecosystem. In fact, the whole gamut of indigenous knowledge involved in shifting cultivation could be adopted as theoretical bases for practical and field-based approaches to more intensive agricultural systems elsewhere. This could provide improvements in such areas as soil conservation and fertility management, management of fallow land, conservation of germplasm and crop diversity, weed control and management, indigenous pest controls and use of bio-pesticides, and development of agro-climatic calendars for seasonal schedules of farming activities. Technologies arising from shifting cultivation could thus have a great bearing on other farming systems.

Incorporating *Macaranga*

Fallow management is a vital practice among the traditional shifting cultivators of Nagaland. As much as any other part of their agricultural system, it has been developed over many centuries. Over time, the *jhumias* of Nagaland – particularly those of the Konyak Naga people of Mon district (Figure 26-2) – have perceived that trees of the *Macaranga* genus, which were a common pioneer

FIGURE 26-2: An elder of the Konyak Nagas, whose farmers have developed a fallow-management system based on *Macaranga* trees.

Photo: Malcolm Cairns.

species in their shifting cultivation fallows and in natural gaps of evergreen forest, had a distinct effect on their farming system. Swiddens cleared from fallows dominated by *Macaranga* trees had richer soil and the harvests were better than those from fields without *Macaranga*. Ultimately, when they began planting the first crops in new swiddens, they retained seedlings of *Macaranga denticulata*, *Macaranga indica* and *Macaranga peltata* that grew naturally. When necessary, they purposely planted trees of the same three species. They nurtured the young trees throughout the cultivation phase as they grew alongside the annual crops, and then they allowed them to dominate the years of fallow recovery.

The three *Macaranga* species are small evergreen trees from the plant family Euphorbiaceae. Growing up to 18 m, their natural range covers Southern China, Indochina, Indonesia (Sumatra and Java), the Malay Peninsula, Thailand, Laos, Cambodia, Myanmar, Sri Lanka and northeastern India. All of these trees grow bolt upright with straight, often tapering, trunks with smooth, pale-gray bark, large, simple leaves, unisexual flowers on different trees and an open crown. Although the primary use of the trees is in fallow enrichment, the wood

Macaranga denticulata (Blume) Müll.Arg. [Euphorbiaceae]

These trees grow rapidly, up to about 18 metres tall, with an open crown. This is often used as a pioneer species when restoring forest land. A dioecious species, both male and female trees are needed to produce fruit and seed. The trees also have medicinal uses: a decoction of the leaves can be used to cleanse wounds.

of mature trees is used for construction or as firewood and the leaves for wrapping purposes – especially, in Nagaland, for wrapping fermented soybean cake.

As *Macaranga* is a pioneer tree in early fallow succession, its role in natural regeneration in later stages of succession may be associated with other long-lived species, such as *Albizia chinensis, Canarium strictum, Castanopsis indica, Duabanga grandiflora, Erythrina arborescens, Schima wallichii, Choerospondias axillaris, Stereospermum chelonoides, Quercus semecarpifolia* and other climax-forest species. As well, various bamboo species also occur in secondary succession in shifting cultivation fallows.

The ability of *Macaranga* species to boost recovery in secondary-forest fallows finds its focus in the soil, among the roots, and a complex symbiotic relationship between the trees and fungus species, known as a mycorrhizal association. Commonly, the fungus species form unique structures, called arbuscules and vesicles, within the root tissue of the host plants as a principal element of this relationship. The trees capture energy from the sun by means of photosynthesis and supply carbon to the fungus, while the fungus helps the trees to capture nutrients such as phosphorus, sulphur, nitrogen and micronutrients from the soil. The host trees' 'investment' in the mycorrhizal relationship may involve transferring up to 20% of their carbon to the fungi (Pfeffer et al., 1999). This represents a significant contribution to the below-ground organic-carbon pool.

Far more than involving only the *Macaranga* species, arbuscular mycorrhizal symbiosis between fungi and plants is found in about 80% of all vascular plant species. This symbiosis is believed to have originated at least 460 million years ago (Simon et al., 1993), and could have played a crucial role in the initial colonization of land by plants, and in the evolution of vascular

Schima wallichii Choisy [Theaceae]

A common species in swidden fallows, this tree is believed to emerge in natural succession from *Macaranga* stands. The tree is valued for its wood, and it is also used medicinally: the astringent inner flower petals are used to treat uterine disorders and hysteria.

plants (Brundrett, 2002), including all flowering plants. Many crop species, such as wheat and rice, also form mycorrhizal associations.

About 29 species of arbuscular mycorrhizal fungi have been recovered from the roots of *Macaranga* trees in the field, and since the trees deliver generous quantities of nutrient-rich leaf litter to the soil, the symbiosis between the fungi and *Macaranga* trees may be playing a major role in reclaiming soil nutrients in swidden fallows (Youpensuk et al., 2004). Upland rice also appears to benefit directly from association with mycorrhizal fungi (Youpensuk et al., 2005), indicating that the fungi play an important role in maintaining soil fertility and upland rice productivity in shifting cultivation by enhancing the accumulation of nutrients. The two species *M. denticulata* and *M. indica* are known to be highly dependent on mycorrhizal fungi for growth and uptake of the nutrients phosphorus, nitrogen and potassium.

This symbiosis highlights the complex interactions that add to high levels of biodiversity in swidden systems, not only in Nagaland, but also in neighbouring countries. H'tin, Khmu and Karen farmers in Thailand believe that *M. denticulata* has fallow enriching properties. Good upland-rice yields of 3.5 t/ha and higher have been reported in short-fallow systems associated with high-density stands of 4200 *M. denticulata* trees per hectare on the poorest of mountain soils, with a pH of 4.0

and just 2 ppm of available phosphorus (Yimyam et al., 2003). In an environment in which fallow periods have been reduced to six years and less, dense growths of *M. denticulata* have maintained a rich diversity of swidden crops over 10 years. Zinke et al. (1978) reported a high level of nutrients in *Macaranga*-dominant biomass at the start of a swidden cycle at Pa Pae village, in Mae Hong Son province, northern Thailand. The nutrient levels were much higher than those previously recorded with seven-year fallow re-growth in a traditional Lawa shifting cultivation system. The *Macaranga*-dominant biomass provided 134 kgN/ha, 16 kgP/ha and 179 kgK/ha after a seven-year fallow.

Studying the root zones (rhizosphere) of *Macaranga*

At the heart of the symbiotic relationship between *Macaranga* trees and fungi species is efficiency in accessing soil nutrients. According to Bolan (1991), mycorrhizas can be much more efficient than plant roots, with an uptake rate of the vital plant nutrient phosphorus reaching up to six times that of root hairs. Less is known about the role of nitrogen nutrition in the arbuscular mycorrhizal system, and while significant advances have been made in explaining this complex interaction, much investigation remains to be done.

A recent study conducted at the Central Agricultural University in Meghalaya, northeastern India (Suting et al., unpublished), sought a clearer understanding of the complex interaction between microscopic fungi species occupying soil around the roots of *Macaranga* trees and living within the root tissues. The study aimed to assess the benefits of cultivating *Macaranga* trees as a dominant fallow species in *jhum* systems by measuring the impacts of the arbuscular mycorrhizal relationship. Soil from around the *Macaranga* roots (rhizospheric soil) was compared with bulk soils from a five-year-old *jhum* cycle at Changki village, in Mokokchung district, to determine levels of colonization by arbuscular mycorrhizal fungi and reveal soil-process indicators (Tables 26-1 and 26-2; Figure 26-3).

TABLE 26-1: Arbuscular mycorrhizal fungal colonization in *Macaranga denticulata*, Changki village, Mokokchung.

Spore counts and colonization in fine roots	Sampling months				Mean	SD
	Jan	Mar	July	Oct		
AMF spores count (per 100 g soil)	475	501	445	410	457.75	±39.2
Hyphal colonization (H) (%)	56.7	46.9	50.7	48.3	50.65	±4.33
Vesicles colonization (V) (%)	18.2	17.6	16.3	15.6	16.925	±1.19
Arbuscles colonization (A) (%)	ND	ND	ND	ND		
Dark septate endophytes (DSE) (%)	10.7	19.7	11.8	9.3	12.875	±4.66
Total colonization (H+V+A+DSE) (%)	85.6	84.2	78.8	73.2	80.45	±5.65

Notes: ND = not detected; SD = Standard deviations.

Source: Suting et al. (unpublished)

TABLE 26-2: Biological and biochemical attributes of *Macaranga* rhizosphere soils versus bulk soils.

Soil properties	*Five-year* jhum *cycle*	
	Macaranga rhizospheric soil	*Bulk soil*
Bacterial counts (colony forming units/g of soil)	6.2×10^6	8.4×10^5
	$(6.79 \pm 0.88b)$	$(5.92 \pm 0.74a)$
Microbial biomass carbon (µg/g of soil)	$2824 \pm 228b$	$1825 \pm 177a$
Potentially mineralizable nitrogen (µg/g of dry soil)	$34.4 \pm 3.5a$	$52.5 \pm 3.2b$
Acid-phosphomonoesterase activity (µg pNP/g soil h)	$1908 \pm 138b$	$840 \pm 98a$
Dehydrogenase activity (µg TPF/g soil h)	$12.4 \pm 2.3b$	$5.8 \pm 1.8a$
Aryl-sulfatase activity (µg pNP/g soil h)	$240 \pm 42b$	$154 \pm 22a$

Notes: Values in parentheses for bacterial counts represent log cfu/g soil; Within a parameter, values that differed significantly are followed by different letters as determined by paired t-test ($P \leq 0.05$).

Source: Suting et al. (unpublished).

Methods

The rhizospheric soils and fine roots of two-year old *Macaranga denticulata* trees were analyzed for spore counts, presence of hyphae (filaments making up fungal mycelium), colonization of vesicles and arbuscles within host roots, and presence of a lesser-known fungal group called dark septate endophytes (Table 26-1). The latter group frequently co-occurs with other mycorrhizal fungi and earlier studies suggest that inoculation with dark septate endophytes can increase total, root and shoot biomass by up to 80% (Newsham, 2011).

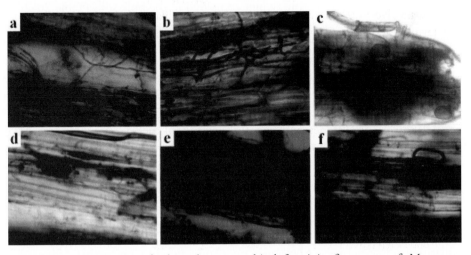

FIGURE 26-3: Structures of arbuscular mycorrhizal fungi in fine roots of *Macaranga denticulata*. Hyphal colonization (a and b); vesicles colonization (c); and coexistence of arbuscular mycorrhizal fungi and dark septate endophytes (d, e and f).

Results

Eight different species of arbuscular mycorrhizal fungi were found in the rhizospheric soil of *Macaranga denticulata*: They were *Acaulospora bireticulata* Rothwell & Trappe; *Acaulospora tuberculata* Janos & Trappe; *Funneliformis fragilistratum* Walker & Schußler; *Funneliformis verruculosum* Walker & Schußler; *Gigaspora decipiens* Hall & Abbott; *Glomus botryoides* Rothwell & Victor; *Glomus claroides* Schenck & Smith; and *Glomus fuegianum* Speg., Trappe & Gerd.

Biological and biochemical activities

The rhizospheric soils of *Macaranga denticulata* and the bulk fallow soils were compared in terms of the population of bacteria, microbial biomass carbon, potentially mineralizable nitrogen and the activity of soil enzymes (dehydrogenase, acid-phosphomonoesterase and aryl-sulfatase). The rhizospheric soils of *Macaranga* contained a significantly higher bacterial population and microbial biomass carbon (Table 26-2). There was also significantly higher biological activity in the *Macaranga* rhizosphere, evident from higher dehydrogenase enzyme activity, and the nitrogen mineralization potential was also significantly higher, compared to that in the bulk soils.

Data on the total percentage of colonization in fine roots (Figure 26-3) and the spore count in rhizospheric soils indicated that the *M. denticulata* plants were heavily colonized by arbuscular mycorrhizal fungi and this made them a good source of inoculum in post-burnt *jhum* soils.

The benefits of *Macaranga* in rejuvenation of burnt soils in *jhum* fallows were also evident in significantly higher activity of acid-phosphomonoesterase and aryl-sulfatase enzymes in the rhizospheric soils, when compared to the bulk soils (Table 26-2). All of these soil-process indicators clearly provided evidence to support the perception of shifting cultivators that *Macaranga* species are potential rejuvenators of degraded *jhum* fields (Suting et al., unpublished).

In the shifting cultivation systems of Naga farmers, *Macaranga dendiculata*, *M. indica* and *M. peltata* are commonly used as dominant fallow species in lower-altitude tropical and subtropical forests, while alder trees (*Alnus nepalensis*) are used as soil improvers at altitudes between 1000 and 2000 metres above sea level (masl) (Cairns and Brookfield, ch. 22, this volume). However, *A. nepalensis* uses a different soil-enriching process. In this case, bacteria of the genus *Frankia* form nodules on the roots of *A. nepalensis* trees, from which they convert atmospheric nitrogen into ammonia via the enzyme nitrogenase, in a process known as nitrogen fixation. As a result of their mutually beneficial relationship with *Frankia*, alder trees improve the fertility of the soils in which they grow (Cairns, 2009) – a process with many similarities to that between *Macaranga* trees and arbuscular mycorrhizal fungi.

Alnus nepalensis D. Don [Betulaceae]

Often regarded as an ultimate fallow species, these trees fix nitrogen from the atmosphere and produce large quantities of leaf litter, enriching soils that are recovering from cropping. While *Macaranga* may be used to enrich fallows at lower altitudes, skilful fallow management of *A. nepalensis* at altitudes from 1000 m to 2000 masl has proven capable of halving fallow periods without significant loss of crop productivity.

Management practices of Konyak Naga farmers

During the cropping year, *Macaranga* seedlings emerge naturally among the rice and other crops (Figure 26-4). The seedlings are not regarded as weeds and are protected during hand weeding, but dense stands may be thinned. *Macaranga* seeds are also broadcast in the *jhum* fields. Germination of *Macaranga* seeds takes place along with the rice, but initially the seedlings are slow to grow. However, with onset of monsoon rains the seedlings grow steadily. At weeding time, they are left to grow in association with the upland rice and other minor crops, and creepers are allowed to climb on the young trees. Very dense stands are thinned and seedlings transplanted into sparsely populated areas. Additional seeds may also be broadcast on to swiddens and even into early fallow growth to bolster tree density. A good stand of *Macaranga* will be shoulder-high by the time of rice harvest (Figure 26-5), and after four to six years, the fallow vegetation has a dominant *Macaranga* canopy (Figure 26-6).

Dense stands that bring the benefit of increased upland-rice yields have about 800 trees/ha, or more. For example, in Sop Moei district, Mae Hong Son province, on Thailand's side of its border with Myanmar, dense six-year-old fallow stands of *Macaranga* with 4200 trees per hectare were followed by upland rice crops with an average grain yield of 3.04 t/ha (Yimyam et al., 2003). This raises the importance of *Macaranga*-dominated fallows as fallow periods are being forced to shorten by increasing population, decreasing availability of

FIGURE 26-4: *Macaranga* seedlings grow naturally among the crops in a new swidden.

Photo: Malcolm Cairns.

land, and official policy. In these circumstances the innovative use of fallow-enriching species such as *Macaranga* can maintain soil quality, even with shorter fallows, and contribute significantly to food security.

'Standing firewood'

Another innovative use of mature *Macaranga* trees by Konyak Naga farmers comes after a *Macaranga*-dominated fallow is re-opened for cultivation. If the trees are felled along with the rest of the secondary forest they will be consumed by fire as the swiddens are burnt, and much valuable firewood will be lost. So, rather than go to the trouble of harvesting and storing the firewood, the farmers simply leave the trees standing, as bare poles that emerge through the flames, to serve first of all as supports for climbing crops, such as beans, yams, gourds, pumpkins and cucumbers (Figure 26-7). Then, as the need for firewood arises, the farmers fell the bare poles and carry the 'standing firewood' home after days spent working in the swidden (Yaden, 2018).

FIGURE 26-5: A Naga farmer stands amid a vigorous growth of upland rice and *Macaranga* trees in his swidden.

Photo: Sapu Changkija

FIGURE 26-6: An eight-year-old fallow stand of *Macaranga* alongside a swidden under cultivation.

Photo: Sapu Changkija

Folklore associated with *Macaranga*

Macaranga has long been a part of the relationship between the farmers of Nagaland and their environment – so much so that the tree has a place in local folklore. The story relates to a witch called 'Longkongla'. This witch stole the lunch of a farmer working in his field, and in her hurry to eat it, choked on some fish bones. She coughed and spat the bones on to the leaves of a nearby *Macaranga* tree, making many small holes in the leaves. The *Macaranga* tree complained to the god 'Lijaba', who dropped down a thin thread and instructed the witch to climb it, to face punishment.

She was warned not to look down, but she ignored this advice and looked down at some children playing in the swidden. The witch slipped, fell and was impaled on the same *Macaranga* tree. The story accounts for the fact that *Macaranga* leaves have many tiny holes and a red-coloured resin oozes from cut stumps of *Macaranga*, said to be the blood of the witch Longkongla.

FIGURE 26-7: 'Standing firewood'. Bare *Macaranga* poles scattered around a swidden, waiting to be cut and carried to the home hearth.

Photo: Malcolm Cairns.

Naga farmer's perceptions of *Macaranga* and its use in shifting cultivation

- *Macaranga* trees are fast growing and their litter decomposes rapidly.
- The trees shed leaves profusely in winter, helping to retain soil moisture, and in the early summer the leaf litter acts as fertilizing mulch.
- *Macaranga* roots spread quickly, decompose early and release nutrients rapidly into the soil.
- *Macaranga* trees yield good quantities of firewood (Figure 26-8).
- The wood is easy to cut and split and is light for carrying in head-loads. Moreover, the wood can be burnt on the day after harvest even when it is not properly dried, and can be burnt throughout the rainy season.
- Mature *Macaranga indica* timber may be used for making furniture and in construction.

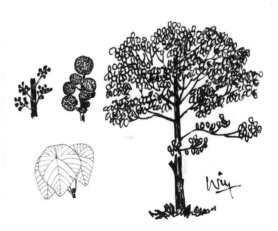

Macaranga peltata (Roxb.) Müll.Arg. [Euphorbiaceae]

A smaller member of the *Macaranga* genus, these trees grow up to 10 metres tall. As well as being noted as pioneer colonizers of disturbed land, their major use in India is producing wood for making pencils and plywood. In the forest, the trees often form symbioses with ants because they have hollow stems that can serve as nesting spaces.

- *Macaranga* stumps are killed easily by burning and release nutrients into *jhum* fields.
- Soil where *Macaranga* grows always gives higher crop yields, especially from rice, maize, millet, Job's tears, potatoes, perilla and pulses.
- In short-fallow shifting cultivation, *Macaranga* fallows maintain the productivity of upland rice at more or less the same levels as when the fallow period was 10 to 15 years.

FIGURE 26-8: Nagaland's former Deputy Conservator of Forests, Social Forestry Division, T. Amenba Yaden, in a stand of mature *Macaranga* trees.

Photo: Malcolm Cairns.

Conclusions

Traditional indigenous systems of shifting cultivation are practised by all of the peoples of Nagaland. However, the Konyak Nagas of Mon district, the Aos of Mokokchung and the Lothas of Whokha district practice a more extensive swiddening system and many farmers incorporate *Macaranga indica, M. peltata* and *M. denticulata* into their swiddens to dominate the fallow phase, in the belief that they will bring higher yields. The Konyak community in Mon district is the most advanced, in terms of managing *Macaranga* in its *jhums*, and this form of shifting cultivation is a striking example of sustainable land-use intensification evolved over centuries of experimentation by upland farmers – without the intervention of outside technologies.

Pressures on upland communities are pushing traditional shifting cultivators in northestern India to intensify their land use with alternative practices introduced by outsiders. At the same time, farmers are adapting shifting cultivation systems into more innovative and intensive forms of agriculture. One of the more significant improvements has been the management of fallow succession as a form of agroforestry, in order to achieve sustainability. Strategies depend upon the available length of the fallow period, and this determines suitable management regimes for fallow succession, i.e., weed succession, bush fallow and secondary forest succession. Despite differences in their ethnic and cultural backgrounds, farmers are responding to the pressures on shifting cultivation by adopting a variety of innovative strategies to modify fallow vegetation to enhance the economic and ecological functions of the fallow.

Naga farmers, especially those of the Konyak people, are struggling to restore the productivity of traditional shifting agriculture to ensure food security for their communities. They have found that use of local pioneer trees of the *Macaranga* genus, particularly *M. denticulata*, can contribute to the sustainability of shifting cultivation through productive regeneration of *Macaranga*-dominated secondary

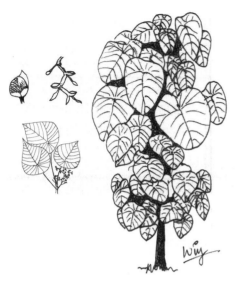

Macaranga indica Wight [Euphorbiaceae]

These trees grow up to 20 m and thrive in full sun as pioneers in forest clearings. A crimson-coloured resin called '*Macaranga* gum' is obtained from this species. Another exudate may be applied as a medicine to treat sores. This species has been under intense investigation for extracts which may be medically beneficial.

forest and nutrient cycling under six years of fallow re-growth. The value of this species has often been ignored. Understanding the scientific processes underpinning the benefits found in managing them could be a key to successful and sustainable land management in Nagaland and the wider subregion. Understanding the role and contribution of *Macaranga* in traditional shifting cultivation could also provide solutions to the restoration of subsistence crop yields in shorter-fallow systems as well as achieving reforestation with a natural process of regeneration.

References

Bolan, N. S. (1991) 'A critical review of the role of mycorrhizal fungi in the uptake of phosphorus by plants', *Plant and Soil* 134 (2), pp.189–207, doi:10.1007/BF00012037

Brundrett, M. C. (2002) 'Coevolution of roots and mycorrhizas of land plants', *New Phytologist* 154(2), pp.275-304, doi:10.1046/j.1469-8137.2002.00397.x

Cairns, M. F. (2009) 'The alder managers: The cultural ecology of a village in Nagaland, N.E. India', PhD dissertation to the Australian National University, Canberra, available at http://www.digitalhimalaya.com/collections/rarebooks/

Newsham, K. K. (2011) 'A meta-analysis of plant responses to dark septate root endophytes', *New Phytologist* 190, pp.783-793

Pfeffer, P., Douds, D., Becard, G. and Shachar-Hill, Y. (1999) 'Carbon uptake and the metabolism and transport of lipids in an arbuscular mycorrhiza,' *Plant Physiology* 120(2), pp.587-598, doi:10.1104/pp.120.2.587

Simon, L., Bousquet, J., Levesque, C. and Lalonde, M. (1993) 'Origin and diversification of endomycorrhizal fungi and coincidence with vascular land plants', *Nature* 363(6424), pp.67-69, doi:10.1038/363067a0

Suting, G. E., Thakuria, D. and Changkija, S. (2019) (unpublished) 'Assessment of arbuscular mycorrhizal fungal association and rhizospheric soil enzyme activity in the *jhum* fallow plant *Macaranga denticulata*', a case study under the Department of Technology, Government of India project: 'Impact assessment of *jhuming* on native plants and soil microbiota and restoration of sustainable jhum agroecosystems in northeast India (sanction order no. DBT-NER/Agri/14/2012 dated 13/10/2012)'.

Yaden, T. A. (2018) Personal communication with T. Amenba Yaden, former Deputy Conservator of Forests, Social Forestry Division, Kohima, Nagaland

Yimyam, N., Rerkasem, K. and Rerkasem, B. (2003) 'Fallow enrichment with *pada* (*Macaranga denticulata* (Blume) Müll. Arg.) trees in rotational shifting cultivation in northern Thailand', *Agroforestry Systems* 57(2), pp.79–86

Youpensuk, S., Lumyong, S., Dell, B. and Rerkasem, B. (2004) 'Arbuscular mycorrhizal fungi in the rhizosphere of *Macaranga denticulata* (Blume) Müll. Arg. and their effect on the host plant', *Agroforestry Systems* 60(3), pp.239–248

Youpensuk, S., Lordkaew, S. and Rerkasem, B. (2005) 'Comparing the effect of arbuscular mycorrhizal fungi on upland rice and *Macaranga denticulata* in soil with different levels of acidity', *ScienceAsia* 32, pp.121–126, doi: 10.2306/scienceasia1513-1874.2006.32.12

Zinke, P., Sabhasri, S. and Kunstadter, P. (1978) 'Soil fertility aspects of the Lua' forest fallow system of shifting cultivation', in P. Kunstadter, E. C. Chapman and S. Sabhasri (eds) *Farmers in the Forest: Economic Development and Marginal Agriculture in Northern Thailand*, The University Press of Hawaii for East-West Center, Honolulu, pp.134–159

27

SENTINELS OF A POLYNESIAN PAST

Biogeography, ethnobotany and conservation status of *koka* (Java cedar, *Bischofia javanica*) in the Pacific Islands and Southeast Asia

*Randolph R. Thaman**

Introduction

Java cedar (*Bischofia javanica*) is one of the most important agroforestry trees in Southeast Asia and the Pacific islands, especially in western Polynesia, Fiji and Vanuatu, where it is widely known as *koka*, and remains a central component in shifting cultivation due to its wide cultural utility and perceived role in soil enrichment. It serves as a sentinel tree in the traditional agroforestry systems of many Pacific islands, where although severely pollarded, ring-barked or coppiced, the tree rarely dies and is allowed to regenerate in gardens and fallows because of its cultural and ecological importance. This practice, although apparently sustainable, is declining in many areas in the process of 'agrodeforestation' – the disappearance of trees from the agricultural landscape due to failure to protect or replant them in agricultural land-use systems (Thaman, 1989, 1992). Factors contributing to agrodeforestation include increasing emphasis on commercial monocultures, mechanization, use of inorganic fertilizers, loss of knowledge related to the ecological, cultural and economic importance of trees such as *koka* and cessation of traditional uses for tree products or their replacement by imported substitutes. Agrodeforestation not only denudes the landscape, but also seriously erodes the ability of indigenous people to build cultural and environmental resilience in the face of environmental change.

This chapter discusses the biogeography, cultural utility, ecology and threatened status of *Bischofia javanica* in agroforestry systems. It highlights the central role that conservation and sustainable management of this tree could play in food, health, energy and livelihood security in the Pacific, in the face of climate, environmental, economic and cultural change. Although there are a number of common names and countless local vernacular names for *Bischofia javanica*, the names *koka* or *B. javanica* will be

* Dr. RANDOLPH R. THAMAN is Emeritus Professor of Pacific Islands Biogeography, the University of the South Pacific, Suva, Fiji.

used in this chapter (sometimes interchangeably) because the main focus is on its importance in western Polynesia, Fiji and Vanuatu, where it is widely known as *koka*.

The findings are based on archival research and field studies by the author in the Pacific islands and parts of Southeast Asia over more than 40 years. The main study was an in-depth year-long survey of all plant communities, including fallow vegetation, on 101 shifting cultivation bush allotments (*api uta*) on Tongatapu, the main island of Tonga, as part of the author's PhD research in 1971 (Thaman, 1975, 1976). A further in-depth survey specifically targeted *koka* in 1988, during which farmers from different areas of Tongatapu were interviewed as part of a questionnaire survey to determine management practices and uses for *koka*. Additional in-depth surveys of agroforestry trees and practices were carried out in Fiji in the mid-1980s (Thaman, 1993a) and in Vanuatu in 1988 (Thaman and Clarke, 1993b), during which practices associated with *koka* were studied as the basis for a book entitled *Pacific Islands Agroforestry Systems for Sustainability* (Clarke and Thaman, 1993a). Further, the author studied trees in agricultural systems in Tonga, Samoa, Kiribati and Tuvalu for the Food and Agriculture Organization of the United Nations (FAO) in 1993 (Thaman and Whistler, 1996). Observations and opportunistic studies were also carried out over the years on other Pacific islands, such as Samoa, the Cook Islands and New Caledonia, as well as in Indonesia, Okinawa, Hong Kong, Hainan, elsewhere in southern China, and Taiwan, where *koka* is also an important cultural tree. Archival research was conducted at Kew Gardens in London in 1987 and subsequently in a number of other locations. More recently, the author's research has used online sources.

Koka: The tree

Bischofia javanica Blume is known as *koka* or its cognate throughout most of its Polynesian and parts of its Melanesian range. Its other common names are Java cedar, bishopwood, Javanese bishopwood, beefwood (Australia), and autumn maple (Hong Kong). It is recognized as a member of the family Phyllanthaceae (Wagner et al., 1990; Wurdack et al., 2004), although it was traditionally placed in the family Euphorbiaceae, with a number of authorities having given it a separate family of its own – Bischofiaceae – because of its somewhat unique characteristics (Airy Shaw, 1965, 1967). The genus *Bischofia* was long thought to be a single-species genus, but a second distinct endemic species, *Bischofia polycarpa* (Leveille) Airy Shaw, known as Chinese bishopwood, was found in China (Airy Shaw, 1972).

B. javanica is a large evergreen or semi-deciduous tree, up to 30 m or more in height, with a short, thick trunk, a wide-spreading crown and alternate compound trifoliate leaves with toothed or notched bright green to bronze-green leaflets, which turn red before falling. The tree is dioecious, having small inconspicuous greenish petal-less unisexual flowers which are born in many-flowered racemose panicles on separate male and female trees. The female flowers ripen into small fleshy yellowish-brown or reddish fruits, up to 1 cm in diameter. Both Fijians and Tongans differentiate male and female trees by name (*koka tagane* and *koka yalewa* in Fiji and *koka tangata* and *koka fefine* in Tonga) (See plates 35 to 37, Coloured Plates section).

The bark is flaky dark grey-brown to dark reddish or purple-brown, shaggy and scaly. The inner bark, the source of a red-brown dye throughout much of its range, is fibrous, spongy and pink, exuding watery or jelly-like red sap. The narrow sapwood is light cream to reddish brown, sharply differentiated from the reddish or purplish-brown to deep brick or meaty-red, coarse-textured and moderately hard to heavy heartwood (Walker, 1954; Yuncker, 1959; Whitmore, 1972; Smith, 1981; Whistler, 2004, 2009).

Biogeography

B. javanica's native range is widely considered to extend from the western Himalayas, Nepal, southwestern India and southern China through the Philippines, Indonesia, New Guinea and northern Australia to Taiwan, Japan's Ryukyu Islands and as far

Bischofia javanica Blume
[Phyllanthaceae]

The focus of this chapter, this tree is known as *koka* in many Pacific island territories, but in its global range (Figure 27-1), it is more widely known as bishopwood. In northern Thailand, it has been planted as a pioneer species in reforestation projects along with other species to produce dense, weed-suppressing crowns and attract wildlife. In the forests of Assam, in northeastern India, it is commonly used by tigers to scratch-mark their territories, while in Taiwan, the indigenous people consider it a sacred tree.

east as Palau in Micronesia and the Cook Islands and Tahiti in Polynesia (Benthall, 1946; Airy Shaw, 1967; Whitmore, 1972, 1979; Fosberg et al., 1979, 1980; Thaman and Whistler, 1996). However, it is argued here that although it was probably indigenous as far east as Fiji and Palau, *B. javanica* was possibly an exotic tree introduced by early Polynesian settlers in the western parts of its range on the smaller, geologically-recent, oceanic islands of Tonga, Samoa, Wallis (Uvea), Futuna, and possibly the Cook Islands and Tahiti (Whistler, 2004, 2009). Moreover, it was introduced to Hawaii, the Ogasawara Islands southeast of Japan, Florida in the United States, Kenya and a number of other areas outside its native range in a more recent post-European-contact expansion. It should also be mentioned that Whitmore and Tantra (1986) reported *B. javanica* as being indigenous as far west as Georgia, in southwestern Europe (Figure 27-1).

The contention that *koka* may be a recent, culturally important introduction by early Polynesians into Samoa, Tonga, the Cook Islands, Tahiti and elsewhere, is based on the fact that only from India to Papua New Guinea and Australia does *B. javanica*

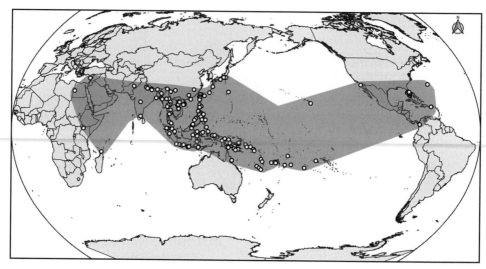

FIGURE 27-1: The natural distribution of *koka* (*Bischofia javanica*), from southwestern India and the western Himalayas, throughout Southeast Asia to Palau in Micronesia, Fiji in Melanesia and to the Cook Islands and Tahiti in Polynesia.

Sources: www.discoverlife.org; Aleen Prasad, University of the South Pacific, 2020.

seem to be found in native forests, mainly in riparian, ravine, swamp or moist lowland forests. Rarely, if ever, (except in Samoa and maybe Rarotonga in the Cook Islands) is it found in undisturbed native forests on the smaller oceanic islands of the Pacific. Instead, it is almost always found in secondary and fallow forests, actively cultivated shifting agriculture gardens and former village sites or towns, which is also characteristic of its occurrence in some areas of Southeast Asia.

In India, *B. javanica* is found up to elevations of 1200 metres above sea level (masl), and in the early 1900s was found mainly in shady ravines and valleys, on stream and river banks and in swamp forests and damper areas (Hooker, 1890; Gamble, 1902; Brandis, 1906). It is reported to be a medium to large tree of hill and plains forests in upper and lower Myanmar (Rodger, 1936). It is also found in Thailand and Vietnam. In Malaysia, where *B. javanica* grows in both mountain forests and along river banks, it is found in the Cameron Highlands, at elevations of almost 1000 m, as well as in lowland Kelantan (Whitmore, 1972). In China, where it is found in evergreen forests and open valley woodlands, it is widely planted on stream banks, in gardens and as a street tree. *B. javanica* is very widespread throughout most of southeastern China, including Hong Kong and Hainan Island (Lee, 1935; Airy Shaw, 1967; FCEC, 2008). Evidence suggests that it was generally found in riparian forests and open habitats (Airy Shaw, 1967). In highly urbanized Hong Kong and Taipei, Taiwan, it is now planted as a street tree, in parks and as an ornamental, but it was a major component of the original forest, which was completely cleared about 300 to 800 years ago (Hau, 2010).

In Indonesia, *B. javanica* is widespread in lower-elevation secondary forests, swamps and moist-area forests, as an agroforestry species and an intercrop or remnant tree in *Cacao* plantations and in settled areas on most of the main islands (Poppenborg and Hölscher, 2009). In Borneo, it is reportedly most common in lowland to mountain forest up to around 1500 masl, as well as being found on riverbanks in riparian forest (Airy Shaw, 1967; Meijer, 1974; Whitmore and Tantra, 1986). In monsoon forests in Timor, it is found in association with *Alstonia scholaris, Cordia monoica* subsp. *subpubescens, Exocarpos latifolius, Ficus saxophila, Tetrameles nudiflora* and *Toona sureni* (ICRAF, 2012).

In the Philippines, where it is found from northern Luzon to Palawan and Mindanao, it is reportedly common in forests along streams at low and medium altitudes, rising to 1500 masl (Merrill, 1923). Paijmans (1976) reported that on the island of New Guinea, *B. javanica* was one of the main canopy trees of widespread mixed-swamp forest. However, it was not reported in mixed-evergreen forest, mixed savannah or grassland associations, did not appear among frequent canopy-tree genera of hill forests, and was not reported to occur in 'man-made vegetation'. He said that although, with prolonged inundation,

Colocasia esculenta (L.) Schott [Araceae]

Taro is a ubiquitous staple food throughout the Pacific and in many other parts of the world, growing strongly in swamps and moist-area forests. While it is valued mainly for its edible starchy corms, the leaves and leaf stems of taro are also eaten as a vegetable. In its raw state, 100 g of taro leaf provides 30% of the daily nutritional needs for vitamin A; 38% of vitamin B2; 63% of vitamin C; 103% of vitamin K; and high levels of manganese, iron and potassium.

mixed-swamp forest 'usually becomes broken up into groups of trees and patches of shrubbery overgrown by rattan and other climbers', certain trees that 'occur sparsely on better drained sites', including *Bischofia javanica*, grow gregariously, increase in numbers, and grade into a swamp forest with its tall structure and dense canopy.

In Australia, where it is known as 'beefwood', *B. javanica* is mainly found in humid tropical North Queensland, from south of Cairns to the Cape York Peninsula (Bailey, 1913; Tracey, 1982). Occurrence records show that it is found from sea level up to an elevation of 560 masl (ALA, 2013). Although reportedly present in 'rainforest and vine thickets', it seems to be found mainly on forest margins, in degraded or logged-over forest, *Melaleuca* and *Eucalyptus woodlands*, pasturelands, near rivers and in rural settlements, but not in primary rainforest.

Li (1963) cited *B. javanica* as an example of a 'relatively recent arrival' on the 36,000-sq.-km island of Taiwan, located only 150 km from the Asian mainland, where it 'generally occurs on the plains and lowlands, especially in secondary forests.' In Japan's continental Ryukyu Islands, about 500 km off the Asian mainland and 100 km east of Taiwan, *B. javanica* is also 'a secondary forest and cultivated tree' (Walker, 1976). However, on the isolated volcanic Ogasawara (Bonin) Islands, *B. javanica* was reportedly introduced from Okinawa before 1905 for forestry purposes and as a source of firewood, but is accused of being the worst example of an exotic plant. It has invaded the floristically poor native vegetation and endangered endemic species, especially in areas of evergreen forest destroyed by a tropical cyclone (Ono et al., 1986; Shimizu, 1999, 2003).

To the east of New Guinea and Australia, although *B. javanica* has been reported throughout New Caledonia (Daniker, 1932; Guillaumin, 1948), listed as present on Bougainville Island, southwest of New Guinea, and also on Malaita, in the Solomon Islands (Whitmore, 1966), its presence has been largely overlooked in forest and vegetation literature from those islands.

Although common in Vanuatu and an 'occasional canopy tree in primary and mature secondary lowland forest', it is most common in 'old garden areas' where it appears to 'regenerate most readily in disturbed situations, following the first pioneers' (Gowers, 1976; Wheatley, 1992). It is also found in regenerating younger fallow forest on Vanuatu, often under encouragement from selective weeding (Thaman and Clarke, 1993b).

The only recorded presence of *B. javanica* in Micronesia is on Palau, where it is presumed to be indigenous, although there is no known local name for it (Fosberg et al., 1980). It is conspicuously absent from Guam (Stone, 1970) and Pohnpei (Fosberg et al., 1979), both larger islands in Micronesia, and from the island of Rotuma (Clarke and Thaman, 1993b; McClatchey et al., 2000), which lies north of Fiji. These are all islands where *koka* would grow well, but do not seem to have developed the associated Polynesian *tapa*-making tradition (see below) that is found in Fiji, Tonga, Wallis and Futuna and Samoa.

In Fiji, where it is apparently native, *koka* is widely distributed throughout the group of islands, from near sea level up to 900 masl, where it is one of 'the commonest trees in forests or on [their] edges, on grassy slopes' (Smith, 1981). However, it is clearly more common in secondary forests and on abandoned or active shifting-agriculture garden areas and old village sites. It is also common along creeks and streams and in ravine forests, and as isolated individuals or groups of trees in grazing areas and grasslands. It is almost never found in relatively undisturbed primary tropical rainforest. Wendy Southern of the Australian National University, who conducted palynological studies in Fiji, reported that '*Bischofia* pollen (which is distinctive and not easily confused with other pollen taxa) is present in the Fiji pollen record from at least 23,000 years ago and is common to the present' (Southern, 1987). This is long before the first record of human occupation, just over 3000 years ago. She commented: '*Bischofia* pollen occurs cyclically in the record, which is suggestive of its role as a tree poised to take advantage of disturbance', and suggested that this related to periodic volcanic activity and landslides. To that I might add human disturbance over the past two to three millennia. On the urbanized Suva peninsula on Fiji's main island of Viti Levu, *koka* is found in protected parklands, privately-owned properties, cemeteries and other protected areas, but is absent from most areas open to uncontrolled urban and peri-urban shifting cultivation and foraging for firewood.

To the east of Fiji, *B. javanica* makes an occasional appearance in secondary, lowland and montane forests from sea level up to 1120 masl on all of the main islands of Samoa. In Tonga, it is restricted to cultivation and secondary vegetation (Whistler, 2004) in the Vavau, Ha'apai and Tongatapu groups, as well as on the island of 'Eua, to the east of Tongatapu (Yuncker, 1959). On 'Eua, which is geologically the oldest island in Tonga, Drake et al. (1996) documented the presence of *B. javanica*, along with *Dendrocnide harveyi* and *Rhus taitensis*, as dominant species in areas that showed evidence of past disturbance, probably due to shifting cultivation of root crops and *kava* (*Piper methysticum*). On Tongatapu, where there is no surface water and only a few very small isolated stands of native broadleaf evergreen lowland forest, *koka* was second only to the ubiquitous coconut in 1971 as the island's most important large tree species, in areas not covered by mangrove forest. This was due to years of shifting cultivation and high population density (Thaman, 1976).

On both Uvea (Wallis) and Futuna Islands, which are west of Samoa and northeast of Fiji, *koka* is reportedly present in hill and lowland forests from 20 to 200 masl, although it is not clear whether this is primary or secondary forest (St. John and Smith, 1971). It is commonly planted in villages as a multipurpose tree (Figure 27-2).

In the absence of pollen records, the possible anthropogenic origin of *B. javanica* in western Polynesia is supported by Whistler (2009), even though he, too, suggests that *koka* appears to be native to Samoa, Tonga and Uvea. The same seems to be the case for at least Rarotonga in the Cook Islands, where *koka* was reported by Wilder (1931) to be particularly common in most valleys and hillsides at lower elevations in the 1920s, by Merlin (1985) in low-elevation disturbed forest, and by Thaman (1993b) on lower mountain slopes and near old irrigated taro terraces. It also appears to be indigenous to Tahiti, where it was reportedly collected by one of Captain James Cook's three expeditions between 1768 and 1779. Despite this, Whistler (2009) says 'it is difficult to determine how far east it [*koka*] is native in Polynesia, since it would have been introduced (aboriginally), if not already found there, because of its usefulness'.

FIGURE 27-2: *Bischofia javanica* planted in a village garden, Uvea Island, Wallis and Futuna.

All Photos in this chapter were taken by the author, Randolph Thaman.

The reason that *koka* might 'seem to be native' in forests in both Samoa and Rarotonga could be because it became rapidly naturalized on these geologically and floristically young volcanic islands, both of which were probably settled via Fiji by Polynesian peoples who had long given *koka* a central place in their cultural array of plants and animals. This is supported by the fact that *koka* – unlike a wide range of indigenous oceanic island plants – is reportedly incapable of long-range natural dispersal. However, once established, it seems to be ideally adapted to medium-range dispersal by birds and subsequent establishment in disturbed sites (Whistler, 2009). This is strongly supported by evidence from Japan's Ogasawara Islands, mentioned earlier, where *B. javanica* was introduced early last century and the islands were left almost uninhabited for almost 25 years following World War Two. Here, *B. javanica*, along with many introduced weeds and escaped crops and trees, such as *Pinus luchuensis*, *Casuarina equisetifolia*, *Acacia confusa*, *Leucaena leucocephala* and *Psidium cattleianum*, invaded the native vegetation in most parts of the main island, including a 'now designated national park' (Ono et al., 1986; Shimizu, 1999, 2003; Hata et al., 2007). Ono et al. (1986) stated that although it preferred 'exposed spaces', it was the 'worst invader' of native vegetation because 'it grows very fast, especially in areas of evergreen forest destroyed by a tropical cyclone … and vigorously, and produces a large amount of fruit that is dispersed by birds'. Shimizu (2002) suggested that the ability of *B. javanica* to invade the native vegetation in such a short time was because of the poor flora and fauna and the geological youth of the Ogasawara Islands' ecosystem, which seemed to have many vacant niches.

The ability of *B. javanica* to become rapidly naturalized outside its native range is supported by evidence from Florida and Hawaii. In Florida, where it was reportedly introduced as an ornamental in 1947 and extensively planted in street landscaping, it became naturalized and by 1974 was considered a 'weed tree'. In south Florida, where it became common in old fields and disturbed wetland sites, it displaced native vegetation and altered the structure of the plant community. Its use in landscaping is now discouraged by some horticulturists in Florida (Morton, 1976, 1985; Broschat and Meerow, 1991). It is also invasive in some areas of Oahu, Hawaii (also a young volcanic island), where Daehler's (2005) 'risk assessment results' rank *koka* as 'high risk', with a score of seven out of 10, based mainly on the fact that it is naturalized and spreading in some areas. However, with the exception of Samoa, there is almost nowhere else in Southeast Asia or the South Pacific where *koka* is regarded as an invasive species. Instead, it exists almost exclusively in riparian, secondary or fallow forests or as a valued agroforestry tree in shifting agriculture areas and pasturelands.

Functional utility

Throughout its traditional Southeast Asian and Pacific range, *B. javanica* has wide functional utility (Thaman et al., 2000). This was particularly true before the recent onset of commercial monocultures, mechanization (including ploughing) and the use of inorganic fertilizers and herbicides that inhibit multispecies fallow regeneration.

Koka's utility has also faltered because imported materials have replaced a wide range of traditional goods originally made from *koka*, including dyes, clothing, fabric, bedding, fertilizer, fuel and wooden items. In Tonga and Fiji, where *koka* is considered one of the most useful of trees, its wide functional utility continues, but is declining. The drive toward monocultural cropping has been more recent in Tonga and Fiji than in places such as Samoa, where the dramatic expansion of coconut and cocoa (*Cacao*) plantations during the German colonial period before World War One converted expansive areas of shifting cultivation land at lower elevations into monocultural plantations. The critical importance of maintaining multi-species fallows, as a basis for livelihood security in the face of rapid expansion of monocultures, is stressed by Delang (2007) in a study of ecological succession of usable plants in an 11-year fallow cycle in northern Laos. Both *B. javanica* and paper mulberry (*Broussonetia papyrifera*), two central components of plant communities in Tonga, Samoa and Fiji, are mentioned in Delang's study.

Table 27-1 shows the frequency that different uses for *B. javanica* were mentioned in the author's 1988 Tonga survey of management practices and uses for *koka*. The most important uses included dying and preserving *tapa* (bark) cloth, firewood, medicine, house posts, improving soil fertility, wood carving, use of the branches in trellising for yams and the use of the small fruit in games. Other uses, certainly more important in the past, include its use as a meat tenderizer, dyes for uses apart from *tapa* cloth, provision of shade and animal food.

Dyes and tannin

Throughout most of its range, *koka* bark, and in some cases the roots, are important sources of dyes and preservative tannin. These are of particular cultural importance

TABLE 27-1: The most important uses of *Bischofia javanica* in a survey of 10 farmers on Tongatapu Island, Tonga, in 1988.

Use	x/10	Parts used
Dying/preserving *tapa* cloth (*koka'anga*)	10	reddish-brown sap from inner bark
Firewood (*fefie*)	10	wood
Medicine (*vai Tonga, faito'o*)	9	bark
Improving soil fertility	8	leaves, roots
House posts (*pou fale*)	8	bole/stems
Woodcarving (*ta tiki*)	7	wood
Games	5	fruit
Yam trellising (*felei*)	5	branches
Cooking/preparing dog	4	wood
Shade/shelter (*faka malumalu*)	4	tree
Meat tenderizer (horsemeat, fish, octopus)	4	leaves
Dying pandanus leaves black	2	roots
Animal food (goats, pigs, birds)	2	leaves
Dying traditional waist sash (*ta'o vala*) reddish-brown	1	roots

in Tonga, Fiji and Samoa, where the inner bark is the source of the reddish-brown dye and preservative for traditional *tapa* cloth (*ngatu*, *masi* and *siapo* in Tonga, Fiji and Samoa, respectively). This is made from paper mulberry (*Broussonetia papyrifera*) and, along with fine pandanus mats, pigs, yams or taro and *kava*, is still among the most important ritual exchange items, forms of wealth and providers of social security in these Pacific societies. In the past, *tapa* cloth was also used as clothing and bedding, and for curtains, mosquito netting, cordage and fishing nets. Ceremonies such as funerals, weddings, installations of chiefs or a coronation (in the case of Tonga) are still not complete without the presentation of such goods, and *tapa* cloth is used to decorate marriage venues, dress the bride and groom, wrap corpses for burial and decorate graves. All of these items are part of an ancient Polynesian culture complex, derived from plants and animals (pigs) that were introduced to the islands by the original Proto-Polynesian Lapita settlers (Whistler, 2009).[1] In Tonga, *koka* is so important to the communal *tapa*-production process, performed by large groups of women, that the process itself is known as *koka anga*.

The process of obtaining the dye starts by using a strong sea shell or, more recently, a tin lid or metal scraper, to scrape off the outer bark of a *koka* tree before abrading the inner bark to collect scrapings, which are then squeezed out by hand, using a coconut-fibre strainer or a sophisticated press made of the plaited bast-fibre of *fau* (*Hibiscus tiliaceus*) (although plastic rice bags are commonly used today). When large quantities were required, work parties of women would go into the bush together, traditionally requiring strict traditions relating to behaviour and personal hygiene. The shavings were then placed in coconut baskets lined and covered with giant taro (*Alocasia macrorrhizos*) or other leaves to minimize evaporation and protect the shavings from rain. More recently, plastic bags are often used. The shavings are then placed in a special wringer, known as *unu*

Alocasia macrorrhizos (L.) G.Don [Araceae]

Giant taro is cultivated in many parts of the Pacific as a food source, although the starchy corms, leaves and stems must be cooked very thoroughly to rid them of Calcium oxalate crystals, which are shaped like tiny needles and irritate the mouth. These fast-growing plants can rise to five metres tall, from seeds, corms, and root suckers, and thrive in a wide range of conditions. They are regarded as invasive in New Zealand, Hawaii, Fiji, French Polynesia, New Caledonia and Palau.

or *tō* in Samoa, which is folded to enclose the shavings, bound with coconut sennit cordage and then suspended from a limb, pipe or beam with a strong pole passed through the lower part of the wringer. Two or more people push down hard on the pole or walk slowly around twisting the wringer, and a wooden bowl is placed below it to collect the sap. Twisting is done in both directions, and it is said that well-wrung shavings will produce as much as 6.8 litres of sap. The red-brown *koka* sap ferments rapidly and must be stored carefully in containers until needed (Pritchard, 1984; Whistler, 2000).

Nothing is added to the *koka* sap, which is then applied directly to the *tapa* cloth in two or three coats. If four or five coats are applied, the *koka* will crack. In Tonga, Samoa and Fiji, the sap of the oriental mangrove, *togo* (*Bruguiera gymnorhiza*), is also used to produce an additional, shinier dye for tapa, but the colour runs and it is not used as much.

B. javanica is listed among plants used on the Indian subcontinent in the tanning of leather, because of their tannin content of 10% or more by dry weight. The twigs and bark of *B. javanica* are listed as containing 16% tannin by dry weight (Howes, 1953; Dekker, 1908). Tannin extracted from the bark was once widely used to coat bamboo huts or treat basket work and fishing nets in Indonesia (Aumeeruddy, 1994).

In Tonga, the roots are reportedly boiled with dried treated pandanus leaves to dye them black for use in mat making and other handicrafts. Whistler (2000) reports that, in Samoa, the brown sap from *B. javanica* is mixed with soot from the smoke emitted by burning candlenut (*Aleurites moluccanus*) kernels, and the resulting black dye is used for traditional tattoos and dying tapa.

Firewood

Koka is highly sought after as a renewable source of high-quality firewood. In Tonga, Samoa, Fiji and Vanuatu, *koka* is considered to be one of the 'hottest' (high heat), slowest-burning, longest-lasting and easiest-to-ignite all-purpose firewoods, even when it is recently cut. This means that a smaller volume of firewood is required (Thaman and Ba, 1979; Thaman, 1984; Wheatley, 1992) (Figure 27-3). This is supported by evidence from Indonesia, where it is protected as a secondary forest tree due to its capacity for coppicing from the base to produce firewood (Aumeeruddy, 1994). This proven sustainability is particularly important in Tonga, where there is limited forest cover

FIGURE 27-3: *Koka* (*Bischofia javanica*) firewood, for sale at Talamahu market, Nuku'alofa, Tonga, most of which is sourced from pollarded, coppiced or felled trees from shifting agricultural areas.

and fewer species to select from, with a greater percentage of firewood coming from active agricultural areas.

In Tonga, it is used for all types of cooking, including firing traditional earthen ovens (*umu*), for roasting (*feitunu*), spit-roasting of pigs (*tunu puaka*), boiling staple foods (*feihaka*), baking or broiling fish (*fakapaku ika*) and, in the past, for singeing and cooking dog (*ta'o kuli*). Dog is a national delicacy that is still eaten occasionally, although surreptitiously. On Pongsu no Tau (Orchid Island), in Taiwan, it is favoured as firewood to smoke and preserve flying fish to protect them from maggot infestations during wet years (Cheng and Yu, 2000).

Medicine

Throughout its range, *B. javanica* is a very important medicinal plant, although its applications and benefits are often contradictory, from place to place (Perry and Metzger, 1980). In India, juice from the leaves is applied to sores (Benthall, 1946; Agarwal, 1986). In China, the leaf is reportedly applied to ulcers and boils; red sap from the stem to sores; the fruit is said to provide a tonic for babies (although another source claims that the fruit are poisonous), and the roots are used to provide a diuretic for night-time urination. Elsewhere, the species is regarded as poisonous, but providing an astringent to be used for burns, coughs, cracked feet, diarrhoea, fever, gastritis, ophthalmia, sores, throat ailments, tonsilitis, toothache, urticaria and wounds (Duke and Wain, 1981; Duke and Ayensu, 1985; Perry and Metzger, 1980).

In Samoa, an infusion of the scraped bark is taken to treat gastrointestinal ailments, mouth infections and sores, often in infants; the sap from the scraped bark is rubbed onto sores, and an infusion of the crushed leaves is taken as a potion. Less commonly, juice from the chewed or crushed leaves is dripped into the eyes to treat pterygium, blurry vision, eyelid infections and eye injuries. In Tonga, juice from the bark is applied to burns and an infusion of the leaves given to infants with mouth infections (Whistler, 1992, 2000). The author's 1988 Tonga survey also indicated that it is used to treat stomach ache and intestinal disorders.

It is also one of Fiji's most widely used medicinal plants. Camby and Ash, in their *Fijian Medicinal Plants* (1994), report that the juice of the stem is given to children who have not walked by two years of age; the grated inner bark, steeped in water, is used as a treatment for 'furred tongue', and in decoctions, as a tonic to treat the relapse of many ailments. The bark is reportedly used to treat diabetes, the inner bark to treat wounds, an infusion of the leaves or oil from the bark to treat tonsilitis, and an infusion of the bark to treat skin irritation caused by the stinging-nettle tree (*Dendrocnide harveyi*). The bark is also used to treat stomach and mouth ulcers, athlete's foot, thrush and bladder infections, and in a decoction with *Syzygium malaccense* and *Spondias dulcis*, to treat diabetes. With the bark of *S. malaccense* and *Allophylus cobbe* it is used to treat arthritis; and with the bark of *Intsia bijuga* and another unidentified tree, to relieve fatigue. All of these tree species are common components of Fiji's fallow-agriculture plant community.

Okabe (1940) reported that in the late 1930s, parts of the tree were used as a baby tonic and a diuretic in Palau; and in East New Britain, Papua New Guinea, crushed leaves were rubbed on an aching stomach (Holdsworth, 1977).

Soil fertility

Most informants in both Tonga and Fiji believe that *koka* improves soil fertility and that crops such as yams, taro, sweet potatoes, bananas and vanilla are much healthier when they are growing near *koka* trees because of the high fertility, 'moisture' and shade. They say the main reason is the considerable leaf fall from *koka*, as a semi-deciduous tree, which keeps the soil 'moist', although some say the roots are also sources of nutrients. In Fiji, one farmer near Suva protected more than 10 trees on four hectares of land because of his belief that during the dry season, *koka* 'spreads water over the land', and that where *koka* grows there is 'fertile, healthy soil'. Similarly, in southwestern Viti Levu, although commercial *kava* farmers often cut most trees irrespective of their value, one or two *koka* trees are always left standing because of their 'fertilizing ability' (Tokalau, 1987).

Tacca leontopetaloides (L.) Kuntze [Dioscoreaceae]

Polynesian arrowroot is an ancient root crop closely related to yams. Valued for its ability to grow on low islands and atolls, the corms produce a long-lasting starch that can be cooked into starchy puddings. It was used as the original "paste" that joined pieces of tapa cloth together in the tapa-making process. It was replaced in this function by special varieties of cassava, and is seldom cultivated today.

There are limited data on the role of *koka* in enhancing soil fertility. However, studies of four different habitats in 'uplifted coral-reef forest' in southern Taiwan found that study plots on limestone soils, where *koka* and *Palaquium* sp. were the dominant indicator trees, had among the highest annual nutrient returns and the highest rates of annual nitrogen input, despite the initial concentrations of N in the leaf litter at all four sites being the same (Liao et al., 2006).

Timber and woodcarving

Despite *koka* timber having only moderate weight, strength and hardness, and being reportedly liable to warp and split, it is widely valued because of its resistance to wear and tear. It is used as a medium-heavy timber for general construction and the production of agricultural implements, planks and thatching work, furniture, containers and woodcarving. In Assam, in northeast India, where it is reportedly common, 'it is esteemed as one of the best timbers'. Because of its ornamental reddish colour, texture and grain, it is commonly used for house construction, railway sleepers, furniture, handicrafts and so on. The timber also seems to have extreme durability in contact with, or when immersed in, water. Consequently, it is highly valued for piles, bridge or boat building, water wheels, and kegs (Troup, 1907; Benthall, 1946; Walker, 1954; Wyatt-Smith and Kochummen, 1979; Wilkinson et al., 2000). On Pongsu no Tau island, in Taiwan, it is the main timber used to divide ancient irrigated terraced taro gardens into family plots (Cheng and Yu, 2000).

In Tonga, Fiji, Samoa and Vanuatu, it is one of the most commonly used trees for foundation posts for traditional houses because of its durability and resistance to rotting in the soil. It is also an important carving wood for making furniture and implements such as *tapa* beaters, bowls and other artifacts and handicrafts (Wheatley, 1992; Whistler, 2009). In Tonga, the ceremonial throne (*taloni*) of the ancient kings of Tonga was made from *koka*.

Yam trellising

One of the most important traditional uses of *koka* in Tonga is an arrangement of cut branches, after pruning and pollarding, for yam-garden trellises called *felei*. The cut, often forked, branches, ranging from about 1 m to 1.5 m long, are placed on top of the yam mounds after planting, so that when the yam vines emerge, they climb on the branches to create a dome-like appearance (see Figure 27-4). The trellises keep the vines off the hot, dark volcanic soil; they create space under the vines so weeds can be hoed without damaging the plants; they reportedly reduce the incidence of Anthracnose wilt (*mahunu*); and because the trellising is low to the ground, the yams are not as susceptible to tropical-cyclone damage as they would be using more erect trellises. Moreover, when the yams are harvested, about six to nine months after planting, the *felei* are

Dioscorea alata L.
[Dioscoreaceae]

In Tonga, yams are the first crops planted in newly cleared shifting cultivation plots. Domesticated within island Southeast Asia and New Guinea, yams are believed to have been spread throughout the Pacific and Indian oceans by Austronesian voyagers. The vast majority of cultivars are sterile, meaning that human agency was vital to their expansion, making them a good indicator of human movement.

used for firewood. Despite the benefits, the use of *felei* is a declining practice because of the loss of tree cover and the use of mechanization (Thaman, 1978).

Food and fodder

B. javanica is listed as an edible plant in *Tanaka's Cyclopedia of Edible Plants of the World* (Tanaka with Nakao, 1976). The leaves are a rich source of vitamin C (Agarwal, 1986) and are used as a condiment in fish dishes in Indochina. *Koka* is also reportedly used as a food in the Ryukyu Islands (Sonohara et al., 1952). Although Tanaka with Nakao (1976) says that the

FIGURE 27-4: A Tongan yam garden, showing the use of *B. javanica* branches as trellising (*felei*) to increase yields, prevent soil-borne disease and facilitate hoeing, Nukunuku, Tongatapu, Tonga.

fruits and seeds are edible, others claim they are poisonous (Perry and Metzger, 1980; Agarwal, 1986). In Tonga and Samoa, the young leaves are sometimes packed into the body cavity of pigs before roasting to improve the flavour of the cooked meat (Whistler, 2009). The liquid from the boiled fruit is reported to be nourishing to infants and of medicinal value (Walker, 1954). The seeds are also eaten by pigeons, doves, starlings and other birds (Whistler, 2004) and the leaves are eaten by goats and occasionally fed to pigs in Tonga.

Conservation value and provision of ecological services

As an agroforest and secondary-forest species, *B. javanica* has considerable conservation value and provides a wide range of ecological services. It provides an important habitat for native birds, reptiles, insects, snails and a range of important vines, epiphytes and understorey species, many of which have medicinal uses or other cultural values, and it is commonly planted in villages as a multipurpose tree. As well as improving soil (discussed above) it also provides many other ecological services, including erosion control, protection from strong winds and salt spray, shade and shelter (Thaman and Clarke, 1993a).

Studies in Kalimantan, on the island of Borneo, by Clough et al. (2009a, 2009b) have shown that the 'boom or bust' conversion of significant areas of forest and species-rich traditional agroforest to monocultural 'full-sun' *Cacao* plantations has played a major role in the loss of functional bird diversity. A number of planted living-fence trees remained following this conversion, such as *Gliricidia sepium* and *Erythrina subumbrans*, but *B. javanica* was one of the most common remaining trees that supported bird diversity, thus underlining its importance as a protected tree within rapidly degraded agroforestry landscapes.

Similarly, comparative surveys in western Kenya of indigenous land-snail fauna in indigenous rainforest and monocultural plantation forests comprising the exotic species *B. javanica* and *Pinus* spp., and a local indigenous species, *Maesopsis eminii*, showed that although many land-snail species were significantly more abundant in rainforest communities, most species were also present in *Bischofia* and *Maesopsis* plantation forests, with significantly lower numbers in *Pinus* plantations (Tattersfield et al., 2001). The authors suggest that this shows that 'plantations hold the potential, at least in some circumstances, to provide alternative habitats for forest molluscs (among the world's most threatened taxa), where indigenous rainforest has been cleared.'

B. javanica is also among the framework species with the highest potential for versatility in reforestation programmes because of its high growth rate, dense crown and its fleshy fruits that attract seed-dispersers (ICRAF, 2012). It was listed among the most promising native plants for forest restoration and slope rehabilitation in Hong Kong (Hau, 2010); was one of five species performing well out of 23 species trialled in 1977 to address forest degradation in the Western Ghats in India (Rai, 1990); and in Fiji, it has been one of the main species propagated for distribution as part of the 2011 'Plant a Million Trees programme'. It is also commonly used as an ornamental, street and shade tree in Southeast Asia and is one of the most common street trees in Taipei and Okinawa (Figure 27-5).

Ecological adaptability and durability

B. javanica is one of the most valuable, adaptable and durable trees in Pacific agroforestry systems. It is fast-growing and easily propagated from seed and possibly from cuttings (Walker, 1954). Available evidence suggests that it makes no special demands as to soil or climate, although its reported occurrence along streams and in ravines or swampy ground seems to imply a liking for a good water supply (Airy Shaw, 1967). In an island environment, *B. javanica* seems to have reasonable resistance to periodic exposure to sea spray, which commonly affects interior areas of smaller islands, often causing serious damage to both coastal and inland vegetation. Similarly, it has the ability to withstand tropical cyclones which 'attack' the Ogasawara islands almost every year (Ono et al., 1986), and which cause frequent widespread devastation to vegetation in Tonga, Fiji, and elsewhere in the western Pacific (Thaman, 1982).

FIGURE 27-5: *Koka* as a main-street tree in Naha, Okinawa, Japan.

B. javanica exhibits a surprising level of ecological adaptability, durability and resilience in its ability to compete with other plants in disturbed sites and to withstand human predation and abuse. It can survive burning, pruning, severe pollarding and even felling, after which it regenerates through root shoots. In Tonga, in the early 1970s, it was second only to the ubiquitous coconut palm as the most common tree in bush garden allotments, and the trees invariably showed evidence of coppicing, pollarding, severe pruning, ring-barking, multiple felling, widespread burning, and associated root shooting. The trees even seemed to have the ability to sprout from the base after ring-barking. Studies in India stress its ability to 'coppice well' (Manjunath, 1948), and surveys by the author in Tonga (Thaman,1989) found that up to one quarter of all trees surveyed had two or more mature trunks, indicating that they probably grew from root shoots, rather than being original trees. When *koka* trees are felled during garden clearance or for use as timber or firewood in parts of Fiji, they are deliberately cut at least 1.3 m above the ground to allow the regrowing trees to compete successfully against regenerating grasses and other fallow plants.

FIGURE 27-6: In 1985, the author visited a very large *B. javanica* tree in Bantarkalang village, southwest-central Java, Indonesia. The tree was about 40 m in height and reportedly hundreds of years old. It was protected in an enclosure as a sacred tree.

If allowed to grow without interference, *koka* trees can reach giant dimensions. Whitmore (1972) reported that in (what was then) Malaya, the species grew to a height of almost 45 m, often with steep 3-m-high buttresses. The author has seen one specimen in Bantarkalang village in southwest-central Java, which is protected in an enclosure as a sacred tree. It is about 40 m or more in height, and is reportedly hundreds of years old (Figure 27-6). The largest tree on Okinawa in the Ryukyu Islands in the 1950s was reportedly a *Bischofia javanica* 'reaching over 25 m in height, but with a short very thick trunk and wide-spreading branches' (Walker, 1954); and a large *akagi* tree (*B. javanica*), reportedly between 200 and 300 years old, stands solemnly at the Uchi-Kanagusuku Utaki shrine in Shuri Kanagusuku, Naha, Okinawa, and has been designated a national natural treasure (NOVOA, 2013). In Fiji, *koka*'s legendary durability and longevity is captured by the expression *qase vā koka* (old as the *koka* tree), which is widely used when referring to something that looks old, but it is still young and full of life. The expression is often used, in a humorous way, to describe older men who are still active.

Although it is easily propagated, *koka* seedlings and young trees are nowadays rarely seen in the Pacific islands. Small trees are almost always root shoots from destroyed 'mother' trees. As suggested in the introduction to this chapter, this may be due to increasing commercialization and the gradual removal or reduction in size of tree groves and areas of fallow forest that were formerly ubiquitous components of traditional shifting cultivation systems. Another cause may be predation of seeds by rats, pigs and other animals. In a study of fruit consumption by introduced black rats (*Rattus rattus*) in Hawaii to determine their role in the dispersal of invasive plant seeds, it was found that the seeds of *Bischofia* generally did not survive ingestion by this voracious species (Shiels and Drake, 2011). The increasing use of herbicides may be a further contributing factor to the disappearance of tree seedlings.

Place in agroforestry and fallow systems

In many areas of Southeast Asia and the western Pacific, *koka* is a dominant arboreal component of traditional agroforestry systems because of its durability and wide utility. At the beginning of the 20th century in Indonesia, it was reportedly always preserved as a secondary forest tree for the tannin extracted from its bark, which was used to coat bamboo huts and to treat basket work and fishing nets. Although now little used for the production of tannin, it is still conserved by some farmers for its many medicinal uses and its great capacity for coppicing from the base to produce firewood (Aumeeruddy, 1994). In Tonga, Fiji and Vanuatu, in particular, *koka* remains one of the dominant, but increasingly threatened, plants in traditional multi-species shifting agroforestry systems and cropping cycles.

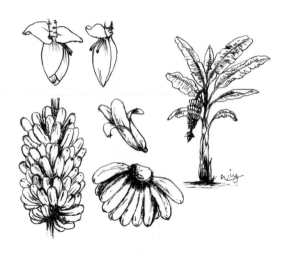

Musa x paradisiaca L. [Musaceae]

Pacific plantains (in broad terms, cooking bananas) belong to the AAB group of cultivars within the complex realm of *Musa* species, hybrids, varieties and cultivars. They are sterile hybrids of *Musa x paradisiaca*, with three sets of chromosomes. Plantains are a major food staple in many tropical countries, and play a starring role in regional cuisines. Compared with 'dessert bananas', plantains are longer, heavier, starchier, harder to peel – and need to be cooked.

In Tonga, for example, the cropping cycle begins with the clearing of secondary vegetation, including many pioneering tree species, while protecting and sometimes pruning, pollarding or coppicing useful species to allow light to reach ground crops. The dried plant remains are then usually burned, and the first crops planted in the ashes. The first crops are often yams (*Dioscorea alata*), usually intercropped with plantains (triploid cultivars of *Musa x paradisiaca*, or *Musa* AAB Group), giant taro (*Alocasia macrorrhizos*), and other crops. After the yams are harvested, true taro (*Colocasia esculenta*), tannia (*Xanthosoma sagittifolium*), sweet potatoes (*Ipomoea batatas*) or cassava (*Manihot esculenta*) are generally the next crops in the succession. Other crops, including bananas, peanuts, paper mulberry (*Broussonetia papyrifera*) or kava (*Piper methysticum*) are also integrated into the cropping sequence and may be planted at any time. If taro has been planted, cassava often follows and is sometimes planted up to four or five times consecutively on a given plot. Then, after the cassava is harvested, the land will be allowed to revert to fallow, although some cassava is commonly left as a component in the bedraggled fallow vegetation (Thaman, 1975, 1976).

Among almost 100 tree species cultivated or protected by Tongans as part of a sophisticated agroforestry system, surveys in 1971 of 101 agricultural allotments, averaging 3.2 hectares each, showed that *Bischofia javanica* was second only to coconut in terms of frequency (Table 27-2). It was present on 88 allotments, ahead of mango (*Mangifera indica*) (79); candlenut (*Aleurites moluccanus*) (78); *Hibiscus tilliaceus* (73); *Citrus* spp. (including *C. sinensis*, *C. reticulata*, *C. hystrix* x *limon*, *C. x aurantium*, *C. aurantiifolia* and *C. maxima*) (72); island litchi (*Allophylus cobbe*) (70); Indian mulberry (*Morinda citrifolia*) (68); breadfruit (*Artocarpus altilis*) (66); *Macaranga harveyana* (66); *Rhus taitensis* (63); Polynesian chestnut (*Inocarpus fagifer*) (55); and guava (*Psidium guajava*) (53). Of the 88 allotments with *koka*, 43 had between 10 to 25 trees each, and two allotments had more than 25 trees each (Thaman, 1975, 1976). All of these trees, except *Hibiscus tilliaceus*, *Morinda citrifolia*, *Macaranga harveyana*, and, at times, *Rhus taitensis*, were generally protected, although sometimes they were severely pruned or pollarded when preparing new garden plots, because they were regarded as useful trees. Other species found in the surveys that were commonly regarded as useful included avocado (*Persea americana*), ylangylang (*Cananga odorata*), beach almond (*Terminalia catappa*), soursop (*Annona muricata*), Polynesian vi-apple (*Spondias dulcis*) and Malay apple (*Syzygium malaccense*). The latter two are early Polynesian introductions that were found on only 13 and 11 allotments respectively.

Almost two decades later, in 1988, as can be seen in Table 27-3, of 17 multipurpose tree species deliberately protected when clearing fallow vegetation for a new cropping cycle, *koka* remained among the top four, along with coconut and breadfruit (two of the staple 'trees of life' in the Pacific), and mango. The remaining deliberately protected trees included a range of other important fruit, medicinal and multipurpose trees with a significant place in the Tongan traditional agroforestry system.

There were other species that, although common in fallow areas in 1971, were no longer deliberately protected in the process of clearing fallows to initiate a new gardening cycle in 1988 (Table 27-4). Of 23 multipurpose trees, shrubs or grasses deliberately removed when clearing fallow vegetation for a new cropping cycle, *koka* was listed only once. This was by a farmer who no longer grew yams and focused

TABLE 27-2: Frequency of occurrence and abundance of the most common useful trees found on 101 bush allotments in Tonga in 1971.

Scientific name	Common name	Tongan Name	Freq x/101	25+ trees	10 to 25 trees
Cocos nucifera	coconut	*niu*	98	81	17
Bischofia javanica	Java cedar	*koka*	88	2	43
Mangifera indica	mango	*mango*	79	9	27
Aleurites moluccanus	candlenut	*tuitui*	78	3	8
Hibiscus tiliaceus	beach hibiscus	*fau*	73	18	14
Citrus spp.	citrus trees	*moli*	72	12	9
Allophylus cobbe	island litchi	*tava*	70	7	27
Morinda citrifolia	Indian mulberry	*nonu*	68	7	9
Artocarpus altilis	breadfruit	*mei*	66	13	19
Macaranga harveyana		*loupata*	66	4	12
Rhus taitensis	Tahitian sumac	*tavahi*	63	7	7
Inocarpus fagifer	Tahitian chestnut	*ifi*	55	1	8
Psidium guajava	guava	*kuava*	53	26	8
Grewia crenata		*fo'ui*	48	2	11
Erythrina variegata	dadap tree	*ngatae*	39	3	12
Persea americana	avocado	*'avoka*	38	0	3
Pandanus spp.	pandanus	*fā*	33	7	2
Glochidion concolor		*malōlō*	32	0	0
Ceiba pentandra	kapok	*vavae*	27	3	2
Santalum yasi	sandalwood	*ahi*	25	0	0
Alphitonia zizyphoides		*toi*	21	0	1
Cananga odorata	ylangylang	*mohokoi*	18	0	0
Terminalia catappa	tropical almond	*telie*	17	2	1
Annona muricata	soursop	*'apele 'initia*	16	0	0
Pittosporum arborescens		*masi 'aukava*	16	0	0
Vavaea amicorum	false sandalwood	*ahivao*	16	3	4

Note: There were at least eight different *Citrus* spp, three of which, orange, mandarin orange and lemon, would have been among the top 15 trees, but they were not assessed for abundance.

Source: Thaman, 1976.

TABLE 27-3: Tree species reportedly protected when clearing and burning fallow land to commence a new cropping cycle in a 1988 survey of 10 farmers on Tongatapu island, Tonga.

Common name	Tongan name	Scientific name	x/10
Coconut palm	*niu*	*Cocos nucifera*	8
Breadfruit	*mei*	*Artocarpus altilis*	8
Mango	*mango*	*Mangifera indica*	8
Java cedar	*koka*	*Bischofia javanica*	7

TABLE 27-3 (cont.): Tree species reportedly protected when clearing and burning fallow land to commence a new cropping cycle in a 1988 survey of 10 farmers on Tongatapu island, Tonga.

Common name	Tongan name	Scientific name	x/10
Oceanic litchi	tava	Allophylus cobbe	5
Avocado	avoka	Persia americana	4
Polynesian vi-apple	vi	Spondias dulcis	4
Tahitian chestnut	ivi	Inocarpus fagifer	4
Citrus trees	moli	Citrus spp.	2
Malay apple	fekika	Syzygium malaccense	2
Pandanus	paongo	Pandanus cultivar	2
Tropical almond	telie	Terminalia catappa	1
Perfume tree	mohokoi	Cananga odorata	1
–	fo'ui	Grewia crenata	1
–	fekika vao	Syzygium dealbatum	1
Macaranga	loupata	Macaranga harveyana	1
–	toi	Alphitonia zizyphoides	1

increasingly on commercial cropping. The remaining most commonly cleared trees included guava (an invasive fruit tree), *Erythrina variegata*, *Hibiscus tilliaceus*, red bead tree (*Adenanthera pavonina*), frangipani (*Plumeria* spp.), macaranga (*Macaranga harveyana*), mango and candlenut, all easy-to-grow trees. There were a number of shrubby invasive species among those deliberately removed from new shifting cultivation fields. They included leucaena (*Leucaena leucocephala*), lantana (*Lantana camara*), *Triumfetta bartramia*, indigo (*Indigofera suffruticosa*), and woolly nightshade (*Solanum mauritianum*). While some of these are important nitrogen-fixing plants, many of them, unlike *koka*, are no longer deliberately protected in the fallow cycle. As stressed by Wolf (1998) in an article on

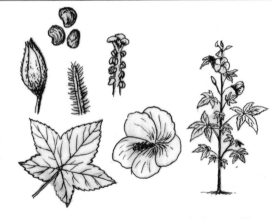

Abelmoschus manihot (L.) Medik. [Malvaceae]

This perennial shrub can grow to three metres in height. It is easily propagated from cuttings and is widely planted either on field borders or as an intercrop in traditional swiddens. Formerly thought to be a *Hibiscus* species, its leaves and shoots are very nutritious, with high levels of vitamins A and C and iron, with 12% protein by dry weight. It also produces a starchy substance that is used in Japan and Korea to make traditional paper.

TABLE 27-4: Fallow species reportedly killed or removed when opening land for a new cropping cycle in a February 1988 survey of 10 farmers on Tongatapu island, Tonga.

Common name	Tongan name	Scientific name	x/10
Guava	kuava	Psidium guajava	9
Leucaena, false tamarind	sialemohemohe	Leucaena leucocephala	8
Lantana	talatala	Lantana camara	6
Beach hibiscus tree	fau	Hibiscus tilliaceus	4
Hibiscus burr	mo'osipo	Triumfetta bartramia	4
Dadap	ngatae	Erythrina variegata	4
Red-bead tree	lopa	Adenanthera pavonina	3
Frangipani	kalosipani	Plumeria spp.	3
Macaranga	loupata	Macaranga harveyana	3
Mango	mango	Mangifera indica	2
Candelnut	tuitui	Aleurites moluccanus	2
Indigo	'akauveli	Indigofera suffruticosa	2
Woolly nightshade	pula	Solanum mauritianum	2
	fo'ui	Grewia crenata	2
Guinea grass	sāfa	Panicum maximum	2
Java cedar	koka	Bischofia javanica	1
Pacific fan palm	piu	Pritchardia pacifica	1
Papaya, pawpaw	lesi	Carica papaya	1
	fekika vao	Syzygium dealbatum	1
Banyan	'ovava	Ficus prolixa	1
	toi	Alphitonia zizyphoides	1
Alexandrian laurel	feta'u	Calophyllum inophyllum	1
Sensitive plant	mateloi	Mimosa pudica	1

deforestation and forest degradation, *koka* would probably have ranked even higher had it not been for increasing monocultural commercial cropping and use of the plough, particularly during rapid expansion of butter-pumpkin (squash) production (*Cucurbita moschata* and *C. maxima* cultivars) for export to Japan in the early 1980s, during which 'trees are often removed as a nuisance' (Wolf, 1998).

Koka was very much a part of the traditional Tongan polycultural shifting cultivation system, which was focused on yam culture and the production of *tapa* cloth as a major form of wealth and exchange. Yam culture required good soil, trellising and an enriched fallow, which depended on the protection and management of *koka*. As people moved away from this labour-intensive and fallow-dependent system to a shorter term, more monocultural, fertilizer- and herbicide-dependent system that was less dependent upon trees, *koka* and other trees began to perceptibly disappear. Few people still plant *koka*, and most remaining trees are remnants that have been protected or have survived over the years. Because of the increasing need for cropping land, the seedlings are no longer protected. Although one respondent in 1988 said he had planted it because it was one of the most useful of trees, most respondents had not heard of people planting it. Many suggested that this was because *koka* was still perceived as common and its abundance taken for granted.

Conclusion: A sentinel of the past offers a foundation for future sustainability

Java cedar (*Bischofia javanica*) – *koka* – is clearly a sentinel of an ancient past – a loyal, resilient, long-serving guardian that has, for millennia, provided protection, enrichment and sustainability to ancient Polynesian, Melanesian and Southeast Asian shifting cultivation systems and cultures. The tree is highly resistant to burning, drought, flooding, tropical cyclones, strong monsoon winds and salt spray and has few significant pests or diseases. It improves soil fertility, is among the most sustainable sources of high-quality firewood and timber and is an important medicinal plant. It provides an extremely important dye and preservative that is central to the manufacture of *tapa* cloth, which, in cultural terms, is one of the most valuable forms of traditional and modern wealth and social security in the western Pacific. As well, it serves an important role in reforestation, wildlife conservation and provision of ecosystem services. Although severely pollarded, ring-barked or coppiced, *koka* is rarely killed and is allowed to regenerate in gardens and fallows because of its cultural and ecological importance. However, this practice is declining in many areas. Increasing emphasis is being given to monoculture cropping, mechanization, and the use of inorganic fertilizers and herbicides, and knowledge about the importance of trees like *koka* is being lost, along with the trees. This process is seriously eroding the ability of traditional agricultural societies to build cultural and environmental resilience in the face of environmental change.

It is argued that the conservation and deliberate replanting of *Bischofia javanica* and other appropriate trees and multipurpose plants should be part of a concerted effort to reverse this process of 'agrodeforestation' and the spread of monoculture cropping. In environmental, cultural and economic terms, this would be one of the most appropriate ways to alleviate poverty, promote soil conservation and achieve food, fuel and livelihood security. Such a campaign would also build resilience against drought, floods, tropical cyclones and climate change and promote the conservation of biodiversity as a basis for sustainability in the face of unprecedented environmental, social and economic change.

Dedication

This study of *koka* is dedicated to the late Alifereti Bogiva, my *taciqu ni kai* ('tree brother', in Fiji's Ra dialect), a dedicated forester, community-based conservation pioneer, and humanitarian. His tireless efforts, over countless hours of consultation and drinking *kava* (*yaqona*), to engage local Fijian communities in the sustainable use of their forests and promote agricultural and marine biodiversity, are legend.

His ancestral totem tree (*icavuti*) was *koka*. More than 30 years ago, while we were attending a Food and Agriculture Organization workshop in Bangkok, he learned of my interest in and knowledge of *koka*. He declared that, henceforth, I would be his *taciqu ni kai*, and this is how we referred to one another thereafter. Alifereti touched countless lives with his knowledge, humour, enthusiasm and commitment to sustainability and the indigenous Fijian people. As we say in Fijian: *vinaka vakalevu taciqu ni kai*! (This is dedicated to you).

References

Agarwal, V. S. (1986) *Economic Plants of India*, Kailash Prakashan, Calcutta

Airy Shaw, H. K. (1965) 'Diagnoses of new families, new names, etc., for the seventh edition of Willis's 'Dictionary', *Kew Bulletin* 18(2), pp.249-254

Airy Shaw, H. K. (1967) 'Notes on the genus *Bischofia* Bl. (*Bischofiaceae*)', *Kew Bulletin* 21(2), pp.327-329

Airy Shaw, H. K. (1972) 'Nomenclatural note on *Bischofia racemosa* Cheng & Chu (*Bischofiaceae*)', *Kew Bulletin* 27(2), pp.271-272

ALA (2013) *Atlas of Living Australia*, available at https://bie.ala.org.au/species/https://id.biodiversity.org.au/node/apni/2904930, accessed 5 September 2021

Aumeeruddy, Y. (1994) 'Local representations and management of agroforests on the periphery of Kerinci Seblat National Park, Sumatra, Indonesia', *People and Plants Working Paper*, Division of Ecological Sciences, United Nations Educational, Scientific and Cultural Organization (UNESCO), Paris

Bailey, F. M. (1913) *Comprehensive Catalogue of Queensland Plants, both Indigenous and Naturalized*, Government Printer, Brisbane

Benthall, A. P. (1946) *The Trees of Calcutta and its Neighbourhood*, Thacker Spink and Co, Calcutta

Brandis, D. (1906) *Indian Trees: An Account of Trees, Shrubs, Woody Climbers, Bamboos, and Palms Indigenous or Commonly Cultivated in the British Indian Empire*, Archibald Constable & Co, London

Broschat, T. K. and Meerow, A. W. (1991) *Betrock's Reference Guide to Florida Landscape Plants*, Betrock Information Systems, Cooper City, Florida

Camby, R. C. and Ash, J. (1994) *Fijian Medicinal Plants*, Commonwealth Scientific and Industrial Research Organization (CSIRO), Melbourne

Cheng, H. W. and Yu, S. Y. (2000) *Botel Tabaco, Yami and Plants*, Lamper Enterprises, Taipei

Clarke, W. C. and Thaman, R. R. (eds) (1993a) *Pacific Island Agroforestry: Systems for Sustainability*, United Nations University Press, Tokyo

Clarke, W. C. and Thaman, R. R. (1993b) 'Rotuma Island, Fiji', in W. C. Clarke and R. R. Thaman (eds) *Pacific Island Agroforestry: Systems for Sustainability*, United Nations University Press, Tokyo, pp.95-98

Clough, Y., Faust, H. and Tscharntke, T. (2009a) 'Cacao boom and bust: Sustainability of agroforests and opportunities for biodiversity conservation', *Conservation Letters* 2 (2009), pp.197-205

Clough, Y., Putra, D. D., Pitopang, R. and Tscharntke, T. (2009b) 'Local and landscape factors determine functional bird diversity in Indonesian cacao agroforestry', *Biological Conservation* 142, pp.1032-1041

Daehler, C. (2005) *Bischofia javanica, Pacific Island Ecosystems at Risk (PIER)*, University of Hawai'i, Honolulu

Daniker, A. U. (1932) '4. Katalogue der Pteridophyta und Embryophyta siphonogama', in Ergebnisse der Reise von Dr A.U. Daniker nach Neu-Caledonien und den Loyalty-Inseln (1924/6) (Results of the journey of Dr A. U. Daniker to New Caledonia and the Loyalty Islands, 1924-1926), *Vierteljahrsschrift der Naturforschenden Gesellschaft in Zürich* (Quarterly Magazine of the Natural History Society in Zurich) 77(19), pp.115-235

Dekker, J. (1908) *De Looistoffen: Botanisch-Chemische Monographie der Tanniden* (The Tannins: Botanical-Chemical Monograph of the Tanniden), J. H. de Bussy, Amsterdam

Delang, C. O. (2007) 'Ecological succession of usable plants in an eleven-year fallow cycle in northern Lao P.D.R.', *Ethnobotany Research and Applications* 5, pp.331–350

Drake, D. R., Whistler, W. A., Motley, T. J. and Imada, C. T. (1996) 'Rain forest vegetation of 'Eua, Kingdom of Tonga', *New Zealand Journal of Botany* 34, pp.65–77

Duke, J. A. and Ayensu, E. S. (1985) *Medicinal Plants of China*, vol. 1, Reference Publications, Algonac, MI

Duke, J. A. and Wain, K. K. (1981) *Medicinal Plants of the World*, computer index with more than 85,000 entries, three vols, 1654pp

FCEC (2008) *Flora of China* 11, 1–622, Flora of China Editorial Committee, Science Press and Missouri Botanical Garden Press, Beijing and St. Louis

Fosberg, F. R., Sachet, M-H. and Oliver, R. (1979) 'A geographical checklist of Micronesian Dicotyledonae", *Micronesica* 15(1-2), pp.41–295

Fosberg, F. R., Otobed, D., Sachet, M-H., Oliver, R. L., Powell, D. A. and Canfield, J. E. (1980) *Vascular Plants of Palau with Vernacular Names*, Department of Botany, The Smithsonian Institution, Washington, DC

Gamble, J. S. (1902) *A Manual of Indian Timbers*, Sampson, Low, Marston and Co, London

Gowers, S. (1976) *Some Common Trees of the New Hebrides and Their Vernacular Names*, Forestry Section, Department of Agriculture, Port Vila, Vanuatu

Guillaumin, A. (1948) *Flore Analytique et Synoptique de la Nouvelle - Caledonie Phanerogames* (Flora and Analytical Overview of the New Caledonian Phanerogams [Spermatophytes]), Editions Larose, Office de la Reserche Scientifique Coloniale, Paris

Hata, K., Suzuki, J-I. and Kachi, N. (2007) 'Effects of an alien shrub species, *Leucaena leucocephala*, on the establishment of native mid-successional tree species after disturbance in the national park in Chichijima Island, a sub-tropical oceanic island', *Tropics* 16(3), pp.283–290

Hau, B. C. H. (2010) 'Using native plants species in forest restoration and slope rehabilitation in Hong Kong, China', PowerPoint presentation at conference on Mainstreaming Native Species-based Restoration, 15-16 July, 2010, University of the Philippines Diliman, Quezon City, Philippines

Holdsworth, D. K. (1977) *Medicinal Plants of Papua New Guinea*, technical paper no. 175, South Pacific Commission, Noumea

Hooker, J. D. (1890) *The Flora of British India*, V. L. Reeve and Co., London

Howes, F. N. (1953) *Vegetable Tanning Materials*, Butterworths Scientific Publications, London

ICRAF (2012) *AgroForestry Tree Database: A Tree Species Selection Guide*, World Agroforestry Centre (ICRAF), Nairobi

Lee, S. C. (1935) *Forest Botany of China*, The Commercial Press, Shanghai

Li, H. L. (1963) *Woody Flora of Taiwan*, Livingston Publishing, Narbeth, PA

Liao, J. H., Wang, H. H., Tsai, C. C. and Hseu, Z. Y. (2006) 'Litter production, decomposition and nutrient return of uplifted coral reef tropical forest', *Forest Ecology and Management* 235(1-3), pp.174–185

McClatchey, W., Thaman, R. and Vodonaivalu, S. (2000) 'A preliminary checklist of the flora of Rotuma and Rotuman names', *Pacific Science* 54(4), pp.345-363

Manjunath, B. L. (ch. ed) (1948) *The Wealth of India: A Dictionary of Indian Raw Materials and Industrial Products*, vol 1, Council of Scientific and Industrial Research, Delhi

Meijer, W. (1974) *Field Guide to Trees of West Malaysia*, Book Store, University of Kentucky, Lexington, KY

Merlin, M. D. (1985) 'Woody vegetation in the upland region of Rarotonga, Cook Islands', *Pacific Sciences* 39(1), pp.81–99

Merrill, E. D. (1923) *An Enumeration of Philippine Flowering Plants*, vol. 2, fascicle 1, Bureau of Science, Department of Interior, Manila

Morton, J. F. (1976) 'Pestiferous spread of many ornamental and fruit species in south Florida', *Proceedings of the Florida State Horticultural Society* 89, pp.348-353

Morton, J. F. (1985) 'Nobody loves the *Bischofia* anymore', *Proceedings of the Florida State Horticulture Society* 97, pp.241-244

NOVOA (2013) *Native Okinawan Village and Omoro Arboretum*, https://churaumi.okinawa/en/area/okipark/nativeokinawanvillage-omoroarboretum/, accessed 5 September 2021

Okabe, M. (1940) 'Investigation of the medicinal plants found on the Palau Islands, their virtues and popular remedies', *Bulletin of Tropical Industry, Palau*, no. 5, Tokyo (translated by Hisayoshi Takeda, December, 1952)

Ono, M., Kobayashi, S. and Kawakubo, N. (1986) 'The present situation of endangered plant species in the Bonin (Ogasawara) Islands', *Ogasawara Research* 12, pp.1-32, Ogasawara Research Committee, Tokyo Metropolitan University, Tokyo

Paijmans, K. (1976) 'Vegetation', part II, in K. Paijmans (ed.) *New Guinea Vegetation*, Elsevier Scientific Publishing and Australian National University Press, Amsterdam and Canberra, pp.23-105

Perry, L. M. and Metzger, J. (1980) *Medicinal Plants of East and Southeast Asia: Attributed Properties and Uses*, The MIT Press, Cambridge, MA

Poppenborg, P. and Hölscher, D. (2009) 'The influence of emergent trees on rainfall distribution in cacao agroforest (Suwavesi, Indonesia)', *Flora: Morphology, Distribution, Functional Ecology of Plants* 204(10), pp.730-736

Pritchard, M. J. (1984) *Siapo: Bark cloth art of Samoa*, American Samoa Council on Culture, Pago Pago, American Samoa

Rai, S. N. (1990) 'Restoration of degraded tropical rain forests of Western Ghats', *Indian Forester* 116(3), pp.179-188

Rodger, A. (1936) *A Handbook of the Forest Products of Burma*, Government Printing and Stationery, Rangoon (Yangon)

St. John, H. and Smith, A. C. (1971) 'The vascular plants of the Horne and Wallis Islands', *Pacific Science* 25, pp.313-348

Shiels, A. B. and Drake, D. R. (2011) 'Are introduced rats (*Rattus rattus*) both seed predators and seed dispersers in Hawaii?' *Biological Invasions* 13, pp.883-894

Shimizu, Y. (1999) 'Natural history of the Ogasawara Islands', *Forest Science* 25, pp.42-45

Shimizu, Y. (2002) 'Biodiversity of the Ogasawara Islands: Three viewpoints', *Forest Science* 34, pp.2-8

Shimizu, Y. (2003) 'The nature of Ogasawara and its conservation', *Global Environmental Research* 7, pp.3-14

Smith, A. C. (1981) *Flora Vitiensis Nova (Spermatophytes only)*, vol. 2, Pacific Tropical Botanical Garden, Lawai, Kavai, Hawaii

Sonohara, S., Tawada, S. and Amano, T., with E. H. Walker (ed.) (1952) *Flora of Okinawa (Okinawa Shokubutsushi)*, preliminary edition, US Civil Administration of the Ryukyu Islands in cooperation with the Ryukyu Forestry Bureau, Government of the Ryukyu Islands and Pacific Science Board, National Academy of Sciences, Washington, DC

Southern, W. (1987) Personal communication between the author and Dr Wendy Southern, former palynologist with the Australian National University, Canberra

Stone, B. C. (1970) 'The flora of Guam', *Micronesica*, vol. 6 (complete), pp.1-659

Tanaka, T., with Nakao, S. (ed.) (1976) *Tanaka's Cyclopedia of Edible Plants of the World*, Keigaku Publishing, Tokyo

Tattersfield, P., Seddon, M. B. and Lange, C. N. (2001) 'Land-snail fauna in indigenous rainforest and commercial forestry plantations in Kakamega forest, western Kenya', *Biodiversity and Conservation* 10(11), pp.1809-1829

Thaman, R. R. (1975) 'Tongan agricultural land use: A study of plant resources' in E. Stokes (ed.) *Proceedings of the International Geographical Union Regional Conference and Eighth New Zealand Geography Conference*, December 1974, Palmerston North, New Zealand, Geographical Society (Inc.), Wellington, pp.153-160

Thaman, R. R. (1976) *The Tongan Agricultural System: With Special Emphasis on Plant Assemblages*, published version of 1975 PhD dissertation to the University of California, Los Angeles, University of the South Pacific, Suva

Thaman, R. R. (1978) 'Cooperative yam gardens: An adaptation of a traditional agricultural system to serve the needs of the developing Tongan market economy', in E. K. Fisk (ed.) *The Adaptation of Traditional Agriculture*, Development Studies Centre Monograph no 11, Australian National University, Canberra, pp.116-130

Thaman, R. R. (1982) 'Hurricane Isaac and Tonga: A natural or cultural disaster?' *Review* 3(8), (quarterly publication of the School of Social and Economic Development, University of the South Pacific, Suva), pp.22-34

Thaman, R. R. (1984) *The Firewood Crisis and Smallholder Fuelwood Systems on Tongatapu Island, Tonga: Present Systems and Development Potential*, PEDP Report, Tonga 85-1, United Nations Pacific Energy Development Programme (UNPEDP), Suva

Thaman, R. R. (1989) 'Fijian agroforestry: Trees, people and sustainable polycultural development', in J. Overton (ed.) *Rural Fiji*, Institute of Pacific Studies, The University of the South Pacific, Suva, pp.31-58

Thaman, R. R. (1992) 'Agrodeforestation as a major threat to sustainable development', Box 19.4 in R. Thistlethwaite and G. Votaw (eds) *Environment and Development: A Pacific Island Perspective*, Asian Development Bank and South Pacific Regional Environment Programme, Manila and Apia, pp.194-195

Thaman, R. R. (1993a) 'Fijian agroforestry at Namosi and Matainasau', in W. C. Clarke and R. R. Thaman (eds) *Pacific Island Agroforestry: Systems for Sustainability*, United Nations University Press, Tokyo, pp.63-84

Thaman, R. R. (1993b) 'Rarotonga and Aitutaki, the Cook Islands', in W. C. Clarke and R. R. Thaman (eds) *Pacific Island Agroforestry: Systems for Sustainability*, United Nations University Press, Tokyo, pp.99-111

Thaman, R. R. and Ba, T. (1979) 'Energy needs and forest resources of small islands', in *Proceedings of the 49th ANZAAS Congress and 10th NZ Geographical Conference*, Auckland, pp.198-204

Thaman, R. R. and Clarke, W. C. (1993a) 'Pacific island agroforestry: Functional and utilitarian diversity', in W. C. Clarke and R. R. Thaman (eds) *Pacific Island Agroforestry: Systems for Sustainability*, United Nations University Press, Tokyo, pp.17-33

Thaman, R. R. and Clarke, W. C. (1993b) 'Agroforestry on Aneityum and Tanna, Vanuatu', in W. C. Clarke and R. R. Thaman (eds) *Pacific Island Agroforestry: Systems for Sustainability*, United Nations University Press, Tokyo, pp.54-63

Thaman, R. R. and Whistler, W. A. (1996) *A Review of Uses and Status of Trees and Forests in Land-Use Systems in Samoa, Tonga, Kiribati and Tuvalu, with Recommendations for Future Action*, working paper 5 (RAS/92/361), South Pacific Forestry Development Programme, Suva

Thaman, R. R., Elevitch, C. R. and Wilkinson, K. M. (2000) 'Multipurpose trees for agroforestry in the Pacific Islands', in C. R. Elevitch and K. M. Wilkinson (eds) *Agroforestry Guides for Pacific Islands*, no. 2, Permanent Agriculture Resources (PAR), Holualoa, Hawaii, pp.24-69

Tokalau, F. (1987) 'Kava in Fiji: A case study of Namosi Village, Fiji', unpublished research paper, School of Social and Economic Development, University of the South Pacific, Suva

Tracey, J. G. (1982) *The Vegetation of the Humid Tropical Region of North Queensland*, Division of Plant Industry, Division of Forest Research, Commonwealth Scientific and Industrial Research Organization (CSIRO), Atherton, Queensland

Troup, R. S. (1907) *Indian Forest Utilization*, Office of the Superintendant of Government Printing, Calcutta

Wagner, W. L., Herbst, D. R. and Sohmer, S. H. (1990) *Manual of the Flowering Plants of Hawai'i*, vol. 1, University of Hawai'i Press, Honolulu

Walker, E. H. (1954) *Important Trees of the Ryukyu Islands*, Special Bulletin no. 3, US Civil Administration of the Ryukyu Islands, Okinawa

Walker, E. H. (1976) *Flora of Okinawa and the Southern Ryukyu Islands*, Smithsonian Institution Press, Washington, DC

Wheatley, J. I. (1992) *A Guide to the Common Trees of Vanuatu with Lists of their Traditional Uses and ni-Vanuatu Names*, Department of Forestry, Port Vila, Vanuatu

C. Extending the window of productivity into the fallow phase (economically improved fallows)...

In northern Lao PDR, shifting cultivators have built upon traditional practices of tapping *Styrax tonkinensis* trees for their benzoin resin (see Vongkhamho and Ingalls, ch. 34, this volume). *S. tonkinensis* seedlings that regenerate naturally in swidden fields are thinned and transplanted, and tapping begins when they are six to seven years old (see plates 62-64 in the Coloured Plates section, this volume). This Khmu farmer was collecting resin from a *Styrax* tree in Samneua district, Houaphan province.

Sketch based on a photo by Simone Vongkhamho on March 9, 2016.

Synthesis

FROM USED TO IMPROVED FALLOWS

*Glenn Hunt and Andreas Heinimann**

The idea of improved fallow systems has been gaining considerable traction since the mid-1980s. Planted or 'improved' fallows can be understood as the targeted use of planted species in order to achieve one or more of the aims of natural fallows within a shorter time or on a smaller area (Prinz, 1986, p.31). Much current literature on improved fallowing of upland shifting cultivation systems cites the necessity of improving fallows due to ongoing stresses of increasing land scarcity and subsequent reduction in fallow periods. This may be seen as a result of government policy, donor- or government-supported 'development', or 'environmental' paradigms that aim to supress rotational cropping and replace it with sedentarized farming practices and/or conservation programmes. In this context, improved fallows are a mechanism to support and hasten natural regeneration processes.

In parallel, economic approaches to improved fallow management aim to enhance the economic benefits flowing to farmers from the manipulation and management of fallow vegetation (Hansen and Sodarak, 1996). Economically improved fallows can manifest in various ways, from age-old practices such as the biochar methods used by Peruvian shifting cultivators during the burning cycle (Coomes and Miltner, 2017), to newly introduced economic activities in fallow areas (Trakansuphakon and Ferrari, ch. 33, this volume).

Many discussions around the economic improvement of fallows refer to the importance of value-chain development (see Chakraborty and Choudhury, ch. 32, and Roshetko et al., ch. 29 , both this volume), as development paradigms pursue the integration of upland shifting cultivators into local and trans-national production chains. However, these shifts to market integration, away from traditional

* GLENN HUNT is a PhD candidate and researcher at the Center for Development and Environment at the University of Bern, Switzerland, and DR. ANDREAS HEINIMANN is head of a Regional Stewardship Hub at the Wyss Academy for Nature, Bern, Switzerland, and an associated scientist of both the Center for Development and Environment and the Institute of Geography at the University of Bern, Switzerland.

management, can pose their own challenges in terms of over-exploitation and the loss of cultural practices vis-à-vis natural resources (Li, 2014; Matias et al., 2018).

Larger socio-political issues surrounding the long-term viability of shifting cultivation systems often underlie discussions about improved shifting cultivation fallows. These discussions typically focus on whether economically improved fallows are part of a longer-term process leading to gradual sedentarization of shifting cultivation, or whether economically improved fallows can help to support the long-term livelihoods of established shifting cultivation communities as part of diversified occupational models. The following chapters present examples from across this spectrum, from the established shifting cultivation communities of Hin Lad Nai in northern Thailand (Trakansuphakon and Ferrari, ch. 33, this volume), to the former shifting cultivation communities of Ha Giang province in northern Vietnam who have moved into sedentarized farming and agroforestry (Tran and Pham, ch. 38, this volume).

The following chapters represent an important contribution to these discussions by presenting a variety of perspectives from South and Southeast Asia. They examine both current shifting cultivators and those who have transitioned, or are in the process of transitioning, to more permanently located agricultural practices. They examine traditional techniques as well as more recent innovations and experimentation through which communities are deriving economic returns from the management of vegetation and resources in fallows. As many chapters demonstrate, there is still much to be done to support both current and past shifting cultivators who are either moving into niche local markets or being integrated into larger global supply chains. Much of the research presented in these chapters predates the covid-19 pandemic, and post-pandemic research will be necessary to reveal how shifting cultivation communities have managed market access during this unprecedented global shutdown. The pandemic has shown as never before the vulnerability of overweighted dependency on international markets, and while much post-pandemic research remains to be done, some preliminary analysis from Nepal shows that farmers who were predominantly engaged in subsistence farming proved to be more resilient than those engaged in commercial farming (Adhikari et al., 2021). Such findings have important implications for shifting cultivators, governments and those actors working with them, as they strive to strike a balance between economic utility and long-term food security.

Various themes permeate through the following chapters. Several of them follow traditional indigenous practices of economically improved fallows, documenting traditional techniques for making productive use of fallow areas. A number of case studies are presented where indigenous knowledge of local resources continues to support economically improved fallows across long-fallow shifting cultivation systems.

Chakraborty and Choudhury (ch. 32, this volume) follow the remarkable case of traditional lac production by Karbi farmers in Assam's Karbi Anglong district in northeast India. They posit that the lucrative lac production of the area has traditionally been intertwined with shifting cultivation fallows, and while lac production has significantly declined over the past 150 years due to a variety of pressures, they follow the innovative measures of two farmers faced with a declining resource base to keep this traditional economically improved fallow practice alive. Similarly, Vongkhamho and Ingalls (ch. 34, this volume) examine the production of benzoin resin in northern Laos. Benzoin resin is a high-value product used in the global perfume industry, and the trees are uniquely suited to long-fallow shifting cultivation systems, where resin tapping takes place when the trees are 8 to 15 years old. The trees are endemic to the area and germinate naturally in shifting cultivation fields, after which they are nurtured by the farmers along with their main rice crop. Yet another successful example of centuries-old indigenous knowledge supporting improved fallows is presented by Supratman et al. (ch. 36, this volume), who describe the success of shifting cultivators in South Sulawesi who plant candlenut trees (*Aleurites moluccanus*) in their long 25-year fallows.

Another theme in the following chapters is sustainable forest management, which is a subject for key debates on shifting cultivation policy in the Asian region, where government policy often directs blame for deforestation on politically marginalized shifting cultivators. Godbole and Sarnaik (ch. 28, this volume) illustrate how improved fallow management can lead to both improved forest management and greater economic benefits. The chapter introduces the traditional forest-management system of *honeyem*, which is part of the traditional shifting cultivation management of the Wancho people of Arunachal Pradesh in India. Through careful selection and nurturing of particular fast-growing tree species in the second year of upland-rice cropping, Wancho communities maintain a sustainable supply of firewood, which reduces pressure on protected community-forest areas. Burgers (ch. 37, this volume) reports on communities in Kerinci district, in Sumatra, Indonesia, that have built upon generations of indigenous knowledge to incorporate cash crops of coffee, cinnamon and vegetables into shifting cultivation fallows, allowing for socially, economically and ecologically sustainable forest-management practices. Roshetko et al. (ch. 29, this volume) detail smallholder teak systems in Central Java, where farmers have used agroforestry techniques to plant teak in old fallow areas, leading to the rejuvenation of heavily degraded areas and restoration of social and economic security for farming families. While the authors say that silvicultural practices and market knowledge are still impediments to the farmers maximizing the potential of these agroforestry systems, they identify ways in which the current system can be built upon to increase benefits to smallholders.

Besides building on existing techniques and local knowledge to economically improve fallows, a number of the following chapters describe examples and experiences involved in introducing new crops and technologies. Trakansuphakon and Ferrari (ch. 33, this volume) report on the shifting cultivators of Hin Lad Nai in Thailand's Chiang Rai province, who have successfully used an endemic bee species to incorporate beekeeping into their shifting cultivation cycle as a new economic activity. Bora et al. (ch. 31, this volume) examine traditional broom grass-harvesting communities in northeast India, and farmers' experimentation with intercropping broom grass and nitrogen-fixing pigeon pea in their fallows. Fox (ch. 35, this volume) provides a comprehensive examination of the multiple uses of the *Borrasus* palm across Southeast Asia.

Finally, Schreer (ch. 30, this volume) presents the other side of economically improved fallows, examining the plight of rattan farmers at Katingan, in Kalimantan, who previously enjoyed booming demand for rattan in the 1970s and 1980s. Present unfavourable terms of trade have led the farmers to gradually abandon their cultural affinity for rattan gardens and turn to wage-labour jobs in oil-palm concessions.

Most of the chapters in this section provide a broad and very valuable contribution to the debate surrounding improved fallow management, with highly contextual examples of economically improved fallow systems. However, one of the key underlying factors in terms of the potential for development of economically improved fallow systems is the widespread tenure insecurity that exists in long-fallow shifting cultivation systems (Albers and Goldbach, 2000; Fox et al., 2009). While mentioned briefly as a side note in many papers, Vongkhamho and Ingalls (ch. 34, this volume) discuss this issue substantially in relation to their example of long-fallow benzoin resin production. The case from northern Laos clearly illustrates the need to ensure long fallow periods in order to maintain economic benefits from niche markets. Not only are farmers less likely to make long-term management changes to their fallow practices if they cannot be assured of security of tenure, but similarly, buyers of niche products, aware of the risk coming from insecure land-tenure systems, are also less likely to feel confident that a certain level of long-term supply can be guaranteed.

Consequently, an orchestrated approach is needed to address the key underlying factor of tenure security, which continues to burden shifting cultivation communities. Ultimately, tenure insecurity is a policy issue, and a concerted effort is needed to address policy change. This could highlight positive examples of communities integrating economically improved fallows and document how multifunctional landscapes go hand-in-hand with broader livelihood improvements. Shifting cultivation provides various ecosystem services as part of a broader ecological system and supports livelihoods. Evidence of the benefits of multifunctional landscapes needs to be compiled and presented in policy processes, advocating the part that shifting cultivation can play in achieving various social and environmental aims. Without policy change toward increased tenure security, shifting cultivators will remain in the most precarious of situations, vulnerable to the loss of their land, and transitional pathways to mutually beneficial outcomes for both nature and people will continue to be hampered.

References

Adhikari, J., Timsina, J., Khadka, S. R., Ghale, Y. and Ojha, H. (2021) 'COVID-19 impacts on agriculture and food systems in Nepal: Implications for SDGs', *Agricultural Systems* 186, 102990, available at https://doi.org/10.1016/j.agsy.2020.102990" https://doi.org/10.1016/j.agsy.2020.102990, accessed 1 February 2022

Albers, H. J. and Goldbach, M. J. (2000) 'Irreversible ecosystem change, species competition, and shifting cultivation', *Resource and Energy Economics* 22(3), pp.261-280, available at https://doi.org/10.1016/S0928-7655(00)00034-8" https://doi.org/10.1016/S0928-7655(00)00034-8, accessed 1 February, 2022

Coomes, O. T. and Miltner, B. C. (2017) 'Indigenous charcoal and biochar production: Potential for soil improvement under shifting cultivation systems', *Land Degradation and Development* 28(3), pp.811-821, available at https://doi.org/10.1002/ldr.2500" https://doi.org/10.1002/ldr.2500, accessed 1 February 2022

Fox, J., Fujita, Y., Ngidang, D., Peluso, N., Potter, L., Sakuntaladewi, N., Sturgeon, J. and Thomas, D. (2009) 'Policies, political-economy, and swidden in Southeast Asia', *Human Ecology* 37(3), pp.305-322, https://doi.org/10.1007/s10745-009-9240-7" https://doi.org/10.1007/s10745-009-9240-7, accessed 1 February 2022

Hansen, P. K. and Sodarak, H. (1996) *Agroforestry Research for Development in Shifting Cultivation Areas of Laos*, Shifting Cultivation Research Sub-programme, Lao Swedish Forestry Programme.

Li, T. (2014) *Land's End: Capitalist Relations on an Indigenous Frontier*, Duke University Press, Durham, NC

Matias, D. M. S., Tambo, J. A., Stellmacher, T., Borgemeister, C. and von Wehrden, H. (2018) 'Commercializing traditional non-timber forest products: An integrated value chain analysis of honey from giant honey bees in Palawan, Philippines', *Forest Policy and Economics* 97, pp.223-231, available at https://doi.org/10.1016/j.forpol.2018.10.009" https://doi.org/10.1016/j.forpol.2018.10.009, accessed 1 February 2022

Prinz, D. (1986) 'Increasing the productivity of smallholder farming systems by introduction of planted fallows', *Plant Research and Development* 24, pp.31-56

28

HOUSEHOLD ENERGY SECURITY THROUGH FALLOW MANAGEMENT

By the Wanchos of Arunachal Pradesh, India

Archana Godbole and Jayant Sarnaik*

Introduction

For many generations, the indigenous knowledge of local communities has provided a variety of options for using and managing forest resources on a sustainable basis. Over the past few years, researchers and policy-makers have come to realize that sustainable use of forest resources will only be successfully achieved if the local communities using and managing those resources are made an integral part of protection and management planning. Local communities in northeast India are a prime example of the use of indigenous-knowledge systems for forest-resources management in today's context.

The significance of indigenous knowledge has been recognized in theory in all recent approaches to development, but the concept of applying indigenous knowledge in community development and conservation has only recently been understood. Our failure to investigate historical, socio-cultural and rational accounts of local knowledge has been the root cause of this poor understanding (Tyagi and Rao, 1997).

The study site

The northeastern state of Arunachal Pradesh forms the eastern corner of India's Himalayan region. Arunachal Pradesh is very rich in natural resources and biodiversity, as well as having a wide diversity of cultures and indigenous people of various origins. More than 80% of the tribal population of Arunachal Pradesh are primarily dependent on shifting cultivation and forest resources for their subsistence needs. Small land holdings yield very meagre crops and villagers gather many plants and animals from surrounding forests to supplement their diet.

---

* Dr. Archana Godbole and Jayant Sarnaik both work for the Applied Environmental Research Foundation (AERF), Pune, India.

This chapter deals with the traditional forest-management practices of the Wancho people of Arunachal Pradesh and their links with conservation. The Wanchos are regarded as being the most underdeveloped of the state's ethnic minorities (Datta, 1990). A thorough survey of the literature indicated the need to document indigenous-knowledge based strategies used by the Wanchos for natural-resource management. The study site was in Tirap district, which was relatively remote and inaccessible, and was not a well-explored area of Arunachal Pradesh, in terms of ethnobiology (Saklani and Jain, 1996). Yet the area was very rich in biological and ethnic diversity. After completion of the study, Tirap was split into two districts in 2012, and the study site became part of the new district, called Longding.

The literature survey and a review of the situation based on preliminary field visits suggested that the forests of the area, which were once very rich (Woodthorpe, 1878), were being depleted very rapidly and the Wanchos were not aware of this problem. It was noticed that there were no written records of the traditions employed by the Wanchos in forest management; nor had there been any analysis of community forest management in the study area in the current context. After a thorough survey of Wancho villages in the district, a village called Zadua was selected for detailed analysis (Figure 28-1). Various criteria for its selection included a long shifting cultivation cycle, better preserved community forests, a strong community organization,

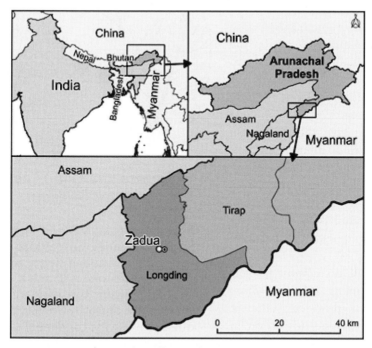

FIGURE 28-1: The study village of Zadua, in Longding district, Arunachal Pradesh. At the time of the study Tirap and Longding made up one large district, called Tirap. Longding district was split from Tirap for administrative purposes in 2012.

leadership by knowledgeable elders and enthusiasm among villagers to take part in the research process.

The work with the Wancho tribal community was an effort to understand the complex mechanisms involved in natural-resource management and to find out how these systems could be used effectively in today's context. A multi-method approach was adopted, including participant observation, informal surveys both along transect walks and in the villages, a biodiversity inventory, resource mapping, and the quantification of ethnobotanical knowledge using the pair-wise ranking method.

Natural-resource management

Various natural-resource management systems used by the Wanchos were closely linked to their understanding of forest resources and their acceptance of the limited availability of these resources. Along with shifting-cultivation fields that were currently cultivated, the Wanchos had fallows at various stages of maturity, along with specifically protected community forests and home gardens as important resource areas.

Zadua village had two main agro-ecosystems: terraced wet-rice cultivation and shifting cultivation, or *jhum*. The terraced cultivation was of recent origin, having begun just 30 years earlier. *Jhum*, on the other hand, had been practised and in a process of development for hundreds of years. These shifting cultivation practices had moulded the lifestyle of the mountain people as well as influencing their ecosystem (Pei, 1994, 1996). Along with *jhum* and wet-rice cultivation, Zadua had home-gardens (*sawat*), plots for opium poppies (*Papaver somniferum*) and community forests (*lings*).

The village's *jhum* farmers were maintaining a 10-year *jhum* cycle, which provided sufficient time for biodiversity to recover in successional-vegetation phases, and for recovery of soil fertility. In the wet-rice terraces, only one crop was produced per year and the terraces

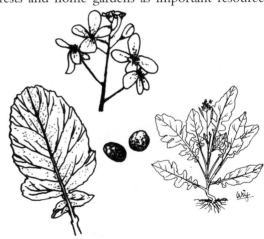

Brassica juncea (L.) Czern.
[Brassicaceae]

The mustard plant is important to many regional cuisines, such as that of the Wanchos. The leaves and stems, which are a rich source of vitamins A, C and K, are eaten as mustard greens. The seeds are ground and mixed with water, vinegar, or other liquids to make the condiment mustard. This plant is also used in phytoremediation, to remove heavy metals, such as lead, from hazardous waste sites by storing the metals in its cells. Such plants are later discarded.

were used as fishponds in the three to four months after rice harvesting. Although opium cultivation was an important economic activity, most of the product was used at household level.

During the documentation of natural resource-management practices that led to this chapter, two traditional ethnoforestry systems called *honeyem* and *loham* were identified as Wancho strategies for sustainable management of fallows, helping to reduce pressures on community forests not currently in use for shifting cultivation (Godbole, 1998). These are discussed below.

Shifting cultivation

Zadua village still maintains a 10-year *jhum* cycle. The village's total resource area used for *jhum* cultivation is divided into nine resource areas. Each *jhum* field is

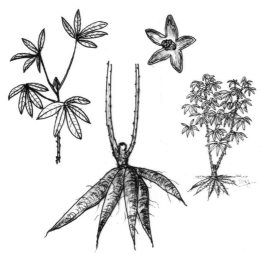

Manihot esculenta Crantz [Euphorbiaceae]

Cassava tubers are the third-largest source of food carbohydrates in the tropics, after rice and maize, and are believed to provide a basic diet for more than half a billion people. The world-wide popularity of this woody shrub also comes from its drought tolerance and ability to thrive on marginal soils. As well as its vital food role, it is increasingly grown as an energy crop, to produce biofuels.

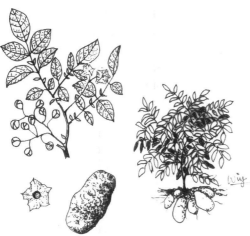

Solanum tuberosum L. [Solanaceae]

Potatoes were domesticated in South America as long as 10,000 years ago and their cultivation spread around the world, to become a vital part of the European diet and a major crop in Asia. China and India are now the world's leading producers, and potatoes are the world's fourth-largest food crop, after maize, wheat and rice. There are now more than 5000 potato varieties or cultivars.

cultivated for two years. The major crops in the first year are millet, *kochu* (taro – *Colocasia esculenta*), cassava (*Manihot esculenta*), beans, leafy vegetables, mustard greens, potatoes and several varieties of rice. After these are harvested at the end of the first year, rice is sown as the sole second-year crop. The field is left fallow in the third year. Every year, each household of shifting cultivators has two *jhum* fields under cultivation.

Jhum fallows

The fallow phase allows biological diversity to be regenerated and maintained through natural restoration and succession; the soil is able to replenish its natural fertility through litter decomposition, deposition of organic matter and nutrient recycling.

Mature *jhum* fallows are important resource areas that provide a continuous supply of livelihood needs, one of the most important of which is firewood. The fallows are also major sources of many non-timber forest products, and with their constant supply of firewood, they help to minimize harvesting pressures on community reserved forests.

In Zadua, decisions regarding the selection of resource areas for *jhum* are made by the community, with the approval of the village's customary chief (*wangham*) and his council. Certain rituals are performed as part of the selection procedure. The criteria for selection are based on the maturity of the fallow, an understanding of previous *jhum* cycles, and the distance of the resource area from the village.

Honeyem and *loham*: traditional fallow-management practices

As mentioned earlier, two traditional ethnoforestry systems called *honeyem* and *loham* were identified as Wancho strategies for sustainable management of fallows. *Honeyem* is a traditional practice aimed at enabling the Wanchos to rely on *jhum* fallows, which are at various stages of forest regeneration, for their non-timber forest product needs and daily firewood requirements. These things cannot be reliably obtained from community forests, which are small patches of forest spread throughout the village resource areas, because they are generally non-exploitable. Only large timber can be harvested, and even then, only with the collective approval of the community.

The *honeyem* system begins in the second year of cultivation in *jhum* fields. The second-year crop is purely rice, but the field is kept empty for a month or two after the first-year crops are harvested. In these empty fields, tree seedlings grow naturally and profusely. Prominent among them are *puak* (*Macaranga denticulata*) and *puakmi* (*Mallotus tetracoccus*). Selected seedlings are kept in the field and allowed to grow. The rice is sown between them, and the seedlings are carefully nurtured during their first year of growth. Later, they have a natural advantage when the field is left fallow. The management system in which these species are allowed to grow alongside the rice is very important because they are the most favoured firewood species among the Wancho villagers at Zadua (Godbole, 2003). Both are fast-growing species capable of yielding firewood within four or five years.

Therefore, the *honeyem* fallow-management system serves three purposes:

a) it increases soil quality;
b) it checks soil erosion; and
c) it yields firewood and timber very quickly, i.e., within four to five years.

The second fallow-management system in Zadua village is called *loham*. It involves transforming the fallow vegetation into palm groves, particularly growing the fan palm *Livistona jenkinsiana*, the leaves of which are used for thatching. Similarly, well-developed bamboo groves are maintained around the village, mainly on privately owned land.

Five community forests are located within the Zadua village territory and despite strict controls imposed by the village council, they represent a major source of timber, firewood, game, and non-timber forest products. Logging and large-scale extraction are not allowed in the community forests, although timber extraction appeared to have occurred prior to the study period. Additionally,

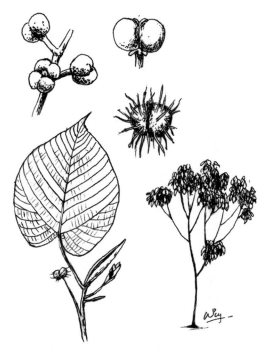

Macaranga denticulata (Blume) Müll.Arg. [Euphorbiaceae]

This medium-sized, fast-growing tree is known as *puak* by the Wanchos. It is best known for producing firewood from the *honeyem* system, since is it always available when needed, growing as a pioneer species in swidden fallows and at forest edges. It is also known as a soil improver, for its symbiotic relationship with fungal species growing in and among its roots (see Changkija et al., ch. 26, this volume).

the community forests are maintained so as to avoid disputes with neighbouring communities, and they are also used as burial grounds. About 12 species of mammals are found in the forests, as well as numerous bird species.

Quantitative analysis of firewood species

The role of women in resource management was explored in the course of this study. Although they are not represented on the village council and have no formal decision-making power, they are much more involved in resource management than the men. Women are solely responsible for all of the activities in *jhum* cultivation, except for cutting trees when opening new swiddens and transporting harvests to the village. Women also do most of the work of cultivation in the wet-rice terraces. However, it is mostly the men who are involved in opium cultivation.

Using the exacting method of pair-wise ranking, we sought to ascertain the women's preferences for specific firewood species and, thus, confirm the validity of *honeyem*. Pair-wise ranking is a widely accepted method of understanding systems based on indigenous knowledge. Only women were involved in this exercise because women alone were engaged in firewood collection, as well as in the various cultivation activities of *honeyem* in *jhum* fields. Women from different age groups were selected.

In preparation, it was found that 42 plant species were used as firewood in Zadua village. Of these, 10 species were mainly used for timber, and only occasionally used as firewood. Fifteen species were most commonly used as firewood, and the pair-wise ranking exercise set out to discover which of these was most preferred by Wancho women for collecting and use as firewood, and how this preference related to the indigenous knowledge employed in *honeyem* fallow management.

In assessing their priorities, each of the 15 species (Table 28-1) was compared, one at a time, with all of the others. A total of 23 women were asked: Which wood do you prefer? Which wood gives more heat? Which wood is more available? Which wood burns slower? Which wood, on burning, produces more light? Responses clearly indicated that overall preference was dependent on availability, rather than heat-producing capacity. This, in turn, indicated the appropriateness of *honeyem*, in its maintenance of *puak* (*Macaranga denticulata*) and *puakmi* (*Mallotus tetracoccus*). There was a preference based on heat-producing capacity in the case of only one species: *hen* (*Stereospermum* sp.).

TABLE 28-1:   Fifteen firewood species tested by pair-wise ranking in Zadua village.

No.	Wancho name	Botanical name
1	puak	Macaranga denticulata
2	puakmi	Mallotus tetracoccus
3	gnut	Dendrocalamus sp.
4	ottan	Saurauia roxburghii
5	phü	Albizia sp.
6	hen	Stereospermum sp.
7	hetpha	Ficus semicordata
8	chabu	Litsea monopetala
9	nyakay	Itea macrophylla
10	zapzan	Eurya acuminata
11	offang	Ficus hirta
12	chicklong	Syzygium cumini
13	zhak	Schima wallichii
14	zhamlau	Cedrela serrata
15	nyapha	Aralia sp.

It was also clear from the pair–wise ranking exercise that the *honeyem* system had been carefully developed by the Wancho community for maintaining fallows and using them to ensure a continuous supply of quality firewood. The simple *honeyem* technique involved the careful selection and nurturing of the most valued firewood species and their dominant role in the fallow phase of Wancho shifting cultivation. Ultimately, the system helped to conserve the community forests by avoiding depletion as a result of demand for firewood.

Several major problems were identified in the course of the study. There was depletion of first-class timber species from Wancho territory due to commercial logging and a high rate of deforestation, presumably due to population growth. The shifting cultivation cycle was growing shorter, from 10

Mallotus tetracoccus (Roxb.) Kurz [Euphorbiaceae]

Called *puakmi* by the Wanchos, this medium-sized tree is one of the species favoured for firewood production in the *honeyem* system. A fast-growing pioneer species, it is more common at forest edges, in clearings, and in secondary forest at altitudes up to 900 m, than in mature forest interiors. The bark and gum of this tree are used medicinally to treat fractures and venereal diseases.

to 12 years down to eight to 10, and the traditional *honeyem* practice was, itself, declining. Overall, the spread of Christianity and tribal-development processes were affecting the Wancho culture and its resource-use systems.

Conclusions

This research provided an opportunity to test simple participatory-quantification methods like pair-wise ranking, and their significance in both short- and long-term research projects.

The quantification exercise showed that the traditional or indigenous-knowledge based natural-resource management of the Wanchos of Zadua village was not only functioning, but was also economically, socially and environmentally sound. Comparative analyses of similar practices are needed, particularly when there is always a shortage of firewood in remote rural areas.

The selection of particular species by the Wanchos for precedence in their regenerating *jhum* fallows not only demonstrated the importance of household energy security, but also highlighted the importance of shifting cultivation as an age-old, sustainable land-use and livelihood system.

29

SMALLHOLDER TEAK SYSTEMS

Indigenous innovations to improve fallow management

James M. Roshetko, Gerhard E. Sabastian, Aulia Perdana,
*Endri Martini, Desy Ekawati and Muhammad A. Fauzi**

Introduction

Teak (*Tectona grandis*) is a tropical timber species native to South Asia and peninsular Southeast Asia, specifically India, Myanmar, Laos and Thailand. The species grows under a wide range of biophysical conditions from the tropics to the subtropics, with rainfall of 500 to 3500 mm and temperatures of 2° to 48°Celsius. Similarly, soil conditions can range from rocky and infertile acidic soil to fertile alluvial soil (Kaosa-ard, 1988). There is currently about 29 million ha of natural teak forest, with nearly half of this in Myanmar (Kollert and Kleine, 2017). Teak timber is durable, strong, easy to work and commonly used to produce furniture, housing materials, crafts, ships and many other products. These traits make the timber highly valuable; for centuries, it has been in high market demand. The first plantations of teak were established in Sri Lanka in 1680 (Perera, 1962) and in Indonesia as early as the 13th century (Troup, 1921; White, 1992; Simatupang, 2000). Commercial teak plantations started in India in the 1840s, Myanmar in 1856, Indonesia in 1880 (Pandey and Brown, 2000), and Thailand in 1906 (Krishnapillay, 2000). Teak plantations spread to Africa and tropical America in the early 20th century; Nigeria in 1902, Ghana in 1905, Trinidad and Tobago in 1913, and Honduras, Panama and Costa Rica from 1927 to 1929 (Pandey and Brown, 2000).

In 2017, the global area of planted teak was estimated to be between 4.35 and 6.89 million ha, with 80% of it in Asia (primarily India, Indonesia and Myanmar); 10% in Africa; and 6% in tropical America (Kollert and Kleine, 2017). Teak is now planted and cultivated in more than 60 tropical and subtropical countries (Jerez-Rico

* Dr. James M. Roshetko, Dr. Gerhard E. Sabastian, Aulia Perdana and Endri Martini are all from the World Agroforestry Centre (ICRAF)'s Southeast Asia Regional Research Programme in Bogor, Indonesia; Desy Ekawati is a forest and environment researcher at the Development and Innovation Agency, Bogor, Indonesia, and Muhammad A. Fauzi is from the Centre for Forest Biotechnology and Tree Improvement in Yogyakarta, Indonesia.

and de Andrade Coutinho, 2017), and while the area of planted teak continues to expand, the demand for its valuable timber continues to exceed the sustainable yield from large-scale plantations and natural forests. This provides an opportunity for smallholder farmers to grow teak as a market crop. Smallholder production is a minor component of the global resource, but it has become an important source of raw material for international and national teak industries, presenting opportunities for farmers to expand their role in teak markets (Roshetko and Perdana, 2017, Midgley et al., 2017). While definitions of smallholder planting vary greatly from region to region, it is nevertheless estimated that smallholder plantings account for 19% of the teak area in Africa and Asia, 31% in Central America, and 34% in South America (Kollert and Cherubini, 2012).

Tectona grandis L.f. [Lamiaceae]

Teak is one of the world's most important timbers, with superior physical properties, including its tolerance of heavy loads and stress. It is native to South and Southeast Asia, but is grown widely in Africa and the Caribbean. It is said to grow up to 40 m tall, but the teak tree that is claimed to be the biggest in the world can be found in Myanmar. It is 8.4 m in girth and 34 m tall.

Teak is not a single-commodity species. Besides timber, teak yields a number of other commercial and household products. Its leaves and buds produce a dye with characteristics similar to henna (Sharma, 1999). It is used to colour cloth (Bhuyan et al., 2004), including Indonesia's iconic batik clothing and crafts (Widiawati, 2009). Teak sawdust is used to produce incense on Java in Indonesia (Roemantyo, 1990). Household uses are many. Villagers in India use the oil extracted from teak leaves and wood to treat skin diseases (Siddiqui et al., 1989, Gupta, V. C. et al., 1997; Gupta, R. et al., 2010). Compresses made from teak leaves are used to hasten the healing of skin wounds (Majumdar, 2005). In Bangladesh, roof thatching is made from dried teak leaves (Chakraborty and Bhattacharjee, 2003), a practice that is common in other countries of the region. Dried leaves are also used as a dry-season feed supplement for goats and sheep, at compositions of 5% to 25% of the animals' diet (Reddy and Reddy, 1984; Anabarasu et al., 2001, 2004). On Java, a caterpillar (*Hyblaeca puera*) that is commonly found on teak is collected to cook as a side dish for home consumption and local sale (Pramono et al., 2010, 2011). Branches and other woody biomass are used for firewood by rural households in many areas.

Despite its many secondary uses, teak is best known for its high-value, high-quality tropical timber, produced in both industrial plantations and smallholder farming systems. It is widely recognized that teak timber makes significant contributions to national and international commerce, local economies, and household incomes. There is another important contribution made by teak that is often unrecognized: its ability to establish and grow on rocky, infertile and even degraded soil, with little or no management intervention, making it an excellent improved fallow crop that yields a valuable product while contributing to environmental rehabilitation and landscape restoration. In its role as a fallow crop, teak is often just one component of a multi-species tree-farming system adopted by smallholders to diversify production, income and risks.

Indonesia is the second-largest global producer of teak, behind India, and smallholder farmers play a substantial role in the national teak sector. Studies over the past decade have found that 80% of raw material used by small- to medium-sized teak-furniture firms comprises small diameter logs, with a diameter at breast height (dbh) equal to or less than 30 cm, that are grown by smallholder farmers (Achdiawan and Puntodewo, 2011). These small- to medium-sized firms account for more than 90% of furniture-making at Jepara, in Central Java, the centre of Indonesia's teak-furniture industry (Yovi et al., 2013). Smallholder teak-farming systems are generally found on marginal to degraded agricultural land, and since the 1960s, teak has made significant contributions to the rehabilitation of these areas.

Yogyakarta is regarded as the centre of smallholder teak cultivation in Indonesia. The Special Administrative Region of Yogyakarta is located on the southern coast of Java, and is surrounded by Central Java province on three sides and the Indian Ocean to the south. Yogyakarta is the historical centre of Javanese culture and the seat of the Javanese Sultanate. The Special Administrative Region contains five administrative units: the City of Yogyakarta and the districts of Bantul, Gunungkidul, Kulon Progo and Sleman. Smallholder teak systems are found in all five administrative units of Yogyakarta, throughout Central and East Java, and parts of eastern Indonesia (Figure 29-1). The smallholder teak-farming systems of Gunungkidul are representative of those found in other parts Indonesia.

Research on smallholder teak production

Since 2007, the World Agroforestry Centre (ICRAF), the Center for International Forestry Research (CIFOR), the Indonesian government's Forestry and Environmental Research, Development and Innovation Agency (FOERDIA), and local partners have conducted research projects in Gunungkidul, Central Java, and Nusa Tenggara, focused on improving production and marketing systems for smallholder timber and non-timber forest products (NTFPs). Smallholder teak systems are a priority of these research activities, support for which has come from the Australian Centre for International Agricultural Research (ACIAR).

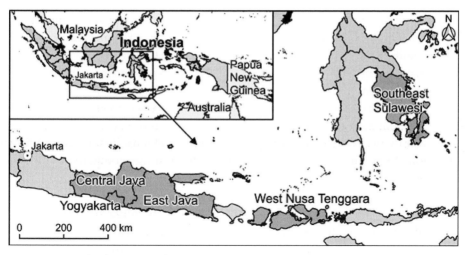

FIGURE 29-1: The locations of major smallholder teak systems in Indonesia, shown in darker shading.

Source: Elissa Dwiyanti in Roshetko and Perdana (2017).

Gunungkidul is located at 7°46' to 8°09' latitude and 110°21' to 110°50' longitude. Its landscape is characterized by hilly terrain, with half the district having slopes of 15% to 40%. The northern zone of the district is hilly with elevations from 200 to 700 metres above sea level (masl); the central zone is primarily flat with some hills of 150 to 200 masl; and the southern zone is characterized by infertile, dry karst (limestone) soils at elevations from sea level to 300 masl. The climate is strongly influenced by the northwest monsoon from November to May and the southeast monsoon from June to October. Annual rainfall ranges from 1500 to 2500 mm, with an average temperature between 24° and 26°C (Sudiharjo and Notohadiprawiro, 2006). The population of the district in 2020 was 747,161, with 18 subdistricts and 144 villages, and 17% of the people lived below the poverty line (BPS Kabupaten Gunungkidul, 2022). The main components of the district economy are agriculture, mining, manufacturing, construction, wholesale and retail (BPS Kabupaten Gunungkidul, 2018). Within the agricultural sector, food crops account for 64% of economic value, followed by forestry (27.3%), livestock (6.3%), plantation crops (1.7%), and fisheries (0.7%). Gross per capita annual income was 22.2 million Indonesia rupiah (US$1563) with growth of 7.16% (BPS Kabupaten Gunungkidul, 2018).

At the beginning of the programme the research team, local government agencies and communities collaborated to select the villages in which the research would be conducted. A baseline study was implemented to identify the socio-economic conditions and farming characteristics of teak-farming families. It involved 275 households living in seven villages representative of Gunungkidul and managing 1074 land parcels covering a total of 276.5 ha. An inventory of 227 teak farms, covering 47.1 ha, was undertaken to document species composition, tree density and management practices. A rapid market appraisal of 293 respondents (the baseline farmers plus 11 traders and seven sawmill owners) was conducted to identify smallholders' teak-marketing practices and related opportunities. Farmer demonstration trials (Roshetko et al., 2005) were designed and established with landowners on six farms to show the advantages of silvicultural management under smallholder conditions. Using the results from these and other trials, computer simulations were applied to evaluate a broader set of intercropping and management options (Khasanah et al., 2015).

Over the course of the research programme, focus-group discussions and interviews with key respondents were undertaken to cross-check survey results, fill information gaps and develop a comprehensive understanding of key issues. These studies have yielded several publications that are cited in this chapter. Similarly, while much of the teak research was conducted in and around Gunungkidul, other related research was conducted across the programme area. Results from those studies, as well as from smallholder teak research in other countries, are included in this chapter where relevant.

A perspective on teak and landscape conversion on Java

Teak was introduced to Java from India in the 3rd century, at the onset of the archipelago's Hindu period. The timber resource quickly became valuable, due to high demand. By the 13th century, teak plantations were established on Java to support ship building (Whitten et al., 1996; Simatupang, 2000). Before 1600, the Javanese sultans claimed the teak and other forests within their realms by royal decree. Community access to and use of the forests was officially denied, but not strictly prevented. During the period of its supremacy, from 1619 to 1796, the Dutch East India Company (Vereenigde Oost-Indische Compagnie, VOC) managed teak as a vital resource to maintain its naval assets for commercial dominance in Asia and around the world. Initially, the company contracted the local nobility to access teak. However, as the company gained power on Java, it forced the sultans and communities to deliver larger quantities of teak on favourable terms. Local access to forests was forbidden and strictly enforced. During the brief English period, from 1811 to 1815, forest control was liberated and the use of teak for other purposes was permitted. The Dutch government resumed direct control over Java in 1816. Teak harvesting was expanded to support the agricultural sector through the construction of factories, warehouses, sheds, housing complexes and railroads for sugar, coffee and tobacco operations, and others. Ship building remained important. Unauthorized teak harvesting was illegal, yet common (Peluso, 1992). By 1856, teak began to be established though the *taungya* system (Wiersum, 1982), in which teak seedlings were intercropped with annual crops for up to three years to improve establishment and early growth.

The agrarian law of 1870 reasserted the central authority of the Dutch government. All unclaimed and forest lands were declared the domain of the state and rural communities lost free access to forests. Subsequent forest laws were passed in 1913,

1927, 1928, 1931 and 1934. Teak forests were defined as land or land parcels on which teak was grown or partially grown. Forest lands on Java and Madura were declared the domain of the state, to which other people or parties had neither rights nor control (Peluso, 1992). Some central control remains today. While individuals and organizations own the teak produced on their land, government permits are required to transport and trade in timber.

The original vegetation of virtually all of Java, including Yogyakarta, consisted of tropical evergreen forests. This forest ecosystem remained relatively undisturbed until the arrival of teak. Teak became a dominant land cover in the lowlands because it was planted on agricultural land that had become marginal or infertile. This was the first historic example of teak being used as a fallow crop in Indonesia. By the 11th century, there may have been as much as 1.5 million hectares of teak in the lowlands of Java (Whitten et al., 1996). Largely native vegetation remained in upland areas of Yogyakarta and other parts of Java, as cultivation of the lowlands with rice, other annual crops and perennials was sufficient to support the small local population, with adequate surpluses for sale and storage. This changed drastically in the 1830s under the Cultivation System (*cultuurstelsel*) of the Dutch East Indies colonial government, which mandated that farmers had to grow commercial commodities, including coffee, sugar, tobacco, rubber, tea and cloves. Called *tanam paksa* (forced planting) in Indonesian, the system resulted in the conversion of the uplands to commodity crops, so farmers could fill their quotas. The production of staple crops declined, resulting in food shortages and social stress. The Cultivation System was stopped in 1870. Private companies were allowed to manage plantation crops, thus continuing the cultivation of the uplands (Whitten et al., 1996).

At the beginning of the 20th century, population pressure resulted in the expansion of intensive semi-permanent agriculture into the uplands of Yogyakarta. To conserve soil and water, farmers began to terrace the uplands, first with tree trunks and later by constructing stone walls. Upland fields were generally cultivated for two years, followed by fallow periods of two to three years. However, the conversion of the uplands to permanent agriculture, even with fallow periods, resulted in soil erosion, declines in soil fertility and water shortages. The population continued to increase, resulting in the conversion of more forests to agricultural production. Agriculture was not the only pressure on the forest. Between 1898 and 1937, large areas of forest were harvested to develop an extensive railroad network to support economic growth. Surging demand for firewood from the growing population continued to contribute to forest degradation and loss. The Japanese occupation in World War Two led to widespread timber harvesting and deforestation. The Indonesian Revolution and post-independence period did not improve the state of Java's forests as central control was weak and the government eager for sources of income (Nibbering, 1999; Whitten et al., 1996). During the 1900s, forest cover on Java steadily declined. Drawing data from various sources, Whitten et al. (1996) estimated that forest cover on Java was 23% before World War Two and it steadily decreased during the 1940s and 1950s to only 11% by 1973 (Donner, 1987, and Seidensticker, 1987, both cited in Whitten et al., 1996).

A perspective on smallholder farming systems in Gunungkidul

Rural communities in Gunungkidul largely mimicked those in Yogyakarta, Central Java and Java as a whole. They were small and largely self-sufficient in terms of agricultural production through the early 1880s. Teak became a common component of lowland vegetation, on marginalized agricultural soils. Rural communities were exploited by the Cultivation System between 1830 and 1870, and by the turn of the 20th century, the valley bottoms were under permanent agriculture and the uplands were cultivated with annual crops, with short fallow periods to rejuvenate the soils. As the population grew, fallow periods were shortened and eventually stopped, and soil-conservation measures were ignored (Wiersum, 1982; Soerianegara and Mansuri, 1994; Nibbering, 1999; Sunkar, 2008). Teak continued to be established using *taungya* systems (Wiersum, 1982). In the 1930s, studies in Gunungkidul found that rural communities were again self-sufficient in food production and the quality of their lives compared favourably with those in neighbouring locations. Land-use pressures accelerated during the following 20 years, with the area of upland cultivation increasing more than three-fold, from 36,000 ha to 119,000 ha (Nibbering, 1999).

Continuous agricultural production in the uplands began to take its toll in the 1950s. Upland soils were eroded and infertile and soil moisture was limited. Ploughing became difficult because severe erosion had left upland soils rocky. Crop yields began to plummet. As a result, farmers changed from cultivating upland rice to growing cassava. The advantage of cassava was its ability to grow and be productive on infertile soils, even in the absence of organic or chemical fertilizers (Van Der Poel and Van Dijk, 1987; Filius, 1997; Sunkar, 2008). While it was capable of providing food, cassava was unable to deliver quality of life or income. At the same time, the fundamental problems of eroded, infertile, dry soils were not being addressed. The soil contained very little organic matter. Demands for firewood, poles, timber and other wood products continued to exert pressure on the remaining forests. At this time, farmers were not planting trees outside of their home gardens because existing trees in home gardens and remnant forests were considered adequate to meet their needs for tree products (Van Der Poel and Van Dijk, 1987; Nibbering, 1999). Little to no interest in planting trees, when forest and tree access is considered adequate, is a common problem across Southeast Asia, even when outsiders observe an urgent need to replenish tree stocks.

By the late 1950s and 1960s, land degradation in Yogyakarta was severe, causing a deep economic and social crisis. The landscape was nearly denuded of trees. Soil erosion was critical, agricultural production was failing and poverty was severe. People survived on a diet of cassava and annual pre-harvest food shortages were common. A number of drought-induced famines occurred. People were malnourished and had to sell their possessions and land to buy food. Many of them moved away to find work to provide their families with basic food and livelihood needs (Van Der Poel and Van Dijk, 1987; Nibbering, 1999).

According to reports from farmers, some progressive individuals in Yogyakarta and surrounding areas began to plant trees on their own land in the 1960s. They developed this innovation because agricultural yields from the infertile soils were uneconomic, even for subsistence farming. They favoured teak, jackfruit (*Artocarpus heterophyllus*), melinjo (*Gnetum gnemon*) and other fruit trees. Teak was grown to generate income; the other species for the household products they yielded. *Lamtoro* (*Leucaena leucocephala*), *gamal* (*Gliricidia sepium*) and *tayuman* (*Piliostigma malabaricum*) were also planted to produce dry-season livestock fodder. According to Filius (1997), off-farm employment opportunities began to increase in Central Java at this same time, lessening the need for rural dwellers to produce all of their own food. Roshetko et al. (2008a) also reported that off-farm employment in Central and East Java enabled smallholders to develop and invest in tree-farming systems. Subsidized chemical fertilizers also become available in the 1960s, increasing yields from more intensively managed farm plots (Filius, 1997; Nibbering, 1999) and enabling diversification with tree crops on other parcels of land.

Leucaena leucocephala (Lam.) de Wit [Leguminosae]

Once hailed as a 'miracle tree' for its multiplicity of uses, this native of Mexico and Central America is also regarded as an aggressively invasive species. Called *Lamtoro* in Indonesia, it is an excellent source of fodder for livestock, is efficient at fixing nitrogen in the soil, provides generous quantities of biomass for green manure, and its wood is used to make paper pulp or as fuel for domestic fires (see Suson and Lasco, ch. 23, this volume).

The spontaneous tree-planting movement by this cadre of progressive farmers in Gunungkidul and surrounding areas was reinforced by government 'tree-planting programmes' in the mid-1970s. These promoted tree planting to regreen (*penghijauan*) private land and reforest (*reboisasi*) public land. As well as revegetation of the landscape, a key objective of these programmes was reversing soil degradation by implementing soil- and water-conservation measures. The *penghijauan* programme provided free tree seedlings and cash inducements for seedling survival (Figure 29-2). Density targets were 400 to 800 trees per ha. Over the course of the programme the species distributed included *lamtoro*, teak, acacia (*Acacia auriculiformis*), *sengon* (*Paraserianthes falcataria*), calliandra (*Calliandra calothyrsus*), mahogany (*Swietenia macrophylla*), jackfruit, coconut (*Cocos nucifera*), and cashews (*Anacardium occidentale*) (Van Der Poel and Van Dijk, 1987; Soerianegara and Mansuri, 1994; Filius, 1997; Nibbering, 1999). The *penghijauan* programme has been criticized for being overly centralized, with 'top-down' planning and implementation. For example, the species that were distributed reflected central decisions, and not local priorities or biophysical conditions. These shortcomings aside, the *penghijauan* programme must be credited with reinforcing spontaneous efforts by progressive farmers to plant trees and engendering a tree-planting culture in the rural communities of Yogyakarta, Java and other parts of Indonesia. National and international research organizations, universities and non-governmental organizations have all conducted research and development activities to analyse, strengthen and expand smallholder tree-farming systems in Yogyakarta and Central Java (Wiersum, 1982; Filius, 1997; Rohadi et al., 2012; Perdana et al., 2012).

In Java, these smallholder tree-planting or agroforestation[1] activities resulted in rehabilitation of communities and farms, restoration of soil fertility, diversification of crop production and improved food security (Van Der Poel and Van Dijk, 1987; Soerianegara and Mansuri, 1994; Filius, 1997; Nibbering, 1999). Gunungkidul was transformed from a nearly treeless condition in the 1950s to having tree cover of 28.1%, covering 41,773 ha, in the first decade of the 21st century. Of this, 8.9% is state forest land and 19.2% is comprised of smallholder agroforestry systems (BPS Kabupaten Gunungkidul, 2008). The development and prominence of these smallholder agroforestry systems were the

FIGURE 29-2: Newly-planted teak seedlings shelter from the sun beneath the fallen leaves of mature trees.

Photo: James M. Roshetko.

Swietenia macrophylla King [Meliaceae]

This species was promoted in the Indonesian government's mid-1970s tree-planting programme. It is one of three species producing genuine mahogany timber. Although a native of South America and Mexico, plantations in many Asian countries are now the major sources of mahogany on world markets. The strong, attractive wood is used to make furniture, musical instruments and ships.

outcome of spontaneous tree planting by farmers, the government's *penghijauan* programme, and related research and development efforts by stakeholders mentioned above.

Smallholder teak production: An improved fallow strategy

In Gunungkidul, the average landholding per family is about 1 ha, ranging from 0.5 to 3.0 ha. Each family owns multiple parcels of land, and almost all families grow teak. It is estimated that 30% to 50% of the land parcels include teak as a component. Farmers' innovations in using teak, along with other species, as a productive fallow crop to regenerate degraded agricultural land has spawned the development of four smallholder teak agroforestry systems: *kitren, tegalan, pekarangan* (home gardens), and line plantings.

Teak agroforestry systems

The following summary of smallholder teak agroforestry systems called *kitren, tegalan, pekarangan* (home gardens) and line planting is supported by the cropping calendar shown in Table 29-1.

- *Kitren* are woodlots dominated by teak and other timber species. They are commonly found 1 to 1.5 km away from the owner's home. Compared to *tegalan*, *kitren* may be found on less accessible sites or have less fertile, rocky soils. They account for 21.9% of land parcels and have an average size of 0.31 ha. As they are timber woodlots, *kitren* have the highest tree density (1532 trees/ha) and the lowest tree diversity (5 species/parcel).
- *Tegalan* are upland systems where trees and annual crops are intercropped (Figure 29-3). Like *kitren*, they are also found 1 to 1.5 km from the owner's home, but the site may be more accessible, with more fertile soils. *Tegalan* are the most common and largest of the teak systems, accounting for 50.6% of smallholder teak-system land parcels, with an average size of 0.47 ha. These systems have tree densities of 1072 trees/ha and tree diversity of 8 species/parcel.
- *Pekarangan* are dominated by tree species, with annual crops commonly cultivated in the understorey. They account for 21.9% of the teak-system land parcels and have an average size of 0.24 ha. They are located adjacent to farmers' homes. The tree density of *pekarangan* is similar to that of *tegalan*, with 1177 trees/ha. As they are 'home garden' systems, *pekarangan* have a high tree diversity averaging 13 species/parcel.
- Border planting can be around or across annual cropping systems or around irrigated rice fields (*sawah*). Border plantings are the least dominant teak production system, accounting for only 4.8% of the total. They occupy the borders of annual cropping systems with an average size of 0.31 ha. As they are line plantings they have low tree density of only 138 trees/ha, with a tree diversity of 7 species/parcel.

TABLE 29-1: Cropping calendar for teak agroforestry systems in Gunungkidul.

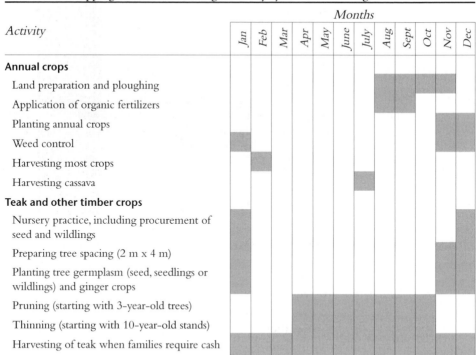

Activity	Jan	Feb	Mar	Apr	May	June	July	Aug	Sept	Oct	Nov	Dec
Annual crops												
Land preparation and ploughing								■	■			
Application of organic fertilizers								■			■	■
Planting annual crops											■	■
Weed control	■											
Harvesting most crops		■										
Harvesting cassava							■					
Teak and other timber crops												
Nursery practice, including procurement of seed and wildlings	■										■	■
Preparing tree spacing (2 m x 4 m)											■	■
Planting tree germplasm (seed, seedlings or wildlings) and ginger crops											■	■
Pruning (starting with 3-year-old trees)					■	■	■					
Thinning (starting with 10-year-old stands)			■	■	■	■	■	■	■	■	■	■
Harvesting of teak when families require cash	■	■	■	■	■	■	■	■	■	■	■	■

The landscape transformation that led to the development of teak agroforestry systems can be summarized as follows: Natural forests were converted to shifting cultivation, or swidden farming, between 1830 (the *tanam paksa* period) and 1900. Land was cultivated for two to three years and fallowed for one or two years. Population pressure resulted in fallows being abandoned around 1930, and by the 1950s, the land and soil had become very degraded. Agricultural production collapsed and there were food shortages and famines. In the 1960s and 1970s, tree farming became common as the only viable option on the degraded soils. Usually, trees were planted in rows along the contours of the land. Annual food crops were grown for three to four years in alleys between the rows of trees. These were *tegalan* systems. In some cases, often when soils were rocky or infertile, trees were established at block spacing without intercropping. These were *kitren* systems. Where soils

FIGURE 29-3: A crop of maize growing in a *tegalan* teak plot.

Photo: Gerhard E. Sabastian.

remained fertile enough to support agricultural production, trees were planted along the boundaries of land parcels and sometimes along contours in the middle of a parcel. These were the border plantings.

As trees begin to dominate, annual-cropping parcels with border and contour plantings could evolve into *tegalan* or even *kitren*. Similarly, *tegalan* parcels could transform into *kitren*, with higher tree density (Figure 29-4). In general, teak is harvested when the trees are between 15 and 50 years old, but younger trees can be harvested when

FIGURE 29-4: A farming couple leave their *kitren*, laden with fodder for domestic animals.

Photo: Gerhard E. Sabastian.

needs occur. The trees are harvested in small numbers, creating canopy openings when they are cut. If soils are sufficiently rehabilitated, a *kitren* parcel may become a *tegalan* or border-planting system. Similarly, *tegalan* may be ready to transform into border plantings, with larger areas of annual crops. *Pekarangan* are determined by location – adjacent to the home. When land is inherited, usually when a child marries, border plantings or *tegalan* may transform into a *pekarangan* when a new home is built at the location. *Kitren* generally do not transform to *pekarangan* (see Figure 29-5). In addition to canopy openings, the main factors that influence the

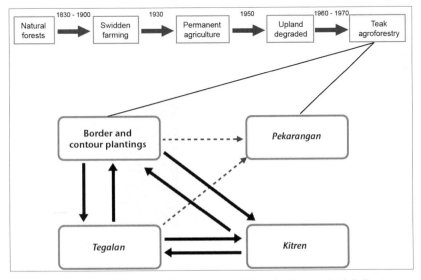

FIGURE 29-5: Landscape and teak–system transitions in Gunungkidul district.

transformation of teak agroforestry systems are soil fertility, soil moisture, availability of labour, and livelihood requirements. Farmers would prefer to move their teak agroforestry systems towards *kitren* status, but the need to produce intercrops for home use or sale, and a lack of other opportunities, results in most teak systems being *tegalan* or *pekarangan*.

General management and intercropping

The management of the smallholder teak systems is shared between men and women. Women are the primary managers of the *pekarangan* systems, particularly the vegetable and other annual crops. Generally, both women and men are responsible for managing the annual and non-timber crops in the *kitren* and *tegalan* systems. While men are responsible for timber-tree management in all systems, women are involved in firewood collection in all systems. Rohadi et al. (2012) estimated that 10% of these teak agroforestry systems are single-purpose *kitren* systems with timber (of one or more species) as the sole product. The other 90% are mixed systems producing a variety of products. Structurally, *tegalan* and *pekarangan* are very similar and are generally managed in the same manner. This similarity between *tegalan* and *pekarangan* was also recognized by Wiersum (1982).

Even after the trees are established, intercropping is common in the teak agroforestry systems. Traditional intercropping is called *tumpangsari*. It is an innovation based on the *taungya* system used for plantation establishment, in which annual crops are planted with the growing seedlings for the first few years. *Tumpangsari* is a farming strategy that aims to diversify farm production, reduce farm risk, produce food and increase farm income. With *tumpangsari*, intercropping is not limited to the tree-establishment phase, but is practised with trees of all ages. Farmers will intercrop when they have available labour and agricultural input, or when market opportunities occur (Roshetko et al., 2013). Annual crops are planted to within 50 cm of trees. Because of competition for light, nutrients and moisture, systems with lower tree density, such as *tegalan* and *pekarangan*, are more favourable to intercropping. However, even *kitren* can be intercropped. A survey of farmers in Yogyakarta found that in the previous year 82% of them intercropped their teak systems with annual crops. Overall, 44% of all teak systems had been intercropped, including 54.4% of *tegalan*, 34.4% of *pekarangan*, and 11.2% of *kitren*. The most common intercrops were cassava (*Manihot esculenta*) in 26.6% of those land parcels cultivated; peanuts (*Arachis hypogaea*), 23.8%; upland rice (*Oryza sativa*), 18%; soybeans (*Glycine max*), 8.1%; and long beans (*Vigna unguiculata* subsp *sesquipedalis*), 2.9% (Roshetko et al., 2013). Other crops that farmers are known to cultivate in their teak agroforestry systems include maize (*Zea mays*), bananas (*Musa* spp), ginger (*Zingiber officinale*), turmeric (*Curcuma longa*), and *temulawak* (*Curcuma zanthorrhiza*). Land preparation, planting and management occur between August and January (Table 29-1). Agricultural crops are usually harvested in February, about 100 days after planting, and cassava is harvested in July, about 9 months after planting.

Tree composition

Teak is the dominant tree crop in Gunungkidul, accounting for 55.9% of all trees in smallholder agroforestry systems (Figure 29-6). Timber species account for 77% of trees; after teak, key timber species include mahogany, acacia, senna (*Senna siamea*) and Indian rosewood (*Dalbergia latifolia*). Fodder and green-manure species account for 15% for the trees; common species are *lamtoro* (*Leucaena leucocephala*), *tayuman* (*Piliostigma malabaricum*), *gamal* (*Gliricidia sepium*) and *turi* (*Sesbania grandiflora*). Spice, nut and condiment species account for only 3.4% of the trees, the most common of which are *melinjo* (*Gnetum gnemon*) and cashews. Fruit species account for only 2.2% of the total, including mangoes (*Mangifera indica*) and coconuts (Figure 29-7). Teak seedlings account for only 47.2% of natural regeneration (Figure 29-8), 9% less than their composition of the overstorey. Farmers remark that they select teak seedlings over other species when managing the understorey of their tree gardens. *Lamtoro*, *acacia*, *gamal*, *turi* and Indian rosewood together comprise 17.3% of the trees in smallholder teak systems. All five are nitrogen-fixing species known for their soil improvement characteristics (Roshetko, 2001). Most farmers are aware of and value those attributes. Besides the value of the timber and fodder products they yield, another reason these species are retained is that they serve as improved-fallow companion crops for teak and other species.

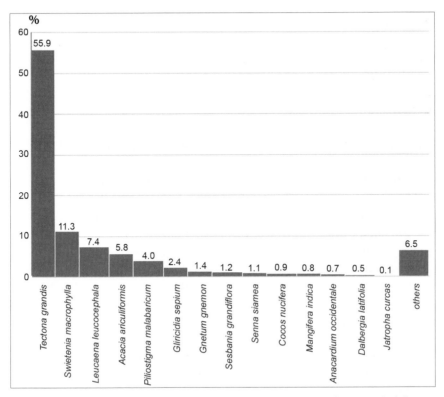

FIGURE 29-6: Tree population in the teak agroforestry systems of Gunungkidul.

It is interesting to compare the species currently found in smallholders' systems, with those reported as prominent in the *penghijauan* programme of the 1970s and farmers' spontaneous tree-planting efforts in the 1960s. The species currently found in smallholder systems that were part of the *penghijauan* programme are teak, *lamtoro* and acacia. Current species that were also planted by farmers in the

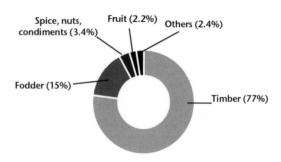

FIGURE 29-7: Tree population in teak agroforestry systems by crop type.

1960s are teak, *lamtoro*, *gamal*, *tayuman* and *melingo*. Species that were promoted in the *penghijauan* programme, but are not present in current smallholder systems, are *sengon* (*Paraserianthes falcataria*), calliandra and jackfruit. As well, coconut and cashews were part of the *penghijauan* programme, but are present in current smallholder systems at only a fraction of 1%.

In an evaluation of the *penghijauan* programme, Soerianegar and Mansuri (1994) found that teak and acacia were the species with the best survival rates and were preferred by farmers. *Sengon* also demonstrated a good survival capacity and fast growth, but

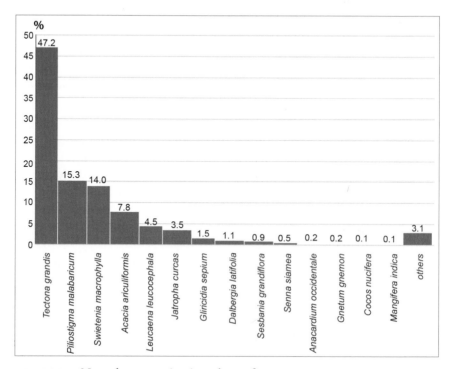

FIGURE 29-8: Natural regeneration in teak agroforestry systems.

after being harvested it was not replanted, indicating it was not preferred by farmers. Mahogany had only a low survival rate, but has persisted. Cashews showed poor survival and were not preferred by farmers because of the low price for the nuts. In another programme analysis, Filius (1997) reported that farmers altered the species composition on their land by neglecting or cutting species that they did not prefer and replanting those that they did – often fruit species. He also commented that the importance of the *penghijauan* programme receded with time, giving the example that only 15% to 24% of the trees planted by farmers came from government agencies. Most seedlings came from relatives, neighbours and local markets and a wide range of species was available, empowering communities with the ability to control the quality and species of seedlings they planted on their farms. These developments were beyond the original scope of

Paraserianthes falcataria (L.) I.C.Nielsen, a synonym of *Falcataria moluccana* (Miq.) Barneby & J.W.Grimes [Leguminosae]

Growing throughout tropical and subtropical regions, this tree has many names, including albizia and madre de cacao. In the study area, it is known as *sengon*. Its soft wood is used for making match sticks, chopsticks and shipping pallets. Although valued as a shade tree for crops like coffee, it has proven to be unpopular with smallholder teak farmers.

the *penghijauan* programme, but should be a part of its legacy in contributing to the evolution of a tree planting culture in rural communities. *Sengon* and calliandra are both nitrogen-fixing species and are well known as soil improvers (Roshetko, 2001). Their inclusion in the *penghijauan* programme certainly contributed to the rehabilitation of degraded agricultural lands. *Sengon* (locally known as *albasiah* or *kalabise*) is used by Baduy farmers at Banten, in West Java, as a fallow crop to improve soil fertility in their culturally important swidden systems. The timber is harvested and sold on four-year rotations (Iskandar and Ellen, 2000, Iskandar, ch. 24, this volume). Calliandra was made famous by Javanese farmers for its soil rehabilitation and intercropping characteristics and its delivery of many products (NRC, 1983). Both species remain common in Yogyakarta, but not as prominent components of teak agroforestry systems. Jackfruit is also common in Gunungkidul, but was not prominent in the inventory of smallholder teak systems.

Silvicultural management

Farmers say that they favour spacing of 2 m x 4 m when planting teak and other timber crops (Figure 29-9). This enables intercropping in the first few years while the seedlings are growing, and even in subsequent years. Field activities found that spacings of 2.5 m x 2.5 m and 3 m x 3 m were common in smallholder teak systems (Roshetko et al., 2013) and observations indicated that spacings from 2 m x 6 m to 2 m x 10 m were used in *tegalan* systems intended for

FIGURE 29-9: A closely-spaced stand of young teak trees in a *kitren* plot.

Photo: James M. Roshetko.

continuous intercropping. Similar wide spacing for teak agroforestry is also reported by Ugalde Arias (2013). About 30% of smallholder teak systems in Yogyakarta have uniform block spacing. The rest have irregular spacing due to undulating slopes, rocky shallow soils, and tree harvesting or mortality having left the residual stand with non-uniform spacing.

When planting teak and other trees, farmers access germplasm from various sources: wildlings from their own farms, neighbours' farms or forest plantations; local seedlings given by family, friends and neighbours; local seedlings bought from traders and nurseries; and 'improved quality' seedlings provided by government programmes and research and developmental projects. Some farmers also operate small backyard nurseries to produce small numbers of seedlings for their own needs or for donation to family and friends. Most of the germplasm used by farmers comes from local sources. A survey in Gunungkidul found that wildlings were the main source of teak seedlings for 72% of farmers. Thirty per cent of farmers also used local teak seedlings and 20% nurtured sprouts growing from stumps. Only 12% of farmers used some improved quality seedlings in their teak systems, representing only a small portion of the on-farm germplasm (Roshetko et al., 2013).

Traditional tree management in Gunungkidul is neither intensive nor proactive. The survey mentioned in the previous paragraph found that 73% of farmers controlled weeds and applied fertilizer to their teak systems, but only in association with annual crops when intercropping. Similarly, 64% of farmers pruned their trees, but primarily to harvest firewood, resulting in 55% of the teak trees owned by smallholders being pruned. Unfortunately, the pruning techniques were often poor, usually retaining 10-to-15-cm branch stubs that were vulnerable to insects and diseases and often caused defects that reduced the quality and value of the timber. The survey found that 43% of smallholder teak stands had been thinned. However, only a few trees were usually removed at a time, primarily to harvest timber, poles or firewood. Farmers generally waited until stands were 10 years old before thinning. While coppiced stumps were often used to regenerate smallholder teak systems, thinning of the new growth was not practised (Roshetko and Manurung, 2009). At the time of the survey, silvicultural management (pruning and thinning) to improve the growth, quality and value of the retained teak stand was an unknown concept. At the initial spacing common in Gunungkidul – 2.5 m x 2.5 m and 3 m x 3 m – heavy early thinning was recommended to retain fast growth, quality and value of the residual stand. Two options were recommended: 60% at 4 years of age or two 25% thinnings at 4 and 5 years of age (Kanninen et al., 2004). Similarly, Zahabu et al. (2015) recommended initial spacing of 2.5 m x 2.5 m or 3 m x 3 m, with thinning before the trees were 5 years old, particularly for trees planted at the closer spacing.

The farmers were supportive of researchers conducting trials on their farms to demonstrate the advantage of silvicultural management. Thinning and pruning demonstration trials were established in *kitren* systems. Thinning demonstration trials on new growth from coppiced stumps were established in border plantings. All systems were even-aged with trees and coppice regrowth about 4 to 6 years old. Two thinning treatments were control and 40%; three pruning treatments were control, 50% of total height, and 60% of total height. There were two treatments for thinning coppice regrowth: control and removing all regrowth except a single best stem (singling). Thinning of 40% was selected to obtain a residual stand density of about 625 trees/ha (4 m x 4 m spacing) (Figure 29-10). Following the trials, average annual increments over a two-year period showed that the combined 'thinning and 60% pruning' treatment increased diameter at breast height (dbh) by 60% and tree height by 124%, compared to the 'no pruning and no thinning' control. The singling treatment on coppice regrowth resulted in 45% greater incremental dbh growth. The trials demonstrated that proper thinning, pruning and singling could improve the dbh and height growth of smallholder teak systems (Roshetko et al., 2013).

The demonstration trials were used as venues for field days and silvicultural training for farmers. Training events included the distribution of silvicultural tools to farmer groups that committed to applying silvicultural practices in their teak agroforestry systems. Within two years, more than 300 farmers and extension agents were trained in silvicultural practices (Rohadi et al., 2012). Additionally, a farmers' teak silvicultural manual was produced in collaboration with farmers, forestry extension officers and researchers (Pramono et al.,

FIGURE 29-10: Harvesting a young tree to create space for others to grow. Farmers usually wait 15 to 50 years before harvesting teak.

Photo: Iwan Kurniawan.

2010, 2011). These efforts were successful in altering farmers' attitudes towards silvicultural management. An impact study conducted by doctoral students from Bogor Agricultural University found that 70% of the farmers in the project area increased their knowledge of silvicultural practices, with 50% adopting silvicultural practices on their own farms and 30% disseminating management practices to other farmers. In areas neighbouring the project, 30% of farmers increased their silvicultural knowledge as a result of project activities, with 20% adopting silvicultural practices and 15% sharing information with others (Rohadi et al., 2012).

Economic and financial aspects

Teak is a dominant component of smallholder farming systems in Yogyakarta. As mentioned earlier, teak accounts for more than 50% of trees on farms and teak systems cover between 30% and 50% of farmers' land. Annual household incomes vary greatly, from zero to 58 million Indonesian rupiah, with an average of 10 million rupiah (US$1125). About 40% of household incomes come from teak agroforestry systems: 25% from annual cropping and livestock systems; 12% from teak; and 3% from other timber species (Figure 29-11). The remaining 60% of household incomes is generated by off-farm activities, casual or skilled labour, shopkeeping, home industries, and services (Rohadi et al., 2012). Off-farm income has dominated household livelihoods since the 1980s (Van Der Poel and Van Dijk, 1987). *Tumpangsari*, the traditional intercropping practice, is an important aspect of smallholder teak systems, enabling farmers to respond to market opportunities for annual crops and bringing in both short- and long-term returns. When and what to cultivate is decided according to existing market prices for crops, availability of household labour, and household capital. Only 12% of the farmers borrow money to support annual-crop production (Perdana et al., 2012). On some occasions, income from timber harvests is used to finance annual cropping (Van Der Poel and Van Dirk. 1987). In either case, intercropping is undertaken in anticipation of profits from the annual crops justifying the loan or the harvest of trees.

Smallholder farmers have limited capital and household labour, and they deploy those resources to generate the best returns, with emphasis on the short-term. Teak is not prioritized when these scarce resources are allocated. Fortunately, the cash

FIGURE 29-11: A harvest where little goes to waste. Teak contributes 12% of household incomes in the study area.

Photo: Iwan Kurniawan.

investment needed to establish and manage teak systems can be low. As discussed above, most germplasm is sourced directly by farmers and some is given free by family, friends, government agencies or research and development projects. Only limited numbers of seedlings are purchased. Investment costs for site preparation, weed control, fertilizers and labour are considered to be investments needed not for trees, but for annual-crop production. Tree planting and management activities are undertaken when opportunity costs for other on- or off-farm activities are low. The limited investment made by farmers in teak and other timber crops is reasonable, considering the long-term nature of the crop and that timber is not a primary source of household income (Perdana et al., 2012). It is also reasonable to argue that farmers' opportunistic management of teak, only when opportunity costs are low, provides good returns on their limited investment. Smallholder teak-management practices minimize risks, diversify farm production and effectively utilize the capacity of households to produce high-value, non-perishable on-farm assets (Roshetko et al., 2013). Evaluating the success of tree-farming activities in Gunungkidul, Soerianegara and Mansuri (1994) concluded that growing teak on rocky infertile soils was a good investment, and analyzing smallholder timber production in Yogyakarta and South Kalimantan, Rohadi et al. (2010) determined that intercropping provided better returns to investment than timber monocultures.

Timber is not the only product that teak provides to generate income for farm families. Teak seed also makes money. A study in Central and East Java found that farming families earned between 275,000 and 795,000 rupiah (US$32 to $94) a year by collecting and processing tree seed from all species for seed dealers and companies. While this amount may seem inconsequential, it represented between 33% and 66% of household income over a three-month period. The seed-collection season coincides with the dry season, when there are few other income-generating opportunities in rural areas. About 20% of the tree seed collected and processed is teak. About 22,500 farmers were involved in tree-seed collecting in the study areas (Roshetko et al., 2008b).

Access to off-farm employment also influences farmers' ability to invest in timber production (Filius, 1997). Urban-based employment opportunities in Central and East Java that involved temporary migration to those areas led to the extensification of tree farming, particularly timber. Under those conditions, farming teak and other tree crops developed as a means of diversifying farm production, reducing risks, building family assets and reallocating labour to lucrative off-farm employment opportunities (Roshetko et al., 2008a). Sabastian et al. (2014) found that larger landholdings and greater on-farm income enabled farmers to intensify the management of their teak crops. It was also found that the development of strong relationships between farmers and extension agents strengthened farmer groups and improved both technical capacity and awareness of government regulations. Moreover, participation in farmer-group activities enhanced the rate of silviculture adoption (Sabastian et al., 2017). Similarly, working in Java, Riau and South Kalimantan, Kallio et al. (2011) found that involvement in farmer groups, larger landholdings, and more total assets made farmers more likely to actively manage their tree crops.

Teak trees are a valuable on-farm asset that serves as a living savings account. The trees are grown to cover specific family expenses, such as weddings, school fees, large medical expenses, periodic social commitments and emergencies. Only 15% of farmers were found to plant teak as a strategy to maximize market opportunities, and only 14% of them harvested trees on the basis of economic maturity. In the teak value chain the role of farmers is limited to that of producers, and they generally have only limited access to market information. To improve their market position, farmers collect information from other farmers who have recently sold teak and offer their trees to more than one trader (Perdana and Roshetko, 2015). However, they usually receive prices below market rates because of their limited access to market information, weak negotiating position and inability to minimize the market transaction costs, including transportation. Because of the nature of the smallholder teak market, traders face high transaction costs and high risks. Traders interact with multiple farmers, each producing small quantities of small diameter trees. Trees often contain hidden defects, such as hollow stems; prices paid by processers fluctuate; and traders may face extra-legal fees during transport. These costs and risks result in lower prices for farmers (Perdana et al., 2012). Studies show that farmers increase their teak income by selling timber to provincial-level traders or direct to industry (Tukan et al., 2004; Anyonge and Roshetko, 2003).

Farmers' perspectives

Farmers began planting teak on infertile agricultural land that could no longer support the economic production of annual crops. Teak and other tree crops were seen as the only viable option for utterly degraded sites. The innovation was more successful than anticipated. Farmers now see teak farming as a system that effectively integrates perennial and annual crop production, yielding diverse products for home use and sale while reducing risks and establishing living assets that serve as a savings account. Farmers recognize that as a growing savings account, teak enables them to bequeath valuable assets to their children and grandchildren. This aspect of teak, as inheritance from

Calliandra calothyrsus Meisn. [Leguminosae]

This small tree was introduced to Indonesia from Central America in 1936 and has since spread throughout Southeast Asia and other parts of the world. It is highly valued as a nitrogen-fixing and soil-stabilizing species that also provides nutritious livestock fodder. It coppices readily, so can be used to produce firewood.

farmers to descendants, was identified by Soerianegara and Mansuri (1994). The importance of teak agroforestry systems to the local population is reflected in the fact that almost all farmers in Yogyakarta grow teak and that teak cultivation is seen as an integral part of the local culture (Perdana et al., 2012).

Farmers also recognize that planting teak on their own land enables them to access state forest land, under agreement with the government, to cultivate annual crops. By intercropping with annual crops, farmers improve the production of perennial crops on state forest land for the benefit of the government. Yields from annual crops support their short-term livelihood needs. Farmers also recognize that the minimal management required by teak after it is established enables family members to pursue more lucrative off-farm employment opportunities in local areas or through temporary migration.

Farmers value teak agroforestry systems highly and acknowledge the environmental and livelihood transformation that has occurred through their widespread adoption. They also know that the systems could be improved. The main shortcoming they see is the extensive management approach taken with teak production. Excluding intercropping activities, which are assessed and undertaken solely on the basis of annual crop production, farmers' main teak-management activities are planting and harvesting. Improving this situation, giving farmers incentives to produce quality timber and meeting other challenges within smallholder teak systems, is a matter of concern.

Improving smallholder teak systems

The key impediments limiting the development of smallholder teak systems include poor silvicultural management, limited market access and policy disincentives. These issues must be addressed if smallholder teak is to achieve its potential as an alternative source of quality timber for commercial teak industries and enhance its contribution to smallholder livelihoods. However, improving smallholder teak systems does not have to alter teak's important function as a social and economic safety net.

Smallholders' current silvicultural management of their teak systems needs to evolve; management should not start with planting seedlings and end with timber harvesting. A key step is the adoption of thinning to enhance incremental diameter growth. In Gunungkidul, most teak systems have spacing of 2.5 m x 2.5 m, 3 m x 3 m, or 2 m x 4 m. When trees are five to six years old, heavy thinning should be conducted to achieve a density of 625 trees per ha. Similarly, pruning to 60% of total height is recommended for five- to six-year-old trees. Subsequent thinning and pruning should be considered every five years. Mature teak trees should be retained until the age of 20 to 30 years, when they reach a diameter size that attracts a lucrative market price. Intercropping, which is currently a common practice, should continue to be encouraged. To encourage the adoption of silvicultural practices, more capacity building and technical support should be provided to smallholders in the form of training, cross-visits, demonstration trials and farmer-oriented extension material.

Some farmers remain reluctant to thin their trees, which they see as a loss of future income rather than a practice that improves the quality and value of the residual stand. To overcome this problem, alternative designs may be used when establishing new teak stands. Trees can be planted in alternating rows of teak and a short-rotation timber species (e.g. acacia, *sengon*, *Gmelina arborea*) (Roshetko et al., 2004; Bertomeu et al., 2011; Sudomo et al., 2021). The short-rotation species can be harvested in five to eight years. Alternatively, trees can be planted at wider densities, 4 m x 4 m or 2 m x 10 m, to reduce the need for thinning. To make new planting as productive as possible, improved quality germplasm of teak and all species should be used.

Limited market access greatly hinders farmers' incentives to produce quality teak. A shared-value model, which is a business

Acacia auriculiformis Benth.
[Leguminosae]

This species remains popular with farmers in the study area following its promotion by the Indonesian government in the mid-1970s. While the trees may be twisted and gnarly; the dense wood is extensively used for paper pulp, for firewood and making charcoal, although it is also suitable for furniture and joinery. This is an outstanding shade tree with spreading roots, making it an effective soil stabilizer.

strategy focused on creating economic value in a way that also creates economic value for farmers, would improve the knowledge of both teak farmers and local traders. These interactions could be further expanded to become farmer-industry partnerships in which farmers produce trees to meet market specifications. Farmers could also engage in group marketing to reduce transaction costs and improve timber supply.

Governments should provide incentives to smallholder teak farmers, and indirectly to the teak industry, by simplifying timber-trade regulations to minimize transaction costs and eliminate extra-legal fees. They should also publish regular market information on teak prices and quality. Governments and support agencies should also provide silvicultural training and extension services to enhance smallholders' technical knowledge and capacity. Moreover, in collaboration with research agencies and industry, they should facilitate farmers' access to sources of quality germplasm (Roshetko and Perdana, 2017).

Smallholder teak in other countries

Indonesia was the first and is the widest-known example of teak production systems by smallholder farmers. Similar systems in other countries also deserve review and discussion. In northern Laos, teak is a key component of integrated systems that reduce risk, diversify production, provide income and increase tree cover. Labour requirements are reduced and can be re-allocated to off-farm employment (Midgley et al., 2007; Newby et al., 2012, 2014). Similarly, smallholder teak systems in Nigeria are credited with rehabilitating soils, improving fallows and restoring tree cover to degraded sites (Osemeobo, 1989). In the central highlands of Vietnam, teak has been used to rehabilitate degraded and infertile soils (Huy et al., 2018). In Thailand, teak is used to diversify production on smallholder farms, while increasing the income of farming families (Mittelman, 2000). In Togo, although teak competes directly with annual crops and food security, it is grown by smallholders to increase household income and establish living assets (Kenny, 2007; Kenny et al., 2014). Farmers in Nigeria cultivate teak in order to join a government programme, diversify their farm production and increase household income (Osemeobo, 1989). Farmers in southern Benin grow teak for poles to diversify farm production and earn income (Aoudji et al., 2011), and high market demand in Panama and Costa Rica enables smallholders to grow teak to diversify and increase family income (De Vriend, 1998; Zanin, 2005).

Unfortunately, poor silvicultural practices by smallholders are reported from most of the countries listed above, and in general from a review of smallholder teak growers (Bhat and Ma, 2004). Midgley et al. (2007) recommended the staging of field demonstrations to show smallholders the benefits of silvicultural management (thinning and pruning). Ling et al. (2018), Newby et al. (2012), Zanin (2005), Bhat and Ma (2004) and De Vriend (1998) all called for more capacity building and technical support to enable smallholder teak farmers to adopt silvicultural management. In some countries, smallholder teak farmers also face policy-based disincentives and compliance issues (Midgley et al., 2017; Smith et al., 2017).

Summary and conclusion

Although exotic to Indonesia, teak was naturalized on Java as long ago as the 3rd century, and by the late 1800s teak plantations were a common land use in the lowlands of Java. Population pressures on shifting cultivation systems in the first half of the 20th century saw the abandonment of fallow periods for soil rejuvenation. Driven by government policy and market demands, the uplands of Yogyakarta and Central Java were then subjected to decades of unsustainable cultivation. The land became thoroughly degraded and could no longer support agriculture, resulting in social and economic crises. In the 1960s, some progressive farmers in Yogyakarta and surrounding areas drew on their indigenous knowledge and began to plant teak on these rocky, infertile degraded sites with the intention of increasing soil organic matter content, improving water infiltration and enhancing water-holding capacity, while producing

Sesbania grandiflora (L.) Pers. [Leguminosae]

This very fast growing tree can reach 8 m tall in less than a year. With an extensive root system, it is valued in soil-reclamation schemes. The wood is used to produce paper pulp, but the tree is perhaps best known as a source of human food. The flowers and the young and tender leaves and pods can be eaten raw or cooked and are popular in many regional cuisines.

a commercial timber crop where agricultural production was no longer possible. Teak, the dominant species in these rehabilitation efforts, was usually cultivated with other tree crops, including nitrogen-fixing species known to farmers for their soil improvement characteristics. Establishing mixed-teak systems on the highly degraded lands enabled farmers to concentrate their agricultural efforts on selected land parcels that retained some fertility and allocate their labour to off-farm employment opportunities. In the 1970s the Indonesian government's *penghijauan* programme promoted tree planting on private land, reinforcing the efforts of those progressive farmers to cultivate teak for land rehabilitation, and engendering a tree planting culture among smallholder farmers.

The intentional use of teak as an improved fallow crop is not the only innovation developed by Javanese farmers. They also improved the *taungya* system, from a plantation-establishment practice to the more dynamic practices of *tumpangsari*, in which intercropping was no longer limited to the tree-establishment phase, but was practised with trees of all ages. Farmers adopted *tumpangsari* to diversify farm production, reduce farm risk, produce food and increase farm incomes. The concurrent impacts of intercropping on tree growth were a positive and welcomed benefit, but were not considered in decisions to cultivate annual crops. Farmers undertook intercropping when they had available labour, capital and germplasm, and when there were clear market opportunities. Another farmer innovation was the four teak agroforestry systems: *kitren, tegalan, pekerangan* and line planting. These types were developed in response to a broad array of household, socio-economic, biophysical and typographical characteristics. Most teak-farming families now cultivate more than one of these teak agroforestry types at any one time, with management practices and agroforestry types shifting as farming families' conditions and characteristics change. Farmers also developed the concept of teak as a living savings account, from which timber is harvested and sold to meet specific family needs, and not to maximize market opportunities. This strategy leads farmers to harvest trees of various ages, some quite small. Under commercial plantation conditions, harvesting small-diameter trees would be considered a loss of future profit, but with smallholders, harvesting small-diameter trees meets a critical need and follows their management strategy. Some of these practices are not unique to either teak or Indonesia, but all represent innovations of teak plantation and smallholder farming management practised on Java.

The teak agroforestry systems that are so common in Yogyakarta, Central Java and East Java today were created by smallholder farmers whose ancestors struggled through the local demise of shifting cultivation, as it was forced well beyond its sustainable capacity. Through the provision of food, tree products and income, the teak agroforestry systems restored social and economic security to communities that had suffered from food shortages, social stress and economic collapse. These systems also achieved environmental restoration through the principles of 'forest landscape restoration', decades before the concept was developed. Establishing mixed teak systems on highly degraded infertile slopes proved to be a best option for the sites, as there was no alternative agricultural use to be made of them. Besides the household and environmental achievements, smallholder teak systems also produced valuable timber and became an important source of raw material for national and international teak industries. However, while smallholder teak agroforestry systems have proven to be successful, with many great impacts, improvements are still needed. Chief among them are better silvicultural management, improved market access and integration, and policy incentives that support farmers and their teak systems. Smallholder teak systems on Java are illustrative of other tree-farming systems across Indonesia and the tropics; their history and examples can be informative and beneficial to other tree farming systems. As global and national demand for teak continues to exceed sustainable supply, it is clear that the significance of smallholder teak agroforestry systems will continue to grow.

Acknowledgements

This chapter arose from the 'Developing and Promoting Market-based Agroforestry Options and Integrated Landscape Management for Smallholder Forestry in Indonesia Project' (FST/2016/141), funded by the Australian Centre for International Agricultural Research and the CGIAR Consortium research programme 'Forests, Trees and Agroforestry: Livelihoods, Landscapes and Governance (CRP6), Flagship 1: Tree Genetic Resources to Bridge Production Gaps and Promote Resilience'.

References

Achdiawan, R. and Puntodewo, A. (2011) 'Livelihood of furniture producers in Jepara', unpublished report from the project 'Mahogany and teak furniture: Action research to improve value chain efficiency and enhance livelihoods (FST/2007/119)', Australian Centre for International Agricultural Research, Canberra

Anabarasu, C., Dutta, N. and Sharma, K. (2001) 'Use leaf meal mixture as a protein supplement in rations of goats fed wheat straw', *Animal Nutrition and Feed Technology* 1(2), pp.113-123

Anabarasu, C., Dutta, N., Sharma, K. and Rawat, M. (2004) 'Response of goats to partial replacement of dietary protein by leaf meal mixture containing *Leucaena leucocephala*, *Morus alba* and *Tectona grandis*', *Small Ruminant Research* 51(1), pp.47-56

Anyonge, C. H. and Roshetko, J. M. (2003) 'Farm-level timber production: Orienting farmers towards the market', *Unaslyva* 212(54), pp.48-56

Aoudji, A. K. N., Adegbidi, A., Ganglo, J. C., Agbo, V., Yevide, A. S. I., De Canneniere, C. and Lebailly, P. (2011) 'Satisfaction across urban consumers of smallholder-produced teak (*Tectona grandis* L.f.) poles in South Benin', *Forest Policy and Economics* 13(8), pp.642-651

Bertomeu, M., Roshetko, J. M. and Rahayu, S. (2011) 'Optimum pruning strategies for reducing crop suppression in a gmelina-maize smallholder agroforestry system in Claveria, Philippines', *Agroforestry Systems* 83, pp.167-180

Bhat, K. M. and Ma, H. O. (2004) 'Teak growers unite!' *ITTO Tropical Forest Update* 14(1), pp.3-5

Bhuyan, R., Saika, C. N. and Das, K. K. (2004) 'Commercially adoptable process for manufacturing natural dyes for cotton', *Natural Product Radiance* 3(1), pp.6-11

BPS Kabupaten Gunungkidul (2008) *Gunungkidul Dalam Angka Tahun 2008* (Gunungkidul in 2008 figures), Badan Pusat Statistik bekerjasama dengan Badan Perencanaan Pembangunan Daerah Kabupaten Gunungkidul, Wonosari, Gunungkidul, Yogyakarta

BPS Kabupaten Gunungkidul (2018) *Gross Regional Domestic Product of Gunungkidul district by Industry 2013-2017*

BPS Kabupaten Gunungkidul (2022) *Buku Saku Indikator Strategis Kabupaten Gunungkidul 2022* (Strategic Indicator Pocket Book, Gunungkidul Regency 2022), available at https://gunungkidulkab. bps.go.id/publication.html?Publikasi%5BtahunJudul%5D=2022&Publikasi%5BkataKunci%5D=ke miskinan&Publikasi%5BcekJudul%5D=0&yt0=Tampilkan, accessed 13 April 2023

Chakraborty, M. K. and Bhattacharjee, A. (2003) 'Plants used for thatching purposes by the tribals of Purulia District, West Bengal, India', *Journal of Economic and Taxonomic Botany* 27(3), pp.571-572

De Vriend, J. (1998) 'Teak: An exploration of market prospects and the outlook for Costa Rican plantations based on indicative growth tables,' thesis to the Subdepartment of Forestry, Department of Environmental Sciences, Wageningen Agricultural University, Wageningen, Netherlands

Donner, W. (1987) *Land Use and Environment in Indonesia*, C. Hurst and Company, London

Filius, A. M. (1997) 'Factors changing farmers' willingness to grow trees in Gunungkidul (Java, Indonesia)', *Netherlands Journal of Agricultural Sciences* 45, pp.329-345

Gupta, R., Vairale, M. G., Deshmukh, R. R., Chaudhary, P. R. and Wale, S. R. (2010) 'Ethnomedical uses of some plants by Goad tribe of Bhandara district, Maharashtra', *Indian Journal of Traditional Knowledge* 9(4), pp.713-717

Gupta, V. C., Hussain, S. J. and Imam, S. (1997) 'Important folk-medicinal plants and traditional knowledge of tribals of Aurangabad and Nasik forest divisions of Maharashtra, India', *Hamdard Medicus* 40(2), pp.59-61

Huy, B., Tri, P. C. and Triet, T. (2018) 'Assessment of enrichment planting of teak (*Tectona grandis*) in degraded dry deciduous dipterocarp forest in the Central Highlands, Vietnam', *Southern Forests: a Journal of Forest Science* 80(1), pp.75-84, DOI: 10.2989/20702620.2017.1286560

Iskandar, J. and Ellen, R. F. (2000) 'The contribution of *Paraserianthes* (Albizia) *falcataria* to sustainable swidden management practices among the Baduy of West Java', *Human Ecology* 28(1), pp.1-17.

Jerez-Rico, M. and de Andrade Coutinho, S. (2017) 'Establishment and management of planted teak forests', in W. Kolbert and M. Kleine (eds) *The Global Teak Study: Analysis, Evaluation and Future Potential of Teak Resources*, IUFRO World Series, vol. 36, International Union of Forest Research Organizations, Vienna, pp.49-65

Kallio, M. H., Kanninen, M. and Rohadi, D. (2011) 'Farmers' tree planting activities in Indonesia: Case studies in the provinces of Java, Riau and South Kalimantan', *Forest, Trees and Livelihoods* 20, pp.191-210

Kanninen, M., Perez, D., Montero, M. and Viquez, E. (2004) 'Intensity and timing of first thinning of *Tectona grandis* plantations in Costa Rica: Results of a thinning trial', *Forest Ecology and Management* 203, pp.89-99

Kaosa-ard, A. (1998) 'Overview of problems in teak plantation establishment', in M. Kashio and K. White (eds) *Teak for the Future: Proceedings of the Second Regional Seminar on Teak*, 29 May - 3 June, 1995, Yangon, Myanmar, FAO Regional Office for Asia and the Pacific (RAP) publication 1998/5, pp.49-60

Kenny, A. L. (2007) 'Optimal land allocation of maize, cassava and teak for small landholders in southern Togo, West Africa', MS Thesis to Michigan Technological University, Houghton, MI

Kenny, A. L., Pickens, J. B. and Orr, B. (2014) 'Land allocation with the introduction of teak: A case study of smallholder farms in Southern Togo', *Journal of Sustainable Forestry* 33(8), pp.776-795, DOI: 10.1080/10549811.2014.925810

Khasanah, N., Perdana, A., Rahmanullah, A., Manurung, G., Roshetko, J. M. and van Noordwijk, M. (2015) 'Intercropping teak (*Tectona grandis*) and maize (*Zea mays*): Bioeconomic trade-off analysis of agroforestry management practices in Gunungkidul, West Java', *Agroforestry Systems* 89(6), pp.1019-1033, DOI: 10.1007/s10457-015-9832-8

Kollert, W. and Cherubini, L. (2012) *Teak resources and market assessment 2010* (*Tectona grandis* Linn. F.), Food and Agriculture Organization of the United Nations, Rome

Kollert, W. and Kleine, M. (2017) 'Introduction', in W. Kolbert and M. Kleine (eds) The Global Teak Study: Analysis, Evaluation and Future Potential of Teak Resources', IUFRO World Series vol. 36, International Union of Forest Research Organizations, Vienna, pp.15-16

Krishnapillay, B (2000) 'Silviculture and management of teak plantations', *Unasylva* 201(51), pp.14-21

Ling, S., Smith, H., Xaysavongsa, L. and Laity, R. (2018) 'The evolution of certified teak grower groups in Luang Prabang, Lao PDR: An action research approach', *Small-scale Forestry* 17, pp.343-360, doi.org/10.1007/s11842-018-9391-8

Majumdar, M. (2005) 'Evaluation of *Tectona grandis* leaves for wound healing activity', *Pakistan Journal of Pharmaceutical Sciences* 20(2), pp.120-124

Midgley, S., Blyth, M., Mounlamai, K., Midgley, D. and Brown, A. (2007) *Towards Improving Profitability of Teak in Integrated Smallholder Farming Systems in Northern Laos*, ACIAR Technical Reports 64, Australian Centre for International Agricultural Research, Canberra

Midgley, S. J., Stevens, P. R. and Arnold, R. J. (2017) 'Hidden assets: Asia's smallholder wood resources and their contribution to supply chains of commercial wood', *Australian Forestry* 80(1), pp.10-25, DOI: 10.1080/00049158.2017.1280750

Mittelman, A. (2000) 'Teak planting by smallholders in Nakhon Sawan, Thailand', *Unasylva* 201(51), pp.62-65

NRC (1983) *Calliandra: A Versatile Small Tree for the Humid Tropics*, National Research Council and National Academy Press, Washington DC

Newby, J. C., Cramb, R. A., Sakanphet, S. and McNamara, S. (2012) 'Smallholder teak and agrarian change in Northern Laos', *Small-scale Forestry* 11(1), pp.27-46

Newby, J. C., Cramb, R. A. and Sakanphet, S. (2014) 'Forest transitions and rural livelihoods: Multiple pathways of smallholder teak expansion in Northern Laos', *Land* 3, pp.482-503, DOI: 10.3390/land3020482

Nibbering, J. W. (1999) 'Tree planting on deforested farmlands, Sewu Hills, Java, Indonesia: Impact of economic and institutional changes', *Agroforestry Systems* 46, pp.65-82

Osemeobo, G. J. (1989) 'An impact and performance evaluation of smallholder participation in tree planting, Nigeria', *Agricultural Systems* 29(2), pp.117-138

Pandey, D. and Brown, C. (2000) 'Teak: A global overview', *Unasylva* 201(51), pp.3-13

Peluso, N. L. (1992) *Rich Forests, Poor People: Resource Control and Resistance in Java*, University of California Press, Berkeley, CA

Perdana, A., Roshetko, J. M. and Kurniawan, I. (2012) 'Forces of competition: Smallholding teak producers in Indonesia', *International Forestry Review* 14(2), pp.238-248

Perdana, A. and Roshetko, J. M. (2015) 'Survival strategy: Traders of smallholder teak in Indonesia', *International Forestry Review* 17(4), pp.461-468

Perera, W. R. H. (1962) 'The development of forest plantations in Ceylon since the seventeenth century', *Ceylon Forester* 5, pp.142-147

Pramono, A. A., Fauzi, M. A., Widyani, N., Heriansyah, I. and Roshetko, J. M. (2010) *Pengelolaan Hutan Jati Rakyat: Panduan Lapang Untuk Petani* (Management of Community Teak Forests: A Field Manual for Farmers), Center for International Forestry Research (CIFOR), Southeast Asia Regional Program, World Agroforestry Centre (ICRAF), Forestry Research and Development Agency, Bogor, Indonesia

Pramono, A. A., Fauzi, M. A., Widyani, N., Heriansyah, I. and Roshetko, J. M. (2011) *Management of Community Teak Forests: A Field Manual for Farmers*, Center for International Forestry Research (CIFOR), Southeast Asia Regional Program, World Agroforestry Centre (ICRAF), Forestry Research and Development Agency, Bogor, Indonesia

Reddy, G. V. N. and Reddy, M. R. (1984) 'Utilization of fallen dry teak leaves (*Tectona grandis*) as roughage source in complete pelleted rations of sheep', *Indian Journal of Animal Sciences* 54(9), pp.843-848

Roemantyo, H. S. (1990) 'Ethnobotany of the Javanese incense', *Economic Botany* 44(3), pp.413-416

Rohadi, D., Kallio, M., Krisnawati, H. and Manalu, P. (2010) 'Economic incentives and household perceptions of smallholder timber plantations: Lessons from case studies in Indonesia', paper presented at an International Community Forestry Conference. Montpellier, France

Rohadi, D., Roshetko, J. M., Perdana, A., Blyth, M., Nuryartono, N., Kusumowardani, N., Pramono, A. A., Widyani, N., Fauzi, A., Sasono, J., Sumardamto, P. and Manalu, P. (2012) *Improving Economic Outcomes for Smallholders Growing Teak in Agroforestry Systems in Indonesia*, Australian Centre for International Agricultural Research, Canberra

Roshetko, J. M. (2001) *Agroforestry Species and Technologies: A Compilation of the Highlights and Factsheets Published by NFTA and FACT Net 1985-1999*, Taiwan Forestry Research Institute and Council of Agriculture, Taiwan, Republic of China and Winrock International, Morrilton, AR

Roshetko, J. M. and Manurung, G. S. (2009) 'Smallholder teak production systems in Gunungkidul, Indonesia', poster presentation at the Second World Congress of Agroforestry, Nairobi

Roshetko, J. M. and Perdana, A. (2017) 'The significance of planted teak for smallholder farmers', in W. Kolbert and M. Kleine (eds) *The Global Teak Study: Analysis, Evaluation and Future Potential of Teak Resources*, IUFRO World Series, vol. 36, International Union of Forest Research Organizations, Vienna, pp.66-70

Roshetko, J. M., Mulawarman and Purnomosidhi, P. (2004) 'Gmelina arborea: A viable species for smallholder tree farming in Indonesia?', *New Forests* 28, pp.207-215

Roshetko, J. M., Purnomosidhi, P. and Mulawarman (2005) 'Farmer demonstration trials (FDTs): Promoting tree planting and farmer innovation in Indonesia', in J. Gonsalves, T. Becker, A. Braun, J. Caminade, D. Campilan, H. De Chavez, E. Fajber, M. Kapiriri and R. Vernooy (eds) *Participatory Research and Development for Sustainable Agriculture and Natural Resource Management: A Sourcebook*, International Potato Center, Laguna, Philippines; International Development Research Centre, Ottawa; and International Fund for Agricultural Development, Rome, pp.384-392

Roshetko, J. M., Lasco, R. D. and De los Angeles, M. D. (2007) 'Smallholder agroforestry systems for carbon storage', *Mitigation and Adaptation Strategies for Global Change* 12, pp.219-242

Roshetko, J. M., Snelder, D. J., Lasco, R. D. and van Noordwijk, M. (2008a) 'Future challenge: A paradigm shift in the forestry sector', in D. J. Snelder and R. Lasco (eds) *Smallholder Tree Growing for Rural Development and Environmental Services: Lessons from Asia*, Springer, the Netherlands, pp.453-485

Roshetko, J. M., Mulawarman and Dianarto, A. (2008b) 'Tree seed procurement-diffusion pathways in Wonogiri and Ponorogo, Java', *Small-scale Forestry* 7, pp.333-352

Roshetko, J. M., Rohadi, D., Perdana, A., Sabastian, G., Nuryartono, N., Pramono, A. A., Widyani, N., Manalu, P., Fauzi, M. A., Sumardamto, P. and Kusumowardhani, N. (2013) 'Teak agroforestry systems for livelihood enhancement, industrial timber production, and environmental rehabilitation', *Forests, Trees, and Livelihoods* 22(4), pp.241-256, DOI: 10.1080/14728028.2013.855150

Sabastian, G., Kanowski, P., Race, D., Williams, E. and Roshetko, J. M. (2014) 'Household and farm attributes affecting adoption of smallholder timber management practices by tree growers in Gunungkidul region, Indonesia', *Agroforestry Systems* 88(1), pp.1-14, DOI 10.1007/s10457-014-9673-x

Sabastian, G. E., Yumn, A., Roshetko, J. M., Manalu, P., Martini, E. and Perdana, A. (2017) 'Adoption of silvicultural practices in smallholder timber and NTFPs production systems in Indonesia', *Agroforestry Systems*, DOI: 10.1007/s10457-017-0155-9

Seidensticker, J. (1987) 'Bearing witness: Observations on the extinction of *Panthera tigris balica and Panthera tigris Sondaica*', in R. L. Tilson and U. S. Seal (eds) *Tigers of the World: The Biology, Biopolitics, Management and Conservation of an Endangered Species*, Noyes, New Jersey, pp.1-8

Sharma, S. K. (1999) 'Plants used as henna dye by the Bhils of southern Rajasthan', *Journal of Economic and Taxonomic Botany* 23(2), p.257

Siddiqui, M. B., Alam, M. M. and Husain, W. (1989) 'Traditional treatment of skin diseases in Uttar Pradesh, India', *Economic Botany* 43, pp.480-486

Simatupang, M. H. (2000) 'Some notes on the origin and establishment of teak forest (*Tectona grandis* L.F.) in Java, Indonesia', in E. B. Hardiyanto (ed.) *Proceedings of the Third Regional Seminar on Teak: Potential and Opportunities in Marketing and Trade of Plantation Teak: Challenges for the New Millennium*, 31 July–4 August, Yogyakarta, Indonesia, pp.91-98

Smith, H. F., Ling, S. and Boer, K. (2017) 'Teak plantation smallholders in Lao PDR: What influences compliance with plantation regulations?', *Australian Forestry* 80(3), pp.178-187, DOI: 10.1080/00049158.2017.1321520

Soerianegara, I. and Mansuri (1994) 'Factors which determine the success of regreening in Gunungkidul, Central Java', *Journal of Tropical Forest Science* 7(1), pp.64-75

Sudiharjo, A. M. and Notohadiprawiro, T. (2006) *Sekuen produktivitas lahan di wilayah karst Karangasem, Kecamatan Ponjong*, Ilmu Tanah Universitas Gadjah Mada, Yogyakarta, Indonesia, Kabupaten Gunung Kidul

Sudomo, A., Maharani, D., Swestiani, D., Sabastian, G. E., Roshetko, J. M., Perdana, A., Prameswari, D. and Fambayun, R. A. (2021) 'Intercropping short rotation timber species with teak: Enabling smallholder silviculture practices', *Forests* 12(12), 1761, available at https://doi.org/10.3390/f12121761, accessed 11 April 2023

Sunkar, A. (2008) 'Deforestation and rocky desertification processes in Gunung Sewu karst landscape', *Media Konservasi* 13(3), pp.1-7

Troup, R. S. (1921) *The silviculture of Indian Trees*, vol. 2, Clarendon Press, Oxford, UK, pp.697-769

Tukan, J. C. M., Yulianti, Roshetko, J. M. and Darusman, D. (2004) 'Pemasaran Kayu dari Lahan Petani di Provinsi Lampung (Marketing Timber from Farmers' Land in Lampung Province)', *Agrivita* 26, pp.131-140

Ugalde Arias, L. A. (2013) *Teak: New Trends in Silviculture, Commercialization and Wood Production*, International Forestry and Agroforestry (INFOA), Cartago, Costa Rica

Van Der Poel, P. and Van Dijk, H. (1987) 'Household economy and tree growing in upland Central Java', *Agroforestry Systems* 5(2), pp.169-184

White, K. J. (1991) *Teak: Some Aspects of Research and Development*, FAO publication 1991/17, Food and Agriculture Organization of the United Nations Regional Office for Asia and Pacific, Bangkok

Whitten, T., Soeriaatmadja, R. E. and Afiff, S. A. (1996) *The Ecology of Java and Bali*, The Ecology of Indonesia Series, vol. 2, Periplus Editions, Hong Kong

Widiawati, D. (2009) 'The revival of the usage of natural fibers and natural dyes in Indonesian textiles', *Journal of Visual Arts and Design* 3(2), pp.115-128

Wiersum, K. F. (1982) 'Tree gardening and taungya on Java: Examples of agroforestry techniques in the humid tropics', *Agroforestry Systems* 1(1), pp.53-70

Yovi, E. Y., Nurrochmat, D. R. and Sidiq, M. (2013) 'Domestic market of Jepara's small scale wooden furniture industries: its potential and barriers', unpublished report from the ACIAR project 'Mahogany and teak furniture: Action research to improve value chain efficiency and enhance livelihoods (FST/2007/119)', Australian Centre for International Agricultural Research, Canberra

Zahabu, E., Raphael, T., Chamshama, S. A. O., Iddi, S. and Malimbwi, R. E. (2015) 'Effect of spacing regimes on growth, yield, and wood properties of *Tectona grandis* at Longuza Forest Planatation, Tanzania', *International Journal of Forestry Research*, available at http://dx.doi.org/10.1155/2015/469760

Zanin, D. K. (2005) 'Feasibility of teak production for smallholders in eastern Panama', MS thesis to the Michigan Technological University, Houghton, MI

Note

1. 'Agroforestation' refers to the establishment of smallholder agroforestry systems, and implies land rehabilitation through the establishment of a tree-based system and intensification of land management (Roshetko et al., 2007).

30

'RATTAN IS SICK'

Exploring the (dis)continuity of Kalimantan's rattan-swidden complex[1]

*Viola Schreer**

Introduction

'These days, rattan is sick', Bapa Edwin** declared, as we sat on the remnants of a fallen tree in his rattan garden in April 2013. Bapa Edwin's garden was a legacy from his ancestors, handed down by his great-grandfather to him and his siblings. We had taken a break from harvesting rattan (locally called *manetes*) to take a sip of coffee and have some biscuits. Bapa Edwin smoked. The cigarette smoke and the smouldering of a small fire protected us from the mosquitos attracted by the sweat of our bodies. It was 10 o'clock in the morning and the heat had started to build up, adding to the travails involved in working rattan. Harvesting the canes, said Bapa Edwin, was *uyuh* – in his language, tedious, painful and exhausting. My back was hurting. My arms were tired from holding the machete and pulling the canes. Feeling small pieces of spiny leaves in my hair and inspecting the scratches on my hands despite the gloves I had worn, I had to agree. Bapa Edwin, who was 52 years old at that time, continued:

> Rattan harvesting is the hardest work on Earth: first, the spines; second, you have to pull hard; third, you have to climb; fourth, you have to peel the skin off; fifth, you have to carry the rattan to the river; sixth, you have to bundle the rattan; seventh, you have to release it into the water and then lift it again; and eighth, there are many mosquitoes and other insects. In fact, harvesting rattan just makes trouble.

* Dr. Viola Schreer is a postdoctoral researcher at the Anthropology Department of Brunel University London, where she explores a community conservation scheme in Central Kalimantan as part of a broader study into the global nexus of orang-utan conservation. Since 2009, she has carried out almost two years of anthropological fieldwork in Central Kalimantan. This study stems from her PhD fieldwork, conducted in 2012 and 2013 with support from the German Academic Scholarship Foundation and the World Agroforestry Centre (ICRAF).
** All names used in this chapter, including those of villages and people, are pseudonyms.

Rattan (from the plant family Arecaceae, subfamily Calamoideae) is a general term for a large and complex group of mostly climbing spiny palms that occur in Old World tropical forests and constitute the world's most important (agro-)forest product (Siebert, 2012, p.1).[2] For centuries, local communities have used rattans for tying, basketry, dying, construction, medicine, food and rituals, and have sold the cane to international markets (Schreer, 2016b).

Despite the pain and frustration I felt throughout that day, I was happy to accompany Bapa Edwin and learn more about working the cane. During 16 months of fieldwork, it remained my first and last-but-one time that I joined in harvesting in Indonesia's self-proclaimed 'Rattan Regency'.

In 2008, Katingan (Figure 30-1) set itself the goal of becoming the production and trade centre of rattan in Indonesia.[3] Supported by the national government and several non-governmental organisations (NGOs), Katingan's government drafted an impressive master plan with the aim of developing the local rattan industry. The Ministry of Forestry decided that, as part of Indonesia's ambitions for so-called 'green development', it would establish a rattan cluster in Katingan as part of a wider government strategy of developing its non-timber-forest-product (NTFP) sector. Diverse initiatives were set up, including an inquiry into rattan stocks, rattan reforestation projects, the formation of farmer cooperatives, rattan certification, handicraft workshops, the resettlement of Javanese furniture-makers and the establishment of a state-owned factory. However, what I found throughout

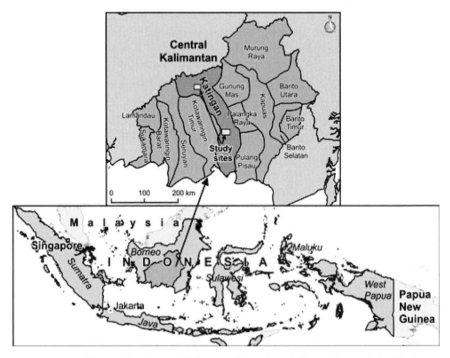

FIGURE 30-1: The study sites at Dahanen (lower) and Sapan (upper) in Katingan regency, Central Kalimantan.

my research in 2012 and 2013 contradicted this vision of a thriving local rattan industry. Rattan gardens were left unmanaged or converted to alternative land uses (e.g. banana plantations, rubber gardens and oil-palm fields), while more and more farmers worked in small-scale gold mining operations or as wage labourers in expanding oil-palm plantations.

'These days, you can no longer live from rattan, not like in the past. Rattan is sick indeed', Bapa Edwin repeated. In the past, rattan harvesting had allowed for livelihood improvement and well-being. But nowadays, Katingan's rattan farmers consider it *pehe* (sick). Not only does rattan no longer guarantee their livelihood, the prices also fail to compensate for the stresses and strains involved in working the cane. Just as rattan is sick, so is the economic situation of the local rattan farmers. Given that there was little chance for his life to improve under present political-economic conditions, I asked Bapa Edwin whether he was considering replacing rattan with rubber, oil palm, or bananas. 'No,' he said without hesitation. 'Rattan has always been there and it will always remain. Rattan won't disappear.'

This chapter asks why so many of Katingan's farmers keep their rattan gardens – if rattan is as sick as they claim. On one hand, people consider that rattan harvesting is a painful labour from which they can no longer make a living, yet, on the other hand, they keep their gardens and remain deeply convinced that they will still exist in the future. How can we make sense of this obvious paradox? By addressing this question, I seek to unravel the factors influencing smallholder decision-making, and thereby contribute to discussions surrounding the decline and resilience of forest-garden systems.

In the light of the rapid transformation of tropical landscapes to annual crops and monocultures of tree crops, scholars have begun to investigate farmer decision-making in the context of these on-going, dynamic land-use changes in an attempt to discover why traditional forest-management systems, such as rattan gardens, disappear or continue to exist. As well as demographic change due to in- and out-migration, which results in land pressure, shorter fallow periods, more intensive-based farming, labour shortages and thus increased production costs (Cramb et al., 2009), politico-economic factors also impact negatively on rattan management. Government policies and resulting unfavourable market conditions, expansion of alternative land uses and infrastructure development all bring pressure to bear on the rattan-swidden complex. Although many factors – both endogenous and exogenous – simultaneously and interactively influence smallholder decision-making, it will be shown that the foremost reasons why rattan farmers keep their gardens are non-economic. Previous analyses of the economic benefits of rattan cultivation in East Kalimantan suggested that rattan gardens were kept mainly for their non-economic values. These included using the flexible harvesting potential of rattan as insurance in case of an urgent need for cash, the low need for labour inputs to maintain the gardens, as a sign of land ownership, spreading the risk of farming, and affective ties to ancestors (Belcher et al., 2004, pp.85-86; Pambudhi et al., 2004, pp.361-362). My study thus reaffirms the findings of previous analyses showing the demise of rattan cultivation in Kalimantan and swiddening in general (e.g. Belcher et al., 2004; Pambudhi et al., 2004; Cramb et al., 2009; Siebert, 2012; Bizard, 2013). However, the following ethnographic account seeks to add to these studies by exploring the material processes involved in rattan management, as well as people's emotional investments in their rattan gardens, in order to detail the non-economic values of rattan gardens and situate them in their socio-cultural and historical context. This chapter makes explicit the fact that rattan gardens are not simply a form of tropical forest management. Rather, rattan gardens are affective spaces allowing people to recall and narrate the past, relate to their forefathers, imagine times of improvement, and thereby experience a sense of security and well-being.

The findings stem from 16 months of anthropological fieldwork carried out in several villages in Katingan regency in 2012 and 2013. I spent most of my fieldwork in the lowland village of Dahanen, a small Muslim settlement with a population of 374 (in October 2012), which was surrounded by swamp and peat-swamp forest, logged and burnt-over forest, shrubland and oil-palm plantations. In the past, people practised swidden farming, complemented by the sale of rattan, the collection of forest products and fishing. With the arrival of logging in the 1970s and a decline in the rattan price in the 1980s, local livelihoods gradually underwent major changes. More and more people left shifting cultivation to work as loggers until the era of logging finally came to an end in 2006. Since then, the residents have sustained their livelihoods mainly by fishing and working as wage labourers in an oil-palm plantation that was established nearby in 2009. While spending most of my time in Dahanen, I also conducted research in neighbouring villages and stayed for several months in an upland village called Sapan. In May 2013, Sapan had a population of 237 people living in 60 nuclear families in 54 households.[4] The people of Sapan practise an animistic religion called *Kaharingan*, involving the diverse yet related indigenous cosmological and ritual traditions of southeast Borneo. In contrast to the people of Dahanen, those at Sapan are active swidden cultivators who

Calamus trachycoleus Becc.
[Arecaceae]

Having been cultivated in Kalimantan for more than a century, the stems of this rattan species grow up to 60 metres long, and are up to 13.5mm in diameter. The canes are light and pliable, and are used as skin peels for weaving; the cores are used for furniture and basket-making.

are able to meet their rice and vegetable needs almost exclusively on a subsistence basis. They also practise small-scale gold mining to obtain cash. The experience and data collected in Sapan served both as a point of comparison and to give a broader overview of the (dis)continuity of the rattan-swidden complex in Katingan.

Rattan smallholders in Indonesian Borneo

The villagers of Dahanen are Ngaju-speaking Dayak, a people who together with other Dayak groups and the (formerly) nomadic Punan, are the native inhabitants of Borneo.[5] Like other Dayak groups, Ngaju were traditionally smallholders practising extensive swidden agriculture. The term 'smallholder' is usually reserved for 'rural cultivators practising intensive, permanent, diversified agriculture on relatively small farms of dense population' (Netting, 1993, p.2). Yet, as noted by Netting (1993, pp.10-15), and in Dove's (2011, p.5) study of Kantu' swidden cultivators in West Kalimantan, the idea of swiddeners living in isolation from market forces has led to a misunderstanding of their economy, history and identity. For Bapa Edwin and his fellow villagers at Dahanen, trading rattan and other forest products is intrinsically tied to their history and identity (Schreer, 2016a).

As Dove (2011) pointed out in *The Banana Tree at the Gate: A History of Marginal People and Global Markets in Borneo*, the island's native inhabitants have been involved for millennia in commodity production for global markets. The same assertion has been emphasized by several other authors (e.g. Wolters, 1967; Padoch and Peluso, 1996; Wadley, 2005). Since at least the 5th century A.D., forest dwellers have collected and traded forest products to supply coastal Malay kingdoms located at river mouths, as part of the commerce between the Indonesian archipelago and China (Wolters, 1967, p.158f). Far from being 'without history' (Wolf, 1982), Bornean societies were enmeshed in global commerce long before the onset of the modern era. In fact, production for the market was, for the majority of the island's population, a complementary pillar to what Dove (2011, pp.13-16) referred to as a 'dual household economy' composed of subsistence-oriented extensive agriculture and market-oriented trade in forest products and cash crops. People's ancestors met their food needs with rice and other food crops from their swiddens, while they gathered forest products, such as gold, beeswax, resin, and particularly rattan, and/or cultivated cash crops, for the market. While they initially collected rattan only from the forest, growing global demand during the mid-19th century led the ancestors of present Dayak groups in Kalimantan to begin cultivating rattan (e.g. van Tuil, 1929, cited in Pambudhi et al., 2004, p.349; Weinstock, 1983, p.60; Knapen, 2001, p.363).

Kalimantan's rattan gardens: vanishing forest-garden systems

Although the specific characteristics of rattan cultivation differ from one grower to another and from place to place, a comparison of studies of the rattan-swidden complex shows that its basic features are similar throughout Kalimantan (e.g.

Weinstock, 1983; Godoy, 1990; Godoy and Feaw, 1991; Fried and Mustofa, 1992; Fried, 2000; Belcher, 2001; Gönner, 2001; Belcher et al., 2004; Arifin, 2003; Matius, 2004; Sasaki, 2007; Bizard, 2013; Schreer, 2016a). In the context of swidden agriculture, people plant rattan seeds – foremost, *Calamus caesius* and *Calamus trachycoleus* – together with rice or other annuals like maize or cassava. Alternatively, farmers raise seedlings in a nursery and then transplant them in the subsequent year. Seedlings may also be collected from productive rattan gardens and transplanted after the rice harvest. The young rattan plants are left to grow with other secondary vegetation during the fallow period, and eventually they become the focal species in a rattan garden. The first harvest is possible about seven to 10 years after planting. Depending on the species, rattan can then be harvested over a span of 30 to 50 years (Belcher et al., 2005, p.247), with large harvests possible every second or third year, and small harvests whenever there is need for immediate income.

Like other so-called 'forest garden systems' (Asbjørnsen et al., 2000, p.9) in

Calamus caesius Blume
[Arecaceae]

One of the main rattan species grown at the study sites, this species can grow canes that climb an astounding 100 metres into the forest canopy.It produces the highest-quality small-diameter canes that are widely used in the local furniture industry as well as for traditional uses such as weaving.

Indonesia, Kalimantan's rattan gardens are part of a broader, multi-species agricultural system that enables diversification, thus spreading risk. Similar to the damar gardens (Michon et al., 2000), coffee gardens (Michon et al., 1986), and cinnamon gardens (Aumeeruddy, 1994) in Sumatra, the fruit gardens of East Kalimantan (Michon and de Foresta, 1999), and rubber gardens found in Sumatra (Joshi et al., 2002; Feintrenie and Levang, 2009) and West Kalimantan (Dove, 2000, 2011), rattan gardens produce one commercially valuable main crop that matures in the medium-term and allows for multiple harvests. Rattan gardens require medium levels of energy, labour and capital input and involve relatively simple equipment, with sophisticated environmental knowledge as a compensation. Access to and control of rattan gardens are defined by customary property regimes that establish long-term use rights or ownership. The gardens also fulfil various ecological functions, including biodiversity conservation, hydrological regulation, soil protection and carbon sequestration (Asbjørnsen et al., 2000; Belcher et al., 2005). Like all smallholder forest gardens, Kalimantan's rattan gardens have contributed significantly to both village economies and national foreign-exchange earnings since at least the middle to the end of the 19th century.

Over the last few years, however, Kalimantan's rattan gardens have been disappearing at an increasing pace. In some areas, in-migration leads to land pressure, shorter fallows and more intensive farming, whereas in others out-migration and lower birth rates result in labour shortages. Increased production costs, as a consequence (Cramb et al., 2009), have been identified as the foremost politico-economic factors impacting negatively on rattan management (e.g. Belcher et al., 2004; Pambudhi et al., 2004; Siebert, 2012; Bizard, 2013; Myers, 2015). Since 1986, government policies have resulted in unfavourable market conditions for rattan smallholders. Expansion of alternative land-uses, particularly oil-palm plantations, and infrastructure development has put the rattan-swidden complex under additional pressure and the disappearance of Katingan's rattan gardens is accelerating, much like it is elsewhere in Kalimantan.

Pain: working the cane in contemporary times

At first glance, harvesting didn't look particularly troublesome (Figure 30-2). Bapa Edwin's body movements assumed a routine aspect. In the middle of the garden, he cleared the vegetation to create an open space, so that he could move freely and gather the canes. Before cutting any of the climbing palms entangled in the surrounding trees from their clumps, Bapa Edwin carefully inspected the rattans, their climbing path and maturity, looking for old, mature canes. It would be a pity to harvest young rattan that was not fully-grown, he said. Mature canes are easier to harvest, as the skin can be removed by knocking the spiny sheath off with the machete. Not only can mature rattan be distinguished from younger parts by the nature of its sheath, but also because – according to Bapa Edwin – fully grown rattan smells like gunpowder.

Holding the rattan with his left hand and with the machete in his right, he knocked off the spiny sheath to be able to grab it more firmly. Then, with both hands he pulled the cane down, hunkering down to swing his weight against the cane. Cutting off four to five spiny leaves, he then ripped off the remaining skin with his machete, followed by hitting the cane strongly, causing the epidermis to fully peel off. Repeating this procedure, Bapa Edwin slowly worked along several canes until pulling was no longer dislodging them. He then had to climb one of the supporting trees to disentangle the spiny palms.

FIGURE 30-2: Bapa Edwin hauls a collection of freshly-harvested cane out of his rattan garden.

Photo: Viola Schreer.

What looked simple was a challenge for a beginner. I was unused to holding a machete, much less using it to effect. The spines got caught in my clothes and hair. Lacking the necessary technique, knocking and ripping off the sheaths was difficult and exhausting. Pulling the cane tired my arms. Inspecting my first trials, Bapa Edwin showed me how to hold the rattan, how and where to cut it, and how to proceed. 'In three days, you will have become proficient', he suggested. I doubted it. We chatted, joked and worked to the sounds of our labour and the surrounding forest: the 'whack' of the machetes, the rustling of the rattan, insects and distant birds. 'Take it easy, if you are tired. Let's take a break', Bapa Edwin said and sat down on the fallen tree, lighting another cigarette.

Our efforts were just the start of a long process. The canes would be bundled and immersed in water until traders collected them. If the rattan was not sold 'wet', but was processed before being sold, the canes needed to be washed, polished, sometimes sulphured, and in any case, dried. Apart from sulphuring, which posed a serious health issue, the processing was often done by women and children (Mulyoutami et al., 2009, p.2058) (Figure 30-3).

Throughout 2012 and 2013, Bapa Edwin was one of the few people harvesting cane in the Dahanen area, despite the widespread occurrence of rattan gardens. In Sapan, people frequently collected rattan from their swiddens to make baskets, but no one harvested rattan for commercial purposes – apart from an elderly couple who harvested cane following a rumour that a trader would visit the village. It remained a rumour; the trader never came.

According to a comprehensive survey conducted across Katingan in 2005, more than half of all the households in the regency owned a rattan garden (TeROPONG and SHK Kaltim, 2005). During my fieldwork in 2012 and 2013, a large majority of households in Dahanen and Sapan still owned such gardens (Figure 30-4). In Dahanen, 81% of households each had an average of 1.48 rattan plots, whereas in Sapan, 79% of households owned, on average, 4.6 plots each.[6] Still, hardly anybody was harvesting. Several interdependent factors help to explain why, in recent years, there has been a demise of rattan management. They include unfavourable terms of trade, ideological prejudices, alternative income opportunities and the laborious difficulty of working rattan.

FIGURE 30-3: Processing rattan for sale. In this case, villagers near the Sebangau National Park in Central Kalimantan debarking rattan in what is a standard process.

Photo: Gerard Persoon.

Unfavourable terms of trade

Farmers in various parts of Katingan say that they have become disinterested in harvesting rattan, because 'there is no price', meaning that the price of rattan no longer justifies the labour input. Following implementation of the export ban on unfinished and semi-finished rattan that came into effect in January 2012, the farm-gate price for rattan fell significantly.[7] In 2012 and 2013, it ranged from 1200 to 1600 rupiah (US$0.09 to $0.12) per kilogram of unprocessed *uei/uwei sigi* (*Calamus caesius*) and 800 to 1200 rupiah ($0.06 to $0.09) per kilogram of unprocessed *uei irit* (*Calamus trachycoleus*). This was significantly lower than the year

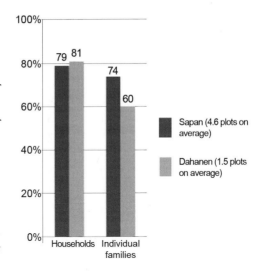

FIGURE 30-4: Local ownership of rattan gardens (%).

before.[8] The export ban imposed in 2012 was not the first of its kind. Over the past three decades, the Indonesian government has implemented several policies that have resulted in rises and falls of farm-gate prices for rattan.[9]

During the 1970s and 1980s, the objective of these interventions was to halt further depletion of natural stocks in the face of serious over-harvesting, as recorded in East Kalimantan by Peluso (1983).[10] In recent decades, policy initiatives have aimed mainly at boosting the national rattan industry, although the most-recent ban was also justified by the threat of over-exploitation. However, there seems little empirical evidence to support this claim (Myers, 2015, p.212). Studies have shown that if properly managed, rattan collection is a sustainable way of extracting forest products (e.g. Salafsky and Wollenberg, 2000; Siebert, 2012).

In his analysis of the impact of the 2012 ban on domestic and international markets, forests, and the livelihoods of rattan collectors, Myers (2015) concludes that the policy mainly serves elite interests; those engaged in smuggling rattan from Indonesia have profited most. The country's rattan-processing industry has benefitted only partially, while demand at local level has been depressed. Therefore, the impact of the ban on local cultivators and extractors has not been much different from the consequences of earlier government interference. Several recent studies have shown that trade regulations, including the latest ban, have depressed demand and raw-material prices and, hence, led to severe economic losses for cultivators and extractors (e.g. Gönner, 2001; Belcher et al., 2004; Belcher, 2007; Bizard, 2013; Myers, 2015). The economic position of cultivators and extractors has been further weakened by world market prices and currency fluctuations (Gönner, 2001, p.141), as well as by their feeble bargaining power vis-à-vis middlemen, and their inability to sell rattan directly to urban traders and manufacturers. The increasing number of middlemen has progressively marginalized collectors (see Dove, 2011, pp.201-206), and together with the government policies, this has added to the sickness that is nowadays afflicting rattan.

Like his fellow villagers, Bapa Edwin sold his rattan to the local middleman, who lived in Dahanen and was, in fact, Bapa Edwin's first cousin. Thanks to the propinquity of their kinship, Bapa Edwin did not suffer a lower price, as was the usual outcome if people harvested on a credit basis. The price was commonly between 5000 to 10,000 rupiah ($0.38 to $0.76) lower per 100 kilograms of wet rattan. Social relations proved to be essential for this particular patron-client arrangement. Apart from impacting on the rattan price that farmers received – sometimes for better, sometimes for worse – the fact that the middleman was kin actually protected Dahanen's residents from the economic loss that rattan harvesters in Sapan had to suffer.

Sapan's residents usually sold their rattan to Banjar traders, who repeatedly asked me how I could voluntarily stay in such an isolated and backward place. Worse, they capitalized on the inhabitants' weak bargaining power. First, the villagers were offered a much lower price because of Sapan's rather remote location. Moreover, the prices offered in the first instance often fell by as much as 50% on the day of collection. On several occasions the traders failed to collect the rattan, resulting in severe losses and frustration for the people of Sapan. Under these unfavourable politico-economic conditions, contemporary rattan harvesting no longer guarantees a stable income. In fact, it may be economically risky. As suggested above and observed by Tsing (2005, p.185), the cultural prejudices of the downstream traders contribute significantly to the economic disadvantages of the villagers. In keeping with the evolutionary logic pervading popular thinking in Indonesia, downstream traders often look down on disadvantaged swidden cultivators living further upstream and seek to profit at their expense. Non-state actors often perpetuate the pejorative connotations attached to swidden agriculturalists by Indonesian development ideologies.

Ideological prejudices and the hidden premises of the NTFP concept

Indonesian development ideologies have long stigmatized shifting cultivation as a primitive form of agriculture with low technology and low returns, and as such in need of replacement with modern land-use systems (e.g. Dove, 1983; Li, 1999). As Asbjørnsen et al. (2000, p.19) wrote, 'A perfect example is found in Indonesia, where the government classified rattan gardens as "degraded forests" and systematically scheduled such lands for conversion to large-scale plantations.' Fried (2000) witnessed such a tragic exercise of state power over rattan gardens in East Kalimantan (see also Belcher et al., 2005, p.250). Even to the present day, the central government of Indonesia does not acknowledge rattan as a cultivated crop, but Forestry Law No. 46/2009 classifies all rattans as non-timber forest products (NTFPs) (*hasil hutan bukan kayu, HHBK*) growing on forest land (Dharma, 2013).[11] While even scholars disagree on what constitutes a non-timber forest product, classifying rattan as an NTFP may make sense with regard to species growing wild in the forest, but it seems

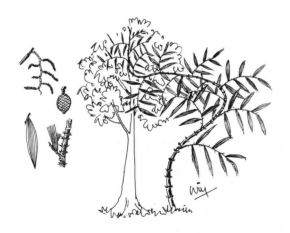

Calamus manan Miq.
[Arecaceae]

Synonymously known as *Calamus giganteus*, this rattan species has a single stem that grows up to 100 metres long and 8 cm in diameter. Growing at higher altitudes, it is said to produce the best large-diameter canes in the genus, with great strength and flexibility. They are hotly sought-after for making furniture.

contradictory in terms of rattan cultivated by farmers in their fields (Belcher, 2003, p.166).

The undifferentiated conceptualization even caused confusion among officials in Katingan, many of whom owned rattan gardens themselves. Nevertheless, the local government strove to implement permits for the extraction of NTFPs (*Izin Pemungutan Hasil Hutan Bukan-Kayu, IPHHBK*). But unless farmers could prove that their rattan gardens were located outside the state forest by means of legal ownership, the permits only allowed people to extract rattan from the 'forest' – that is, from their rattan gardens – without acknowledging their ownership. When discussing with Katingan's farmers the government's conceptualization of rattan as an NTFP rather than as a cultivated crop, they usually became emotional. Bapa Edwin, in a raised voice, told of a discussion he and other farmers once had with officials of the local forestry office:

> They don't consider rattan as a garden product but as a forest product, although rattan has been planted ever since the time of our ancestors. *Uei sigi* doesn't grow in the forest, only here along the river. It doesn't survive, unless it is cultivated and maintained.

The Ministry of Forestry's denial of the existence of people's rattan gardens was a mystery not only to Bapa Edwin, but to many other farmers who voiced their discontent with the official position:

> According to them, a garden has (to look) like an oil-palm garden (or) a rubber garden, although in the case of rattan it's impossible to establish a garden like that. Rattan needs trees.

> They really deny it and consider rattan to be just the accidental result of birds. That's what we discussed with them. Whereas all of the people of Katingan, in essence Central Kalimantan, say that it is not like they say, but (that rattan) really has been planted since the time of our ancestors

For the farmers, the government's position implied more than a lack of comprehension of rattan ecology, ignorance of local environmental knowledge and denial of ownership rights; it meant a complete disregard for the 'drops of sweat' spent by the farmers, and especially their ancestors, in labouring to prepare and maintain the gardens. Failing to pay tribute to the 'results of their ancestors' toil' – as one person termed it – was not only unacceptable, but as I argue and will further show below, it was also a disregard for the emotive relationships between the people and their rattan gardens. In fact, it was a complete denial of their history (Schreer, 2016a).

As aptly argued by Dove (2011, pp.211-212), while the non-timber-forest-product concept appears to be 'a politically neutral botanical reference to a slice of tropical forest resources, it has hidden premises' with 'far-reaching implications for power and equity'.[12] Revealing the logic behind the official construction of rattan as an NTFP shows the tensions between local and state conceptualizations of landscape, natural resources and their 'proper' management. Moreover, it uncovers aspects of the politico-economic environment of contemporary rattan management that disadvantage and disempower. The case of rattan exemplifies on-going conflicts over rights of access to, and ownership and management of, Indonesia's forests and, by association, it reveals the risk to local people of land capture, dispossession and disempowerment, often in the interests of large-scale plantation development. While such politico-economic factors may lead to an understanding of why rattan is now 'sick', it is important to emphasize that plantation and infrastructure development, accompanied by market expansion, also provide more lucrative alternative economic opportunities for Kalimantan's rattan farmers (Belcher et al., 2004; Pambudhi et al., 2004; Bizard, 2013).

Alternative income opportunities

Following the establishment of an oil-palm plantation on Dahanen's village grounds, many local people chose to work on the plantation as temporary or daily-wage labourers, since this guaranteed them a higher and, more importantly, a stable income (Schreer, 2016a). As has been observed elsewhere (e.g. Belcher et al., 2004, 2005; Cramb et al., 2009), younger generations, in particular, prefer plantation work to agricultural labour, not least because of changes in values and aspirations for modern lifestyles as a consequence of school attendance away from the village.

For elders like Bapa Edwin, plantation work was hardly an option. He couldn't stand the heat in the open fields, and like many villagers, he rejected the prospect of working under someone else's authority (Schreer, 2016a). 'If we harvest rattan, we are the ones deciding. You don't have to follow the rules of others', he said. Even though many agreed with him, that harvesting rattan allowed a self-determined working routine, the income it provided was unstable at best, so they were drawn into wage labour. Increasing involvement in plantation work led to a local shortage of labour, and this made it even more difficult to find people to sharecrop rattan gardens. It therefore contributed to the demise of rattan management.

However, the move from rattan and other livelihood activities into oil-palm wage labour was not necessarily an either–or choice. In line with the Dayak appreciation of independence and flexible adaptation to economic opportunities, the residents of Dahanen who temporarily engaged in wage labour were not implying a permanent abandonment of rattan harvesting, but rather a diversification of their household strategies. If no other alternatives were available, they could switch back to harvesting rattan, provided that the price was attractive and weather conditions were suitable. During the rainy season, harvesting is not feasible. It is too dangerous to climb trees, and the rattan gardens in Dahanen become flooded. Rattan farmers' decision-making is thus not predicated solely on economic considerations; it may also be influenced by individual

Elaeis guineensis Jacq. [Arecaceae]

Oil-palm plantations near the study site of Dahanen have offered rattan gardeners an alternative means of livelihood. The drift to wage-labouring in the plantations is leading to the disappearance of rattan gardens. The loss of biodiversity as a consequence is collateral damage from the expansion of oil palm.

preferences, age, values, climatic and ecological conditions, access to alternatives and risk management (see Belcher et al., 2004, p.S78). Farmers operate within a complex web of endogenous and exogenous factors, including how they feel about different kinds of labour.

The sensory experience of rattan harvesting

Having felt the pain of engaging with rattan myself, I can readily understand why the villagers of Dahanen and Sapan consider rattan harvesting, first and foremost, as painful (Figure 30-5). The analysis of my interlocutors' statements clearly shows the stresses and strains involved in rattan harvesting. Most of my local interlocutors mentioned terms such as *pehe* (painful), *uyuh* (exhausting) and *are duhi* (many spines) when describing how they felt about harvesting. Given the physical pain of working rattan, the expected income per person from a day's difficult labour – about 52,000 rupiah ($3.95) – does not justify the effort required. This further explains why people prefer engaging in physically less painful and more profitable work: oil-palm wage labour or fishing in Dahanen and mining for gold in Sapan.[13]

Against the background of this interdependent set of factors, including ecological factors such as forest fires (see Gönner, 2001; Pambudhi et al., 2004), it is possible to understand the 'sickness' that has overwhelmed rattan. Having outlined various factors that help to explain the recent demise of rattan management in Kalimantan generally (Belcher et al., 2004; Pambudhi et al., 2004; Bizard, 2013), the rest of this chapter explores the conundrum of why people nevertheless keep their gardens, even though there is an observable shift to alternative land uses.

FIGURE 30-5: Feeling the cane – experiencing rattan harvesting.

The (dis)continuity of rattan gardens

In some parts of Katingan, farmers have converted their rattan gardens into other land-use systems, mostly rubber, bananas and oil-palm gardens. Such land-use changes have been observed in Sapan, where 46% of the households have sold or converted single rattan plots, with 74% of them replacing rattan with rubber, 42% selling their gardens, and 26% converting them into gold mines (Figure 30-6). In the latter case, people began mining their own gardens or allowed others to mine there. As compensation, the garden owner received either a 10% share of the profits or an installation payment. By comparison, just 7% of people in Dahanen had ever sold or converted a rattan garden.

The picture becomes more complicated when rattan-planting activities in the two villages are taken into account. Between 2010 and 2013, only 5% of Dahanen's households planted rattan, all of which was aimed at rejuvenating existing gardens. By contrast, 46% of all households in Sapan planted rattan. Of these, 63% aimed to establish new rattan gardens following their swidden-rice harvests and 37% intended to rejuvenate old gardens. When rattan-planting activities in and before 2009 were taken into consideration,

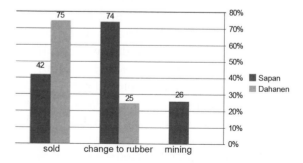

FIGURE 30-6: Conversion of rattan gardens into other land uses (%).

it became evident that people in Sapan had been more active in managing rattan than people in Dahanen, where 75% of all households had never (trans) planted rattan (Figure 30-7).

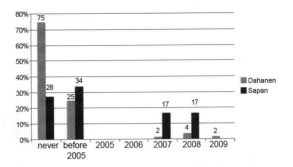

FIGURE 30-7: Planting of rattan before 2009 (%).

How can this paradox be explained? Why should there be a high rate of sale and conversion of rattan gardens together with more active planting activities on one hand, and a low rate of sale and conversion of rattan gardens together with a low rate of planting activities on the other?

In Dahanen, ecological conditions have become unfavourable for agricultural activities. The shallow alluvial soils along riverbanks, where rattan gardens are located, have become prone to flooding as a consequence of logging activities. Therefore, people have established new rubber gardens on less fertile, but higher, flood-proof ground, instead of converting rattan gardens to rubber and taking the risk that the young rubber trees may die due to flooding.

In contrast to the people of Dahanen, residents of Sapan are active swidden cultivators. Since 2009, only 10% of all households in Dahanen have attempted to open swiddens. All of them failed because of flooding. In fact, most of Dahanen's households (61%) have never engaged in growing swidden-based rice. In Sapan, 92% of all households opened a swidden between 2011 and 2013, with rice self-sufficiency in the village reaching as high as 79% in the 2012-2013 season (Schreer, 2016a). For the people of Sapan, planting and replanting in a rotational system of forest management and modification of the landscape are natural ways of engaging with their environment. The swidden cycle usually involves the cultivation of rattan, or more recently, rubber, following the rice harvest. In 2013, 85% of Sapan's households that planted cash crops following the rice harvest planted rubber, and 95% of those intending to plant something in their swidden fallows and empty land said that rubber was their first priority.[14] Soil fertility and distance to the hamlet were decisive criteria for field selection. The villagers of Sapan prefer the easily accessible, flat and fertile areas alongside rivers for their agricultural activities, which might imply the conversion of previously established rattan swiddens.

Given the high mean number of 4.6 rattan plots per household in Sapan, replacing rattan with rubber did not inevitably mean the complete abandonment of rattan gardens. Rather, establishing a rubber garden was a way for Sapan's residents to diversify their household economies, predicated on the hope for future well-being. Both children and adults of different age and sex expected that rubber would become the major income activity, and that this would guarantee a stable income and allow them to improve their lives.[15] As well as the hope for a better future, there was another psychological motive driving people's decisions to plant rubber: by joining the trend of establishing rubber gardens, the villagers stuck to the local principle of 'following the festivities of others', meaning that they did not want to miss the projected prosperity arising from the hoped-for rubber boom. Thus, future visions shaped resource-management practices just as firmly as past experiences, as we shall see below.

In addition to the above motives, villagers believed that, compared to harvesting rattan, tapping rubber was 'relaxing'. This supported my claim that the physical experience of different kinds of labour played an important role in people's decision-making. Moreover, local rules of inheritance and ownership rights had to be taken into account. As can be seen in Figure 30-8, most of Dahanen's rattan gardens were inherited. In Sapan, by contrast, most farmers established their gardens themselves, although some were inherited or received as bride wealth. However, a decisive

Hevea brasiliensis (Willd. ex A.Juss.) Müll.Arg. [Euphorbiaceae]

Much like rattan, but in more recent decades, rubber has been associated with shifting cultivation as 'rubber gardens' in fallow vegetation. More recently, rubber has brought pressure to bear on the continued existence of rattan gardens because of the commercial 'sickness' of rattan and farmers' desire to diversify.

difference was that 37% of all inherited rattan gardens in Dahanen were the collective property of siblings. Collectively managed gardens were absent in Sapan, where, under the observance of cognatic kinship, rattan gardens were usually divided on an equal basis among male and female children.[16] Whereas individual ownership meant that decisions were made at the level of an individual family, collective ownership constrained conversion of rattan gardens insofar as common agreement had to be reached among siblings.

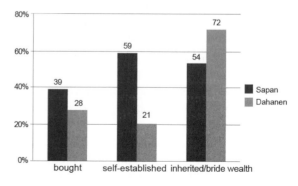

FIGURE 30-8: How rattan gardens were acquired (%).

As the previous sections show, ecological constraints – many of which are human-induced – economic diversification, and rules of ownership help to explain why rattan gardens have been converted in Sapan but not in Dahanen. It seems that in the future, rattan gardens may disappear in Sapan, not least because some villagers are uncertain whether their children will still have a rattan garden and many households express interest in further land-use change. Still, the large majority of all residents, in both Dahanen and Sapan, follow Bapa Edwin's claim that rattan gardens have always been there and will always exist in the future.

Sayang – rattan gardens as affective spaces

When I asked Bapa Edwin whether he had any plans to convert his garden, he replied: 'It would be a shame'. The Indonesian word he used was *sayang*. The same word was used by most of my interlocutors in spontaneous reactions to the same question. There is no single translation of *sayang*. Depending on context, it connotes meanings such as 'affection', 'dear', 'love', 'compassion', or 'sadness', and despite its multiple meanings, it conveys a sense of feeling. I suggest, therefore, that *sayang* expresses the affective connections between the people and their ancestors, entangled in their rattan gardens.

I asked Bapa Edwin: 'Why would it be sad?' He explained, in his words, *warisan datu hiang* – the garden was a legacy from the ancestors – *akan anak-esu hindai* – and it was intended to be passed on to his children and grandchildren. His reasoning was the same as that of many others to whom I put the same question (see Figure 30-9): future generations, both young and old, would continue to manage the legacy.[17] Thus, images of the future that are not necessarily detached, but are imbued with the past, shape people's present-day decisions.

As revealed in Figure 30-9, parents hoped that their children and grandchildren would still own a rattan plot. Although wishing that their children would have more successful jobs, thanks to higher education, and would no longer be the ones to do the harvesting, the gardens would still serve their male children as bride-wealth (in Sapan), indicate land ownership and especially support their children in times of economic hardship (see also Belcher et al., 2004, pp.85-86; Pambudhi et al., 2004, pp.361-362). Among the elders who explained to me that rattan had always served as a safeguard in times of need, there was a group of women:

> *Indu Refan:* Because, how shall I say (…), just don't let rattan disappear, because rattan is our last resort. It has never misguided us, but we could always still look for rice…
>
> *Indu Lia:* We really need rattan, really need it indeed…you get sick, if there is no rattan…
>
> *Indu Tina:* It's possible as livelihood, the last resort…

FIGURE 30-9: Reasoning behind the continuity of rattan in Sapan (left) and Dahanen (right).

Indu Lia: If you go (harvesting) for one to two days, you get a sack of rice...

Indu Tina: The last resort...

Rattan gardens thus convey a sense of stability and security, when feeling 'safe' is extremely valuable under conditions of increasing economic and environmental uncertainty. Even though rattan no longer provides a stable income, it serves people as a last resort. However, many farmers are deeply convinced that this is a temporary situation. Having become accustomed to fluctuations in commodity prices over recent decades, many people predict that rattan prices will rise again and harvesting rattan will become a major form of livelihood once again – at least in Sapan (Figure 30-9). Keeping the gardens, which requires only a little labour for maintenance, is thus seen not only as a secure investment, but also as sticking to a diversified household economy that is considered a necessity in anticipation of possible future price shocks and a way to spread the risk of economic hardship (see also Belcher et al., 2004, p.85; Pambudhi et al., 2004, p.361). Even though future visions influence people's choices, 'present and future refuse to offer stable hinges on which to build a semblance of certainty' (Pelkmans, 2013, p.5). It is therefore of little surprise, Pelkmans continues, that the past not only becomes a source of experience, but 'a reservoir of nostalgia'. People's hopes for the continuity and resurgence of rattan are nourished by their memories of rattan's heyday.

At the beginning of the 1970s, the growing popularity of rattan mats in Japan and wicker furniture in North America and Europe led to an expansion of the international rattan trade (Fried, 2000, p.208; Sasaki, 2007, p.464).[18] Throughout the two following decades, Indonesia became the leading exporter of raw materials. Not least because of the rapid development of the rattan-mats industry in South Kalimantan, small-diameter canes became highly sought after (Pambudhi et al., 2004, p.356), and this benefitted local collectors and cultivators. People in Dahanen recalled that 'rattan harvesting was *rami-rami* at that time', meaning that people were busily working with rattan. Buyers came to the village every second day, offering cash and consumer goods in advance to secure rattan stocks. Worldwide demand gave farmers much greater bargaining power than they have at present. Rattan was processed at village level, guaranteeing labour, higher prices and a decent life. To quote the people of Dahanen:

> Try to imagine counting rattan's value at that time. For 100 kilograms of (finished) rattan we could buy 15 grams of gold. Hence, people said: 'Working one day is enough for eating one week', eating from rattan's yields, even longer...[19]

> In the past, the rich people had a rattan garden. If you owned a rattan garden, life was already pleasant, because rattan was the primary livelihood.

In the past, rattan provided a steady income, food security, prosperity and improvement. Envisioning rattan's future by recalling its glorious past allows people to (re-)imagine themselves into times of well-being.

However, there was a challenge to the common narrative of a comfortable past. 'No garden, no harvesting', an old woman declared. 'If you did not have a rattan garden, life was sick, difficult.' So not everyone benefitted from the rattan boom. Just as prosperity and well-being are part of people's rattan memories, inequality and economic hardship are also a part of that local past. Multiple, and even conflicting, pasts converge, notwithstanding the dominant narrative of rattan enabling a decent life.

As we sat on the remnants of the fallen tree, Bapa Edwin's gaze wandered around the garden. An expert eye could spot the ordered lines of rattan clumps once planted by his forefathers. Lighting another cigarette, he became lost in childhood memories. Each day after school, he said, he would accompany his parents to the family's rattan gardens. He recalled how his father taught him how to hold the machete properly; how to move it; how to cut the cane; how to select seeds and seedlings; how to plant, transplant and manage rattan. As a child, he had crawled between the spiny clumps, sitting on his haunches and fighting the mosquitos, waiting for his father to shout from the top of the trees: 'Now! Cut the cane'. His eyes alight with the memories, Bapa Edwin told of how frightened he had been when an old person (a spirit), appeared to him at the back of the garden one day. But smiling whimsically, he remembered spending most of the time simply relaxing on the forest floor, watching his parents, sleeping, and nibbling snacks. For Bapa Edwin, just as for other villagers, rattan gardens were imbued with memories, most of them of childhood, evoking mixed feelings of pain, boredom, anxiety, relaxation and joy. Rattan gardens were spaces rich in sentiment, which created a sense of belonging and responsibility.

In Sapan, emotive ties to the surrounding landscape were created not only through the inheritance of rattan gardens, but through the active engagement of people in reworking the land. In Dahanen, however, such feelings could relate to a single rattan stem once planted by forefathers. Replacing rattan clumps without particular reason amounted to disregard for their ancestors' drops of sweat; it was *sayang*, connoting a notion of sadness. 'If we give up our rattan gardens, we forget our parents', one person reasoned. Keeping rattan gardens and thereby showing respect to their ancestors was not only seen as a moral obligation (see also Belcher et al., 2004, p.S86), but, metaphorically speaking, rattan gardens also bound families together. 'Since the ancestors in the past, it has just been rattan. It can't be broken,' Bapa Edwin concluded as he rose from the fallen tree. 'If we give up our gardens, we lose our history.'

As I have shown elsewhere, this history tells of the village's formation, local kinship, improvement, resource control, trade, material affluence, social cohesion and conviviality (Schreer, 2016a). In short, it is a chronicle of prosperity and well-being, thanks to their ancestors' success in trading rattan as far as Singapore from the middle to the end of the 19th century. For the villagers of Dahanen, rattan gardens were part of their identity; the gardens allowed them to relate to their ancestors. Bapa Edwin and his siblings' garden once belonged to their great-grandfather, who had accompanied the village's founding father to Singapore to sell the local rattan in international trade. For Bapa Edwin and his fellow villagers, keeping their gardens was an assertion of who they were. Beyond that, it was an assertion of what they and their life could be, if rattan was able to recover from its sickness.

Conclusion: 'Rattan identity?'

In his study of the dual economy of Kantu' smallholders, composed of subsistence rice cultivation and rubber cultivation for global markets, Dove (2011, p.15) argues that the 'cultural niches' of the two crops 'are fundamentally different', insofar as they belong to separate 'transactional orders'. Whereas rice cultivation centres on long-term reproduction of social and cosmological order, the market-oriented production of rubber focuses on individual short-term benefit. He sees these two different transactional orders as having different places in the moral hierarchy, with rice taking supremacy. Dove's analysis clearly holds true for the local economy in Sapan, where people place rice cultivation above all other livelihood pursuits, given its importance for the local economy as well as its sacred character (Schreer, 2016a). His analysis becomes less relevant in places such as Dahanen, where traditional swidden farming no longer exists. As has been observed elsewhere, the demise of rice cultivation can lead to 'redefinitions of local identity' and cause, for instance, the appearance of a Dayak 'pepper identity' (Cramb et al., 2009, p.332)

While rattan cultivation is geared towards market production and thus short-term maximization of individual benefit, this chapter has shown that rattan gardens are, at the same time, of great non-economic significance. In the near absence of swidden farming, rattan gardens seem to have taken moral supremacy in places like Dahanen, although purchased rice remains the primary ingredient in ritual conduct and fishing is the primary livelihood activity. As one person stated, 'rattan is identical with our culture'. Rattan gardens are more than a patch of forest managed by means of elaborate environmental knowledge. They are spaces that evoke memories and associated feelings; that allow people to (re-)imagine themselves into times of well-being, recall and narrate the past, and thus relate to their forefathers. As a material manifestation of people and their ancestors' labour, rattan gardens not only reflect the continuous interaction between humans and their environment, but they also symbolize an unbroken legacy, conceived to endure for an indefinite time.

King (1993, p.167) claims that 'the traditional view of Borneo natives is that natural resources are held in trust for future generations'. This, he contends, is an expression of 'traditional adaptation to the natural environment' and as such, a sign of intrinsic sustainable resource management.[20] The case of Katingan's rattan gardens shows that resources are indeed held for future generations. The reasons behind this are far more complex than suggested by King's materialist approach. Despite a multitude of endogenous and exogenous factors simultaneously and interactively influencing the decision-making of rattan smallholders, the expectation that rattan gardens will be managed by future generations reflects their hope for their children's security and the continuity and recognition of their history. An understanding of why farmers maintain their gardens and imagine that they have a future comes not least through an elaboration of people's emotive relation to the gardens.

The government's construction of farmers collecting 'wild' rattan from the forest, together with the fantasies of non-governmental organizations seeking to clear and replace the 'untidy' vegetation with 'orderly' rattan plantations implies more than disregard for people's rattan gardens. It is a complete misinterpretation of rattan ecology, people's environmental knowledge and their history and identity, which is entangled with rattan.[21] Writing about the threat of appropriation of 'the banana tree at the gate' (a metaphor for the rich but vulnerable natural-resource wealth of Borneo's peoples), Dove (2011, p.258, emphasis in original) holds that it is necessary to recognize that this is 'not an image of a community that *needs* a resource that it doesn't have; rather, it is an image of a community that cannot afford to *lose* a resource that it does have.' The case of Katingan's rattan gardens makes this more than explicit.

Epilogue: About one year after I sat with Bapa Edwin on the remnants of the fallen tree in his rattan garden, I returned to Dahanen. Nobody was harvesting rattan, not even Bapa Edwin. He was working as a wage labourer in the adjacent oil-palm plantation. These days, rattan is sick indeed.

References

Achdiawan, R. (2014) Personal communication between the author and R. Achdianwan, expert for non-timber forest products, Center for International Forestry Research (CIFOR), Bogor, August 2014

Arifin, Y. F. (2003) *Traditionelle Rattangärten in Zentralkalimantan, Indonesien*, Cuvillier Verlag, Göttingen, Germany

Asbjørnsen, H., Angelsen, A., Belcher, B., Michon, G. and Ruiz-Perez, M. (eds) (2000) *Cultivating in Tropical Forests? The Evolution and Sustainability of Systems of Management between Extractivism and Plantations*, Proceedings of a workshop, 28 June to 1 July 2000, Lofoten, Norway, available at http://www.gip-ecofor.org/doc/drupal/etfrn_series_3.pdf, accessed 12 February 2011

Aumeeruddy, Y. (1994) *Local Representations and Management of Agroforests on the Periphery of Kerinci Seblat National Park Sumatra, Indonesia*, People and Plants Working Paper no. 3, United Nations Educational, Scientific and Cultural Organization (UNESCO), Paris

Belcher, B. (2001) 'Rattan cultivation and livelihoods: The changing scenario in Kalimantan', *Unasylva* 205

Belcher, B. (2003) 'What isn't an NTFP?', *International Forestry Review* 5(2), pp.161-168

Belcher, B. (2007) 'The feasibility of rattan cultivation within shifting cultivation systems: The role of policy and market institutions', in M. F. Cairns (ed.) *Voices From The Forest: Integrating Indigenous Knowledge Into Sustainable Upland Farming*, Resources for the Future Press, Washington, DC, pp.729-742

Belcher, B., Imang, N. and Achdiawan, R. (2004) 'Rattan, rubber, or palm oil: Cultural and financial considerations for farmers in Kalimantan', *Economic Botany* 58, pp.77-87

Belcher, B., Michon, G., Angelsen, A., Ruiz Pérez, M. and Asbjørnsen, H. (2005) 'The socioeconomic conditions determining the development, persistence, and decline of forest garden systems', *Economic Botany* 59(3), pp.245-253

Bizard, V. (2013) *Rattan Futures in Katingan: Why do Smallholders Abandon or Keep their Gardens in Indonesia's 'Rattan District'?*, Working Paper no. 175, World Agroforestry Centre (ICRAF), Bogor, Indonesia

Casson, A. (2001) *Decentralisation of Policies Affecting Forests and Estate Crops in Kotawaringin Timur District, Central Kalimantan*, Center for International Forestry Research (CIFOR), Bogor, Indonesia

Cramb, R. A., Pierce Colfer, C. J., Dressler, W., Laungaramsri, P., Le, Q. T., Mulyoutami, E., Peluso, N. L. and Wadley, R. L. (2009) 'Swidden transformations and rural livelihoods in Southeast Asia', *Human Ecology* 37, pp.323-346

Dharma, K. (2013) personal communication between the author and K. Dharma, Head of the Division for the Management of Forest Products at the Forestry Office, Katingan Regency, November 2013

Dove, M. R. (1983) 'Theories of swidden agriculture and the political economy of ignorance', *Agroforestry Systems* 1, pp.85-99

Dove, M. R. (1998) 'Living rubber, dead land, and persisting systems in Borneo: Indigenous representations of sustainability', *Bijdragen tot de Taal-, Land- en Volkenkunde* 154(1), pp.20-54

Dove, M. R. (2000) 'The life-cycle of indigenous knowledge and the case of natural rubber production', in R. Ellen, P. Parkes and A. Bicker (eds) *Indigenous Environmental Knowledge and its Transformations: Critical Anthropological Perspectives*, Harwood Academic Publishers, Amsterdam, pp.213-252

Dove, M. R. (2011) *The Banana Tree at the Gate: A History of Marginal Peoples and Global Markets in Borneo*, Yale University Press, New Haven, CT

Dransfield, J. and Manokaran N. (eds) (1994) *Rattans, Plant Resources of Southeast Asia (PROSEA) 6*, Pudoc, Wageningen, The Netherlands

Dransfield, J., Uhl, N. W., Asmussen, C. B., Baker, W. J., Harley, M. M. and Lewis, C. E. (2008) *Genera Palmarum. The Evolution and Classification of Palms*, Kew Publishing, Kew Royal Botanical Gardens, UK

Feintrenie, L. and Levang, P. (2009) 'Sumatra's rubber agroforests: Advent, rise and fall of a sustainable cropping system', *Small-scale Forestry* 8(3), pp.323-335

Fried, S. G. (2000) 'Tropical forests forever? A contextual ecology of Bentian agroforestry systems', in C. Zerner (ed.) *People, Plants, and Justice: The Politics of Nature Conservation*, Columbia University Press, New York, pp.204-233

Fried, S. G. and Mustofa, A. S. (1992) 'Social and economic aspects of rattan production, middle Mahakam region: a preliminary survey', *GFG Report* 21, pp.63-72

Godoy, R. A. (1990) 'The economics of traditional rattan cultivation', *Agroforestry Systems* 12, pp.163-172

Godoy, R. A. and Feaw, T. C. (1991) 'Agricultural diversification among smallholder rattan cultivators in Central Kalimantan, Indonesia', *Agroforestry Systems* 13, pp.27-40

Gönner, C. (2001) *Muster und Strategien der Ressourcennutzung: Eine Fallstudie aus einem Dayak Benuaq Dorf in Ost-Kalimantan, Indonesien*. Forstwissenschaft-liche Beiträge der Professur Forstpolitik und Forstökonomie, Eidgenössische Technische Hochschule Zürich, Switzerland

Haug, M. (2010) *Poverty and Decentralisation in East Kalimantan: The Impact of Regional Autonomy on Dayak Benuaq Wellbeing*, Centaurus, Freiburg, Germany

Joshi, L., Wibawa, G., Vincent, G., Boutin, D., Akiefnawati, R., Manurung, G., van Noordwijk, M. and Williams, S. (2002) *Jungle Rubber: A Traditional Agroforestry System under Pressure*, World Agroforestry Centre (ICRAF), Bogor, Indonesia

King, V. T. (1993) *The Peoples of Borneo*, Blackwell, Oxford, UK

Knapen, H. (2001) *Forests of Fortune? The Environmental History of Southeast Borneo, 1600-1880*, KITLV Press, Leiden, the Netherlands

Li, T. M. (1999) 'Marginality, power and production: Analysing upland transformations', in T. M. Li (ed.) *Transforming the Indonesian Uplands: Marginality, Power and Production*, Harwood Publishers, Amsterdam, pp.1-44

Matius, P. (2004) 'Plant diversity and utilization of rattan gardens', *Freiburg Forestry Research* 28, Silviculture Institute, Freiburg, Germany

Michon, G., Mary, F. and Bompard, J. (1986) 'Multistoried agroforestry garden system in West Sumatra, Indonesia', *Agroforestry Systems* 4(4), pp.315-338

Michon, G. and de Foresta, H. (1999) 'Agro-forests: Incorporating a forest vision in agroforestry', in L. Buck, J. Lassoie and E. Fernandez (eds) *Agroforestry in Sustainable Agricultural Systems*, CRC Press, New York, pp.381-406

Michon, G., de Foresta, H., Kusworo, A. and Levang, P. (2000) 'The Damar agroforests of Krui: Justice for forest farmers', in C. Zerner (ed.) *People, Plants, and Justice: The Politics of Nature Conservation*, Columbia University Press, New York, pp.159-203

Mulyoutami, E., Rismawan, R. and Joshi, L. (2009) 'Local knowledge and management of *simpukng* (forest gardens) among the Dayak people in East Kalimantan, Indonesia', *Forest Ecology and Management* 257, pp.2054-2061

Myers, R. (2015) 'What the Indonesian rattan export ban means for domestic and international markets, forests, and the livelihoods of rattan collectors', *Forest Policy and Economics* 50, pp.210-219

Netting, R. McC. (1993) *Smallholders, Householders: Farm Families and the Ecology of Intensive, Sustainable Agriculture*, Stanford University Press, Stanford, CA

Padoch, C. and Peluso, N. (eds) (1996) *Borneo in Transition: People, Forests, Conservation, and Development*, Oxford University Press, Kuala Lumpur

Pambudhi, F. P., Belcher, B., Levang, P. and Dewi, S. (2004) 'Rattan (*Calamus* spp.) gardens of Kalimantan: Resilience and evolution in a managed non-timber forest product system', in K. Kusters and B. Belcher (eds) *Forest Products, Livelihoods and Conservation: Case studies of Non-Timber Forest Product Systems*, Center for International Forestry Research, Bogor, Indonesia, pp.347-365

Pelkmans, M. (2013) 'Ruins of hope in a Kyrgyz post-industrial wasteland', *Anthropology Today* 29(5), pp.17-21

Peluso, N. L. (1983) 'Networking in the commons: A tragedy for rattan?', *Indonesia* 35, pp95-108

Persoon, G. A. and van Est, D. M. E. (2000) 'The study of the future in anthropology in relation to the sustainability debate', *Focaal* 35, pp.7-28

Salafsky, N. and Wollenberg, E. (2000) 'Linking livelihoods and conservation: A conceptual framework and scale for assessing the integration of human needs and biodiversity', *World Development* 28(8), pp.1421-1438

Sasaki, H. (2007) 'Innovations in swidden-based rattan cultivation by Benuaq farmers in East Kalimantan, Indonesia', in M. F. Cairns (ed.) *Voices from the Forest: Integrating Indigenous Knowledge into Sustainable Upland Farming*, Resources for the Future Press, Washington, DC, pp.459-470

Schreer, V. (2016a) 'Longing for prosperity in Indonesian Borneo', PhD dissertation to the School of Anthropology and Conservation, University of Kent, Canterbury, UK (unpublished)

Schreer, V. (2016b) 'Learning knowledge about rattan (Calamoideae, Arecaceae) and its uses amongst Ngaju Dayak in Indonesian Borneo', Special Issue 'Botanical Ontologies', *Journal of Ethnobiology* 36(1), pp.125-146

Siebert, S. F. (2012) *The Nature and Culture of Rattan: Reflections on Vanishing Life in the Forests of Southeast Asia*, University of Hawai'i Press, Honolulu

Sunderland, T. C. H. and Dransfield, J. (2002) 'Species profiles', in J. Dransfield, F. O. Tesoro and N. Manokaran (eds) *Rattan: Current Research Issues and Prospects for Conservation and Sustainable Development*, Food and Agriculture Organization of the United Nations, Rome, pp.9-22

TeROPONG and Yayasan SHK Kaltim (2005) 'Data Dasar Rotan Kabupaten Katingan', an unpublished report by two non-governmental organizations

Tsing, A. L. (2005) *Friction: An Ethnography of Global Connection*, Princeton University Press, Princeton, NJ

Vantomme, P. (2003) 'Growing imbalance between supply and demand for rattan?', *Journal of Bamboo and Rattan* 2(4), pp.407-415

van Tuil, J. H. (1929) 'Handel en cultuur van rotan in de zuiderenooster afdeeling van Borneo (Trade and culture of rattan in the southeastern section of Borneo)', *Tectona* 22, pp.695-717

Wadley, R. L. (ed.) (2005) *Histories of the Borneo Environment. Economic, Political and Social Dimensions of Change and Continuity*, KITLV Press, Leiden, the Netherlands

Weinstock, J. A. (1983) 'Kaharingan and the Luangan Dayaks: Religion and identity in Central East Borneo', PhD dissertation to Cornell University, Ithaca, New York

Wolf, E. R. (1982) *Europe and the People without History*, University of California Press, Berkeley and Los Angeles, CA

Wolters, O. W. (1967) *Early Indonesian Commerce: A Study of the Origins of Srivijaya*, Cornell University Press, Ithaca, New York

Notes

1. Parts of this chapter have been published previously as a working paper (Bizard, 2013).
2. Two readily observable and distinctive features (apomorphies) arising from the evolution of Calamoideae palms are spines and fruits covered by reflexed scales. Rattans are found in the tropical forests of Equatorial Africa, South and Southeast Asia, northern Australia and Fiji (Sunderland and Dransfield, 2002, p.10). The broad range of ecological niches in which rattans grow not only explains the wide geographic distribution and the species richness of rattan, but also the high endemism occurring within this palm group (Siebert, 2012, p.10). Even though rattans are normally described as climbing palms, some species do not climb. For more details on rattan ecology see Dransfield et al. (2008, pp.141-207).
3. Apart from being a political strategy to capture the votes of local rattan farmers in the local elections of 2003 and 2008, the policy was part of a macro-political vision of an economy of added value that arose in the aftermath of the Asian financial crisis in 1997-1998 (Bizard, 2013, p.1). The initiative involved, among other growth measures, a ban on raw-material exports, including rattan raw material and semi-finished rattan products. It came into effect in January 2012.
4. While Indonesian government data is based on nuclear families, this paper takes the household as its unit of analysis.
5. The Ngaju Dayak are the most numerous and dominant group in southeast Borneo (King, 1993, p.53; Knapen, 2001, p.89). The majority of them live along the middle and lower reaches of Central Kalimantan's waterways. The region is also inhabited by different Muslim groups, the most numerous of which are Malay, Buginese, Banjarese, Javanese and Madurese (Casson, 2001, p.1). However, like elsewhere in Borneo, ethnic and religious boundaries are far from clear-cut and people's self-conception is in flux (Schreer, 2016a).
6. People did not measure the size of their gardens in hectares, but spoke of 'plots' – in their language, *lembar*. The size of a plot ranged between 0.5 and 2 hectares, which constrained an accurate analysis. As can be seen in Figure 30-4, when ownership was examined in terms of individual families the numbers changed, meaning that each nuclear family that was part of a larger household did not necessarily possess its own garden.
7. Regulation of Trade Minister no. 35/2011.
8. In 2011, the farm-gate price for one kilogram of wet (unprocessed) *uei/uwei sigi* reached 2000 rupiah ($0.15), which was about the average minimum price that people would have expected, had they been tempted back into rattan harvesting. Some claimed that rattan harvesters in Sulawesi had also experienced reduced volumes, but the latest ban had not resulted in a price change there (Myers, 2015, p.218). This might have been due to the fact that large-diameter canes collected in

the 'wild', foremost in Sulawesi, were in demand, whereas demand for small-diameter rattans, such as those cultivated by farmers, had fallen (Achdiawan, 2014).

9. These policies were bans on the export of raw rattan in 1986 and semi-finished rattan in 1989. These bans were lifted with the introduction of export quotas in 1998 and further policy modifications in 2004, 2005, 2007, 2008 and 2009, allowing the raw material price to slowly recover (Haug, 2010, p.43).

10. Peluso (1983) provides a detailed analysis of the explosion of collection from wild stocks and the threat to the rattan commons in East Kalimantan during the 1970s and 1980s.

11. Regulation of the Minister of Forestry No. 46/2009, about the Issuing of Extraction Permits for Timber and Non-Timber Products.

12. Dove (2012, p.211) holds that these premises are that NTFPs are resources that local people may be allowed to exploit, but that they are resources that no one but local people would want to exploit, rendering them non-valuable forest products.

13. This figure is based on a rattan price of 1500 rupiah ($0.11) per kilogram and farmers' estimation that, on average, a couple can harvest 70 kilograms of rattan per day. However, yields vary depending on individual skills and the condition of the rattan.

14. In 2013, while 85% of all households in Sapan owned a rubber garden, only 5% of them were able to tap the young trees. However, in Dahanen, 52% of all households owned a rubber garden, and in 22% of them, the trees could already be tapped. These gardens mainly stemmed from the 1970s and were frequently mixed with rattan. More recently, further rubber gardens were set up following a ban on logging in 2006, and the World Wide Fund for Nature supported the promotion of rubber as an alternative livelihood option. In Sapan, people received seedlings through a project of the local forestry office.

15. Many of the children's drawings in Sapan showed anticipation of a future with rubber gardens.

16. In Dahanen, kinship is likewise recognised bilaterally, but given the influence of Muslim tradition, male children frequently receive a larger share than their sisters.

17. While 'future generations' refers to unspecified generations in unspecified places and times (Persoon and van Est, 2000, p.18), the notion of *anak-esu* (children and grandchildren) is likewise neither temporarily nor spatially bounded to the two following generations of children and grandchildren.

18. Increased demand for rattan over past decades has not been without consequences for the world's rattan resources. Wild rattan stocks, which meet about 90% of demand from the international rattan industry, are dwindling at an increasing rate (Dransfield and Manokaran, 1994, p.11). In his analysis of the status of rattan resources, with special emphasis on Asia due to its market share, Vantomme (2003, p.414) maintained that scarcity of quality cane was a problem in different parts of Asia and that the difficulty of resource supply had never been more evident. In view of decreasing wild rattan resources, local people's traditional rattan cultivation and the knowledge associated with it are becoming increasingly important, not solely for meeting global demand, but also for developing rattan as a sustainable plantation crop (Siebert, 2012, p.85).

19. In 2013, 100 kilograms of finished rattan was worth less than 2 grams of gold.

20. King's assertion has been criticised for not showing on which 'empirical or authentic ideological evidence' the assumed environmental wisdom is predicated (Persoon and van Est, 2000, p.18). With reference to Dove's (1998) account of emic conceptions of sustainability, Persoon and van Est (2000, p.18) further remark that the term 'sustainability' is not used by those concerned or may be conceptualised differently.

21. It is important to stress that not all NGOs engaged in the revitalization of rattan as a sustainable livelihood pursued the plan to replace existing gardens with rattan plantations.

31

FALLOW MANAGEMENT THROUGH CULTIVATION OF BROOM GRASS

A potential cash crop in northeast India

Indrani P. Bora, Kuntala N. Barua and Pawan K. Kaushik*

Introduction

Shifting cultivation, locally known as *jhum,* is the most widespread cropping system in mountainous regions of the tropics. The value of this system comes from an age-old process that makes use of indigenous knowledge for efficient management of natural resources. It is regarded as having been one of the first steps in humanity's transition from food gathering to food production (Sharma, 1976). There are ethnically and culturally distinct forms of shifting cultivation, according to agro–climatic locations. Mountain dwellers practice various forms of shifting cultivation that maintain harmony with their environments. The basic system, which begins with slashing and burning of forests on hill slopes followed by phases of growing intermixed crops, can be considered as an early form of agroforestry. The cultivation is shifted from one plot to another in a regular cycle in which the process is repeated. The calendar of activities associated with shifting cultivation also varies considerably according to the crop selection and ritual activities of the farmers. One of the most remarkable features of shifting cultivation systems is community or clan ownership of the land, which maintains equity within a community in terms of economic status. However, shifting cultivation has been a primary recipient of blame for forest conversion, deforestation, depletion of resources, loss of soil fertility, soil erosion and lowered productivity, with no scope for application of improved agriculture technologies.

India's northeast region, which covers an area of about 300,000 square kilometres, forms a highly complex mosaic landscape. Forest covers 66.29% of the area, or a total of 173,800 sq. km. However, this constitutes less than 25% of the country's total

* Dr. Indrani P. Bora, Scientist 'C', Shifting Cultivation Division; Dr. Kuntala N. Barua, Chief Technical Officer, Biodiversity and Conservation Division, both from the Rainforest Research Institute, Indian Council of Forestry Research and Education, Jorhat, Assam; and Shri Pawan K. Kaushik, Scientist 'E' and Regional Director, Centre for Forest-based Livelihood and Extension, Agartala, Tripura, India.

forest cover (Ministry of Environment and Forest, 2009). The same ministry reported in 2011 that 12 Indian states, including the northeastern states, showed a decrease of 367 sq. km in forest cover over the previous two years. The major blame for this fell upon shifting cultivation, although the practice remains the main source of livelihood for many hill communities. It is still part and parcel of the linguistic and cultural practices – the socio-cultural life – of more than 200 different tribal groups. Generally, shifting cultivation extends from the moist deciduous forest foothills up to the subtropical and even temperate forests, with slopes of 30° to 40° in many parts. The annual rainfall is always higher than 1000 mm and may go beyond 5000 mm.

The extent of shifting cultivation in India's northeast

Shifting cultivation is accepted as a practice springing from the early stages of agricultural evolution. For millennia, it has provided sustenance for a large section of humanity, and it is still widely practised in tropical hilly areas of Southeast Asia, the Pacific, Latin America, the Caribbean and Africa. It has been estimated that about 200 million people – 7% of mankind – still practise this type of cultivation on about 300 million hectares of land (Craswell et al., 1997; Thomaz, 2009). In India, around two million tribal people still depend upon it, using as much as 11 million hectares of land. Shifting cultivation has long been the main source of livelihood for most tribes of India's northeastern hills and more than 100 tribal ethnic minorities still depend upon it exclusively for their survival.

According to the Ministry of Agriculture's Task Force on Shifting Cultivation, the area under shifting cultivation in northeast India was 3.81 million hectares in 1983 and an estimated 443,336 households earned their livelihood from it. Some studies even show an increase in shifting cultivation. Satellite data studied by the Assam Remote Sensing Application Centre showed that in the Karbi Anglong district of Assam, the area under *jhum* increased from 13,583 ha, or 1.302% of the total geographical area, in 1986-1987 to 69,125 ha, or 6.62% of the total geographical area, in 1993-1994 (ASTE Council, 1996). In Meghalaya, 52,290 families were reportedly practising shifting cultivation on 0.26 million ha of land in 1983 (Task Force on Shifting Cultivation, 1983). However, the Forest Survey of India (1997) found that the average area under shifting cultivation in Meghalaya during the period from 1987 to 1997 was 0.18 million ha, or about 180 square kilometres, representing a decline in the traditional practice (Department of Environment and Forest, Meghalaya, 2005)

Land transformation

Originally, shifting cultivation was the only agricultural practice supporting the livelihoods of indigenous communities in northeast India. Over recent decades a rapidly increasing population and limits on the productive capacity of the land have forced many changes in local farming systems. These changes have mostly been made because of economic pressures rather than ecological concerns. Since swiddening systems were no longer providing sufficient support for farming families, shifting cultivators began to adjust their land-use practices by introducing new crops. The adoption of alternative land-use patterns and promotion of cash-crop cultivation sometimes destabilized practices that were ecologically and economically well adapted. Traditionally, the people of northeast India had grown as many as 35 different crops in their fields (Tiwari, 2001). However, the shortening of fallow periods in shifting cultivation, along with many other reasons, led to the mixed-cropping culture being replaced by a monoculture system. Some non-traditional crops have also entered the system. The sustainability of these different land-use systems depends mainly on the capacity of the land. Some groups are now using longer cropping periods in their swidden cycles because their crops include, for example, broom grass, pineapples, oranges and areca nuts. Some of the cash crops have arisen from the livelihood patterns of the cultivators and have been found to be ecologically more sustainable than shifting cultivation. Hence, swiddening has become a supplementary form of crop production. This diversification of farming systems has resulted in a shift away from purely seasonal cropping to the adoption of perennial crop rotations, which may be referred to as commercial shifting cultivation. Food crops are rotated together with valuable trees, with long-term income from the trees being added to cash benefits from the annual food crops. One of the guiding factors leading to adoption of alternatives to traditional shifting cultivation is the development of transportation networks and marketing facilities. Nevertheless, switching over from food-crop to cash-crop production sometimes has adverse effects on the food security of small, resource-poor landholders.

Shifting cultivation has lately moved closer and closer to what we know as agroforestry. As well as the demographic, economic and biophysical pressures forcing these changes, successive governments over the past half-century have been consistent in their opposition to shifting cultivation, hoping for the adoption of alternative agricultural systems or at least the achievement of a measure of control over slash-and-burn. However, the efforts of policy-makers have been unsuccessful in halting a practice which over countless generations has become a deeply ingrained part of the socio-cultural life of communities and is essential to various ritual activities. Other forces of change have been inexorable; land use has intensified considerably and fallow management has become more refined. The final factor has delivered the ultimate force of change: there is no longer enough land on which to practise traditional rotational shifting cultivation.

The role of *Thysanolaena latifolia* (Roxb. ex Hornem.) Honda (broom grass) as a fallow management crop

While shifting cultivation remains the principal farming system in the mountainous terrain of northeast India, the practices of most *jhumias* of the region are evolving towards some kind of more permanent land use. There is a current need for better approaches and methods to manage this transition; to enable farming systems to become more productive. Improved fallow management is one of the options in this regard, and broom grass (*Thysanolaena latifolia*) can play an important role. This species is distributed across sub-tropical northern India from Kumaon to the Eastern Khasi Hills, growing at altitudes up to about 2000 m. It is a huge tufted perennial reed-like grass that grows to 3 m tall and is commonly found on shady hill slopes. It grows well in a wide range of habitats on silt, sandy and clay-based, quick-draining soils with pH ranging from 5.3 to 9.3; moisture from 11.6% to 37.6%; and carbon from 0.4% to 2.7%. The culm of broom grass is solid, smooth and terete; the leaf-sheaths are tight; the nodes glabrous and the margins bearing some short stiff hairs towards the throat. Leaf blades are lanceolate-acuminate, resembling bamboo. Broom grass has a dense, terminal, foxtail-like panicle that is 60 cm to 90 cm long, wider during the flowering. Spikelets are numerous, often in pairs on a common peduncle and each pedicel is distinct. One to

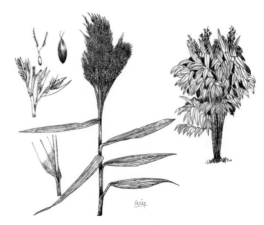

Thysanolaena latifolia (Roxb. ex Hornem.) Honda [Poaceae]

Often known as tiger grass, this species is rapidly becoming an important economic plant across much of upland Southeast Asia. It is especially favoured by women, not only providing income and brooms, but the stems and leaves relieve the drudgery of collecting firewood and fodder for animals, respectively. The leaves are also used as food wrappers and are said to have wound-healing and antibacterial properties.

one-and-a-half-year-old healthy plants begin producing panicles from October to March. In the hilly areas of northeast India, broom grass grows naturally in degraded *jhum* fallows and contributes substantially to the cash income of the *jhumias* (Figure 31-1). The mature inflorescence of *T. latifolia* is used for making household brooms. The brooms are transported annually to other parts of the country, attracting high prices. But despite the demand, supplies of broom grass to date have come from naturally growing stock. Hence, wild stocks of broom grass are depleting rapidly and there is an increasing need to cultivate broom grass to meet escalating demand (Ahlawat and Singh, 1998).

Uses

The panicles of broom grass are used for making soft brooms. These brooms are more durable than those made from other plants. As well, the tips and tender leaves of *T. latifolia* may be used as livestock fodder, although the palatability of broom grass leaves is low compared to other grasses. The grass is usually available in large quantities during the months of July to February and is used for cattle feed in periods of fodder

FIGURE 31-1: A hillside of broom grass growing naturally in northeastern India.

Photo: P. K. Kaushik.

scarcity. After harvest, the broomsticks (culms) are used as fencing or wall-building material. The stalks of the panicle are used for the manufacture of handicrafts and the fibrous roots are very effective in checking soil erosion on steep slopes. A decoction of broom grass roots is also used as a mouthwash for patients with fever.

Cultivation practices

Broom grass grows well on wasteland or degraded land. It can be planted by seeds or by rhizomes. Farmers collect planting stock from wild clumps, or it can be propagated artificially.

Seeds from the inflorescences are collected in March. Farmers broadcast 5 to10 grams of seed on to a bed measuring 2 m x 1 m. The bed is covered with a thin layer of sand and thatch grass. Proper watering is essential. Seeds start to germinate after two to three weeks. The young plants are transplanted into plastic bags or new beds four to six weeks after germination. They are watered regularly and the seedlings are planted into fields during the following rainy season.

Propagation from rhizomes involves digging the rhizomes from wild sources after the panicles are harvested. The culms are trimmed, leaving the rhizome with bud sprouts. These are raised in nursery beds or planted directly into fields. Nursery beds are prepared by mixing soil, sand and farmyard manure in a ratio of 1:2:1. Fields are cleared by slashing the vegetation before March and removing the debris. Pits are dug into the field, measuring 30 cm to 50 cm in diameter, and spaced at intervals of about 2.5 m. They are left for one month for weathering. Sprouted rhizomes are then transplanted into the pits within three months, at the onset of the monsoon. Weeding is needed two to three times in the first year, after which minimal care is required.

Yield and economics

In the months of January through March, broom grass inflorescences turn brownish in colour and are then ready for harvesting. In the first year, the yield is negligible. Maximum yield can be expected in the third and fourth years after planting, due to the need for the plant to reach its highest increment of culms. Selection of appropriate planting materials is essential for obtaining the highest yield. According to Bisht and Ahlawat (1998), the average yield of culms per tussock in a one hectare area is 225 kg and 180 kg in the third and fourth years of cultivation, respectively. Cultivation of broom grass can generate 10,000 to 20,000 rupees (US$135 to $270) per hectare per year. Total economics of the system depends on labour efficiency, cultivation practices and market prices.

Selection of study site and study methods

Study villages were located in the Nilip and Rongmongwe blocks of Karbi Anglong district in Assam, and in the Nongpoh and Dawki areas of Meghalaya (Figure 31-2). A rapid rural appraisal exercise was conducted in different villages of the selected sites, along with surveys using questionnaires. Selection of villages was based solely on the existence of *T. latifolia* plantations in the area. Farmer interviews were combined with on-site observations. Data were generated on existing farming systems, farmer's perceptions of *T. latifolia* plantations, their productivity and economics, comparisons with traditional *jhum* cultivation, seasonal calendars, market possibilities and reasons for planting *T. latifolia* in degraded fallows.

Study site 1: Karbi Anglong district, Assam

Extensive surveys were conducted in Nilip Block (Silonijan, Kailamati, Johnar Sinar and Bordeka Ingti villages) and Rongmongwe block (Jilangso, Rongtara and Dolomora villages) in Karbi Anglong district, Assam.

Climate

Due to variations in the topography, this hill zone experiences diverse climates. The winter commences in the month of October and continues until February. The area receives an annual rainfall of approximately 1235.59 mm. The highest monthly rainfall (214.58 mm) is recorded in July and the lowest (10.92 mm) in January (Sarma et al., 1996).

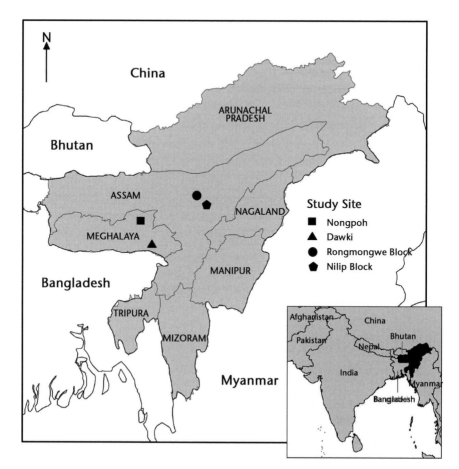

FIGURE 31-2: The study sites in northeastern India.

Temperature

Due to the southwest monsoon circulating over low-lying hills and the absence of any streamlined movement of wind, summer temperatures remain comparatively high. The average maximum temperature ranges from 28.65°C to 31.24°C and the minimum temperature from 14.67°C to 19.38°C. Relative humidity is highest (90%) in the month of August.

Soil

Soils in this zone are distinct red laterite types, mostly old alluvial. The area is within the 'medium hill' zone, with an altitude of 180 m to 900 m above sea level. The range of basic soil characteristics are as follows: pH, 4.98 to 5.45; conductivity, 0.50 to 0.54 μmhos/cm; organic carbon, 1.16 to 2%; phosphorus, 20.34 to 24.54 kg/ha; potassium, 41.11 to 360.56 kg/ha; nitrogen, 0.132 to 0.154 kg/ha; calcium, 2.9 to 3.5 meq/100 g; and magnesium, 1.3 to 1.9 meq/100 g.

Jhum is the traditional system of cultivation and is deeply rooted in the socio-economic and religious life of the communities in Karbi Anglong district. The hill-slope plots are slashed and burnt in January or February and the farmers follow their own respective cropping calendars to cultivate mixed crops. The principal crops are rice and maize. Along with these, sesame, chillies, pumpkins, *Colocasia* sp. (taro), cucumbers, beans, yams, tapioca, pointed gourds, ash gourds and others are grown as mixed crops. Seeds are sown at the onset of the monsoon by dibbling. Harvesting of rice occurs in the months of August and September. Cultivation of a single plot is continued for a couple of years before it is left fallow. The fallow is an intrinsic part of shifting cultivation with many functions that are commonly overlooked. In the past, when population pressures were low and land was available, shifting cultivators cleared new plots of land every year for cultivation and old cropping plots were left fallow. Thus, they created many fallow fields of different ages and natural vegetation succession took place in various stages.

The significance of the fallow can be summarized as follows:

- It restores fertility and rejuvenates the soil;
- It provides forage for livestock;
- It is a source of domesticated, semi-domesticated and wild food plants;
- It is a source of herbal medicine; and
- It provides raw materials for a wide range of crafts.

Recently, as population pressures have driven intensification of the agricultural system, fallow periods have shrunk to as little as two years. This does not allow the fallow to rejuvenate the soil or to restore its fertility, so the productivity of crops diminishes. Under these conditions of considerably intensified land use, a more refined fallow-management concept has arisen. Gradually, farmers have begun to cultivate bamboo, broom grass, pineapples, oranges and other cash crops on their fallowed land, as a subsidiary source of income. Typically, the fallow land has low fertility and diminished water-holding capacity, so it is not suitable for all kinds of crops. However, broom grass is a species that can tolerate harsh environmental conditions, including drought or high rainfall, and can grow in shallow soil on steep rocky mountain slopes. It is suitable for cultivation on wastelands as well as on the undulating landscape of degraded *jhum* fallows. Minimal inputs are required, and with vast areas of degraded *jhum* land available, cultivation of broom grass has a high potential to uplift the socio-economic conditions of *jhumias*. Broom grass has long grown abundantly on the mountain slopes of this hill district, sufficient in the past to meet demands for raw materials with which to make brooms. However, the local people have been ruthless in their exploitation of wild sources of broom grass, for their own requirements as well as for cash sales. As a result, natural stocks have been depleted at an alarming rate. To a certain extent, urbanization and industrialization have exacerbated this depletion.

During the survey from which this paper was produced, we noticed that broom grass was growing spontaneously in *jhum* fallows after three years or longer. Farmers were leaving the fallow succession to occur naturally and in this way were creating an environment suitable for *T. latifolia*. They managed naturally-occurring broom grass as well as supplementing it by planting rhizomes collected from nearby natural sources. They were driven by the ever-increasing national demand for brooms made from *T. latifolia*, and had turned their full attention towards cultivation of broom grass as a cash crop in their *jhum* fallows.

For the cultivation of broom grass, healthy clumps are selected from natural stock. The rhizomes are uprooted and the green stems trimmed to about 12 cm to 15 cm in length. The rhizomes are planted during the onset of the rainy season, in holes about 30 cm to 50 cm in diameter and spaced at intervals of about two metres. Generally, fertilizer is not applied. In some cases rice, ginger and turmeric are intercropped with broom grass during the first year. Panicle initiation takes place from the month of October onwards. Grading of the product depends upon the softness and greenness of the panicles, and yields from the plants depend mainly on the quality of the planting materials and site conditions. Highest yields are obtained from 3- to 4-year-old plants. Harvesting of the panicles generally occurs during January and February. The harvested panicles are dried in the sun for a few days, while the freshly harvested plants are burnt to encourage the sprouting of new shoots. Over time, an area planted to broom grass increases and becomes a permanent plot giving regular annual cash returns.

Bhuyan et al (2007) made a comparison of investment and returns from growing rice and broom grass, and concluded that broom grass gave a cash return 8 to 10 times greater than rice (Table 31-1). Properly managed broom grass can give profitable yields for up to 10 years. However, such perennial cultivation demands long-term ownership of the land. In the land-holding system of Karbi Anglong district, *jhum*-cultivated land belongs to the community. There is no form of permanent ownership.

TABLE 31-1: Broom grass compared with rice productivity and returns.

Rice		Broom grass	
Cost of seed, man days in plot preparation, weeding, harvesting, etc.	Rs15,300	Cost of collecting rhizomes, man days in plot preparation, pit digging, weeding, harvesting, etc.	Rs2200 (US$41)/ha
Productivity of rice	2100 kg/ha	Productivity of broom grass	6900 kg/ha
Price for paddy	Rs6800 (US$125)/ton	Price for broom grass (wet)	Rs8000 (US$147)/ton

Study site 2: Nongpoh (Re Bhoi district) and Dawki (West Khasi Hills district), Meghalaya

Climate

The climate of this area is humid with a small seasonal water deficiency. However, there is great variation in rainfall across the sites, with some readings as high as 7196 mm per year.

Temperature

Summer temperatures rise up to 26°C and winter temperatures fall down to 9°C.

Soil

The area is an ancient pre–Cambrian plateau raised to a height of 600 m to 1800 m above sea level. The area is covered with limestone and has typical karst topography. The soil is loamy skeletal to fine and greatly drained, representing an erosion hazard.

The forest cover in this area is mostly tropical moist deciduous to evergreen. Shifting cultivation provides a diversified cropping system and is well adapted to the conditions in the moist hilly tracts of Meghalaya. People living in remote upland villages of East and West Khasi Hills, the Garo Hills and part of the Jaintia Hills are entirely dependent on shifting cultivation (Mipun and Saikia, 2003). According to the Indian Council for Agricultural Research (1983), the annual loss of topsoil in Meghalaya due to shifting cultivation was 40,900 kg/ha (organic carbon, 702.9 kg; P_2O_5, 0.15 kg; and K_2O, 7.5 kg).

The land belongs to the village council – commonly known as the *Dorbar Shnong* – and the council allots land to farmers for cultivation. Rice and maize are the major food crops. Besides these, some fruit crops such as oranges, pineapples, guavas, bananas, jackfruit, papaya and *jamun* are also

Areca catechu L. [Arecaceae]

Betel-nut palms are grown as a perennial crop – often along with broom grass – in parts of Meghalaya where shifting cultivation is intensifying to more settled agriculture. The palms produce intoxicating and slightly addictive seeds that are chewed along with a leaf of the vine *Piper betle*. The habit, pursued by millions of people, is a known cancer risk and is regarded as a neglected public-health emergency.

cultivated. Pumpkins, ash gourds, cucumbers and chillies are important vegetable crops cultivated in swiddens along with the major crop. Areca nuts, ginger, turmeric, betel leaf and black pepper are the main commercial crops grown in *jhum* fields. This mixed cropping system mimics nature in terms of diversity (Ministry of Home Affairs, 1995). However, shifting cultivation in this area offers very little in economic return and is, in fact, incapable of meeting the increasing livelihood demands of a growing population. Gradually, the local farmers have been changing their cropping patterns to a more 'settled' form of agriculture. Initially they cultivated annual cash crops such as turmeric, ginger, pineapples and so on, but have since changed to perennial crops like areca nut, broom grass, papaya, guavas and oranges, because they offer long-term benefits.

With the growth of population and reduction of the area available for *jhum* cultivation, the swidden cycle in Meghalaya has been reduced to between one and four years, and the length of the cycle tends to be site specific. As a result, the fertility of the soil and the productivity of *jhums* is in constant decline, making the pursuit of shifting cultivation increasingly hazardous. The extensive degradation of land is causing an imbalance in the socio–economic structure of village communities. Along with *jhum* cultivation, the ethnic communities of Meghalaya practice terrace or bun cultivation in valleys and foothills. This is a settled type of cultivation that has been adopted as an improved production system. However, a complete change or switch from shifting cultivation is neither possible nor advisable because of the physical geography of the land and the prominent role that swiddening plays in the cultural life of the people. Therefore, rather than changing *jhum* cultivation, efforts should be made to improve it.

Our survey of Nongkhrah village, in Re-Bhoi district, near Nongpoh, revealed that a large percentage of the 300 households in the village grew broom grass in their *jhum* fallows, having started the practice as long as 25 years ago. In the beginning, two or three families began planting broom grass in their fallows and, over the course of time, more and more families joined them. By the year 2000, half of the farmers in Nongkhrah village were growing broom grass, and recently, the figure reached 70% of all village households. These farmers practice both *jhum* cultivation and broom-grass plantation, on a rotational basis. They grow pineapples, bananas and guavas as major plantation crops in their *jhum* fields. Wet rice is grown only in some low lying areas of the foothills.

For the cultivation of broom grass, farmers first clear the forest land and burn the field. Rhizomes are collected, either from natural stock or using a barter system. The field is prepared by digging 12 cm- to 15 cm-wide pits in which to plant the rhizomes, with space of 60 cm to 90 cm between them. Farmers intercrop yams, turmeric and ginger with the broom grass in the first year. Sometimes bananas are also grown with broom grass, but as the fertility of the land decreases in subsequent years of cropping, it becomes unsuitable for cultivation of agricultural crops. As a result, the plot gradually becomes a permanent broom-grass plantation. The grass thrives in nutrient- and moisture-deficient soils and needs very little maintenance.

It gives maximum productivity for up to five years. After that the yields gradually decrease and after 10 years, the plot is left fallow for rejuvenation.

During our survey, two types of broom grass – red and white – were observed in the area. The red type was found to be more durable because of its stiffness. Farmers sell the broomsticks to vendors for Rs35 (US$0.64) per kg (dry weight) from the field. It was reported that cultivation of broom grass could generate a benefit of Rs110,500 for farmers within four years of cultivation (Table 31-2). Hence, over recent decades, there has been increasing interest from *jhum* farmers in planting broom grass in their fallows.

Extensive surveys were conducted in Sharngan village near Pynurslea, and at Ponshutia, Panlum and Madanhakro villages in the Dawki area of Meghalaya. In these areas, broom grass grows naturally in five-year-old fallows, along with *Dendrocalamus* species. *jhum* cultivators have long been in the habit of collecting the grass from wild stocks for their day-to-day needs. However, the broom grass that grows spontaneously has proven insufficient to meet gradually increasing market demand as well as domestic needs. Therefore, the farmers have begun to cultivate and manage broom grass in their fields.

Despite the demand, cultivation of broom grass is not a community pursuit in these areas, but a practice followed by individual families. Farmers collect planting stock (rhizomes) from nearby natural forest and plant it in a mixed-cropping system with yams, turmeric and ginger. Perennial crops like *Cinnamomum tamala* (bay leaf) and *Areca catechu* (areca nut or betel nut) are preferred as intercrops with broom grass by many farmers. The *T. latifolia* gives maximum yields for up to five years and thereafter, productivity dwindles. A field with 10-year-old broom grass is totally unproductive and the entire crop needs to be uprooted. Harvested broomsticks are dried and stacked in bundles weighing 50 to 70 kg. A recent market price in a nearby city was Rs35 to Rs40 for 1 kg of dry broom grass. Initially, maximum profits went to the city vendors, but farmers have now learned the realities of the market and have been astute in securing for themselves the benefits of broom grass as a commercial crop.

TABLE 31-2: Broom grass: productivity and income.

Year	Productivity (quintals/acre)	Total return (@Rs35/kg dry weight)
1	Negligible	
2	6	21,000
3	10	35,000
4	16	56,000
5	8	28,000
Total return		140,000
Net profit (Rs140,000 minus production cost of Rs29,500)		110,500

Notes: In September 2012, Rs1000 equalled US$18.47; 1 quintal = 100 g; 1 acre = 0.405 ha.

Fallow management involving *T. latifolia* and pigeon pea (*Cajanus cajan*)

A field experiment was conducted at Silonijan, in Assam's Karbi Anglong district, during the years 2007 through 2010, to assess the growth and productivity of broom grass when intercropped with nitrogen-fixing pigeon pea (*Cajanus cajan*). Broom-grass-growing areas of Assam were surveyed and rhizomes were collected on the basis of the number of culms per tussock, the height of the tussock and inflorescence length of the parent plants. Thus gathered, the rhizomes were planted in plastic bags for multiplication.

One-hectare areas of degraded fallow land were cleared and burned during February and March. Pits with a capacity of 30 cubic centimetres were then dug and left for one month for weathering. At the onset of the monsoon in April and May, the multiplied broom-grass seedlings were planted in three spacing regimes: 1 m, 2 m and 2.5 m. *Cajanas cajan* (Pigeon pea) was grown as a nitrogen-fixing species along with the broom grass, following standard statistical design.

The experimental plot was divided into two equal halves, with a path four metres wide. In one part, *T. latifolia* was planted as a monoculture and in the other part, it was intercropped with *C. cajan*. The experiment was laid out in randomized block design of 10 m x 10 m plots with spacing of 1 m. Rhizomes were planted in three spacing trials (1 m, 2 m and 2.5 m) with five replications. The treatment combinations were as follows: T1 = *T. latifolia* + *C. cajan* in 1 m spacing; T2 = *T. latifolia* + *C. cajan* in 2 m spacing; T3 = *T. latifolia* + *C. cajan* in 2.5 m spacing; T4 = *T. latifolia* in 1 m spacing; T5 = *T. latifolia* in 2 m spacing; and T6 = *T. latifolia* in 2.5 m spacing.

Growth data, i.e., plant height, culm diameter and number of culms per tussock, were recorded every three months. At harvesting stage, the number of panicles per tussock, the length of the panicles and the total yield of panicles were recorded every year (Tables 31–3 and 31–4). Input, output and total income were calculated on the basis of harvest data. The plots were burnt to stimulate the initiation of new shoots in pre-monsoon showers after the first year's harvest.

Cajanus cajan (L.) Millsp. [Leguminosae]

The nitrogen-fixing abilities of pigeon pea may obscure the fact that it is a major source of protein for people throughout India, Africa and Central America. As well as producing dried peas, flour, or green beans as human food, it is also a fodder and cover crop. It can be grown on marginal land and is very drought resistant.

TABLE 31-3: Progressive height (cm) and emergence of culms/tussock of *T. latifolia* under different treatments.

Treatment	3		6		9		12		15		18		21	
	A	B	A	B	A	B	A	B	A	B	A	B	A	B
T$_1$	28.4	4	52.4	13	108.5	24	43.5	10	148.0	34	176.8	54	185.7	54
T$_2$	28.2	4	58.0	22	149.2	28	58.5	12	159.6	36	182.6	63	196.5	63
T$_3$	29.0	5	63.6	24	164.8	32	61.8	13	176.4	45	210.7	66	225.4	69
T$_4$	27.0	3	48.6	9	92.5	18	46.7	4	72.8	5	92.6	21	97.5	21
T$_5$	25.4	5	53.5	18	115.2	23	32.6	4	84.9	6	96.4	36	104.2	36
T$_6$	26.6	4	59.7	21	126.4	28	49.8	4	92.6	6	105.6	54	121.4	57
SE±	0.03	0.02	0.15	0.16	0.74	0.13	0.29	0.11	1.22	0.5	1.43	0.47	1.41	0.5
CD (5%)	0.08	0.046	0.06	0.35	1.65	0.3	0.65	0.26	2.79	1.13	3.2	1.06	3.16	1.12

Note: T1 = *T. latifolia* + *C. cajan* in 1 m spacing; T2 = *T. latifolia* + *C. cajan* in 2 m spacing; T3 = *T. latifolia* + *C. cajan* in 2.5 m spacing; T4 = *T. latifolia* in 1 m spacing; T5 = *T. latifolia* in 2 m spacing; T6 = *T. latifolia* in 2.5 m spacing; A = progressive height; B = emergence of culms/tussock; SE (standard error) is the standard deviation of the sampling distribution; CD (critical difference) is the smallest difference between sequential laboratory results.

TABLE 31-4: Productivity and total return of *T. latifolia* under different treatments.

Treatment	Total yield of panicles (kg/10 sq. m dry weight)		Total return/ha (@Rs25 per kg dry weight)	
	1st year	2nd year	1st year	2nd year
T$_1$	7.00	35.40	1750	8850.00
T$_2$	12.96	45.58	3240	11395.00
T$_3$	9.80	43.25	2450	10812.50
T$_4$	7.00	30.40	1750	7615.50
T$_5$	11.85	36.40	2963	9110.00
T$_6$	9.50	35.04	2375	8762.50
SE±	0.067	0.157	17	39.18
CD (5%)	0.157	0.366	39.50	91.20

Notes: T1 = *T. latifolia* + *C. cajan* in 1 m spacing; T2 = *T. latifolia* + *C. cajan* in 2 m spacing; T3 = *T. latifolia* + *C. cajan* in 2.5 m spacing; T4 = *T. latifolia* in 1 m spacing; T5 = *T. latifolia* in 2 m spacing; T6 = *T. latifolia* in 2.5 m spacing; Rs1000 equals about US$18.78; SE (standard error) is the standard deviation of the sampling distribution; CD (critical difference) is the smallest difference between sequential laboratory results.

C. cajan plays a significant role in improving organic-matter content in soil. Its profuse growth of leafy foliage and heavy litterfall during the rainy season decomposes to contribute significantly to the carbon content of the soil during the second year of cultivation. As *C. cajan* is a nitrogen-fixing plant, its growth along with broom grass has a remarkable influence on the soil nitrogen content. The highest increment of total nitrogen in the sprouting and growing stage in successive years may be due to the combined contribution from root nodules and decomposing biomass. Favourable soil conditions following the growth of a nitrogen-fixing plant in degraded soil also helps in the mineralization of soil nitrogen, leading to a build-up of available nitrogen.

Harvesting of *T. latifolia* was performed just before maturity, with care to avoid damaging newly sprouted shoots that appeared during the winter months up to mid-February. The panicles were removed from the culms and dried in the field. It was observed during the harvesting stage of both years in the study that plant growth was significantly less in the 1 m spacing regime than in the other two spacings. This could mean that 1m between tussocks was insufficient for optimum vegetative growth. Plants with space of 2.5 m recorded comparatively more panicles, while panicle length was more or less the same at 2 m and 2.5 m spacing. This indicates that 2.5 m spacing may be appropriate for optimum growth of *T. latifolia*.

Productivity of *T. latifolia* depends mainly on site conditions, the quality of planting materials and maintenance of plantations. The yield in the second year was high due to the increment of productive culms per tussock. Maximum yields were observed at 2 m spacing in a plot shared with *C. cajan*. This spacing allowed an increase in rhizome and root biomass, resulting in the development of a higher number of sprouting culms.

T. latifolia can be a profitable enterprise when cultivated in degraded *jhum* fallows. This experiment demonstrated that cultivation of broom grass along with pigeon pea (*Cajanus cajan*), as a nitrogen-fixing plant, can bring immediate benefit to local inhabitants in the shortest possible time, while improving the habitat and uplifting the local economy (Figure 31-3).

In order to gather greater insights into fallow management using broom grass cultivation in Karbi Anglong, Assam and Meghalaya, a SWOT (strengths, weaknesses, opportunities and threats) analysis was used to establish the potential benefits from the system. First, a participatory rural appraisal (PRA) analysis was organized so that facts and experiences could be shared by farmers and the research team, based on the four aspects of SWOT. After preparing the final list, a set of six combinations was made, within the four aspects. The principal eigenvectors of the two positive aspects (strengths and opportunities) were compared with the two negative aspects (weaknesses and threats) to identify the level of sustainability (Table 31-5). The lists were as follows:

FIGURE 31-3: Bundles of dry broom grass being prepared for market.

Photo: P. K. Kaushik.

Strengths

- Earns more cash income,
- Less labour intensity,
- Minimum effort for maintenance,
- Grows in harsh conditions,
- Fertiliser not required,
- Good soil binder,
- Increases the seasonal flexibility of the rural livelihood,
- Multiple uses with great demand.

Weaknesses

- Reduces food security,
- Not suitable for intercropping after the first year.

Opportunities

- Alleviation of poverty,
- Helps permanent settlement,
- Easily adjustable to modern changes,
- Market opportunities already exist,
- Provides feed and fodder for domestic animals.

Threats

- Reduces biodiversity,
- Reduces soil moisture content,
- Soil becomes unproductive for a long time after abandonment.

TABLE 31-5: SWOT scores on broom grass cultivation.

Scale	S-W	S-O	S-T	W-O	W-T	O-T
Mean	8.00	5.50	8.00	0.03	4.00	6.50
Reciprocal	0.13	0.18	0.13	36.00	0.25	0.15

Adaptability vector (Influence out of 100)	
Strengths	35.96
Weaknesses	5.87
Opportunities	32.09
Threats	26.08
Sum	100.00

λmax = 23.29708

Consistency Index = 6.432359

Consistency Ratio = 7.1471

The score of positive aspects was higher in terms of sustainability of the crop, making it a suitable fallow management option to lift the economic status of farmers. The system, which is a recent innovation for ethnic farmers in mountainous northeastern India, has been successfully applied in adverse environmental conditions and offers good prospects from both ecological and economic points of view.

Acknowledgements

The authors are grateful to the Director of the Rain Forest Research Institute, Jorhat, for providing facilities and constant encouragement during the study. The authors also gratefully acknowledge the help of Mr K. C. Momin, the retired Director of the Department of Soil Conservation in Meghalaya, who also provided valuable information. It would not have been possible to carry out our field activities without the participatory approach of the villagers of Nilip and Rongmongwe blocks, Karbi Anglong, Assam and the Dawki and Nonghpoh areas of Meghalaya.

References

ASTE Council (1996) *Report*, Assam Science, Technology and Environmental Council, Department of Science and Technology, Government of Assam, India

Ahlawat, S. P. and Singh, U.V. (1998) 'Cultivation of broom grass (*Thysanolaena maxima*) on degraded and *jhum* lands in North East India', in R. C. Sundriyal, U. Shankar and T. C. Upreti (eds) *Perspective for Planning and Development in NE India*, U. P. Himavikas Occasional Publication No. 11, Geographically-Based Programme, India, Kosi-Katarmal, Almora, India, pp.239-245

Bhuyan, P., Pator, C., Pfoze, L. and Jimo, V. P. (2007) 'Study on shifting cultivation and food security', in *Project Village of Karbi Anglong District, Assam*, Karbi Anglong Community Resource Management Society, North Eastern Region Community Resource Management Project (NERCORMP) and International Fund for Agricultural Development (IFAD)

Bisht, N. S. and Ahlawat, S. P. (1998) *Broom Grass*, State Forest Research Institute, Department of Environment and Forests, Government of Arunachal Pradesh, Itanagar, India

Craswell, E. T., Sajjapongse, A., Howlett, D. J. B. and Dowling, A. J. (1997) 'Agroforestry in the management of sloping lands in Asia and the Pacific', *Agroforestry Systems* 38, pp.121-137

Department of Environment and Forest, Meghalaya (2005) *State of the Environment Report*, Government of Meghalaya, Shillong

Forest Survey of India (1997) *State of Forest Report*, Ministry of Environment and Forests, Government of India, Dehradun, India

Indian Council of Agricultural Research (1983) *Shifting Cultivation in Northeastern India*, ICAR Publication Unit, New Delhi

Ministry of Environment and Forest (2009) *State of Environment Report, India*, Ministry of Environment and Forests, New Delhi

Ministry of Environment and Forest (2011) *State of Environment Report, India*, Ministry of Environment and Forests, New Delhi

Ministry of Home Affairs (1995) *Basic Statistics of North Eastern Region*, North Eastern Council Secretariat, Ministry of Home Affairs, Government of India, Shillong

Mipun, B. S. and Saikia, Anup. (2003) 'Landuse/landcover detection mapping and identification of shifting cultivation areas of Tirap district', in S. Singh, D. K. Nayak and B. S. Mipun (eds) *Environment, Locational Decisions and Regional Planning, The Geographical* , The Geographical Society of North-Eastern Hill Region, North Eastern Hills University, Shillong, pp.167-178

Sarma, N. N., Paul, S. R. and Sarma, D (1996) 'Rainfall patterns and rainfall based cropping system for the hill zone of Assam', *Annals of Agricultural Research* 3, pp.223-229

Sharma, T. C. (1976) 'The Pre-historic background of shifting cultivation', in *Proceedings of a Seminar on Shifting Cultivation in North-East India*, North East Indian Council for Social Science Research, Shillong, Meghalaya, pp.1–4

Task Force on Shifting Cultivation (1983) *Report*, Ministry of Agriculture, Government of India, New Delhi

Thomaz, E. L. (2009) 'The influence of traditional steep land agricultural practices on runoff and soil loss', *Agriculture, Ecosystems & Environment* 130, pp.23–30

Tiwari, B. K. (2001) 'Ecological and socio-economic impacts of modified shifting cultivation in Northeast India', paper delivered to a workshop on Shifting Cultivation: Towards Sustainability and Resource Conservation in Asia, August 2000, International Institute for Rural Reconstruction, Silang, Cavite, Philippines

32

DRAWING ON PAST PRACTICES TO SECURE THE FUTURE

Innovative applications of traditional lac rearing in the fallows of Karbi farmers in Assam, northeast India

*Sanat K. Chakraborty and Dhrupad Choudhury**

Introduction

In recent decades, state-sponsored programmes have combined with pervasive market forces to initiate a process of change that is sweeping across the uplands of northeast India, fuelling expectations among poorer, marginalized communities of their assimilation into modern mainstream development. These aspirations are fanned by the lure of consumerism, delivered by satellite television into the homes of the better-off in areas that are still difficult to access, and they trickle down to stir the dreams and hopes of people in the lower strata of rural life. These are shifting cultivators, belonging to the diverse ethnic groups that have long inhabited the uplands of the region.

The promised transformations have fallen far short of expectations. State-sponsored schemes aimed at replacing shifting cultivation by introducing alternatives have selectively benefited some, while the vast majority remain without opportunities or benefits. The main problems are the inability of poorer people to access either resources or the relevant programmes, and the failure of the programmes to ensure equitable access and a minimum degree of inclusion (Leduc and Choudhury, 2012). However, while poor farmers have been disheartened by their failure to benefit from state programmes, it has not eroded their determination.

Having witnessed the changed fortunes of those fortunate enough to have taken part in the government schemes – in particular their enhanced purchasing power – the resolve of many of those left out has driven them to independently develop options with the potential to raise their incomes and improve their livelihoods. These include the innovative application of traditional approaches to present-day challenges

* SANAT K. CHAKRABORTY is an independent journalist and researcher and DR. DHRUPAD CHOUDHURY is Chief of Scaling Operations at the International Centre for Integrated Mountain Development (ICIMOD), Kathmandu, Nepal.

and the adaptation of age-old practices to take advantage of market opportunities. These innovations provide them with opportunities to adapt the traditional practice of shifting cultivation.

While such innovations are not exactly rare in the uplands of northeast India (or elsewhere – see Cairns, 2007), a particularly appealing example is that of lac culture, as practised by Karbi shifting cultivators in India's northeastern state of Assam. Lac culture involves the management and 'farming' of microscopic insects. They live gregariously in colonies, feeding on the succulent parts of host plants by sucking the plant juices. The insects release a resinous material from their bodies that accumulates around the colony. It hardens on contact with air, thus providing a protective covering for the colony, excluding parasites and predators. The encrusted lac resin on the twigs of host plants is called stick lac or raw lac, and is harvested for both commercial and domestic uses. As the lac resin is natural, biodegradable and non-toxic, it finds applications in food, dyes, textiles and pharmaceutical industries and is used by goldsmiths in making jewellery. It is also used as a surface coating in electrical and other fields. Before the advent of magnetic tapes and digital technology, lac was a principal raw material for making gramophone records and hence, was essential for the music industry.

This chapter describes two case studies based on a survey of five villages in the Hamren subdivision of Assam's Karbi Anglong district.[1] It provides an account of initiatives taken by Karbi farmers to continue a tradition practised by their forefathers, notwithstanding various challenges, including declining markets, a rapidly depleting resource base and deteriorating production systems.

The case studies highlight the determination of farmers to find innovative ways to revive and continue a tradition that seems, elsewhere, to be dying. In the first instance, farmers have drawn on fundamental principles of resource management inherent in shifting cultivation to manage their lac colonies sustainably while supplementing their incomes with complementary activities. In the second, a farmer revives the use of a leguminous pulse as a host for his lac colonies, providing a profitable bridge for the transition from cultivation to early fallow regeneration in his severely shortened shifting cultivation system. In doing so, he adopts an innovative approach to diversifying and enhancing his income opportunities while ensuring the restoration of soil fertility in his severely distorted practice of shifting cultivation. The latter provides an unusual approach to fallow enrichment.

The poignancy of the case studies lies in the fact that they involve the innovations of individual farmers striving to secure their livelihoods and translate their families' aspirations into reality in an environment that is rapidly changing. State-sponsored support eludes them. Ironically, policy-makers promote options that have succeeded elsewhere, but which are inappropriate, inadequate and often ineffective in the context of shifting cultivation in northeast India.

The people and the land: The Karbis (or the erstwhile Mikirs)

The Karbis, believed to be of Tibeto-Burman origin, are said to have migrated from the Chindwin river valley of western Myanmar. Early ethnographical accounts, particularly from the time of British rule in India, refer to the Karbis as the Mikirs (Hunter, 1879). They suggest that the Karbis (or Mikirs) were relatively widespread across the northeastern region, as in the following passage:

> The Mikirs are undoubtedly the most peaceful and industrious of the hill tribes....The tract of country inhabited by them is stated to be about sixty miles in length from east to west, and thirty-five or forty miles in a straight line from north to south. Tradition states that the tribe was originally located in that tract of country lying between the Dhaneswari, the Jamuna, and the Kapili rivers, known as "Tularam Senapati's country"; but that being invaded and conquered by a Cachari prince, they were driven by his oppressions to take refuge in the Jaintia Hills. ......They afterwards emigrated from Jaintia, some of the tribe going north-east towards Kamrup, and others northwards into Nowgong and the Naga Hills, where they appear now to have finally settled down. The Mikirs generally inhabit the interior portion of the hills, but most of their villages are within a day's journey of the plains.... (Hunter, 1879).

Nowadays, the Karbis primarily inhabit the district of Karbi Anglong, a 'hill district' of southern Assam in much the same area as that described by Hunter. The district is bounded by the Ri-bhoi, East Khasi Hills and Jaintia hills districts of Meghalaya in the west, the Dima Hasao district of Assam to the south, Dimapur and Kohima districts of Nagaland in the east and Morigaon, Nagaon (erstwhile Nowgong) and Golaghat districts of Assam in the north (Figure 32-1). The district is dominated by rolling hills and is dissected by several rivers, the most important being the Kopili, a northward-flowing tributary of the Brahmaputra.

Traditionally, the Karbis lived in and around forests, clearing patches for their shifting cultivation (*jhum*), in which they grew rice and cotton along with numerous minor crops. They moved to a fresh location after every two or three years, as Hunter (1879) wrote:

> The people live in solitary huts or in small hamlets, with from five to thirty individuals in each house......many families herd together in the same house, in order to avoid paying the house tax [tax was levied on a household basis]. The villages are generally situated in the midst of dense jungles, their sites being changed every two or three years, when fresh land is taken up for cultivation.... (Hunter, 1879).

The Karbis drew on forest resources for both sustenance and trade. Many of them subsisted by collecting and trading jungle products, primarily consisting of beeswax, several dyes, a variety of cinnamon and several kinds of fibre (Hunter, 1879). In

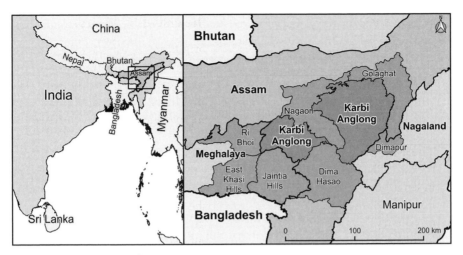

FIGURE 32-1: The study location.

Leea indica (Burm. f.) Merr.
[Vitaceae]

This shrub or small tree, called *sopleple* by the Karbis, is perhaps the best host plant for lac insects. However, it falls victim to grazing cattle (below). It is commonly gathered from the wild for local use as a food and medicine. The tender young shoots are eaten as a vegetable and are chewed to relieve a severe cough. The roots are believed to combat fever and induce perspiration.

contemporary times, the Karbis continue to practise shifting cultivation, although they no longer shift their settlements as they did in the past. Rice remains their staple food, supplemented by several varieties of millet, maize, tubers and vegetables grown in *jhum*. The Karbis maintain strong bonds with nature and natural events. The intensity of this linkage is ingrained in their lives and activities and is reflected prominently in their folklore.

Karbi ethno-botanist and folklorist, Robindra Teron, who has been documenting the ancient knowledge systems of his people, says that Karbi farmers are keen observers of nature and have woven such observations into an intricate calendar of major activities. Their annual calendar is based on observations of plant, animal and bird behaviour in their immediate surroundings. Thus, falling leaves, emergence of new leaves, flowering and fruiting and animal behaviour, including courtship, mating, egg laying and hatching, all enable farmers to understand the underlying changes in weather conditions. Biological indicators determine when farming plots are selected, forest is cleared, vegetation is slashed, dried and burnt, and seeds are sown.

Karbi farmers explain: 'When the *vosobiko* (a certain bird) sings, it signals the arrival of the rains'. This indicates that the time is ripe to burn the slashed debris in cleared *jhum* fields and, immediately afterwards, to sow crops before the arrival of the rain. Failure to heed nature's cue may result in the slashed debris being soaked, making burning difficult and ineffective. This would condemn the household to food insecurity for a year and the need to incur debt to keep meagre supplies of food on the table.

The Karbis also describe the lunar calendar months by making elaborate references to natural events. Thus, the description for the month of June, called *Vosik,* includes a variety of biological indicators: '..as *henup kardik* (or the search for young bamboo shoots) suggests their first appearance; *laha aso,* (lac insects) signifies both the emergence of lac insects and the sprouting of young leaves of *sopleple* (*Leea indica)*, an important host plant for the lac insects. The wild fowl, *vorek* and *vohar,* begin to hatch and the chicks feed on *laha aso.* The appearance of *laha aso* is also the indicator for the hatching of the jungle fowl. All these are characteristics of the month of June (Teron, 2012).

Although this makes it clear that the Karbis were keen observers of nature and well acquainted with the life cycle of lac insects, history does not relate whether they actively reared lac insects in the past. Early ethnographic accounts do not suggest anything to the contrary.

The history of the lac trade in northeast India

The colonial period

Accounts of trade and commerce in Assam during the British period suggest that lac was an important item of trade (Hunter, 1879; Khadria, 1990; Handique, 2006). Handique, citing Hamilton, asserts that Assam used to export a considerable quantity of lac that was of 'highly superior quality'. Hamilton's accounts, as cited by Handique, put the exports of stick lac at 10,000 maunds (1 maund equals 40 kgs), valued at Rs35,000, in the first decade of the 19th century. Hunter (1879), providing details of produce from British-administered districts, mentions lac collection in Goalpara, Garo Hills, Khasi and Jaintia Hills and Sylhet. Lac was one of the principal 'jungle products' from Goalpara (although the district's total volume of 'jungle products' was considered 'insignificant'). However, the bulk of it was said to come from the Garo, Khasi and Jaintia Hills. While the product was collected by Garos in considerable quantities and sold by them at frontier market villages, in the Jaintia Hills it was reported to have been gathered by the Mikirs – 'but not to any great extent'.

In his section on commerce and trade from Khasi and Jaintia Hills, Hunter provides a table of the estimated exports from the district in the year from 1876 to 1877. This table suggests that the trade in stick lac for that year was significant, estimated at 5000 maunds with a value of Rs100,000, being priced at Rs20 per maund. The trade in stick lac made the product fourth in importance, after potatoes, limestone and cotton (Hunter, 1879). However, Hunter's accounts do not suggest or even indicate that active culturing of lac insects was taking place among any of the ethnic communities. His only mention of cultivation refers to Sylhet, where lac was cultured and maintained on banyan trees (*Ficus cordifolia*). Broods were reportedly initiated in March and April and the produce collected the following October and November. Hunter also reported that lac could be produced on the same tree only once in every three years.

Further insights into the trade from Assam in that period were provided by Khadria (1990), who listed Goalpara, Kamrup and Darrang as the chief lac producing and exporting areas in 1870 (This offers a possible explanation of why lac collections by the Karbis in Jaintia Hills was regarded as 'insignificant'. Most of their collection was probably traded through Kamrup, rather than Jaintia Hills). Encouraged by growing exports and increased earnings, the British established plantations of *Ficus* in the foothills of the Garo Hills in 1885 to promote and expand the trade. While the initial results were promising, the widespread incidence of malaria and *kala-azar* (leishmaniasis) brought the venture to an end. The trade in Darrang also diminished as a consequence of widespread epidemic diseases that plagued the lac colonies.

Khadria (1990) noted that although stick lac topped the list of exports from Assam during the late 19th century, processing of lac took place in the districts of Kamrup and Sibsagar only, and even then, on an insignificant scale. By the turn of the century, this had become limited to the district of Sibsagar alone. In Khadria's words: 'Apart from natural calamities contributing to the failure of lac culture in the valley [Brahmaputra], there is no other reference to attempts at developing a home industry from this major item of export'. It would seem paradoxical that a trade which provided such promise and generated considerable revenue should slowly wither away instead of flourishing. However, a coherent understanding of historical events in northeast India is often constrained by the paucity of literature. While there are no direct accounts describing the trajectory of lac culture or trade during the period up to independence, the authors offer some insights on factors that possibly contributed to the gradual decline of the lac trade in colonial Assam.

The emergence of tea and decline of lac

Central to the gradual decline of lac trade seems to be the discovery of tea (*Camellia sinensis*) in 1823 in parts of upper Assam under the administration of Ahom kings, and its later domestication for tea production by the British. After successful experimental tea production around 1837 and realization that tea offered tremendous economic opportunities for the imperial economy, all of upper Assam was immediately brought under direct British control. These areas were opened up to Europeans for tea plantations and planters were encouraged to expand their gardens.

To aid the process, the Wasteland Rules were introduced in 1838, allowing the leasing of 'wastelands' to planters at nominal fees, thereby encouraging the procurement of land for establishing and expanding tea gardens. Fuelling the process further, leasing conditions were liberalized by Lord Canning in 1861, to further encourage the process. Forest areas were rapidly converted to agricultural use, in particular to tea gardens. With increased revenue coming from extraction of forest resources and minerals and the trade in tea, Assam, hitherto part of the Bengal principality, was elevated to the status of a fully-fledged province in 1874 under the control of a Chief Commissioner based in Shillong. A provincial forest department was established which facilitated the generation of revenue through extraction of forest resources and the conversion of primary forests into plantations. The process was further boosted by the introduction of the Assam Land Revenue Settlement of 1886 and the Assam Forest Regulation of 1891. Both helped to hasten transformation of forest land to agriculture (Saikia, 2005).

The discovery of coal and gasfields in upper Assam by 1880 provided a source of energy for fuelling imperial enterprises. Railways were soon established to transport the coal to tea factories, and construction of a rail network required railway sleepers, leading to massive extraction of timber from Assam's forests. Estimates made by Guha (2006) suggest that 860 sleepers were required for every mile of railway lines laid; with a network covering 400 miles, that translated to 344,000 sleepers. The resulting destruction of forests was substantial.

With a railway network connecting to river ports and the development of steamer services to transport goods down the Brahmaputra river to markets as far away as Calcutta, the tea industry flourished. In his studies of Assam, Guha (2006) estimated that about 283,280 hectares of land were occupied by tea planters – about one seventh of the entire settled area of the Assam plains. Along with the growth of the tea industry grew demand for tea chests in which to package and transport the product over long distances. Saw mills were established and the exploitation of Assam's forests intensified. Much of the timber came from forests under British control, one of which was the Nambor Reserve in the Dhaleswari, Jamuna and Dayang valley (Saikia, 2008). Combined with extensive conversion for agricultural use, forests were rapidly depleted. Tucker (1988) estimated that 607,072 hectares of forest land was converted to agriculture in just two decades, between 1930 and 1950. A further impetus resulted from Britain's 'Grow More Food' campaign during the 1940s.

Therefore, as more lucrative resources such as timber, minerals and tea started replacing lac as major export products and revenue earners, their growth and development accelerated the process of deforestation and the destruction of pristine habitats. In the process, most of the habitats, hosts and resident colonies of lac insects were wiped out forever, and lac rearing became confined to a few pockets in Assam. These were mainly areas where tea had not penetrated, or where the easy extraction of timber was constrained by lack of navigable waterways. These were also areas where lac had a domestic value and was used by the local people. Such areas, not surprisingly, were confined mostly to the hills. Often, they were 'excluded areas', outside direct British administration, so they escaped the onslaught of large-scale extraction and ended up playing an unexpected role in the conservation and continuity of the lac trade.

The pre-independence period

Experts familiar with lac production in the region,[2] along with our review of literature, have led us to believe that the aforementioned areas with difficult access, where tea plantations were not established, later became the centres for lac production and trade in northeast India in the early decades of the 20th century and the pre-Independence period. Erstwhile Assam (then including what are now the northeastern states of Meghalaya, Mizoram and Nagaland) was an important centre for lac production. The state produced between 1100 and 1850 tons of raw lac every year until the 1930s, and Assam remained one of the country's major lac-growing states up to the late 1950s.

Within Assam, Karbi Hills was one of the principal centres of lac production until the 1930s. But over subsequent decades, production declined drastically. This was due mainly to a steady decline in demand and fluctuating prices for raw lac, leading to a loss of markets and consequently, to erosion of farmers' interest in lac cultivation. Once abandoned, restarting lac cultivation and expanding it to desired levels was a very difficult and time-consuming process (Sharma, 2012). Add to these factors indiscriminate logging in the Karbi Hills and the government's policy for the eradication of shifting cultivation. Despite these odds, lac cultivation, practised primarily by the Karbis, continues in a few pockets of Assam's Karbi Anglong district and in neighbouring Meghalaya, in particular, along the entire Umswai valley in Hamren subdivision, and in neighbouring Ri-bhoi district.

Lac production in contemporary India

Until recently, India was the largest lac-producing country and exporter of processed lac in the world, accounting for about 50% to 60% of global production. Annual production of raw lac, as reported by Sharma et al. (2006), was estimated at 20,000 tonnes, with foreign-exchange earnings of about US$28 million to $30 million. The major lac-producing states are Jharkhand (57% of the country's production), Chhattisgarh (23%) and West Bengal (12%), while Orissa, Gujarat, Maharashtra, Uttar Pradesh, Andhra Pradesh and Assam are minor producers. More than three million ethnic households in these states are engaged in lac cultivation and the lac insects and their host plants play an important role in their livelihoods. For example, about 20% to 38% of the total agricultural income of tribal growers in Jharkhand comes from lac. Nearly one million person-days of work per annum are generated by industries engaged in post-harvest processing of lac (Sharma et al., 2006).

According to the Indian Lac Research Institute (ILRI) - now renamed the Indian Institute of Natural Resins and Gums - about 100 to 150 tons of lac are still produced annually by indigenous farmers in Assam (Karbi Anglong district) and neighbouring

Meghalaya (Ri-Bhoi district and pockets of the Garo Hills district) (Sharma, 2012). They therefore contribute significantly to keeping the region on India's map of lac-producing areas.

Biological relationships underlying lac production

Lac insects are gregarious sap-sucking insects, living in colonies protected by a resinous secretion exuded by the females (Figure 32-2). They are closely related to scale insects and mealy bugs. Of the nine genera and 87 species of lac insects reported globally, only two genera and 19 species are found in India. Of these, the only genus that is commercially important is *Kerria*, while species belonging to the genus *Paratachardina* are often regarded as pests of important horticultural and commercial crops such as lychee (*Litchi chinensis*), mango (*Mangifera indica*), *ber* or jujube (*Ziziphus jujuba*) and sandalwood (*Santalum album*). *Paratachardina* species have also been studied for their potential role as biocontrol agents for managing weeds (Sharma et al., 2006). *Kerria lacca* is the principal species exploited for commercial production of lac in the major lac-growing states of India. However, in the northeast of the country, where this study was based, the commercially important species are *Kerria chinensis* and *Kerria rangoonensis* (Sharma et al., 2006; Sharma, 2012). Both species are also common in the southern Chinese province of Yunnan as well as in Thailand, Myanmar and Vietnam.

Lac insects are sessile herbivores, surviving and multiplying by sucking the phloem sap of their host plants. The soft-bodied insects produce a resinous secretion that protects them from predators and adverse environmental conditions. The vital and valuable ingredients in the secreted resin are the dye *erythrolaccin* and the pigment *laccaic acid*, which is present in the insects' haemolymph (equivalent to blood). The commercial viability (and value) of the lac industry is intrinsically influenced by the relationship between the insects and their host plant. Various biological attributes of the lac insect - such as fecundity, sex ratio, body colour and yield of resin – are significantly influenced by variations in the host plant. The characteristics of the host plant also affect both the quality and quantity of lac-insect dye. Similarly, inter- and intra-specific variations of the insect species also determine the quality and quantity of dye produced. These influences can result in the insects' body colour (and consequently that of the dye) ranging from crimson to yellow, albino and cream. The

FIGURE 32-2: Lac colonies on the branch of a host tree.

quality and quantity of lac dye present in the resin is dependent in equal measure upon both the lac insect and the host plant on which it is reared. This attribute is of crucial importance to the lac industry (Sharma et al., 2006).

The lac insect's ecosystem is a complex multitrophic web of flora and fauna. Besides lac insects and their host plants, it includes several predators as well as parasites of lac insects, microbes and a variety of pests of the host plants. Ecologically, the lac host plants constitute the first trophic level in the food chain, while pests of the host plants and the lac insects constitute the second level; lac predators, along with their primary parasites and parasitoids, make up the third and fourth trophic levels. Lac host plants can be divided into various categories, according to the preference of the lac insects and the abundance and quality of lac obtained (Sharma, 2012). More than 400 plant species have been reported as hosts to various lac insects around the world. While many of the insects are highly host specific, others are known to have a wider tolerance and can thrive on a broad range of host plants.

Sharma et al. (2006) provide a list of host plants exploited for lac cultivation in the prominent lac growing areas of India. Significantly, except for *Ficus* and *Zizyphus* spp., most of the lac host plants recorded in the northeast region are very different from those reported elsewhere in India. Thus, just as the lac insects in the northeast are different from their counterparts elsewhere in India, so too are their hosts (Table 32-1).

The Lac farmers of Karbi Anglong

While Hunter's account suggests that Assam's lac trade during the British period was based solely on the collection of stick lac from natural forests (Hunter, 1879),

TABLE 32-1: Lac host plants of northeast India.

	Host species	Family
	Major hosts	
1	*Cajanus cajan (indicus)*	Leguminosae
2	*Ficus semicordata*	Moraceae
3	*Grewia multiflora, G. serrulata*	Malvaceae
4	*Leea macrophylla, L. asiatica, L. indica*	Vitaceae
	Minor hosts	
5	*Albizia lucidior*	Leguminosae
6	*Engelhardia spicata*	Juglandaceae
7	*Ficus benghalensis*	Moraceae
8	*Ficus benjamina*	Moraceae
9	*Ficus infectoria*	Moraceae
10	*Ficus religiosa*	Moraceae
11	*Ficus rumphii*	Moraceae
12	*Flemingia macrophylla*	Leguminosae
13	*Kydia calycina*	Malvaceae
14	*Zizyphus mauritiana*	Rhamnaceae

Source: Based on Sharma, 2012

interactions with Karbi villagers indicate otherwise. The villagers say that for several generations, the traditional Karbi practice of rearing lac insects has always taken place in their *jhum* plots and the subsequent regenerating fallows. The insects were reared for the resin they secreted, and while this was traditionally used for fixing farming implements and as a dye, it was sold to traders as stick lac from the time it gained a market value – a practice that continues to the present day.

A possible explanation for the contradictory views could be that the early traders, predominantly from the plains, were unfamiliar with the shifting cultivation cycle and believed fallowed land was natural forest. They assumed that collections were made from naturally occurring insect colonies and therefore categorized lac as a 'jungle product', whereas in reality, the lac colonies were actively nurtured on trees in regenerating

Cajanus cajan (L.) Millsp.
[Leguminosae]

Pigeon pea was domesticated on the Indian subcontinent about 3500 years ago. As well as producing an annual food crop and fixing nitrogen to help enrich the soil, it also hosts the lac insect. It grows in marginal soil, is drought-resistant, and is a staple food and a major source of protein for the population of India. It is now grown throughout the world.

fallows. This misplaced assumption continued quite naturally with the British, and hence official gazetteers such as Hunter had no reasons to document otherwise. They reinforced the judgmental view of the traders and for generations deprived the Karbis of credit for being active lac rearers, rather than mere collectors of a jungle product.

Current lac-rearing practices of the Karbis continue in much the same way, although on a much-reduced scale and with certain modifications that have evolved over the years in response to rapidly changing conditions. Presently, lac is harvested twice a year – once in May or June and again around October and November – coinciding with the lac secretion cycle, which is completed every six months. Within a fortnight after harvesting, the Karbis hang brood pouches containing the lac insects on the branches of host plants to allow the insects to gradually spread on the host trees and re-establish their colonies. Despite the two-week deadline observed by Karbi farmers, Sharma et al. (2006) wrote that brood lac (akin to seeds in crops) should not be stored for more than a week. This is probably because the insects will escape from the brood pouch in search of food if they are not given a timely introduction to a host plant. Any delay in the process of inoculation – technically, introducing an infective agent into a culture medium – is seen as increasing the risk of brood dispersing, and since the availability of fresh brood can be a major constraint to continued lac cultivation, every effort is made to introduce the brood to host trees in a timely fashion (Sharma, 2012). However, current recolonization practices among the Karbis involves a longer period than one week between harvesting and inoculation. This could be because the lac species reared in northeast India is different to those reared elsewhere, or because the Karbis, through generations of empirical experience, have developed brood-storage conditions that are more conducive to brood survival, and they have been able to master this aspect of lac rearing better than the technical experts.

During field visits, Karbi farmers identified at least six lac host plants. They are known locally as *hot chiri* and *chirijangphong* (both *Ficus* spp.), *sopleple* or *soplipli*

Ficus religiosa L. [Moraceae]

The sacred fig, or Bodhi tree, is the tree beneath which the Buddha is believed to have achieved enlightenment. The tree grows up to 30 m, with a trunk diameter up to 3 m. It has a very long lifespan, averaging 900 to 1500 years, with some believed to have been living for more than 3000 years. This is one of the Ficus species named *hot chiri* by Karbi farmers and is popular as a lac host tree.

(*Leea indica*), *thekek* (arhar, or pigeon pea – *Cajanus cajan*), *inghok*, *thengrong ku* and *senam longdak*. While some of these tally with the list provided by Sharma (2012) (Table 32-1), a few remain unidentified, amid suspicions that they could be *Grewia* sp and *Albizia* or *Schleichera* sp.

In all five study villages, most lac farmers preferred *hot chiri* (*Ficus* spp.) as host trees (Figure 32-3). However, a few of them favoured *thekek* (arhar, or pigeon pea) planted in fallow plots. Traditionally, the Karbis preferred *sopleple* (*soplipli*) shrubs growing in their *jhum* fallows for rearing lac; but as cattle raising became an important aspect of villagers' livelihoods, it became increasingly difficult to protect the young *sopleple* saplings from free-grazing cattle. Consequently, some households were forced to give up lac rearing for want of sufficient host plants. Others took up the cultivation on alternative hosts such as *hot chiri* and *thekek*.

Sharma et al. (2006) observed that cattle grazing and the use of twigs and boughs from lac host trees as fodder, particularly during times of adverse conditions and fodder scarcity, severely hampered the multiplication of lac colonies in production areas such as Rajasthan and Gujarat, and this contributed to the gradual decline of the lac trade in these states. Commenting on the factors that have contributed to the decline and gradual eradication of the lac trade from many areas of India, Sharma et al. (2006) blamed wide-scale deforestation and conversion of forests to non-forest uses, as much as volatile markets and price fluctuations. This, the authors said, was further accentuated by clearing the

FIGURE 32-3: Lac colonies growing in a *hot chiri* (*Ficus* spp.) host tree.

natural habitats of lac host trees for promotion of agriculture or commercially important crops and the replacement of known lac hosts (such as *ber - Ziziphus jujuba*) with more lucrative commercial crops.

In the northeast region, concerted efforts by governments to replace shifting cultivation with settled agriculture, particularly through the promotion of cash crops, have seen large areas of regenerating fallows permanently replaced by crops such as citronella, rubber, coffee, cashews and tea, notwithstanding the fact that many of these introduced crops have been a total failure. While this has resulted in permanent conversion of shifting cultivation fallows, it has also contributed directly to a drastic reduction of *jhum* cycles, with the result that *jhum* fallows have been prevented from developing into secondary forests and are covered with scrub vegetation.

The consequences for lac growers such as the Karbis have been the permanent and irreversible reduction of the area that can be used for lac cultivation and the drastic depletion of host plants, thus affecting both the volume and quality of lac production and incomes that accrue from it. Further, the severe reduction in *jhum* cycles has ruled out the fallow regeneration of host plants with long maturation periods, drastically narrowing the choice of hosts that lac producers can employ. Even in cases where mature individuals of preferred host species are retained in fallows as relicts, increased exposure to burning often destroys both the insect colonies and the individual hosts. To compound matters, as pockets of *jhum* cultivation become fragmented and interspersed between plantations and settlements, managing lac colonies and host plants becomes more labour demanding and time consuming, discouraging many Karbi households from continuing with lac rearing and forcing them to explore other options, even if they are less lucrative.

Innovative applications of traditional approaches

Despite such daunting constraints, many Karbi farmers continue with lac-rearing, which is often their only means of generating income. The government, meanwhile, continues to promise to wean farmers away from shifting cultivation, but many official initiatives seem to be doomed to failure – like the much-hyped introduction of citronella growing in the Karbi Anglong area in the 1990s. As transformations driven by policy and market forces have swept through the uplands, intensifying the challenges to traditional rural subsistence even in the most insular villages, the lac-farmers of Karbi Anglong have responded by developing ingenious coping strategies. The following section presents two examples of the ways in which Karbi lac growers have drawn on traditional principles of resource management to overcome the rapid depletion of host plants and adopt a common leguminous lac host to profitably bridge the transition from *jhum* cultivation to fallow, with drastically shortened cycles. The innovations blend lac rearing into management of shifting cultivation fallows, creating options for income generation and rejuvenating the productive capacity of the short-duration *jhum* systems that are fast becoming common in the region.

Sarbura Phungcho of Putsari

Mimicking jhum *practices by rotating lac colonies*

At Putsari village, 40-year old Sarbura Phungcho has as many as 20 *hot chiri* (*Ficus* spp.) trees scattered around the fallow forests he inherited from his father. A shifting cultivator, Sarbura manages his share of the family plot to support his eight-member household, which includes his six children. His production of rice, the staple crop, has been declining year by year because of the shortened *jhum* cycle. The increasing difficulty of making a living from shifting cultivation alone has led Sarbura, along with many other shifting cultivators in the area, to grow broom grass (*Thysanolaena latifolia*) and betel leaves (*Piper betle* - the leaves to wrap *Areca catechu* nuts to make *paan*) as fallow crops to enhance his cash-generating options, while continuing with lac rearing.

Shifting cultivation at Putsari village follows a five-year cycle with a cropping season of only one year. After the harvest of rice, millet, maize, vegetables and spices from the *jhum* plot, Sarbura plants broom grass and betel leaves, which trail over coppiced trees that have been retained in his fallows, as well as rearing lac colonies on *hot chiri* trees. Stick lac and broom grass are his biggest money-earners. Farmers begin to harvest broom grass within a year of planting and continue to do so every year for the following three years. By the fifth year, the entire fallow regains its vegetative cover and is cleared again to resume the *jhum* cycle. Thus, the shifting cultivators of Putsari, like Sarbura, have diversified their cash-generating options and 'commodified' the traditional practice of shifting cultivation, even when the cycle is drastically shortened, in their pursuit of livelihood security.

Lac is Sarbura's most lucrative option, despite a highly volatile market. Stick lac earns anything between Rs250 and Rs500 (US\$4.06 to \$8.12) per kilogram,[3] depending on the market, while the same quantity of dried broom grass would earn a maximum of Rs30 to Rs35. The returns from lac are high in years of good production, fetching around Rs500 per kilogram, but prices plummet to about half of that when production is low. This seems to defy conventional market logic, but may be because of high procurement costs for traders in years of low production. If this is the case, it suggests that profits for traders are higher if lac can be procured in large volumes (hence the tendency to pay higher prices) – an indication that market demand for lac is favourable. Not surprisingly, Sarbura values his lac colonies and nurtures them diligently. In earlier times, most farmers in Putsari reared lac on *sopleple* shrubs, which were abundant in their *jhum* fallows. Sarbura was no exception, explaining that wherever *sopleple* grew, the soil was soft and moist and delivered good crops. Production of lac was also good from these plants. However, as families began to rear cattle and buffaloes, it became increasingly difficult to protect *sopleple* saplings growing in swidden fallows from the depredations of free-grazing cattle.

While many farmers gave up lac cultivation, Sarbura shifted his lac rearing to *hot chiri* trees, which are bigger, have many branches and cannot be destroyed by cattle. However, diminishing forests and constant conversion of fallows into plantations and settlements have made *hot chiri* trees increasingly rare. Sarbura was forced to develop a simple solution that drew on the traditional principles of shifting cultivation. He knew that for generations, *jhum* fields had been rotated across the landscape to allow the land to recover under fallows, with vegetation rejuvenating and gradually maturing into secondary forests. Drawing on this knowledge, Sarbura began to rotate his lac colonies within the host trees. Brood lac has to be reintroduced to the host trees within two weeks of harvesting, and since *hot chiri* trees are large and have many branches, he found that the brood lac could be rotated: if lac is harvested in May or June from one side of a tree, the new lac colony can be raised on another side of the same tree for the October-November harvest. He explains that in this way, site rotation can continue on a single *hot chiri* tree just as *jhum* fields are rotated around the landscape for cropping; the tree is not drained of its vitality just as *jhum* fields do not lose their fertility.

Sarbura has, thus, found an innovative solution to the problem of a rapidly depleting host-plant base and the looming possibility of yield decline. The yield from his 20 *hot chiri* trees in 2011 was 3.5 quintals (350 kg) – an impressive production for a shifting cultivator, and one of considerable value. Priced at even the lowest value of Rs250 per kilogram, Sarbura would have earned at least Rs87,500; in the best of years, this might have risen to Rs175,000 (from US$1421 to $2842) – in either case, a considerable sum for a marginal upland farmer. Discussions with farmers in nearby villages suggest that Sarbura's case is not an isolated one, and others are also adopting his innovative practice.

Mansing Bey of Belangso

The lac-pigeon pea option

For most shifting cultivators, income opportunities from nascent fallows immediately following crop harvests are normally non-existent, or at best, notoriously insignificant

(Cairns, 2007). The Karbis seem to have been more resourceful than many, having introduced options such as betel leaves and broom grass to ensure continuity of income from their *jhum* systems, even during the early land-recovery phase. However, the volume and value of returns from such fallows, particularly during the first year, are not very impressive, even by Karbi standards.

The Karbis claim it is one of their farming traditions to cultivate pigeon peas (*arhar* or *thekek* – *Cajanus cajan*) in their *jhum* fields immediately following crop harvests, especially as host plants for lac rearing. This combination would presumably have offered income from both pigeon peas and lac, while the nitrogen-fixing ability of the pigeon peas would simultaneously have contributed to enhancement of soil fertility – a crucial input for the rejuvenation of any *jhum* system, let alone those with a shortened cycle. The combination should, therefore, have been an extremely attractive option for farmers faced with the challenge of sustainably managing resources and income opportunities in challenging circumstances. However, discussions with villagers indicated that *thekek* cultivation, particularly as a lac host, had not retained its past appeal, at least in the five study villages. In fact, it was difficult to find a farmer who preferred *thekek* as a lac host and even harder to find a villager who was using pigeon peas for lac rearing. It was some relief, therefore, to meet Mansing Bey at Belangso village – a farmer who had been experimenting with the lac-*thekek* combination, notwithstanding the fact that most of his fellow villagers were more comfortable in rearing lac on *hot chiri* trees.

Mansing explained that he found lac cultivation on big trees problematic, especially for those who could not climb trees. Rearing lac on trees such as *hot chiri* requires climbing high up the tree or venturing towards the end of boughs in order to inoculate the lac insects on new growth sufficiently succulent for the insects to feed on and establish their colonies. At the end of the season, more such risks must be taken to harvest the raw lac. Further, both inoculation and harvesting must be repeated twice a year, given the two seasons for harvesting. Accidents are not uncommon, involving broken bones and even fatalities. Moreover, these tasks are virtually impossible for women, Mansing said, and compound the difficulty of finding *hot chiri* trees in fallows. Mansing, therefore, prefers *thetek* (pigeon peas) as lac host plants.

Showing us his fallow land where he planted *thekek* experimentally the previous year for lac production, Mansing explained that his parents and grandparents had planted *thekek* in their *jhum* fallows for lac cultivation after harvesting their rice crop. Although he didn't understand why, his parents had explained that *thekek* was good for the soil. Mansing said he expected that over the following three years, he would harvest *dhal* (pulses) from the *thekek* plants and at the same time, would extract lac twice a year from the same plants. He hoped to earn a good income.

Mansing's reasoning merits serious consideration and support, particularly in the context of managing severely distorted, short-cycled *jhum* systems. Most Karbi farmers rear lac on *hot chiri* trees, but these are slow-growing, requiring several years before they are ready to support lac production in a volume sufficient to be marketable. Moreover, the chances of finding sufficient numbers of suitable host trees diminish with each passing year. *Thekek* or *arhar* is a fast-growing shrub that is well suited to shifting cultivation systems. It is ready for lac inoculation within a year and a small plot of land can support sufficient host plants to ensure a marketable volume of lac. As *thekek* plants are small, introducing the insects and harvesting the raw lac is easy. Brood lac, contained in bamboo pouches, is tied to the *thekek* plants and the tiny insects soon crawl out of the pouch and colonise young branches (Figure 32-4). As they feed on the sap, they begin to secrete the resinous liquid that subsequently forms the stick lac. Lac harvesting from *thekek* shrubs is not as accident-prone as that from tree species such as *hot chiri*. Moreover, *thekek* bushes enable women to take part in lac cultivation.

FIGURE 32-4:   Brood lac is attached to a *Cajanus cajan* (pigeon pea) host plant in a small bamboo basket.

Lac hosts such as *hot chiri* have a growing period of at least a year, and often more, before they can support lac production. In contrast, lac rearing on *thekek* plants allows first harvesting in the second half of the first year, thus supporting uninterrupted generation of household income. In the six-year *jhum* cycle practised in Belangso, Mansing would be able to take harvests from five of his fallow *jhum* plots from the first year of fallow onwards, giving him a steady and stable income each year. Other farmers using *hot chiri* or other host plants would be able to harvest raw lac only from those fallows where the hosts had attained sufficient maturity to allow a satisfactory harvest from lac colonies, effectively limiting their production and income.

Thekek is a fast growing leguminous plant that fixes atmospheric nitrogen, thus contributing to the improvement of soil fertility. For short-cycle *jhum* systems, this is a crucial attribute that helps in faster soil recovery following cropping. The use of legumes such as pigeon peas, soybeans and others as food crops and a means to sustainably maintain soil fertility (and, therefore, subsequent productivity) is reportedly a widespread practice in shifting cultivation systems in many parts of Asia (Cairns, 2007). The use of pigeon peas as a food crop and later, as a soil-fertility enhancer, is common at Ngibat, Kalinga, in the Cordillera region of the Philippines (Daguitan and Tauli, 2007). The additional utility of pigeon peas as a lac host is also reported among the Khmu of northern Laos. A lac-host species that has a crop value and offers both an income opportunity as well as a soil-fertility enhancement attribute and is easily incorporated into *jhum* systems would seem a logical choice, especially for marginalized, short-cycle shifting cultivation systems.

It is difficult, therefore, to understand why the majority of Karbis have rejected the lac-*thekek* system and have adopted the lac-*hot chiri* system in its place, allowing the former practice to decline gradually over the years. The underlying rationale requires detailed investigation, although the reasoning would probably be strong. Preliminary indications suggest that the reasons include grazing damage and the susceptibility of lac colonies to strong winds, although the latter would presumably be true in all other hosts. In the context of marginalized shifting cultivation, particularly that with drastically shortened cycles and a limited capacity for recovery, the rapid depletion of resources and the urgent need to find viable income-generating options, Mansing Bey's experiment gains relevance and stands out as an innovative approach worthy of promotion among shifting cultivators within the region and elsewhere.

Does lac rearing offer a commercially viable option for the shifting cultivators of Karbi Anglong?

The critical challenge facing the marginalized upland farmers of Karbi Anglong (and the majority of their counterparts elsewhere in northeast India) is to find viable options that allow them to earn a decent living within the present-day framework of a monetized economy. Ensuring a cash income sufficient to meet household needs and to address the growing aspirations of their children is a fundamental requirement. But given the constraints of access to government schemes and programmes, the poor links between most rural households and markets and the near absence of opportunities for off-farm enterprises, their choices are extremely limited.

Not surprisingly, Karbi farmers are eager to embrace any opportunity that offers to raise their incomes. They willingly experiment with every crop that offers a potential business opportunity and regularly add new crops to their fields. Over the last decade, they have added broom grass and betel leaves, and recently, ginger, turmeric and rubber (although the latter is being tried mainly by the better-off). Non-performing crops are replaced with the new, with the underlying hope that a few of them will prove commercially viable. However, while experimenting with new opportunities, Karbi shifting cultivators have been careful not to neglect or abandon their steady sources of income. Many of them persevere with the age-old practice of lac rearing while they experiment with new crops, showing an uncanny knack for risk management and a fairly sound business sense.

Meanwhile, the emergence of new industrial uses and diversified applications has given impetus to global demand for processed lac. Sharma et al. (2006) point out that *laccaic* acid or lac dye – the pigment present in the insect haemolymph – is a natural product that is both biodegradable and non-toxic. Therefore, it has numerous applications in food, textiles and pharmaceutical industries. Uses for lac have also diversified into other industrial sectors, such as surface coatings (paint, varnish and furniture polish), electrical goods and handicrafts (bangles and other ornaments), and most recently, lac-based products have emerged as a coating for fruit and vegetables to enhance their shelf-life (Sharma, 2012). Even in India, which used to be the world's leading producer and exporter of lac, imports of processed lac have increased steadily to meet the growing demands of domestic industries. New opportunities for lac production appear to be brightening up, with a growing potential for resurgence of the traditional industry.

While these developments appear to offer a bright future for the lac trade, they are of no immediate value for Karbi farmers because of their poor integration into the industry. The neglect of governments and development agencies in the region is reflected in the fact that lac remains unlisted as a forest product by Forest Departments and the Forest Development Corporations of most states in northeast India. There are no plans to develop lac production, nor investments earmarked; lac remains excluded from the various schemes and programmes of different agencies. This neglect is further reflected by the absence of efforts to establish any kind of processing facilities within the region, whether by the government or the private sector - a neglect continuing from colonial times (see Khadria, 1990). Concerted efforts are needed in order to change the attitudes and perceptions of planners, so a home-grown lac industry can develop in northeast India.

A more practical approach is needed to the integration of Karbi farmers into the lac value chain and hence to the trade and industry, if lac rearing is to provide a viable solution to their immediate livelihood-security needs, and to those of other upland shifting cultivators. Lac production should be linked with the most promising local users, thereby adding value to the product and shortening the value chain, whether locally or within the region. The most promising of options seems to be a linkage with traditional weaving.

For generations, indigenous communities in both Karbi Anglong and Ri bhoi have been using lac for dyeing traditional garments and costumes. Karbi women are highly skilled weavers, much like the women of other ethnic groups in the region. The Karbis weave their own clothes and have always used natural dyes, prepared from lac and wild plants. Although this practice is rapidly being replaced by commercially available yarns, traditional garments are still woven for use on special occasions, or to fill outside orders. These products have a good market, given their unique motifs and striking colour combinations, and are much sought after both locally and further afield.

Weaving and lac are intrinsically linked with the lives and livelihood of women in Karbi Anglong and neighbouring Meghalaya. There are about 3500 weavers in the neighbourhood of Ri bhoi district of Meghalaya who still use lac as a natural dye, and at least 50 nearby villages are reported to produce lac for their use. Many of these weavers use lac together with wild plants to colour the silk yarns that are used to make traditional *Jaintia* shawls for women. These are unique products that of late have been used by fashion designers to develop high-value designer garments.

Despite the latent business potential of such products, large-scale commercialization is constrained by a lack of credit for expansion, poor institutional support for promotion of these high-value niche products and the absence of training opportunities for designers. Lack of support has led to the gradual depletion of traditional knowledge related to dye-making, and this not only threatens its continuance, but also places earnings and livelihoods in peril. Therefore, Karbi Anglong's shifting cultivators are not the only ones who would benefit from the revival of lac rearing. The livelihoods of many others are intricately dependent on this practice, a situation demanding immediate attention and action at government level.

Lac as a pivot for livelihoods and land-use transformation in northeast India

Policies that promote the lac trade and industry can dramatically change lives and livelihoods in the northeastern states, while simultaneously providing a pragmatic solution to transforming the region's shifting cultivation areas. The obscure upland farmers whose innovations were presented as examples in this chapter remain unaware of the diverse opportunities opening up in the lac trade and industry. And while they may seem inconsequential to government policy-makers, their innovations and tenacity provide a significant pulse to business opportunities that are latent even in their challenging conditions. They therefore offer decision-makers potentially attractive options in the task of stimulating the local economy and bringing change to the uplands.

A vision of local economic growth being driven by a lac-centred transformation is neither far-fetched nor unrealistic. But it requires a systematic approach, with clear milestones and the linking of seemingly unrelated elements of the development scenario. The first step would involve the convergence and effective coordination of presently isolated schemes and activities being implemented by different departments and agencies. Transformation of shifting cultivation areas and communities cannot be achieved by insular and isolated approaches with little coordination; efforts by all concerned departments – agriculture, horticulture, soil conservation, forests, rural development, cottage industry, commerce and trade, and revenue – must converge if transformation is to be acceptable to the target communities and therefore effective. Such efforts must also involve, and be synchronized with, the activities of the banking sector and private industry at large. The result must be a synergized orchestration for the achievement of the ultimate goal.

For this to happen, simple changes need to be incorporated into the approaches of concerned departments, under a common framework. In areas with a drastically shortened *jhum* cycle, the promotion of *Cajanus cajan* (or similar shrubby legumes) should be encouraged, providing not only the introduction of a cash crop, but also enhancement of soil fertility (and hence, stabilization of distorted forms of shifting cultivation), while simultaneously providing a host for the rearing of lac. This would offer farmers a means of diversifying and enhancing their incomes. Sharma et al. (2006) suggested the integration of lac culture with agriculture and the inclusion of multi-purpose lac host plants in social-forestry programmes in states of India where lac cultivation is still practised. Agroforestry and silvipastoral approaches should be central to projects launched by agriculture and horticulture departments in these states, with encouragement of incorporating lac hosts. This is especially pertinent in programmes promoted for watershed management and afforestation in shifting cultivation areas. Silvipastoral approaches are needed if forage and fodder requirements are to be met in support of animal husbandry. This would address the problem of free grazing and allow the resurgence of highly preferred lac hosts such as *sopleple*, among others. These approaches should stabilize incomes, after which careful application of land-management techniques could allow the gradual lengthening of *jhum* fallows as the need to pursue current shifting cultivation rotations diminished.

Forest departments have an important role in strengthening this process. In the relevant states, the forest department and Forest Development Corporation need to recognize the potential opportunities offered by products such as lac, in addition to their conventional menu of medicinal and aromatic plants (MAPs), non-timber forest products and commercially important timber species. As Sharma et al. (2006) pointed out: while afforestation for purposes of producing firewood and timber requires a long waiting period before revenue can be generated, returns from lac can start within a few months of adopting the initiative. Moreover, while revenue may be generated from a firewood and timber forest once every several years, that from lac can commence within one or two years and provide bi-annual returns thereafter – an option that farmers find very attractive (Sharma et al., 2006). This could substantially enhance the revenue-generating capacities of both households and the forest department.

As these agencies turn to promote different species of lac hosts as well as their conventional MAPs, NTFPs and timber species, this could lead to a gradual increase in tree cover and the subsequent regeneration of forests. In the process, this would contribute to the conservation of biodiversity in forested areas, reversing the current practice of promoting monocultures, and restore the habitat of lac insects, along with their associated flora and fauna (Sharma et al., 2006).

While such efforts would positively contribute to an increase in lac production and could result in the expansion of lac growing areas, this alone would not succeed in enhancing incomes or ensuring the integration of Karbi lac-producers into the trade and industry. Success will only follow the creation of a sufficiently large local demand for processed lac. This demand could be triggered quite quickly by supporting both women weavers in the area and the activities of numerous cooperatives and women's groups. In the long term, it would also require an environment that enabled the development, promotion and marketing of the niche products that were being created. Further, steps would need to be taken to gradually increase the capacity to absorb these products within the region, by encouraging the growth of enterprises with diversified uses for processed lac.

Evidence suggests that both the knowledge required and the practical exercise of extracting dye from stick lac still exist in the indigenous communities around Karbi Anglong and Ri Bhoi. It would seem logical to harness this knowledge and ensure its continuity by establishing microenterprises that process local stick lac for the manufacture of natural dyes. This objective could be pursued by involving women's groups and cooperatives in the area, along with concerned departments, cooperatives and district industrial centres.

Sharma (2012) points out that *Kerria chinensis*, the lac species reared in the northeast region, is very productive and a rich source of dye that is darker in colour than that derived from the *Kerria lacca* species cultivated in mainland India. This suggests the possibility of developing niche products that could offer a comparative advantage to the northeast lac producers, if sufficiently large volumes were produced. The establishment of dye-processing units, run by women's cooperatives and supported by relevant departments, could contribute to the development of a value-added product for use by local weavers, thus triggering the absorption of locally produced lac. With improvements in quality and volume, the processed dye could enjoy a comparative advantage when marketed throughout the country.

As there is already a growing demand from high-end buyers for traditional Karbi garments and *Jaintia endi* silk, a linkage with the private sector – in particular, fashion designers and branded-garment manufacturers – would ensure a steady market and high-end demand. Such a partnership could provide the stimulus for projecting the unique traditional woven garments and fabrics of northeast India into niche markets, thus benefiting many women weavers in the uplands. It could also pave the way for integration of Karbi lac growers into the lac value chain and link them to larger markets, providing much-needed impetus for the transformations sought by upland communities and development planners alike.

Acknowledgements

The authors are extremely grateful to the upland farmers in the study villages for sharing their knowledge and experiences. The untiring efforts of John Rongpi of the Rural Areas Development Society in organizing meetings and interviews and providing logistic support was invaluable. Without his selfless contribution, this chapter would not have been possible. We also gratefully acknowledge the invaluable contributions of Robindra Teron and Dr. K. K. Sharma. Financial support for this study was extended by the International Centre for Integrated Mountain Development (ICIMOD), Kathmandu, through *Adapt* Himal, a regional programme funded by the International Fund for Agricultural Development, Rome.

References

Cairns, M. F. (ed) (2007) *Voices from the Forest: Integrating Indigenous Knowledge into Sustainable Upland Farming*, Resources for the Future Press, Washington, DC, USA

Daguitan, F. M. and Tauli, M. (2007) 'Indigenous fallow management systems in selected areas of the Cordillera, the Philippines', in M. F. Cairns (ed) *Voices from the Forest: Integrating Indigenous Knowledge into Sustainable Upland Farming*, Resources for the Future Press, Washington, DC, USA, pp.673-678

Guha, A. (2006) *Planter Raj to Swaraj: Freedom Struggle and Electoral Politics in Assam, 1826-1947*, Tulika, Delhi

Hunter, W. W. (1879) *A Statistical Account of Assam*, vol I and II, Truber & Company, London, reprinted by B R Publishing Corporation, Delhi

Handique, R. (2006) *British Forest Policy in Assam*, Concept Publishing Company, New Delhi

Khadria, N. (1990) 'Traditional crafts and occupational structure of Assamese rural society in the 19th Century', *Social Scientist* 18(11-12), pp.36-63

Leduc, B. and Choudhury, D. (2012) 'Agricultural transformations in shifting cultivation areas of northeast India: Implications for land management, gender and institutions', in D. Nathan and V. Xaxa (eds) *Social Exclusion and Adverse Inclusion: Development and Deprivation of Adivasis in India,* Oxford University Press, New Delhi, pp.237-258

Saikia, A. (2005) *Jungles, Reserves, Wildlife: A History of Forests in Assam,* Wildlife Areas Development and Welfare Trust, Guwahati, Assam

Saikia, A. (2008) 'Forest land and peasant struggles in Assam, 2002-2007', *Journal of Peasant Studies* 35, pp.39-59

Sharma, K. K. (2012) Personal communication between author Sanat Chakraborty and Dr. K. K. Sharma, head of the Lac Production Division of the Indian Institute of Natural Resins and Gums

Sharma, K. K., Jaiswal, A. K. and Kumar, K. K. (2006) 'Role of lac culture in biodiversity conservation: Issues at stake and conservation strategy', *Current Science* 91(7), pp.894-898

Teron, R. (2012) Personal communication between author Sanat Chakraborty and Karbi ethno-botanist and folklorist, Robindra Teron

Tucker, R. P. (1988) 'The depletion of India's forests under British imperialism: Planters, foresters and peasants in Assam and Kerela', in D. Woster (ed.) *The Ends of the Earth: Perspectives on Modern Environmental History,* Cambridge University Press, Cambridge, UK, pp.118-140

Notes

1. The villages were Langparpam, Putsari, Kolbari, Doloiarong and Belangso.
2. As well as interacting with lac farmers, the authors interviewed a number of experts by email or telephone, including officials of the Karbi Anglong Autonomous District Council. The most prominent of these experts were Karbi folklorist and ethno-botany researcher Robindra Teron and the head of the Lac Production Division of the Indian Institute of Natural Resins and Gums (formerly the Indian Lac Research Institute – ILRI), Dr. K. K. Sharma. Dr. Sharma has worked extensively in the major lac-producing belts of India. His co-authored paper of 2006 provides valuable information on the diversity of lac insects in India and identifies the important lac-producing regions of the country in contemporary times (Sharma et al., 2006). Dr. Sharma also provided vital advice on the biological aspects of lac insects.
3. Calculated at a rupee-dollar exchange rate of Rs61.57 to US$1, as existed in January 2014.

33

BEE PASTURES NEED NO FENCES

Honey production from forest fallows by a *Pgaz K' Nyau* (Karen) village in northern Thailand

Prasert Trakansuphakon and Maurizio Farhan Ferrari[*]

Introduction: Hin Lad Nai village

The Hin Lad Nai community is located between a national forest reserve and the Khun Chae National Park, in Wiang Pa Pao district, Chiang Rai province, northern Thailand, about 130 km south-southwest of the city of Chiang Rai (Figure 33-1). The community's land is hilly forest, through which flow more than 14 small streams. Community forest covers 3110 ha, while agricultural land occupies about 570 ha. The community comprises four settlements: Hin Lad Nai, Pha Yuang and Hin Lad Nok, populated by *Pgaz K' Nyau* (Karen) people, and Sai Khao, occupied by Lahu people. The total population of the four hamlets is about 350. Hin Lad Nai has 20 households with a total population of 105. The community's main natural-resource and agricultural systems are focused on a constant cycle of shifting cultivation, practised and innovated upon by the *Pgaz K' Nyau* indigenous people over countless generations.

The villagers of Hin Lad Nai believe in both animism and Buddhism. Their ancestors came from Mae Chaem, a district in the west of Chiang Mai province, and settled at the present location more than 150 years ago. In the years before the move, the community had shifted about 11 times, in search of access to the best sites for shifting cultivation. The present site was chosen on that basis (Trakansuphakon et al., 2016). At Hin Lad Nai, the shifting cultivation cycle is based on short cultivation and long fallow: the farmers cultivate for one year and allow the area to rest and regenerate for seven to 10 years in a long fallow (Trakansuphakon and Kamphonkun, 2008).

The main crop cultivated by the Hin Lad Nai community is rice. In addition to traditional upland rice varieties grown in swiddens, the community has terraced fields

[*] Dr. Prasert Trakansuphakon is Director of the Pgakenyaw Association for Sustainable Development (PASD), Chiang Mai, Thailand; Dr. Maurizio Farhan Ferrari is Senior Policy Adviser on Environmental Governance to the Forest Peoples' Programme (FPP), Moreton-in-Marsh, England.

for wet rice. Work on the terraces began about 60 years ago, after village farmers observed wet-rice production in lowland areas. The terraced paddy fields provide the community with additional sources of rice, but swidden cropping systems still provide most of the food and bring a cultural focus to Hin Lad Nai's agricultural calendar, much as they have for many years. The community also grows fruit and other crops, gathers forest products such as wild tea, bamboo shoots and forest honey, and breeds pigs, chickens, ducks, cattle and buffaloes. The shifting cultivation system supports an exceptionally rich biodiversity of edible cultivars and semi-domesticated crops, which, together with grain from the wet-rice fields, provide the

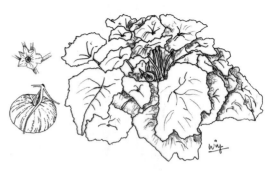

Cucurbita pepo L. [Cucurbitaceae]

This is one of the oldest domesticated species in the world, with a vast range of subspecies and cultivars growing throughout tropical and temperate regions. Variously known as pumpkin, squash, marrow, gourd, zucchini, scallop or crookneck, the species appears in various forms in Hin Lad Nai's swiddens. As well as the fruit, the flowers and tender tips are also eaten.

community with its food security. Hin Lad Nai is, thus, regarded as inspirational, since it produces more than 70% of its own food (NDF, 2009). In recognition of the community's efforts and achievements, Hin Lad Nai was a recipient of the international Green Globe Award for 1999-2009. It has also received a number of local and national awards over the past 20 years.

For more than a century, shifting cultivation has maintained a controlling influence over all aspects and characteristics of life for the Hin Lad Nai people. However, the farmers have introduced many innovations over the years in response to environmental and social changes. One of the most recent innovations is keeping bees for the honey they produce. This addition to agricultural tradition is reported in this chapter, following a brief background account of *Pgaz K' Nyau* shifting cultivation.

Shifting cultivation as the backbone of a sustainable and resilient biological and cultural (or socio-ecological) system

The shifting cultivation system practised in Hin Lad Nai is a central and vital component of a natural-resource management strategy developed by the *Pgaz K' Nyau* people. It draws upon the full range of their knowledge and wisdom, cosmology, spirituality and technical knowledge for conservation and sustainable use of natural resources. It is central to their cultural identity and social well-being. For example, most rituals align with the shifting cultivation farming system and it defines when

FIGURE 33-1: The location of Hin Lad Nai, in Chiang Rai province, northern Thailand.

and where different events take place in the course of a calendar year. It has provided the *Pgaz K'Nyau* people with stories and tales of their culture and many metaphors and teachings in their language. It is crucial for the protection and transmission of the local spiritual and cultural patrimony.

The traditional occupation of swidden farming is known as *quv*. It lies at the heart of the way the *Pgaz K'Nyau* people manage natural resources, as expressed in the *hta* (a poem or poetic expression), '*Auf hti k'tau hti, auf kauj k' tauz kauj*' (Use water, take care of river; use land, take care of forest). The poem reflects the understanding of the *Pgaz K'Nyau* and their practical process of sustainable management based on the concept of coexistence between humans and nature.

The shifting cultivation cycle of use, regeneration and new growth is linked to the *Pgaz K'Nyau* worldview of change and flow, and belief that the past generates new life.

The *Pgaz K'Nyau* also have deeply embedded personal and collective memories of fallow fields, providing a strong basis for community well-being. Fallow fields and their related ecosystems, more than any other local space, have value in the form of memories – of human links to their own fallow land; as a space in which human-to-human connections were made, as well as connections to animals and plants. These heartfelt memories are expressed through *hta* story telling, which recalls 'feeling good about fallow land and the community'. The kinds of memories conjured by thoughts of fallows include relationships between people as they once worked together side by side, sharing labour, talking and smiling; good collective times and close personal relationships. Fallow land is, therefore, symbolic of a happy life. It creates a kind of love in the *Pgaz K'Nyau* people for their swiddens and the fallows that are resting; for the various kinds of food that the swiddens provide and, more broadly, a love of the forest and its natural surroundings. This makes people want to care for these resources and to manage them sustainably. This cultural devotion has its wider benefits: a Thai forest academic recently noted that during the seven- to 10-year fallow period, fallow fields were always green, and helped to reflect sunshine back into space, thereby mitigating climate change (Trakansuphakon et al., 2016).

Research carried out in 2004 by members of the Hin Lad Nai community in collaboration with academics documented the succession of vegetation and wildlife that took place year-by-year in a forest fallow. It showed that the shifting cultivation system enhances both food security and biodiversity, given that changes in vegetation attract a wide variety of animals. This suggests that both humans and animals rely on shifting cultivation for food: humans use *quv* (the year of cultivation) to supply their main staple food (rice), while fallow fields provide food (as well as shelter and habitats) for wild and domesticated animals. For people, fallows also provide spices, wild vegetables, herbal medicines and building materials. The study found that there were no fewer than 207 edible plant species and varieties of food in Hin Lad Nai's shifting cultivation system (Ganjanapan et al., 2004), providing the basis for a rich, healthy and tasty diet. The *Pgaz K' Nyau* therefore consider swidden fields, under both cultivation and fallow, as supermarkets and food centres for people, wildlife and semi-domesticated animals (Figure 33-2).

These and other positive aspects of shifting cultivation, including the resilience of its socio-ecological systems, have now begun to be highlighted at an international level (Forest Peoples Programme et al., 2016).

The concept behind Hin Lad Nai's community governance system stems from the traditional philosophies of the *Pgaz K' Nyau* people and the teachings of the grandfather of the present official leader of Hin Lad Nai, Chaiprasert Phokha. 'The land and forest never end if we know how to take care of them and use them', the old man said. This implies that the community's governance system is not only about preserving and protecting the forest, but also about how it should be used, for food security and to obtain a sustainable income, while at the same time conserving it. This was the philosophy that led the Hin Lad Nai villagers, about 30 years ago, to recover and revitalize their forests.

In the 1970s and 1980s, a logging company severely affected Hin Lad Nai's forest and environment, until its activities were stopped by a government logging ban in 1989. Most of the large trees were taken and wildlife nearly disappeared, leaving the forest full of dry branches and leaves, making it very vulnerable to fires. However, the Hin Lad Nai community turned the crisis into an opportunity by gradually

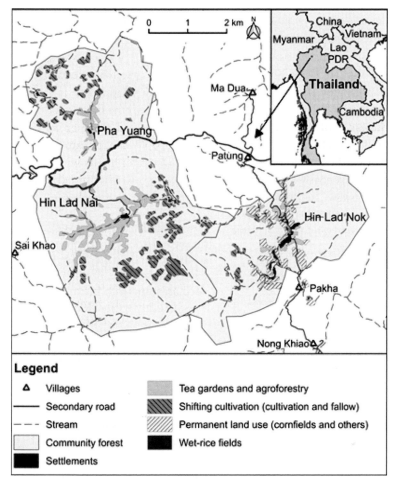

FIGURE 33-2: The land uses in Hin Lad Nai's community forest, showing the extent of shifting cultivation land, most of which would be in fallow at any given time, and other agricultural uses.

and steadily recovering their forest and natural resources. They used a combination of traditional knowledge, livelihood options and modern knowledge from 'outside'. This included making fire breaks along the boundaries of their territory, planting diverse varieties of new trees along with native species, classifying types of forest land use and developing community-based Geographical Information System (GIS) mapping. Over the past 30 years, Hin Lad Nai's forest environment has become very rich again and the livelihood options of the people have diversified and increased.

Forest honey

Around 2010, a new venture gained popularity in the Hin Lad Nai community. Various people gained an interest in natural forest bee-keeping and established beehives in their agroforestry areas, as well as in the tea gardens, within the community forest. Thus began a learning process that has so far developed a very successful new enterprise.

At first, the villagers contacted a commercial honey producer at Wiang Pa Pao, near Hin Lad Nai, and took bees from that source into their forest environment. The new bees had no established food sources, so they were fed with sugar. Within one month, all the bees had died, apparent victims of an inability to adapt to the natural environment.

The villagers then went to another community where people had been keeping bees by making wooden boxes and allowing bees to inhabit them naturally. The Hin Lad Nai villagers adopted this idea and set about making it work in their own environment. Some had

Apis cerana [Apidae]

The Asian honey bee is a close relative of the Western, or European honey bee. It has a vast natural range, from Russia south to eastern Indonesia and from Japan west to Afghanistan. With human intervention, it has recently expanded into New Guinea and Australia. Colonies of 6000 to 7000 individuals normally build multiple combs within some kind of cavity.

previous experience of harvesting honey from colonies of bees living naturally in the forest, and they had learnt that if they spread beeswax on a tree, bees would come to stay there. This principle was adapted to attract bees to boxes made from sawn wooden planks, the trunks of palm trees and other materials. Not only were the boxes spread with beeswax, but the wax was also burned to produce an attractive smell (Figure 33-3). The bees soon began to arrive.

There are several types of wild bees in Hin Lad Nai's forest, but only three of them are relevant to this chapter. The first is a larger bee (sometimes referred to as the 'giant' honey bee) called *ke nae* by the villagers or *phung luang* in Thai (*Apis dorsata*). These bees build single honey combs beneath thick branches high up in trees, and their honey is traditionally gathered by climbing 15 to 18 metres high into the trees (Raygorodetsky, 2017). The second is a medium-sized bee called *kwae do* by the people of Hin Lad Nai, or *phung phrong* or *phung kron* by Thai speakers (*Apis cerana*). This is the Eastern or Asiatic Honey Bee, colonies of which build nests of multiple honey combs in underground cavities or in old trees. They are similar in most respects to Western, or European honey bees (*Apis mellifera*), but whereas a colony of *A. mellifera* may comprise 50,000 individuals or more, colonies of *A. cerana* are much smaller, consisting of 6000 to 7000 workers. The third species is known as *kwae pho* by the Hin Lad Nai villagers and is a small, stingless species known as *chanarong* in Thai (*Apidae Meliponini*). These bees often build their combs in bamboo trunks.

Many *Pgaz K'Nyau* families now raise the medium-sized *kwae do* in beehives along forest trails and in sheltered forest areas. Increasingly, they are also raising the small stingless *kwae pho* bees. Having grown accustomed to the habits of their insect 'livestock', the *Pgaz K' Nyau* have words for the bee 'seasons'. The bees are called *kwae do* or *kwaiv dof* when they are in the beehives constructed of sawn wood or palm tree trunks, and *hsuv hpauv* when they freely roam the forest and find temporary shelter along tree branches in the wet season.

The villagers began in a forest area, where they put the beehives in sheltered spots, such as under big trees or big rocks (Figure 33-4), on top of sheltered rocks, in protected niches along river banks and under small cliffs. Following their observation that natural beehives were most often found in hollow trees, they proceeded on the basis that any well-sheltered place was good for the man-made wooden beehives (Figure 33-5). They found that the bees did not like wet areas, hot places in direct sunshine, or places where they would be disturbed by enemies. Wasps, bears, some species of palm civet, hawks and insect-eating birds are their worst enemies, and if ants, spiders or cockroaches enter the beehives, the bees are likely to leave immediately. For these reasons, the Hin Lad Nai farmers have learnt to take care of the beehives before the bees arrive, and even after

FIGURE 33-3: A round beehive, its entrance smeared liberally with beeswax, awaits the arrival of bees in the shelter of a hollow forest tree at Hin Lad Nai.

Photo: Boonsri Chalakanok.

Tectona grandis L.f. [Lamiaceae]

The importance of its timber eclipses the fact that teak is an attractive tree to forest-dwelling honey bees. Teak bears profuse clusters of tiny fragrant white flowers at the ends of its branches from June to August, and they fall to cover the forest floor like snow. The flowers are pollinated by insects; prominent among them, bees.

they have settled in. One farmer, S' lex hpo Papa, explained: 'I need to go every week to clean the beehives before the bees come and even when bees are already in the beehive I need to come and see what has happened to them because they may be disturbed or eaten by wasps, birds or other animals. Abnormal weather is also an enemy if it is too hot, too cold, or it changes too quickly'.

Of utmost danger to beehives is forest fire, so farmers who keep bees need to take special care of their forest, ensuring that fires are prevented. The Hin Lad Nai people work hard throughout the long summers, clearing fire breaks around their settlements and the bee-keeping areas. While bee-keeping may look easy, in reality there are many tasks that must be performed to take care of and protect the bees and their environment.

After they found that the bees liked the boxes they made, the Hin Lad Nai farmers increased the number of beehives from 5 to 10, from 10 to 20, and now some farming families have almost 100 boxes. At the end of the third year, they had large amounts of honey from their beehives, but could not market or sell it. At that point, the village head, Chaiprasert Phokha, contacted the *Pgaz K' Nyau* Association for Sustainable Development (PASD), which provided training in packaging, branding and marketing. Since then, all of

FIGURE 33-4: Two bee boxes shelter in the lee of a giant forest tree.

Photo: Maurizio Farhan Ferrari.

FIGURE 33-5: Two bee hives in a forest hut at Had Lin Nai.

Photo: Maurizio Farhan Ferrari.

the honey has been sold and the farmers are very keen to expand their bee-keeping enterprise. Moreover, the word has passed around: surrounding communities of indigenous people in other districts and even in other provinces have begun to follow the example of the bee-keepers of Hin Lad Nai.

Bee-keeping as a promising innovation for shifting cultivators

Over the past three to four years, building on their success with forest bee-keeping, a group of farmers at Hin Lad Nai has been experimenting with bee-keeping in shifting cultivation fallows. In 2014 and 2015, some farmers began placing hives in old field huts on fallow land, where the forest was regenerating following a year of swidden cultivation. One of the farmers explained that bees liked to make their hives in shady, dry and cool areas that were safe from their natural enemies. Therefore, old field huts were very suitable, and finding that the bees seemed to thrive in this environment, they began increasing the number of hives in or under the huts (Figure 33-6).

FIGURE 33-6: Bee boxes both inside and under an old field hut in a swidden fallow at Hin Lad Nai.

Photo: Boonsri Chalakanok.

Nine families have been involved in this experiment, with very positive results so far. Table 33-1 shows the location of bee boxes and honey production from fallow fields as part of a trial-and-error process to assess the viability of bee-keeping as a complement to shifting cultivation.

Further studies in 2018 are beginning to illustrate the innovative aspects of bee-keeping as part of a shifting cultivation system, and the main findings continue to be encouraging.

Importantly, bees seem to like the environment in fallow fields. This may be due to the fact that the farmers put their bee boxes either in old field huts located in

TABLE 33-1: Bee-keeping in the first three years of fallow in shifting cultivation land at Hin Lad Nai, 2015 to 2017.

Bee-keeper's name	Location of bee boxes	Fallow years	No. of boxes [*]	Honey harvested (kg)
Nipon Pookhrae	Under fallow hut	First year	14	26
Nivet Siri	Under fallow hut	First year	2	8
Chaiprasert Phokha	Under fallow hut	First and second years	2	15
Somsak Praiwanhtoon	1 box under hut and 1 in hut	First year	2	2
Somchai Boonyong	In fallow hut	First year	1	1
Duangdee Siri	Under fallow hut	First year	1	6
Kasem Haedoo	Under fallow hut	First year	2	8
Somboon Prawankul	In fallow hut	First, second and third years	4	16
Sura Phokha	In fallow hut	First year	2	9
Totals (9 families)			30	91

Note: [*]All bee boxes recorded in this table were placed in fallow areas by farmers and subsequently occupied by bees.

fallow fields or under the elevated floor of the huts. Both of these places are safe, dry and clean; the hives do not get wet and there are no attacks from large animals. Perhaps even more important is the fact that during the early years of fallow, the surrounding fields offer a very high variety of flowers and plants suitable for bees and other pollinators. As the seasons change, the bees can find different kinds of flowers to feed on, while their presence contributes to the process of pollination, increasing the diversity of flora and fauna in the fallow fields. Interviews with farmers have highlighted the diversity of plants and flowers in fallow regrowth, both in the course of just one-year and within the full 7- to 10-year fallow phase of the swidden cycle.

In May, June and July, the main bee-forage species include *Amaryllis* sp., flowers of lettuce and maize, tuberoses (*Polianthes tuberosa*), *Mirabilis jalapa*, and many kinds of grass flowers. Flowers on nearby crops and plants in August, September and October include rice flowers and blooms on pumpkins, squash, eggplants, cowpeas, bitter and dishcloth gourds and lettuce. From November to January there are cockscombs (*Celosia* sp.), marigolds, sesame flowers and sunflowers (*Helianthus annuus*). Farmers report that their bees produce good-quality honey from nectar gathered from flowers found during the first year of fallow.

However, significant changes take place in the successional process as the fallow matures. In the first to third years of fallow (*hsgif bauf* in the *Pgaz K'Nyau* language, or 'young fallow period'), many young trees sprout from old stumps and numerous varieties of grass and shrubs produce flowers suitable for bees. The early species include *Chromolaena odorata*, *Blumea balsamifera*, *Ageratum conyzoides* and *Melastoma malabathricum*, as well as many grass and vine species. In the third and fourth years, grasses decrease or disappear and young trees rise above the lower vegetation (a stage called *hsgif loov htauf* by the local farmers, or 'stand up of fallow land'). Some species flower profusely, including Siamese pom-pom trees (*Mallotus barbatus*), crepe mytle (*Lagerstroemia* sp.), wild sugar cane or kans grass (*Saccharum*

Celosia cristata L., a synonym of *Celosia argentea* L. (Amaranthaceae]

Cockscomb is mainly cultivated as an ornamental plant because of its spectacular flowers. It offers rich bee fodder in recently fallowed swiddens. The leaves and flowers are also eaten as vegetables, particularly in Africa and India. A compound produced by the plants has also been found to rid cereal crops of root parasites when they are sown in the same fields.

spontaneum) and others that persist from the preceding period. In years five and six (a stage called *hsgif yauv ploj* at Hin Lad Nai), young trees grow to maturity and flower – notably Siamese pom-pom trees. This is the time for bees to do their pollination work among the flowering trees. Some trees will bear fruit, bringing many kinds of animals to the young forest in search of food. In the final period (known as *doo lax*, or old fallow), the land has gathered sufficient nutrients and the soil is ready for new cultivation. Flowers during fallow years five to 10 bring together the blossoms of all trees that have grown to maturity in the fallow fields. The flowering reaches a peak during November to April, which is when the Hin Lad Nai farmers place their bee boxes, because it is around these months that the bees will be looking for places to create new hives.

Life cycle of the Asiatic honey bee

According to Seeley (2009), the Asiatic honey bee (*Apis cerana*) must collect and store around a third of its nectar in a concentrated form – honey – in order to ensure an adequate supply of food during the time of year when nectar sources are scarce, or the weather prohibits foraging. It follows that if this cycle is interrupted by human honey harvesters, the colony will simply die of starvation.

If the colony survives the 'hungry' months, it will quickly build up its strength, and before rice-harvest time, it will have attained full size, with a single queen, 6000 to 7000 female worker bees and several hundred male drone bees. The colony will then begin to reproduce. It rears several new queens, and when they are close to hatching, the old queen will swarm away from the hive with about half of the workers. This is the opportunity the bee-keepers of Had Lin Nai depend upon.

The swarm settles 20 to 30 metres away from the old hive, and sends out scouts to search for a suitable place to build a new home. When the scouts return, having found an empty bee box or similar suitable cavity, the

Mallotus barbatus Müll.Arg. [Euphorbiaceae]

This evergreen tree grows to 12 m and is commonly found in forests or in disturbed or burnt areas. Called the Siamese pom-pom tree, it flowers profusely in three- to four-year-old fallows. The roots and fruit are used as a medicine to treat muscle stiffness and the seeds yield a fatty oil that is used for making candles.

swarm leaves *en masse* for the new location. During the rest of the dry season, the new colony will build combs, rear brood and gather food to quickly rebuild its population and food reserves prior to the approaching 'hungry' period.

As well as reproductive swarming, *A. cerana* has migration and absconding behaviour, abandoning the current nest and settling at a new location where an abundant supply of nectar and pollen is available. These bees usually do not store great amounts of honey, so they are more vulnerable to starvation if a prolonged shortage of nectar and pollen occurs. Absconding will start when not enough pollen and nectar are available at an existing site (Seeley, 2009).

At Hin Lad Nai, the bee-keeping calendar starts after rice harvesting, when surviving colonies of bees have already been building their strength and numbers for some weeks (Figure 33-7). Having left the bees to their own devices during the months from June through November, the farmers notice large numbers of them appearing again when the swidden rice is flowering. In September and October, there are many vegetable flowers in the cultivated swiddens to help boost the bee numbers, and trees such as *Litsea glutinosa* and crepe myrtle (*Lagerstroemia* sp.), and clumps of greater galangal (*Alpinia galanga*) are in full blossom in more mature fallows. Then, in November and December, wild sunflowers (*Tithonia diversifolia*) create a vivid display on the edges of rice fields and in the first- and second-year fallows. The constant hum of bees around the swiddens and fallows suggests that strong colonies remain, hidden in the fallows, forest trees and other places, but in most cases, the beehives of the *Pgaz K'Nyau* are empty as the landscape emerges from the wet

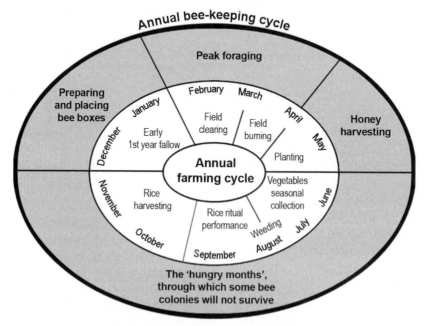

FIGURE 33-7: The 'calendars' of swidden cultivation (inner portion) and bee-keeping show how the two activities are complementary.

season. However, the farmers believe there is a 'trend' for the bees to stay on in their bee boxes on fallow land, even during the rains.

After the swidden harvest, in December and January, the farmers place the wooden beehives they have made for the bees in old rice-field huts, amid the rice stubble and burgeoning plant growth. The freshly fallowed swiddens soon have many flowers to provide pollen and nectar, and the bees are attracted to the sudden abundance. Behind the scenes, the colonies that survived the wet season are now strong and ready for the reproductive cycle of swarming to expand their numbers in new locations. The wooden beehives, spread with fragrant wax and secure in the old huts, are prime places for the early swarms.

The first swarms arrive in December, about two months after the start of the tropical dry season (winter), but the greatest number come to the wooden hives in January and February. Many of the large forest trees are in bloom

Oryza sativa L. [Poaceae]

Rice is a grass species that is usually regarded as a wind-pollinated crop. However, honey bees are attracted to the copious production of pollen on flowering rice, and have been known to carry rice pollen more than 500 m from its source. At Hin Lad Nai, one of the first harbingers of a new honey season is the hum of bees in the swidden rice.

between January and April, with species such as black plum (*Sygygium cumini*), teak (*Tectona grandis*), bishop wood (*Bischofia javanica*), mangosteen (*Garcinia* sp.), *Erythrina subumbrans* and *Duabanga grandiflora* making this time of year a peak period for foraging bees. The bee-keepers of Had Lin Nai expect to harvest honey from their bee boxes at the end of April and beginning of May (Figures 33-8 and 33-9) – before the rainy season comes again. The shifting cultivation and bee-keeping calendars complement one another well. With honey collecting in April and May, the farmers are able to take harvests from their two intertwined production systems at different times of the year. The farmers are also able to make use of a part of the dry season, following the harvest of their agricultural products, to take care of their bee-keeping procedures.

Fallow versus forest

In their bee-keeping experiments, the farmers at Hin Lad Nai have discovered that it is much easier to look after beehives in fallow fields rather than in the forest. In the fallows, they have to take care of the bee boxes only two or three times per year, while in the forest they need to check the hives as often as every week, depending on the time of year and various factors such as wild animals, rain and potential forest fires. The following are farmers' comments on bee-keeping, taken from interviews:

FIGURE 33-8: A square beehive laid on its side with the top removed, to show the bounty within.

Photo: Chalit Saphaphak.

> I have taken care of bees on fallow land only two times a year: once when I brought the box into the hut and another time when I harvested honey. It is safer than bee-keeping in the forest area.

> I take care of my bees in the fallow area two or three times a year, when I pass-by for other purposes, mainly just to check whether the bees are actively using the box. If the bees have not come into the box, I need to change the box to a more suitable position.

> In fallow land it is easier, and is not hard work like bee-keeping in the natural forest area because the bees are already safe from rain, sunshine, animals, and so on. This is another reason why people are more interested in keeping bees on fallow land.

FIGURE 33-9: The top of a round beehive removed to show the honeycombs that hang beneath it.

Photo: Chalit Saphaphak.

> We keep bees on fallow land because we hope to increase the amount of good honey. It is also easy to harvest honey and to take care of the beehives. It is easy for us in our daily work because the fallow land where we keep bees is in the same direction as the fields we are farming, so we can easily also take care of the beehives.

There is another important argument for fallows over the forest: In the three-year study, bees not only occupied all of the boxes located in fallow land, but also returned to use them in the second and the third years. In forest areas, not all boxes were regularly occupied, suggesting that fallows seem to provide a more stable environment and shelter for bees.

Whence comes the flavor?

The Hin Lad Nai bee-keepers have found that honey from fallow land has a different taste, smell, colour and viscosity (stickiness) to that from the forest (Figures 33-10 and 33-11). Fallow honey is darker and more viscous than honey from forest areas, and this has sparked their interest. They would like to see the honey analysed, to see which is superior, in terms of nutrition and medicinal value. Perhaps this arises from the copious flowering of a shrub that they call *hpau pgaj laj*, in the second year of fallow. It is identified as *Blumea balsamifera*, which is recognized in many parts of the Asian tropics for its medicinal properties, particularly as a diuretic and a treatment for infected wounds. As the Hin Lad Nai village leader, Chaiprasert Phokha, said: 'Honey from the second year of fallow is produced from the special flowers of *hpau pgaj laj*, which gives it a special smell and more stickiness. It tastes soft and sweet and may have special medicinal qualities'.

It has also been noticed that differences in honey depend on the time the bee boxes are placed in the field (and therefore the time over which the bees can forage for nectar). When the boxes are placed early in the season, the honey will be stickier and stored in greater quantities. But as their fascination with bee-keeping gathers expertise, the farmers are convinced the flavour and sweetness depends mainly on the type of flowers and tree species from which the bees collect nectar.

Chaiprasert points to the sheer diversity of flowering plants and trees on fallow land, giving bees a greater foraging range that includes both fallows and nearby forest trees. Depending on the time of year, particular trees and shrubs that flower profusely have a clear influence on the colour, taste, herbal content and viscosity of fallow honey, making it different to that harvested from the forest.

Mirabilis jalapa L.
[Nyctaginaceae]

The so-called 'four o'clock flower' is widely known for its spectacular shows of strongly fragrant flowers that open in the late afternoon and bloom throughout the night. Flowers with different colours – yellow, red, magenta, pink or white – can grow simultaneously on the same plant. This species appears in early fallow vegetation at Hin Lad Nai.

Closer to the land

Interviews of farmers at Hin Lad Nai suggest very strongly that the growing importance of bee-keeping as a complement to their traditional shifting cultivation system has led to an increasingly positive relationship between the farmers and their environment. They report that this extends from good feelings and spirituality to concern for better care and management of fallowed land. In particular, they have become acutely watchful for wildfires that can race through fallows after a lightning strike. The farmers also claim to have seen an increase in the diversity of flowering and medicinal plants in their shifting cultivation systems as they have encouraged the growth of plants suitable for pollinators and honey production. This, they say, has made the already rich biodiversity found in shifting cultivation systems even richer.

FIGURE 33-10: Hin Lad Nai village leader Chaiprasert Phokha harvests honey from a beehive in an old field hut, in the centre of a young swidden fallow.

Photo: Chalit Saphaphak.

As a result of the recent years of experimentation, Hin Lad Nai's farmers are increasingly interested in keeping bees in swidden fallows. This has important potential for the community's long-term food security and well-being, as the extra care lavished on the shifting cultivation system may yield benefits beyond providing honey for food and household income, in that it creates an entrepreneurial spirit in which farmers are exploring other kinds of food processing and income generation from their

FIGURE 33-11: A honey-laden comb is lifted from another hive in a fallow field hut.

Photo: Chalit Saphaphak.

swiddens. This may also contribute to a long-term increase in biodiversity as shifting cultivation areas become even richer from the constant and protected work of bees and other pollinators.

Table 33-2 shows the broad scope of the Hin Lad Nai bee-keeping venture, including the locations, shapes, sizes and materials used in the farmers' bee boxes. It also points to a seamless progression from forest to fallow honey production.

What does the future hold for Hin Lad Nai's fallow honey?

Table 33-2 shows that 14 families, or most of the families in the Hin Lad Nai community, have been involved in bee-keeping. However, despite the increasing popularity of the activity, the community's experimentation is far from finished.

TABLE 33-2: The experiences of Hin Lad Nai's farmers with bee-keeping in natural forest, shifting cultivation fallows and in settlement areas.

Farmer	No. of boxes	Harvest	Under big trees	Under cliffs or on rocks	Swidden fallows	wet rice fields	Village area	Square	Round	Materials for box	Small	Medium	Large
Nipon Pookhrae	Total 25 Bees 20 No bees 5		√	√	√	√	√	√	√	Melia azedarach Livistona speciosa Thysanolaena latifolia		√	√
Poolong Kham	Total 21 Bees 13 No bees 8		√	√				√	√	Paper mulberry (Broussonetia papyrifera)		√	√
Mano Phokha	Total 26 Bees 19 No bees 7		√	√	√			√	√	Small pieces of wood Livistona speciosa		√	√
Hsaf htoo paj	Total 43 Bees 32 No bees 11	70 kg		√	√	√	√			Livistona speciosa			√
Nivet Siri	Total 28 Bees 21 No bees 7	90 kg	√	√	√	√	√	√	√	Michelia alba Livistona speciosa Shorea roxburghii Star gooseberry		√	√
Chaiprasert Phokha	Total 40 Bees 32 No bees 8	91 kg	√	√	√	√	√	√	√	Livistona speciosa Thysanolaena latifolia Shorea roxburghii		√	√
Somchai	Total 5 Bees 3 No bees 2	4 kg	√			√		√		Livistona speciosa Thysanolaena latifolia		√	
Achira	Total 40 Bees 32 No bees 8	91 kg	√	√	√	√	√	√	√	Thysanolaena latifolia		√	√
Sukhawong	Total 6 Bees 2 No bees 4	4.5 kg	√			√		√	√	Livistona speciosa Thysanolaena latifolia		√	
Duangdee Siri	Total 19 Bees 5 No bees 14	3 kg	√	√				√		Livistona speciosa		√	√
Somboon	Total 15 Bees 5 No bees 10	15 kg	√	√			√	√	√	Livistona speciosa Thysanolaena latifolia		√	√

TABLE 33-2 (cont): The experiences of Hin Lad Nai's farmers with bee-keeping in natural forest, shifting cultivation fallows and in settlement areas.

Farmer	No. of boxes	Harvest	Location					Box shape		Materials for box	Size		
			Under big trees	Under cliffs or on rocks	Swidden fallows	wet rice fields	Village area	Square	Round		Small	Medium	Large
Name withheld	Total 26 Bees 23 No bees 3	102 kg	√	√	√			√	√	*Livistona speciosa* *Thysanolaena latifolia*			√
Name withheld	Total 64 Bees 47 No bees 17	90 kg	√	√						*Melia azedarach* *Dipterocarpus obtusifolius* *Erythrina variegata*			
Name withheld	Total 15 Bees 5 No bees 10	12 kg	√	√	√					*Dipterocarpus alatus* *Livistona speciosa*			

Note: *All bee boxes recorded in this table were placed in fallow areas by farmers and subsequently occupied by bees.

From the very beginnings of their experiment in 2010, the Hin Lad Nai farmers' curiosity about their insect colonies has led to intensive observation of bee behaviour. In the early years, they were keen to learn where the bees went in the wet season, after the honey harvest, when the bees abandoned their boxes. Careful observation

Saccharum spontaneum L. [Poaceae]

Wild sugar cane is a perennial grass species native to the Indian subcontinent. It grows up to three metres high, in dense swards with spreading rhizomatous roots and masses of flowering heads. Its ability to colonize disturbed soil has made it an invasive species that takes over croplands. In some places it is harvested to thatch roofs and make garden fences.

has suggested that the bees find their food over short distances from their hives; that there is no migration and the bee colonies continue to exist in the same forest, as long as they have sufficient honey to feed themselves during the 'hungry' months when the weather is bad and nectar is scarce. This revelation has sparked a process of refinement in the farmers' techniques for managing their bees.

During the honey harvest, the farmers will no longer collect honey from all of their hives, but will allow some of the bees to continue living, with their honey stores intact, so that the hives will not be unoccupied. Farmers must also observe each hive closely enough to know precisely the right time in the harvest season to take its honey. They point to indicators such as bees crowding the entrance to hives, and numbers of worker bees 'fanning' their wings at the hive entrance to maintain a passage of cooling air into a crowded hive. If the combs in a hive are not yet full, the farmers should not harvest the honey, but let the bees increase their stores until the hive is full. This should lead to swarming behaviour, with about half of the bees leaving the old hive to start a new colony. This is one way of increasing the number of bees.

New approaches to management will also dissuade farmers from harvesting honey outside of the seasonal harvest period, even if the combs are full. The Hin Lad Nai bee-keepers say that unseasonal harvesting stops the bees from swarming and expanding the number of colonies. In order to swarm, the bees need to gorge themselves on honey before taking flight, so they are prepared for the demanding tasks of setting up a new home. The farmers see this as letting nature take its course, in the belief that if they look after nature, then nature will look after them. There are also the long-term benefits arising from natural swarming, of expanding the number of hives and enhanced biodiversity because of more intensive pollination.

In future, the bee hives that remain occupied after the honey harvest will be left alone, while those that are no longer occupied will be routinely gathered and carried back to the village, to be stored in the months from about June to November. Continuing occupation occurs mostly in the hives placed in fallows, particularly those in old fallow huts. The empty hives will be returned to places in the fallows and forests around December or January, ready for the first swarms of the season and the peak period for nectar foraging.

Conclusion

After a logging company severely damaged Hin Lad Nai's forest and environment in the 1970s, the village's recent history includes a strong component of recovery and revitalization. The community turned the crisis into an opportunity by gradually and steadily recovering the forest and their natural resources, by combining traditional knowledge and customary natural-resource management with innovative livelihood options and modern knowledge and technology.

At the heart of their socio-ecological landscape is a traditional shifting cultivation system that sustains an exceptionally rich biodiversity of edible cultivars and semi

Melastoma malabathricum L. [Melastomataceae]

This erect, free-flowering shrub is among the contenders for space in recently fallowed swiddens. It grows on a wide range of soils, and has been declared a noxious weed in the United States. This species can accumulate high levels of aluminium that would be toxic to other plants, so it can be used to clean soils of harmful contamination.

domesticated crops. It is driven by a full range of the traditional knowledge and wisdom of the *Pgaz K'Nyau* (Karen) people, backed by centuries of experience in conservation and sustainable use of natural resources. Together with the village's wet-rice fields, the swidden system provides the community with its food security.

While bee-keeping in fallow fields is a relatively new innovation in Hin Lad Nai, it is already clear that farmers are keen to further develop and expand it. Already, the product from Hin Lad Nai's bee boxes is appearing on the shelves of domestic markets (Figure 33-12). The village's bee-keeping venture has thus become an important model for shifting cultivation communities elsewhere, that are striving to develop solutions and good practices to boost incomes and improve livelihoods while enriching the environment.

The village head of Hin Lad Nai, Chaiprasert Phokha, himself one of the bee-keeping innovators, provided a fitting summary in an interview:

> Bee-keeping is a natural process that is linked very closely to preservation; to taking care of and benefiting from nature; to managing natural resources in a balanced and sustainable way. It also helps to answer questions about how much farmers can benefit from a shifting cultivation system, because fallow land is not just land that we are allowing to regenerate. We also benefit from fallow land, so we take care of it and contribute to it; make it richer by increasing the number and species of plants and enhance

FIGURE 33-12: A jar of Hin Lad Nai honey, professionally packaged and ready for the supermarket.

Photo: Nutdanai Trakansuphakon.

biodiversity through the pollination process. Perhaps most importantly, this bee-keeping process creates income and a social enterprise for the local community, and this practical and visible income process creates inspiration for young people. They see that they can have a good life in the village through sustainable and innovative occupations based on good relations between nature and humans.

The children of Hin Lad Nai's farmers recently supported Chaiprasert's philosophy (Figure 33-13). They gave their views in January 2019 at

FIGURE 33-13: Eighteen-year-old Niraphaw Japhaw explains to an international forum in northern Thailand how the innovative farming systems at Hin Lad Nai have created a brighter future for village youngsters.

Photo: Maurizio Farhan Ferrari.

an international forum entitled 'Dialogue across Indigenous, Local and Scientific Knowledge Systems, Reflecting on the Intergovernmental Platform on Biodiversity and Ecosystem Services (IPBES) Assessment on Pollinators, Pollination and Food Production'. The village's youth representatives said that they now preferred to live in the village and earn income from bee-keeping and other forest-related activities, rather than moving to cities in search of work. They were therefore proud that their community was providing opportunities to live and develop in a healthy environment, following and building upon the knowledge and efforts of their parents and earlier generations.

References

Forest Peoples Programme, the International Indigenous Forum on Biodiversity and the Secretariat of the Convention on Biological Diversity (2016) *Local Biodiversity Outlooks. Indigenous Peoples' and Local Communities' Contributions to the Implementation of the Strategic Plan for Biodiversity 2011-2020. A complement to the fourth edition of the Global Biodiversity Outlook*, Moreton-in-Marsh, UK, pp.63-64 and 113-114

Ganjanapan, A., Pinkaew, L. A., Tawit, J., Paiboon, H., Artcharam, R., Wichien, A., Sakunee, N., Monthol, C., Phasutha, S. and Surin, O. (2004) *Rotational Farming System: Status and Trends*, no. 1, Faculty of Social Science, Chiang Mai University, Chiang Mai (Thai language)

NDF (2009) *Climate Change, Trees and Livelihoods: A Case Study of the Carbon Footprint of Pgaz K' Nyau Community in Northern Thailand*, Northern Development Foundation and Asia Indigenous Peoples Pact, Chiang Mai, Thailand

Raygorodetsky, G. (2017) 'Swidden Honey', in *Archipelago of Hope – Wisdom and Resilience from the Edge of Climate Change*, Pegasus Books, New York and London, pp.169-204

Seeley, T. D. (2009) *The Wisdom of the Hive: The Social Physiology of Honey Bee Colonies*, Harvard University Press, Cambridge, MA

Trakansuphakon, P. and Kamphonkun, T. (2008) *Rotational Farming System Knowledge and Practice of Pgaz K'Nyau People in Northern Thailand*, Indigenous Knowledge and Peoples, Chiang Mai, Thailand

Trakansuphakon, P., Sudsearee, N., Trakansuphakon, N., Phokha, C., Siri, Preecha, Siri, N., Vetchakit, P., Siri, D., Siri, Prasit and Sri-uangdoi, P. (2016) *Mobilizing Traditional Knowledge, Innovations and Practices in Rotational Farming for Sustainable Development*, The Community of Hin Lad Nai and Pgakenyaw Association for Sustainable Development, Thailand, and Swedbio at Stockholm Resilience Centre, Sweden

34

NEGOTIATING THE FOREST-FALLOW INTERFACE

Benzoin trees in the multifunctional shifting cultivation landscapes of Lao PDR

*Simone Vongkhamho and Micah L. Ingalls**

Introduction

As the Lao People's Democratic Republic (Lao PDR, or Laos) strives to balance the demands of rural development, poverty alleviation and environmental sustainability, the fate of the Lao uplands – the largely rural, mountainous areas along Laos's north and east, comprising nearly 70% of the national territory – have been thrust on to centre stage. These uplands are generally poorer than lowland areas along the Mekong and its tributaries. This is due in large part to their steeply-sloping terrain and limited fertile flat land for cultivation, as well as their inaccessibility, which has hindered agricultural investment and the penetration of development assistance. Shifting cultivation remains a dominant livelihood strategy in these areas. As of the last agricultural census (2011), about 1.4 million people, or 29% of the agricultural population, are engaged in shifting cultivation. They are distributed across half of all villages in Laos (Epprecht et al., 2018). Shifting cultivation is not only a leading livelihood strategy in the Lao uplands, it is also the largest agricultural land use. About 212,000 hectares are planted annually with upland rice, or 17% of the country's total rice-production land. But when the associated fallow areas are added to this figure, the total area involved in shifting cultivation amounts to as much as 6.5 million ha (Messerli et al. 2009), more than six times the total production area of wet rice, the country's second-largest agricultural land use. Shifting cultivation has a long history in the Lao uplands, dating back centuries, if not millennia, resulting in a complex landscape of fields, fallows and forests. This mosaic of land uses, habitats and associated species produces a unique, multifunctional socioecological system (Fox et al., 2000; Ingalls and Dwyer, 2016).

---

* Dr. Simone Vongkhamho is a senior researcher at the Forestry Research Centre, National Agriculture and Forestry Research Institute, Lao PDR; Dr. Micah L. Ingalls is a senior scientist at the Center for Development and Environment, University of Bern, Switzerland.

However, the persistence of shifting cultivation is perceived by many development agencies and policymakers as an obstacle to both development and environmental sustainability. Land-sparing models that emphasize intensification and commercialization within existing agricultural land, alongside enhanced forest and environmental conservation, have become increasingly dominant. Within the land-sparing concept, shifting cultivation is a misfit; it is a livelihood strategy that is neither apparently intensive nor considered compatible with forest conservation. As a consequence, development and conservation policies have long sought to eradicate or stabilize shifting cultivation. These efforts have recently redoubled as Laos strives to achieve the United Nations' Sustainable Development Goals; to lift itself from Least Developed Country status and to reach the government's target of achieving 70% forest cover by 2020.

Development practitioners, conservationists and state foresters have long sought viable alternatives to shifting cultivation in the Lao uplands, but these have generally remained elusive. Nevertheless, restrictive policies continue to diminish the land available for shifting cultivators, resulting in shorter fallow periods (Fujita and Phanvilay, 2008; Kenney-Lazar, 2010). The shortening of shifting cultivation fallows has made way for the expansion of large commercial plantations and land concessions, alongside increased forest areas, all to the detriment of poor shifting cultivation households. Declining yields, soil fertility and fallow health and the replacement of natural vegetation with non-native monocultures has, in the main, been disastrous for both shifting cultivators and, perhaps paradoxically, the natural environment (Foppes and Ketphanh, 2004; Rerkasem et al., 2009; Mertz et al., 2009). This has often undermined efforts to achieve either rural development or environmental sustainability.

Lao PDR is at a crossroads, faced with disparate possible futures. On one hand, it may follow a land-sparing development pathway that emphasizes intensification through commercialization, investment in commodity-oriented exports and increased foreign direct investment alongside the modernization of forest-conservation measures. On the other, there is the possibility of land-sharing for development and conservation, building on the unique strengths of its agricultural and ecological base and the indigenous cultural practices of the country's multifunctional uplands. Depending on the direction it follows, shifting cultivation faces an uncertain and increasingly threatened future in Lao PDR.

The success of the multifunctional, land-sharing development and conservation pathway depends on a number of inter-related factors. The most important of these is the ecological and socio-economic viability of shifting cultivation systems and the degree to which rural households are able to secure access to and control over their resources.

The viability of shifting cultivation relates in some measure to the management of fallows as an intermediate land use between cropped lands and forests. This is due to the importance of fallows for:

- their ability to regenerate soil fertility and reduce pests and weed pressures during the cropping stage;
- their abundance of wild, semi-domesticated and domesticated species, including both non-timber forest products (NTFPs) and planted crops, and the role these play in household consumption and revenue generation; and
- the various other ecosystem services that depend on them, including water provisioning, carbon sequestration and wild biodiversity.

In general, the value of fallows increases with fallow length (de Rouw, 1995; Mertz, 2002; Xu et al., 2009; Cramb et al., 2009). In addition to the direct benefits for agricultural production, fallow length is positively associated with the value of other ecosystem services relating to wild biodiversity, water provisioning and climate regulation (van Vliet et al., 2012; Fox et al., 2014; Ingalls and Dwyer, 2016).

Without adequate security of upland resources, local communities will fail to realize the benefits of long fallows. While formal tenure security in Laos is low (Broegaard et al., 2017; Ingalls et al., 2018), shifting cultivation areas – which are typically managed as communal commons – are particularly insecure and under current laws, are generally ineligible for titling or other types of formal tenure recognition.

In the light of these threatening conditions, this chapter focuses on two inter-related innovations in the Lao uplands that provide some measure of hope. First, we highlight *Styrax tonkinensis*, a species that produces benzoin resin. It is highly adapted to long-fallow shifting cultivation systems and has enabled local communities to benefit from increased fallow lengths and enhanced overall system viability. Second, we examine the ways in which community-led land-use planning – best exemplified in the participatory approach to forest and agricultural land-use planning and management (FALUPAM) – has enabled benzoin producers to demonstrate customary-resource claims and negotiate a politically acceptable space for long fallows.

Stryrax tonkinensis (Lao Benzoin or, locally, *yarn*), is a tree species native to Laos and Vietnam. It is closely related to other species found throughout Southeast Asia, Asia Minor and South America. It produces a resin that has been used in and exported from Indochina since the 16th century. Benzoin resin is used in incense, cosmetics and inhalants, as well as in Chinese traditional medicine. The composition of Lao Benzoin – comprising about 65% Coniferyl benzoate, 10% Benzoic acid and other aromatic esters – is particularly suitable for producing high-value perfumes and cosmetic products in the European Union, especially in France, where benzoin from Laos claims about 70% of the market (FAO, 2014).

S. tonkinensis, which is a scarce species in global terms, but abundant locally, is emblematic of a group of endemic species and cultivars in Laos that are highly adapted to shifting cultivation systems. It is a pioneering species that occupies forest breaks and newly-cleared land. It also appears to be semi-pyrophilic, as burning seems to invigorate germination and promote seedling emergence. As a medium-term woody species, *S. tonkinensis* has specific agroecological needs that make it particularly suited to long-fallow systems. It thrives across the entire spectrum of successional stages in the shifting cultivation landscape, from cropping to fallow, to forest.

In the uplands of Laos, *S. tonkinensis* is self-generated within rice fields and managed fallows, along with more than one thousand other NTFPs that together comprise the complex food-production systems of Laos's shifting cultivation landscapes. These NTFPs, which are integral to local livelihoods and the area's complex biotic interactions, play a fundamental role in the social-ecological systems of the Lao uplands, providing food and medicine, household income and regulatory benefits. These services are especially valuable for low-income households that rely on these resources most directly, and that also tend to lack other resources to buffer shocks and disturbances such as those resulting from climate change. In the uplands of Laos, benzoin and other NTFPs provide as much as 45% of cash incomes and nearly 50% of non-cash resources for upland families (Weyerhaeuser et al., 2010). However, these benefits are not restricted to upland communities. Nationally, trade in NTFPs comprises 19% of all exports from Laos (see Wiemann et al., 2009).

Styrax tonkinensis Craib ex Hartwich [Styracaceae]

A fast-growing, light-demanding tree that invades forest gaps and disturbed land. The resin tapped from its trunk is thick and brownish-yellow, with a sweet balsamic odour and a hint of vanilla. Its many uses range from medicinal – as an inhalant and an antiseptic; to culinary – as a food flavouring; and cosmetic – in creating fragrances. Elsewhere, the species is also an important pulpwood crop.

Despite their significance, NTFPs – and perhaps benzoin in particular – are under threat. The expansion of commercial markets and the related boom in export-oriented commodity crops in the uplands, the reduction in land available for shifting cultivation, poor management and overharvesting have had a significant impact on NTFP diversity and abundance. More than 15% of NTFP species are at risk of extinction, while 80% of all NTFP species have shown significant decline (NAFRI, NUoL and SNV, 2007; Ketphanh and Vongkhamho, 2008). The obligate relationship of *S. tonkinensis* with long-fallow systems presents acute obstacles to its future as fallow periods continue to decline. Despite this, many communities have used the benefits of benzoin to enhance returns from long-fallow systems.

Understanding of the agroecological, production and marketing needs of *S. tonkinensis* and its benzoin product is critically limited. Likewise, there is little knowledge of the ways in which traditional *S. tonkinensis* cultivation intersects with national development and conservation policies and ongoing efforts to secure land tenure in Laos's uplands. Our research, which was carried out in partnership with producing communities, private-sector partners,[1] government agencies and The Agrobiodiversity Initiative of the Lao PDR (TABI), addresses this need. In the sections that follow, we provide some context of the benzoin-production landscape in the Lao PDR and describe key attributes of the benzoin-production system. We also offer a comparative analysis of alternative land uses and explore some ways in which participatory land-use planning has functioned to support both benzoin production and long fallowing, while providing a measure of tenure security for producing communities. We close by drawing out some implications relating to national policies on development and conservation.

Assessment of benzoin production practices and producing communities

The initial stages of the research process involved the collation and review of existing documentation, reports and market information relating to the agroecological dynamics of *S. tonkinensis* and the benzoin value chain, along with relevant policy domains that influence them. Published information was sourced through Internet and library resources, especially from the archival resources of the Lao National Agriculture and Forestry Research Institute (NAFRI), and through extensive canvassing of stakeholders involved in all stages of the benzoin supply chain, from producers to processors, exporting companies and government officials. The results of this review served as a baseline upon which structured field observations, participatory research and data collection were carried out between 2015 and 2018.

Research was conducted in five upland provinces of northeastern Laos, across the known range of benzoin: Xiengkhouang, Houaphan, Luang Prabang, Oudomxay and Phongsaly. In these provinces, the research team conducted extensive field studies and interviews of experts across 17 districts (Figure 34-1), aimed at investigating the distribution of *S. tonkinensis* trees. There was a greater in-depth analysis of key benzoin-producing areas in 11 focal districts (Table 34-1). In total, field interviews involved 128 individuals, including government authorities, and focused particularly on local experts in producing villages.

After completion of the surveys, the team mapped the distribution of *S. tonkinensis* and made an inventory of the current benzoin-production areas with their corresponding yields. Data was cross-analyzed with regard to agroecological conditions, production practices and local knowledge, as well as the effectiveness of community-led land-use planning, using the participatory forest and agricultural land-use planning and management approach. We also conducted a brief comparative analysis of costs and benefits associated with two common livelihood strategies in benzoin-producing areas: the production of rice and maize. Analysis of government programmes and policies was carried out in order to identify key gaps and potential leverage points.

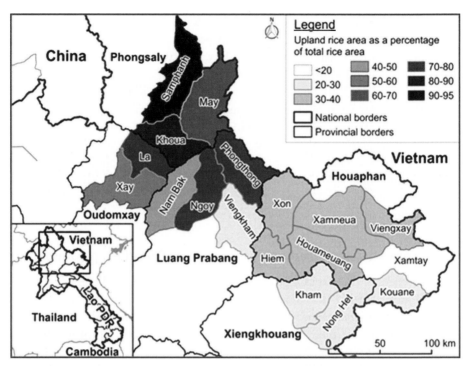

FIGURE 34-1: Benzoin-producing districts and shifting cultivation of rice in Lao PDR.

Data source: Epprecht et al., 2018.

TABLE 34-1: Key benzoin-production areas in Lao PDR.

Province	District	No. of villages	Households (hh)	Area (ha)	Average (ha/hh)
Houaphan		**69**	**1397**	**2432**	**1.74**
	Houameuang	3	13	9	0.69
	Xamneua	39	794	1541	1.94
	Viengxay	7	150	379	2.53
	Xamtay	11	219	423	1.93
	Kouane	9	221	80	0.36
Luang Prabang		**34**	**730**	**1528**	**2.09**
	Nam Bak	12	285	536	1.88
	Ngoy	15	271	578	2.13
	Phonthong	7	174	414	2.38
Phongsaly		**81**	**1158**	**1347**	**1.16**
	Khoua	41	623	350	0.56
	May	9	167	120	0.72
	Samphanh	31	368	877	2.38
Total		**184 villages**	**3285 hh**	**5307 ha**	**1.62 ha**

Sources: Field surveys, 2015, 2016 and 2018.

The context of benzoin production in Lao PDR

As mentioned earlier, benzoin production has a long history in the uplands of Laos and is rooted in traditional practices and local socioecological systems. In the recent past, exports of benzoin resin amounted to about 50 tons per year, but records were not systematically maintained nor were production areas officially registered. Between 1996 and 1998, the Food and Agriculture Organization of the United Nations (FAO) carried out a survey of benzoin production areas in Luang Prabang, covering about 1200 ha. At that time, there was no information about *S. tonkinensis* production in other provinces. According to the data collected by the FAO, benzoin production was generally on the decline. Key causal factors in this decline were identified as (1) livelihood transitions away from benzoin and shifting cultivation towards maize production, livestock raising, cardamom cultivation and others, and (2) the migration of young people to urban areas, leading to a reduction in the rural workforce. Similar dynamics remain relevant today, but have intensified due to policy measures promoting agricultural intensification and the restriction of shifting cultivation within state forests.

Our benzoin survey results indicate that there are at least 184 producing villages across the three focal provinces of Houaphan, Luang Prabang and Phongsaly, involving 3285 families (or about 19,700 individuals) and 5307 ha currently under production. Average smallholding stands of *S. tonkinensis* cover about 1.62 ha per family (Table 34-1). Of the three provinces, Houaphan has the largest area of benzoin resin production (2432 ha), amounting to 46% of the three-province production area, involving 1397 families in 69 villages, with an average of 1.74 ha/household. Within this province, Xamneua district has the largest concentration. Luang Prabang province has the second-largest area of production, followed by Phongsaly. There are known but unmeasured production areas in Oudomxay and Xiengkhouang provinces. Benzoin-producing communities and households typically have low to very-low harvests of wet rice and are instead primarily dependent on shifting cultivation of upland rice. In benzoin-producing districts, the area of upland-rice production, as a percentage of total rice-production areas, ranges from just below 20% to more than 90% (Figure 34-1). Households engaged in benzoin production tend to be poorer and more remote from urban centres.

Benzoin production has a long historical association with the ethnic Khmu,[2] who cultivate around 90% of all *S. tonkinensis*. For many Khmu communities, benzoin production is an integral element of traditional practices and beliefs that are highly-adapted to remote areas. By providing a product that has a low volume-to-value ratio, a long shelf-life, and a strong market demand, benzoin production is suited to the dynamic, mobile livelihoods of the remote Lao uplands (cf. Scott, 2009). While the Lao ethnic majority comprises only 1% of benzoin-producing households, they are disproportionately involved in processing and exports, thus capturing an outsized portion of the total value of the benzoin trade. For all producing communities and households, benzoin supplements livelihood strategies that are dominated by rice, providing cash income for expenses and a safety-net during periods of shortfall.

Silvicultural techniques for cultivating *Styrax tonkinensis*

S. tonkinensis grows well in acidic soil with a pH level lower than 4.5 and organic matter content higher than 2%. Higher resin yields generally come from areas with higher potassium (K) soil content, and where clay loam soils predominate in the upper soil layers – with good drainage, as the trees are intolerant of waterlogging. While *S. tonkinensis* plantations are found as low as 750 m above sea-level (masl) within the research area, the majority of high-yielding stands occur above 1000 masl.

S. tonkinensis is a pioneering species that establishes in upland fields following vegetation clearance and burning. Most cultivated trees occur naturally, emerging from seeds and rootstocks within newly-cleared fields. However, some farmers systematize the silviculture of *S. tonkinensis* by planting seeds in freshly cleared swiddens between February and early March. About 15 days after slashing the fallow vegetation and leaving it to dry, farmers who plant supplementary *S. tonkinensis* push small holes in the ground with a dibble stick and place five to eight seeds in each hole. Spacing is 5 m by 10 to 15 m, resulting in between 130 and 200 holes per hectare. The cut vegetation is left to dry for about one month. Then, depending on weather conditions, it is burnt to clear the field for planting rice and other vegetables and herbs, and to trigger the germination of the *S. tonkinensis* seeds. When the rains begin (in late April or May), the farmers plant rice by dropping seeds into holes spaced about 25 cm by 25 cm apart. The rice and tree seeds usually germinate together in late May, although some *S. tonkinensis* seedlings emerge later.

Over the following year, the farmers cultivate the rice while managing the emerging *S. tonkinensis* seedlings. Depending on the length of the previous fallow, the field is weeded two or three times, first in June and July and then again in September and October. During the first weeding, the *S. tonkinensis* saplings are handled carefully. They may be about 20 cm tall, and at this stage the farmers select the strongest seedlings with the highest potential for future resin production. Preferred characteristics include dark green bark and dark green, curling leaves. The farmers say that saplings with straight trunks, light green bark and straight leaves yield lower quantities of resin. Selected seedlings are retained, while crowded and weaker seedlings are weeded out. The farmers generally prefer to keep one or two seedlings per clump during this initial thinning, leaving between 130 and 200 seedlings per hectare or, in fields where they prefer a denser stand, as many as 400 seedlings per ha. In some cases, healthy seedlings found in dense clumps are transplanted for spacing, but these have a high casualty rate. The seedlings are maintained and tended during the second year of rice cultivation, when healthy saplings reach a height of two to three metres. Farmers then cut off the top shoot to promote lateral branching, a practice that enhances future tapping potential and overall resin yield.

Rice and other indigenous species and perennials such as broom grass (*Thysanolaena latifolia*) are commonly intercropped with the growing *S. tonkinensis* trees through the second cropping year, after which canopy closure precludes rice cultivation. Shade-tolerant species and NTFPs such as galangal (*Alpinia* spp.), mushrooms (*Russula* spp., *Lentinus* spp., *Termitomyces* spp., etc.), cardamom (*Amomum villosum*), and bamboo

(*Bambusa* spp., *Indosasa* spp., *Phyllostachys* spp. etc.) continue to be grown within the maturing stands of *S. tonkinensis*.

Tapping and harvesting

Benzoin producers generally begin tapping *S. tonkinensis* trees when they are about eight years old, when the bark is thick enough to produce a large volume of high-quality resin while ensuring the longevity of the tree. Producers continue to tap until the tree dies, normally sometime after the 15th year, although some trees continue to provide resin until after their 20th year. Ideally, tapping begins between May and June, when the trees are in flower and there is a higher volume of resin. However, the availability of labour is lowest during these months, when households are commonly involved in rice planting and weeding. Consequently, many households do not begin tapping until after August, when the second weeding has been done, up until November, prior to rice harvest. Field surveys across all producing areas indicate that August is the most intensive tapping period on average, involving 38% of producers.

The main tool used in tapping is a sharp knife; farmers cut the bark deeply to reach a small portion of the xylem and then push down about 8 to 10 cm to make the shape of a valve to receive the resin (Figure 34-2). Tapping is divided into portions, depending on the health of the tree and the skill of the tapper. Lower tappings are done within the first two metres of the trunk, from 20 cm to 1 m from the base of the tree (Figure 34-3). The second set of tappings is done between two and four metres above the base, and a third set of tappings is done between four and six metres up the trunk (Figure 34-4). In each group, the tree is tapped on two to four faces, with about three tappings per face (thus there are six to 12 taps in each set of tappings). The tapper lashes a wooden step to the tree with rope, and stands on this to reach the higher tapping positions. In traditional practice, an

FIGURE 34-2: Tappers cut away the bark to reach the xylem, then make a shape to receive the resin.

FIGURE 34-3: Lower tappings are up to one metre from the base of the tree.

Photos: Simone Vongkhamho.

individual tree will be tapped only in alternating years, to maintain yields and prevent premature tree die-off. In cases where trees are tapped only in the lower position, one tapper can handle six to eight trees per day, but difficulties in accessing the second and third positions limits the worker to about four trees per day, on average.

The trees are left for some months, allowing the resin to collect and dry at each tap position. The main harvesting period is from November to April, outside of the rice season when labour is available. The most intensive month of harvesting is December, following the rice harvest. During the harvesting of the resin, farmers use a knife or another sharp tool to cut off the bark and carefully pick the hardened resin off the tapped trees (Figure 34-5). As in the tapping procedure, rope and wooden steps are used to reach the higher tapping positions. The resin is kept in baskets until it is sold to traders.

FIGURE 34-4: Tapping the trees at higher levels requires ropes and a pole to stand on.

Photo: Simone Vongkhamho.

Yield and tree death

According to participants in the study, resin yield per tree varies considerably, depending on the health and age of the tree and the experience and technique of the tapper. Lower yields range around 270 grams/tree to higher-level yields of around 540 g/tree. Younger trees – those between five and seven years old – produce substantially less. In Houaphan, for example, younger trees produced 36% less than trees eight years old and over, and 47% less than trees older than 11 years. In other areas, where tree-stand densities are greater, yield variations between age classes are less pronounced, possibly due to crowding. Total annual yield per family also varies significantly, depending on the availability of labour, tree numbers and stand age, ranging from 12.54 kg to 24.23 kg. Wide variations in yield suggest the need for further assessment.

While it was observed that tree mortality rates increase considerably following 15 years of age, many trees survive for more than 20

FIGURE 34-5: After some months, the dried resin is picked off the trees into baskets.

Photo: Keooudone Souvannakhoummane.

years. However, the main constraint on the tapping age of trees is the length of the shifting cultivation cycle, with most farmers clearing and burning their stands of *S. tonkinensis* at a relatively young age (a point we return to below).

The benzoin value chain: Processing and pricing

Benzoin is commonly sold to agents of exporting companies in producing villages. These agents usually transport the resin to the Laotian capital, Vientiane, for processing, during which around 11.4% of the purchase weight is lost in cleaning. Processed benzoin is exported to several countries, although the formal market is dominated by exports to European Union countries and the United States. The benzoin market in Lao PDR is dominated by two companies, which together command nearly 90% of formal trade. As well as the registered companies, there is a large but unregulated trade carried out by independent traders from China and Vietnam, who purchase raw benzoin at village level and export it 'unofficially'. Administrative practices relating to quotas, taxation and exports are highly variable and weakly regulated, creating market uncertainties for producer villages and companies throughout the value chain.

While Fischer et al. (2007) observed that the price of benzoin had declined considerably from historic levels, systematic production and trade data was generally lacking at that time. They reported that low and uncertain benzoin pricing had led to overall reductions in benzoin production across the country. However, since that time the price of benzoin has increased considerably, with average farm–gate prices currently at around US$17/kg.

Despite consistent increases in the market price of benzoin, available data indicates that production rates have been variable; there was a sharp increase between 2009 and 2011, but production has fallen considerably since then (Figure 34-6). The reasons are numerous. Recent years have seen a reduction in production area as *S. tonkinensis* forests have been squeezed by the expansion of boom crops – the success of which are partly attributable to favourable policies and promotion by state agencies – alongside enhanced forest conservation efforts that have restricted shifting cultivation in the Lao uplands (see below). However, anecdotal evidence suggests that there has also been a rise in informal trading, chiefly through

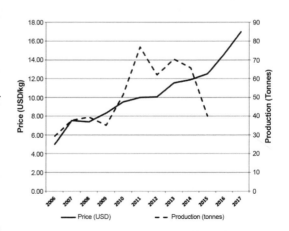

FIGURE 34-6: Benzoin pricing and production in the Lao PDR, 2006–2017.

Source: Field surveys, 2015 and 2016.

Vietnamese and Chinese middlemen, and a growing illicit cross-border trade in benzoin. It is unclear to what degree this informal trade affects production statistics, as these record only formal production and trade.

Benzoin in the short(ening) fallow

As mentioned above, the annual and total yield of benzoin is considerably higher in long-fallow shifting cultivation regimes. Figure 34-7 shows the observed relationship between tree age and bark thickness (the most predictive factors for yield rates) over time, compared with harvesting periods under short- and long-fallow swidden systems. Historically, producers did not tap trees younger than eight years old, due to low rates of yield and higher risks of premature tree death. However, fallow lengths have recently decreased considerably, as has the total area of upland-rice production. This largely accounts for the drop in overall production and per-household benefits from benzoin production, despite rising prices. Many producer communities now tap trees at year five, and continue to tap for only one or two years more before the field is cleared for fresh rice production. Moreover, many farmers tap trees annually, rather than in alternating years, due to the limited time available for tapping under short-fallow regimes. In 2015, for example, 88% of trees that were tapped were younger than 10 years old.

Short-fallow shifting cultivation systems have become the norm in the uplands of Lao PDR. Crucial to this transition has been the perception that shifting cultivation is unsustainable, environmentally destructive and, fundamentally, a 'backward' agricultural practice inconsistent with a vision of modernization, economic development and national identity that is largely rooted in lowland Lao cultural values (Baird and Shoemaker, 2007). Policies and state programmes have endeavoured to eradicate or stabilize shifting cultivation through semi-voluntary and involuntary resettlement (primarily before the early 2000s), various shifting cultivation bans, subsidization and promotion of alternatives (such as growing maize, cassava, rubber and other commodity crops produced primarily for export markets) and forest-conservation measures (Fujita and Phanvilay, 2008, Ingalls

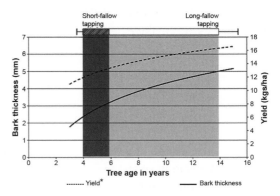

FIGURE 34-7: *S. tonkinensis* bark thickness and yield, over time, in short- and long-fallow systems.

Note: * Yield varied greatly in sampled fields. Yield values here are based on Luang Prabang, where an average of 167 trees were tapped per hectare. The figure depicts the logarithmic trend of averaged values per age–class

Source: Field surveys, 2016.

and Dwyer 2016). These drivers intersect with national policies focused on the concept of 'turning land into capital', a policy direction that seeks to use Laos's large land-resource base to attract (principally foreign) investments, principally (but not solely) through the granting of large-scale land concessions and commercial operations. As of 2017, land concessions covered 1.1 million ha of land (Hett et al., 2019), creating a leading driver of land-use change in the Lao uplands. While population growth in the uplands is commonly blamed for the reduction of land available for shifting cultivation, this is largely untrue. Between 2005 and 2015, the rural population of Laos generally remained stable, at around 4 million people. The agricultural censuses of 1999 and 2011 – the only systematic sources of data specifically on the shifting cultivator population – indicate that the number of households involved in these rice-production systems actually declined, from around 260,000 households to 240,000.

Croton cascarilloides Raeusch. [Euphorbiaceae]

A shrub growing up to two metres, this species is called *pao ngern* in the benzoin-producing forests, and is harvested in the wild for its medicinal properties. The bark and roots are antipyretic, and are used to prevent or reduce fever. It flowers and bears fruit all year round.

Assessing the alternatives: A comparative analysis

Land-sparing approaches to development and conservation in Laos have focused not only on restricting access to forest areas and reducing shifting cultivation, but also on the promotion of alternatives. To assess the implications of these alternatives we carried out a limited comparative analysis between the production of benzoin with shifting cultivation of rice and the production of maize, a common alternative. Systematic assessment was carried out with participating farmers within the research area to quantify and compare the costs and benefits associated with these alternatives. An analysis then focused on returns per labour and per hectare, both of which are limiting input factors in upland cultivation systems. Costs associated with purchased inputs were deducted from revenue under each scenario to allow for direct comparisons. Yield rates under each scenario were averaged across production years for simplification (Table 34–2).

TABLE 34-2: Comparison of cost benefits between benzoin resin and shifting cultivation of rice and maize.

Product	Labour costs (Person days/ ha)	Yield (kg/ha)	Unit value (US$/kg)	Input costs (US$/ha)	Benefit (land) (US$/ha)	Benefit (labour) (US$/person/ day)
Benzoin resin	35	41.8	18	6.96	745	21
Upland rice	271	2090	0.36	19.92	732	3
Maize	113	4200	0.144	59.16	546	5

Sources: Field surveys, 2016.

Labour requirements for benzoin, rice and maize vary considerably. Benzoin tapping has the lowest labour requirements (35 person days/ha/yr), followed by maize (113 person days/ha/yr) and upland rice (271 person days/ha/yr). While the volume of yield for benzoin is much lower, a high price-per-volume ratio means that benzoin compares favourably with maize and rice, particularly with regard to returns per unit of labour, where return differential is four- and seven-times higher, respectively. Returns per unit of land are also enhanced under traditional co-production of rice and benzoin. Maize is not suitable for intercropping with benzoin due to the shade-intolerance of *S. tonkinensis* during its early seedling stages. Benzoin production thus shows particular potential for lower-income families with limited labour resources and capital inputs.

The comparative analysis above does not include revenue generated from intercropped species and NTFPs. In the Lao uplands, numerous vegetables and other products are grown within cultivated fields and fallows, including *S. tonkinensis* fallows, contributing a significant proportion of household consumption and income. While NTFPs are commonly thought to be primarily associated with forests, this is not typically the case; the abundance and diversity of such species are far higher in fallows. An assessment of the contribution of NTFPs to household income across 225 villages in Laos's northern

Litsea glutinosa (Lour.) C.B.Rob. [Lauraceae]

Called *mee tho* in northern Laos, this evergreen tree has multiple uses. The fruit has a sweet creamy edible pulp; the roots, bark and leaves are used medicinally to reduce fever or swelling and to treat diarrhoea; the timber can be used to make furniture; and the roots yield a fibre that is used to make rope.

uplands under the TABI project indicated that 48% of the value of NTFPs was derived from shifting cultivation fields and fallows, versus only 10% from mature forests. In the surveyed villages, NTFPs contributed as much as 33% of total household income (Ingalls and Roth, 2018).

The comparison of costs and benefits above also excludes key values not easily captured, such as those producing non-monetary environmental and social values. Maize cultivation is representative of the alternative land-use systems promoted for intensification under the land-sparing paradigm. Maize cultivation entails high-levels of external inputs, including herbicide and insecticide applications and inorganic fertilizers, as well as modern technical expertise. The short-cycle annual production system of maize also presents particular risks related to soil erosion. While no systematic comparative assessment was done in this study with regard to these factors, some reasonable inferences can be drawn. The known impacts of pesticide application and intensive fertilizer regimes are well-documented, particularly within the context of the Lao PDR, where poor regulation of such inputs and inadequate management have been shown to present substantial and far-reaching risks to local communities and natural ecosystems. These, together with the typically low biodiversity of monocultural plantations of non-native species like maize, suggest that *S. tonkinensis* cultivation – which employs local knowledge and a long-fallow system in which chemical inputs are not typical – compares favourably beyond economic returns alone.

Benzoin at the fallow-forest interface

Upland shifting cultivation systems are a poor fit within current policies that emphasize land-sparing (intensification of agriculture within limited areas alongside strict forest protection and environmental conservation) over land-sharing (extensive integration of human-environment systems across multiple land uses). Yet the latter systems – and particularly those with long-fallows – are a critical precondition for the benzoin production cycle as well as numerous other environment services that characterize the multifunctional uplands of Lao PDR. This places benzoin production in a precarious position vis-à-vis the directions of national development and conservation. A very practical issue relates to the ways in which forest and agricultural land are defined, and the ways in which these definitions intersect with applicable legislation. The official definition of forest in Lao PDR – the areas to which forest-protection measures ostensibly apply - includes areas larger than half a hectare, wherein canopy closure is 20% or greater, tree height is 5 m or more and tree diameter at breast height (dbh) is 10 cm or more. As stated above, the success of benzoin production depends on fallow length, relating to tree maturation and thus yields. Ideally, *S. tonkinensis* trees are not tapped until they reach eight years old. By age 15, resin production declines while rates of tree death increase rapidly, requiring clearance and burning to begin the swiddening cycle anew. However, even by the fifth year of growth, *S. tonkinensis* stands may already have reached the thresholds of the forest definition

(Figure 34-8). The clearance and burning of *S. tonkinensis* stands, which is necessary to ensure continued production, is in conflict with forest-protection legislation and is thus forbidden, particularly under recently-enhanced regulatory approaches to conservation.

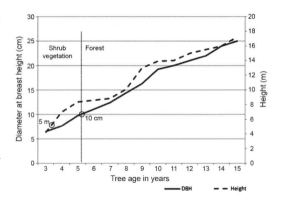

FIGURE 34-8: *S. tonkinensis* growth rates and land-cover classifications.

Source: Field data, 2016.

In practice, the enforcement of forest-conservation laws as they apply to shifting cultivation and *S. tonkinensis* cultivation has been mixed. Lack of clarity with regard to legal provisions, contradictory policies and the local political consequences of implementation have often chilled enthusiasm for strict enforcement. Until now, this has resulted in the tenuous persistence of shifting cultivation, albeit under increasingly restrictive conditions. Nevertheless, this study identified forest-conservation policies as a leading reason for the reduction in fallow periods and the abandonment of *S. tonkinensis* cultivation, consistent with a trend that is widely acknowledged across the country (Fujita and Phanvilay, 2008; Schmidt-Vogt et al., 2009).

Despite the historic ambiguity that has allowed the persistence of shifting cultivation and benzoin-production systems, recent policy shifts and a renewed intensification of forest-conservation efforts provide ample reason for concern with regard to the fate of shifting cultivation in general, and long-fallow systems in particular. The Politburo's *Resolution on the Enhancement of Land Management and Development in New Period* (2017) reiterated the Government's long-standing goal of achieving 70% forest cover by 2020. Despite current forest cover being only about 63.5% (DoF, 2018), the 70% goal is also a precondition for Laos's bid to graduate from Least Developed Country status by that same year. Without a radical redefinition of what constitutes forest, it can be argued that this is impossible to achieve while shifting cultivation continues to occupy such a dominant role in the Lao uplands. The eradication of shifting cultivation and the succession of fallows to forest is widely viewed as the 'low-hanging fruit' for the achievement of national forest-cover goals (Dwyer and Ingalls, 2015; cf. DoF, 2018). High-level endorsements (such as the Politburo resolution) have galvanized those forest-conservation advocates who are opposed to shifting cultivation. This is reflected in recent revisions of the Forestry Law and the National Land-use Master Plan, each of which envision the near-term replacement of shifting cultivation with intensive 'sedentary' alternatives.

There may be another way to achieve these national-development and forest-conservation goals, one that builds on (rather than undermines) indigenous livelihood systems rooted in Laos's multifunctional landscapes and agroecological diversity. *S. tonkinensis* cultivation in long-fallow systems is representative of such an alternative option. Benzoin production within long-fallow shifting cultivation has all the hallmarks of the plantation-forestry alternatives proposed by forest-conservation advocates. Like rubber (*Hevea brasiliensis*), which is regularly touted as an alternative for upland systems, *S. tonkinensis* is a fast-growing species that rapidly matures to forest, produces a high-value resin with a strong global market demand and has a low labour-to-area ratio suited to the low labour density of the Lao uplands. Old *S. tonkinensis* trees whose resin-yields have declined have also shown some promise as sources of pulp and plywood in Vietnam (Fischer et al., 2007). This

Castanopsis hystrix Hook. f. & Thomson ex A. DC. [Fagaceae]

This evergreen broadleaf tree grows to 30 metres and is commonly found in the forests of northern Laos. The seeds are edible either raw or cooked. Such is the value of its hard wood that southern China has plantations of *C. hystrix* covering about 5000 hectares. The timber is used for both construction and making furniture.

is similar to pulpwood species such as eucalypts (*Eucalyptus* spp.) and acacia (*Acacia mangium*), both of which are other proposed alternatives for reforesting the Lao uplands.[3] However, unlike rubber and eucalyptus, *S. tonkinensis* has three distinct advantages: (1) it is suited to intercropping with rice and other species; (2) it has deep biocultural associations with upland communities and uses local agroecological knowledge (versus external knowledge and resources) in its cultivation and use; and (3) it is an endemic species embedded within local agroecosystems, constituting an integral element in multifunctional upland landscapes.

While policies to protect and promote *S. tonkinensis* and other endemic fallow species and the shifting cultivation landscape more generally are a necessary precondition to realizing their potential for livelihood and ecosystems benefits, such policies alone would not be enough. There must also be adequate recognition of customary use and tenure over these resources. As we have said, tenure security within the uplands of Lao PDR is weak, particularly for communally-managed shifting cultivation fallows. In recent years, innovative approaches such as participatory forest and agricultural land-use planning and management have emerged that embrace indigenous land-use systems and traditional livelihood practices and act to demonstrate and defend customary land-use claims within these areas. In so doing, such approaches create the enabling conditions for multifunctional, long-rotation shifting cultivation systems and sustainable benzoin production. By using participatory forest and agricultural land-use planning and management, local communities have been able to document customary land uses and demonstrate tenure claims in a way that has proven largely acceptable to government authorities. At the time of writing, participatory planning for collective rotational shifting cultivation within existing fallows has been used by more than 200 communities to increase both fallow lengths and forest cover under village management (Dwyer et al., 2018). At the same time, this has generated a measure of formal recognition of customary claims over land and forest resources.

While 'win-win' is certainly a fraught term, we cautiously suggest that cultivation of *S. tonkinensis* within long-fallow shifting cultivation systems, made possible by community-led land-use planning, may be just that: local innovations that enable both enhanced forest cover during the long period of mature tree management and tapping, and the creation of space for the return of long-fallow rice production, together with the numerous environmental and social benefits that accompany them. Achieving such a 'win-win' would, however, require a more enabling policy environmentally grounded in land-sharing (rather than land-sparing) paradigms in multifunctional landscape systems, within which long-term tenure security was ensured. At present, despite the partial security afforded by land-use plans, stands of *S. tonkinensis* in shifting cultivation fallows remain largely insecure in terms of local tenure claims because they have yet to receive official validation, similar to that of the alternative sedentary crops actively promoted by state agencies. While tree plantations of eucalypts, rubber and other commercial crops receive such protection, *S. tonkinensis* continues to be seen as part and parcel of the shifting cultivation system demonized by many forest-conservation advocates, particularly since *S. tonkinensis* stands are cleared and burned, a practice with all the trappings of 'slash-and-burn' farming.

Formal recognition of both customary tenure and culturally embedded management regimes is essential if a sustainable future for Laos's multifunctional upland landscapes is to be ensured. This is currently lacking. While formal collective land titling has been piloted in Laos on a limited basis, it is unclear whether it will be systematically employed and, if it is, national coverage remains a long way off. Enhanced formal recognition of participatory land-use plans as documentation of local tenure claims is a practical and proven innovation to address this gap.

Conclusions

The sustainability of the Lao uplands depends on policy directions that remain contested and uncertain; policies that will ultimately decide the fate of upland communities, their multifunctional landscapes and the complex interplay between social and ecological dynamics that have shaped their character. While contemporary discourses have framed these choices as zero-sum, there are legitimate alternatives. It remains possible to employ robust systems of local knowledge, innovation and agroecological diversity to achieve key national goals for both rural improvement and environmental sustainability. Our research suggests that benzoin production in long-fallow shifting cultivation systems in the Lao uplands is a preferred alternative to either a transition toward sedentary commercial cultivation of commodity crops, on one hand, or strict forest protection on the other. But a better enabling environment is needed, one that values these systems and provides tenure security for upland communities. As we have said, Laos is at a crossroads. The fate of multifunctional upland landscapes hangs in the balance and, with them, the future of benzoin production and the broader agroecological systems of which it is a part.

Acknowledgements

We wish to thank the benzoin producers and other participants too numerous to name who took an active part in this research, as well as Michael Victor and F. Chagnaud for their helpful contributions and reviews of the draft manuscript.

References

Baird, I. G. and Shoemaker, B. (2007) 'Unsettling experiences: Internal resettlement and international aid agencies in Laos', *Development and Change* 38(5), pp.865-888

Broegaard, R. B., Vongvisouk, T. and Mertz, O. (2017) 'Contradictory land use plans and policies in Laos: Tenure security and the threat of exclusion', *World Development* 89, pp.170-183

Cramb, R. A., Colfer, C. J. P., Dressler, W., Laungaramsri, P., Le, Q. T., Mulyoutami, E., Peluso, N. L. and Wadley, R. L. (2009) 'Swidden transformations and rural livelihoods in Southeast Asia', *Human Ecology* 37(3), pp.323-346

de Rouw, A. (1995) 'The fallow period as in shifting cultivation (tropical wet forests)', *Agriculture, Ecosystems and Environment* 54, pp.31-43

DoF (2018) *Emissions Reduction Programme Document*, Department of Forestry, Ministry of Agriculture and Forestry, Vientiane

Dwyer, M. B. and Ingalls, M. (2015) *REDD+ at the Crossroads: Choices and Tradeoffs for 2015–2020 in Laos*, CIFOR working paper no. 179, Center for International Forestry Research, Bogor, Indonesia

Dwyer, M. B., Vongvisouk, T., Ingalls, M. L. and Bounmixay, L. (2018) *Tenure without Title? Land-Use Planning as a Possible Pathway to Village Tenure Security in Laos*, report to The Mekong Region Land Governance Project, Vientiane

Epprecht, M., Weber, A-K., Bernhard, R., Keoka, K., Saphanthong, T., Manivong, V., Ingxay, P., Vongsamphanh, P., Bosoni, N., Hanephom, S., Vanmeexay, P., Kaungbounhieng, A., Sisouvan, H., Khounthikoumman, S., Xaichounorxoa, P., Ingalls, M., Nanhthavong, V., Liu, J., Norasingh, I., Wiesmann, U. and Breu, R. (2018) *Atlas of Agriculture in the Lao PDR: Patterns and Trends between 1999 and 2011*, Centre for Development and Environment, University of Bern, Switzerland and the Lao Ministry of Agriculture and Forestry, Open Bern Publishing, Vientiane

FAO (2014) *Chemical and Technical Assessment for Benzoe Tonkinensis*, Joint FAO/WHO Expert Committee on Food Additives (JECFA), Food and Agriculture Organization of the United Nations, Rome

Fischer, M., Savathvong, S. and Pinyopusarerk, K. (2007) 'Upland fallow management with *Styrax tonkinensis* for benzoin production in northern Lao PDR', in M. F. Cairns (ed.) *Voices from the Forest: Integrating Indigenous Knowledge into Sustainable Upland Farming*, Resources for the Future Press, Washington, DC, pp.564-570

Foppes, J. and Ketphanh, S. (2004) 'Non-timber forest products for poverty reduction and shifting cultivation stabilization in the uplands of Lao PDR', in *Poverty Reduction and Shifting Cultivation*

Stabilization in the Uplands of Lao PDR: Technologies, Approaches and Methods for Improving Upland Livelihoods, Proceedings of a Workshop held in Luang Prabang, Lao PDR, pp.27-30

Fox, J., Truong, D. M., Rambo, A. T., Tuyen, N. P., Cuc, L. T., and Leisz, S. (2000) 'Shifting cultivation: A new old paradigm for managing tropical forests', *BioScience* 50(6), pp.521-528

Fox, J., Castella, J. C. and Ziegler, A. D. (2014) 'Swidden, rubber and carbon: Can REDD+ work for people and the environment in Montane Mainland Southeast Asia?', *Global Environmental Change* 29, pp.318-326

Fujita, Y. and Phanvilay, K. (2008) 'Land and forest allocation in Lao People's Democratic Republic: Comparison of case studies from community-based natural resource management research', *Society and Natural Resources* 21(2), pp.120-133

Hett, C., Nanhthavong, V., Hanephom, S., Phouangphet, K., Phommachanh, A., Sidavong, B., Epprecht, M., Heinimann, A., Ingalls, M., Shattuck, A. and Lu, J. (2019) *Targeting Land Deals in the Lao PDR: A Characterization of Investments in Land and their Impacts*, Ministry of Natural Resources and Environment, Lao PDR, and the Centre for Development and Environment, University of Bern, Vientiane

Ingalls, M. L. and Dwyer, M. B. (2016) 'Missing the forest for the trees? Navigating the trade-offs between mitigation and adaptation under REDD', *Climatic Change* 136(2), pp.353-366

Ingalls, M. L. and Roth, V. (2018) 'Insights from the northern uplands: Informing the multifunctional/ monofunctional debate', paper presented at the Lao Uplands Initiative Conference, 12-14 March 2018, Luang Prabang, Lao PDR

Ingalls, M. L., Diepart, J-C., Truong, N., Hayward, D., Niel, T., Sem, T., Phomphakdy, M., Bernhard, R., Fogarizzu, S., Epprecht, M., Nanthavong, V., Vo, D. H., Nguyen, D., Nguyen, P. A., Saphanthong, T., Inthavong, C., Hett, C. and Tagliarino, N. (2018) *The Mekong State of Land*, Bern Open Publishing, Centre for Development and Environment, University of Bern, Switzerland and Mekong Region Land Governance

Kenney-Lazar, M. (2010) *Land Concessions, Land Tenure and Livelihood Change: Plantation Development in Attapeu Province, Southern Laos*, Fulbright and the National University of Laos, Vientiane

Ketphanh, S. and Vongkhamho, S. (2008) *Review of Existing NTFP Quotas and Marketing Systems at Local Level in Laos*, Forestry Research Center, National Agriculture and Forestry Research Institute, Vientiane

Mertz, O. (2002) 'The relationship between length of fallow and crop yields in shifting cultivation: A rethinking', *Agroforestry Systems* 55(2), pp.149-159

Mertz, O., Padoch, C., Fox, J., Cramb, R. A., Leisz, S. J., Lam, N. T. and Vien, T. D. (2009) 'Swidden change in Southeast Asia: Understanding causes and consequences', *Human Ecology* 37(3), pp.259-264

Messerli, P., Heinimann, A. and Epprecht, M. (2009) 'Finding homogeneity in heterogeneity: A new approach to quantifying landscape mosaics developed for the Lao PDR', *Human Ecology* 37(3), pp.291-304

NAFRI, NUoL and SNV (2007) *Non-Timber Forest Products in the Lao PDR. A Manual of 100 Commercial and Traditional Products*, The National Agriculture and Forestry Research Institute, the National University of Laos and SNV Netherlands Development Organization, Vientiane

Rerkasem, K., Lawrence, D., Padoch, C., Schmidt-Vogt, D., Ziegler, A. D. and Bruun, T. B. (2009) 'Consequences of swidden transitions for crop and fallow biodiversity in Southeast Asia', *Human Ecology* 37(3), pp.347-360

Roder, W. (2001) *Slash-and-burn Rice Systems in the Hills of Northern Lao PDR: Description, Challenges and Opportunities*, International Rice Research Institute, Los Baños, Philippines

Schmidt-Vogt, D., Leisz, S. J., Mertz, O., Heinimann, A., Thiha, T., Messerli, P., Epprecht, M., Van Cu, P., Chi, V. K., Hardiono, M. and Dao, T. M. (2009) 'An assessment of trends in the extent of swidden in Southeast Asia', *Human Ecology* 37(3), p.269

Scott, J. C. (2009) *The Art of Not Being Governed: An Anarchist History of Southeast Asia*, Yale University Press, New Haven, CT

Van Vliet, N., Mertz, O., Heinimann, A., Langanke, T., Pascual, U., Schmook, B., Adams, C., Schmidt-Vogt, D., Messerli, P., Leisz, S. and Castella, J. C., Jorgensen, L., Birch-Thomsen, T., Hett, C., Bruun, T. B., Ickowitz, A., Vu, K. C., Yasuyuki, K., Fox, J. M., Dressler, W., Padoch, C. and Ziegler, A. D. (2012) 'Trends, drivers and impacts of changes in swidden cultivation in tropical forest-agriculture frontiers: A global assessment', *Global Environmental Change* 22(2), pp.418-429

Weyerhaeuser, H., Bertomeu, M., Wilkes, A. and Mei, Y. (2010) *Cross-border NTFP value chains: Laos, China*, technical report, the National Agriculture and Forestry Research Institute and the World Agroforestry Centre (ICRAF), Vientiane

Wiemann, J., Ashoff, V., Grad, M., Meyer, A. K., Ruff, S. and Staiger, T. (2009) 'Laos on its way to WTO membership: Challenges and opportunities for developing high-value agricultural exports', draft report presented in a workshop, 30th April 2009, Vientiane

Xu, J., Lebel, L. and Sturgeon, J. (2009) 'Functional links between biodiversity, livelihoods and culture in a Hani swidden landscape in southwest China', *Ecology and Society* 14(2), p.20

Notes

1. Especially the Agroforex Company of Vientiane.
2. This name of this ethnic group is variously transliterated as Khmu, Kh'hmu, or Khamu.
3. The definition of forest in the Lao PDR, consistent with that of the Food and Agriculture Organization of the United Nations, includes monocultural plantations of non-native species as 'forest', despite the obvious and substantial differences with regard to ecological functions.

35

BORASSUS PALM UTILIZATON

As a complementary mode of livelihood in dry land cultivation

James J. Fox*

Introduction

In his masterwork *The Ambonese Herbal,* written in the 17th century but published decades later in 1741, Georg Eberhard Rumphius devoted considerable attention to a discussion of the many remarkable qualities of the lontar palm.[1] He described the lontar as the second most-important palm tree to be found in Asia after the coconut, and the first among all juice-yielding palms. He said he made his full estimation of the lontar's importance and his extended exposition of its significance 'because of its multiple uses'. Although he listed dozens of other names for this palm, and for his *Herbal* he formally nominated it as the *jager* (or *jaggery*) tree (*jagerboom*), from the outset of his discussion, he referred to it specifically as the 'lontar tree'. In his day, and to the present, this designation derived from Malay was, and is, the most common term used to identify the tall, fan-leafed *Borassus* palm in Indonesia.

Rumphius began his description of the lontar with an extended comparison with the coconut palm (in his nomenclature: *calappus*). In this comparison, he noted a critical ecological difference between these – to his mind – similar palms: 'It is amazing that these two nursemaids of the Indies, to wit the *calappus* and the *jager* tree, harbour such jealousy and hatred towards one another that they will not grow together in the same landscape or in the same field...' In detailing the different areas in which these two palms flourished, Rumphius implicitly suggested the reasons for this 'jealousy' towards one another.[2]

Whereas the coconut flourishes under a regime of heavy rainfall, the lontar flourishes in much drier conditions. The 'lontar landscapes' that Rumphius outlined in his *Herbal* were all notably hot, dry areas; semi-arid fringe country, with neutral or

---

* Dr. JAMES J. FOX is Emeritus Professor of Anthropology in the College of Asia and the Pacific at the Australian National University. He continues to conduct research in Indonesia, in both Java and the Timor region.

slightly alkaline soils and extended dry seasons where the palm's capacity for drought resistance was paramount. Thus the lontar occurred in large numbers in parts of south India rather than in the north; on the east coast – the Coromandel coast – and into Odisha, as well as in Jaffna, rather than on the west coast of India, and in Sri Lanka;[3] on Java, primarily in the east and on the island of Madura, in parts of Sulawesi, especially along the Mandar coast, and on the islands up to Timor. Unknown to Rumphius, there were similar environments in mainland Southeast Asia, from Pagan in Myanmar to Angkor Wat, with large numbers of lontar palms in the 'dry zone' of central Myanmar as well as in scattered parts of Cambodia and Vietnam (Figure 35-1) (see Lubeigt, 1979, pp.9-15 for a careful delineation of these areas).

Given its considerable distribution, the *Borassus* palm is known by many names. Its most common English name is palmyra palm, although, more colloquially, it is often referred to as the toddy tree. In India, it is known by various cognates of *tal* or *tala*, which have given rise to similar terms in much of Indonesia, such as *tal*, *ental* and *lon-tar*. In Myanmar, it is *tanbin*; in Thailand, *tôn tan tanôt*; in Cambodia, *thnôt*; and in Vietnam, *cay thôt lot*. In Javanese it is the *siwalan*; among the Rotenese and many of the populations of the Timor area, it is the *tua*.

The botanical designation for the palm is *Borassus flabellifer* L. An older classification was *Borassus flabelliformis*, which, in the 19th century, was used as a single designation for *Borassus* palms from West Africa to New Guinea. This classification was overturned by the distinguished Italian botanist Odoardo Beccari, who recognized seven species of *Borassus* (1914). In his classification, he distinguished the *Borassus* palms of India, Ceylon (Sri Lanka) and Indochina as a separate species, *Borassus flabellifer*; those of the Indonesian archipelago he designated as *Borassus*

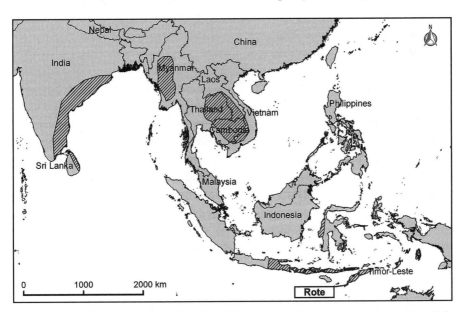

FIGURE 35-1: The main areas of South and Southeast Asia in which *Borassus flabellifer* is found, showing the Indonesian island of Rote.

sundaicus. However, much of the general literature on the Asian *Borassus* (see Heyne, 1927, vol 1, p.324; Burkill, 1966, pp.350-351) seems to have remained unconvinced by Beccari's differentiation and has persisted in the use of *Borassus flabellifer* for both species. In this paper, I will use the term lontar in my specific discussions of this *Borassus* palm in Indonesia or the term *Borassus* as a general term for this palm throughout Asia.

The 'multiple uses' of the *Borassus*

Much of Rumphius's exposition on the lontar palm was taken up with describing its 'multiple uses' and hence its value as a livelihood resource. He was preceded in this description by the Tamil poet, Atunachalam, who enumerated 801 uses of the palm in his *Tala Vilasam: The Glory of the Palm.* In addition to its prolific juice production, which can be made into jaggery sugar, allowed to ferment to make toddy, or distilled to make *arak,* the tree yields edible fruit and offers a durable timber. Its leaves can be used for thatch and plaited into a wide variety of baskets and utensils. For centuries, the leaves served as a medium for writing. Thus, to this day, the historical records of the Bugis and Makassarese are known collectively as *Lontara.* Most accounts of this palm since Rumphius have continued the tradition of enumerating its usefulness (Ferguson, 1850; Corner, 1966, pp.284-286; Fox, 1977, pp.23-29; Lubeigt, 1979; Dalibard, 1999; Depommier, 2003). The capacity of this palm to produce substantial livelihoods; the diversity of these livelihoods in different parts of Asia, and the flexibility of these alternative livelihoods to complement other economic pursuits is what gives this palm its great social and economic importance. Given its wide distribution, the question to be asked is how the specialized economic pursuits associated with the *Borassus* palm relate to agriculture in general and to swidden cultivation in particular. (It is perhaps worth noting that African *Borassus* palms, particularly *Borassus aethiopum,* have many of the same botanical characteristics as the Asian *Borassus* and are also of major importance for local livelihoods.)

Modes of *Borassus* utilization

Rumphius's long exposition on the '*jager* tree' is more than a botanical description of the palm, its habitats and its physical features: trunk, crown, leaves, fruit, inflorescences and the differences between male and female trees. It is also a detailed account of the uses made of this palm by different populations from India's Coromandel coast to Timor. In doing this, he established a Dutch tradition of economic botany that continued and was elaborated in later works such as Heyne's three-volume 1927 study, *De nuttige planten van Nederlandsch-Indië.* It is instructive to compare Rumphius's botanical descriptions and his account of the use of the *Borassus.* His description of the inflorescences of the male lontar is a notable attempt to provide a graphic but accurate description of the plant for which a scientific lexicon had not yet been developed:

'The male lontar is similar to the female, and grows by itself, and differs only in that it does not bear fruit but instead some long, round tails like the flowers of the *soccum*, or cattails of our country, but coarser and almost scaly, a hand or more long, with same colour, usually two, rarely three, hang down together, and these spikes have a particular or unique bloom, being little white flowers, fashioned from three spoon-shaped leaflets, with five little threads inside, on top of one of the spike's scales, parallel in slanted rows, with a lovely scent and appearing very slowly' (Rumphius, 2010).

In tapping the lontar, these inflorescences are deliberately wounded by repeated squeezing and their tips are cut to initiate a steady flow of juice. The female lontar can also be tapped, but its inflorescences must be squeezed before they begin to develop fruit.

Rumphius described various methods for climbing the palm and gathering its juice. He then wrote about its processing to sugar and sugar 'loafs', drawing on a combination of Indian and Javanese methods:

'The *sura* is first boiled over a slow fire until it is a thick syrup, called *carpune*, whereafter it is poured into small oblong baskets, and dried in the smoke, and from this comes a brown tree-sugar, which is what is really called *jagara*. The Javanese pour this syrup into small plates, *tampurongs*, or bamboos, and let it dry until small loafs of sugar, then they wrap the hardest ones in leaves and bring these to the market…' (Rumphius, 2010).

***Borassus flabellifer* L.
[Arecaceae]**

The lontar palm: 'second most-important palm in Asia, but first among all juice-yielding palms'. A species with potential to complement swidden systems.

Rumphius recognized basic similarities in the exploitation of *Borassus*, but made it abundantly clear that there was a great variety in these modes of utilization. In fact, much of his exposition was a collection of comparisons of different uses. Thus, for example, he wrote:

'On Ceylon and on the Coromandel coast, they do more with the fruits than on Java and in the Eastern Islands… The Macassarese are much easier about preparing this fruit, taking much less trouble, they pour the juice that was pressed out in large platters, mix some rice-meal with it, and make all sorts of food with it. They also have a use for the stones or kernels, which are otherwise thrown away [on Ceylon and the Coromandel coast]... The people on Java, Bali, Timor and Rotthe use the sura they tap from the tree more to make sugar than for drink, boiling it first down to a syrup (called *gula*) which they let dry in little baskets or plates ...' (Rumphius, 2010).

Thus, in his exposition, Rumphius highlighted one of the significant features of *Borassus* exploitation: its diverse products – timber, fruit, leaves and juice – have given rise to a variety of methods of utilization, some more specialized, some more intensive, but all with recognizable similarities.

Borassus palm succession

One of the remarkable features of the *Borassus* palm is its ability to resist drought and to flourish in areas with extended dry seasons. Its distribution in South and Southeast Asia makes it one of the key signature trees of the region's arid zones and enclaves. Another of its remarkable features is its ability to withstand fire and to tolerate herbivore grazing. Like other palms (see Rull, 1999 for the *Mauritia* palm), *Borassus* palms have the capacity to colonize habitats created by disturbances, particularly recurrent burning.

The *Borassus* palm is dioecious, with distinct male and female plants. Where the demographic strategies of the *Borassus* have been studied in detail, there is evidence of different spatial clustering for male and female trees on nutrient-rich patches, with seedlings closer to maternal trees, but with eventual survival and greater dispersal dependent on deep-root foraging (see Barot et al, 1999 for *Borassus aethiopum*). Over time, this can lead to partial enclaves and eventually to extended palm savannah. Evidence suggests that many of the areas of *Borassus* concentration were produced or promoted by earlier and repeated swidden burning. Rumphius's 17th-century delineation of areas of *Borassus* concentration gives some indication of where swidden activities gave rise to high palm densities.

One such area of *Borassus* concentration noted by Rumphius was on Timor and Rote. At the time, in the 17th and 18th centuries, the Dutch controlled only a small enclave around the Dutch East India Company settlement at Kupang. Rumphius's reference was thus specifically to this area at the western end of Timor, including the off-shore island of Rote, further to the southwest. These southernmost islands of the archipelago are subject to pronounced dry seasons and recurrent El Niño droughts, combined with a local agricultural regime of regular burning to clear and prepare the land for planting. In his *Malay Archipelago*, Alfred Russel Wallace provided a succinct picture of this area outside of Kupang. He visited Kupang twice, once for a day in 1857 and then again for two weeks in May 1859. Although he was visiting at a time when the dry season had only just begun, he presented a stark portrait of the landscape:

'The vegetation is everywhere scanty and scrubby...and the whole country has a parched and desolate appearance... The most conspicuous feature of the vegetation was the abundance of fine fan-leaved palms (*Borassus flabelliformis*)...' (Wallace, 1869, vol I, p289).

Wallace then goes on, as commentators have done both before and after him, to enumerate the 'multiple uses' of *Borassus*. However, his report is deficient in one critical aspect. Although he noted the abundance of *Borassus*, he failed to mention that *Borassus* palms often occur in close and mixed proximity to another major fan-leafed palm, *Corypha elata* Roxb. (In the Dutch literature on this palm, it is sometimes designated as *Corypha utan* Lamarck or *Corypha gebang*). Although *Borassus* is the palm that is most exploited, *Corypha* is also put to a variety of uses. Together, these two palms – lontar and *gebang* – offer a major source of livelihood for the local populations that exploit them.

Borassus utilization as a complement to cultivation

It is more than simply the existence of palm enclaves, and in some areas, extensive palm savannah, that fosters the utilization of *Borassus* as a complement to certain systems of swidden cultivation – especially in areas of dry land cultivation. It is also more than the potential 'multiple uses' of *Borassus* palm. It is the seasonality of its productivity – the cycle of its principal flow of juice – that is critical to its role in complementing swidden.

The timing of dry-land swidden and related gardening are critically dependent on seasonal rainfall. In effect, this requires that field preparation be completed prior to the onset of the first appreciable rains. Harvesting is invariably done after the rains, and thereafter in the semi-arid tropics, there is a relatively long period of little or no rain. *Borassus* production follows a different seasonality: initial juice production occurs after the rains cease, with production rising to a peak at the height of the dry season (Figure 35-2). This means that *Borassus* palms can be tapped and their juice rendered to syrup (and other products such as alcohol) during the hiatus in swidden or other cultivation activities. As a consequence, even the most intensive *Borassus* utilization can complement seasonal cultivation.

To illustrate this complementarity, it is useful to consider specific examples.

Borassus utilization on the island of Rote

Rumphius, who was the first to document the uses of the *Borassus* palm, referred specifically to the use of the palm on the island of Rote, and described at length the dependence of the local population on the juice they obtained from tapping the tree. He wrote:

'...on Timor and Roti it [the tapped juice from the palm] is used more for a liquor, which they call *tua*, and of which they are uncommonly fond, saying

FIGURE 35-2: A stand of *Borrasus* palms on dry land near a settlement on the island of Rote.

Photo: James Fox.

that it serves them as both food and drink; for daily use they mix it with some water and they become hale and fat from it; they keep the same in vats made from *Saribu* leaves, wherewith they also hang them in the front part of their houses. Anyone who enters can partake thereof freely, and quench his thirst, even before he has greeted someone, which is a courtesy and custom of that country' (Rumphius, 2010).

This late 17th-century description comes close to describing Rotenese practices to this day. The juice (*tua matak*) of the Rotenese *Borassus* or lontar palm (*tua*) is consumed daily. It is hung in buckets (*haik*) shaped from the leaves of the palm and is freely available, in its fresh state, to any visitor during the tapping season. At other times of the year, cooked lontar syrup (*tua nasu*) is mixed with water to produce a drink called *tua hopo*. Either way, lontar juice is part of the daily diet for most Rotenese.

On Rote (and the neighbouring island of Savu) as well as in parts of Timor, *Borassus* utilization has been well-established for centuries. This historical continuity makes Rote an illustrative case for consideration.

Rote is a small low-lying island off the western tip of Timor with soil that is predominantly a jumble of limestone, raised coral debris and marl, ideally suited to *Borassus*. The island has a number of small rivers, all of which run dry in the dry season, and is dotted with fresh-water springs, many of which continue to yield precious water throughout the year.

Rote experiences a short period of monsoon rains whose irregular onset is in late November or early December. These rains become heavier in January and February and then tail off in March. By April, the dry season has come and this continues until the rains resume towards the end of the year. Under these conditions, the lontar palm can be tapped initially for a short period in April and May (*tua timu*). However, the main tapping season (*tua fanuk*) occurs later in the dry season, in September and October.

At the time that Rumphius was writing, swidden cultivation on Rote centred on the cultivation of millet and sorghum; maize planting began at the end of the 17th century and it gradually became a major swidden component. Other crops in the swidden mix were Job's tears, sesame and a variety of beans. Squash was introduced into this mix, probably in the 18th century, and sweet potato arrived even later. Swidden cultivation was, and still is, combined with the cultivation of house gardens that generally involve a more intensive cultivation of one or another of the swidden crops or, later in the dry season, a combination of tobacco and onions. Rain-dependent or spring-fed rice, or rice that is irrigated by channels drawn from rivers in the wet season, is a valued crop that is grown separately, wherever possible. The productivity of any and all of these crops varies considerably from year to year in a region that is prone to a marked El Niño-La Niña climatic cycle.

By contrast, the lontar palm is consistently productive and provides a dependable level of food security. Tapping varies from household to household, depending on the individual requirements of each and their specific success in the cropping season. Some households may tap only a few trees, while others may tap 20 or more. Early season tapping is generally used as a daily drink, whereas later in the year, during the period of intensive tapping, most lontar juice is cooked into syrup for storage.

The leaves of the palm are indispensable to *Borassus* utilization, and are used for an enormous range of products. They can be shaped to make hats, rolled to make a kind of cigarette paper or laid down in sheets as roof thatch. They can be woven into mats or baskets, and folded to form durable buckets. These buckets, made from different-sized leaves, come in many sizes and are used for different purposes. The largest are used for carrying water, watering gardens or holding quantities of palm juice; smaller buckets make simple receptacles for every-day drinking.

These easily made, serviceable palm-leaf buckets are vital to Rotenese palm tapping. To keep palm juice fresh and drinkable, it is essential to reduce the inevitable rapid fermentation that occurs once the juice has left the tree. Unlike most *Borassus*-palm tappers, who add a fermentation retardant to the receptacles that catch the juice, the Rotenese limit fermentation by alternating the use of lontar-leaf buckets (Figure 35-3).

The trees are tapped twice each day, in the early morning and again late in the afternoon. The tapper climbs the tree carrying a large leaf bucket (*hai kuneuk*) to collect the juice from various smaller leaf buckets (*hai seseluk*) in the crown of the palm. He also carries, strapped to a belt, a gourd of fresh water that he uses to clean each of the smaller leaf buckets once they have been emptied. The number of leaf buckets depends on the number of tapped inflorescences – at least three to five per palm. The tapper uses a small brush, itself made from the frayed end of a palm-leaf stalk, to clean each leaf bucket; he then exchanges leaf buckets, leaving one to dry in the sun while the other is used to catch the dripping juice. To limit the effects of direct sunlight and disturbance by birds or bats, each leaf bucket is set inside an individual protective leaf basket (Figure 35-4). On his descent, the tapper pours the juice he has collected into an even larger leaf bucket (*hai sasalik*) before setting out to climb the next tree. This simple reliance on a combination of lontar-leaf buckets, with a considerable effort made to ensure that all receptacles are clean, and by encasing those in the crown of the palm in a separate lontar basket, allows the palm juice to remain fresh for longer – ready either for drinking or for cooking.

The cooking process itself is dependent on the palm. In the high tapping season, the Rotenese fashion special ovens for syrup cooking. They hollow out narrow tunnel-like cavities in the ground; then they cover these with earth, leaving a row of holes – usually from three to five – into which cooking pots can be tightly placed so that each pot is half submerged

FIGURE 35-3: A Rotenese tapper climbing a *Borassus* Palm. Note the leaf bucket swinging from his waist.

Photo: James Fox.

FIGURE 35-4: A tapper in the crown of a *Borassus* palm, showing the leaf buckets and covering baskets in which smaller leaf buckets are secured to receive the juice. A tied and slivered male inflorescence can be seen near the tapper's foot.

Photo: James Fox.

in the earth. This simple but highly efficient earthen oven uses little firewood. Instead, leaf stalks of the palm (and other plant detritus) are inserted and burned in the narrow shaft beneath the pots (Figure 35-5).

At the productive peak of its second flowering season, the *Borassus* palm produces a prodigious quantity of juice. Although this juice, either as a fresh drink or as a syrup mixed with water, is a basic human nutrient, it can also be used as a feed source for pigs. This allows the Rotenese to convert the surplus of their *Borassus* juice to useful protein.

A generous portion of *Borassus* syrup is also converted to a sweet mash and then distilled to produce an *arak* or gin. The Rotenese claim to have learnt the art of distilling from the Dutch and they recount how they simultaneously learnt of both Christianity and distilling in their early dealings with the Dutch East India Company. Rotenese gin stills are simple constructions: the fermented lontar mash is boiled in an earthen pot capped with fresh lontar fruit and attached to a pipe that extends through a hollowed lontar log filled with cooling water, leading to a receiving bottle at the other end. The distillate is a delicious gin that, when it is well crafted, resembles a good Dutch *genever*.

Rotenese settlements are generally in the close vicinity of various stands of *Borassus* palms. Many of the houses are thatched with lontar leaves, lontar buckets are hung prominently around these houses and discarded leaves are found scattered in and around homes, gardens and cooking areas. Discarded leaves have a further special use. They can be laid out to carpet a house garden and then burnt to a fine ash just before planting – a kind of limited burning, probably adopted in imitation of the more extensive burning of swidden fields. This has the effect of sterilizing the ground and may possibly add nutrients to the soil.

FIGURE 35-5:   A woman cooking *Borassus* juice on Rote.

Photo: James Fox.

Borassus utilization in central Myanmar

A comparable case study to that of the Rotenese of eastern Indonesia is that of the Burmese cultivator-palm tappers of central Myanmar. The *Borassus* palm is found in large numbers in what is generally referred to as Burma's 'dry zone' – a broad area that extends from Shewbo, through Monywa, Prome and Taungdwingyi to Mandalay. This is an area that receives on average about 600 to 800 mm of rain per year. Reliance on the use of the *Borassus* palm, in combination with dry-land cultivation, can be historically documented for more than a millennium. Inscriptions dating from the Pagan period, from the 9th to the 11th centuries, refer to the donation of *Borassus* palms to Buddhist monasteries of the period. Other traditions link the palm to even earlier dynasties and populations, especially in relation to the establishment of monasteries. The fact that the dried leaves of the palm were used as precious writing material enhanced this association.

Guy Lubeigt has documented this *Borassus* utilization in considerable detail in his monograph, *Le Palmier a Sucre (Borassus flabellifer) en Birmanie Centrale* (1979). His research, based on extensive fieldwork undertaken between 1968 and 1975, indicated varying patterns of palm utilization in relation to dry land cultivation. Lubeigt estimated that in the 1970s, several hundred thousand Burmese were dependent on the palm. An aerial survey of the region revealed a clear contrast in these patterns. In some areas there was what Lubeigt termed an 'anarchic' distribution of trees, whereas elsewhere there was a clear 'alignment' pattern of palms, indicating the strategic planting of palms either to delimit cultivated fields or to form regular dense plantations. These patterns reflected both contemporary and past land use and indicated different intensities of utilization.

Lubeigt noted a higher percentage of young trees in the Pagan-Kyaukpadaung area than in Swebo. Young trees were considered to be more productive and were certainly easier to climb than older, taller trees. Consequently, by Lubeigt's calculations, cultivator-tappers in Kyaukpadaung concentrated more of their efforts on exploiting the *Borassus*, tapping from 30 to 70 trees and therefore allocating less of their efforts to their fields. By contrast, at Swebo, cultivator-tappers put more effort into their fields, but still managed to tap from 25 to 40 palm trees. The ability of cultivator-tappers to shift their efforts between *Borassus* tapping and dry-land cultivation, depending on a variety of factors, was one of the features that made *Borassus* utilization complementary and mutually supportive of cultivation efforts. One of the main reasons for switching from one to the other was the effect of seasonal rainfall on cropping success.

Differences in tapping technique are pertinent to the exploitation of the palm. On Rote, most of the equipment used for tapping is derived from the palm itself. Although in some parts of Rote, the trunk of the lontar is notched for climbing, in other areas, inverted leaf stalks are tied to the tree as steps. The tapper carries a leaf bucket to the crown of the palm to collect the juice, which has dripped into smaller leaf buckets. He uses a frayed leaf stalk as his cleaning brush and encloses the leaf-bucket receptacle in a protective palm-leaf basket. By contrast, in Burma's dry zone, a long ladder is used to mount the palm and earthenware pots are used as receptacles. The climber carries a large pot up into the tree and gathers the juice from smaller pots that hang in the crown. As a result, the contamination from open pots is appreciable and rapid fermentation of the juice is a problem. To retard fermentation, the bark of the *thitya* tree (*Shorea obtrusa*) is placed in the pots.[4]

In the dry zone of central Myanmar, there are roughly two tapping seasons for the *Borassus* palms. The lesser season (*wazo*) occurs during the months of June and July, the greater season (*tabaung*) occurs during the months of February and March. However, where reliance on the palm is heavier, tapping can be extended over many months. Tapping in the greater season can be continued through April to May, linking with the lesser season in which tapping may be continued through August to September. Most farmers in the region combine tapping with the cultivation of maize, sesame and groundnuts in the period from October to January. In some areas, where water for irrigation is available, this cultivation may include rice.

Most villagers are cultivator-tappers, but the balance between cultivating fields and tapping palms varies according to local and seasonal conditions. Landowners with a large number of trees can exploit some of their palms and rent out other palms to be tapped by villagers without trees. Villagers without land may combine work as palm tappers with other occupations.

In his monograph, Lubeigt provides a variety of vignettes – short sketches of various cultivator-tappers in different areas of the dry zone. Each presents a variation on the exploitation of *Borassus* palms. The versatility of the palm allows various levels of utilization and combinations.

There is no mention of the daily consumption of fresh juice on the scale that occurs among the Rotenese. In fact, the use of a fermentation retardant affects the flavour of the juice. On the other hand, fermented juice – a palm wine or toddy – is a regularly consumed product of the palm. Similarly, the fruit of the palm is collected and sold, but preference for tapping female palms limits the commercial production of this product. The main product of the *Borassus* is cooked syrup or jaggery, much of which is used in the production of *arak*. This distilled product is much sought after. Although its distillation is supposed to be strictly government-controlled, *arak* is easily produced and is illegally available throughout Myanmar.

As elsewhere, the cooking of juice into syrup is an intensive process and must be done on a daily basis. A major impediment to the cooking process is obtaining enough firewood for the ovens, and firewood is relatively expensive. From Lubeigt's sketches, the ovens used for cooking the juice are not built into the ground as on Rote, but are raised brick structures with chimney vents at one end. Nor is there any indication that either juice or syrup is used as an animal feed, such as it is on Rote, where it supports intensive pig rearing.

Borassus utilization in Myanmar is not unlike that in Cambodia (see Romera, 1968). In the 1990s, drawing on evidence from Rote of the use of *Borassus* juice as a pig feed, and from Cuba of the similar use of molasses, a project was launched by the Food and Agriculture Organization of the United Nations (FAO), to promote the use of *Borassus* juice as pig feed in Cambodia (see Dalibard, 1996; Khieu Borin, 1996). The project was successful and led to a notable increase in palm tapping (Khieu Borin, 1998). However, there was one significant difference. Whereas on Rote both juice and cooked syrup mixed with water were used as pig feed, only the *Borassus* juice was used in Cambodia because, as in Burma, the lack of firewood and consequent high cost of cooking juice into syrup made its use uneconomic. The Rotenese cooking methods that relied on *Borassus* leaf-stalks were not introduced along with the livestock-feeding practices.

Thus, in modern-development parlance, the *Borassus* was recognized as having a critical – indeed innovative – role to play in the creation of an integrated farming system. However, its full potential was never properly utilized.

The vulnerability of *Borassus* utilization

Despite the versatility and productivity of *Borassus* palms, their use as a means of livelihood remains precarious. Similarly, despite frequent references to the *Borassus* as a means of increasing food security and alleviating poverty, the *Borassus* remains the 'poor man's tree'. There are many reasons for this. For one, the *Borassus* is found mainly in geographical areas of marginal agricultural productivity; areas of relatively low rainfall and generally poor soils. As far as can be determined, the distribution of *Borassus* covers regions where evidence suggests that there may earlier have been more various and more productive modes of cropping. The tenacious capacity of the *Borassus* to withstand the repeated and extensive burning that devastates other vegetation leaves the palm a sturdy sentinel survivor.

The versatility of the *Borassus* also contributes to its vulnerability. The trunk of the palm offers a durable timber for construction. It is claimed that the palm can grow for 100 years and then its timber can be used for another 100 years. To many, this timber is more valuable than its other products. Thus, for example, the abundance of palms remarked upon by Alfred Russel Wallace in the palm savannah forests outside of the town of Kupang on Timor have now been largely cleared to supply a voracious appetite for house construction in the rapid expansion of the town.

Borassus leaves are also a coveted product. During the Dutch colonial period, from the 1920s onward, discerning European buyers insisted that all Java coffee be packed in *Borassus*-leaf matting and baskets. This produced such an alarming demand for *Borassus* leaves that the Dutch colonial government was forced to conduct an inquiry into the future survival of *Borassus* utilization on the island of Madura (Gebuis and Kadir, 1928-29).

Perhaps the biggest stigma on the use of the *Borassus* palm as a means of livelihood is the fact that its juice can be fermented into toddy and distilled to produce gin. Everywhere that the *Borassus* is utilized, its tappers are suspected (not without reason) of illegal alcohol production. This stigma has often contributed to the dubious social status of tappers. This is certainly the case for the Nadar caste of *Borassus* tappers of South India (Hardgrave, 1969). Even without the imposition of a caste system, the suspicion of being illegal *arak* producers colours the status of the *Borassus* tappers of central Burma. In Muslim Madura, even mild toddy production, not to mention the distillation of *arak*, carries with it religious sanctions. And on Rote, where most of the population prides itself in its special 'water of words' (Fox et al, 1983), attempts have been made to impose prohibition.

Looking back to Rumphius's early account of the remarkable qualities of the *Borassus* palm, we can conclude, as he recognized, that the palm's versatility held great potential for the populations of South and Southeast Asia. However, in a changing world, not all of the palm's potential has proven to be advantageous.

References

Barot, S., Gignoux, J. and Menaut, J-C. (1999) 'Demography of a savanna palm tree: Predictions from comprehensive spatial analyses', *Ecology* 80(6), pp.1987-2005

Beccari, O. (1914) 'Studia Sui Borassus' (On studying *Borassus*), *Webbia* 4, pp.295-385

Burkill, I. H. (1966) *A Dictionary of the Economic Products of the Malay Peninsula*, vol 1 (A-H), Ministry of Agriculture and Cooperatives, Kuala Lumpur, Malaysia

Corner, E. J. H. (1966) *The Natural History of Palms*, Weidenfeld and Nicolson, London

Dalibard, C. (1996) 'The potential of tapping palm trees for animal production', in the Second FAO Electronic Conference on Tropical Feeds: https://wgbis.ces.iisc.ac.in/envis/tapdoc1012.html, accessed 21 April 2023

Dalibard, C. (1999) 'Overall view on the tradition of tapping palm trees and prospects for animal production', *Livestock Research for Rural Development* 11(1)

Depommier, D. (2003) 'The tree behind the forest: Ecological and economic importance of traditional agroforestry systems and multiple uses of trees in India', *Tropical Ecology* 44(1), pp.62-71

Ferguson, W. (1850) *The Palmyra Palm*, Observer Press, Colombo, Ceylon

Fox, J. J. (1977) *Harvest of the Palm: Ecological Change in Eastern Indonesia*, Harvard University Press, Cambridge, MA

Fox, J. J., Asch, T. and Asch, P. (1983) *The Water of Words: A Cultural Ecology of an Eastern Indonesian Island*, (documentary film, 16mm colour, 30m duration), Documentary Educational Resources, Watertown, MA

Gebuis, Ir. L. and Kadir, R. A. (1928-29) 'Enkele gegevens omtrent den Siwalan op Madoera', *Landbouw* 4, pp.304-321

Hardgrave, R. I. (1969) *The Nadars of Tamiland*, University of California Press, Berkeley, CA

Heyne, K. (1927) *De nuttige planten van Nederlandsch-Indië*, (three volumes), Gedrukt bij Ruygrok and Company

Khieu Borin (1996) 'The sugar palm tree as the basis of integrated farming systems in Cambodia', in the Second FAO Electronic Conference on Tropical Feeds, https://www.semanticscholar.org/paper/The-Sugar-Palm-Tree-As-the-Basis-of-Integrated-in-Borin/c5c9331bf8eb6327a6cdfa9bfb94581311ba4766, accessed 21 April 2023

Khieu Borin (1998) 'Sugar palm (*Borassus flabellifer*): Potential feed resource for livestock in small-scale farming systems', *World Animal Review* 91(2), www.fao.org/docrep/W9980T/W9980T00.htm, accessed on 23 November 2012

Lubeigt, G. (1979) *Le Palmier a Sucre* (Borassus flabellifer) *en Birmanie Centrale*, Département de Géographie, l' Université de Paris-Sorbonne, Paris

Ramzan Ali (1969) 'Process development on production of sugar palm juice', *Union of Burma Journal of Science and Technology* 11(1), pp.173-180

Romera, J. P. (1968) 'Le Borassus et le sucre de palme au Cambodge', *L'Agronomie Tropicale* 8, pp.801-843

Rull, V. (1999) 'A palynological record of a secondary succession after fire in the Gran Sabana, Venezuela', *Journal of Quaternary Science* 14, pp.137-152

Rumphius, G. E. (2010) *The Ambonese Herbal*, translated and annotated by E. M. Beckman (six volumes), Yale University Press, New Haven, CT

Wallace, A. R. (1869) *The Malay Archipelago*, (two volumes), Macmillan and Co, London

Notes

1. Rumphius's *Herbarium Amboinense* was written in two languages: Latin and Dutch, and was published with the two languages side by side, in two columns. E. M Beckman has done an enormous service to scholarship by introducing, translating and annotating Rumphius's study in a six-volume edition covering more than 3300 pages, published by Yale University Press (Rumphius, G. E., 2010, *The Ambonese Herbal*).

2. Although Rumphius was able to recognize the different environments where these two palms flourished, his explanation of the differences between them was more theological than ecological. He commented that the locational differences between the two palms could 'be ascribed to the wisdom of the Creator, who will not suffer these two profitable and necessary trees for humankind to be in one and the same country'.

3. Throughout most of his time on Ambon, Rumphius carried out an extensive correspondence to gather a wealth of information that he incorporated into his various writings. For his detailed information on the uses of the lontar palm on the Coromandel Coast, he relied upon and properly acknowledged a fellow officer of the Dutch East India Company, Herbert d'Jager, a man of prodigious linguistic ability whom the company dispatched to a number of its trading posts, from Persia to the Far East. D'Jager was stationed for 10 years (1670 to 1680) on the Coromandel coast, where he studied Sanskrit as well as Tamil and Telugu, and supplied Rumphius with any botanical information that he sought.

4. Retardants from related *Shorea* species are used in Thailand (*Shorea talura*) and Cambodia (*Shorea cochinchinensis*). However, according to a study cited by Lubeigt, such retardants have no apparent effect in lessening fermentation (Ramzan Ali, 1969). On the Indonesian island of Madura, the common retardant used in palm tapping is derived from the *kusambi* tree (*Schleichera oleosa* (Lour.) Merr.). In India, lime is often added to the juice receptacle to retard fermentation. All of these retardants affect the flavour of the fresh juice from the palm.

36

INNOVATIONS IN *KEMIRI* SHIFTING CULTIVATION

Household strategies, land institutions, and a hedge against official policy in South Sulawesi

*Supratman, Muhammad Alif K. Sahide and Micah R. Fisher**

Introduction

Just a short distance from the sprawling metropolitan area of Makassar, the provincial capital of South Sulawesi, in Indonesia, a cluster of villages continues to practise shifting cultivation based around the candlenut tree (*Aleurites moluccanus*, locally called *kemiri*). The villages, which together cover about 15,000 hectares, are in the Camba region (see Figure 36-1). Nuts from the *kemiri* trees are harvested mainly for the oil they contain, amounting to 15 to 20% of their weight. However, they are also an essential part of regional cuisines. The *kemiri* shifting cultivation practice emerged from centuries of intermixing between the technical knowledge of regional migrants and that of local upland communities (Yusran, 2005). Over time, a number of agrarian practices developed that were based on *kemiri* production. This was driven by subsistence needs and other household-income strategies tied to regional markets, the integration of new cash crops with changing market possibilities and adaptation to development pressures.

The swidden fields are defined by concentric circles that ripple outwards from settlement areas. Within these circles, land-management plots are patched together in a honeycomb-shaped arrangement that follows the *kemiri* life cycle. The first circular area extending away from villages is occupied by terraced wet-rice fields (*sawah*). Where sources of irrigation are available, the communities prioritize wet-rice cultivation. Beyond the rice, each household manages a two-hectare site. These sites are further divided into eight hexagonal quarter-hectare plots that fit together like a honeycomb, the cultivation and management of which follow a 25-year

* PROFESSOR SUPRATMAN, Department of Forestry, Universitas Hasanuddin, Makassar, Indonesia; DR. MUHAMMAD ALIF K. SAHIDE, Associate Professor, Department of Forestry, Universitas Hasanuddin, Makassar, Indonesia; and MICAH R. FISHER, Research Fellow, East-West Center, Honolulu, Hawaii, and Department of Forestry, Universitas Hasanuddin, Makassar, Indonesia.

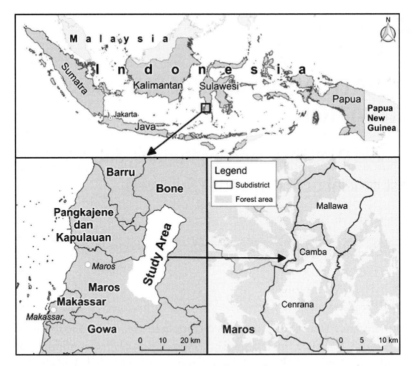

FIGURE 36-1: The study site: *kemiri* shifting cultivation areas within state forest boundaries in the Camba region of South Sulawesi, including the subdistricts of Camba, Cenrana and Mallawa.

Note: this map highlights the extent of *kemiri* shifting cultivation systems only within state forest areas. Maps of the entire system are unavailable, but the broader extent of the system is estimated at 15,000 hectares (Bakosurtanal, 1999).

rotation dictated by the *kemiri* life cycle. The cultivation process begins in the first quarter-hectare plot with peanuts (*canggoreng – Arachis hypogaea*), maize (*ba'do – Zea mays*), upland rice, tubers, and hardwood trees to establish natural fencing. At the same time, *kemiri* is established in this first sub-plot, along with other tree crops, such as cacao (*Theobroma cacao*) and sugar palm (*Arenga pinnata*). As the *kemiri* forest matures, land management shifts from one hexagonal plot to the next in a sequential system that is discussed in detail below. After 25 years, *kemiri* is no longer productive, at which stage the timber is harvested. The quarter-hectare plot is cleared and burnt, and cultivation begins anew, repeating the same cycle around the eight plots.

In this chapter, we focus on three main themes that describe the origins, adaptations and continued resilience of this swidden system. After a brief historical section, we explain how it works in detail, highlighting land relations, cultivation, harvesting and market institutions. We describe the formal and informal institutional mechanisms that continue to support the system. Thereafter, we highlight the main forces that threaten its integrity, by applying a framework introduced by Fox et al. (2009).

Fox et al. showed that the decline of swiddening across Southeast Asia was due to a variety of factors. They provided six reasons for this decline, and these were used to form the heuristic framework that we apply in this chapter. They included:

> (1) classifying swiddeners as ethnic minorities within nation–states; (2) dividing the landscape into forest and permanent agriculture; (3) expansion of forest departments and the rise of conservation; (4) resettlement; (5) privatization and commoditization of land and land-based production; and (6) expansion of markets, roads, and other infrastructure and the promotion of industrial agriculture (Fox et al., 2009, p.305).

In this chapter we will engage with these broader themes through the historical political economy and land relations surrounding *kemiri*-based shifting cultivation in the Camba region. To some degree, each of the threats listed by Fox et al. is present in the Camba region. However, strong land institutions and alliances with certain state policies have helped to protect *kemiri*-based shifting cultivation there.

We focus on how communities have adapted to these conditions by drawing on sustained research which began in 1999, when the Ford Foundation initiated grants in cooperation with Hasanuddin University in Makassar to study community forestry in the Camba region, alongside comparative work in the nearby regions of Toraja and Bulukumba. These studies yielded a series of theses and doctoral dissertations that engaged with the landscapes and livelihoods of *kemiri*-based shifting cultivation systems. In 2006, state forest schemes for community forestry were proposed, and further studies were conducted to develop appropriate institutional mechanisms for implementation of a forest-management unit for community forestry based on *kemiri* shifting cultivation. From 2008 to 2010, new community forestry programmes were proposed for the Camba region, this time supported by the Center for People and Forests (RECOFTC), a non-governmental organization, which launched additional studies. Finally, Hasanuddin University's Faculty of Forestry conducted studies of the Camba region, which is adjacent to the faculty's teaching forests, as a priority site in its master plan. The research that shapes this chapter therefore draws on sustained engagement and continued research involving local communities in the Camba region.

This chapter examines the adaptive practices in the Camba region by focusing on existing land institutions and describing the key features that continue to sustain *kemiri* shifting cultivation in the region. Land relations are premised around socio-cultural systems that allow *kemiri* to continue to evolve alongside pressures from the state, commoditization, resettlement dynamics and development programmes. This contrasts Camba with neighbouring regions that once had similar cultivation systems based on teak and other tree crops. These neighbouring systems have long disappeared as a result of a cacao boom in Sulawesi and the expansion of state development authority.

The fluctuating history of *kemiri* shifting cultivation in Camba

In 2018, the plantation agency of Maros district (see Figure 36-1) noted steady improvement in the agriculture commodity sector. The most dominant commodity was *kemiri*, with total production of 4400 tons per year, accounting for about 70% of total production from the district's 11 plantation commodities (Pemkab Maros, 2018) (Figure 36-2). There is a unique history behind the origins of *kemiri* as the dominant crop in Maros. Early communities that established continued residence first migrated to the Camba region in 1826, led by Isossong Lamappaselling – the

17th crown prince of Bone (Muspida, 2008). Isossong Lamappaselling established a kingdom in the area called *Baholiang*, meaning the palace beneath the caves, due to the region's karst landscape and hilly terrain. The main impetus for the migration was to establish *kemiri* groves. The trees grew well in dry regions and provided a lucrative crop for export to the Netherlands, involving the Dutch East India Company.

Isossong Lamappaselling required all of his followers to plant *kemiri* on any hilly or mountainous terrain (Muspida, 2008; Supratman, 2007a, 2007b). Flat terrain was set aside for wet–rice (*sawah*) cultivation for household consumption. The agroforestry systems initially began by developing ownership rights to lands based on cultivation and the opening of new forest areas. Seasonal crops initiated the land-use process in the first three or four years, including intensive cultivation systems of rice, maize, peanuts,

FIGURE 36-2:

Aleurites moluccanus (L.) Willd. [Euphorbiaceae]

Candlenut, known in Indonesia as *kemiri*, grows to a height of 25 m over a productive lifetime of 25 years. Its life cycle lies at the heart of a complex shifting cultivation system; its nuts harvested mainly for their oil.

Photos: All photographs in this chapter were taken by Muhammad Alif K. Sahide.

chillies, tomatoes, vegetables, bananas, cassava and other roots and tubers. The long-term cultivation crops were then planted in a staging process detailed below. This land management system revolved around the main crop of *kemiri*, but also included rows of teak, acting as boundary markers between plots, signifying ownership and providing an important resource for construction of the typical homes of the region. When the *kemiri* stands became unproductive, farmers harvested the timber, which also yielded profits, and began the cycle of replanting once again. Although *kemiri* was the central cash crop, it was susceptible to several waves of change that redefined the early swidden groves and eventually gave rise to the current system.

Beyond the systems that were first established by Isossong Lamappaselling, the 1920s saw significant expansion of *kemiri*, which gained renewed interest in international markets. Farmers organized themselves into groups and sought agreements from the village chiefs of the time to act as indigenous leaders in negotiations with the Dutch colonial government. At that time the Dutch also set up enforcement mechanisms to patrol areas designated as forest areas. Forest rangers were assigned to delineate and patrol state forest borders, and the Camba communities called them *tuan pa'*. These officers ensured that activities within areas designated as the Forest Estate (*Bosch Wezen*) were 'allowable'. Nowadays, swiddening communities in the Camba region, especially their elders, often refer to the legitimating authority of the *Bosch Wezen* forest boundaries that were designated by the *tuan pa'* because they reshaped the trajectory of local land relations. The legitimating authority of these historical boundaries has become pronounced because the State has subsequently sought to expand national forest boundaries by enclosing community forest lands and only providing limited consultation with local communities when doing so.

The state-administered forest concept that was applied in Camba during the late colonial period also had the effect of dividing forest areas between those managed by communities (*kemiri* groves) and areas preserved as state forests. Such designations followed geographic and topographical features. For example, areas in the uplands with steeper slopes were prohibited for cultivation and designated as state forests. Pathways (*balatu*) constructed by the *tuan pa'* were once designed to highlight the formal divisions between community forests and natural forests.

Populations in the region began to increase in the 1940s, requiring new innovations for increased food production. *Kemiri* forests nearby settlement areas and adjacent to the main thoroughfare (the Makassar-Bone highway) that had relatively flat land that was easy to irrigate were converted to wet-rice (*sawah*) cultivation. However, the political situation in the 1950s and 1960s, characterized by Islamist rebellions across South Sulawesi, had a stifling effect. Communities sought to avoid the violence and were afraid of getting stuck between the military forces of the State and the *Darul Islam* rebels. They retreated deeper into inaccessible areas and opened up new areas of *kemiri*-based shifting cultivation, thereby expanding the extent of the region's *kemiri* groves. After the violence subsided in the mid-1960s, the farmers returned to areas with greater access to markets, but they also continued to work in the new swiddens that they had opened during the years of unrest.

In the 1970s, state-supported silviculture programmes transformed much of the uplands in South Sulawesi by introducing pine (*Pinus merkusii*) plantations. The fast-growing pines were planted on land that was considered 'critical' – a designation that covered any upland areas without tree cover. Some parts of the Camba region were affected by the silviculture programmes, but not to the extent of other regions in South Sulawesi, where conversion to pine plantations was accompanied by an erasure of tenurial designations. In much of Camba, there was no need for the pines because *kemiri* already provided tree cover, so the extent of *kemiri* was especially important for maintaining local land relations in the Camba region. However, there was one unintended consequence of the pine forests: unlike *kemiri*, the pines were vulnerable to fires.

In the 1980s, there was significant pressure for change in *kemiri* shifting cultivation. State programmes promoted tree crops and expanded efforts to commoditize land, which disrupted swiddening institutions. Swidden farmers began to incorporate new elements into their agroforestry systems, to

Pinus merkusii Jungh. & de Vriese [Pinaceae]

Sumatran pines – the only pine species to occur naturally in the southern hemisphere – were introduced to South Sulawesi to reforest areas without tree cover. There was no need for them in *kemiri*-growing areas because of the generous canopy provided by *kemiri* stands.

benefit from price increases in global markets for specific commodities. Plantation extension officers also made chemical inputs more readily available, along with courses on how to apply them. Attempts to incorporate coffee into shifting cultivation systems under the shade of *kemiri* endured for some time,[2] but slowly dwindled as coffee prices began to fall in the 1990s. The 1980s also saw the introduction of cacao (*Theobroma cacao*) trials, with households first growing the trees in the yards around their homes. Cacao was introduced by local people who returned from working on cacao plantations in Malaysia. Using the experience they gained while working on cacao plantations abroad, local farmers slowly began to integrate cacao into local farming systems in the hope of gaining additional profits. The national government also prioritized the province of South Sulawesi for cacao development, and extension programmes were offered to those interested in converting land to cacao (Limbongan, 2016).

Theobroma cacao L. [Malvaceae]

Cacao produces seeds from which chocolate and cocoa are made. After its introduction to South Sulawesi it threatened to supplant *kemiri* cultivation, but *kemiri* persisted, particularly in more remote areas.

The 1998 Asian financial crisis also had a major impact on rural livelihood strategies across Indonesia. As the value of the Indonesian rupiah (IDR) plummeted, cacao prices in Camba rose from IDR 2000 per kilogram to 14,000, a sharp increase compared to other plantation crops in a time of desperate need for cash flow. Numerous farmers expanded their commodity crops, electing to convert *kemiri* stands to cacao, or planting cacao in the understorey of more productive *kemiri* stands. However, this mixed cropping system did not last long. After about six years, cacao began losing sunlight under the increasing density of the *kemiri* canopy, and farmers were forced to make a choice. The case for cacao was affected by a perception that canopy cover caused fungal disease in cacao, so the choice between *kemiri* and cacao was a difficult one. Prices had stabilized, so in many places *kemiri* stands became more irregular as they made way for cacao. At a limited number of sites, cacao completely took over. Those farmers that were wary of the fungal disease rumours, or had personally experienced its effects, opted to continue *kemiri*-based shifting cultivation for its robustness against potential disease outbreaks.

In 1999, as part of broader political changes taking place across Indonesia, timber production in the region also fell sharply, resulting in increased demand for timber as a building material. New local markets emerged for *kemiri* wood, which had never previously been used as a building material. Upon harvesting *kemiri* timber, farmers also began to reconsider their options for switching to more intensive cultivation of seasonal crops, or more intensive agroforestry systems focused around cacao. *Kemiri* timber sales were only viable in areas close to roads that were accessible by four-wheeled vehicles. If a harvest was further inland, *kemiri* timber did not have the same value because of high transportation costs. These dynamics continued to reshape the geographic distribution of *kemiri* forests.

Finally, in the early 2000s, demand for chillies and tomatoes began to rise because of a growing inter-island market for these commodities. The sea route from Mamuju, on the western coast of Sulawesi, across the Makassar Strait to Balikpapan, in Kalimantan, had opened up and emerging markets in both places were connected. This took place alongside periods of high demand for tomatoes and chillies. High prices resulted in land-use changes in Camba, with some *kemiri* areas being converted to perennial ventures. One village, Sawaru, switched completely to intensive farming

Capsicum annuum L. [Solanaceae]

Inter-island trade and high prices for chillies in the early 2000s led to a surge away from *kemiri* shifting cultivation into permanent cropping, buoyed by the prospect of regular weekly incomes. *Kemiri* cultivation survived the onslaught, with chillies as a regular swidden crop among the younger trees.

with chemical fertilizers and pest control, resulting in an increased demand for labour. Cultivation of these commodities was attractive because of high prices and constant weekly income from chilli and tomato sales. Areas with road access began to follow these approaches as farmers elected to convert their *kemiri* groves and pursue the successful ventures emerging around them.

In summary, a centuries-old cultivation system underwent various ebbs and flows, expanding under some political-economic situations and decreasing its area of influence during others. However, there was one distinct change differentiating the periods before and after the 1990s. Before the 1990s, an integrated network of *kemiri*-producing communities in Camba had been established, and the network sold to a reliable market, so *kemiri* production was driven by stable prices. The systems that maintained *kemiri* shifting cultivation were also a thoroughly ingrained part of the social fabric, and it was seen as an insurance crop that could sustain farmers beyond their years of productive labour. Unique social relations also emerged alongside the *kemiri* cultivation system, characterized by a determination to protect land-tenure claims arising from local indigenous decision-making processes. After the 1990s, this stability came under heavy pressure from more intense commoditization and development processes. Road access and more integrated markets provided new opportunities for cacao and horticultural crops, and timber shortages also led farmers to consider alternatives to the production value of *kemiri*. As a result, this period initiated intense changes. *Kemiri* shifting cultivation underwent significant conversion, at a time when prices were stagnating, productivity was falling and competitive crops were booming. Yet *kemiri* shifting cultivation endured, and the system remains widespread and intact to this day. As we will show in following sections, its survival was due to the deep socio-cultural relations surrounding both land and *kemiri*-based shifting cultivation, as well as its continued robustness and economic viability. Next, we turn to the intricacies of the cultivation system in its current form, before raising some recent challenges threatening its resilience.

Solanum lycopersicum L. [Solanaceae]

Tomatoes were a part of the chilli boom in the early 2000s, leading to more land in the Camba region being converted to annual cropping. A native of South America, the tomato is among the world's most popular vegetables. However, the fruit is highly perishable and postharvest losses can be a challenge.

Kemiri landscapes, land relations, cultivation and markets

Understanding the landscape

Land use throughout the Camba region follows a distinct set of patterns based on physical and socio-economic attributes. The key physical attributes include soil fertility, slope and availability of continuous water. Wet rice is a critical component of local livelihoods, so areas without a steep grade that also have access to continuous supplies of water throughout the year are almost always set aside for *sawah* cultivation. Where the land is fertile, but it has steeper slopes and lacks a continuous water supply, it is likely to be designated for agricultural fields. However, where land is both infertile and steeply sloped, it is likely to be cultivated with *kemiri*, timber groves, other cash crops and fruit trees. Very steep lands are left to remain as permanent forest. As described in the previous section, each of these broad categories is also influenced by socio-political considerations. For example, road access, designation as a state forest and other key factors play an important role in defining the landscape.

Figure 36-3 provides a cross section of these broad land classifications. The topographical land-use model features a diametral line through outward concentric circles rippling away from a settlement. Settlement areas are determined by their proximity to water sources, both for daily household requirements and for their regular association with flat, fertile lands cultivable as wet-rice paddies. Labour requirements for *sawah* cultivation also demand that villages are nearby and often help to define the distribution of settlements. The boundaries of *sawah* areas are marked by timber stands that not only provide a valuable resource, but also indicate ownership of the more intensive crop-cultivation lands. The range of cash crops grown on these lands is likely to depend on proximity to roads. For example, road access creates a higher likelihood that horticultural varieties such as chillies and tomatoes will be grown, along with tree crops like cacao. In areas beyond these intensive croplands,

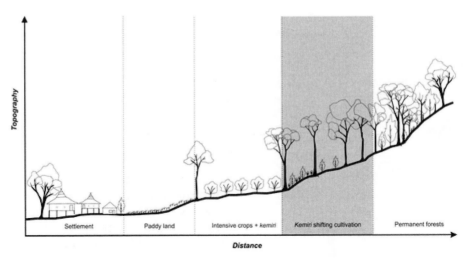

Settlement   Paddy land   Intensive crops + kemiri   Kemiri shifting cultivation   Permanent forests

Distance

FIGURE 36-3:   Topographical model of land management categories in the Camba region.

kemiri-based shifting cultivation becomes the dominant land use, and in steeper areas with more difficult access, *kemiri* stands give way to permanent state forests. This chapter is focused on the land uses in *kemiri*-based shifting cultivation areas, marked by the shaded zone in Figure 36-3. In the sections below, we detail the rotational land management and tenure schemes within the *kemiri* shifting cultivation zone, as well as supply chains and local markets. However, it is important to point out that although *kemiri* is a central feature of the landscape, as shown in Figure 36-3, *kemiri*-based shifting cultivation is only one part of what shapes livelihoods in the Camba region.

A model cycle of kemiri shifting cultivation

A unique tenurial dynamic governs land management in the *kemiri* shifting cultivation zone. This can be seen as a honeycomb pattern of eight hexagonal plots each covering one quarter of a hectare. This amounts to a total of about two hectares managed by a given household. Figure 36-4 highlights how the land-management stages rotate from one quarter-hectare plot to the next, covering all eight plots in the course of a 25-year cycle. Although this model provides a generalized description, cultivation areas also depend on the ability of a household to open and manage land while pursuing other economic opportunities.

Within the *kemiri* shifting cultivation system, land management shifts from one hexagonal plot to the next in a sequential system. Table 36-1 provides a more detailed description of crop cultivation and income generated from each hexagonal plot. It is important to note that although Table 36-1 provides detailed economic data from cultivation of *kemiri* and other crops, there are also numerous other cultivation activities taking place within the plots for household and subsistence uses.

The sequence of the eight plotted areas occurs as follows. In the first phase, farmers plant seasonal crops to make a quick return on profits and to open up lands for initial *kemiri* plantings. Often, landowners enter into contracts with those willing

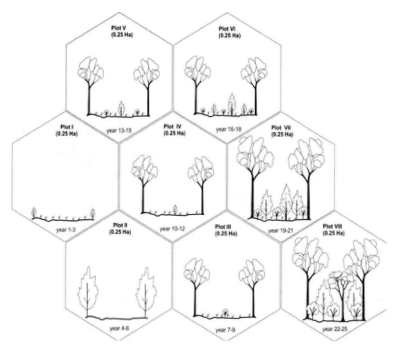

FIGURE 36-4:   The honeycomb rotation pattern involving eight quarter-hectare plots over the 25-year *kemiri* shifting cultivation cycle.

to work the land and split the earnings through sharecropping arrangements. *Kemiri* is thus introduced to plot 1; seedlings are planted at spacing of 5 x 5 metres. The fast-growing seasonal crops are intended to recoup initial investments. The types, varieties and economic value of the seasonal crops vary from year to year, depending on seedling availability, experience and market prices. The income figures from seasonal crops in Table 36-1 are discussed further in the section on household strategies, below. In the fourth year, the farmers move to plot 2 with initial plantings of seasonal crops and *kemiri* seedlings. Thus, the rotation system of the broader honeycomb begins to appear. In years 4 to 6, limited *kemiri* production becomes available from plot 1 as the trees have a limited harvest potential at five years. In years 7 to 9 *kemiri* production increases in plot 1, while harvests begin in plot 2 and seasonal crops are initiated in plot 3, and so the system progresses in a cyclical manner.

The most productive ages of *kemiri* are from 15 to 25 years. In years 16 to 18, *kemiri* production increases markedly through to the end of the 25-year cycle, at which point production begins to decline and farmers elect to harvest the timber, which can also yield cash benefits. In the 26th year, the cycle begins anew as plot 1 is cleared for seasonal crops and the planting of *kemiri* seedlings. Subsequent plots are reopened every three years. According to Supratman (2007a, 2007b) the total area of these typical two-hectare *kemiri* plots in Maros regency amounted to about 15,000 hectares, of which 9299 hectares were located within state forests. The same studies also showed that 9404 families managed these shifting cultivation lands in and around the state forests.

Arachis hypogaea L. [Leguminosae]

Peanuts are a popular annual crop in the first years of the *kemiri* shifting cultivation cycle, when the trees are still small. Unlike most other crop plants, peanut pods develop under the ground. As legumes, they also fix nitrogen, which enriches the soil. Peanuts are rich in essential nutrients and food products derived from peanuts are often used in famine relief.

From an economic point of view, Table 36-1 highlights the different ways in which income is generated from the *kemiri* shifting cultivation system. Initial income is obtained through what we describe as seasonal crops, which are often cultivated in sharecropping arrangements. Seasonal crops change with availability, price and household needs (Figure 36-5). For example, upland rice and maize are grown for ritual and subsistence purposes alongside peanuts, shallots and ginger. On the basis of comprehensive research on livelihoods in 2011, which calculated earnings from peanuts and maize, annual revenue from these crops averaged IDR862,815 per year (about US$100) (Alam 2007b; Supratman, 2007a). Seasonal crops tend to change according to market prices, networks, and availability of seedlings. Therefore, the potential earnings from this aspect of the shifting cultivation system, and its composition, are also subject to change.

The economic value of *kemiri* is based on different stages of its production potential. At five years old, the trees begin producing an average of half a kilogram of fruit per tree per year (Figure 36-6). This level of productivity continues through to about the 14th year, at which point *kemiri* has doubled its earlier fruit production, to about one kilogram per tree per year, and this level is maintained from years 15 to 25. The entire *kemiri* cultivation system, with all quarter-hectare intercropping areas involved sequentially over one complete 25-year rotation, is presented in Table 36-1. This forest-management system yields between 50 and 550 kg of fruit per year. Assuming an average price of IDR7000 per kg, the income from *kemiri* was estimated in 2011 to range from IDR350,000 to 3,850,000 per year (at that time between US$40 and $447) over the course of an entire rotation. As the end of the rotation sweeps sequentially around the eight plots, the farmers earn additional income from harvested logs as they prepare each plot for fresh cultivation. In 2011, *kemiri* timber was sold by individuals at the roadside at a price of IDR300,000 (about US$35) per cubic metre, or in the form of standing trees at an average price of IDR40,000 (about US$5) per tree. Most elect to sell standing trees because farmers can gain

TABLE 36-1:   Land cover and harvest potential from the 25-year *kemiri* shifting cultivation cycle, including seasonal crop cultivation.

Shifting area (0.25 ha)	Income sources	Periodic dynamic crop cover (year)								
		1 to 3	4 to 6	7 to 9	10 to 12	13 to 15	16 to 18	19 to 21	22 to 24	25
I	Seasonal crops	2,588,445								862,815.00
	Kemiri fruit		700,000	1,050,000	1,050,000	1,050,000	2,100,000	2,100,000	2,100,000	4,000,000
	Kemiri wood									
	Total	**2,588,445**	**700,000**	**1,050,000**	**1,050,000**	**1,050,000**	**2,100,000**	**2,100,000**	**2,100,000**	**4,862,815**
II	Seasonal crops		2,588,445							
	Kemiri fruit			700,000	1,050,000	1,050,000	1,050,000	2,100,000	2,100,000	700,000
	Kemiri wood									
	Total		**2,588,445**	**700,000**	**1,050,000**	**1,050,000**	**1,050,000**	**2,100,000**	**2,100,000**	**700,000**
III	Seasonal crops			2,588,445						
	Kemiri fruit				700,000	1,050,000	1,050,000	1,050,000	2,100,000	700,000
	Kemiri wood									
	Total			**2,588,445**	**700,000**	**1,050,000**	**1,050,000**	**1,050,000**	**2,100,000**	**700,000**
IV	Seasonal crops				2,588,445					
	Kemiri fruit					700,000	1,050,000	1,050,000	1,050,000	350,000
	Kemiri wood									
	Total				**2,588,445**	**700,000**	**1,050,000**	**1,050,000**	**1,050,000**	**350,000**
V	Seasonal crops					2,588,445				
	Kemiri fruit						700,000	1,050,000	1,050,000	350,000
	Kemiri wood									
	Total					**2,588,445**	**700,000**	**1,050,000**	**1,050,000**	**350,000**
VI	Seasonal crops						2,588,445			
	Kemiri fruit							700,000	1,050,000	350,000
	Kemiri wood									
	Total						**2,588,445**	**700,000**	**1,050,000**	**350,000**
VII	Seasonal crops							2,588,445		
	Kemiri fruit								700,000	350,000
	Kemiri wood									
	Total							**2,588,445**	**700,000**	**350,000**
VIII	Seasonal crops								2,588,445	
	Kemiri fruit									350,000
	Kemiri wood									
	Total								**2,588,445**	**350,000**
	Periodic income	2,588,445	3,288,445	4,338,445	5,388,445	6,438,445	8,538,445	10,638,445	10,150,000	8,012,815
	Annual income	862,815	1,096,148	1,446,148	1,796,148	2,146,148	2,846,148	3,546,148	3,383,333	

immediate net income, as the cost of felling and hauling the timber to the village roadside is borne by the buyer. However, depending on the availability of labour and proximity to road access, landowners may choose to harvest the trees themselves. Income from *kemiri* timber harvests can reach IDR4 million per year (about $464 in 2011), if about 100 trees are harvested from each quarter-hectare plot. After the 25th year, the rotational system yields a new timber harvest every three years, as the system moves from one plot to the next. Labour dynamics are discussed in the following sections.

Land-management institutions

The *kemiri* shifting cultivation landscape is managed under formal institutional property rules, but also under local indigenous notions of access to land, property and harvests. Overall, the Ministry of Environment and Forestry has the largest formal authority

FIGURE 36-5: The 25-year rotation begins with fast-growing seasonal crops – in this case, maize – interplanted with *kemiri*. The quarter-hectare plot is 'fenced' with hardwood trees.

FIGURE 36-6 A stand of young *kemiri* trees about the age when they first begin producing candlenuts.

in the Camba region, with control over 85% of the total land area. The remainder is governed under private property legal provisions of the Basic Agrarian Law.

The formal designation of the Camba region as an area under the authority of the Ministry of Environment and Forestry took place in several ways. The high percentage of tree cover as a result of *kemiri* shifting cultivation helped to identify the region as ideal for forestry authority under assessments made by the central government in the early 1980s. This was done by remapping the land through surveys called new forest plan agreements (*Tata guna hutan kesepakatan*, or TGHK). There was not much local consultation involved in TGHK designations, as administrators from afar used a scoring system based on physical characteristics, such as soil type, slope and rainfall to bring areas under forest-management authority. The Camba region was easily designated as it was a rural region without substantial claims of local administrative authority, it had an inland and contoured landscape and a large extent of tree cover. Within the forestry designation system, the Camba region was set aside as 'protection forest', a zoning category that especially restricts land uses. These restrictions became significant for current *kemiri* shifting cultivation practices and will be discussed later.

Although formalized under state authority, local indigenous land institutions had long been established to govern these lands. The local intricacies were much more complex than the state categories, including various layers of private property and sharecropping claims; *sanra'* (land held as security over debt); *teseng* (sharecropping land, but with a differentiation between *makkoko* for agroforestry systems and *maddare* for seasonal crops; and *mana'* (ancestral land). These land institutions are the foundation of socio-cultural relations in the Camba region, and are upheld in both rural and more densely populated settlement areas. In subsequent sections we will describe the institutions that govern cultivation and harvests on these lands.

The study area for this chapter has a population density of about one family per hectare. This is despite the model of shifting cultivation described in Figures 36-3, 36-4 and Table 36-1, showing a management area of about two hectares per family. The ratio of family-to-area land management has changed with population growth, creating new land-management dynamics. More generally, however, land relations begin through the inherited transfer of *mana'*, or the passing down of land rights to children, transferred as inheritance during marriage ceremonies. *Mana'* transfers are not gender specific, as sons and daughters have equal access to inherited land. *Sanra'* is a land category that was developed as a lending institution. In other words, if someone wants to borrow money, they can *sanra'* their land as security over the debt, and *kemiri* harvests are provided to the lender until the debtor can repay the loan. *Sanra'* is also a mechanism to ensure that ownership of land does not change, particularly by falling under the influence of outsiders. *Teseng* (*makkoko*) are split ventures between landowners and cultivators. The landowner creates a *teseng* agreement by jointly agreeing with the cultivator to investment terms and labour arrangements, with the intention of splitting the eventual harvests. These land ownership and access arrangements also take on new dynamics during cultivation and harvests.

In other areas of Sulawesi, where there has been a boom in cacao plantations, there have been significant indications of land accumulation by local elites, who have disrupted historical indigenous land relations (e.g. Li, 2014). In the Camba region, the stability of land relations in *kemiri*-based shifting cultivation areas has been due to several factors, but is largely because land areas cannot accumulate to levels that exceed the capacity of owners to work them. Furthermore, *sanra'* also acts as a safety net for debtors, to ensure that land will not be sold outside of families in the region. As landowners approach the end of their productive labour years, they can be certain that predetermined *teseng* arrangements covering a final 25-year cycle will sustain them by recouping the value of labour invested at a shifting cultivation site.

Kemiri *land-cultivation institutions*

Land-cultivation institutions govern the way that plots are cleared and prepared for planting, as well as providing guidance for the planting process. There are also different arrangements for sales of *kemiri* timber, depending on whether the landowner wishes to personally harvest the trees or to contract the harvest to a potential buyer to avoid transportation problems. Once a plot is cleared for planting, a land owner may either manage his or her land on an individual household basis, or enter into contracts with farmers who wish to manage only the cultivation of seasonal crops, or only the cultivation of tree crops. These arrangements are called *pakkoko* or *paddare*, the former describing tree-crop cultivation arrangements, and the latter, seasonal crops. In the *pakokko* system, farmers return the *kemiri* harvests to the landowners and split the profits according to their sharecropping arrangements. The early phase of crop cultivation also requires collaboration across households.

After the timber is felled and collected, the quarter-hectare plot is cleared and burnt. *Kemiri* farmers are careful in their burning systems. Buffer zones are cleared, firebreaks are made and pathways are provided to both coordinate the burning and prevent outbreaks of fire on adjacent family plots or those of their neighbours. The fires are followed by planting both seasonal crops and *kemiri* seedlings. Pigs are the biggest concern among farmers when they begin to cultivate seasonal crops, and during this phase they work together and stay close to their fields to protect early seedlings. After the fourth year, crops are more established and resistant to pests.

Kemiri *harvest institutions*

The institutions that regulate *kemiri* harvests are called *hompong*, and these involve a prohibition against collecting fruit on land owned by another person throughout a given period, or until the landowner cancels the *hompong*. White cloths are wrapped around selected *kemiri* trees to indicate ownership and to symbolize and apply a *hompong* as harvests draw near. If an outside harvester should enter an area in which *hompong*

Arenga pinnata (Wurmb) Merr.
[Arecaceae]

Sugar palm is another economic species that is often grown along with *kemiri*. Its sap is tapped to make sugar or a drink and can also be fermented to make vinegar or palm wine (see Fox, ch. 35, this volume, for similar uses of the *Borassus flabellifer* palm). The immature fruit is widely consumed and fibre from the trunk is made into cordage or roofing material.

remains current, the act of harvesting will be regarded as theft, and people throughout the Camba region believe that karmic retribution will strike those who take from *hompong* sites. *Hompong* is thus a sacred and spiritual institution that helps to ensure that the social customs of harvesting rights are upheld (Figure 36-7).

A *kemiri* harvest is undertaken in three stages, known as *makkampiri*, *mabbali* and *makkalice*. The first stage, *makkampiri*, involves collection efforts led by landowners who have coordinated with other farmers to provide labour support. These *kemiri* foragers are called *pakkampiri*. However, as populations have increased along with household demands for cash income, the *makkampiri* institutions have shifted, and nowadays *pakkampiri* are often rewarded through wages. The second stage, *mabbali*, is a collection process conducted by landowners or others after the *makkampiri* has ended. The foragers during this stage are called *pabbali*. They perform a second wave of harvesting that seeks to gather any missed or additional fruit following the earlier *makkampiri* stage. The fruits collected during the *mabbali* are divided between the landowner and the foragers (the *pabbali*).

The third stage is the *makkalice*. The *hompong* cloths are taken down prior to the *makkalice*, providing access to anyone who wishes to freely collect remaining fruit. In this final stage, there is no need for foragers to seek the permission of the landowner. Unlike the previous stages, the foragers, called *pakkalice*, may keep their harvests without providing any portion to the landowner. The *makkalice* institution serves a socio-cultural function in that it provides a safety net for community members who would like to generate extra income.

Upon completion of the first two stages of the harvest (*makkampiri* and *mabbali*), post-harvest institutions are organized. The first, called *maddepa*, gathers people to peel the shells of *kemiri* fruit by using a very simple tool made from woven rattan (Figure 36-8). These practices depend on the customs of individual households. Women are usually less involved in the various stages of cultivation, but are more active in foraging parties during harvests, and in the preparation of fruit for market. Women are especially involved in the first post-collection process, in which they are known as *paddepa* – those who crack and peel the *kemiri* nuts (Figure 36-9). The *maddepa* stage also involves a division of the harvest yield among the *padeppa*. Previously, large organizations were attracted to the *maddepa*, but this has become less common as *paddepa* are increasingly organized within individual households and family networks. Women also play an important role in price negotiations for the market-

FIGURE 36-7:   A man gathers candlenuts in a stand of mature *kemiri* trees.

ready fruit, as bargaining is seen as taboo among men. Household finances in the Camba region are generally controlled by women and negotiations of sale are approached with this in mind.

From harvest to market

Harvests may be taken to market in three ways. The first is when the farmer sells to local traders. This is a typical procedure when harvests are less than one *karung*. A *karung* is a sack that is often used for packaging fertilizer, and carries between 30 and 40 kg of *kemiri* fruit. The traders gather a number of harvests until the quantity justifies transport to subdistrict markets or to larger exporting companies in Makassar (Figure 36-10).

The second approach is for farmers to arrange their own transport to take harvests to subdistrict markets. This is a normal procedure if a harvest yields between one and three *karung*. The local markets are supported by the district government, which has an interest in facilitating *kemiri* commodity production. In this scenario, farmers like not only to deliver their *kemiri* harvests to markets, but also to make purchases for household and farming needs while they are there. The weight of a harvest is ideal because it does not require additional transport costs and the *karung* sacks can be re-used for the purchase of basic needs. The Camba market, which is the largest in the region, has about 20 regular buyers of *kemiri* fruit on

FIGURE 36-8: *Kemiri* fruit and simple tools made from woven rattan that are used to crack and peel the nuts for market.

FIGURE 36-9: Freshly cracked and peeled candlenuts.

FIGURE 36-10: Candlenuts packed in *karung*, ready for market.

given market days. The farmers tend to sell successive harvests to the same buyers. The local-market buyers then coordinate sales to exporters located in the industrial areas of Makassar. There are a total of seven such companies that buy from local markets.

The third approach is for the farmers to sell directly to the larger-scale exporting merchants in Makassar. In this scenario, the farmers must have more than three *karung* to make the transport costs worthwhile, so they often work together to fill a truckload. This approach is used when farmers need large amounts of cash, especially leading up to family events like weddings and other ceremonies, or to cover home-improvement costs.

Prices for *kemiri* fruit are determined by negotiations between village traders and regional traders at subdistrict markets. They will agree upon prices paid by local traders to farmers at the village level. If these prices exceed the cost of transport, farmers will elect to sell directly to the markets or bypass the markets by going directly to exporters. At the level of the exporters, prices are not negotiable. Therefore, price differentiation is determined solely by transport costs, which may be different depending on the location of certain villages. The farmers, village buyers and regional traders at subdistrict markets all sell to the large exporters, and the choices available to farmers in selling their harvests are determined by established relationships between buyers and sellers. Such relationships may be disrupted if farmers and local traders feel unfairly treated, or *kemiri* is bought at levels below prices at trader levels, but this is rare.

Between threats and innovations

Applying a framework to kemiri *shifting cultivation in Camba region*

Given the empirical detail we have provided, we now return to the heuristic framework presented by Fox et al. (2009), which highlights six broad reasons for the rapid pace of decline among swiddening systems across Southeast Asia. In examining these reasons for decline, we ask: what are the factors that continue to support the resilience of *kemiri*-based shifting cultivation systems, such that they defy this pan-regional decline? We argue that these factors may be found in consideration of the historical and evolving land relations in the Camba region, and particularly in the institutional cultivation-to-market arrangements.

First, the framework of Fox et al. (2009) describes the common State practice of labelling swiddeners as ethnic minorities and describing the practitioners of shifting cultivation as primitive and undeveloped. Indeed, Indonesia's official policy throughout the 1990s labelled upland communities as isolated [*terasing*] and backward [*terbelakang*]. In the Camba region however, the cross-section of ethnicities and languages are perceived as part of a proud history that is still associated with the dominant cultural influence of the Buginese people, and the history of royal bloodlines that settled the region. Therefore, the Camba region fits into the networks and narratives of the leading regional aristocracies, and the connection to Isossong Lamapaselling is part of the lineage of dominant kingdoms. Furthermore, since *kemiri* maintains the highest volume and earnings of all commodities produced in Maros district, shifting cultivation practices have generated a positive perception among regional administrators and thus there has been no forceful undermining of the status of rural communities, such as has occurred elsewhere in Indonesia, or in Southeast Asia more broadly.

However, this does not mean that connotations of primitiveness have been altogether absent in reference to *kemiri* shifting cultivation in Camba. Enclosures of state protection forests are increasingly considered to play an important conservation function, and the illegality of cultivation in these areas, particularly regenerative practices such as burning, have come into question by state administrators. Although local officials have shied away from enforcement because of the economic benefits from *kemiri*, targeted enforcement of burning practices could come into play as state forestry monitoring systems are modernized with new technologies. Outside of designated forest boundaries, state development priorities also encourage intensive monocropping. Agricultural offices and extension workers are of the same opinion, that multiple-cropping agroforestry systems are in need of modernization. For the time being, the existence of wet-rice systems, the stability of *kemiri* prices, and the primacy of the commodity in district revenue streams all help to protect *kemiri*-based shifting cultivation as a viable socio-economic system.

Second, Fox et al. describe how nation-states have divided landscapes between the administrative authorities of forest land and permanent agriculture, and this comes alongside certain beliefs about what constitutes proper uses of such lands. Indonesia is a classic example of this dual forest-agriculture system of state land authority, where permanent agriculture and forests are divided into two land categories of forest and non-forest on the basis of key land features (Sahide and Giessen, 2015). However, there is a more dynamic approach to land use at the community level, with numerous variations that are not fixed. In Camba, for example, lands that are difficult to access and have steeper slopes will be defined as permanent forests, while lower down, if there is a shift in rivers, streams, or irrigation potential, wet-rice areas may be relocated. The rigidity of state systems and the enforcement of new institutional rules have been incompatible with many shifting cultivation systems. However, in Camba such policies have had unique outcomes. Because 85% of *kemiri*-based shifting cultivation is in state protection forest areas, these are the areas that have largely continued to support and maintain it. This is because the regional government has a strong interest in maintaining *kemiri* due to its primacy for revenue generation. Indeed, areas outside the protection forests that are designated as permanent agriculture lands have been more susceptible to conversion, especially to the booming monocrop, cacao (Alam, 2007a). Therefore, as long as tree-cover continues to be maintained in the protection forest areas due to the canopy of *kemiri* forests, law–enforcement officials are willing to overlook what are institutionally illicit activities.

That having been said, there are other potential factors that emerge from state designations of allowable land uses. Indeed, Fox et al.'s third point highlights rising interest in conservation. *Kemiri* farmers in the Camba region have begun to refer to increasing pressure from forestry administrators to do away with burning. Anti-burning campaigns in Indonesia are part of a broader policy imperative to maintain protection forests as areas free from burning.[3] Even social-forestry designations, in which management authority is vested in local communities, have provisions on fire usage in state forest areas. With increasing prohibitions on burning, the balance of the *kemiri* system may be affected. It remains to be seen if stricter enforcement of fire bans results in new directions for land use in the Camba region.

The fourth element of Fox et al.'s framework describes how resettlement has crippled the land-management practices of swiddening communities, particularly the ability of farmers to access and manage their land. Indeed, the establishment of *kemiri*-based shifting cultivation was a product of resettlement when, in the 19th century, Isossong Lamapaselling migrated into the region and established a viable trading network for *kemiri* cultivation. In more recent times, resettlement in the Camba region has taken place in three ways. The first involved long-term government resettlement programmes aimed at formalizing communities into village structures. The second was a consequence of turbulent and violent times during the Islamic rebellion of the 1950s and 1960s, when communities were forced to retreat deeper into the forest. The final form of resettlement is a part of Fox et al.'s fifth element: voluntary resettlement to get closer to infrastructure and markets.

The fifth and sixth elements of the heuristic framework relate to privatization, commoditization, land-based production and expansion of infrastructure, particularly in support of industrial agricultural development. In settlement areas with roads, state programmes encourage monocropping systems by introducing single crops. In addition, private-sector ventures seek to incorporate rural communities into contract-farming arrangements, while smallholders establish their own ventures to benefit from greater access to global price increases. Development programmes also support intensive planting systems by providing subsidized fertilizer and seedlings along with methodologies that regard both shifting cultivation and diverse agroforestry systems as inappropriate. Such formal state-development viewpoints and programming also reinforce the first point of Fox et al.'s heuristic framework by describing centuries of innovation in shifting cultivation as primitive. Thus, village heads, extension officers and other officials are compelled to support programmes in which success is measured by the extent to which intensive monocropping systems are established. However, formal government-sanctioned programmes also require clarity, especially in their administration of land. In this way, the designation of Camba as 85% protected forest helps to maintain *kemiri* shifting cultivation, not only because of its advantages and productivity, but also because converting it to agricultural and more intensive uses would be illegal. This has also hindered road development into the region and limited development to narrow corridors.

Conclusions

Although it is located close to the large metropolitan area of Makassar, *kemiri*-based shifting cultivation in the Camba region persists for three main reasons. The first is that a centuries-old system based on a convergence of uplander and migrant histories, with networks to markets, became a deeply ingrained part of socio-cultural land relations. With connections to the proud ruling identities of South Sulawesi – the Bugis – these shifting cultivators were not easily relegated to 'primitive' status. As we have shown, *kemiri*-based shifting cultivation in the Camba region has weathered and adapted to impacts arising from enclosures by colonial authorities, Islamic rebellions, state development and resettlement programmes and narrow definitions of allowable land uses in protection forests.

The second reason that *kemiri* shifting cultivation continues to thrive is economic. The Maros district government, as the leading administrative authority in the Camba region, has a vested interest in supporting high-volume *kemiri* production. As the highest-yielding plantation crop in the district, it props up local revenue streams. As a result, a third supporting factor has emerged: although national policies seek to delineate state forests and remove communities from these lands, regional forestry administrators have set aside their enforcement responsibilities in the case of *kemiri*-based shifting cultivation, especially with regard to resettlement and burning. Thus, unexpectedly, state enclosures designating forest lands in the region have had an opposite effect to that originally intended: they enable *kemiri* farmers to weather aggressive state programmes pushing intensive agriculture; keep road development and other larger infrastructure programmes at bay; and provide a buffer against plantation–crop commoditization, compelling smallholders to continue practising shifting cultivation. Indeed, the areas of *kemiri*-based shifting cultivation that are experiencing the greatest land-use change are those located outside of state forests, covering about 6000 hectares of agricultural land. These areas have been under pressure from agricultural commoditization processes brought on by the cacao boom, and have experienced increasing infrastructure development.

Overall, the threats to the *kemiri* system still follow the heuristic elements identified in the Fox et al. framework. These include heightened programmes that undermine shifting cultivation as a primitive and 'unintensive' agricultural system and new state forest policies encouraging aggressive enforcement of allowable activities in protection forests, such as cracking down on burning practices or resettlement. There is also the threat that prices for *kemiri* could lose their lustre in the face of new incentives for change, such as those seen in the case of cacao and other horticultural crops in the past.

References

Alam, S. (2007a) 'Analisis deskriptif pola konversi hutan kemiri rakyat (HKR) di Kabupaten Maros (A descriptive analysis of forest management in community candlenut forests in the District of Maros)', *Jurnal Hutan dan Masyarakat* 2(1), pp.136-144

Alam, S. (2007b) Nilai manfaat dan pola konversi hutan kemiri rakyat di Kabupaten Maros Provinsi Sulawesi Selatan (Benefits from candlenut land management approaches from the District of Maros, South Sulawesi)', PhD dissertation to the Graduate Programme, Hasabuddin University, Makassar (unpublished)

Bakosurtanal (1999) *Peta Rupa Bumi* (Base Map) 1:50.000, *Badan Koordinasi Survei dan Pemetaan Nasional* (National Survey and Mapping Coordination Board), Jakarta

Fox, J., Fujita, Y., Ngidang, D., Peluso, N., Potter, L., Sakuntaladewi, N., Sturgeon, J, and Thomas, D. (2009) 'Policies, political-economy, and swidden in Southeast Asia', *Human Ecology* 37(3), pp.305-322

Islam, M. S., Hui Pei, Y. and Mangharam, S. (2016) 'Trans-boundary haze pollution in Southeast Asia: Sustainability through plural environmental governance', *Sustainability* 8(5), p.499

Li, T. M. (2014) *Land's End: Capitalist Relations on an Indigenous Frontier*, Duke University Press, Durham, NC

Limbongan, J. (2016) 'Kesiapan penerapan teknologi sambung samping untuk mendukung program rehabilitasi tanaman kakao', *Jurnal Penelitian dan Pengembangan Pertanian* 30(4), pp.156-163

Muspida (2008) 'Kearifan lokal dalam pengelolaan hutan kemiri rakyat di Kabupaten Maros Sulawesi Selatan (Local wisdom and candlenut community forest management in the District of Maros, South Sulawesi)', *Jurnal Hutan dan Masyarakat* 3(2)

Pemkab Maros (2018) *Potensi Perkebunan* (Plantation Figures), available at https://maroskab.go.id/potensi-perkebunan/, accessed 11 June 2018

Sahide, M. A. K., and Giessen, L. (2015) 'The fragmented land use administration in Indonesia: Analysing bureaucratic responsibilities influencing tropical rainforest transformation systems', *Land Use Policy* 43, pp.96-110

Supratman, S. (2007a) 'Analisis struktur wilayah dan kinerja pengelolaan hutan di Kabupaten Maros, Provinsi Sulawesi Selatan (An analysis of regional structure and forest management in the District of Maros, South Sulawesi)', *Jurnal Hutan dan Masyarakat* 2(2),

Supratman, S. (2007b) 'Hubungan spesialisasi wilayah pedesaan hutan dengan degradasi hutan di Sulawesi Selatan (The relationship of specializing village commodities with forest degradation in South Sulawesi)', *Jurnal Ilmu Pengetahuan Sosial* 8(3)

Yusran (2005) 'Mengembalikan kejayaan hutan kemiri rakyat (Restoring the glory of community *kemiri* forests)', Governance Brief June no. 10, under the programme Building Leadership for Forest Reforms of the Future, Center for International Forestry Research and the Ford Foundation

Notes

1. The Camba region of South Sulawesi consists of several communities defined by their linguistic and ethnic characteristics. They include the Bugis-Camba, Dentong, Bugis-Mallaha, Bugis-Bone and Makassar. During the colonial era, the Dutch administered the region as the district of Camba. However, Camba is nowadays split into three administrative subdistricts (*kecamatan*): Camba, Cenrana and Mallawa. These communities are quite diverse in their backgrounds, but are united in their common practice of cultivating *kemiri*. In the past, the region was well known as an area of great wealth, and was called the *Haji Kemiri* due its lucrative *kemiri* harvests that long ago provided sufficient income for local people to make Hajj pilgrimages to Mecca.

2. The Toraja *Coffea arabica* plantations are located within the same province of South Sulawesi, and these produce one of the most famous varieties of coffee in the world.

3. Indonesia recently suffered from one of the worst fire-related crises in living memory: the so-called 'transboundary haze' problem. It resulted in significant respiratory-tract disorders across the region and losses amounting to US$20.1 billion (Islam et al., 2016). However, in this case the smoke haze came from peatland fires that smouldered underground for a long time, and were significantly different to the regenerative burning of small patches of vegetation undertaken by *kemiri* farmers. Nevertheless, international pressure to curb burning across Indonesia led to a severe crackdown on any burning activities, led by state rangers and foresters.

37

ROTATING AGROFORESTS

Using shifting cultivation practices to construct a sustainable livelihood

*Paul Burgers**

Introduction

For many generations, the inaccessible inland agricultural areas of the island of Sumatra consisted of villages of rather autonomous people with closed economies. They practised subsistence farming, based on shifting cultivation of rice. This practice was known for its rotation of fields, rather than crops. Yearly use rights for a certain plot of forest land were allocated to individual households, and after their cropping was finished, the land was returned to the village for the restoration of soil fertility in long forest fallows. To make these systems sustainable, a community organization at village level was required to prevent unsustainable forms of natural-resource use.

The introduction of pepper cultivation by Islamic traders from India in the 13th century in Aceh, the northernmost part of Sumatra, marked the beginning of the penetration of more commercialized, market-oriented forms of agriculture. From the 16th and 17th centuries onwards, the Dutch took control in West Sumatra, interrupted by a short period of British rule at the end of the 18th century.

In the late 1800s, the Dutch found substantial areas of arabica coffee (*Coffea arabica*) growing on a voluntary basis in the Padang highlands of West Sumatra.[1] In most cases, these plantations had arisen from the planting of coffee trees in the fallow vegetation of shifting cultivation systems. Judging from the thickness of the stems, many of them appeared to be a few hundred years old. The old age of the trees, combined with the local use of coffee – drinking an extract from the dried and smoked coffee leaves – indicated that coffee must also have been introduced in the 1400s by Islamic traders, who prepared their coffee in a similar way. The coffee variety 'robusta' (*Coffea canephora*) was introduced in about 1915. Its easy management, lesser need for labour and compatibility with shifting cultivation systems led to widespread on-farm coffee

* Dr. Paul Burgers, formerly of the Faculty of Geosciences, Utrecht University, the Netherlands, now of CO2 Operate BV, Woerden, the Netherlands.

production among small-scale producers in the forest margins. To this day, the robusta variety still dominates coffee production by small farming households in Sumatra. When the ethical policies of the Dutch administration enabled rural people to move from shifting cultivation to more intensive wet–rice cultivation in the early 1900s, the rather autonomous villages became integrated into a wider, supra-local and national political system. Moreover, with growing overseas markets for perennial cash crops, including coffee, the cash income from tree crops was used to purchase food, putting an end to possible constraints in household food security. This caused an unexpected boom in economic development in Sumatra. The expansion of highly intensive irrigated rice-farming areas both absorbed the growing population and largely solved constraints associated with agricultural-land

Coffea robusta L. Linden, a synonym of *Coffea canephora* Pierre ex A. Froehner [Rubiaceae]

This is one of two primary species providing the world's coffee production. The other is *Coffea arabica*, of which *Coffea canephora* is actually a parent plant. Robusta coffee makes up less than half of world production, and is often regarded as being of lesser quality than Arabica. However, it is easier to grow and is more tolerant of disease and drought. Moreover, it has nearly double the caffeine of Arabica. Blends of the two are commonplace, especially for espresso.

extension. Farming systems were largely transformed from shifting cultivation of rice for subsistence needs to an integrated system of food cropping on irrigated ricefields alongside commercial tree-based production. Between 1880 and 1950, there was a rapid decline in dependency on shifting cultivation (Figure 37-1).

These new crops brought new opportunities and with them, different farming techniques. However, farming communities in Kerinci district, West Sumatra, continue to use accumulated ecological knowledge and management of forest-based natural resources, developed over many generations, to make their agricultural-cropping systems sustainable. They have developed agro-ecosystems that imitate forest regeneration during the fallow period of shifting cultivation.

The setting and the study sites

Kerinci has always been known for its favourable agricultural conditions. At an altitude of about 500 metres above sea level (masl), the study sites of Selampaung and Masgo are situated in the Gunung Raya subdistrict in the southern part of

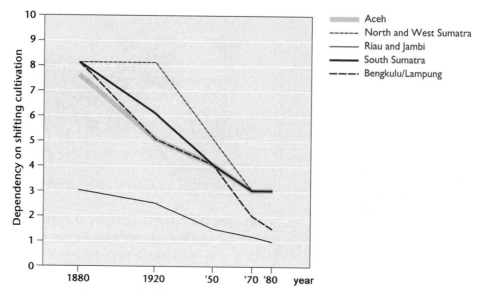

FIGURE 37-1: Decreasing livelihood dependency on shifting cultivation per province, Sumatra (1880 to 1980).

Source: based on Richards and Flint, 1994.

Kerinci district (Figure 37-2). Whereas most of the upland fields are privately owned, rice cultivation in Selampaung and Masgo is largely structured by kinship organizations, based on local regulations under the *adat* system of customary law, originating from the mainly ethnic descendants of the early settled Minangkabau people. A specific matrilineal organization of regulations for use of natural resources continues to prevail. An indigenous system of communal ownership and temporary individual land-use rights for wet-rice cultivation is said to have evolved when the villagers switched their agricultural practices from shifting cultivation of upland rice to irrigated rice farming (Scholz, 1988). Instead of private ownership, yearly access rights to rice-cropping land may be obtained by the female members of the family on a rotational basis (known as *gilir ganti*). A ricefield managed under these arrangements is known as a *sawah giliran*. Within the boundaries set by *adat* regulations, people can manoeuvre between the exploitation rights and the allocation of plots, not only directly under the *giliran* system, but also on land belonging to others through sharecropping (the share stipulated by *adat* is 50:50). Specialized rice farming continues to be an important fallback mechanism to secure food needs. In most parts of the district, access to upland areas can be secured, where perennial crops are usually grown.

As stated before, enrichment planting of useful products or promotion of certain products (e.g., coffee) through the manipulation of forest vegetation has for many years added to the stability of livelihoods in Kerinci district. Since households no longer depend solely on the forest for their livelihood, more intensified management of patches of forest and secondary fallow vegetation in upland areas has resulted in

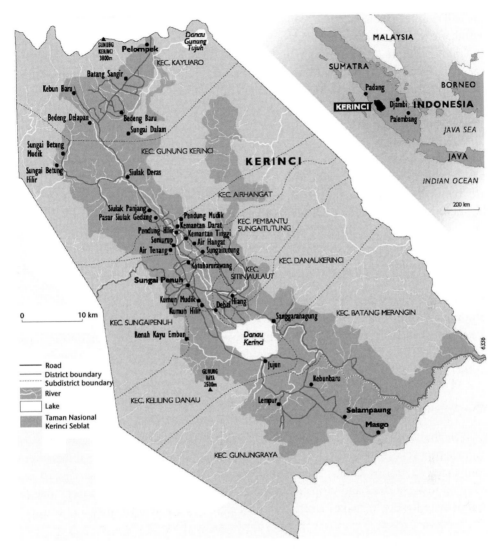

FIGURE 37-2: Kerinci district, in Sumatra, Indonesia, and the research villages.

their complete conversion into agricultural fields. However, in order to protect the forest functions that are important for the sustainability of their livelihood in general and for rice cultivation in particular, tree-based systems or so-called multi-strata agroforests have evolved, imitating the shifting cultivation cycle of a cropping phase and a period where tree cover takes over.

Multi-strata agroforests in Selampaung and Masgo

Largely by imitating the processes of natural vegetation succession, as occurs during the establishment of fallow vegetation in shifting cultivation systems, a highly efficient and productive agroforest system has evolved in upland areas surrounding the villages and their ricefields. In various configurations, annual crops are grown simultaneously or in sequence with coffee and cinnamon trees. The cinnamon species is *Cinnamomum burmanni*, which under the Dutch colonial administration was called *korintje*, the Dutch word for Kerinci. Depending on the major crop present in the field, i.e., either annual crops or tree crops, different names are used. A *ladang* refers to a field on which annual crops are most important, although dispersed trees are generally present as well. Where tree crops are dominant, the term *kebun* is used.

When vegetables are the major crop (i.e. on a *ladang*), this represents a temporary stage, waiting for the time when the *ladang* transforms into a *kebun*, and the maturing tree crops take over as the main crop. Chillies, tobacco, and to a lesser extent, peanuts are the most common annual crops cultivated. The combination of these exotic and local crops and tree species has secured farming households in the research villages, giving them both economic resilience and ecological sustainability. The differing growth rates and functions of the mixture of species and varieties on the farm provide multiple socio-economic benefits at different times, and in a flexible manner. This implies higher resilience against socio-economic and environmental shocks and stresses. Moreover, the system largely substitutes for the ecological functions of a natural forest by safeguarding hydrological functions through shading and root systems that regulate water flows and minimise erosion. These are among the most important means of maintaining agronomic sustainability in the irrigated rice systems in the valleys (Scholz, 1977).

The Agroforest: a winning team

One of the farmers in Selampaung, Pak Rustam, explained that the characteristics of an agroforest that enabled economic, social and ecological sustainability were comparable to those of an 'unbeatable' soccer team (Box 37-1). He said that a winning team of trees and crops consisted of flexible components that were able to compensate for the 'mistakes' of the others (i.e., they were able to work as a team).

BOX 37-1: Agroforests and winning football teams

According to Pak Rustam, of Selampaung, a good soccer team has players that can be relied upon. They maintain interest in the team. First, there are attackers who regularly try 'to score goals' and attract our attention. These are the annual crops, like peanuts, chillies and potatoes. These are planted when development of the agroforest begins, and they provide us with an immediate and regular source of income; they 'score goals' at regular intervals.

But there must also be a reliable defence, for when the 'game' starts getting difficult. This is coffee. Although coffee cannot be harvested too often, it usually gives a higher income. There are times though, when this defence is not enough and the 'team' must rely on its goalkeeper, the cinnamon tree. Cinnamon bark gives a sure, flexible income. If small amounts of cash are needed, just a few trees or branches are harvested. If the need is great, all of the cinnamon trees can be cut down at once. This is just as important as a goalkeeper stopping a penalty.

Source: Burgers and Wiliam, 2000.

If this principle is applied to an agroforest, the individual components need to be able to compensate for certain pitfalls like disappointing yields or low prices from other crops. Flexibility in choosing the various crops (the players) and techniques for 'renewing' the perennial species, allows households to weigh the costs and the speed of growth for each crop against its economic advantages and the sufficiency of their resource base at the time of perennial renewal (i.e., they can create a winning team at any point in time). In the research villages, households have not only capitalized on the cultivation of various crops, they have also capitalized on certain management strategies that best fit their socio-economic circumstances.

This is done by choosing different crops and the different abilities of the perennial species to either recover from harvesting or grow from seedlings; the entire process representing an agroforest cycle. If a farmer is dealing with an existing agroforest, renewal can be achieved by harvesting a complete stand of cinnamon trees and removing the rest of the vegetation. The tree crops can be 'renewed' in several ways: either by uprooting the trees and planting new seedlings, coppicing the trees to ground level, or pollarding the trees, leaving a trunk about one metre high. The latter procedures aim to encourage regrowth. There is also a combination of these practices, and farmers are largely influenced by prices in how they proceed. When the trees are pollarded, cinnamon is reduced to a one-metre-high trunk; coffee to about 50 cm high. When seedlings are used for the establishment of a new agroforest, coffee seedlings are usually planted at the same time as annual crops. Because of their relatively fast growth, cinnamon seedlings are not incorporated into the system until the coffee is about two to three years old, i.e., when it starts producing coffee beans. This is done to avoid early closure of the cinnamon–tree canopy, which causes the other crops to stop producing. Since cinnamon is planted mainly to serve as a 'savings bank' for problematic times when large amounts of cash are required, a uniform age structure is crucial in order that an accurate estimate can be made of the

Cinnamomum burmanni (Nees & T.Nees) Blume [Lauraceae]

The 'goalkeeper of the agroforest football team', providing a sure, flexible income for Kerinci district farmers. The popular spice cinnamon comes from the inner bark, which is peeled, cleaned and dried into 'sticks' or 'quills'. The trees are often pollarded at harvest time, and they grow back with new, tender branches providing high-quality spice. Despite their utility in Indonesia, these trees have become invasive in Hawaii, where they were introduced as a cultivated crop.

value of a stand, based on age and quality. Respondents said that when cinnamon seedlings were used to establish a new agroforest, two seedlings were planted in each place where they expected only one tree to grow to maturity. This not only made sure that at least one would reach maturity, but also provided additional cash when the stand was thinned. When the trees are pollarded and left to regrow, the canopy of coffee and/or cinnamon closes more rapidly, because regrowth from pollarding is faster than seedling growth, and the trees reach harvestable size in less time.

The pollarding of cinnamon and coffee is a common practice in the study villages of Selampaung and Masgo. Since regrowth from pollarding is faster, this approach is preferred to planting new seedlings when long-term prospects for the tree crops are better than those for annual crops. The pollarded tree crops grow more rapidly, the canopy closes earlier and vegetable cultivation is limited to one year. Normally, however, this practice is not favoured. When resources are sufficient to allow alternatives, such as uprooting the trees and replacing them with seedlings, this is usually preferred. In most cases, cinnamon trees can only be pollarded (or coppiced) three times, because of a declining rate of regrowth after harvesting. The trees are usually replanted after four harvests (Suyanto et al., 2001). However, the farmers' needs for cash can override the routine harvests. They commonly harvest individual branches to cover household needs. Hence, pollarding cinnamon trees is not common, when compared to the practice of pollarding coffee. The farmers say that robusta coffee trees respond vigorously to pollarding, developing a thick bush of branches. This means more beans and higher production. During the research, some coffee trees in upland fields were estimated to be more than 50 years old.

Management phases of the agroforest

The growth rates of the different species point to variations in the duration of production and management of the various crops. Three different stages were identified:

- An intensive-management phase, when vegetables are cultivated and sold;
- An intermediate-management phase, when the focus is entirely on the harvesting of coffee beans and their sale; and
- A phase of extensive management, once the cinnamon canopy has closed.

In Selampaung and Masgo, annual crops of commercial vegetables are always cultivated at the beginning of a new cycle. The remains of the previous vegetation are used either as a green manure or are brought together in scattered heaps in the field and burned. Entire fields are never burned, especially when trees have been pollarded, as this would destroy their ability to regrow. The common approach, in which burning of entire fields is avoided, also provides another important advantage, although the farmers may not have been aware of it. Ketterings (1999) found in her research among rubber farmers in Jambi that burning entire plots – a common practice there – can cause severe damage to the soil and its nutrients. In particular, when the fire is too hot, it may lead to soil oxidation and compaction. This can happen when there is a lot of easily combustible material in the fire, such bushes that often occur when fallow periods are short, or there is limited natural regeneration. By burning vegetation in small heaps, most of the field is not exposed to fire and does not suffer this damage. After burning, the ashes from the various patches of burnt biomass are spread out in the field so that nutrients are available for crop cultivation. This burning process also allows seeds and other plant remains in the soil, which are not tolerant of burning, to re-establish once the system moves into the extensive-management phase. This final phase favours the establishment of natural vegetation.

The initial years of establishing an agroforest, when commercial vegetables are cultivated and perennial crops need to come to maturity, is a time of high resource investment and management. This is therefore referred to as the intensive-management phase (similar to the rice-cultivation phase in shifting cultivation). Many respondents said that if the previous land cover had been forest, tobacco would be the first crop cultivated in this phase. If, on the other hand, there had been one or several agroforests on the plot before, the soil properties would no longer be suitable for tobacco cultivation without the use of external inputs (in particular, fertilizers). The relatively

high investments that must be made to maintain tobacco production at profitable levels are usually not worth the risk for the majority of resource-constrained households in the study villages. Many households prefer to cultivate peanuts (*Arachis hypogaea*), instead. Besides the fact that peanuts serve as a food and cash crop, they provide quick ground cover that minimises soil erosion, while leafy material remaining after the harvest is said to be good quality green manure for the next crop, which is usually chillies.

Arachis hypogaea L. [Leguminosae]

The shading effects of the perennial crops largely determine the time at which the agricultural system progresses into the next management phase. Through careful planning, coffee trees begin shading out vegetables at the time when their berries are ready to be harvested. This timing varies, depending on whether seedlings have been planted, trees have

Peanuts are regularly grown as a food and cash crop in the first phase of the Kerinci agroforests, when the coffee and cinnamon trees are still establishing. Peanuts have dense foliage that protects the soil from erosion and later becomes a rich green manure. As they grow, they also fix nitrogen to boost the levels of N available to plants in the soil. Peanuts are classified as both a grain legume and, because of their high oil content, an oil crop.

been pollarded or coffee trees have been left standing after replacing the cinnamon trees. When coffee trees are left standing or are left to regrow after pollarding, this second phase may begin after one year. In the case of seedlings, it takes about two to three years to reach this stage. However, once the coffee trees start producing, the system moves into the intermediate-management phase. Household labour resources alone are used for the coffee harvest, and this continues until the canopy of the cinnamon trees closes. At this point, the coffee trees can no longer produce coffee beans because of the absence of sufficient sunlight. As might be expected, the timing of the closure of the cinnamon-tree canopy also varies, depending on whether the trees were pollarded or replaced with fresh seedlings. When cinnamon trees are left to regrow after pollarding, the intermediate-management phase lasts for about one year, but it extends for up to three years in the case of growth from seedlings.

With the closure of the cinnamon canopy, the agroforest has reached its third and final stage: the extensive-management phase. This can be compared to the mature fallow stage in shifting cultivation. The field is abandoned and natural vegetation is allowed to re-establish as a third layer under the canopy of cinnamon and coffee trees. Management is restricted to occasional thinning or harvesting of branches for daily or weekly cash needs, while the regenerating natural vegetation may be slashed once or twice a year. However, a number of respondents said they might manipulate and manage the natural-succession regrowth at this stage by promoting and even planting species that they believe will have positive effects on soil fertility and provide green manure for the cultivation of vegetables in the next intensive-management phase, when the system's cycle begins again. A number of respondents pointed out that they collect seeds of the pioneer species *Austroeupatorium inulaefolium* from other fields and broadcast them in the *kebun*. They say that *A. inulaefolium*, with its vigorous growth, has beneficial effects on soil fertility because of its large amounts of biomass, which easily decomposes.

Balancing household needs, resources and vegetation

The previous section has shown that management contributes significantly to the way these systems protect or enhance a certain degree of vegetative cover. The state of a household's resource base will usually decide whether the system can be managed in a comprehensive way. When the original shifting-cultivation system was still in use, temporary-use rights automatically turned into long forest fallows to restore soil fertility. Nowadays, a heterogeneous pattern of management capabilities and constraints have brought huge variations to the way the vegetative cover is manipulated, and hence the agroforestry system is able to protect soil fertility and enhance the recovery of vegetation.

The ideal situation is for each household to have (or rotate around) at least five plots of land. Cultivating food crops for up to two years during the intensive-management phase on one plot will allow it to remain in the medium- and extensive-management phases for at least eight years. Respondents said this would be long enough to open the cinnamon/coffee stand again and grow food crops without the use of external inputs. By rotating the system on five different plots of land, a farming household always has one plot in the intensive-management phase (equivalent to the rice-cultivation phase in shifting cultivation). Favourable *adat* regulations in the study villages allow anyone to obtain temporary-use rights to upland fields. This attracts many migrants in search of a satisfactory livelihood, from as far afield as Java. As individual ownership prevails, rather than communal natural-resource management, it is clear that the socio-economic circumstances of individual households are important in their ability to gain access to five different plots of land. We were able to identify three major groups of households managing the upland fields, each with its own needs, priorities, opportunities and constraints. The first group consists of rich absentee families

Austroeupatorium inulaefolium (Kunth) R.M.King & H.Rob. [Asteraceae]

This highly invasive woody shrub can grow up to two metres and colonizes disturbed areas such as roadsides and fields prepared for planting (see plate 43 in Colour Plates section II, this volume). However, its vigorous growth, high biomass production and heavy leaf litter have made it a valuable green manure and soil improver (Cairns, 2007). In some of the Kerinci agroforests it is allowed to grow under the coffee and cinnamon canopy to boost the soil quality ahead of the annual-cropping phase.

owning large tracts of land, whose livelihoods are well above survival needs. Since the accumulation of wealth in cinnamon trees is their foremost strategy, and their resource base is large enough to hire labour, they make use of sharecroppers to establish plantations for them.

Sharecroppers constitute the second group of upland 'managers'. They work full time in agroforests usually owned by the first group, the rich absentee families, and usually hold a low socio-economic position. Quite often they are migrants, although local villagers also engage in sharecropping deals in order to access the 'required' five plots of land. The third category consists of local households who must combine the cultivation of their ricefields with upland farming, using their own resources and family labour. Depending on their resource base, they can be found along the entire spectrum of low to high socio-economic positions. As mentioned, these people sometimes act as sharecroppers to accumulate more access to land. These differences in resource base have caused variations in the way the agroforests are able to support both environmental and socio-economic functions. Figure 37-3 shows that five different systems can be identified. Systems B, C and D were the most commonly practised, by 46% of all respondents. A further 33% followed management system E, while the rest used system A.

The variations in these management systems provide ingenious ways to balance short-term survival needs with long-term sustainability. The size and extent of the vegetation and benefits in terms of biodiversity are greatest the sooner the canopy closes – or the sooner the system moves into the extensive-management phase – and the longer the (mature) tree components are left to grow in the field. Short periods of annual-crop cultivation followed by long phases of perennial cropping not only support an early build up of natural vegetation, but also allow it to remain in the field during long 'fallow periods', when the cinnamon trees are left long enough to achieve the highest-quality bark (Figure 37-3, system E). Early closure of the canopy is achieved when both coffee and cinnamon trees are pollarded, which limits the intensive- and intermediate-management phases to one year each. After two years, the system progresses into the extensive-management phase and natural vegetation is allowed to regenerate from the third year onwards. The duration of the extensive-management phase is a second major factor contributing to the persistence of, or re-colonization by, native plant and animal species. In some cases this final stage of the cycle lasts at least 10 to 12 years, or until the trees are 12 to 14 years old. This duration of the extensive-management phase may occur when the cinnamon trees are considered a wealth-accumulation strategy and the owners do not depend on the land for their survival. Clearly, this applies to rich absentee households. In reality, the duration of the extensive-management phase is rarely this long and appears to fluctuate between two and 10 years, ending when the cinnamon trees are between five and 15 years old. Only rich absentee households are able to wait for long periods before the cinnamon trees are harvested. Since this group aims at income maximization, they will also use sharecroppers to provide a long phase of intensive management at the start of the next cycle (Figure 37-3, system A).

Discussing Figure 37-3 with the villagers, it became clear that management system E is followed mainly under conditions of severe resource constraint. This type of management is not preferred, because these constraints limit a more intensive and more profitable management of the upland fields. This shows, however, that agroforest management largely depends on the constraints and capabilities of individual owners or households. As a household continuously weighs the advantages of one crop against the benefits of another for livelihood stability, it may opt to cut down its cinnamon trees, for instance, if prices for cinnamon bark are low compared to those for vegetables and/or coffee. In the research villages, cinnamon trees are cut down after an average of eight to 10 years. This applies to both trees that start from seedlings and those that are pollarded and regrow. Usually, those households that choose to pollard their cinnamon trees are the same ones that say their trees might be harvested at an earlier age (at an average age of six to eight years).

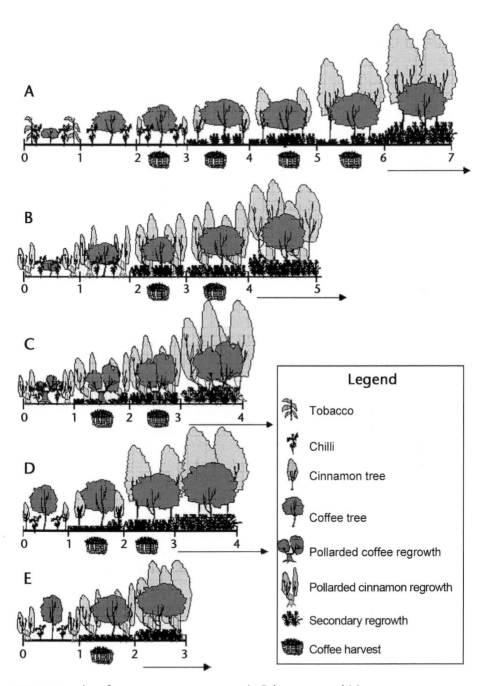

FIGURE 37-3: Agroforest management systems in Selampaung and Masgo.

In terms of livelihood security and possibly accumulation of cash, management system A in Figure 37-3 is most preferred from a socio-economic point of view, as it maximises income-generating opportunities in the intensive- and intermediate-management phases. However, this system has obstacles to the enhancement of vegetation in the early years, so it delays the build-up of perennial vegetation. Intensive management also has high demands for both finance and labour. Usually, labour in these systems is deployed on a full-time basis, so management system A is restricted almost entirely to sharecroppers and households with enough labour and adequate financial resources. With regard to long-term recolonization by native plants and animal species, the extensive-management phase of system A may contribute significantly because the livelihoods of rich absentee land owners do not depend heavily on the agroforest, so they usually opt for the longest extensive-management phases. This is consistent with the fact that planting cinnamon trees is, for these land owners, a wealth-accumulation strategy. The same holds true for households managing systems A, B, C and D by themselves.

Conclusions

The various agroforestry-management strategies identified in this chapter have a number of factors that explain the extent to which they protect a certain degree of vegetative cover – with its associated biodiversity – while sustaining the livelihoods of their practitioners. This commercialized type of shifting cultivation forces a household to continuously weigh the advantages of one crop against the benefits of another for livelihood stability. The household may opt, for instance, to cut down its cinnamon trees if prices for cinnamon bark are low compared to those for vegetables and/or coffee. The same factors force the adoption of different combinations of practices for renewing the systems at the end of each cycle, along with crop choices for the cycle ahead. A farming household must focus on the size and composition of the resource base at its disposal at any given point in time, and decide to manage its agroforests in a certain way, and within the constraints of its important assets and capabilities. Each individual management system varies in its ability to maximize income from the various cash crops and to maintain its contribution to protecting a certain level of vegetative cover, which is important for the sustainability of these agricultural practices. Each management system, with its specific combination of renewal practices and crop choices, has strengths and weaknesses in contributing to livelihood security while safeguarding the most important function of a 'fallow' – restoring soil fertility. The ecological knowledge of these people, inherited from generations of shifting cultivation practices, seems to enable them to maintain the social, economic and ecological sustainability of their rotational agroforestry systems. The practices of shifting cultivation communities with long experience of farming in the forest margins provide important insights into how more sustainable and more profitable forms of agriculture can be built up by using the processes of natural vegetation succession. Making use of the ecological knowledge gleaned from generations of farming in the forest margins represents a huge challenge for researchers and policy-makers, particularly in the context of volatile global market prices. Combining these adapted forms of shifting cultivation with poverty-reduction programmes may have much to offer if pricing policies are able to help to stabilize export products from tree crops.

References

Burgers, P. and Wiliam, D. (2000) 'Indigenous strategies: Options for sustainable agriculture in the forest margins', *ILEIA Newsletter* 16(3), pp.16-18

Cairns, M. F. (2007) 'Management of Fallows based on *Austroeupatorium inulifolium* by Minangkabau Farmers in Sumatra, Indonesia', in M. F. Cairns (ed.) *Voices from the Forest: Integrating Indigenous Knowledge into Sustainable Upland Farming*, Resources for the Future Press, Washington, DC, pp.153-184

Ketterings, Q.M. (1999) 'Fire as a land management tool in Sepunggur, Sumatra, Indonesia. Can farmers do without it?' PhD dissertation, graduate school of Ohio State University, Columbus OH

Richards, J. F. and Flint, E. P. (1994) *Historic Land Use and Carbon Estimates for South and Southeast Asia, 1880-1980*, Global Change Research Program, Environmental Sciences Division, Office of Health and Environmental Research, US Department of Energy, available at https://cdiac.ess-dive.lbl.gov/ftp/ndp046/ndp046.pdf, accessed 23 April 2023

Scholz, Ulrich (1977) *Minangkabau: die Agrarstuktur in West Sumatra und Moglichkeiten ihrer Entwicklung*, Giessener Geograpische Schriften, Heft 41, Selbstverlag des Geographischen Instituts der Justus Liebig-Universitat Giessen

Scholz, Ulrich (1988) *Agrargeographie von Wet Sumatra*, Giessener Geograpische Schriften 63, Selbstverlag des Geograpischen Instituts der Justus Liebig-universitat Giessen

Suyanto, S., Tomich, T.P. and Otsuka K. (2001) 'Land tenure and farm management efficiency: the case of paddy and cinnamon production in customary land areas of Sumatra', *The Australian Journal of Agricultural and Resource Economics* 45(3), pp.411-436

Note

1. Coffee plants are commonly regarded as shrubs or small trees. They can grow to 10 m, but when pollarded, they are smaller. However, in this chapter, coffee plants are referred to as trees.

38

FARMING MEDICINAL PLANTS UNDER A FOREST CANOPY

Innovations by shifting cultivators in the uplands of north Vietnam

*Tran Thi Thu Ha and Pham Van Dien**

Introduction

Ever since the introduction of the *Doi Moi* economic reforms in Vietnam in the mid–1980s, common land has no longer been available for local communities or tribal groups to practise shifting cultivation. As a result, shifting cultivators have had to adapt to new cultivation systems, not only to meet the government's legal requirements, but also to provide income and enhance the quality of their lives.

Shifting cultivation was popular in Vietnam during the French colonial period, when more than 3.5 million ha of land was used by 50 ethnic minority groups for various stages of the shifting cultivation cycle. The percentage of swidden practitioners in the population of each ethnic group varied largely, depending on the group, geographical zones and development periods. However, soon after the introduction of *Doi Moi*, the rate in the 1990s was 7 per cent for the Tay, 16 per cent for the Nung, 45 per cent for the Thai and 100 per cent for the H'mong and Dao (Do, 1994). Swidden farming was said to be an excellent agricultural practice in Vietnam's uplands, given low population density and a high forest cover. However, growing population pressure and shrinking forest cover led to a gradual shortening of fallow periods, from 15 to 20 years down to only 4 to 5 years, and as a result of forest loss, soil erosion caused a rapid reduction in soil fertility and hence crop productivity. These were the inevitable consequences of the fallow period in swidden agriculture being shortened and the forest being robbed of its ability to recover its fertility (Do, 1994; Khong, 2002; Hoang, 2013). Given the high rates of population growth and increasing deforestation, new policies were issued to control the practice of shifting cultivation (Lundberg, 1996; Le et al., 1996; Thai and Nguyen, 2002).

* Associate Professor Dr. Tran Thi Thu Ha, Director, Institute of Forestry and Sustainable Development, Thai Nguyen University of Agriculture and Forestry, Vietnam; and Professor Dr. Pham Van Dien, Ministry of Agriculture and Forestry, Vietnam.

The Government of Vietnam's *Doi Moi* policy ushered in radical economic reforms that turned a centrally-planned economy into a market-oriented one. The policy reached the country's upland regions in the early 1990s. Its most important effects were transferring use rights from state to real users (households, individuals, organizations, companies, and so on), and decentralizing the management of forest and forest land from central state agencies to local authorities and non-state bodies, along with socio-economic reforms (Fforde and Goldstone, 1995; Dang, 2004; Kerkvliet, 2005). The purpose of the new policy was to achieve good use of forest and forest land by applying appropriate cultivation systems and strategies to harmonize conservation and development goals – in other words, minimizing shifting cultivation and promoting sedentary or permanent cultivation in order to raise the incomes of upland people, secure timber and non-timber products, and protect ecological systems for their environmental benefits. In short, achieving sustainable development.

Unfortunately, after more than 30 years of implementation, these objectives of *Doi Moi* have not been met (Nguyen and Gilmour, 1999; De Koninck, 1999; Le and Rambo, 2001; Chu, 2002; World Bank, 2005; GoV, 2005; Tran and Pham, 2017). Even today, negative impacts such as flooding, landslides and drought are occurring more often and they directly threaten the livelihoods of upland people who are still practising shifting cultivation. The maintenance and restoration of upland forests for their environmental functions is critical, but these realities are not adequately realized. On the contrary, fast-growing industrial crops have been introduced to the uplands, and these not only appear to be unsuitable for local conditions, but they also lead to greater losses of natural forest. Thus, the uplands of Vietnam remain characterized by continuing forest loss, a rapidly degrading environment, and poverty (Nguyen and Gilmour, 1999; Tran, 2007; Tran and Hoang, 2014).

The aim of this chapter is to reveal how upland shifting cultivators have responded to this changing situation by adopting the best practices on their land to meet both objectives of securing income and maintaining sustainable micro-environments in a way that provides safe and secure livelihoods.

Research area and methodology

This study was conducted on the Dong Van karst plateau of Ha Giang province, Vietnam's northernmost province. Field trips and surveys were carried out in remote areas where ethnic minority groups have been practising traditional cultivation for centuries. Four main ethnic groups were selected for this research: the H'mong, Dao, Nung and Tay.

The Dong Van karst plateau has a total area of 2400 sq. km, spread across four districts of Ha Giang province: Dong Van, Meo Vac, Yen Minh and Quan Ba (Figure 38-1). It is a frontier geographic area of the country, sharing a border of more than 100 km with China's Guangxi and Yunnan provinces. Most of the area consists of limestone mountains that are home to 17 ethnic minority groups, the largest of which is the H'mong, claiming 80% of the local population. Ha Giang has the

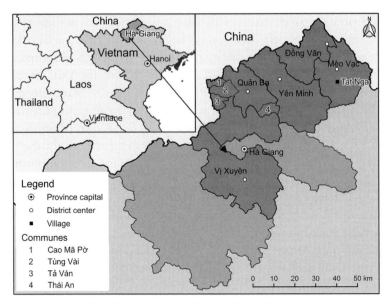

FIGURE 38-1: Ha Giang province in northern Vietnam, showing the study sites discussed in this chapter.

highest concentration of H'mong people in Vietnam. In the past, most people lived a subsistent lifestyle, but more recently, the Dong Van plateau has become well known as a place of geological heritage, indigenous cultures and endemic biodiversity. It is also recognized by the United Nations Educational, Scientific and Cultural Organization (UNESCO) as a Global Karst Plateau Geopark – a unified area with a geological heritage of international significance. Human survival in its harsh natural conditions has been a rare feat, but the plateau's unique conditions have provided the potential for eco-tourism and historical and cultural tourism at both national and international levels. For the region's farming communities, environmental conservation is strict, demanding that the local people change their traditional shifting cultivation systems for new, innovative agricultural practices.

This study was conducted from June 2017 to August 2018. Many field trips were made to 40 villages belonging to four districts on the Dong Van karst plateau. Interviews and field surveys were carried out throughout four seasons of the year. The study team lived and worked with local people to learn how their culture and indigenous knowledge have helped them to survive in the harsh mountainous conditions

Findings and discussions

Shifting cultivation and environmental impacts in Ha Giang

It is commonly acknowledged that there are two types of shifting cultivation in Ha Giang province: *pioneer* shifting cultivation and *rotational* shifting cultivation. Pioneer shifting cultivation usually uses a swidden until the soil is exhausted and harvests fall below demand. The land is then abandoned without any concern for its recovery since the cultivators have no intention of ever returning to cultivate that plot. This type of shifting cultivation is mainly practised by H'mong people, who are accustomed to a nomadic lifestyle at high elevations of at least 700 masl, with steep slopes and poor soil. When land formerly covered by dense forest becomes exhausted by repeated cultivation, the farmers move entire villages to find fresh forest land to clear. Thus, the practice of pioneer shifting cultivation obviously threatens natural forests, but it is not the sole cause; there are other reasons for the reduction of forest areas at a national scale. Influxes of new settlers to areas normally used for shifting cultivation have led to abruptly shorter fallow periods in mountainous areas as farmers strive to produce more food from less land. The growing number of mouths to feed from a steadily shrinking area of available land has resulted in the shifting cultivation fallow period – if a fallow period was initiated at all, with the intention of soil and nutrient recovery – falling from 10 to 15 years down to just three or four years by late 1980s or early 1990s.

Rotational shifting cultivation has been practised by most other ethnic minorities in the study area, such as the Dao, Nung and Tay. Upland rice, maize and beans are cropped for two to four years, after which the land is left fallow for six to eight years while the farmer shifts to cultivate other plots. Under this system, crop yields are reportedly declining and fallow periods are being forced downward as land available for new swiddens becomes harder and harder to find. Farmers whose main source of food is shifting cultivation are being forced into a nomadic lifestyle in order to feed their families. Nowadays, most rotational shifting cultivation fields have been transformed into sedentary or permanent farms, with very short fallow periods, or none at all.

Over recent decades, the breakdown of traditional shifting cultivation systems in Ha Giang, when farmers were denied the ability to exercise astute stewardship of the land, led to the destruction of thousands of hectares of natural forests, causing negative impacts in both upstream and downstream

Coix lacryma-jobi L. [Poaceae]

Wild varieties of Job's tears produce hard grains that have been used for thousands of years, and in many cultures, as beads. Domesticated plants produce softer grain that is widely used as human food and a traditional medicine. When grown as a medicinal species in the study sites, the seeds are used to treat inflammation of the intestines, prolonged diarrhoea in children, oedema, kidney stones, numbness, aches and pains.

areas. Consequently, various policies and regulations banning shifting cultivation were applied by local governments over the entire province. As indigenous ethnic groups had fewer and fewer opportunities to practise shifting cultivation in the style of their forefathers, ambitious programmes were introduced to permanently resettle shifting cultivators and their families.

Despite these programmes, it has proven difficult to achieve a strict ban on an agricultural system which, despite its relatively low crop yields, has been sustainable for centuries. Its traditions have been passed down through many generations, and have provided people living in remote and inaccessible areas with year-round sustenance and food security. Shifting cultivation has also helped to conserve soil moisture, enrich the soil structure and sustain a good crop cover, thanks to its mixed-cropping system. Moreover, for most shifting cultivation communities, it was not simply a means of livelihood, but was deeply ingrained in their cultures and traditions. Although shifting cultivation was still practised after the turn of the millennium, there were few practitioners. Reduced land resources and natural forests were unable to justify a wider perseverance. However, market opportunities beckoned, and former swidden farmers were able to apply their indigenous knowledge to adapt high-value crops, such as medicinal and herbal plants, to local conditions and grow them in permanent farming systems.

Growing medicinal plants under natural forests

Our survey found that growing medicinal plants was an expanding enterprise in four districts of Dong Van Karst Plateau Park, with a particular concentration in Quan Ba district, which has conditions conducive to growing high-quality medicinal plants.

Quan Ba is the gateway to the Dong Van UNESCO Geopark, with an elevation of more than 1000 masl (see Figure 38-1). The terrain is serrated, with many river systems, high mountains and deep valleys. The district has 13 communes and towns, of which nine communes are ranked as poor. In recent years, local shifting cultivators have changed their crops of upland rice and

Aconitum carmichaelii var. *truppelianum* (Ulbr.) W. T. Wang & P. K. Hsiao [Ranunculaceae]

This perennial grows to more than 1 m tall, has large leathery leaves and dense panicles of blue flowers in late summer and autumn. Native to China, Vietnam and the Korean peninsula, it is valued as a garden species. However, all of its parts are extremely toxic and have been used to produce poison for arrows. When prepared by a trained person, the tuberous roots can be used to treat limb pain and numbness.

corn for medicinal plants, since the government no longer allows the practice of shifting cultivation.

Swidden farmers living within natural forests grow medicinal plants that live only under a natural forest canopy, at altitudes around 1000 masl. Some of these medicinal plants, which are grown in a sustainable system, make a significant contribution to the eradication of hunger and alleviation of poverty. There is substantial market demand for two species in particular, cardamom (*Amomum tsao-ko*) and *Lysimachia foenum-graecum*, which have been selected as major species for economic development in Quan Ba district. Currently, the district has 2850 hectares of cardamom and *L. foenum-graecum* concentrated in several communes, such as Tung Vai, Thai An, Cao Ma Po and Ta Van (see Figure 38-1).

The practice of growing medicinal plants under the forest canopy has existed for many years, but has previously involved only small areas and a few farmers (Figure 38-2). Most present-day crops are planted in old fallowed fields that are now reforested, because old-growth forests are usually classified as protection or special-use forests, which cannot be cleared for any reason. In degraded forests or former swiddens where forest vegetation has regrown, local farmers plant medicinal species under trees without cutting them at canopy level. This is an indigenous practice developed by local farmers, although it is not officially accepted by government authorities, who continue to associate both farmers and their crops with clearing forests for shifting cultivation.

The Quan Ba district authority has introduced an agriculture and forestry development structure based on the comparative advantages of each sub-region or commune, to promote economic development and increase the incomes of local households. In practice, the people grow those plant species that bring them the highest benefits, and the economic value of these plants has positively influenced their awareness of the need to protect forests, not only in terms of conserving the remaining primary forests, but also to improve the forest cover and actively maintain growing conditions for medicinal and herbal species that can increase their income.

In 2018, a survey of 560 households in seven villages in Cao Ma Po, Tung Vai and Ta Van communes in Quan Ba district found that 352 of them were growing medicinal species under forest canopies (Table 38-1). The surveyed households belonged to the H'mong, Dao, Nung, Han, Tay, Pu Y and Giay ethnic groups.

FIGURE 38-2: Farmer Vang A Chinh, in his traditional H'mong costume, stands among a crop of medicinal plants (*Strobilanthes cusia*) growing under his forest canopy

Photo: Tran Thi Thu Ha.

Table 38-1 shows that most villagers living adjacent to forests in Tung Vai, Cao Ma Po and Ta Van communes are engaged in forest understorey cultivation of two herbs – *Amomum tsao-ko* (cardamom) and *Lysimachia foenum-graecum*, a low-growing, yellow-flowered herb of the primrose family. Returns from these crops are lucrative and make an important contribution to incomes for many households.

Cardamom cultivation has now largely stabilized in the forests of the study communes, since all suitable forests have, for several years, been fully occupied at ground level by cardamom plantations. Close to 80 per cent of households in Tung Vai commune are now engaged in growing cardamom. In the past, cardamom was the only species being cultivated under forest cover, but recently *Lysimachia foenum-graecum* began to be intercropped with cardamom among the forest trees. Our survey in early 2018 showed that *Lysimachia* cultivation was expanding, with about 70 per cent of forests in Tung Vai, Cao Ma Po and Ta Van communes being used to cultivate both plants.

It is estimated that cardamom and *Lysimachia* plantations cover 159 ha in

Lysimachia foenum-graecum Hance [Primulaceae]

This flowering perennial grows up to 30 cm, and is commonly found in humus-rich soils in mountain valleys between 800 and 1700 masl. It has been used in traditional medicine for its anti-inflammatory properties, and the whole plant is used in the study area as a treatment for rheumatoid arthritis and migraine. Recent studies have found that it has potential as an anti-obesity treatment. The root has also been used to treat halitosis.

Tung Vai commune. The average productivity was reportedly about 2500 kg per ha, although in the year before the survey, the plantations produced only 1200 kg/ha due to hail storms. The total harvest is equivalent to 191 tonnes of dried products. In Cao Ma Po commune, cardamom and *Lysimachia* cultivation occupies an area of 468 ha. As a result, many H'mong households have become richer thanks to income from the cardamom harvest. The yield from both *Lysimachia* and cardamom depends on the winter weather, which can be harsh in the mountains of Ha Giang province. In 2013 and 2016, for instance, cold weather meant most of the families in the area went without income because of low yields and low prices.

Use of medicinal plants for treatment of illness and injuries, as well as food processing, reaches back for centuries in both the study area and neighbouring China. Government health care was inaccessible, so herbal medicines assumed an

TABLE 38-1: Medicinal species grown in seven villages of Quan Ba district in 2018.

Commune	Villages	Ethnic groups	Households growing medicinal plants/ Total households	Main medicinal species	Other medicinal plants
Tung Vai	Ban Thang	Tay, Nung, H'mong, Giay, Han, Dao, Pu Y	71/129	Amomum tsao-ko, Lysimachia foenum-graecum	Panax pseudoginseng, Paris polyphylla, Anoectochilus setaceus, Aconitum carmichaelii var. truppelianum
	Tung Vai Phin	H'mong	63/104	Amomum tsao-ko, Lysimachia foenum-graecum	Codonopsis sp., Eucommia ulmoides, Ardisia silvestris
Cao Ma Po	Va Thang	Dao, Nung	37/57	Amomum tsao-ko, Angelica sinensis	Lysimachia foenum-graecum, Panax pseudoginseng, Paris polyphylla, Eucommia ulmoides
	Vang Cha Phin	Han, Dao	25/46	Amomum tsao-ko, Lysimachia foenum-graecum	Lonicera japonica, Aconitum carmichaelii var. truppelianum, Zingber zerumbet
Ta Van	Ta Van	H'mong	43/62	Amomum tsao-ko, Lysimachia foenum-graecum	Amomum villosum, Anoectochilus setaceus, Aconitum carmichaelii var. truppelianum
	Chung Trai	H'mong	46/70	Amomum tsao-ko	Lysimachia foenum-graecum, Fibraurea tinctoria, Panax pseudoginseng, Aconitum carmichaelii var. truppelianum,
	Lo Suoi Tung	H'mong, Nung	67/92	Amomum tsao-ko, Lysimachia foenum-graecum	Coix lacryma-jobi, Panax pseudoginseng, Zingber zerumbet

important position in traditional healing. The indigenous people had a significant pharmacopoeia, and only recently has continuing demand for traditional medicines prompted the cultivation of the most sought-after species in both local and Chinese markets (Table 38-2).

TABLE 38-2: Medicinal plants at the study sites and their uses.

Species Local name/ botanical names	Sources	Plant parts harvested for medicinal purposes	Ailments for which plants are used	Harvest season	Process after harvest	Main markets
Thao qua (cardamom) (Amomum tsao-ko)	Local natural forest	Fruit	Abdominal pain; mouth odour; malaria; diarrhoea		Drying	China, Europe, local markets
Huong thao (Lysimachia foenum-graecum)	Local species	whole plants	Rheumatoid arthritis; migraine.	Any time, dry	Drying	Domestic, China
Tam that (Panax pseudoginseng)	Local natural forest	Flower buds, root tubers	Angina; poor heart condition; acute and chronic leukemia; lower back pain; general health effects.	5 to 7 years after planting	Drying	Domestic
Bay la mot hoa (Paris polyphylla)	Local natural forest	tubers, roots, stems, leaves	General health effects; snake bite.		Keep fresh	Domestic, China
Lan Kim tuyen (Anoectochilus setaceus)	Local natural forest	Whole plants	Pulmonary tuberculosis; cough due to heat exhaustion; rheumatism; general health; blood circulation; antibacterial.		Keep fresh	China, Domestic
Au tau (Aconitum carmichaelii var. truppelianum)	Local species	Tubers	Limb pain; numbness.	3 years after planting	Keep fresh	Local market
Dang sam (Codonopsis javanica)	Local species	Roots	Blood pressure; anorexia due to defective taste; general health.	3 years after planting	Keep fresh	China, domestic
Do trong (Eucommia ulmoides)	Local species	Bark	Rheumatism; numbness; lower back pain; knee pain; preventing miscarriage and hypotension; kidney ailments; osteoarthritis.	Any time	Drying	Domestic
La khoi (Ardisia silvestris)	Local natural forests	Leaves	Soothing ulcers; stimulating the skin; healing wounds; peptic ulcers	Any time	Keep fresh	Domestic
Kim ngan hoa (Lonicera japonica)	Native species	Flowers	Acne and boils; itchy skin; acute inflammatory bowel disease; Sore throat; mumps; influenza; measles.	March to June	Keep fresh and drying	China, domestic

TABLE 38-2 (cont.): Medicinal plants at the study sites and their uses.

Species Local name/ botanical names	Sources	Plant parts harvested for medicinal purposes	Ailments for which plants are used	Harvest season	Process after harvest	Main markets
Gung gio (*Zingber zerumbet*)	Local species	Tubers	Cold symptoms caused by rain; middle-aged men with fat in the blood; painful joints, dyspeptic symptoms	2 years after planting	Keep fresh	Domestic
Sa nhan (*Amomum villosum*)	Local natural forests	Fruit	Abdominal cold; flatulence; urinary care in pregnant women; diarrhoea; toothache; numbness.	June to September	Drying	China, Local market
Hoang dang (*Fibraurea tinctoria*)	Native species	Bark, leaves, fruit	Colitis; deadly toxoplasmosis	Any time	Keep fresh	Local market
Y di (*Coix lacryma-jobi*)	Local species	Seeds	Inflammation of the intestines; prolonged diarrhoea in children; oedema, kidney stones, numbness, aches and pains	August to October	Drying	Local market
Giao co lam (*Gynostemma pentaphyllum*)	Local natural forests	Whole plant	High cholesterol; high blood pressure; Boost and stimulate the immune system; liver protection	Any time	Keep fresh	Domestic (pharmacy company)
Duong quy (*Angelica sinensis*)	Local species	Root tubers	Anaemia, depression of the body; rheumatism; bone pain	12 months after planting, November and December	Keep fresh	Domestic (pharmacy company)

In the past, most local people used these plants for sickness or to protect their health because they were isolated from government health-care systems. Nowadays, they cultivate them to generate cash income. They use traditional indigenous knowledge for propagation, harvesting and simple post-harvest processing. They also know how to control problems related to the survival of these plants by growing them under the canopy of big trees to protect them against the cold and frost of winter.

Cardamom and *Lysimachia* thrive in conditions of forest-canopy shade, high humidity, and well-drained soil, with moderate temperatures between 16°C and 24°C. Most households prepare their own seedlings and apply traditional cultivation techniques handed down over many generations, without observance of modern techniques for forest protection. They do not apply fertilizers, weed by clearing grass and creepers, and tend to surrounding stumps in the regenerating forests by dressing them. Woody trees are maintained only for shading purposes.

Although this may cause loss of biodiversity and degradation of forest vegetation, some households claim their plantations control soil erosion. As one local farmer commented: 'We stop cutting trees and only clear grass and herbs so our cardamom and *Lysimachia* can grow. The forest canopy protects them during winter and prevents soil erosion during the rainy season, and this secures a harvest every year'.

When cardamom and *Lysimachia* plantations are established away from the shade of the forest, the farmers have difficulty managing harvest outcomes. Harvesting of unripe fruit can result in a drop in yield of 20% to 30%, and this translates into a loss of income amounting to 10%

Lonicera japonica Thunb.
[Caprifoliaceae]

Japanese honeysuckle is a twining vine that can climb up to 10 m in trees. It has an important place in folk medicine. The flowers are used to treat acne and boils, itchy skin, acute inflammatory bowel disease, sore throat, mumps, influenza and measles. Importantly, computational studies of a compound called loniflavone, found in this plant, have demonstrated antiviral action that could be an early step towards developing a drug to combat the virus that causes Covid-19 (Kadioglu et al., 2021).

or 15%. Low quality harvests, with small-sized fruit of dark colour that deteriorate rapidly, demand additional firewood consumption in the drying process. It normally requires between three and five cubic metres of firewood to produce 1000 kg of dried products, so the farmers are constantly concerned to minimize their need for firewood for drying cardamom.

H'mong farmers' knowledge about cultivation of cardamom under a forest canopy

Our study found that H'mong farmers had a wealth of experience in growing cardamom in both Quan Ba and Yen Minh districts. They advised that cardamom was a special plant that favoured shade and moisture. It can be planted only under a forest with 40% to 60% canopy cover, at altitudes of 1300 to 2200 masl, where the average temperature is 13°C to 15°C and annual rainfall is between 3500 and 3800 mm. Flowers begin to appear at the base of the stems from late-April to mid-May, and by November, the cardamom seeds are mature.

If the seeds are required for new planting, they are first dried and then stored in moist sand for three to four months. Before planting, they are soaked in warm water at about 45⁰C – or, in the words of the farmers, three boiling water, two cold – for about eight hours. The seeds are then transferred to a moist base and incubated until they crack, before being planted in nursery beds containing a mixture of soil and organic fertilizer. The seeds are planted one or two centimetres deep and are spaced at 10 cm x 20 cm, under pergolas with a cover level of 70% to 90%. These beds are given adequate water and are periodically weeded. After 12 to 18 months, the seedlings with a height of 60 cm to 80 cm can be planted out beneath the forest between December and March.

When preparing understorey plots for cardamom seedlings, all vegetation, including vines and shrubs, is chopped and spread on the ground at least one month beforehand, so that the slash decomposes into humus under the forest canopy. The seedlings are planted in holes dug deep enough to accommodate the roots, and close attention is paid to cleaning the forest floor so that all the invasive trees and plants around the root zone are removed. It is not necessary to fertilize newly cleared forest land, but looking ahead, fertilizer and ash should be applied to the cardamom after each harvest. The usual harvest time extends from September to the end of November. A knife is used to cut whole bunches of seed pods from the base of the plant, and these are taken directly to be dried. When drying cardamom, the bunches of seed pods should be left intact, to create ventilation.

Anoectochilus setaceus Blume
[Orchidaceae]

One of the so-called jewel orchids, this plant is often cultivated for its unusual leaf patterns. It is commonly found in humid primary forests in moist soils as a small terrestrial plant that produces erect inflorescences of small white flowers. The whole plant is used to treat tuberculosis, coughs, rheumatism, and circulation issues. It is also said to be a general health tonic with antibacterial qualities.

The effects of cardamom cultivation

There are two different opinions on the impacts of cardamom cultivation. Provincial authorities, researchers and conservationists believe that it leads to forest degradation. As mentioned earlier, reducing the forest cover leads to reduced cardamom yields and unstable quality. However, they say that most households 'over–open' the forest cover, by clearing most of the forest vegetation under the main forest trees. This adversely affects the forest structure and eliminates seedling generation, resulting in loss of biodiversity in areas planted to cardamom. It also reduces the capacity to protect forests. In addition, the need for firewood to dry the cardamom harvest is another cause of deforestation.

The opposing view, expressed in interviews with cardamom-farming families, claims that the economic value of cardamom has positively influenced the attitude of people towards forest protection. In order to maintain the forest cover in old, natural protection forests, the farmers proactively plant new species under the forest canopy to ensure an optimum environment for the cardamom plants that have eradicated their hunger and reduced poverty. The farmers point out that planting cardamom has brought many environmental benefits, including the prevention of soil erosion during the rainy season, control of forest fires in the dry season, and the promotion of water conservation. Cardamom also improves soil fertility and increases the efficient and sustainable use of forest land.

Medicinal plants: The perspective of shifting cultivators

The first priority of shifting cultivators in the study sites of Dong Van Plateau Park is maintaining the cultivation of medicinal plants so that the cash income of local people is secure and their various cultures preserved. However, their care for the forests is a high consideration and the farmers are striving to optimize the efficiency of their use of forest land. They have implemented forest-care models that include the use of indigenous species to replace ageing and dead trees, and intercropping with fast-growing species to be used as a main source of firewood for drying harvests.

Stimulation of forest growth is aimed at providing a favourable environment for plantations of medicinal plants, which affects budding, rooting, blooming and bearing of fruit, hence increasing the productivity of the plantation and shortening of time to harvest. As

Paris polyphylla Sm. [Melanthiaceae]

This plant, which grows from a rhizome to about 1 m tall, thrives in moist conditions beneath a forest canopy, in full or partial shade. It is listed as vulnerable and, in some places, is threatened by overharvesting. It has long been valued in Nepal, where it is used to treat fevers and headaches, burns and wounds. The tubers, roots, stems and leaves are used in the study sites for general health remedies and snakebite.

well as cardamom and *Lysimachia foenum-graecum*, market demand for other medicinal species, such as *Anoectochilus setaceus*, *Aconitum carmichaelii* var. *truppelianum*, *Gynostemma pentaphyllum*, *Paris polyphylla* and *Codonopsis javanica*, has seen the establishment of new plantations in Dong Van Plateau Park. Most households have their own nurseries, in which high quality seedlings are propagated to expand the cultivation area.

There is also a movement that is gradually gathering strength for the growing of medicinal plants outside of natural forests, in areas where forest concerns are critical or very critical. Intensive cultivation of cardamom and controls to ensure sustainable yields in these circumstances are aimed at maintaining incomes for poverty reduction while protecting forests and improving biodiversity.

The governing authority of Ha Giang province has recently

Codonopsis javanica (Blume) Hook.f. & Thomson [Campanulaceae]

This perennial climber is commonly found in damp fertile areas at the edges of mountain forests. The root contains an essential oil, glucose, fatty substances, and a glucoside called scutellarin, which is the focus of research and development for some experimental medical therapies. From the study sites, it is marketed locally and in China for the treatment of blood pressure and anorexia, and as a general health tonic.

recognized that growing medicinal plants under natural forests is linked with economic development from forest land, and provided it does not negatively affect biodiversity, it can gradually generate a sustainable income for local forest-dependent people.

Surveys elsewhere on the Dong Van karst plateau

In addition to our studies in Quan Ba district, we surveyed 180 households in six villages in another three districts of the Dong Van karst plateau. They were Phieng Tra and Sang Lung villages in Yen Minh district; Lo Lo Chai and Xa Lung villages in Dong Van district; and Lung Chin and Khau Vai villages in Meo Vac district (see also Box 38-1, for details of another medicinal-plant enterprise). We found that farmers in these locations also placed a high priority on growing medicinal plants for both their own consumption and for sale at the border with China. These areas are characterized by difficult topography, narrow roads and relative isolation, so planting medicinal plants was regarded as a best choice, and incomes from this source were regarded as important in terms of returns for labour and provision of food and livestock fodder. However, not every household on the Dong Van plateau has the

BOX 38-1: Elsewhere in Ha Giang

In a related case, a community of the Dao minority ethnic group is growing the medicinal species *Millettia speciosa* in their 'forest gardens'. Tan Binh village, in Ngoc Minh commune, in Ha Giang province's Vi Xuyen district (see Figure 38-1), was once a community of shifting cultivators with swiddens at about 600 masl. Under pressure from government bans on shifting cultivation on forest land, they began planting spice-producing trees of the *Cinnamomum* genus, along with vines of the medicinal legume *Millettia speciosa*, in sunny locations at forest edges.

The roots of *M. speciosa* are used to make a general health tonic, or specifically for the treatment of headaches or urinary retention. Alternatively, the young roots can be used as human food. As household food, the roots can be harvested after 18 months' growth, but for medicinal use they are harvested after three years. With high market demand across the district's border with China, the farmers of Tan Binh claim the income from *M. speciosa* is five to 10 times what they used to expect from swidden crops of upland rice and maize. Moreover, the income is spread over the short and long term.

FIGURE 38-3: A group of Tan Binh villagers carry baskets of *M. speciosa* seedlings to their 'forest garden'.

FIGURES 38-4 and 38-5:

Left: Ha Thi Lien, a woman of Tay ethnicity, plants *M. speciosa* in a forest garden along with others from her Dao husband's village.
Right: Chang Thi Song, in her traditional Dao costume, plants an *M. speciosa* seedling in the forest garden of Tan Binh village.

All photos: Mai Van Que.

conditions to grow these species. Different geographic conditions, such as soil and weather, make some places unsuitable. For example, most medicinal plants can grow in Quan Ba and Yen Minh districts, but very few species can survive the dry, cold conditions of Dong Van and Meo Vac districts.

In Phieng Tra village, Yen Minh district, 24 of the village's 90 households had been resettled from shifting cultivation sites in natural forests, and under the village's natural forest, several medicinal species, including *Curcuma longa*, *Salvia miltiorrhiza*, *Zingber zerumbet*, *Gynostemma pentaphyllum* and *Ardisia silvestris* were growing well. Other popular medicinal species such as *Angelica sinensis*, *Salvia miltiorrhiza* and *Codonopsis* sp. were growing on fallow land. These species have a short rotation of 18 to 24 months, so the local people were able to obtain income in a short term. One farmer, Mr. A Pao, told us his household received a net income of 380-million Vietnamese dong (about US$16,443) per hectare after growing *Angelica sinensis* for 18 months. This compared to 80-million dong (US$3461) that he previously received from growing two crops of maize on the same land.

In the cold, dry conditions of Dong Van and Meo Vac districts, several species such as cardamom (*Amomum tsao-ko*), *Anoectochilus setaceus*, *Aconitum carmichaelii* var. *truppelianum*, *Amomum villosum* and *Ophiopogon japonicus* were able to grow under the forest canopy. Shifting cultivators there recognized that growing medicinal plants under natural forests could bring them more income. The top layers of their forests were trees for collecting bark, oil and fruit, and the lower layers comprised herbs for harvesting roots, stems, flowers and leaves. Under these conditions, they said their water resources would be safe during the dry season for the rest of their lives.

Forest management in Meo Vac district: the case of Tat Nga

Unlike many indigenous communities in Vietnam's northern mountainous region, the people of Meo Vac district can enjoy a beautiful landscape. During the first village meeting to discuss forest degradation and livelihood problems, a village elder spoke out:

> Primary forests have formed the traditional culture of our indigenous communities for many generations. Our ancestors lived in primary forests for hundreds of years. Forests provide everything we need, such as food, shelter, furniture, hunting equipment, musical instruments, cooking and eating utensils, ceremonial artifacts, and hundreds of other practical and spiritual items. The life of the indigenous people has lasted for centuries and will continue for centuries, if we do not destroy [the forests].

Tat Nga village is located about 20 km from the administrative centre of Meo Vac district (see Figure 38-1). The remote village is connected with the outside world by a dirt track accessible only by motorcycles and carts. It was first established in the early 1930s with only five households. The number of households increased to 17 in

1965, 36 in 1990 and 77 at present, with a population of 387 people. The majority (84%) belong to the Giay ethnic group. The rest are H'mong.

The total area of Tat Nga village is about 679 ha, of which 615 ha are classified as forest land and the rest as agricultural land, including 25 ha of wet-rice fields. A watershed forest covering about 400 ha, now classified as protection forest, has been managed by the local community since 1961. Within this forest are 10 ha of 'sacred forest', classified as old forest, which has been protected by the Tat Nga community for many years.

The remaining 260 ha of mostly degraded forest land covering lower slopes were allocated to individual households for long-term use in 1998, as part of a government forest-land reform.

During the French colonial period until 1954, the villagers of Tat Nga lived under feudal institutional arrangements. A significant change occurred after the country gained independence in 1954, with the establishment of the 'cooperative' or 'collective' model. This principle was applied to every household in each village, or cooperative. The members of every household who were capable of working as labourers worked in the wet-rice and agricultural fields as a cooperative, and received food based on the number of household members. The cooperative model was abolished in 1986, to be replaced by the *Doi Moi* economic reforms.

After independence in 1954, almost all of the old forests, both inside and outside Tat Nga village, were cleared for shifting cultivation as a result of policies encouraging food self-sufficiency. Downstream, the village's 25 ha of wet-rice fields were severely affected by an irregular flow of water. An inadequate cover of vegetation in the watershed forest areas also caused a lack of water in streams and creeks during the dry season. Animal hunting for food became difficult and some non-timber-forest-product species were lost. The livelihoods of Tat Nga's forest-dependent villagers were severely affected. The majority of households experienced, for the first time, food shortages that lasted for many months in each year. They recognized the negative impacts of forest loss and also realized that they could neither rely upon, nor wait

Eucommia ulmoides Oliv.
[Eucommiaceae]

This tree, which grows up to 15 m tall, is one of the 50 fundamental herbs of Chinese herbology. It produces a latex that makes an elastic material capable of substituting for natural rubber, with qualities exceeding those of rubber. In traditional medicine, the bark is used to treat rheumatism, numbness and lower back and knee pain; also to prevent miscarriages and hypotension, kidney ailments and osteoarthritis.

for, outside help. The villagers held several meetings to discuss their problems, and agreed to work together to reverse the situation.

In 1962, a simply structured 'community forest management committee' was established at the initiative of the Tat Nga villagers. The committee consisted of a chairman (also the village head) and two forest-protection teams of seven members each. Other members were volunteers or nominated by villagers to work voluntarily without payment. The immediate task of the committee was to formulate village rules for the protection and conservation of the watershed forest within the village boundaries. The simple rules that they established, which have been in force for nearly half a century, were never put in writing, but they are well respected and enforced by all Tat Nga households. They are as follows:

- All forests and forest land on the top of mountains are strictly protected and belong to the village community.
- No form of agricultural cultivation is permitted in the protected area.
- Harvesting of timber and non-timber forest products in the protected area is strictly limited to non-commercial quantities and is permitted only in designated areas. Use of timber from such harvests is restricted to house construction and for community purposes.
- Each household, including those newly setting up for residence, is permitted one timber harvest only to build one house to last for their entire life-time. Additional small harvests will be considered for house repairs, but not for house extension. No fee payments are required. Households must submit a harvest plan to the committee chairman for consideration and approval at a community meeting. Only the agreed number of trees can be cut down and harvesters must ensure minimum damage to surrounding trees.
- All households have a duty to protect the community forest and report forest encroachment to the committee.

The village's total area of protected forest is about 400 ha. Each of the forest-protection teams works on a two-week roster to patrol the forest and prevent trees being cut by people from other villages. Other household members contribute free labour to community work, such as maintaining the village roads and firebreaks. Women are not expected to work as forest guards, but can take part in emergency tasks such as helping to extinguish forest fires.

Although neighbouring villagers still cut trees in Tat Nga's forests, these violations occur mainly in private household forests because the community forests are strictly protected by frequent patrols. In most cases, conflicts are resolved at village level with sanctions agreed by village heads rather than being forwarded to the Commune People's Committee. However, cross-border forest-management continues to be a big challenge, not only at village level but also at higher administrative levels in Vietnam. High demand and high prices for wood encourage tree-cutting by people from outside Tat Nga. Forest guards have to work hard and make random inspections of the protected forest at least twice a week. The security of Tat Nga's forests is considered 'good enough', with very few incidents of theft.

Villagers can harvest non-timber forest products (NTFPs) such as bamboo, bamboo shoots, banana stems and leaves, mushrooms, forest vegetables and cardamom from the community forest for household consumption only, provided that no permanent damage is done to the forest resources. Sustainable harvesting is applied to avoid overexploitation, young plants are kept for the future and collecting dead branches is allowed only for firewood. From interviews conducted during the study, poorer people go to the community forest to collect NTFPs more often than those who are better-off. All households are allowed to graze their cattle in the community forest area. This usually takes place for a few months during the dry and cold seasons, from December to February.

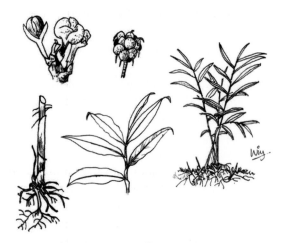

Amomum villosum Lour.
[Zingiberaceae]

This vigorous member of the ginger family grows to 3 m tall, commonly in wet forest clearings at altitudes from 600 m to 800 m. In traditional medicine, the fruit is used to treat colds, flatulence, diarrhoea, toothache, numbness, and for urinary care in pregnant women. Scientific studies have also found it effective in the prevention and treatment of nonalcoholic fatty liver disease (Lu et al., 2018) and inflammatory bowel disease (Chen et al., 2018).

In 2001, 360 ha of Tat Nga's protected watershed forest were formally classified as 'protection forest' and qualified for payments of about US$5 to $10 per ha for protection tasks. Since then, the Tat Nga village committee has signed forest-protection contracts with local authorities and has received payments of $2000 to $3000 each year. The forest-protection teams also receive small payments of $2 per person each time they patrol the forest. At the end of each year, the village committee gives each household about $20 from money saved from the government contract. It is a New Year gift in recognition of the household's contribution to protection of the community forest. Any remaining money is kept as a village development fund. At the time this chapter was written, the development fund held 220 million Vietnamese dong (about $10,000). Poor households are able to borrow up to one million dong ($45) from the fund.

Tat Nga's villagers believe strongly that the protection and natural regeneration of the protected watershed forest has improved and stabilized the entire local ecosystem. Soil erosion in the protected forest is no longer an issue. Water continues to flow from the mountains into the village streams and creeks throughout the year, even in the dry season, enabling cultivation of cash crops and vegetables for household use and sale.

Most interviewed households said they had better crop yields since the degraded forest or bare hills were replaced by regenerating forests. This has encouraged them to protect upland fields in their privately owned patches of degraded forest to reduce soil erosion after heavy rain. The forest canopy also provides shade and reduces temperatures in the upland fields, creating a favourable microclimate for crash crops. Most of villagers agreed that without the primary forests in community-forest areas, their wet-rice fields could not be cultivated.

A recent investigation of nine sample plots (40 m × 25 m each) laid out along transects of the Tat Nga community forest found that the primary forest had a medium biomass level with a total volume of 248 cu. m/ha. Trees with a diameter of 50 cm to 60 cm were dominant, and there were some individuals with a diameter larger than one metre. Since the primary forests were a mixture of limestone and mountain forest, they had many tree species with high economic value, including *Syzygium aromaticum*, *Madhuca pasquieri* and *Burretiodendron hsienmu*. There

were not many high-value species in naturally regenerated forest on hill tops, and biomass was still low there because, before it was protected, the area had been burned many times by uncontrolled fires from an adjacent village. From field observations and interviews of key stakeholders at village, commune, district and provincial levels, it was clear that Tat Nga's forests had better management than that in other villages. The main reason was Tat Nga's devotion of time to protecting their forest from outsiders who would cut down trees, collect non-timber products and allow their livestock to graze. Tat Nga has also used a selective thinning approach to forest management that is considered to be 'conservative harvesting'. For example, if a household needs three trees, they are allowed to cut exactly three trees and must do so without damaging adjacent trees.

Zingiber zerumbet (L.) Roscoe ex Sm. [Zingiberaceae]

A close relative of ginger, this perennial has leaves that grow from a rhizome to more than 1 m tall, are fragrant, and are used to enhance flavours in food cooked in earthen ovens in Polynesia. A clear fragrant juice in the pinecone-shaped flower heads is also used as a shampoo and hair conditioner. In the study sites, the rhizomes are used to treat cold symptoms, to counter high triglyceride levels, and for painful joints and dyspepsia.

Tat Nga's community forests comprise closed seasonal tropical evergreen forests with a high diversity of species. Data collected from the sample plots showed particular high species diversity within the primary forest. There were 80 species, ranging from ground cover to high trees, in the primary forest and 48 species in seven-year-old regenerated forests. Natural regeneration has also improved biodiversity. Mixed forests are rich in species and produce an abundance of resources, including valuable timber, medicinal plants, resins and oils, and other forest products. The old-growth forests provide habitats for birds and wildlife. Since hunting was banned in the community forest seven years ago, the villagers have seen large wildlife return, including formerly common species like monkeys and other small mammals.

Many high-value non-timber-forest-product plants, such as *Stephania rotunda*, *Dioscorea hamiltonii*, wild orchids and cardamom are also found in Tat Nga's community forests. Some more common NTFPs, such as mushrooms, bamboo shoots and herbs, are still available and these are sometimes harvested by people from other villages.

Tat Nga village has two well-known traditional handicrafts: knitting hats known as *dan non*, and producing woven cloth. These attract tourists to the village who are interested in learning of the culture of the Giay ethnic minority. Both handicrafts require the collection of raw materials, such as leaves of particular plants to produce natural dyes. The community's sustainable exploitation regulations allow the collection of NTFPs to maintain traditional handicrafts, but they are strict in forbidding the sale of raw materials. Therefore, the forests have the appearance of botanical gardens, or genetic conservation parks.

Conclusions and recommendations

A number of valuable lessons were learned in the course of our case studies on the Dong Van karst plateau of Ha Giang province, concerning both best cultivation practices and forest protection.

First, when it was no longer possible for them to practice traditional shifting cultivation, local farmers were quite nimble in changing from clearing forests to planting different crops directly under the canopies of intact forests. For reasons of both productivity and ecological care, these farmers now have increasing incentives for protecting remaining forests. Former shifting cultivators who have lived their entire lives in harsh conditions know well how to perform the best cultivation of natural forests and medicinal plants under the forest canopy. The medicinal plants play the role of cash crops for annual income, and they return a higher reward than the upland rice and maize crops of tradition. This creates the necessity for local farmers to protect natural forests as a vital environment for growing medicinal plants with generous yields and high quality. This is a convincing example of farmer innovation that reverses their reputation as destroyers of forests and makes them forest defenders.

Second, upland farmers who belong to mountain-dwelling ethnic minorities often face hardships in farming and generating a livelihood. However, their harsh environment offers a competitive advantage in producing specialty products such as medicinal plants that are highly valued in lowland markets. Their recent circumstances prompt the question: What might have happened if the indigenous population had been forced to abandon their traditional homelands because of bans on shifting cultivation? Their indigenous knowledge might have been irretrievably lost, and this new farming sector based on medicinal plants might never have emerged without their traditional expertise in the medicinal properties of local plants, and their cultivation.

Third, the case study of community-forest protection at Tat Nga village gave us another important lesson on how local people can protect their forests for multipurpose use in a sustainable way. In practice, the indigenous knowledge of ethnic minority groups enables them to protect forest resources in harsh natural conditions better than might be achieved with the application of 'advanced knowledge' or 'technical solutions' from outside. Moreover, forest protection enables local people to maintain their ethnic cultures, since they are deeply connected to the natural environment.

In addition, urban dwellers and modern technology have recently made their mark on most remote areas. While the people of Dong Van karst plateau, in Ha Giang province, embrace their uniqueness by holding on to their traditional knowledge and lifestyle, they create an inimitable bond with the beauty of their upland environment. For outsiders, the combination is fascinating, luring many of them to visit and explore remote parts of our country.

References

Chen, Z., Ni, W., Yang, C., Zhang, T., Lu, S., Zhao, R., Mao, X. and Yu, J. (2018) 'Therapeutic effect of *Amomum villosum* on inflammatory bowel disease in rats', *Frontiers in Pharmacology*, 20 June 2018, 9:639, doi: 10.3389/fphar.2018.00639

Chu, H. Q. (2002) 'Ten years of social-economic development in the uplands' in Le Trong Cuc and Chu Huu Quy (eds) *Sustainable Development in the Upland Region: A 10-year Review and Beyond*, Agriculture House Publishing, Hanoi, pp.38-97

Dang, P. (2004) 'Stages on the road to Doi Moi of Vietnamese economy: An historical perspective', in M. Beresford and A. N. Tran (eds) *Reaching for the Dream: Challenges of Sustainable Development in Vietnam*, NIAS Press, Copenhagen, pp.19-50

Do D. S. (1994) *Shifting Cultivation in Vietnam: Its Social, Economic and Environmental Values Relative to Alternative Land Use*, IIED Forestry and Land Use Series, International Institute for Environment and Development, London

De Koninck, R. (1999) *Deforestation in Vietnam*, International Development Research Centre (IDRC), Ottawa, Canada, available at https://idl-bnc-idrc.dspacedirect.org/handle/10625/24496, accessed 26 April 2023

Fforde, A. and Goldstone, A. (1995) *Vietnam to 2005: Advancing on all Fronts*, The Economist Intelligence Unit (E.I.U), London

GoV (2005) *National Report to the Fifth Session of the United Nations Forum on Forests*, Government of Vietnam, Hanoi, available at https://dokumen.tips/documents/national-report-to-the-fifth-session-of-the-united-nations-forum-on-.html?page=1, accessed 26 April 2023.

Hoang, H. T. M (2013) *Discourse of Vietnam Forest Land Policy and Local Livelihood Strategy: A Case Study of Shifting Cultivation*, Chiang Mai University, Thailand

Kadioglu, O., Saeed, M., Greten, H. J. and Efferth, T. (2021) 'Identification of novel compounds against three targets of SARS CoV-2 coronavirus by combined virtual screening and supervised machine learning', *Computers in Biology and Medicine* 133, 104359, doi:10.1016/j.compbiomed.2021.104359. pmc 8008812

Kerkvliet, B. J. T. (2005) *The Power of Every Day Politics: How Vietnamese Peasants Transformed National Policy*, Cornell University Press, Ithaca, New York

Khong, D. (2002) 'Some issues related to population and development on minority ethnic groups and mountainous regions in Vietnam', in Le Trong Cuc and Chu Huu Quy (eds) *Sustainable Development in the Upland Region: A 10-year Review and Beyond*, Agricultural House Publishing, Hanoi, pp.140–159

Le, T. C., Rambo, A. T., Fahrney, K., Vien, T. D. and Dang, T. S. (1996) *Red Books, Green Hills: The Impact of Economic Reform on Restoration Ecology in the Midlands of Northern Vietnam*, Center for Natural Resources and Environmental Studies, Hanoi University; Southeast Asian Universities' Agroecosystem Network; Program on Environment, East-West Center, Honolulu; University of California at Berkeley

Le, T. C. and Rambo, A. T. (2001) *Bright Peaks, Dark Valley: A Comparative Analysis of Environmental and Conditions and Development Trend in Five Communities in Vietnam's Northern Mountain Region*, Upland Working Group, Centre for Resources and Environment Studies in Vietnam (CRES), Vietnam National University, Hanoi

Lu, S., Zhang, T., Gu, W., Yang, X., Lu, J., Zhao, R. and Yu, J. (2018) 'Volatile oil of *Amomum villosum* inhibits nonalcoholic fatty liver disease via the gut-liver axis', BioMed Research International 2018, article 3589874, https://doi.org/10.1155/2018/3589874

Lundberg, M. (1996) *Ethnic Minorities and the State: Conflicting Interest Between Shifting Cultivators and the Governments in Peru and Vietnam*, research report no 7, Environmental Policy and Society (EPOS), Linkoping University, Sweden

Nguyen, V. S. and Gilmour, D. (1999) 'Forest rehabilitation policy and practice in Vietnam', in *Forest Rehabilitation Policy and Practice in Vietnam*, proceedings of a National Workshop, Hoa Binh, Vietnam, 4-5 November 1999, International Union for Conservation of Nature (IUCN) in Vietnam, Hanoi, pp.4–34

Thai, P. and Nguyen, T. S. (2002) *Sustainable Use of Uplands and Mountainous Areas*, Hanoi Agricultural Publishing, Hanoi

Tran, T. T. H. (2007) 'The impacts of the "*doi moi*" process on communities and forest land management in the Northern Mountainous Region of Vietnam', PhD dissertation to the Australian National University, Canberra

Tran, T. T. H and Hoang, V. C. (2014) 'Climate change and livelihoods of people living in protected areas: A critical perspective', in *The 5th Engaging with Vietnam: An Interdisciplinary Dialogue Conference. Integrating Knowledge: The multiple Ways of Knowing Vietnam*, Thai Nguyen University, Vietnam; The University of Hawaii at Manoa, Vietnam National University, Hanoi; and Monash University, Melbourne, Australia, pp.278-284

Tran, T. T. H and Pham, V. D. (2017) 'Vietnam's 'renovation' policies: Impacts on upland communities and sustainable forest management', in M. F. Cairns (ed.) *Shifting Cultivation Policies: Balancing Environmental and Social Sustainability*, CABI, Wallingford, UK, pp.732-750

World Bank (2005) *Taking Stock: An Update on Vietnam's Economic Developments and Reforms, by the World Bank in Vietnam*, Consultative Group Meeting for Vietnam, The World Bank in Vietnam, Hanoi

The innovation of 'jungle rubber' in Indonesia is one of the best-known and most widely practised examples of farmers repurposing their fallow period to produce a commercial product, latex rubber (see Yonariza, ch. A18, in the supplementary chapters to this volume). The rubber seedlings are '*taungya*-planted' into the swidden during the cropping phase, and later develop into a rubber plantation during what would otherwise have been the fallow phase (see plates 51 to 53 in Coloured Plates section II, this volume).

Sketch based on a photo by Meine van Noordwijk on October 29, 2011.

D. Asteraceae shrubs as a preferred fallow and biological tool to smother out *Imperata*...

A farmer in Mindanao, the Philippines, slashes a thick stand of *Tithonia diversifolia* ('wild sunflower') to add a green manure to his soil. Many of the woody stems that fall to the ground will take root and develop into new plants, making the stand thicker and more effective in performing fallow functions (see van Noordwijk et al., ch. 39, and Daguitan et al., ch. 40, both this volume).

Sketch based on a photo by Malcolm Cairns, ca. 1995.

Synthesis

WHEN EXOTIC PLANT SPECIES HELP SHIFTING CULTIVATORS

Hubert de Foresta[*]

Exotic species have a bad reputation, not only in biodiversity-conservation circles, but also in the media, and thus more generally in the public at large. However, as long as they are confined to their place of introduction, such as a botanical garden for ornamental plants, or as long as they conform to the function(s) for which they were introduced, such as providing food or feed, exotic species are well accepted. Moreover, when they gain importance for the livelihood and economy of a country in which they thrive, their exotic origin may be forgotten with time. In central Africa for instance, shifting cultivators are surprised to learn that cassava (*Manihot esculenta*), their staple food, was introduced from South America and is thus an exotic species. Similarly, in Southeast Asia, the rubber tree (*Hevea brasiliensis*), also introduced from South America, has become a very important economic species, not only cultivated in large monoculture plantations, but also in swiddens, where they have become the backbone of rubber agroforests established by shifting cultivators who introduced the trees by planting them amid their crops of upland rice (Dove, 1993; Gouyon et al., 1993).

Therefore, not all exotic plant species have a bad reputation. Those that are poorly regarded have found such favourable growing conditions in their new environments that they reproduce and establish new populations in natural and semi-natural habitats. In other words, they invade, or threaten to invade, ecosystems in which they do not belong. These species are now called 'invasive alien species' (IAS) and are considered one of the major challenges for biodiversity worldwide. They have been identified by the Intergovernmental Science-Policy Platform on Biodiversity and Ecosystem Services in its *Thematic Assessment of Invasive Alien Species and Their Control* (IPBES, 2019).

[*] Dr. Hubert de Foresta is Editor of the journal *Forests, Trees and Livelihoods* and a Senior Scientist, AMAP, IRD, CNRS, INRA, Université Montpellier, Montpellier, France.

Tithonia diversifolia, *Chromolaena odorata* and *Austroeupatorium inulaefolium*, the three Asteraceae species dealt with in this section, belong to this invasive alien species group, and are still considered as major problems by the biodiversity-conservation community (see, for example, Yang et al., 2012, and Kritikos and Kritikos, 2021, for *T. diversifolia*; and Shao et al., 2018 for *C. odorata*). However, over time, many IAS, including the three species cited above, have found new roles in ecosystem dynamics and new functions in the livelihoods of local people. Recent studies have begun to consider how people have adapted to IAS (e.g. Howard, 2019) and to assess not only the risks, but also the benefits, of IAS (e.g. Shackleton et al., 2019), to go beyond the current paradigm that regards all IAS as threats that must be combatted by any means.

I first encountered one of these Asteraceae species (*C. odorata*) in the Congo People's Republic in 1986, while studying forest regeneration after swidden cultivation. *C. odorata* was considered as no more or less than a noxious invasive weed by all agronomists I met at the time, and I was convinced it was – at least, to begin with. The species was so vigorous that it was blocking natural forest regeneration: while swiddens on slashed-and-burned old forest reverted to forest fallows, those after secondary-forest fallows were invaded by a mix of pioneer trees and *C. odorata*, while swiddens cleared from a mixture of pioneer trees and *C. odorata*, or from *C. odorata* fallows, evolved into almost monospecific and very vigorous *C. odorata* fallows. However, while farmers who were still able to open old growth forest or forest fallows clearly had a negative opinion of *C. odorata*, farmers who had to clear new swiddens from two- to three-years-old *Chromolaena* fallows claimed that yields from mixed peanut, maize, and cassava crops were similar to those from 10- to 15-years-old forest fallows. Their opinion was clearly very positive, and it made me realize the importance that such Asteraceae thickets held for the improvement of fallows in shifting cultivation systems (de Foresta and Schwartz, 1991).

In Southeast Asia, shifting cultivators have been confronted by *T. diversifolia*, *C. odorata* and *A. inulaefolium* since these species first appeared in their environment more than a century ago. Since then, farmers have observed that when their swiddens are abandoned to fallow, these invasive species self-establish themselves quickly, and are strong enough to prevent the establishment of the local pioneer trees that made up their forest fallows since time immemorial. There should have been a period of uncertainty, in which to answer the question: how will this new type of fallow vegetation, once slashed and burned, compare to a forest fallow, in terms of swidden crop yields and labour required? Fortunately, these new bush fallows compared quite well. They were similar to forest fallows in terms of yields and labour, but they achieved it in a much shorter period of fallow. Shifting cultivators soon found that they were able to adapt their management of the fallow phase of the swidden cycle and considerably improve the efficiency of their land use.

As in Central and West Africa, many Southeast Asian landscapes, where swidden cultivation has not been replaced by various forms of permanent agriculture, are now dominated by Asteraceae fallows instead of forest fallows. Agricultural researchers have for long looked for ways to improve fallows, but they have considered the three Asteraceae species only as weeds that were detrimental to agriculture and in need of control. This paradigm took time to evolve towards a more balanced approach, as illustrated by the chronology of research on *C. odorata*. At the first International Workshop on Biological Control of *Chromolaena odorata*, in 1988, there were 16 presentations focused exclusively on its distribution and control. The second international workshop, in 1991, had 21 presentations focused on the same themes, but two presentations (one from Nigeria and mine from Congo) focused on the positive role of *C. odorata* in fallows. The third workshop, in 1993, saw a real change of paradigm: 24 presentations focused on distribution and control, but six discussed the role of *C. odorata* in fallows, and a major recommendation of the workshop was 'to assess the role of *C. odorata* in short-term fallows of West and Central Africa cropping systems' (Prasad et al., 1996). In the case of *T. diversifolia*, the perception of the species solely as a noxious weed began to change to recognition of a useful fertilizer plant with the first experiments conducted in the 1990s by World Agroforestry Centre (ICRAF) researchers in Kenya, on the use of *T. diversifolia* biomass as a green manure (Jama et al., 2002), and as a fallow species (Smestad et al., 2002). These experiments came in support of indigenous innovations, including the use of *C. odorata*, *T. diversifolia* and *A. inulifolium* in Southeast Asia (Burgers et al., 2000).

The two chapters in this section both focus on the potential benefits of the three Asteraceae species for the livelihoods of shifting cultivators. Thus, they are a part of the research effort aimed at better understanding the positive side of the three species. Van Noordwijk et al. (ch. 39, this volume), after reviewing the roles acquired by the three species in fallows, rightly circumscribe their utility to an intermediate stage of land-use intensification, when time available for fallowing land decreases but permanent agriculture is not yet possible nor needed. It proposes a pathway for research, beginning with the study of practices that have been developed by swidden cultivators to adapt to the new conditions imposed by the Asteraceae invaders, and then tackling important related questions, such as, 'are these species simply indicators of more fertile soils or are they true improvers?' It also presents relatively old, but original observational and experimental results suggesting that these species are more 'indicators' than 'improvers'. Daguitan et al. (ch. 40, this volume) present an original and interesting agronomic experiment carried out by a group of shifting cultivators, in which crops of *Vigna unguiculata* were grown after a two-year planted fallow of *T. diversifolia*. The results were much appreciated by the farmers, as they clearly suggested a substantial improvement in yield compared to that from a control plot. The improvement largely compensated the farmers for the additional labour involved in planting the fallow and managing the crop by pruning the *Tithonia* and using it as green manure.

There are many recently published studies on the properties, agricultural uses and benefits of the three Asteraceae species. This shows that they remain at the core of a vibrant dynamic research field. However, the questions raised by Van Noordwijk et al. (ch. 39, this volume) are still not unequivocally answered.

Although field observations and experimental data reported by Van Noordwijk et al. (ch. 39, this volume) give, according to the authors, 'a clear indication of an "indicator" value and little evidence of soil improvement during the fallow period', other studies point to clear soil improvement in *Chromolaena* fallows in Lamto, Côte d'Ivoire (e.g. Tondoh et al., 2013), and to its durability over time (Kassi et al., 2017). Indeed, the question is simple, but the methods required are anything but trivial challenges. In observational studies, the first challenge is to choose the right ecosystem of reference, and in this case that is, first and foremost, the forest fallow that established before the IAS settled in. This is often not an easy task. In addition, other fallow types may be compared, as was done in observational studies by van Noordwijk et al. (ch. 39, this volume). The second challenge is to ensure that all parameters of soil fertility are taken into account, which is far from an easy task. In any case, yield after fallow has to be assessed, in parallel to soil properties, as it is the most integrative proxy of soil fertility. Again, this is not easy because there may be some variations in material and management practices that have to be taken into account. In experimental planted-fallow studies, the main challenge is to ensure that the planted fallow has the same characteristics as a self-established fallow (which was not the case in the *Chromolaena* experiment reported by van Noordwijk et al. (ch. 39, this volume), especially in terms of density and vigour, because we know that the large quantity of quality biomass produced by these Asteraceae species is a key factor in soil improvement. This being said, it would be quite a paradox if the three Asteraceae species were proven to be merely indicators of more fertile soils. First, how can we explain that they establish themselves everywhere in the landscape, including on soils that are considered to be of low fertility, when they invade? Moreover, how can we explain that swidden cultivators have to wait for 10 to 15 years to clear a forest fallow and get reasonable swidden yields, while with the Asteraceae as dominant fallow species, they have to wait only two to three years to obtain similar yields? Could the improvement perceived by shifting cultivators be an artefact? Or is science limited in its analytical methods?

Another important question is: if the three Asteraceae species are improvers, is it possible to use their capacity to improve soil fertility in contexts where they are not yet present? A first way is to use their biomass as mulch or green manure in permanent fields. Some recent studies report success in testing *C. odorata* and *T. diversifolia* biomass as mulch in various crops, sometimes added to mineral fertilizers, and recommend propagation of this practice (e.g. Tshinyangu et al., 2017; Agbede et al., 2014). A second way is through planted fallows, but the challenge is the same as explained above: ensuring that the planted fallow has the same characteristics as a self-established one. There are also species-specific differences: *T. diversifolia* has good vegetative propagation characteristics and the cuttings are relatively easy to plant and manage so that they soon build up vigorous biomass. *C. odorata* and *A. inulaefolium* have extremely efficient seed-propagation qualities, but their cuttings seem to be much weaker than those of *T. diversifolia*. As they are highly invasive, it seems preferable to allow these two species to invade the fallow and adopt management practices favourable to them, such as those described by van Noordwijk et al. (ch. 39, this volume), rather than trying to establish fallows through planting cuttings. Another challenge that should be remembered is the current status of the three species as noxious weeds, which may prevent their voluntary dispersion in areas where they are not yet present.

I hope that the two chapters in this section will be taken into account when the benefits and risks of the three invasive alien Asteraceae species are assessed. Indeed, these species are a clear demonstration that the weed status of a given plant species is a relative concept: in the same region, *C. odorata* may be a noxious weed in forest or tree crop plantations, but a 'miracle' species in nearby fallows, allowing swidden cultivators to considerably improve their land-use efficiency at almost no cost (de Foresta, 1996)!

I also hope that these chapters will stimulate new, more holistic, research to narrow the gap between the divergent perceptions of swidden cultivators and scientists with regards to the soil-fertility improvement capacity of the three Asteraceae species. The two chapters make clear that swidden cultivators can benefit (van Noordwijk et al., ch. 39, this volume) or actually benefit (Daguitan et al., ch. 40, this volume) from Asteraceae fallows through a substantial reduction of the fallow time compared to forest fallows. This is probably the most important and interesting point: while farmers could only crop once every 10 to 15 years when the fallow was forest, they have been able to crop once every three years since the Asteraceae invasion, with no decrease in crop yields. This means that in the 10 to 15 years needed for the previous forest fallow, their swiddens now produce three to five times more. Therefore, besides the various scientific debates, we should acknowledge that the three invasive alien Asteraceae species dealt with in this section have promoted a true 'green revolution' for swidden agriculture, allowing shifting cultivators *in fine* to multiply three to five times their swidden yields compared to those from the previous forest fallows.

References

Agbede, T. M., Adekiya, A. O., and Ogeh, J. S. (2014) 'Response of soil properties and yam yield to *Chromolaena odorata* (Asteraceae) and *Tithonia diversifolia* (Asteraceae) mulches', *Archives of agronomy and soil science* 60(2), pp.209-224

Burgers, P., Hairiah, K., and Cairns, M. F. (2000) *Indigenous Fallow Management*, International Center for the Research in Agroforestry, Bogor, Indonesia

Dove, M. R. (1993) 'Smallholder rubber and swidden agriculture in Borneo: A sustainable adaptation to the ecology and economy of the tropical forest', *Economic Botany* 47(2), pp.136-147

de Foresta, H. (1996) '*Chromolaena odorata*: Calamité ou chance pour l'Afrique tropicale', in Prasad, U. K., Muniappan, R., Ferrar, P., Aeschliman, J. P., and de Foresta, H. (eds) *Distribution, Ecology and Management of* Chromolaena odorata. *Proceedings of the Third International Chromolaena Workshop, Abidjan, Côte d'Ivoire*, publication no. 202, Agricultural Experiment Station, University of Guam, Mangilao, Guam, pp.45-53

de Foresta, H. and Schwartz, D. (1991) '*Chromolaena odorata* and disturbance of natural succession after shifting cultivation: An example from Mayombe, Congo, Central Africa', in R. Muniappan and P. Ferrar (eds) *Ecology and Management of* Chromolaena odorata, BIOTROP Special Publication No 44, pp.23-41

Gouyon, A., De Foresta, H., and Levang, P. (1993) 'Does "jungle rubber" deserve its name? An analysis of rubber agroforestry systems in southeast Sumatra', *Agroforestry Systems* 22(3), pp.181-206

Howard, P. L. (2019) 'Human adaptation to invasive species: A conceptual framework based on a case study metasynthesis', *Ambio* 48(12), pp.1401-1430

IPBES (2019) *Thematic Assessment of Invasive Alien Species and Their Control*, Intergovernmental Science-Policy Platform on Biodiversity and Ecosystem Services, Bonn, Germany, available at https://ipbes.net/invasive-alien-species-assessment, accessed 22 January 2022

Jama, B., Palm, C. A., Buresh, R. J., Niang, A., Gachengo, C., Nziguheba, G. and Amadalo, B. (2000) '*Tithonia diversifolia* as a green manure for soil fertility improvement in western Kenya: A review', Agroforestry Systems 49, pp.201-221

Kassi, S. P. A., Koné, A. W., Tondoh, J. E., and Koffi, B. Y. (2017) '*Chromoleana odorata* fallow-cropping cycles maintain soil carbon stocks and yam yields 40 years after conversion of native-to farmland: Implications for forest conservation', *Agriculture, Ecosystems and Environment* 247, pp.298-307

Kriticos, J. M. and Kriticos, D. J. (2021) 'Pretty (and) invasive: The potential global distribution of *Tithonia diversifolia* under current and future climates', *Invasive Plant Science and Management*, pp. 205-213, doi: 10.1017/inp.2021.29

Prasad, U. K., Muniappan, R., Ferrar, P., Aeschliman, J. P. and de Foresta, H. (eds) (1996) 'Distribution, ecology and management of *Chromolaena odorata*', *Proceedings of the Third International* Chromolaena *Workshop, Abidjan, Côte d'Ivoire,* publication no. 202, Agricultural Experiment Station, University of Guam, Mangilao, Guam

Shackleton, R. T., Shackleton, C. M. and Kull, C. A. (2019) 'The role of invasive alien species in shaping local livelihoods and human well-being: A review', *Journal of Environmental Management* 229, pp.145-157

Shao, X., Li, Q., Lin, L. and He, T. (2018) 'On the origin and genetic variability of the two invasive biotypes of *Chromolaena odorata*', *Biological Invasions* 20(8), pp.2033-2046

Sonke, D. (1997) '*Tithonia* weed: A potential green manure crop', *Echo Development Notes* 57, pp.1-2

Smestad, B. T., Tiessen, H. and Buresh, R. J. (2002) 'Short fallows of *Tithonia diversifolia* and *Crotalaria grahamiana* for soil fertility improvement in western Kenya', *Agroforestry Systems* 55(3), pp.181-194

Tshinyangu, K. A., Mutombo, T. J. M., Kayombo, M. A., Nkongolo, M. M., Yalombe, N. G. and Cibanda, M. J. (2017) 'Effet comparé de *Chromolaena odorata* King et HE Robins, et *Tithonia diversifolia* A. Gray sur la culture du Maïs (Zea mays L) à Mbujimayi (RD. Congo)', *Journal of Applied Biosciences* 112, pp.10996-11001

Tondoh, J. E., Koné, A. W., N'Dri, J. K., Tamene, L. and Brunet, D. (2013) 'Changes in soil quality after subsequent establishment of *Chromolaena odorata* fallows in humid savannahs, Ivory Coast', *Catena* 101, pp.99-107

Yang, J., Tang, L., Guan, Y. L. and Sun, W. B. (2012) 'Genetic diversity of an alien invasive plant, Mexican sunflower (*Tithonia diversifolia*) in China, Weed Science 60(4), pp.552-557

39

RESEARCHING *TITHONIA* AND OTHER DAISIES

As fallow species in Southeast Asia[1]

*Meine van Noordwijk, Malcolm Cairns and Kurniatun Hairiah**

Introduction

As mounting land-use pressures in the uplands of Southeast Asia force a progressive shortening of fallow periods in rotational cultivation systems, there is a serious erosion of the fallow's ability to either provide useful products or perform its intended ecological services. Forest regeneration is impeded as longer cropping periods and more intense tillage kill tree stumps and roots. Fallows are periodically swept by fires that retard growth of emerging tree seedlings and enhance the overwhelming competitive ability of the rhizomatous grass, *Imperata cylindrica*. In this way, once-forested slopes are rapidly transformed into treeless, degraded grasslands where farmers ruefully observe, '*there is nothing left to burn*'. This is the precarious intermediate stage in land-use intensification where traditional long-fallow rotations are no longer viable, but land is not yet so scarce that permanent cropping is necessary. In this context, a number of 'fallow species' belonging to the daisy family, Asteraceae (also known as Compositae) have caught the attention of farmers and researchers, and will be reviewed in this chapter.

Southeast Asia's 'economic miracle' has spurred the pace of change in the region's uplands, bringing market access to previously isolated villages, along with outside influences and competing demands on their resource bases. The land scarcity that has accompanied these broad changes is a relatively new phenomenon, often first encountered by the current generation of swidden farmers. Their indigenous technologies, developed under conditions of land abundance, have had little application in solving the many new problems, and many swidden communities

* DR. MEINE VAN NOORDWIJK, former Chief Science Advisor, World Agroforestry Centre (ICRAF), Bogor, Indonesia; DR. MALCOLM CAIRNS, former Associate Scientist at ICRAF-Southeast Asia, Bogor, Indonesia and more recently a Research Fellow at the Centre for Southeast Asian Studies at Kyoto University; and DR. KURNIATUN HAIRIAH, Professor of Soil Biology, Brawijaya University, Malang, Indonesia.

continue to search for means of escaping the ubiquitous 'swidden–degradation syndrome'. Given this context, farmers are in urgent need of technologies to enhance and accelerate fallow functions, to provide the same level of ecological benefits in a shorter time than that needed by traditional natural bush fallows. This is most likely to be accomplished by introducing improved fallow species with superior abilities to conserve soil, regenerate fertility and control weeds. A critical caveat though, is that any proposed technology must have limited demands for labour and capital to have any chance of wide adoption.

In traditional shifting cultivation systems, fallow or secondary-forest vegetation has direct value to the farmer, in terms of household uses or cash sales from fallow products (including firewood or other burnable biomass). As well, fallow vegetation has an indirect value, gained from enrichment of the soil and elimination of weeds, which is exploited during the following cropping phase. Intensified management of fallows can be aimed at both direct and indirect benefits. However, increasing the direct value may make fallows so lucrative that farmers will extend their duration, until any thought of annual cropping is abandoned and the fallows become long-term agroforests (van Noordwijk et al., 2014). On the other hand, increasing the indirect value of fallows may lead to shorter fallows and a higher intensity of annual food cropping (Figure 39-1).

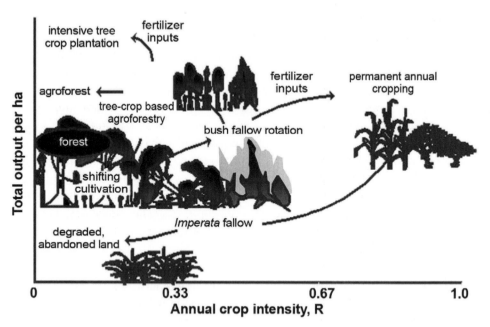

FIGURE 39-1: Pathways for intensification of shifting cultivation systems and their effect on annual crop intensity, R (that fraction of time in which land is used for annual crops) and total output per ha. Whereas initially, shifting cultivation systems can be intensified by increasing both the value of the fallow and the frequency of cropping, at some point a choice has to be made between an annual food-crop based pathway or one leading to agroforests and tree-crop plantations (van Noordwijk, 1999).

Figure 39-2 indicates three pathways for intensification of undifferentiated, multi-purpose fallows: one leading to systems where the value is increasingly derived from tree products; one where it is based on fodder value for grazing livestock; and one leading to increased cropping intensity. The first step in farmer management is usually the retention or promotion of certain plant species which appear spontaneously (either from native vegetation or as neophytes) in the fallow. In many cases, these are commonly regarded as weeds.

Weed competition is frequently cited as one of the most serious problems in collapsing swidden systems. In cases of shortened fallows, weed competition can accelerate beyond the ability of farming households to control. However, in the eyes of farmers, not all 'weeds' are created equal. Some species associated with soil-building properties are not only retained, but also actively managed as beneficial components of cropping systems. Studies from northeast India, for example, have shown that partial weeding (retaining up to 20% of the weed population *in situ*) is a deliberate farmer strategy to conserve nutrients during the early phase of crop establishment (Ramakrishnan, 1993, p.155). Other authors have described careful management of weeds as effective cover crops (Oyen, 1995; Pandeya, 1995). Within swidden fallows, farmers recognize some succession species as valuable soil-improvement plants, while others are considered soil-degrading, or having no value either way. Thus, the efficiency of a fallow, as measured by the time required for effective site recovery, is strongly associated with the plant species involved. There is, therefore, the possibility that farmers can move beyond passive recognition of indicator species and begin to actively manipulate fallow successions to ensure domination by preferred species. Tapping into the agronomic properties of beneficial pioneer shrubs provides a basis for converting weedy fallows into self-sown green-manure crops that perform the desired fallow functions in a shorter time. Southeast Asian examples of such 'shrub-based accelerated fallows' are based primarily on nutrient-scavenging members of the Asteraceae family – in colloquial terms, the 'daisy family' – and nitrogen-fixing *Mimosa* species (see plate 23 in the Coloured Plates section of *Voices from the Forest* (Cairns, 2007)).

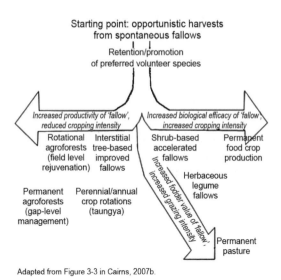

Adapted from Figure 3-3 in Cairns, 2007b.

FIGURE 39-2: Spectrum of indigenous approaches to modifying fallow vegetation in Southeast Asia, emphasizing the tree, fodder or subsequent crop production as a main function.

Asteraceae forbs and shrubs: Overlooked fallow plants[2]

Although most agronomic literature refers to Asteraceae species in the context of weeds,[3] a growing body of studies has documented farmers' appreciation of *Chromolaena odorata*[4] (see plate 44 in Coloured Plates section II, this volume, and plate 15 in the Coloured Plates section of *Voices from the Forest* (Cairns, 2007)) in Southeast Asia and elsewhere in the tropics (Dove, 1986; Agbim, 1987; de Foresta and Schwartz, 1991; Field, 1991; de Foresta, 1993; Slaats, 1993; Roder et al., 1995, 2007; Zachariades et al., 1999), *Austroeupatorium inulaefolium*[5] (Stoutjesdijk, 1935; Cairns, 1994, 1997a), and *Tithonia diversifolia* (van de Goor, 1953; Verliere, 1966; Nagarajah and Nizar, 1982; Derpsch, 1993) for their efficiency in performing fallow functions.[6] Interestingly, these are all exotics originating from Central and South America and, lacking the pests or diseases that co-evolved with them,[7] are probably

Chromolaena odorata (L.) R.M.King & H.Rob. [Asteraceae]

Reputedly one of the world's worst weeds, this fast-growing shrub can reach two metres high in six months after germination, forming dense thickets that preclude all other plant growth (see Figure 39-13 and Plate 44, in Coloured Plates section II, this volume). One hectare of this daisy species produces about one billion seeds, many of which are dispersed by the wind and can survive in the soil for up to six years. As well, branches that touch the ground grow roots and produce new plants that bolster the density of thickets.

more aggressive and dominating in Southeast Asia and tropical Africa than in their area of evolutionary origin.[8] As light-demanding species, they thrive in swidden environments and aggressively colonize disturbed forest areas.

A number of cases have been documented (below) in which farmers prefer Asteraceae fallows, even though the direct value of such fallows is limited. Their direct values are as sources of honey, as a tourist attraction (postcards featuring mountainsides ablaze with blooming *T. diversifolia* claim to illustrate typical 'native' Thai landscapes) and as fuel. The dried stems burn very quickly but generate considerable heat, making them useful in lime kilns, such as those found in south Malang district, East Java, Indonesia.

Properties of Asteraceae as efficient fallow plants

Evidence suggests that these spontaneous invaders share an ability to sponge up labile soil nutrients that might otherwise be lost through runoff or leaching, and store them in their rapidly accumulating biomass. Work done in Sumatra, for example, provides quantitative evidence of the superior growth and nutrient uptake capacity of *A. inulaefolium* (see plate 43 in Coloured Plates section II, this volume), as compared to adjacent fallows populated by Pteridophyta species (ferns) and *Imperata cylindrica*[9] (Cairns, 2007a) (Figures 39-3 and 39-4). Under conditions of heavy monsoon rains and highly erodible slopes, both of which characterize Southeast Asia's uplands, this rapid shift of the nutrient pool from the soil to the regenerating vegetation appears to play an important role in nutrient conservation in disturbed swidden environments, especially where the development of woody biomass is slow. Fast development of a dense Asteraceae canopy further assists nutrient retention by filling in the gap left after crop harvest and shields the soil from erosion and leaching (Roder et al., 2007). As the stand matures, copious quantities of litter decompose quickly, releasing nutrients that are recaptured by the extensive lateral roots of Asteraceae species, mimicking the tight nutrient cycling of natural-forest ecosystems. This greater flux of nutrients pumped through the vegetation and redeposited on the soil surface may partially account for preliminary data from the Philippines that show strikingly higher nutrient stocks in soils under *T. diversifolia* fallows than under fallows of similar ages dominated by *Imperata cylindrica* (locally known as *cogon*) and *Pteridium aquilinum* (see below). There is also growing interest in the apparent ability of *T. diversifolia*, probably in association

FIGURES 39-3 and 39-4: A research assistant (left) collects the above-ground biomass of an *Austroeupatorium inulaefolium* fallow in West Sumatra, Indonesia (see Cairns, 2007a). This fallow was about two years old. It compares (above) with the biomass collected from a nearby fallow of similar age dominated by the rhizomatous grass, *Imperata cylindrica*.

All photographs in this chapter were taken by
Malcolm Cairns.

with mycorrhiza, to mobilize and accumulate soil phosphorus (P) (Scrase et al., 2019). These combined attributes thus appear to speed up the nutrient-accumulation process and contribute to a steeper soil-fertility-restoration curve. When the fallow is later reopened, nutrients immobilized in the standing Asteraceae biomass are directed to crop production as ash or decomposing mulch.

This rather optimistic appraisal of the fertility-restoring capacity of Asteraceae shrubs needs to be tempered with some caution. There is, first and foremost, a need for careful interpretation of existing data. Although sampling of farmers' fields demonstrates clear trends of superior biomass and nutrient accumulation in these shrub fallows, it leaves unanswered the question of whether Asteraceae species actually improve the soil or simply tend to colonize the most fertile sites, i.e., the 'indicator' vs. 'improver' debate. Careful field trials are needed to filter out the effects of topography, parent material, cultivation history and other confounding variables and provide a clearer assessment of the degree to which these differences can be attributed to the fallow species. Secondly, if Asteraceae-based

Austroeupatorium inulaefolium (Kunth) R.M.King & H.Rob. [Asteraceae]

A native of Central and South America, this perennial scrambling shrub aggressively colonizes agricultural fields, fallow fields, waste lands and roadsides, growing to 5 m in very dense thickets (see plate 43 in Coloured Plates section II, this volume) and spreading by wind-borne seeds. Naturalized territories include Indonesia and the Philippines. In the latter case, it has become a serious weed in rubber and tea plantations, upland rice fields and clearings in secondary forest.

fallows do not, in contrast to species fixing atmospheric nitrogen, bring external nutrients into the system, but are simply efficient 'nutrient sponges', can short-term fallows based on these shrubs be sustainable, or are we simply enabling more efficient mining of the soil to lower fertility levels? This suggests that Asteraceae fallows might best be viewed as transitional systems that can play a valuable bridging role in mitigating the deterioration of swidden agroecosystems as the cycle intensifies into more permanent forms of cultivation.

C. odorata, *A. inulaefolium* and *T. diversifolia* display additional attributes that recommend them as effective fallow species. Their rapid growth, early canopy closure and formation into dense, almost monospecific thickets prevents invasion by *Imperata* and other persistent grasses, thus reducing weeding labour during the subsequent cropping phase. Allelopathic inhibitions may also play a role in bolstering *C. odorata*'s weed-suppression abilities (Hu and Zhang, 2013). Both *C. odorata*[10] and *T. diversifolia*[11] (Figure 39-5) are reported to have the capacity to kill microscopic plant-parasitic nematodes, and this may help to prevent the build-up of soil-borne diseases, even in short fallows. Farmers say the abundance of litter and leaves improves soil tilth so that seeds can be dibbled directly without any need for tillage. In very young fallows, the initial populations of these plants are high, but they thin with age, leaving intermittent clumps that can easily be cleared with a machete. Any Asteraceae seedlings regenerating during the cropping period are easily removed and are not a serious weed threat.

FIGURE 39-5: This Tala-andig farmer in the Manupali River watershed, Bukidnon province, Mindanao, shows how large *Tithonia diversifolia* can grow, even in a relatively short fallow. Aside from the high levels of biomass, very few weeds are likely to survive in this thicket.

Among its wide variety of uses in Africa and Asia, *T. diversifolia* is regarded as a medicinal plant, with various extracts and parts used to treat many maladies, including diabetes, malaria, snake bite, measles, gastric ulcers, menstrual pains, and wounds (Ajao and Moteetee, 2017). Further clinical and pharmacological investigations are suggested to explore the potential of the plant in ethnomedicine. *T. diversifolia* also produces major active defensive substances known as sesquiterpenes (Tagne et al., 2018) that can also act as allelopathic substances, inhibiting the germination and growth of neighbouring plants and contributing to the invasiveness of the species (Kato-Noguchi, 2020). Oyerinde et al. (2009) found that the allelopathic effects of *T. diversifolia* on maize germination were inconclusive but positive, including growth stimulation. As a forage supplement for goats (see Figure 39-18), a 30% content of *T. diversifolia* leaf meal in a guinea-grass diet has produced no deleterious effect (Wambui et al., 2006; Odediri and Oloidi, 2014).

These combined properties of Asteraceae shrubs may provide the biological basis for accelerating fallow functions and intensifying the swidden cycle without it sliding into ecological decline. Given that grazing by free-ranging livestock has been a persistent cause of failure in attempts to promote improved fallow models in Southeast Asia's uplands, it is significant to note that these Asteraceae species are generally unpalatable to cattle and buffaloes and do not require protective fencing.

Considerable research in Africa has quantified the use of *Tithonia* and *Chromolaena* in mulch-transfer systems to fertilize cropped soils. When directly compared, *Tithonia* was superior to *Chromolaena* in N, P and K content and had positive effects on crop growth comparable to applications of cattle or chicken manure (Olabode et al., 2007). Tshinyangu et al. (2017) found a similar ranking for these two species after experiments with maize in DR Congo. Partey et al. (2011) confirmed significantly high N, P and K concentrations in *T. diversifolia*, comparable to levels recorded for *Senna spectabilis*, *Gliricidia sepium*, *Leucaena leucocephala* and *Acacia auriculiformis*, which are commonly used in biomass-transfer systems. However,

FIGURE 39-6: Near Barangay Songco, in Bukidnon province, these farmers were burying *Tithonia diversifolia* biomass to create a valuable compost. They were keen to demonstrate their composting method, although it did not appear to be widely used.

Tithonia decomposed and released its nutrients faster than the other species (Figure 39-6). Nitrogen supplies from *T. diversifolia* biomass, when added to maize-crop residues, alleviated N-immobilization that otherwise could be detrimental to young stages of subsequent crops (Partey et al., 2014). Although applications of *T.diversifolia* green manure to soils were found to increase ammonia and nitrous oxide emissions, the application of biochar counteracted this process under field conditions for multiple years (Fungo et al., 2019). Moreover, Ikerra et al. (2006) found that *T. diversifolia*

Tithonia diversifolia (Hemsl.) A.Gray [Asteraceae]

This native of Mexico and Central America has become virtually pantropical as an introduced species. Like other Asteraceae 'weeds', it spreads by prolific seed production and grows in dense, barely penetrable thickets up to 5 m tall. While showing great potential as an improver of depleted soils, it is also known for its dazzling floral displays that may cover entire mountainsides with yellow blooms (see Plates 40, 41 and 42, in Coloured Plates section II, this volume). It is the official flower of Thailand's Mae Hong Son province, where an annual festival coincides with its blooming in November every year.

led to significant increases in soil pH, exchangeable Ca and labile and moderately-labile inorganic P, while reducing exchangeable Al and P sorption in an acid soil and enhancing the effectiveness of rock–phosphate added to such soils.

A study in western Kenya concluded that a *Tithonia*-biomass transfer system could improve maize yields in the short term, but had limitations because of the large amount of biomass needed and the associated labour requirements (Gachengo et al., 1999). Another study in western Kenya found that P and K concentrations in *T. diversifolia* leaves were higher when the leaves were collected from naturalized hedges (3.2 g P/kg and 35 g K/kg), rather than from unfertilized fields (2.2 g P/kg and 23 g K/kg). The critical level of 2.5 g P/kg for net P mineralization was exceeded by > 90% of the leaves from hedges, but by only 14% of those from unfertilized fields (George et al., 2001).

Studying daisy fallows

Research on improving fallows by using these Asteraceae species generally proceeds along the following path:

1. First, farmers' practices, perceptions and preferences must be understood. When it is clear that farmers are interested in developing daisy fallows on their farms, further steps will require answers to a series of questions.
2. Indicator or improver? Do daisy fallows actually improve soil fertility, or are they simply indicators of favourable conditions? The answer to this question will depend on what is seen as the main crop-growth limiting factor that fallows can improve. Is it to improve N, P or cation supply, enhance soil structure and infiltration, or suppress soil-borne diseases or weeds?
3. Bearing in mind these critical aspects of fallow function, can we identify a mechanism by which specific types of fallow vegetation have the desired effect? Can knowledge of this mechanism give direction to the search for further improvements of the system?
4. The ease of establishing the fallow vegetation and the vigour of weed growth during the cropping phase must be determined. Is specific management required to establish the fallow in each rotational cycle, or can it regrow from a seedbank or stumps maintained in the field? If so, is the fallow vegetation a weed itself during cropping? Is there a risk of it spreading beyond control?
5. Can further improvements be made by incorporating other species or adjusting the management regime? Is there any advantage in combining fertilizer use with daisy fallows?
6. Based on an improved understanding of the biophysical mechanisms and the socio-economic context of this type of fallow management, are there communities of farmers elsewhere that might benefit from the lessons learned?

Tithonia research that began in the mid-1990s in East Africa and Southeast Asia had progressed to phase 6 and phase 3 of this scheme, respectively, when assessed in 2010. Since then, progress on fallow research appears to have been limited, but syntheses on research into folkloric medicinal uses continued (Ajao and Moteetee, 2017; Tagne et al., 2018), as did interest in using daisy fallows for phytoremediation of soils with high concentrations of heavy metals (Tanee and Jude, 2020; Jampasri et al., 2021). The East African soil-fertility experience has been synthesized by Grierson et al. (2004), Oberson et al. (2011) and Chivenga et al. (2011), so we will summarize here the largely unpublished Southeast Asian experiences. These are grouped as descriptions of farmer practices (phase 1); analyses from surveys and field experiments involving both *Tithonia* in Mindanao, Philippines, and *Chromolaena* in Lampung, Sumatra, Indonesia (see Figure 39-7), to identify their role as indicator or improver (phase 2); and the results of pot experiments with *Tithonia* and *Chromolaena* aimed at revealing underlying processes and mechanisms (phase 3).

Phase 1: Farmers' practices

The most compelling indicator that the three Asteraceae species in question are useful fallow plants and worthy of further research attention is evidence that farmers themselves are investing in their management. Field observations confirm that swidden farmers are vigilant of the floristic composition of fallow successions and intervene to ensure post-cropping domination by these valued shrubs (Figure 39-8). These interventions

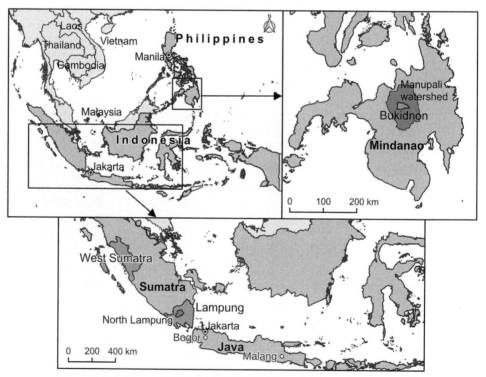

FIGURE 39-7: Shaded areas in Indonesia and the Philippines show the location of study sites for research into Asteraceae fallows described in this chapter.

range from subtle manipulation of succession dynamics to, where necessary, active planting.

A critical attraction of Asteraceae-based fallows is their potential to be largely spontaneous, requiring minimal human intervention. *C. odorata* and *A. inulaefolium* produce large quantities of wind-dispersed seeds, enabling both species to rapidly colonize disturbed land. The seeds of *T. diversifolia* are heavier and lack a pappus – the tuft of hair that can assist wind dispersal – so they are less invasive and require birds, running water or other mechanical means to disburse. This restricted mobility means that a nearby source of *T. diversifolia* propagules is essential if it is to self-seed into adjacent fallowed plots. This usually takes the form of 'mother shrubs' maintained on field boundaries, along footpaths, in rock piles or other pockets of marginal land within or nearby the swiddens.[12] In a particularly effective practice in Bukidnon, the Philippines, *T. diversifolia* poles erected as support for climbing beans are allowed to take root and provide an automatic reversion to fallow after the beans are harvested, without further labour inputs.[13] (Figure 39-9). This is less critical for *C. odorata* and *A. inulaefolium*, whose lightweight seeds can be blown in from long distances. Regardless of their method, farmers intervene to encourage recruitment of the Asteraceae species they want. In Laos, for example, swidden farmers jump-start fallow establishment by deliberately refraining from weeding out *C. odorata* seedlings during the latter part of the cropping cycle, promoting rapid ground cover and its domination of the fallow vegetation.

Farmers' tinkering with succession dynamics often

FIGURE 39-8: This swidden has been carved from a hillside covered by *T. diversifolia* in Lantapan, Bukidnon province, Mindanao. The larger plants at the field perimeter suggest that these may have been retained as an on-going seed source through successive swidden cycles.

FIGURE 39-9: This vegetable grower in Barangay Songco, Bukidnon province, Mindanao, used *T. diversifolia* stems to create a trellis for his climbing beans. As the bean crop grew, the trellis took root, and after harvesting his bean crop, the farmer had a pre-established *T. diversifolia* fallow or source of green manure.

continues well into the fallow period, aimed at enhancing the competitive edge of the desired Asteraceae species within floristically mixed succession communities. The pivotal successional struggle of importance to swidden farmers is that between the beneficial Asteraceae and detrimental *Imperata* and other course grasses. Farmer knowledge of the ecophysiological differences between these two camps provides a scientifically sound basis for deflecting fallow succession in the desired direction. Repeated burning is widely recognized as favouring *Imperata* and ferns at the expense of Asteraceae shrubs, so fire control is often a vital part of farmer strategies.[14] Minangkabau swiddenists in West Sumatra, Indonesia, stress the importance of fallowing their land before excessive exhaustion, to ensure *A. inulaefolium's* domination over *Imperata* in the ensuing fallow (Cairns, 2007a). This observation may be grounded in the fact that C4 grasses are most competitive on nutrient-depleted soils because of their higher efficiency in the use of nutrients, whereas the less efficient C3 dicotyledons prevail in more fertile soils (Ramakrishnan, 1993, pp.302–303).[15] Banjarese shifting cultivators in Kalimantan make spot applications of the herbicide glyphosate on *Imperata* clumps in new fallows, specifically to favour the expansion of *C. odorata* (Garrity, 1994). Lao swidden farmers in Luang Prabang reportedly use grazing cattle for the same purpose – again to selectively target *Imperata* and encourage the spread of *C. odorata* (Hansen, 1997). More innovative managers in Mindanao slash their *T. diversifolia* fallows every three to four months, so that fallen stems take root and thicken the stand (see Figure 39-10). Additional advantages of this procedure include preventing the aerial biomass from becoming excessively woody (see Figure 39-11) and enhanced nutrient cycling through regular applications of young and easily decomposed green manure (author's field notes).[16] While none of the above constitute planted fallows, they are all deliberately planned and achieve the same ends with little or no additional investment of labour.

In other circumstances, benefits seen by farmers from establishing Asteraceae species in their swidden fields may lead them to actively plant them, usually by vegetative cuttings. In the case of *A. inulaefolium* (Stoutjesdijk, 1935) and *T. diversifolia* (author's field notes), farmers have been shown to engage in ecological restoration by planting stem cuttings into *Imperata* swards (see plate 39 in the Coloured Plates section of volume 1 (Cairns, 2015b)). They are thus used as biological tools to rehabilitate degraded grasslands and bring the land back into productive cultivation.[17] Labour investment in planting the Asteraceae shrubs is much less than would otherwise be required to eradicate the *Imperata* manually. Farmers in the Cordillera Region of the northern Philippines often plant *T. diversifolia* as growing fertilizer banks (Daguitan and Tauli, 2007). In one variation of this theme, 0.5 m-long stem cuttings of *T. diversifolia* are planted on the steepest inclines of new swiddens. After a vigorous *T. diversifolia* thicket is established, its regular lopping and application as a green manure to the remaining cropped portion of the field becomes a routine part of swidden maintenance. During the subsequent fallow period, these thickets expand to occupy the remainder of the field and smother out troublesome weeds. Some farmers in the region are reported to accelerate this process by planting vegetative

FIGURES 39-10 and 39-11: Extremes in *Tithonia diversifolia* fallows: Left, a farmer near Lantapan municipality, Bukidnon, Mindanao, slashes a young stand of *T. diversifolia*, probably to thicken the stand by dropping many stems to the ground that will take root and grow into new plants. Right, also near Lantapan, an older *T. diversifolia* fallow has grown so woody that the farmer has chosen an axe to open it. This fallow sparked special interest because a lower strata of the wild viny legume, *Calopogonium mucunoides*, covered the soil, adding a N-fixing component to the mix.

cuttings throughout the swidden at the time of fallowing (Daguitan and Tauli, 2007).[18] Finally, *T. diversifolia* has become a favourite contour hedgerow species in Mindanao (see Figure 40-5 in ch. 40, this volume), due to its abundance, ease of propagation by vegetative cuttings and rapid growth. Instead of applying loppings directly to the soil, some farmers instead feed them to confined sheep and goats in a cut-and-carry system, and then return the manure to the alleyways. This promising system combines benefits of livestock production, efficient nutrient cycling and soil conservation (Cairns, 1997b). Although fields with contour hedgerows are usually permanently cropped, and therefore outside the realm of this study, data from the Philippines reveal that a surprisingly high percentage of these fields continue to be periodically fallowed (Suson et al., 2007). It is worth noting that *T. diversifolia*, with its prolific seeding and aggressive growth, is adept at colonizing the inter-rows to create a solid fallow thicket within a very short time. When later re-opening the fallow, the original configuration is regained by simply trimming back the still discernable hedgerows and clearing only the alleyways for cropping. *T. diversifolia* thus appears ideally adapted for straddling this farmer hybridization of fallow and contour-hedgerow technologies.

These examples clearly illustrate that the restorative properties of Asteraceae pioneers are widely recognized and used by shifting cultivators to enhance the biological efficiency of their fallows and intensify their land use.

Phase 2a: Indicator or improver? observational analysis

A quantitative analysis was made in 1995 of different types of fallow vegetation in the upper Manupali River watershed in Mindanao, the Philippines (see Figure 39-7), in the course of assessing the sustainability of land use and exploring new methods of natural resource management.

Three main types of fallow vegetation were distinguished in the field, dominated by *Imperata*, bracken (*Pteridium*, a fern) and *Tithonia diversifolia*, respectively. Nearly 90 fallow fields were sampled, spanning a fallow age range of 0.5 to 5 years. Farmer interviews accompanied the collection of soil and above-ground biomass samples. Soil samples were taken at depths of 0-5 cm and 5-15 cm. Above-ground biomass was sorted by plant species and plants from species contributing more than about 5% to the total biomass were taken for chemical analysis. Analysis of soil and plant samples was carried out at the laboratory of the International Rice Research Institute (IRRI) at Los Baños in the Philippines, using conventional methods.

Macronutrient concentrations in different components of fallow vegetation

Data on nutrient concentrations in above-ground biomass were grouped into the following categories (Table 39-1):

- *Tithonia diversifolia* (n=57 samples; partly split into stems and leaves. The results here are presented as a 1:1 average of the two tissues.)
- Other plants from the Asteraceae family (n=37; *Bidens pilosa, Ageratum conyzoides, Chromolaena odorata, Conyza canadensis*)
- Legumes (n=12; *Mimosa invisa, Crotalaria mucronata*)
- Grasses (n=52; *Imperata cylindrica, Paspalum conjugatum, Pennisetum polystachion, Panicum repens, Saccharum* sp.)
- Others (n=67; *Stachytarpheta jamaicensis, Sida rhombifolia, Elephantopus tomentosus, Commelina diffusa, Corchorus olitorius, Hyptis suaveolens, Crassocephalum crepidioides*)
- Bracken (n=88; *Pteridium aquilinum, Pteridium* spp.)

TABLE 39-1: Average macronutrient concentrations in above-ground biomass of plant samples taken from fallows of different ages in Mindanao, the Philippines.

Plant group	N	N%	NP ratio	P%	K%	Ca%	Mg%
Tithonia diversifolia	57	1.81	6.56	0.234	2.59	0.741	0.287
Other Asteraceae	37	0.88	5.72	0.163	2.32	0.691	0.227
Legumes	12	1.46	15.71	0.094	1.07	0.455	0.220
Grasses	52	0.66	6.99	0.097	1.35	0.288	0.217
Other Angiosperms	67	0.90	7.18	0.142	1.47	0.791	0.250
Bracken	88	1.23	10.37	0.128	1.51	0.338	0.183
F-probability		<0.001	<0.001	<0.001	<0.001	<0.001	0.001
s.e.d		0.24	0.92	0.026	0.306	0.107	0.040

Note: F-probability indicates the probability that the differences observed can be observed by chance under the null-hypothesis of no differences between the populations tested; s.e.d.= standard error of difference.

Source: Cairns, unpublished.

Tithonia has a remarkably high P concentration in its tissues, as well as a high P content (Table 39-2), but does not differ from other plants in N:P ratio; the N:P ratio for legumes is high, as may be expected for plants which are primarily limited by P supply. The K concentration is high in *Tithonia*, as well as in the other Asteraceae samples, while Ca was low in the ferns and grasses.

Soil-fertility parameters of different fallow types and ages

The data were grouped into three main types of fallow vegetation, dominated by *Imperata cylindrica* (cogon) (n=21); bracken (*Pteridium*) (n=37); and *Tithonia diversifolia* (n=28), respectively.

The cation-exchange capacity (CEC) of the soils did not differ, indicating that there were no major differences in texture or mineralogy. The organic C content was highest in the top layer of *Imperata* fallows, followed by fern and *Tithonia* fallows, but the organic N content did not differ. The C/N ratio was thus lowest for the *Tithonia* fallows, which could lead to a higher N mineralization after clearing for food crops (Table 39-3).

Fallows dominated by *Tithonia* had a significantly higher cation (K, Ca and Mg) status than the other two fallow types. Exchangeable acidity (H and Al) was highest on the fern-dominated sites, in line with their lower pH (H_2O). The relation between pH and cation status, or pH and exchangeable acidity did not differ between the fallow types (Table 39-4).

Some insight into whether these differences are simply differences between the sites that have led to the development of different vegetation – so that vegetation is an indicator of soil fertility – or if they are in fact an effect of the fallow vegetation, can be obtained by considering a regression of soil fertility for the top 5 cm on the age of the fallow, as reported by the farmer. Results of regression analysis (Table 39-5) show very little evidence that soil-fertility parameters change with fallow age. The only slopes that are significantly different from zero indicate an increase of C_{org}, CEC and P_{Bray}, with time, for *Imperata* fallows.

TABLE 39-2: Average macronutrient concentrations in *Tithonia* stems and leaves.

Part of plant	N	N%	N:P ratio	P%	K%	Ca%	Mg%
Leaves	26	3.55	9.29	0.388	4.33	1.42	0.48
Stems	31	0.35	4.28	0.104	1.12	0.17	0.13
F-probability		<0.001	<0.001	<0.001	<0.001	<0.001	<0.001
s.e.d.		0.113	0.656	0.023	0.169	0.080	0.031

Notes: F-probability indicates the probability that the differences observed can be observed by chance under the null-hypothesis of no differences between the populations tested; s.e.d.= standard error of difference.

Source: Cairns, unpublished.

TABLE 39-3: Average soil fertility parameters for the top 5 cm of fallows dominated by three types of vegetation in Mindanao, the Philippines.

Vegetation type	N	pH (H_2O)	C_{org} (%)	N (%)	CN ratio	P_{Bray} mg/kg	CEC cmol/kg
Imperata cylindrica	21	5.00	6.50	0.46	17.2	6.62	20.9
Bracken	37	4.78	5.85	0.42	14.9	5.36	18.6
Tithonia diversifolia	28	5.06	4.95	0.48	11.0	8.80	20.9
F-probability		<0.001	0.005	0.226	0.053	0.053	0.105
s.e.d.		0.078	0.447	0.039	1.53	1.53	1.32

Vegetation type	N	K $cmol_e$ kg	Na $cmol_e$ kg	Mg $cmol_e$ kg	Ca $cmol_e$ kg	H $cmol_e$ kg	Al $cmol_e$ kg
Imperata cylindrica	21	0.495	0.0423	1.36	3.65	0.753	0.535
Bracken	37	0.429	0.0454	1.30	2.96	1.081	0.813
Tithonia diversifolia	28	1.052	0.0531	2.81	7.47	0.427	0.261
F-probability		<0.001	0.190	<0.001	<0.001	<0.001	<0.001
s.e.d.		0.1364	0.00594	0.281	0.982	0.1217	0.1059

Notes: F-probability indicates the probability that the differences observed can be observed by chance under the null-hypothesis of no differences between the populations tested; s.e.d.= standard error of difference.

Source: Cairns, unpublished.

TABLE 39-4: Results of regression analysis of exchangeable cation content on the pH (H_2O) in the top 5 cm of soils under different fallow types.

Vegetation type	(K+Ca+Mg) = intercept + slope x pH						(Al+H) = intercept + slope x pH					
	n	r^2	intercept	s.e.	slope	s.e.	n	r^2	intercept	s.e.	slope	s.e.
Imperata	21	16.9	4.78	0.092	0.0417	0.0152	21	14.3	5.16	0.072	-0.117	0.047
Tithonia	28	75.5	4.62	0.055	0.0395	0.0035	28	60.9	5.40	0.049	-0.473	0.059
Bracken	37	27.9	4.59	0.057	0.0426	0.00962	37	66.1	5.14	0.039	-0.194	0.020

Note: s.e. = standard error of preceding parameter estimate.

In the Philippines survey, data of plant-nutrient contents were analysed for relatedness to the soil-fertility indicators measured. For five plant-nutrient concentrations of eight types of plant material (*Tithonia* leaf, *Tithonia* stem, *Imperata*, other grasses, fern, other *Asteraceae*, *Leguminosae*, other dicotyledons) the correlation (r value) with 11 soil parameters was calculated. Only six out of these 440 r values were larger than 0.6 and two were smaller than -0.6 (K in legumes vs Al_{exch} and H_{exch}). These relatively strong plant-soil relations were restricted to legumes (K plant with Ca$_{exch}$ and Mg$_{exch}$, Ca plant with Na$_{exch}$), other grasses (K plant vs Mg$_{exch}$, P plant vs P$_{Bray}$) and other Asteraceae (P plant vs P$_{Bray}$).

Only 129 of the 440 r values were larger than +0.3 or less than -0.3. The P concentration in plant tissue was most often correlated (outside the -0.3 to +0.3 range) with soil parameters (43 out of 88 combinations tested all had positive r values, except for those with Al$_{exch}$ and H$_{exch}$, which were always negative), while the N concentration showed the least relatedness to soil parameters (only 13 out of 88 combinations were outside the -0.3 to +0.3 range) and the K, Ca and Mg

TABLE 39-5: Results of regression analysis to test changes in soil fertility parameters for the top 5 cm of fallows of three different vegetation types (<0 means that the regression did not explain any part of data variance).

Type		pH = intercept + slope x time						$C_{organic}$ = intercept + slope x time				
	n	r^2	intercept	s.e.	slope	s.e.	n	r^2	intercept	s.e.	slope	s.e.
Imperata	21	16.9	4.78	0.092	0.0417	0.0152	21	14.3	5.16	0.072	-0.117	0.047
Tithonia	28	75.5	4.62	0.055	0.0395	0.0035	28	60.9	5.40	0.049	-0.473	0.059
Bracken	37	27.9	4.59	0.057	0.0426	0.00962	37	66.1	5.14	0.039	-0.194	0.020
		N = intercept + slope x time						C/N ratio = intercept + slope x time				
Imperata	21	<0	0.388	0.061	0.0119	0.0177	21	1.1	16.31	2.32	0.786	0.672
Tithonia	28	<0	0.476	0.020	-0.0046	0.0069	28	0.7	10.18	1.45	0.553	0.486
Bracken	37	<0	0.426	0.025	-0.0013	0.0088	37	<0	15.45	1.19	-0.405	0.421
		P = intercept + slope x time						CEC = intercept + slope x time				
Imperata	21	15.2	3.3	1.35	1.021	0.393	21	26	17.6	1.33	1.351	0.386
Tithonia	28	<0	12.5	2.6	-0.578	0.874	28	<0	20.9	1.76	0.468	0.591
Bracken	37	3.2	6.19	0.86	-0.487	0.303	37	<0	19.2	0.796	-0.155	0.281
		K = intercept + slope x time						Na = intercept + slope x time				
Imperata	21	<0	0.58	0.116	-0.009	0.034	21	3.4	0.052	0.0079	-0.003	0.002
Tithonia	28	<0	1.18	0.206	0.009	0.069	28	<0	0.055	0.0045	-0.0007	0.001
Bracken	37	<0	0.42	0.072	0.008	0.026	37	2.5	0.042	0.0066	0.003	0.002
		Mg = intercept + slope x time						Ca = intercept + slope x time				
Imperata	21	<0	1.31	0.226	0.0269	0.066	21	6.3	2.92	0.458	0.236	0.133
Tithonia	28	<0	2.85	0.408	0.111	0.137	28	<0	8.13	1.69	0.314	0.567
Bracken	37	5.8	1.14	0.183	0.1288	0.066	37	2.8	2.70	0.475	0.259	0.168
		H = intercept + slope x time						Al = intercept + slope x time				
Imperata	21	<0	0.741	0.13	0.0104	0.038	21	<0	0.524	0.118	0.011	0.034
Tithonia	28	<0	0.345	0.092	0.0001	0.031	28	<0	0.192	0.075	0.001	0.025
Bracken	37	<0	1.092	0.124	-0.0436	0.044	37	0.2	0.829	0.108	-0.040	0.038

concentrations were intermediate in this respect (20 to 29 r values were outside the
–0.3 to +0.3 range). Clear differences between the plant categories emerged, with 32
out of 55 correlations tested for the other grasses category being outside the –0.3 to
+0.3 range, and only 6 to 11 out of 55 for the *Imperata, Tithonia* leaf and *Tithonia*
stem categories. In *Tithonia*-leaf material, only 1 positive r value was found outside
of the –0.3 to +0.3 range (Ca in plant vs P_{Bray}). Plant P concentrations for *Tithonia*
leaf and other Asteraceae showed 0 and 1 relatedness, respectively, while other grasses
and other Asteraceae had 10 and 9 relatedness, respectively, outside of the –0.3 to
+0.3 range. Plant Ca concentration was correlated with soil parameters, especially in
the other grasses, other Asteraceae and other dicotyledon categories; while plant K
concentration responded, especially in the Leguminosae and other grasses categories.

When grouped by soil parameters, P_{Bray}, K_{exch} and Ca_{exch} had the highest number
of correlates (17, 16 and 15 out of 40, respectively), with one negative r value (with
Mg concentration in *Tithonia* leaf) in each case. The Na_{exch} and C_{org} content of the
soil had the lowest number of correlates (7 and 8 out of 40, respectively, outside the
–0.3 to +0.3 range).

Further analysis of the relatedness between plant-tissue P concentration and the
P_{Bray} value, as an indicator of P availability in the soil, showed similar slopes of the
regression line for all plant categories, except for the *Tithonia* leaf and *Tithonia* stem
tissue. Data for all non–*Tithonia* plant samples were thus combined. The combined
model (separate slopes and intercepts for *Tithonia* stems, *Tithonia* leaves and all other
plants, seen in Figure 39-12) accounted for 66.8% of the variance in the data set. The
combined model shows that for the 'other plant' categories, the tissue-P concentration
increases by 0.05% if P_{Bray} increases by 10mg/kg, and that a P_{Bray} of 20mg/kg is needed

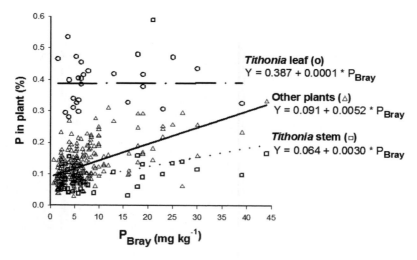

FIGURE 39-12:   Relatedness between plant-tissue concentrations of P and the
P_{Bray} value of the soil, as an indicator of P availability, for three types of plant
tissue in a range of fallow types and ages. The overall regression model has an
adjusted r^2 of 0.668.

for an average tissue-P concentration of 0.2%, while *Tithonia* leaves have a tissue-P concentration of 0.39 at P_{Bray} values of less than 5 mg/kg. Somehow, *Tithonia* is able to achieve such tissue-P concentrations on soils where other fallow species cannot reach their normal (lower) tissue-P concentrations. It is similar in this respect to *Brassica* species, which can take up more P than is considered to be 'available', by a process of selective rhizosphere acidification (Hedley et al., 1983).

Thus, we tentatively conclude that differences showing an indicator value are more important than those demonstrating fallow effects.[19] *Tithonia* fallows are associated with high soil-cation status, relatively high pH and low C content. Only in *Imperata* fallows can we find indications of soil-fertility improvement (at least for C_{org} and cation buffer capacity or CEC) during fallow growth. Fern fallows dominate on the most acid soils with the highest probability of Al toxicity to crops and there is no indication of soil improvement with increasing age for this fallow type.

In a field survey in western Kenya, George et al. (2001) found that *Tithonia* leaf nutrient concentrations were related to soil-nutrient status. Leaf P and K concentrations were higher for plants growing in hedges (3.2 g P/kg and 35 g K/kg) than in unfertilized fields (2.2 g P/kg and 23 g K/kg).

Phase 2b: Indicator or improver analysis in field experiments

Chromolaena odorata was introduced into Indonesia at the beginning of the 20th century as a soil-improving cover crop. The plant, a perennial herb with a shrub–like appearance (Figure 39-13), has spread rapidly and is now a common constituent of fallow vegetation on any abandoned crop land. It is considered to be a weed that must be eradicated by biological control (Obatolu and Agboola, 1993; Zachariades et al, 1999), but farmers generally appreciate its soil-improving characteristics and definitely prefer it over *Imperata* grasslands, which it may replace under certain conditions. Internationally coordinated programmes for biological pest control have been active in selecting insects for release. In Indonesia, permission was given for an initial release, but the wisdom of doing so has been questioned, in view of the presumed benefits of *Chromolaena* to smallholders. In the ensuing debate, the issue of 'improver' versus 'indicator' found a new meaning: if farmers' preference is based merely on an 'indicator' value, then they will

FIGURE 39-13:   A Batak villager in Tina village, in the uplands of Palawan, the Philippines, hacked a pathway into this *Chromolaena odorata* fallow so that one of the authors could gain access to take a soil sample. A camera bag was left on the trail to provide scale.

not suffer serious losses if the indicator disappears. However, if it is a real 'improver', they will suffer real losses.

Although many references indicate that land cleared from *Chromolaena* delivers relatively good crop yields, no conclusive evidence exists that *Chromolaena* actually improves soil fertility, and if so, by what mechanism. The correlation between relatively high soil-cation contents and *Chromolaena* abundance on acid soils might be based on selective growth of *Chromolaena* on good sites, as well as on actual improvement of soil conditions, possibly based on *Chromolaena*'s relatively deep root system and its high biomass and litter production. Experiments are needed in which there is a controlled introduction of *Chromolaena* in some plots and not in others, to test its effects on the soil. As a comparison, *Imperata* grassland may be used, along with the leguminous cover crop *Pueraria*, which is known to enrich soil organic-matter content and to supply N to subsequent crops, and the local non-N-fixing leguminous tree *Peltophorum dasyrrhachis*,[20] known to be deep rooted.

The following hypotheses were tested in trials at Lampung in Indonesia:

- A two-year fallow of *Chromolaena*, *Pueraria* or *Peltophorum* will lead to higher subsequent crop yields than a fallow consisting of *Imperata*;
- Soil improvement by *Chromolaena* and *Peltophorum* is based on high litter inputs and on topsoil enrichment with cations taken up from deeper soil layers; and
- Soil improvement by *Pueraria* is primarily based on a higher soil-N content.

Methods

Experiment 1 was conducted at the Biological Management of Soil Fertility project site in north Lampung, Indonesia (see Figure 39-7), in an existing *Imperata* grassland on an acid ultisol of low fertility. This plot had been used for maize experiments for three years after being opened by slash-and-burn from logged-over forest and had been invaded by *Imperata* for two to three years.

Four types of fallow (*Imperata cylindrica, Chromolaena odorata, Pueraria phaseoloides* – a viny

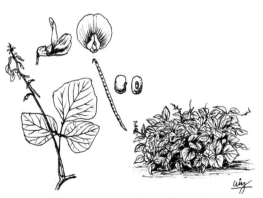

Pueraria phaseoloides (Roxb.) Benth. [Leguminosae]

This native of East and Southeast Asia is a forage or green manure crop known widely as *kudzu*. It is a deep-rooted, vigorous perennial vine that can grow as much as 30 cm in a day, its longer stems reaching 20 m. Vines that touch the ground grow new plants from nodes. It is popular as a nitrogen-fixing cover crop in tree plantations, as a grazed forage crop, a green manure in crop rotations, and to protect soil against erosion.

legume cover crop (van Noordwijk et al., 1995) – and the local tree *Peltophorum dasyrrhachis*) were planted from local wildings and maintained for two years, from September 1994 to September 1996, and were subsequently cropped for three maize seasons, between December 1996 and March 1998. Treatments were arranged in a randomized block design with four replications and a plot size of 9.5 m x 12 m. Root trenches were dug around the tree plots to prevent or reduce interference.

During the fallow period, above- and below-ground development of biomass was monitored on a three-to-six-monthly basis, while litterfall was recorded in the *Chromolaena* and *Peltophorum* plots and analyzed for nutrient content and quality (lignin and polyphenolics).

In the cropping stage, all plots were split into four parts: receiving no fertilizer, N only, P only, or N + P (30+30 kg N/ha at planting and 30 days after planting, respectively, and 20 kg P/ha, respectively). Maize (cv *Arjuna*) was planted at 0.4 m x 1 m spacing and harvested from a 2.5 m x 4 m core area in each sub-plot.

Results

Results obtained during the fallow period were summarized by Hairiah et al. (1996). Initial growth of *Chromoleana* from stem cuttings was satisfactory, but after six months plant development slowed down. Plant-nutrient analysis indicated that contents of P and cations in the experimental fallow were below those of vigorous growing plants on sites that had been burned but not cropped. Root development of the plants grown from cuttings appeared to be similar to that of seed-derived plants. Litterfall rates of about 4 Mg/ha/year peaked during the early dry seasons, but rapid decomposition meant that there was little surface litter at most times. The *Peltophorum* fallow plots had a slow start, but by the end of the first year they had a complete canopy cover and during the second year produced a steady litterfall of more than 8 Mg/ha/year, which accumulated on the soil

Peltophorum dasyrrhachis (Miq.)
Kurz [Leguminosae]

This deciduous tree is a pioneer species native to Southeast Asia that quickly invades open areas and fallowed agricultural land and grows to 30 m tall. Its leaf litter decomposes slowly, allowing a humus layer to build up in the soil, reducing erosion and contributing to the suppression of weeds. Due to a deep rooting system, the trees are drought tolerant, can be severely pruned, and can be managed as hedges to minimize shading of crops.

surface. *Pueraria* had a rather slow start but produced a thick mat of green biomass on the surface, with a gradually accumulating litter layer underneath.

Maize growth was poor, amounting to less than 2 Mg/ha, on all four fallow types in the absence of fertilizer. *Pueraria* + P gave a good first crop of maize, while for the other fallow types, only the crops with both N and P were satisfactory. There were no statistically significant differences in maize yield between the *Imperata*, *Chromolaena* and *Peltophorum* fallows.

It is clear that hypotheses 1 and 2 were not confirmed, while hypothesis 3 survived the test. On this site of low soil fertility, where *Chromolaena* could invade into *Imperata* only very slowly in the absence of fire (as was evident on nearby plots, outside the experiment) and developed slowly from stem cuttings, *Chromolaena's* performance was probably limited by its nutrient-acquisition abilities. With the low rate of litterfall, there was no improvement in fallow function compared to the *Imperata* fallow, which was used as a control, and which, under these conditions, was the default fallow type.

Phase 3: Underlying processes and mechanisms studied in pot experiments

Methodology

Pot experiments were carried out in a glasshouse of the Faculty of Agriculture, Universitas Brawijaya, Malang, East Java (see Figure 39-7), from February through July, 1998. Soil was collected from the Biological Management of Soil Fertility research site in north Lampung (Hairiah et al., 1996) from a depth of 0 to 20 cm. It was air-dried and sieved through 2 mm mesh. Green biomass of *Imperata cylindrica*, *Chromolaena odorata* and *Tithonia diversifolia* were used as sources of organic matter. *Imperata* and *Chromolaena* were collected from Lampung and *Tithonia* was collected from Malang. The green biomass was oven dried at 80°C for 48 hours and ground for chemical analyses, including total C, P, concentration of polyphenolic, lignin (for methods, see Handayanto, 1994).

Three series of experiments were conducted:

- Incubation experiments to test the rate of P-mineralization of different organic-matter inputs;
- Pot experiments to test the enhancement of P-mineralization by adding organic materials and inorganic-P (unlabelled) fertilizer, and its effect on P-uptake by maize; and
- Pot experiments to test the enhancement of P-mineralization by adding organic materials and inorganic-P (^{32}P) fertilizer, and its effect on P-uptake by maize.

Treatments were arranged in randomized block design, with three replications. Inorganic P fertilizer was used at rates of 0 or 75 kg/ha. Organic-matter inputs were either none; *Imperata* of low quality, having a C:P ratio of 497; *Chromolaena* of high quality, with a C:P ratio of 138; or *Tithonia* of high quality, with a C:P ratio of 121.

Experiment 2

Organic matter was added to 0.25 kg of soil at a rate of about 10 Mg/ha. It was watered at about 80% of field capacity and placed in the laboratory at room temperature. Soil moisture was controlled every day by weighing it. Soil samples were collected at 0, 5, 10, 15, 22, and 30 days after incubation and analysed for P-organic concentration, microbial biomass-P and P_{Bray1}.

Experiment 3

Organic matter was added to 1.5 kg of soil at a rate of about 10 Mg/ha. Fertilizers were applied to every pot at rates equivalent to 100 kg N/ha; 25 kg K/ha; and 2.5 kg Zn/ha. To avoid an extreme P deficiency, 10 mg of P was applied to every pot. Soil moisture content was maintained at about 80% of field capacity by weighing, and the pots were kept in the laboratory at room temperature.

Soil samples were collected at harvesting time, 30 days after incubation. Soil samples were analysed for soil pH, microbial biomass P and P_{Bray1}. Maize was harvested four weeks after planting, oven dried at 80°C for 48 hours, and ground for analysis of total C and P concentrations.

Experiment 4

Soil preparations were the same as those for experiment 3. The ^{32}P-labelled fertilizer was kindly provided by the National Nuclear Energy Agency of Indonesia (BATAN) as NaH_2PO_4 and was applied to the soil before planting.

Soil samples were collected at harvesting time, 30 days after incubation, and analysed for soil pH, microbial biomass P and P_{Bray1}. Maize was harvested four weeks after planting, oven dried at 80°C for 48 hours and ground for analysis of total C and P concentrations.

Extraction methods

Analytical methods for P fractionation followed a process outlined by Hedley et al. (1982):

- P-inorganic: extraction with 1M H_2SO_4
- P-total: extraction with 1M H_2SO_4 at 550° C
- P-organic: (Total-P minus P-inorganic)
- P_{Bray1} = P-available. Extraction: 0.03M NH_4F + 0.025 M HCl
- P_{resin} = 24-hour sorption to resin strips in a soil paste
- Microbial biomass P = difference in P-resin with and without chloroform fumigation, corrected for recovery of added inorganic P

Results

Incubation experiment

Organic P sourced from *Chromolaena* and *Tithonia* was mineralized at a much higher rate than that from *Imperata* material or the background soil organic matter (Table 39-6). Maize growth and P uptake were improved with *Chromolaena* and *Tithonia* inputs, and were equivalent to that on the +P control. A combination of daisy inputs and P gave the best performance, with some indication of more–than–additive effects of *Tithonia*.

P balance

Data indicated a substantial decrease (compared to the no-P control) of total organic P for all pots where daisy or *Imperata* biomass + P were used (Figure 39-14). Measured inorganic available P (P_{Bray}) was clearly increased by P fertilizer use, but was increased over the control with *Tithonia* litter as well (again with a more–than–additive effect in the combined treatment). Microbial P was reduced below the control by all organic and inorganic additions.

Labelled P experiment

The labelled P experiment gave more direct evidence that the increased P uptake, where daisy biomass had been used, exceeded the amount of P supplied with the litter and that a substantial part of the effect was probably due to increased mineralization of organic P sources present in the control soil, but unavailable to the maize plant as such (Table 39-7). This experiment should be treated as a first step in this direction, and further tests of this hypothesis will be needed.

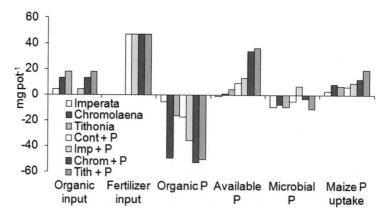

FIGURE 39-14:   Differences in components of the P balance between soil with fertilizer and/or organic inputs and the control system without inputs.

TABLE 39-6: Decomposition constants (k values/day) for total soil organic P during incubation of Lampung soil with and without inorganic and organic inputs.

	Without fertilizer P	*With fertilizer P*
Control	0.011	0.020
Imperata	0.019	0.025
Chromolaena	0.045	0.047
Tithonia	0.033	0.044

TABLE 39-7: P uptake by 30-days-old maize, showing differences in P derived from inorganic P sources in the soil and that from organic inputs and/or increased mineralization of organic soil P (all figures mg/pot).

	P addition with biomass	*P uptake by maize*	*^{32}P-labelled fertilizer inputs in plant*	*P derived from soil*	*P derived from biomass and/ or increased mineralization*
Control soil	0	24.2	0.40	23.8	–
Imperata	4.0	27.1	0.16	10.0	17.0
Chromolaena	12.9	38.4	0.17	9.9	28.3
Tithonia	17.9	45.5	0.14	8.2	37.2

Further exploration of hypotheses regarding underlying mechanisms

Research by the *Tithonia* team in Malang confirmed that farmers use *Tithonia* – which is available on nearby public-access land – as green manure under specific circumstances in clear preference over other types of biomass. There are two phenomena to be considered: (A) *Tithonia* has a high P, N and cation content and must, itself, be an efficient scavenger of nutrients, and (B) it also appears that the nutrient transfer to subsequent crops is very effective, especially for P. Underlying mechanisms should be understood for both phenomena, and Table 39-8 lists the hypotheses that have been formulated.

Tithonia *P uptake*

Research elsewhere in the *Tithonia* project funded by the (UK) Department for International Development is focussed on hypothesis A1 and on testing variations between *Tithonia* provenances in this respect. Hypotheses A1, A2 and A3 have in common the assumption that *Tithonia*'s P uptake exploits specific pools of organic P. This aspect can be tested by applying a modified Hedley fractionation scheme to soils before and after *Tithonia* growth. Research in Vietnam showed that the P organic amount in the NaOH extract was specifically reduced by *Tephrosia*, in contrast to other fallow species and crops (Hoang et al., 2002).

TABLE 39-8: Hypotheses regarding underlying mechanisms in *Tithonia*'s ability to scavenge for and transfer nutrients.

Hypotheses explaining phenomenon A	Hypotheses explaining phenomenon B
A1. *Tithonia* has efficient mycorrhizal partnerships, allowing it to utilize organic P sources in the soil not accessible by other vegetation.	B1. The beneficial effect of *Tithonia* on subsequent crops in situ is based on stimulated mycorrhizal development of the crops.
A2. *Tithonia* is able to take up additional P from poor soils by an effective rhizosphere modification mechanism, based on the release of organic acids.	B2. *Tithonia* biomass (including the green parts) contains organic acids that when released into the soil, compete for P sorption sites and thus mobilize P in the soil.
A3. *Tithonia* can mobilize organic P sources by the release of phosphatases.	B3. *Tithonia* roots leave a modified rhizosphere behind which allows subsequent crops to benefit from a residual P-mobilizing effect.
–	B4. *Tithonia* biomass contains other substances, which when released into the soil, stimulate crop growth.

Benefits for food crops

Hypotheses B1 and B3 (Table 39-8) do not account for the positive effects of *Tithonia* mulch transfer. Hypothesis B4 is still very general, and would deserve attention if hypothesis B2 was not supported by subsequent experiments. To test hypothesis B2, several approaches could be tried.

- Measuring and identifying the organic acids in fresh *Tithonia* biomass, e.g. in comparison to those of less-effective green manures.
- Testing the impact of the same type and amount of organic acids on P dynamics in a soil, both at soil test and pot experiment levels. As certain organic acids are quickly metabolized by microbes, a 'slow release' effect of litter transfer may be compared to a one-off addition of organics.
- Comparing the effects of *Tithonia* biomass in soils that already contain a lot of organic matter (and in which the additional impact may therefore be small), with those on a soil with a low organic content (where the impacts of adding organics will be high). This could be achieved by comparing topsoil and subsoil from the same site, by comparing profiles in different stages of pedogenesis, or by mild heat treatment of the soils (which is known to increase P sorption, probably by reducing organic-acid loading of P sorption sites).

Hypotheses A1 through B3 can all be represented by modifying parameters in the WaNuLCAS model of water, nutrient and light capture in agroforestry systems (van Noordwijk et al., 2011) (Figure 39-15), and as the hypotheses are not mutually exclusive, it may be relevant to try to set all parameters at a 'guesstimate' level and

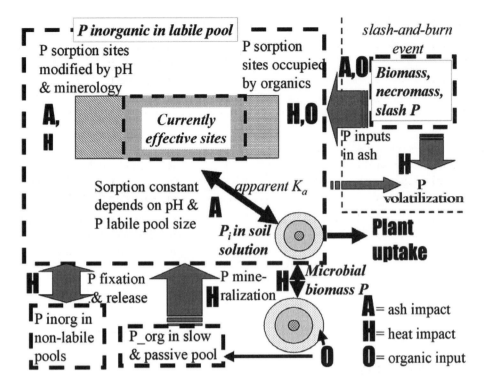

FIGURE 39-15: Conceptual scheme of P pools in the soil as represented in the WaNuLCAS model and potential impacts of ash (A), heat (H) or addition of organics (O).

perform a sensitivity analysis of the model around these values to identify the parameters in greatest need of quantification.

A pot experiment conducted at Lampung subsequent to the studies detailed above tested two soil types, andisols of volcanic origin from East Java and an ultisol from Lampung. There was some evidence of a positive interaction between *Tithonia* root or shoot material and the rate at which rock phosphate solubilizes on the andisol, but the ultisol showed a negative interaction (Figure 39-16). For a maize test crop, the *Tithonia* root material was an additive with rock phosphate (neither positive nor negative interaction) in both soils, but the amount of shoot material was too high to test interactions (Figure 39-17).

Discussion: priority issues for further research

Grierson et al. (2004) reviewed the literature on phosphorus dynamics and mobilization by *Tithonia diversifolia* following the initial enthusiasm generated by Sanchez et al. (1997), and reported that *Tithonia* was superior to other common organic materials as a source of N for maize (Buresh and Niang, 1997; Jama et al., 2000). *Tithonia diversifolia* is a good host to root-knot nematodes, although it does not show the typical root-gall symptoms (Desaeger et al., 2004), so a directly biological explanation of positive fallow effects via pest control seems unlikely.

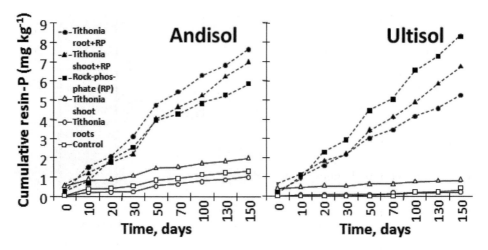

FIGURE 39-16: P availability in the soil (expressed as cumulative extractions of resin P) on an andisol and ultisol for various combinations of *Tithonia* shoot and root material and rock–phosphate.

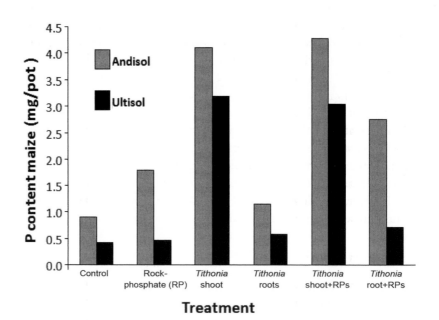

FIGURE 39-17: Maize P content for different combinations of *Tithonia* shoot and root material and rock phosphate on an andisol and an ultisol.

There appeared initially to be synergies between *Tithonia* and added inorganic fertilizers (Gachengo et al., 1999). However, after several years of study and testing, the conclusion emerged that *Tithonia* was neither the unique plant some claimed it to be, nor was it the 'useless myth based only on hype', as claimed by others. In an experiment where *Tithonia* was superior to urea as a source of N, the site's K deficiency was addressed by the organic and not by the inorganic source. However, maize yields were generally higher when *Tithonia* was applied as a green manure than when equivalent inorganic N-P-K fertilizers were used in P-limited soils (Nziguheba et al., 2002). When *Tithonia* manure was applied in conjunction with rock phosphate, Savini (1999) showed that *Tithonia* seemed to depress rather than enhance phosphate-rock solubility, although the effect was temporary (Savini, 1999; Mutuo et al., 1999; Savini et al., 2006).

FIGURE 39-18: In Indonesia, this farmer, near Malang, in East Java, found his goats were fond of *Tithonia* leaves as a fresh forage – adding to the list of farmers' uses for *T. diversifolia.*

Overall, 'daisy fallows' seem to invade slightly better sites, rather than causing an improvement to those sites. Nevertheless, *Tithonia* and other Asteraceae often have large quantities of nutrients in their tissues. Even so, composts derived from *Tithonia* (see Figure 39-6) have a high Ca content, which limits their potential use in promoting the dissolution of phosphate rock, because Ca inhibits the dissolution process (Smithson, 1999; Wong et al., 2004). Yet, Ikerra et al. (2006) found on a chromic acrisol in Tanzania that *Tithonia* increased crop availability of rock phosphate in the second year, probably due the effects of P sorption and Al toxicity in the test site. Oberson et al. (2011) and Chivenga et al. (2011) carried out a meta-analysis of reported field studies of interacting organic and inorganic nutrient sources, and found that interaction effects were the exception rather than the rule. However, on specific sites and with low quantities of fertilizer, some positive interactions had been found. Overall, biological access to recalcitrant inorganic and organic P forms remains the critical issue for further research.

Current understandings

While daisy fallows, and in particular *Tithonia*, are preferred by farmers, the field data give a clear indication of an 'indicator' value and little evidence of soil improvement during the fallow period.[21] *Tithonia* and to a lesser extent, *Chromolaena*, are effective nutrient scavengers, and there is little doubt that they have effective mycorrhizal associations (this was confirmed for the poorly performing *Chromolaena* in the experiment in Lampung). Scrase et al. (2019) provided direct evidence of the effectiveness of mycorrhizal hyphae on *T. diversifolia* roots in P uptake from occluded sources.

Where *Chromolaena* was planted in an experiment just on the edge of its normal occurrence, it did not perform well and was of no value to a subsequent maize crop.

Tithonia leaf litter and, to a lesser extent, that of *Chromolaena*, appears to have a positive effect on the P nutrition of maize on P-limited soils in pot experiments – beyond what can be expected from the P content of the litter itself. The indications suggest that somehow these litters stimulate the mineralization of P and breakdown of soil organic P ('robbing the bank').

The evidence we have in hand on the 'induced P-mineralization hypothesis' is not yet sufficiently clear and further experiments are needed to verify the effect and clarify the underlying mechanisms. These might include a release of organic acids from the litter or a stimulation of soil biological activity (see plate 42 in Coloured Plates section II, this volume), as similar effects (increased mineralization and reduction of microbial biomass at a potentially higher turnover rate and activity) can be obtained when bacterial grazers are introduced into soils.

Current indications are that part of the positive effect of *Tithonia* on crops is based on accelerated breakdown of soil organic resources. This may be welcomed in the short term, as it activates 'non-performing' soil nutrient capital stocks, but it may also give reason for serious concerns over the sustainability and long-term prospects for these systems. If *Tithonia* fallows are viewed as an intermediate stage in the evolution of farming systems towards permanent cropping, accelerated mineralization may help provide the resources to invest in further intensification. Evidence so far indicates that *Tithonia* may actually increase the returns on moderate or small P fertilizer applications.

Acknowledgements

To support and encourage research interest in farmer-developed strategies for intensifying shifting cultivation systems, ICRAF-Southeast Asia coordinates the Indigenous Fallow Management network in Southeast Asia, which links research and development agencies to innovations in the field. The IFM Network has received generous support from the International Development Research Centre (IDRC – Canada). The *Tithonia* survey in Mindanao described in this chapter was funded under the Sustainable Agriculture and Natural Resource Management (SANREM) programme. Special thanks are owed specifically to Delia Delfin and Rolly Yam-oc, for supporting author Cairns as field assistants and interpreters during his work in Bukidnon. They played a large role in the collection of fallow biomass analyzed in this chapter. Back at ICRAF's regional office in Bogor, Indonesia, Ms Subekti Rahayu helped in the statistical analysis of the data.

References

Agbim, N. N. (1987) 'Carbon cycling under *Chromolaena odorata* (L.) K.&R. canopy', *Biological Agriculture and Horticulture* 4, pp.203-212

Ajao, A. A. and Moteetee, A. N. (2017) '*Tithonia diversifolia* (Hemsl) A. Gray. (Asteraceae: Heliantheae), an invasive plant of significant ethnopharmacological importance: A review', *South African Journal of Botany* 113, pp.396-403

Buresh, J. R. and Niang, I. A. (1997) '*Tithonia diversifolia* as a green manure: Awareness, expectations and realities', *Agroforestry Forum* 8, pp.29-30

Cairns, M. F. (1994) 'Comparative analysis of agroecosystems of three case study villages in the buffer zone of Kerinci-Seblat National Park', chapter 2 in Master of Environmental Science dissertation to the Faculty of Environmental Studies, York University, Canada, pp.73-208

Cairns, M. F. (1997a) 'Indigenous fallow management (IFM) in Southeast Asia: New research exploring the promise of farmer-generated technologies to stabilize and intensify stressed swidden systems', paper presented at an international workshop on 'Green-Manure Cover-Crop Systems for Smallholders in Tropical and Subtropical Regions', 6-12 April 1997, Chapeco, Santa Catarina, Brazil

Cairns, M. F. (1997b) 'Fallows, fodder and fences: The critical elements of integrating livestock into swidden systems', paper prepared for presentation at a workshop on 'Upland Farming Systems in the Lao P.D.R.: Problems and Opportunities for Livestock', 19-23 May 1997, Vientiane, Lao P.D.R.

Cairns, M. F. (Ed.) (2007) (ed.) *Voices from the Forest: Integrating Indigenous Knowledge into Sustainable Upland Farming.* Proceedings of regional workshop held in Bogor, Indonesia on June 23 to 27, 1997. Resources for the Future Press, Washington, DC, USA.

Cairns, M. F. (2007a) 'Management of fallows based on *Austroeupatorium inulaefolium* by Minangkabau farmers in Sumatra, Indonesia', in M. F. Cairns (ed.) *Voices from the Forest: Integrating Indigenous Knowledge into Sustainable Upland Farming*, Resources for the Future Press, Washington, DC, pp.153-184

Cairns, M. F. (2007b) 'Conceptualizing indigenous approaches to fallow management: A road map to this volume', in M. F. Cairns (ed.) *Voices from the Forest: Integrating Indigenous Knowledge into Sustainable Upland Farming*, Resources for the Future Press, Washington, DC, pp.16-36

Cairns, M. F. (2015a) 'Ancestral domain and national park protection: A logical union?' in M. F. Cairns (ed.) *Shifting Cultivation and Environmental Change: Indigenous People, Agriculture and Forest Conservation*, Earthscan from Routledge, London, pp.597-634

Cairns, M. F. (ed.) (2015b) *Shifting Cultivation and Environmental Change: Indigenous People, Agriculture and Forest Conservation*, Earthscan from Routledge, London

Capundag, O. S. (1989) 'Effect of different levels of sunflower-leaf meal in the meat quality of broilers', thesis for Bachelor of Science in Agriculture, Central Mindanao University, Musuan, Bukidnon, Philippines

Chivenga, P., Vanlauwe, B. and Six, J. (2011) 'Does the combined application of organic and mineral nutrient sources influence maize productivity? A meta-analysis', *Plant and Soil* 342, pp.1-30

Chung, C. S. (1994) Personal communication between author Cairns and Dr. Chin See Chung, later Director of the Singapore Botanic Gardens

Daguitan, F. M. and Tauli, M. A. (2007) 'Indigenous fallow management systems in selected areas of the Cordillera, the Philippines', in M. F. Cairns (ed.) *Voices from the Forest: Integrating Indigenous Knowledge into Sustainable Upland Farming*, Resources for the Future Press, Washington, DC, pp.673-691

de Foresta, H. (1993) '*Chromolaena odorata*: Calamité ou chance pour l'Afrique Tropicale?' paper presented at the third International Workshop on Biological Control and Management of *Chromolaena odorata*, 15-19 November 1993, Abidjan, Cote d'lvoire

de Foresta, H. and Schwartz, D. (1991) '*Chromolaena odorata* and disturbance of natural succession after shifting cultivation: An example from Mayombe, Congo, Central Africa', in R. Nuniappau and P. Ferrar (eds) *Ecology and Management of* Chromolaena odorata, BIOTROP Special Publication No 44, pp.23-41

Derpsch, R. (1993) 'Descripcion de un sistema de produccion sostenible, basado en la siembra directa, en laderas con 100% de declive, dessarrolado por un pequeno agricultor del Sur de Honduras', paper presented at a Latin-American meeting on 'Plantio Direto Na Pequena Propriedade (Tilling a small property)' 22-26 November 1993, Ponta Crossa, Brazil

Desaeger, J., Rao, M. R. and Bridge, J. (2004) 'Nematodes and other soil-borne pathogens in agroforestry', in M. van Noordwijk, G. Cadisch and C. K. Ong (eds) *Belowground Interactions in Tropical Agroecosystems*, CAB International, Wallingford, UK, pp.262-284

Dove, M. R. (1986) 'The practical reason of weeds in Indonesia: Peasant vs state views of *Imperata* and *Chromolaena*', *Human Ecology* 14(2), pp.163-190

Field, S. P. (1991) '*Chromolaena odorata*: Friend or foe for resource poor farmers?' Chromolaena odorata *Newsletter* 4, pp.4-7

Fungo, B., Lehmann, J., Kalbitz, K., Thiongo, M., Tenywa, M., Okeyo, I. and Neufeldt, H. (2019) 'Ammonia and nitrous oxide emissions from a field Ultisol amended with *Tithonia* green manure, urea, and biochar', *Biology and Fertility of Soils* 55(2), pp.135-148

Gachengo, C. N., Palm, C. A., Jama, B. and Othieno, C. (1999) '*Tithonia* and *Senna* green manures and inorganic fertilizers as phosphorus sources for maize in Western Kenya', *Agroforestry Systems* 44, pp.21-36

Garrity, D. (1994) Personal communication between author Cairns and Dr. Dennis Garrity, then Regional Director of the Southeast Asia Programme, ICRAF, Bogor, Indonesia

Geilfus, F. (1994) 'El arbol al servicio del agricultor: Manual de agroforesteria para el desarrollo rural (The service-tree farmer: An agroforestry manual for rural development)', in *Tomo: Guia de especies* (*Volume: Guide to Species*), second edition, Enda-CATIE, Santa Domingo, Dominican Republic, pp.632-633

George, T. S., Gregory, P. J., Robinson, J. S., Buresh, R. J. and Jama, B. A. (2001) '*Tithonia diversifolia*: Variations in leaf nutrient concentration and implications for biomass transfer', *Agroforestry Systems* 52, pp.199-205

Grierson, P. F., Smithson, P., Nziguheba, G., Radersma, S. and Comerford, N. B. (2004) 'Phosphorus dynamics and mobilization by plants', in M. van Noordwijk, G. Cadisch and C. K. Ong (eds) *Belowground Interactions in Tropical Agroecosystems*, CAB International, Wallingford, UK, pp.127-143

Hairiah, K., Kasniari, D. N., Van Noordwijk, M., de Foresta, H. and Syekhfani, M. S. (1996) 'Soil properties, litter fall, above- and below-ground biomass during a *Chromolaena odorata* fallow', *Agrivita* 19, pp.184-192

Hairiah, K., Van Noordwijk, M. and Cadisch, G. (2000) 'Quantification of biological N2 fixation of hedgerow trees in Northern Lampung', *NJAS-Wageningen Journal of Life Sciences* 48(1), pp.47-59

Handayanto, E. (1994) 'Nitrogen mineralization from legume tree prunings of different quality', PhD dissertation to the University of London, UK

Hansen, P. (1997) Personal communication between author Cairns and Dr. Peter Hansen, then Land-use Policy and Planning Advisor, Policy and Planning Division, Ministry of Agriculture, Thimphu, Bhutan

Hedley, M. J., Stewart, J. W. B. and Chauhan, B. S. (1982) 'Changes in inorganic and organic soil phosphorus fractions induced by cultivation practices and by laboratory incubations', *Soil Science Society of America* 46, pp.970-976

Hedley, M. J., Nye, P. H. and White, R. E. (1983) 'Plant-induced changes in the rhizosphere of rape (*Brassica napus* var. *emerald*) seedlings, IV: The effect of rhizosphere phosphorus status on the pH, phosphatase activity and depletion of soil phosphorus fractions in the rhizosphere and on the cation-anion balance in the plants', *New Phytologist* 95, pp.69-82

Hoang, M. H., Nilsson, S. I., van Noordwijk, M., Phien, T., Olsson, M., Hansson, A. and Svensson, C. (2002) 'Does *Tephrosia candida* as fallow species, hedgerow or mulch improve nutrient cycling and prevent nutrient losses by erosion on slopes in northern Vietnam?' *Agriculture, Ecosystems and Environment* 90, pp.291-304

Hu, G. and Zhang, Z. (2013) 'Allelopathic effects of *Chromolaena odorata* on native and non-native invasive herbs', *Journal of Food, Agriculture and Environment* 11(1), pp.878-882

Ikerra, S. T., Semu, E. and Mrema, J. P. (2006) 'Combining *Tithonia diversifolia* and *minjingu* phosphate rock for improvement of P availability and maize grain yields on a chromic acrisol in Morogoro, Tanzania', *Nutrient Cycling in Agroecosystems* 76, pp.249-260

Jama, B., Palm, C. A., Buresh, R. J., Niang, A., Gachengo, C., Nziguheba, G. and Amadalo, B. (2000) '*Tithonia diversifolia* as a green manure for soil fertility improvement in western Kenya: A review', *Agroforestry Systems* 49, pp.201-221.

Jampasri, K., Saeng-Ngam, S., Larpkern, P., Jantasorn, A. and Kruatrachue, M. (2021) 'Phytoremediation potential of *Chromolaena odorata*, *Impatiens patula*, and *Gynura pseudochina* grown in cadmium-polluted soils', *International Journal of Phytoremediation* 23, pp.1061-1066

Kato-Noguchi, H. (2020) 'Involvement of allelopathy in the invasive potential of *Tithonia diversifolia*', *Plants* 9(6), p.766

Mutuo, P. K., Smithson, P. C., Buresh, R. J. and Okalebo, R. J. (1999) 'Comparison of phosphate rock and triple superphosphate on a phosphorus deficient Kenyan soil', *Communications in Soil Science and Plant Analysis* 30(7-8), pp.1091-1103

Nagarajah, S. and Nizar, B. M. (1982) 'Wild sunflower as a green manure for rice in the mid-country wet zone', *Tropical Agriculturist* 138, pp.68-80

Nziguheba, G., Merckx, R., Palm, C. A. and Mutuo, P. (2002) 'Combining *Tithonia diversifolia* and fertilizers for maize production in phosphorus-deficient soil in Kenya', *Agroforestry Systems* 55, pp.165-174

Obatolu, C. R. and Agboola, A. A. (1993) *The Potential of Siam weed* (Chromolaena odorata) *as a Source of Organic Matter for Soils in the Humid Tropics, monograph*, Soils and Chemistry Division, Cocoa Research Institute of Nigeria, Ibadan, Nigeria

Oberson, A., Pypers, P., Bünemann, E. K. and Frossard, E. (2011) 'Management impacts on biological phosphorus cycling in cropped soils', *Soil Biology* 100, pp.431-458

Odediri, J. A. and Oloidi, F. F. (2014) 'Feeding wild sunflower (*Tithonia diversifolia* Hemsl., A. Gray) to West African dwarf goats as a dry season forage supplement', *World Journal of Agricultural Research* 2(6), pp.280-284

Olabode, O. S., Sola, O., Akanbi, W. B., Adesina, G. O. and Babajide, P. A. (2007) 'Evaluation of *Tithonia diversifolia* (Hemsl.) A.Gray for soil improvement', *World Journal of Agricultural Sciences* 3(4), pp.503-507

Oyen, L. (1995) 'Aggressive colonizers work for the farmers', *ILEIA Newsletter* 11(3), pp.10-11

Oyerinde, R. O., Otusanya, O. O. and Akpor, O. B. (2009) 'Allelopathic effect of *Tithonia diversifolia* on the germination, growth and chlorophyll contents of maize (*Zea mays* L.)', *Scientific Research and Essays* 4(12), pp.1553-1558

Pandeya, C. N. (1995) 'We love and protect it', *ILEIA Newsletter* 11(3), p.8

Partey, S.T., Quashie-Sam, S. J., Thevathasan, N.V. and Gordon, A. M. (2011) 'Decomposition and nutrient release patterns of the leaf biomass of the wild sunflower (*Tithonia diversifolia*): A comparative study with four leguminous agroforestry species', *Agroforestry Systems* 81(2), pp.123-134

Partey, S. T., Preziosi, R. F. and Robson, G. D. (2014) 'Improving maize residue use in soil fertility restoration by mixing with residues of low C-to-N ratio: Effects on C and N mineralization and soil microbial biomass', *Journal of Soil Science and Plant Nutrition* 14(3), pp.518-531

Potter, L. and Lee, J. (2007) 'Managing *Imperata* grasslands in Indonesia and Laos', in M. F. Cairns (ed.) *Voices from the Forest: Integrating Indigenous Knowledge into Sustainable Upland Farming*, Resources for the Future Press, Washington, DC, pp.113-121

Ramakrishnan, P. S. (1993) *Shifting Agriculture and Sustainable Development: An Interdisciplinary Study from North-Eastern India*, Oxford University Press, Delhi

Rios, C. (1994) 'Botton de *Tithonia diversifolia* (Hemsl.) Gray', in *Arboles y Arbustos Forrajeros Utilizados en la Alimentacion Animal*, Foundation CIPAV, Cali, Colombia, pp.115-126

Rios, C. and Salazar, A. (1995) '*Tithonia diversifolia* (Hemsl.) Gray: A protein source for the tropics', *Livestock Research for Rural Development* 1995, 6(3)

Roder, W., Keoboualapha, B. And Manivanh, V. (1995) 'Teak (*Tectona grandis*), fruit trees and other perennials used by hill farmers of northern Laos', *Agroforestry Systems* 29, pp.47-60

Roder, W., Maniphone, S., Keoboualapha, B. and Fahrney, K. (2007) 'Fallow improvement with *Chromolaena odorata* in upland rice systems of northern Laos', in M. F. Cairns (ed.) *Voices from the Forest: Integrating Indigenous Knowledge into Sustainable Upland Farming*, Resources for the Future Press, Washington, DC, pp.142-152

Sanchez, P. A., Shepherd, K. D., Soule, M. J., Place, F. M., Buresh, R. J. and Izac, A. M. (1997) 'Soil fertility replenishment in Africa: An investment in natural resource capital', in R. J. Buresh, P. A. Sanchez and F. Calhoun (eds) *Replenishing Soil Fertility in Africa*, Soil Science Society of America special publication 51, ASA and SSSA Madison, WI, pp.1-46

Savini, I. (1999) 'The effect of organic and inorganic amendments on phosphorus release and availability from two phosphate rocks and triple superphosphate in phosphorus-fixing soils, MSc thesis to the University of Nairobi, Nairobi, Kenya

Savini, I., Smithson, P. C., Karanja, N. K. and Yamasaki, H. (2006) 'Influence of *Tithonia diversifolia* and triple superphosphate on dissolution and effectiveness of phosphate rock in acidic soil', *Journal of Plant Nutrition and Soil Science* 169(5), pp.593-604

Scrase, F. M., Sinclair, F. L., Farrar, J. F., Pavinato, P. S. and Jones, D. L. (2019) 'Mycorrhizas improve the absorption of non-available phosphorus by the green manure *Tithonia diversifolia* in poor soils', *Rhizosphere* 9, pp.27-33

Slaats, J. J. P. (1993) 'The use of *Chromolaena odorata* as fallow in a semi-permanent cropping system in S.W. Cote d'Ivoire', paper presented at the Third International Workshop on Biological Control and Management of *Chromolaena odorata*, 15-19 November 1993, Abidjan, Cote d'Ivoire

Smithson, P. C. (1999) 'Interactions of organic materials with phosphate rocks and triple superphosphate', *Agroforestry Forum* 9, pp.37-40

Stoutjesdijk, J. A. J. H. (1935) '*Eupatorium pallescens* DC op Sumatra's westkust' (*Eupatorium pallescens* DC on the West Coast of Sumatra), *Tectona* 28, pp.919-926.

Susana, I. V. R. and Tangendjaja, B. (1988) 'Leaf protein concentrate (LPC) from *Calopogonium caeruleum* and *Tithonia diversifolia*', in *Proceedings of the Second National Seminar on Animal Husbandry and Farmers' Forum*, 18-20 July 1988, Bogor, Indonesia, pp.190-202

Suson, P. D., Garrity, D. P. and Lasco, R. D. (2007) 'Pruned-tree hedgerow fallow systems in Mindanao, the Philippines', in M. F. Cairns (ed.) *Voices from the Forest: Integrating Indigenous Knowledge into Sustainable Upland Farming*, Resources for the Future Press, Washington, DC, pp.403-413

Tagne, A. M., Marino, F. and Cosentino, M. (2018) '*Tithonia diversifolia* (Hemsl.) A. Gray as a medicinal plant: A comprehensive review of its ethnopharmacology, phytochemistry, pharmacotoxicology and clinical relevance', *Journal of Ethnopharmacology* 220, pp.94-116

Tanee, F. B. G. and Jude, K. (2020) 'Stimulating the rate of phytoremediation of *Chromolaena odorata* in crude oil contaminated soil using inorganic urea', *International Journal of Plant and Soil Science* 32, pp.13-25

Tshinyangu, K. A., Mutombo, T. J. M., Kayombo, M. A., Nkongolo, M. M., Yalombe, N. G. and Cibanda, M. J. (2017) 'Effet comparé de *Chromolaena odorata* King et HE Robins, et *Tithonia diversifolia* A. Gray sur la culture du Maïs (Zea mays L) à Mbujimayi (R.D. Congo)', *Journal of Applied Biosciences* 112, pp.10996-11001

van de Goor, G. A. W. (1953) 'An investigation of the relative value of various legumes and non-leguminous plants as green manure', *Teknik Pertanian* 3

van Noordwijk, M. (1999) 'Productivity of intensified crop fallow rotations in the Trenbath model', *Agroforestry Systems* 47, pp.223-237

van Noordwijk, M., Sitompul, S. M., Hairiah, K., Listyarini, E. and Syekhfani, M. S. (1995) 'Nitrogen supply from rotational or spatially zoned inclusion of Leguminosae for sustainable maize production on an acid soil in Indonesia', in R.A. Date, N. J. Grundon, G.E. Rayment and M.E. Probert (eds) *Plant-Soil Interactions at Low pH: Principles and Management*, Springer, Dordrecht, pp.779-784

van Noordwijk, M., Lusiana, B., Khasanah, N. and Mulia, R., (2011) *WaNuLCAS version 4.0: Background on a Model of Water Nutrient and Light Capture in Agroforestry Systems*, World Agroforestry Centre (ICRAF), Bogor, Indonesia

van Noordwijk, M., Bizard, V., Wangpakapattanawong, P., Tata, H. L., Villamor, G. B. and Leimona, B. (2014) 'Tree cover transitions and food security in Southeast Asia', *Global Food Security* 3(3-4), pp.200-208

Verliere, G. (1966) 'Valeur fertilisante de deux plantes utilisees dans les essais de paillage du cafaier: *Tithonia diversifolia* et *Flemingia congesta*', *The Café Cacao* X(3)

Voss, J. (1995) Personal communication between author Cairns and Dr. Joachim Voss, then Manager, Research Division, International Development Research Centre (IDRC), Ottawa, Canada

Wambui, C. C., Abdulrazak, S. A. and Noordin, Q. (2006) 'The effect of supplementing urea treated maize stover with *Tithonia, Calliandra* and *Sesbania* to growing goats', *Development* 18, pp.5

Wilson, G. and Kang, B. T. (1981) 'Developing stable and productive biological cropping systems for the humid tropics', in B. Stonehouse (ed.) *Biological Husbandry: A Scientific Approach to Organic Farming*, Butterworth, London, pp.193-203

Wong, M. T. F., Hairiah, K. and Alegre, J. (2004) 'Managing soil acidity and aluminium toxicity in tree-based agroecosystems', in M. van Noordwijk, G. Cadisch and C. K. Ong (eds) *Belowground Interactions in Tropical Agroecosystems*, CAB International, Wallingford, UK, pp.143-156

Yuksel, N., Acetpah, A., and Imo (P. Burgers (ed.)) (1999) *The Amarasi Model: An Example of Indigenous Natural Resource Management in Timor, Indonesia*, Indigenous Fallow Management Network, Southeast Asia Programme, World Agroforestry (ICRAF), Bogor, Indonesia

Zachariades, C., Strathie-Korrûbel, L. W. and Kluge, R. L. (1999) 'The South African programme on the biological control of *Chromolaena odorata* (L.) King & Robinson (Asteraceae) using insects', *African Entomology Memoir* 1, pp.89-102

Notes:

1. This chapter is an updated version of a paper prepared for a (UK) Department for International Development Forestry Research Programme planning meeting on *Tithonia* at the International Center for Tropical Agriculture (CIAT), in Cali, Colombia on 2-4 September 1998. The original paper was authored by Malcolm Cairns, Meine van Noordwijk, Parwi, Eko Handayanto, Sugeng Priyono, Kurniatun Hairiah and Dennis P. Garrity. Authors Cairns, van Noordwijk and Garrity represented ICRAF-Southeast Asia and the others, Brawijaya University, Malang, Indonesia.

2. Forbs include biennial and perennial plants that are not grasses, sedges or rushes, and have a storage root for survival of adverse seasons when they have little or no woody tissues surviving aboveground.

3. The previous family name, Compositae, is still in use.

4. *Eupatorium odoratum* is a synonym of *Chromolaena odorata*.

5. *Eupatorium inulaefolium* and *Eupatorium pallescens* are synonyms of *Austroeupatorium inulaefolium*.

6. *Ageratum conyzoides* is another member of the Asteraceae family that is widely appreciated by shifting cultivators as a soil-fertility indicator and a cover crop that establishes rapidly, protecting vulnerable soils after the harvest of swidden crops (author's field notes; Oyen, 1995; Pandeya, 1995). However, it is reported to act as an intermediate host for the fungal rice blast disease, and may be problematic in areas where upland rice is grown as a staple food (Roder et al., 2007).

   Work in northeast India shows that *Mikania micrantha*, yet another Asteraceae, is also an important component of early fallow successions and plays a similar role in nutrient sequestration, particularly potassium (K) (Ramakrishnan, 1993, p.260). *Mikania* species are aggressive creepers, however, and become a major weed problem when farmers attempt enrichment planting of economic trees into swidden fallows. Even trees several metres tall can be pulled to the ground and smothered by *Mikania*.

   The semi-woody Asteraceae, *Blumea balsamifera*, is reported to produce an abundance of wind-dispersed seeds and colonize swidden fallows in almost pure stands in Borneo (Chung, 1994). It remains unknown whether *B. balsamifera* has similar properties to the other Asteraceae species highlighted in this paper that would qualify it as an efficient fallow species.

7. Interestingly, juice extracted from all three species is often used in farmers' recipes for botanical insecticides. The underlying reasoning is that their notable freedom from pest attack is evidence that they contain compounds with insect-repelling properties.

8. A popular notion suggests that incorporation of indigenous flora into managed fallows will ensure a more robust and sustainable system and conversely, exotic species will be more vulnerable to ecological collapse. As a point of argument, Asteraceae originating from Central and South America appear to prosper in Southeast Asian environments precisely because they were introduced and there is consequently an absence of insects, diseases and herbivores that would otherwise have co-evolved with them.

9. It is noteworthy that in this dataset, there was virtually no nitrogen accumulation in either the *Imperata* or *Pteridophyta* fallows. This tends to substantiate farmers' claims that if fallows are dominated by fern or *Imperata* successions, then regardless of the fallow length, the land will remain infertile (Cairns, 2007a, 2015a).

10. Work in Laos indicates that *C. odorata* improves rice yields, in part by displacing *Ageratum conyzoides*, which acts as an intermediate host for the root-knot nematode (*Meloidogine graminicola*). Therefore, by reducing the *A. conyzoides* population, the nematode density is automatically diminished (Roder et al., 2007).

11. In Bukidnon province, Mindanao, Philippines, vegetable growers have reported marked decreases in disease problems after short *T. diversifolia* fallows – often only two to three months in duration – compared to alternative floristically-mixed fallow compositions of similar length (author's field notes).

12. On a landscape level, the resulting mosaic of *T. diversifolia* fallows and boundary hedgerows dispersed throughout cropped hillsides may play an important role in acting as filter strips or sinks, trapping soil and nutrients washed away from cropped areas and thus minimizing erosion, as is evident, for example, in northern Mindanao.

13. This is essentially the same concept as that described by Wilson and Kang in 1981 (p.197): 'In the area around Ibadun, *Gliricidia sepium*, established from stakes used for yam support, has become the recognized fallow species. Farmers in the area claim that only two years' resting under *Gliricidia* fallow is required for restoring soil productivity to economic levels'.

14. This is the reverse strategy of farmers on the islands of Muna and Bali, in Indonesia, who carefully nurture *Imperata cylindrica* in fallows for use in thatching. They describe annual burning, minimizing tillage and weeding out invading shrubs as important steps in maintaining the *Imperata* 'crop' and preventing its invasion by *C. odorata* 'weeds' (Potter and Lee, 2007).

15. This is reflected in the visible shift of weed populations from C3 herbs to C4 grasses as soil-nutrient status declines during the second or third cropping year in swidden plots.

16. While they get increasingly woody with age, the rapid decomposition and nutrient release by Asteraceae species makes them strong candidates for 'slash-and-mulch' systems (*in situ* use), offering superior soil protection, accumulation of organic matter and reduced nutrient loss due to volatilization, runoff and leaching. This avoids the less efficient nutrient flush associated with burning, and may replace it with a better synchrony between crop nutrient demand and mulch nutrient release. Some farmers express fears that the convenient habitat provided by mulch might encourage an increase in rats and other pests. For example, farmers practising the *Leucaena leucocephala*-based Amarasi system in West Timor, Indonesia, insist that burning is necessary to control a snail pest (Yuksel et al., 1999). Derpsch (1993) describes an innovative system in southern Honduras in which *T. diversifolia* fallow biomass is slashed-and-mulched in the production of common beans (*Phaseolus vulgaris*) and maize. Some farmers in West Sumatra, Indonesia, have developed mulching practices to manage *A. inulaefolium* biomass, as a deliberate strategy to avoid burning because it encourages intrusion by fire-adapted *Imperata* and ferns (Cairns, 2007a).

    The ability of Asteraceae species to accumulate large quantities of nutrients also makes them well suited for the *ex situ* application of biomass-transfer technologies, in which marginal areas of a farm may be converted into special green-manure source plots and the biomass used to manure cropped areas. Roder et al. (2007) refer to such practices with *C. odorata* in Laos, while Daguitan and Tauli (2007) reported widespread transfer of *T. diversifolia* biomass in the Cordillera Region, Philippines.

17. Farmers interviewed in Bukidnon Province, the Philippines, contended that after planting stem cuttings into cogon (*Imperata cylindrica*) (see plate 18 in the Coloured Plates section of *Voices from the Forest* (Cairns, 2007)), *T. diversifolia* shaded out *Imperata* after the first year, and by the end of the second year, the soil had been sufficiently rejuvenated that the fallow could be reopened and a good crop grown without nutrient inputs. Others described such improvements in soil physical properties that ploughing was no longer necessary and soil erosion was reduced.

18. Daguitan and Tauli (2007) reported the equally purposeful incorporation of *T. diversifolia* into permanent cropping systems in Northern Luzon. Bontok farmers are known to plant it along streams and river banks on the upper reaches of watersheds. The harvested biomass is then placed in

the water for swift decomposition and the nutrients carried downstream to irrigated rice fields. It is popularly planted in hedgerows, often near irrigated rice fields or along the boundaries of sweet potato fields for use as a green manure or a component of compost (see Figure 39-6). It is so highly valued as a green manure that disputes have been known to break out between neighbouring villages over harvesting rights to *T. diversifolia* biomass growing naturally on state land (author's field notes).

In addition to its agronomic benefits, *T. diversifolia* is commonly protected in Northern Luzon as an important nectar source in apiculture. In a study from northern Thailand, Trakansuphakon and Ferrari (ch. 33, this volume) similarly found that Karen shifting cultivators who were using their fallows as 'bee pastures' to produce honey also identified *T. diversifolia* as one of their key sources of nectar. Farmers in northern Luzon also describe broadcasting its leaves in ponds for tilapia to browse (Voss, 1995). Studies in China and Latin America (Geilfus, 1994; Rios, 1994; Rios and Salazar, 1995) confirm the fodder potential of *T. diversifolia* for sheep, goats (see Figure 39-18), rabbits and even pigs, as well as its utility in honey production. Further feeding trials in Indonesia (Susana and Tangendjaja, 1988) and the Philippines (Capundag, 1989) showed promising results from incorporating *T. diversifolia* leaf meal as a component of poultry feed.

19. For farmers searching for a promising site to open a swidden, the main criterion is that it is sufficiently fertile to yield a good crop. As an important part of their indigenous knowledge, shifting cultivators are known to be acutely aware of which fallow plants indicate soil fertility, and search for them when choosing a site to open. The key issue to them is the soil fertility of the site. The question of whether the *Tithonia* is acting as an actual soil improver or just an indicator is largely immaterial to them, at least in the short term.

    Conversely, they try to avoid sites that are colonized by *Imperata*, ferns or other plants that they associate with degraded soils (see Figure 39-4).

20. The lead author confirms that examination of the roots of many *Peltophorum dasyrrhachis* trees has never found any nodulation, and that followup work with 15N isotopes confirmed that it derives all its nitrogen from the soil (Hairiah et al., 2000). It is, so to speak, from the poor branch of the Leguminosae family.

21. Beyond the fertility question, there are of course other reasons why shifting cultivators value *T. diversifolia* as a fallow species. *Tithonia*'s large leaves create a dense fallow canopy that will shade out any light-demanding weeds and assure the farmer of reduced weeding labour in the next cropping phase.

    In addition to good fertility and fewer weeds, farmers also prefer the softer humus soils that they say they usually find below *Tithonia* fallows (see plate 42 in Coloured Plates section II, this volume). In contrast to the hard and dry soils typically found under *Imperata* or fern fallows, a *T. diversifolia* fallow offers soft soils that can be dibbled directly without any need for tillage.

40

THE EFFECTIVENESS OF WILD SUNFLOWER

As a fallow species in *uma* cultivation

*Florence Daguitan, Montañosa Research and Development Center, and Tomiangan Unity for Progress and Action**

Introduction

The Cordillera Administrative Region in the Philippines is the watershed of seven major river systems in the northern part of the island of Luzon (Figure 40-1). It supports a substantial indigenous population. However, as early as the start of the 20th century, extractive and exploitative development projects, exemplified by large-scale mining and logging and the advancement of commercial vegetable production, began to erode the livelihood of the people, leaving the region in a precarious state.

Over the years, social, economic and political pressures have taken their toll on indigenous subsistence agriculture in the region, with the result that the mountain ecosystem has become inherently fragile. In the past, swidden (*uma*) farmers left their fields fallow for five to seven years, and even as long as 10 years. This period, which is appropriate for forest recovery, has shrunk dramatically because of the decreasing area available for their practice of shifting cultivation. Now, farmers' swidden fields lie fallow for just two to four years, and relentless population pressure is expected to exacerbate the region's ecological imbalance in the near future.

The study on which this chapter is based was an attempt to contribute to the stabilization of the Cordillera region's *uma* cultivation system by proposing a fallow crop – wild sunflower (*Tithonia diversifolia*) – as a species capable of enhancing the functions of the fallow, thus making the brief fallow period more effective in achieving soil recovery.

---

* FLORENCE DAGUITAN is Program Coordinator for the Indigenous Peoples and Biodiversity Program Tebtebba, Baguio City, Philippines; MONTAÑOSA RESEARCH AND DEVELOPMENT CENTER (MRDC) is a non-governmental organization based in Sagada, Mountain province, Philippines, with a principal aim of promoting sustainable agriculture and other appropriate technologies; and TOMIANGAN UNITY FOR PROGRESS AND ACTION (TUPAC) is a local people's organization based at Tomiangan, in Kalinga province. Philippines.

FIGURE 40-1: The Cordillera Administrative Region in the Philippines, showing locations for the studies discussed in this chapter.

T. diversifolia grows abundantly in the Cordillera region. In Mountain Province, one of the region's six provinces, it is widely used as a green fertilizer to support almost all kinds of crops. It is intentionally maintained on the peripheries of rice fields and swiddens as a source of green manure.

The study was originally proposed by a people's organization, Tomiangan Unity for Progress and Action (TUPAC), whose members in Kalinga province were prepared to supply the labour required. They felt compelled to pursue the study because they were aware that Bontok people living in the Cordillera region, along with other mountain communities, commonly used wild sunflower as a fertilizer, and the more industrious of them also planted it for use as a green manure.

Preparations for the trial involved the collection of sunflower cuttings for the experimental plots. As the materials were collected, some of the participant farmers took the opportunity to gather planting materials for their own farms. Some announced plans to plant sunflower in idle land near their rice fields and swiddens.

Two farmers who helped to collect sunflower cuttings for the trial later used sunflower biomass as the main fertilizer in their rice fields, and did away with inorganic fertilizers. Observers claimed they harvested as much rice as those who used commercial inorganic fertilizer.

Scope and limitations of the study

The project aimed to determine the effectiveness of wild sunflower as a fallow crop and as a green manure to (a) suppress the noxious grass *Imperata cylindrica*, the presence of which indicates poor soil quality; and (b) to improve crop performance.

Three members of TUPAC were tasked with planting and maintaining the plots, but more than 15 people took part in all of the other processes, while others became regular observers and actively participated in discussions.

This chapter is based solely on that research related to the effectiveness of wild sunflower as a fallow crop, to improve crop performance. This part of the project had a closer focus on the following objectives:

FIGURE 40-2: The vibrant yellow blooms of wild sunflower (*Tithonia diversifolia*) can transform whole mountainsides (see plates 40, 41 and 42 in Coloured Plates section II, this volume. See also Figure 40-3 for an example of the ground-covering capacity of *T. diversifolia*).

- to determine the capacity of wild sunflower as a fallow crop to improve the biophysical and nutrient properties of the soil;
- to determine the capacity of wild sunflower as a fallow crop to increase crop yields, using bush *sitao*, also known as yard-long beans (*Vigna unguiculata* subsp. *sesquipedalis*) as a test crop;
- to determine the optimum planting distance of sunflower vis-à-vis labour input; and
- to make recommendations with regard to the use of sunflower in developing the *uma* system.

The experiment had the following limitations:

- A plan to compare the growth and yield of the test crop and soil improvement after a one-year fallow did not materialize. The experimental plots were planted after two years of fallow.
- Soil analyses were conducted on samples taken (a) before the establishment of wild sunflower as fallow plants; (b) after more than a year of fallow growth; and (c) just before the wild sunflower was slashed and burnt. There was no soil analysis after the slashed biomass had been burnt.
- The study did not benefit from a review of the literature for reference to the use of *T. diversifolia* as a fallow crop. There were many studies related to the use of wild sunflower as a fertilizer, focusing on its use as a green manure or compost material, but none were found on the use of sunflower as a fallow crop.

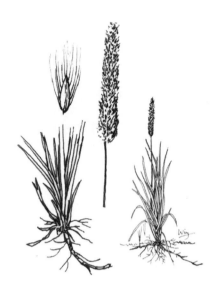

Imperata cylindrica ((L.) Raeusch. [Poaceae]

This troublesome weed has motivated research into the ability of species such as *Tithonia diversifolia* to overcome its dominance. Locally known as *cogon*, it establishes dense swathes on a tangled mat of tough rhizomes, typically on soil depleted of nutrients, and produces countless wind-dispersed seeds. A highly flammable species, it feeds the flames of seasonal fires that kill other plants, after which it rises from its rhizomes to reassert its dominance.

Procedures and methods

A site measuring 2250 square metres was prepared and divided into three blocks. Each block was subdivided into five plots, each measuring 10 m x 15 m. The experiment used randomized block design. Treatments are shown in Table 40-1.

Soil samples were collected for laboratory analysis three times during the experiment. The first was before the wild sunflower cuttings were planted (November 1998); the second in May 2000 (but was incomplete) and the third in June 2001.

A total of 2250 wild sunflower cuttings were collected, each about two feet long. Of these, 750 were planted in plots with one–metre spacing (D1) and 1500 were planted in plots with 50 cm spacing (D2).

The cuttings were planted simply by making a hole in the soil with a crowbar (used as a simple digging tool) and burying them four to six inches deep. Due to a long drought, the majority of these first cuttings did not survive, and the plots were replanted six months later, in May 1999.

The sunflower plants were pruned in May 2000. One year later, in May 2001, the sunflowers and other weeds were slashed, dried and burnt. *Sitao* (*Vigna unguiculata* subsp. *Sesquipedalis*) seeds were planted at a rate of two seeds per hill on June 10 and harvested on August 14, 2001. Twice during the vegetative stage of crop development, the sunflower regrowth was slashed and laid on the ground.

Data gathered included (a) changes in the soil; (b) labour input; and (c) *sitao* yield.

TABLE 40-1: Plot identity and treatment.

Plot code	Treatment of T. diversifolia
T0	No activity, no sunflower planted
D1F2	Sunflower planted at one-metre spacing; two years of fallow
D2F2	Sunflower planted at 50 cm spacing; two years of fallow

Note: T0 = control.

Vigna unguiculata subsp. *sesquipedalis* (L.) Verdc. [Leguminosae]

This vigorous annual vine is known by many names, including *sitao* in the Philippines and asparagus bean, yard-long beans or snake beans elsewhere. The long immature pods are used in much the same way as common beans and are highly nutritious. The plants are also able to fix nitrogen and their tangled foliage is often the basis for a green manure cover crop that protects the soil and supplies valuable fodder for livestock.

Results and discussion

The experimental site

The experiment was conducted in *barangay* Dupag, in the Tabuk municipality of Kalinga province (see Figure 40-1).

The site had been used from the 1960s to the late 1970s for shifting cultivation, and had been planted with cassava, *camote* (Sweet potato, *Ipomoea batatas*), legumes (especially black beans), squash and other vegetables. It was very accessible to a nearby house and had been heavily and continually cropped. There had been no fallowing of the site in all of the time it was used as a swidden.

In 1984 it had been abandoned as a *uma* site and converted to pastureland, but it was abandoned after more than 10 years because of overgrazing. At the time it was opened as the experimental site for this study, it was dominated by kikuyu grass (*Pennisetum clandestinum*), a few patches of *cogon* (*Imperata cylindrica*), *makahiya* (*Mimosa pudica*) and a few stands of *Lantana camara*.

Analysis of soil samples collected in November 1998, before the wild sunflower was planted, revealed the soil characteristics shown in Table 40-2.

TABLE 40-2: Soil characteristics before planting *T. diversifolia*.

Parameter	Result
Soil Texture	Loam
OM Content (%)	1.66
Soil pH	5.7
Water Retention Capacity	62.63

Note: OM = organic matter.

In May 2001, all sunflowers and weeds on the plots were cleared and burnt, and on June 2001 the plots were planted to maize, *sitao* and vegetables. Traditional processes of *uma* cultivation were then followed, except for uniform planting of sitao and maize, which were supposed to serve as test crops, i.e. the same spacing per hill and same number of seeds planted in one plot. However, the maize crop failed in some plots, so it was not considered in the final analysis.

Changes in the soil

Water-holding capacity

As shown in Table 40-3, there was a significant increase in moisture-retention capacity in D1F2, from 62.63 to 69.3, an increase of 6.67 points. Importantly, while the sunflower fallow on plot D2F2 had twice the volume of that on D1F2, this difference was not reflected in increased water-retention capacity. This capacity in D2F2 was

Pennisetum clandestinum Hochst. ex Chiov. [Poaceae]

Kikuyu grass is a native of Kenya that has become naturalized in many countries and is widely used as a lawn and golf-course species in Asia, Australia, The United States and the Pacific. It uses rhizomes below the ground and long runners on the soil surface to rapidly form dense mats that suppress other species and choke ponds and waterways. It is regarded as an invasive weed in many of its naturalized territories.

only 0.11 points higher that D1F2, and this suggested that the treatment of D1F2, i.e. sunflower planted at one-metre spacing, was most ideal when the labour requirement for the closer spacing of D2F2 was taken into account.

Organic-matter content

After two years of fallow, there was a significant difference in organic-matter (OM) content between the control plot (T0), and the plots planted to wild sunflower. The control plot showed an increase of only 0.49% in OM content. The highest

TABLE 40-3: Changes in moisture-retention capacity.

	November 1998	*August 2001*	*Difference*
T0	62.63	64.08	1.44
D1F2	62.63	69.30	6.67
D2F2	62.63	69.41	6.78

OM-content change was an increase of 1.47% in those plots planted to sunflower with 50 cm spacing, followed by 1.33% in plots with one-metre spacing (Table 40-4). There was not much difference between the two wild-sunflower treatments in terms of OM-content increase. The result of the experiment upheld the findings of Bulwayan (1993), that use of wild sunflower has a significant effect on the OM content of the soil.

FIGURE 40-3: Very shortly after this field was left fallow, *T. diversifolia* was already engulfing the abandoned field hut and rejuvenating the field for its next round of cultivation. While this chapter refers to the north of the Philippines, this photo was taken in the Manupali river watershed in Bukidnon province, Mindanao, in the south.

Photo: Malcolm Cairns, 1995.

Soil pH

As shown in Table 40-5, the control plot became more acidic after two years. It registered a pH of 5.3, falling from the original 5.7. Both plots planted to sunflower retained a more favourable pH with a slight increase of 0.1. The decrease in the control plot may be attributed to the leaching of elements in the soil, making it more acidic. The mere presence of sunflower, with its numerous rootlets, and the increase in OM content may have prevented or minimized leaching, or the organic matter may have replaced leached elements, maintaining a more favourable pH for crop growth.

TABLE 40-4: Changes in organic-matter content.

Field identity	Organic-matter (OM) content (%)		Difference
	November 1998	August 2001	
T0	1.66	2.15	+ .49
D1F2	1.66	2.99	+ 1.33
D2F2	1.66	3.13	+ 1.47

TABLE 40-5: Changes in soil pH.

Field identity	Results of pH analysis		Difference
	November 1998	August 2001	
T0	5.7	5.3	– 0.4
D1F2	5.7	5.8	+ 0.1
D2F2	5.7	5.8	+ 0.1

Changes in trace elements

The trace elements copper (Cu), zinc (Zn), iron (Fe) and manganese (Mn) all increased after two years of fallow beneath wild sunflower.

However, as shown in Table 40-6, changes in copper and zinc were not consistent. Copper showed a slight increase in D1F2, but decreased by a similar figure in D2F2. Zinc decreased in both the second year of fallow and in the overall trial in D1F2, by -0.44 and -0.33 parts-per-million (ppm) respectively. In D2F2 there was a marked decrease of 1.38 in zinc in the second year of fallow, but over the entire trial D2F2's zinc level rose by an even larger figure to finally show an increase of 0.15 ppm over the original analysis.

After two years of fallow, there was a consistent increase in iron, rising by 46.85 ppm in D1F2 and by 46.85 ppm in D2F2.

The trace element manganese increased consistently in D2F2 – by 7.19 ppm in the second year of fallow and by 7.76 ppm overall – and in D1F2, where it rose by 5.42 ppm over the entire trial period.

Effectiveness of sunflower as a fallow crop in terms of production

Average number of bush sitao pods and their weight

The spacing between the wild sunflower plants had a slight influence over the number of pods produced by the bush sitao crop. The highest average number of pods (256 per square metre) was recorded in plots where sunflower had been planted with a 50 cm spacing (D2F2). In D1F2, where the sunflowers were planted with a spacing of

TABLE 40-6: Changes in trace elements (parts per million).

Treatment	Year 1	Year 2	Year 3	Difference between: Year 2 and Year 3	Year 1 and Year 3
			Copper		
T0	2.39	2.27	2.99	+ 0.72	+ 0.60
D1F2	2.39	2.27	2.99	+0.72	+ 0.60
D2F2	2.39	2.82	1.85	- 0.97	- 0.54
			Zinc		
T0	1.35	1.46	1.02	- 0.44	- 0.33
D1F2	1.35	1.46	1.02	+ 0.44	+ 0.33
D2F2	1.35	2.88	1.50	- 1.38	+ 0.15
			Iron		
T0	61.75	83.41	108.6	+ 25.19	+ 46.85
D1F2	61.75	83.41	108.6	+ 25.19	+ 46.85
D2F2	61.75	88.17	129.1	+ 40.93	+ 67.38
			Manganese		
T0	58.23	52.83	63.65	+ 10.82	+ 5.42
D1F2	58.23	52.83	63.65	+ 10.82	+ 5.42
D2F2	58.23	58.80	65.99	+ 7.19	+ 7.76

one metre, the average harvest was 240 pods per square metre. The control plot (T0) produced the lowest average – 208 pods per square metre. This finding showed that the closer spacing of sunflowers produced a slight increase in the number of pods harvested per square metre.

Table 40-7 also reveals a significant difference in the weight of bush *sitao* pods harvested from the former wild sunflower plots. Plants from D2F2 produced the highest mean weight of 689.9 gm per 100 pods, followed by a mean weight of 603.7 gm per 100 pods from D1F2. The control (T0) yielded the lowest mean weight of 431.2 gm per 100 pods.

Per-hectare yield

On a one-hectare basis, the highest mean weight of 17,661.4 kg of bush *sitao* was taken from D2F2, followed by D1F2, with a mean harvest weight of 14,488.8 kg.

Lantana camara L. [Verbenaceae]

This native of Central and South America typically grows in sprawling thickets up to two metres tall. It is regarded as an invasive weed in many parts of Africa, Asia and Oceania, where it readily establishes in agricultural areas, reducing the productivity of pastures, and in secondary forests, where it crowds out native species and reduces biodiversity. It is toxic to livestock and produces chemicals that inhibit germination and root development in other plants.

As Table 40-8 shows, while there was not much difference in the increase in OM content between D1F2 and D2F2 (Table 40-4), the increase in harvest weight of 61.9% and 96.9% respectively over the control plot was very significant. This is attributed to the effect of the cutting and laying of young sunflower prunings on the ground.

TABLE 40-7: Average number and weight of bush *sitao* pods.

Field identity	No. of plants/sq m	Ave no. of pods/sq m	Ave no. of pods/plant	Ave weight of 100 pods (gm)	Weight increase vs. control (gm)	Increase in weight
T0	16	208	13	431.3		
D1F2	16	240	15	603.7	172.5	40%
D2F2	16	256	16	689.9	258.7	60.23%

TABLE 40-8: Average yield per hectare.

Field identity	No. of pods/ sq m	Ave weight of 100 pods (gm)	Weight (gm/sq m)	Yield (kg/ha)	Increase
T0	208	431.2	896.896	8,968.96	
D1F2	240	603.7	1,448.88	14,488.80	61.90%
D2F2	256	689.9	1,7661.40	17,661.40	96.91%

Labour input

Total labour involved in establishing the wild sunflower in the experimental plots was 31 person–days – 12 days for D1F2 and 19 days for D2F2.

Table 40-9 shows the number of days spent in the procurement, planting and pruning of the sunflower.

The monetary value of the labour required for D2F2 – the field with denser 50 cm sunflower spacing – was 58.33% higher than that for D1F2.

Cost benefit on per hectare basis

Table 40-10 shows the increase in yield value of bush *sitao* grown after a wild sunflower fallow, along with benefits from the various planting treatments after the deduction of costs. In terms of yield value, there was a highly significant 96.91% difference between the control plot and D2F2. However, after deducting costs, the increase in benefit was reduced to 56.56%.

TABLE 40-9: Labour requirement for establishing the sunflower fallow.

Field identity	Labour Input (person-days)			Total	Rate/day	Labour cost
	Procurement	Planting	Pruning			
D1F2	5.5	3.5	3	12	100.00	1,200.00
D2F2	8.0	7.0	4	19	100.00	1,900.00
T0	0	0	0	0	100.00	0

Notes: Currency used is Philippine pesos. Exchange rate at the time of the trial: 25 pesos = US$1.

TABLE 40-10: Cost-benefit calculation for bush sitao grown after wild sunflower fallows.

Field identity	Yield value	Increase over control	Cost	Benefit	Increase over control
T0	62,782.72		Nil	62,782.72	
D1F2	101,421.60	61.54%	16,000.00	85,421.60	36.06%
D2F2	123,627.80	96.91%	25,333.33	98,294.47	56.56%

Notes: Labour costs include only that work related to sunflower. Other work done in the production process was not considered. Values, costs and benefits expressed in Philippine pesos. Average farmgate price of *sitao* (4th qtr, 2001) = 7 pesos/kg. Exchange rate at the time of the trial: 25 pesos = US$1.

Farmers' perceptions and assessment of the trial

After a presentation of results and findings from the experiment, both positive and negative views were expressed by farmers. Foremost, the local trial had been positive for the farmers because they were able to discuss maximizing the use of a plant that was both locally available and easily grown.

In the eyes of local farmers, the trial proved the advantages and disadvantages of using wild sunflower as a fallow crop (Figure

FIGURE 40-4:   An innovation dawns.

40-4). They were impressed by the marked increase in yield from plots planted with sunflowers. Other advantages they observed were:

a)   *T. diversifolia* grows abundantly in patches within their community, so planting materials are readily available; and
b)   it produces large quantities of biomass in a short time, prevents some weed growth by shading the soil, and makes the soil porous so that weeds that do grow are easily uprooted. The noxious grass *Imperata cylindica* rarely grows under a wild sunflower canopy.

On the other hand, they observed the following disadvantages:

a)   Establishing sunflower fallows by planting cuttings is labour intensive; and
b)   its use is limited to erosion control (Figure 40-5), organic fertilizer and food for goats.

After planting food crops of maize and pole (climbing) *sitao* in association with wild sunflower, farmers made the following observations:

Even after the wild sunflower was slashed and burnt, the

FIGURE 40-5:   To farmers in the south of the Philippines, *T. diversifolia* is not only a preferred fallow species, but it also grows into strong contour hedgerows that help to reduce soil erosion. These hedgerows would have been easily established by using vegetative cuttings. This field was photographed outside the Mt. Kitanglad Nature Park in Bukidnon province, Mindanao.

Photo: Malcolm Cairns, 1995.

fallow crop regrew faster than the food crops. Additional labour was then required to prevent the sunflower from competing with the crops for sunlight and nutrients. The sunflowers had to be pruned twice during the vegetative stage of crop growth and once during pod formation. However, some farmers said this made weed management easier. The regrowing sunflowers meant that there was continuous shading of the soil, which minimized weed growth and made the soil more porous. Other weeds were easily uprooted and the slashed sunflowers served as mulch. Moreover, the slashed sunflower directly fertilized the soil as it decomposed rapidly in the warm climate with regular rain.

Farmers' testimonies

Following the trial, there was wider use of wild sunflower among the swidden and wet-rice farmers in the Cordillera region. Some found wider benefits from using sunflower in their fields. The following are excerpts from interviews with farmers in Kalinga province:

> Sunflower is very good. When I see signs of tungro[1] in the paddy fields, I collect sunflower leaves and lay them as mulch on the newly planted rice, putting more on the tungro sites. After a month, the plants have fully recovered and they grow with more vigour. However, there are times when complete plant health is not restored if sunflower leaves are applied after the onset of tungro. It is best to put in the sunflower leaves a week after transplanting, as it will effectively prevent the occurrence of tungro. My husband and I do the sunflower cutting, hauling and field application. But I am the one who does the check-ups of the farm.

> It not only prevents tungro, but also *lisao*[2], and kills small worms that disturb plant growth.

> Before the sunflower experiment and these personal experiences, I didn't bother about sunflowers. But now, after transplanting, and when we proceed to shifting cultivation areas, I make an extra effort to look for *uma* land with sunflowers growing nearby. I cut the stems and leave the roots in the ground. When the stems resprout, I cut them and lay them on the ground as mulch, and when I harvest, I leave the sunflower to grow again – Julia Wanagon, of Tomiangan, Tabuk, Kalinga.

> I plant sunflowers in all available spaces on the periphery of my rice fields. During land preparation, I cut the leaves and young stems and incorporate these into the soil. Our rice fields are small, but with sunflower, the yield is good without us buying fertilizers – Willam Banutan, of Tomiangan, Tabuk, Kalinga.

> My husband is the one mainly in-charge of our swidden farm, where we have established sunflower to bring into our rice fields for green manure. While he

is not very conscious of the effect of the sunflower, I have observed it closely. Sunflowers are very effective in preventing rice tungro in the paddy fields and producing good yields from our swiddens and rice fields. It has also significantly lessened labour in weeding, both in the paddy and swidden farms – Hilda Banatao, of Tomiangan, Tabuk, Kalinga.

Mimosa pudica L. [Leguminosae]

Sometimes known as sensitive plant, this species folds its compound leaves and allows its stems to droop when touched or disturbed. The foliage also closes in darkness and reopens in light. Assumed to be a defence against herbivores, the movements cost the plant in energy and interfere with the process of photosynthesis. This prickly native of Central and South America is a pantropical weed primarily found on soils depleted of nutrients.

Conclusion and recommendations

The study demonstrated that the establishment of wild sunflower as a fallow crop for two years significantly contributed to the improvement of the soil by increasing the organic-matter content and maintaining a more favourable pH value.

Changes in the trace elements may have been affected by other factors like rainfall and may not have been attributable solely to the presence or absence of the sunflower fallow.

After cutting and burning the sunflower, there was no significant improvement in the soil in the plots planted to sunflower, i.e. D1F2 and D2F2, on a per-hectare basis. The increase in yield was more attributable to continued cutting and laying down of prunings on the ground, which acted as a green manure during the vegetative growth of the *sitao* crop.

While there were observed disadvantages of establishing wild sunflower as a fallow crop, the advantages far outweighed the negative aspects – the main one of which was the added labour needed for fallow establishment and subsequent control of the sunflowers' regrowth. On the positive side, as well as improving the soil, the sustained supply of biomass as an organic fertilizer was ensured, soil erosion could be controlled and growth of noxious weeds was prevented, thus increasing crop yield.

Following these findings, the following recommendations are made with regard to optimizing the benefits of wild sunflower fallows in shifting cultivation:

- In establishing sunflower fallows, the 1.5 m to 2 m spacing of sunflower plants may be employed, and the number of times the plants are pruned should be increased;
- Burning should be discouraged to avoid loss of nutrients contained in the biomass. Green sunflower prunings should simply be piled or scattered on farming plots;
- Planting of wild sunflower in contour rows or hedgerows is also recommended. In this manner, with the correct distance between plants, farming systems can be made more permanent.

Further research

- Further studies should be undertaken to verify the data arising from this study, since it was pioneer research. Specifically, the effects of wild sunflower fallows on the major physical soil elements should be determined in other agro-economic zones of the Cordillera region;
- Farmers should be encouraged to take the initiative in finding better ways of propagating sunflower that involve less labour; and
- Farmers should be encouraged to experiment with the use of wild sunflower as a means of reclaiming grassland for crop production, where this is feasible.

References

Bulwayan, W. B. (1993) 'Effect of different rates of wild sunflower and inoculation on the growth and yield of snap beans', unpublished BSc Thesis, Benguet State University, La Trinindad, Benguet, Philippines

Notes

1. Rice tungro bacilliform virus (RTBV) is a plant pararetrovirus of the family Caulimoviridae. Tungro means 'degenerated growth' and was first observed in the Philippines. Tungro and *lisao* both cause degenerated growth, but with distinct symptoms. Rice plants afflicted with tungro first become yellow and whole fields are affected. *Lisao*, on the other hand, affects only patches of rice fields.
2. *Lisao* also causes 'degenerated growth' and the eventual death of newly transplanted rice seedlings. Informants claim it can be prevented if rice fields are drained and kept dry before land preparation or if powder from dry cell batteries is applied to paddy fields.

This farmer in Bukidnon, the Philippines, judged that his *Tithonia diversifolia* fallow had grown so woody that an axe was the tool best suited to cut it. As an understorey to the wild sunflower, a wild viny legume, *Calopogonium mucunoides,* fixes nitrogen and adds to the biological value of this fallow (see plates 42 to 44 in Coloured Plates section II, this volume).

Sketch based on a photo by Malcolm Cairns, ca, 1995.

V. CONCLUSIONS

A Batak farmer in Palawan, the Philippines, weaves a basket while her baby sleeps in her lap. This volume concludes by focusing on the next generation of swidden farmers who are inheriting centuries of agricultural tradition. Can they learn from their forefathers, and from shifting cultivators elsewhere, to improve their farming methods, their standards of living and their resource management?

Sketch based on a photo by Malcolm Cairns in 1993.

41

SWIDDENING IN THE 21ST CENTURY AND BEYOND

History and diversity between regional and global perspectives

*Roy Ellen**

Introduction

This chapter, and the volume of which it is a part, brings to a conclusion a remarkable publishing project that has simultaneously celebrated and documented the diversity, resilience and ingenuity of an often-despised mode of agriculture and way of life. The project was initiated through the collected papers of a regional workshop (Cairns, 2007), followed by three further volumes (Cairns, 2015, 2017, and this volume). The overall achievement is distinguished not only by its scale and thoroughness, but in adopting an approach in which communities convincingly speak for themselves through participatory research, decades of professional experience, collective authorship, local experts and the personal testimony of family involvement. The whole is accompanied by some attractive artwork, all of which serves to humanize what might easily have become an arid technical appraisal. Cumulatively, all volumes furnish a wealth of quantitative data on crop combinations, fallow options, anthropic soils, ecological implications and socio-economics, and provide a near-exhaustive bibliography of the literature, supplementing the landmark compilation of Conklin (1961). Through this body of work, we might hope to finally demolish a caricature and provide a strong evidential base for the re-assessment of a cultural practice that is still important.

As the 21st century beds down, many will be surprised to learn that what we call 'shifting cultivation' or 'swiddening'[1] continues to thrive and adapt. But because it does so is reason enough to regard this volume and its companions as a serious basis for pragmatic engagement by governments and other agencies that encounter it. Given the impressive collaboration between authors drawn from upwards of 22 different countries, the 226 chapters and more than 3.970 pages of the three volumes

* PROFESSOR ROY ELLEN is Emeritus Professor of Anthropology and Human Ecology at the University of Kent, and attached to its Centre for Biocultural Diversity within the School of Anthropology and Conservation.

and prequel together, as well as 'overflows' of about 80 further chapters unable to be accommodated in the volumes and published in associated online addenda, it might be thought that there is little more left to be said. However, like all forms of biocultural activity, swiddening is typified by its endless ability to adapt, regenerate and diversify; so it may be worth emphasizing just a few themes and dimensions that tie this volume together, and which in turn connect it to those that have preceded it, and to other attempts to synthesize the literature.

The focus of the present volume is farmer innovation and best practice. Previous volumes have focused on indigenous knowledge and sustainability, and environmental change and policy. So we might reasonably expect much overlap with what has gone before, as well as attention to themes not previously prioritized. The first main theme is how farmers manage different factors of production innovatively: field selection and preparation, tools, selecting what to plant (including balancing cash against subsistence cropping), maintaining soil health and weed management. The remaining themes range widely through food security, social and land tenure innovations, fireless swiddening, innovation in fallow management, and so on. I do not propose to provide an in-depth reflection on all of these themes, nor a systematic synthesis (see e.g. Found, 2017), or even less, a meta-synthesis of the series. I am in awe of the collective knowledge reflected overall, and acknowledge the danger of needlessly repeating what has been well-said by other contributors to this and earlier volumes, as summaries themselves are summarized, and as we try to make sense of a vast compendium of research. Indeed, in sketching out a plausible plan, I found time-after-time that points I had intended to make had already been well made, perfectly clearly, by others. I did, therefore, wonder whether there was really anything left to say. In the end I have decided that there is virtue in a broad overview involving the strategic re-emphasis of a few main points, with selectively developing themes that either have not been previously explored and that emphasize matters otherwise submerged in the wealth of data, or which include insights from comparative studies based on work conducted outside the focal geographic area of the main project. These build on observations I have made in a previous review of the literature (Ellen, 2012). Finally, given the large number of chapters, I have avoided making references to named authors, except in a few very specific cases, usually where there is a direct quotation.

The geography of swiddening

It is reckoned that 200 to 500 million people worldwide rely on swidden agriculture in some form, perhaps involving 280 million hectares of land (Heinimann et al., 2017). The focus of the present volume is Asia-Pacific, with contributions on India (especially the northeast), China (especially Yunnan), Indonesia, Thailand, Malaysia, Cambodia, Laos, the Philippines, Vietnam, Myanmar, Bangladesh, Papua New Guinea, Nepal and the Pacific islands. Previous volumes in the series have covered similar ground, adding Micronesia and Bhutan, Sri Lanka and Timor-Leste.

FIGURE 41-1: More 'crowded' than normal, this swiddening scene in Ban Lung district, Ratanakiri province, northeast Cambodia, resulted from two villages being crowded on to the territory of a third village, after being displaced by the establishment of a new provincial capital.

Sketch based on a photograph by Jeremy Ironside, in late 1996.

Combined, these works supply a wealth of case-study material for a part of the world where swiddening remains an important component of agricultural patterns, especially in marginal and undeveloped areas. Within the region, states with extensive mountainous areas tend to have larger tracts of swidden cultivation, with the distribution of newly-opened swiddens more concentrated in the northern borderlands of mainland Southeast Asian countries. Because of the importance of swiddening in Southeast Asia (and especially given the focus of the Cairns volumes) it is likely that we know more about these systems than about those in any other part of the world.

Swiddening is nevertheless a global phenomenon, and remains economically important in tropical South America, subtropical Central America and the Caribbean, and in sub-Saharan Africa. While predominantly a feature of tropical and subtropical landscapes, it has historically been important in high altitude (Ekvall, 1968), temperate and even subarctic zones as well, especially as a pioneering frontier technology in places where humans have been extending agriculture into previously forested areas. Even in late 19th-century Scandinavia, Finland and northern Russia, fire-assisted swidden agriculture was a feature of remote areas with poor populations, and of course the derivation of the term 'swidden' from the Middle English 'swithen' (Izikovitz, 1951) suggests the importance of this form of farming as English settlers in the British Isles moved into previously forested areas (Darby, 1956, p.186). Thus, while the Asia–Pacific region provides much material with which to conduct a comparative assessment, and probably exhibits more diversity than any other part of the world, an examination of global distribution, looked at historically (as well as archaeologically), is an opportunity to explore those features that Southeast Asian swiddening has in common with systems elsewhere, and to note any features that are regionally distinctive.

Swiddening as a subject of investigation

Swiddening was not taken seriously as a subject of scientific examination until well into the second half of the 20th century. Up until that time it was widely regarded as a 'primitive' form of farming of interest only to colonial administrators and similar agents, who were mainly concerned to account for its deficiencies and advise governments on how it might be eliminated and replaced with more efficient methods of agriculture. Among academics, it was often placed between hunter-gathering and permanent agriculture in a single linear timeline (what Pollock in ch. A25, in supplementary chapters to this volume calls 'the standard narrative'), culminating in the agricultural revolution and industrial grain production. Even within the contemporary homelands of swiddening, where there were traditional controlling polities with powerful economic centres engaged in intensive modes of production, swiddening – typically a technique of the periphery – was mythologized and looked down upon, 'othered' as a kind of 'barbarism' in the ancient Roman sense (Dove, 1983a, 1983b, 1986).

In the scientific study of swiddening, we can identify several themes situated in the prevailing preoccupations of their times. The earliest studies from the 1950s (e.g. Bartlett, 1956) were concerned to restore its reputation as a legitimate form of farming through systematic description and to refute claims that it destroyed primary rainforest on a large scale. They also sought to address the issue of how it might overcome the limitations of what were assumed to be dangerously low carrying capacities. Hence, there was a bias in anthropological and geographical studies in favour of exploring the technicalities of carrying-capacity computations (e.g. Carneiro, 1960), and with predictions that swidden intensification would be held responsible for the creation of unproductive anthropogenic grasslands (Conklin, 1959) and even the collapse of entire civilizations (Santley, 1986). Close scrutiny, however, often showed swiddening to possess an underlying rationality. This rationale was often a response to conditions where labour rather than land was in short supply (Boserup, 1965) and was generally rooted in a deep appreciation of local ecological particularities and comprised techniques that were carefully considered and elaborated, and in which – for example – primary rainforest represented only a small percentage of vegetative cover that was destroyed (Leach, 1950, p.338, 1957). Even so, definitive ethnographic monographs of swiddening, such as that by Freeman (1955) on the Iban, still promoted simplistic views of its wastefulness, and this invited the defence that such wastefulness was not necessarily so (Padoch, 1982). Successively, swidden agriculture became re-understood as 'forest farming' (Kunstadter et al., 1978), as ecology, in the context of promoting local (ethnoecological) knowledge systems (Geertz, 1963), in terms of the politics of sustainability, peripherality and indigeneity, and as a form of landscape management and biodiversity conservation. Along the way, earlier scientific skirmishes were re-enacted as the idea of swiddening as primitive farming re-asserted itself, for instance as a force of environmental degradation in the context of forest fires blamed on swiddening, but associated with (often corrupt) corporate clearance for oil palm and other plantations in Indonesia (e.g. Ellen and Watson, 1997). But as Street (1969) and others noticed quite early in the burgeoning of swidden literature, this mode of cultivation was hardly intrinsically harmonious in its function, and there was a danger that those who studied it became over-enamoured with the subject of enquiry. Therefore, swiddening has, in a sense, been reinvented in the political and academic imagination numerous times, under different conditions, and the objections of earlier generations have often been revisited.

The problem of swiddening as a category

Part of the problem in defending the reputation of swiddening is knowing how first to define it. It is not really a single agricultural form and seldom occurs by itself as a subsistence technique; it is commonly either subsidizing, or being subsidized, by other food-procuring measures or sources of income. In many cases, it is part of a composite agroecosystem that includes, amongst other strategies, home gardens, wet-rice terraces and commercial plantations or agroforests. While it has been shown that swiddening can be robust and with inbuilt flexibility, so, too, are the composite systems of which it is often a part. Rarely static, it is constantly changing, and this is why – as I argue below – we need not only a historical approach that can explain the circumstances in which particular practices evolved, but also a history of the study of swiddening that reminds us that many of the issues for state administrations and for the professional development community (agriculturalists, foresters, health workers) have not only been previously encountered, but often examined in some depth. In an early study, Conklin (1961) distinguished between integral and insipient or pioneer systems: the first productive, regulated, sustainable, embedded in socio-cultural norms and practices and involving the application of traditional ecological knowledge to management; the second more opportunistic, intensive and destructive, often a short-term survival strategy for settlers with little accumulated experience. Since then, others have proposed more complex typologies, of which the continuum model distinguishing long-fallow from short-fallow systems and forest pioneer farming has proven to be particularly instructive (e.g. Brown and Schreckenberg, 1998). What we have come to recognize as swiddening or shifting cultivation arises, therefore, in circumstances where there is a coming together of particular factors and techniques in situations of (usually) low population density and ecological marginality.

One constant variable influencing the particular form and ecology of all swiddening systems is the crops grown and their combination. Where swiddens are the main source of starch for a population, they tend towards a low variety of cultigens; where they are supplementary, they are more likely to have a high variety of crop types. Rice and similar grain crops have very different production requirements and demands on ecology compared with root crops such as cassava or yams. What crops are planted has different implications for group social relations and for the 'transactional orders' between exchanging populations.

This volume and its companions provide many examples of the consequences of growing different crops and crop combinations, but even then, they hardly exhaust the range of possibilities. Consider, for example, swiddening in areas where most edible starch is derived from sago palms, such as *Metroxylon sagu*: these areas are often, but not always, swampland. A case in point is the island of Siberut, off the coast of Sumatra, discussed here by Darmanto and Persoon (ch. 10, this volume). But equally striking cases can be found at the eastern end of the Indonesian archipelago, and in New Guinea. Where dietary sago has a prominent role and is culturally salient, as amongst the Nuaulu of Seram, in Indonesia's Maluku islands (Ellen, 2006), swiddens are often supplementary, providing backup alternative starch and a diverse combination of vegetables, leaves and fruits. Since the beginning of the 20th century, however, the status of swiddens has become more significant, rather than less. This has occurred through the introduction of cassava and increasing pressure to grow plantation crops for sale.

Swidden cultivation in history

It is sometimes claimed that swiddening was the first form of agriculture. Quite apart from the fact that its archaeological study is problematic, this seems unlikely for a number of reasons. It is certainly not the view of most of those investigating the origins of agriculture around the world, in which the category 'swidden' seems irrelevant (e.g. Bellwood, 2005). Where we have evidence of the adoption of swiddening among hunter-gatherers, such as the Agta of Luzon, in the Philippines, it is often an adjunct to obtaining food otherwise sourced through the market. Swiddening has probably been practised opportunistically by Agta over many centuries, but in the modern period it has more generally become an adaptation to rapid socio-economic change (Minter and Headland, ch. 8, this volume). By comparison, for the Penan of Sarawak, Malaysia, rather than a means of preserving a preferred way of life, swidden cultivation has served as a hallmark and route to modernity (Mashman, ch. 9, this volume).

The early historical appearance of swiddening was most likely circumstantial, important in some places (such as the humid tropics) rather than others (the Levant). There is nothing about it that is intrinsically 'primitive', and as scores of chapters in the Cairns volumes testify, it is anything but. By contrast, it is often a specialist adaptation in complex economies with extensive divisions of labour. There is increasing evidence of its antiquity and its transformative impact, particularly on the lowland forest ecology of the tropics (e.g. Arroyo-Kalin, 2012; Hunt and Rabett, 2014): not only influencing species composition, but also the formation of 'dark earths', in lowland South America, West Africa and Borneo (e.g. Hecht, 2003; Sheil et al., 2012). The cumulative effects of these ecological changes in deep history represent some of the oldest evidence for what we now recognize as the Anthropocene epoch. And in the present day, it isn't just that fallow is intrinsic to the effective working of the swidden system, or simply a consequence or a means of re-establishing conditions for further short-term agriculture, but rather swiddening becomes the instrument for producing resource-rich fallows. It is a way of managing forest and making a patchwork of successional habitats, whether, for example, it is focused on *Bactris gasipaes* in the upper Amazon (Rival, 2002, pp.80-84), *Gnetum gnemon* in lowland New Guinea (Belharte, 2011), or in combination with *Metroxylon sagu* in the Moluccas (Ellen, 2007).

Early descriptions of swiddening portray it as a timeless and unchanging system, but this is an historical illusion based on ignorance of non-modern farming systems. Many swidden populations have been part of wider systems of exchange, connected to markets, on the remote peripheries of centralized political entities, for centuries, perhaps millennia (Dove, 2011). The conditions of European colonialism also encouraged specialized forms of swiddening, promoting commercial incentives, markets and infrastructures, for example focused on candlenut harvesting in south Sulawesi (see Supratman et al., ch. 36, this volume) or the farming of lac scale insects in Assam (see Chakraborty and Choudhury, ch. 32, this volume). Similarly, in the Kerinci region of south Sumatra, from the 1800s, coffee and cinnamon began to supplant pepper in swidden forest-fallows in a rotating agroforestry system of coppicing and replanting (Burgers, ch. 37, this volume). Rattan production has existed as part of traditional forest management and exchange systems in Borneo since at least the 19th century, and the demise of rattan gardening has accompanied the demise of swiddening, with the growth of the logging industry and the opening-up of employment on oil-palm plantations. Novellino (2011) has shown how, for the Batak of Palawan, 'non-indigenous' swiddens have a complex history, where productivity and significance has fluctuated depending on cycles determined by external political and other factors. During the 19th century, preferred combinations of crops and trees in agroforestry systems in central Java were influenced by successive changes in colonial land-use policy, while in Cebu, in the Philippines, population replacement led to a process of experimentation which resulted in the deliberate planting of *Leucaena leucocephala* (Suson and Lasco, ch. 23, this volume).

Another variable, the implications of which only really become clear through the historical record, is movement. From the early work on carrying-capacity computations, we know that swiddening can be sustainable without settlements having to shift. Indeed, most peoples engaged in swiddening are surprisingly sedentary, although we need to recognize that the conditions of sedentarism are, in many cases, imposed by governments that seek to control local populations.

Some patterns evidently involve movement of settlements in addition to rotation of fields. In Southeast Asia we have the classic description of the H'mong Gar by Condominas (1957), and in the Amazon some peoples establish swiddens in one place and then move on, returning later to harvest a crop, such as the fruit of the palm *Bactris gasipaes*, or cassava (e.g. Rival, 2002, pp.80-88). More dramatic cases of linear movement and expansion are to be found in the 19th century migrations of the Iban through northern Borneo (Vayda, 1961; Padoch, 1982), and the H'mong (Geddes, 1976) and other groups, from Yunnan in China, southwards. The main economic interface between these latter peripheral peoples and centralized states and trading centres during much of the 19th and 20th centuries was opium production.

Where swiddening occurs in frontier areas, along national peripheries, on the edges of states, it becomes a kind of 'escape agriculture' (Scott, 2009, p.187), its precariousness and lack of permanence evading effective control from political centres. And this indeed has often been the complaint of those same centres: that until swidden populations are conveniently settled, they cannot be effectively administered. It has been argued that to see swiddening, as Scott does, as a political choice made in historic times, is to restore a degree of agency to 'hill peoples', to 'rescue' them from assumptions that they have a set of essential characteristics that make them what they are (Lieberman, 2010, p.336).

Swiddening as a situated knowledge system and a way of life

The emergence and subsequent viability of the various historic production systems involving swiddening and forest fallow depended almost entirely on local ecological knowledge. It is important to recognize that this was not limited to a fragmented knowledge of the growing conditions of particular plants, but involved knowledge of system properties and of complex inter-species relationships, including not only those of plants, but of the life cycles of insects serving as pollinators, or bee-plants for honey production, or – as with lac – the best species for encouraging the rotation of colonial hosts between different trees. Among the more impressive studies of local insect knowledge in subsistence systems focused on swiddening is Posey's (2002) study of Brazilian Kayapó land management, and how forest patches are selectively retained to deliberately ensure a supply of beneficial insects. Knowledge of this kind, although not always linguistically explicit, is embedded in local languages and technical practices, with classifications of habitats, ecotypes or secondary biodiversity linked to an understanding of the ecological consequences of historical disturbances (e.g. Ellen, 2007, 2010).

There is sometimes a danger in writings on swiddening to confine its description and assessment to a set of physical techniques, plus the kind of accompanying ecological knowledge discussed in the previous paragraph. But 'integral' swiddening is generally a distinctive way of life, where the identity of those who practise it is often severely challenged, and where, if languages disappear, the traditional knowledge embedded in them also disappears. This is partly because the technical knowledge of forest variation and change is linked to social knowledge of how people have used the forest in the past, and imprinted cultural experience, not only of forest use, but also of how it may sometimes have been contested. For many swiddening peoples, the rituals and festivals that punctuate their lives, their cosmography and cultural aesthetic, are geared to and profoundly grounded in the cyclical character of swiddening, and these fundamental elements lose their meaning when swiddening ends. We can see this in the Brunei Dusun, for whom in the mid-1990s the material need for extensive annual swiddens disappeared, yet the swiddens were maintained in residual form to sustain meaning for rituals and to supply glutinous rice varieties otherwise not available for Dusun religious ceremonies called *temarok* (Ellen, 2021, p.25).

With in-migration of non-swiddening peoples and remaining swidden communities gathered in marginal enclaves, the livelihood techniques themselves may become even more emblematic of identity, as in the case of the Baduy of upland west Java (see Iskandar, ch. 24, this volume). In such circumstances, the defence of swiddening against the state and against commercial intrusion can become the defence of indigenous rights. As with the Baduy and nearby Kasepuhan (see Laforge, ch. A32, in supplementary chapters to this volume), there are often complex reasons for maintaining swiddening, and in both these cases the communities are within wider more-encompassing traditional-political entities of which they are an historically intrinsic part.

Integral swiddening has consequences for how people conceptualize the relationship between humans and the natural realm. It blurs the conceptual distinction between nature and culture, by generating geographic patchiness within forest, and by the way in which old gardens merge into forest through fallow. Where mulching is employed exclusively (as in Siberut), these same boundaries are even more fuzzy and a hard semantic binary more difficult to correlate with environmental experience. Swiddening, therefore, may provide 'an integrated representation of the world, rather than a mere technique' (Bourdier, ch. A4, in supplementary chapters to this volume), a kind of 'cultural backbone' (Trakansuphakon and Ferrari, ch. 33, this volume) through the alignment of community rituals with the agricultural calendar and with fallows embodying structured memories, while spatial taboos and 'ritual gates' (Baird, ch. A3, in supplementary chapters to this volume) reinforce agreed conventions regarding the juxtaposition of different kinds of land and the order in which they are created; where particular practices, such as composting and mounding, may have cultural as well as agricultural significance, constituting part of an identity that serves to reinforce those practices and make them more effective.

Mulching and fireless swiddening

One apparent incongruity pertinent to the themes covered in this volume arises where forest-fallow systems do not involve the use of fire (practices sometimes called 'slash-and-mulch'). In many definitions of swiddening, fire is seen as integral, but there is plenty of evidence that it is not essential, coming from Southeast Asia (e.g. from Siberut (Darmanto and Persoon, ch. 10)), Papua New Guinea (Bourke and Allen, ch. 12) and parts of northeast India (Singh et al., ch. 11, all this volume), from Amazonia, e.g. the Huaorani people of Ecuador (Rival, 2002, p.89), and from Africa. Firing and mulching regimes can be alternative strategies in the soil-management repertoire of some peoples, found in the same local systems, in Southeast Asia and further afield (e.g. Posey, 2002, p.172-80). These 'no-burning' systems tend to evolve in marginal environments where rainfall is high and soil fertility low, some of them swampy. High rainfall, humidity and temperatures accelerate vegetal decomposition, contributing significant amounts of nutrients to growing crops.

The ecological ramifications of mulching versus fire can be considerable. Mulching is the response of least effort where high rainfall and microclimatic moisture make effective firing difficult. Fire is preferable for farmers where a completely clear field is required, as with certain types of grain, but the kinds of fires used by most swiddeners do not destroy all seed and root systems, and controlled firing can be very beneficial. A consequence of mulching is that preceding biodiversity and incipient structure is preserved, while it also serves to retain moisture, soil micro-organisms and microclimates, and from a global perspective, reduces atmospheric carbon emissions. The gradual transition of forest to agroforest, as found on Siberut, by cutting, mulching and planting, provides protection for growing trees; while the cutting and mulching of generally smaller areas creates a patchier micro-mosaic structure than we find with the clearing and burning of one-hectare plots. Complete burning, while rapidly releasing nutrients in the soil for immediate absorption by growing crops, also destroys the physical structures that prevent topsoil erosion, run-off and nutrient leaching. Depending on the crop, open-fired swiddens generate a dependency on weeding regimes at different stages in the cultivation cycle and have historically encouraged invasive grasses such as *Imperata cylindrica*.

Some systems have been developed that avoid burning while simultaneously encouraging high yields (including double-cropping) and preventing erosion on steep slopes, as achieved with foliage from fallows of *Leucaena leucocephala* used as green manure along contour ridges, or other kinds of foliage piled along embankments. On poor land, fireless swiddening avoids nutrient loss following burning, when the process is complicated by wind and water erosion, by encouraging biomass decomposition, microbial activity and the preservation of other living matter, including the underground parts of plants and seeds. In places, dried debris is gathered into ridges and furrows, covered in soil, and then burned (see Tiwari, ch. A42, in supplementary chapters to this volume). Elsewhere, fireless composting practices alternate with slow burning. However, in considering the range of fireless forms of swiddening, we should distinguish between low labour-input systems such as those on Siberut and the high labour-input systems that have developed in upland areas such as Meghalaya and Manipur.

Calculating the carbon balance: burning and the conservation of biodiversity

There can be little doubt that swidden burning releases carbon, causing atmospheric pollution, which in turn contributes to climate change. But the accusation that it releases significant quantities of carbon dioxide and other greenhouse gases, when compared to other burning, is contested. In the context of global warming, low-intensity agriculture releases fewer carbon emissions. Long-fallow systems with fast tree biomass recovery do not necessarily cause net carbon emissions at all, but are rather carbon neutral or carbon positive, storing or sequestering more than they emit, making them comparable to old-growth forests and better than timber plantations. Where swiddening is combined with reliance on sago (*Metroxylon sagu*), there is less burning, and the swamp ecosystems in which sago grows provide effective carbon sinks. Such scenarios make natural regeneration a cost-effective restoration strategy, and at a policy level, the evidence suggests that swiddening is compliant with mitigation schemes such as the United Nations' Reducing Emissions from Deforestation and Forest Degradation programme (REDD). Unfortunately, swidden burning continues to be confused (sometimes maliciously) with clearance for oil-palm plantations and logging (Bamba and Munandar, ch. 15, this volume). In West Kalimantan and Sumatra, since at least 1979, this has been accompanied by forest fires and smog contributing to year-on-year temperature increases, and which itself has led to widespread loss of swiddening and fallow. These problems have been exacerbated by the effects of El Niño climate patterns.

In the context of debates around global environmental change, there has been much discussion about the politics of environmentalism and conservation and the ecological virtues of swidden burning, beyond the clearing of ground and carbon balance. Shortened fallow periods under intensified regimes can lead to poor vegetation recovery and reduced soil fertility from less ash, while nutrients may deplete through volatilization from burning fields, the removal of crops, erosion and leaching during prolonged cropping. It is argued that targeted burning is beneficial in destroying pathogens and suppressing self-propagating weed seeds in top layers of soil. It may even be a deliberate strategy for tackling invasive species such as *Lantana camara* (e.g. Puri, 2015). Burning, therefore, often facilitates forest-fallow restoration through concentrating nutrients in ash, encouraging growth-enriching agrobiodiversity, and invigorating germination and seedling emergence, while retention of 'relict emergent' species generally indicates sustainability. Without fire there is no charcoal, which can contribute to the formation of dark earths.

Compared with carbon sequestration, there is less contestation regarding the role of swiddening in reproducing and even increasing biodiversity, although there is still much official ignorance. Ecological studies of vegetation conditions and soil fertility under forest fallows have been used to clarify and validate site-selection practices in communities where intensification is taking place. Swidden cultivation has an important role in the making of soil, through the introduction of charcoal and rich biomass and encouraging micro-ecology by planting nitrogen-fixing plants. By moving around and selecting germplasm (including between forest and swidden) and managing pollinators, local species diversity and diversity below the species level - *alpha diversity* – both increase, with swiddens serving as reservoirs for the conservation of landrace diversity. There is also a strong correlation between swidden agrodiversity, biodiversity, dietary diversity and food sovereignty, which regulates seasonal availability and buffers against periods of hunger. Through reliance on small plots, diverse planting and fallow-management regimes, swiddening increases patchiness – and therefore *beta diversity*. These interconnections have provided resistance against climatic stress (including El Niño disturbances), and constitute not only a way of managing tropical and sub-tropical forest, but are also responsible, in many parts of the world, for making the forest as we know it.

In many cases, a rich agrobiodiversity developed in swidden fallows involves deliberate planting and protection of particular species, which grow amongst crops and enrich soil by using mycorrhizal fungi and nitrogen-fixing bacteria. Many of these are leguminous, but many are not (Cairns and Brookfield, ch. 22, this volume). Different species used in this way are illustrated throughout this book, with selection varying according to local ecology, biogeographic availability and economic and functional considerations. These are often trees, but also include bamboo, shrubs and grasses such as *Thysanolaena latifolia*. Many local systems described here involve the use of several interacting or complementary species, some deliberately planted and harvested, some that can be wild-harvested, some of which have commercial potential, and some of which provide mulch. In addition, forest-management regimes involving swiddening provide all manner of 'ecosystem services', including protection against erosion and an integrated approach to watershed maintenance.

Innovation in swiddening

Once we see swiddening as part of a fluid historical dynamic, as an ever-changing set of interacting conditions responding to new situations, the early but persistent caricatures of it as static and mechanical become absurd. It is true that until recently, studying farmer innovation and its consequences in a swiddening context has not always been realistic, given the lack of longitudinal studies of particular systems. Hitherto, it has been widely assumed that swiddening systems are stable over long periods (that they have existed '... since time immemorial...' and so on), and that farming strategies were conservative techniques that had proven to be adaptive over the long term. Such assumptions were encouraged by the absence of historical detail and were the consequence of a certain kind of ethnographic description. While this may be true for some systems in some places, it clearly does not work everywhere. In my experience, farmer innovation is intrinsic to the system, given the precariousness of material conditions and local ecological variation, and may often entail the expansion of long-insipient existing practices. No wonder swiddens are so often characterized as 'laboratories for innovation', but no wonder, also, that the responses found for similar problems in widely separated spatio-temporal populations are fundamentally the same, differing mainly due to local ecology and to the biocultural resources available in a particular place. Farmer innovations should interest us less because individual innovations can provide solutions that others might emulate, but because they serve to remind us that swidden farmers are always active agents and that this agency is always to be protected and encouraged.

But the literature on innovation is conceptually fraught. What constitutes an innovation? Many ideas lie dormant for years before being developed rather than being a spontaneous response. How can we identify something that is really new? How do we distinguish an internal from an external innovation, whether through an official agency or from some other source? What do we mean by 'sustainable innovation'? The innovations that count are really only those that are successful. Most attempts to innovate generally fail, and do not spread within or beyond a local population. And among innovations most likely to fail are those introduced under pressure through governments or other bodies, including through persuasion to adopt certain economically-valuable crops, where there is official ignorance of the context and knock-on effects, or restrictions on nuanced knowledge amongst outside operatives working in the field. Moreover, successful innovations are not just technical events, but depend on effective social delivery, often linked to other innovations in subsystems of practice that can only be understood in a wider sense – an entire political economy of knowledge transfer.

Farmer innovations and experimentation occur in a variety of different ways. Little purpose would be served here by providing a simple catalogue of types covered in this volume, though were we to do so, Conklin's (1961) aide-memoire would serve as a useful organizational point of departure. The present volume provides diverse examples across many types of innovation: in tools, plants selected, firing strategies and control measures, soil management, rotation and fallowing, as well as in terms of the social organization of production, including land tenure. There is generally a tendency to prioritize the introduction and development of new material equipment and technical procedures when considering innovation in agricultural systems, such as the use of sprays to control pests and weeds, replacement of bush knives by felling axes, and then by chainsaws and brush cutters, or dibblesticks by hoes. But while these have certainly impacted on swiddening, they are generally less important than changes in crops and auxiliary species, or soil, fallow and forest management.

New crop types and combinations may be deliberately introduced and developed to increase productivity and adapt to different conditions, as improved food types, or as cash crops driven by the changing market economy. And wherever there is evidence of earlier crop innovation this will have had consequences for system management as a whole. We can see this in the introduction of New World species into Old World swiddens. Sweet potato had an impact on Southeast Asian swiddening as early as the 17th century, and cassava a much greater impact – but much later. The development of new cultivars is often a response to micro-climatic change and loss of forest cover (for example cassava in the Kei Islands, in Ellen et al., 2012). As a result, swiddens and subsequent fallows have become more specialized, or have undergone changes in structure and ecological organization, for example in the role of shade and of

FIGURE 41-2:   Karen shifting cultivators in Mae Hong Son province, northern Thailand, work together to plant their upland rice. The men generally wield the dibble sticks to poke planting holes in the soil, and the women follow closely behind to drop rice seeds into the holes.

Sketch based on a photograph by Peter Kunstadter on April 24, 1964.

juxtaposed plant types in retaining water moisture. In certain situations, the cycle is fundamentally diverted and altered by the permanent planting of tree crops grown for the market.

Other innovations relate to soil and fallow conservation: ways of retaining fallow length or introducing short fallows using fast-growing and soil rejuvenating species, mulch lines, cover crops and green manures, contoured vegetation strips and log bunding (Lotha et al., ch. A21, in supplementary chapters to this volume), bamboo retaining walls, improved long fallows incorporating key crops and non-timber forest products, intercropping tree species, soil erosion and nitrogen-fixing achieved by planting a single species or multifunctional perennial crops, planting in termite mounds, and management of fire-resistant features through use of trees that can tolerate burning regimes.

There is a wide range of different trees that people have discovered to optimize the effectiveness of swiddening systems, often in combination with other species, rotating plots of different fallow lengths, and rotating different kinds and sizes of plot. However, innovation is rarely about self-contained technical changes with specific outcomes, but rather about combinations of changes with a range of consequences. Different species may individually contribute to different outcomes, but combinations collectively improve the quality of swiddening as a whole. Across Assam and Manipur, a basic system of swiddening persists, modified locally into diversified and specialist land-use subsystems involving different crop combinations in response to possibilities and perceived needs in particular areas (Darlong et al., ch. A17, in supplementary chapters to this volum4e). Where swiddening has been abandoned under pressure to plant monocultures and agro-industrial deforestation, where food security has suffered from reliance on chemical inputs or a decline in the marketability of 'boom crops', there are instances where people are attempting to restore traditional systems and revert to self-sufficient organic practices (Bamba and Munandar, ch. 15, and Huy, ch. 18, this volume, and Gypmantasiri and Limnirankul, ch. A14, in supplementary chapters to this volume).

No technical innovation is without social correlates or consequences: some may explicitly derive from participation in particular kinds of social network, and new social arrangements are often necessary to manage successful innovations. Some social innovations have permitted more flexibility and control: recognition of tenure, drawing up of boundaries, gender-inclusive development committees, participatory forest mapping, agricultural land-use planning, and institutional support for the retention of traditional knowledge. Some social changes may be spontaneous within communities, but many derive from the experience of other communities, involving collaborative decision-making with various kinds of non-governmental organizations (NGOs) – community-based, local, national and international – or with government agencies.

Local responses to problems are often hybrid, combining various technical and social arrangements. Thus, *Imperata cylindrica* infestation in northern Laos has been effectively addressed by combining changes in weeding, fallowing, chemical application, labour exchange, crops and off-farm employment (Keoboualapha et al., ch. A23, in supplementary chapters to this volume); while fallow management is transforming systems in Nepal through a combination of shortening fallow periods to a minimum - partly because of requirements of cardamom cultivation - and using forest and fallows as sources of diverse income opportunities, including harvesting of medicinal plants (Aryal et al., ch. 13, this volume).

The future of swiddening in the 21st century

One recent estimate is that about 7% of humankind is involved in shifting cultivation or swiddening, approximately 200 million people using 300 million hectares of land. In places (such as in Assam) it continues to expand, yet on a global scale it is in decline. Satellite imagery of its distribution and rapid and dramatic transformation across mainland montane Southeast Asia over the period from 1988 to 2016 (Li et al., ch. 6, this volume) provides clear evidence of the stress factors encouraging this decline across the region. They are reflected in reduced fallow lengths and the changing intensity of swidden farming: continued expansion during the 1990s, a slow decline in the 2000s, and a sharp reduction in the 2010s.

Various frameworks have been suggested through which we might organize and understand the causes of this trend (e.g. Fox et al., 2009). For one thing, the way swiddening is socially constructed as a category by governments and agencies often makes it invisible. Swiddens are seldom marked cartographically, are ambiguous in land-use classifications, falling between official divisions of the landscape into forest and permanent agriculture, and between the division of responsibility of ministries and the legal constraints under which they operate, such as the expansion of forestry departments and the rise of conservation as an international concern. Moreover, there are moral challenges to the viability of swiddening, especially given the link between 'ethnic minorities' or 'backward peoples', and resistance in the face of resettlement policies and resettlement practice, privatization and commoditization of land and produce.

While we need to acknowledge that swiddening sometimes leads to accelerated soil erosion, runoff and nutrient loss, all factors combine as markets expand to further erode its viability. Commercialization of agriculture has impacted on customary laws, leading to land alienation, accentuating class distinctions and gender injustice and aggravating ethnic conflict. As swiddening degrades further, due to factors its practitioners cannot easily control, governments are, at the same time, under additional pressure to restrict it, to intensify cultivation and expand conservation forest. Many governments still attempt to eradicate it, and where this is not deliberate, they have nevertheless been responsible for actions that shorten fallows, degrade land and result in poor forest-regeneration outcomes, made worse by resettlement and plantation-agriculture schemes. Swiddening peoples are caught in the slipstream between environmental campaigning (including reaction to the advent of the UN-REDD+ programme) and the dominance of agribusiness and neoliberal policies at both national and international levels.

The impact of oil palm in equatorial Southeast Asia in this regard is relentless, making it more-or-less impossible for local people to sustainably maintain a traditional way of life. And while NGOs such as the Centre for International Forestry Research (CIFOR) produce briefing documentation on all kinds of sustainable options, governments and agribusiness seem to continue regardless of warnings. Oil-palm expansion has been accompanied by criminalization of local farmers (seen as the main contributors to burning), land conflict and theft, corruption (impeding prosecution of companies), impacts on agriculture, and an increase in plant pests despite high levels of insecticide use. The whole comprises a veritable non-virtuous downward spiral in socio-environmental conditions typified by ecological simplification.

A specific cause and consequence of ecological simplification in once high-diversity systems is genetic erosion. In Yunnan (Yin et al., ch. A28, in supplementary chapters to this volume) temperature increases have led to flooding, rainstorms, drought, landslides and pest-infestations, and these have impacted on agrobiodiversity, the agricultural calendar and techniques. Most traditional rice varieties have been abandoned and there is concentration on a few drought-resistant varieties. With a reduction in diversity, in Yunnan as elsewhere, it is now difficult to fine-tune the composition of fields by adjusting landrace combinations. Thus, with the move away from swiddens and their simplification, there is also a simplification of diet. At the same time, cash-crop prioritization reduces both diversity and labour available for swiddening, with labour also impacted by off-farm employment.

Nevertheless, while a further decline of swiddening may be inevitable for both ecological and economic reasons, this book bears witness to how predictions of its early demise have been exaggerated. In some areas, swiddening remains in resilient enclaves, where populations have been able to resist external factors of deforestation and incursions from growing outsider populations. Forest loss makes it less sustainable in its integral form, but pioneer swiddening thrives on the margins, where, in Indonesia for example, voluntary migrants extend the settlement frontier. If present rates of decline continue, and if swiddening remains a part of local economies, then local innovations and best practices may help to further sustain its productivity and resilience.

Several things that we have learned should specifically inform policy. These have already been addressed in a previous volume:

- The main factor sustaining swiddening is undeniably poverty, linked to an inability to raise capital and land shortage, sometimes exacerbated by labour shortages and ageing populations.
- The precariousness of swiddening populations is aggravated by unrecognized land tenure systems, labour shortages and ageing populations, marginal topography and geomorphology. These are the same pressures that can lead to innovation under contemporary socio-economic conditions.
- Swiddening should not and cannot be eradicated by government edict. It will always be around, wherever environmental, social and economic conditions make it the most efficacious way of farming.
- If we accept its inevitability, or even its preference in particular areas, it is difficult to legislate and administer for its 'improvement'. Certainly, government reforms relating to recognition of customary forest are helpful. Farmers may take advantage of information and resources brought to their attention by outsiders, but are unlikely to accept changes unless they have worked out for themselves that the prospects on offer are in their best interests.
- Successful recognition of the rights of local swiddening groups provides effective resistance in some places, including land-tenure security and encouragement of women's voices, increasingly amplified through local NGOs and interest groups.
- There are few prescriptions when it comes to swiddening policy, only proscriptions on how not to handle it, and possible scenarios about how to encourage 'best practices'.

Despite the continuing ignorance of some official decision-makers, after Cairns it is impossible to claim that published research on swiddening is inadequate, or that its versatility and diversity still need to be demonstrated, or that in many places it cannot play a positive developmental role.

This volume has sought to emphasize 'best practices'. This is usually understood as a body of procedures accepted as being correct or most effective in a particular context. Although a concept of increasing importance in regulatory bureaucracies, it may need adjustment here. What may be a technical best practice in a particular domain of knowledge, its implementation may not be a best practice overall in a system of conflicting technical and social desirabilities, and may vary in terms of the different actors involved, be they swiddeners, environmental managers, forestry departments, development agencies, or indeed those who analyse swiddening systems. We need always to ask, therefore, who defines what best practices might be?

Those outside authorities seeking to introduce new practices to farmers should at all times compare their performance against existing indigenous practices to test if they offer any advantages. Farmers often fail to adopt innovations because they have more relevant knowledge and experience than those advising them, both on technical matters and on socio-economic context. Governments tend to think linearly rather than holistically and outside experts doing things for other people seldom see the complete picture. As this entire series confirms, there are many different kinds of swidden system, functioning in situation-specific contexts, exhibiting constant change and adaptation. The market economy has penetrated into even the most remote mountainous areas, and traditional subsistence crops abandoned in favour of cash crops. But cash crops in swidden economies come and go, mirroring the competing demands for food security and cash income. Heavy investment in particular cash crops may lead to over-dependence and lack of flexible response in the face of a changing economy and environment, but there are numerous mitigations to protect farmers from the perturbations of a cash economy.

As Pollock suggests (ch. A25, in supplementary chapters to this volume), swiddening may be part of the solution and not the problem in a world where lack of ecological diversity is the issue. In many cases, swiddening cannot be replaced by anything 'permanent' that is nearly as efficient ecologically, and which can retain food self-sufficiency, resilience and sustainability. There is no single way of doing this. Swiddening continues to be a realistic option for farmers and not merely a safety net. In a world where there is still a lack of joined-up thinking when it comes to policy, we might be advised to avoid 'solutions to problems' and opt for best practices in particular circumstances. This must be grounded in recognition of the essential role of indigenous and local perspectives, and in acknowledgment that understanding and progress only comes through the shared enterprise of swiddeners themselves and those who engage with them as researchers, or in the political domain. For Malcolm Cairns, it is only when we take the long and overarching view, and accept swiddening practices as forms of 'permanent' land use, that we will be able to find a political space for their continued existence in state structures that view them negatively. I agree.

References

Arroyo-Kalin, M. (2012) 'Slash-burn-and-churn: Landscape history and crop cultivation in Pre-Columbian Amazonia', *Quaternary International* 249, pp.4-18

Bartlett, H. H. (1956) 'Fire, primitive agriculture, and grazing in the tropics', in W. L. Thomas (ed.) *Man's Role in Changing the Face of the Earth*, University of Chicago Press, Chicago, pp.692-720

Belharte, S. (2011) 'The ecological view: Management of woody perennials and the transition to vegeculture in Southeast Asia and the Pacific', in G. Barker and M. Janowski (eds) *Why Cultivate? Archaeological and Anthropological Approaches to Foraging/Farming Transitions in Southeast Asia*, McDonald Institute for Archaeological Research, Cambridge, pp.27-46

Bellwood, P. (2005) *First Farmers: The Origins of Agricultural Societies*, Blackwell, Oxford

Boserup, E. (1965) *The Conditions of Agricultural Growth: The Economics of Agrarian Change under Population Pressure*, Allen and Unwin, London

Brown, D. and Schreckenberg, K. (1998) 'Shifting cultivators as agents of deforestation: Assessing the evidence', *ODI Natural Resource Perspectives* 29, pp.1-4

Cairns, M. F. (ed.) (2007) *Voices from the Forest: Integrating Indigenous Knowledge into Sustainable Upland Farming*, Proceedings of a regional workshop held in Bogor, Indonesia, 23-27 June 1997, Resources for the Future Press, Washington DC

Cairns, M. F. (ed.) (2015) *Shifting Cultivation and Environmental Change: Indigenous People, Agriculture and Forest Conservation*, Earthscan, London

Cairns, M. F. (ed.) (2017) *Shifting Cultivation Policies: Balancing Environmental and Social Sustainability*, CAB International, Wallingford, UK

Carneiro, R. L. (1960) 'Slash-and-burn agriculture: A closer look at its implications for settlement patterns', in A. F. C. Wallace (ed.) *Men and Cultures: Selected Papers of the Fifth International Congress of Anthropological and Ethnological Sciences* 1956, University of Pennsylvania Press, Philadelphia, pp.229-234

Condominas, G. (1957) *Nous Avons Mangé la Forêt de la Pierre-génie Gôo*, Mercure de France, Paris

Conklin, H. C. (1959) 'Shifting cultivation and the succession to grassland', *Proceedings of the Ninth Pacific Science Congress* 1957 (7), pp.60-62

Conklin, H. C. (1961) 'The study of shifting cultivation', *Current Anthropology* 2, pp.27-61

Darby, H. C. (1956) 'The clearing of the woodland in Europe', in W. L. Thomas Jr. (ed.), *Man's Role in Changing the Face of the Earth*, University of Chicago Press, Chicago, pp.183-216

Dove, M. R. (1983a) 'The agroecological mythology of the Javanese and the political economy of Indonesia', *Indonesia* 39, pp.1-36

Dove, M. R. (1983b) 'Theories of swidden agriculture and the political economy of ignorance', *Agroforestry Systems* 1(3), pp.85-99

Dove, M. R. (1986) 'The practical reasons of weeds in Indonesia: Peasant vs state views of *Imperata* and *Chromolaena*', *Human Ecology* 14(2), pp.163-190

Dove, M. R. (2011) *The Banana Tree at the Gate: A History of Marginal Peoples and Global Markets in Borneo*, Yale University Press, New Haven, CT

Ekvall, R. B. (1968) *Fields on the Hoof: Nexus of Tibetan Nomadic Pastoralism*, Holt, Rinehart and Winston, New York

Ellen, R. (2006) 'Local knowledge and management of sago palm (*Metroxylon sagu* Rottboell) diversity in South Central Seram, Maluku, eastern Indonesia', *Journal of Ethnobiology* 26(2), pp.83-123

Ellen, R. (2007) 'Plots, typologies and ethnoecology: Local and scientific understandings of forest diversity on Seram', in P. Sillitoe (ed.) *Global vs Local Knowledge*, Berghahn, Oxford, pp.41-74

Ellen, R. (2010) 'Why aren't the Nuaulu like the Matsigenka? Knowledge and categorization of forest diversity on Seram, eastern Indonesia', in L. M. Johnson and E. S. Hunn (eds) *Landscape Ethnoecology: Concepts of Biotic and Physical Space*, Berghahn, New York, pp.116-140

Ellen, R. (2012) 'Studies of swidden agriculture in Southeast Asia since 1960: An overview and commentary on recent research and syntheses', *Asia Pacific World* 3(1), pp.18-38

Ellen, R. (2021) 'Thinking through nostalgia in anthropologies of the environment and ethnographies of landscape', in O. Angé and D. Berliner (eds) *Ecological Nostalgias: Memory, Affect and Creativity in Times of Ecological Upheavals*, Berghahn, New York, pp.17-38

Ellen, R. and Watson, C. W. (1997) 'Fire, forest, people and the Southeast Asian smog', *APFT Briefing* 4 (December), pp.1-2

Ellen, R., Soselisa, H. L. and Wulandari, A. P. (2012) 'The biocultural history of Manihot esculenta in the Moluccan islands of eastern Indonesia: Assessing the evidence for the movement and selection of cassava germplasm', *Journal of Ethnobiology* 32(2), pp.157-184

Found, W. C. (2017) 'Lessons learned from the identification and implementation of policies affecting shifting cultivation in the Asia-Pacific region', in M. F. Cairns (ed.) *Shifting Cultivation Policies: Balancing Environmental and Social Sustainability*, CAB International, Wallingford, UK, pp.952-970

Fox, J., Fujita, Y., Ngidang, D., Peluso, N., Potter, L., Sakuntaladewi, N., Sturgeon, J. and Thomas, D. (2009) 'Policies, political-economy, and swidden in Southeast Asia', *Human Ecology* 37, pp.305-322

Freeman, D. (1955) *Iban Agriculture: A Report on the Shifting Cultivation of Hill Rice by the Iban of Sarawak*, H. M. Stationery Office, London

Geddes, W. R. (1976) *Migrants of the Mountains: The Cultural Ecology of the Blue Miao (Hmong Njua) of Thailand*, Clarendon Press, Oxford

Geertz, C. (1963) *Agricultural Involution: The Processes of Ecological Change in Indonesia*, University of California Press, Berkeley and Los Angeles

Hecht, S. (2003) 'Indigenous soil management and the creation of Amazonian dark earths: Implications of KayapÃ³ practices', in J. Lehmann, D. C. Kern, B. Glaser and W. I. Woods (eds) *Amazonian Dark Earths: Origins, Properties, and Management*, Kluwer, Dordrecht, The Netherlands, pp.355-372

Heinimann, A., Mertz, O., Frolking, S., Egelund Christensen, A., Hurni, K., Sedano, F., Chini, L. P., Sahajpal, R., Hansen, M. and Hurtt, G. (2017) 'A global view of shifting cultivation: Recent, current, and future extent', *PLoS ONE* 12(9): e0184479. https://doi.org/10.1371/journal.pone.0184479

Hunt, C. O. and Rabett, R. J. (2014) 'Holocene landscape intervention and plant food production strategies in island and mainland Southeast Asia', *Journal of Archaeological Science* 51, pp.22-33

Izikovitz, K. G. (1979 [1951]) *Lamet: Hill Peasants in French Indochina* [Etnologiska Studier 17, Göteborg], AMS Press, New York

Kunstadter, P., Chapman, E. C. and Sabhasri, S. (eds) (1978) *Farmers in the Forest: Economic Development and Marginal Agriculture in Northern Thailand*. East-West Centre, Honolulu

Leach, E. R. (1950) *Social Science Research in Sarawak, A Report on the Possibilities of a Social Economic Survey of Sarawak presented to the Colonial Social Science Research Council*, H.M. Stationery Office for the Colonial Office, London

Leach, E. R. (1957) 'Some economic advantages of shifting cultivation', *Proceedings of the Ninth Pacific Science Congress*, Bangkok

Lieberman, V. (2010) 'Review article: A zone of refuge in Southeast Asia? Reconceptualising interior spaces', *Journal of Global History* 5(2), pp.333-346

Novellino, D. (2011) 'Cycles of politics and cycles of nature: Permanent crisis in the uplands of Palawan', in R. Ellen (ed.) *Modern Crises and Traditional Strategies: Local Ecological Knowledge in Southeast Asia*, Berghahn, Oxford, pp.185-219

Padoch, C. (1982) *Migration and its Alternatives among the Iban of Sarawak*, Nijhoff, The Hague

Posey, D. A. (2002) *Kayapó ethnoecology and culture*, Routledge, London and New York

Puri, R. K. (2015) 'The uniqueness of the everyday: Herders and invasive species in India', in J. Barnes and M. R. Dove (eds) *Climate Cultures: Anthropological Perspectives on Climate Change*, Yale University Press, New Haven, CT, pp.249-272

Rival, L. (2002) *Trekking Through History: The Huaorani of Amazonian Ecuador*, Columbia University Press, New York

Santley, R. S. (1986) 'On the Maya collapse', *Journal of Anthropological Research* 42(2), pp.123-159

Scott, J. C. (2009) *The Art of not Being Governed: An Anarchic History of Upland Southeast Asia*, Yale University Press, New Haven, CT

Sheil, D., Basuki, I., German, L., Kuyper, T. W., Limberg, G., Puri, R. K., Sellato, B., van Noordwijk, M. and Wollenberg, E. (2012) 'Do anthropogenic dark Earths occur in the interior of Borneo? Some initial observations from East Kalimantan', *Forests* 3(2), pp.207-229, doi:10.3390/f3020207

Street, J. M. (1969) 'An evaluation of the concept of carrying capacity', *The Professional Geographer* 21(2), pp.104-107, doi: 10.1111/j.0033-0124.1969.00104.x

Vayda, A. P. (1961) 'Expansion and warfare among swidden agriculturalists', *American Anthropologist* 63, pp.346-358

Note

1. The term 'shifting cultivation' appears in the titles and editorial matter of volumes in this series, although 'swiddening' is also employed throughout by Cairns and other authors. My preference here will be clear, but it is less that I am persuaded by *sui generis* etymological arguments than by the fact that the adjective 'shifting' was always semantically troubling, while 'swiddening' - though obscure - was, at least for most readers, arbitrary and concise.

As the sun sets behind the hills of northern Thailand, it silhouettes a Lahu Nyi mother and child enjoying the evening light in the village of Ban Pang Farn, in the hills above Phrao valley (see Walker, ch. A13, in the supplementary chapters to this volume). Some observers have long predicted that 'the sun is also setting' on the era in which shifting cultivation remains a viable farming practice – but, as documented by this volume, shifting cultivators have been impressively adept at changing with the times, and innovating better ways to manage their resources.

Sketch based on a photo by Anthony Walker in 1968.

POSTSCRIPT: THE UNIMAGINED COUNTRY

The past and present future of swidden agriculture

Michael R. Dove[*]

Introduction

This volume, in examining 'innovation' in swidden cultivation practices, represents a radical departure from a century and more of work on swidden. To innovate is to change; to differentiate past, present and future. This is not the way that people – except for swiddeners themselves – have thought about swiddening; the only change that has been imagined has been its disappearance. Swidden agriculture, perhaps unique among systems of agriculture, was long perceived as having a past and present so troubled as to prejudice any possibility of its having a future. The cultivation of swiddens has been seen as quintessentially unsustainable.

The future is sometimes called the 'imagined country'. Since swidden has always been seen as having no future, it is thus the 'unimagined country'. The question in the historic, global discourse pertaining to swidden cultivation has always been, 'How long can it last?' And the answer to this question has almost always been, regardless of era or latitude, 'not long'. This teleological view of swidden – as something that is barreling toward its demise and will not exist in the future – has been integral to its popular representation and discussion. This teleology has been one of the most ubiquitous and influential dimensions of the very idea of swidden, which compels us to ask, why is this *the* swidden question?

The Dogma versus the Ethnography

Across diverse time periods, cultures and political structures, there has been remarkable homogeneity in the way that outside authorities have viewed swidden

[*] Professor Michael R. Dove is Margaret K. Musser Professor of Social Ecology at the Yale School of the Environment, Professor in the Department of Anthropology, and Curator of the Peabody Museum of Natural History, Yale University, New Haven, CT.

agriculture; so much so, indeed, that it can be said to have attained the status of a global discourse. According to this discourse, the technology is primitive, and its practitioners are backward and uncivilized, if not criminal or subversive. The burning of the forest (albeit not universal – see Section III, this volume) is viewed with disdain, if not horror; and the resulting fields are derided as having a disorderly appearance. The harvests from these fields are thought to barely meet subsistence needs – keeping people in poverty – much less producing anything for the market, with its practitioners often seen as living in a pre-monetary, barter economy (Tsing, 1999). After cultivation, the fields are thought to be abandoned, whereupon their cultivators, often called 'nomadic', are supposed to move on to new territories. The burning and then perceived abandonment of the forest combined to produce a picture of destructive, profligate behaviour; the colonial Dutch referred to swidden agriculture in Indonesia as *roofbouw* (exhaustive or robber agriculture) (Jansonius, 1950, p.1245). Given this perceived wanton use of natural resources, coupled with the perceived intemperate fertility of backward people, swidden communities have been seen as uniquely vulnerable to the Malthusian curse: the population must outgrow and degrade its resource base, rendering swidden no longer possible. Hence, it is the system of agriculture without a future.

This global discourse of swidden represented one of the most widespread examples – if not *the* most widespread example – of unproven dogma in conservation and development thinking in the modern era. The abundance and energy devoted to denouncing swidden was never matched by its empirical study: indeed, the former seemed to erase the need for the latter. Given the former ubiquity of swidden throughout the tropics, this dogma biased views of rural peoples across huge swathes of the world. It is no accident that swidden agriculture became, in consequence, a focus of much of the pioneering work in the mid-20th century by anthropologists and geographers in the non-Western world (e.g. Izikowitz, 1951; Freeman, 1970/1955; De Schlippe, 1956; Condominas, 1977/1957; Conklin, 1957; Geertz, 1963).

These scholars directly challenged the popular dogma regarding swidden. Indeed, Conklin (1954) begins one of his classic papers on swidden by listing ten of 'the most frequent and problematic statements and assumptions' regarding swidden. Research by Conklin, among others, demonstrated that swidden peoples often have unparalleled knowledge of their tropical forest environments; they practice complex systems of intercropping in their swiddens; and the return to labour in swiddens often exceeds that of permanent-field agriculture. Moreover, the relatively low demand on labour enables them to engage in commodity production or trade in forest products (see Section VI in the supplementary chapters to this volume); they are not nomadic, they don't abandon but rather fallow their fields, and they manage the fallow vegetation (see Section IV, this volume), which is often highly biodiverse; and they draw an energy 'subsidy' (Alcorn, 1993) from the natural dynamics of regrowth. Overall, this research demonstrated that in the absence of external disturbances that undermine traditional tenure, create indebtedness, or otherwise disrupt the system, swidden agriculture is one of the most sustainable production systems ever devised for the tropical forest.[1] The strengths and attractions of this system of agriculture are reflected in its extraordinary tenacity through the colonial and into the post-colonial era, in the face of not merely a lack of the research and extension support enjoyed by other types of agriculture, but open hostility accompanied by prohibition, land appropriation and other punitive measures.

Politics and culture

Given the overwhelming weight of this scholarly evidence regarding the sustainability of swidden, whence comes the discourse of a system of agriculture with no future? How can this long-lived diatribe against swidden be explained, given the lack of evidence to support it, the abundance of evidence to the contrary and the sheer persistence of the practice? To paraphrase Mark Twain, swidden has many times outlived the writing of its obituary.[2]

Politics is at least part of the answer. Throughout the modern era, swidden has been ill-regarded by central political authorities. Even in 18th-century Sweden, Linnaeus was hard-pressed by government authorities for his defence of swiddening or 'burn beating' (Dove, 2015). This near-universal state enmity towards swidden is due to the fact that swidden is characteristically ill-suited to the imperatives of central governments to extract surplus products from agriculture and to control the producers themselves. More subtly, swidden systems are quintessential local systems of resource exploitation, incomprehensible to outside authorities, who must deprecate the expertise involved in order to privilege the metropolitan systems of knowledge and practice that accompany centralized rule. This helps to explain why swidden has been seen as unsustainable, in stark contrast to many systems of manifestly less sustainable permanent-field agriculture – involving high inputs of chemicals and so-called 'high-yielding' varieties, or expansive monocrops like oil palm and soybean – that have not been subjected to the same critique. Even when these latter systems of cropping are critiqued, it is not said that 'they have no future'; they retain their future regardless.

The ubiquity and tenacity of the critique of swidden suggests that there is more at work here than politics. The anthropologists who did the pioneering work on swidden spent much of their energy grappling with how swiddeners manage not just land or space, but also time. The elaborate graphs and charts produced by Conklin (e.g. 1957, tables 12,13, figures 10,11,12,13) are testament to the close management of both intra-annual and inter-annual swidden time. Condominas (1977/1957, p.xvii) famously wrote that the Montagnard swiddeners of Cambodia reckon historical time as the year that they farmed such-and-such an area of forest: '*Hii saa brii* . . . (We ate the forest of ...), followed by a place name, is how the Mnong Gar, or *Phii Brêe* ("The Men of the Forest") designate this or that year.' This represents an unusual conflation of time and space: the passage of time is in effect perceived as the movement among forest plots. Absent this movement, time does not pass.

The swiddener's view of time is embedded in a world view that is often at odds with the Western, Cartesian divide between nature and culture. Many of the early ethnographers of swidden remarked upon the extent to which swidden fields resembled the natural world more than they did the cultural world. They were impressed by the fact that the guiding principle in swidden seemed to be to mimic or reproduce the natural forest, not transform it. As Geertz (1963, p.25) wrote in his classic study:

> With respect to degree of generalization (diversity), to proportion of total system resources storied in living forms, to closed-cover protection of an already weakened soil against the direct impact of rain and sun, the swidden plot is not a 'field' at all in the proper sense, but a miniaturized tropical forest, composed mainly of food-producing and other useful cultivates.

Geertz was saying that, through mimicry, the swiddeners seek not to dominate natural dynamics, but rather to utilize them, to extract the subsidy mentioned by Alcorn (1993). As Geertz (1963, p.16) further noted in his comparison of irrigated and swidden agriculture in Indonesia:

> Any form of agriculture represents an effort to alter a given ecosystem in such a way as to increase the flow of energy to man: but a wet-rice terrace accomplishes this through a bold reworking of the natural landscape; a swidden through a canny imitation of it.

Ironically, given its reputation for destruction, swidden does not radically alter the landscape; it does not bring about a permanent shift from a natural to an anthropogenic landscape. Swiddeners see time pass, and history unfold, within this landscape that is constantly being recycled between cultivation and forest fallow; but outsiders with a Cartesian mindset do not. The latter see no change, no progress, no history – and thus neither a past nor a future. This is an example of the way that time is used to construct the 'other', by denying that the other – the swiddener – is in the same time as ourselves, a time that has both a past and a future (Fabian, 1983).

The current situation

The 20th-century ethnographic studies of swidden led to waning predictions of its demise, although state disparagement of swidden continued, for example, in Indochina. Ironically, if predictably, just as the colonial discourse of swidden's demise was put to rest, its actual demise began to unfold in certain parts of the world, especially in Southeast Asia (Mertz et al., 2009; Padoch et al., 2010, Heinimann et al., 2017). This is due not to the hoary Malthusian curse, but to market forces; it is due not to demographics, but to economics and politics in the form of 'land grabbing' and an explosion in estate cropping (Li, 2018). Importantly, it is not factors internal to swidden systems that are undermining the conditions of its reproduction, but rather external factors, which do not impugn its inherent sustainability.

The recent retrenchment of swidden in certain parts of the world has been accompanied by statements of appreciation that are almost without historical precedent beyond the anthropological community. And they focus not on swidden's ability to adapt to 19th and 20th century worlds, but on what it has to offer society in the 21st century.[3] Thus, its ability to support biodiversity is extolled, as is its importance in the survival strategies of rural peoples hard-pressed by environmental change (see Sections VI and VII, both in the supplementary chapters to this volume), and also its ability to help sequester carbon (van Vliet et al., 2012; Mertz et al., 2013; Section IV, this volume).

There has been no more important contribution to the current reappraisal of swidden than the series of volumes, including this one, from Malcolm Cairns and his coeditors (Cairns, 2007, 2015, 2017, and this volume). The sheer breadth of these volumes, in both topic and geography, testifies to the continued tenacity of swidden. And the diversity of practices depicted in these volumes, especially the innovative practices that feature in the current volume, testify to the ability of swidden to change, to evolve, to have a history, to have a future. The depiction of swidden as a system without a future was an effort to take that future away from swidden cultivators; the present volume is a praiseworthy effort to restore it.

References

Alcorn, J. (1993) 'Indigenous Peoples and Conservation', *Conservation Biology* 7(2), pp.424-426

Cairns, M. F. (ed.) (2007) *Voices from the Forest: Integrating Indigenous Knowledge into Sustainable Upland Farming*, Resources for the Future, Washington, DC

Cairns, M. F. (ed.) (2015) *Shifting Cultivation and Environmental Change: Indigenous People, Agriculture and Forest Conservation*, Earthscan, London

Cairns, M. F. (ed.) (2017) *Shifting Cultivation Policies: Balancing Environmental and Social Sustainability*, CAB International, Wallingford, UK

Condominas, G. (1977) *We Have Eaten the Forest: The Story of a Montagnard Village in the Central Highlands of Vietnam*, translated by Adrienne Foulke (original: 1957) *Nous Avons Mangé la Forêt de la Pierre-Génie Gôo*. Mercure de France, Paris), Hill and Wang, New York

Conklin, H. C. (1954) 'An ethnoecological approach to shifting agriculture', *Transactions of the New York Academy of Sciences*, Ser. 2, 17, pp.133-142

Conklin, H. C. (1975) *Hanunóo Agriculture: A Report on an Integral System of Shifting Cultivation in the Philippines*, (original: (1957) Food and Agriculture Organization of the United Nations, Rome), Elliot's Books, Northford, CT

Dove, M. R. (2015) 'Linnaeus' study of Swedish swidden cultivation: Pioneering ethnographic work on the "economy of nature"', *Ambio* 44(3), pp.239-248

Fabian, J. (1983) *Time and the Other: How Anthropology Makes its Object*, Columbia University Press, New York

Freeman, J. D. (1970) *Report on the Iban*, London School of Economics Monographs on Social Anthropology No.41, (original: (1955) *Iban Agriculture: A Report on the Shifting Cultivation of Hill Rice by the Iban of Sarawak*, Her Majesty's Stationery Office, London), The Athlone Press, New York

Geertz, C. (1963) *Agricultural Involution: The Process of Ecological Change in Indonesia*, University of California Press, Berkeley, CA

Heinimann, A., Mertz, O., Frolking, S., Christensen, A. E., Hurni, K., Sedano, F., Chini L. P., Sahajpal, R., Hansen, M. and Hurtt, G. (2017) 'A global view of shifting cultivation: Recent, current, and future extent', *PLoS One* 12(9), e0184479, https://doi.org/10.1371/journal.pone.0184479

Izikowitz, K. G. (1951) *Lamet: Hill Peasants in French Indochina*, Etnografiska Museet, Göteborg, Sweden

Jansonius, H. (1950) *Groot Nederlands-Engels Woordenboek voor Studie en Practijk*, Nederlandsche Uitgeversmij, Leiden, The Netherlands

Kunnas, J. (2005) 'A dense and sickly mist from thousands of bog fires: An attempt to compare the energy consumption in slash-and-burn cultivation and burning cultivation of peatlands in Finland in 1820-1920', *Environment and History* 11, pp.431-446

Li, T. M. (2018) 'After the land grab: Infrastructural violence and the "Mafia System" in Indonesia's oil palm plantation zones', *Geoforum* 96, pp.328-337

Mertz, O., Padoch, C., Fox, J., Cramb, R. A., Leisz, S. J., Lam, N. T. and Vien, T. D. (2009) 'Swidden change in Southeast Asia: Understanding causes and consequences', *Human Ecology* 37(3), pp.259-264

Mertz, O., Egay, K., Bruun, T. B. and Colding, T. S. (2013) 'The last swiddens of Sarawak, Malaysia', *Human Ecology* 41(1), pp.109-118

Myllyntaus, T. and Mattila, T. (2002) 'Decline or increase? The standing timber stock in Finland, 1800–1997', *Ecological Economics* 41(2), pp.271-288

Padoch, C., Coffey, K., Mertz, O., Leisz, S. J., Fox, J. and Wadley, R. L. (2010) 'The demise of swidden in Southeast Asia? Local realities and regional ambiguities', *Geografisk Tidsskrift, Danish Journal of Geography* 107(1), pp.29-41

De Schlippe, P. (1956) *Shifting Cultivation in Africa*, Routledge and Kegan Paul, London

Sigaut, F. (1979) 'Swidden cultivation in Europe: A question for tropical anthropologists', *Social Science Information* 18(4/5), pp.679-694

Tsing, A. L. (1999) 'Becoming a tribal elder and other green development fantasies', in T. M. Li (ed.) *Transforming the Indonesian Uplands: Marginality, Power and Production*, Berg, London, pp.159-202

van Vliet, N., Mertz, O., Heinimann, A., Langanke, T., Pascual, U., Schmook, B., Adams, C., Schmidt-Vogt, D., Messerli, P., Leisz, S. J., Castella, J. C., Jørgensen, L., Birch-Thomsen, T., Hett, C., Bech-Bruun, T., Ickowitz, A., Vu, K. C., Yasuyuki, K., Fox, J., Padoch, C., Dressler, W. and Ziegler, A. D. (2012) 'Trends, drivers and impacts of changes in swidden cultivation in tropical forest-agriculture frontiers: A global assessment', *Global Environmental Change* 22(2), pp.418-429

Weimarck, G. (1968) *Ulfshult: Investigations Concerning the Use of Soil and Forest in Ulfshult*, Parish of Örkened, During the Last 250 Years, C. W. K. Gleerup, Lund, Sweden

Notes

1. Swidden's historic success was not confined to tropical latitudes. The 20th-century studies in the tropics eventually led to the rediscovery of the surprisingly recent history of swidden in Europe and North America (Weimarck, 1968; Sigaut, 1979; Myllyntaus et al., 2002; Kunnas, 2005; Dove, 2015).
2. On May 31, 1897, when Mark Twain was in London, in response to an inquiry from the New York Journal as to whether he was ill or dead, Twain replied, 'The report of my death was an exaggeration.'
3. Recent efforts in Scandinavia to re-enact swidden farming in some national parks make no such claims, but depict swidden as part of a past that should be remembered and respected.

BOTANICAL INDEX

Note: Page numbers in bold type refer to figures or sketches
    Page numbers in italic type refer to tables
    Page numbers followed by 'n' refer to end of chapter notes
    Page numbers followed by 'box' refer to a box that appears on that page.

ETHNIC GROUP INDEX

Note: Page numbers in bold type refer to figures or sketches
Page numbers in italic type refer to tables
Page numbers followed by 'n' refer to end of chapter notes
Page numbers followed by 'box' refer to a box that appears on that page.

GENERAL SUBJECT INDEX

Note: Page numbers in bold type refer to figures or sketches
Page numbers in italic type refer to tables
Page numbers followed by 'n' refer to end-of-chapter notes (endnotes)
Page numbers followed by 'box' refer to a box that appears on that page.